Behrouz Abedian

BUOYANCY-INDUCED FLOWS AND TRANSPORT

BUOYANCY-INDUCED FLOWS AND TRANSPORT

Textbook Edition

Benjamin Gebhart
Samuel Landis Gabel Professor of Mechanical Engineering
University of Pennsylvania
Philadelphia, Pennsylvania

Yogesh Jaluria
Rutgers University
New Brunswick, New Jersey

Roop L. Mahajan
AT&T Engineering Research Center
Princeton, New Jersey

Bahgat Sammakia
IBM Corporation
Endicott, New York

HEMISPHERE PUBLISHING CORPORATION, New York
A subsidiary of Harper & Row, Publishers, Inc.

Cambridge Philadelphia San Francisco Washington
London Mexico City São Paulo Singapore Sydney

BUOYANCY-INDUCED FLOWS AND TRANSPORT: Textbook Edition

1 2 3 4 5 6 7 8 9 0 BCBC 8 9 8 7

Cover art from W. C. Grigull and W. Hauf, *Proceedings of the Third International Heat Transfer Conference,* Chicago, vol. 2, p. 194. Reproduced by permission of the American Institute of Chemical Engineers.

This book was set in Times Roman by Edwards Brothers, Inc. The editors were Sandra Tamburrino and Mary Prescott.
BookCrafters, Inc. was printer and binder.

Library of Congress Cataloging in Publication Data

Buoyancy-induced flows and transport.

Bibliography: p.
Includes indexes.
1. Fluid dynamics. I. Gebhart, Benjamin.
QA911.B86 1988 532′.05 87-23703
ISBN 0-89116-728-5 Hemisphere Publishing Corporation

CONTENTS

PREFACE

Buoyancy-induced flows abound in nature and in our enclosures and devices. They arise through the interaction of a body force, such as gravity, with a density difference. This difference usually is caused by the diffusion of thermal energy and/or chemical species. Early pioneering insights concerning such transport arose over a century ago. Data, correlations, and analyses have accumulated at an increasing rate ever since.

Increasingly intensive study in the past 30 years has further clarified our conceptual understanding of many more fundamental aspects. A single generation of researchers has explained in detail many of the mechanisms of flow and transport, both in extensive and quiescent ambient media and within enclosed bodies of fluid. The most subtle kinds of transport, those that combine the features of both external and internal flows, have also received careful attention.

The diversity of commonly occurring buoyancy-driven transport is immense. It arises in single- and in multiple-phase bodies of fluid. Interesting and important effects are seen at every scale, from the microscopic in living systems, through that of meters in our immediate lives and in technology and that of hundreds of kilometers in the atmosphere and in terrestrial waters, to immense scales in extraterrestrial circulations.

At every scale a body of fluid may be subject to buoyancy effects arising singly or multiply from many different kinds and combinations of physical processes. The buoyancy force may arise from density differences in a body force field, following from both thermal and mass transfer density effects. The force-generating thermal and mass transport may arise, in turn, from many diverse mechanisms. For example, even the apparently simple buoyancy effect around a leaf of corn in the sun is quite complicated. Sun warms the leaf, which may also transpire water vapor, for thermal regulation. In photosynthesis the leaf stomata draw in CO_2 from the air and expel O_2 in return. Thus, one thermal effect and three mass transfer effects operate together to produce net buoyancy. These processes are coupled with radiant transfer. Another example is the loss of metabolic heat by mammals over the extent of their bodies. Their warmth results in thermally driven transport near the surface. However, perspiration is often a comparable effect. Surface vaporization humidifies the adjacent air. Thus, two upward buoyancy force components arise.

This book has been written to draw together current knowledge over a considerable and representative part of this broad and diverse field. Stress is placed on the fundamental transport process mechanisms. Rather than complete coverage of all the detailed information that has appeared related to the many specific applications thus far studied, emphasis is concentrated on the subfields in which intense study has unraveled and formulated the processes and in which transport formulations of great practical value have arisen. The buoyancy-generated motions considered in the bulk of the book are those that arise internal to a single fluid phase due to thermal and mass diffusion.

Emphasis throughout is in terms of physical mechanisms, their general formulation, the resulting analysis and general results, and the data related to these matters. Less emphasis is given to the detailed characteristics of particular technological and environmental applications. In most of the treatment, the aim has been to bring certain fields up to date to help both those who use such information and those attempting to extend it.

The first seven chapters concern the simplest fundamental mechanisms that arise in steady and transient external flows driven by thermal and mass diffusion. Chapters 8 and 9 concern the next level of complexity, the effects of important and of anomalous fluid property variations. Chapter 10 treats mixed convection in both external and internal flows. Chapters 11 and 12 are an account of instability, transition, and turbulent transport for external flows. Chapter 13 treats unstable stratified fluid layers, leading up to transport in enclosures and partial enclosures in Chapter 14. In Chapter 15 both external and internal flows in porous media are discussed. In Chapter 16 the effects of non-Newtonian behavior are formulated and presented. Finally, Chapter 17 collects information on rotational and other force fields, random and radiative effects, and the treatment of conjugate effects and of entropy generation.

Problems relative to a wide sample of the book material are presented at the end of each chapter. Common notation is used throughout the book instead of the many different usages that have arisen in the literature. This notation is given in the Nomenclature following Chapter 17. Fluid, thermal, and transport property information is appended. Author and Subject Indexes are included for the location of specific material in this diverse coverage.

The reference edition of this book also includes a listing of important additional references, with a separate author index. Most of these publications are very recent results. References containing material that could not be conveniently incorporated into the format of the book are also included.

Given the form of this book, with the problems and appended material, a variety of one-semester courses may be covered. Chapters 1–3, 5–7, 10, 13, and 14 are the basic material, sufficient as background to the rest. For a course concentrating on fundamentals, Chapters 4, 8, and 9, along with selected material in Chapter 11, may be added. On the other hand, Chapters 12 and 15–17 would broaden the conceptual base and perhaps the immediate usefulness of the course for some users having interests in different and emerging fields.

This book reports the wide-ranging and often complete work that has been done. We acknowledge the truly deep understanding and insight of so many researchers, which collectively has made possible this account of our field. We also acknowledge another direct contribution in terms of specific book content. Dr. Zafar Qureshi, at Johns Hopkins University, collected and organized several components of the contents. Several of us have used portions of this material in lectures and courses. We acknowledge the responses over the years that have contributed to both the clarity and suitability of much of the material.

The authors, in turn, acknowledge the substantial comfort and support in this effort arising from association with and support from the University of Pennsylvania and the Samuel Landis Gabel Fund established by Richard Gabel, President of Superior Tube Company; Rutgers University; AT&T, the Engineering Research Center in Princeton, New Jersey; and the IBM Corporation in Endicott, New York. We can only acknowledge in a general way the excellent clerical efforts of many helpers during the years of manuscript preparation. Of great importance to each of us in our effort, and to the quality of the result, was the forbearance and cooperation of our families and others close to us. The long hours of absence, in work and consultation together, were matched by their understanding and generosity.

Benjamin Gebhart
Yogesh Jaluria
Roop L. Mahajan
Bahgat Sammakia

BUOYANCY-INDUCED FLOWS AND TRANSPORT

CHAPTER

ONE

INTRODUCTION

1.1 INCIDENCE AND DIFFERENT CHARACTERISTICS

Most of the fluid motions and transport that affect our lives, our immediate surroundings, and our near and far environment are induced by buoyancy. Such flows are found in the air circulation around our bodies, in the enclosures we frequent, in cooking, in processing, in pools of water, and in atmospheric, lake, and oceanic circulation at every scale. They are found inside planetary bodies and are presumed to occur in and around celestial ones. The buoyancy force arises from motive density differences resulting from inhomogeneities in temperature, differences in concentration of chemical species, changes in material phase, and many other effects. A great many different kinds of buoyancy-induced flows are found because of the various separate effects and their combinations, the occurrence of many different geometric configurations, and the different bounding conditions and force fields that arise.

Yet all such occurrences have similarities and are very different from transport processes driven by such familiar forcing conditions as an imposed motion or a fan or pump. A signal difference is that little is known a priori about the resulting buoyancy-driven flow. The flow and temperature fields are invariably completely coupled and must be considered together, and the flows are relatively weak. That is, the velocities are relatively small and the momentum and viscous effects are commonly of the same order.

The group of all such processes is subdivided into two useful categories. Flows arising in a very extensive medium due to a local inhomogeneity in density or to imposed temperature or energy conditions at a location or at a surface are called external flows. Such a flow is seen in Fig. 1.1.1, generated adjacent to a surface that dissipates a uniform heat flux, over both sides. Flows arising in a body of

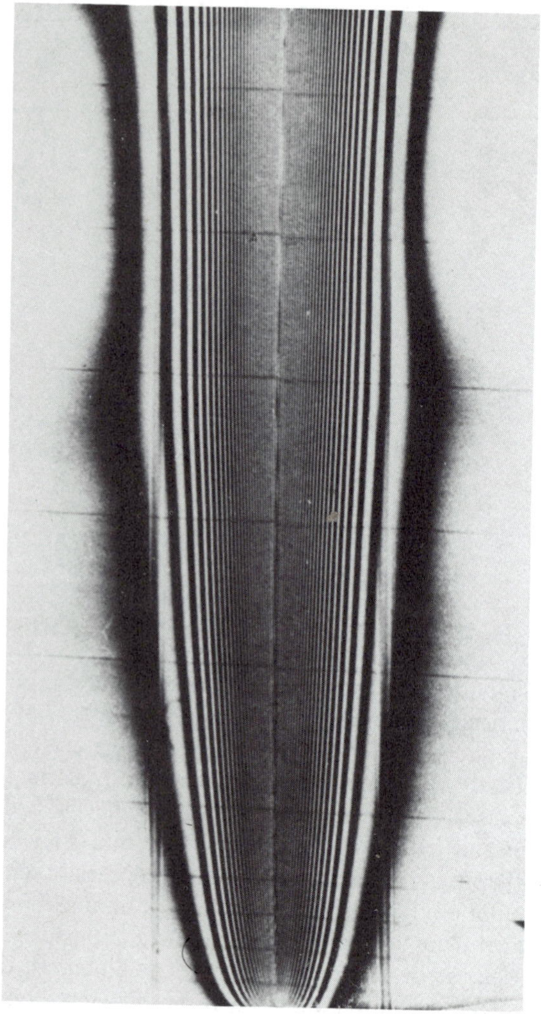

Figure 1.1.1 Interferogram of a vertical flow in nitrogen gas generated adjacent to a stretched and vertical metal foil. An electrical current dissipates energy, which is convected away on both sides. The horizontal scale is enlarged through an anamorphic lens system by a factor of 6. *(From Polymeropoulos and Gebhart, 1967).*

fluid contained in a cavity or otherwise completely bounded by surfaces are called internal flows. Internal flows, not completely bounded, as in a cavity with openings, are called partial enclosure flows. Some completely external flows, such as plumes, thermals, and jets, do not arise in conjunction with any solid boundary, and these are called free-boundary flows. A plane plume is seen in Fig. 1.1.2, rising in a plane flow above a heated wire.

This classification of flows is not controversial. However, the terminology applied to the flows is. One body of usage calls external flows "free" convection;

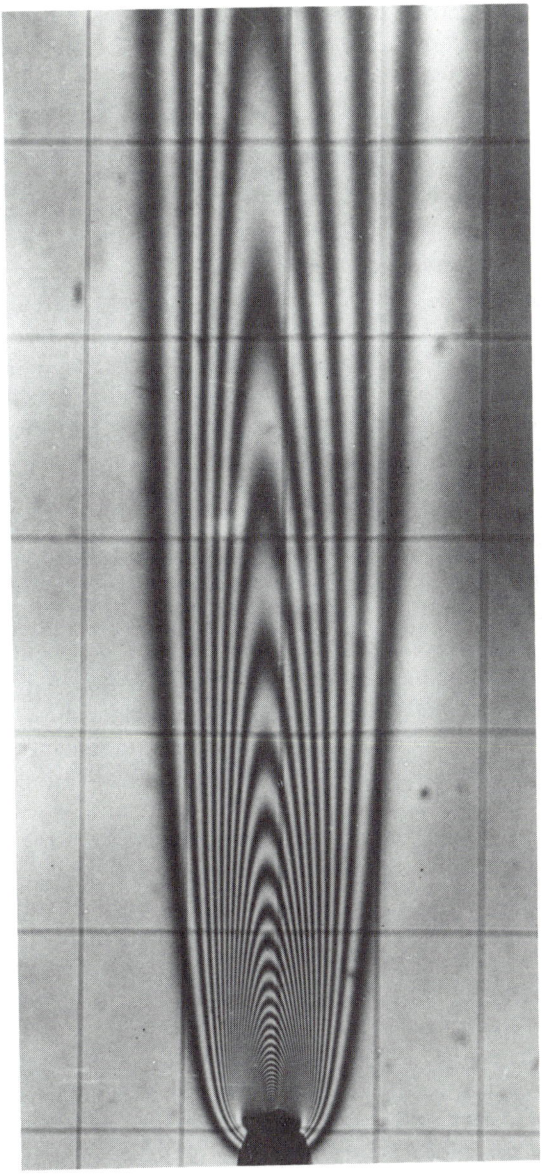

Figure 1.1.2 Transport arising above a horizontal electrically heated wire, the plane thermal plume. *(Reprinted with permission from* Int. J. Heat Mass Transfer, *vol. 13, p. 161, B. Gebhart et al. Copyright © 1970, Pergamon Journals Ltd.)*

another uses this term for both external and internal flows. Still another calls all such flows "natural" convection. There is some sentiment that neither term, natural or free, is very appropriate and that something more descriptive, such as "buoyancy convection" or "gravity-induced flow," should be used. This divergence of usage clearly stems from the absence of any obviously suitable and simple term. The only position taken here in this matter is the use of a single general term for all processes: buoyancy-induced flows.

In recent years there has been a rapid increase in the intensity of research in this field. Some of this effort represents a shift in interest from several conventional fields of fluid mechanics of diminished relative importance. However, most of it is due to a growing demand for detailed quantitative information concerning buoyancy-induced motions in the atmosphere, bodies of water, quasi-solid bodies such as the earth, enclosures, and devices and process equipment. The result has been a rapid expansion of knowledge in areas little considered only a few years ago.

In assembling a treatise of reasonable size on buoyancy-induced flows, only a fraction of what is known can be covered. The field is vast. The catalogue of studied mechanisms is very large and rapidly growing. Further, such transport is seldom simple and is often extremely subtle in both geometry and mechanisms. Single and multiphase motions occur. Many different body force fields such as gravitation and rotation exist to drive flows through density differences. The density differences arise from phase differences, thermal transport, and/or single or multiple diffusion of chemically distinct species. The resulting motion may often be correctly anticipated to be highly organized, as in a rising thermal plume. It may also be spatially somewhat fixed by the presence of a stable phase interface, as on both sides of a horizontal air-water interface. On the other hand, any motion that arises may have initially unknown spatial characteristics. This occurs in a body of fluid with an unstable density stratification, in the gravitational field, as in a pan of water heating on a stove. A detailed initial surmise concerning the pattern of motion cannot be made confidently. There are also many flow configurations that arise through a combination of these and/or other effects.

Common buoyancy-induced flows are also very different in physical extent. They increase in scale from that around a small sun-heated particle in the atmosphere, through that arising around our bodies to carry away the metabolic heat, to the circulations in the atmosphere, oceans, and earth's mantle, and finally to the presumed motions interior to stars.

Buoyancy-induced motions are initially laminar at a small scale. However, the vigorous flows at larger scales are inevitably turbulent. Laminar and turbulent transport characteristics usually differ widely in buoyancy-induced flows, as in forced ones. It is interesting that the flows in water and in atmospheric air, at the human size scale, are initially laminar, before perhaps beginning a transition toward downstream turbulence. Laminar transport characteristically persists to a distance of the order of 1 m, more or less, depending on bounding conditions. Therefore, the many flows of smaller scale are treated as laminar. For those at a much larger scale, the initial laminar portion may often be ignored and the whole flow field treated as turbulent. At scales between these there are important questions of when and how a laminar flow becomes unstable and eventually proceeds to turbulence. Inevitably, laminar transport is much better understood at this early time in the study of buoyancy-induced flows. However, the results for laminar transport will always be generally instructive; as the initial estimates of characteristic quantities such as velocity levels, as estimates of the lower limits of transport, in establishing bounds between different kinds of processes, and in determining the initial geometries of various flows.

Thus, among buoyancy-induced flows there are great diversities in mechanism, physical extent, and resultant motions. However, few of the possible combinations of these diverse factors are well understood, even among those of greatest immediate and practical importance. Much present knowledge concerns laminar processes in a single fluid phase driven by the interaction of a gravitational force with thermal transport effects on density. Effects on density of diffusion of chemical species at low concentration levels are easily incorporated in equally simple analyses. In most species' diffusion processes in the atmosphere and in terrestrial water, the absolute concentration level of commonly diffusing species is very low. Atmospheric humidity and CO_2 levels are around 1 and 0.04%, respectively. The oceanic salinity is around 3.5%.

Therefore, a considerable portion of the material included here treats laminar flow arising through the interaction of a uniform gravitational field with thermal and species transport effects on the density level in a single fluid phase. Most of the treatment concerns steady flows, although some of the much smaller body of available information concerning time-dependent flows is also included.

For a few steady laminar flows, a considerable amount is known about how they become initially unstable. For fewer yet, it is known how they progress through the transition to a more or less developed turbulence. Such information is summarized both for its specific value and as a guide to general concepts concerning instability and transition.

A number of common and largely turbulent processes at a larger size scale have been treated approximately by making some simple assumptions. These include vertical plumes arising from a steady concentrated input of energy and non-buoyant and buoyant jets resulting from a steady momentum and buoyancy input. Simple analysis also results for the upward propagation of a single input event of a pulse of vertical momentum, a puff, and for the upward flow resulting from a single input event of buoyancy, a thermal, as from a concentrated explosion. Thermals are included here as an informative and interesting generalization for commonly occurring flows. These formulations are based on very little information but on good intuition.

Phase interfaces are present in many processes. In technology they occur with liquid films and droplets in vapors and gases, with bubbles in liquids, and so forth. They also occur at the many interfaces between the atmosphere and terrestrial waters. Motions in the presence of phase interfaces are complicated by wave action and, in the atmosphere and large bodies of water, by the additional rotational effects. Even in a single phase, wavelike motions arise in connection with rapid vertical density changes, as in thermoclines in bodies of water. These matters are not considered here, except for several much simpler single-phase configurations with rotational effects.

There has been considerable study of important additional effects, even within the constraints of thermally induced laminar flows. In applications, the imposed temperature conditions of, for example, a surface and the quiescent ambient medium in which it is immersed are often sufficiently different that large variations in the fluid viscosity and thermal conductivity occur across the thermal diffusion region. Guidelines are set forth for treating these effects, as well as those arising from density stratification of the ambient medium, which may result from a vertical temperature variation. Such stratification has important effects on transport.

The effects of an imposed velocity of the ambient medium have also been assessed for many simple surface geometries. The inverse problem concerns the effects of buoyancy forces due to temperature gradients in vigorous forced flows of an ambient medium over a surface embedded in it. The latter situation arises, for example, in internal forced flows in a fluid with temperature gradients flowing through a tube. These processes, generally called either mixed or combined convection—a combination of forced and buoyancy-induced flow effects—are treated here.

Another important application concerns buoyancy-induced flows in water at low temperatures. The density maximum of pure water at 1 atm, which occurs at about 4°C, persists to higher pressure and salinity levels. If the temperature field in cold water spans the density maximum, there is a buoyancy force reversal. If this reversal is sufficiently large, local flow reversals occur that have large effects on transport. In some circumstances there is a complete change in net flow direction, called a convective inversion. These complicated processes commonly occur in ice freezing or melting in both pure and saline water. Findings concerning such processes are summarized in the light of current understanding.

A summary is also given of the mechanisms and transport in a quite different kind of buoyancy situation in which induced flow and transport may arise. An extensive horizontal layer of fluid may be unstably stratified. That is, the fluid density may increase upward, as in a pan of water on a stove. Then heavier fluid overlies lighter fluid. Any local tendency to motion may cause the heavier fluid to fall, causing the lighter fluid to rise elsewhere. Such motions would be opposed by viscosity. This is a potentially unstable circumstance, and the instability is usually called thermal instability.

Such unstable stratification is common in air spaces in enclosures, in solar collectors, within stored liquids, in processing equipment, in the atmosphere, and in bodies of water. The first matters of importance are the conditions of thickness, viscosity, and so forth under which such layers first become unstable to ever-present disturbances. Then the concern is what kind of motion first arises. Finally, the matter of interest is how the net thermal transport across such a layer differs from the pure conduction heat transfer that takes place when the fluid in the layer is stable and stationary.

Another kind of buoyancy-induced flow of interest is that resulting from energy transport in a porous solid saturated with a fluid, such as sand containing water. Temperature differences cause a buoyancy force and the fluid circulates through the porous medium. Such flows also occur in melting dense ice-slush in water, in geothermal deposits of water in porous rock, and as air circulation in fibrous or granular building insulation. Buoyancy-induced velocities are often very low because of large viscous effects in the small flow passages. This permits an important simplification of the analysis of transport, at some loss in physical realism in some configurations. Some of the results of greatest value are given.

Finally, a group of somewhat special yet interesting and important kinds of transport are treated in the last chapter. The information included in this book, therefore, provides a broad and deep coverage of this field. It is especially related to flows at the small scales common to most technological applications.

Reference is made here to a group of past summaries of available and relevant information concerning buoyancy-induced flows. Among these are Ostrach (1964), Ede (1967), Gebhart (1969, 1971, 1973), Turner (1973), and Spalding and Afgan (1977). More recent surveys and treatments of more particular areas and recent results include Jaluria (1980), Platten and Legros (1984), Kakac et al. (1985), and Raithby and Hollands (1985). Reference to the first group will often provide a more complete account of earlier work. The second group relates to more particular areas and to some specific present concerns.

1.2 THE BUOYANCY FORCE

Consider a fluid whose density ρ is dependent on temperature t, on the concentration C of some chemical component, and on static pressure p; that is, $\rho(t, C, p)$. In a quiescent ambient environment of such a fluid in a gravitational field of strength $\mathbf{g}$, a local region of lower density produces an upward buoyancy force $\mathbf{B}$, written as a vector. This force will result in motion. The situation is seen in Fig. 1.2.1 for a density field, which may be described in two space coordinates x and y. The gravitational force is vertical and acts downward. Motion arises because the pressure, or hydrostatic gradient in the ambient environment, governed by $dp_a/dx = -\rho_a g$, is not the same as that at various locations in the region of density variation from ρ_a, that is, $dp/dx = -\rho g$. Here ρ is the instantaneous local density. This difference in pressure gradient drives the motion through the buoyancy force $\mathbf{B}$. Its magnitude B is calculated as $B = B(x, y) = g(\rho_a - \rho)$. Thus, B is the difference between the two body forces. It is taken as positive if it, and therefore the resulting flow, is upward.

The local magnitude of this buoyancy force depends on the local temperature and/or concentration levels. This force is written as a balance of viscous forces and momentum. A mass continuity equation arises as well. An energy balance equation sums the diffusion and convection of thermal energy, along with all other energy effects. Finally, for each chemical species there is a balance of diffusion, convection, and any rate of gain or loss due to chemical reactions. Since the thermal and chemical transport equations are in terms of t and C, they are coupled directly to force-momentum through the appearance of $\mathbf{B}$ in the latter equation. This is the primary source of the complexity of the mechanisms involved in buoyancy-induced flows. The same kind of complexity arises with other fields in which

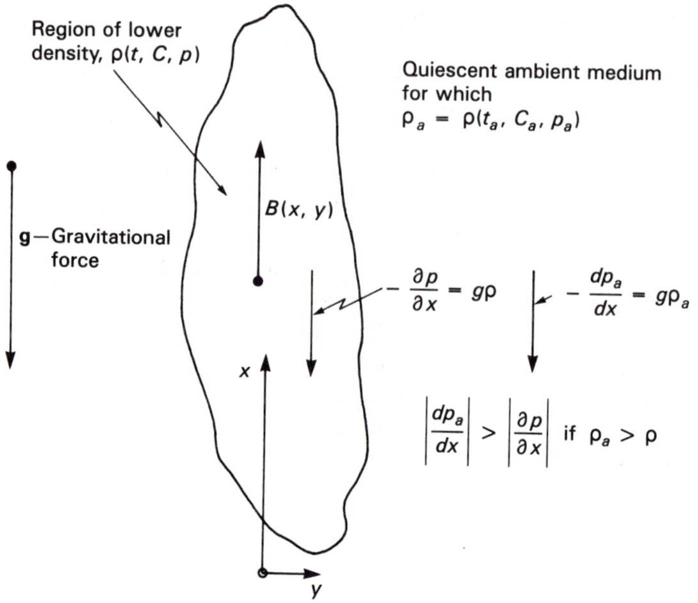

Figure 1.2.1 Density disturbance in an extensive and quiescent ambient environment and the resulting transport. The buoyancy force is $\mathbf{B} = B(x, y) = g(\rho_a - \rho)$, where $\mathbf{g} = -g$.

the force depends on the local density level, for example, in a rotating body of fluid.

Buoyancy forces also arise in bodies of fluid of limited extent, such as heated water in a pan and air in a heated enclosure. These are called internal flows, internal to boundaries. A variation over the volume of the temperature and/or the concentration of one or more chemical components also results in a tendency to motion. The local buoyancy force then is calculated from the local instantaneous gravitational force at any point, $-\rho g$. The buoyancy force B is obtained by subtracting from $-\rho g$ some average or representative body force $-g\rho_r$. Thus, $B = g(\rho_r - \rho)$.

The choice of reference density ρ_r depends on the internal flow under consideration. A simple example is a horizontal fluid layer between upper and lower bounding surfaces at different temperatures t_1 and t_2, respectively. If $t_2 > t_1$, the fluid layer is unstably stratified if the density of the fluid in question decreases with temperature. Buoyancy-induced motion results under certain conditions. For such a flow, the local reference density at different elevations, ρ_r, is calculated from the linear temperature distribution that would apply upward through the layer of fluid in the absence of fluid motion, that is, for pure thermal conduction. The procedures for choosing the reference density variation in other and more complicated internal flow configurations and/or body force fields are frequently less simple than in the foregoing example. The differing circumstances will be discussed in later chapters where such flows are considered.

1.3 FLOW AND TRANSPORT

The gravitational buoyancy force $g(\rho_r - \rho)$ drives the flow. For analysis, it enters into the general vector balance of forces and momentum. The other balances are mass continuity and the balance applying to whatever transport process causes the density variation. Thus, there are always at least three coupled equations governing the flow variables: velocity, pressure, and temperature or concentration. In addition, several property equations are needed, notably $\rho(t, C, p)$. Also appearing are the values of the molecular diffusion coefficients: viscosity μ for a Newtonian fluid, fluid thermal conductivity k, Fickian species diffusion D, and any additional ones that may arise in special circumstances.

These relations are then applied to any physical circumstance of interest. The particular circumstance is characterized by the conditions it imposes on the operating mechanisms in terms of geometry, surface conditions, ambient medium condition, and so forth. For example, the flow will be vertical if confined by or attached to vertical surfaces. It will also be vertical in a quiescent ambient medium if due to an energy source not associated with an extensive surface. The flow may be inclined if induced by a condition on an inclined or curved surface. Horizontal flows may even arise along horizontal surfaces.

The quantities to be determined for any flow, by either analysis or experimentation, are usually the heat transfer rate and the velocity field. The heat transfer coefficient h from a surface to a fluid is defined in terms of the heat flow rate Q and the imposed temperature conditions t_0 and t_∞ as follows. The temperatures t_0 and t_∞ are those of the surface and the ambient medium, respectively, and A is the surface area:

$$Q = hA(t_0 - t_\infty) \tag{1.3.1}$$

The heat transfer or convection coefficient is then written as a Nusselt number

$$\mathrm{Nu} = \frac{hL}{k} \tag{1.3.2}$$

where L is a characteristic dimension of the surface.

The velocity u generated by an external flow at elevation L in a region of density disturbance $\rho_r - \rho = \rho_\infty - \rho = \Delta\rho$ may be estimated by equating the kinetic energy produced, $\rho u^2/2$ per unit volume, to the work input of the buoyancy force in distance L, $gL\,\Delta\rho$, also per unit volume. Thus,

$$\frac{\rho u^2}{2} \simeq gL\,\Delta\rho \qquad \text{or} \qquad u = O\left(\sqrt{\frac{gL\,\Delta\rho}{\rho}}\right)$$

neglecting viscous forces and some other usually yet smaller effects. The small velocities arising in buoyancy-induced flows are calculated from this estimate. For example, for standard gravity, an L of 50 cm, and a 2% density difference, the maximum value of u at $x = L = 50$ cm would be 44 cm/s. Including the inevitable viscous and entrainment effects that oppose the flow, the actual values are about 34, 17, and 8 cm/s in liquid mercury, air, and water, respectively, when generated adjacent to a warm vertical surface.

Writing u as a Reynolds number Re yields

$$\mathrm{Re} = \frac{uL}{\nu} \propto \sqrt{\frac{gL^3\,\Delta\rho}{\nu^2\rho}} = \sqrt{\left(\frac{gL^3}{\nu^2}\right)\left(\frac{\Delta\rho}{\rho}\right)} = \sqrt{\mathrm{Gr}} \tag{1.3.3}$$

where ν is the kinematic viscosity, μ/ρ. Thus Gr, the Grashof number, replaces the Reynolds number of forced flows as the "vigor" parameter of buoyancy-induced flows. This number is the ratio of the buoyancy force and/or the resulting velocity to the viscous effect. It was apparently first called the Grashof number by Groeber (1921) to honor an eminent engineer; see Jakob (1949).

The Grashof number in Eq. (1.3.3) is seen to be gL^3/ν^2, a unit Grashof number, times $\Delta\rho/\rho$, an estimate of the units of buoyancy. This will sometimes prove to be a valuable distinction. Both groups are individually dimensionless, as is Gr.

In fluids generally, the transport, as Nu, depends on the vigor of the flow. It also depends on the Prandtl number $\mathrm{Pr} = \nu/\alpha$, where $\alpha = k/\rho c_p$ is the thermal diffusivity and c_p is the specific heat. This is the ratio of the molecular diffusion of momentum to that of thermal energy. Thus,

$$\mathrm{Nu} = f(\mathrm{Gr}, \mathrm{Pr}, \text{others}) \tag{1.3.4}$$

where "others" indicates the role of additional effects. The following chapter will set forth the governing equations for buoyancy-induced fluid motions. Succeeding chapters will treat many configurations and circumstances to determine specific kinds of transport information.

REFERENCES

Ede, A. J. (1967). *Adv. Heat Transfer 4,* 1.

Gebhart, B. (1969). *Appl. Mech. Rev. 22,* 691.

Gebhart, B. (1971). *Heat Transfer,* 2d ed., Chap. 8, McGraw-Hill, New York.

Gebhart, B. (1973). *Adv. Heat Transfer 9,* 273.

Gebhart, B., Pera, L., and Schorr, A. W. (1970). *Int. J. Heat Mass Transfer 13,* 161.

Groeber, H. (1921). *Die Grundgestze der Waermeleitung und des Waermeueberganges,* Julius Springer, Berlin.

Jakob, M. (1949). *Heat Transfer,* vol. 1, p. 487, Wiley, New York.

Jaluria, Y. (1980). *Natural Convection Heat and Mass Transfer,* Pergamon, Oxford, U.K.

Kakac, S., Aung, W., and Viskanta, R., Eds. (1985). *Natural Convection Fundamentals and Applications,* Hemisphere, Washington, D.C.

Ostrach, S. (1964). *High Speed Aerodynamics and Jet Propulsion,* vol. 4, Sec. F, Princeton Univ. Press, Princeton, N.J.

Platten, J. K., and Legros, J. C. (1984). *Convection in Liquids,* Springer-Verlag, Berlin.

Polymeropoulos, C. E., and Gebhart, B. (1967). *J. Fluid Mech. 30,* 225.

Raithby, G. D., and Hollands, K. G. T. (1985). *Handbook of Heat Transfer Fundamentals,* 2d ed., Chap. 6, McGraw-Hill, New York.

Spalding, D. B., and Afgan, N. (1977). *Heat Transfer and Turbulent Buoyant Convection,* vols. 1 and 2, Hemisphere, Washington, D.C.

Turner, J. S. (1973). *Buoyancy Effects in Fluids,* Cambridge Univ. Press, Cambridge, U.K.

CHAPTER
TWO

GENERAL FORMULATION OF BUOYANCY–INDUCED FLOWS

2.1 MECHANISMS OF TRANSPORT

The general equations of fluid transport for buoyancy-induced flows are the same as those for fluid mechanics generally. The specific difference is that the driving mechanism is the changing body force $\rho\mathbf{g}$ per unit volume. Actually, the net body force or buoyancy force $\mathbf{B} = \mathbf{g}(\rho - \rho_r)$ will arise in the force-momentum balance eventually used for calculations. The driving mechanism in forced flows is usually an imposed pressure gradient or velocity level. Here, the velocity and pressure effects jointly result from the interaction of buoyancy with other flow effects such as viscous forces. The equations of motion induced by thermal transport are

$$\frac{D\rho}{D\tau} = -\rho\nabla\cdot\mathbf{V} \qquad \text{or} \qquad -\frac{\partial\rho}{\partial\tau} = \rho\nabla\cdot\mathbf{V} + \mathbf{V}\cdot\nabla\rho \tag{2.1.1}$$

$$\rho\frac{D\mathbf{V}}{D\tau} = \rho\mathbf{g} - \nabla p + \mu\nabla^2\mathbf{V} + \frac{1}{3}\mu\nabla(\nabla\cdot\mathbf{V}) \tag{2.1.2}$$

$$\rho c_p\frac{Dt}{D\tau} = \nabla\cdot k\nabla t + \beta T\frac{Dp}{D\tau} + \mu\Phi + q''' \tag{2.1.3}$$

In Eqs. (2.1.1)–(2.1.3) the following local quantities arise: velocity $\mathbf{V} = (u, v, w)$; the conventional and absolute temperatures t and T; the gradient of static pressure ∇p; the local body force due to gravity $\rho\mathbf{g}$; the viscous dissipation energy effect $\mu\Phi$; the volumetric energy generation rate q'''; the local fluid properties ρ, c_p, and the coefficient of thermal expansion β; the thermal conductivity k; and the time τ. For brevity, the viscosity μ has been taken as uniform and constant. Its variation will be accommodated later in more specialized equations. The equations above apply in general equally for laminar, unsteady, and turbulent transport in any coordinate system. A simple formulation for chemical species diffusion appears at the end of this section.

The first equation is the mass continuity condition, a scalar equation relating the rate of density change of the fluid instantaneously at a point in the fluid field, through its particle derivative $D/D\tau$, to the local rate of expansion or compression resulting from the velocity field $\nabla \cdot \mathbf{V}$. The second is a vector relation equating local acceleration to the sum of the local body force, pressure gradient, and viscous forces per unit volume for a Newtonian fluid. The third equation, also a scalar, results from an energy balance. It equates the rate of temperature increase to the sum of several effects. The first is the volumetric gain of energy by conduction, expressed by the Fourier law of conduction. The second is a pressure term from the general stress tensor; this pressure is identified with the usual thermodynamically significant equilibrium pressure level, as an approximation. The volumetric strength of any distributed energy source internal to the fluid material arises as q''', which may be a function of location, temperature, and so forth. The viscous dissipation term, $\mu\Phi$, is the volumetric rate of flow energy dissipation into thermal energy. This term is approximately the difference between the total mechanical power input by the stress system and the smaller amount of the total power input that produces thermodynamically reversible effects, such as increases in potential and kinetic energy. The difference is that dissipated as thermal energy by viscous effects. The dissipation function is

$$\Phi = 2\left[\left(\frac{\partial u}{\partial x}\right)^2 + \left(\frac{\partial v}{\partial y}\right)^2 + \left(\frac{\partial w}{\partial z}\right)^2\right] + \left(\frac{\partial v}{\partial x} + \frac{\partial u}{\partial y}\right)^2 + \left(\frac{\partial w}{\partial y} + \frac{\partial v}{\partial z}\right)^2 + \left(\frac{\partial u}{\partial z} + \frac{\partial w}{\partial x}\right)^2 - \frac{2}{3}(\nabla \cdot \mathbf{V})^2 \tag{2.1.4}$$

Detailed developments of these equations may be found in Schlichting (1968), Howarth (1956), Gebhart (1971), and Yih (1969).

In the above relations the body force per unit volume is written as $\rho\mathbf{g}$, where ρ is the local density and $\mathbf{g}$ the gravitational force per unit mass. The usual approximation is made for bulk viscosity (e.g., see Schlichting, 1968). Here the shear viscosity μ is taken as constant. The form that Eq. (2.1.2) takes when μ is variable is given in vector form by Milne-Thompson (1960) and Pao (1967) and in indicial notation in Cartesian coordinates by Schlichting (1968) as follows:

$$\rho\left(\frac{\partial v_i}{\partial \tau} + v_j\frac{\partial v_i}{\partial x_j}\right) = \rho g_i - \frac{\partial p}{\partial x_i} + \frac{\partial}{\partial x_j}\left[\mu\left(\frac{\partial v_i}{\partial x_j} + \frac{\partial v_j}{\partial x_i}\right) - \frac{2}{3}\delta_{ij}\frac{\partial v_k}{\partial x_k}\right] \tag{2.1.5}$$

where $i, j, k = 1, 2, 3$, the volumetric coefficient of thermal expansion is β, and k is the thermal conductivity, taken as variable here. No force-producing electrical, magnetic, or rotational effects are included in Eq. (2.1.2).

The complexity and coupling inherent in natural-convection processes are apparent in this set of equations. Motion results because the local density ρ in Eq. (2.1.2) is variable due to variable t. The variable density in Eqs. (2.1.1) and (2.1.2) is known only by taking into account the "temperature" equation. This equation, in turn, inevitably involves velocity. Thus, the distributions of ρ, p, $\mathbf{V}$, and t in space and perhaps also in time τ must be found simultaneously from five scalar equations and a property relation for ρ.

In spite of these difficulties, the study of natural-convection flows via the governing equations has produced a large amount of information of importance in understanding and predicting transport behavior. Most of this information comes from the simpler forms of the governing equations that are applicable in most physical circumstances. We will consider various approximations made to simplify the equations and the limitations of these approximations.

The principal complexities in the above equations result primarily from the possible variation of the transport properties μ and k and of the density ρ. These are separate aspects for the vast majority of interesting flows. Since μ and k are dependent primarily on temperature, an important variation occurs in processes involving large temperature differences. These properties may often be taken as constant. However, the density variation must always be taken into account, to provide motion. Nevertheless, for processes not involving large temperature differences, density differences may often be approximated in a way that greatly simplifies the equations.

Gradients in the concentration of chemical species in a fluid often cause buoyancy-induced flows in a gravitational force field. Common examples are humidity in air and solutes in water. For example, the various components of oceanic salinity diffuse when ice melts in seawater. Diffusing water vapor makes air lighter, and diffusing salinity makes water heavier.

Therefore, the body force $\mathbf{g}$ in the force-momentum balance, Eq. (2.1.2), must often include the effect of varying species concentration C on density. In fact, it is the only driving force in many important circumstances. As a result, C would appear in Eq. (2.1.2), as does temperature for a thermally induced flow. An additional conservation equation, similar to Eq. (2.1.3) for temperature, is necessary to relate the convection and diffusion of the chemical species. Several such equations are required if there is simultaneous diffusion of several distinct chemical species. An example occurs in largely quiescent air immediately adjacent to a sun-heated leaf. Temperature regulation is accomplished by thermal transport and by water vapor formation and diffusion away from the surface. However, photosynthesis requires that CO_2 diffuse to the surface from the atmospheric background reservoir, at a concentration of about 350 ppm. Also, O_2 is released and diffuses away. Thus, there are three actively diffusing species: H_2O vapor, CO_2, and O_2. Each is driven by a very small but different concentration difference $C_0 - C_\infty$. These processes are superimposed on the other constituents of air, largely N_2 and the background level of O_2.

Most species diffusion in the atmosphere and in terrestrial water occurs at low concentration differences. This is also true in many occurrences in processing and manufacture. In this case, a simple formulation follows from Fick's law of diffusion:

$$m'' = -D\frac{dC}{dy} \tag{2.1.6}$$

or, in general,

$$\mathbf{m}'' = -D\nabla C \tag{2.1.7}$$

where $\mathbf{m}''$ is, for example, the mass flux rate of species C, and D is the diffusivity coefficient. The dimensions of D are meters squared per second in SI units. A balance of convection with the diffusion effects and with any chemical reaction yields the scalar equation

$$\frac{DC}{D\tau} = \nabla \cdot D\nabla C + c''' \tag{2.1.8}$$

where c''' is the local rate of production of species C per unit volume. This rate may depend on concentration, temperature, and so forth. Equation (2.1.8) then supplements Eqs. (2.1.1)–(2.1.3). There are two additional effects, the Dufour and Sorét effects, which are ignored in Eqs. (2.1.3) and (2.1.8). Combined buoyancy mode transport is considered further in Chapter 6, where these effects are discussed.

2.2 A SIMPLE CALCULATION

The general equations for thermally buoyant flows in the preceding section are very complicated. They must be to account for the diverse physical mechanisms encountered. However, both the study and the increasing understanding of buoyancy-induced transport began with flows for which much simpler formulations apply, at least approximately. The development in this book will also progress from the simplest insights, through gradually increasing complexity, to the most recent results.

The first detailed calculation of buoyancy-induced transport was by Lorenz (1881). The simplest kind of circumstance was considered: the steady laminar transport adjacent to a flat vertical surface maintained at a uniform temperature t_0 while immersed in a quiescent ambient gas uniformly at $t_r = t_\infty$; see Fig. 2.2.1. Assume that an energy source maintains the surface temperature at $t_0 > t_\infty$. Then if the density of the fluid decreases with temperature, which is most common, the buoyancy force $B(x, y) = g(\rho_\infty - \rho)$ is positive and flow is up, as in Fig. 2.2.1*a*. If $t_0 < t_\infty$ the flow is down and is taken instead as increasing downward from the top edge of the surface, as in Fig. 2.2.1*b*. The top edge is then the leading edge for the flow.

For a surface of wide span in the z direction, the transport is described in the two Cartesian coordinates shown, x and y. The full equations are (2.1.1)–(2.1.3). Steady laminar flow is a simple kind of transport. Further, in many flows the density differences are sufficiently small that they are important only in causing a buoyancy force, as shown in Fig. 1.2.1. Then $\nabla \cdot \mathbf{V}$ in Eqs. (2.1.1) and (2.1.2)

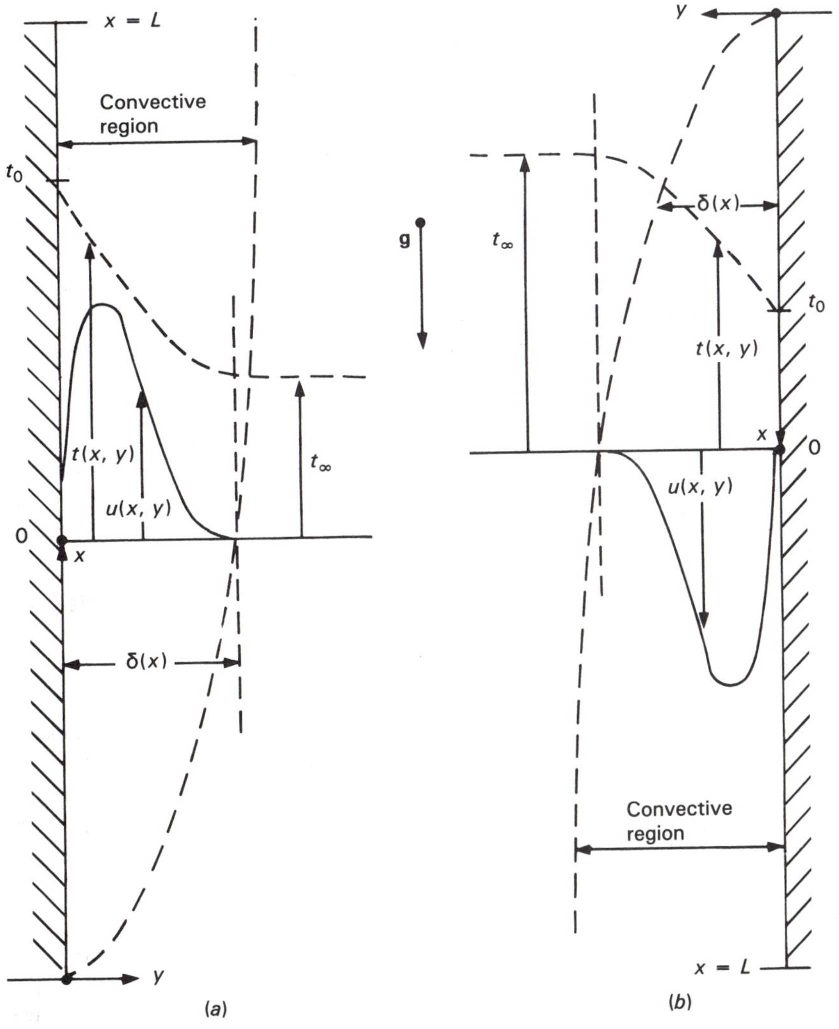

Figure 2.2.1 Vertical thermally induced flow. (*a*) $t_0 > t_\infty$, upflow; x is taken positive upward from the lower edge. (*b*) $t_0 < t_\infty$, downflow; x is positive downward from the upper edge.

may then be taken as zero. Thermal effects arising in Eq. (2.1.3) from the pressure field, from distributed sources q''', and from $\mu\Phi$ are often very small. Also, the difference between the body force and the pressure force in Eq. (2.1.2), $\mathbf{g}\rho - \nabla p$, is frequently merely $\mathbf{g}(\rho - \rho_\infty)$ to a good approximation, as will be seen in the next two sections. Therefore, Eqs. (2.1.1)–(2.1.3) become, for the flow in Fig. 2.2.1,

$$\nabla \cdot \mathbf{V} = \frac{\partial u}{\partial x} + \frac{\partial v}{\partial y} = 0 \tag{2.2.1}$$

$$\rho\left(u\frac{\partial u}{\partial x}+v\frac{\partial u}{\partial y}\right)=\mu\left(\frac{\partial^2 u}{\partial x^2}+\frac{\partial^2 u}{\partial y^2}\right)+g(\rho_\infty-\rho) \tag{2.2.2}$$

$$\rho\left(u\frac{\partial v}{\partial x}+v\frac{\partial v}{\partial y}\right)=\mu\left(\frac{\partial^2 v}{\partial x^2}+\frac{\partial^2 v}{\partial y^2}\right) \tag{2.2.3}$$

$$\rho c_p\left(u\frac{\partial t}{\partial x}+v\frac{\partial t}{\partial y}\right)=k\left(\frac{\partial^2 t}{\partial x^2}+\frac{\partial^2 t}{\partial y^2}\right) \tag{2.2.4}$$

The distributions $u(x, y)$, $v(x, y)$, $\rho(x, y) = \rho(t)$, and $t(x, y)$ are expressed in terms of x and y. The properties μ, k, and c_p also appear.

Very early experiments with several simple flows of this kind showed a peculiar result. It was found that the heat transfer rate Q per unit span in the z direction was not proportional to $t_0 - t_\infty = \Delta t$, as had been conjectured in Newton's law of cooling, that is, $Q \propto \Delta t$. This law, in turn, was supported by measurements in forced and nonbuoyant flows. In fact, for a number of buoyancy-induced flows, Q had been found to be approximately proportional to $\Delta t(1 + K\,\Delta t)$ or to a noninteger power of Δt, $(t_0 - t_\infty)^n$, where n depended on specific conditions or K was some constant.

Then Lorenz (1881) applied a keen intuition to guide a very simple analysis of this vertical flow in air adjacent to an isothermal surface of height L. The resulting heat transfer prediction is within a few percent of much later experimental data and calculations. Schmidt et al. (1930) experimentally demonstrated boundary layer behavior and gave a similarity solution for the heat transfer, also in air. This agreement may now be thought somewhat fortuitous. Yet the additional approximations and method of calculation of Lorenz are interesting and instructive. This analysis will be given, starting from the full equations above.

In this complete formulation, in Eqs. (2.2.1)–(2.2.4), the velocity component normal to the surface $v(x, y)$ is taken as zero. This results in $u(x, y) = u(y)$ from Eq. (2.2.1). Thus, Eqs. (2.2.1) and (2.2.3) disappear and Eqs. (2.2.2) and (2.2.4) are simplified. The remaining momentum convection term in Eq. (2.2.2), $u\,\partial u/\partial x$, is omitted since $u = u(y)$. As some justification of this, flow velocities are often small. In addition, as in subsequent boundary region analysis, viscous and conduction effects arising through downstream gradients are neglected. Finally, $\rho(t)$ was taken as a linear function of temperature:

$$\rho(t) = \rho(t_\infty)[1 - \beta(t - t_\infty)] \tag{2.2.5}$$

where $\beta = -(1/\rho)\,(\partial\rho/\partial t)_p$ is the coefficient of thermal expansion of the fluid. For an ideal gas, $p = \rho RT$, $\beta = 1/T$.

Then the remaining Eqs. (2.2.2) and (2.2.4) in $u(y)$ and $t(x, y)$ become

$$0 = \mu\frac{\partial^2 u}{\partial y^2} + g\rho\beta(t - t_\infty) \tag{2.2.6}$$

$$\rho c_p u\frac{\partial t}{\partial x} = k\frac{\partial^2 t}{\partial y^2} \tag{2.2.7}$$

These are still partial differential equations because of $\partial t/\partial x$ in Eq. (2.2.7). Since the temperature level in the thermal diffusion region changes from t_∞ at $x = 0$ to $t(x, y)$ at $x = L$, the quantity $\partial t/\partial x$ is approximated as $(t - t_\infty)/L$. The x dependence of t is thereby lost and $t(x, y) = t(y)$. Finally, expressing $t(y)$ as $\phi = [t(y) - t_\infty]/(t_0 - t_\infty)$, the now ordinary differential equations in coordinate y become

$$\frac{d^2u}{dy^2} = -\frac{g\rho\beta(t_0 - t_\infty)\phi}{\mu} \tag{2.2.8}$$

$$u\phi = \frac{kL}{\rho c_p}\frac{d^2\phi}{dy^2} \tag{2.2.9}$$

The boundary conditions for $u(y)$ and $\phi(y)$ in a quiescent ambient medium are

$$u(0) = u(\infty) = \phi(\infty) = 1 - \phi(0) = 0 \tag{2.2.10}$$

where ∞ means some still indefinite distance y_∞ out from the surface. It is the location in the quiescent ambient medium where the effects of thermal transport and of the buoyancy-induced flow have largely disappeared.

Equations (2.2.8) and (2.2.9) are next converted to nondimensional form by the following transformations of y and U:

$$Y = y\left[\frac{gc_p\rho^2\beta(t_0 - t_\infty)}{\mu kL}\right]^{1/4} \quad \text{and} \quad U = u\left[\frac{\mu c_p}{kg\beta L(t_0 - t_\infty)}\right]^{1/2} \tag{2.2.11}$$

The equations and boundary conditions become

$$\frac{d^2U}{dY^2} = -\phi \quad \text{or} \quad U'' = -\phi \tag{2.2.12}$$

$$U\phi = \frac{d^2\phi}{dY^2} \quad \text{or} \quad U\phi = \phi'' \tag{2.2.13}$$

$$U(0) = U(\infty) = \phi(\infty) = 1 - \phi(0) = 0 \tag{2.2.14}$$

where the primes indicate differentiation.

The numerical calculation techniques routinely used today for such problems were not available to Lorenz. Instead, a convenient and accurate approximate formulation was used. Transforming the coordinate Y to s as

$$Y = \ln[1/(1 - s)] \tag{2.2.15}$$

yields the following relations for the distributions $U(s)$ and $\phi(s)$:

$$(1 - s)\frac{dU}{ds} - \frac{d^2U}{ds^2} = \phi \tag{2.2.16}$$

$$\phi U = (s - 1)\frac{d\phi}{ds} + \frac{d^2\phi}{ds^2} \tag{2.2.17}$$

$$U(0) = U(1) = \phi(1) = 1 - \phi(0) = 0 \tag{2.2.18}$$

Note that the unknown and troublesome y_∞, or Y_∞, has become $s = 1$ in the transformation, Eq. (2.2.15). Now $U(s)$ and $\phi(s)$ are approximated by polynomials as

$$U(s) = a_1 s + a_2 s^2 + \cdots \tag{2.2.19}$$

$$\phi(s) = 1 + b_1 s + b_2 s^2 + \cdots \tag{2.2.20}$$

where the a_1, a_2, ... and b_1, b_2, ... remain to be determined to satisfy Eqs. (2.2.16)–(2.2.18). However, the most important quantity in the analysis is the rate of heat transfer from the surface to the fluid, that is, the surface heat flux q''. This is calculated from the Fourier law of conduction at $y = 0 = s$ to be

$$q'' = -k \left.\frac{dt}{dy}\right)_0 = -k(t_0 - t_\infty)\frac{d\phi}{dy} = -\,k(t_0 - t_\infty)\frac{d\phi}{ds}\frac{ds}{dY}\frac{dY}{dy} \tag{2.2.21}$$

This is rewritten as a convection coefficient h and Nusselt number hL/k as follows:

$$\frac{q''}{(t_0 - t_\infty)}\frac{L}{k} = \frac{hL}{k} = \mathrm{Nu} = L\left[\left(-\frac{d\phi}{ds}\right)\frac{ds}{dY}\frac{dY}{dy}\right]_{s=0} \tag{2.2.22}$$

Note from Eq. (2.2.20) that $d\phi/ds = b_1$ at $s = 0$. Lorenz found that b_1 was approximated by 0.548. Also $ds/dY = 1$ at $s = 0$ and dY/dy is found from Eq. (2.2.11). Therefore, the Lorenz estimate of what has since been defined as the Nusselt number is

$$\mathrm{Nu} = L(-b_1)\frac{dY}{dy} = 0.548 \sqrt[4]{\frac{g c_p \rho^2 \beta L^3 (t_0 - t_\infty)}{\mu k}} \tag{2.2.23}$$

or, equivalently, in terms of additional dimensional parameters that have since been defined

$$\begin{aligned} \mathrm{Nu} &= 0.548 \sqrt[4]{\mathrm{Ra}} \\ &= 0.548 \sqrt[4]{\frac{gL^3}{\nu^2}\beta(t_0 - t_\infty)} \sqrt[4]{\frac{\mu c_p}{k}} \\ &= 0.548 \sqrt[4]{\mathrm{Gr}} \sqrt[4]{\mathrm{Pr}} \end{aligned} \tag{2.2.24}$$

Thus, the result may be expressed in terms of both the Grashof and Prandtl numbers. However, the Lorenz result shows them combined in a single parameter, the Rayleigh number. The Rayleigh number often suffices for approximate formulations of transport in buoyancy-induced flows in many different transport configurations.

The full significance of this pioneering result is seen in several ways. First, Eq. (2.2.24) yields

$$Q = q''L = 0.548\, kL(t_0 - t_\infty)\left[\frac{gc_p\rho^2\beta(t_0 - t_\infty)}{\mu kL}\right]^{1/4}$$
$$\propto x(t_0 - t_\infty)^{5/4} \tag{2.2.25}$$

This explains the failure of Newton's law of cooling. Further, Eqs. (2.2.23) and (2.2.24) are in very good agreement with measured thermal transport in fluids having Prandtl numbers around 1.0, when the experimental value of Gr is large and yet the flow is still laminar. Finally, an exact treatment of the full Eqs. (2.2.1)–(2.2.4) is possible through the boundary region approximations applicable for large Gr. The solution, given in Section 3.4, results in the following prediction for Pr = 1:

$$\text{Nu} = 0.535\sqrt[4]{\text{Gr}} \tag{2.2.26}$$

This is only 2% smaller than the result of Lorenz.

Successful though the Lorenz analysis and results are, they are misleading in a number of more particular matters. For example, it will be seen in Section 3.4 that the surface heat flux, predicted to be independent of vertical location x in Eq. (2.2.21), is calculated by boundary region theory to vary as $x^{-1/4}$. Then, for Pr << 1, Eq. (2.2.24) is a very inaccurate estimate of the Prandtl number effect. Further, both u and t vary in important ways with the vertical distance x. Finally, the normal component of velocity, v, is not zero. It is the flow component normal to the surface that entrains ambient fluid into the transport region. It supplies the fluid to an increasing flow downstream, that is, toward larger x.

Nevertheless, the Lorenz result was of great importance. Almost a half-century passed before boundary region theory was first applied to buoyancy-induced transport, by Schmidt et al. (1930). This theory has since been applied to a diversity of both internal and external flows. Succeeding chapters detail many such results.

2.3 BUOYANCY FORCE AND MOTION PRESSURE

The Lorenz analysis proceeds directly, through intuitive notions and reasonable postulates, to a quite accurate quantitative result. On the other hand, the formulations that are applied to varied and to more complicated configurations usually follow from relations that simultaneously embody the numerous general mechanisms. These are expressed in Eqs. (2.1.1)–(2.1.3). However, initially rather obscure additional steps must be made in Eq. (2.1.2) so that the buoyancy force **B** appears explicitly there. Consider the body force and pressure field terms in Eq. (2.1.2)

$$\rho\mathbf{g} - \nabla p$$

where $\mathbf{g} = (g_x, g_y, g_z)$ for the Cartesian coordinate system, is the gravitational force vector per unit mass. Here p is the local static pressure in the fluid. This is the pressure that could be measured, for example, by a device moving with the fluid. It is related to the stagnation pressure by the dynamic pressure. Recall that the local reference density is ρ_r, or $\rho_a = \rho_\infty$ in an external flow in an extensive ambient medium. This was discussed in Section 1.2 in evaluating the buoyancy force as

$$B = g(\rho_r - \rho) \qquad \text{or} \qquad g(\rho_\infty - \rho) \tag{2.3.1}$$

More generally, the vector buoyancy force per unit volume is given below. Recall that $\mathbf{g} = -g$ for the geometry and coordinates in Fig. 1.2.1.

$$\mathbf{B} = \mathbf{g}(\rho - \rho_r) \qquad \text{or} \qquad \mathbf{g}(\rho - \rho_\infty)$$

Now consider the hydrostatic pressure field p_h that would result in a fluid everywhere at rest but locally at the reference density chosen, ρ_r or ρ_∞. The relation that determines changes in this postulated hydrostatic pressure field p_h is

$$\nabla p_h = \mathbf{g}\rho_r$$

This is the consequence of Eq. (2.1.2) with $\mathbf{V}$ taken as zero. In Fig. 1.2.1, for example, $\mathbf{g} = (g_x, g_y, g_z) = (-g, 0, 0)$. Then

$$\nabla p_h = \left(\frac{dp_h}{dx}, 0, 0\right) = -g\rho_r \qquad \text{or} \qquad -g\rho_\infty \tag{2.3.2}$$

However, in general, we write

$$\begin{aligned} \rho\mathbf{g} - \nabla p &= \rho\mathbf{g} - \nabla p_h - \nabla(p - p_h) \\ &= \mathbf{g}(\rho - \rho_r) - \nabla(p - p_h) \\ &= \mathbf{B} - \nabla(p - p_h) \\ &= \mathbf{B} - \nabla p_m \end{aligned} \tag{2.3.3}$$

Thus, the first of the resulting two terms is the buoyancy force $\mathbf{B}$. The second term is the gradient of the difference between two pressures: p, the actual static pressure at any location, and p_h, the pressure that would pertain at the same point, given the definition or choice of ρ_r, in the absence of any motion that would result in the departure of the pressure field from the hydrostatic variation imposed by gravity. With buoyancy force and motion, the difference between these two, $p - p_h$, is the pressure change that arises through fluid motion. It is due to acceleration, viscous forces, and the buoyancy force. The difference $p - p_h$ is called the "motion" pressure field p_m. That is, the actual static pressure p is decomposed into p_h and p_m, as $p = p_h + p_m$. Motion pressure may also be looked on as the impetus to motion caused by the buoyancy force. Consider Eq. (2.1.2) when $\rho\mathbf{g} - \nabla p$ is replaced as in Eq. (2.3.3) and $\mathbf{V}$ is taken as zero. Then $\nabla p_m = \mathbf{B}$.

In an external flow, as in Figs. 1.1.1 and 1.1.2, ρ_r would be chosen as $\rho_a = \rho_\infty$. Then p_h is given by

$$\frac{dp_h}{dx} = -g\rho_\infty \tag{2.3.4}$$

and, denoting the x-direction unit vextor as $\mathbf{i}$,

$$\mathbf{g}\rho - \nabla p = \mathbf{i}g(\rho_\infty - \rho) - \nabla p_m \tag{2.3.5}$$

However, the last result in Eq. (2.3.3) is the proper general relation to substitute into Eq. (2.1.2), which in turn becomes

$$\rho \frac{D\mathbf{V}}{D\tau} = \mathbf{g}(\rho - \rho_r) - \nabla p_m + \mu\nabla^2\mathbf{V} + \frac{1}{3}\mu\nabla(\nabla \cdot \mathbf{V}) \tag{2.3.6}$$

In an external flow it might be a reasonable surmise that $p_m \leq 0$ everywhere in the fluid because the motion of the ambient fluid entrained by the flow is induced from the quiescent ambient by decreasing pressures. Also, viscous forces further oppose that motion. However, this is not a safe general assumption for all flow configurations and choices of ρ_r.

2.4 SOME CHARACTERISTIC QUANTITIES

Consider a flow due to density differences resulting from thermal and chemical species transport. The local buoyancy force $\mathbf{B} = \mathbf{g}(\rho - \rho_r)$ is calculated in terms of the local temperature and species concentration C. The effect of local pressure must also be accounted for if its variation in the domain of the flow has an important effect on density. A density relation for the fluid, $\rho(t, C, p)$, is necessary.

In many processes of great interest the resulting density difference is very small. This is the case in the atmosphere and in almost all naturally occurring terrestrial waters (see Section 1.1). An example is the so-called human boundary layer. It is driven upward by the surface dissipation of the thermal energy released metabolically in the body. This energy is carried away by rising currents of air around the body. The mechanisms are thermal transport and, simultaneously, transport of the water vapor released by perspiration. For a typical skin temperature, 33°C, in surrounding air at 25°C, the thermal contribution to $\Delta\rho/\rho$ is only about 0.03 or 3%. The enhanced humidity level due to perspiration produces a comparable additional contribution to $\Delta\rho/\rho$.

A purely thermally buoyant flow will be taken as an example to assess various characteristic flow quantities. It is the flow generated in air adjacent to a warm vertical surface, as in Fig. 2.2.1a. In Section 1.3 the maximum velocity at a height of 0.5 m in a flow in which $\Delta\rho/\rho = 0.02$ was given as $u = 0.17$ m/s. From this, the dynamic pressure may be calculated as $p_d = \rho u^2/2 = 1.44 \times 10^{-2}$ N/m^2 (pascal, Pa). This is 1.7×10^{-7} bar, where 1 atm = 1.0132 bars. This is very small. For example, it is equal to the change in hydrostatic pressure in an elevation difference of only 0.0015 m. However, this change in dynamic pressure was generated over a vertical distance of 50 cm for an average gradient in p_d, $\partial p_d/\partial x$, of only 3.5×10^{-2} N/m^3 (or Pa/m).

Another way to consider the magnitude of motion-associated pressure effects is in terms of the magnitude of the buoyancy force and what that implies about motion pressure and its variation in the induced flow field. The maximum buoyancy force per unit volume in the present example occurs at the surface and is damped by viscous forces. Nevertheless, its value is 0.25 N/m^3. In the absence of motion, as at the beginning of flow, $\mathbf{V} = 0$. Then it is seen from Eqs. (2.1.2) and (2.3.3), for this configuration, that

$$B = g(\rho_\infty - \rho) = g\,\Delta\rho = \frac{dp_m}{dx} = \frac{\Delta p_m}{\Delta x} \tag{2.4.1}$$

Since $p_m = 0$ below the region of thermal diffusion and, therefore, of buoyancy, the value of p_m at $x = 0.5$ m may be calculated as 0.13 N/m^2. This is again a very small quantity compared to the hydrostatic variation, which is 4.9 N/m^2. Yet it is about 10 times as large as the dynamic pressure generated at this location in the eventual steady flow, as calculated above. Thus, the potential of the buoyancy force to generate motion has been greatly reduced by viscous forces and by the necessity to accelerate fluid downstream, including that continually entrained from the largely quiescent ambient medium.

These matters are very important in determining the principal functions of the buoyancy force and assessing the effects of flow-caused pressure differences and motion pressure. Such weak flows are subject to subtle effects. Yet various small effects may be ignored in many flows. It is usually necessary to be very careful in such matters.

2.5 DETERMINATION OF DENSITY EFFECTS

In a thermally buoyant flow the local density appears in the force-momentum balance, Eq. (2.3.6), in the driving force as $\mathbf{g}(\rho - \rho_r)$. The local density ρ also appears in four other places in the total formulation of transport. The local level of the density appears as a coefficient in the mass, force-momentum, and energy balances, Eqs. (2.1.1)–(2.1.3). It also appears as a temporally and spatially changing quantity on the left-hand side of Eq. (2.1.1).

For forced flows, driven by large velocities or pressure differences, the level of density may vary widely over the flow field. It often becomes another principal variable, like p, $\mathbf{V}$, and t. Its distribution is then determined by Eqs. (2.1.1)–(2.1.3) in conjunction with a given density equation of state $\rho(t, p)$. A strong driving effect results in a nonuniform pressure distribution and so forth.

However, in most terrestrial buoyancy-induced flows the only driving force, $\mathbf{B}$, is relatively very small. The resulting variation in the density level, $\Delta\rho$, is also often relatively very small compared to ρ. Recall the results of Section 2.4. The present section will first evaluate how the density difference in the buoyancy force is to be related to the temperature and pressure fields. Then the effects of the density variation as it appears elsewhere in the equations will be considered.

In Section 2.3 the buoyancy force $\mathbf{B}$ was determined from the body force $\rho\mathbf{g}$ and pressure force ∇p. The pressure gradient was rewritten in terms of motion pressure $p_m = p - p_h$ to obtain

$$\mathbf{B} = \mathbf{g}(\rho - \rho_r) = -g(\rho_r - \rho) \tag{2.5.1}$$

Once a reference condition has been defined to fix ρ_r, an equation of state for the density of the fluid is necessary, in principle, to determine **B**. Both thermal and/or any species transport effects on density must be accounted for.

For an external flow in an extensive ambient medium, ρ_r at each elevation may be taken as the local ρ_∞. Then a simple and convenient way to express the density difference with combined thermal and concentration buoyancy modes would be

$$(\rho_r - \rho) = (\rho_\infty - \rho) = \rho'\beta(t - t_\infty) + \rho'\beta^*(C - C_\infty) = \Delta\rho = \Delta\rho_t + \Delta\rho_C \tag{2.5.2}$$

where $\rho_\infty = \rho(t_\infty, C_\infty)$, ρ' is some suitable density level, and β and β^* are the values necessary to make Eq. (2.5.2) correct for the choice of ρ'. If density varies linearly with both t and C over the ranges encountered in the transport process, β and β^* are simply

$$\beta = -\frac{1}{\rho}\left(\frac{\partial\rho}{\partial t}\right)_{p,C} \tag{2.5.3}$$

$$\beta^* = -\frac{1}{\rho}\left(\frac{\partial\rho}{\partial C}\right)_{p,t} \tag{2.5.4}$$

where both β and β^* may be evaluated anywhere in the ranges $t_0 - t_\infty$ and $C_0 - C_\infty$, where t_0 and C_0 are the other bounding conditions on the flow. If it is assumed that motion pressure p_m is sufficiently small to have a negligible effect on density, then the expansion coefficients β and β^* may be evaluated at $p_h = p_\infty$. The discussion in Section 2.4 is a guide for this approximation.

Density approximations. For many actual fluids and flow conditions Eqs. (2.5.2)–(2.5.4) are good approximations, especially for $t_0 - t_\infty$ and $C_0 - C_\infty$ small. The approximation for the thermal buoyancy effect in Eqs. (2.5.2) and (2.5.3) is one of the Boussinesq approximations (Boussinesq, 1903). Reference is also made to the earlier detailed treatment by Oberbeck (1879); other recent considerations of these approximations are referred to later. However, if the density variation is substantially nonlinear in t and C over the ranges of their values in the buoyancy region, then the expressions for β and β^* must in general be much more complicated to yield an accurate representation in Eq. (2.5.2). This occurs for large temperature differences in any fluid. It also may arise, for example, in thermally driven motions in cold water. Recall that a density extremum occurs at about 4°C. If $t_0 = 0$°C and $t_\infty = 6$°C, the density variation is not even approximately linear. The value of $\beta(t)$ is positive at 6°C, zero somewhere around 4°C, and negative at 0°C. No approximation such as Eq. (2.5.2) would suffice. A density equation must be used.

The frequent suitability of Eqs. (2.5.2) and (2.5.3) as an approximation for analysis in a thermally driven flow, when $\rho = \rho(t, p)$, will be assessed in terms of flow parameters. The density difference in Eq. (2.5.2) at any given elevation, $\rho_\infty(t_\infty, p_\infty) - \rho(t, p)$, is written below as an expansion in terms of $(t_\infty - t)$ and $(p_\infty - p) = p_m$.

$$\rho_\infty - \rho = \left(\frac{\partial \rho}{\partial t}\right)_p (t_\infty - t) + \frac{1}{2!}\left(\frac{\partial^2 \rho}{\partial t^2}\right)_p (t_\infty - t)^2 + \cdots$$
$$+ \left(\frac{\partial \rho}{\partial p}\right)_t (p_\infty - p) + \frac{1}{2!}\left(\frac{\partial^2 \rho}{\partial p^2}\right)_t (p_\infty - p)^2 + \cdots$$
$$+ \frac{\partial^2 \rho}{\partial t\, \partial p}(t_\infty - t)(p_\infty - p) + \cdots \tag{2.5.5}$$

First, considering the fluid to be an ideal gas, that is, $p = \rho RT$ and $\beta = 1/T$, Eq. (2.5.5) becomes

$$\rho_\infty - \rho = \rho\beta(t - t_\infty)\,[1 + \beta(t - t_\infty) + \beta^2(t - t_\infty)^2 + \cdots]$$
$$+ \frac{p_m}{RT} + \left[\frac{(t - t_\infty)(p_\infty - p)}{RT^2} + \text{ACT}\right]$$
$$= \frac{\rho\beta(t - t_\infty)}{1 - \beta(t - t_\infty)} + \frac{\beta p_m}{R}\,[1 + \beta(t - t_\infty) + \text{ACT}] \tag{2.5.6}$$

where ACT means additional cross terms of the expansion. Recall that the motion pressure gradient was estimated in Section 2.4 as $g(\rho_\infty - \rho)$. Thus, p_m at some height $x = L$ is merely $gL(\rho_\infty - \rho)$ from Eq. (2.4.1). Then Eq. (2.5.6) may be rewritten as

$$\frac{\rho_\infty - \rho}{\rho}\left\{1 - \frac{g\beta L}{R}\,[1 + \beta(t - t_\infty)]\right\} = \frac{\beta(t - t_\infty)}{1 - \beta(t - t_\infty)} + \text{ACT} \tag{2.5.7}$$

Therefore, the accuracy of the approximation of $\rho_\infty - \rho$ in Eq. (2.5.2) for the temperature effect on density depends on the magnitude of the following quantities, from Eq. (2.5.7), compared to 1. The first quantity below, R_0', is the motion pressure effect on ρ. The second quantity, R_1, is the first term in the approximation for the temperature effect in Eq. (2.5.5) on $\rho_\infty - \rho$, divided by ρ'.

$$R_0' = \frac{g\beta L}{R} = \frac{g\beta L}{c_p}\,\frac{\gamma}{\gamma - 1} = R_0\,\frac{\gamma}{\gamma - 1} \tag{2.5.8}$$

where $\gamma = c_p/c_v$ and R_0' is transformed to R_0 for later convenience.

$$R_1 = \beta(t - t_\infty) = \frac{1}{\rho'}\left(\frac{\partial \rho}{\partial t}\right)_p (t - t_\infty) \tag{2.5.9}$$

where R_1 becomes $(t - t_\infty)/T \simeq (t_0 - t_\infty)/T$ for an ideal gas. However, in general, R_1 is written as

$$R_1 = \frac{\rho_\infty - \rho}{\rho'} \le \frac{\rho_\infty - \rho_0}{\rho'} = \frac{\Delta\rho}{\rho'} \tag{2.5.10}$$

where $\Delta\rho$ is the characteristic density difference and ρ' is a density level.

For terrestrial gravity and for air at $T = 300$ K, $R_0 = 1.2 \times 10^{-4} L(\text{m})$. Then for $t_0 - t_\infty = 10°\text{C}$, $R_1 \leq 0.03$. Clearly, $R_0 << R_1$ are both small compared to 1 for many common buoyancy-induced flows in gases. Thus **B** may often be written with sufficient accuracy simply as

$$\mathbf{B} = -\mathbf{g}(\rho_\infty - \rho) = -\mathbf{g}\rho'\beta(t - t_\infty) \tag{2.5.11}$$

The expansion in Eq. (2.5.5) also applies for liquids. However, the relative magnitudes of the various terms may not be accurately assessed in a general way. There is no widely applicable density equation for liquids that can be used to evaluate the temperature and pressure effects on density. However, we may compare the first pressure and first temperature term in Eq. (2.5.5) with each other as follows:

$$R_0' = \frac{\partial\rho/\partial p)_t\,(p_\infty - p)}{\partial\rho/\partial t)_p\,(t_\infty - t)} \simeq \frac{gL(\rho_\infty - \rho)\,\partial\rho/\partial p)_t}{\rho\beta(t - t_\infty)} \simeq gL\left.\frac{\partial\rho}{\partial p}\right)_t \tag{2.5.12}$$

Clearly R_0' becomes R_0' in Eq. (2.5.8) for an ideal gas. However, R_0' for a liquid is much less than for a gas since the pressure effect on density, $\partial\rho/\partial p)_t$, is very much less. Thus, the purely pressure effects on density may be neglected in calculating $\rho_\infty - \rho$, except perhaps in very extreme circumstances, as near the thermodynamic critical point of a vapor-liquid system. Further arguments show that both the higher-order temperature and pressure terms in Eq. (2.5.5) and all cross terms may also be neglected to a good approximation, for liquids, when $\beta(t - t_\infty) = R_1$ is much less than 1. Therefore, Eq. (2.5.11) also applies widely for buoyancy-induced motions in liquids. It is interesting to note that R_1 for a liquid is commonly much less than R_1 for a gas for given bounding temperatures t_0 and t_∞.

Another observation at this point is that the Boussinesq approximation for the temperature effect on density also applies widely to the transport of a chemical species at a low concentration difference $C_0 - C_\infty$. This measure was postulated in Eq. (2.5.2) in the evaluation of $\Delta\rho_C$. Arguments similar to those for $\Delta\rho_t$ also apply to make Eq. (2.5.2) frequently an equally good approximation. This is discussed further in Chapter 6.

The next most intricate appearance of density is in the mass balance Eq. (2.1.1), which is written in Cartesian coordinates as

$$\frac{\partial\rho}{\partial\tau} + u\frac{\partial\rho}{\partial x} + v\frac{\partial\rho}{\partial y} + w\frac{\partial\rho}{\partial z} + \rho\frac{\partial u}{\partial x} + \rho\frac{\partial v}{\partial y} + \rho\frac{\partial w}{\partial z} = 0 \tag{2.1.1}$$

Consider, for example, the flow of Fig. 2.2.1. The ratio R_2 written below expresses the relation between two principal terms in Eq. (2.1.1), in a steady laminar flow in this geometry. This ratio will be evaluated approximately from foregoing estimates, including a possible density stratification in the ambient medium, that is, $\partial\rho_\infty/\partial x \neq 0$.

$$R_2 = \frac{u\,\partial\rho/\partial x}{\rho\,\partial u/\partial x} \simeq \frac{u\,\partial\rho/\partial x}{\rho(u/x)} = \frac{x}{\rho}\frac{\partial\rho}{\partial x} \tag{2.5.13}$$

This is rewritten, allowing for stratification and evaluating $\rho_\infty - \rho$ as in Eq. (2.5.11):

$$R_2 = -\frac{x}{\rho}\frac{\partial}{\partial x}[(\rho_\infty - \rho) - \rho_\infty] = -\frac{x}{\rho}\frac{\partial}{\partial x}[\rho\beta(t - t_\infty)] + \frac{x}{\rho}\frac{\partial \rho_\infty}{\partial x}$$

$$= -\frac{x}{\rho}R_1\frac{\partial \rho}{\partial x} - x\frac{\partial R_1}{\partial x} + \frac{x}{\rho}\frac{\partial \rho_\infty}{\partial x} \qquad (2.5.14)$$

Since it is implicit that $(\rho_\infty - \rho)/\rho$ is much less than 1, the two $1/\rho$ coefficients may be taken as $1/\rho_\infty$ and Eq. (2.5.14) may be rewritten as

$$R_2 = -O\left(R_1\frac{x}{H}\right) - O\left(R_1\frac{x}{H'}\right) + O\left(\frac{x}{H}\right) \qquad (2.5.15)$$

where O indicates the order of magnitude of each term and H' and H are the stratification lengths of $R_1 = \beta(t - t_\infty) \leq \beta(t_0 - t_\infty)$ and ρ_∞, respectively:

$$H' = \left(\frac{1}{R_1}\frac{\partial R_1}{\partial x}\right)^{-1} \quad \text{and} \quad H = \left(\frac{1}{\rho_\infty}\frac{\partial \rho_\infty}{\partial x}\right)^{-1} \qquad (2.5.16)$$

The magnitude of H' depends entirely on how $t_0 - t_\infty$ varies downstream, based on t_∞ stratification and/or changing t_0, which is the other imposed bounding condition. On the other hand, H expresses only the degree of ambient medium density stratification. The imposed stratification may be large or small. However, it must be stable for quiescence of the ambient medium. It is small—that is, H is very large compared to $x = L$—in many circumstances of interest. For example, it will later be seen that $H = H_a = c_pRT/c_vg = \gamma RT/g$ in an adiabatically stratified ideal gas. Thus, x/H (or L/H) is extremely small as seen below, where

$$\frac{L}{H_a} = \frac{Lg}{\gamma RT} = \frac{R_0'}{\gamma} = \frac{1.2 \times 10^{-4}\,L}{\gamma} \simeq 1 \times 10^{-4}L$$

using the estimate of R_0' above.

In summary, both x/H' and x/H are very small except for extreme stratifications of R_1 and ρ_∞. For moderate or no stratifications, the continuity Eq. (2.1.1) becomes

$$\frac{\partial \rho}{\partial \tau} + \rho\nabla \cdot \mathbf{V} = 0 \qquad (2.5.17)$$

The remaining question concerns the relative magnitude of the transient term. This would arise as a local density change due to transient imposed boundary conditions. Circumstances may arise in which this term is locally very large. In steady laminar flow this term is zero. Its relative magnitude when transient bounding conditions are imposed may be assessed.

A further consideration concerns density level effects in the coefficients in Eqs. (2.1.2) and (2.1.3). This is a complicated matter that depends on the kind of transport process considered as well as the particular fluid. These questions will be addressed in Chapter 8, which considers the effects of variations of c_p, μ, and k over the region of transport. It is sufficient here to note that the total variation of a ρ across the region is $\rho_\infty - \rho_0 \simeq \rho'\beta(t_0 - t_\infty)$. Thus, the fractional change in the density as a coefficient is only

$$\frac{\rho_\infty - \rho_0}{\rho'} = \beta(t_0 - t_\infty) \geq R_1 \tag{2.5.18}$$

For R_1 much less than 1, this effect is assumed to be small. However, its magnitude will be estimated later when the effects of variable properties are considered.

Considering density as a constant when it appears as a coefficient and approximating the full continuity equation, Eq. (2.1.1), as in Eq. (2.5.17), are additional steps in the total set of Boussinesq approximations. The first step was in Eq. (2.5.11) in calculating the buoyancy force in terms of $t - t_\infty$. Finally, although the above assessments of the Boussinesq approximations are based on a particular configuration of flow, they are widely applicable.

2.6 SOME OTHER PHYSICAL EFFECTS

The foregoing evaluations of the effects of density variation have permitted conditional simplifications of the equations from their general forms given in Section 2.1. Some further simplifications are also admissible in many of the most important buoyancy-induced flows. They concern the pressure and viscous dissipation terms in Eq. (2.1.3), which resulted from an energy balance. The magnitude of each will be assessed by comparison to other terms in Eq. (2.1.3) that are known to be important thermal transport effects in reasonably vigorous flows. These are the thermal convection and conduction terms. Again, consider a steady laminar flow like that in Fig. 2.2.1 as a convenient example, although the results are not limited to this geometry or flow.

A principal component of the pressure term in Eq. (2.1.3) is

$$\beta T u \frac{\partial p}{\partial x} \qquad \text{from} \qquad \beta T \frac{Dp}{D\tau}$$

A principal convection term is

$$\rho c_p u \frac{\partial t}{\partial x} \qquad \text{from} \qquad \rho c_p \frac{Dt}{D\tau}$$

The magnitude of the ratio of the two, R_3, is assessed:

$$R_3 = \frac{\beta T u \; \partial p/\partial x}{\rho c_p u \; \partial t/\partial x} = \frac{\beta T}{\rho c_p} \frac{\partial p_h/\partial x + \partial p_m/\partial x}{\partial t/\partial x} \simeq \beta T \frac{R_0}{R_1} \tag{2.6.1}$$

where $R_0 = g\beta L/c_p$, p_m is estimated as in Eq. (2.4.1), and $\Delta\rho$ is still considered very much smaller than ρ and $\partial t/\partial x$ is estimated as $(t_0 - t_\infty)/L$.

The product βT is 1 for an ideal gas and much less than 1 for most conditions with liquids. Therefore, the pressure terms may be neglected in Eq. (2.1.3) for $R_0/R_1 << 1$ or for

$$R_0 << R_1 \tag{2.6.2}$$

This is not a generous condition, since previous arguments have shown that R_1 is commonly much less than 1. However, R_0, following Eq. (2.5.8), was seen to be very small for terrestrial gravity.

Analogously, the viscous dissipation effect $\mu\Phi$ in Eq. (2.1.3) is compared with conduction $\nabla \cdot k\nabla t = k\nabla^2 t$ for k uniform throughout the fluid. Principal terms in each effect are identified and their ratio R_4 is formed and evaluated as done above for R_3:

$$\mu\left(\frac{\partial u}{\partial y}\right)^2 \quad \text{from} \quad \mu\Phi$$

$$k\frac{\partial^2 t}{\partial y^2} \quad \text{from} \quad k\nabla^2 t$$

$$R_4 = \frac{\mu(\partial u/\partial y)^2}{k\,\partial^2 t/\partial y^2} = O\left[\frac{\mu u^2}{k(t_0 - t_\infty)}\right] \tag{2.6.3a}$$

$$= O\left(\frac{\mu c_p}{k}\frac{g\beta L}{c_p}\right) = O(\text{Pr}R_0) \tag{2.6.3b}$$

recalling from Section 1.3 that in general, for $\text{Pr} \leq 1$

$$\frac{\rho u^2}{2} \simeq gL(\rho_\infty - \rho) \simeq \rho g L\beta(t_0 - t_\infty) \tag{2.6.4}$$

Thus, viscous dissipation may be neglected for $\text{Pr}R_0 << 1$ for $\text{Pr} \leq 1$. However, this conclusion also holds for large Prandtl number fluids as well when $R_0 << 1$, since Eq. (2.6.4) is a substantial overestimate of u^2 in fluids of high viscosity. This will be shown in Section 3.8. From Eq. (3.8.12) developed there for $\text{Pr} >> 1$,

$$\rho\frac{u^2}{2} \approx \rho g\beta L(t_0 - t_\infty)\text{Pr}^{-1} \tag{2.6.5}$$

Therefore, from Eq. (2.6.3*a*),

$$R_4 = O(R_0) \tag{2.6.6}$$

Thus, even for $\text{Pr} >> 1$, R_4 is independent of Pr and the viscous dissipation effects may be neglected when $R_0 << 1$. It can also be shown (Mahajan and Gebhart, 1988) that for moderate values of Pr, R_4 is at most $O(R_0)$. Combining this result with Eq. (2.6.3*b*), it is clear that for the whole range of Pr, $(R_4)_{\max} = O(R_0)$. Viscous dissipation effects may, therefore, always be neglected if $R_0 << 1$.

Next, the principal viscous dissipation and pressure terms are compared directly with each other, as R_5:

$$R_5 = \frac{\mu(\partial u/\partial y)^2}{\beta T u \,\partial p/\partial x} \simeq \frac{R_4}{R_3} = \frac{\Pr R_1}{\beta T} \tag{2.6.7}$$

For ideal gases $\beta T = 1$ and Pr is around 1. Therefore, viscous dissipation effects are much smaller than pressure effects if R_1 is much less than 1. For liquids, βT is usually substantially less than 1. However, since Pr is very much less than 1 for liquid metals, viscous dissipation effects are again much smaller than pressure effects. For other liquids, the Prandtl number ranges from greater than 1 to very large values for highly viscous liquids.

However, since for these liquids $R_4 = O(R_0)$, it follows from Eq. (2.6.7) that R_5 is at the most of the order of $R_1/\beta T = (t - t_\infty)/T \simeq (t_0 - t_\infty)/T$. For most applications the temperature difference $t_0 - t_\infty$ is small compared to T. Then viscous dissipation effects are again smaller than pressure effects. Thus, R_5 indicates that viscous dissipation may not consistently be retained in the equation if the pressure term is omitted.

The useful simplifications that result if R_1 is much less than 1 and if $R_0 << 1$ are that both the pressure and dissipation terms may generally be omitted from the energy equation. These approximations and those of the preceding section are applied in the last section of this chapter to set forth the equations that apply to almost all of the terrestrial buoyancy-induced flows encountered.

2.7 USEFUL EQUATIONS AND BOUNDING CONDITIONS

The first section of this chapter set forth the conventional general equations for transport in fluids, Eqs. (2.1.1)–(2.1.4) and (2.1.8). These results form the conservation laws for mass, force-momentum, energy, and a single diffusing chemically distinct species at a low concentration difference. In subsequent sections, characteristic quantities were determined for vertical laminar flows as a basis for estimating suitably accurate simplifications of the full equations. A number of opportunities were seen to arise. They are summarized in Table 2.7.1. Many other permissible simplifying approximations arise. However, most of them apply to more limited circumstances and will be considered when their use is appropriate.

Among the foregoing approximations, different ones apply to different common kinds of transport of wide interest. The sets of resulting equations are collected below for reference for a number of common kinds of processes.

The most notable and helpful approximation is that which results if $R_0 = g\beta L/c_p$ and $R_1 = \Delta\rho/\rho$ are much less than 1. It greatly simplifies the treatment of thermal and/or species transport effects on the buoyancy force. Density must also be sufficiently linearly dependent on t and/or C. The resulting equations for constant and uniform viscosity when the molecular transport parameters k and D, as well as c_p, may be variable are

Table 2.7.1 Approximations in external vertical laminar buoyancy-induced flows (at terrestrial gravity)

Small parameter	Other requirements	General result
$R_0 = \dfrac{g\beta L}{c_p} \propto R_0' = \dfrac{g\beta L}{R}$		Ignore motion pressure effect on density, Eq. (2.5.6)
$R_1 = \dfrac{\Delta\rho}{\rho}$	Density variation sufficiently linear with both t and C	Linear dependence of $\rho_\infty - \rho$ on both t and C, Eq. (2.5.2). The coefficients ρ in Eqs. (2.1.1)–(2.1.3) taken as a suitable constant value
R_2; ratio of largest terms in $\mathbf{V}\cdot\nabla\rho$ and $\rho\nabla\cdot\mathbf{V}$	No extreme vertical variation of either R_1 or ρ_∞	Neglect $\mathbf{V}\cdot\nabla\rho$ in Eq. (2.1.1)
$R_3 \propto R_0/R_1$; ratio of pressure and convection energy effects	$R_0 << R_1$	Neglect $\beta T \dfrac{Dp}{D\tau}$ in Eq. (2.1.3)
$R_4 = O(\mathrm{Pr}R_0)$; ratio of viscous dissipation and conduction energy effects	$R_0 << 1$ for Pr >> 1	Neglect $\mu\Phi$ in Eq. (2.1.3)
$R_5 = O(R_4/R_3)$; ratio of viscous dissipation and pressure energy effects	For all Pr	$\mu\Phi$ is smaller than $\beta T \dfrac{Dp}{D\tau}$

$$\frac{D\rho}{D\tau} = -\rho\nabla\cdot\mathbf{V} \qquad \text{or} \; -\frac{\partial\rho}{\partial\tau} = \mathbf{V}\cdot\nabla\rho + \rho\nabla\cdot\mathbf{V} \tag{2.7.1}$$

$$\rho\frac{D\mathbf{V}}{D\tau} = \mathbf{B} - \nabla p_m + \mu\nabla^2\mathbf{V} + \frac{1}{3}\mu\nabla(\nabla\cdot\mathbf{V}) \tag{2.7.2}$$

where

$$\mathbf{B} = -\mathbf{g}\rho[\beta(t - t_\infty) + \beta^*(C - C_\infty)] \tag{2.7.3}$$

$$\rho c_p\frac{Dt}{D\tau} = \nabla\cdot k\nabla t + \beta T\frac{Dp}{D\tau} + \mu\Phi + q''' \tag{2.7.4}$$

$$\frac{DC}{D\tau} = \nabla\cdot D\nabla C + c''' \tag{2.7.5}$$

Recall that the coefficient ρ in Eqs. (2.7.2)–(2.7.4) may be taken as a suitable constant value for R_1 small.

If the parameter related to the vertical stratification, R_2, is small and transient effects are negligible, Eq. (2.7.1) becomes simply $\nabla \cdot \mathbf{V} = 0$. If in addition R_3 and R_4 are small, the equations are

$$\nabla \cdot \mathbf{V} = 0 \tag{2.7.6}$$

$$\rho \frac{D\mathbf{V}}{D\tau} = -\mathbf{g}\rho[\beta(t - t_\infty) + \beta^*(C - C_\infty)] - \nabla p_m + \mu \nabla^2 \mathbf{V} \tag{2.7.7}$$

$$\rho c_p \frac{Dt}{D\tau} = \nabla \cdot k \nabla t + q''' \tag{2.7.8}$$

$$\frac{DC}{D\tau} = \nabla \cdot D \nabla C + c''' \tag{2.7.9}$$

For a two-dimensional steady laminar vertical plane flow, as in Fig. 2.2.1, $-\mathbf{g} = (g, 0, 0)$ and the above equations become, when viscosity μ is considered variable,

$$\frac{\partial u}{\partial x} + \frac{\partial v}{\partial y} = 0 \tag{2.7.10}$$

$$\rho\left(u \frac{\partial u}{\partial x} + v \frac{\partial u}{\partial y}\right) = -\frac{\partial p_m}{\partial x} + 2\frac{\partial}{\partial x}\left(\mu \frac{\partial u}{\partial x}\right) + \frac{\partial}{\partial y}\left[\mu\left(\frac{\partial u}{\partial y} + \frac{\partial v}{\partial x}\right)\right] + g\rho[\beta(t - t_\infty) + \beta^*(C - C_\infty)] \tag{2.7.11}$$

$$\rho\left(u \frac{\partial v}{\partial x} + v \frac{\partial v}{\partial y}\right) = -\frac{\partial p_m}{\partial y} + 2\frac{\partial}{\partial y}\left(\mu \frac{\partial v}{\partial y}\right) + \frac{\partial}{\partial x}\left[\mu\left(\frac{\partial u}{\partial y} + \frac{\partial v}{\partial x}\right)\right] \tag{2.7.12}$$

$$\rho c_p\left(u \frac{\partial t}{\partial x} + v \frac{\partial t}{\partial y}\right) = \frac{\partial}{\partial x}\left(k \frac{\partial t}{\partial x}\right) + \frac{\partial}{\partial y}\left(k \frac{\partial t}{\partial y}\right) + q''' \tag{2.7.13}$$

$$u \frac{\partial C}{\partial x} + v \frac{\partial C}{\partial y} = \frac{\partial}{\partial x}\left(D \frac{\partial C}{\partial x}\right) + \frac{\partial}{\partial y}\left(D \frac{\partial C}{\partial y}\right) + c''' \tag{2.7.14}$$

If the molecular diffusion quantities μ, k, and D are uniform and constant over the field of transport, these equations become

$$\frac{\partial u}{\partial x} + \frac{\partial v}{\partial y} = 0 \tag{2.7.10}$$

$$u \frac{\partial u}{\partial x} + v \frac{\partial u}{\partial y} = \nu\left(\frac{\partial^2 u}{\partial x^2} + \frac{\partial^2 u}{\partial y^2}\right) - \frac{1}{\rho}\frac{\partial p_m}{\partial x} + g\beta(t - t_\infty) + g\beta^*(C - C_\infty) \tag{2.7.15}$$

$$u \frac{\partial v}{\partial x} + v \frac{\partial v}{\partial y} = \nu\left(\frac{\partial^2 v}{\partial x^2} + \frac{\partial^2 v}{\partial y^2}\right) - \frac{1}{\rho}\frac{\partial p_m}{\partial y} \tag{2.7.16}$$

$$u \frac{\partial t}{\partial x} + v \frac{\partial t}{\partial y} = \alpha\left(\frac{\partial^2 t}{\partial x^2} + \frac{\partial^2 t}{\partial y^2}\right) + \frac{q'''}{\rho c_p} \tag{2.7.17}$$

$$u \frac{\partial C}{\partial x} + v \frac{\partial C}{\partial y} = D\left(\frac{\partial^2 C}{\partial x^2} + \frac{\partial^2 C}{\partial y^2}\right) + c''' \tag{2.7.18}$$

The general equations for μ, k, and D uniform are

$$\nabla \cdot \mathbf{V} = 0 \tag{2.7.19}$$

$$\frac{D\mathbf{V}}{D\tau} = -\mathbf{g}[\beta(t - t_\infty) + \beta^*(C - C_\infty)] - \frac{1}{\rho} \nabla p_m + \nu\nabla^2\mathbf{V} \tag{2.7.20}$$

$$\frac{Dt}{D\tau} = \alpha\nabla^2 t + \frac{q'''}{\rho c_p} \tag{2.7.21}$$

$$\frac{DC}{D\tau} = D\nabla^2 C + c''' \tag{2.7.22}$$

where α and ν are the thermal diffusivity and kinematic viscosity. Recall that $\mathrm{Pr} = \nu/\alpha$, the Schmidt number $\mathrm{Sc} = \nu/D$, and the Lewis number $\mathrm{Le} = \mathrm{Sc}/\mathrm{Pr} = \alpha/D$.

Each of the above sets of equations determines the distributions of $\mathbf{V}$, t, C, and p_m, over the region of transport. However, admissible distributions must conform to limiting bounding conditions characteristic of the particular circumstance. Some such conditions are general; others are particular. Among general ones for an external flow in an extensive quiescent ambient medium is that velocity $\mathbf{V}$ and sometimes also p_m approach zero far out in the medium, away from the source of the buoyancy-causing density disturbance. The differences $t - t_\infty$ and $C - C_\infty$ also go to zero there. In general, $\mathbf{V}$ is zero at any impervious surface because of a zero normal component and zero tangential component due to the no-slip condition. Imposed conditions at a surface either fix t_0 and/or C_0 or amount to a condition on thermal or species flux there. For external flows independent of a surface, such as a free-rising plume, analogous conditions may arise from considerations of symmetry and so forth.

Succeeding chapters will consider a wide range of kinds of transport. Many additional and more specific convenient approximations arise to facilitate many of the formulations. Such analysis always begins with specific simple bounding conditions characteristic of the particular transport process.

2.8 PARAMETERS OF TRANSPORT

The dimensionless parameters characteristic of buoyancy-induced flows are in some particulars different from those of forced flow transfer. The reason is that the characteristic velocity, which we call U_c, does not arise directly from an externally imposed forcing condition. Rather, it arises from other than imposed velocities or pressure differences. For example, it may be a consequence of imposed temperature conditions interacting with gravity. It is, therefore, expressed in terms of such conditions.

A general way to demonstrate the nature and origins of important parameters is to nondimensionalize the full set of applicable and relevant equations in terms of the characteristic quantities of any particular flow, for example, that in Fig. 2.8.1. The procedure is to determine the parameters on which transport depends. For example, a matter of interest in calculation is to determine the resulting convection coefficient h or Nusselt number $\text{Nu} = hL/k$. This would be done in a calculation by solving for $t(x, y, z, \tau)$, from the equations and boundary conditions, and then calculating the heat flux to the fluid at the fluid-surface interface. The flux would then be integrated over the surface A to determine the total heat flow rate Q.

The determination of the relevant parameters will start with the differential equations, Eqs. (2.7.19)–(2.7.22), for uniform μ, k, and D. However, the pressure and viscous dissipation terms in the energy balance will be included to determine any additional parameters on which they might depend. Recall that the particle derivative operator $D/D\tau$ may be written [e.g., see Eq. (2.1.1)] as

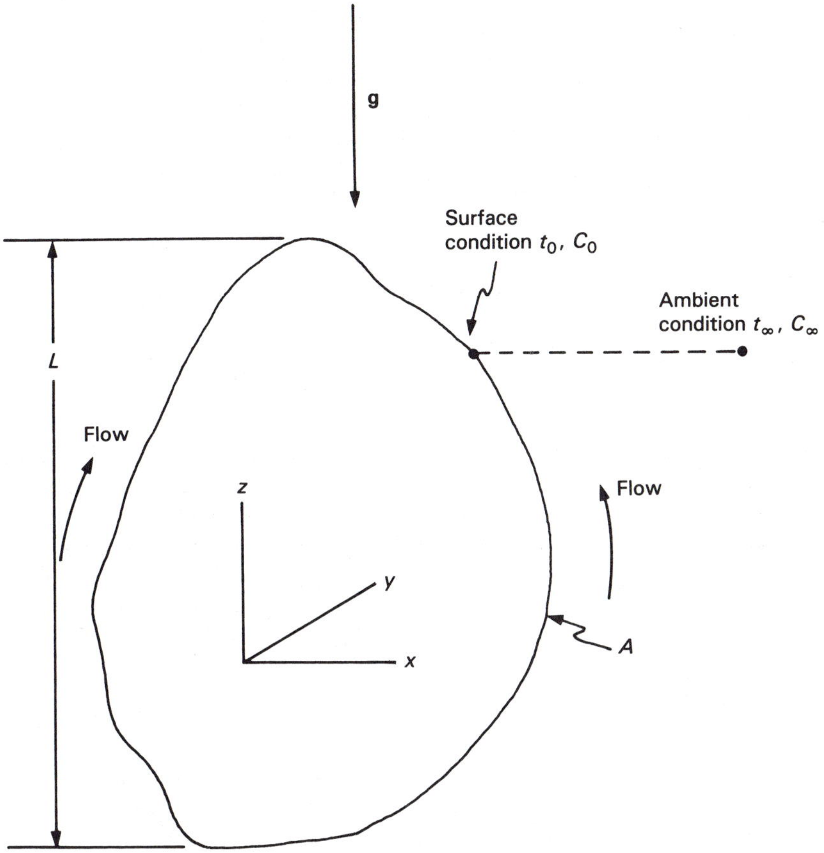

Figure 2.8.1 A buoyancy-induced flow generated adjacent to a solid body of surface area A.

$$\frac{D}{D\tau} = \frac{\partial}{\partial \tau} + (\mathbf{V} \cdot \boldsymbol{\nabla}) \tag{2.8.1}$$

The equations and boundary conditions for the conditions specified are then

$$\boldsymbol{\nabla} \cdot \mathbf{V} = 0 \tag{2.8.2}$$

$$\frac{\partial \mathbf{V}}{\partial \tau} + (\mathbf{V} \cdot \boldsymbol{\nabla})\mathbf{V} = -\mathbf{g}[\beta(t - t_\infty) + \beta^*(C - C_\infty)] - \frac{1}{\rho} \boldsymbol{\nabla} p_m + \nu \nabla^2 \mathbf{V} \tag{2.8.3}$$

$$\frac{\partial t}{\partial \tau} + (\mathbf{V} \cdot \boldsymbol{\nabla})t = \alpha \nabla^2 t + \frac{\beta T}{\rho c_p} \left[\frac{\partial p_m}{\partial \tau} + (\mathbf{V} \cdot \boldsymbol{\nabla})p \right] + \mu \Phi + \frac{q'''}{\rho c_p} \tag{2.8.4}$$

$$\frac{\partial C}{\partial \tau} + (\mathbf{V} \cdot \boldsymbol{\nabla})C = D\nabla^2 C + c''' \tag{2.8.5}$$

Note that $\partial p_h / \partial \tau = 0$ in Eq. (2.8.4), from the definition of p_h in Section 2.3. The boundary conditions on $\mathbf{V}$, t, and C at the surface, over the extent of A, are

$$\mathbf{V} = 0 \qquad t = t_0 \qquad C = C_0 \tag{2.8.6}$$

Far out in the quiescent ambient medium they are

$$\mathbf{V} \rightarrow 0 \qquad t \rightarrow t_\infty \qquad C \rightarrow C_\infty \tag{2.8.7}$$

The local heat flux q'' and total heat transfer rate Q would be calculated from $t(x, y, z, \tau)$ as

$$q'' = -k \left.\frac{\partial t}{\partial n}\right)_{n=0} \tag{2.8.8}$$

$$Q = \int_A q'' \, dA = -k \int_A \left.\frac{\partial t}{\partial n}\right)_{n=0} dA \tag{2.8.9}$$

where n is the local normal coordinate to the surface and $n = 0$ at the surface.

In finding the parameters on which Q, Nu, and so forth depend, the calculation will not be considered in detail. It will be done only in principle to see which parameters arise. This is done by writing the physical variables x, y, z, n, τ, $\mathbf{V}$, p_m, t, and C in terms of their characteristic values. These are L for the apparent length scale, τ_0 the time constant of any transient effect, and U_c, $t_0 - t_\infty$, $C_0 - C_\infty$, and A for the other characteristic values. The new variables are

$$X, Y, Z, N = \frac{x, y, z, n}{L} \qquad \tau' = \frac{\tau}{\tau_0} \qquad \mathbf{W} = \frac{\mathbf{V}}{U_c} \qquad P_m = \frac{p_m}{\rho U_c^2} \tag{2.8.10}$$

$$P_h = \frac{p_h}{\rho U_c^2} \qquad dA' = \frac{dA}{A} \qquad \phi = \frac{t - t_\infty}{t_0 - t_\infty} \qquad \text{and} \qquad \tilde{C} = \frac{C - C_\infty}{C_0 - C_\infty} \tag{2.8.11}$$

where τ_0 is the time constant for a starting transient that ends in a steady state.

For example, the transient might result from an internal heat source that increases to a value that is constant thereafter. Eventually, a steady state is achieved.

Substituting x, y, z, etc. from Eqs. (2.8.10) and (2.8.11) into the formulation, Eqs. (2.8.2)–(2.8.5), yields

$$\frac{U_c}{L}\nabla\cdot\mathbf{W} = 0$$

$$\frac{U_c}{\tau_0}\frac{\partial\mathbf{W}}{\partial\tau'} + \frac{U_c^2}{L}(\mathbf{W}\cdot\nabla)\mathbf{W} = -\mathbf{g}[\beta(t_0 - t_\infty)\phi + \beta^*(C_0 - C_\infty)\tilde{C}] - \frac{U_c^2}{L}\nabla P_m + \frac{\nu U_c}{L^2}\nabla^2\mathbf{W}$$

$$\frac{t_0 - t_\infty}{\tau_0}\frac{\partial\phi}{\partial\tau'} + \frac{U_c(t_0 - t_\infty)}{L}(\mathbf{W}\cdot\nabla)\phi = \frac{\alpha(t_0 - t_\infty)}{L^2}\nabla^2\phi + \frac{\beta T}{\rho c_p}\left[\frac{\rho U_c^2}{\tau_0}\frac{\partial P_m}{\partial\tau'} + \frac{\rho U_c^3}{L}(\mathbf{W}\cdot\nabla)(P_h + P_m)\right] + \frac{\mu U_c^2}{\rho c_p L^2}\Phi + \frac{q'''}{\rho c_p}$$

$$\frac{C_0 - C_\infty}{\tau_0}\frac{\partial\tilde{C}}{\partial\tau'} + \frac{U_c(C_0 - C_\infty)}{L}(\mathbf{W}\cdot\nabla)\tilde{C} = \frac{D(C_0 - C_\infty)}{L^2}\nabla^2\tilde{C} + c'''$$

In these equations, for example, $\nabla^2 = \partial^2/\partial X^2 + \partial^2/\partial Y^2 + \partial^2/\partial Z^2$ for a cartesian coordinate system.

The first equation is next divided by U_c/L. Each of the other three is divided by the coefficient of its respective molecular diffusion term on the right-hand side. Then the four differential equations and their transformed boundary conditions become

$$\nabla\cdot\mathbf{W} = 0 \tag{2.8.12}$$

$$\frac{L^2}{\nu\tau_0}\frac{\partial\mathbf{W}}{\partial\tau'} + \frac{U_c L}{\nu}(\mathbf{W}\cdot\nabla)\mathbf{W} = \frac{gL^2\beta(t_0 - t_\infty)}{\nu U_c}\phi + \frac{gL^2\beta^*(C_0 - C_\infty)}{U_c}\tilde{C} - \frac{U_c L}{\nu}\nabla P_m + \nabla^2\mathbf{W} \tag{2.8.13}$$

$$\frac{L^2}{\alpha\tau_0}\frac{\partial\phi}{\partial\tau'} + \frac{U_c L}{\alpha}(\mathbf{W}\cdot\nabla)\phi = \nabla^2\phi + \frac{\beta T U_c^2}{c_p(t_0 - t_\infty)}\left[\frac{L^2}{\alpha\tau_0}\frac{\partial P_m}{\partial\tau'} + \frac{U_c L}{\alpha}(\mathbf{W}\cdot\nabla)(P_h + P_m)\right] + \frac{\Pr U_c^2}{c_p(t_0 - t_\infty)}\Phi + \frac{q''' L^2}{k(t_0 - t_\infty)} \tag{2.8.14}$$

$$\frac{L^2}{D\tau_0}\frac{\partial \tilde{C}}{\partial \tau'} + \frac{U_c L}{D}(\mathbf{W}\cdot\nabla)\tilde{C} = \nabla^2\tilde{C} + \frac{c'''L^2}{D(C_0 - C_\infty)} \tag{2.8.15}$$

At the surface, for $\tau' \geq 1$,

$$\mathbf{W} = 0 \qquad \phi = 1 \qquad \tilde{C} = 1 \tag{2.8.16}$$

Far out in the quiescent ambient medium

$$\mathbf{W} \rightarrow 0 \qquad \phi \rightarrow 0 \qquad \tilde{C} \rightarrow 0 \tag{2.8.17}$$

Also, the local heat flux and total heat transfer rate now become

$$q'' = -\frac{k(t_0 - t_\infty)}{L}\left.\frac{\partial\phi}{\partial N}\right)_{N=0} \tag{2.8.18}$$

$$\frac{Q}{A(t_0 - t_\infty)}\frac{L}{k} = \frac{hL}{k} = \text{Nu} = \int_1 \left.\frac{\partial\phi}{\partial N}\right)_{N=0} dA' \tag{2.8.19}$$

The Sherwood number for mass transfer, Sh, is determined analogously below, where Q_C is the integral of m'' over the surface

$$m'' = -\frac{D(C_0 - C_\infty)}{L}\left.\frac{\partial\tilde{C}}{\partial N}\right)_{N=0}$$

$$\frac{Q_C}{A(C_0 - C_\infty)}\frac{L}{D} = \frac{h_c L}{D} = \text{Sh} = \int_1 \left.\left(\frac{\partial\tilde{C}}{\partial N}\right)\right)_{N=0} dA' \tag{2.8.20}$$

The apparently complicated Eqs. (2.8.12)–(2.8.15) are, in reality, quite simple. The same terms are present as before. However, each component other than the molecular diffusion terms $\nabla^2\mathbf{W}$, $\nabla^2\phi$, and $\nabla^2\tilde{C}$ is divided into two components, the coefficient and the generalized term. However, neither component alone reliably indicates the order of magnitude of the physical importance of the term. For example, ϕ in Eq. (2.8.13) is of order 1, whereas $\nabla^2\mathbf{W}$ is very large for vigorous flows. It is the product of the two components that indicates the relative magnitude of the terms in the equation. Nevertheless, the coefficient of the thermal buoyancy force in Eq. (2.8.13) is sometimes regarded as the ratio of the thermal buoyancy and the viscous forces. It will be seen later that it is accurately that ratio for many flows.

Recalling that $U_c^2 = O(gL\,\Delta\rho/\rho) = O[gL\beta(t_0 - t_\infty)]$ for a relatively vigorous and purely thermally driven flow of a fluid having a Prandtl number of about 1 or less, the coefficient of ϕ in Eq. (2.8.13) becomes

$$\frac{gL^2\beta(t_0 - t_\infty)}{\nu U_c} = O\left[\frac{gL^2\beta(t_0 - t_\infty)}{\nu\sqrt{gL\beta(t_0 - t_\infty)}}\right]$$

Thus, this coefficient becomes approximately the square root of the thermal Grashof number:

$$\sqrt{\frac{gL^3}{\nu^2}\beta(t_0 - t_\infty)} = \sqrt{\text{Gr}} \tag{2.8.21}$$

This is the unit Grashof number gL^3/ν^2 times the units of buoyancy $\beta(t_0 - t_\infty) \simeq \Delta\rho/\rho = R_1$. If the buoyancy force is instead entirely due to species diffusion, the second component of the buoyancy force in Eq. (2.8.13), in $\tilde{C}$, is taken. Then the Grashof number Gr_C for a Schmidt number of about 1 or less becomes

$$\text{Gr}_C = \frac{gL^3}{\nu^2}\beta^*(C_0 - C_\infty) \tag{2.8.22}$$

If buoyancy arises from both effects, then there are two supplementing values simultaneously, Gr and Gr_C.

Examining the other coefficients, the three for convection $(\mathbf{W}\cdot\boldsymbol{\nabla})$ operating on $\mathbf{W}$, ϕ, and $\tilde{C}$ are, respectively,

$$\frac{U_cL}{\nu} = O(\sqrt{\text{Gr}}) \tag{2.8.23}$$

$$\frac{U_cL}{\alpha} = O(\sqrt{\text{Gr}}\,\text{Pr}) \tag{2.8.24}$$

$$\frac{U_cL}{D} = O(\sqrt{\text{Gr}_C}\,\text{Sc}) \tag{2.8.25}$$

The coefficient of the transient temperature term in the energy equation is

$$\frac{L^2}{\alpha\tau_0} = \frac{1}{\text{Fo}} \tag{2.8.26}$$

where $\text{Fo} = \alpha\tau_0/L^2$ is analogous to the familiar conduction Fourier number. The transient flow and concentration coefficients are

$$\frac{L^2}{\nu\tau_0} = \frac{1}{\text{PrFo}} \quad \text{and} \quad \frac{L^2}{D\tau_0} = \frac{\text{Sc}}{\text{PrFo}} = \frac{\text{Le}}{\text{Fo}}$$

Considering the three pressure terms in the energy equation, the coefficient of the transient term $\partial P_m/\partial\tau'$ becomes

$$(\beta T)\frac{g\beta L}{c_p}\frac{L^2}{\alpha\tau_0} = \frac{(\beta T)R_0}{\text{Fo}} \tag{2.8.27}$$

The coefficient of the two convection terms in P_h and P_m becomes

$$(\beta T)\sqrt{\text{Gr}}\,\text{Pr}\frac{g\beta L}{c_p} = (\beta T)R_0\text{Pr}\sqrt{\text{Gr}} \tag{2.8.28}$$

Recall, however, that the term in P_m is usually very much smaller than that in P_h. The coefficient of the viscous dissipation term is

$$\frac{\text{Pr}U_c^2}{c_p(t_0 - t_\infty)} = R_0\text{Pr} \tag{2.8.29}$$

The two parameters characteristically associated with the distributed sources q''' and c''' are also apparent in Eqs. (2.8.14) and (2.8.15).

The total collection of dimensionless parameters that have arisen from the full equations is as follows:

Gr and/or Gr_C
Pr and/or Sc
Fo for transient effects
(βT)
$R_0 = g\beta L/c_p$

Now consider solving Eqs. (2.8.12)–(2.8.15) with Eqs. (2.8.16) and (2.8.17), for example, for ϕ. The resulting function $\phi(X, Y, Z, \tau')$ would also depend on the specific value of each of the dimensionless parameters above, since each appears in the equations. Therefore, since the calculated heat or mass transfer parameters in Eqs. (2.1.19) and (2.1.20) depend on ϕ (and $\tilde{C}$), they would also depend on these parameters. Note that the X, Y, and Z dependence disappears in integrating Eqs. (2.8.19) and (2.18.20) and τ' remains. As a result, for heat transfer the Nusselt number (and the Sherwood number also) has the following dependence

$$\text{Nu} = F(\tau', \text{Gr}, \text{Gr}_C, \text{Pr}, \text{Sc}, \text{Fo}, \beta T, R_0) \tag{2.8.30}$$

In steady flow, τ' and Fo disappear. In addition, βT and R_0 arise only in terms that are often of much smaller order than the principal ones of convection and molecular diffusion (Section 2.6). Therefore, the common dependence of Nu is much simpler, as

$$\text{Nu} = F(\text{Gr}, \text{Gr}_C, \text{Pr}, \text{Sc}) \tag{2.8.31}$$

or in purely thermal buoyant flows

$$\text{Nu} = F(\text{Gr}, \text{Pr}) \tag{2.8.32}$$

Experimental data would be cast in terms of these three parameters as a first step in organizing the data into a more compact and useful result. Note that the simple calculation of Section 2.2 resulted in this same form.

REFERENCES

Boussinesq, J. (1903). *Theorie Analytique de la Chaleur*, vol. 2, Gauthier-Villars, Paris.
Gebhart, B. (1971). *Heat Transfer*, 2d ed., McGraw-Hill, New York.
Howarth, L., Ed. (1956). *Modern Development in Fluid Mechanics, High Speed Flow*, Oxford Univ. Press, New York.

Lorenz, L. (1881). *Ann. Phys. 13*, 582.
Mahajan, R. L., and Gebhart, B. (1988). In preparation.
Milne-Thomson, L. M. (1960). *Theoretical Hydrodynamics*, 4th ed., Macmillan, New York.
Oberbeck, A. (1879). *Ann. Phys. Chem. 7*, 21.
Pao, R. H. (1967). *Fluid Dynamics*, Merrill, Columbus, Ohio.
Schlichting, H. (1968). *Boundary Layer Theory*, 6th ed., McGraw-Hill, New York.
Schmidt, E., and Beckmann, W., with E. Pohlhausen (1930). *Tech. Mech. Thermodyn. 1*, 341–391.
Yih, C. S. (1969). *Fluid Mechanics*, McGraw-Hill, New York.

PROBLEMS

2.1 A wide flat vertical plate of height 30 cm, at 20°C in a quiescent ambient fluid at 20°C, is warmed by internal heaters at a rate of 0.1°C/s for 50 s. Compare the values of various general parameters if the fluid is air or water.

(a) Calculate the steady-state "convection velocities" and Grashof numbers. Compare the two sets of results.

(b) Calculate the Fourier numbers. Are transient effects equally important in each fluid?

(c) Calculate the Prandtl numbers and βT.

(d) Determine the values of R_0, R_1, R_2, R_3, R_4, and R_5. Note any systematic differences between the results for the two fluids.

2.2 A water droplet is settling in dry air. The diameter of the droplet is 0.001 in. Reduce the general form of the Navier-Stokes and other diffusion equations to a form adequate for the analysis of this problem.

2.3 The general equations governing transport for μ, k, and D uniform, and neglecting pressure and viscous dissipation energy effects, are Eqs. (2.7.19)–(2.7.22). Write these in scalar form for:

(a) Two-dimensional plane flow in cartesian coordinates x and y.

(b) Two-dimensional flow in polar coordinates r and θ.

2.4 For the surface and conditions in Problem 2.1, calculate the convection coefficient h, the Nusselt number, and the heat transfer rate, per side and unit width, for both air and water ambients.

2.5 The density difference causing the buoyancy force, $\rho_\infty - \rho$, is written for a perfect gas in Eq. (2.5.5). To assess the relative magnitude of the terms depending only on $t_\infty - t$, calculate the ratio of the second and third terms to the first term. Use $T_\infty = 300$ K, $t_\infty - t = 10$°C, and a pressure of 1 atm.

CHAPTER

THREE

EXTERNAL VERTICAL THERMALLY INDUCED FLOWS

3.1 THERMALLY BUOYANT FLOWS

Buoyancy-induced flows caused by thermal transport alone are common. Many kinds arise as a heating or cooling condition is applied in an extensive body of fluid in a single phase. Many different geometric circumstances are important in applications. The resulting flow may be laminar, in transition, or turbulent. It may be either steady, periodic, or transient. The flow may be vigorous when of large vertical extent L, that is, when the Grashof number $gL^3\beta(t_0 - t_\infty)/\nu^2$ is very large. Often, however, the flow is very weak. These collective possibilities are considered in this and in following chapters.

The present chapter treats vigorous vertical and steady laminar flows. These arise in a quiescent medium in the absence of any nonvertical constraint, such as an inclined or curved surface. Two such flows are seen in Figs. 1.1.1 and 1.1.2. The additional geometric qualification in this chapter is that the flows be of wide horizontal extent, such as adjacent to a wide heated vertical surface or above a long horizontal electrically heated wire. Such a flow may be called a plane flow. It is described in two cartesian coordinates, x in the vertical direction and y in the horizontal direction, normal to the principal flow direction. See Fig. 2.2.1 and the calculation of Section 2.2. For upflow, x increases upward; for downflow, it increases downward.

For such vigorous flows, both the thermal and momentum transport are accomplished in a vertical region whose thicknesses δ_t and δ are often very small compared to the height L of the flow considered. Referring back to the calculations in Section 2.2, $\delta/L << 1$ is seen to be an implicit assumption followed in making some of the approximations that simplifed the full equations, (2.2.1)–(2.2.4), to Eqs. (2.2.6) and (2.2.7). This procedure is the boundary layer concept of Prandtl (1904) for forced flows, applied to buoyancy-induced flows.

However, it was not until the careful measurements of Schmidt and Beckmann (1930), in air adjacent to a heated vertical surface, that it was known that such thin-layer mechanisms also occurred in vigorous thermally buoyant flows. The boundary region transport equations resulting from these measurements are simpler than Eqs. (2.2.1)–(2.2.4). However, they retain much more reality than Eqs. (2.2.6) and (2.2.7). A similarity solution of the equations was found immediately by E. Pohlhausen, as given in the paper by Schmidt and Beckmann. A calculation was given for air, Pr = 0.733. In a similarity solution the calculation is simplified in that x and y may be replaced by a single space coordinate $\eta = \eta(x, y)$. The partial differential equations become ordinary differential equations. During recent decades there has been a vast expansion in the use of the boundary region concept in calculating many different kinds of buoyancy-induced transport. Many of the most general of these calculations are found in this book.

In the following section the permissible boundary region approximations are used to obtain the simplest equations. Then a general transformation is made to find the kinds of transport that have similarity solutions. Such results are compared with measured transport. A number of further and more special aspects, such as transport in a fluid having an extreme value of the Prandtl number and the effects of ambient medium stratification, are considered.

3.2 BOUNDARY REGION APPROXIMATIONS

The simplest vertical thermally buoyant flows are those in which the transport properties $\mu(t)$ and $k(t)$ do not vary appreciably across the transport region. In practice, this occurs if both μ and k are weak functions of temperature or if the temperature difference is sufficiently small. The relevant plane flow equations are written as in Section 2.7, but including in the energy equation the appropriate pressure and viscous dissipation terms, the latter from Eq. (2.1.4), as below. The Boussinesq approximations for $\Delta\rho/\rho$ and R_0 much less than 1 have been applied.

$$\frac{\partial u}{\partial x} + \frac{\partial v}{\partial y} = 0 \tag{3.2.1}$$

$$u\frac{\partial u}{\partial x} + v\frac{\partial u}{\partial y} = \nu\left(\frac{\partial^2 u}{\partial x^2} + \frac{\partial^2 u}{\partial y^2}\right) - \frac{1}{\rho}\frac{\partial p_m}{\partial x} + g\beta(t - t_\infty) \tag{3.2.2}$$

$$u\frac{\partial v}{\partial x} + v\frac{\partial v}{\partial y} = \nu\left(\frac{\partial^2 v}{\partial x^2} + \frac{\partial^2 v}{\partial y^2}\right) - \frac{1}{\rho}\frac{\partial p_m}{\partial y} \tag{3.2.3}$$

$$u\frac{\partial t}{\partial x} + v\frac{\partial t}{\partial y} = \alpha\left(\frac{\partial^2 t}{\partial x^2} + \frac{\partial^2 t}{\partial y^2}\right) + \frac{\beta T}{\rho c_p}\left(u\frac{\partial p_h}{\partial x} + u\frac{\partial p_m}{\partial x} + v\frac{\partial p_m}{\partial y}\right)$$

$$+\frac{\mu}{\rho c_p}\Phi + \frac{q'''}{\rho c_p} \tag{3.2.4}$$

where

$$\Phi = 2\left(\frac{\partial u}{\partial x}\right)^2 + 2\left(\frac{\partial v}{\partial y}\right)^2 + \left(\frac{\partial v}{\partial x} + \frac{\partial u}{\partial y}\right)^2$$

These relations are applied to a general vigorous flow situation arising from an energy source, as shown in Fig. 3.2.1. An upflow is assumed to begin at the leading edge at $x = 0$ and to proceed downstream in the positive x direction. For $t_0 < t_\infty$, a downflow begins at $x = 0$ and proceeds in the negative x direction. The regions of thermal and momentum transport are concentrated near $y = 0$. Their local thicknesses $\delta_t(x)$ and $\delta(x)$ are sketched. These regions are not necessarily of equal thickness if $\text{Pr} \neq 1$. For example, it will be shown that $\delta_t(x) < \delta(x)$ if $\text{Pr} > 1$. Both t_0 and t_∞ may vary with x. The latter dependence reflects any ambient medium stratification. This stratification must be stable for the treatment that follows since the ambient medium is considered quiescent. The conditions at $y = 0$, partially represented in Fig. 3.2.1 by $t_0(x)$, might be due to a surface there. Other and various attendant boundary conditions may also arise. There might also be a plane of symmetry of transport, as in the plane plume seen in Fig. 1.1.2.

Diverse circumstances and bounding conditions are considered in succeeding sections. However, it is necessary first to obtain the boundary region approximations of the above equations. These will apply in vigorous flows, when δ_t/L

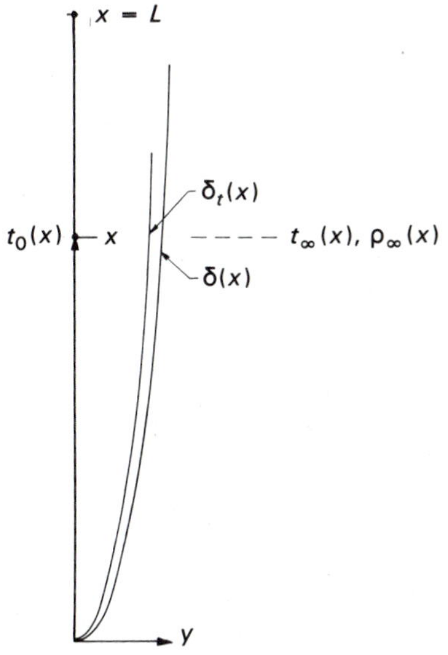

Figure 3.2.1 General flow situation arising from an energy source.

and δ/L are both much less than 1. To evaluate the relative magnitude of the various terms in Eqs. (3.2.1)–(3.2.4), all variables are scaled by characteristic quantities, as was done in Section 2.8 to find relevant parameters. The new variables are

$$X, Y = \frac{x, y}{L} \qquad \Delta_t, \Delta = \frac{\delta_t, \delta}{L} \qquad U, V = \frac{u, v}{U_c} \qquad P = \frac{p}{\rho U_c^2} \tag{3.2.5}$$

where

$$U_c = O[\sqrt{gL\beta(t_0 - t_\infty)}] \qquad \text{or}$$

$$\frac{U_c L}{\nu} = O(\sqrt{\mathrm{Gr}}) \qquad \text{for Pr around 1 or less} \tag{3.2.6}$$

Also,

$$\phi = \frac{t - t_\infty}{t_0 - t_\infty} \tag{3.2.7}$$

where $t_0 = t(x, 0)$ is the temperature at $y = 0$. It is taken as constant in this analysis.

The above transformation is introduced into the governing equations, (3.2.1)–(3.2.4). The coefficients are suitably simplified by division to yield the following equations in $\phi(X, Y)$, $U(X, Y)$, $V(X, Y)$, and $P(X, Y)$.

$$\frac{\partial U}{\partial X} + \frac{\partial V}{\partial Y} = 0 \tag{3.2.8}$$

$$1 \qquad 1$$

$$U\frac{\partial U}{\partial X} + V\frac{\partial U}{\partial Y} = \frac{1}{\sqrt{\mathrm{Gr}}}\left(\frac{\partial^2 U}{\partial X^2} + \frac{\partial^2 U}{\partial Y^2}\right) - \frac{\partial P_m}{\partial X} + \phi \tag{3.2.9}$$

$$1 \quad 1 \qquad \Delta \; \Delta^{-1} \qquad\qquad 1 \qquad \Delta^{-2} \qquad\qquad 1$$

$$U\frac{\partial V}{\partial X} + V\frac{\partial V}{\partial Y} = \frac{1}{\sqrt{\mathrm{Gr}}}\left(\frac{\partial^2 V}{\partial X^2} + \frac{\partial^2 V}{\partial Y^2}\right) - \frac{\partial P_m}{\partial Y} \tag{3.2.10}$$

$$1 \quad \Delta \qquad \Delta \quad 1 \qquad\qquad \Delta \qquad \Delta^{-1}$$

$$U\frac{\partial \phi}{\partial X} + V\frac{\partial \phi}{\partial Y} = \frac{1}{\mathrm{Pr}\sqrt{\mathrm{Gr}}}\left(\frac{\partial^2 \phi}{\partial X^2} + \frac{\partial^2 \phi}{\partial Y^2}\right)$$

$$1 \quad 1 \qquad \Delta \; \Delta_t^{-1} \qquad\qquad 1 \qquad \Delta_t^{-2}$$

$$+ (\beta T)\frac{g\beta L}{c_p}\left(U\frac{\partial P_h}{\partial X} + U\frac{\partial P_m}{\partial X} + V\frac{\partial P_m}{\partial Y}\right)$$

$$1 \qquad\quad 1 \qquad\quad \Delta$$

$$+\frac{g\beta L}{c_p\sqrt{\text{Gr}}}\left[2\left(\frac{\partial U}{\partial X}\right)^2+2\left(\frac{\partial V}{\partial Y}\right)^2+\left(\frac{\partial V}{\partial X}+\frac{\partial U}{\partial Y}\right)^2\right]$$

$$\qquad\qquad 1 \qquad\qquad 1 \qquad\qquad \Delta \qquad \Delta^{-2}$$

$$+\frac{q'''L^2}{\mu c_p(t_0-t_\infty)\sqrt{\text{Gr}}} \tag{3.2.11}$$

The procedure now is to estimate the relative magnitudes in this diverse array of physical effects. This is done here through physical reasoning based on reasonable surmises concerning the behavior of U, V, and so forth. An alternative to this procedure frequently invoked in making such kinds of approximations is an expansion technique. However, a less formal method is perhaps more instructive initially and certainly is of wider application.

The first observation concerning magnitudes is that U changes from 0 to order 1 as X changes from 0 to 1; see Eq. (3.2.5) and Fig. 3.2.1. Therefore, $\partial U/\partial X$ is of order 1, as, therefore, is $\partial V/\partial Y$ by Eq. (3.2.8). Now $V(X, 0) = 0$ for an impermeable or symmetry condition at $y = 0$. However, $V(X, Y) \neq 0$ in the flow region at distances of order Δ out. Therefore, V is of order Δ. Similarly, $\partial^2V/\partial Y^2 = O(1/\Delta)$. Since U decreases from about 1 to zero in distance Δ, $\partial U/\partial Y = O(1/\Delta)$ and $\partial^2U/\partial Y^2 = O(1/\Delta^2)$. Analogously, $\partial V/\partial X$ and $\partial^2V/\partial X^2 = O(\Delta)$. Then, since $\phi(X, 0) = 1$ and $\phi(X, \Delta_t) = 0$, $\partial\phi/\partial Y = O(\Delta_t^{-1})$ and $\partial^2\phi/\partial Y^2 = O(\Delta_t^{-2})$. Since $\phi = 0$ for $X < 0$ and $\phi \simeq 1$ downstream in the thermal region, $\partial\phi/\partial X = 1$ and $\partial^2\phi/\partial X^2 = 1$.

These estimates are written below the relevant terms in the equations above, with very interesting consequences. First, in Eq. (3.2.9), $\partial^2U/\partial Y^2 = O(\Delta^{-2})$ in the viscous force term is a very large term compared to other determined magnitudes. Yet it is known that the buoyancy force $\phi = O(1)$ is important in driving momentum convection in the outer part of the thermal diffusion layer. This, in turn, is seen to be of order 1 in Eq. (3.2.9). Also, ϕ must balance viscous forces farther in, where acceleration is often less. Therefore,

$$\frac{1}{\sqrt{\text{Gr}}}\left(\frac{\partial^2U}{\partial X^2}+\frac{\partial^2U}{\partial Y^2}\right)=O\left(\frac{1}{\sqrt{\text{Gr}}}\frac{1}{\Delta^2}\right)=O(1)$$

and

$$\frac{1}{\sqrt{\text{Gr}}}=O(\Delta^2) \qquad \text{or} \qquad \Delta=\frac{\delta}{L}=O\left(\frac{1}{\sqrt[4]{\text{Gr}}}\right) \tag{3.2.12}$$

Thus, $\Delta << 1$, for very large Gr.

This conclusion applied to Eq. (3.2.10) shows that the convection and viscous terms are of order Δ or less. Therefore, $\partial P_m/\partial Y = O(\Delta)$ and Eq. (3.2.10) may be neglected entirely compared to Eq. (3.2.9). That is, the y-direction force-momentum balance may be disregarded compared to the x-direction one to effects of order Δ but not to order Δ^2. This result also shows that $P_m = O(\Delta^2)$. Then

$\partial P_m/\partial X = O(\Delta^2)$, and this term disappears from Eq. (3.2.9). Therefore, along with the loss of one equation, (3.2.10), one of the distributions to be determined, $P_m(X, Y)$, has also disappeared.

In the energy balance, a principal convection term is of order 1. Since this must be the order of the largest conduction term,

$$\frac{1}{\Pr\sqrt{\mathrm{Gr}}} = O(\Delta_t^2) \qquad \text{or} \qquad \frac{\delta_t}{L} = O\left(\frac{1}{\sqrt[4]{\mathrm{Gr}}\sqrt{\Pr}}\right) \tag{3.2.13}$$

and

$$\frac{\delta}{\delta_t} = O(\sqrt{\Pr}) = \frac{\Delta}{\Delta_t} \tag{3.2.14}$$

Therefore, the two convection terms and the largest of the two conduction terms are retained.

Turning to the pressure term in Eq. (3.2.11), the first contribution is evaluated as

$$U\frac{\partial P_h}{\partial X} = \frac{UL}{\rho U_c^2}\frac{\partial p_h}{\partial X} = O\left[U\frac{\rho}{\rho\beta(t_0 - t_\infty)}\right] = O\left(U\frac{1}{R_1}\right)$$

Recalling that $\partial P_m/\partial X = O(\Delta^2)$ and $\partial P_m/\partial Y = O(\Delta)$, the hydrostatic term is the largest. This term is of the following magnitude:

$$(\beta T)R_0/R_1$$

Recall that R_0 was considered small in evaluating the buoyancy force $B = g(\rho_\infty - \rho)$ in Section 2.5, as written in Eq. (3.2.2). However, R_1 was also taken as small. Therefore, this leading pressure term will be retained for circumstances in which R_0 may not actually be very small compared to R_1.

Finally, the viscous dissipation term $(\partial U/\partial Y)^2 = O(\Delta^{-1})$ is large compared to the other three contributions. Thus, the principal dissipation effect is

$$\frac{g\beta L}{c_p}\frac{\Delta^2}{\sqrt{\mathrm{Gr}}} = O\left(\frac{g\beta L}{c_p}\right) = O(R_0)$$

Again, R_0 was considered small in evaluating B. It expresses the effect of the vertical hydrostatic pressure variation on B. If this viscous dissipation term will be retained, additional questions are raised concerning the evaluation of B.

This collection of conclusions concerning relative magnitudes results in the following boundary region equations for thermally driven vertical transport:

$$\frac{\partial u}{\partial x} + \frac{\partial v}{\partial y} = 0 \tag{3.2.15}$$

$$u\frac{\partial u}{\partial x} + v\frac{\partial u}{\partial y} = \nu\frac{\partial^2 u}{\partial y^2} + g\beta(t - t_\infty) \tag{3.2.16}$$

$$u\frac{\partial t}{\partial x} + v\frac{\partial t}{\partial y} = \alpha\frac{\partial^2 t}{\partial y^2} + \frac{\beta T}{\rho c_p}u\frac{\partial p_h}{\partial x} + \frac{\mu}{\rho c_p}\left(\frac{\partial u}{\partial y}\right)^2 + \frac{q'''}{\rho c_p} \tag{3.2.17}$$

A similar boundary region equation arises from Eq. (2.7.18) for chemical species diffusion. This will be determined in a later chapter.

The above equations are reduced in order from the initial full equations. The loss of any contribution from Eq. (3.2.3) to the order of approximation of $\Delta = O(1/\sqrt[4]{\text{Gr}})$ has eliminated one distribution, p_m. Another result of the reduced complexity in Eqs. (3.2.15)–(3.2.17) is that no boundary condition may be specified for $v(x, y)$ at the outer edge of the momentum region. This is the same result as in a forced-flow boundary layer, for example, the Blasius flow.

The clear boundary conditions for the above equations as the ambient medium is approached are

$$u = t - t_\infty = 0$$

The only other general one that arises is $t(x, 0) = t_0$. Yet t_0 is in turn determined from the particular physical condition imposed at $y = 0$. Conditions on $u(x, 0)$ and $v(x, 0)$ arise in the same way. The following sections, which set forth similarity solutions, will consider appropriate conditions characteristic of important applications.

3.3 SIMILARITY SOLUTION FOR AN ISOTHERMAL SURFACE

The boundary region equations determined in the preceding section apply to a wide diversity of thermal transport circumstances of interest as well as of practical utility. However, the more complicated general treatment will be preceded by the calculation of transport in perhaps the simplest of all thermally induced motions. It is the vertical isothermal surface at t_0 in a quiescent ambient medium at t_∞, including only buoyancy, convection, and thermal diffusion. This was analyzed by Pohlhausen in the paper reporting the experimental results of Schmidt and Beckmann (1930), using the boundary region equations and boundary conditions established as below.

$$\frac{\partial u}{\partial x} + \frac{\partial v}{\partial y} = 0 \tag{3.3.1}$$

$$u\frac{\partial u}{\partial x} + v\frac{\partial u}{\partial y} = \nu\frac{\partial^2 u}{\partial y^2} + g\beta(t - t_\infty) \tag{3.3.2}$$

$$u\frac{\partial t}{\partial x} + v\frac{\partial t}{\partial y} = \alpha\frac{\partial^2 t}{\partial y^2} \tag{3.3.3}$$

$$u(x, 0) = v(x, 0) = u(x, \infty) = 0 \tag{3.3.4}$$

$$t(x, 0) = t_0 \qquad t(x, \infty) = t_\infty \tag{3.3.5}$$

The first two conditions, on u and v at $x = 0$, result from no fluid slip at the fluid-solid interface and an impervious surface or symmetry condition there.

A similar solution for this transport circumstance is sought by postulating that x and y may be combined into a single space coordinate $\eta(x, y) = b(x)y$, where $b(x)$ is to be bounded for $x > 0$ and is to be found. A two-dimensional stream function $\psi(x, y)$, defined below, satisfies and replaces Eq. (3.3.1). It is then transformed into another stream function f through $c(x)$. The temperature is transformed into ϕ.

$$\eta = b(x)y \tag{3.3.6}$$

$$u = \psi_y \qquad v = -\psi_x \qquad \psi(x, y) = \nu c(x) f(x, y) \tag{3.3.7}$$

$$\phi(x, y) = \frac{t(x, y) - t_\infty}{t_0 - t_\infty} \tag{3.3.8}$$

The question of similarity is whether or not functions $b(x)$ and $c(x)$ may be found such that ϕ and f will depend only on η while simultaneously satisfying all conditions of the transport, Eqs. (3.3.1)–(3.3.5). This is assessed by assuming that $f(x, y) = f(\eta)$ and $\phi(x, y) = \phi(\eta)$, and substituting the transformation, Eqs. (3.3.6)–(3.3.8), into the equations and boundary conditions. Note that

$$u = \psi_y = \nu c(x) \frac{\partial f}{\partial \eta} \frac{\partial \eta}{\partial y} = \nu c(x) b(x) f' = \nu c b f' \tag{3.3.9}$$

$$v = -\psi_x = -\nu \left[c(x) f' \frac{\partial \eta}{\partial x} + f \frac{dc}{dx} \right] = -\nu (c f' y b_x + f c_x) \tag{3.3.10}$$

where the prime and the subscript x indicate differentiation with respect to η and x, respectively. Other quantities are simply calculated and the equations and boundary conditions become

$$f''' + \frac{c_x}{b} f f'' - \left(\frac{c_x}{b} + \frac{c b_x}{b^2} \right) f'^2 + \frac{t_0 - t_\infty}{c b^3} \frac{g\beta}{\nu^2} \phi = 0 \tag{3.3.11}$$

$$\frac{\phi''}{\mathrm{Pr}} + \frac{c_x}{b} f \phi' = 0 \tag{3.3.12}$$

$$f'(0) = f(0) = f'(\infty) = 1 - \phi(0) = \phi(\infty) = 0 \tag{3.3.13}$$

The f''' and ϕ terms in Eq. (3.3.11) are the viscous and buoyancy forces. The term ϕ'' in Eq. (3.3.12) in thermal conduction. The other terms in both equations are convection.

Since x remains in the formulation only in the functions $b(x)$ and $c(x)$, f and ϕ will not depend on x if all of the apparently x-dependent coefficients are actually constants:

$$\frac{c_x}{b} = C_1 \qquad \frac{c b_x}{b^2} = C_2 \qquad \frac{t_0 - t_\infty}{c b^3} \frac{g\beta}{\nu^2} = C_3 K_1 \tag{3.3.14}$$

where $K_1 = g\beta/\nu^2$. Note that $c \propto x^{3/4}$ and $b \propto x^{-1/4}$ satisfy all these relations. Equations (3.3.14) are now used to determine b and c. First, b is obtained by eliminating c in the second and third parts of Eq. (3.3.14) as follows:

$$c = C_2 \frac{b^2}{b_x} = \frac{t_0 - t_\infty}{C_3 b^3}$$

The result is a differential equation in b

$$\frac{b_x}{b^5} = \frac{C_2 C_3}{t_0 - t_\infty}$$

This is integrated, taking the constant that arises to be zero, to become

$$b = \left(- \frac{t_0 - t_\infty}{4K_1 C_2 C_3 x} \right)^{1/4} = \frac{K}{\sqrt[4]{x}} \tag{3.3.15}$$

Then the first relation in Eq. (3.3.14) evaluates c as

$$c_x = \frac{C_1 K}{\sqrt[4]{x}} \qquad c = C_1 K \frac{4}{3} x^{3/4}$$

again taking the integration constant $c(0)$ as zero. But the second relation in Eq. (3.3.14) also gives c, as

$$c = \frac{b^2 C_2}{b_x} = -4KC_2 x^{3/4} \tag{3.3.16}$$

Therefore, $C_1 = -3C_2$. Now the simplest result for the last part of Eq. (3.3.14) would be that $C_3 K_1 = 1$. Then the buoyancy force in Eq. (3.3.11), $C_3 K_1 \phi$, becomes simply ϕ. That is, $K_1 = 1/C_3$. From this consideration, along with Eq. (3.3.16), c is

$$c = -4C_2 \left[- \frac{gx^3}{4\nu^2} \frac{\beta(t_0 - t_\infty)}{C_2} \right]^{1/4}$$

Choosing $C_2 = -1$, then $C_1 = 3$ and c and d become

$$c(x) = 4 \left[\frac{gx^3}{4\nu^2} \beta(t_0 - t_\infty) \right]^{1/4} = 4 \left(\frac{\text{Gr}_x}{4} \right)^{1/4} = G \tag{3.3.17}$$

$$b(x) = \frac{1}{x} \left[\frac{gx^3}{4\nu^2} \beta(t_0 - t_\infty) \right]^{1/4} = \frac{1}{x} \left(\frac{\text{Gr}_x}{4} \right)^{1/4} = \frac{G}{4x} \tag{3.3.18}$$

where the local Grashof number Gr_x expresses the local vigor of the flow at downstream location x. Note that if $t_0 - t_\infty < 0$, downflow results and x is increasingly negative downward from its origin. Thus, Gr_x remains positive.

Therefore, if b and c are chosen as above, Eqs. (3.3.11) and (3.3.12) are ordinary but nonlinear differential equations for f and ϕ. They are entirely in terms of η, as are the boundary conditions. The formulation is

$$f''' + 3ff'' - 2f'^2 + \phi = 0 \tag{3.3.19}$$

$$\phi'' + 3\mathrm{Pr} f\phi' = 0 \tag{3.3.20}$$

$$f'(0) = f(0) = f'(\infty) = 1 - \phi(0) = \phi(\infty) = 0 \tag{3.3.21}$$

This is the similarity formulation found by Pohlhausen for the boundary layer equations verified by the experiments of Schmidt and Beckman (1930). The Prandtl number is the only parameter that appears. That is, the solution of these equations as a function of Pr covers all such transport circumstances that may arise. However, it may be anticipated from Eq. (3.3.20) that both extremely small and large values of Pr may require or permit special treatment. Such solutions are determined later. Solutions and other aspects of these results are discussed in the following section.

3.4 TRANSPORT AND SOLUTIONS

The similarity formulation in the preceding section, for an isothermal surface condition t_0, was solved numerically for Pr = 0.733 by Pohlhausen in the paper by Schmidt and Beckmann (1930). Then Schuh (1948) reported calculated results for Pr = 10, 100, and 1000, along with solutions for the plane plume flow of Fig. 1.1.2 and the axisymmetric plume, both for Pr = 0.72. Plumes are considered later. Many aspects of such results are of interest and of practical value.

The solutions of Eqs. (3.3.19)–(3.3.21) for any value of the Prandtl number gives the velocity field $u(x, y)$ and $v(x, y)$ as follows:

$$u = \frac{\nu}{4x} G^2 f'(\eta) \tag{3.4.1}$$

$$v = \frac{\nu}{4x} G(\eta f' - 3f) \tag{3.4.2}$$

Since $f(\infty) \neq 0$, $v(x, \infty)$ is not zero. This is called the entrainment velocity. It provides the continuing downstream induction of ambient fluid into the increasing flow. Assuming that the active transport region extends from $\eta = 0$ to $\eta = \eta_\delta$, where $y = \delta$, Eq. (3.3.6) indicates that

$$\eta_\delta = b(x)\,\delta(x) \qquad \text{or} \qquad \delta(x) = \frac{\eta_\delta}{b(x)}$$

If $\delta(0)$ is to be zero, then $b(0)$ must be unbounded. That is the condition used in obtaining Eq. (3.3.15), where the constant of integration $[1/4b(0)]^4$ was taken as zero.

The local flux $q''(x)$ and total heat transfer $Q(x)$ are calculated from Eqs. (3.3.6), (3.3.8), and (3.3.18).

$$q''(x) = -k \left.\frac{\partial t}{\partial y}\right)_{y=0} = [-\phi'(0)]k(t_0 - t_\infty)b(x) \tag{3.4.3}$$

The local Nusselt number becomes

$$\frac{q''(x)}{(t_0 - t_\infty)}\frac{x}{k} = \frac{h_x x}{k} = \text{Nu}_x = \frac{[-\phi'(0)]}{\sqrt{2}}\sqrt[4]{\text{Gr}_x} = F(\text{Pr})\sqrt[4]{\text{Gr}_x} \tag{3.4.4}$$

Note that $h_x \propto x^{-1/4}$.

$$Q(x) = \int_0^x q''(x)\,dx = \frac{4}{3}[-\phi'(0)]k(t_0 - t_\infty)xb \tag{3.4.5}$$

The average convection coefficient to x, $\bar{h}_x$, is then

$$\frac{Q(x)}{x(t_0 - t_\infty)}\frac{x}{k} = \frac{\bar{h}_x x}{k} = \frac{4}{3}\frac{[-\phi'(0)]}{\sqrt{2}}\sqrt[4]{\text{Gr}_x} = \frac{4}{3}F(\text{Pr})\sqrt[4]{\text{Gr}_x} \tag{3.4.6}$$

Comparing Eqs. (3.4.4) and (3.4.6), $\bar{h}_x = 4/3\ h_x$. This results from higher values of h_x at smaller x. The boundary layer estimate of h_x for a surface of height L would similarly be $h_L = \bar{h}_x$. Later, h_L is simply written as h. The surface shear stress τ and resulting drag coefficient C_D are easily found in terms of $f''(0)$ and $f'(\eta)$ as

$$\tau(x) = \mu \left.\frac{\partial u}{\partial y}\right)_{y=0} = \mu\nu c b^2 f''(0)$$
$$C_D = \frac{\tau(x)}{\rho u^2/2} = \frac{2f''(0)/[f'(\eta)]^2}{G} \tag{3.4.7}$$

where perhaps $f'(\eta)$ is chosen as its maximum value across the region, $f'_{\max}$.

Simple solutions for f and ϕ are not available. However, it is possible to write a solution formally. From Eq. (3.3.20)

$$\frac{\phi''}{\phi'} = \frac{d}{d\eta}\ln\phi' = -3\text{Pr}f$$

yields

$$\phi'(\eta) = \phi'(0)\exp\left(-3\text{Pr}\int_0^\eta f\,d\eta\right)$$

Then

$$\phi(\eta) - 1 = \phi'(0)\int_0^\eta \exp\left(-3\text{Pr}\int_0^\eta f\,d\eta\right)d\eta$$

Noting that $\phi = 0$ at $\eta = \infty$, $\phi'(0)$ is evaluated and $\phi(\eta)$ becomes

$$\phi(\eta) = 1 - \frac{\displaystyle\int_0^\eta \exp\left(-3\mathrm{Pr}\int_0^\eta f\,d\eta\right)d\eta}{\displaystyle\int_0^\infty \exp\left(-3\mathrm{Pr}\int_0^\infty f\,d\eta\right)d\eta} \tag{3.4.8}$$

This may be put into Eq. (3.3.19) to obtain an integro-differential equation for $f'(\eta)$.

However, this is not the common procedure for such calculations. The functions f and ϕ and their necessary derivatives are determined by numerical integration. Three of the boundary conditions in Eq. (3.3.21) are at $\eta = 0$, and the other two are for f' and ϕ to be asymptotic to zero at increasing η. Therefore, a shooting method from $\eta = 0$, with initial guesses of $f''(0)$ and $\phi'(0)$, proceeds to find values of f' and ϕ at large η. These values, in a correction scheme, provide new guesses for the next calculation, and so forth (Ostrach, 1953). Alternatively, asymptotic solutions of f and ϕ at large η are obtained from Eqs. (3.3.19) and (3.3.20). These then become the targets of the shooting method. Other methods have also been used.

Resulting distributions of $\phi(\eta)$ and $f'(\eta) \propto u$ are shown in Fig. 3.4.1 for the Prandtl number range 0.01 to 100. Many effects are seen. First, the maximum value of f' across the flow region, f'_{max}, sharply decreases with increasing Pr. Another interesting result is that the ratio δ_t/δ is very small for Pr = 100 but increases to about 1 as Pr decreases to about 1. It remains about 1 even for a

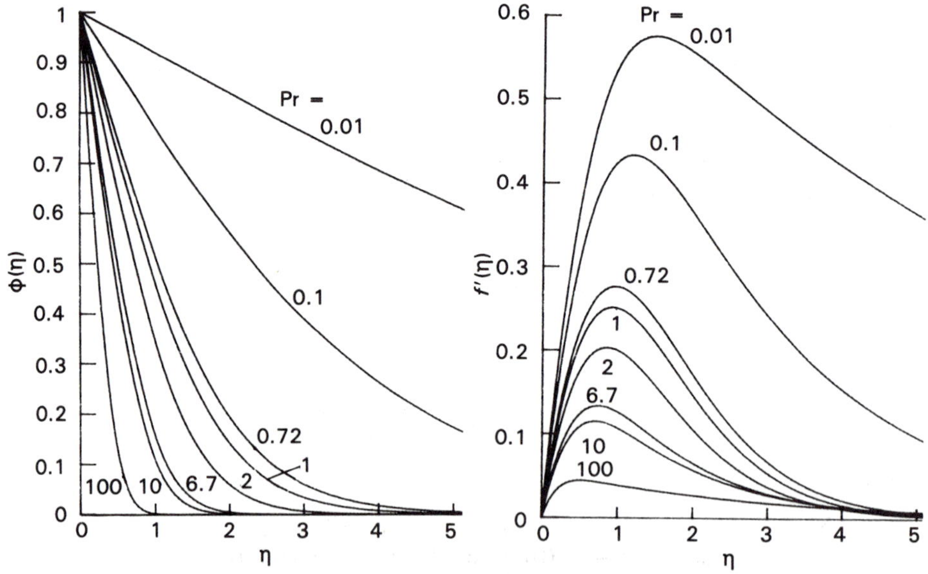

Figure 3.4.1 Boundary region velocity and temperature distributions for thermally buoyant laminar flows adjacent to a vertical isothermal surface. *(Prepared by Ramesh Krishnamurthy.)*

further 100-fold decrease of Pr to 0.01. Recall that for Pr >> 1, $\nu >> \alpha$. Therefore, the thermal layer δ_t lies buried deeply in the velocity layer δ and next to the surface. Buoyancy force is largely consumed in producing a fluid shear layer that drives the outer flow layer. This layer is uniformly at t_∞. However, decreasing the Prandtl number to Pr << 1 does not produce the inverse effect, that is, $\delta_t > \delta$. Buoyancy penetrates out as far as does the temperature and causes motion. Thus δ_t and δ remain locked together as δ_t increases. Note that this does not occur in forced flows. This behavior is also apparent in Table 3.4.1.

Concerning heat transfer, the parameter is seen in Eqs. (3.4.4) and (3.4.6) to be $[-\phi'(0)]/\sqrt{2}$ for both the local heat transfer coefficient h_x and the average value $\bar{h}_x$ from 0 to x. This quantity is found in Table 3.4.1 and the values are plotted against Pr in Fig. 3.4.2, along with those of $f''(0)$, $f'_{\max}$, and $f(\infty)$, the entrainment parameter. There are two trends of $[-\phi'(0)]$ vs. Pr. That is, $[-\phi'(0)] \propto \sqrt{\text{Pr}}$ for Pr << 1 and $[-\phi'(0)] \propto \sqrt[4]{\text{Pr}}$ for Pr >> 1. These will be seen later to correspond to the calculated asymptotic behaviors of $[-\phi'(0)]$ for Pr << 1 and Pr >> 1, respectively.

The other quantities tabulated show additional aspects of the Prandtl number dependence. These are of interest and value in assessing other effects of a wide variation in the relative values of ν and α.

Table 3.4.1 Values of the transport quantities for thermally buoyant laminar flows adjacent to a vertical isothermal surface[a]

Pr	$-\phi'(0)$	$F(\text{Pr}) = \frac{[-\phi'(0)]}{\sqrt{2}}$	$f''(0)$	$f'_{\max}$ $[\eta(f'_{\max})]$	$f(\infty)$	η_{δ_t}[b]	η_δ[c]
0.01	0.0807	0.0571	0.9873	0.5761 [1.5]	4.536	32.5	29.25
0.1	0.2301	0.1627	0.8591	0.4335 [1.21]	1.512	11	11.5
0.72	0.5046	0.3568	0.6760	0.2762 [0.97]	0.5988	4.42	5.5
1.0	0.5671	0.4010	0.6422	0.2513 [0.92]	0.5230	3.8	5.29
2.0	0.7165	0.5066	0.5713	02028 [0.87]	0.4046	2.82	5.75
5.0	0.9540	0.6750	0.4818	0.1484 [0.78]	0.3031	1.98	6
6.7	1.0408	0.7360	0.4548	0.1335 [0.73]	0.2786	1.80	6.91
10.0	1.1693	0.8268	0.4192	0.1149 [0.71]	0.2492	1.57	7.5
100.0	2.1913	1.5495	0.2517	0.0442 [0.51]	0.1365	0.78	12.2

[a]Prepared by Ramesh Krishnamurthy.
[b]Corresponds to the value of η where $\phi = 0.01$.
[c]Corresponds to the value of η where $f' = 0.01(f'_{\max})$.

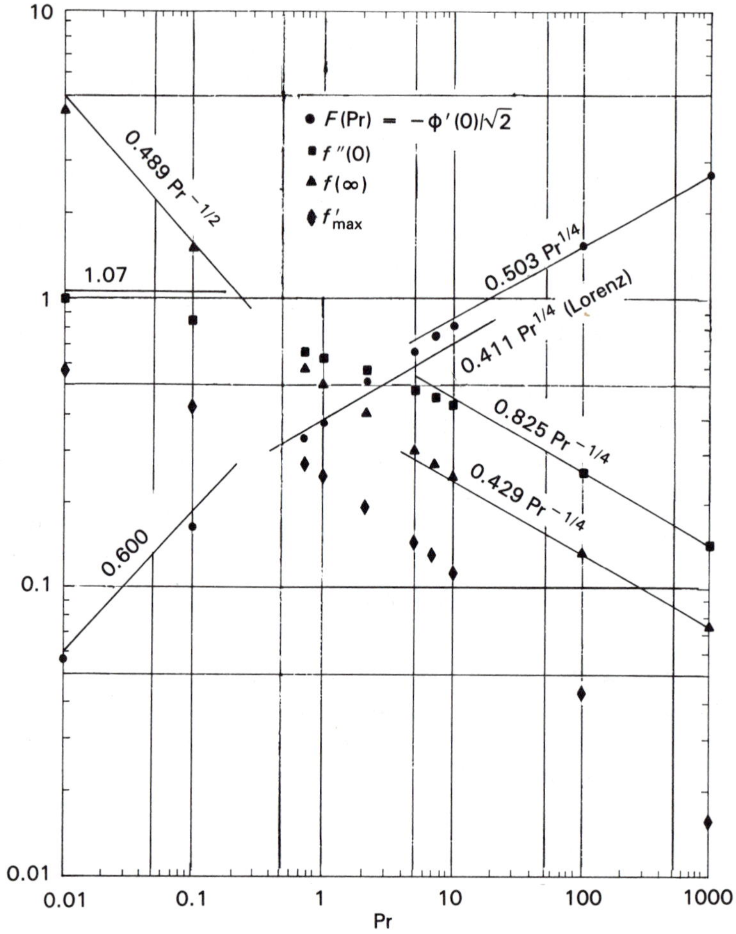

Figure 3.4.2 Heat transfer and flow characteristics for thermally buoyant laminar flow adjacent to a vertical isothermal surface. *(Prepared by Ramesh Krishnamurthy.)*

3.5 GENERAL SIMILARITY FORMULATION

Many of the general characteristics of thermally buoyant flows are apparent in the preceding two sections. However, the collection of important applications includes many kinds of bounding conditions at $y = 0$ that are not those of an isothermal surface, for example, the flows seen in Figs. 1.1.1 and 1.1.2. The thermal condition at $y = 0$ for the first is $q''(x) \propto (\partial t/\partial y)_{y=0} =$ const. and for the second $q''(x) \propto \partial t/\partial y)_{y=0} = 0$. Early studies indicated that similarity also resulted for other kinds of surface conditions, taking into account the transport and convection effects in the preceding simple formulation, in Section 3.3, that is, in Eqs. (3.3.1)–(3.3.5). Sparrow and Gregg (1956) showed that similarity still resulted if a heat flux input q'', uniform in x, was imposed by the surface on the adjacent fluid.

Sparrow and Gregg (1958) also found similarity to result if the surface temperature t_0 varied downstream as x^n or e^{mx}. Yang (1960) found similarity for a transient convective response for certain surface temperature and flux variations, as discussed in Chapter 7.

The general analysis below will recover these kinds of conditions. It will also consider similarity for the many other effects ignored in using the simplest energy equation (3.3.3) and in assuming uniform t_0 and/or t_∞. The full equations are listed below with the same transformation as before. Also postulated are the variations of $t_0(x)$ and $t_\infty(x)$, $d(x)$ and $j(x)$, along with the invariant boundary conditions. The formulation is expressed as in the analysis given in Gebhart (1971).

$$\frac{\partial u}{\partial x} + \frac{\partial v}{\partial y} = 0 \tag{3.5.1}$$

$$u\frac{\partial u}{\partial x} + v\frac{\partial u}{\partial y} = \nu\frac{\partial^2 u}{\partial y^2} + g\beta(t - t_\infty) \tag{3.5.2}$$

$$u\frac{\partial t}{\partial x} + v\frac{\partial t}{\partial y} = \alpha\frac{\partial^2 t}{\partial y^2} + \frac{\beta T}{\rho c_p}u\frac{\partial p_h}{\partial x} + \frac{\mu}{\rho c_p}\left(\frac{\partial u}{\partial y}\right)^2 + \frac{q'''}{\rho c_p} \tag{3.5.3}$$

$$\eta = b(x)y \qquad u = \psi_y \qquad v = -\psi_x \qquad \psi(x, y) = \nu c(x)f(\eta) \tag{3.5.4}$$

$$\phi(\eta) = \frac{t - t_\infty}{t_0 - t_\infty} = \frac{t - t_\infty}{d(x)} \tag{3.5.5}$$

$$t_0 - t_\infty = d(x) \qquad \text{and} \qquad t_\infty - t_r = j(x) \tag{3.5.6}$$

$$u(x, \infty) = 0 \qquad t(x, 0) = t_0 \qquad t(x, \infty) = t_\infty \tag{3.5.7}$$

In Eq. (3.5.6) t_r is a reference temperature, perhaps $t_\infty(0)$. Applying the definitions and transformation, Eqs. (3.5.1)–(3.5.3) and (3.5.7) become

$$f''' + \frac{c_x}{b}ff'' - \left(\frac{c_x}{b} + \frac{cb_x}{b^2}\right)f'^2 + \frac{d}{cb^3}\frac{g\beta}{\nu^2}\phi = 0 \tag{3.5.8}$$

$$\frac{\phi''}{\mathrm{Pr}} + \frac{c_x}{b}f\phi' - \frac{cd_x}{bd}f'\phi - \frac{cj_x}{bd}f' - \frac{c}{bd}\frac{g}{c_p}(\beta T)f'$$

$$+ \frac{b^2c^2}{d}\frac{\nu^2}{c_p}f''^2 + \frac{1}{db^2}\frac{1}{\mathrm{Pr}}\frac{q'''}{k} = 0 \tag{3.5.9}$$

$$f'(\infty) = 1 - \phi(0) = \phi(\infty) = 0 \tag{3.5.10}$$

Equation (3.5.8) differs from Eq. (3.3.11) only in that $d(x)$ appears above, where $t_0 - t_\infty$ was constant before. However, Eq. (3.5.9) is very different from Eq. (3.3.12) and each difference is an added physical effect. The term in d_x is the effect of the changing level of t_0 on convection. The term in j_x is a similar effect from changing t_∞. The following three terms are pressure, viscous dissipation, and distributed source effects.

Similarity will again result if each apparently x-dependent coefficient and quantity that must appear in properly modeling any transport process is a constant and if the relevant boundary conditions do not contain x and/or y, except as η. Then f and ϕ depend only on η.

All the parameters in Eqs. (3.5.8) and (3.5.9) are listed below, with an indication of the physical effect to which they are related. The values of the constants C_1, C_2, and C_3 below, determined in Section 3.3 for $t_0 - t_\infty$ constant, were 3, −1, and $1/K_1$.

$$\left.\begin{aligned} \frac{c_x}{b} &= C_1 \qquad (3.5.11a) \\ \frac{cb_x}{b^2} &= C_2 \qquad (3.5.11b) \end{aligned}\right\} \text{momentum and thermal convection}$$

$$\frac{d}{cb^3}\frac{g\beta}{\nu^2} = C_3K_1 = 1 \qquad \text{thermal buoyancy} \tag{3.5.11c}$$

$$\frac{cd_x}{bd} = C_4 \qquad \text{nonuniform } t_0 \tag{3.5.11d}$$

$$\frac{cj_x}{bd} = C_5 \qquad \text{temperature stratification, } t_\infty \tag{3.5.11e}$$

$$\frac{c}{bd}\frac{g(\beta T)}{c_p} = \frac{c}{bd}K_2 = C_6K_2 \qquad \text{pressure energy effect} \tag{3.5.11f}$$

$$\frac{b^2c^2}{d}\frac{\nu^2}{c_p} = \frac{b^2c^2}{d}K_3 = C_7K_3 \qquad \text{viscous dissipation} \tag{3.5.11g}$$

$$\frac{1}{b^2d}\frac{1}{\Pr}\frac{q'''}{k} = F_1 = F_1(\eta) \qquad \text{distributed energy source} \tag{3.5.11h}$$

Conditions for similarity. The minimum necessary terms to be retained in Eqs. (3.5.8) and (3.5.9) for any flow are molecular diffusion f''' and ϕ'', the buoyancy force ϕ, and the convection terms. These resulted in similarity in Section 3.3 for d constant. Additional effects will be added to these one at a time to find additional conditions under which similarity results. However, the first step is to allow only for varying $t_0 - t_\infty = d(x)$ with x. The collection of relevant coefficients for only this additional effect is

$$\frac{c_x}{b} = C_1 \qquad \frac{cb_x}{b^2} = C_2 \qquad \frac{d}{cb^3} = C_3K_1 = 1 \qquad \frac{cd_x}{bd} = C_4 \tag{3.5.12}$$

These four conditions determine the convection parameters b and c, the constants, and the permissible x dependence of $d(x)$. Through the elimination of $c(x)$ from the first two conditions, the following differential equation in $b(x)$ is obtained:

$$\frac{b_{xx}}{b_x} = \left(2 - \frac{C_1}{C_2}\right)\frac{b_x}{b} \tag{3.5.13}$$

Integration yields

$$\frac{b_x}{b^{2-C_1/C_2}} = A = \text{const.} \tag{3.5.14}$$

For $C_1 \neq C_2$ the following power law form results from the above:

$$b(x) = \left[\left(\frac{C_1 - C_2}{C_2}\right)(Ax + B)\right]^{C_2/(C_1-C_2)} \tag{3.5.15}$$

Then c is calculated from the C_2 condition as

$$c(x) = \frac{C_2}{A} b^{C_1/C_2} \tag{3.5.16}$$

However, for $C_1 = C_2$ the left-hand side of Eq. (3.5.14) is the derivative of ln b. The resulting x dependence of b is exponential as follows:

$$b(x) = Be^{Ax} \tag{3.5.17}$$

The function c becomes

$$c(x) = \frac{C_2 B}{A} e^{Ax} \tag{3.5.18}$$

The resulting x dependence of the temperature difference, $d(x)$, is determined for each result above through the third condition in Eq. (3.5.12), as follows:

$$d(x) = cb^3 \frac{g\beta}{\nu^2} = \frac{cb^3}{K_1} \tag{3.5.19}$$

$$= \frac{C_2}{K_1 A}\left[\frac{A(C_1 - C_2)}{C_2}\left(x + \frac{B}{A}\right)\right]^{(C_1+3C_2)/(C_1-C_2)}$$

$$= N(x + C)^n \qquad \text{power law} \tag{3.5.20}$$

$$= \frac{C_2 A^2 B}{K_1} e^{4Ax} = Me^{mx} \qquad \text{exponential} \tag{3.5.21}$$

where each of the forms of $d(x)$ above, which admit similarity solutions, is also written in simplest forms. The last condition in Eq. (3.5.12) relates C_4 to the other constants.

The power law form in Eq. (3.5.20) has a peculiar consequence when $C \neq 0$. The total heat transfer $Q(x)$, from $x = 0$ downstream to location x, is obtained by integrating $q''(x)$, the local heat flux. See, for example, Eq. (3.4.5). The result here is

$$Q(x) \propto (x + C)^{(5n+3)/4} \qquad \text{or} \qquad Q(0) \propto C^{(5n+3)/4}$$

Thus, Q at the leading edge, $x = 0$, is greater than zero. The value of $u(x, y) = \psi_y$ also is not zero at $x = 0$. This means that the flow into the convective region from below is not induced from a quiescent ambient medium at t_∞. Therefore, taking $C \neq 0$ means that the origin $x = 0$ was not placed at the proper location. It should be moved to $x = -C$. Nevertheless, use has been made of this formulation, for instance, by Cheesewright (1967), to model circulation and stratification in the ambient medium. Such questions are deferred to a latter section. Here $C = 0$ and

$$d(x) = Nx^n \tag{3.5.22}$$

Similarity forms. The simplest constants C_1, C_2, C_3, and C_4 are determined for the power law and exponential variations, Eqs. (3.5.22) and (3.5.21), respectively. The resulting values of b and c, the relevant Grashof numbers, the differential equations, and the boundary conditions for $v(x, 0) = 0$ are summarized below.

For the power law variation:

$$C_1 = n + 3 \qquad C_2 = n - 1 \qquad C_3 = \frac{1}{K_1} \qquad C_4 = 4n \tag{3.5.23}$$

$$b = \frac{1}{x}\sqrt[4]{\frac{gx^3}{4\nu^2}\beta d} = \frac{G}{4x} \qquad c = G \qquad d = Nx^n \tag{3.5.24}$$

$$\mathrm{Gr}_x = \frac{gx^3}{\nu^2}\beta d \qquad \text{local Grashof number} \tag{3.5.25}$$

$$f''' + (n + 3)ff'' - (2n + 2)f'^2 + \phi = 0 \tag{3.5.26}$$

$$\phi'' + \Pr[(n + 3)f\phi' - 4nf'\phi] = 0 \tag{3.5.27}$$

$$f'(0) = f(0) = 1 - \phi(0) = f'(\infty) = \phi(\infty) = 0 \tag{3.5.28}$$

This result is very similar to that in Section 3.3 for $t_0 - t_\infty$ constant, Eqs. (3.3.17)–(3.3.21). In fact, those relations are recovered above for $n = 0$.

For the exponential variation:

$$C_1 = 1 \qquad C_2 = 1 \qquad C_3 = 1/K_1 \qquad C_4 = 4 \tag{3.5.29}$$

$$\eta = b(x)y \qquad \psi(x, y) = \nu c(x)f \qquad \phi = \frac{t - t_\infty}{d(x)} \tag{3.5.30}$$

$$b = m\sqrt[4]{\frac{\mathrm{Gr}_{x,m}}{4}} = \frac{m}{4}G_m \qquad c = 4\sqrt[4]{\frac{\mathrm{Gr}_{x,m}}{4}} = G_m \tag{3.5.31}$$

$$\mathrm{Gr}_{x,m} = \frac{g}{m^3\nu^2}\beta d \qquad d = Me^{mx} \tag{3.5.32}$$

$$f''' + ff'' - 2f'^2 + \phi = 0 \tag{3.5.33}$$

$$\phi'' + \Pr(f\phi' - 4f'\phi) = 0 \tag{3.5.34}$$

$$f'(0) = f(0) = f'(\infty) = 1 - \phi(0) = \phi(\infty) = 0 \tag{3.5.35}$$

The transformation in terms of $b(x)$ and $c(x)$ is as before. However, here the unit Grashof number, analogous to gx^3/ν^2 before, becomes $g/m^3\nu^2$. The distance measure x is replaced by $1/m$. A surprising result is that the solution of Eqs. (3.5.33)–(3.5.35) is independent of the value of m, in contrast to the dependence of Eqs. (3.5.26)–(3.5.28) on n.

The usefulness and practicality of the above two forms of $d(x) = t_0 - t_\infty$ remain to be determined. Questions concern what values of N, n and M, m correspond to what realistic and/or important circumstances. These matters are considered first by plotting $d(x)$ against x in Fig. 3.5.1 to indicate the available variations. Both N and M are taken as positive. Later in this section, quantities like $u(x, y)$, $\delta(x)$, $q''(x)$, and $Q(x)$ will be evaluated to find the limits that physical realism applies to these variations. However, some observations are useful here. When $n > 0$ in Fig. 3.5.1*a*, $t_0 - t_\infty = 0$ at $x = 0$. The distributions for n very near 0 are also close to the isothermal condition of $n = 0$. The unboundedness for $n < 0$ at $x = 0$ initially appears an unacceptable model for any actual circumstance. However, dismissing such variations at this point would be premature, as will be seen later. Continuations of $d(x)$ are shown to $x < 0$, taking $d = N|x|^n$.

With the exponential variations shown in Fig. 3.5.1*b* for $M > 0$, there is an increase of $t_0 - t_\infty$ downstream for $m > 0$ and a decrease for $m < 0$. For $m = 0$ there is no result; see Eq. (3.5.32). Note that d remains of the same sign for $x < 0$ and that mx changes sign. Therefore, a growth with $x > 0$ becomes a decay in the negative x direction.

An interesting aspect of these distributions is seen in considering the two forms of the Grashof number for the power law and exponential forms:

$$\mathrm{Gr}_x = \frac{gx^3}{\nu^2}\beta N x^n \qquad \text{and} \qquad \mathrm{Gr}_{x,m} = \frac{g}{m^3\nu^2}\beta M e^{mx} \tag{3.5.36}$$

Taking $\mathbf{g} = (-g, 0, 0)$, assumes x increasing positive upward and negative downward from its origin. The foregoing discussions have been largely in terms of N and M positive, that is, upflow. However, for N and M negative, both buoyancy forces in Eq. (3.5.36) become negative and flow is down from $x = 0$. For $d(x) = Nx^n$, x in Gr_x changes sign, Gr_x remains positive, and x^n is taken as $|x|^n$. Alternatively, flows known to be downward may be modeled by taking x increasing positive downward from its origin, considering $N > 0$. An interesting and important exceptional circumstance arises for $n = 1$. Then $d = Nx^n = Nx$ is positive in one direction and negative in the other, depending on the sign of N. For $n = 1$, although the sign of Gr_x is determined entirely by N, it must be taken as positive to calculate G in Eq. (3.5.24). Flow direction is known from the sign of N.

For an exponential variation Me^{mx}, the direction of flow is determined entirely

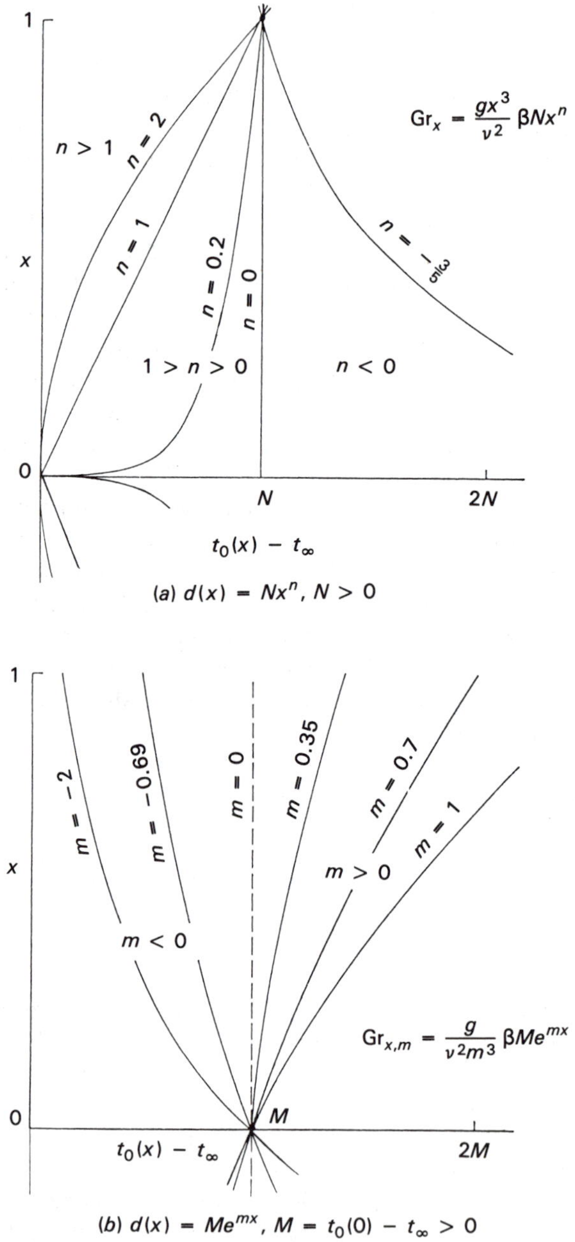

Figure 3.5.1 Variations of $t_0 - t_\infty$ for which similarity solutions result. (*a*) $t_0 - t_\infty = Nx^n$; (*b*) $t_0 - t_\infty = Me^{mx}$.

by the sign of M. On the other hand, the sign of $\mathrm{Gr}_{x,m}$ as written in Eq. (3.5.36) depends on M/m^3. However, it will be shown later in this section that $m < 0$ leads to unreasonable consequences, for example, for any upflow that begins at $x = 0$. Similarly, $m > 0$ is unreasonable for downflow. Clearly, M must be taken positive to calculate G_m in Eq. (3.5.31) from $\mathrm{Gr}_{x,m}$. Therefore, flow direction is simply known from the sign of M.

Note, in addition, that the solutions of the equations and boundary conditions for both the power law and exponential variations, Eqs. (3.5.26)–(3.5.28) and (3.5.33)–(3.5.35), are independent of whether x and N or M are positive or negative. Under some conditions it is only a convention to interpret the solution as being either an upflow or a downflow, separately. The solution is actually made simultaneously for flows in both the positive and negative x directions.

Other limitations. The limitations on reasonable downstream variations of $t_0 - t_\infty = d(x)$ determined above are not the only ones that arise from physical considerations. For the power law variation, the local heat flux $q''(x)$, the total convected energy at x, $Q(x)$, the tangential component of velocity u, and the boundary region thickness $\delta(x)$ are calculated as

$$q''(x) = -k\left.\frac{\partial t}{\partial y}\right)_{y=0} = \frac{[-\phi'(0)]}{\sqrt{2}}\frac{kd}{x}\sqrt[4]{\mathrm{Gr}_x} \propto N[-\phi'(0)]x^{(5n-1)/4} \qquad (3.5.37)$$

$$Q(x) = \int_0^\infty c_p(t - t_\infty)\rho u\, dy$$

$$= c_p\mu\, dc\int_0^\infty \phi f'\, d\eta \propto N\, x^{(3+5n)/4}\int_0^\infty \phi f'\, d\eta \qquad (3.5.38)$$

$$u(x, y) = \psi_y = \nu cbf'(\eta) \propto f'(\eta)x^{(n+1)/2} \qquad (3.5.39)$$

$$\delta(x) = \frac{\eta_\delta}{b} \propto x^{(1-n)/4} \qquad (3.5.40)$$

where $Q(x)$ is as above in the present approximations, but in the absence of stratification.

These relations are examined to determine values of n for which unreasonable consequences would follow. For example, unless $n \geq -1$, $u(x, y)$ in Eq. (3.5.39) is unbounded at the leading edge ($x = 0$). On the other hand, Eq. (3.5.40) indicates that $\delta(0)$ is unbounded unless $n \leq 1$. This leaves the range $1 - \leq n \leq 1$ as reasonable. However, another more restrictive lower limit arises from Eq. (3.5.38). For $N > 0$, $t_0 > t_\infty$ and $q''(x)$ is expected to be positive or at least zero. Therefore, $Q(x)$ must at least be constant or must increase with x. This results for $n \geq -\frac{3}{5}$. Note that for $n = -\frac{3}{5}$, $Q(x) = Q(0)$ for all x. It is a constant. This suggests a horizontal line source at $x = 0$ in an otherwise adiabatic flow. Note in Eq. (3.5.37) that this requires $[-\phi'(0)] = 0$. This is the plane plume or a wall plume, which are considered in Section 3.7. Note that for $n < -\frac{3}{5}$, $Q(0)$ is unbounded, yet its derivative with respect to x is negative. This amounts to an unbounded line source at $x = 0$. The downstream flow loses energy in both directions, at $y = 0$ to the surface and outward into entraining ambient fluid at t_∞.

The reasonable limit then is $-\frac{3}{5} \leq n \leq 1$. This range includes the isothermal condition $n = 0$. It also includes another important special circumstance, first analyzed by Sparrow and Gregg (1956), the uniform imposed surface flux condition. See Fig. 1.1.1 for an example. From Eq. (3.5.37), this condition is seen to result for $n = \frac{1}{5}$, or

$$t_0 - t_\infty = N_{q''} x^{1/5} \tag{3.5.41}$$

where $N_{q''}$ is calculated from Eq. (3.5.37) as

$$q''(x) = q'' = \frac{[-\phi'(0)]}{\sqrt{2}} k N_{q''} x^{-4/5} \sqrt[4]{\frac{gx^3}{\nu^2} \beta N_{q''} x^{1/5}}$$

or

$$N_{q''} = \left\{ \frac{q''}{k[-\phi'(0)]} \right\}^{4/5} \left(\frac{4\nu^2}{g\beta} \right)^{1/5} \tag{3.5.42}$$

The equations and boundary conditions for uniform surface flux transport are merely Eqs. (3.5.26)–(3.5.28) with $n = \frac{1}{5}$. The Prandtl number is the only parameter in the equations. Calculated results are summarized in Table 3.5.1.

This formulation is different from that of Sparrow and Gregg (1956), where a special transformation was used. Sparrow and Gregg introduced a modified Grashof number $\mathrm{Gr}_x^* = g\beta q'' x^4 / k\nu^2$ and defined the similarity transformation in terms of this parameter as $\eta = (y/5x)\, G^*$, $\psi = \nu G^* f(\eta)$, $\phi(\eta) = (t - t_\infty)/(5xq''/kG^*)$ with $G^* = 5(\mathrm{Gr}_x^*/5)^{1/5}$. The governing equations with the boundary conditions become

$$f''' + 4ff'' - 3f'^2 + \phi = 0$$

$$\phi'' + \mathrm{Pr}(4f\phi' - \phi f') = 0$$

$$f(0) = f'(0) = f'(\infty) = \phi(\infty) = \phi'(0) + 1 = 0$$

For the numerical results, see Sparrow and Gregg (1956).

Over the range of n, the Nusselt number for a surface at $y = 0$ is calculated from Eq. (3.5.37) to be, as before in Eq. (3.4.4),

$$\frac{q''(x)}{d} \frac{x}{k} = \frac{h_x x}{k} = \mathrm{Nu}_x = \frac{[-\phi'(0)]}{\sqrt{2}} \sqrt[4]{\mathrm{Gr}_x}$$

$$= \frac{[-\phi'(0)]}{\sqrt{2}} \sqrt[4]{\frac{gx^3}{\nu^2} \beta d}$$

$$= F(n, \mathrm{Pr}) \sqrt[4]{\mathrm{Gr}_x} \tag{3.5.43}$$

An average convection coefficient for the region 0 to x may again be calculated,

Table 3.5.1 Values of transport quantities for laminar flows generated adjacent to a vertical surface dissipating a uniform heat flux q''[a]

Pr	$-\phi'(0)$	$F(\text{Pr}) = \frac{[-\phi'(0)]}{\sqrt{2}}$	$f''(0)$	$f'_{\max}$ $[\eta(f'_{\max})]$	$f(\infty)$	η_{δ_t}[b]	η_{δ}[c]
0.01	0.0947	0.0669	0.9351	0.5274 [1.46]	4.189	30.4	27.3
0.1	0.2669	0.1887	0.8135	0.3966 [1.17]	1.403	11	11.5
0.72	0.5756	0.4070	0.6389	0.2514 [0.97]	0.538	4.3	5.2
1.0	0.6453	0.4563	0.6069	0.2288 [0.92]	0.473	3.73	5.12
2.0	0.8116	0.5739	0.5394	0.1843 [0.82]	0.365	2.8	5.13
5.0	1.0759	0.7608	0.4543	0.1345 [0.77]	0.268	1.94	5.36
6.7	1.1726	0.8291	0.4286	0.1209 [0.7]	0.244	1.8	5.4
10.0	1.3164	0.9309	0.3951	0.1042 [0.67]	0.227	1.5	7.5
100.0	2.4584	1.7380	0.2367	0.040 [0.51]	0.125	0.8	12.7

[a]Prepared by Ramesh Krishnamurthy.
[b]Corresponds to the value of η where $\phi = 0.01$.
[c]Corresponds to the value of η where $f' = 0.01\ (f'_{\max})$.

as in Eq. (3.4.6) for an isothermal surface. It would be used as below to calculate total heat transfer from 0 to x as

$$Q(x) = \bar{h}_x x(\overline{t_0 - t_\infty}) = \int_0^x q''(x)\, dx \tag{3.5.44}$$

where $\overline{t_0 - t_\infty}$ might be taken as the average value of $d(x)$ from 0 to x:

$$\overline{t_0 - t_\infty} = \frac{1}{x}\int_0^x d(x)\, dx = \frac{N}{n+1} x^n \qquad \text{for } n \geq 0 \tag{3.5.45}$$

The resulting value of $\bar{h}_x$ for $n \geq -\frac{3}{5}$ is obtained from Eq. (3.5.43) as

$$\bar{h}_x = 2\sqrt{2}\, k\, \sqrt[4]{N}\, [-\phi'(0)]\, \frac{n+1}{5n+3}\, x^{(n-1)/4} \left(\frac{g\beta}{\nu^2}\right)^{1/4} \tag{3.5.46}$$

Considerations similar to the above are now applied to the exponential form $d(x) = Me^{mx}$. The principal anomaly of this form, shown in Gebhart and Mollendorf (1969), is seen in calculating $Q(x)$ and also $M(x)$, the local level of momentum in the boundary region, at distance x from the leading edge:

$$Q(x) = \int_0^\infty \rho c_p(t - t_\infty)u\, dy \propto e^{5mx/4} \int_0^\infty \phi f'\, d\eta$$

$$M(x) = \int_0^\infty \rho u^2\, dy \propto e^{3mx/4} \int_0^\infty f'^2\, d\eta$$

Neither of these is zero at $x = 0$. Forming the ratio of local values $Q(x)$ and $M(x)$ at $x = L$ to their values at zero, $Q(0)$ and $M(0)$, yields

$$\frac{Q(L)}{Q(0)} = e^{5mL/4} \tag{3.5.47}$$

$$\frac{M(L)}{M(0)} = e^{3mL/4} \tag{3.5.48}$$

Clearly, m must be greater than zero, as seen above. Also, mL must be large in order that the onflow of energy and momentum from below, at $x = 0$, has a small net effect in the region $x = 0$ to L. These onflow effects at $x = 0$ are actually the convection generated adjacent to a surface from $x = 0$ to $-\infty$ along which $d = Me^{mx}$, where $mx < 0$. This is the decaying continuation of d for $x \geq 0$ to negative x.

Local transfer quantities and the boundary region thickness are found to be

$$q''(x) = -k\left.\frac{\partial t}{\partial y}\right)_{y=0} = \frac{[-\phi'(0)]}{\sqrt{2}}\, mdk\, \sqrt[4]{\mathrm{Gr}_m} \tag{3.5.49}$$

$$\mathrm{Nu}_m = \frac{h}{mk} = \frac{q''(x)}{d}\frac{1}{mk} = \frac{[-\phi'(0)]}{\sqrt{2}} \sqrt[4]{\mathrm{Gr}_m} \tag{3.5.50}$$

$$\delta(x) = \frac{\eta_\delta}{b} = \frac{\sqrt{2}\,\eta_\delta}{m\sqrt[4]{\mathrm{Gr}_m}} \tag{3.5.51}$$

The characteristic length in Nu_m is $1/m$ and x appears only in $d(x)$ in Gr_m. The value of $\delta(0)$ is greater than zero, recalling that $m > 0$.

3.6 SIMILARITY WITH ADDITIONAL TRANSPORT EFFECTS

The foregoing determinations of admissible variations of the local downstream motive temperature difference $t_0 - t_\infty = d(x)$ have extended the range of the similarity formulation. It remains to determine, under the constraint of power law and exponential downstream variations of $d(x)$, the conditions under which other physical effects may be retained while preserving similarity. Among these other effects in thermally buoyant flows is ambient medium stratification. Then there are the other three remaining physical effects in the energy equation: the pressure, viscous dissipation, and distributed energy source terms in Eqs. (3.5.9) and (3.5.11).

Finally to be considered here is the effect of the addition or removal of the same fluid at the location $y = 0$, where t_0 is applied. This may arise, for example, at a porous surface at $y = 0$ through which the fluid also flows. This affects the velocity boundary condition there.

3.6.1 Ambient Medium Stratification

This is expressed in the present formulation as an x dependence of $t_\infty(x) - t_r = j(x)$ in Eq. (3.5.6) and, therefore, of density ρ_∞. The term in j_x in Eqs. (3.5.9) and (3.5.11e) results from the thermal convection term. The coefficient that must be constant for similarity is

$$\frac{cj_x}{bd} = C_5 \tag{3.5.11e}$$

This is the same as the term that arose from the d variation downstream, Eq. (3.5.11d), except that d_x there is replaced by j_x above. Therefore, given the permissible variations of $d(x)$ determined in Section 3.5 along with the resulting values of $b(x)$ and $c(x)$, $j(x)$ must vary as $d(x)$. That is,

$$j = t_\infty - t_r = \frac{C_5}{4n} Nx^n = N_\infty x^n \qquad C_5 = \frac{4nN_\infty}{N} \tag{3.6.1}$$

and

$$j = t_\infty - t_r = \frac{C_5}{4} Me^{mx} = M_\infty e^{mx} \qquad C_5 = \frac{4M_\infty}{M} \tag{3.6.2}$$

Thus, j may only be some positive or negative multiple of d, for each of the variations, as in Fig. 3.5.1. There would then be two varying differences between t_0 and the constant reference temperature t_r, $d(x)$ and $j(x)$, respectively. The actual resulting vertical variation of t_0 is

$$t_0 - t_r = t_0 - t_\infty + t_\infty - t_r = Nx^n\left(1 + \frac{C_5}{4n}\right) = (N + N_\infty)x^n \tag{3.6.3}$$

$$\text{or} \qquad (M + M_\infty)e^{mx} \tag{3.6.4}$$

Another possibility is that $d = t_0 - t_\infty$ is constant. Then from Eq. (3.5.11e) $j(x) \propto \ln x$, which becomes singular at the leading edge. This variation will be set aside as unreasonable.

Strict limits on permissible stratification arise to ensure a quiescent ambient medium. Since the physical extent and configuration of the medium are not to be specified or limited in any way here, it must be stable when of infinite extent. A simple and approximate initial estimate of stability might be that ρ_∞ may not increase upward. A considerable increase would lead to instability and cause circulations in the ambient medium. A simple condition that implies this kind of stability is

$$\frac{d}{dx}\,\rho_\infty(t_\infty, p_\infty) \le 0 \tag{3.6.5}$$

If density is essentially a function only of temperature, $\rho_\infty = \rho_\infty(t_\infty)$, and $\beta > 0$, this amounts to

$$\frac{dt_\infty}{dx} \ge 0 \tag{3.6.6}$$

These simple rules apply to a good approximation for most states of most liquids. However, they are not conservative estimates for all fluids. The general condition for stable stratification is that arising from thermodynamic considerations. For stability, the actual vertical gradient in density $d\rho_\infty/dx$ must be equal to, less than, or more negative than the adiabatic one $(\partial\rho_\infty/\partial x)_S$, where S means constant entropy. That is,

$$\frac{d\rho_\infty}{dx} \le \left.\frac{\partial\rho_\infty}{\partial x}\right)_S < 0 \qquad \text{note that } \left.\frac{\partial\rho}{\partial p}\right)_S > 0 \tag{3.6.7}$$

The requirement is that the actual vertical distribution of ρ_∞ must decrease at the same rate as or more rapidly than the adiabatic one. The adiabatic decrease, in turn, is that which would occur if an amount of the fluid were moved upward, expanding reversibly and adiabatically at the increasingly lower hydrostatic pressure levels. For $\rho_\infty = \rho_\infty(p_\infty, t_\infty)$, the above condition may also be written as

$$\frac{dt_\infty}{dx} = j_x \ge \left.\frac{\partial t}{\partial x}\right)_S \qquad \text{note that } \left.\frac{\partial t}{\partial x}\right)_S < 0 \tag{3.6.8}$$

A general hydrostatic variation of p_∞ and vertical variations of ρ_S and t_S are seen in Fig. 3.6.1. Also shown are stable and unstable vertical variations of both ρ_∞ and t_∞.

The above general limit, Eq. (3.6.7), may be expressed in terms of j_x and thermodynamic behavior as follows. The hydrostatic pressure field is determined by

$$\frac{dp}{dx} = -g\rho_\infty \tag{3.6.9}$$

For $\rho = \rho(t, p)$

$$\frac{d\rho}{dx} = \left.\frac{\partial\rho}{\partial t}\right)_p \frac{dt}{dx} + \left.\frac{\partial\rho}{\partial p}\right)_t \frac{dp}{dx}$$

Then

$$\frac{d\rho_\infty}{dx} = \left.\frac{\partial\rho}{\partial t}\right)_p j_x - g\rho_\infty \left.\frac{\partial\rho}{\partial p}\right)_t$$

and

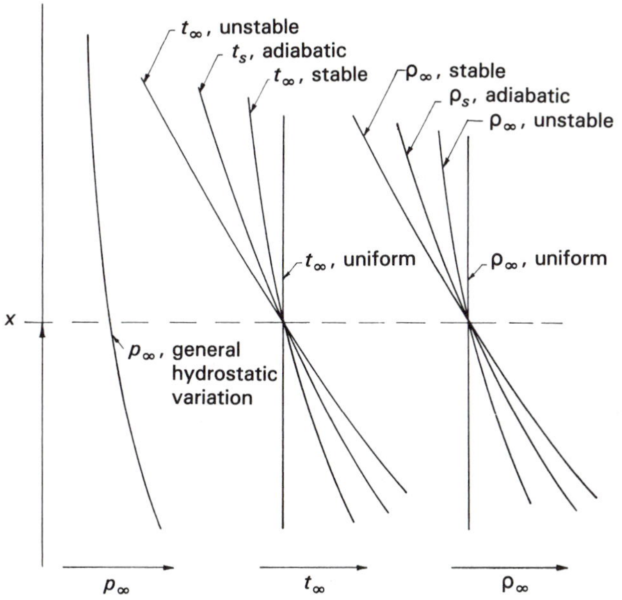

Figure 3.6.1 Adiabatic, stable, and unstable vertical variations of t_∞, p_∞, and $\rho_\infty(t, p)$.

$$\left.\frac{d\rho_\infty}{dx}\right)_S = \left.\frac{\partial\rho}{\partial t}\right)_p \left.\frac{dt}{dx}\right)_S - g\rho_\infty \left.\frac{\partial\rho}{\partial p}\right)_t$$

Therefore, from Eq. (3.6.7)

$$\left.\frac{\partial\rho}{\partial t}\right)_p j_x \leq \left.\frac{\partial\rho}{\partial t}\right)_p \left.\frac{dt}{dx}\right)_S \tag{3.6.10}$$

For $\beta > 0$ this becomes

$$j_x \geq \left.\frac{\partial t}{\partial x}\right)_S = \left(\frac{\partial t}{\partial p}\right)_S \frac{dp}{dx} = -g\rho_\infty \left.\frac{\partial t}{\partial p}\right)_S \tag{3.6.11}$$

The right side in Eq. (3.6.11), the vertical gradient in adiabatic temperature t_S, is negative. Therefore, j_x, the gradient of t_∞, may also be negative. However, t_∞ may not decrease more rapidly than t_S.

For ideal gases the gradient of t_S may be simply calculated:

$$\left(\frac{\partial t}{\partial p}\right)_S = \frac{\gamma - 1}{\gamma}\frac{T}{p} = \frac{1}{c_p\rho}$$

Thus

$$j_x \geq -\frac{g}{c_p} \quad \text{for ideal gases} \tag{3.6.12}$$

This is a very small quantity for terrestrial gravity, about 0.009°C/m for air. For liquids $(\partial t/\partial p)_S = \beta T/\rho c_p$ is much smaller and is, therefore, the lower limit of j_x in Eq. (3.6.11); see Jaluria (1980). The limits for a power law variation of j, Eq. (3.6.1), are then

$$j_x = C_5 \frac{N}{4} x^{n-1} \geq -g\rho \left.\frac{\partial t}{\partial p}\right)_S = -\frac{g}{c_p} \text{ (ideal gases)} \simeq 0 \text{ (liquids)} \quad (3.6.13)$$

Using Eq. (3.6.1), this relation is written as a general limit on the value of C_5:

$$C_5 \geq -\frac{4g\rho}{N} \left.\frac{\partial t}{\partial p}\right)_S x^{1-n} \quad \text{or} \quad -\frac{4g}{Nc_p} x^{1-n} \quad \text{for gases} \quad (3.6.14)$$

There are several limitations on this result. First, it is apparent from Eq. (3.6.1) that $n < 0$ results in t_∞ unbounded as x goes either to zero or to large values, depending on the sign of C_5/n. Therefore, the permissible range of n for stratification for reasonable flow modeling is narrowed to $0 \leq n \leq 1$. For $n = 1$ a single nonzero lower limit results for C_5. However, for $0 < n < 1$, the lower limit of C_5 is set as zero as x goes to zero. Taking this into account, the general lower limits regarding stratification, in terms of C_5 and t_∞, for $0 < n < 1$ and $N > 0$ are

$$C_5 \geq 0 \quad \text{or} \quad \frac{dt_\infty}{dx} \geq 0 \quad (3.6.15)$$

These apply if not even small local regions of unstable stratification are to arise. This means in effect that t_∞ is constant or increasing upward.

However, another interesting circumstance arises for $n = 1$. Then

$$C_5 = \frac{4}{N} j_x \geq -\frac{4g\rho}{N} \left.\frac{\partial t}{\partial p}\right)_S \quad \text{or} \quad -\frac{4g}{Nc_p} \quad \text{for gases} \quad (3.6.16)$$

The limiting stratification is adiabatic and the resulting distribution, from Eqs. (3.6.13) and (3.6.16), is j_S, where

$$j_S = t_\infty - t_r = -\frac{g}{c_p} x = N_\infty x \quad (3.6.17)$$

Another use of the result in Eq. (3.6.16), when $n = 1$, is to study the effects of stratification on transport. For example, take $N = g/c_p > 0$, $C_5/4n = -1$, and $t_0 - t_r = 0$. Thus, an isothermal condition arises even with stratification and $j_x \propto d_x$. Then $t_r = t_0$, $t_0 - t_\infty = Nx$, and $t_\infty - t_0 = -N_\infty x$. Then $C_5 = -4$ and the term in the energy balance equation, (3.5.9), $-C_5 f'$ becomes simply $+4f'$. This is similar to the one-dimensional analysis of Gill (1966) considered in Section 3.11.

The relative magnitude of the effect of ambient medium stratification is assessed by comparing the term $C_5 f'$ with a principal convection term

$$R' = \frac{C_5 f'}{C_1 f\phi'} = O\left(\frac{C_5}{C_1}\right) \tag{3.6.18}$$

With conditions for similarity satisfied, this ratio becomes

$$\frac{C_5}{C_1} = \frac{cj_x}{dc_x} = \frac{2x^{1-n}}{N(n+3)} j_x \tag{3.6.19}$$

As an example, an adiabatically stratified gas results in $N_\infty = -g/c_p$ and $n = 1$; therefore,

$$\frac{C_5}{C_1} = -\frac{g}{c_p N} \tag{3.6.20}$$

For $N >> g/c_p$, this is a very small effect. However, for N smaller, comparable effects may arise.

3.6.2 The Pressure Energy Effect

The pressure term in the energy equation (3.5.9) under the transformation in terms of b and c becomes $-(\beta T)(c/bd)(g/c_p)f'$. The coefficient of f' must be independent of x for this effect to be included in a similar solution. This coefficient becomes

$$(\beta T)\frac{c}{bd}\frac{g}{c_p} = (\beta T)\frac{4g}{Nc_p}x^{1-n} \qquad \text{for } d = Nx^n \tag{3.6.21}$$

$$= (\beta T)\frac{4g}{mM}e^{-mx} \qquad \text{for } d = Me^{mx} \tag{3.6.22}$$

The explicit x dependence above disappears only for the power law form and then only for $n = 1$. The implicit nonsimilarity effect in βT disappears for ideal gases since $\beta T = 1$. For other fluids generally it is small since the variation of T across the transport region is very small if $R_1 \simeq \beta(t - t_\infty)$ is much less than 1, as required for this formulation.

Considering $n = 1$ and an ideal gas, this pressure term is

$$-\beta T C_{10} K_4 f' = -\frac{4g}{Nc_p}f' \tag{3.6.23}$$

Recall that the stratification term for an adiabatically stratified ideal gas became

$$-C_5 f' = \frac{4g}{Nc_p}f' \tag{3.6.24}$$

The stratification and pressure terms thus cancel each other. Therefore, neglecting both at the outset is justified.

3.6.3 Viscous Dissipation

The viscous dissipation term in the energy equation is $C_6 K_3 f''^2$. This becomes

$$+\frac{b^2c^2}{d}\frac{\nu^2}{c_p}f''^2 = \frac{4g\beta x}{c_p}f''^2 \qquad \text{for } d = Nx^n \tag{3.6.25}$$

$$= \frac{4g\beta}{mc_p}f''^2 \qquad \text{for } d = Me^{mx} \tag{3.6.26}$$

For the power law form an inevitably x-dependent coefficient arises, $4g\beta x/c_p$. It is called the dissipation parameter. Similarity does not result, that is, $f = f(\eta, x)$ and $\phi = \phi(\eta, x)$. However, this parameter is usually small, and solutions were given by Gebhart (1962) using the following asymptotic expansion in terms of $\epsilon(x) = g\beta x/c_p$:

$$\psi(x, y) = \nu c f(\eta, x) = \nu c[f_0(\eta) + \epsilon(x) f_1(\eta) + \cdots] \tag{3.6.27}$$

$$\phi(\eta, x) = \phi_0(\eta) + \epsilon(x)\phi_1(\eta) + \cdots \tag{3.6.28}$$

where f_0, ϕ_0 are solutions of Eqs. (3.5.26)–(3.5.28), which neglect viscous dissipation. Then f_1, ϕ_1 are the first correction for the effects of viscous dissipation, and so forth. Results were calculated for Prandtl numbers from 10^{-2} to 10^4. Roy (1969) extended the range to $\mathrm{Pr} \to \infty$.

For the exponential form, the coefficient $g\beta/mc_p$ is a constant and a similar solution results; see Gebhart and Mollendorf (1969). This new dissipation parameter appears as above in the formulation of Eq. (3.5.9), in addition to the Prandtl number, as in Eq. (3.5.34). Extensive calculations of this effect were given for Prandtl numbers from 0.7 to 100 and for $4g\beta/mc_p$ from 0 to 2.

In connection with viscous dissipation, recall the discussion in Section 2.6 concerning its magnitude with respect to the pressure energy effect. In general, the latter effect is the larger one (Ackroyd, 1974; Mahajan and Gebhart, 1988).

3.6.4 A Distributed Energy Source

The next effect considered is the presence of a distributed energy source in the fluid. The term is

$$\frac{1}{b^2 d}\frac{1}{\mathrm{Pr}}\frac{q'''}{k} = \sqrt{\frac{\nu^2}{g\beta N}}\,\frac{x^{(1-3n)/2}}{\mathrm{Pr}}\,\frac{q'''}{Nk} \equiv \frac{Q'''}{\mathrm{Pr}}x^{(1-3n)/2} \tag{3.6.29}$$

The nature of this term results from the dependence of $Q'''\,x^{(1-3n)/2}$. If Q''' were a simple constant, then buoyancy would arise throughout the fluid medium. That would not be boundary region transport. On the other hand, the condition might be $Q''' = Q'''(x, y)$. Similarity would result only if $Q'''(x, y)x^{(1-3n)/2}$ were η-dependent, as $F_1(\eta)$, and if $F_1(\eta)$ decayed rapidly enough out from $\eta = 0$. This would arise, for example, if $F_1(\eta)$ depended only on ϕ.

Similarity is seen to be both possible and of practical value for certain restricted conditions for each of the four added physical effects considered above. Permissible stratification variations are Eqs. (3.6.1) and (3.6.2), subject to the ambient medium stability limitations set forth. Considering hereafter only the power law variation of $d(x)$, similarity results, including the pressure term, only for $n = 1$ and if βT may be approximated as 1. Viscous dissipation effects invariably give rise to the parameter $4g\beta x/c_p = 4R_0$. Finally, the distributed source, for $n = 1$, results in similarity if $F_1 = F_1(\eta)$ decays sufficiently rapidly with η. The resulting equations that govern the transport are

$$f''' + (n+3)ff'' - (2n+2)f'^2 + \phi = 0 \tag{3.6.30}$$

$$\phi'' + \Pr[(n+3)f\phi' - 4nf'\phi] - \frac{4nN_\infty}{N}f' + (\beta T)\frac{4g}{Nc_p}x^{1-n}$$

$$+ \frac{4g\beta x}{c_p}f''^2 + \frac{Q'''}{\Pr}x^{(1-3n)/2} = 0 \tag{3.6.31}$$

The added complexities in Eq. (3.6.31) affect Eq. (3.6.30) only in the coupling through f and ϕ. It is apparent that the particular circumstance $n = 1$ has considerable utility in assessing the influence of two of these effects.

3.6.5 Fluid Addition or Removal at the Bounding Surface

The last condition considered is fluid addition or removal at $y = 0$, as through a porous surface. Similarity considerations concerning this effect are independent of those considered above in this section. For an impervious surface $v(x, 0) = 0$. That is, $f(0) = 0$ in Eqs. (3.5.28) and (3.5.35). This condition results from the following calculation of $v(x, 0)$ from $\psi = \nu cf(\eta)$:

$$v(x,0) = -\psi_x = -\nu[c_x f + cf'yb_x] = -\nu c_x f(0) \tag{3.6.32}$$

Thus, $v(x, 0) = 0$ results in $f(0) = 0$. In general, $f(0)$, often called the blowing parameter, becomes, from Eq. (3.6.32),

$$f(0) = -\frac{v(x,0)}{\nu c_x} = -\frac{\sqrt{2}\,v(x,0)x^{(1-n)/4}}{(3+n)\sqrt[4]{g\beta N\nu^2}} \quad \text{for } d(x) = Nx^n \tag{3.6.33}$$

$$= -\frac{\sqrt{2}\,v(x,0)e^{mx/4}}{\sqrt[4]{gm\beta M\nu^2}} \quad \text{for } d(x) = Me^{mx} \tag{3.6.34}$$

Thus, similarity results only if the imposed normal velocity component at $y = 0$ varies with x as

$$v(x,0) \propto x^{(n-1)/4} \quad \text{power law} \tag{3.6.35}$$

$$\propto e^{-mx/4} \quad \text{exponential} \tag{3.6.36}$$

The condition $n = 1$ permits a similarity solution for $v(x, 0)$ uniform. Then $f(0)$ is a positive constant for fluid removal at $y = 0$ and negative for fluid addition.

For n other than 1, $v(x, 0)$ must be x-dependent. Solutions for various blowing and suction conditions are given in Section 3.12.

3.7 THERMAL PLUMES

The general similarity formulation of Section 3.5 permits considerable flexibility in the kinds of temperature conditions and collection of physical effects that may be simply modeled to calculate, for example, $u(x, y)$ and $t(x, y)$. An important kind of flow included among these is that of plane thermal plumes.

Free plume. The example of a free plume seen in Fig. 1.1.2 arises from a horizontal electrically heated wire of small diameter D. The thermal transport region seen around the wire, partly obscured by the wire support, is not of boundary region form. However, farther downstream, for $x >> D$, the condition $\delta(x)/x << 1$ is a much better approximation, and boundary region simplifications apply. The nature of local downstream transport is sketched in Fig. 3.7.1. The conditions at $y = 0$ are completely symmetric. At $y = 0$ there is no shear stress, no cross-flow thermal flux, and no y-direction velocity component. That is, at $y = 0$, $\partial u/\partial y = \partial t/\partial y = 0 = v(x, 0)$. In terms of f and ϕ these will become $f''(0) = \phi'(0) = f(0) = 0$, with similarity. Note that an isolated spherical or point

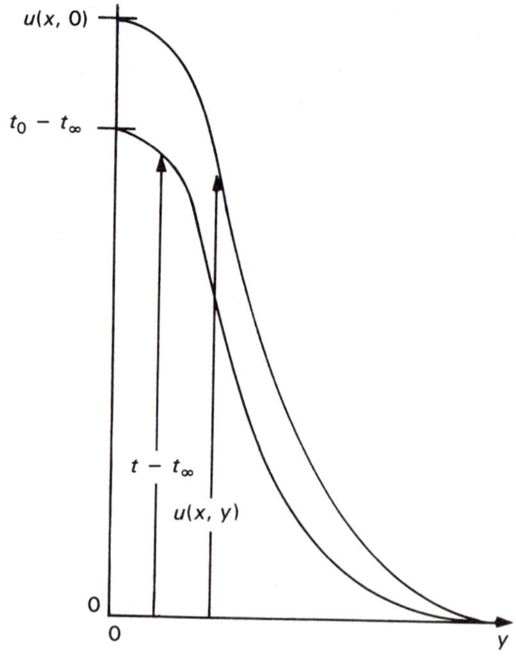

Figure 3.7.1 Distributions of temperature and vertical velocity downstream of a horizontal concentrated line source of energy of strength Q per unit span, the plane thermal plume.

source of energy Q also produces a plume downstream. The resulting axisymmetric flow is treated in detail in Section 4.4.

Wall plume. Another similar plane flow results from a horizontal line of energy sources, such as a row of electronic components, embedded at the surface of a vertical slab of insulating material. These thermal energy sources might be idealized as a horizontal line source of strength Q embedded at the surface. Transport is visualized in Fig. 3.7.2. The source produces a buoyancy force in the fluid. The heated fluid rises, carrying Q away downstream. This flow upward along the surface is again a different kind of circumstance than those with heat flux at the solid-fluid interface. It was called a wall plume by Zimin and Lyakhov (1970).

The surface does not exchange energy with the rising fluid in steady state. Although there is shear at the interface, there is no heat transfer. Thus, the distribution of $t - t_\infty$ has zero slope at $y = 0$. That is, $\phi' = 0$ there. The energy Q is simply diffusing out into the entrained ambient fluid. A principal question here is how $t_0(x)$ varies with downstream location, which might be relevant in the positioning and permissible thermal loading of other dissipating components to be placed downstream along the surface.

Plume behavior. Downstream temperature decay for plumes is determined by calculating the condition under which $Q(x)$ in Eq. (3.5.38) is equal to a con-

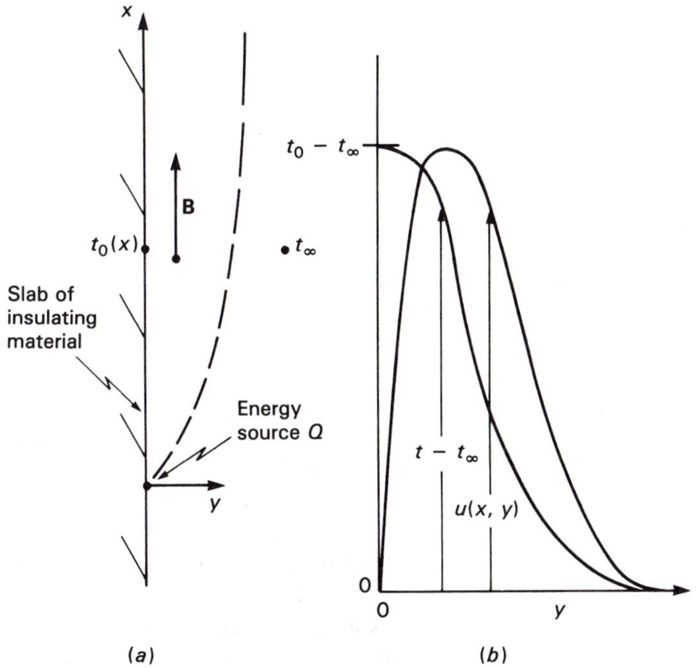

Figure 3.7.2 Wall plume arising adjacent to an insulating surface downstream of a concentrated horizontal energy source, of strength Q per unit span, embedded at the surface. (*a*) Wall plume; (*b*) distributions downstream.

stant Q. This is half the input per unit length by the source for the free plume and the total input for the wall plume.

$$Q(x) = \sqrt[4]{4^3 g\beta N^5 c_p^4 \rho^2 \mu^2}\, x^{(5n+3)/4} \int_0^\infty \phi f'\, d\eta = Q \qquad (3.7.1)$$

Thus, $n = -3/5$ and N is found from Eq. (3.7.1) as follows, where I is to be determined from the solution f, ϕ:

$$t_0 - t_\infty = Nx^{-3/5} = \left(\frac{Q^4}{4^3 g\beta c_p^4 \rho^2 \mu^2 I^4}\right)^{1/5} x^{-3/5} \qquad (3.7.2)$$

where

$$I = \int_0^\infty \phi f'\, d\eta = I(\mathrm{Pr}) \qquad (3.7.3)$$

The velocity then becomes

$$u(x, y) = \psi_y = \nu c b f' = \left(\frac{2g\beta Q}{c_p I\sqrt{\mu\rho}}\right)^{2/5} x^{1/5} f' \qquad (3.7.4)$$

Velocity is seen to increase with x. Of course, this prediction does not continue to apply as the flow becomes turbulent downstream. The above results apply for both the free plume and the wall plume, which are considered in turn below. Note that a downward-flowing plume arises for a sink of strength Q. Then Q may still be taken as positive in Eq. (3.7.2) and x taken as increasing positive downward.

3.7.1 Characteristics of Free Plumes

Zeldovich (1937) first described thermally buoyant plumes arising from a point and from a horizontal line source of heat. The similarity methods used by Tollmien (1926), to solve for the turbulent flow velocity in two-dimensional and axisymmetric jets, and by Schlichting (1933), to solve for laminar flow velocities, were employed. Buoyancy and a similarity form of temperature distribution were included. The treatment by Zeldovich does not permit a velocity component normal to the plane of symmetry of the plume. However, using the conditions that all the terms of the x-momentum equation are of the same order of magnitude and that the heat produced by the source crosses each horizontal plane, expressions were given for the velocity and temperature distributions for both the two-dimensional and axisymmetric cases for both laminar and turbulent flow.

Schmidt (1941) investigated the behavior of natural convection in a turbulent plume above a line and point source of heat. A similarity technique was used. The governing flow equations were solved by assuming a series solution in terms of the similarity variable. Experiments included measurements of the temperature and velocity above an electrically heated wire.

Schuh (1948) reported an analysis of plume flows above line and axisymmetric sources, assuming the form of the similarity variable as originally proposed by Prandtl. Numerical solutions of the coupled equations were given for Pr = 0.7. Yih (1952) found a transformation that results in closed-form solutions for the temperature and velocity distribution for flow above a line source of heat for Prandtl numbers of $\frac{5}{9}$ and 2. Measurements of velocity and temperature distributions above a line of small gas flames designed to simulate a line source of heat were made by Rouse et al. (1952). A solution was given by Sevruk (1958) in a power series. Crane (1959) analyzed the boundary layer equations for a plume in a gas, taking the viscosity and thermal conductivity as dependent directly on the absolute temperature. A conventional transformation was used and a series solution determined for a Prandtl number of $\frac{5}{9}$. An asymptotic Prandtl number solution was given by Spalding and Cruddace (1961), after simplifying the differential equations as appropriate, in a medium of very high Prandtl number, $\text{Pr} \to \infty$.

A very thorough treatment of both the plane and the axisymmetric plume appears in Fujii (1963). Closed-form boundary layer and numerical solutions were given for Pr = 2 and for Pr = 0.01, 0.7, and 10, respectively. A closed-form solution for Pr = $\frac{5}{9}$ and numerical calculations for other values were also given by Brand and Lahey (1967).

In summary, the closed-form solutions for Pr = $\frac{5}{9}$ and 2, although of considerable interest, arise in a range that excludes many of the most important fluids. Thus, numerical calculations are very widely used to determine transport in both plane and axisymmetric plumes. The following formulation, by Gebhart et al. (1970), sets forth the equations and boundary conditions in a convenient and efficient form for calculation.

The momentum and thermal transport equations for $n = -\frac{3}{5}$, written in terms of the transformations in Eqs. (3.5.24) and (3.5.25), are

$$f''' + \frac{12}{5} ff'' - \frac{4}{5} f'^2 + \phi = 0 \tag{3.7.5}$$

$$\phi'' + \frac{12\text{Pr}}{5}(f\phi' + f'\phi) = \phi'' + \frac{12\text{Pr}}{5}(f\phi)' = \phi'' + K(f\phi)' = 0 \tag{3.7.6}$$

where similarity, that is, $f(\eta)$ and $\phi(\eta)$, is assured if the collection of boundary conditions may be expressed in terms of η above. The midplane conditions at $y = 0$, $\partial t/\partial y = 0$, $\partial u/\partial y = 0$, $v(x, 0) = 0$, and $t(x, 0) = t_0$, yield

$$\phi'(0) = f''(0) = f(0) = 1 - \phi(0) = 0 \tag{3.7.7}$$

The conditions at large η, in the quiescent ambient medium at t_∞, are

$$f'(\infty) = \phi(\infty) = 0 \tag{3.7.8}$$

Further, since $Q(x)$ is constant, one might invoke the integral in Eq. (3.7.3) as an additional condition on the distributions f and ϕ, using a transformation of the variables that results in a constant value of I, say $I = 1$.

Thus, six conditions appear in Eqs. (3.7.7) and (3.7.8), plus the integral, whereas only five are required for the integration of Eqs. (3.7.5) and (3.7.6). Some of these conditions have been used in earlier calculations. Considering the six conditions in Eqs. (3.7.7) and (3.7.8), it has been shown by Gebhart et al. (1970) that only five are independent, as follows. Since the energy equation (3.7.6) is a perfect differential, it is integrated to give

$$\phi' + \frac{12}{5} f\phi \mathrm{Pr} = C_1 = 0 \tag{3.7.9}$$

Since $\phi'(0)$ and $f(0)$ are zero, $C_1 = 0$. Integrating again from 0 to η yields

$$\phi(\eta) = \phi(0) \exp\left[-\left(\frac{12}{5}\right)\mathrm{Pr} \int_0^{\eta} f\, d\eta\right] = \exp\left[-\left(\frac{12}{5}\right)\mathrm{Pr} \int_0^{\eta} f\, d\eta\right] \tag{3.7.10}$$

Since f is positive and becomes constant at large η—that is, $f(\infty)$ is the entrainment velocity from the ambient medium—the integral in Eq. (3.7.10) is unbounded. Therefore,

$$\lim_{\eta\to\infty} \phi(\eta) = 0 \tag{3.7.11}$$

Thus, $\phi(\infty) = 0$ is not an independent condition; it results from the other conditions $\phi'(0) = f(0) = 1 - \phi(0) = 0$ used to evaluate Eq. (3.7.10). Therefore, there is some choice among the boundary conditions.

Since integration of the equations is a two-point boundary value problem, numerical efficiency is achieved by having as many as possible of the five conditions apply at one boundary. Using the following, as in Gebhart et al. (1970), four arise at $\eta = 0$ and only one applies at large η:

$$\phi'(0) = f''(0) = f(0) = 1 - \phi(0) = f'(\infty) = 0 \tag{3.7.12}$$

Finally, this simplicity results from the formulation of ϕ, where $t_0 - t_\infty$ is as expressed in Eqs. (3.7.2) and (3.7.3).

Calculated downstream plume transport is seen in Figs. 3.7.3–3.7.5 for the Prandtl number range 0.01–100. Various parameters are also given in Table 3.7.1. The boundary conditions at $\eta = 0$ are seen to be satisfied. The Prandtl number effect is similar to that seen by Schuh (1948) for flows adjacent to an isothermal surface. That is, δ_t/δ decreases from around 1.0 at Pr $\simeq$ 1 to much lower values as Pr becomes large.

Experimental study of laminar plumes. Laminar plumes rising above long electrically heated wires of small diameter have been studied. Brodowicz and Kierkus (1966) measured velocity and temperature distributions in air above a wire having a length-to-diameter ratio L/D of 3330. Temperatures were measured in air above a wire of $L/D = 250$ by Forstrom and Sparrow (1967). Temperatures were determined by Schorr and Gebhart (1970) in an interferometric study of a plume, as in Fig. 1.1.2, in a light silicone oil (Pr = 6.7) above a wire of 1.27×10^{-4} m diameter and 0.1524 m length, $L/D = 1200$. The data of all three

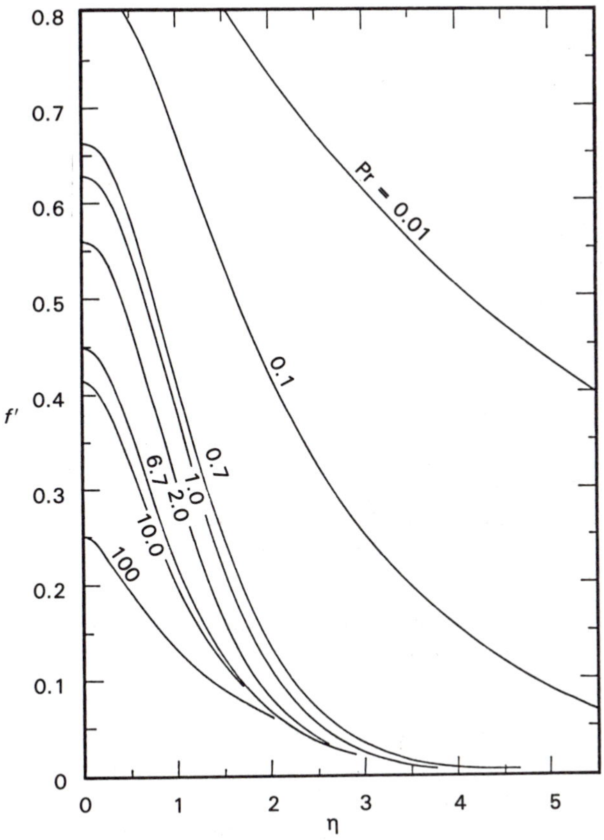

Figure 3.7.3 Plume velocity distributions across the boundary region as a function of Prandtl number. *(Reprinted with permission from* Int. J. Heat Mass Transfer, *vol. 13, p. 161, B. Gebhart et al. Copyright © 1970, Pergamon Journals Ltd.)*

investigations indicated plume temperatures significantly below the analytical predictions. The generalized midplane temperatures from all three studies at different heating rates are shown in Fig. 3.7.6 plotted against the local downstream Grashof number in the plume. The results have the correct slope but are uniformly low by approximately 15%. Measurements by Fujii (1973) also found low temperature levels, except at smaller Gr_x. Laser Doppler measurements downstream in water by Nawoj and Hickman (1977) showed 20–25% deficiencies in velocity level. Similar instrumentation used in air by Goodman et al. (1974) showed deficiencies of about 8%.

Schlieren studies of plumes in air and in water, in a direction normal to the plane of the plume, clearly show large end effects, that is, necking of the plume, an L/D effect. Schorr and Gebhart (1970) showed that allowing for finite wire size and nonsimilar flow near the wire by finding a "virtual" line source does not remove the systematic disagreement between theoretical results and data. Also, a

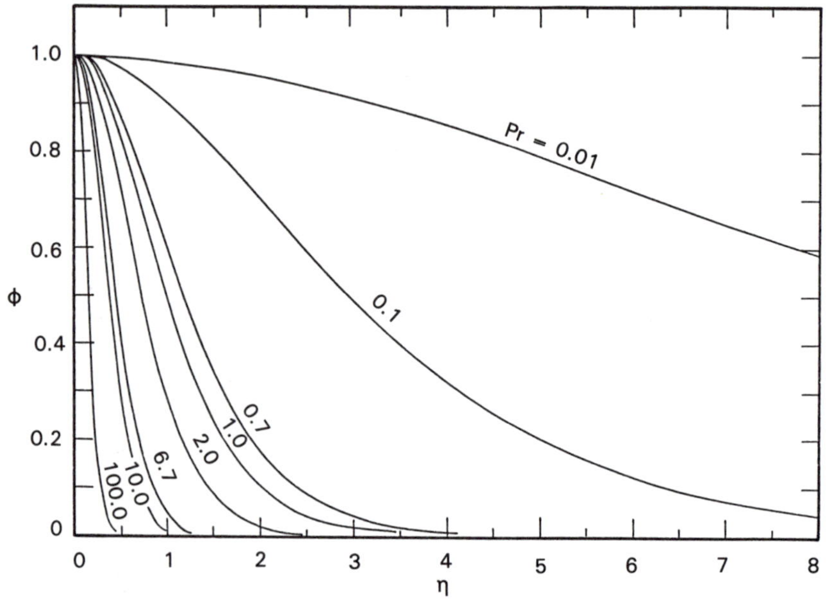

Figure 3.7.4 Plume temperature distributions. *(Reprinted with permission from* Int. J. Heat Mass Transfer, *vol. 13, p. 161, B. Gebhart et al. Copyright © 1970, Pergamon Journals Ltd.)*

"conduction" end effect was found to be inadequate as an explanation for the temperature deficiency, compared with two-dimensional theory.

The discrepancy seen in Fig. 3.7.6 was resolved in measurements in air and water by Lyakhov (1970). The region below the energy source, an electrically heated wire, was closed off by a horizontal insulating surface at about $x = 0$. Upflow from below the energy source was thus prevented and $u(0, y)$ was actually zero as envisioned in the foregoing formulation [see Eq. (3.7.4)]. The resulting measured downstream midplane temperatures were within a few percent of calculations.

This experiment clearly demonstrated an easily overlooked and frequent unrealistic result in the boundary region modeling of actual buoyancy-induced flows. The model results in $u(0, y) = 0$ [Eq. (3.5.39)] for the permissible range of n. That is, the exponent of x is positive. This means that in the model entrainment is accomplished only by horizontal flow, $v(x, \infty) \propto f(\infty)$. However, in actual flows of many configurations, the decreased motion pressure in a buoyant region also induces entrainment upward from below. The relative effect of this on local flow downstream might be expected to decrease with x or Gr_x. Such matters are discussed further in Section 3.10.

Concerning thermal plume calculations, it was mentioned above that closed-form solutions may be derived for certain fixed values of the Prandtl number, starting from the perfect differential, Eq. (3.7.6), in conjunction with Eq. (3.7.5) and the boundary conditions, Eq. (3.7.12). Fujii (1963) and Brand and Lahey

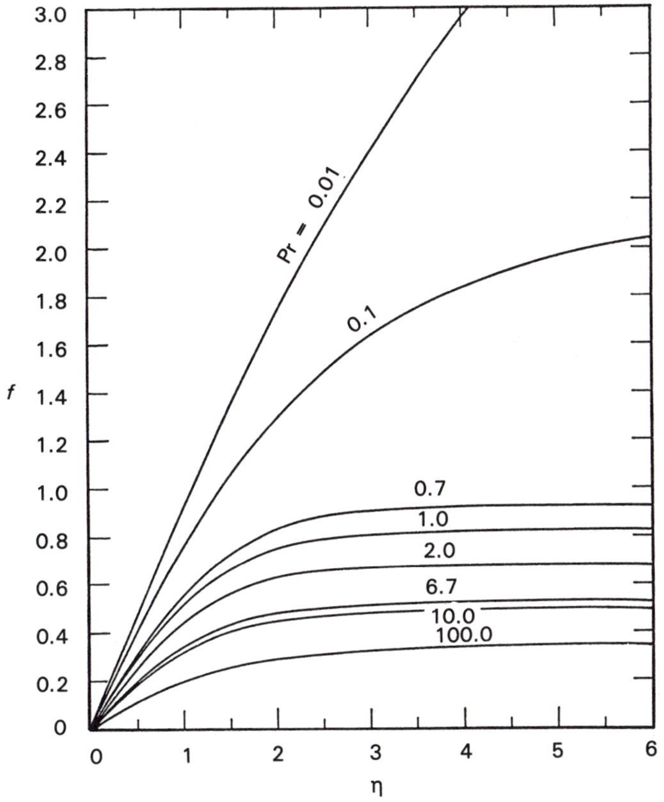

Figure 3.7.5 Distributions of f across the boundary region. Recall that $f(\infty)$ is proportional to the entrainment velocity. *(Reprinted with permission from* Int. J. Heat Mass Transfer, *vol. 13, p. 161, B. Gebhart et al. Copyright © 1970, Pergamon Journals Ltd.)*

(1967) presented closed-form solutions for Pr = 2 and $\frac{5}{9}$. A nondimensionalization different from that given above was used:

$$\tilde{\eta} = \frac{y}{x}\,\mathrm{Gr}_x^{1/5} \qquad \psi = \nu \mathrm{Gr}_x^{1/5}\tilde{f}(\eta) \qquad t - t_\infty = \mathrm{Gr}_x^{-1/5}\left(\frac{Q}{Ik\mathrm{Pr}}\right)\tilde{\phi} \tag{3.7.13}$$

with

$$\mathrm{Gr}_x = g\beta\left(\frac{Q}{Ik\mathrm{Pr}}\right)x^3/\nu^2 \qquad \text{and} \qquad I = \int_{-\infty}^{\infty} \tilde{f}'\tilde{\phi}\, d\tilde{\eta}$$

where $\tilde{\eta}$, $\tilde{f}$, and $\tilde{\phi}$ are the similarity variables. Then for Pr = 2 the solution is

$$\tilde{f} = \left(\frac{10a}{3}\right)^{1/2} \tanh\left[\left(\frac{3a}{10}\right)^{1/2} \tilde{\eta}\right] \tag{3.7.14}$$

Table 3.7.1 Numerical values of computed parameters for plane plumes

Pr	0.01	0.1	0.7	1.0	2.0	6.7	10.0	100.0
$f'(0)$	0.9751	0.8408	0.6618	0.6265	0.5590	0.4480	0.4139	0.2505
$\phi = 0.01$ at $\eta =$	—	11.0	3.9	3.2	2.2	1.2	1.0	0.4
$\frac{f'}{f'(0)} = 0.01$ at $\eta =$	14.6	9.3	4.1	3.8	3.7	4.1	4.3	5.6
I	—	3.090	1.245	1.053	0.756	0.407	0.328	—
J	—	4.316	1.896	1.685	1.393	1.094	1.024	—

Source: From Gebhart et al. (1970).

$$\tilde{\phi} = \frac{4}{5} a^2 \operatorname{sech}^4 \left[\left(\frac{3a}{10} \right)^{1/2} \tilde{\eta} \right] \tag{3.7.15}$$

Here a is the nondimensional vertical velocity component $\tilde{f}'(0)$ at $\tilde{\eta} = 0$. For Pr = 2 it was determined as 0.837. Similarly, for Pr = $\frac{5}{9}$ the closed-form solution is

$$\tilde{f}' = \left(\frac{675}{2048} \right)^{1/5} \operatorname{sech}^2 \left[\left(\frac{5}{768} \right)^{1/5} \tilde{\eta} \right] \tag{3.7.16}$$

$$\tilde{\phi} = \left(\frac{3}{640} \right)^{1/5} \operatorname{sech}^2 \left[\left(\frac{5}{768} \right)^{1/5} \tilde{\eta} \right] \tag{3.7.17}$$

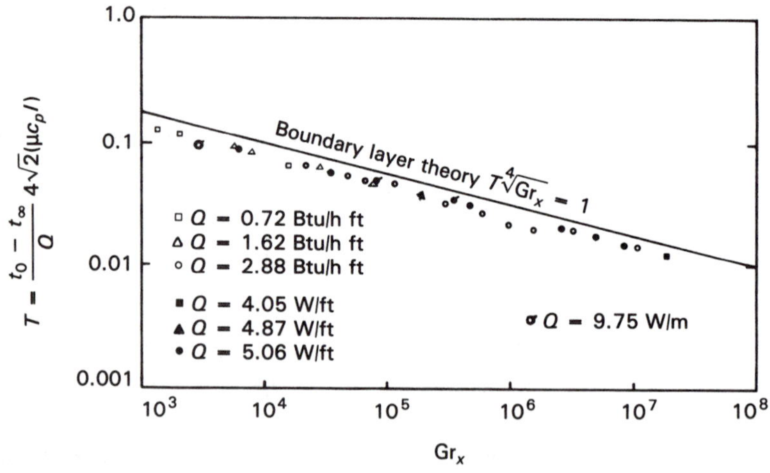

Figure 3.7.6 Measured midplane temperatures in plumes rising from horizontal heated wires. Winged points are from Brodowicz and Kierkus (1966), filled ones from Forstrom and Sparrow (1967), and open ones from Schorr and Gebhart (1970).

Although both the numerical and closed-form solutions are for thermal plumes, the introduction of a different chemical species from a long concentrated or "line" source may be treated similarly if buoyancy results. For example, the introduction of methane (CH_4) or CO_2 into ambient air would produce upflowing and downflowing plumes, respectively. The analysis, very similar to that given above, will be referred to in Chapter 6.

For free thermal plumes other conditions of thermal input would also result in similarity solutions. The energy input might occur along the plane at $y = 0$ for $x \geq 0$, as with an upward-facing slit-radiation source at $x = 0$. The value of n in $t_0 - t_\infty = Nx^n$ for any such plume would be found either from the expression for $q''(x)$, Eq. (3.5.37), or from $Q(x)$, Eq. (3.5.38). For example, for $q''(x)$ increasing linearly with x, the value of n is seen to be 1. The differential equations would be Eqs. (3.5.26) and (3.5.27). The boundary condition $\phi'(0) = 0$ would not apply. Instead, the conditions are

$$f''(0) = f(0) = 1 - \phi(0) = f'(\infty) = \phi(\infty) = 0 \tag{3.7.18}$$

3.7.2 Characteristics of Wall Plumes

Such a plume is shown in Fig. 3.7.2. The formulation, calculations for Pr = 7.0, and experimental results for water are reported by Zimin and Lyakhov (1970). The calculations and measurements were found to have only qualitative agreement.

The formulation used here for calculations will again be that of Section 3.5. As seen in Eq. (3.7.1), n is again $-\frac{3}{5}$. Thus, the equations are still Eqs. (3.7.5) and (3.7.6). However, the boundary conditions must permit shear stress and zero tangential velocity, as at a surface. Five conditions are

$$f(0) = f'(0) = \phi'(0) = 1 - \phi(0) = f'(\infty) = 0 \tag{3.7.19}$$

The additional apparent condition $\phi(\infty) = 0$ is not independent but results from those above. Thus $f''(0) = 0$, for a free plume, is replaced by $f'(0) = 0$, for a wall plume. Recall that $Q(x)$, $t_0 - t_\infty = Nx^n$, I, and $u(x, y)$ are still as given in Eqs. (3.7.1)–(3.7.4). However, the value of $I(\mathrm{Pr})$ will be different from that for the free plume above.

Calculations were reported by Liburdy and Faeth (1975) and Jaluria and Gebhart (1977). The latter gave results for the Prandtl number range 0.01–100. The temperature and velocity distributions ϕ and f' are plotted as the solid curves in Figs. 3.7.7 and 3.7.8 for Pr = 0.7 and 6.7. Comparable free plume distributions, drawn as dashed lines, show the effect of viscous stress at the surface. The plume is thickened and the velocity levels decreased. Also shown are the distributions of f' for an isothermal surface. They are similar to those for the wall plume. Their lower level results from the difference in the formulation between $n = 1$ and $n = -\frac{3}{5}$.

A more detailed comparison of the wall and free plumes is seen in Fig. 3.7.9.

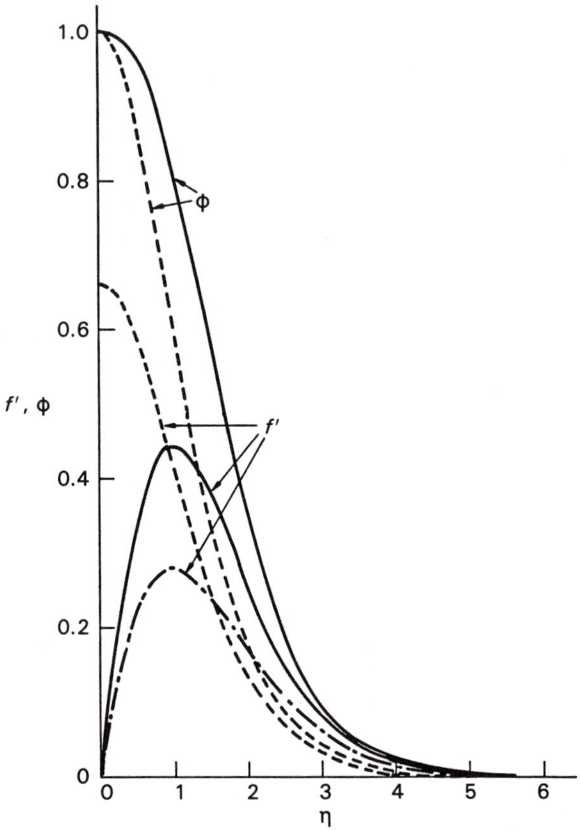

Figure 3.7.7 Temperature and velocity distributions for the wall plume (———) compared with those for the free plume (– – –), Pr = 0.7. Velocity results (– – – – –) are also shown for an isothermal surface at $y = 0$. *(Reprinted with permission from* Int. J. Heat Mass Transfer, *vol. 20, p. 153, Y. Jaluria and B. Gebhart. Copyright © 1977, Pergamon Journals Ltd.)*

Wall plume quantities are plotted, normalized by the corresponding free plume quantities calculated at the same Prandtl number. Here

$$F(x) = \int_0^\infty u\,dy = \nu G \int_0^\infty f'\,d\eta = \nu G f(\infty) \tag{3.7.20}$$

is the local volume flow rate and u_{max} is the maximum value of $u(x, y)$ locally at x. All measures of flow are seen to be lower for the wall plume, due to the viscous force arising at the surface. The larger N, on the other hand, means that downstream temperature decay is less rapid. This is a consequence of the decrease in the vigor of the flow and of entrainment. The actual variation of several important flow parameters across the Prandtl number range is shown in Fig. 3.7.10. The relative increase of viscous effects with increasing Prandtl number is apparent from the trends.

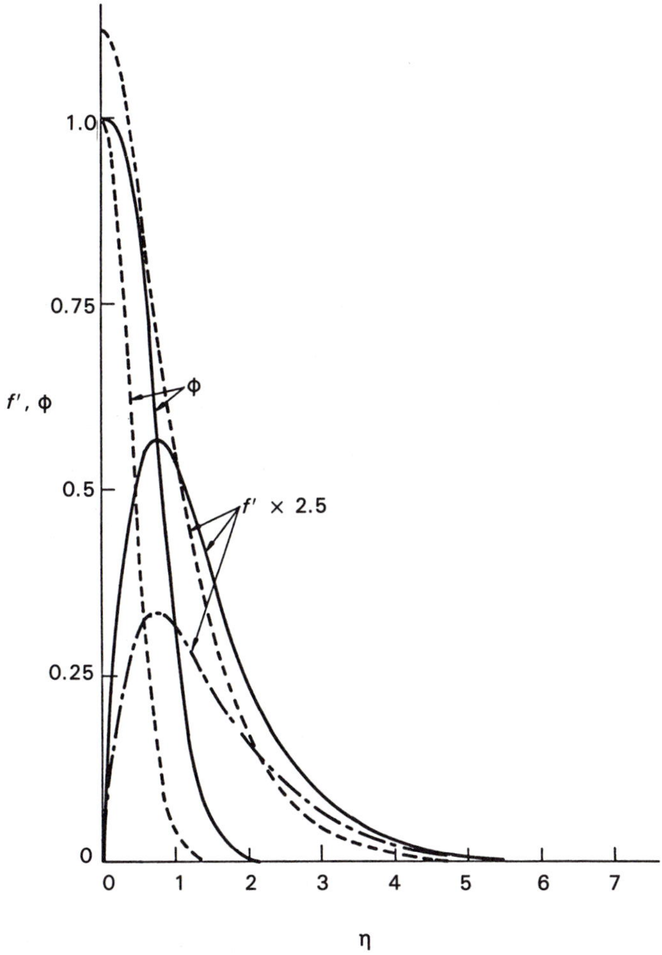

Figure 3.7.8 Temperature and velocity distributions for the wall plume (———) compared with those for the free plume (– – –), Pr = 6.7. Velocity results (- - - - -) are also shown for an isothermal surface at $y = 0$. *(Reprinted with permission from* Int. J. Heat Mass Transfer, *vol. 20, p. 153, Y. Jaluria and B. Gebhart. Copyright © 1977, Pergamon Journals Ltd.)*

3.8 TRANSPORT IN EXTREME PRANDTL NUMBER FLUIDS

Several applications of technological significance use fluids having either very large or extremely small Prandtl numbers. Large Prandtl number hydrocarbon fuels and silicone polymers are being used increasingly in processing. Low Prandtl number fluids, such as liquid sodium, are used as coolants in the fast breeder reactor. Several other liquid metals have been proposed as working fluids in space environments. Transport in such fluids is also of theoretical interest. For example, for laminar boundary layer flows, one would like to know if the Nusselt number in

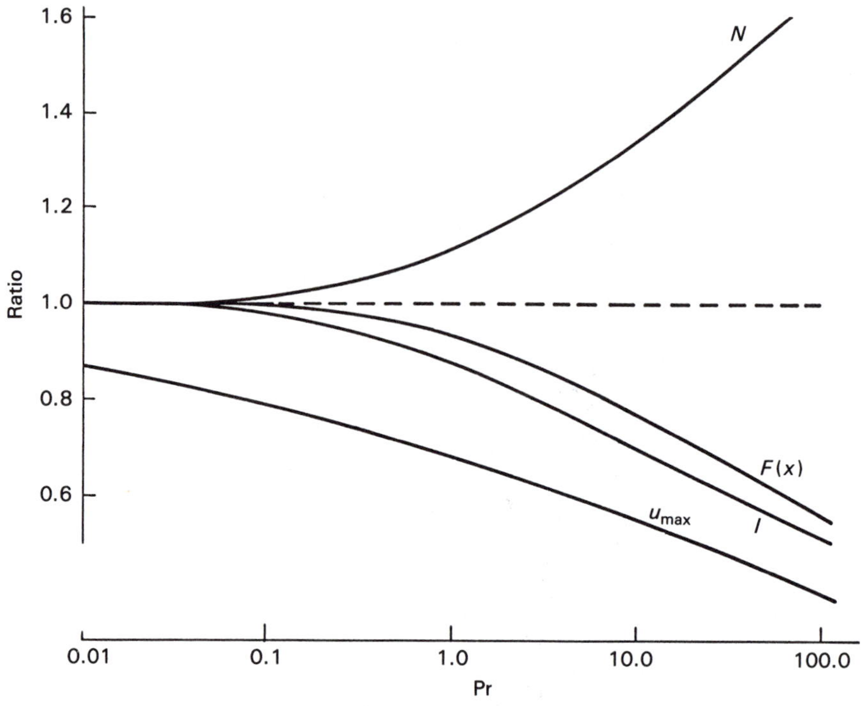

Figure 3.7.9 Ratio of the quantity indicated calculated for the wall plume to that calculated for the free plume. *(Reprinted with permission from* Int. J. Heat Mass Transfer, *vol. 20, p. 153, Y. Jaluria and B. Gebhart. Copyright © 1977, Pergamon Journals Ltd.)*

Eq. (3.4.4) reaches an asymptotic dependence, through $F(\mathrm{Pr})$, for extreme values of Pr.

The first such study appears to have been by LeFevre (1956) for a vertical isothermal surface. The governing equations are Eqs. (3.3.19)–(3.3.21). To determine the limiting behavior of these equations, the independent variable ζ and dependent variables $\bar{f}(\zeta)$ and $H(\zeta)$ are introduced as follows:

$$\zeta = \sqrt[4]{\frac{3\mathrm{Pr}^2}{1+\mathrm{Pr}}}\,\eta \qquad \bar{f}(\zeta) = \sqrt[4]{3^3\mathrm{Pr}^2(1+\mathrm{Pr})}\,f(\eta) \qquad H(\zeta) = \phi(\eta) \quad (3.8.1)$$

where η, f, and ϕ are as defined in Eqs. (3.3.6)–(3.3.8). Using Eq. (3.8.1), the governing equations (3.3.19)–(3.3.21) become

$$\mathrm{Pr}(\bar{f}''' + H) + (\bar{f}\bar{f}'' - \tfrac{2}{3}\bar{f}'^2 + H) = 0 \quad (3.8.2)$$

$$H'' + H'\bar{f} = 0 \quad (3.8.3)$$

$$\bar{f}(0) = \bar{f}'(0) = \bar{f}'(\infty) = H(0) - 1 = H(\infty) = 0 \quad (3.8.4)$$

Recalling that $F(\mathrm{Pr}) = -\phi'(0)/\sqrt{2}$ and using the above transformations,

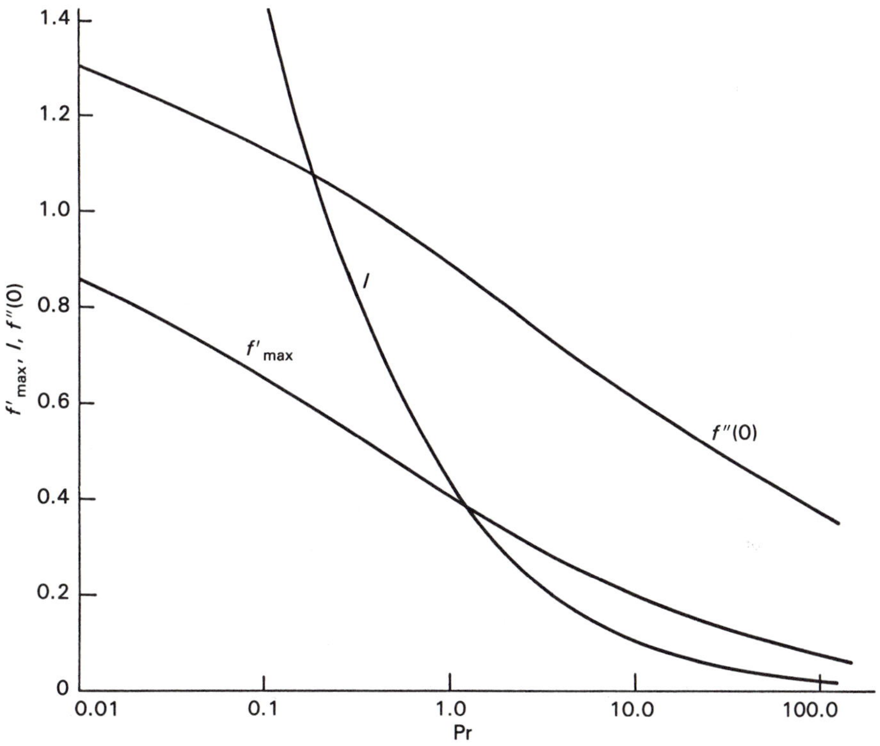

Figure 3.7.10 Variation of wall plume flow parameters with Prandtl number. *(Reprinted with permission from* Int. J. Heat Mass Transfer, *vol. 20, p. 153, Y. Jaluria and B. Gebhart. Copyright © 1977, Pergamon Journals Ltd.)*

$$F(\text{Pr}) = -\frac{H'(0)}{\sqrt{2}}\left(\frac{3\text{Pr}^2}{1+\text{Pr}}\right)^{1/4} \tag{3.8.5}$$

The simplicity of Eq. (3.8.2) at extreme values of Pr is apparent. At each extreme, one of the terms in this equation becomes negligible. First, for Pr $\rightarrow \infty$ Eqs. (3.8.2)–(3.8.4) reduce to

$$\tilde{f}''' + H = 0 \tag{3.8.6}$$

$$H'' + H'\tilde{f} = 0 \tag{3.8.7}$$

$$\tilde{f}(0) = \tilde{f}'(0) = \tilde{f}''(\infty) = H(0) - 1 = H(\infty) = 0 \tag{3.8.8}$$

Note that $\tilde{f}'(\infty) = 0$ in Eq. (3.8.4) has been replaced by $\tilde{f}''(\infty) = 0$ in Eq. (3.8.8). A later, more rigorous analysis justifies this.

These equations were solved numerically by LeFevre with the following results:

$$-H'(0) = 0.5402348$$

$$\tilde{f}''(0) = 1.0851246$$

Substituting this value of $-H'(0)$ in Eq. (3.8.5), the result in the limit $\mathrm{Pr} \to \infty$ is

$$F(\mathrm{Pr}) = -\frac{\phi'(0)}{\sqrt{2}} = 0.5027\mathrm{Pr}^{1/4} \tag{3.8.9}$$

This asymptotic behavior is sketched in Fig. 3.4.2. In the original variables η, f, and ϕ, $f''(0)$ reduces to

$$f''(0) = 0.8245\mathrm{Pr}^{-1/4} \tag{3.8.10}$$

Thus, in the limit $\mathrm{Pr} \to \infty$ the Nusselt number varies as $\mathrm{Pr}^{1/4}$, while the shear stress varies as $\mathrm{Pr}^{-1/4}$.

An estimate of u is obtained next. Recalling from Eq. (3.5.4) that $u = \psi_y = \nu cbf'(\eta) = \nu G^2 f'(\eta)/4x$ and using Eq. (3.8.1), we get

$$\frac{u}{\nu cb} = \tilde{f}'(\zeta)3^{-1/2}\mathrm{Pr}^{-1/2}\left(1 + \frac{1}{\mathrm{Pr}}\right)^{-1/2} \tag{3.8.11}$$

In the limit $\mathrm{Pr} \to \infty$, $u/\nu cb \simeq \tilde{f}'(\zeta)3^{-1/2}\mathrm{Pr}^{-1/2}$. Therefore,

$$\frac{\rho u^2}{2} \simeq \rho g \beta (t_0 - t_\infty) L \mathrm{Pr}^{-1} \tag{3.8.12}$$

This is the estimate used in Eq. (2.6.5) in Section 2.6. As pointed out there, this relationship clearly shows that for $\mathrm{Pr} >> 1$, Eq. (2.6.4) is a substantial overestimate of u^2.

For $\mathrm{Pr} \to 0$, Eqs. (3.8.2)–(3.8.4) reduce to

$$\tilde{f}''\tilde{f} - \tfrac{2}{3}\tilde{f}'^2 + H = 0 \tag{3.8.13a}$$

$$H'' + \tilde{f}H' = 0 \tag{3.8.13b}$$

$$\tilde{f}(0) = \tilde{f}'(\infty) = H(\infty) = H(0) - 1 = 0 \tag{3.8.13c}$$

Arguing that for $\mathrm{Pr} \to 0$ the fluid may be treated as essentially inviscid, the boundary condition $\tilde{f}'(0) = 0$ is omitted. These equations were solved numerically and the value of $-H'(0)$ was given as $3^{-1/4}(0.849105)$. Substitution of this in Eq. (3.8.5) gives

$$F(\mathrm{Pr}) = \frac{[-\phi'(0)]}{\sqrt{2}} = 0.6004\mathrm{Pr}^{1/2} \tag{3.8.14}$$

This asymptotic relation is also shown in Fig. 3.4.2.

3.8.1 Matched Asymptotic Solution

A more thorough analysis follows from the method of matched asymptotic expansions; see Van Dyke (1964) for a description of the general technique. The method is also discussed in some detail in Section 3.10, where it is used to obtain higher-order corrections to the classical boundary layer solution adjacent to a vertical surface. For the analysis of heat transfer from a vertical surface in extreme Pr fluids, the technique has been applied by Stewartson and Jones (1957), Kuiken (1968a), and Roy (1969, 1973) for high Pr fluids and by Kuiken (1969) for low Pr fluids. These and other relevant studies are discussed briefly next.

High Pr fluids. The flow region is divided into two layers: the thin layer close to the surface, called the inner layer, where the temperature difference is brought to zero, and the outer layer, where the velocity parallel to the surface is brought to zero. The thicknesses of these two layers are determined to scale as $O(\text{Pr}^{-1/4})$ and $O(\text{Pr}^{1/4})$, respectively. Suitable variables ζ and ξ, as defined below, are then sought for these two layers.

Inner layer:

$$\zeta = (3\text{Pr})^{1/4}\eta \qquad \tilde{f}(\zeta) = (3\text{Pr})^{3/4}f(\eta) \qquad H(\zeta) = \phi(\eta) \tag{3.8.15}$$

Outer layer:

$$\xi = A(3\text{Pr})^{-1/4}\eta \qquad \tilde{G}(\xi) = (3\text{Pr})^{1/4}f(\eta)/A \qquad \phi(\eta) = 0 \tag{3.8.16}$$

The numerical constant A is to be determined from matching conditions. Using these transformations, Eqs. (3.3.19) and (3.3.20) give the following governing equations for the functions $\tilde{f}$, H, and $\tilde{G}$:

$$\tilde{f}''' + H + \frac{1}{3\text{Pr}}(3\tilde{f}\tilde{f}'' - 2\tilde{f}'^2) = 0 \tag{3.8.17}$$

$$H'' + \tilde{f}H' = 0 \tag{3.8.18}$$

$$\tilde{G}''' + 3\tilde{G}\tilde{G}'' - 2\tilde{G}'^2 = 0 \tag{3.8.19}$$

where a prime denotes differentiation with respect to the appropriate variable ζ or ξ.

The solution to these equations is now sought in the form of series expansions:

$$\begin{aligned} \tilde{f} &= \tilde{f}_0 + \epsilon\tilde{f}_1 + \cdots \\ H &= H_0 + \epsilon H_1 + \cdots \\ \tilde{G} &= G_0 + \epsilon G_1 + \cdots \end{aligned} \tag{3.8.20}$$

The perturbation parameter is $\epsilon = (3\text{Pr})^{-1/2}$. Substituting Eq. (3.8.20) into Eqs. (3.8.17)–(3.8.19) and collecting terms of the equal order of ϵ, the governing equations in $\tilde{f}_i$, H_i, and $\tilde{G}_i$, where $i = 0, 1$, are obtained. The relevant boundary conditions up to first order are

$$\left.\begin{aligned} &\tilde{f}_0(0) = \tilde{f}_0'(0) = \tilde{f}_0''(\infty) = H_0(0) - 1 = H_0(\infty) = 0 \\ &\tilde{G}_0(0) = \tilde{G}_0'(\infty) = 0 \qquad \tilde{G}'(0) = 1 \end{aligned}\right\} \tag{3.8.21}$$

$$\left.\begin{aligned} &\tilde{f}_1(0) = 0 = \tilde{f}_1'(0) = H_1(0) = H_1(\infty) \qquad \tilde{f}_1''(\infty) = A^3 G_0''(0) \\ &\tilde{G}_1(0) = \frac{1}{A}\tilde{f}_0(\infty) - A\zeta_\infty \qquad \tilde{G}_1'(0) = \frac{1}{A^2}\tilde{f}_1'(\infty) - \zeta_\infty \tilde{f}_1''(\infty) \\ &\tilde{G}_1'(\infty) = 0 \end{aligned}\right\} \tag{3.8.22}$$

where $A^2 = \tilde{f}_0'(\infty)$ and ζ_∞ is the value of ζ at the boundary of the inner layer. The boundary conditions on $\zeta = \infty$ and $\xi = 0$ are determined by matching the inner solutions for large values of ζ with the outer solutions for small values of ξ. Note that the matching condition results in $\tilde{f}_0''(\infty) = 0$. Replacement of $f'(\infty) = 0$ by $f''(\infty) = 0$, as done by LeFevre (1956), is therefore justified.

The numerical solution to the zeroth-order equations subject to boundary conditions (3.8.21) has been obtained by Stewartson and Jones (1957), while the first-order equations have been solved by Roy (1973). From these results,

$$\begin{aligned} &\tilde{f}_0''(0) = 1.08506 \qquad H_0'(0) = -0.54023 \qquad \tilde{f}_0'(\infty) = 0.88425 \\ &\tilde{f}_1''(0) = -0.70013 \qquad H_1(0) = 0.24542 \qquad \tilde{f}_1(\infty) = -5.40819 \\ &\tilde{G}_0''(0) = -1.54079 \qquad \tilde{G}_1''(0) = -1.65437 \end{aligned} \tag{3.8.23}$$

Combining these results with Eqs. (3.8.15) and (3.8.16) yields

$$-\phi'(0)/\sqrt{2} = F(\mathrm{Pr}) = 0.5027\mathrm{Pr}^{1/4} - 0.1319\mathrm{Pr}^{-1/4} + \cdots \tag{3.8.24}$$

$$f''(0) = 0.8245\mathrm{Pr}^{-1/4} - 0.3071\mathrm{Pr}^{-3/4} - \cdots \tag{3.8.25}$$

Note that the zeroth-order solutions, that is, the first terms in Eqs. (3.8.24)–(3.8.25), are the same as those obtained by LeFevre (1956). Further, the inclusion of first-order terms in the above equations also allows their use for moderate Prandtl number fluids. For example, the values of $F(\mathrm{Pr})$ calculated from Eq. (3.8.24) are close to those given in Table 3.4.1 for $\mathrm{Pr} = 100$. For completeness, it is noted that this case has also been solved by Kuiken (1968a), using matched expansions, and by Eckert (1949), using the momentum integral method.

Roy (1969) extended the isothermal surface analysis of Stewartson and Jones to the uniform flux surface condition. It was found that the inner and outer layers then have thicknesses of $O(\mathrm{Pr}^{-1/5})$ and $O(\mathrm{Pr}^{3/10})$, respectively, and the appropriate transformations for the inner and outer layer are

Inner layer:

$$\zeta = \mathrm{Pr}^{1/5}\eta \qquad \tilde{f}(\zeta) = \mathrm{Pr}^{4/5}f(\eta) \qquad H(\zeta) = \mathrm{Pr}^{1/5}\phi(\eta) \tag{3.8.26}$$

Outer layer:

$$\xi = A\mathrm{Pr}^{-3/10}\eta \qquad \tilde{G}(\xi) = \frac{1}{A}\mathrm{Pr}^{3/10}f(\eta) \qquad \phi(\eta) = 0 \tag{3.8.27}$$

Proceeding in the same manner, the following expression for the local Nusselt number is obtained:

$$\mathrm{Nu}_x(1.1476 + 0.2268\mathrm{Pr}^{-1/2} - 0.0304\mathrm{Pr}^{-1}) = \mathrm{Pr}^{1/5}\left(\frac{\mathrm{Gr}_x^*}{5}\right)^{1/5} \tag{3.8.28}$$

The equivalent problem of laminar natural convection mass transfer at a vertical surface with uniform flux condition has been solved by Selman and Newman (1971).

The asymptotic behavior of $F(\mathrm{Pr})$ for $\mathrm{Pr} \to \infty$ for a uniform flux surface ($n = \frac{1}{5}$) is given as $F(\mathrm{Pr}) = 0.563\mathrm{Pr}^{1/4}$; see Gebhart (1971). That is,

$$\mathrm{Nu}_x = 0.563\mathrm{Pr}^{1/4}\mathrm{Gr}_x^{1/4} \tag{3.8.29}$$

It can easily be shown by use of the relationship

$$\mathrm{Gr}_x = \left[\frac{\mathrm{Gr}_x^*}{F(\mathrm{Pr})}\right]^{4/5} \tag{3.8.30}$$

that Eq. (3.8.29) and the leading term in Eq. (3.8.28) are identical. Equation (3.8.30) is a simplified form of the relation given by Jaluria and Gebhart (1974). The function $F(\mathrm{Pr})$ is given in Table 3.5.1 for $n = \frac{1}{5}$.

The thermal plume solution at high Prandtl number has also been determined. Kuiken and Rotem (1971) solved the perturbation equations up to second order while Spalding and Cruddace (1961) obtained the first term of the outer solution on a semi-intuitive basis.

Low Pr fluids. Using matched asymptotic expansions, the flow is again divided into two distinct regions. Near the surface there is a thin inner layer of large velocity gradient due to viscous effects. In the much wider outer layer, the gradients are smaller than those near the wall and viscous effects arise at higher order. The appropriate inner and outer region expansions are

Inner layer:

$$f = f_0 + \epsilon f_1 + \epsilon^2 f_2 + \cdots \tag{3.8.31}$$

$$\phi = 1 + \epsilon\phi_1 + \epsilon^2\phi_2^2 + \cdots \tag{3.8.32}$$

Outer layer:

$$\tilde{F}(\xi) = \tilde{F}_0(\xi) + \epsilon\tilde{F}_1(\xi) + \epsilon^2\tilde{F}_2(\xi) + \cdots \tag{3.8.33}$$

$$H(\xi) = H_0(\xi) + \epsilon H_1(\xi) + \epsilon^2 H_2(\xi) + \cdots \tag{3.8.34}$$

where

$$\xi = 3(18)^{-1/4}\mathrm{Pr}^{1/2}\eta \tag{3.8.35}$$

$$\tilde{F}(\xi) = 18^{1/4}\mathrm{Pr}^{1/2}f(\eta) \tag{3.8.36}$$

$$H(\xi) = \phi(\eta) \tag{3.8.37}$$

The perturbation parameter is $\epsilon = \mathrm{Pr}^{1/2}$.

The choice of leading term in Eq. (3.8.32), as $\phi_0 = 1$, follows from a simple observation. If $\text{Pr} \to 0$ is assumed to result from infinitely large conductivity k, the surface temperature t_0 penetrates far into the fluid. Therefore, near the wall, $\phi = 1$. This also follows from the generalized solution of Eq. (3.3.20), which for $\text{Pr} \to 0$ is given by

$$\phi = C_1 + C_2\eta \tag{3.8.38}$$

For $C_1 = 1$ and $C_2 = 0$, $\phi = 1$.

Substituting Eqs. (3.8.31)–(3.8.35) in Eqs. (3.3.19)–(3.3.20) and collecting terms of like powers of ϵ, governing equations on f_i, ϕ_i, $\tilde{F}_i$, and H_i, where $i = 0, 1, 2$, are obtained. The boundary conditions, Eq. (3.3.21), at the surface apply to the inner functions, while the outer functions satisfy the ambient conditions. The other boundary conditions at $\eta = \infty$ and $\xi = 0$ are determined by matching the inner solution for large values of η with the outer solution for small of values of ξ.

Using the numerical results of the perturbation equations obtained in the manner cited above, Kuiken (1969) gave the following expressions for local Nusselt number and $f''(0)$:

$$\text{Nu}_x = \text{Gr}_x^{1/4}\text{Pr}^{1/2}[0.6004 - 0.32385\text{Pr}^{1/2} + O(\text{Pr}) + \cdots] \tag{3.8.39}$$

$$f''(0) = 1.0699496 - 1.001023\text{Pr}^{1/2} + O(\text{Pr})^{2/3} \tag{3.8.40}$$

Comparison of Eq. (3.8.39) with the results for $\text{Pr} = 0.01$ and 0.1 in Table 3.4.1 and with those of Sparrow and Gregg (1958) indicates good agreement for low Prandtl numbers. For $\text{Pr} \leq 0.02$ this expression is accurate up to at least the second significant digit.

The perturbation solution for a thermal plume at low Prandtl number has also been obtained by Kuiken and Rotem (1971). No such solution appears to have been reported for a uniform flux vertical surface, although numerical solutions are available, for example, in Chang et al. (1964) and the data in Table 3.5.1. Chang et al. extended Sparrow and Gregg's similarity solution of 1956 to Prandtl number values of 0.01 and 0.03 for the uniform flux condition and then obtained first-order perturbation improvements to the boundary layer analysis at these Prandtl numbers. The data in Table 3.5.1 were obtained from numerical solutions of Eqs. (3.5.26)–(3.5.28) for $n = \frac{1}{5}$ where $\text{Gr}_x = g\beta N x^n x^3/\nu^2$.

Other studies of transport at low Prandtl numbers used the integral method of Von Karman and Pohlhausen. See, for example, Cygan and Richardson (1968), Braun and Heighway (1960), Eckert (1950), and Saunders (1939). Although these calculations are relatively straightforward and result in explicit functions of the Prandtl number, inherent errors are introduced through the assumed profiles.

Experimental results. The analytical predictions of the Nusselt number discussed above have been fairly well-corroborated by the few known experimental studies in low Prandtl number fluids; see Saunders (1939), Julian and Akins (1969), Colwell and Welty (1973), and Humphreys and Welty (1975). The fluid used in all these studies was mercury at $\text{Pr} = 0.023$, and the vertical surface had a uniform flux condition.

Saunders compared the temperature data with an approximate theoretical treatment and reported good agreement. Julian and Akins gave the following experimental correlation in terms of $Gr_x^* = g\beta q'' x^4/k\nu^2$:

$$Nu_x = 0.196(Gr_x^*)^{0.188} \qquad 10^4 < Gr_x^* < 10^9 \tag{3.8.41}$$

This expression correlated the data of Humphreys and Welty (1975) up to $Gr_x^* = 10^{11}$. Colwell and Welty obtained the following correlation of data at lower Gr_x^*:

$$Nu_x = 0.230(Gr_x^*)^{0.18} \qquad 10^4 < Gr_x^* < 10^9 \tag{3.8.42}$$

The interpolated value of $F(\text{Pr})$ for $\text{Pr} = 0.023$ from Table 3.5.1 is 0.1004. Using Eq. (3.8.30), the heat transfer in terms of Gr_x^* is then

$$Nu_x = 0.159(Gr_x^*)^{0.2} \tag{3.8.43}$$

The difference between the exponents 0.18 and 0.2 is small. Therefore, for the purpose of direct comparison, the fit of experimental data may be forced to have an exponent of 0.2, using a data regression, to obtain a new value of the coefficient. This was done by Humphreys and Welty, and the resulting correlation became

$$Nu_x = 0.150(Gr_x^*)^{0.2} \tag{3.8.44}$$

This result is only about 6% lower than the analytical result, Eq. (3.8.43). Similar conclusions follow from the experimental results of Colwell and Welty (1973) and Julian and Akins (1969).

It is interesting to note, however, that Eqs. (3.8.41) and (3.8.42), Nu_{41} and Nu_{42}, when used to calculate Nusselt numbers at different Grashof numbers, indicate a systematic deviation from Eq. (3.8.43), called Nu_{43}, as seen in Table 3.8.1. The discrepancies are larger at lower Grashof numbers, particularly at $Gr_x^* = 10^4$ and 10^5. At these values of Gr_x^*, the terms neglected in the boundary layer analysis used in obtaining the numerical results in Eq. (3.8.43) are not negligible, and this equation may be increasingly inaccurate at decreasing Grashof numbers. Further, the experimental temperature and velocity profiles also show deviations from the boundary layer similarity results; see Julian and Akins (1969)

Table 3.8.1 Comparison of theoretical and experimental predictions of Nusselt number, Nu_{41}/Nu_{43} and Nu_{42}/Nu_{43}

Gr_x^*	Nu_{41}/Nu_{43}	Nu_{42}/Nu_{43}
10^4	1.104	1.203
10^5	1.074	1.149
10^6	1.044	1.097
10^7	1.016	1.048
10^8	0.988	1.001
10^9	0.961	0.956

and Humphreys and Welty (1975). The trends of these deviations are in agreement with those predicted by the first-order perturbation analysis of Chang et al. (1964). These observations suggest that higher-order approximations are necessary for better predictions at low Pr.

3.9 DATA AND CORRELATIONS

The similarity analyses discussed in earlier sections give transport results in close agreement with the experimental data. However, these analyses are valid only for a certain range of Rayleigh or Grashof numbers. For a vertical heated surface the boundary layer approximations become locally invalid with decreasing values of Grashof numbers ($\mathrm{Gr}_x < 10^4$). This corresponds to the regions near a leading edge. Higher-order approximations are applied to describe the transport in this region, as discussed in Section 3.10.

Another limitation of the similarity analyses is that they fail to predict the transport quantities in downstream regions in which the flow undergoes transition and eventually becomes turbulent. In these regimes the transport rates are much higher than those predicted by laminar analysis. Typically, the laminar experimental data show good agreement with the analysis for $10^5 < \mathrm{Ra}_x < 10^9$, where $\mathrm{Ra}_x = \mathrm{Gr}_x\mathrm{Pr}$ and where Ra_x is the local Rayleigh number.

Many experimental studies have reported natural convection heat transfer rates from vertical heated surfaces, with either a uniform surface temperature or uniform surface heat flux condition. To assess the effect of fluid Prandtl number, various fluids were used; air and water were the fluids most frequently used. Several difficulties arise in reproducing the idealized conditions assumed in a theoretical analysis. Ideally the fluid motions should be due only to the effect of the heated surface. However, structural vibrations, flow disturbances in the ambient medium, and fluid circulations and stratification due to a finite ambient medium may influence the data. Another major question is how precisely the thermal boundary condition at the surface is attained experimentally. The effect of variable properties is also often quite large across the boundary region. Various approaches have been suggested by different investigators to account for these effects.

The experimental data and the transport correlations presented below apply for the two common surface conditions, the uniform temperature and flux conditions.

Uniform surface temperature. Following the boundary layer analysis, the local Nusselt number is given as

$$\mathrm{Nu}_x = F(\mathrm{Pr})(\mathrm{Gr}_x)^{1/4} \tag{3.9.1}$$

where $\mathrm{Nu}_x = hx/k$ and $\mathrm{Gr}_x = g\beta(t_0 - t_\infty)x^3/\nu^2$. For $\mathrm{Pr} >> 1$, $F(\mathrm{Pr}) \simeq 0.5027(\mathrm{Pr})^{1/4}$. Then the Rayleigh number ($\mathrm{Ra}_x = \mathrm{Gr}_x\mathrm{Pr}$) is a better choice to represent the heat transfer correlations. However, this parameter does not correlate the Prandtl number effect well over the whole range. Therefore, one may write

$$\text{Nu}_x = F_1(\text{Pr})\text{Ra}_x \tag{3.9.2}$$

Most correlations in the literature are of this form. Ede (1967) reviewed all the known experimental data and surmised that the similarity solution agrees reasonably well for Rayleigh numbers between 10^6 and 10^8. For lower and higher Rayleigh numbers the experimental data indicate higher heat transfer rates.

Churchill and Usagi (1972) analyzed various experimental results and suggested that the effect of the Prandtl number may be accurately incorporated in terms of the asymptotic trends for $\text{Pr} \to 0$ and $\text{Pr} \to \infty$. The following correlation was given:

$$\text{Nu}_x = \frac{0.503\text{Ra}_x^{1/4}}{[1 + (0.492/\text{Pr})^{9/16}]^{4/9}} \tag{3.9.3}$$

Equation (3.9.3) is taken to apply for $\text{Pr} = 0$ to $\text{Pr} = \infty$ and $10^5 < \text{Ra}_x < 10^9$. This equation is integrated over $x = 0$ to L to find the average Nusselt number as

$$\text{Nu} = \frac{hL}{k} = \frac{0.670\text{Ra}^{1/4}}{[1 + (0.492/\text{Pr})^{9/16}]^{4/9}} \tag{3.9.4}$$

Churchill and Chu (1975) considered the available experimental data over a wider range of Rayleigh number. The following correlation was suggested for $\text{Ra} < 10^{12}$ and for any Prandtl number:

$$\text{Nu}^{1/2} = 0.825 + \frac{0.387\text{Ra}^{1/6}}{[1 + (0.492/\text{Pr})^{9/16}]^{8/27}} \tag{3.9.5}$$

This equation also correlates the limited data in the transition and turbulent regime available for an isothermal surface.

Uniform surface heat flux. Churchill and Ozoe (1973) reviewed the solutions obtained by Sparrow and Gregg (1956), Chang et al. (1964), Churchill and Usagi (1972), and Selman and Newman (1971). The following correlation was given for Nu_x, in the laminar flow range, for all values of Pr:

$$\text{Nu}_x = \frac{0.563\text{Ra}_x^{1/4}}{[1 + (0.437/\text{Pr})^{9/16}]^{4/9}} \tag{3.9.6}$$

Generally, it is more convenient for a uniform heat flux surface to give the correlations in terms of the modified local Grashof number $\text{Gr}_x^* = g\beta q''x^4/k\nu^2$. This is based on the known surface heat flux instead of the unknown surface temperature excess. It may be shown from the laminar analysis of Sparrow and Gregg (1956) that

$$\text{Gr}_x^* = \text{Gr}_x\text{Nu}_x \tag{3.9.7}$$

where Gr_x is based on the local surface temperature excess, as in Eq. (3.5.25).

The laminar boundary layer solutions for a vertical surface with uniform heat flux at various Prandtl numbers indicate that the variation of local Nusselt number is

$$\mathrm{Nu}_x = F_2(\mathrm{Pr})(\mathrm{Gr}_x^*\mathrm{Pr})^{1/5} \tag{3.9.8}$$

Fujii and Fujii (1976) reviewed the solutions obtained by Sparrow and Gregg (1956), Gebhart (1962), Fujii and Uehara (1970), Kuiken (1968a, 1968b), and Churchill and Ozoe (1973) and suggested the following relationship to calculate $F_2(\mathrm{Pr})$:

$$F_2(\mathrm{Pr}) = \left(\frac{\mathrm{Pr}}{4 + 9\mathrm{Pr}^{1/2} + 10\mathrm{Pr}}\right)^{1/5} \tag{3.9.9}$$

This relationship is said to apply to all Prandtl number values and $10^5 < \mathrm{Gr}_x^* < 10^{12}$.

Churchill and Chu (1975) correlated the experimental data of Dotson (1954), Vliet and Liu (1969), Chang and Akins (1972), and Julian and Akins (1969). The following correlation was suggested for computing the average Nusselt number for a wide range of GrPr, from laminar to turbulent flow:

$$\mathrm{Nu}^{1/2} = \left(\frac{hL}{k}\right)^{1/2} = 0.825 + \frac{0.387(\mathrm{GrPr})^{1/6}}{[1 + (0.437/\mathrm{Pr})^{9/16}]^{8/27}} \tag{3.9.10}$$

This is said to be applicable for $1 < \mathrm{GrPr} < 10^{11}$ and for any Prandtl number. The correlation is given in terms of Grashof number instead of the modified Grashof number Gr*. However, once the modified Grashof number is computed, the average Nusselt number may be determined implicitly from Eq. (3.9.10) by replacing Gr by Gr*/Nu, enabling use of this equation for the uniform surface heat flux condition as well.

3.10 HIGHER–ORDER APPROXIMATIONS

The similarity solutions discussed in Section 3.2 are based on laminar boundary layer equations derived from the complete Navier-Stokes, continuity, and energy equations by neglecting terms of $O(\mathrm{Gr}^{-1/4})$ and higher. See Eqs. (3.2.8)–(3.2.11) and recall that $\Delta = O(\mathrm{Gr}^{-1/4})$. Therefore, these solutions apply for large Grashof numbers only. For moderate Grashof number flows, refinements to the boundary layer solutions are necessary. This has been done by several investigators by use of perturbation techniques, with the classical boundary layer solution serving as a first step in a scheme of successive approximations.

Yang and Jerger (1964) obtained the first-order correction to the boundary layer solution for an isothermal finite plate. The procedure followed was identical to that used by Kuo (1953) in the Blasius forced flow past a semi-infinite plate. For the same surface condition, Kadambi (1969) and Riley and Drake (1975) used the method of matched asymptotic expansions and presented perturbation solutions to the second order. However, there are certain uncertainties in these analyses. Yang and Jerger calculated the correction to the Nusselt number as negative. This, as pointed out by Gebhart in a comment following that paper, is at variance with the experimental data at low Grashof numbers, which indicate Nusselt number values higher than predicted by the boundary layer theory. In the second-order correction obtained by Kadambi and Riley and Drake, there is an error due to improper matching. These discrepancies have been resolved by Hieber (1974). This analysis is discussed next in detail to illustrate the principles involved in obtaining such improvements. The results for the uniform flux condition, obtained by Mahajan and Gebhart (1978), are then presented.

3.10.1 The Isothermal Condition

If the pressure and viscous dissipation terms are assumed negligible and there is no distributed heat source q''' in the energy equation, the full governing equations for two-dimensional plane flow induced adjacent to a semi-infinite vertical surface are

$$\frac{\partial u}{\partial x} + \frac{\partial v}{\partial y} = 0 \tag{3.10.1}$$

$$u\frac{\partial u}{\partial x} + v\frac{\partial u}{\partial y} = \nu\left(\frac{\partial^2 u}{\partial x^2} + \frac{\partial^2 u}{\partial y^2}\right) - \frac{1}{\rho}\frac{\partial p_m}{\partial x} + g\beta(t - t_\infty) \tag{3.10.2}$$

$$u\frac{\partial v}{\partial x} + v\frac{\partial v}{\partial y} = \nu\left(\frac{\partial^2 v}{\partial x^2} + \frac{\partial^2 v}{\partial y^2}\right) - \frac{1}{\rho}\frac{\partial p_m}{\partial y} \tag{3.10.3}$$

$$u\frac{\partial t}{\partial x} + v\frac{\partial t}{\partial y} = \alpha\left(\frac{\partial^2 t}{\partial x^2} + \frac{\partial^2 t}{\partial y^2}\right) \tag{3.10.4}$$

The associated boundary conditions are

$$u = 0 = v \qquad t = t_0 \qquad \text{at } y = 0,\ x > 0 \tag{3.10.5}$$

$$\frac{\partial u}{\partial y} = 0 = v \qquad \frac{\partial t}{\partial y} = 0 \qquad \text{at } y = 0,\ x < 0 \tag{3.10.6}$$

$$u = v \sim 0 \qquad t \sim t_\infty \qquad p \sim p_\infty \qquad \text{as } y \to \infty \text{ and upstream} \tag{3.10.7}$$

These equations are now scaled by characteristic quantities defined in Eqs. (3.2.5)–(3.2.7) and a stream function ψ is defined as $U = \psi_Y$ and $V = -\psi_X$ so that the continuity equation is satisfied. By eliminating pressure between Eqs. (3.10.2) and (3.10.3), Eqs. (3.10.1)–(3.10.4), when written in terms of dimensionless variables, reduce to the following equations in ψ and ϕ:

$$\left(\psi_Y\frac{\partial}{\partial X} - \psi_X\frac{\partial}{\partial Y}\right)\nabla^2\psi = \mathrm{Gr}^{-1/2}\nabla^4\psi + \phi_Y \tag{3.10.8}$$

$$\psi_Y \phi_X - \psi_X \phi_Y = \frac{\mathrm{Gr}^{-1/2}}{\mathrm{Pr}} \nabla^2 \phi \tag{3.10.9}$$

When $\mathrm{Gr} \to \infty$ the highest-order derivative terms are lost and no solution can then satisfy the boundary conditions. This is a classical singular perturbation problem that may be treated by the method of matched asymptotic expansions. For a complete discussion of the method, the reader is referred to Van Dyke (1964).

In the method of matched asymptotic expansions, the perturbation solution to Eqs. (3.10.1)–(3.10.7) is obtained by constructing inner and outer expansions. The appropriate inner expansions valid in the boundary region close to the surface are

$$\begin{aligned}\psi &= \psi_0 + \epsilon\psi_1 + \epsilon^2\psi_2 + C_n\epsilon^{\lambda_n}\psi_n = \nu c f(\eta, \epsilon)\\ &= \nu c[f_0(\eta) + \epsilon f_1(\eta) + \epsilon^2 f_2(\eta) + C_n\epsilon^{\lambda_n}F_n(\eta) + \cdots]\end{aligned} \tag{3.10.10}$$

$$\begin{aligned}t - t_\infty = \Delta t\, \phi(\eta, \epsilon) &= \Delta t[\phi_0(\eta) + \epsilon\phi_1(\eta) + \epsilon^2\phi_2(\eta)\\ &\quad + C_n\epsilon^{\lambda_n}H_n(\eta) + \cdots]\end{aligned} \tag{3.10.11}$$

$$p - p_\infty = \rho U^2 P(\eta, \epsilon) = \rho U^2[\epsilon^2 P_2(\eta) + \cdots] \tag{3.10.12}$$

where $c = G = 4\sqrt[4]{\mathrm{Gr}_x/4}$ and $\epsilon = 4/G$ is the perturbation parameter. Also, f_0 and ϕ_0 correspond to the boundary layer solutions obtained in Section 3.2, and λ_n is the eigenvalue associated with eigenfunctions $C_n\epsilon^{\lambda_n}F_n$ and $C_n\epsilon^{\lambda_n}H_n$, which identically satisfy the boundary conditions at $\eta = 0$ and $\eta = \infty$. The multiplicative constant is C_n, which Stewartson (1964) found to be associated with the stream function upstream.

The outer expansions, in the outer inviscid region, are

$$\psi = \psi_0^o + \psi_1^o + \psi_2^o + \cdots \tag{3.10.13}$$

$$t - t_\infty = \phi_0^o + \phi_1^o + \phi_2^o + \cdots \tag{3.10.14}$$

$$p - p_\infty = P_0^o + P_1^o + P_2^o + \cdots \tag{3.10.15}$$

where superscript o refers to the outer region defined by $\theta \neq 0$, $r > 0$. Here θ is the angular coordinate measured away from the surface, r is the polar radial coordinate, and L is the characteristic length, $\frac{1}{4}(\nu^2/g\beta\,\Delta t)^{1/3}$. The perturbation parameter $\epsilon = \delta/L = 4/G$ is determined by consideration of the order of magnitude of the normal velocity component at the outer edge of the boundary region. Recalling that $u = \psi_y$ and $v = -\psi_x$, Eqs. (3.10.2)–(3.10.4) are evaluated in terms of variables defined in Eqs. (3.10.10)–(3.10.12). Then the perturbation equations are obtained by collecting terms with like powers of ϵ. Thus, the following successive approximations are obtained. It may be noted that the variables c, ϵ, η used here differ from those used by Hieber (1974) due to the numerical constants chosen. As a result, the perturbation equations derived below are also different. However, one set may be transformed into the other.

Zeroth order. Collecting terms of order ϵ^0, the classical boundary layer Eqs.

(3.3.19)–(3.3.21) are recovered. The asymptotic forms of the functions f_0 and ϕ_0 as $\eta \to \infty$ are obtained as

$$f_0(\eta) \sim A_0 + \text{EST}$$

$$\phi_0(\eta) \sim \text{EST}$$

where A_0 is a constant for a given Prandtl number and EST denotes exponentially small terms. In the outer region it can easily be shown (e.g., Kadambi, 1969) that $\psi_0^o = 0 = \phi_0^o = P_0^o$.

First order. The first-order inviscid solution ψ_1^o, as shown by Yang and Jerger (1964) and by Kadambi, is governed by

$$\nabla^2\psi_1^o = 0 \qquad \psi_1^o = 0 \qquad x < 0 \tag{3.10.16}$$

supplemented by

$$\psi_1^o = \nu c f_0(\infty) \qquad y = 0, x > 0 \tag{3.10.17}$$

Equation (3.10.17) is obtained by matching the zeroth-order boundary layer solution ψ_0 as $y \to \infty$ to the inviscid solution ψ_1^o at $y = 0$. In terms of polar conditions, Eqs. (3.10.16)–(3.10.17) reduce to

$$\nabla^2\psi_1^o = 0 \qquad \psi_{1(\theta=0)}^o = \nu f_0(\infty)\left(\frac{r}{L}\right)^{3/4} \qquad \psi_{1(\theta=\pi)}^o = 0 \tag{3.10.18}$$

The solution to Eq. (3.10.18) is given by

$$\psi_1^o = \frac{-\nu f_0(\infty)}{\sin(3\pi/4)}\left(\frac{r}{L}\right)^{3/4} \sin\frac{3}{4}(\theta - \pi) \tag{3.10.19}$$

Using Bernoulli's theorem in the outer inviscid region,

$$P_1^o = -\frac{1}{2}\rho\mathbf{V}^2 = -\rho\left[\frac{3}{4}f_0(\infty)\right]^2\left(\frac{\nu}{L}\right)^2\left(\frac{L}{r}\right)^{1/2} \tag{3.10.20}$$

Further, for the outer isothermal region, $\phi_1^o \equiv 0$.

To obtain the first-order equations for the inner region, it is first necessary to determine the boundary condition of $f_i'(\infty)$. This is obtained by matching the two-term inner expansion of the u component as $r \to \infty$, $\theta \neq 0$, with the concomitant two-term outer expansion as $\theta \to \infty$. Noting that, as $\theta \to 0$, $r = x[1 + O(\epsilon^2\eta^2)]$ and $\theta = \epsilon\eta + O(\epsilon^3\eta^3)$, the behavior of ψ_1^o in the matching region is determined to be

$$\psi_{1(\theta\to 0)}^o = \nu f_0(\infty)\left(\frac{x}{L}\right)^{3/4}\left[1 + \frac{3}{4}\epsilon\eta + O(\epsilon^2\eta^2)\right] \tag{3.10.21}$$

From the matching consideration discussed above

$$f_1'(\infty) = \frac{3}{4}f_0(\infty) \tag{3.10.22}$$

Then collecting terms of power ϵ, as before, the following perturbation equations in f_1 and ϕ_1 are obtained:

$$f_1''' + 3f_0f_1'' - f_0'f_1' + \phi_1 = 0 \tag{3.10.23}$$

$$\phi_1'' + 3\Pr(f_0\phi_1' + f_0'\phi_1) = 0 \tag{3.10.24}$$

$$f_1(0) = f_1'(0) = 0 = \phi_1(0) = \phi_1(\infty) \qquad f_1'(\infty) = \frac{3}{4}f_0(\infty) = \frac{3}{4}A_0 \tag{3.10.25}$$

In particular, $f_1(\eta) = \frac{3}{4}f_0(\infty)\eta + A_1$ as $\eta \to \infty$, where A_1 is a numerically determined constant.

Second order. Knowing the asymptotic behavior of f_1 and matching the two-term inner and outer stream functions, it is easily deduced that

$$\nabla^2\psi_2^o = 0 \qquad \psi_{2(\theta=0)}^o = 4\nu A_1 \qquad \psi_{2(\theta=\pi)}^o = 0 \tag{3.10.26}$$

The solution to Eq. (3.10.26) is

$$\psi_2^o = 4\nu A_1\left(1 - \frac{\theta}{\pi}\right) \tag{3.10.27}$$

As before, Bernoulli's equation gives

$$P_2^o = \frac{3\sqrt{2}}{\pi}\rho f_0(\infty)A_1\left(\frac{\nu}{L}\right)^2\left(\frac{L}{r}\right)^{5/4}\cos\frac{3}{4}(\theta - \pi) \tag{3.10.28}$$

and

$$\phi_2^o \equiv 0 \tag{3.10.29}$$

Finally, terms of order ϵ^2 defining the second-order inner problem are obtained as follows:

$$f_2''' - 3f_2f_0'' + 2f_0'f_2' + 3f_0f_2'' + \phi_2$$
$$= \frac{f_0'}{4} - \frac{1}{16}\eta f_0'' - \frac{\eta^2}{16}f_0''' - f_1'^2 - 2P_2 - P_2'\eta \tag{3.10.30}$$

$$\frac{\phi_2''}{\Pr} + 6f_0'\phi_2 + 3f_0\phi_2' = 3f_2\phi' - \phi_0''\frac{\eta^2}{16} - \frac{5}{16}\eta\phi_0' \tag{3.10.31}$$

$$P_2' = f_0'^2\frac{\eta}{16} + \frac{3}{16}\eta f f_0'' - \frac{9}{16}ff_0' - \frac{f_0''}{16} + \frac{\eta}{16}f_0''' \tag{3.10.32}$$

$$f_2(0) = f_2'(0) = 0 = \phi_2(0) = \phi_2(\infty) \qquad P_2(\infty) = -[\tfrac{3}{4}f_0(\infty)]^2 \tag{3.10.33}$$

The condition on $P_2(\infty)$ is obtained through the matching considerations. Note that in the region $\theta \to 0$,

$$P_{1(\theta\to 0)}^o = -\frac{9}{16}f_0^2(\infty)\rho\epsilon^2U^2[1 + O(\epsilon^2)] \tag{3.10.34}$$

Three-term matching of inner and outer u components gives

$$f_2'(\infty) = -\frac{A_1}{\pi} - \frac{3}{8} f_0(\infty)\eta \tag{3.10.35}$$

so that

$$f_2(\eta) \sim -\frac{A_1}{\pi}\eta - \frac{3}{16} f_0(\infty)\eta^2 + A_2 \tag{3.10.36}$$

As first pointed out by Hieber (1974), the boundary conditions $f_2''(\infty) = 0 = P_2(\infty)$ used by Kadambi (1969) are inappropriate. The nonzero values of $P_2(\infty)$ and $f_2''(\infty)$ arise from the behavior of P_1^o in Eq. (3.10.34) and of the $O(\epsilon^2)$ terms in Eq. (3.10.21), respectively.

To complete the analysis, the eigenfunctions in Eqs. (3.10.10) and (3.10.11) must be determined to assess the order of their contributions. To do so the terms of order ϵ^{λ_n} are collected from Eqs. (3.10.2)–(3.10.4) after substituting the expansions, Eqs. (3.10.10) and (3.10.11). The following linear homogeneous equations in F_n and H_n are obtained:

$$F_n''' + H_n = (-3\lambda_n + 4) f_0' F_n' + 3F_n f_0''(\lambda_n - 1) - 3 f_0 F_n'' \tag{3.10.37}$$

$$\frac{H_n''}{\text{Pr}} = -3 f_0 H_n' + 3\lambda_n f_0' H_n - 3F_n \phi_0'(1 - \lambda_n) \tag{3.10.38}$$

subject to the homogeneous boundary conditions

$$F_n(0) = F_n'(0) = F_n'(\infty) = H_n'(0) = H_n(\infty) = 0 \tag{3.10.39}$$

Nontrivial solutions of these equations exist only for particular values of λ_n. The smallest such eigenvalue of λ_n is $\frac{4}{3}$, and the associated eigenfunctions are

$$F_1 = C_1(3f_0 - \eta f_0') \tag{3.10.40}$$

$$H_1 = C_1(-\eta\phi') \tag{3.10.41}$$

where C_1 is an indeterminate constant. Thus, the contributions due to these eigenfunctions appear in expansions before the second-order terms. Other eigenvalues found by numerical solution of Eqs. (3.10.37)–(3.10.39) have values greater than 2. As noted by Hieber, the terms $\epsilon^{4/3}\nu c F_1$ and $\epsilon^{4/3}\,\Delta t\, H_1$ are proportional to the x derivative of the zeroth-order stream function and temperature. The constant C_1 is therefore related to an apparent shift in the location of the leading edge as seen by the downstream boundary region; see Stewartson (1964).

The inner region perturbation equations, as well as the equations governing the eigenfunctions, have been solved numerically for Pr = 0.72. The results are summarized in Table 3.10.1. The missing values of $F_1''(0)$ and $H_1'(0)$, corresponding to the first eigenvalue $\lambda_1 = \frac{4}{3}$, were determined as 0.08383 and 0.06256, respectively. Based on these results, the local surface heat flux and shear stress are given by

Table 3.10.1 Numerical data for higher-order approximations for isothermal and uniform flux surfaces

Isothermal surfaces, Pr = 0.72 (Hieber, 1974)			
	$f_i''(0)$	$\phi_i'(0)$	A_i
$i = 0$	0.67602	−0.50463	0.59888
$i = 1$	0.19818	0.0	−0.63449
$i = 2$	−0.24872	−0.63036	0.26519

Uniform flux surfaces (Mahajan and Gebhart, 1978)						
	$f_i''(0)$		$\phi_i(0)$		A_i	
	Pr = 0.733	Pr = 6.7	Pr = 0.733	Pr = 6.7	Pr = 0.733	Pr = 6.7
$i = 0$	0.808931	0.356332	1.479807	0.841702	0.507505	0.205823
$i = 1$	−0.083596	−0.006691	−0.361412	−0.082575	−1.109141	−0.537460
$i = 2$	0.4461	−0.001106	−2.3230	−0.2488	2.6414	0.14178

$$q'' = -k\frac{\Delta t}{\delta}[\phi_0'(0) + C_1\epsilon^{4/3}H_1'(0) + \epsilon^2\phi_2'(0) + \text{HOT}] \quad (3.10.42)$$

$$\tau = \mu\left.\frac{\partial u}{\partial y}\right)_{y=0} = \mu\frac{U}{\delta}[f_0''(0) + \epsilon f_1''(0) + C_1\epsilon^{4/3}F_1''(0)$$
$$+ \epsilon^2 f_2''(0) + \text{HOT}] \quad (3.10.43)$$

where HOT represents higher-order terms.

Other physical quantities of interest are the total surface heat transfer Q and total drag D between the leading edge and any x. These would be evaluated from

$$\int_0^x q''\,dx \qquad \text{and} \qquad \int_0^x \tau\,dx$$

where q'' and τ are given by Eqs. (3.10.42) and (3.10.43). However, the improved estimates of q'' and τ are valid only in the region for $\epsilon < O(1)$, that is, $x > O(L)$. In the region of the leading edge, these expressions are inapplicable. This is indicated by the fact that some of the higher-order terms are nonintegrable at $x = 0$. An alternative procedure that avoids this difficulty is to obtain the total heat transfer rate and the drag from considerations of global (boundary region plus surroundings) energy and momentum, respectively. This general technique was used by Imai (1958) for forced flows and by Heiber (1974) and Mahajan and Gebhart (1978) for buoyancy-driven flows.

The global heat transfer rate is obtained by calculating the total thermal convection across the boundary layer (BL) at any x. That is,

$$Q = \rho c_p \int_{\mathrm{BL}} u(t - t_\infty)\, dy - k \int_{\mathrm{BL}} \frac{\partial t}{\partial x}\, dy$$

$$= \frac{k\, \Delta t}{\delta} x(a_0 + a_1 \epsilon + a_{4/3} C_1 \epsilon^{4/3} + a_2 \epsilon^2 + \cdots) \qquad (3.10.44)$$

where

$$\Delta t = t_0 - t_\infty$$

$$a_0 = 4\mathrm{Pr} \int_0^\infty (f_0' \phi_0)\, d\eta \qquad a_1 = 4\mathrm{Pr} \int_0^\infty f_1' \phi_0\, d\eta$$

$$a_{4/3} = 4\mathrm{Pr} \int_0^\infty (F_1' \phi_0 + H_1 f_0')\, d\eta \qquad (3.10.45)$$

$$a_2 = 4\mathrm{Pr} \int_0^\infty \left(f_2' \phi_0 + \phi_2 f_0' - \frac{\phi_0}{16} \right) d\eta$$

for $\mathrm{Pr} = 0.72$, $a_0 = 0.67284$, $a_1 = 0.62480$, $a_{4/3} = 0.25024$, and $a_2 = -0.84048$.

Comparison of Eqs. (3.10.42) and (3.10.44) indicates that, although the $O(\epsilon)$ correction to q'' is zero in the boundary layer region, the $O(\epsilon)$ correction to Q is nonzero. The additional term $a_1 k\, \Delta t$ represents a convection of the zeroth-order temperature by the first-order velocity and is independent of x. Hieber argued that this represents the leading edge effect on the global heat transfer rate.

Based on Eqs. (3.10.42) and (3.10.44) and the computed results for Pr = 0.72, the local and average Nusselt numbers are expressed as

$$\mathrm{Nu}_x = \frac{q''}{\Delta t} \frac{x}{k} = \frac{1}{\sqrt{2}} (\mathrm{Gr}_x)^{1/4} (0.50463 - 0.06256 C_1 \epsilon^{4/3} + 0.63036 \epsilon^2 + \mathrm{HOT}) \qquad (3.10.46)$$

$$\mathrm{Nu} = \frac{1}{\sqrt{2}} (\mathrm{Gr})^{1/4} (0.67284 + 0.62480 \epsilon + 0.25024 C_1 \epsilon^{4/3} - 0.84048 \epsilon^2 + \mathrm{HOT}) \qquad (3.10.47)$$

Thus, although the leading correction to the local Nusselt number is indeterminate, the leading correction to the average Nusselt number is known explicitly and is positive for Pr = 0.72.

The above expressions are for a semi-infinite surface. For a rigorous comparison with experimental results obtained with surfaces of finite length, the effects of the trailing edge and the wake region on Nu must be included in the analysis. The wake effect has been estimated by Yang and Jerger (1964) and in more detail by Hieber (1974). The trailing edge effects have not been investigated so far. However, the wake results seem to indicate corrections in the right direction. Positive contributions to the heat transfer results arise at moderate values of the Grashof number, in agreement with experimental results.

As mentioned before, the total drag D can be obtained from considerations of the global momentum and buoyancy A global force balance in the vertical direction gives

$$D = \mu \frac{U}{\delta} x \left[\frac{4}{5} f_0''(0) + 2\epsilon f_1''(0) + 4C_1 \epsilon^{4/3} F_1''(0) + O(\epsilon^{5/3}) - 4\,\epsilon^2 f_2''(0) + \cdots \right] \tag{3.10.48}$$

or

$$D = \frac{\rho \nu^2}{L} (4)^{-1/3} \left[\frac{4}{5} f_0''(0)\epsilon^{-5/3} + 2 f_1''(0)\epsilon^{-2/3} + 4C_1 F_1''(0)\epsilon^{-1} + O(1) - 4\epsilon^{1/3} f_2''(0) + \text{HOT} \right] \tag{3.10.49}$$

where the lower-order $O(1)$ contribution is indeterminate and is due to the global buoyant force acting throughout the leading edge region. The contribution by the leading eigenfunction $4C_1 F_1''(0)\epsilon^{-1/3}$ is seen to be unbounded as $x/L \to \infty$. Such an unbounded contribution to D by an eigenfunction appears to be peculiar to natural convection boundary layers.

3.10.2 The Uniform Flux Condition

A perturbation analysis of higher-order boundary layer effects in natural convection flow adjacent to a semi-infinite vertical uniform flux surface has been given by Mahajan and Gebhart (1978). The analysis is similar to that above. It will be discussed briefly, emphasizing the differences between the two analyses.

Inner and outer expansions, Eqs. (3.10.10)–(3.10.15), are used again, except that the perturbation parameter ϵ is now given by $\epsilon = 5/G^*$, where $G^* = 5(\text{Gr}_x^*/5)^{1/5}$. Proceeding exactly as above, the perturbation equations and the equations governing the eigenfunctions are easily obtained. The only eigenfunction appearing up to second order is determined to be that corresponding to the smallest eigenvalue of λ_n, $\frac{5}{4}$. The associated eigenfunctions are

$$F_1 = C_1(4f_0 - \eta f_0') \tag{3.10.50}$$

and

$$H_1 = C_1(\phi_0 - \eta\phi_0') \tag{3.10.51}$$

However, in contrast to the isothermal condition, C_1 is not indeterminate. It is shown to be identically zero. The sum of the total convected thermal energy $Q(x)$ and the heat conducted downstream is equated to the total heat transferred from the surface to the flow, $q''x$:

$$\rho c_p \int_0^\infty u(t - t_\infty)\, dy - k \int_0^\infty \frac{\partial t}{\partial x}\, dy = q''x \tag{3.10.52}$$

Substitution results in

$$\begin{aligned}\int_0^\infty & [(f_0'\phi_0) + \epsilon(f_1'\phi_0 + f_0'\phi_1) + C_1\epsilon^{5/4}(F_1'\phi_0 + f_0'H_1) \\ & + \epsilon^2(f_0'\phi_2 + f_1'\phi_1 + f_2'\phi_0) \\ & + \text{HOT}]\, d\eta = \frac{1}{5\text{Pr}} - \frac{1}{25\text{Pr}}\epsilon^2 \int_0^\infty [(\phi_0'\eta - \phi_0) + \text{HOT}]\, d\eta \end{aligned} \tag{3.10.53}$$

Equating like powers of ϵ on both sides,

ϵ^0:

$$\int_0^\infty f_0'\phi_0\, d\eta = \frac{1}{5\text{Pr}} \tag{3.10.54}$$

ϵ^1:

$$\int_0^\infty (f_0'\phi_1 + f_1'\phi_0)\, d\eta = 0 \tag{3.10.55}$$

ϵ^2:

$$\int_0^\infty (f_0'\phi_2 + f_1'\phi_1 + f_2'\phi_0)\, d\eta = \frac{2}{25\text{Pr}} \int_0^\infty \phi_0\, d\eta \tag{3.10.56}$$

$\epsilon^{5/4}$:

$$\begin{gathered} C_1 \int_0^\infty (F_1'\phi_0 + f_0'H_1)\, d\eta = 0 \\ C_1 \int_0^\infty [(4f_0' - f_0' - \eta f_0'')\phi_0 + f_0'(\phi_0 - \eta\phi_0')]\, d\eta = 0 \\ 5C_1 \int_0^\infty (f_0'\phi_0)\, d\eta = 0 \end{gathered} \tag{3.10.57}$$

From Eq. (3.10.54), $\int_0^\infty f_0'\phi_0\, d\eta \neq 0$. Therefore, Eq. (3.10.57) implies that $C_1 \equiv 0$. For the uniform flux condition, therefore, the boundary layer expansions, Eqs. (3.10.10)–(3.10.12), are appropriate up to $O(\epsilon^2)$.

The numerical data of interest for both Pr = 0.733 and 6.7, as calculated by Mahajan and Gebhart (1978), are also listed in Table 3.10.1. The calculated velocity and temperature profiles for Pr = 0.733 are shown in Figs. 3.10.1 and 3.10.2, respectively. The temperature difference $t_0 - t_\infty$ across the boundary layer is

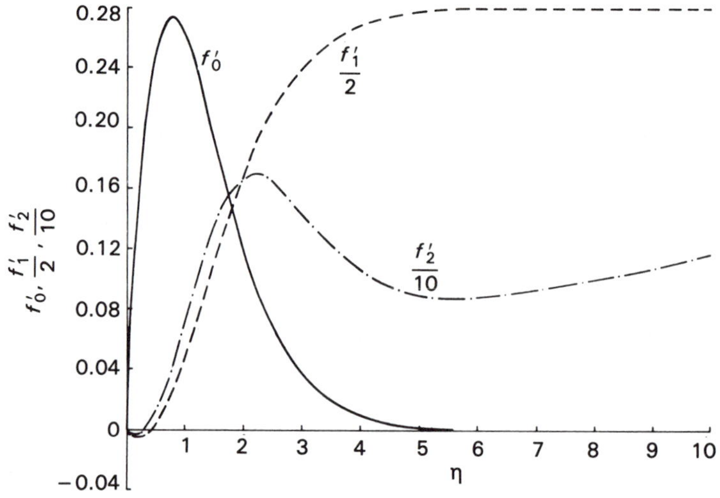

Figure 3.10.1 Velocity function distributions for Pr = 0.733. *(Reprinted with permission from* Int. J. Heat Mass Transfer, *vol. 21, p. 549, R. L. Mahajan and B. Gebhart. Copyright © 1978, Pergamon Journals Ltd.)*

$$\Delta t = t_0 - t_\infty = \frac{q''}{k}[\phi_0(0) + \epsilon\phi_1(0) + \epsilon^2\phi_2(0)] \tag{3.10.58}$$

$$= \frac{5q''}{k}L^{4/5}\phi_0(0)x^{1/5}\left[1 + \epsilon\frac{\phi_1(0)}{\phi_0(0)} + \epsilon^2\frac{\phi_2(0)}{\phi_0(0)}\right] \tag{3.10.59}$$

where $L = \frac{1}{5}(kv^2/g\beta q'')^{1/4}$. This result shows how the boundary layer surface temperature variation downstream, $\Delta t \propto x^{1/5}$, is modified by the higher-order corrections. In particular, for Pr = 0.733 and 6.7, from the values in Table 3.10.1, the boundary layer theory overpredicts the local value of $t_0 - t_\infty$. The effect is larger for the lower Prandtl number.

The improved local Nusselt number is

$$\text{Nu}_x = \frac{q''}{(t_0 - t_\infty)}\frac{x}{k} = \frac{1}{\epsilon\phi_0(0)}\left\{1 - \epsilon\frac{\phi_1(0)}{\phi_0(0)} - \epsilon^2\left[\frac{\phi_2(0)}{\phi_0(0)} - \frac{\phi_1^2(0)}{\phi_0^2(0)}\right]\right\} \tag{3.10.60}$$

The numerical values of $\phi_i(0)$ yield

$$\frac{\text{Nu}_x}{\text{Nu}_{x,0}} = 1 + 0.22423\epsilon + 1.6294\epsilon^2 \qquad \text{for Pr} = 0.733 \tag{3.10.61}$$

$$= 1 + 0.098105\epsilon + 0.3052\epsilon^2 \qquad \text{for Pr} = 6.7 \tag{3.10.62}$$

where subscript 0 refers to the zeroth-order boundary layer results. Clearly, the predicted corrections to the local Nusselt number are positive.

The improved shear stress τ is again Eq. (3.10.41) with $C_1 = 0$. It has a

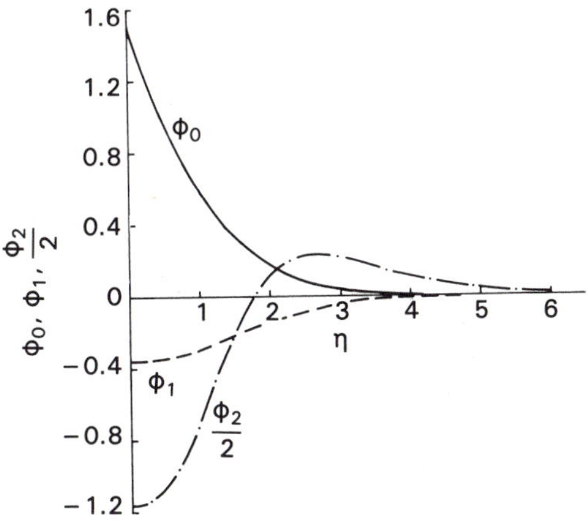

Figure 3.10.2 Temperature function distributions for Pr = 0.733. *(Reprinted with permission from* Int. J. Heat Mass Transfer, *vol. 21, p. 549, R. L. Mahajan and B. Gebhart. Copyright © 1978, Pergamon Journals Ltd.)*

nonintegrable singularity in the third term at $x = 0$. The total drag D therefore cannot be obtained from $D = \int_0^x \tau \, dx$. However, as before, the global momentum analysis can be applied to obtain the following expression for drag:

$$D = \frac{\nu^2 \rho}{L} 5^{3/4} \left[\frac{1}{7} \epsilon^{-3/4} f_0''(0) + \frac{1}{3} \epsilon^{-3/4} f_1''(0) + O(1) - \epsilon^{1/4} f_2''(0) + \cdots \right] \tag{3.10.63}$$

where the lower-order $O(1)$ contribution is indeterminate and again represents the leading edge effect on the total drag. However, there is no unbounded term due to the first eigenfunction, as found for the isothermal condition.

The solutions above are valid everywhere except in a small region around the leading edge, where $x = O(L)$. Although the consideration of global energy and momentum has resulted in an estimate of the contributions of the leading edge effects on total heat transfer rate and total drag, no rigorous analysis has appeared for the transport in the leading edge region. An associated problem is the pattern of inflow into the boundary region, that is, entrainment. Brodowicz (1968) found that this inflow may be unsteady and that it influences boundary region flow, particularly in the region of the leading edge, where the boundary region velocities are very low. These questions deserve further consideration.

3.11 AMBIENT MEDIUM DENSITY STRATIFICATION

In nature and in industry, buoyancy-induced flows occurring in a density-stratified environment are frequently encountered. This effect is of particular concern in energy and material rejection into the environment, in energy storage systems such as solar ponds, and in heat transfer from bodies in enclosed regions. In these circumstances it is important to determine the effect of the ambient medium density stratification on the flow and on the heat and mass transfer from a given body. In Section 3.6 the stability of a stratified environment was considered and it was shown that, for a fluid whose density decreases with an increase in temperature, the ambient temperature must decrease at a rate lower than in an adiabatically stratified medium for the stratification to be stable. This condition is satisfied when the ambient temperature increases with height. This is the circumstance that has been studied most extensively.

3.11.1 Flat Surfaces in Stratified Media

Among the earliest investigations of buoyancy-driven flows in thermally stratified media was that by Prandtl (1952), who considered an infinite surface in a linearly stratified environment. It was assumed that an infinite inclined surface has a constant temperature excess $\Delta t = t_0 - t_\infty$ over the local ambient temperature t_∞. If N is the constant vertical temperature gradient, the temperature and velocity distributions for the resulting one-dimensional parallel flow were calculated to be

$$\frac{t - t_\infty}{\Delta t} = e^{-\tilde{y}} \cos \tilde{y} \qquad \text{with } t_\infty - t_r = Nx \tag{3.11.1}$$

$$u = \left(\frac{g\beta\alpha}{\nu N}\right)^{1/2} \Delta t \, e^{-\tilde{y}} \sin \tilde{y} \tag{3.11.2}$$

where $\tilde{y} = y/l$ and l is the following characteristic length scale for a surface inclined at an angle θ with the vertical:

$$l = \left(\frac{4\nu\alpha}{g\beta N \cos^2 \theta}\right)^{1/4} \tag{3.11.3}$$

Similar flows over infinite surfaces immersed in linearly stratified media have also been studied by Gill (1966), Gill and Davey (1969), Iyer (1973), and Jischke (1977). In these studies the thermal stratification does not vary with time. However, in any actual circumstance, vertical thermal diffusion would lead to a gradual decay of the stratification, as studied by Jaluria and Gupta (1982). The analysis for either a finite or a semi-infinite surface assumes a slow decay of the stratification, compared to the characteristic time for the flow. Otherwise, the transient effects would have to be included. Jaluria and Gebhart (1974) discuss the implications of this consideration.

The natural convection flow over a semi-infinite vertical surface (i.e., with a leading edge) immersed in a thermally stratified environment was first analyzed by Cheesewright (1967). Similarity solutions were obtained for isothermal and nonisothermal vertical surfaces. Results for a stably stratified environment indicated an increase in the local heat transfer coefficient and a decrease in velocity and buoyancy levels. Eichhorn (1969) studied the flow over an isothermal vertical surface in a thermally stratified medium and obtained a series solution. Chen and Eichhorn (1976) used the local nonsimilarity method for the same problem and obtained good agreement between analysis and experiment. Yang et al. (1972), Eichhorn et al. (1974), Takeuchi et al. (1974), Singh (1977), and Raithby and Hollands (1978) considered natural convection flow with various geometries in a stably stratified environment.

Following the similarity analysis given in Section 3.6, Jaluria and Gebhart (1974) and Jaluria (1979) calculated the transport from a heated vertical surface in a thermally stratified environment. The governing boundary layer equations are obtained as

$$f''' + (n + 3)ff'' - 2(n + 1)f'^2 + \phi = 0 \tag{3.11.4}$$

$$\frac{\phi''}{\mathrm{Pr}} + (n + 3)f\phi' - 4nf'\phi - Sf' = 0 \tag{3.11.5}$$

with the boundary conditions

$$f(0) = f'(0) = 1 - \phi(0) = f'(\infty) = \phi(\infty) = 0 \tag{3.11.6}$$

where S, the stratification parameter, is simply the constant C_5 defined in Eq. (3.6.1) for the power law variation. Therefore, the relevant governing parameters of the problem are Pr, S, and n. The unstratified ambient solution, for $S = 0$, is discussed in detail in Section 3.5. It was also shown in Section 3.6 that the ambient temperature variation must be of the same form as the variation of the surface temperature excess $t_0 - t_\infty$ for similarity to arise. Note that the reverse problem of the downflow due to a cooled surface, $t_0 < t_\infty$, in a stably stratified medium may be treated similarly by taking x downward from the top end of the surface.

Figures 3.11.1 and 3.11.2 show the velocity and temperature profiles obtained by Jaluria and Gebhart (1974) for Pr values of 0.733 and 6.7, respectively, and various values of S. The value of n for these curves is 0.2, which corresponds to the uniform surface heat flux. Results for other values of n and S are given by Jaluria (1979). The figures show that an increasing stratification rate results in a decreasing buoyancy level downstream. This results in a lower velocity level. The temperature gradient steepens at the surface, accompanied by a decrease in the temperature in the outer boundary region. The temperature defect that arises at larger values of S results as follows.

At a downstream location x in the flow coming up from below, the outer region tends to have a temperature lower than the local ambient temperature. If the ambient temperature increase with height is rapid enough, the temperature in this region is unable to attain the local ambient temperature, even though the physical temperature does increase due to heat transfer mechanisms. Therefore, a temperature defect arises, as shown, in the nondimensional temperature distribution.

The ambient temperature increase with height also results in a lowering of the local buoyancy. This is seen in terms of the decrease in the relative temper-

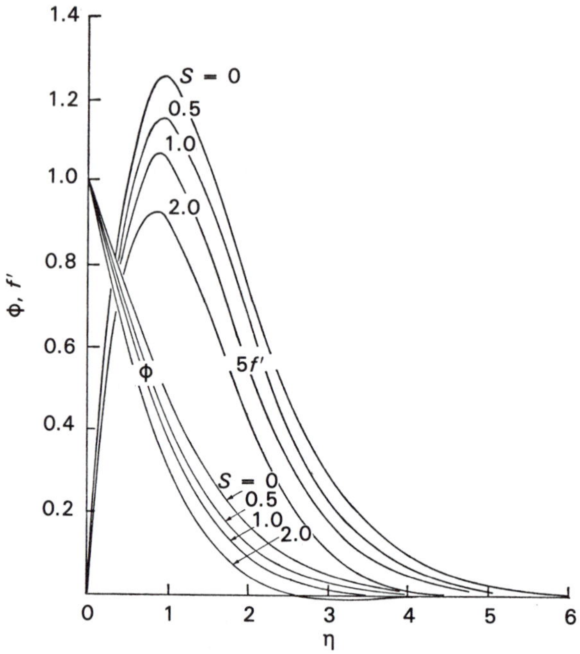

Figure 3.11.1 Velocity and temperature distributions in the boundary layer flow over a uniform heat flux surface immersed in a stably stratified medium for Pr = 0.733. *(Reprinted with permission from Y. Jaluria and B. Gebhart,* J. Fluid Mech., *vol. 66, p. 593. Copyright © 1974, Cambridge University Press.)*

ature level and the consequent lowering of the velocity level in the boundary layer. Flow reversal is also seen to arise, although it is found to be a very small effect over the range of S shown. Such reversals would probably be very unstable. The other studies mentioned above also found both a temperature defect and flow reversal at large stratification levels. Boundary layer theory does not, in general, deal with flow reversal satisfactorily since the vertical velocity component decreases through zero. The implications of the full equations must be considered.

Figure 3.11.3 shows the effect of stratification on the heat transfer and skin friction parameters $[-\phi'(0)]$ and $f''(0)$, respectively. As expected, the former increases with S, the effect being larger at smaller Pr, and the latter decreases with an increase in S. This implies that the convective heat transfer coefficient increases due to thermal stratification. For a given temperature difference Nx^n, the heat transfer q'' from the surface increases due to stratification, as is evident from

$$q'' = [-\phi'(0)]\,\frac{kNx^n}{x}\,\frac{G}{4} \tag{3.11.7}$$

Since the temperature difference is kept unchanged with an increase in S, this implies that the local surface temperature must also be increased to keep pace

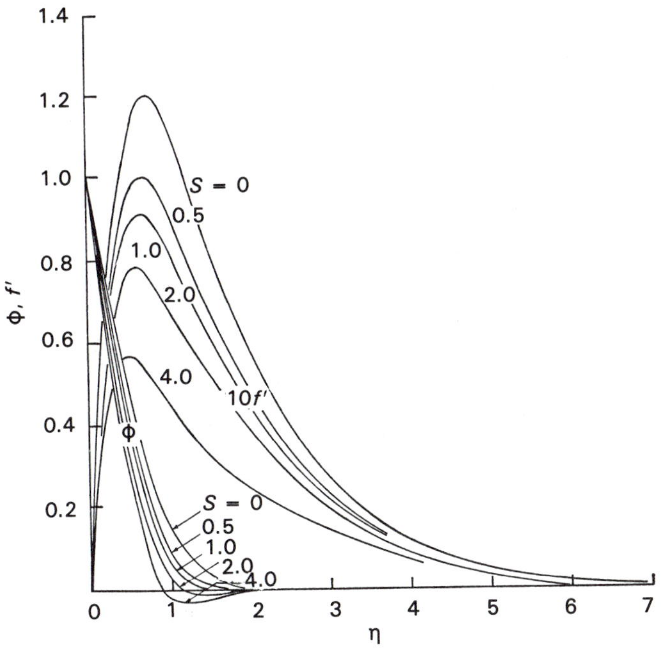

Figure 3.11.2 Velocity and temperature distributions in the boundary layer flow over a uniform heat flux surface immersed in a stably stratified medium for Pr = 6.7. *(Reprinted with permission from Y. Jaluria and B. Gebhart,* J. Fluid Mech., *vol. 66, p. 593. Copyright © 1974, Cambridge University Press.)*

with the ambient temperature increase. However, a more practical circumstance is one in which the local surface temperature is held constant instead. Then the temperature difference $t_0 - t_\infty$ decreases and the heat transfer from the surface is found to decrease, despite the increase in the heat transfer coefficient. This result is expected since stratification decreases the buoyancy level and the velocity and must result in lower heat transfer from a vertical surface at a given temperature (Jaluria, 1980).

The heat transfer from an isothermal vertical surface to a thermally stratified fluid has been studied in detail, experimentally and analytically, by Chen and Eichhorn (1976) and Takeuchi et al. (1974). The ratio of the overall heat transfer Q in a stratified fluid to that in an isothermal fluid Q_{iso} was calculated by Chen and Eichhorn (1976), using the local nonsimilarity method of Sparrow and Yu (1971) discussed in Section 3.13, as

$$\frac{Q}{Q_{\text{iso}}} = 1.343\left[\frac{g\beta(t_0 - t_{\infty,0})x^3}{4\nu^2}\right]^{1/4}(1 - 0.3836\xi - 0.02883\xi^2) \qquad (3.11.8)$$

where $t_{\infty,0}$ is the ambient temperature at the leading edge of the vertical surface and ξ is a dimensionless vertical distance given by $\xi = x(dt_\infty/dx)/(t_0 - t_{\infty,0})$. Thus, ξ can be considered a local stratification parameter.

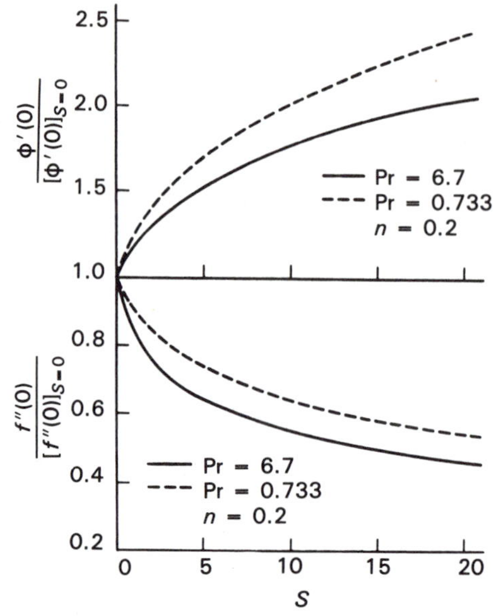

Figure 3.11.3 Effect of stable thermal stratification on the heat transfer and skin friction parameters. *(From Jaluria, 1979.)*

The overall heat transfer rate Q in Eq. (3.11.8) is obtained by integrating the local heat transfer over x. Therefore, a reduction in heat transfer results for an isothermal surface immersed in a stratified environment due to the stratification. The Nusselt number may be based on the temperature difference at the leading edge $(t_0 - t_{\infty,0})$ or on the mean value over the length ξ. Figure 3.11.4 shows the experimental and analytical results of Chen and Eichhorn (1976) and the approximate analytical results of Raithby and Hollands (1978). The Nusselt numbers are based on the mean temperature difference and the stratification parameter S is defined as $S = L(dt_\infty/dx)/\Delta t_m$, where Δt_m is the mean temperature difference between the surface and the fluid $(t_0 - t_\infty)$ and L is the plate height. Figure 3.11.4 shows the expected increase in the heat transfer coefficient with an increase in the stratification level and the agreement between theory and experiment. The Nusselt number for a heated surface in isothermal surroundings is denoted by $\mathrm{Nu}_{\mathrm{iso}}$.

3.11.2 Free Boundary Flows

Buoyancy-induced flow in thermal plumes and in buoyant jets may also occur in a stratified environment. This problem is of particular interest in heat rejection to the environment and has therefore been considered extensively for turbulent flows. Axisymmetric flows have received more attention than plane plumes and jets because of their more frequent occurrence in practice. Of particular interest in these flows is the height to which the flow rises in a stably stratified environment, as

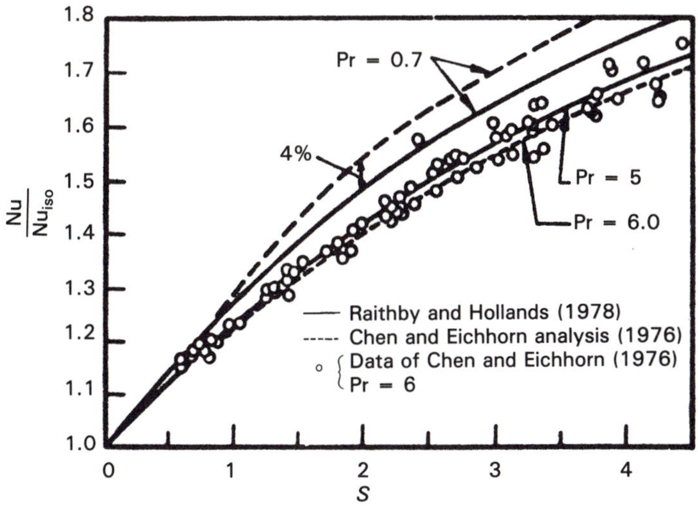

Figure 3.11.4 Effect of ambient medium stratification on the heat transfer coefficient for an isothermal surface. *(Reprinted with permission from G. D. Raithby and K. G. T. Hollands,* J. Heat Transfer, *vol. 100, p. 378. Copyright © 1978, ASME.)*

discussed in Chapter 4 for laminar flows and in Chapter 12 for turbulent flows. Heat transfer from bodies immersed in media that are stably stratified due to chemical species diffusion is also of considerable interest and importance. The corresponding combined heat and mass transfer situation is considered in Chapter 6.

Similarity is not obtained for a plane plume rising in a stably stratified environment because of the additional term $[-(cj_x/bd)f']$ that arises in the energy equation; see Eq. (3.5.9). As shown in Section 3.6, similarity requires that $j(x)$ vary as $d(x)$. Since the centerline temperature in a thermal plume decreases with height due to entrainment, this condition implies that t_∞ must also decrease with height for similarity to arise. For most practical circumstances, this requirement results in an unstable thermal stratification for which the analysis discussed here is not applicable (see Chapter 13). In an adiabatically stratified environment, t_∞ decreases with height. However, the variation with x of the plume centerline temperature excess is not of the same form as that in the ambient fluid. Thus, similarity is not obtained.

Free boundary flows in a stratified environment may be studied numerically, as done by Jaluria and Himasekhar (1983), in considering different ambient temperature distributions for Pr = 0.7 and 6.7. They considered long thermal sources of finite size in the numerical simulation of plumes rising in stably stratified media. Numerical results were obtained at various values of the local stratification parameter S, defined similarly to that given before, as

$$S = \frac{1}{\Delta t}\frac{dt_{\infty,x}}{dX} \tag{3.11.9}$$

where $\Delta t = t_0 - t_{\infty,0}$, $X = x(g\beta\, \Delta t/\nu^2)^{1/3}$, and $t_{\infty,x} - t_{\infty,0} = N_\infty x^n$. For a linearly stratified medium, S is a constant. Here t_0 is the temperature at the surface of a long horizontal source of small finite width that generates the flow, and $t_{\infty,0}$ is the ambient temperature at $x = 0$. The dimensionless temperature ϕ, velocity U, and transverse coordinate Y are given by

$$\phi = \frac{t - t_{\infty,x}}{t_0 - t_{\infty,0}} \qquad U = \frac{u}{(\nu g\beta\, \Delta t)^{1/3}} \qquad Y = y\left(\frac{g\beta\, \Delta t}{\nu^2}\right)^{1/3} \tag{3.11.10}$$

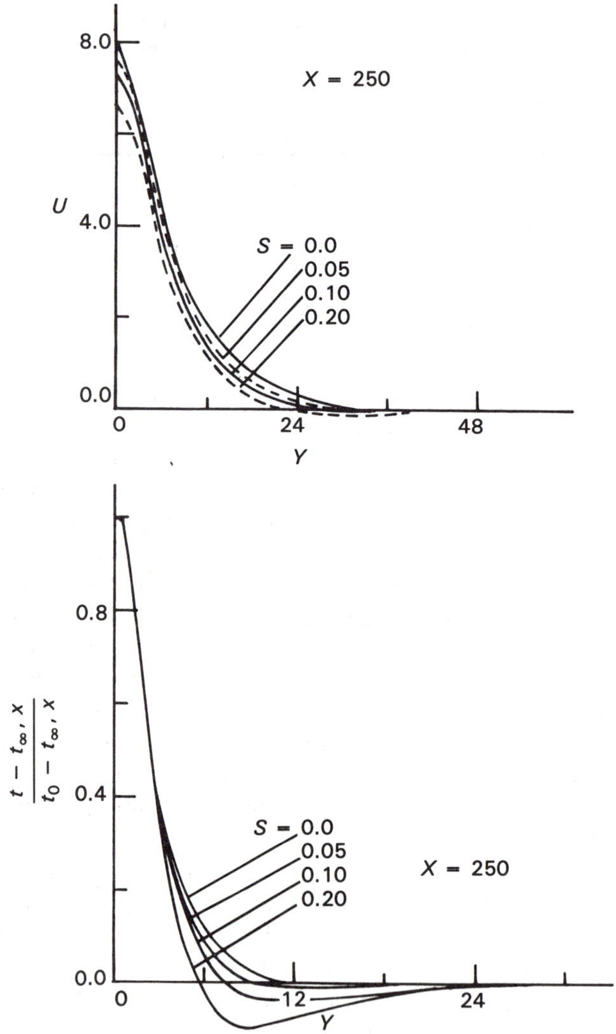

Figure 3.11.5 Velocity and temperature profiles in a plane plume rising in a linearly stratified environment at Pr = 6.7. *(Reprinted with permission from* Comput. Fluids, *vol. 11, p. 39, Y. Jaluria and K. Himasekhar. Copyright © 1983, Pergamon Journals Ltd.)*

Figure 3.11.5 shows the computed velocity and temperature profiles. The observed trends are very similar to those obtained for a vertical surface in a thermally stratified environment. A temperature defect arises in the outer region of the flow. This effect increases with an increase in the stratification level and also as the flow proceeds downstream. The centerline velocity decreases with an increase in S, as expected. Figure 3.11.6 shows the downstream variation of the centerline velocity and temperature. For an isothermal ambient medium, $S = 0$, these vary as $x^{1/5}$ and $x^{-3/5}$, respectively, far downstream of the source, as predicted by similarity analysis. For $S > 0$ the buoyancy level decreases more rapidly downstream due to the increasing ambient temperature, ultimately leading to a negative buoyancy level. This results in a decrease in the centerline velocity to zero, followed by flow reversal. Some of these effects are discussed again for axisymmetric flows in Chapter 4.

Numerical results for a heated isothermal or uniform heat flux surface in an

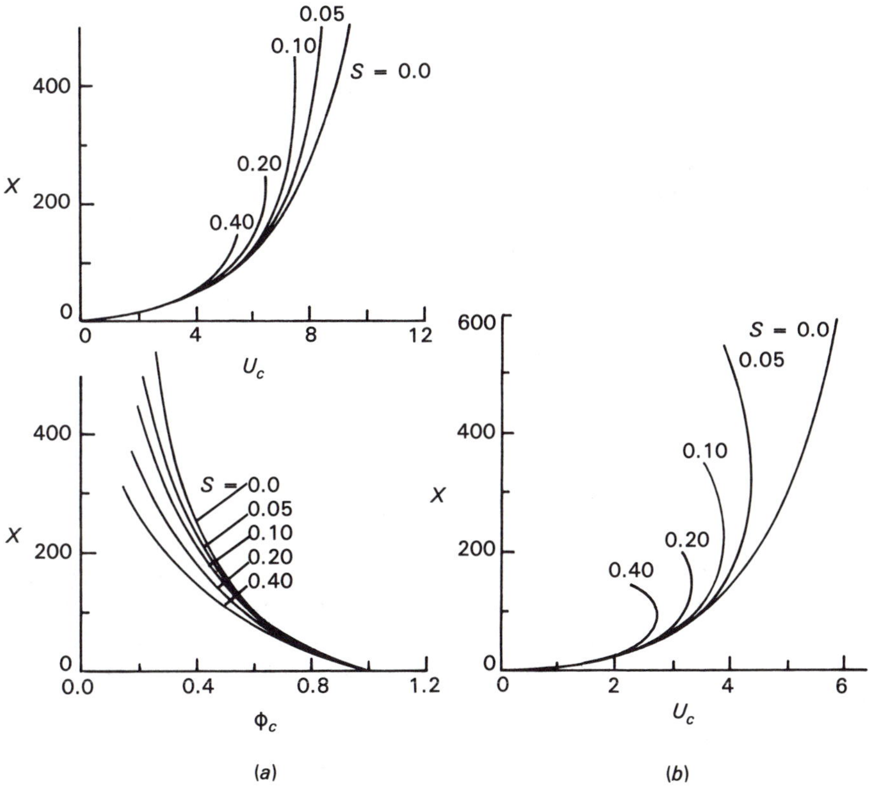

Figure 3.11.6 Downstream variation of the centerline velocity in a plane plume rising in a linearly stratified environment at (*a*) Pr = 6.7 and (*b*) Pr = 0.7. Also shown is the temperature variation for Pr = 6.7. *(Reprinted with permission from* Comput. Fluids, *vol. 11, p. 39, Y. Jaluria and K. Himasekhar. Copyright © 1983, Pergamon Journals Ltd.)*

arbitrarily stratified medium are also presented by Jaluria and Himasekhar (1983). An experimental study of these flows was carried out and the results were found to agree with the theoretical predictions as discussed by Himasekhar (1980).

Experimental studies of external natural convection are generally carried out in fluid regions of finite extent. Therefore, as time elapses, the heat transfer mechanisms are expected to lead to thermal stratification of the medium. A distant recirculating flow also arises to compensate for the buoyancy-induced flow. It is therefore necessary to limit the time for experimentation so that the effects of thermal stratification and recirculation are negligible. The dimensions of the experimental system used are also determined by these considerations so that an adequate ambient fluid region results. A detailed study of these aspects is very involved since it is a transient internal natural convection process. Similar considerations also arise in several problems of practical interest, such as the cooling of electronic components located in an enclosed region. Some of the relevant related work is outlined in Chapter 14.

3.12 OTHER BOUNDARY CONDITIONS

Boundary conditions other than those discussed in the preceding sections have also been considered because of their relevance to practical applications. Often, similarity does not arise and solutions have been obtained by series, approximate, and numerical methods. One important nonsimilar flow arises for a surface temperature or heat flux imposed only over a limited height of a vertical surface. This occurs in many practical circumstances, such as electronic circuitry cooling. The energy-dissipating devices are idealized as thermal sources located on vertical adiabatic surfaces. The wall plume, due to a line thermal source on a vertical adiabatic surface, was considered in Section 3.7. The interaction of wakes due to multiple heated surface elements is discussed in Section 5.7. The natural convection wake above a finite heated vertical surface and the flow adjacent to a vertical surface with a step discontinuity in the wall temperature have also been studied.

Wakes. Yang (1964) used the asymptotic series expansion technique to study the velocity and temperature profiles in the wake region immediately above the trailing edge of an isothermal vertical surface. The calculations were continued downstream by means of an integral solution. Hardwick and Levy (1973) employed finite-difference methods to solve the full elliptic wake region equations for an isothermal vertical surface. The solution was compared with experimental results for air, indicating fairly good agreement. Figure 3.12.1 shows the computed velocity and temperature profiles in the wake. The centerline velocity at $y = 0$ rises sharply beyond $x = L$ due to the absence of wall shear downstream. Within a short distance, the Gaussian profile characteristic of a thermal plume is obtained.

Sparrow et al. (1978) numerically calculated the development of wall and free plumes. The wall plume generated by a finite-size heated region located on a

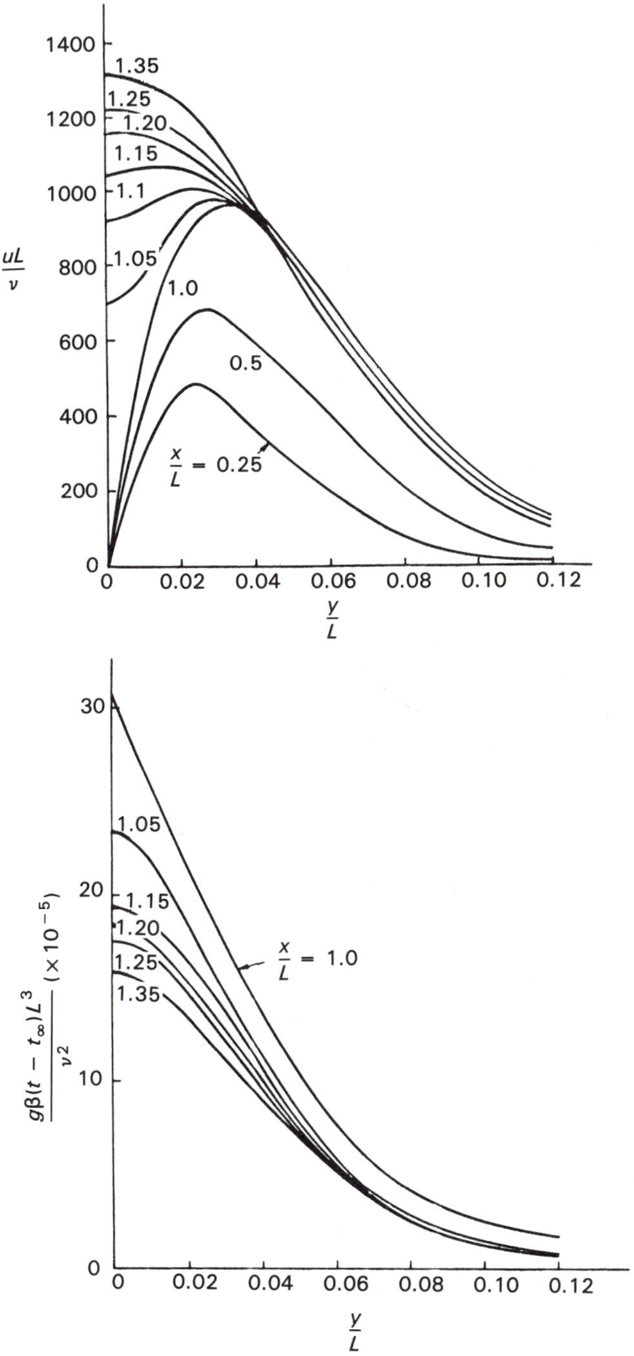

Figure 3.12.1 Velocity and temperature profiles in the wake region downstream of a heated vertical surface at Gr = 3.07×10^6. *(Reprinted with permission from N. E. Hardwick and E. K. Levy,* J. Heat Transfer, *vol. 95, p. 289. Copyright © 1973, ASME.)*

vertical adiabatic surface and the free plume above a heated isothermal vertical surface of given height L were studied by boundary layer analysis. Finite-difference methods were used to solve the governing equations. As seen in Fig. 3.12.1, the sharp discontinuity at the trailing edge of the surface gives rise to large gradients in both x and y directions. Therefore, the region adjacent to the discontinuity demands a solution of the full conservation equations. However, this zone in which the complete equations are needed is relatively small. The boundary layer solution is then valid farther downstream. Consequently, a solution of the boundary layer equations is expected to apply in the region away from the trailing edge. Figure 3.12.2 shows the calculated downstream variation of the maximum temperature and velocity in the wake above a finite heated vertical surface. A rapid approach to the characteristics of a thermal plume is observed downstream of the trailing edge. The results agreed with the findings of the earlier studies on wakes and on wall plumes. A numerical and experimental study of this problem, with the coupling due to heat conduction and thermal radiation, was carried out by Kishinami and Seki (1983), confirming trends observed in earlier work.

Discontinuous heating. The natural convection flow over a vertical surface with a step discontinuity in the wall temperature level has been studied by several investigators. Schetz and Eichhorn (1964) carried out a detailed experimental study of the natural convection boundary layer flow over such a vertical surface. The temperature field and the heat transfer rate from the surface were measured. The upper portion of the vertical surface was maintained at a temperature different

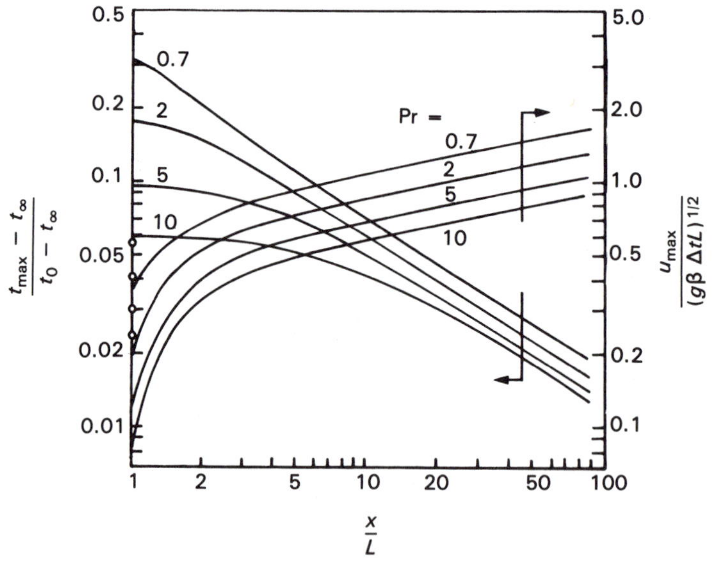

Figure 3.12.2 Downstream variation of the maximum temperature and velocity in the wake above a heated surface at various values of the Prandtl number. *(Reprinted with permission from E. M. Sparrow et al.,* J. Heat Transfer, *vol. 100, p. 184. Copyright © 1978, ASME.)*

from that of the lower portion, as shown in Fig. 3.12.3. An important parameter that arises is the wall temperature difference ratio e, given by

$$e = \frac{t_{01} - t_\infty}{t_{02} - t_\infty} \tag{3.12.1}$$

where t_{01} and t_{02} are the lower and upper wall temperatures, respectively. For a semi-infinite vertical surface, the height L of the lower portion, at temperature t_{01}, arises as the characteristic dimension, and therefore there is no similar solution.

The circumstances where $t_{02} = t_\infty$ or $t_{02} < t_\infty$ were also considered, the latter condition being studied by visualization. Then, a downflow is generated in the upper region and an upflow in the lower region. These two opposing streams meet and separate from the surface at a location dependent on the temperature levels and the relative heights of the two portions. This configuration is also of interest in buoyancy-driven flows in thermally stratified media, where the upper portion of the surface may be cooler than the ambient medium and the lower portion warmer. This situation arises in flow over the walls of a room in which a fire is located, since a hot layer overlying a much cooler lower layer is generated. This was considered by Jaluria and Steckler (1982) and is discussed further in Section 14.8. Chen and Eichhorn (1976) have also considered this configuration for an isothermal surface in a linearly stratified air or water environment.

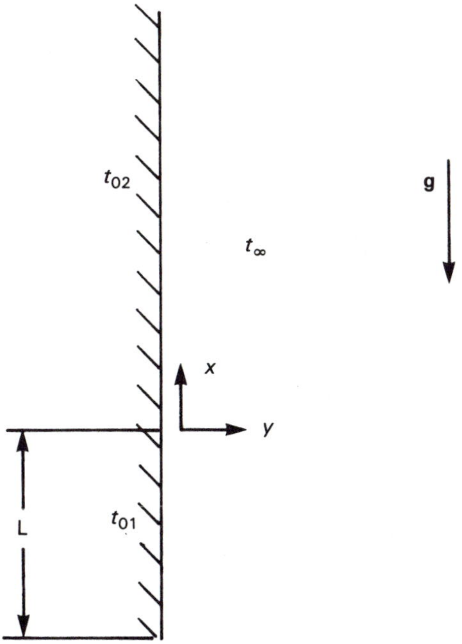

Figure 3.12.3 Coordinate system for a vertical surface with a step discontinuity in surface temperature.

Hayday et al. (1967) made numerical calculations of the air flow over a vertical surface with a step discontinuity in the surface temperature. Kelleher (1971) employed an asymptotic series method to study the velocity and temperature profiles in the region immediately above the discontinuity. A comparison of the heat flux results obtained by Kelleher (1971) with the experimental data of Schetz and Eichhorn (1964) and the numerical results of Hayday et al. (1967) is seen in Fig. 3.12.4. Good agreement is seen. In Fig. 3.12.4 q_{02} is the actual local heat transfer rate at x and q_{01} is the local heat transfer rate that would have resulted if the surface were maintained at t_{01}. Also, x is measured downstream from the discontinuity. For $e > 1.0$ the lower portion of the surface is hotter than the upper one, and $e \to \infty$ corresponds to the upper portion being maintained at t_∞. For a hotter upper portion $q_{02}/q_{01} > 1.0$ and for a colder upper portion $q_{02}/q_{01} < 1.0$, as expected. In the latter case, heat transfer occurs to the surface near the discon-

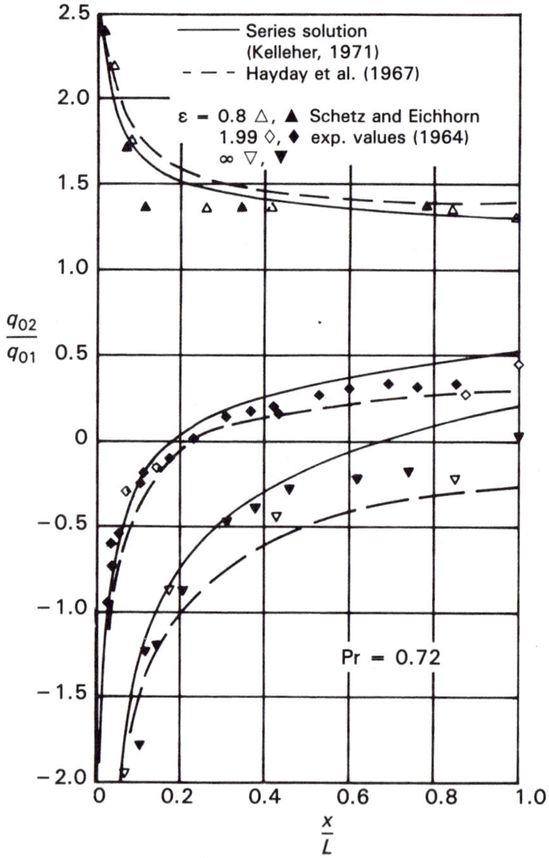

Figure 3.12.4 Comparison of calculated surface heat flux variation with experimental results for a vertical surface with a step discontinuity in the wall temperature. *(Reprinted with permission from M. Kelleher,* J. Heat Transfer, *vol. 93, p. 349. Copyright © 1971, ASME.)*

tinuity at $x = 0$, since the fluid coming from below is hotter than the surface. Such processes clearly require consideration of the full equations in this region. Finite-difference methods provide a convenient and satisfactory solution procedure there.

Most studies have treated these flows as a boundary layer for simplicity. However, such results are not valid in the immediate vicinity of any discontinuity. For the opposing flow that arises when the upper portion is colder than the ambient medium and the lower one is hotter, the solution of the full governing equations is particularly necessary since reverse flow and large x-direction gradients arise.

Blowing and suction. Another mechanism of interest in some applications concerns the effect of blowing and suction in a natural convection flow. This occurs, for instance, if fluid is added to or removed from the natural convection flow at a porous surface. The required velocity conditions for similarity in the flow over a vertical plate, with blowing or suction at the surface, were determined in Section 3.6.5. A nonzero horizontal velocity component $v(x, 0)$ arises at the surface. Similarity requires that $f(0)$ be a constant, being zero for an impervious surface. This requires a variation of $v(x, 0)$ as $x^{(n-1)/4}$ for the power law surface temperature distribution and as $e^{mx/4}$ for the exponential variation. For an isothermal surface, $n = 0$, this condition results in a variation of $v(x, 0)$ as $x^{-1/4}$. Suction corresponds to a negative value of $v(x, 0)$. This results in a positive value of $f(0)$ from Eq. (3.6.33). Similarly, blowing is obtained for $f(0) < 0$.

Eichhorn (1960) considered this flow. For very large suction rates, the vertical velocity component becomes much smaller than the horizontal component. Then vertical convection may be neglected to obtain the energy equation, for large positive values of $f(0)$, as

$$\phi'' + \Pr[(n + 3)f(0)\phi'] = 0 \tag{3.12.2}$$

with

$$\phi(0) = 1 \text{ and } \phi(\infty) = 0 \tag{3.12.3}$$

The resulting temperature distribution and Nusselt number are

$$\phi = e^{-(n+3)\Pr f(0)\eta} \tag{3.12.4}$$

and

$$\text{Nu} = (n + 3)\Pr f(0)(\text{Gr}/4)^{1/4} \tag{3.12.5}$$

The velocity and temperature profiles for various values of $f(0)$ were also calculated by Eichhorn (1960) for $\Pr = 0.733$. Figure 3.12.5 shows some of the results obtained for an isothermal surface. There is a significant effect on the heat transfer, but only a small effect on the surface shear. Heat transfer from the surface was found to be essentially zero at $f(0) = -1.0$. For large $f(0)$ the above asymptotic solution was approached. With increasing blowing velocity, the maximum vertical velocity increases and occurs farther from the surface, as expected. It also results in a reduction of the heat transfer rate due to increased boundary layer thickness. The reverse occurs for suction.

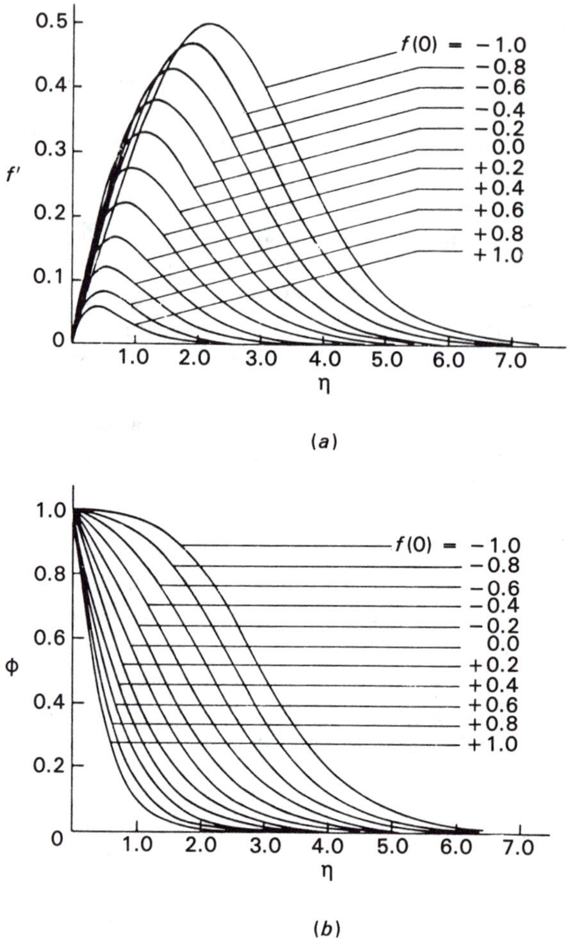

Figure 3.12.5 Effect of blowing and suction on (*a*) velocity and (*b*) temperature profiles for natural convection from a vertical isothermal surface. *(Reprinted with permission from R. Eichhorn,* J. Heat Transfer, *vol. 82, p. 260. Copyright © 1960, ASME.)*

Sparrow and Cess (1961) considered a uniform blowing or suction velocity $v(x, 0)$ for a vertical isothermal surface. Since similarity does not arise, a perturbation solution with the following perturbation parameter ϵ was obtained:

$$\epsilon = \frac{v(x, 0)x}{4\nu(\mathrm{Gr}_x/4)^{1/4}} \tag{3.12.6}$$

Numerical results were presented for Pr = 0.72. Again, blowing was found to decrease the heat transfer and suction to increase it. Merkin (1972) obtained asymptotic expansions as $x \to \infty$ for velocity and temperature. Clarke (1973) obtained the next order of approximation to the solution of the governing equations for large Grashof number without employing the Boussinesq approximations. Merkin (1975) presented solutions for a horizontal cylinder and other body shapes that give similarity. Experimental work on this problem has been done by Brdlik and Mochalov (1966), who used interferometry for small magnitudes of blowing and suction, and by Parikh et al. (1974), who presented temperature profiles obtained with an interferometer. Close agreement between theory and experiment was obtained.

3.13 APPROXIMATE METHODS OF ANALYSIS

In several natural convection boundary layer flows, an exact solution of the governing equations by similarity analysis is not possible. One may then resort to perturbation and local similarity methods, if applicable, or to numerical solution by finite-difference and finite-element methods. Since most of these methods are quite involved, integral methods, which provide simpler approximate solutions to the boundary layer equations, may be used as an alternative approach of reasonable accuracy.

3.13.1 Integral Method

This method is based on the overall balances of mass, momentum, and energy and on the use of approximate representations of the local velocity and temperature fields. In many practical problems one is interested largely in the overall flow rate, momentum flow rate, and energy transfer. The integral analysis then provides a convenient, though approximate, method for determining the desired transport quantities.

Following the integral analysis developed by von Karman (1921), the method was first used by Squire (1938) for the analysis of the buoyancy-driven boundary layer flow adjacent to a heated isothermal vertical surface. This solution is outlined here, followed by a discussion of the extension of the method to other boundary conditions. The governing integral equations may be obtained by considering a differential control volume, as shown in Fig. 3.13.1, and writing the mass, momentum, and energy balance for the control volume. These equations may also be obtained by integrating the governing differential equations across the boundary layer and using the corresponding boundary conditions. If the velocity and thermal boundary layer thickness are taken as equal and denoted by δ, the integral momentum and energy equations are obtained as

$$\frac{d}{dx}\left[\int_0^\delta u^2\,dy\right] = -\nu\left.\frac{\partial u}{\partial y}\right)_{y=0} - g\beta\left[\int_0^\delta (t - t_\infty)\,dy\right] \tag{3.13.1}$$

and

$$\frac{d}{dx}\left[\int_0^\delta u(t - t_\infty)\,dy\right] = -\alpha\left.\frac{\partial t}{\partial y}\right)_{y=0} \tag{3.13.2}$$

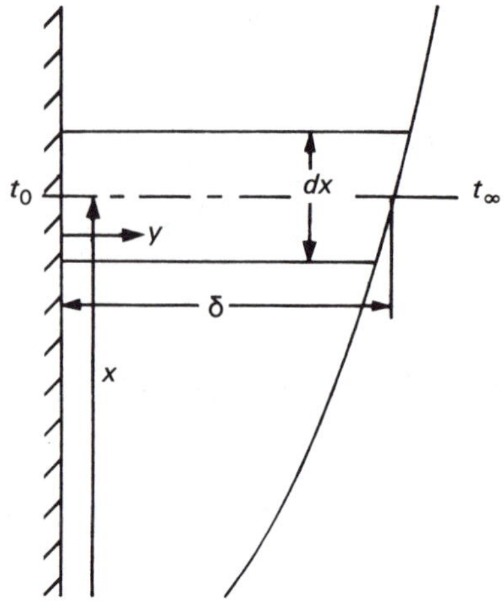

Figure 3.13.1 Control volume for derivation of the integral boundary layer equations.

where the terms on the left represent convective transport, the first term on the right the surface shear or heat flux, and the second term on the right in Eq. (3.13.1) the buoyancy effect.

The velocity and temperature profiles are then chosen to satisfy the boundary conditions at the surface and in the ambient medium. If a second-order polynomial is assumed, as used by Squire (1938), the distributions for velocity and temperature are obtained as

$$u = U\frac{y}{\delta}\left(1 - \frac{y}{\delta}\right)^2 \qquad \text{and} \qquad t - t_\infty = (t_0 - t_\infty)\left(1 - \frac{y}{\delta}\right)^2 \tag{3.13.3}$$

where U and δ are functions of x and are to be obtained from the governing integral equations. These distributions obviously satisfy the conditions of zero vertical velocity at the surface and in the ambient medium and of the temperature being t_0 and t_∞ at these locations. In addition, the velocity and temperature gradients are zero at the edge of the boundary layer, $y = \delta$. Higher-order polynomials may also be chosen. Then the additional constants are determined by requiring that higher-order derivatives of u and t be zero at $y = \delta$. For the quadratic distribution, the maximum velocity is $(4/27)U$ and occurs at $y = \delta/3$. However, as found in forced flow and discussed by Schlichting (1968), the eventual results are usually relatively insensitive to the order of the polynomial chosen.

If the assumed profiles, Eq. (3.13.3), are substituted into the integral equations, Eqs. (3.13.1) and (3.13.2), we obtain the following ordinary differential equations for $U(x)$ and $\delta(x)$.

$$\frac{d}{dx}\left(\frac{U^2\delta}{105}\right) = \frac{g\beta(t_0 - t_\infty)\delta}{3} - \frac{\nu U}{\delta} \tag{3.13.4}$$

$$\frac{d}{dx}\left(\frac{U\delta}{30}\right) = \frac{2\alpha}{\delta} \tag{3.13.5}$$

These nonlinear coupled equations may be solved simply by assuming a power law variation, in x, of U and δ and then determining the relevant constants. Thus, if

$$U = C_1 x^m \qquad \text{and} \qquad \delta = C_2 x^n$$

the constants C_1, C_2, m, and n may be determined by substituting these expressions in Eqs. (3.13.4) and (3.13.5) and equating the indices and the coefficients. This yields $m = \frac{1}{2}$ and $n = \frac{1}{4}$. The expressions for U and δ are

$$U = 5.17\nu\left(\text{Pr} + \frac{20}{21}\right)^{-1/2}\left[\frac{g\beta(t_0 - t_\infty)}{\nu^2}\right]^{1/2} x^{1/2} \tag{3.13.6}$$

$$\delta = 3.93\text{Pr}^{-1/2}\left(\text{Pr} + \frac{20}{21}\right)^{1/4}\left[\frac{g\beta(t_0 - t_\infty)}{\nu^2}\right]^{-1/4} x^{1/4} \tag{3.13.7}$$

The variation of U and δ with $t_0 - t_\infty$ and x is found to be of the same form as that obtained from the similarity analysis in Section 3.3. A comparison of the u and ϕ profiles with those from the exact solution is shown in Fig. 3.13.2 for $\text{Pr} = 0.733$. Fairly good agreement is observed. Since the thermal and velocity boundary layer thicknesses are taken as equal, the solution is expected to apply to Prandtl numbers near unity. However, the analysis has been found to be quite satisfactory even for larger Pr. Sugawara and Michiyoshi (1951) considered different thermal and velocity boundary layer thicknesses. Ede (1967) has summarized the integral analyses carried out by several other investigators, such as Merk and Prins (1953–1954) and Fujii (1959).

From the expressions for U and δ given above, the heat flux and the corresponding Nusselt number are

$$q'' = -k\left.\frac{\partial t}{\partial y}\right)_{y=0} = k\frac{2}{\delta}(t_0 - t_\infty)$$

and

$$\text{Nu}_x = \frac{q''}{(t_0 - t_\infty)}\frac{x}{k} = \frac{2x}{\delta}$$

$$= 0.508\text{Pr}^{1/2}\left(\text{Pr} + \frac{20}{21}\right)^{-1/4}\text{Gr}_x^{1/4} \tag{3.13.8}$$

Therefore,

$$\text{Nu} = \frac{hL}{k} = 0.68\text{Pr}^{1/2}\left(\text{Pr} + \frac{20}{21}\right)^{-1/4}\text{Gr}^{1/4} \tag{3.13.9}$$

A comparison of the results obtained from this expression with those from the

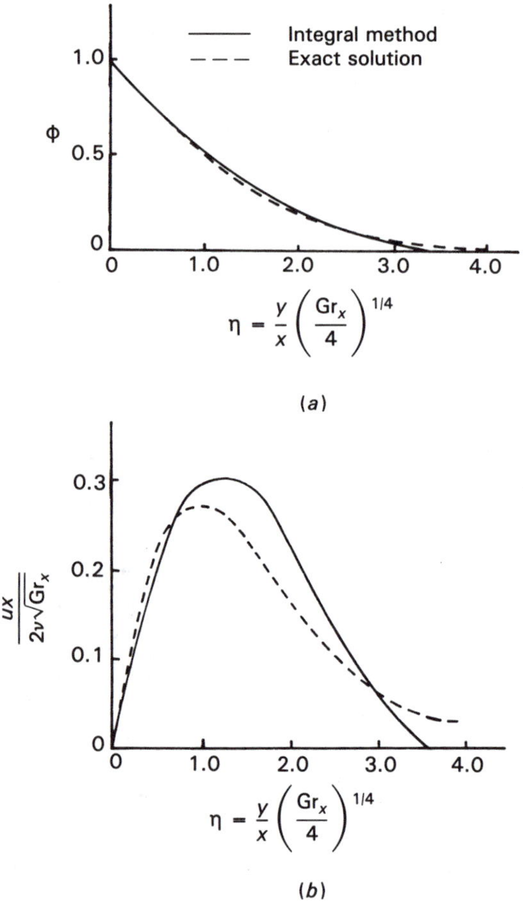

Figure 3.13.2 Comparison of results from the approximate integral analysis and those from the similarity solution for flow over a vertical isothermal surface at Pr = 0.733. *(Reprinted with permission from S. Goldstein, Ed.,* Modern Developments in Fluid Dynamics, *H. B. Squire. Copyright © 1938, Oxford University Press.)*

similarity solution indicates a difference of the order of 5% for Pr $\approx$ 1. The mass, momentum, and thermal energy flow rates in the boundary layer may also be obtained by suitable integration of the profiles given in Eq. (3.13.3). The corresponding calculation with the similarity formulation requires the numerical values of the relevant integrals involving the similarity variables. Frequently, these integrals are not tabulated and a numerical solution of the similarity equations is needed to determine them.

The integral method analysis may be extended to transport adjacent to nonisothermal vertical surfaces. Sparrow (1955) considered power law variations of the heat flux and of the surface temperature. Tribus (1958) considered an arbitrary variation of the heat flux $q''(x)$ and obtained the following differential equations for δ and U, corresponding to the profiles of Eq. (3.13.3):

$$\frac{d}{dX}\left(\frac{\hat{U}^2\hat{\delta}}{105}\right) = \frac{q''\hat{\delta}^2}{6q_0''} - \frac{\hat{U}}{\hat{\delta}} \tag{3.13.10}$$

$$\frac{d}{dX}\left(\frac{\hat{U}\hat{\delta}^2 q''}{30}\right) = \frac{2q''}{\text{Pr}} \tag{3.13.11}$$

where q_0'' is the heat flux at $x = 0$. The velocity parameter $\hat{U}$, distance X, and the boundary layer thickness $\hat{\delta}$ are dimensionless variables, defined in terms of U, δ, and x in Eq. (3.13.3), as

$$\hat{U} = \frac{U(x)}{(g\beta q_0''\nu^2/k)^{1/4}} \qquad X = \left(\frac{g\beta q_0''}{k\nu^2}\right)^{1/4} x \qquad \hat{\delta} = \left(\frac{g\beta q_0''}{k\nu^2}\right)^{1/4} \delta(x) \tag{3.13.12}$$

Equations (3.13.10) and (3.13.11) may be solved for U and δ. For $q'' \propto x^m$, the Nusselt number was obtained as

$$\text{Nu} = \frac{(\frac{7}{27}\text{Pr})^{1/4}(m + 1)^{1/2}(\text{GrPr})^{1/4}}{[(1 + 35\text{Pr}/12)(m + 1) + 4/3]^{1/4}} \tag{3.13.13}$$

The agreement of this result with that from the similarity solution is found to be good. For $m < 0$ a singularity arises at $x = 0$. Thus, the integral method analysis may be used to obtain an approximate solution in several circumstances. It is particularly useful in cases where similarity does not arise. It is also useful for turbulent flows, as shown in Chapters 11 and 12.

3.13.2 Other Approximate Methods

In several cases, similarity does not arise and the integral approach given above does not yield results of desired accuracy. In problems where an additional effect is to be studied, such as the effect of an external induced flow on the natural convection flow over a vertical surface or the effect of inclination with the vertical of a flat surface on the transport, the analysis can often be approached as a perturbation for small magnitudes of the additional term. This method again gives rise to ordinary differential equations, the zero-order equations being those for the problem without the additional term, for which the solution is known. On the basis of this known solution, the first-, second-, and higher-order equations are solved, each order being dependent on the solution of the lower order. This technique has been used in several problems, for instance, to study the higher-order approximations to a boundary layer flow (Section 3.10).

Two other methods that have received considerable attention in recent years for solving several natural convection flows in which similarity does not arise are the local similarity and the local nonsimilarity methods. Mixed convection flows, axisymmetric flows over vertical cylinders, and flows in stratified media are among those studied by these methods. The basic approach for both of the methods is based on neglecting the nonsimilar terms at various levels of truncation to obtain progressively higher levels of accuracy. The first level of truncation is known as the local similarity method, which converts the governing partial differential equations into ordinary differential equations, applicable locally, by neglecting the nonsimilar terms in the conservation equations. These equations are then solved at various locations and the overall heat transfer rates determined. The solution at any downstream location is independent of the solutions at upstream locations. However, this method often has uncertain accuracy because of the uncertainty regarding the effect of the terms involving downstream derivatives, which are neglected to obtain the local similarity, on the final results.

Higher levels of truncation are known as local nonsimilarity methods. These also give rise to such ordinary differential equations and locally independent solutions. However, the nonsimilar terms are retained in the conservation equations. Eventually, terms are selectively neglected in the derived subsidiary equations to simplify these equations. A variable ξ, which is similar to the similarity variable η and is dependent on the downstream location x, still appears in the equations and is treated as a parameter in the numerical solution. The accuracy of the method improves with the level of truncation and, thus, provides a method for the assessment of accuracy. Sparrow et al. (1970) and Sparrow and Yu (1971) discuss the use of these methods in natural convection. Recall that the results of Chen and Eichhorn (1976) were obtained by this approach (Section 3.11). Further details on this method are given in Section 5.2. Also, results obtained by this approach for several flows are presented in the following chapters.

Other similar methods may be derived to obtain the equations in the ordinary differential equation form. The solution is then obtained by one of the several methods available for solving ordinary differential equations. However, it must be noted that all these methods often have an uncertain accuracy and generally require validation by other analytical, numerical, or experimental results on the flows being considered. For instance, see the work of Kumari et al. (1985), who compared finite-difference results with those from the local nonsimilarity method and found the comparison to be good at small values of ξ and to become quite poor as ξ increases beyond 1.0.

In many natural convection flows the techniques outlined in this chapter, to obtain the governing equations as ordinary differential equations, cannot be applied. Then one may resort to the solution of the governing partial differential equations by numerical techniques such as finite-difference methods. Finite-element methods may also be employed for very complex geometries. These techniques are very well documented, and only a brief outline is given here.

For parabolic partial differential equations, which apply for a boundary layer flow, the various terms in the equations are written in their finite-difference form at each point in the grid into which the flow region is subdivided. The solution is obtained by marching in the downstream direction, starting at a location where the boundary conditions are given, and determining the values of the velocity components and the temperature at the various grid points in the flow region.

Similarly, elliptic equations, which arise, say, near a leading edge or in the wake of a vertical surface, may be solved by finite differences. Iterative schemes are often employed to get convergence, as indicated by a small variation from one iteration to another. This is generally a very time-consuming process and is employed when the other simpler methods cannot be employed.

In many problems of practical interest, flows are often generated by bodies of arbitrary shape, the boundary layer approximations may not apply, and there may be interactions with other flows and bodies. In such cases, one usually has to resort to finite-difference methods for obtaining the solution. Hardwick and Levy (1973) employed this method for a detailed study of wakes above an isothermal vertical plate, as discussed in Section 3.12. For details on these methods, see, for instance, the books by Roache (1976), Spalding (1977), Patankar (1980), and Jaluria and Torrance (1986).

REFERENCES

Ackroyd, J. A. D. (1974). *J. Fluid Mech. 62*, 677.
Brand, R. S., and Lahey, F. J. (1967). *J. Fluid Mech. 29*, 305.
Braun, W. H., and Heighway, J. E. (1960). *NASA Tech. Note D-292*, 4.
Brdlik, P. M., and Mochalov, V. A. (1966). *J. Eng. Phys. (Inzh.-Fiz. Z.) 10*, 3.
Brodowicz, K. (1968). *Int. J. Heat Mass Transfer 11*, 201.
Brodowicz, K., and Kierkus, W. T. (1966). *Int. J. Heat Mass Transfer 9*, 81.
Chang, B. H., and Akins, R. G. (1972). *Int. J. Heat Mass Transfer 15*, 513.
Chang, K. S., Akins, R. G., Burris, L., and Bankoff, S. G. (1964). Argonne National Laboratory Rept. ANLG835.
Chen, C. C., and Eichhorn, R. (1976). *J. Heat Transfer 98*, 446.
Cheesewright, R. (1967). *Int. J. Heat Mass Transfer 10*, 1847.
Churchill, S. W., and Chu, H. S. (1975). *Int. J. Heat Mass Transfer 18*, 1323.
Churchill, S. W., and Ozoe, H. (1973). *J. Heat Transfer 95*, 540.
Churchill, S. W., and Usagi, R. (1972). *AIChE J. 18*, 1121.
Clarke, J. F. (1973). *J. Fluid Mech. 57*, 45.
Colwell, R. G., and Welty, J. R. (1973). ASME Paper 73-HT-52.
Crane, L. J. (1959). *Z. Angew. Math. Phys. 10*, 453.
Cygan, D. A., and Richardson, P.D. (1968). *Can. J. Chem. Eng. 46*, 321.
Dotson, J. P. (1954). M.S. Thesis, Purdue Univ., Lafayette, Ind.
Eckert, E. R. G. (1949). *Einfuhring in den Warme und Stoffaustausch*, p. 116, Springer-Verlag, Berlin.
Eckert, E. R. G. (1950). *Introduction to the Transfer of Heat and Mass*, McGraw-Hill, New York.
Ede, A. J. (1967). *Adv. Heat Transfer 4*, 1.
Eichhorn, R. (1960). *J. Heat Transfer 82*, 260.
Eichhorn, R. (1969). *Prog. Heat Mass Transfer 2*, 41.
Eichhorn, R., Lienhard, J. H., and Chen, C. C. (1974). *Proc. 5th Int. Heat Transfer Conf.*, Tokyo, Paper NC 1.3.
Forstrom, R. J., and Sparrow, E. M. (1967). *Int. J. Heat Mass Transfer 10*, 321.
Fujii, T. (1959). *Bull. Jpn. Soc. Mech. Eng. 2*, 365.
Fujii, T. (1963). *Int. J. Heat Mass Transfer 6*, 597.
Fujii, T. (1973). *Int. J. Heat Mass Transfer 16*, 755.
Fujii, T., and Fujii, M. (1976). *Int. J. Heat Mass Transfer 19*, 121.
Fujii, T., and Uehara, H. (1970). *Int. J. Heat Mass Transfer 13*, 607.
Gebhart, B. (1962). *J. Fluid Mech. 14*, 225.

Gebhart, B. (1971). *Heat Transfer*, 2d ed., McGraw-Hill, New York.
Gebhart, B., and Mollendorf, J. (1969). *J. Fluid Mech. 38*, 97.
Gebhart, B., Pera, L., and Schorr, A. W. (1970). *Int. J. Heat Mass Transfer 13*, 161.
Gill, A. E. (1966). *J. Fluid Mech. 26*, 515.
Gill, A. E., and Davey, A. (1969). *J. Fluid Mech. 35*, 775.
Goodman, J., Abauf, N., and Laufer, G. (1974). *Isr. J. Technol. 12*, 198.
Hardwick, N. E., and Levy, E. K. (1973). *J. Heat Transfer 95*, 289.
Hayday, A. A., Bowlus, D. A., and McGraw, R. A. (1967). *J. Heat Transfer 89*, 244.
Hieber, C. A. (1974). *Int. J. Heat Mass Transfer 17*, 785.
Himasekhar, K. (1980). An Analytical and Experimental Study of Laminar Free Boundary Layer Flows in a Stratified Medium, M. S. Thesis, I.I.T., Kanpur, India.
Humphreys, W. W., and Welty, J. R. (1975). *AIChE J. 2*, 268.
Imai, I. (1958). *J. Aerosp. Sci. 24*, 155.
Iyer, P. A. (1973). *Boundary Layer Meterol. 5*, 53.
Jaluria, Y. (1979). *Proceedings of International Conference on Numerical Methods in Thermal Problems*, Swansea, Wales, p. 443, Pineridge Press, U.K.
Jaluria, Y. (1980). *Natural Convection Heat and Mass Transfer*, Pergamon, Oxford, U.K.
Jaluria, Y., and Gebhart, B. (1974). *J. Fluid Mech. 66*, 593.
Jaluria, Y., and Gebhart, B. (1977). *Int. J. Heat Mass Transfer 20*, 153.
Jaluria, Y., and Gupta, S. K. (1982). *Solar Energy 28*, 137.
Jaluria, Y., and Himasekhar, K. (1983), *Comput. Fluids 11*, 39.
Jaluria, Y., and Steckler, K. D. (1982), Proceedings of Technical Meeting Eastern Section, Atlantic City, N.J., Combustion Institute, Paper 46.
Jaluria, Y., and Torrance, K. E. (1986). *Computational Heat Transfer*, Hemisphere, Washington, D.C.
Jischke, M. C. (1977). *J. Heat Transfer 99*, 676.
Julian, D. V., and Akins, R. G. (1969). *Ind. Eng. Chem. Fundam. 8*, 641.
Kadambi, V. (1969). *Waerme Stoffuebertrag. 2*, 99.
Kelleher, M. (1971). *J. Heat Transfer 93*, 349.
Kishinami, K., and Seki, N. (1983). *J. Heat Transfer 105*, 759.
Kuiken, H. K. (1968a). *J. Eng. Math. 2*, 95.
Kuiken, H. K. (1968b). *Int. J. Heat Mass Transfer 11*, 1141.
Kuiken, H. K. (1969). *J. Fluid Mech. 37*, 785.
Kuiken, H. K., and Rotem, Z. (1971). *J. Fluid Mech. 45*, 585.
Kumari, M., Pop, I., and Nath, G. (1985). *Int. J. Heat Mass Transfer 28*, 2171.
Kuo, Y. H. (1953). *J. Math. Phys. 32*, 83.
LeFevre, E. J. (1956). *Proc. 9th Int. Cong. Appl. Mech. 4*, 168.
Liburdy, J. A., and Faeth, G. M. (1975). *Lett. Heat Mass Transfer 2*, 407.
Lyakhov, Y. N. (1970). *J. Appl. Mech. Tech. Phys. 11*, 355 (translated 1973).
Mahajan, R. L., and Gebhart, B. (1978). *Int. J. Heat Mass Transfer 21*, 549.
Mahajan, R. L., and Gebhart, B. (1988). In preparation.
Merk, H. J., and Prins, J. A. (1953–1954). *Appl. Sci. Res. A4*, 11, 195, 207.
Merkin, J. H. (1972). *Int. J. Heat Mass Transfer 15*, 989.
Merkin, J. H. (1975). *Int. J. Heat Mass Transfer 18*, 237.
Nawoj, H. J., and Hickman, R. S. (1977). *J. Heat Transfer 99*, 609.
Ostrach, S. (1953). NACA Rep. 1111, 63.
Parikh, P. G., Moffat, R. J., Kays, W. A., and Bershader, D. (1974). *Int. J. Heat Mass Transfer 17*, 1465.
Patankar, S. V. (1980). *Numerical Heat Transfer and Fluid Flow*, Hemisphere, Washington, D.C.
Pohlhausen, E. (1930). See Schmidt and Beckmann (1930).
Prandtl, L. (1904). *Proc. 3rd Int. Math. Cong.*, Heidelberg, p. 484.
Prandtl, L. (1952). *Essentials of Fluid Dynamics*, Hafner, New York.
Raithby, G. D., and Hollands, K. G. T. (1978). *J. Heat Transfer 100*, 378.
Riley, D. S., and Drake, D. G. (1975). *Appl. Sci. Res. 30*, 193.

Roache, P. J. (1976). *Computational Fluid Dynamics*, rev. printing, Hermosa, Albuquerque, N.M.
Rouse, H., Yih, C. S., and Humphreys, H. W. (1952). *Tellus 4*, 201.
Roy, S. (1969). *Int. J. Heat Mass Transfer 12*, 239.
Roy, S. (1973). *J. Heat Transfer 95*, 124.
Saunders, O. A. (1939). *Proc. R. Soc. London Ser. A 172*, 55.
Schetz, J. A., and Eichhorn, R. (1964). *J. Fluid Mech. 18*, 167.
Schlichting, H. (1933). *Z. Angew. Math. Mech. 13*, 260.
Schlichting, H. (1968). *Boundary Layer Theory*, 6th ed., McGraw-Hill, New York.
Schmidt, E., and Beckmann, W., with Pohlhausen, E. (1930). *Tech. Mech. Thermodyn. 1*, 341, 391.
Schmidt, W. (1941). *Z. Angew. Math. Mech. 21*, 351.
Schorr, A. W., and Gebhart, B. (1970). *Int. J. Heat Mass Transfer 13*, 557.
Schuh, H. (1948). Boundary Layers of Temperature, in *Boundary Layers*, W. Tollmein, Ed., Sec. B6, British Ministry of Supply, German Doc. Cent. Ref. 3220T.
Selman, J. R., and Newman, J. (1971). *J. Heat Transfer 93*, 465.
Sevruk, I. G. (1958). *Appl. Math. Mech. 22*, 807.
Singh, S. N. (1977). *Int. J. Heat Mass Transfer 20*, 1155.
Spalding, D. B. (1977). *GENMIX: A General Computer Program for Two-Dimensional Parabolic Phenomena,* Pergamon, Oxford, U.K.
Spalding, D. B., and Cruddace, R. G. (1961). *Int. J. Heat Mass Transfer 3*, 55.
Sparrow, E. M. (1955). NACA-TN-3508.
Sparrow, E. M., and Cess, R. G. (1961). *J. Heat Transfer 83*, 387.
Sparrow, E. M., and Gregg, J. L. (1956). *Trans. ASME 78*, 435.
Sparrow, E. M., and Gregg, J. L. (1958). *Trans. ASME 80*, 379.
Sparrow, E. M., and Yu, H. S. (1971). *J. Heat Transfer 93*, 328.
Sparrow, E. M., Patankar, S. V., and Abdel-Wahed, R. M. (1978). *J. Heat Transfer 100*, 184.
Sparrow, E. M., Quack, H., and Boerner, C. J. (1970). *AIAA J. 8*, 1936.
Squire, H. B. (1938). In *Modern Developments in Fluid Dynamics,* by S. Goldstein, Oxford Univ. Press, New York.
Stewartson, K. (1964). *The Theory of Laminar Boundary Layers in Compressible Fluids*, Sec. 3.5, Oxford Univ. Press, New York.
Stewartson, K., and Jones, L. T. (1957). *J. Aerosp. Sci. 24*, 379.
Sugawara, S., and Michiyoshi, I. (1951). *Trans. Jpn. Soc. Mech. Eng. 17*, 109.
Takeuchi, I., Ota, M. Y., and Tanaka, Y. (1974). *Trans. Jpn. Soc. Mech. Eng. 40*, 1046.
Tollmien, W. (1926). *Z. Angew. Math. Mech. 6*, 468.
Tribus, M. (1958). *Trans. ASME 80*, 1180.
Van Dyke, M. (1964). *Perturbation Methods in Fluid Mechanics*, Academic Press, New York.
Vliet, G. C., and Liu, C. K. (1969). *J. Heat Transfer 91*, 517.
Von Karman, T. (1921). *Z. Angew. Math. Mech. 1*, 233; also NACA-TM-1092.
Yang, K. T. (1960). *J. Appl. Mech. 27*, 230.
Yang, K. T. (1964). *J. Appl. Mech. 31*, 131.
Yang, K. T., and Jerger, E. W. (1964). *J. Heat Transfer 86*, 107.
Yang, K. T., Novotny, J. L., and Cheng, Y. S. (1972). *Int. J. Heat Mass Transfer 15*, 1097.
Yih, C. S. (1952). *Trans. Am. Geophys. Union 33*, 669.
Zeldovich, Y. B. (1937). *Zh. Eksp. Teor. Fiz. 12*, 1463.
Zimin, V. D., and Lyakhov, Y. N. (1970). *J. Appl. Mech. Tech. Phys. 11*, 159 (translated 1973).

PROBLEMS

3.1 Consider the flows and transport in the eventual steady states that would be achieved for the conditions in Problem 2.1.

(a) Calculate the magnitude of all terms in the full two-dimensional plane flow equations in generalized form as in Eqs. (3.2.5)–(3.2.11) that apply. Assume that convection and diffusion are comparable to obtain initial estimates.

(b) Compare the resulting magnitudes for air and water.

3.2 A heated vertical surface 30 cm high is at a temperature 20°C above the ambient air temperature of 30°C. Evaluate the air flow rate at the top of the surface and the overall heat transfer rate. Also determine the maximum velocity that occurs. Compare its value with what you would obtain if the fluid were water under the same conditions. Assume laminar flow.

3.3 For natural convection from a vertical surface at 60°C in water at 20°C, find the maximum velocity and the total heat transfer per side if the plate is 30 cm high.

3.4 For a vertical isothermal heated surface, plot the variation of the local maximum velocity and the location of this maximum point against the downstream distance. In terms of the physical distance from the vertical surface, at what rate downstream does the location of maximum velocity shift outward?

3.5 Determine the heat transfer from a vertical surface at 100°C placed in stagnant air at 15°C if the surface is 1 m high and 0.5 m wide.

3.6 A plate 10 cm high in air at 20°C is heated electrically to obtain a uniform heat flux of 100 W/m^2 in air. Find the maximum and average surface temperatures in natural convection.

3.7 On a vertical surface the temperature excess over the ambient temperature, which is taken as constant, varies as $1/x$, where x is the distance along the surface from the leading edge. Obtain the variation with x of the following physical variables in the natural convection flow; q'', Q, δ, u_{max}, and $v(\infty)$. Is this circumstance obtainable physically? If so, under what conditions would this flow arise?

3.8 For natural convection over a vertical flat surface whose temperature varies as $t_0 - t_\infty = 10\sqrt{x}$°C, determine the heat transfer coefficient and the mass flow rate into the boundary layer at $x = 1$ m. What is the local heat transfer flux? Obtain an expression for the mass flow rate as a function of x. Take $f(\infty) = 0.2$ and $[-\phi'(0)] = 1.0$; $\beta = 1/T$; $T_\infty = 300$ K.

3.9 Consider a semi-infinite vertical surface generating a buoyancy-induced flow. Neglect all but the simplest energy effects. We have seen that boundary layer similarity results for $d = Nx^n$ and also for a concentrated line source along the leading edge of an adiabatic surface. Does similarity exist for the general heat flow or $t_0 - t_\infty = Nx^n$ variation maintained on the surface simultaneously with a concentrated line source at the leading edge? Specify any such conditions completely.

3.10 For natural convection flow over a 1-m-high heated vertical surface, the surface temperature varies as $A/x^{1/3}$. Determine the variation of the boundary layer thickness, velocity level, heat flux, and convected thermal energy. Is this circumstance physically possible? Explain your answer briefly.

3.11 We have seen that $d = Nx^n$ leads to a similar solution for natural convection from a vertical surface in a quiescent ambient medium. Consider the distribution $d = Nx^n + Px^p$, where P and p are constants. Find the conditions on the values of N, n, P, and p such that similarity also results, where both c and b are still in terms of Gr_x and $\mathrm{Gr}_x = g\beta x^3\, d/\nu^2$.

3.12 A thin stainless steel foil 0.02 mm thick, 15 cm wide, and 60 cm high is heated electrically to obtain a uniform heat flux input of 200 W/m^2. Determine the required current and the temperature and maximum velocity at a height of 30 cm along the surface for laminar flow in water at 30°C.

3.13 In Problem 3.12 determine the maximum boundary layer thickness, velocity, and temperature. Also determine the same quantities if the heat flux is halved and the height doubled. Which effect, flux or height, is larger? Compare the boundary layer thicknesses, calculated with the expected dependence on Pr, in air and water.

3.14 Compare the orders of magnitude of the viscous dissipation and pressure terms in the energy equation and show that the latter is generally of greater significance than the former. Determine the physical circumstances, if any, under which the viscous dissipation effect may be much larger than the pressure term.

3.15 Consider laminar vertical natural convection with a power law temperature variation [i.e., $d(x) \equiv t_0(x) - t_\infty = Nx^n$] in a fluid at t_∞ with volumetric energy generation, $q''' = A(t - t_\infty)P$, where A and P are constant.

(a) Determine the condition on P such that a similarity solution exists.

(b) Write the similar form of the energy equation and circle the term that represents volumetric energy generation.

Neglect stratification, viscous dissipation, and compression work.

3.16 Consider a buoyancy-induced flow generated adjacent to a vertical isothermal surface in a quiescent ambient medium of air at $t_\infty = 40°F$. The pressure level is 1 atm. Calculate the gravitational level g and any other conditions such that viscous dissipation will have an appreciable effect on transport, for example, on Nu_x, at say $x = 1$ ft.

3.17 If the pressure term is added as a perturbation in the governing equations for a vertical isothermal surface,

(a) Determine the perturbation parameter for the power law variation of $t_0 - t_\infty$.

(b) Write the equations for the stream and temperature function expansions in terms of this parameter.

3.18 Carry out a perturbation analysis for a small amount of blowing at a vertical porous surface. Determine the perturbation parameter and the governing equations. Up to what value of this parameter is the effect on the overall heat transfer rate less than 5%?

3.19 For an isothermal vertical surface, the surface blowing velocity is given as $5x^{-1/4}$ (cm/s). If the surface is at 100°C, in air at 25°C, determine the maximum velocity in the boundary layer and the heat transfer at $x = 1$ m. Compare these values with those without blowing.

3.20 For a vertical surface with an exponential surface temperature variation, evaluate the effect on the heat transfer due to viscous dissipation of $4g\beta/mc_p = 2.0$ and $Pr = 10.0$. If m is 1 m^{-1}, find the necessary value of g for this magnitude of dissipation. Compare, qualitatively, the expected effect due to the pressure term.

3.21 A horizontal heated wire has a thermal input of 50 W/m in air at 20°C. Find the maximum velocity and temperature 2 cm above the wire.

3.22 Consider a plume arising from a long line energy source in air at 1 atm and 70°F. There is a limitation on the applicability of boundary layer theory near the source (small x) because the relative boundary region thickness (δ/x) becomes too large. Estimate this lower limit for boundary layer theory if the upper limit on δ/x is 0.2 for boundary layer theory to apply. Express the limit in terms of Gr_x.

3.23 Consider the flow and transport in a freely rising plane generated by a uniform plane source of thermal energy q''. Determine the x dependence of the temperature, vertical velocity component, horizontal velocity component, and boundary layer thickness.

3.24 A long, thin horizontal nichrome wire of diameter 0.001 cm is heated electrically by a current of 15 A in air. In the thermal plume that arises, determine the centerline velocity and temperature at a height of 5 cm above the wire. Also determine the entrainment velocity and its dependence on height and on energy input. Repeat this calculation for water as the fluid. Take the resistivity of nichrome as 10^{-6} ohm cm.

3.25 A thin horizontal wire of length 20 cm has a resistance of 2 ohms. A current of 20 A flows through the wire. If the wire is located in an extensive air medium, find the centerline temperature and velocity at a height of 10 cm above the wire.

3.26 A resistor located on a ceramic panel may be taken as a line thermal source, dissipating 5 W/m, at the leading edge of a vertical adiabatic surface. Find the surface temperature 5 cm above the source and compare it with that for a plume above a line source dissipating 10 W/m. Also compare the velocity levels at this location.

3.27 There is a great deal of emphasis on boundary layer formulations because of the ease of analysis. How do the results compare with measured characteristics of flow and of transport? Take two examples of laminar boundary layer analysis in this chapter and compare some aspect of the resulting predictions with *measured behavior*, as found in the book.

3.28 A vertical plate is at 30°C above the ambient air medium at 20°C. If its surface emissivity is 0.8, determine whether the radiative heat transfer is comparable to the convective heat transfer. Repeat the calculation if the temperature excess is 150°C. Assume the ambient to be black and at 0 K.

3.29 Consider a tall, flat, vertical surface placed in an ambient fluid. The surface must dissipate heat uniformly by natural convection. The ambient fluid is stratified.

(a) For t_∞ increasing upward, find a condition for which a similar solution may be found for the resulting flow.

(b) Find how h_x and t_0 vary with x for the condition found in (a).

(c) Is there a similarity solution for the uniform flux surface condition when the ambient medium is a gas stratified at the condition of neutral stability, that is, adiabatically?

3.30 A heated vertical plate is in a stationary stratified water medium. If the surface has a uniform heat flux input, write the governing similarity equations and determine (from the similarity solution)

(a) The maximum velocity at $x = 10$ cm.

(b) The local heat transfer coefficient at $x = 10$ cm.

3.31 A vertical isothermal plate is at 100°C and is immersed in stratified air whose temperature varies linearly from 0°C at $x = 0$ to 100°C at $x = L$. Here, L is the height of the plate and is given as 1 m. Obtain the governing equations in terms of the similarity variables. Does similarity arise? If $[-\phi'(0)]$ is 1.5 at $x = 1$ m, find the local heat transfer coefficient. For air, take $c_p = 10^3$, $\nu = 1.6 \times 10^{-5}$, $k = 0.026$ SI units.

3.32 A vertical surface dissipates 200 W/m^2 uniformly and is immersed in a thermally stratified medium such that the stratification parameter $S = 4.0$. Determine the ambient temperature distribution and the maximum flow velocity if the surface is 0.5 m high.

3.33 A flat plate is placed vertically in air at 80°F and 1 atm. The surface temperature is uniformly 120°F. The plate is of such a height that the Grashof number (at L) is 10^8.

(a) Calculate the "convection" velocity.

(b) Estimate the relative magnitude of all of the terms in the full two-dimensional equations.

(c) Repeat (b), calculating magnitudes from specific numerical boundary layer results.

3.34 A flat vertical surface 8 ft high in water having a uniform temperature stratification of 10°F/ft, increasing from $t_\infty = 70$°F at $x = 0$, is dissipating energy at 800 W/ft^2 side. Estimate the plate temperature variation and the average plate temperature. Assume that the surface temperature also increases linearly with x.

CHAPTER

FOUR

VERTICAL AXISYMMETRIC FLOWS

4.1 AXISYMMETRIC FLOWS AND TRANSPORT EQUATIONS

In many buoyancy-induced flows, a circumferential symmetry arises because the surface or the body that gives rise to the flow is symmetric about a vertical axis. Such axisymmetric flows frequently arise, for example, adjacent to a long vertical cylinder and around a vertical cone or sphere, when the energy input is uniform circumferentially. Boundary layer approximations, similar to those for two-dimensional vertical flows considered in the preceding chapter, may often be made. Axisymmetric free boundary flows, such as plumes and buoyant jets, also are of considerable interest, particularly in the environmental considerations of energy and material rejection. In this chapter we consider laminar vertical axisymmetric boundary region flows arising only from thermal buoyancy, leaving the turbulent and multiply buoyant flows for later chapters.

Consider the free-boundary flow arising from an axisymmetric heat source as well as the flow over a vertical axisymmetric surface—for example, the surface of a vertical cylinder as shown in Fig. 4.1.1*b*. The governing scalar equations for an axisymmetric flow may be derived from the vector form of the equations, given in Chapter 2. These are written in terms of the vertical coordinate x, the radial coordinate y measured from the axis, and the corresponding velocity components u and v. For a vertical axisymmetric flow, the boundary layer approximations may again be made if the boundary layer thickness δ is small compared to the vertical distance x. The resulting equations may be written as follows, using the Boussinesq approximations for the density variation, assuming the remaining

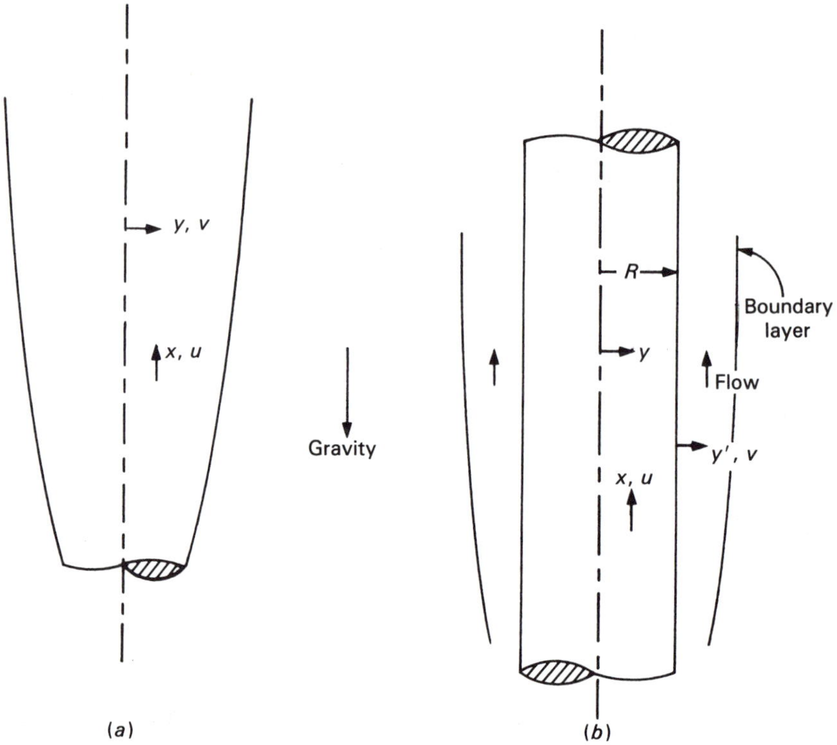

Figure 4.1.1 (*a*) Axisymmetric free boundary flow due to a thermal energy input. (*b*) Boundary layer flow over a vertical axisymmetric surface.

properties to be constant, and neglecting the viscous dissipation and pressure work effects:

$$\frac{\partial(yu)}{\partial x} + \frac{\partial(yv)}{\partial y} = 0 \tag{4.1.1}$$

$$u\frac{\partial u}{\partial x} + v\frac{\partial u}{\partial y} = g\beta(t - t_\infty) + \frac{\nu}{y}\frac{\partial}{\partial y}\left(y\frac{\partial u}{\partial y}\right) \tag{4.1.2}$$

$$u\frac{\partial t}{\partial x} + v\frac{\partial t}{\partial y} = \frac{\alpha}{y}\frac{\partial}{\partial y}\left(y\frac{\partial t}{\partial y}\right) \tag{4.1.3}$$

These equations apply for the boundary region flow over vertical axisymmetric surfaces and the flow in axisymmetric plumes and jets. The coupled equations, with the corresponding boundary conditions, are solved for the velocity and temperature fields. For flow over a vertical cylinder of radius R, shown in Fig. 4.1.1*b*, the treatment is often, instead, in terms of the radial coordinate y' measured outward from the surface. Then $y' = y - R$, $y' = 0$ represents the surface, and $y' \to \infty$ is the distant ambient medium.

The apparent boundary conditions for an axisymmetric free-boundary, or plume, flow arise from symmetry and ambient conditions. For the flow in Fig. 4.1.1*a*, they are

$$\begin{aligned} &\text{at } y = 0: \quad v = \frac{\partial u}{\partial y} = \frac{\partial t}{\partial y} = 0 \qquad t = t_0(x) \\ &\text{as } y \to \infty: \quad u \to 0 \qquad t \to t_\infty \end{aligned} \tag{4.1.4}$$

where $t_0(x)$ is the temperature along the axis. This centerline temperature is determined from the input conditions at $x = 0$, using the conservation of the convected thermal energy in the flow. For flow adjacent to a vertical axisymmetric surface, in Fig. 4.1.1*b*, the boundary conditions are

$$\begin{aligned} &\text{at } y' = 0: \quad u = v = 0 \qquad t = t_0 \\ &\text{as } y' \to \infty: \quad u \to 0 \qquad t \to t_\infty \end{aligned} \tag{4.1.5}$$

Here t_0 is the surface temperature. It is constant for an isothermal surface but may be a function of the vertical position x for other conditions of interest, such as imposed surface heat flux conditions.

4.2 SIMILARITY FORMULATION

The similarity formulation for vertical axisymmetric boundary layer flows is discussed. The governing equations and the relevant boundary conditions for flows that give rise to similarity are obtained. These include axisymmetric plumes, non-buoyant jets, and vertical cylinders and needles.

Axisymmetric plumes. These arise from an isolated concentrated thermal source, as idealized in Fig. 4.2.1 for the buoyancy-induced wake arising above a heated body. The flow is governed by Eqs. (4.1.1)–(4.1.3). The continuity equation is satisfied by the Stokes stream function $\psi(x, y)$, defined as follows:

$$u = \frac{1}{y}\frac{\partial \psi}{\partial y} \qquad v = -\frac{1}{y}\frac{\partial \psi}{\partial x} \tag{4.2.1}$$

To determine any plume similar solutions of the boundary layer equations, the procedure of Chapter 3 is followed. The similarity variable η, dimensionless stream function $f(\eta)$, and dimensionless temperature $\phi(\eta)$ are again postulated as

$$\begin{aligned} &\eta(x, y) = yb(x) \qquad \psi(x, y) = \nu c(x) f(\eta) \\ &\phi(\eta) = \frac{t - t_\infty}{t_0 - t_\infty} = \frac{t - t_\infty}{d(x)} \end{aligned} \tag{4.2.2}$$

where t_0 is the temperature at the axis of symmetry, $y = 0$.

It may again be shown, following the treatment given by Gebhart (1973), that similarity arises for the power law and the exponential variations of the centerline temperature. For the power law variation, $t_0 - t_\infty = Nx^n$, Mollendorf and Gebhart

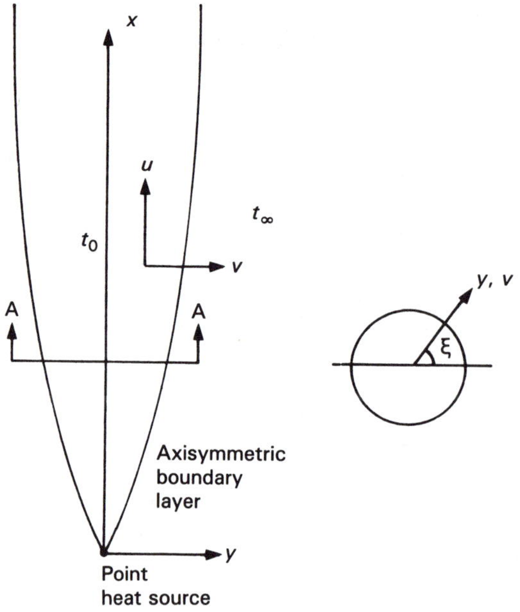

Figure 4.2.1 Axisymmetric plume due to a point heat source.

(1974) carried out the above substitution to obtain the governing equations in terms of the similarity variables as

$$f''' + \frac{g\beta}{\nu^2}(yb)\frac{d}{b^4c}\phi + \frac{c_x}{yb}ff'' - \left(\frac{c_x}{yb} + \frac{1}{yb}\frac{2cb_x}{b}\right)f'^2$$

$$-\frac{c_x}{(yb)^2}ff' - \frac{1}{yb}f'' + \frac{1}{(yb)^2}f' = 0 \qquad (4.2.3)$$

and

$$\frac{\phi''}{\mathrm{Pr}} + \frac{c_x}{yb}f\phi' - \frac{1}{yb}\frac{d_xc}{d}f'\phi + \frac{1}{yb}\frac{1}{\mathrm{Pr}}\phi' = 0 \qquad (4.2.4)$$

For similarity to arise in these equations, the coefficients of the terms must be constant or functions of η alone. Since $yb = \eta$, similarity will result in these equations if

$$\frac{d}{b^4c} = B_1 \qquad c_x = B_2 \qquad \frac{cb_x}{b} = B_3 \qquad \frac{d_xc}{d} = B_4$$

where B_1, B_2, B_3, and B_4 are constants or functions of η. For the power law variation, $d(x) = Nx^n$, the above equations yield $c(x) \propto x$ and $b(x) \propto x^{(n-1)/4}$. These variations may be written in terms of the usual local Grashof number Gr_x as

$$\eta = \frac{y}{x}\sqrt[4]{\mathrm{Gr}_x} \qquad \psi = \nu x f(\eta) \qquad (4.2.5a)$$

where

$$\mathrm{Gr}_x = \frac{g\beta x^3(t_0 - t_\infty)}{\nu^2} \tag{4.2.5b}$$

With these definitions, the governing boundary layer equations become

$$f''' + (f-1)\left(\frac{f'}{\eta}\right)' - \left(\frac{1+n}{2}\right)\frac{f'^2}{\eta} + \eta\phi = 0 \tag{4.2.6}$$

$$(\eta\phi')' + \mathrm{Pr}(f\phi' - nf'\phi) = 0 \tag{4.2.7}$$

The velocity components are obtained as

$$u = \frac{\nu}{x}\sqrt{\mathrm{Gr}_x}\left(\frac{f'}{\eta}\right) \qquad v = -\frac{\nu}{x}(\mathrm{Gr}_x)^{1/4}\left(\frac{f}{\eta} - \frac{f'}{2}\right) \tag{4.2.8}$$

Therefore, the necessary boundary conditions are written for symmetry at the centerline and for an asymptotic approach to the ambient conditions as $\eta \to \infty$. As discussed in Section 4.4, these are

$$f(0) = f'(0) = 1 - \phi(0) = \phi'(0) = \frac{f'}{\eta}(\infty) = 0 \tag{4.2.9}$$

The numerical solutions of the similarity equations are discussed in detail in Section 4.4. It may be mentioned here that the exponential variation of $d(x)$ is not acceptable since it does not yield a constant thermal convection flux $Q(x)$ downstream.

Nonbuoyant jets. These arise from a point vertical momentum source. The above procedure may again be followed to obtain the similarity formulation. The flow is governed by the continuity and momentum equations, Eqs. (4.1.1) and (4.1.2). Schlichting (1933) obtained the similarity variables as

$$\eta = \frac{y}{x\sqrt{\gamma}} \qquad \psi = \nu x f(\eta) \tag{4.2.10}$$

where γ is a constant that depends on the input momentum flux J. The corresponding similarity equation is

$$f''' + (f-1)\left(\frac{f'}{\eta}\right)' + \frac{f'^2}{\eta} = 0 \tag{4.2.11}$$

The solution obtained by Schlichting for this flow is discussed later, along with the corresponding buoyant jet circumstance.

Vertical cylinders. Another interesting similarity solution corresponds to the axisymmetric boundary layer flow adjacent to an infinite vertical cylinder, shown in Fig. 4.1.1*b*. Millsaps and Pohlhausen (1956, 1958) considered this case and showed that a similarity solution may be obtained if the surface temperature excess, over the unstratified ambient temperature, varies linearly, that is, $t_0 - t_\infty = Nx$, where N is a constant. Using the stream function ψ given in Eq. (4.2.1) along with the transformations given by $\psi = Xf(r)$ and $\text{Gr}_R = X\phi(r)/(e^r)^4$, where $X = x/R$, $(y - R)/R = e^r$, $\text{Gr}_R = g\beta R^3(t - t_\infty)/\nu^2$, and R is the radius of the cylinder, Eqs. (4.1.1)–(4.1.3) become

$$f''' + (f - 4)f'' + (4 - 2f - f')f' + \phi = 0 \tag{4.2.12}$$

$$\phi'' + (\text{Pr}f - 8)\phi' + (16 - 4\text{Pr}f - \text{Pr}f')\phi = 0 \tag{4.2.13}$$

where the primes denote derivatives with respect to the new independent variable r. Note that these equations retain no x dependence. Thus, both the heat transfer coefficient h_x and the boundary layer thickness $\delta(x)$ are obtained independent of x. This is expected behavior for an infinite surface, since the origin at $x = 0$ may be located at any position along the surface.

The boundary conditions are obtained as

$$f(0) = f'(0) = f'(\infty) = \phi(\infty) = 0 \qquad \phi(0) = \text{const.} \tag{4.2.14}$$

where the last boundary condition, $\phi(0) = \text{const.}$, requires that the surface temperature excess $t_0 - t_\infty$ vary linearly with x. Therefore, similar solutions may be obtained. The numerical results obtained by Millsaps and Pohlhausen (1956, 1958) for $t_0 - t_\infty = Nx$ are discussed in the next section.

For other cylinder surface temperature distributions and for cylinders with a leading edge (Fig. 4.2.2) similarity solutions are not obtained since the partial differential equations do not reduce to ordinary differential equations. Such solutions must therefore be obtained by approximate analytical methods, such as the local similarity method and numerical techniques. However, Nagendra et al. (1969) used a transformation that reduces the governing equations to ordinary differential equations except for an additional parameter η_w given by

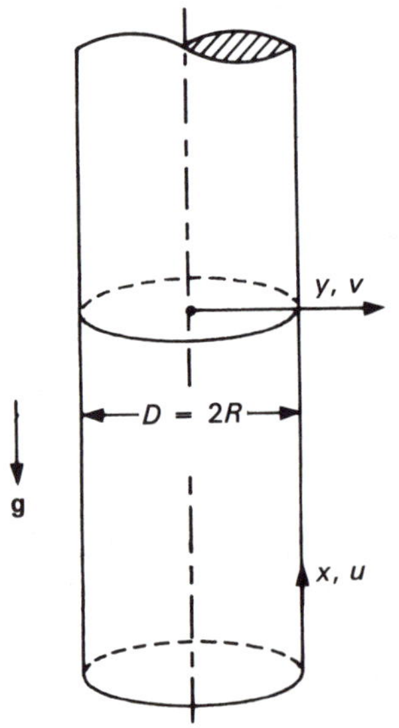

4.2.2 Coordinate system for natural convection flow adjacent to a semi-infinite vertical cylinder with a leading edge.

$\eta_w = 1/[16(\mathrm{Gr}_D D/2x)]^{1/2}$, where D is the diameter of the cylinder and Gr_D the Grashof number based on D. The parameter η_w represents the radial location of the cylindrical surface, locally, instead of the usual $\eta = 0$ employed above. Numerical results may thus be obtained with η_w as a parameter. Good agreement between such results and earlier analytical and experimental results was demonstrated by Nagendra et al. (1969) for various values of the Prandtl number, considering both the isothermal and uniform heat flux conditions. The natural convection flow over vertical needles has also been shown to yield similarity solutions and is discussed in the following section.

4.3 FLOWS ADJACENT TO SURFACES

Buoyancy-induced vertical axisymmetric boundary region flows adjacent to various kinds of surfaces of limited height have been of interest because of their many industrial applications. Transport from vertical cylinders, needles, and wires has received considerable attention. Such transport, sketched in Fig. 4.2.2, is governed by Eqs. (4.1.1)–(4.1.3). If the boundary region thickness δ is very much less than the radius of the cylinder R, the coordinate distance may be taken as $y' = y - R << R$. This implies that $y \simeq R$ across the flow region. Then, as an approximation, y may be taken out of the parentheses in the continuity equation, (4.1.1), and in the diffusion terms of the momentum and energy equations, (4.1.2) and (4.1.3). The resulting governing equations for vertical boundary layer flow are then

$$\frac{\partial u}{\partial x} + \frac{\partial v}{\partial y} = 0 \tag{4.3.1}$$

$$u\frac{\partial u}{\partial x} + v\frac{\partial u}{\partial y} = g\beta(t - t_\infty) + \nu\frac{\partial^2 u}{\partial y^2} \tag{4.3.2}$$

$$u\frac{\partial t}{\partial x} + v\frac{\partial t}{\partial y} = \alpha\frac{\partial^2 t}{\partial y^2} \tag{4.3.3}$$

These equations are the same as those for two-dimensional boundary region flow over a flat vertical surface, Eqs. (3.3.1)–(3.3.3). Therefore, if the boundary layer thickness δ is much smaller than the radius of curvature R, the influence of the radial curvature on the axisymmetric boundary layer flow is neglected. This condition may be justified quantitatively by considering the variation of δ downstream with the local Grashof number.

For a vertical surface of height L, the maximum boundary layer thickness arises at $x = L$ and its variation with L is given by $\delta \propto L/(\mathrm{Gr})^{1/4}$, where Gr is the Grashof number based on the height L. Therefore, $\delta << D$ if

$$\frac{L}{\mathrm{Gr}^{1/4}} << D$$

where D is the diameter of the cylinder. Sparrow and Gregg (1956) used a series method to analyze the transport from an isothermal vertical cylinder. The results were compared with those for plane flow. For Pr $\simeq$ 1.0 it was found that a difference in the heat transfer rate of less than 5% arises if

$$\frac{D}{L} \geq \frac{35}{(\mathrm{Gr})^{1/4}} \tag{4.3.4}$$

Ede (1967) made a detailed comparison of the experimental transport results for flows over a vertical cylinder with those over a flat vertical surface. A criterion was also given for neglecting the radial curvature effects for a vertical cylinder, that is, $(\mathrm{GrPr})^{1/4}D/L \geq M$, where M is suitably chosen from the available information. If such criteria are satisfied, the boundary layer transport becomes that for a flat vertical surface of width πD. The local and overall heat transfer coefficients may then be determined from the results given in Chapter 3.

Appreciable curvature effects. If these effects are not negligible, Eqs. (4.1.1)–(4.1.3) must be solved for boundary layer flows. Several studies have considered this problem. Elenbaas (1948) postulated a conduction film around the vertical cylinder and obtained the Nusselt number as a function of $\mathrm{Ra}_D D/L$, where $\mathrm{Ra}_D = \mathrm{Gr}_D\mathrm{Pr}$ is the Rayleigh number based on the diameter D. The following expression resulted for the Nusselt number Nu_D based on diameter:

$$\mathrm{Nu}_D \exp\left(-\frac{2}{\mathrm{Nu}_D}\right) = 0.6\left(\frac{D}{L}\right)^{1/4} \mathrm{Ra}_D^{1/4} \tag{4.3.5}$$

Le Fevre and Ede (1956) used the integral method to solve the governing equations and gave the following expression for Nusselt number Nu based on the height L of the cylinder (Ede, 1967):

$$\mathrm{Nu} = \frac{4}{3}\left[\frac{7\mathrm{GrPr}^2}{5(20 + 21\mathrm{Pr})}\right]^{1/4} + \frac{4(272 + 315\mathrm{Pr})L}{35(64 + 63\mathrm{Pr})D} \tag{4.3.6}$$

Millsaps and Pohlhausen (1956, 1958) obtained a similarity solution for $t_0 - t_\infty = Nx$, as outlined earlier, and presented numerical results for Pr = 0.733, 1, 10, and 100. The local heat transfer coefficient h_x is independent of the local position x. The Nusselt number Nu_D may then be expressed as a function of the Prandtl number and a Grashof number Gr′, which depends on the temperature gradient N and is defined as $\mathrm{Gr}' = g\beta D^4N/\nu^2$. The numerical results were found to be well correlated by the equation

$$\mathrm{Nu}_D = \frac{hD}{k} = 1.058\left(\frac{\mathrm{Pr}^2\mathrm{Gr}'}{4 + 7\mathrm{Pr}}\right)^{1/4} \tag{4.3.7}$$

Yang (1960) also obtained similarity solutions for a surface temperature varying linearly with x.

The perturbation analysis of Sparrow and Gregg (1956) for an isothermal surface condition employed the axisymmetric boundary layer equations and the variables

$$\eta = C_1 \frac{y^2 - R^2}{x^{1/4}} \qquad \xi = C_2 x^{1/4}$$
$$\psi(x, y) = C_3 x^{3/4} f(\eta, \xi) \qquad \phi = \frac{t - t_\infty}{t_0 - t_\infty} \tag{4.3.8}$$

Here C_1, C_2, and C_3 are parameters that depend on the Grashof number Gr_R, where $\mathrm{Gr}_R = g\beta(t_0 - t_\infty)R^3/\nu^2$, and the cylinder radius R as

$$C_1 = \frac{\mathrm{Gr}_R^{1/4} R^{-7/4}}{2^{3/2}} \qquad C_2 = \mathrm{Gr}_R^{-1/4} R^{-1/4} 2^{3/2}$$
$$C_3 = \frac{\mathrm{Gr}_R^{-3/4} 2^{3/2}}{\nu R^{3/4}} \tag{4.3.9}$$

The functions f and ϕ are expanded as power series in ξ with the coefficients taken as functions of η. Then the zeroth-order set of equations gives the flat surface solution and the higher-order solutions, the influence of radial curvature. Figure 4.3.1 shows the computed Nusselt number for the cylindrical geometry, compared to corresponding results for a flat surface for $\mathrm{Pr} = 0.72$ and 1.0. The deviation from the flat surface results increases downstream, with ξ. This is expected since the boundary layer thickness δ increases and the ratio δ/R decreases, resulting in an increasing effect of curvature on the heat transfer as the flow proceeds downstream.

Nonisothermal surface temperature conditions, $t_0 - t_\infty = Nx^n$, were considered by Kuiken (1968) for vertical cylinders and closed-bottom cones. A power series solution was obtained for various values of the Prandtl number. The uniform surface heat flux condition was also considered. It was found that a lower surface

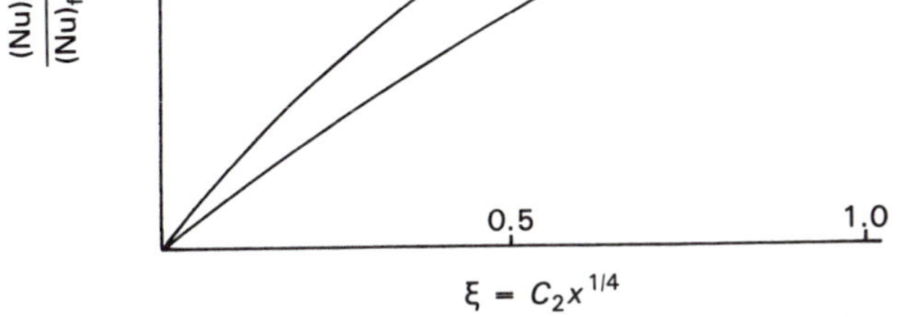

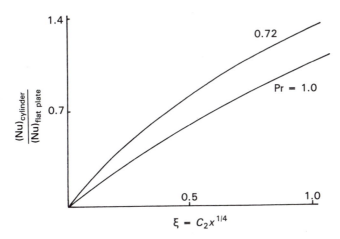

Figure 4.3.1 Variation of the Nusselt number for flow over a vertical cylinder, compared to that for flow over a flat vertical surface, in terms of the parameter ξ, which indicates the downstream location. *(Reprinted with permission from E. M. Sparrow and J. L. Gregg,* J. Heat Transfer, *vol. 78, p. 1823. Copyright © 1956, ASME.)*

temperature arises for a cylinder than for a flat surface, indicating a higher heat transfer coefficient. The surface temperature excess varies as Nx^n. The uniform heat flux condition again arises for $n = 0.2$. Figure 4.3.2 shows the velocity and temperature profiles for Pr = 0.7 at two values of the downstream variable $\xi \propto x^{(1-n)/4} = x^{0.2}$. This general parameter is similar to the ξ defined in Eq. (4.3.8) for the isothermal condition. The effect of curvature was again found to increase downstream. An increase in the heat transfer coefficient was also seen from the decrease in surface temperature downstream.

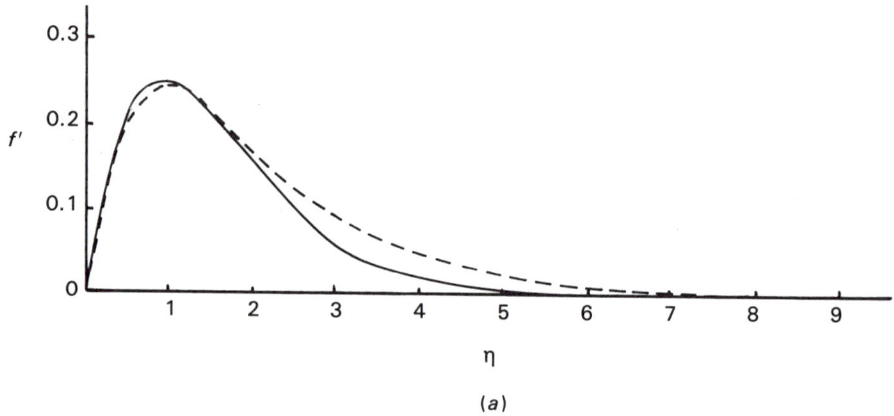

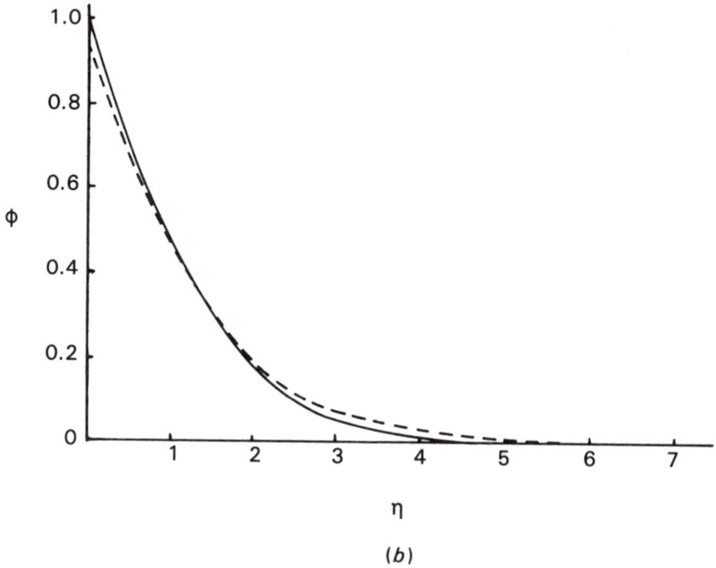

Figure 4.3.2 Velocity and temperature profiles for a vertical cylinder with uniform surface heat flux at Pr = 0.7. (——) $\xi = 0$; (– – –) $\xi = 0.2$. Here $\xi = Cx^{(1-n)/4}$, where C is a constant. *(Reprinted with permission from* Int. J. Heat Mass Transfer, *vol. 11, p. 1141, H. K. Kuiken. Copyright © 1968, Pergamon Journals Ltd.)*

Minkowycz and Sparrow (1974) reconsidered the isothermal surface condition using a local nonsimilarity method to obtain the velocity and temperature distributions in flows in which large deviations from the flat-surface results are expected. The governing equations become ordinary differential equations upon omission of the nonsimilarity terms in the higher-order equations. Trends similar to those discussed above were observed and good agreement with the earlier results of Sparrow and Gregg (1956) was obtained for Pr = 0.72, as mentioned earlier.

Vertical needles. The laminar natural convection heat transfer from vertical needles has been considered by several investigators. The governing equations, (4.1.1)–(4.1.3), were reduced to ordinary differential equations by Cebeci and Na (1969, 1970), using a similarity analysis, for an isothermal vertical needle. Narain and Uberoi (1972) analyzed the uniform heat flux condition and also corrected a few errors in the earlier studies. The transformations used were

$$\psi = xf(\eta) \qquad t - t_\infty = x^{1/5}\phi(\eta) \qquad \eta = \frac{y^2}{x^{2/5}} \tag{4.3.10}$$

The equations thus obtained for the uniform heat flux condition are

$$8\eta f''' + 8f'' + 4ff'' - \frac{12}{5}f'^2 + \phi = 0 \tag{4.3.11}$$

$$\eta\phi'' + \left(1 + \frac{1}{2}\Pr f\right)\phi' - \frac{1}{10}\Pr\phi f' = 0 \tag{4.3.12}$$

The surface of the needles was taken to correspond to $\eta = C$ in Eq. (4.3.10), where C is a constant. Thus, the needle diameter increases downstream as $x^{1/5}$. These equations were solved numerically and the computed velocity and temperature profiles presented. The expression for the Nusselt number Nu as a function of the Grashof number Gr, where both Nu and Gr are based on a reference length L, was given for an isothermal needle as

$$\text{Nu} = -\frac{8}{3}(C)^{1/2}\phi'(C)\text{Gr}^{1/4} \qquad \text{where Gr} = \frac{g\beta q''L^4}{k\nu^2} \tag{4.3.13}$$

A similar configuration was also considered by Van Dyke (1967). Raithby and Hollands (1976) carried out an analysis of natural convection from prolate spheroids (Section 5.4.4). When the ratio of the minor to the major axis approaches zero, a thin vertical needle tapered from the center toward both ends is obtained. Oosthuizen (1979) considered the transport from long wires with an axial curvature in the vertical plane.

Several experimental studies of transport from vertical cylinders have been reported. The results of Eigenson (1940) are for vertical cylinders in air and those of Mueller (1942) for fine vertical wires in air. Kyte et al. (1953) reported measurements for vertical wires in various gases. Hama and Recesso (1958) carried out an integral method analysis, which indicated a correlation for the Nusselt

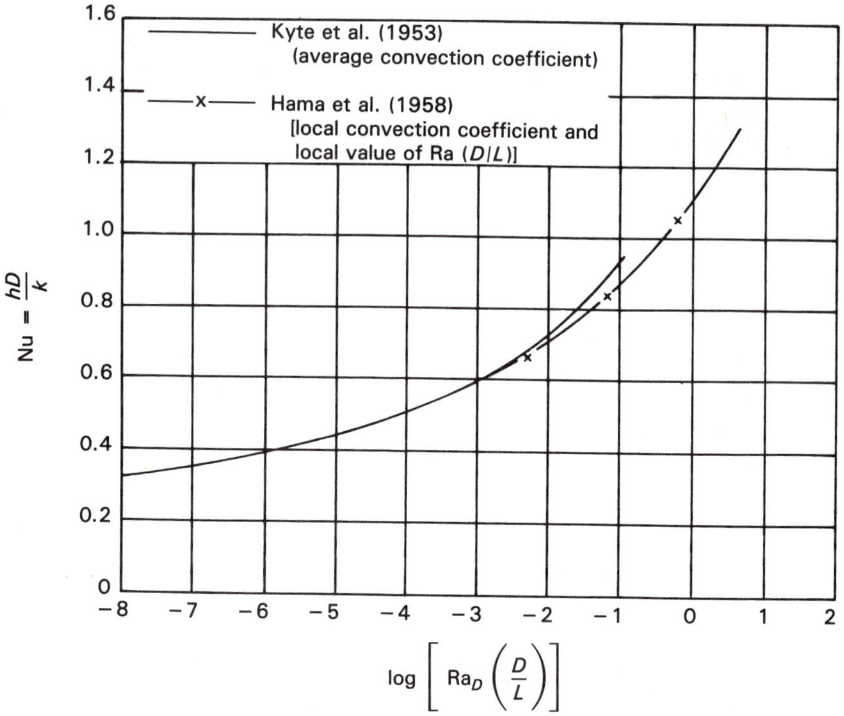

Figure 4.3.3 Correlations for natural convection flow over a vertical cylinder. *(Reprinted with permission from B. Gebhart,* Heat Transfer, *2d ed. Copyright © 1971, McGraw-Hill Book Company.)*

number in terms of the parameter $Gr_D(D/L)$, and also reported measurements in air. As mentioned earlier, Elenbaas (1948) also obtained the Nusselt number as a function of $Ra_D(D/L)$. Figure 4.3.3 summarizes some of these results as compiled by Gebhart (1971). The average and local values diverge at large $Ra_D(D/L)$. This arises because of increasing downstream effects on heat transfer at large Ra_D.

In this section we have discussed the axisymmetric flow over vertical cylinders and needles. Nonvertical axisymmetric flows that arise, for example, over spheres, spheroids, vertical cones, and various arbitrary axisymmetric contours are discussed in the next chapter.

4.4 PLUMES AND BUOYANT JETS

This section discusses the results obtained for laminar axisymmetric free-boundary flows such as plumes and jets. The basic characteristics of these flows are outlined and related to the underlying physical mechanisms.

Axisymmetric plume. The transport calculated from a concentrated source of thermal energy is important since the developing natural convection flow downstream of an actual finite-size source approaches axisymmetric plume behavior. The thermal wake in the region close to the surface or body, which generates the flow, does reflect the mechanisms that give rise to the wake. However, these characteristics decay rapidly as the flow proceeds downstream. The energy content remains of principal importance as the flow eventually approaches the behavior of an axisymmetric plume arising from a point heat source.

Yang (1964), Hardwick and Levy (1973), and Sparrow et al. (1978) considered wakes above heated vertical surfaces. Pera and Gebhart (1972), Jaluria and Gebhart (1975), and Jaluria (1976) also studied the wake generation over finite-size bodies. As demonstrated by Bill and Gebhart (1975), even the two-dimensional plume flow due to a horizontal line source of finite length approaches axisymmetric plume behavior far downstream. This occurs because entrainment, from the sides, also sweeps the flow together. The axisymmetric flow is also often more stable. Therefore, the axisymmetric plume usually represents all wakes sufficiently far downstream. See also the discussion on separation and wake formation in Sections 5.7 and 5.8.

In the first study concerning the axisymmetric plume arising from a point heat source, Schuh (1948) numerically integrated the equations for Pr = 0.7 and obtained the downstream velocity and temperature fields. Then Yih (1952) gave closed-form solutions for Pr = 1 and 2. Fujii (1963) obtained numerical results over a wide range of Pr and gave corrected closed-form solutions for Pr = 1 and 2. Mollendorf and Gebhart (1974) also presented numerical results for various values of the Prandtl number, as discussed later in this section. The analyses of the flow in these studies are quite similar and follow the similarity formulation given in Section 4.2, with some differences in posing the boundary conditions, as considered later.

Since there is no continuing input of thermal energy downstream of the source, the total convected thermal energy $Q(x)$ at any vertical location x must be constant and equal to the energy input at the source Q. This assumes that both viscous dissipation and pressure energy effects are negligible. Thus

$$Q(x) = \int_0^\infty \rho c_p (t - t_\infty) u 2\pi y \, dy = Q \tag{4.4.1}$$

By using the similarity formulation in Eq. (4.2.2), this equation may be written as

$$2\pi\rho c_p \nu (t_0 - t_\infty) x \int_0^\infty f'\phi \, d\eta = 2\pi\rho c_p \nu N x^{n+1} \int_0^\infty f'\phi \, d\eta = Q \tag{4.4.2}$$

Thus, $n = -1$, since the x dependence of $Q(x)$ must disappear. The centerline temperature, therefore, varies as

$$t_0 - t_\infty = \frac{N}{x} \tag{4.4.3}$$

where $N = Q/2\pi\rho c_p \nu I$ and $I = \int_0^\infty f'(\eta)\phi(\eta) \, d\eta$.

The governing boundary region equations, (4.2.6) and (4.2.7), for $n = -1$ become

$$f''' + (f - 1)\left(\frac{f'}{\eta}\right)' + \eta\phi = 0 \tag{4.4.4}$$

$$(\eta\phi')' + \Pr(f\phi)' = 0 \tag{4.4.5}$$

where Eq. (4.4.5) may be integrated to give

$$\eta\phi' + \Pr f\phi = \text{const.} = C$$

The apparent boundary conditions in the physical variables are

$$\text{at } y = 0: \qquad \frac{\partial u}{\partial y} = \frac{\partial t}{\partial y} = v = 0 \qquad t = t_0$$

and

$$\text{as } y \to \infty: \qquad u \to 0 \qquad t \to t_\infty$$

These will be converted into generalized forms in the manner set forth by Gebhart et al. (1970) for the plane plume.

Using the relationships between the velocity components and the dimensionless stream function, Eq. (4.2.8), and the definition of ϕ, the boundary conditions may be written as

$$\text{at } \eta = 0: \qquad \left(\frac{f'}{\eta}\right)' = \phi' = \frac{f}{\eta} - \frac{f'}{2} = 0 \qquad \phi(0) = 1$$

and

$$\text{as } \eta \to \infty: \qquad \frac{f'}{\eta} \to 0 \qquad \phi \to 0$$

Thus, f'/η approaches zero as $\eta \to 0$. Therefore, $f'(0) = 0$. The condition on v implies that $f(0) = 0$ since f/η approaches $f'/2$ as $\eta \to 0$; see Appalaswamy and Jaluria (1980). Also, $f'(\infty)$ is bounded. The boundary conditions may then be written concisely as

$$f(0) = f'(0) = 1 - \phi(0) = \phi'(0) = 0 \qquad f'(\infty) \text{ bounded} \tag{4.4.6}$$

The sixth condition, $\phi(\infty) = 0$, follows from these and is therefore not independent, as shown next. Equation (4.4.5) becomes

$$\eta\phi' + \Pr f\phi = 0$$

since the constant of integration C is zero because of the boundary conditions, Eq. (4.4.6). The above equation is integrated from $\eta = 0$ to η to give

$$\phi(\eta) = \phi(0) \exp\left(-\Pr \int_0^\eta \frac{f}{\eta}\, d\eta\right) \tag{4.4.7}$$

Because of entrainment into the flow, f/η has a nonzero value as $\eta \to \infty$. This follows from $f'/\eta \to 0$ as $\eta \to \infty$ and from Eq. (4.2.8). Therefore, $\phi \to 0$ as $\eta \to \infty$. Solution of the governing differential equations with the boundary conditions given in Eq. (4.4.6) yields the velocity and temperature distributions for axisymmetric plumes.

The calculated profiles for Pr = 0.7, 2.0, and 7.0 are shown in Figs. 4.4.1 and 4.4.2, as obtained by Mollendorf and Gebhart (1974). The maximum velocity and temperature occur at zero slope at the centerline, as expected from the symmetry conditions there, implying zero-shear and adiabatic conditions. The centerline temperature, given by Eq. (4.4.3), varies linearly with the heat input Q and inversely with the vertical distance x from the source. The centerline velocity $u(x, 0)$ is given by Eq. (4.2.8) as

$$u(x, 0) = \sqrt{g\beta N}\left(\frac{f'}{\eta}\right)_0 = \sqrt{\frac{g\beta Q}{2\pi\rho c_p \nu I}}\left(\frac{f'}{\eta}\right)_0 \tag{4.4.8}$$

where $(f'/\eta)_0$ is the intercept in Fig. 4.4.1. Therefore, the centerline velocity is predicted to remain constant downstream. This interesting result arises because the accelerating effects of buoyancy are balanced by the decelerating effects of entrainment.

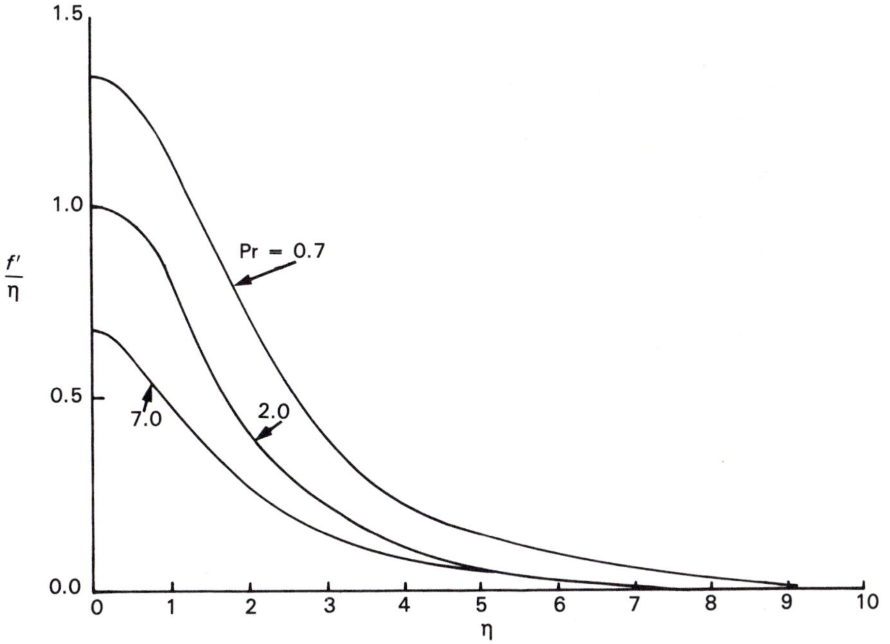

Figure 4.4.1 Velocity profiles in an axisymmetric plume. *(From Mollendorf and Gebhart, 1974.)*

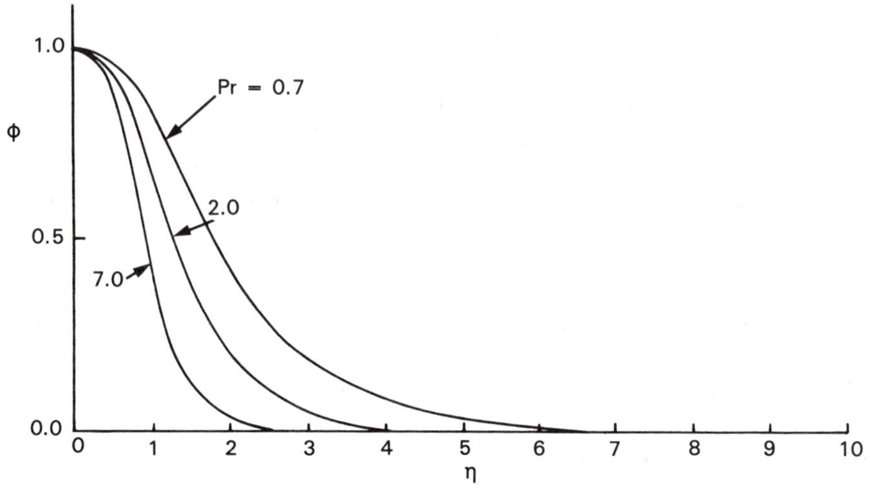

Figure 4.4.2 Temperature profiles in an axisymmetric plume. *(From Mollendorf and Gebhart, 1974.)*

For the buoyancy-induced wake downstream of a finite-size axisymmetric body, the centerline velocity has also been found to approach a constant value downstream, as the initial geometry effects decay; see Jaluria and Gebhart (1975). The velocity boundary layer thickness $\delta(x)$ for an axisymmetric plume varies as

$$\delta(x) = \eta_\delta \left(\frac{2\pi c_p \mu \nu^2 I}{g\beta Q} \right)^{1/4} x^{1/2} \tag{4.4.9}$$

where η_δ is the value of η at the edge of the boundary layer, defined as the location where the dimensionless velocity has dropped to 1% of the centerline value. Similarly, the thermal boundary layer thickness $\delta_T(x)$ may be calculated by replacing η_δ in the above equation by the value of η at which ϕ is 0.01.

Also, other physical variables such as the volume flow rate $\dot{m}$, the momentum flow rate M, and the transverse velocity v may be determined. These are given in terms of the similarity variables as

$$\dot{m} = 2\pi\mu x f(\infty) \qquad M = 2\pi\mu\nu\sqrt{\mathrm{Gr}_x}\, I_M \qquad \text{where } I_M = \int_0^\infty f'^2 \frac{d\eta}{\eta}$$

and

$$v = -\frac{\nu}{x}(\mathrm{Gr}_x)^{1/4}\left(\frac{f}{\eta} - \frac{f'}{2}\right) \tag{4.4.10}$$

The numerical values of $f''(0)$, $f(\infty)$, I, and I_M are given by Mollendorf and Gebhart (1974) as 1.351, 7.91, 2.074, and 3.406 for Pr = 0.7. The corresponding values for Pr = 7.0 are 0.6683, 3.08, 0.2497, and 0.502.

Fujii (1963), using a somewhat different formulation, did not take $\phi(0) = 1$ as a boundary condition. The normalization $\int_0^\infty f'(\eta)\phi(\eta)\, d\eta = 1$ was used instead.

The similarity variables are also defined differently. The numerical results obtained are similar to those presented above and may be converted to the variables of the similarity formulation discussed in Section 4.2. Fujii (1963) gave the centerline temperature in terms of $\tilde{\phi}(0)$, where $\tilde{\phi}$ is the dimensionless temperature employed in that formulation, as

$$t_0 - t_\infty = \tilde{\phi}(0) \frac{Q}{2\pi\rho c_p \nu x} \tag{4.4.11a}$$

where $\tilde{\phi}(0)/\mathrm{Pr}$ is given as 0.759, 0.687, 0.667, 0.625, 0.561, and 0.5 for Pr = 0.01, 0.7, 1.0, 2.0, 10, and ∞, respectively. As seen from Eq. (4.4.3), $\tilde{\phi}(0) = 1/I$. For Pr = 0.7, I is obtained from these results as 2.079. This is in close agreement with the value obtained by Mollendorf and Gebhart (1974).

The closed-form solutions for Pr = 1 and 2 may be obtained from the investigations of Yih (1952), Fujii (1963), and Brand and Lahey (1967). The dimensionless stream function $\tilde{f}$ and temperature $\tilde{\phi}$ were given in terms of the similarity variable $\tilde{\eta}$ by Fujii (1963) as

$$\tilde{f} = \frac{\tilde{\eta}^2}{2(1 + \frac{1}{12}\tilde{\eta}^2)} \qquad \tilde{\phi} = \frac{2}{3(1 + \frac{1}{12}\tilde{\eta}^2)^3} \qquad \text{for Pr} = 1 \tag{4.4.11b}$$

$$\tilde{f} = \frac{C\tilde{\eta}^2}{1 + (C/4)\tilde{\eta}^2} \qquad \tilde{\phi} = \frac{4C^2}{[1 + (C/4)\tilde{\eta}^2]^4} \qquad \text{for Pr} = 2 \tag{4.4.11c}$$

where the tilde indicates the formulation used by Fujii. Here C is a constant, given as 0.559.

Vertical axisymmetric buoyant jets. For a jet issuing into a quiescent ambient of the same density, the momentum flux $J(x)$ remains constant downstream. That is,

$$J(x) = \int_0^\infty \rho u^2 2\pi y \, dy = \text{const.} = J \tag{4.4.12}$$

where J is the input momentum flux at the nozzle. The governing momentum equation is Eq. (4.2.11), whose solution was given by Schlichting (1933) in terms of the following similarity variables:

$$\psi = \nu x f(\eta) \qquad \eta = \frac{y}{x\sqrt{\gamma}} \tag{4.4.13}$$

where $\gamma = (16\pi/3)\rho\nu^2/J$. The dimensionless stream function f and the velocity u were calculated as

$$f(\eta) = \frac{\eta^2}{1 + (\eta^2/4)} \qquad u = \frac{2\nu}{\gamma x} \frac{1}{[1 + (\eta^2/4)]^2} \tag{4.4.14}$$

The temperature distribution, in the absence of buoyancy, was obtained by Squire (1951) and may be written in terms of the variables of Eq. (4.4.13) as

$$\phi = \frac{t - t_\infty}{t_0 - t_\infty} = \left(1 + \frac{\eta^2}{4}\right)^{-2\text{Pr}} \tag{4.4.15}$$

In a buoyant jet the convected thermal energy $Q(x)$ in the flow remains constant downstream in x, being equal to the energy input at the source Q, while $J(x)$ increases due to buoyancy. Thus,

$$\int_0^\infty \rho c_p(t - t_\infty)u2\pi y\, dy = Q \tag{4.4.16}$$

Buoyancy produces additional momentum downstream, and Eqs. (4.4.14) and (4.4.15) do not apply. When the buoyancy term in Eq. (4.1.2) is retained, the equations become coupled and similarity is not found. Mollendorf and Gebhart (1973) considered this coupled problem for small buoyancy levels. An asymptotic expansion in f and ϕ was used that applies in the region close to the nozzle exit. The perturbation parameter $\epsilon(x)$ is

$$\epsilon(x) = \frac{\text{Gr}_x}{\text{Re}^4} \tag{4.4.17a}$$

where

$$\text{Gr}_x = \frac{g\beta x^3(t_0 - t_\infty)}{\nu^2} \quad \text{and} \quad \text{Re} = \frac{UD}{\tilde{s}\nu} \tag{4.4.17b}$$

The Reynolds number Re is defined in terms of the nozzle diameter D and the exit velocity U. The parameter $\tilde{s}$ depends on the velocity profile at the exit. It is 4.62 for a flat or top-hat profile and 4.0 for a parabolic profile.

Figure 4.4.3 shows the effect of buoyancy on the downstream velocity and temperature profiles, including the first approximation. A positive value of ϵ, $t_0 > t_\infty$, is seen to increase the centerline velocity and a negative value, $t_0 < t_\infty$, to decrease it, as expected from the corresponding aiding and opposing buoyancy effects in the two cases. The latter is a negatively buoyant jet. The effect on the temperature profile was found to be very small for $\epsilon < 1$. This flow was also considered by Schneider and Potsch (1979), who obtained the first approximation for the effect of weak buoyancy forces on the flow by the method of matched asymptotic expansions for $\frac{1}{2} < \text{Pr} < \frac{3}{2}$. It was also shown that the analysis of Mollendorf and Gebhart (1973) is not valid in an outer region of the jet if $\text{Pr} < \frac{3}{2}$. It breaks down if $\text{Pr} \leq \frac{1}{2}$. Figure 4.4.4 shows the results for Prandtl numbers between $\frac{1}{2}$ and $\frac{3}{2}$ and for nonbuoyant and positively and negatively buoyant jets. Expected trends are seen. It is noted that the flow is always dominated by buoyancy far downstream. That is, given enough distance, buoyancy always becomes preeminent.

For relatively large buoyancy effects, which inevitably arise in regions far downstream, the perturbation analyses referred to above do not apply. The coupled boundary layer equations must be considered directly, as outlined below. However, most such jet flows of practical interest are turbulent. A considerable amount of work has been done concerning these, as set forth in Chapter 12.

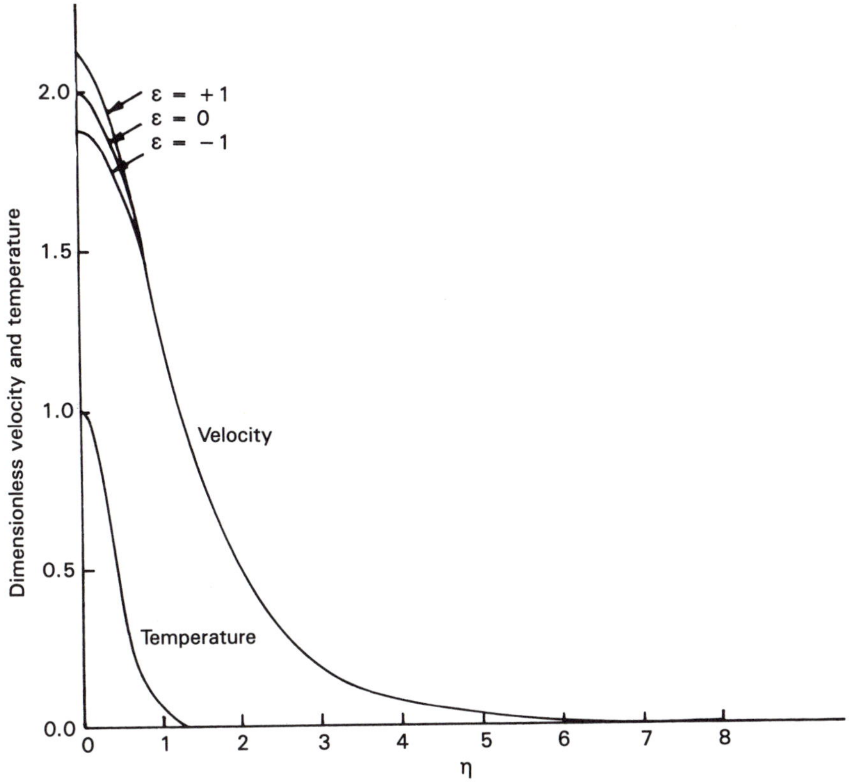

Figure 4.4.3 Velocity and temperature profiles, f'/η and ϕ, in a buoyant jet at small levels of thermal buoyancy for Pr = 6.7. *(Reprinted with permission from* Int. J. Heat Mass Transfer, *vol. 16, p. 735, J. C. Mollendorf and B. Gebhart. Copyright © 1973, Pergamon Journals Ltd.)*

Nevertheless, laminar buoyant jets remain of interest in some technological applications, such as thermal discharge into small enclosed fluid regions for energy storage and extraction. Laminar flow arises in the region close to the source. However, the assumption of boundary layer behavior often is not permissible in such applications and the full elliptic equations must be solved. Far from the source, where a boundary layer treatment is valid, the flow may already be in transition or turbulent. Some of these considerations have been studied by Jaluria and Gebhart (1975) and Jaluria (1976) with respect to the generation of a wake above a heated body.

Consider the laminar flow far enough downstream of a jet exit to be a boundary region flow. The governing equations are nonsimilar because of the buoyancy. They may be solved by numerical techniques, as done by Himasekhar and Jaluria (1982). The equations are nondimensionalized with u_0 and t_0, the jet exit velocity and temperature, and the vertical distance x as the characteristic quantities. The dimensionless velocity components U and V, the coordinate distances X and Y, and temperature ϕ are defined as

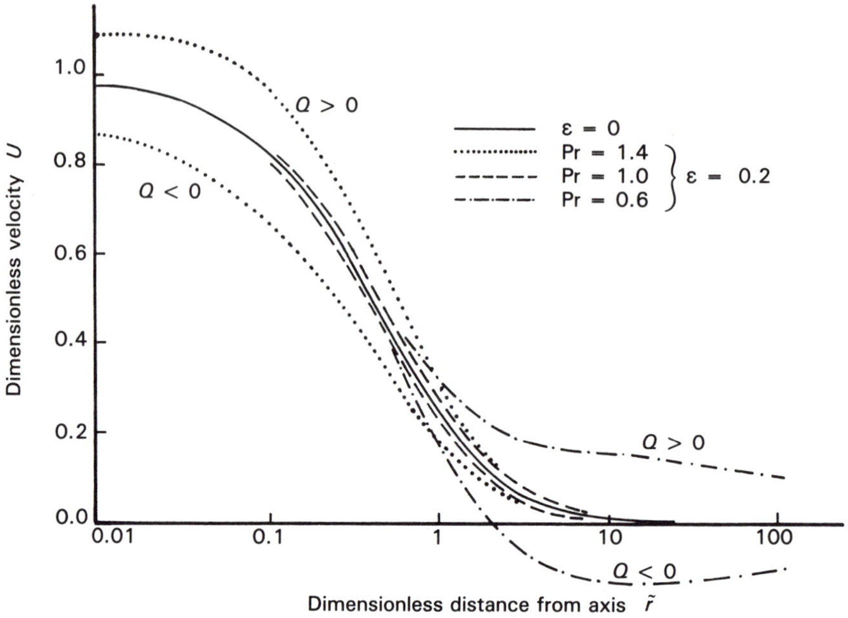

Figure 4.4.4 Vertical velocity distributions in a nonbuoyant ($Q = 0$), a buoyant ($Q > 0$), and a negatively buoyant ($Q < 0$) jet at small levels of thermal buoyancy. The dimensionless velocity $U = \frac{4}{3}\,\nu x/M_o$ and distance $\tilde{r} = \frac{3}{32}\,(M_o/\nu^2)(r/x)^2$, where $2\pi\rho M_o$ is the momentum flux at the jet inlet. *(From Schneider and Potsch, 1979.)*

$$U = \frac{u}{u_0} \qquad V = \frac{v}{u_0} \qquad X = \frac{x'}{x} \qquad Y = \frac{y}{x}\sqrt{\mathrm{Re}_x} \qquad \phi = \frac{t - t_\infty}{t_0 - t_\infty}$$

and

$$\mathrm{Re}_x = \frac{u_0 x}{\nu} \tag{4.4.18}$$

Here x' is the local coordinate distance, which varies from 0 at the source to an arbitrary vertical distance x. The dimensionless governing equations become

$$\frac{\partial(YU)}{\partial X} + \frac{\partial(YV)}{\partial Y} = 0 \tag{4.4.19}$$

$$U\frac{\partial U}{\partial X} + V\frac{\partial U}{\partial Y} = \frac{\mathrm{Gr}_x}{\mathrm{Re}_x^2}\phi + \frac{1}{Y}\frac{\partial}{\partial Y}\left(Y\frac{\partial U}{\partial Y}\right) \tag{4.4.20}$$

$$U\frac{\partial \phi}{\partial X} + V\frac{\partial \phi}{\partial Y} = \frac{1}{\mathrm{Pr}}\frac{1}{Y}\frac{\partial}{\partial Y}\left(Y\frac{\partial \phi}{\partial Y}\right) \tag{4.4.21}$$

with Gr_x defined as in Eq. (4.4.17*b*).

Thus, $\mathrm{Gr}_x/\mathrm{Re}_x^2$ arises as a governing parameter that determines the effect of thermal buoyancy in the flow. For small values of $\mathrm{Gr}_x/\mathrm{Re}_x^2 \propto x$, a perturbation solution may be obtained as outlined earlier. However, the boundary layer far downstream of the nozzle, as formulated above, is numerically determined, with the input thermal energy and momentum specified at the jet exit at $x = 0$. Himasekhar and Jaluria (1982) considered this flow in isothermal and stably stratified environments, using finite-difference marching procedures for the governing parabolic equations. They considered both plumes and buoyant jets. Both flows were found to approach the characteristics of an axisymmetric plume far downstream, as generated by a concentrated thermal source. As thermal buoyancy dominates, a thermal plume behavior results. See the reviews by List (1982) and Jaluria (1985).

4.5 FLOW IN A STABLY STRATIFIED ENVIRONMENT

The rise of axisymmetric plumes and buoyant jets in a stably stratified environment is an important concern in environmental studies. Consequently, several investigators have considered these flows and have determined the effect of stratification on the basic characteristics of the flow and the height to which the flow rises. Most of these studies have considered turbulent flow because of its greater relevance to practical problems. Chapter 12 discusses experimental and analytical results obtained for turbulent free-boundary flows.

Some work has also been done on the flow of laminar plumes and jets in stably stratified media. An experimental study of laminar axisymmetric vertical jets in a salt-stratified water medium was carried out by Tenner and Gebhart (1971). A toroidal cell was found to arise around the jet, with upward flow in the inner region and a downward flow due to negative buoyancy in the outer region. A further discussion of these mechanisms is given in Chapter 6 in terms of the underlying combined thermal and mass transport processes.

Entrainment models have been used extensively for an integral analysis of turbulent flows, as discussed in Chapter 12. Morton (1967) developed a similar model for laminar jets, plumes, and wakes. The entrainment flux scale was obtained by order-of-magnitude arguments, and the model was used to study the ascent of laminar plumes in a stably stratified environment. Further study was carried out by Wirtz and Chiu (1974). The governing integral equations for the conservation of mass, momentum, and energy were obtained for a laminar axisymmetric plume as

$$\int_0^\infty 2\pi\rho u y\, dy = \begin{cases} 2\pi\alpha_e \rho \nu x & \text{for Pr} \geq 1 \\ 2\pi\alpha_e \rho \alpha x & \text{for Pr} \leq 1 \end{cases} \tag{4.5.1}$$

$$\frac{d}{dx}\int_0^\infty y u^2(x, y)\, dy = \int_0^\infty g\beta y \tilde{\theta}(x, y)\, dy \tag{4.5.2}$$

$$\frac{d}{dx}\int_0^\infty y u(x, y)\tilde{\theta}(x, y)\, dy = -\frac{dt_\infty}{dx}\int_0^\infty y u(x, y)\, dy \tag{4.5.3}$$

where $\tilde{\theta} = t - t_\infty$. Here the entrainment coefficient α_e, which is the ratio of the radial inflow velocity to the mean centerline velocity, as defined in Section 12.2 is taken as a constant.

These equations were solved analytically to obtain the height at which the buoyancy becomes zero and that at which the velocity vanishes. The flow is predicted to rise to a definite height x_{max}, where the vertical velocity u becomes zero, and then to drop to an equilibrium level x_{min}, where the excess temperature vanishes. Figure 4.5.1 shows these trends in terms of dimensionless temperature excess $\tilde{\theta}_0$, centerline velocity $\tilde{U}_0$, and plume half-width $\tilde{\delta}$, as defined by Wirtz and Chiu (1974) in terms of the fluid properties, α_e, and the heat input.

They also considered finite-size heat sources and related the downstream flow to that generated by point energy sources. Figure 4.5.2 sketches the relationship between an actual heat source and a virtual (point) source. The vertical distance x_v between the two was determined by equating the width of the point source plume at the height x_v to the width of the given source. Experimental results were also obtained, which gave the value of α_e as about 1.0, for air as the stratified ambient fluid. For further details, see Jaluria (1980).

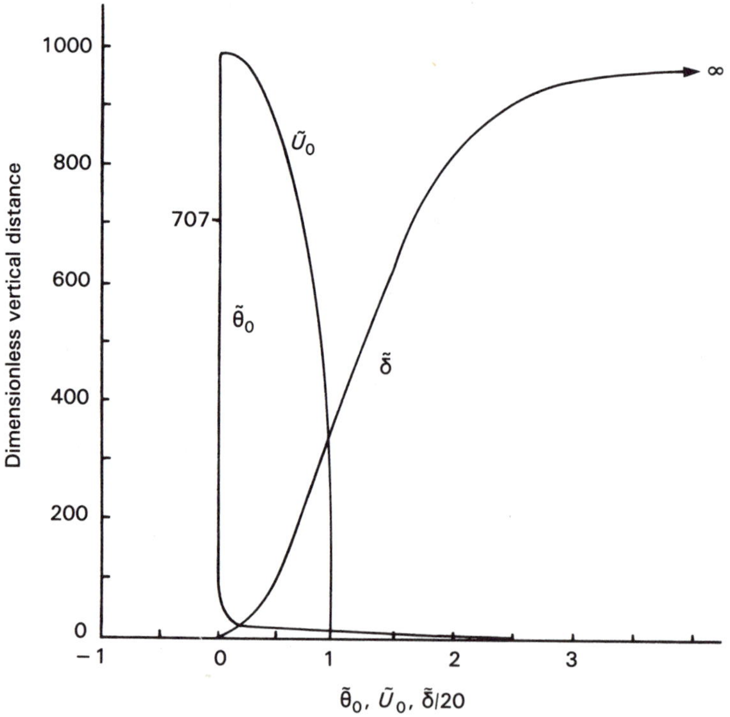

Figure 4.5.1 Variation of the dimensionless centerline temperature $\tilde{\theta}_0$, centerline velocity $\tilde{U}_0$, and plume half-width $\tilde{\delta}$ for a laminar axisymmetric plume in a linearly stratified environment. *(Reprinted with permission from* Int. J. Heat Mass Transfer, *vol. 17, p. 323, R. A. Wirtz and C. M. Chiu. Copyright © 1974, Pergamon Journals Ltd.)*

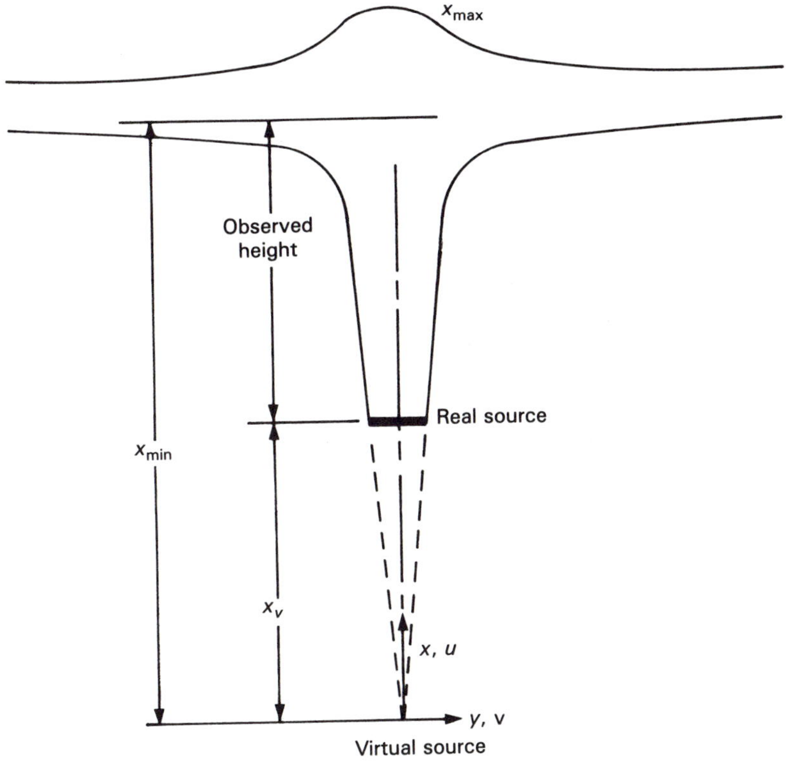

Figure 4.5.2 Relationship between the actual heat source location and the corresponding point source for plume flow in a stratified medium. *(Reprinted with permission from* Int. J. Heat Mass Transfer, *vol. 17, p. 323, R. A. Wirtz and C. M. Chiu. Copyright © 1974, Pergamon Journals Ltd.)*

Himasekhar and Jaluria (1982) carried out a numerical study of laminar axisymmetric free-boundary flows in thermally stratified media. Figures 4.5.3 and 4.5.4 show the downstream variation of the centerline temperature ϕ_0 and the centerline velocity U_0, respectively, for a thermal plume generated by a small disk at temperature t_0. The stratification parameter S and the dimensionless quantities are the same as those defined in Eqs. (3.11.9) and (3.11.10).

The stable thermal stratification is found to result in a faster decrease, than in an isothermal medium, in the buoyancy level downstream and consequently in the velocity level. A temperature defect is found to arise in the outer region of the flow. This defect increases with X and with an increase in the stratification level. The buoyancy force ultimately becomes zero and then negative, leading to a maximum height of plume rise and flow reversal. The region where flow reversals occur is not satisfactorily treated in terms of the boundary layer formulation. The full elliptic equations must be considered.

From the results of Himasekhar and Jaluria (1982), the basic trends in this flow are indicated. It is interesting to note from Fig. 4.5.4 that for $S = 0$ the

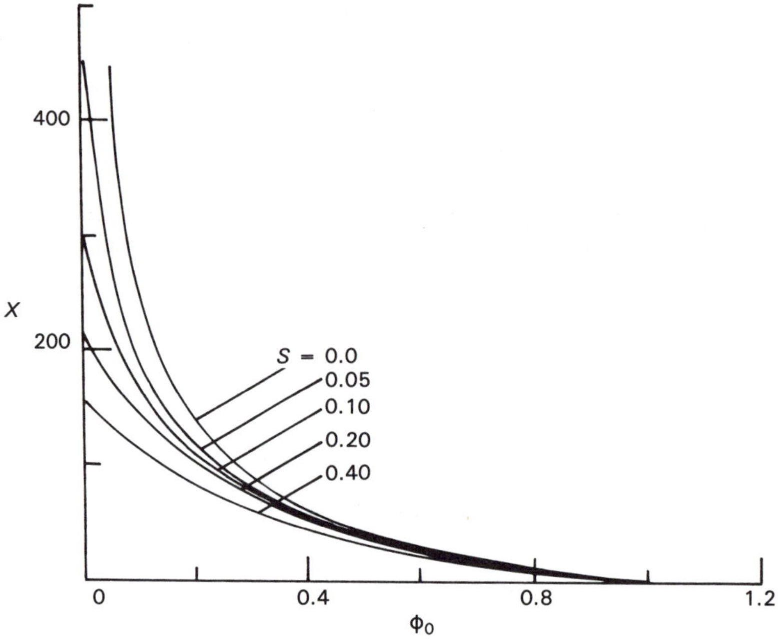

Figure 4.5.3 Downstream centerline temperature variation in an axisymmetric plume rising in a linearly stratified medium at Pr = 7.0 and at various values of the stratification parameter S. *(Reprinted with permission from* Int. J. Heat Mass Transfer, *vol. 25, p. 213, K. Himasekhar and Y. Jaluria. Copyright © 1982, Pergamon Journals Ltd.)*

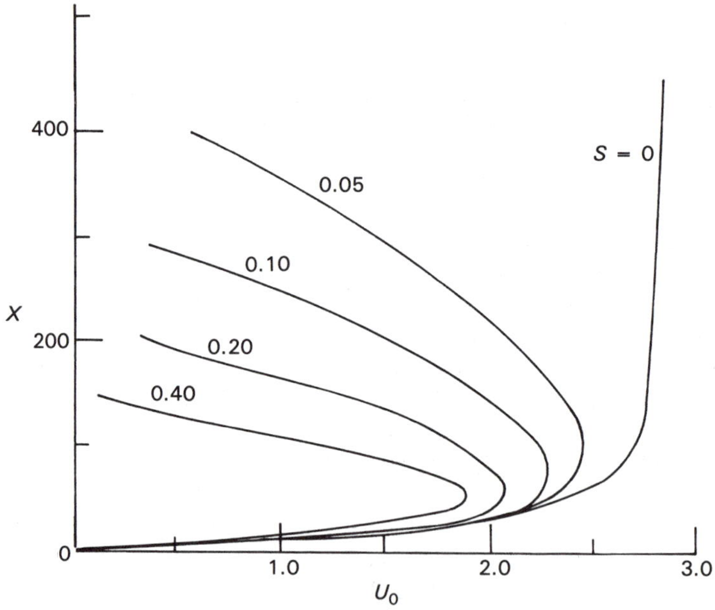

Figure 4.5.4 Downstream variation of the centerline velocity in an axisymmetric plume rising in a linearly stratified medium at Pr = 7.0 and at various values of S. *(Reprinted with permission from* Int. J. Heat Mass Transfer, *vol. 25, p. 213, K. Himasekhar and Y. Jaluria. Copyright © 1982, Pergamon Journals Ltd.)*

constant centerline velocity predicted by similarity analysis is obtained downstream of the source. Also, the temperature excess decreases more gradually than that in a stratified medium. It asymptotically approaches zero as $X \to \infty$. For a stratified ambient, $S > 0$, the height at which the centerline temperature equals the ambient temperature may be inferred to be the height to which the flow rises.

REFERENCES

Appalaswamy, A. V., and Jaluria, Y. (1980). *J. Appl. Mech. 47,* 667.
Bill, R. G., and Gebhart, B. (1975). *Int. J. Heat Mass Transfer 18,* 513.
Brand, R. S., and Lahey, F. J. (1967). *J. Fluid Mech. 29,* 305.
Cebeci, T., and Na, T. Y. (1969, 1970). *Phys. Fluids 12,* 463; *13,* 536.
Ede, A. J. (1967). *Adv. Heat Transfer 4,* 1.
Eigenson, L. S. (1940). *C. R. Acad. Sci. U.S.S.R. 26,* 440.
Elenbaas, W. (1948). *J. Appl. Phys. 19,* 1148.
Fujii, T. (1963). *Int. J. Heat Mass Transfer 6,* 597.
Gebhart, B. (1971). *Heat Transfer,* 2d ed., McGraw-Hill, New York.
Gebhart, B. (1973). *Adv. Heat Transfer 9,* 273.
Gebhart, B., Pera, L., and Schorr, A. W. (1970). *Int. J. Heat Mass Transfer 13,* 161.
Hama, F. R., and Recesso, J. V. (1958). Univ. of Maryland Tech. Note BN-116. See also Hama, F. R., and Christiaens, J. (1958). Univ. of Maryland Tech. Note BN-138.
Hardwick, N. E., and Levy, E. K. (1973). *J. Heat Transfer 95,* 289.
Himasekhar, K., and Jaluria, Y. (1982). *Int. J. Heat Mass Transfer 25,* 213.
Jaluria, Y. (1976). *Lett. Heat Mass Transfer 3,* 457.
Jaluria, Y. (1980). *Natural Convection Heat and Mass Transfer,* chap. 6, Pergamon, Oxford, U.K.
Jaluria, Y. (1985). Hydrodynamics of Laminar Buoyant Jets, in *Encyclopaedia of Fluid Mechanics,* N. P. Cheremisinoff, Ed., vol. 2, p. 317, Gulf Publishing, Houston, Texas.
Jaluria, Y., and Gebhart, B. (1975). *Int. J. Heat Mass Transfer 18,* 415.
Kuiken, H. K. (1968). *Int. J. Heat Mass Transfer 11,* 1141.
Kyte, J. R., Madden, A. J., and Piret, E. L. (1953). *Chem. Eng. Prog. 49,* 653.
Le Fevre, E. J., and Ede, A. J. (1956). *Proc. 9th Int. Cong. Appl. Mech.,* Brussels, vol. 4, p. 175.
List, E. J. (1982). *Annu. Rev. Fluid Mech. 14,* 189.
Millsaps, K., and Pohlhausen, K. (1956). *J. Aerosp. Sci. 23,* 381.
Millsaps, K., and Pohlhausen, K. (1958). *J. Aerosp. Sci. 25,* 357.
Minkowycz, W. J., and Sparrow, E. M. (1974). *J. Heat Transfer 96,* 178.
Mollendorf, J. C., and Gebhart, B. (1973). *Int. J. Heat Mass Transfer 16,* 735.
Mollendorf, J. C., and Gebhart, B. (1974). *Proc. 5th Int. Heat Transfer Conf.,* Tokyo.
Morton, B. R. (1967). *Phys. Fluids 10,* 2120.
Mueller, A. C. (1942). *Trans. AIChE 38,* 613.
Nagendra, H. R., Tirunarayanan, M. A., and Ramachandran, A. (1969). ASME Paper 69-HT-G (1969); also *J. Heat Transfer 92,* 191 (1970).
Narain, J. P., and Uberoi, M. S. (1972). *Phys. Fluids 15,* 928.
Oosthuizen, P. H. (1979). *Numer. Heat Transfer 2,* 477.
Pera, L., and Gebhart, B. (1972). *Int. J. Heat Mass Transfer 15,* 175.
Raithby, G. D., and Hollands, K. G. T. (1976). *J. Heat Transfer 98,* 452, 522.
Schuh, H. (1948). Boundary Layers of Temperature, in *Boundary Layers,* W. Tollmien, Ed., Sec. B6, British Ministry of Supply, German Doc. Cent. Ref. 3220T.
Schlichting, H. (1933). *Z. Angew. Math. Mech. 13,* 260.
Schneider, W., and Potsch, K. (1979). Weak Buoyancy in Laminar Vertical Jets, in *Recent Developments in Theoretical and Experimental Fluid Mechanics,* V. Muller, K. G. Roesner, and B. Schmidt, Eds., p. 501, Springer-Verlag, Berlin.
Sparrow, E. M., and Gregg, J. L. (1956). *J. Heat Transfer 78,* 1823.

Sparrow, E. M., Patankar, S. V., and Abdel-Wahed, R. M. (1978). *J. Heat Transfer 100,* 184.
Squire, H. B. (1951). *Q. J. Mech. 4,* 321.
Tenner, A. R., and Gebhart, B. (1971). *Int. J. Heat Mass Transfer 14,* 2051.
Van Dyke, M. (1967). Free Convection from a Vertical Needle, Sedov Anniversary Volume, Moscow.
Wirtz, R. A., and Chiu, C. M. (1974). *Int. J. Heat Mass Transfer 17,* 323.
Yang, K. T. (1960). *J. Appl. Mech. 27,* 230.
Yang, K. T. (1964). *J. Appl. Mech. 31,* 131.
Yih, C. S. (1952). *Proc. 1st U.S. Natl. Cong. Appl. Mech.,* p. 941.

PROBLEMS

4.1 (a) Show that the energy differential equation for axisymmetric vertical flow transforms to the energy equation given by Eq. (4.2.4).

(b) Find whether there are any possibilities other than Eq. (4.2.5), for the definition of Gr_x, for similarity to be preserved.

(c) Taking the axisymmetric nonbuoyant jet similarity formulation in Eqs. (4.2.10) and (4.2.11), determine

(1) Whether the momentum is actually conserved downstream.
(2) The rate of decay of centerline velocity.
(3) The change in entrainment velocity.
(4) The change in jet size.

4.2 A 3-mm-diameter vertical cylinder 2 m long is in air at 20°C. The cylinder temperature is 20°C at the bottom and increases linearly to 40°C at the top. Calculate the convection coefficient and the total convective heat transfer.

4.3 For the calculation of convective losses, a human being may be treated as a vertical cylinder 35 cm in diameter and 170 cm in height. At a normal surface temperature of 31°C, find the energy lost by natural convection to air at 10°C. How does this result compare with those for a flat vertical surface of equal area and with a rest metabolic heat production for an average person?

4.4 For natural convection flow over a vertical cylinder, write the boundary layer equations, starting with the equations in polar cylindrical coordinates. Show that as the cylinder radius R becomes large compared to the boundary layer thickness δ, the flow approaches that over a flat vertical surface.

4.5 A point source in water dissipates 10 W. Determine the temperature 5 cm above the source and the maximum vertical velocity there. Take Pr = 7.0. Repeat for 10 cm above the source.

4.6 A thermal plume is generated by a concentrated (point) source in an extensive unstratified ambient medium. For the flow still laminar but far downstream from the source, find how the entrainment velocity, mass flow rate, and radius of the plume material vary with distance from the source.

4.7 Consider the thermal plume generated from a vertical line heat source of uniform strength q', per unit length, beginning at $x = 0$. There is no surface present. Find how both the temperature field and centerline velocity vary downstream in x.

4.8 Obtain the independent boundary conditions for an axisymmetric plume. Repeat this for the flow arising from a fine vertical wire of diameter R, whose radius is very much smaller than the resulting plume diameter.

4.9 Consider the plane and axisymmetric plume flows. If the pressure term is retained in the temperature differential equation, the residual term of the equation, after boundary layer approximations, is $\beta T u \, dp_h/dx$. Find whether this effect may be retained and still yield similarity for the two plume flows.

4.10 A jet of air emerges vertically from a nozzle taken as a point source. Determine the variation of the centerline temperature and velocity and of the jet width with height if the flow is buoyant. If the flow is negatively buoyant, that is, the exit temperature is less than the ambient temperature, what do you expect? How would you calculate the transport and what are the results obtained? Take the flow as laminar.

4.11 For axisymmetric free-boundary flow, a similar solution of the boundary layer equations results for $t_0 - t_\infty = Nx^n$. Determine the range in n for which physically reasonable formulations arise for transport of boundary region form.

4.12 Consider a vertical cylinder of height 20 cm in air at 20°C. The outer surface of the cylinder is maintained at 150°C. Calculate the heat transfer rates from the cylinder for diameters 2, 1, 0.5, 0.1, and 0.01 cm and compare the results with those from the corresponding flat-plate solution. Comment on the difference in view of Eq. (4.3.4).

4.13 Assuming Gaussian profiles for the velocity and temperature distributions needed for Eqs. (4.5.1)–(4.5.3), obtain the ordinary differential equations whose solution would yield the velocity and temperature levels and the plume thickness as functions of height. Note that the plume thickness will appear in the assumed profiles.

4.14 For an axisymmetric plume rising in linearly and stably stratified environment, write the governing equations, along with the relevant boundary conditions, that would apply in the region where significant flow reversal occurs. Recall that the problem may not be treated as a boundary region flow there.

CHAPTER
FIVE

OTHER THAN VERTICAL FLOWS

5.1 NONVERTICAL FLOWS AND TRANSPORT EQUATIONS

In many important applications, the surface generating buoyancy may be curved or inclined at an angle θ with the direction of gravity, where θ may vary along the surface; see Figs. 5.1.1 and 5.1.2. The simplest case is an inclined flat surface (Fig. 5.1.1), a plane flow. The angle θ is then constant downstream. For $\theta = \pi/2$ the flow is adjacent to a horizontal surface. For curved surfaces the angle θ varies downstream, as seen in Figs. 5.1.2*a* and 5.1.2*b* for symmetric two-dimensional planar (wedgelike) and axisymmetric (conelike) bodies, respectively. The bodies can be of any arbitrary shape. However, many practical shapes, such as cylinders and spheres, curve back to close. These two geometries are very frequently encountered in applications. For example, the cylindrical configuration is important for many problems such as heat transfer from pipes, tubes, and cables, while the spherical configuration arises in many manufacturing systems as in packed beds of spherical bodies for heat transfer and in many electronic components that are nearly spherical.

Considering the transport equations governing inclined flows, the buoyancy force **B** now has both tangential and normal components, B_t and B_n. If B denotes the magnitude, $B_t = B \cos \theta$ and $B_n = B \sin \theta$. Component B_t drives the tangential motion u. Simultaneously, B_n generates cross-flow effects. This is first felt as a motion pressure gradient across the flow. These effects depend on the angle θ. For larger θ, B_n increases relative to B_t. These effects generally lead to theoretical analyses more complicated than those for vertical surfaces. However, analytical solutions have been obtained for many of the important inclined flow configurations. These are examined in this chapter in considerable detail.

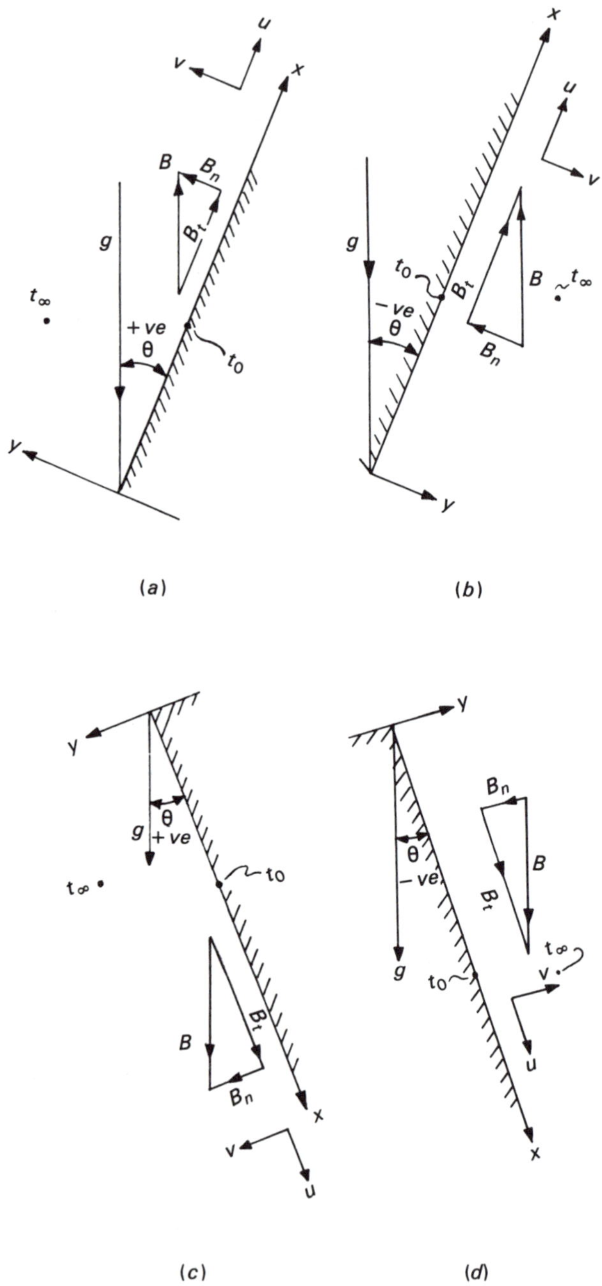

Figure 5.1.1 Coordinate system for flow adjacent to flat inclined surfaces. (*a*) $t_0 > t_\infty$, warm surface facing up; (*b*) $t_0 > t_\infty$, warm surface facing down; (*c*) $t_0 < t_\infty$, cold surface facing down; (*d*) $t_0 < t_\infty$, cold surface facing up.

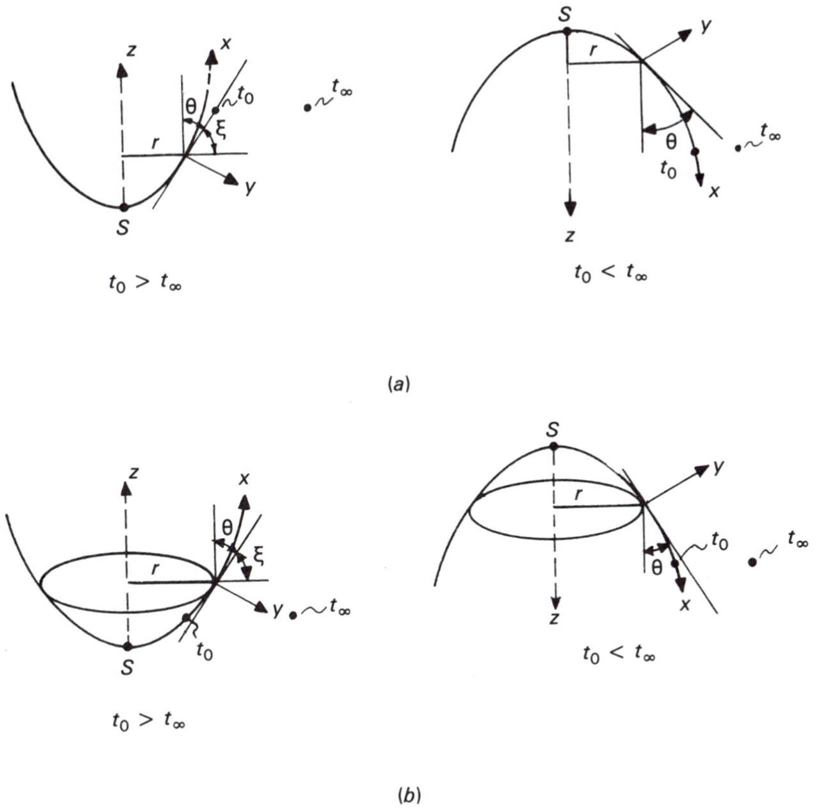

Figure 5.1.2 Coordinate system for flow adjacent to curved surfaces. (*a*) Two-dimensional planar body; (*b*) axisymmetric body.

The coordinate systems used for both $t_0 > t_\infty$ and $t_0 < t_\infty$ are shown in Figs. 5.1.1 and 5.1.2. If the surface temperature t_0 is higher than the ambient medium temperature t_∞, upflow arises and x is measured from the bottom leading edge or the lower stagnation point, as applicable. On the other hand, for $t_0 < t_\infty$, downflow arises and x is measured from the top leading edge or the upper stagnation point. The normal coordinate for both $t_0 > t_\infty$ and $t_0 < t_\infty$ is y, outward from the surface. The angle of inclination θ is taken as positive above a surface and negative below a surface for $t_0 > t_\infty$ and vice versa for $t_0 < t_\infty$. When the coordinate systems are taken in this manner, the two flow formulations are identical. The analyses will be presented as for warm surfaces with the understanding that the results apply to cold surfaces as well.

It is noted that the choice of sign of θ is not arbitrary, as it sometimes appears from the multiplicity of notations used in literature. The sign of θ adopted here correctly reflects, as it should, the physics of the flow for the region above and below the surface and for both the heating and cooling conditions.

Consider Fig. 5.1.1. For a warm surface, $t_0 > t_\infty$, facing upward (Fig. 5.1.1*a*), θ by notation is positive, and B_n is directed away from the surface. That is, $B_n > 0$. The effect is to increase the normal component v away from the surface and to cause the fluid to tend to lift off the surface. This, as will be seen later, is evidenced in the force-momentum balance as a positive motion pressure gradient in the vertical direction, $\partial p_m/\partial y$; see Eq. (5.1.6). For a warm surface facing down, that is, for the region below the surface (Fig. 5.1.1*b*), θ is negative and $B_n < 0$ and is now directed toward the surface. From Eq. (5.1.6), $\partial p_m/\partial y$ is negative and the effect is to increase v toward the surface.

For a surface maintained at $t_0 < t_\infty$, θ is taken as positive for the region below a surface. This again results in B_n directed away from a cold surface (Fig. 5.1.1*c*). Above a cold surface $\theta < 0$ and B_n is directed toward the surface (Fig. 5.1.1*d*). The effect of B_n, then, is to increase v toward the cold surface in the region above it and to increase v away from the surface in the region below the surface. This is just the reverse of the effects for a warm surface.

A clear understanding of the association of sign of θ with B_n is helpful in understanding the physics of inclined flows. From the above discussion, a simple rule follows. For $\theta > 0$ for both warm and cold surfaces, $B_n > 0$; that is, B_n acts away from the surface. For $\theta < 0$, $B_n < 0$ and is directed toward the surface. This rule also applies to curved surfaces, shown in Fig. 5.1.2, as well as horizontal surfaces. Note that for $\theta = \pi/2$, $B_t = 0$ and only the B_n effect remains to cause flow; see Eq. (5.1.6). Such a flow is called an "indirect drive" flow since it is driven indirectly by B_n, through p_m, rather than by B_t. This is discussed in detail in Section 5.3. For other inclined flows in general, B_n and B_t operate together.

With this understanding of the physical mechanisms that interact in inclined flows, transport equations are presented next. The first sets are for inclined plane flows. The boundary layer equations are then determined for both inclined and horizontal flows. Finally, the boundary layer equations are written for curved surfaces of both the wedge and cone geometries and cylinders.

Flow adjacent to flat inclined surfaces, $\theta = const$. This is seen in Fig. 5.1.1, where $-\pi/2 < \theta < \pi/2$. With the Boussinesq approximation and constant fluid properties, the buoyancy force components B_t and B_n in the x and y directions are $g\beta(t - t_\infty)\cos\theta$ and $g\beta(t - t_\infty)\sin\theta$, respectively. Thus, the tangential and normal force-momentum balances for plane flow, Eqs. (3.2.2) and (3.2.3) before, are modified by these two buoyancy force components as below:

$$u\frac{\partial u}{\partial x} + v\frac{\partial u}{\partial y} = \nu\left(\frac{\partial^2 u}{\partial x^2} + \frac{\partial^2 u}{\partial y^2}\right) - \frac{1}{\rho}\frac{\partial p_m}{\partial x} + g\beta(t - t_\infty)\cos\theta \tag{5.1.1}$$

$$u\frac{\partial v}{\partial x} + v\frac{\partial v}{\partial y} = \nu\left(\frac{\partial^2 v}{\partial x^2} + \frac{\partial^2 v}{\partial y^2}\right) - \frac{1}{\rho}\frac{\partial p_m}{\partial y} + g\beta(t - t_\infty)\sin\theta \tag{5.1.2}$$

The continuity and energy equations remain as given by Eqs. (3.2.1) and (3.2.4).

Simplifications again result when the boundary region approximations are applied. This is done here, following the arguments and procedures in Section 3.2.

Equations (5.1.1) and (5.1.2) are generalized in terms of the nondimensional variables defined there. The following equations result, where $\Delta = \delta/L$:

$$U\frac{\partial U}{\partial X} + V\frac{\partial U}{\partial Y} = \frac{1}{\sqrt{\text{Gr}}}\left(\frac{\partial^2 U}{\partial X^2} + \frac{\partial^2 U}{\partial Y^2}\right) - \frac{\partial P_m}{\partial X} + \phi\cos\theta \tag{5.1.3}$$

$$1 \quad 1 \qquad \Delta \quad \Delta^{-1} \qquad\qquad 1 \qquad \Delta^{-2} \qquad\qquad 1$$

$$U\frac{\partial V}{\partial X} + V\frac{\partial V}{\partial Y} = \frac{1}{\sqrt{\text{Gr}}}\left(\frac{\partial^2 V}{\partial X^2} + \frac{\partial^2 V}{\partial Y^2}\right) - \frac{\partial P_m}{\partial Y} + \phi\sin\theta \tag{5.1.4}$$

$$1 \quad \Delta \qquad \Delta \quad 1 \qquad\qquad \Delta \qquad \Delta^{-1} \qquad\qquad 1$$

The known order of each term is written below it in terms of 1 and powers of Δ. Again neglecting terms of order Δ and higher, it is evident from Eq. (5.1.4) that the pressure term must be of $O(\sin\theta)$. Thus, P_m must be of $O(\Delta\sin\theta)$. In the tangential momentum equation the pressure term is therefore of order $\Delta\sin\theta$ since the value of X is of order 1. Its relative magnitude, compared to the buoyancy term in Eq. (5.1.3), is therefore $\Delta\tan\theta$.

To obtain boundary layer approximations, quantities of order Δ and higher are neglected in these equations. Consistent with this approximation, the pressure term in Eq. (5.1.3) may be neglected compared to the buoyancy term if $\Delta\tan\theta$ is of order Δ or smaller. Clearly, this is justified for small and moderate values of θ. For θ close to $\pi/2$, however, the pressure term may not be neglected in Eq. (5.1.3). Since the pressure and buoyancy terms in Eq. (5.1.4) are of $O(\sin\theta)$, this equation may be neglected when these terms are of order Δ or smaller. Again, at small values of θ this is possible. However, over most of the range of θ these terms are to be retained.

In presenting the boundary region equations below for the whole range $-\pi/2 < \theta < \pi/2$, the pressure terms in Eqs. (5.1.3) and (5.1.4), along with the buoyancy term in Eq. (5.1.4), are retained, with the understanding that these terms may be neglected for some values of θ.

$$u\frac{\partial u}{\partial x} + v\frac{\partial u}{\partial y} = \nu\frac{\partial^2 u}{\partial y^2} + g\beta(t - t_\infty)\cos\theta - \frac{1}{\rho}\frac{\partial p_m}{\partial x} \tag{5.1.5}$$

$$0 = g\beta(t - t_\infty)\sin\theta - \frac{1}{\rho}\frac{\partial p_m}{\partial y} \tag{5.1.6}$$

The continuity and the reduced energy equations, with the usual assumptions, are

$$\frac{\partial u}{\partial x} + \frac{\partial v}{\partial y} = 0 \tag{5.1.7}$$

$$u\frac{\partial t}{\partial x} + v\frac{\partial t}{\partial y} = \alpha\frac{\partial^2 t}{\partial y^2} \tag{5.1.8}$$

Flow adjacent to a horizontal surface, $\theta = \pm\pi/2$. The buoyancy force B_t in Eq. (5.1.3) vanishes. The only buoyancy force effect now is B_n, directed outward normal to the surface for a warm surface facing upward, $\theta = \pi/2$, and toward the surface for a warm surface facing downward, $\theta = -\pi/2$. The consequence is a vertical motion pressure gradient, $\partial p_m/\partial y$, which then indirectly drives the flow through its x component $\partial p_m/\partial x$. The resulting flows are systematically weaker than vertical flows. Therefore, correct boundary layer equations for such flows may not be recovered from Eqs. (5.1.3) and (5.1.4) since they are based on characteristic quantities representative of vertical flows. It is instructive, however, to carry out this analysis in detail since horizontal flows are not a special case of inclined flows.

In Eq. (5.1.4), the normal buoyancy force B_n is of order 1. Neglecting terms to order Δ and higher, the conclusion from Eq. (5.1.4) is that P_m is $O(\Delta)$. The pressure term in Eq. (5.1.3), therefore, is of order Δ. However, this term, although of $O(\Delta)$, must be retained in the boundary layer approximations of Eq. (5.1.1) since any flow is now entirely driven by this motion pressure gradient. This is a very different circumstance. In the x direction, a pressure term of $O(\Delta)$ generates the flow, with acceleration terms of $O(1)$. This oddity, as mentioned earlier, is a consequence of using inappropriate characteristic quantities to nondimensionalize the governing equations. A more realistic approach is as follows.

U_c^* is taken as the characteristic velocity for horizontal flows and $U, V = u, v/U_c^*$, $P_m = p_m/\rho U_c^{*2}$. As before, let $X, Y = (x, y)/L$, $\Delta = \delta/L$. Equations (5.1.1) and (5.1.2) for $\theta = \pi/2$ then become

$$U\frac{\partial U}{\partial X} + V\frac{\partial U}{\partial Y} = \frac{\nu}{LU_c^*}\left(\frac{\partial^2 U}{\partial X^2} + \frac{\partial^2 U}{\partial Y^2}\right) - \frac{\partial P_m}{\partial X} \tag{5.1.9a}$$

$$1 \quad 1 \qquad \Delta \quad \Delta^{-1} \qquad\qquad 1 \qquad \Delta^{-2}$$

$$U\frac{\partial V}{\partial X} + V\frac{\partial V}{\partial Y} = \frac{\nu}{U_c^* L}\left(\frac{\partial^2 V}{\partial X^2} + \frac{\partial^2 V}{\partial Y^2}\right) - \frac{\partial P_m}{\partial Y} + \phi\frac{\mathrm{Gr}\nu^2}{U_c^{*2}L^2} \tag{5.1.9b}$$

$$1 \quad \Delta \qquad \Delta \quad 1 \qquad\qquad \Delta \qquad \Delta^{-1} \qquad\qquad\qquad 1$$

An estimate of U_c^* is not known a priori. However, it may be determined from order of magnitude arguments as follows. Since $\partial P_m/\partial X$ drives the motion in x direction, it must be $O(1)$. Therefore, the largest viscous term must also be $O(1)$ and $\Delta = O[U_c^*(L/\nu)]^{-1/2}$. Since P_m is $O(1)$, $\partial P_m/\partial y$ in Eq. (5.1.9b) is of $O(\Delta^{-1})$. Neglecting terms to order Δ and higher, the pressure term in Eq. (5.1.9b) must remain to balance the buoyancy force. Therefore, $\mathrm{Gr}\nu^2/(L^2U_c^{*2}) = O(\Delta^{-1})$. That is, $\Delta = O(\mathrm{Gr}^{-1/5})$. Then, $U_c^*(L/\nu) = O(\mathrm{Gr}^{2/5})$. Recall that in vertical flows, $U_c(L/\nu) = O(\mathrm{Gr}^{1/2})$. Thus, as expected, horizontal flows are weaker than vertical flows.

The boundary layer equations for flat horizontal surfaces are thus recovered from Eqs. (5.1.9a) and (5.1.9b). Neglecting all terms to $O(\Delta)$ or smaller, these are

$$u\frac{\partial u}{\partial x} + v\frac{\partial u}{\partial y} = \nu\frac{\partial^2 u}{\partial y^2} - \frac{1}{\rho}\frac{\partial p_m}{\partial x} \tag{5.1.10a}$$

$$0 = g\beta(t - t_\infty) - \frac{1}{\rho}\frac{\partial p_m}{\partial y} \qquad \text{for } \theta = \frac{\pi}{2} \tag{5.1.10b}$$

$$0 = -g\beta(t - t_\infty) - \frac{1}{\rho}\frac{\partial p_m}{\partial y} \qquad \text{for } \theta = -\frac{\pi}{2} \tag{5.1.10c}$$

The continuity and energy equations remain as before, Eqs. (5.1.7) and (5.1.8). Note that the gradient of p_m is negative for $\theta = -\pi/2$. This does not result in an onflow at the leading edge and a developing boundary region downstream, as discussed in Section 5.3.

Curved surfaces. The conservation equations governing the steady laminar nonbuoyant boundary layer flow over the two-dimensional planar and axisymmetric bodies shown in Figs. 5.1.2*a* and 5.1.2*b* are well known; see, for example, Schlichting (1968) for the derivation. Neglecting the normal component of the buoyancy force B_n and the motion pressure field, these equations with the inclusion of buoyancy are

$$\frac{\partial(r^a u)}{\partial x} + \frac{\partial(r^a v)}{\partial y} = 0 \tag{5.1.11}$$

$$u\frac{\partial u}{\partial x} + v\frac{\partial u}{\partial y} = g\beta(t - t_\infty)\cos\theta + \nu\frac{\partial^2 u}{\partial y^2} \tag{5.1.12}$$

$$u\frac{\partial t}{\partial x} + v\frac{\partial t}{\partial y} = \alpha\frac{\partial^2 t}{\partial y^2} \tag{5.1.13}$$

where $a = 0$ for two-dimensional planar bodies and $a = 1$ for axisymmetric bodies.

These equations are derived under the usual assumptions of constant-property flow, the Boussinesq approximations, and the omission of compression, dissipative, and generation effects in the energy equation. The pressure variation across the boundary layer is omitted, since B_n is not included, along with a normal direction force-momentum balance. In addition, it is assumed that the boundary layer thickness is small compared to the local radius of the curvature of the surface (see Section 4.3). Some of these assumptions are not valid over the whole possible range of ξ, where $\xi = \pi/2 - \theta$. For example, for larger values of ξ the boundary layer may be thick enough that curvature and normal buoyancy effects must be included in the momentum and energy equations. This is discussed in Section 5.4.

The boundary layer equations above may again be shown to admit similarity solutions for some configurations. These and other solutions are discussed in the next three sections. The flat inclined surfaces, $-\pi/2 < \theta < \pi/2$, are discussed in Section 5.2. The horizontal flows are described in Section 5.3. The flows adjacent to symmetric two-dimensional and axisymmetric bodies are dealt with in the following section, including horizontal circular cylinders and spheres. Three-dimensional flows are also reviewed briefly. Correlations based on the experimental data are given in Section 5.5. This is followed by a consideration of the effect of density stratification of the ambient medium on the flow and transport characteristics. In many circumstances of interest, multiple flows occur and interact or a flow interacts with surfaces. These interactive flows are discussed in Section 5.7. The last section describes the mechanism of flow separation observed in horizontal and inclined flows.

5.2 FLAT INCLINED SURFACES

It was pointed out in Section 5.1 that the pressure term in the x-momentum equation may be neglected in obtaining boundary layer solutions for such flows if

$$\Delta \tan \theta << 1 \tag{5.2.1}$$

Then u, v, and t may be found from Eqs. (5.1.5), (5.1.7), and (5.1.8) only. Equation (5.1.6) may then be used to calculate the distribution of p_m, if desired. The solution is therefore identical to that for a vertical surface except that the buoyancy force is merely multiplied by cos θ locally. This result is sometimes referred to as the equivalent vertical plate solution. Thus, all the similarity solutions obtained in Chapter 3 for vertical surfaces may also be applied to inclined surfaces simply by replacing g in the Grashof number by $g \cos \theta$. That is, the Grashof number now is $\mathrm{Gr}_{x,\theta} = \mathrm{Gr}_x \cos \theta = (gx^3/\nu^2)\beta(t_0 - t_\infty) \cos \theta$.

This procedure seems to have been suggested first by Rich (1953), from experimental data taken at inclinations between 0° and 40°. These data are in agreement with the theoretical predictions for vertical flow when cos θ is introduced. Analytical and experimental findings of later investigators also confirm this procedure for calculating heat transfer in laminar flow for small and moderate values of θ.

There is, however, one serious shortcoming in this procedure: it does not distinguish between the flow behavior above and below an inclined surface. This difference arises in the sign of the buoyancy force B_n and the $\partial p_m/\partial y$ terms in the y direction, Eq. (5.1.6). For example, consider upper and lower surface flows for $t_0 > t_\infty$. Then B_n in Eq. (5.1.6) is directed away from the surface on an upper side and directed toward the surface on a lower side. That is, B_n tends to lift warmer fluid off an upper side but pushes it toward the surface on a lower side. This suggests higher flow velocities on the upper side than on the lower. For $t_0 < t_\infty$ the roles are reversed. Therefore, at larger values of θ, pressure terms in Eqs. (5.1.5) and (5.1.6) must be retained. The complete equations, Eqs. (5.1.5)–(5.1.8), must then be solved simultaneously to calculate the u, v, t, and p_m distributions. However, these equations then do not admit similarity solutions. Different and more complicated techniques must be used to obtain solutions.

Perturbation solutions. Kierkus (1968) sought an improved solution through a perturbation analysis of Eqs. (5.1.1) and (5.1.2) in conjunction with Eqs. (5.1.7) and (5.1.8). The flow quantities were expanded in powers of $(\mathrm{Gr}_x \cos \theta)^{1/4}$, with the simplest inclined boundary layer solution, for small values of θ, taken as the zeroth-order approximation. The velocity and temperature fields were calculated in detail for Pr = 0.7 for isothermal surfaces inclined up to 60°, both upward- and downward-facing surfaces, for $t_0 > t_\infty$. The temperature field was not found to be affected by the first-order perturbation correction. This supports Rich's procedure for determining the heat transfer rate. The effect on the velocity field, however, is significant. The main feature of the flow, when compared with the zeroth-order boundary layer solution, is the asymmetry above and below the surface due to B_n. The solution indicates a pressure gradient directed away from the surface above the plate and toward the surface below. This in turn induces a higher flow velocity above the plate than below. The experimental data of Kierkus for air, both above and below a warm surface, for angles of inclination up to 45° show good agreement with these analytical findings.

Despite this agreement, there is a discrepancy in the results of Kierkus. This was first pointed out by Riley (1975). In essence, Kierkus attempted to model the effects with a finite surface in the outer solution while retaining the solution for a semi-infinite surface in the inner region. As a result, the first-order inner and outer solutions are incompatible. Riley corrected the solution for the velocity field by reference to the work of Clarke (1973). Again, an asymmetry in the boundary layer flow arose between the upper and lower flow results. However, the results also supported the conclusion of Kierkus that the first-order perturbation terms have no effect on the heat transfer.

Lee and Lock (1972) solved the boundary layer equations numerically, for air, for $-30 \leq \theta \leq 90°$. For θ up to ±30°, the inclination was shown to have a very slight effect on both the velocity and temperature profiles.

Local nonsimilarity solution. Hasan and Eichhorn (1979) used the method of local nonsimilarity to obtain solutions for an inclined isothermal surface, varying θ from 0° to near horizontal, for a range of Prandtl numbers. For a detailed discussion of the general method, see Sparrow et al. (1970) and Sparrow and Yu (1971). The governing equations are (5.1.5)–(5.1.8) and were taken subject to the following boundary conditions:

$$\begin{aligned} &\text{at } y = 0: && u = v = 0 \qquad t = t_0 \\ &\text{as } y \to \infty: && u = \frac{\partial u}{\partial y} = t - t_\infty = 0 \end{aligned} \tag{5.2.2}$$

As a first step in obtaining a solution to these equations, the pressure term is eliminated in the x- and y-momentum equations through cross-differentiation to reduce Eqs. (5.1.5) and (5.1.6) to the following single equation:

$$u \frac{\partial^2 u}{\partial x \partial y} + v \frac{\partial^2 u}{\partial y^2} = \nu \frac{\partial^3 u}{\partial y^3} + g \cos\theta \frac{\partial t}{\partial y} - g\beta \sin\theta \frac{\partial t}{\partial x} \tag{5.2.3}$$

The stream function is defined by $u = \psi_y$ and $v = -\psi_x$ so that the continuity Eq. (5.1.7) is automatically satisfied. Equations (5.1.8) and (5.2.3) are then transformed from the x,y coordinate system to a ζ,η system, in terms of the stream function $f(\zeta,\eta)$ and dimensionless temperature $\phi(\zeta,\eta)$, where

$$\psi(x, y) = \nu c(x) f(\zeta, \eta) \tag{5.2.4}$$

$$t(x, y) = (t_0 - t_\infty)\phi(\zeta, \eta) + t_\infty \tag{5.2.5}$$

$$\eta(x, y) = b(x)y \qquad \zeta = \zeta(x) \tag{5.2.6}$$

The scaling functions $c(x)$, $b(x)$, and $\zeta(x)$ are determined from the requirement that x not appear explicitly in either the transformed equations or the boundary conditions. In addition, η is chosen so that it reduces to the true similarity variable for boundary region flows that are similar. For an inclined surface, η in Eq. (5.2.6) should reduce to the similarity variable for near-vertical flows, θ small, since the flow is then similar. Using these requirements, $c(x)$, $b(x)$, and $\zeta(x)$ are determined to be

$$c(x) = 4\left(\frac{\mathrm{Gr}_{x,\theta}}{4}\right)^{1/4} \qquad \mathrm{Gr}_{x,\theta} = \frac{gx^3}{\nu^2}\beta(t_0 - t_\infty)\cos\theta \tag{5.2.7}$$

$$b(x) = \frac{c}{4x} \tag{5.2.8}$$

$$\zeta(x) = \frac{\tan\theta}{(4\mathrm{Gr}_{x,\theta}/4)^{1/4}} = \frac{\tan\theta}{G_\theta} \tag{5.2.9}$$

Applying Eqs. (5.2.4)–(5.2.9) to Eq. (5.2.3) and integrating the transformed equation once with respect to η, the following equation in $f(\zeta, \eta)$ results:

$$f''' + 3ff'' - 2f'^2 + (1 + \eta\zeta)\phi - 3\zeta(f''f_\zeta - f'f'_\zeta)$$
$$- \zeta\left(\int_0^\eta \phi\, d\eta - \int_0^\infty \phi\, d\eta\right) + 3\zeta^2\left(\int_0^\eta \phi_\zeta\, d\eta - \int_0^\infty \phi_\zeta\, d\eta\right) = 0 \tag{5.2.10}$$

Similarly, the energy Eq. (5.1.8) and the boundary conditions, Eqs. (5.2.2), transform to

$$\phi'' + 3\mathrm{Pr} f\phi' = 3\mathrm{Pr}\zeta(\phi' f_\zeta - f'\phi_\zeta) \tag{5.2.11}$$

$$f(\zeta, 0) = f'(\zeta, 0) = \phi(\zeta, 0) - 1 = f'(\zeta, \infty) = \phi(\zeta, \infty) = 0 \tag{5.2.12}$$

If the inclination angle is zero, $\zeta = 0$, Eqs. (5.2.10)–(5.2.12) reduce to the similarity equations for a vertical isothermal surface. If $G_\theta >> \tan\theta$, for θ less than $\pm\pi/2$, ζ also approaches zero. Then the boundary layer equations used for near-vertical flows are recovered.

The solutions for nonzero values of ζ are obtained through a sequence of successive approximations. To the first approximation of inclination effects, the transformed Eqs. (5.2.10) and (5.2.11) are retained exactly, including the unknown terms $\partial f/\partial\zeta$ and $\partial\phi/\partial\zeta$. An additional pair of auxiliary equations is then developed to provide estimates for these terms. These are generated by simply differentiating Eqs. (5.2.10) and (5.2.11) with respect to ζ. To close the system of equations at this level of approximation, terms involving higher-order derivatives $\partial^2 f/\partial\zeta^2$ and $\partial^2\phi/\partial\zeta^2$ are neglected from these auxiliary equations. The transformed Eqs. (5.2.10) and (5.2.11) along with the auxiliary equations are then solved as ordinary differential equations with the variable ζ taking on the role of a parameter. If it is desired to obtain the solutions to the next order approximation, the same procedure is repeated. The terms involving $\partial^3 f/\partial\zeta^3$ and $\partial^3\phi/\partial\zeta^3$ are neglected from the second pair of auxiliary equations developed, to obtain an estimate of $\partial^2 f/\partial\zeta^2$ and $\partial^2\phi/\partial\zeta^2$. Generally, the solution obtained to first-order approximation, including only $\partial f/\partial\zeta$ and $\partial\phi/\partial\zeta$ terms, is quite accurate; see, for example, Hasan and Eichhorn (1979), Sparrow and Yu (1971), and Minkowycz and Cheng (1976).

Following this procedure and requiring a solution to the first level of approximation only, the auxiliary equations in $F = \partial f/\partial\zeta$ and $H = \partial\phi/\partial\zeta$, with the exclusion of $\partial^2 f/\partial\zeta^2$ and $\partial^2\phi/\partial\zeta^2$ terms, are

$$F''' + 3fF'' - f'F' + H(1 + \zeta_\eta) + \eta\phi - 3\zeta(FF'' - F'^2)$$
$$- \left(\int_0^\eta \phi\, d\eta - \int_0^\infty \phi\, d\eta\right) + 5\zeta\left(\int_0^\eta H\, d\eta - \int_0^\infty H\, d\eta\right) = 0 \qquad (5.2.13)$$

$$H'' + 3\text{Pr}(fH' + f'H) + 3\text{Pr}\zeta(F'H - FH') = 0 \qquad (5.2.14)$$

The boundary conditions for the complete set are

$$\begin{gathered} f(\zeta, 0) = f'(\zeta, 0) = F(\zeta, 0) = F'(\zeta, 0) = H(\zeta, 0) = \phi(\zeta, 0) - 1 = 0 \\ f'(\zeta, \infty) = F'(\zeta, \infty) = \phi(\zeta, \infty) = H(\zeta, \infty) = 0 \end{gathered} \qquad (5.2.15)$$

Equation (5.2.14) is a homogeneous equation linear in H with homogeneous boundary conditions. Therefore, the solution is trivial, that is, $H(\zeta, \eta) \equiv 0$. Equations (5.2.10), (5.2.11), and (5.2.13) with the boundary conditions, Eq. (5.2.15), are then solved by the method of successive approximation, as described in detail by Hasan (1978), Minkowycz and Cheng (1976), and Eckert et al. (1958). Hasan and Eichhorn (1979) obtained these solutions for Pr = 0.1, 0.7, 6.0, and 275 for various values of ζ. The numerical values of $f''(\zeta, 0)$ and $\phi'(\zeta, 0)$ are presented in Table 5.2.1. Tabulated values of ζ end at the maximum values at which convergent solutions could be obtained, for each value of the Prandtl number.

The $-\phi'(\zeta, 0)$ results in the table confirm again that the effect of ζ on heat transfer is negligible. Therefore, as mentioned before, the local Nusselt number for inclined surfaces, facing either upward or downward, may be correlated by the usual vertical plate formula by using $\text{Gr}_x \cos\theta$ instead of Gr_x. This is not true for very large angles of inclination, very close to the horizontal, or for extremely low Prandtl numbers. This is shown in Fig. 5.2.1 as a plot of $\phi'(\zeta, 0)/\phi'(0, 0)$ versus ζ. The calculated temperature profiles showed similar trends of difference.

The $f''(\zeta, 0)$ results and the computed velocity profiles, on the other hand, indicate appreciable effects of ζ on the flow field. For the upward-facing surface, for $B_n > 0$, the downstream velocity u was found to be higher than that for the equivalent vertical surface, that is, when g is replaced by $g \cos\theta$. For the downward-facing surface, for $B_n < 0$, u was lower. See Fig. 5.2.2 for typical velocity profiles for Pr = 6.0 and 0.7. Recall that for an equivalent vertical surface, there

Table 5.2.1 Local nonsimilarity solutions for inclined isothermal surfaces

		Upward-facing surface		Downward-facing surface	
Pr	ζ	$-\phi'(\zeta, 0)$	$f''(\zeta, 0)$	$-\phi'(\zeta, 0)$	$f''(\zeta, 0)$
0.1	0.00	0.2304	0.85914	0.2304	0.85914
	0.05	0.22875	1.08479	0.22965	0.67582
	0.10			0.22755	0.50756
	0.15			0.22428	0.34425
	0.20			0.21969	0.17865
0.7	0.00	0.49951	0.67891	0.49951	0.67891
	0.05	0.49924	0.74683	0.49928	0.61393
	0.10	0.49831	0.81934	0.49866	0.55096
	0.15	0.49633	0.89976	0.49768	0.48934
	0.20	0.49203	0.99839	0.49633	0.42860
	0.30			0.49285	0.30831
	0.40			0.48798	0.18740
6.0	0.00	1.00740	0.46842	1.00740	0.46842
	0.10	1.00721	0.50712	1.00723	0.42329
	0.20	1.00663	0.55093	1.00677	0.38288
	0.30	1.00555	0.59508	1.00602	0.34164
	0.40	1.00378	0.64185	1.00501	0.30126
275	0.00	2.83580	0.19679	2.83580	0.19670
	0.10	2.83580	0.20290	2.83580	0.19068
	0.20	2.83578	0.20902	2.83578	0.18458
	0.30	2.83574	0.21514	2.83574	0.17848
	0.40	2.83571	0.22126	2.83571	0.17238
	0.60	2.83559	0.23352	2.83559	0.16018
	1.00	2.83521	0.25809	2.83523	0.13582

Source: From Hasan and Eichhorn (1979).

is no difference in transport above and below the surface. The difference actually shown in Fig. 5.2.2 again arises from the change in direction of dynamic pressure in the y-momentum equation, for upward- and downward-facing surfaces.

Also shown in Fig. 5.2.2 is a flow reversal in the outer portion of the boundary layer observed for a downward-facing heated surface for large values of ζ. With an increase in the value of ζ, or equivalently the angle of inclination θ, the region of reversed flow moves closer to the wall. This flow reversal was observed only at lower values of Pr, 0.1 and 0.7.

The direction of flow in the boundary layer depends to a large extent on the direction of the net buoyancy force across the region. From Eq. (5.2.10) the net buoyancy force is $\phi(1 + \eta\zeta) - \zeta(\int_0^\eta \phi \, d\eta - \int_0^\infty \phi \, d\eta)$, where, for a downward-facing heated surface, ζ is negative. Near the heated surface, the buoyancy force is positive. Then, depending on the value of ζ, it goes to zero at some larger value of η. Thereafter it is negative, approaching zero as $\eta \rightarrow \infty$. The velocity profiles are determined by the buoyancy force distribution, modulated by inertia and shear

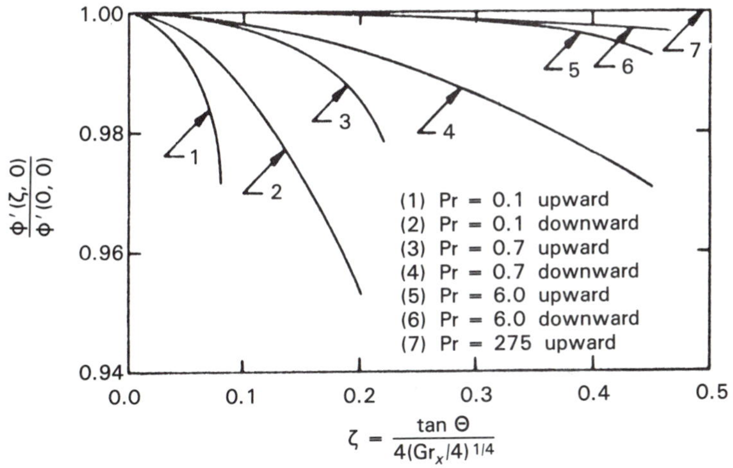

Figure 5.2.1 Ratio of local Nusselt number for the inclined surface to that for the equivalent surface. *(Reprinted with permission from M. M. Hasan and R. Eichhorn,* J. Heat Transfer, *vol. 101, p. 642. Copyright © 1979, ASME.)*

effects. Whether the flow reversal will occur also depends on Pr. The thermal boundary layer is much thicker for low Pr than for high Pr and contributes much more to the negative buoyancy force term in Eq. (5.2.10). At high Pr the velocity profile is much thicker than the temperature profile and dominates the flow.

Another interesting feature of these results is that the effect of the parameter ζ on the heat transfer rate and wall shear stress decreases with increasing Prandtl number. In the limiting condition $\Pr \to \infty$, the transformations $\lambda = \Pr^{1/4}\eta$, $f_1 = \Pr^{3/4} f$, $F_1 = \Pr^{3/4} F$ indicate that Eq. (5.2.13) reduces to $F_1''' = 0$. The boundary conditions are $F_1(\zeta, 0) = F_1'(\zeta, 0) = F_1'(\zeta, \infty) = 0$. This gives $F_1 = 0$, and Eqs. (5.3.10) and (5.3.11) reduce to

$$f_1''' + \phi = 0 \tag{5.2.16}$$

$$\phi'' + 3f_1\phi' = 0 \tag{5.2.17}$$

In this transformation the dependence on Pr is the same as that used in Section 3.8 in the matched asymptotic expansions for vertical flows for $\Pr \to \infty$. Equations (5.2.16) and (5.2.17) do not contain ζ. Therefore, there is no ζ effect on either the flow or the thermal field. These equations are the same as the zeroth-order equations, as given in Section 3.8 for a vertical surface for $\Pr \to \infty$, except for a different constant in Eq. (5.2.17), due to the transformation. The solutions for the temperature and flow fields, therefore, are the same as those in Section 3.8, when Gr_x is replaced by $\mathrm{Gr}_x \cos\theta$.

Other analytical and experimental studies. Other approximate solutions for inclined surface transport are due to Sugawara and Michiyoshi (1955), Fussey and Warneford (1976), and Zariffeh and Daguenet (1981). Sugawara and Michiyoshi solved pertinent boundary layer equations around a long horizontal narrow strip

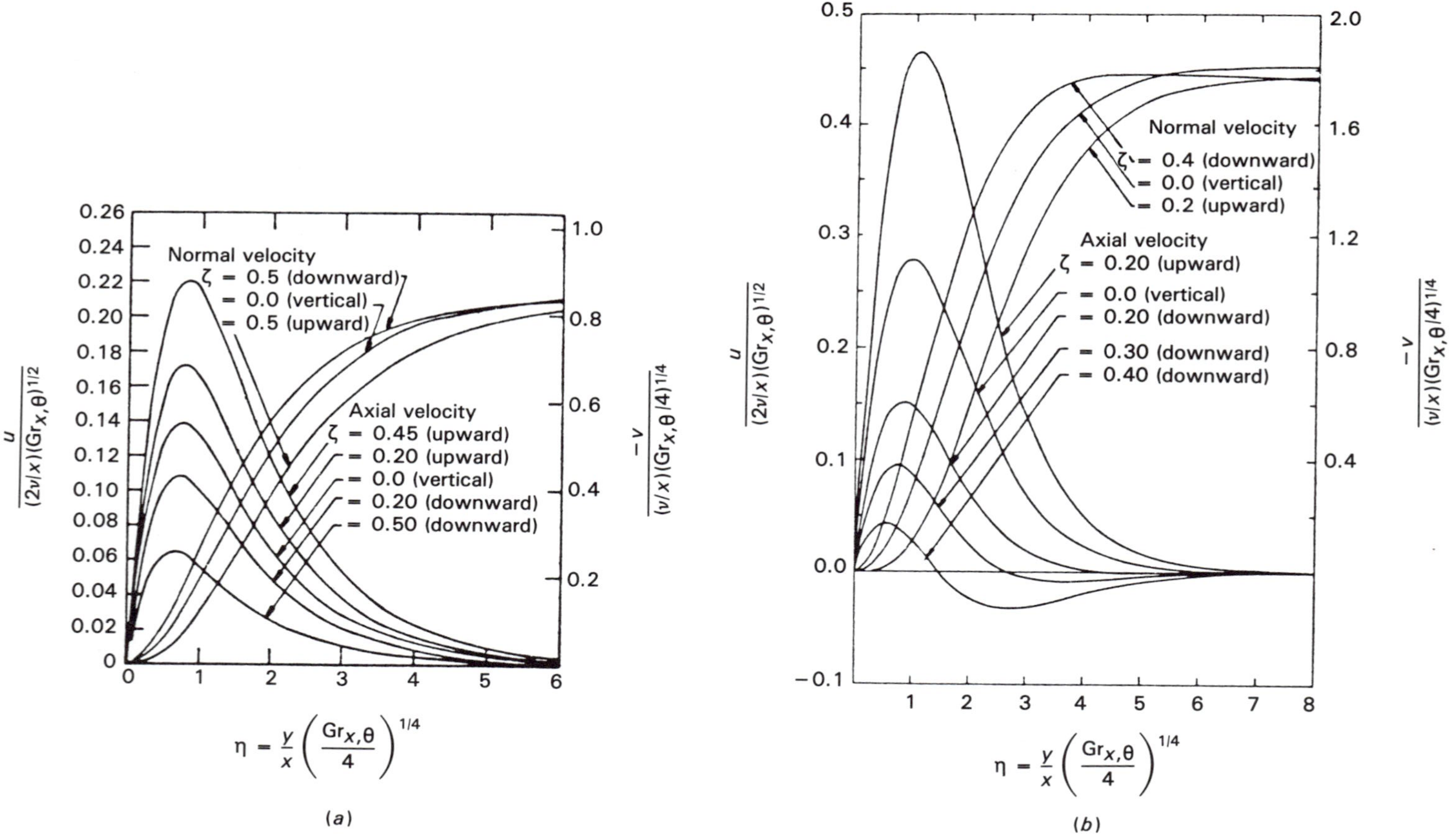

Figure 5.2.2 Dimensionless axial and normal velocity profiles for an inclined surface. (*a*) Pr = 6.0; (*b*) Pr = 0.7. *(Reprinted with permission from M. M. Hasan and R. Eichhorn,* J. Heat Transfer, *vol. 101, p. 642. Copyright © 1979, ASME.)*

inclined from the vertical. It was approximated by a flat elliptical cylinder. Heat transfer coefficients were found to be larger than those measured by Rich (1953) at values of $\theta > 75°$. Fussey and Warneford used an integral analysis for transport above an inclined uniform heat flux plate. Zariffeh and Daguenet proposed a new nonsimilar method to analyze inclined surface transport with a prescribed heat flux. The transformed boundary region equations were solved by a series expansion. However, the authors pointed out that the normal direction momentum equation had been omitted, along with the pressure term in the x-momentum equation. Thus, the applicability of this solution to larger angles of inclination appears questionable.

There have also been several experimental investigations in addition to those of Rich (1953) and Kierkus (1968) mentioned before. Hassan and Mohamed (1970) measured the local heat transfer rates from an inclined isothermal plate in air for both $t_0 > t_\infty$ and $t_0 < t_\infty$ and for $-\pi/2 \leq \theta \leq \pi/2$. The data were in good agreement with the equivalent vertical surface results for θ from 0 to $-75°$, $B_n < 0$, and up to 60° for positive inclinations, $B_n > 0$. The measurements by Fussey and Warneford (1978) for water adjacent to a uniform heat flux surface, for θ from 0 to $-86.5°$, indicated that the equivalent vertical surface result applied to $\theta = -70°$.

The detailed measurements of Fujii and Imura (1972) for water with $\mathrm{Pr} \approx 5$, on the other hand, suggest that equivalent vertical plate correlations can be applied when B_n is toward the surface for θ up to $-85°$. For B_n away from the surface, flow separation occurred downstream, and the extent of the plate over which the flow remained laminar decreased with increasing angle of inclination. However, the transport in the attached region again followed the equivalent vertical plate results.

Some laminar flow data are also available from the studies of turbulent natural convection by Vliet (1969) for air and water, Vliet and Ross (1975) for air, and Black and Norris (1975) for air, and from the study of downstream transition by Skaukatullah and Gebhart (1978) for water. The collective conclusion that may be drawn from the laminar heat transfer data in these studies is that the equivalent vertical plate correlations are applicable for inclined surfaces, with both upward and downward heating, for θ up to $\pm 60°$ in air and water.

Transport data for inclined surfaces in high Prandtl number fluids have been reported by Emery et al. (1976). The experiments were for uniform heat flux upward- and downward-facing inclined surfaces, using water-Pluracol solutions with Prandtl numbers of 270 to 1020. In the range of θ from 0 to $\pm 30°$, their data show an insignificant effect of inclination on the temperature field, in contrast to a noticeable effect on the velocity field. The results are in good agreement with the high Prandtl number solutions of Hasan and Eichhorn (1979).

Mass transfer measurements in inclined flows have been reported by Lloyd et al. (1972) and Moran and Lloyd (1975). In the former work, the data were taken for an upward-facing surface with $0 \leq \theta \leq 20°$, $B_n < 0$. In the latter, θ was varied up to 60° for a downward-facing surface, $B_n > 0$. An electrochemical technique, using an aqueous solution of reagent grade cupric sulfate in reagent grade sulfuric acid, was employed. The Schmidt number was approximately 2000. The

Sherwood number data in both studies again supported the use of vertical surface results with Gr_x replaced by $Gr_x \cos \theta$.

Summary. The analyses and experimental results indicate that Rich's procedure for near-vertical flows, that is, replacement of g by $g \cos \theta$ in the Grashof number, is appropriate for calculating heat transfer rates for even large angles of inclination from the vertical, for B_n both toward and away from the surface. The limit appears to be at least $\theta = 70°$ when B_n is toward the surface. At larger inclinations, with B_n directed outward, flow separation has been found to occur at some distance downstream from the leading edge. This distance is smaller at larger inclinations. Separation and related matters are further discussed in Section 5.8. However, the heat transfer coefficient up to flow separation may apparently again be correlated by the equivalent vertical surface correlations, to θ values of at least 60°.

The flow field is not, however, correctly predicted by Rich's approximation at larger values of θ. The local nonsimilarity solutions of Hasan and Eichhorn (1979) may then be used.

The Prandtl number seems to have an effect on the extent to which the temperature and flow field for inclined surfaces deviate from the equivalent vertical surface solutions. With an increase in Prandtl number, the differences decrease. In the limit $Pr \to \infty$, the angle of inclination affects the heat transfer rate and flow field only as $g \cos \theta$ and equivalent vertical surface solutions then apply.

5.3 HORIZONTAL AND SLIGHTLY INCLINED FLOWS

In studying horizontal flows, there is a temptation to consider them as merely the limiting mechanisms in the general formulation of inclined flows in Section 5.2. However, as discussed in Section 5.1, this is not a viable approach. For $\theta = 90°$, B_t vanishes in the x-momentum equation. The driving force for the motion is then entirely due to B_n. It results in only a vertical motion or dynamic pressure gradient, $\partial p_m/\partial y$. This indirectly drives the flow through the x component $\partial p_m/\partial x$. This, in turn, is the only motive force that drives the flow. This is the basic difference between directly and such indirectly driven flows.

This peculiarity has resulted in extensive investigations of flows adjacent to horizontal surfaces. As for inclined surfaces, two distinct flow configurations arise: (1) a warm surface facing upward or equivalently a cold surface facing downward, and (2) a cold surface facing upward or a warm surface facing downward. For (1), $\theta > 0$ and B_n acts away from the surface while for (2), $\theta < 0$ and B_n is toward the surface. These two cases are discussed next. For (1), the flow adjacent to surfaces slightly inclined from the horizontal is also discussed.

5.3.1 Semi-Infinite Horizontal Flows, B_n Away from the Surface

This idealized flow pattern for $t_0 > t_\infty$ is shown in Fig. 5.3.1a. The surface heats the fluid, which, being higher, then tends to rise. This results in a positive normal pressure gradient $\partial p_m/\partial y$. See Eq. (5.1.6). Since $p_m \simeq 0$ in the quiescent me-

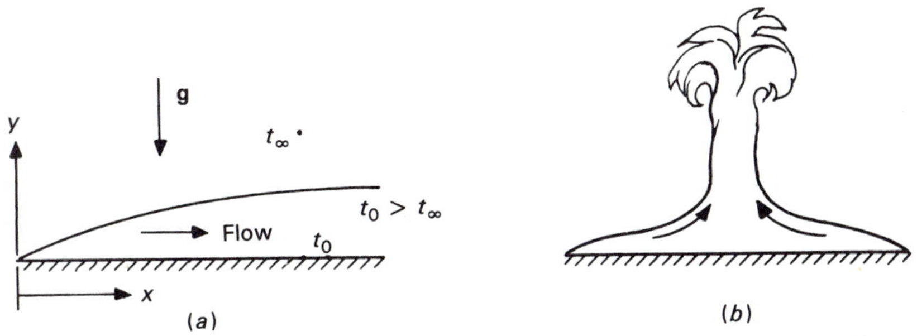

Figure 5.3.1 Idealized natural convection flow over hot horizontal surfaces facing upward. (*a*) Semi-infinite surface; (*b*) finite surface.

dium, a negative horizontal gradient $\partial p_m/\partial x$ therefore arises in the buoyant layer adjacent to the surface. This in turn drives the developing boundary layer flow along the surface, starting from the leading edge. Downstream, perhaps at some critical value of the Grashof number, the flow begins to separate away from the surface. A longitudinal roll mechanism was seen by Pera and Gebhart (1973b).

That such flow patterns exist has been confirmed by several visual observations. The early experimental investigations of Schmidt (1932) and Weise (1935) in air showed that boundary layer flow might be expected near the leading edge. Later, the observations of Rotem and Claassen (1969) with a semifocusing Schlieren system and of Pera and Gebhart (1973a) with an interferometer clearly indicated the existence of a laminar boundary layer near the leading edge, followed by flow separation. An interferogram of Pera and Gebhart for a horizontal surface is shown in Fig. 5.3.2*a*. In this experiment there were walls on both lateral sides of the heated surface to prevent any incoming flow. This ensured essentially two-dimensional flow, conforming to the idealized pattern envisioned in Fig. 5.3.1*a*.

The first analysis of laminar boundary layer transport over a semi-infinite horizontal surface was by Stewartson (1958). A sign mistake in the analysis led to the erroneous conclusion that the boundary layer similarity formulation applied for B_n toward the surface. Later, Gill et al. (1965) and then Rotem (1967) showed that the leading edge onflow is generated only over a warm surface facing upward or a cold surface facing downward, that is, for B_n away from the surface. It is the $\partial p_m/\partial x < 0$ that indirectly drives an onflow. Rotem and Claassen (1969) obtained similarity equations for this flow case for a power law surface temperature variation. Results were presented for the horizontal surface at uniform temperature, for some specific values of Prandtl number. The limits for $\Pr \to 0$ and $\Pr \to \infty$ were also given. Fatt (1976) used an integral method to determine the local Nusselt number.

The similarity analysis is like that given in Chapter 3 for vertical flows. In addition to the stream and temperature functions, defined in Eqs. (3.3.7) and (3.3.8), $\psi = \nu c(x)f(\eta)$, $t - t_\infty = \phi(\eta)\, d(x)$, a pressure function also arises:

$$p_m = P(\eta)a(x) \tag{5.3.1}$$

where $P(\eta)$ is in terms of η. Proceeding exactly as in Chapter 3, similarity was

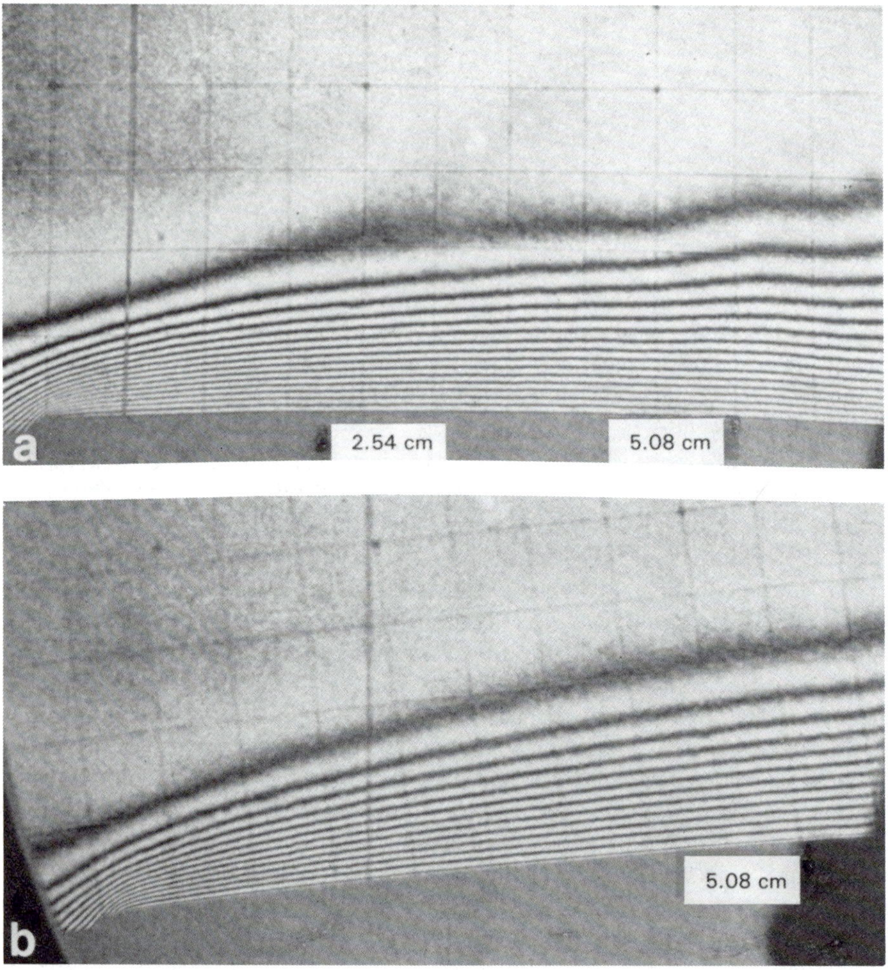

Figure 5.3.2 Interferograms of the boundary region over uniform-temperature horizontal and slightly inclined surfaces. (*a*) $\xi = 0°$, $\Delta t = 11.8°C$; (*b*) $\xi = 6°$, $\Delta t = 14.2°C$. *(Reprinted with permission from* Int. J. Heat Mass Transfer, *vol. 16, p. 1131, L. Pera and B. Gebhart. Copyright © 1973, Pergamon Journals Ltd.)*

shown for the power law surface temperature variation $t_0 - t_\infty = d(x) = Nx^n$. Then

$$c = 5\left(\frac{\mathrm{Gr}_x}{5}\right)^{1/5} \qquad b = \frac{c}{x} \qquad \text{and} \qquad a = 5\frac{\nu^2\rho}{x}\left(\frac{\mathrm{Gr}_x}{5}\right)^{4/5} \tag{5.3.2}$$

The following set of ordinary differential equations and boundary conditions in f, ϕ, and P result.

$$f''' + (n+3)ff'' - (2n+1)f'^2 - \frac{4n+2}{5}P - \frac{n-2}{5}\eta P' = 0 \tag{5.3.3}$$

$$P' = \phi \tag{5.3.4}$$

$$\phi'' + \text{Pr}[(n + 3)f\phi' - 5nf'\phi] = 0 \tag{5.3.5}$$

$$f(0) = f'(0) = 1 - \phi(0) = f'(\infty) = \phi(\infty) = P(\infty) = 0 \tag{5.3.6}$$

For an isothermal surface, $n = 0$. The uniform flux surface condition corresponds to $n = \frac{1}{3}$. This follows from the consideration of the total rate of thermal energy $Q(x)$ convected in the boundary layer at any location x,

$$Q(x) = \int_0^\infty \rho c_p u(t - t_\infty)\, dy = 5\rho c_p \nu N I \left[\frac{g\beta N}{5\nu^2}\right]^{1/3} x^{(6n+3)/5} \tag{5.3.7}$$

where $I = \int_0^\infty \phi f'\, d\eta$. This is a function of Pr and n. For $n = \frac{1}{3}$, Q increases linearly with x, which implies a uniform flux surface. In general, Eq. (5.3.7) can be used to determine the value of exponent n for the imposed condition in which Q may vary with x. For example, the condition that Q be independent of x requires that $n = -\frac{1}{2}$. This is the case when the surface is adiabatic and all the convected energy is released by a line source at the leading edge. This formulation is likely to be physically unrealistic since a boundary layer flow along the surface may not arise.

Numerical solutions of Eqs. (5.3.3)–(5.3.6) for a uniform temperature horizontal surface, $n = 0$, have been obtained by Rotem and Claassen (1969). Those for a uniform flux surface, $n = \frac{1}{3}$, were obtained by Pera and Gebhart (1973a). Prandtl number ranged from 0.1 to 10. The velocity, temperature, and pressure functions were computed for these Prandtl numbers and are shown in Figs. 5.3.3–5.3.5. As for vertical flows, the curves indicate a decrease in velocity and thinning of the thermal layer with increasing Prandtl number. However, the decrease in velocity arises indirectly, as a result of decrease in the motion-driving pressure gradient seen in Fig. 5.3.5.

The local Nusselt number Nu_x is obtained from

$$\text{Nu}_x = \frac{q''(x)}{t_0 - t_\infty}\frac{x}{k} = [-\phi'(0)]\frac{c(x)}{5} = [-\phi'(0)]\left(\frac{\text{Gr}_x}{5}\right)^{1/5} = F(\text{Pr})\text{Gr}_x^{1/5} \tag{5.3.8}$$

The values of $-\phi'(0)$ for $n = 0$ and $n = \frac{1}{3}$ and Prandtl numbers from 0.1 to 10 are given in Table 5.3.1. Values of the other important functions, $f''(0)$, $f(\infty)$, and $P(0)$ are also listed.

On the basis of the heat transfer results in Table 5.3.1, Pera and Gebhart (1973a) gave the following two correlations for Prandtl numbers around 1.0:

$$\text{Nu}_x = 0.501(\text{Gr}_x)^{1/5}(\text{Pr})^{1/4} \quad \text{for a uniform flux surface} \tag{5.3.9}$$

$$\text{Nu}_x = 0.394(\text{Gr}_x)^{1/5}(\text{Pr})^{1/4} \quad \text{for an isothermal surface} \tag{5.3.10}$$

Equations (5.3.9) and (5.3.10) indicate the relative weakness of horizontal flows; the Grashof number appears as the fifth root as opposed to the fourth root for vertical flows. Thus, the heat transfer to the boundary region adjacent to a horizontal semi-infinite surface is lower than that from a vertical surface for the same fluid and temperature conditions. For example, the heat transfer at the location

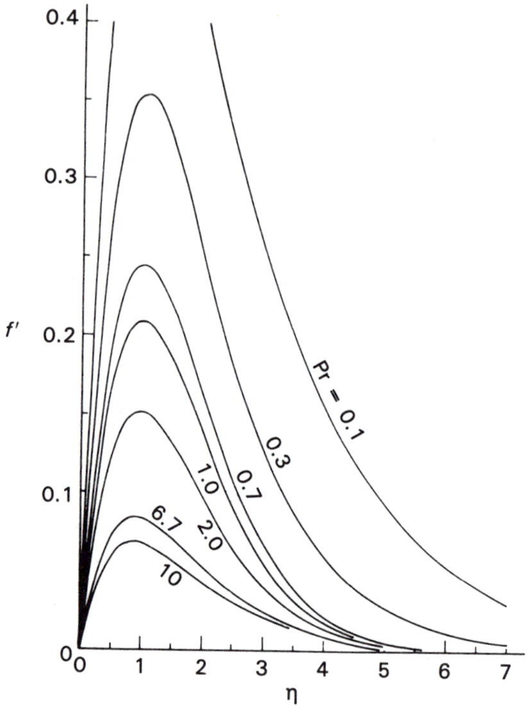

Figure 5.3.3 Computed velocity function for a uniform flux horizontal surface for a range of Prandtl numbers. *(Reprinted with permission from* Int. J. Heat Mass Transfer, *vol. 16, p. 1131, L. Pera and B. Gebhart. Copyright © 1973, Pergamon Journals Ltd.)*

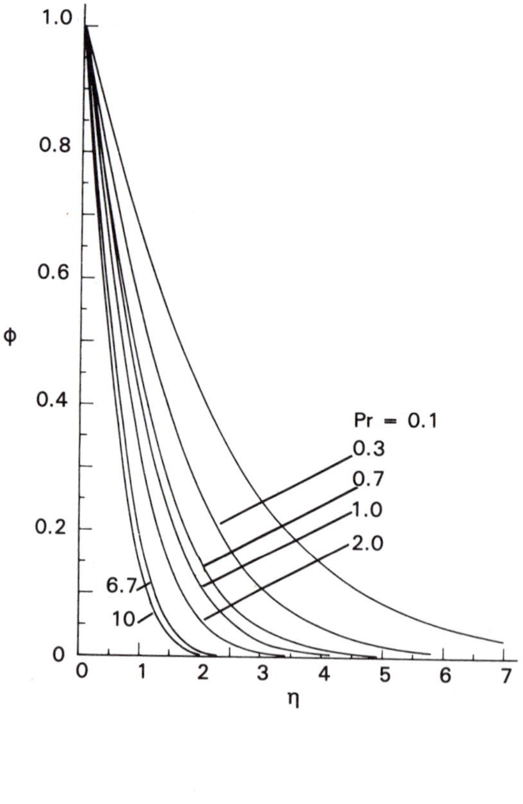

Figure 5.3.4 Computed temperature function for a uniform flux horizontal surface, for a range of Prandtl numbers. *(Reprinted with permission from* Int. J. Heat Mass Transfer, *vol. 16, p. 1131, L. Pera and B. Gebhart. Copyright © 1973, Pergamon Journals Ltd.)*

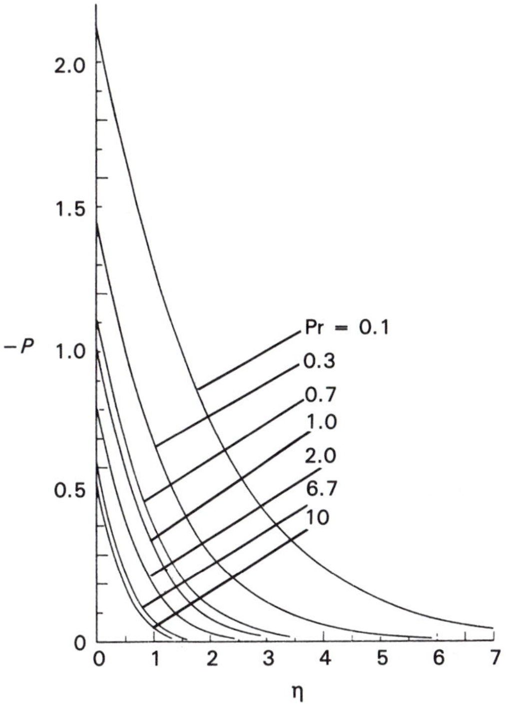

Figure 5.3.5 Computed pressure function for a uniform flux horizontal surface for a range of Prandtl numbers. *(Reprinted with permission from* Int. J. Heat Mass Transfer, *vol. 16, p. 1131, L. Pera and B. Gebhart. Copyright © 1973, Pergamon Journals Ltd.)*

where Gr_x is 10^5 is about 80% higher for a vertical surface than for a horizontal one.

The isothermal results in Table 5.3.1, along with Eq. (5.3.10), are plotted in Fig. 5.3.6. Also shown is the asymptotic behavior for very large and very small Prandtl numbers, determined by Rotem and Claassen (1969).

The similar solutions discussed above have also been extended to include the effects of blowing or transpiration through the surface. For an imposed injection velocity $v(x, 0)$ at the surface, similarity results for a power law surface temperature variation if

$$v(x, 0) \propto c_x \propto x^{(n-2)/5}$$

The choice of n in $d(x)$ fixes the distribution of $v(x, 0)$ in x, required for similarity. Clarke and Riley (1975) considered blowing for $n = 0$, assuming sufficiently weak blowing velocities so that the boundary layer assumptions remain applicable. The Boussinesq approximation was removed in the continuity equation through use of the Howarth transformation. Variable gas properties were taken, assuming that thermal conductivity and viscosity are proportional to absolute temperature. Flows for which the blowing velocity is comparable to the maximum buoyancy-induced velocity throughout the field were analyzed by Smith and Riley (1979) for horizontal as well as inclined and vertical surfaces, for $n = 0$. Similar solutions do not exist for these flows.

Finite-difference solutions for a heated strip have been obtained by Goldstein and Lau (1983). No boundary layer assumptions were made. For $40 < \mathrm{Ra} < 8 \times 10^3$ and $\mathrm{Pr} = 0.7$, the heat transfer rate was correlated by $\mathrm{Nu} = 0.621\mathrm{Ra}^{1/5}$. The calculated value is only about 4% lower than that given by the similarity analysis of Pera and Gebhart (1973a).

Horizontal disk flows. These are radially symmetric flows generated over extensive horizontal surfaces due to radially symmetric surface conditions. They are also called axisymmetric horizontal flows. The first analysis of such flows was by Lugt and Schwiderski (1966). The infinite surface was locally heated or cooled in an axisymmetric manner; see Fig. 5.3.7*a*. The resulting buoyant flow was assumed to vanish asymptotically far upstream, that is, out from the origin. For $t_0 > t_\infty$, the flow is directed inward and rises as a plume near the axis of symmetry. This is referred to here as onflow. Streamlines are shown in Fig. 5.3.7*b*. For $t_0 < t_\infty$, $B_n < 0$ and the motion near the axis is directed toward the surface. Farther out from the axis, the flow is horizontal and outward. This is off-flow. Using boundary layer approximations and local series expansion around the axis of symmetry, results were presented for velocity, pressure, temperature profiles, and heat transfer for $\mathrm{Pr} = 0.7$ and 7.0. For off-flows, the convective flow diminishes in spatial extent with increasing Grashof number but increases in strength, a behavior typical of boundary layer flows. Off-flows always yield stable motions. For onflows, on the other hand, the convective flow grows in strength and extent with an increase in Grashof number. Beyond a certain value of the Grashof number, the flow is thermally unstable. Also, onflows do not remain of boundary region form at smaller r.

A similarity formulation for these disklike boundary layer flows was first given by Rotem and Claassen (1970), for the surface maintained at a uniform temperature. Only off-flows were considered. However, no numerical results were given. Blanc and Gebhart (1974) considered these flows for a generalized surface temperature variation. The boundary layer equations were shown to admit a similarity solution for a power law surface temperature variation $t_0 - t_\infty = Nx^n$. However, the physically realistic solutions lie in the range $-\frac{1}{2} \leq n \leq 2$ for $t_0 > t_\infty$ and $-\frac{4}{3} \leq n \leq -\frac{1}{2}$ for $t_0 < t_\infty$. Exact solutions for several disk and plane flows were also discussed.

Qualitative observations of the flow above a heated finite disk, $t_0 > t_\infty$, indicated that the above considerations for physically realistic horizontal radial flows on an extensive surface may not apply to small disks. Both radially inward flow and flow instability were observed. Leading edge effects and thermal instability mechanisms for unstable stratification contribute to these observed phenomena. For $t_0 < t_\infty$, stable and outward flows are expected even for small disks.

5.3.2 Finite Horizontal Surfaces, B_n Away from the Surface

Stewartson (1958) extended the analysis for a semi-infinite surface to include the flow over a finite surface (Fig. 5.3.1*b*). The analysis is also applicable to a rect-

Table 5.3.1 Numerical results for an isothermal and a uniform flux horizontal surface

	$f(\infty)$		$f''(0)$		$P(0)$		$\phi'(0)$	
Pr	$n = 0$	$n = \frac{1}{3}$	$n = 0$	$n = \frac{1}{3}$	$n = 0$	$n = \frac{1}{3}$	$n = 0$	$n = \frac{1}{3}$
0.1	1.94307	1.66641	1.06644	1.16319	−2.4387	−2.1265	−0.27154	−0.36075
0.3	1.04146	0.89999	0.71535	0.76709	−1.6626	−1.4464	−0.38450	−0.50250
0.7	0.64613	0.54740	0.51910	0.54990	−1.26831	−1.10162	−0.48905	−0.63091
1.0	0.54599	0.46634	0.45497	0.47875	−1.1349	−0.99008	−0.54091	−0.69184
2.0	0.39715	0.33804	0.34994	0.36528	−0.9300	−0.81516	−0.64711	−0.81963
6.7		0.18825		0.22895		−0.60234		−1.08201
10.0	0.21916	0.15750	0.19246	0.19615	−0.6227	−0.54842	−0.95295	−1.18237

Source: From Rotem and Claassen (1969), for isothermal surface, and from Pera and Gebhart (1973a), for uniform flux surface.

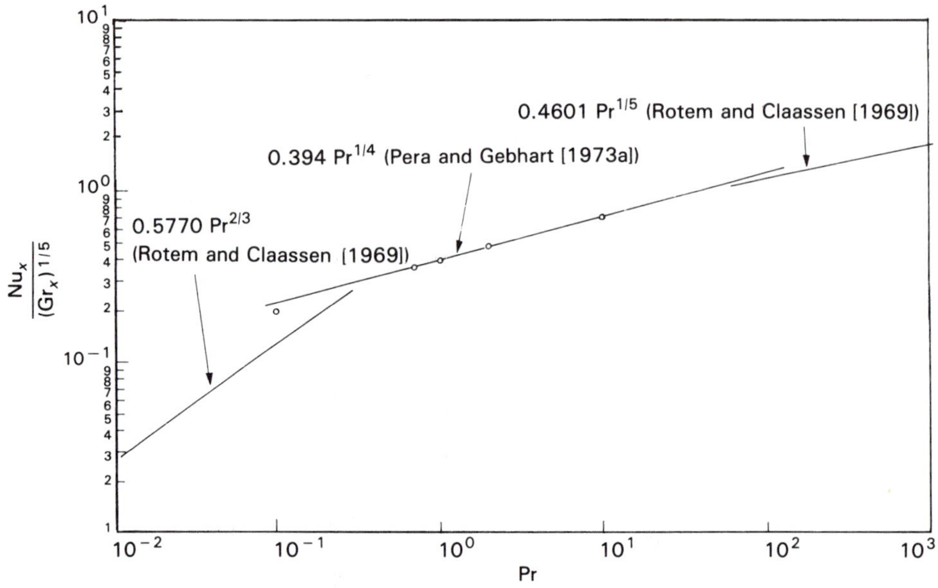

Figure 5.3.6 Heat transfer in laminar natural convection from an isothermal horizontal surface.

angular surface of large planform ratio as shown in Fig. 5.3.8. The aspect ratio is L_2/L_1. The width may be assumed large to exclude two of the edge side flows. Two boundary layers then originate at opposite edges of the plate, somewhat independent of each other. These proceed toward the center, where the two streams merge to form a buoyant plane plume, rising above the surface. Except for the

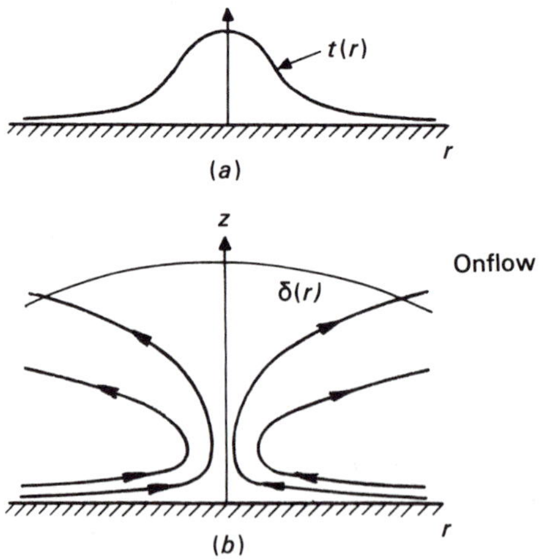

Figure 5.3.7 Convection flow caused by local heating of a surface, around the origin. (*a*) Temperature distribution; (*b*) streamlines. The direction of motion is reversed for a cooled surface. *(Reprinted with permission from H. J. Lugt and E. W. Schwiderski,* J. Atmos. Sci., *vol. 23, p. 54. Copyright © 1966, Cambridge University Press.)*

center region, the inflows are essentially the same as over a semi-infinite surface. The analytical results discussed above may then be applied.

However, for smaller aspect ratio L_2/L_1 such analysis becomes inaccurate; see, for example, Goldstein et al. (1973). A comparison between predictions of boundary layer theory and several sets of experimental data in air, for square, rectangular, and circular plates, showed large discrepancies. Actual heat transfer was consistently above the calculated values. Also, the experimental Nusselt numbers were found to vary with one-fourth power of the Rayleigh number, whereas boundary layer theory predicts the one-fifth power; see Eq. (5.3.8). In the low Rayleigh number range, $10 < \text{Ra} < 500$, an exponent of one-sixth was found to correlate the experimental data better. Later experiments of Goldstein and Lau (1983) supported the use of the one-fifth power law in calculating heat transfer rates at low Ra.

These discrepancies are thought to be due to the effects of the additional edge flows, neglecting fluid property variations in the boundary region, interaction of the flow in the middle of the plate, and the exclusion of higher-order effects in the analysis. Ackroyd (1976) evaluated the first two of these for horizontal plates of rectangular planform. First, boundary layer analysis for a semi-infinite surface was extended to include variable property effects. Detailed calculations were presented for air and water flows. The analysis was then formulated to calculate the heat transfer rates on horizontal surfaces of rectangular planform, as in Fig. 5.3.8. The assumed model of the boundary region flow is in accord with the flow visualization experiments of Husar and Sparrow (1968) above heated horizontal surfaces of various planforms in water. Four independent boundary region flows are postulated, coming in from the four edges of the plate. Confluence of these flows is assumed to occur along the lines *AB*, *BC*, *DE*, *EF*, and *BE*. The flows are

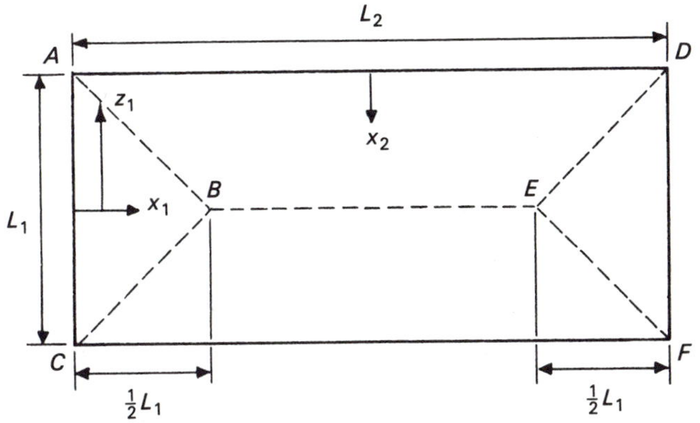

Figure 5.3.8 Model of boundary region flow on a rectangular plate. *(Reprinted with permission from J. A. D. Ackroyd,* Proc. R. Soc. London Ser. A, *vol. 352, p. 249. Copyright © 1976, The Royal Society.)*

assumed to separate along these lines and to rise. If $\bar{q}''$ denotes the average heat transfer rate per unit area of the top surface of the plate, then

$$\bar{q}'' = \frac{4\int_0^{L_1/2} 2q''z_1\,dx_1 + 2(L_2 - L_1)\int_0^{L_2/2} q''\,dx_2}{L_1L_2} \tag{5.3.11}$$

where $q'' = \mathrm{Nu}_x(t_0 - t_\infty)k/x$ is the local surface heat transfer rate for a semi-infinite horizontal surface. When property variations across the boundary region are negligible, the value of Nu_x is given by Eqs. (5.3.9) and (5.3.10) or Table 5.3.1. With variable property effects included, $\mathrm{Nu}_x\mathrm{Gr}_x^{-1/5}$ values may be obtained for air and water from the tabulated results of Ackroyd.

Defining average Nusselt number and Grashof number on the length scale L_1 as $\mathrm{Nu}_{L_1} = \bar{q}''L_1/k(t_0 - t_\infty)$ and $\mathrm{Gr}_{L_1} = gL_1^3\beta(t_0 - t_\infty)/\nu^2$, Eq. (5.3.11) yields the following relationship between the average and local values of Nusselt number:

$$\frac{\mathrm{Nu}_{L_1}}{\mathrm{Nu}_x} = \frac{5}{3}2^{2/5}\left(1 + \frac{1}{4}\frac{L_1}{L_2}\right)\left(\frac{\mathrm{Gr}_{L_1}}{\mathrm{Gr}_x}\right)^{1/5} \tag{5.3.12}$$

For $L_1/L_2 \to 0$ this reduces to the result given by Stewartson (1958) for a surface with only two boundary layer leading edges, that is, a long strip. However, for a square surface, $\mathrm{Nu}_{L_1}\mathrm{Gr}_{L_1}^{-1/5}$ is 25% above the long-strip result.

Zakerullah and Ackroyd (1979) report a similar analysis for isothermal horizontal circular disks. Near the disk periphery, the boundary layer development is two dimensional. As the center of the disk is approached, the boundary layer is increasingly influenced by an axially symmetric squeezing of the flow. Close to the center, boundary layer analysis breaks down. The flow turns upward to form the root of a buoyant plume. The overall surface heat transfer rate, however, depends mainly on the high heat transfer rates near the disk periphery, where the boundary layer model is applicable. The expression for Nu_D, the Nusselt number based on diameter ($D = 2a$) of the disk, is

$$\mathrm{Nu}_D\mathrm{Gr}_D^{-1/5} = -\frac{25}{12}2^{2/5}C^{-2/5}\frac{\rho_0k_0}{\rho_rk_r}\left[\phi_0'(0) + \sum_{n=1}^{\infty}\frac{24\phi_n'(0)}{(5n+3)(5n+8)}\right] \tag{5.3.13}$$

where subscripts 0 and r refer to the conditions at the disk surface and the chosen reference temperature, respectively, $\mathrm{Gr}_D = [(\rho_0 - \rho_r)/\rho_0]gD^3/\nu_r^2$ is the Grashof number based on disk diameter D. C is a constant taken as unity except when the Chapman-Rubesin (1949) viscosity-temperature relationship $\mu/\mu_r = CT/T_r$ is used for gases. Then $(1/C)\rho\mu/\rho_r\mu_r = (1/C)\rho k/\rho_r k_r = c_{pr}/c_p = 1$. Other symbols have the same meaning as before. For such a gas, the calculated values of $-\phi_i'(0)$, where $i = 0$, 1, and 2, are given in Table 5.3.2 for different values of $\Delta t/T_r$, where $\Delta t = t_0 - t_r$.

When the contribution of the summation terms is ignored in Eq. (5.3.13), the expression for Nusselt number reduces to that obtained from Eq. (5.3.12) for a

Table 5.3.2 Computed values of $-\phi_i'(0)$ for flow adjacent to a horizontal circular disk in air

i	$\Delta T/T_r$	$-\phi_i'(0)$
0	0	0.35741
0	0.25	0.36810
0	0.50	0.37767
0	0.75	0.38634
0	1	0.39430
1	0	−0.05006
1	0.25	−0.05115
1	0.50	−0.05217
1	0.75	−0.05312
1	1	−0.05401
2	0	−0.03285
2	0.25	−0.03333
2	0.50	−0.03382
2	0.75	−0.03430
2	1	−0.03476

Source: From Zakerullah and Ackroyd (1979).

square plate, $L_1 = L_2$. The values of $-\phi_i'(0)$ in Table 5.3.2 indicate that $-\phi_1'(0)$ and $-\phi_2'(0)$ are small compared to $-\phi_0'(0)$ and have successively decreasing values. It seems reasonable to assume that the contribution of all the summation terms to Nu_D is small compared to that of $-\phi_0'(0)$. Therefore, the Nusselt number results from boundary layer predictions for circular disks may be taken as very close to those for square plates. This is in agreement with the experimental data of Al-Arabi and El-Reidy (1976) for isothermal horizontal square, rectangular, and circular surfaces in air.

These heat transfer calculations, with the property variations taken into account, are in better agreement with the experimental data of Fishenden and Saunders (1950, 1961) for square plates, of Wragg and Loomba (1970) for circular disks, and of Goldstein et al. (1973) for square and circular plates. At higher Ra, the calculated slope of Nusselt number vs. Rayleigh number was shown to lie closer to the experimental value of one-fourth. However, at low Ra, the data of Goldstein et al. (1973), Bandrowski and Rybski (1976), and Goldstein and Lau (1983) suggest a one-fifth power law correlation.

In conclusion, the heat transfer results for a horizontal circular disk are very close to those for a square surface. The experimental data of Al-Arabi and El-Reidy (1976) for isothermal horizontal surfaces of different shapes in air also showed that the average heat transfer rate from a circular plate is very close to that from a square plate with side length equal to diameter of the circular plate.

5.3.3 Slightly Inclined Horizontal Surfaces, B_n Away from the Surface

For a horizontal surface there is no component of buoyancy force parallel to the surface. The flow is driven entirely by the normal motion pressure gradient. However, for even slightly inclined surfaces B_t arises and has an appreciable effect. The governing equations are again Eqs. (5.1.5)–(5.1.8) and, as pointed out earlier, they do not admit a similarity solution. The solution can, however, be estimated by perturbation techniques, as done by Pera and Gebhart (1973a) and Jones (1973).

For presentation of the analysis, it is convenient to express Eqs. (5.1.5)–(5.1.8) in terms of angle $\xi = 90° - \theta$. Thus, ξ is the angle with respect to the horizontal. The recast equations are

$$\frac{\partial u}{\partial x} + \frac{\partial v}{\partial y} = 0 \tag{5.3.14}$$

$$u\frac{\partial u}{\partial x} + v\frac{\partial u}{\partial y} = \nu\frac{\partial^2 u}{\partial y^2} + g\beta(t - t_\infty)\sin\xi - \frac{1}{\rho}\frac{\partial p_m}{\partial x} \tag{5.3.15}$$

$$0 = g\beta(t - t_\infty)\cos\xi - \frac{1}{\rho}\frac{\partial p_m}{\partial y} \tag{5.3.16}$$

$$u\frac{\partial t}{\partial x} + v\frac{\partial t}{\partial y} = \alpha\frac{\partial^2 t}{\partial y^2} \tag{5.3.17}$$

For very small values of ξ, the structure of the boundary layer would be similar to that adjacent to a horizontal surface. It is reasonable, therefore, to use the same similarity variables f, ϕ, P as in those flows, except that they are now expanded in terms of a perturbation parameter $\epsilon(x)$ to account for departures from similarity as follows:

$$\psi = \nu c f(\eta, x) = 5\nu\left(\frac{\mathrm{Gr}_{x,\xi}}{5}\right)^{1/5}[f_0(\eta) + \epsilon(x)f_1(\eta) + \cdots] \tag{5.3.18}$$

$$\phi(\eta, x) = \phi_0(\eta) + \epsilon(x)\phi_1(\eta) + \cdots \tag{5.3.19}$$

$$p = \frac{5\nu^2\rho}{x}\left(\frac{\mathrm{Gr}_{x,\xi}}{5}\right)^{1/5}[P_0(\eta) + \epsilon(x)P_1(\eta) + \cdots] \tag{5.3.20}$$

where $\epsilon(x) = (\mathrm{Gr}_{x,\xi}/5)^{1/5}\tan\xi$ and $\mathrm{Gr}_{x,\xi} = g\beta(t_0 - t_\infty)x^3\cos\xi/\nu^2$. The value of ϵ is determined so that the buoyancy force and the pressure gradient along the plate are of the same order of magnitude. Noting that $u = \psi_y$ and $v = -\psi_x$, the continuity equation, Eq. (5.3.14), is satisfied identically. The remaining equations are then evaluated in terms of variables defined in Eqs. (5.3.18)–(5.3.20). Collecting terms of the order ϵ^0, Eqs. (5.3.3)–(5.3.6) for a horizontal surface are recovered. The first-order equations in f_1, ϕ_1, and P_1, obtained by collecting terms of $O(\epsilon)$, are

$$f_1''' + (3 + n)f_0 f_1'' - 5(n + 1)f_0' f_1' + 2(n + 3)f_1 f_0''$$

$$- (n + 1)P_1 - \frac{n - 2}{5} P_1' \eta + \phi_0 = 0 \qquad (5.3.21)$$

$$P_1' = \phi_1 \qquad (5.3.22)$$

$$\phi_1'' + \Pr[(n + 3)f_0 \phi_1' - (3 + 6n)f_0' \phi_1 - 5nf_1' \phi_0 + 2(n + 3)f_1 \phi_0'] = 0 \qquad (5.3.23)$$

Similarly, the second-order equations in f_2, P_2, and ϕ_2 are

$$f_2''' + (n + 3)f_2'' f_0 - 2(3n + 4)f_2' f_0' + 3(n + 3)f_2 f_0'' + 2(n + 3)f_1'' f_1$$

$$- (3n + 4)f_1'^2 - \frac{6n + 8}{5} P_2 - \frac{n - 2}{5} P_2' \eta + \phi_1 = 0 \qquad (5.3.24)$$

$$P_2' = \phi_2 \qquad (5.3.25)$$

$$\phi_2'' + \Pr[(n + 3)f_0 \phi_2' - (7n + 6)f_0' \phi_2 - (6n + 3)f_1' \phi_1$$

$$+ 2(n + 3)f_1 \phi_1' + 3(n + 3)\phi_0' f_2 - 5n\phi_0 f_2'] = 0 \qquad (5.3.26)$$

The boundary conditions for a power law surface temperature variation for zeroth order are given by Eq. (5.3.6). For the next two higher orders, these are

$$f_i(0) = f_i'(0) = \phi_i(0) = f_i'(\infty) = \phi_i(\infty) = P_i(\infty) = 0 \qquad i = 1, 2 \qquad (5.3.27)$$

The higher-order equations subject to these boundary conditions have been solved by Pera and Gebhart (1973a) for Pr = 0.7 for an isothermal surface condition ($n = 0$). From these calculations:

$$f_1''(0) = 0.37674 \qquad P_1(0) = 0.23560 \qquad \phi_1'(0) = -0.10426 \qquad (5.3.28)$$

$$f_2''(0) = -0.01032 \qquad P_2(0) = -0.06321 \qquad \phi_2'(0) = 0.013609 \qquad (5.3.29)$$

The velocity and temperature profiles associated with the above solutions are shown in Fig. 5.3.9. Very small angles of inclination are seen to have a large effect on the velocity distribution but a relatively small influence on the temperature distribution. The effect on heat transfer rate, however, is not negligible. Defining the local value of Nusselt number for the inclined surface as $\mathrm{Nu}_{x,\xi}$, we obtain from the numerical data in Eqs. (5.3.28) and (5.3.29) and Table 5.3.1,

$$\frac{\mathrm{Nu}_{x,\xi}}{\mathrm{Nu}_{x,0}} = 1 + 0.21319\epsilon - 0.02783\epsilon^2 + \cdots \qquad \text{for Pr} = 0.7 \qquad (5.3.30)$$

For $\xi = 3°$ and $\mathrm{Gr}_{x,\xi} = 5 \times 10^5$, ϵ is 0.524 and the increase in heat transfer due to inclination is 10.4%. For $\xi = 6°$ at the same $\mathrm{Gr}_{x,\xi}$, the increase is 19%.

If it is desired to analyze the flow when B_n is toward the surface, that is, $\theta < 0$, the governing equations in terms of ξ are to be obtained from Eqs. (5.1.5)–(5.1.8), using $\xi = 90° - (\theta)$, rather than simply replacing ξ by $-\xi$ in Eqs. (5.3.14)–(5.3.17). Also, see Problem 5.15.

Jones (1973) also analyzed the slightly inclined isothermal surface in terms

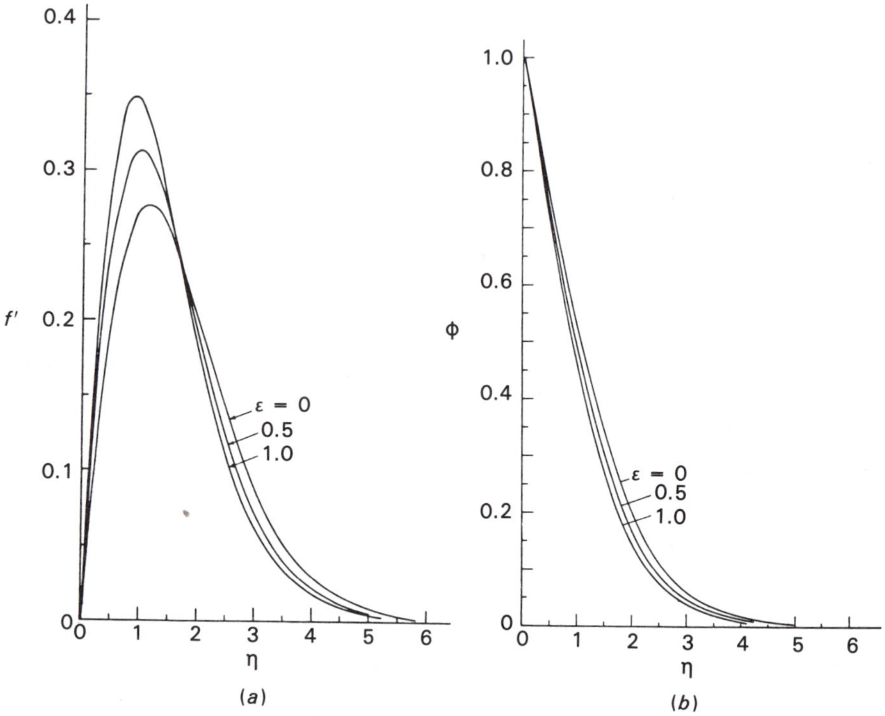

Figure 5.3.9 Effect of inclination from the horizontal on (*a*) the velocity function and (*b*) the temperature function, for an isothermal surface; Pr = 0.7. *(Reprinted with permission from* Int. J. Heat Mass Transfer, *vol. 16, p. 1131, L. Pera and B. Gebhart. Copyright © 1973, Pergamon Journals Ltd.)*

of series solutions, one to apply in the region of the leading edge and the other at large distances downstream. In the upstream region the buoyancy-induced pressure field, due to the normal direction buoyancy component, is taken to predominate in driving the motion. This is called an indirect drive. Farther downstream, the tangential buoyancy force becomes large and predominates. This is called a direct drive. Since for an isothermal surface the pressure gradient and the buoyancy force are proportional to $x^{-3/5}$ and $x^{2/5}$, respectively, this appears reasonable. However, the early separation observed by Pera and Gebhart (1973b), as well as the general agreement of measured attached profiles with those of a single solution that incorporates both driving effects, leads to the question of whether a largely buoyancy-driven direct-drive flow is ever actually found downstream.

The first measurements are those of Pera and Gebhart (1973a). The trends in these measurements, taken in air for a range of surface inclination from 0 to 6°, are in good general agreement with the theoretical predictions. However, there are some differences. At $Gr_x < 10^4$, the measured heat transfer rates are lower than calculated. As for vertical flows, at lower Gr_x, the higher-order buoyancy layer effects arise. These effects, for horizontal surfaces in air, have been treated by Mahajan and Gebhart (1980a) in terms of a perturbation parameter $\epsilon_1 = (\mathrm{Gr}_x)^{-1/5}$, which is related to $\epsilon = (\mathrm{Gr}_{x,\xi}/5)^{1/5} \tan \xi$ as follows:

$$\frac{\epsilon}{\epsilon_1} = \left(\frac{\mathrm{Gr}_{x,\xi}}{5^{1/2}}\right)^{2/5} \tan \xi$$

For a given ξ, the effects of surface inclination are larger than the first boundary layer correction if $\epsilon/\epsilon_1 >> 1$. For $\mathrm{Gr}_x = 10^4$ and $\xi = 2$, 4, and 6°, the value of this ratio is greater than 1. Therefore, the correction due to higher-order boundary layer approximations must simultaneously be considered for small Gr_x.

At larger values of Gr_x, the data were consistently above the calculations. This, as suggested by Pera and Gebhart, may be due to an ambiguity in the location of the effective leading edge. Or it may be due to an additional driving mechanism that may arise in such flows, as follows. Yet farther downstream, the boundary layer was found to separate from the surface, eventually into a rising plume, and produce a large region of vertical buoyancy. This may result in motion pressure levels below that estimated downstream for a continuously attached boundary region. It would also decrease the pressure levels in the upstream attached flow and result in a thinner boundary layer and higher heat transfer rate than calculated.

Using a Mach-Zehnder interferometer, Yousef et al. (1982) measured the local and average heat transfer coefficients and the temperature distributions in air from isothermal heated surfaces in air. The trends in the data were in general agreement with those of Pera and Gebhart (1973a). However, it was reported that the random changes in the local values of Nusselt number could reach +45% and −35% of the mean values. These were attributed to the periodic instabilities that arise.

There is a need for more detailed experiments to further clarify these matters. In particular, detailed velocity measurements will be useful since for such flows the angle of inclination has a strong effect on the velocity field.

5.3.4 Horizontal Downward-Facing Warm Surfaces

Two-dimensional laminar natural convection flow adjacent to a horizontal infinite strip, when the warm surface faces down or the cold surface faces up, has also been investigated experimentally and analytically. Then $B_n < 0$ is toward the surface. Experimental evidence indicates that boundary layer flow results for this case also; see, for example, Singh and Birkebak (1969) and Aihara et al. (1972). In the flow visualization experiment of Singh and Birkebak, the flow field adjacent to a heated horizontal flat strip facing downward was observed by using 10–40-μm neutral density plastic particles in water. The plate was 3 in. wide, 10 in. long, and 1 in. thick. A photograph of the flow field is shown in Fig. 5.3.10*a*. Outward flow boundary layers are indicated by streaklines. In the experiments of Aihara et al. in air, a similar boundary layer flow was observed. The flow in the boundary region, particularly in the vicinity of the free edge, was found to be quite stable, although in the whole region the presence of even a small disturbance such as a little slope of the test plate affected the symmetry of the flow significantly.

Referring to Fig. 5.3.10*a*, as the fluid near the surface is heated by the warm surface, it becomes lighter than the surrounding fluid and tends to accumulate on the surface. The only driving force is again the pressure field induced by $B_n < 0$. Still, $\partial p/\partial x$ is negative because of the thinning layer, with increasing x. Driven by this pressure gradient, this fluid flows outward and off the edges. Under steady-state conditions, a boundary layer is established that has maximum depth at the plate's center and decreases toward the edges. This is quite different from the leading edge onflow and the developing boundary region downstream that results for a horizontal warm surface facing upward. Here $\delta \neq 0$ at $x = 0$ and both $d\delta/dx$ and B_n are <0.

As indicated by Gill et al. (1965) and Clifton (1967), a similarity solution does not exist for this flow. Approximate solutions by integral methods have been reported. Wagner (1956) applied the momentum integral method of Levy (1955) to this flow. The fluid was assumed to flow from the center of the plate to the edges and the boundary layer assumed to have zero thickness at the edges. Singh et al. (1969) extended the analysis to circular and square plates.

The assumption of vanishing boundary layer thickness at the edges is unrealistic. The observations of Fishenden and Saunders (1961), Abdulkadir (1968), Singh and Birkebak (1969), and Birkebak and Abdulkadir (1970) indicate an appreciable off-flow layer thickness at the edges of the plate, as sketched in Fig. 5.3.10*a*.

Singh and Birkebak (1969) presented a two-dimensional solution that includes a finite boundary layer thickness at the edges. The flow configuration is shown in Fig. 5.3.10*b*—a long strip, resulting in a plane flow. The edges are assumed to be attached to two adiabatic vertical walls to inhibit side effects. The width of the plate, $2a$, is assumed large compared to the boundary layer thickness, allowing approximations may be made. The cold fluid is entrained into the boundary region over its whole outer edge. It is heated near the surface and flows outward and upward around the edges.

The governing equations are Eqs. (5.1.7), (5.1.8), (5.1.10*a*), and (5.1.10*c*) developed for $\theta = -90°$. The relevant boundary conditions for an isothermal surface are

$$\text{at } y = 0: \qquad u = v = 0 \qquad t = t_0$$

$$\text{at } y = \delta: \qquad u = 0 = t - t_\infty$$

Using the momentum integral method, Singh and Birkebak obtained the solutions for boundary layer thickness and local heat transfer for various values of the Prandtl number. Assuming parabolic temperature and velocity distributions in the boundary layer, the expression for local Nusselt number is

$$\mathrm{Nu}_a = -\frac{a}{t_0 - t_\infty}\left.\frac{\partial t}{\partial y}\right)_{y=0} = \frac{2a}{\delta} = \frac{2}{y_2}\,\mathrm{Ra}_a^{-1/5} \tag{5.3.31}$$

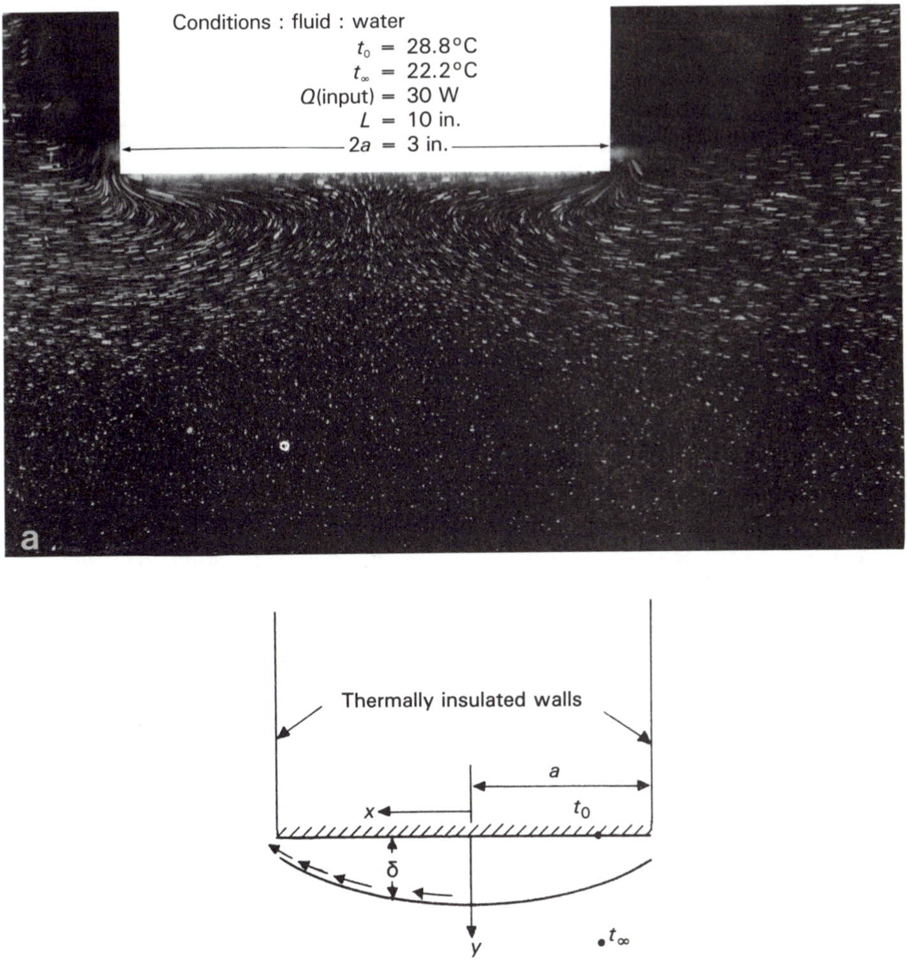

Figure 5.3.10 Flow over a warm horizontal surface facing downward. (*a*) Flow streaklines; (*b*) coordinate system. *(Reprinted with permission from S. N. Singh and R. C. Birkebak,* Z. Angew. Math. Phys., *vol. 20, p. 454. Copyright © 1969, Birkhaeuser Publishers Ltd.)*

where a is the half-width of the plate, δ the boundary layer thickness, $y_2 = (\delta/a)\mathrm{Ra}_a^{-1/5}$, and Ra_a is the Rayleigh number based on dimension a. The values of y_2 at $x' = 0$ and 1, where $x' = x/a$, for various values of Prandtl number are given below:

Pr	0.025	0.1	0.7	1.0	5.0	10	50	∞
y_2 at $x' = 0$	8.77	6.23	4.87	4.61	3.98	3.75	3.55	3.43
y_2 at $x' = 1$	4.99	3.46	2.56	2.30	1.59	1.19	0.68	0

Plots of Nu_a and y_2 as functions of x' are shown in Fig. 5.3.11. Wagner's solution for $Pr = 1$ is also shown. The agreement is seen to be reasonable for $x' < 0.8$. Near the edges, Wagner's solution overpredicts heat transfer because of the assumption of zero boundary layer thickness in that analysis.

Clifton and Chapman (1969) also analyzed this transport by the integral method. The boundary layer depth at the edges was set equal to a critical depth predicted by analogy to open channel water flow. In the center of the surface, the resulting predictions of the heat transfer coefficient are in agreement with those of Singh and Birkebak. Near the edges, significant discrepancies exist. This, again, is due to different boundary layer thicknesses at the edges.

There is only a qualitative agreement between these analyses and the available experimental data; see Aihara et al. (1972), Restrepo and Glicksman (1974), and Hatfield and Edwards (1981). In general, the predictions agree fairly well with the data in the central region of the surface. Around the edges, the heat transfer data are higher than the calculated values. These discrepancies are partly due to the inability of the analyses to accurately describe the flow near the edges. As the flow approaches the edges, it is constrained to be horizontal by the surface. However, it then immediately turns upward to form a vertical plume. The analysis includes no allowance for the entrainment and resulting changes in the motion pressure field. Further, for surfaces of finite span, the aspect ratio of the surface is another variable. The other two edges also influence the flow and heat transfer.

Other studies. Before concluding the discussion on horizontal surfaces, it is noted that the analogous mechanism of natural convection mass transfer from horizontal surfaces has also been investigated analytically and experimentally. Some

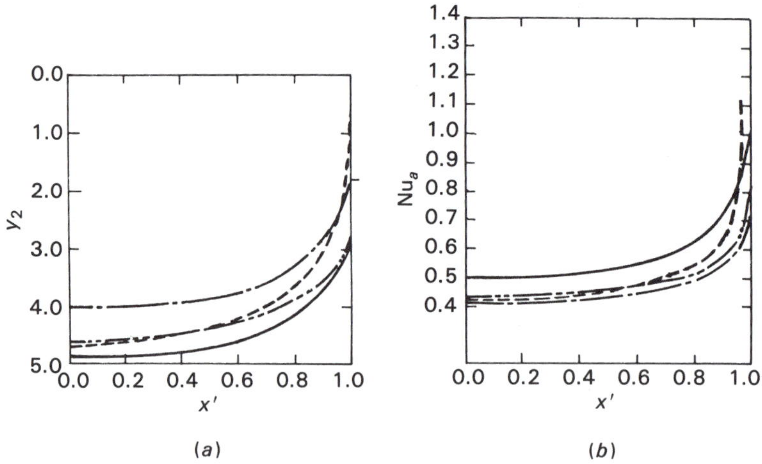

Figure 5.3.11 (*a*) Boundary region thickness and (*b*) Nusselt number vs. x' for a warm horizontal surface facing downward, for different Prandtl numbers. (——) Pr = 5; (–··–) Pr = 1; (–·–) Pr = 0.7 (Singh and Birkebak, 1969); (– –) Pr = 1 (Wagner, 1956). *(Reprinted with permission from S. N. Singh and R. C. Birkebak,* Z. Angew. Math. Phys., *vol. 20, p. 454. Copyright © 1969, Birkhaeuser Publishers Ltd.)*

of the experimental data are due to Wragg (1968), Wragg and Loomba (1970), Goldstein et al. (1973), Lloyd and Moran (1974), and Goldstein and Lau (1983). Bandrowski and Rybski (1976) analyzed the mass transfer from the active surface of an infinite strip for two conditions: (1) the surface facing downward, with the net mass flow directed away from the surface, and (2) the surface facing upward, with the net mass flow directed toward the surface. As for the analogous heat transfer problem (warm surface facing upward), a similarity analysis was obtained for the first condition. An integral method was used to treat the mass transfer from the upward orientation of the active surface (analogous to the warm surface facing downward). Numerical results were presented for Schmidt number values in the range $1 \leq \text{Sc} \leq 2500$.

5.4 CURVED SURFACES

In many natural convection heat transfer problems of practical interest, the surface generating the flow is curved; see Figs. 5.1.2*a* and 5.1.2*b*. The angle θ between the tangent to the surface and the direction of gravity varies downstream along the surface from its upstream beginning. When the curvature of the surface is small, it may be locally approximated as flat. Then the solutions discussed earlier for vertical and inclined flat surfaces may be used. For greater curvature, this approximation may not be accurate. The buoyancy force components B_n and B_t vary with θ, and a general analysis is much more difficult. However, similarity solutions have been obtained for a special class of such surfaces. Where these solutions are not possible, approximate techniques have been used to develop general relations for the transport quantities of interest. Some of these results are reviewed in this section.

5.4.1 Similar Solutions

The two most extensively studied curved surfaces are symmetric two-dimensional planar (wedgelike) bodies as in Fig. 5.1.2*a* and axisymmetric (conelike) bodies as in Fig. 5.1.2*b*. In analyses for these surfaces it is generally assumed that the boundary region thickness is small compared to the local radius of curvature. Also, the normal component of buoyancy B_n and the motion pressure p_m are ignored. With boundary region and these approximations, Eqs. (5.1.11)–(5.1.13) then apply.

These equations are similar to those for the vertical flows discussed in Chapters 3 and 4, except that the buoyancy force term is now modified by $\cos\theta$, which varies downstream. For similar solutions, therefore, the added condition to be satisfied is that the buoyancy force dependence on $\cos\theta$ be included in terms of similarity variables. Such conditions were first discussed by Schuh (1948). The flow over a two-dimensional planar body having finite curvature everywhere and a boundary layer of constant thickness was discussed. Later, Merk and Prins (1953, 1954) developed general relations for similarity solutions for two-dimensional planar and axisymmetric bodies. The horizontal circular cylinder and sphere were shown not to permit similarity solutions. The cone is one axisymmetric body that was shown to have a similarity flow.

Two-dimensional planar flows. Braun et al. (1961) identified a family of two-dimensional body shapes that permit similarity. The requirements for similarity for such surfaces can again be determined following the procedure in Chapter 3. Equations (3.3.6)–(3.3.8) are introduced to transform Eqs. (5.1.11)–(5.1.13) to yield

$$f''' + \phi\left(\frac{g\beta d \cos\theta}{\nu^2 c b^3}\right) = \left(\frac{c_x}{b} + \frac{c b_x}{b^2}\right) f'^2 - \frac{c_x}{b} f f'' \tag{5.4.1}$$

$$\frac{\phi''}{\Pr} = \left(\frac{c}{b}\frac{d_x}{d}\right)\phi f' - \frac{c_x}{b} f \phi' \tag{5.4.2}$$

where, in Eq. (5.1.11), $a = 0$. It may be shown that for an isothermal surface, the above equations have a similarity solution if

$$c \propto x^{m/(m+1)} \qquad b \propto x^{-1/(m+1)} \qquad \text{and} \qquad \cos\theta = \left(\frac{m+1}{m} X\right)^{(m-3)/(m+1)} \tag{5.4.3}$$

where m is a body shape parameter and $X = x/L$, as before, is the dimensionless streamwise coordinate. Also, in terms of Fig. 5.1.2a, let $\tilde{r} = r/L$ and $Z = z/L$ be the local dimensionless body radius and axial coordinate, respectively, where L is some characteristic length of the body. From Fig. 5.1.2a, it can be seen from geometric considerations that $d\tilde{r}/dX = \sin\theta$ and $dZ/dX = \cos\theta$. Then, from the value of $\cos\theta$ in Eq. (5.4.3),

$$\tilde{r} = \int_0^X \left[1 - \left(\frac{m+1}{m} X\right)^{2(m-3)/(m+1)}\right]^{1/2} dX \tag{5.4.4}$$

$$Z = \frac{m}{2(m-1)} \left(\frac{m+1}{m} X\right)^{2(m-1)/(m+1)} \tag{5.4.5}$$

These equations are identical to those derived by Braun et al. (1961) for an isothermal surface condition. They describe the two-dimensional planar bodies with closed lower ends shown in Fig. 5.4.1. As shown by Braun et al., the requirement that $\tilde{r}$ be real restricts the values of m and X to

$$0 \le \left(\frac{m+1}{m} X\right) \le 1 \tag{5.4.6a}$$

where all values of m are permissible except

$$-1 \le m < 3 \tag{5.4.6b}$$

Then $m = 3$ is simply the vertical flat surface and $m = -1$ is the horizontal surface. Similar solutions are also possible for power law surface temperature variations, $d = Nx^n$, for some values of n; see Braun et al. (1961).

Using boundary layer techniques, heat transfer results were computed for these isothermal surfaces for a range of Prandtl numbers. Note, however, that the boundary layer thickness δ for these bodies varies as $\delta = \{[(m+1)/m]X\}^{1/(m+1)}$. This means that for the flat-bottomed contours in Fig. 5.4.1, $m < -1$. Therefore, δ is unbounded as the origin is approached. The thin boundary layer assumption is not valid there. Also, in the region close to the origin, the normal buoyancy force B_n and motion pressure p_m, omitted in Eqs. (5.1.11)–(5.1.13), are very important. In fact, B_n and p_m are the driving forces for the flow. Therefore, these solutions apply only sufficiently far from the origin.

Blowing and suction. Merkin (1975) considered these effects for boundary layer flows over two-dimensional bodies and showed that similar solutions are still possible for certain body shapes and transpiration velocities. Let $\pm V_0\gamma(X)$ be the transpiration velocity at the surface, where $\gamma(X)$ is the nondimensional variation in velocity and the $\pm$ signs are for blowing and suction, respectively. The governing equations are still Eqs. (5.1.11)–(5.1.13), with $a = 0$. The boundary condition $v = 0$ at $y = 0$, for flow without transpiration, is now replaced by $v = \pm V_0\gamma(X)$. It was shown that a similarity solution is permissible provided $\gamma(X) = \{[(m + 1)/m]X\}^{1/(m+1)}$. For constant transpiration velocity, the only body shape for which a similar solution is possible corresponds to $m \to \infty$ in Fig. 5.4.1.

Axisymmetric flows. For such flows, two more shapes, as well as the cone discussed below, were found by Braun et al. (1961) to have similar solutions. These are parabolic-nosed and flat-nosed bodies and a vertical cone, a more familiar configuration. The cone has been investigated in detail by Hering and Grosh (1962) and Hering (1965). This flow is discussed next.

Vertical cone. The laminar boundary layer equations governing the flow are Eqs. (5.1.11)–(5.1.13), with $a = 1$. Now θ is half the apex angle and $r = x \sin\theta$; see Fig. 5.4.2*b*. The equations and the boundary conditions are as below. Again B_n is neglected.

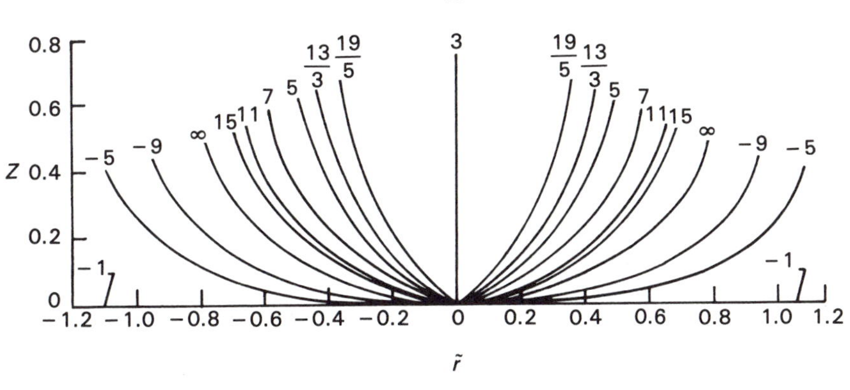

Figure 5.4.1 Two-dimensional similar geometries *(Reprinted with permission from* Int. J. Heat Mass Transfer, *vol. 2, p. 121, W. H. Braun et al. Copyright © 1961, Pergamon Journals Ltd.)*

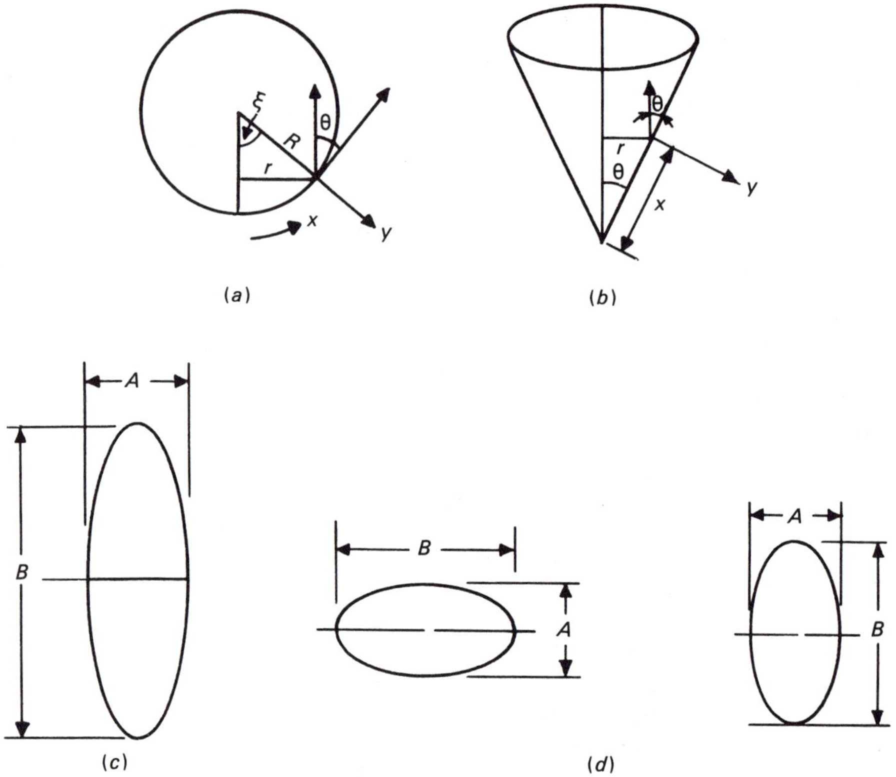

Figure 5.4.2 Coordinate systems for (*a*) horizontal circular cylinder or sphere, (*b*) right circular cone, (*c*) elliptic cylinder, and (*d*) spheroids (left, oblate spheroid, *C/B* = 0.5; right, prolate spheroid, C/B = 0.5).

$$\frac{\partial}{\partial x}(ur) + \frac{\partial}{\partial y}(vr) = 0 \tag{5.4.7}$$

$$u\frac{\partial u}{\partial x} + v\frac{\partial u}{\partial y} = \nu\frac{\partial^2 u}{\partial y^2} + g\beta(t - t_\infty)\cos\theta \tag{5.4.8}$$

$$u\frac{\partial t}{\partial x} + v\frac{\partial t}{\partial y} = \alpha\frac{\partial^2 t}{\partial y^2} \tag{5.4.9}$$

$$\begin{aligned} &\text{at } y = 0: \quad && u = 0 = v \quad && t = t_0 \\ &\text{as } y \to \infty: \quad && u \to 0 \quad && t \to t_\infty \end{aligned} \tag{5.4.10}$$

It is worth examining again the validity of the assumptions made in deriving the above equations. In Eqs. (5.4.7)–(5.4.9), the boundary layer is assumed thin compared to the cone radius. This allows the local radial distance to a point in the layer to be replaced by the local cone radius $r(x)$. Clearly, this condition is not satisfied near the cone tip. Also, since the pressure gradient in the y direction is not accounted for, the equations are strictly applicable to cones of small apex angles.

To investigate the conditions for similarity, the stream function ψ is defined as $u = (1/r)\psi_y$, $v = -(1/r)\psi_x$. Then the continuity equation is automatically satisfied. Following the procedure in Chapter 3, the similarity variable η, dimensionless temperature $\phi(\eta)$, and reduced stream function $f(\eta)$ are introduced as follows:

$$\eta = b(x)y \qquad \phi = \frac{t - t_\infty}{t_0 - t_\infty} = \frac{t - t_\infty}{d(x)} \qquad \psi = \nu c(x) r f(\eta) \tag{5.4.11}$$

These are then introduced into Eqs. (5.4.8) and (5.4.9) to yield

$$f''' + \phi\left(\frac{g\beta d \cos\theta}{\nu^2 c b^3}\right) = \left(\frac{c_x}{b} + \frac{c b_x}{b^2}\right) f'^2 - \left(\frac{c}{bx} + \frac{c_x}{b}\right) f f'' \tag{5.4.12}$$

$$\frac{\phi''}{\text{Pr}} = \left(\frac{c}{b}\frac{d_x}{d}\right)\phi f' - f\phi'\left(\frac{c_x}{b} + \frac{c}{bx}\right) \tag{5.4.13}$$

A condition for similarity is that the quantities in parentheses be independent of x. They may be functions of η. For the surface temperature varying according to a power law $d = t_0 - t_\infty = Nx^n$, it may be shown that similarity requirements are met if c and b are proportional to $x^{(n+3)/4}$ and $x^{(n-1)/4}$, respectively. Defining the local Grashof number as $\text{Gr}_{x,\theta} = g\beta x^3 d \cos\theta/\nu^2$, the functions $c(x)$ and $b(x)$ are written in terms of $\text{Gr}_{x,\theta}$ as $c = \text{Gr}_{x,\theta}^{1/4}$ and $b = c/x$. With those values, Eqs. (5.4.12) and (5.4.13) are as below:

$$f''' + \frac{n+7}{4} f f'' - \frac{n+1}{2} f'^2 + \phi = 0 \tag{5.4.14}$$

$$\phi'' + \text{Pr}\left(\frac{n+7}{4} f\phi' - n f'\phi\right) = 0 \tag{5.4.15}$$

The boundary conditions are

$$f(0) = f'(0) = \phi(0) - 1 = 0 \tag{5.4.16}$$

$$f'(\infty) = \phi(\infty) = 0 \tag{5.4.17}$$

These equations were solved numerically for an isothermal surface, $n = 0$, and a linear surface temperature variation, $n = 1$, for $\text{Pr} = 0.7$, by Hering and Grosh (1962). The computed values of local heat transfer were

$$\text{Nu}_x = 0.45110\,\text{Gr}_{x,\theta}^{1/4} \qquad \text{for } n = 0 \tag{5.4.18}$$

$$\text{Nu}_x = 0.56699\,\text{Gr}_{x,\theta}^{1/4} \qquad \text{for } n = 1 \tag{5.4.19}$$

The analysis was later extended by Hering (1965) to low Prandtl number fluids. The computed local and average heat transfer results, for various values of n in $d = Nx^n$, to low Prandtl numbers are shown in Fig. 5.4.3.

The effect of the radial curvature, neglected in the similarity solutions above, was considered by Kuiken (1968), Oosthuizen and Donaldson (1972), and Alamgir (1979). A common conclusion is that the curvature effects result in higher heat transfer rates.

5.4.2 Flow around a Horizontal Circular Cylinder

As mentioned before, this commonly encountered surface does not admit a similarity solution. However, this shape is of great practical importance and its transport characteristics have been studied in considerable detail. These are discussed next, first for laminar flow when boundary layer assumptions are valid and then for moderate and low values of Grashof numbers when boundary layer solutions may not be sufficiently accurate.

Boundary layer flow. The governing equations around a horizontal cylinder of radius R, again omitting B_n, in terms of Fig. 5.4.2*a*, are

$$\frac{\partial u}{\partial x} + \frac{\partial v}{\partial y} = 0 \qquad (5.4.20)$$

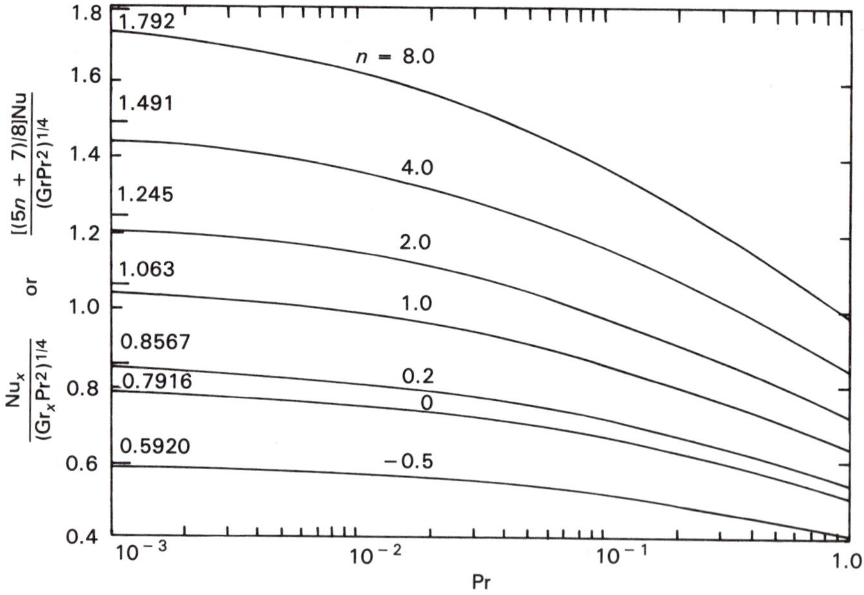

Figure 5.4.3 Local and average heat transfer for a vertical cone for various values of n in $d(x) = Nx^n$. *(Reprinted with permission from* Int. J. Heat Mass Transfer, *vol. 8, p. 1333, R. G. Hering. Copyright © 1965, Pergamon Journals Ltd.)*

$$u \frac{\partial u}{\partial x} + v \frac{\partial u}{\partial y} = \nu \frac{\partial^2 u}{\partial y^2} + g\beta(t - t_\infty) \sin \frac{x}{R} \tag{5.4.21}$$

$$u \frac{\partial t}{\partial x} + v \frac{\partial t}{\partial y} = \alpha \frac{\partial^2 t}{\partial y^2} \tag{5.4.22}$$

The relevant boundary conditions are

$$\begin{aligned} &y = 0 \qquad u = v = 0 \qquad t = t_0(x) \quad \text{or} \quad -k \frac{\partial t}{\partial y} = q''(x) \\ &y \to \infty \qquad u \to 0 \qquad t \to t_\infty \end{aligned} \tag{5.4.23}$$

where $q''(x)$ is a prescribed surface heat flux. This formulation does not cover the whole range $0 \le \theta \le \pi$. That is, the mechanisms arising in any downstream separation are not provided for.

Solutions for an isothermal cylinder. In one of the earliest works, Hermann (1954) modified Pohlhausen's similar solution for a vertical isothermal flat surface to obtain solutions for a horizontal cylinder at Pr = 0.733. The boundary layer thickness at different angles around the cylinder was obtained by multiplying the flat-plate boundary layer thickness by a parameter that is a function of the azimuthal angle $\xi = x/R$. It will be noted later that this solution is strictly accurate only around $\xi = \pi/2$, where $\theta = 0$.

A more accurate solution for an isothermal cylinder was obtained by Chiang and Kaye (1962) through the use of a Blasius series. Equations (5.4.20)–(5.4.22) were first transformed by use of the following variables:

$$\eta = (\text{Gr}_R)^{1/4} \frac{y}{R} \qquad \xi = \frac{x}{R} \qquad \psi = \nu \text{Gr}_R^{1/4} f(\eta, \xi) \qquad \phi = \frac{t - t_\infty}{t_0 - t_\infty} \tag{5.4.24}$$

where $\text{Gr}_R = g\beta(t_0 - t_\infty)R^3/\nu^2$, $u = \psi_y$, and $v = -\psi_x$. The transformed equations become

$$f''' + f'' \frac{\partial f}{\partial \xi} - f' \frac{\partial^2 f}{\partial \eta \partial \xi} + \phi \sin \xi = 0 \tag{5.4.25}$$

$$\frac{\phi''}{\text{Pr}} + \phi' \frac{\partial f}{\partial \xi} - f' \frac{\partial \phi}{\partial \xi} = 0 \tag{5.4.26}$$

$$f(0) = f'(0) = f_\xi(0) = \phi(0) - 1 = f'(\infty) = \phi(\infty) = 0 \tag{5.4.27}$$

Next, f, ϕ, and $\sin \xi$ are expanded in the following series:

$$f = \xi f_0(\eta) + \xi^3 f_1(\eta) + \xi^5 f_2(\eta) + \cdots \tag{5.4.28}$$

$$\phi = \phi_0(\eta) + \xi^2 \phi_1(\eta) + \xi^4 \phi_2(\eta) + \cdots \tag{5.4.29}$$

$$\sin \xi = \xi - \frac{\xi^3}{6} + \frac{\xi^5}{120} + \cdots \tag{5.4.30}$$

Substituting Eqs. (5.4.28)–(5.4.30) into Eqs. (5.4.25)–(5.4.27) and collecting terms of like powers of ξ, the following ordinary differential equations and boundary conditions are obtained:

$$f_0''' + f_0''f_0 - f_0'^2 + \phi_0 = 0 \tag{5.4.31}$$

$$\frac{\phi_0''}{\Pr} + f_0\phi_0' = 0 \tag{5.4.32}$$

$$f_1''' + f_0f_1'' - 4f_0'f_1' + 3f_0''f_1 + \phi_1 - \frac{\phi_0}{6} = 0 \tag{5.4.33}$$

$$\frac{\phi_1''}{\Pr} + f_0\phi_1' - 2f_0'\phi_1 + 3f_1\phi_0' = 0 \tag{5.4.34}$$

$$f_2''' + f_0f_2'' - 6f_0'f_2' + 5f_0''f_2 + 3f_1f_1'' - 3f_1'^2 + \phi_2 - \frac{\phi_1}{6} + \frac{\phi_0}{120} = 0 \tag{5.4.35}$$

$$\frac{\phi_2''}{\Pr} + f_0\phi_2' - 4f_0'\phi_2 + 3f_1\phi_1' - 2f_1'\phi_1 + 5f_2\phi_0' = 0 \tag{5.4.36}$$

$$f_i(0) = f_i'(0) = f_i'(\infty) = \phi_i(\infty) = 0 \qquad i = 0, 1, 2,$$

$$\phi_0(0) = 1, \ \phi_1(0) = 0 = \phi_2(0) \tag{5.4.37}$$

From the numerical results of Chiang and Kaye (1962),

$$\frac{\mathrm{Nu}_R(x)}{\mathrm{Gr}_R^{1/4}} = 0.37023 - 0.01609\xi^2 - 0.00009\xi^4 \qquad \text{for } \Pr = 0.7 \tag{5.4.38}$$

$$= 0.42143 - 0.01861\xi^2 - 0.00011\xi^4 \qquad \text{for } \Pr = 1 \tag{5.4.39}$$

where the local Nusselt number $\mathrm{Nu}_R(x) = h_xR/k$. The velocity and temperature profiles at the bottom of a heated cylinder for $\Pr = 0.7$ are plotted as the dashed curves in Fig. 5.4.4.

Series other than the Blasius have also been used for this cylinder calculation. Saville and Churchill (1967) used a Görtler-type series that converged faster than the Blasius series (Schlichting, 1968). Lin and Chao (1974) used a Merk-type series to obtain solutions for various two-dimensional planar and axisymmetric bodies. The horizontal circular cylinder was a special case.

Other analyses are due to Merk and Prins (1953, 1954) and Muntasser and Mulligan (1978). Merk and Prins used an integral method and gave the following expressions for the average Nusselt number Nu_D, where Nusselt number and Grashof number are based on cylinder diameter D:

$$\mathrm{Nu}_D = \frac{2}{\pi}\int_0^{\pi/2} \mathrm{Nu}_D\left(\frac{x}{D}\right) d\left(\frac{x}{D}\right) = F(\Pr)(\mathrm{Gr}_D\Pr)^{1/4} \tag{5.4.40}$$

where $F(\Pr)$, a function of Prandtl number, was calculated as 0.436, 0.456, 0.520, 0.523, and 0.523 for Pr values of 0.7, 1, 10, 100, and ∞, respectively. In terms of Nu_R and Gr_R, this relation becomes

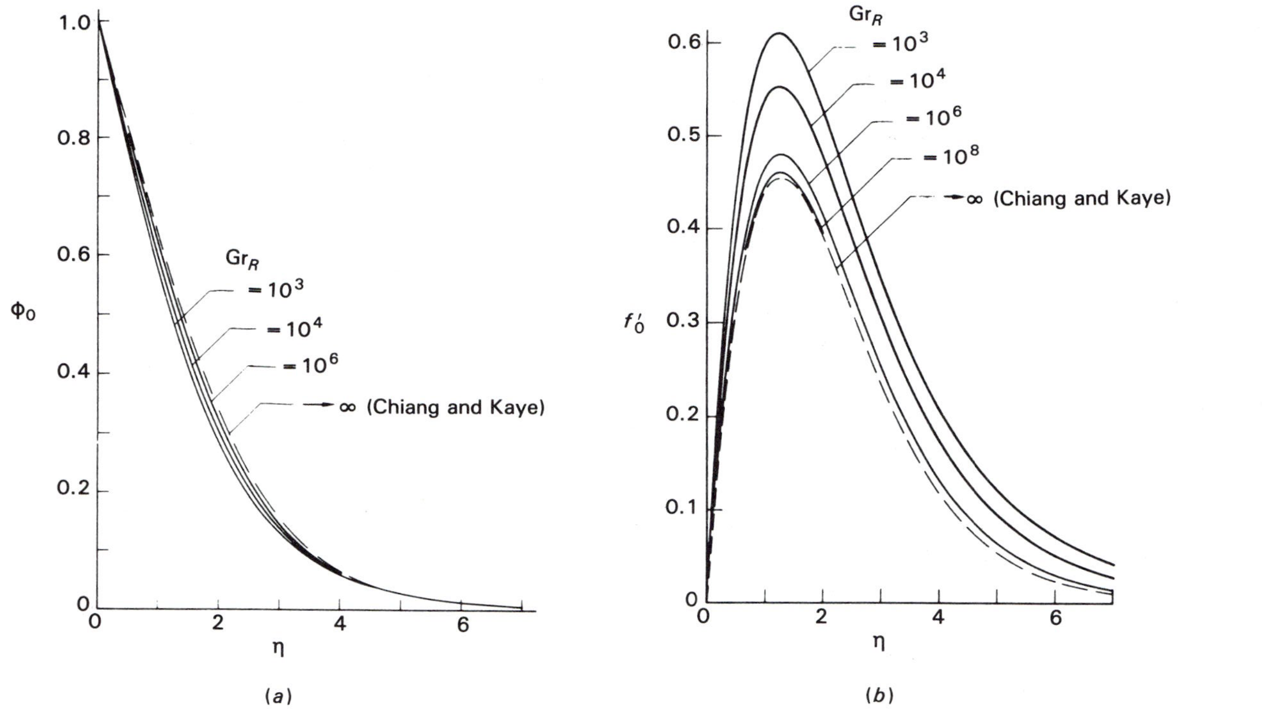

Figure 5.4.4 (*a*) Temperature and (*b*) velocity profiles at the bottom of a horizontal circular cylinder at different values of Grashof number, Pr = 0.7. *(Reprinted with permission from* Int. J. Heat Mass Transfer, *vol. 12, p. 749, J. A. Peterka and P. D. Richardson. Copyright © 1969, Pergamon Journals Ltd.)*

$$\mathrm{Nu}_R = (2)^{-1/4} F(\mathrm{Pr})(\mathrm{Gr}_R \mathrm{Pr})^{1/4} \tag{5.4.41}$$

where Nu_R is the average value of the Nusselt number based on radius R.

Muntasser and Mulligan used the local nonsimilarity method, discussed in Section 5.2. Local solutions were given for different values of the Prandtl number as summarized in Table 5.4.1.

These predictions were compared with results from other studies discussed above and also with the finite-difference results of Merkin (1976). This comparison for Pr = 1.0 is shown in Fig. 5.4.5. In the figure the results of Hermann (1954) for Pr = 0.733 were corrected to Pr = 1, using interpolation on the results of Merk and Prins. The agreement between the local nonsimilarity and finite-difference numerical results is good. Of the series solutions, the Blasius series results are close to these two. The integral method results of Merk and Prins are significantly different, while Hermann's classical solution is close around $\xi = \pi/2$ but diverges over the rest of the ξ domain.

While making these comparisons, it should be kept in mind that at larger values of ξ, the boundary layer equations are an inadequate formulation of the transport mechanisms. At large angles, the two flows converge to form a rising plume. As the boundary layer thickens, the curvature, streamwise effects, and motion pressure neglected in the boundary layer analysis are no longer negligible. The limit of ξ up to which the boundary layer solutions may be applicable is sometimes as low as $\xi = 130°$, as discussed later.

Other boundary conditions. Laminar transport from a nonisothermal cylinder has also been analyzed. Using the Blasius series of the form discussed above and assuming that the surface temperature is varied as

$$\phi = \frac{t - t_\infty}{t_0 - t_\infty} = 1 + a_1\xi^2 + a_2\xi^4 \tag{5.4.42}$$

where a_1 and a_2 are arbitrary constants, Koh and Price (1965) obtained the additional contributions to the Nusselt number due to these effects. The heat transfer parameter in Eqs. (5.4.38) and (5.4.39) for the isothermal condition in terms of a_1 and a_2 now becomes

Table 5.4.1 Values of $\mathrm{Nu}_R(x)/\mathrm{Gr}_R^{1/4}$ for an isothermal horizontal circular cylinder at various azimuthal angles and Prandtl numbers

	ξ						Arithmetic mean from
Pr	0	30	60	90	120	150	ξ = 0 to 180°
0.733	0.3770	0.3724	0.3590	0.3365	0.3039	0.2611	0.3350
1.0	0.4214	0.4162	0.4010	0.3753	0.3379	0.2853	0.3729
5.0	0.7142	0.7051	0.6790	0.6325	0.5617	0.4485	0.6235
10.0	0.8822	0.8699	0.8350	0.7753	0.6877	0.5435	0.7656

Source: From Muntasser and Mulligan (1978).

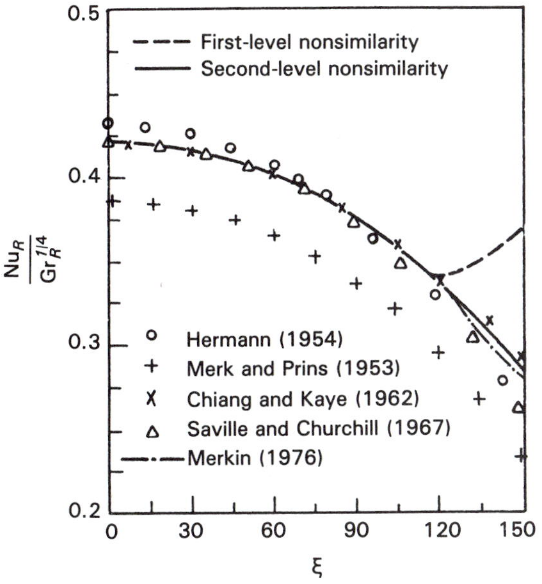

Figure 5.4.5 Peripheral variation of the local convective heat transfer from a horizontal cylinder for Pr = 1.0. *(Reprinted with permission from M. A. Muntasser and J. C. Mulligan, J. Heat Transfer, vol. 100, p. 165. Copyright © 1978, ASME.)*

$$\frac{\mathrm{Nu}_R(\xi)}{\mathrm{Gr}_R^{1/4}} = 0.37023 + (0.75688a_1 - 0.01609)\xi^2 + (0.92847a_2 + 0.09471a_1^2 - 0.02885a_1 - 0.00009)\xi^4 \qquad \text{for Pr} = 0.7 \qquad (5.4.43)$$

$$= 0.42143 + (0.85200a_1 - 0.01861)\xi^2 + (1.0411a_2 + 0.10702a_1^2 - 0.03305a_1 - 0.00011)\xi^4 \qquad \text{for Pr} = 1.0 \qquad (5.4.44)$$

Lin and Chern (1981) extended the solution method of Lin and Chao (1974), using a Merk-type series, to calculate local heat transfer rates over a nonisothermal surface. The results were presented for a horizontal circular cylinder, again for ϕ given in Eq. (5.4.42), where $a_1 = -0.25$ and $a_2 = 0.125$, for Pr = 0.72 and 100. The computations showed that the series behaved well up to $\xi = 130°$.

Koh (1964) presented results for an assigned variable surface heat flux q'' prescribed as follows, where q''_0 is q'' at $x = 0$:

$$\frac{q''}{q''_0} = 1 + a_1\xi^2 + a_2\xi^4 \qquad (5.4.45)$$

The uniform flux condition corresponds to $a_1 = a_2 = 0$. The local Nusselt number is

$$\mathrm{Nu}_R(\xi) = \frac{q''}{t_0 - t_\infty}\frac{R}{k} \qquad (5.4.46)$$

The resulting local temperature difference is

$$t_0 - t_\infty = \frac{Rq_0''}{k(\mathrm{Gr}_R^*)^{1/5}} [2.2142 + (0.04708 + 1.0831a_1)\xi^2 + (0.00157$$

$$+ 0.02896a_1 - 0.05405a_1^2 + 0.88294a_2)\xi^4] \qquad \text{for Pr} = 0.7 \qquad (5.4.47)$$

$$= \frac{Rq_0''}{k(\mathrm{Gr}_R^*)^{1/5}} [1.9963 + (0.04361 + 0.98737a_1)\xi^2 + (0.00150$$

$$+ 0.02691a_1 - 0.05020a_1^2 + 0.80806a_2)\xi^4] \qquad \text{for Pr} = 1.0 \qquad (5.4.48)$$

Here $\mathrm{Gr}_R^* = g\beta q_0''R^4/k\nu^2$. The surface condition of uniform flux has also been analyzed with a Görtler series by Wilks (1972). The solutions for Nusselt number and temperature and velocity fields are presented for Pr = 0.1, 0.7, 1, 10, and 100.

Heat transfer at moderate and low Grashof numbers. There are many applications and experiments in which these conditions arise. Then the boundary layer formulation does not apply. The predominant effects neglected in the boundary layer analysis that become most important at lower Grashof numbers are those due to boundary layer curvature. A more detailed solution of the complete two-dimensional Navier-Stokes equations, coupled with the energy equation, must then be obtained. See Peterka and Richardson (1969) for one such numerical calculation for Pr = 0.72 for Gr_R from 10^8 to moderate values of 10^3. A Blasius-type transformation was used—see Eqs. (5.4.24) and (5.4.28)–(5.4.30)—and the leading equations in f_0 and ϕ_0 were solved numerically. The computed temperature and velocity profiles for various values of Gr_R are shown in Fig. 5.4.4. With decreasing Grashof number, the temperature profiles appear to be relatively unaffected. However, the values of $\phi_0'(0)$ tabulated below show significant variation with Gr_R:

Gr_R	10^3	3×10^3	6×10^3	10^4	3×10^4	10^5	10^6	10^7	10^8	∞
$-\phi_0'(0)$	0.4729	0.4504	0.4385	0.4311	0.4179	0.4068	0.3927	0.3846	0.3800	0.3741

Clearly, the effect of decreasing Grashof number is to increase the heat transfer rate. These $\phi_0'(0)$ values were correlated within 3% as follows:

$$\phi_0'(0) = -0.3741(1 + 1.33\mathrm{Gr}_R^{-1/4} - 0.357\mathrm{Gr}_R^{-1/2} + \cdots) \qquad (5.4.49)$$

The limited Schlieren interferometer data of Peterka and Richardson showed reasonable agreement with the calculations.

The effects of curvature on transport from a cylinder have been analyzed by Akagi (1965) through a perturbation analysis. These effects were found to be small when Ra > 10^5, around Pr = 1. It was indicated that for Pr << 1 and Pr >> 1, the curvature effects exist even at very large Grashof numbers. Curvature effects in unsteady natural convection, during the initial transient period for a suddenly heated circular cylinder, have also been calculated by Gupta and Pop (1977). It was shown that the curvature leads to an increase in skin friction as well as heat transfer rate from the cylinder.

Heat transfer from a circular cylinder at extremely small Grashof numbers has also been analyzed; see Nakai and Okazaki (1975). The temperature field is divided into near and far fields. In the near field around the cylinder, conduction is dominant in comparison with convection. In the far field, at a large distance above the cylinder in the plume, convection is dominant. The temperature field in the near region is obtained from continuity and energy equations only. In the far field the similar solution discussed in Section 3.7 for plumes is obtained. The circumferential average temperatures $\bar{\phi}$ from these two solutions are then joined to determine the heat transfer, where $\bar{\phi} = (1/2\pi)\int_0^{2\pi}\phi \, d\xi$. The following relation was obtained:

$$\frac{1}{\text{Nu}_R} = \frac{1}{3}\ln E - \frac{1}{3}\ln(\text{Nu}_R\text{Gr}_R) \tag{5.4.50}$$

where $E = 3.1(\text{Pr} + 9.4)^{1/2}\text{Pr}^{-2}$, $\text{Pr}^2\text{Gr} \leq 10^{-3}$.

A comparison with the experimental data of Collis and Williams (1954) for $\text{Pr} = 0.72$, Gebhart and Pera (1970) for $\text{Pr} = 6.3$, and Nakai and Okazaki for $\text{Pr} = 0.7$ showed good agreement with this relation. It was concluded that Eq. (5.4.50) is accurate for Grashof numbers extending from extremely low values of $O(10^{-10})$ up to values of $O(10^{-1})$ for all values of the Prandtl number.

The studies of Peterka and Richardson (1969) and Nakai and Okazaki (1975) do not cover the range $10^{-1} < \text{Gr}_R < 10^3$. Numerical solutions by Fujii et al. (1979), Kuehn and Goldstein (1980), and Farouk and Güceri (1981) include this range. Fujii et al. improved the boundary layer solutions by including the curvature terms in the x-momentum and energy equations. The momentum equation in the radial direction and the pressure term in the x-momentum equation were neglected. Based on the numerical results, the average convection coefficient for $\text{Pr} = 0.7$, 10, and 100, in the range $10^{-4} \leq \text{Gr}_D \leq 10^4$, was shown to be given by the following relation to within an accuracy of 1%:

$$\frac{2}{\text{Nu}_D} = \ln\left[1 + \frac{4.065}{F(\text{Pr})\text{Ra}_D^m}\right] \tag{5.4.51}$$

where $m = \frac{1}{4} + 1/(10 + 4\text{Ra}_D^{1/8})$, $\text{Gr}_D = g\beta(t_0 - t_\infty)D^3/\nu^2$, $\text{Ra}_D = \text{Gr}_D\text{Pr}$, and $F(\text{Pr})$ is a function of Pr to be taken from vertical flat surface results such as those given in Table 3.4.1. A comparison with the experimental data of others, for instance, Collis and Williams (1954), Gebhart and Pera (1970), and Fand et al. (1977), at different Pr values showed satisfactory agreement with Eq. (5.4.51) in the range $10^{-10} \leq \text{Ra}_D \leq 10^7$. When compared with Eq. (5.4.50) for $\text{Pr} = 0.7$, 10, and 100, the agreement between the two was very good for $\text{Pr} = 0.7$ but less good at higher Pr.

Kuehn and Goldstein (1980) numerically integrated the complete Navier-Stokes and energy equations for an isothermal local, at ξ, and the average horizontal circular cylinder for values in the range $10^0 \leq \text{Ra}_D \leq 10^7$. The local, at ξ, and the average heat transfer coefficients were computed for $\text{Pr} = 0.01$, 0.1, 0.7, 1, 5, and 10. These results are in good agreement with those in a later study by Farouk and Güceri (1981). Streamlines and isotherms were computed at different values of the Rayleigh number. Some results are seen in Fig. 5.4.6. These results show several interesting features of flow and suggest the bounds of validity of various calculation procedures. At $Ra_D = 10^0$ the flow is close to vertical over the field, convecting heat above the cylinder in a well-defined plume. The temperature distribution resembles that expected near a line heat source.

Similar characteristics are seen at $Ra_D = 10^2$. However, at much higher Rayleigh numbers a boundary layer mode of transport arises. At $Ra_D = 10^4$ the layer thickness is about equal to the cylinder radius and non-boundary layer effects are still very important. At $Ra_D = 10^6$, however, the thickness is much less and the boundary layer solutions might be expected to be reasonably accurate.

Kuehn and Goldstein (1980) also computed the angular and radial velocities and temperature profiles at $Ra_D = 10^5$ for $Pr = 0.7$ at various angles downstream, as shown in Figs. 5.4.7–5.4.9. The tangential velocity component distributions, shown in Fig. 5.4.7, for $30 \leq \xi \leq 150°$ are very similar to those predicted by boundary layer analysis. However, for $\xi > 150°$ the separation effect begins to develop, as seen in the decrease in u, which must be zero at $\xi = 180°$.

The radial velocity plots in Fig. 5.4.8 also indicate separation and plume formation. For $0 \leq \xi \leq 120°$ the radial velocities are fairly small and uniform in the outer region. The entrained flow is moving toward the cylinder. However, between about $\xi = 150$ and $160°$ the flow changes from inflow to outflow, indicating turning to separate into plume flow. This result is in agreement with the numerical calculations of Ingham (1978), which dealt with the formation of a buoyant plume due to colliding boundary layers at ξ near 180°. In Fig. 5.4.8 the separation region outflow velocities are seen to be about an order of magnitude higher than the inflow velocities in the upstream boundary layer. Also, near the cylinder the vertical velocity in the center of the plume at $\xi = 180°$ is less than at $\xi = 175°$. These observations are characteristic of developing plume flow, as found experimentally by Jodlbauer (1933).

The radial temperature distributions (Fig. 5.4.9) follow similar trends. For $0 \leq \xi \leq 120°$ they are similar to boundary layer solutions. At larger angles the turning of the flow to form a plume significantly alters the temperature distribution and the thermal boundary layer thickness.

Separation. The general question of flow separation has been dealt with in detail in the experimental study by Pera and Gebhart (1972) for flow over a cylindrical surface and by Jaluria and Gebhart (1975) for flow over a spherical surface. The results are described in Section 5.8. Here it is sufficient to note that flow separation does not occur in the conventional sense in forced flows. Cross-flow vortices have not been observed. Instead, as the upper stagnation point is reached, the heated fluid coming from the converging boundary region flow simply realigns itself upward and rises into a steady developing plume above the surface. This realignment process begins well ahead of the top stagnation point and is accompanied by an abrupt growth in boundary layer thickness. The flow velocity and entrainment increase significantly and result in an increased heat transfer rate, despite the rapid increase in boundary layer thickness.

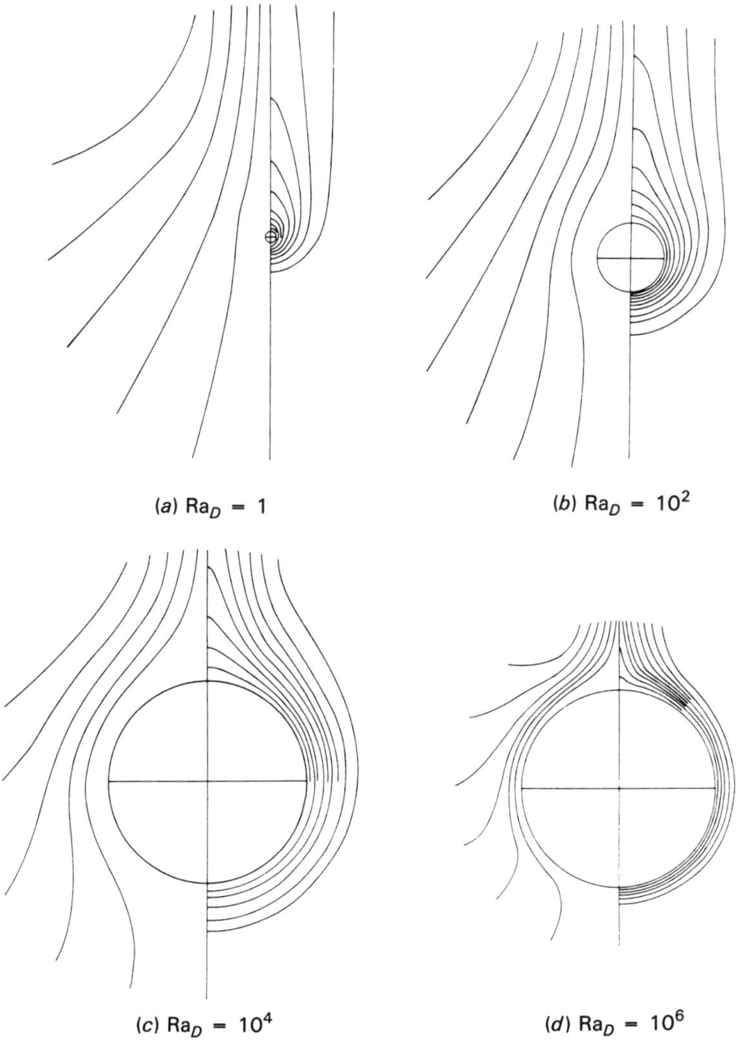

Figure 5.4.6 Streamlines and isotherms for a circular cylinder for Pr = 0.7 at four values of the Rayleigh number; $Ra_D = 10^0(a)$, $10^2(b)$, $10^4(c)$, and $10^6(d)$. *(Reprinted with permission from* Int. J. Heat Mass Transfer, *vol. 23, p. 971, T. H. Kuehn and R. J. Goldstein. Copyright © 1980, Pergamon Journals Ltd.)*

The foregoing studies clearly demonstrate that for values of ξ between 120 and 150°, the first effects of separation into a plume flow arise. Thereafter, the flow and temperature fields depart entirely from boundary layer forms. Such effects increase at larger ξ. The boundary layer formulation becomes inapplicable. Kuehn and Goldstein suggest a value of $\xi = 130°$ for the limit of reasonable accuracy.

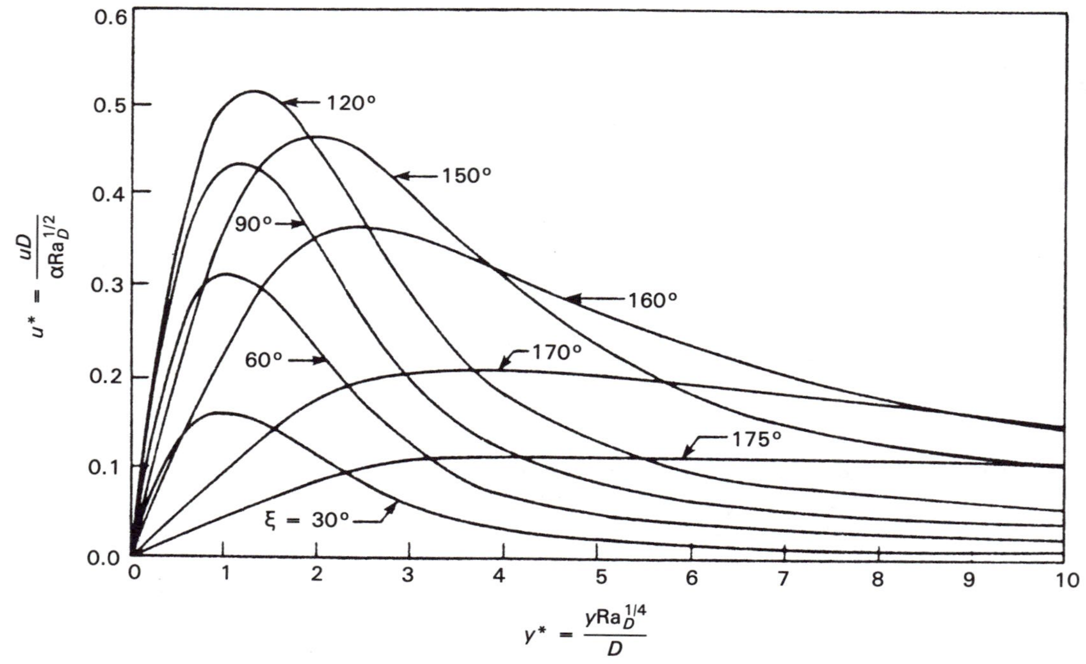

Figure 5.4.7 Tangential velocity distributions at $Ra_D = 10^5$, Pr = 0.7, for a circular cylinder. *(Reprinted with permission from* Int. J. Heat Mass Transfer, *vol. 23, p. 971, T. H. Kuehn and R. J. Goldstein. Copyright © 1980, Pergamon Journals Ltd.)*

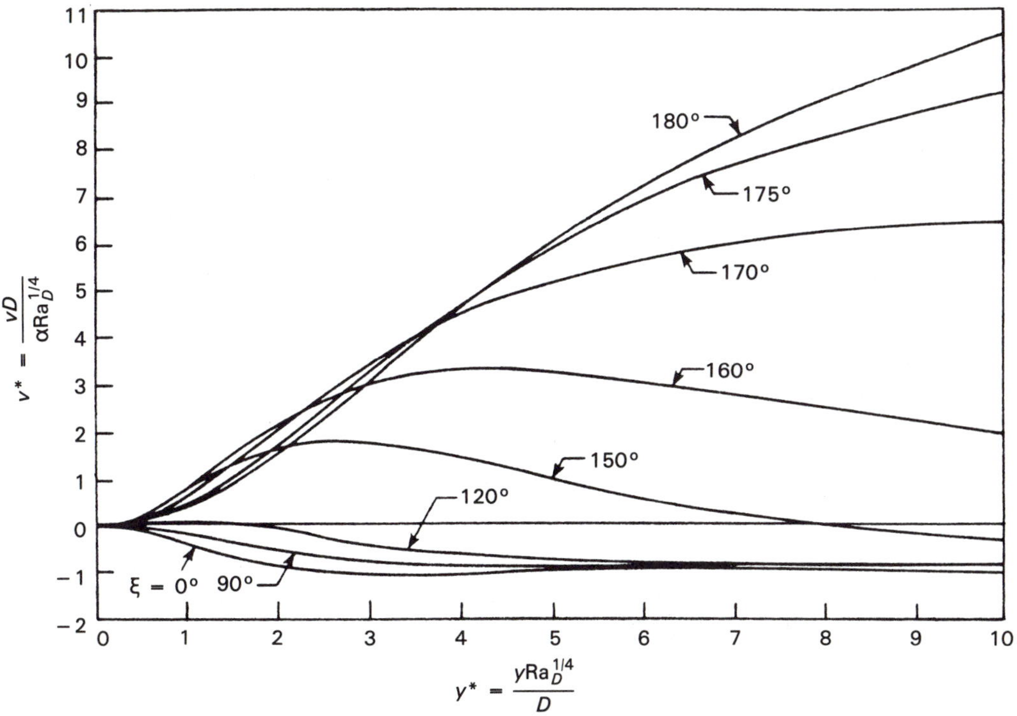

Figure 5.4.8 Radial velocity distributions at $Ra_D = 10^5$, Pr = 0.7, for a circular cylinder. *(Reprinted with permission from* Int. J. Heat Mass Transfer, *vol. 23, p. 971, T. H. Kuehn and R. J. Goldstein. Copyright © 1980, Pergamon Journals Ltd.)*

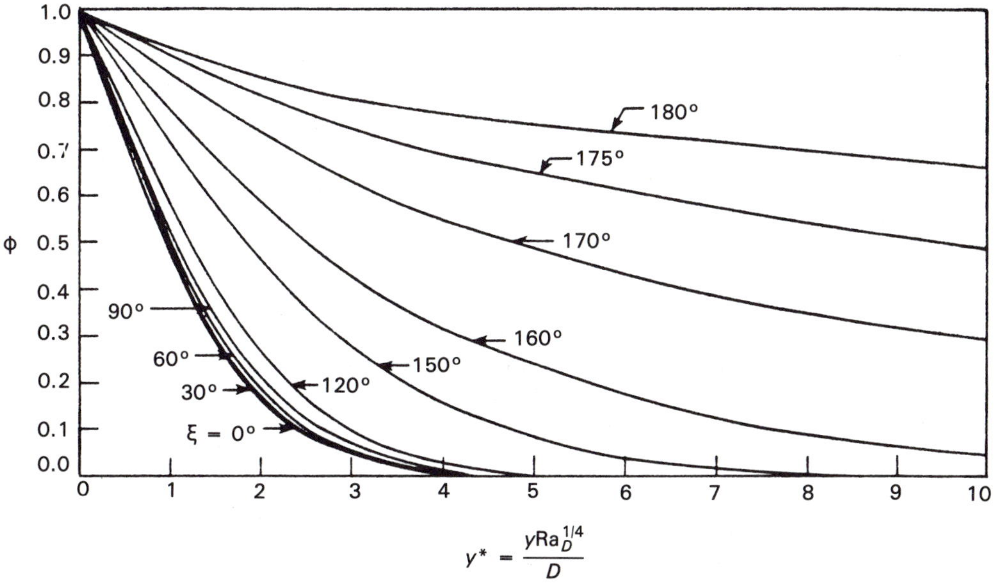

Figure 5.4.9 Temperature distributions at $Ra_D = 10^5$, Pr = 0.7, for a circular cylinder. *(Reprinted with permission from* Int. J. Heat Mass Transfer, *vol. 23, p. 971, T. H. Kuehn and R. J. Goldstein. Copyright © 1980, Pergamon Journals Ltd.)*

5.4.3 Laminar Natural Convection around a Sphere

As for a cylinder, a similarity solution does not exist for the sphere geometry. This geometry has, however, been widely studied because of its importance. Several of the studies are related to large Grashof number transport, at which the boundary region is much thinner than the radius of curvature. These are discussed first, followed by a brief description of the transport at low and moderate Grashof numbers.

Transport at large Grashof number. The governing boundary layer equations in terms of Fig. 5.4.2*a* are Eqs. (5.4.20)–(5.4.22), except that the continuity Eq. (5.4.20) is now replaced by

$$\frac{\partial}{\partial x}(ur) + \frac{\partial}{\partial y}(vr) = 0 \tag{5.4.52}$$

where r is the radial distance from the axis of symmetry. Equations (5.4.52), (5.4.21), and (5.4.22) appear to have been treated first by Yamagata (1943). A transformation similar to that of Hermann (1954) for a horizontal cylinder was used to obtain the azimuthal function concerning the effect of the shape of the body on boundary layer thickness.

Merk and Prins (1954) used an integral method to obtain the following heat transfer results for an isothermal sphere:

$$\mathrm{Nu}_D = F(\mathrm{Pr})(\mathrm{Gr}_D\mathrm{Pr})^{1/4} \tag{5.4.53}$$

where $F(\mathrm{Pr})$ was determined to be 0.474, 0.497, 0.576, 0.592, and 0.595 for Pr values of 0.7, 1.0, 10.0, 100.0, and ∞, respectively. Acrivos (1960) analyzed transport for large values of the Prandtl number. The inertia terms were omitted in the equations, and an exact form for the leading term in the asymptotic expansion of the solution as $\mathrm{Pr} \to \infty$ was derived. The mean Nusselt number for the whole sphere was determined as

$$\mathrm{Nu}_D = 0.583(\mathrm{Gr}_D\mathrm{Pr})^{1/4} \tag{5.4.54}$$

A more accurate solution for an arbitrary value of Pr was proposed by Chiang et al. (1964). They extended the use of a series solution method, discussed earlier for a horizontal cylinder, to the analysis for a sphere. Using Eqs. (5.4.23) and (5.4.28)–(5.4.30) and noting that for a sphere

$$u = \frac{1}{r}\frac{\partial}{\partial y}(\psi r) \qquad v = -\frac{1}{r}\frac{\partial}{\partial x}(\psi r) \qquad r = R \sin \xi \tag{5.4.55}$$

they obtained the solution for the local Nusselt number $\mathrm{Nu}_R(\xi)$ for an isothermal sphere, for $\mathrm{Pr} = 0.7$, where $\mathrm{Nu}_R(\xi) = h_\xi R/k$.

$$\text{Nu}_R(\xi) = (0.4576 - 0.03402\xi^2)\text{Gr}_R^{1/4} \tag{5.4.56}$$

For the uniform flux condition, the surface temperature distribution was determined as

$$\phi = \frac{(t_0 - t_\infty)k\text{Gr}_R^{*1/5}}{q''R} = 1.8691 + 0.08127\xi^2 \tag{5.4.57}$$

where $\text{Gr}_R^* = g\beta q''R^4/k\nu^2$.

The computed temperature and velocity profiles, along with the local Nusselt number results at various angular positions, are shown in Figs. 5.4.10–5.4.12 for the isothermal temperature condition. The results were presented up to 120° since it was expected that the analysis may be invalid beyond that.

Numerical solutions have also been obtained by Potter and Riley (1980), taking into account the eruption of the fluid from the boundary layer to form the plume above the sphere. In this region the results support the flow separation characteristics observed by Jaluria and Gebhart (1975) mentioned earlier in this section. The local Nusselt number results are compared in Fig. 5.4.12 with the solution of Chiang et al. The agreement in the range of overlap of the two solutions is quite good.

Although Potter and Riley gave results for angular positions all the way up to 180°, they indicated that those at larger angular positions might be in error. Again, this results from other effects arising at large angular position. Based on the data of Jaluria and Gebhart (1975), curvature effects must be taken into account quite far upstream of the upper stagnation point for accurate estimates of the heat transfer. The data of Kranse and Schenk (1965) and Amato and Tien (1972) support this.

A first-order correction to account for such effects was obtained by Elenbass (1942), who determined a semiempirical correlation between Nu_D and Gr_D. Raithby and Hollands (1975) used an approximate analysis to consider the effects of boundary layer curvature. A brief description of this technique is given later in this section. The following expression for average heat transfer was obtained:

$$\text{Nu}_D = 2 + C_s\text{Ra}_D^{1/4} \tag{5.4.58}$$

where

$$C_s = 0.56\left(\frac{\text{Pr}}{0.846 + \text{Pr}}\right)^{1/4} \tag{5.4.59}$$

A comparison of this relation for Pr = 0.72 with the data of Elenbass (1942) for air and with the recommended correlation equations of Kyte et al. (1953) and Amato and Tien (1972) showed good agreement.

Transport at small Grashof numbers. Heat transfer from an isothermal sphere at small values of the Grashof number ($0 < \text{Gr}_R < 1$) has also been investigated

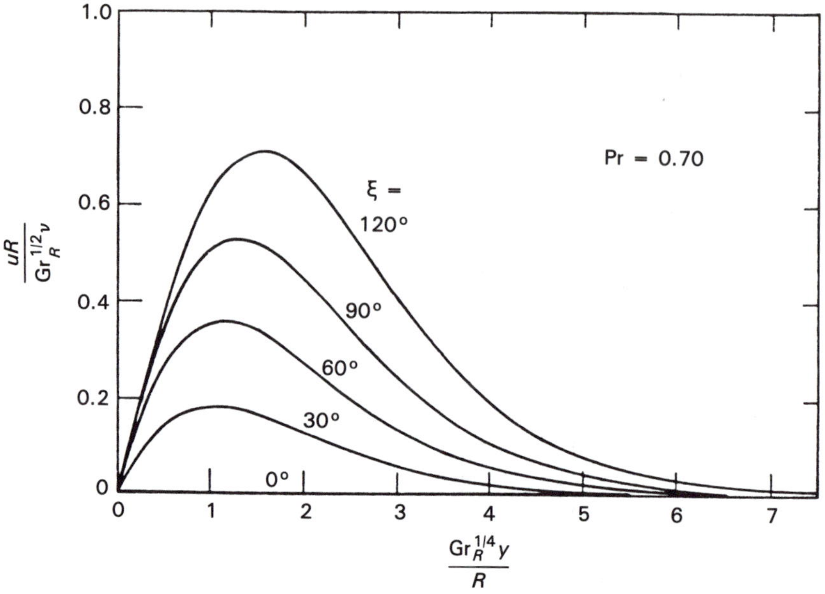

Figure 5.4.10 Dimensionless velocity profiles for an isothermal sphere. *(Reprinted with permission from T. Chiang et al.,* J. Heat Transfer, *vol. 86, p. 537. Copyright © 1964, ASME.)*

theoretically; see Mahony (1956) and Hossain and Gebhart (1970). Mahony solved the natural convection flow around a sphere. It was shown that the pure conduction solution, which might be expected to be valid for very small values of Grashof number, applies only over a distance a out from the sphere, where $a = r/R \leq O(\mathrm{Gr}_R)^{-1/2}$. Farther out, the inertia and convection terms must be considered. Hossain and Gebhart used an asymptotic expansion technique to calculate transport. Solutions to the governing differential equations were expressed in a series

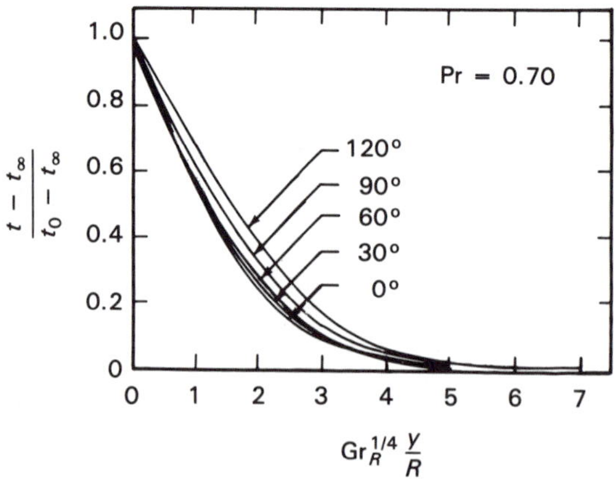

Figure 5.4.11 Dimensionless temperature profile for an isothermal sphere. *(Reprinted with permission from T. Chiang et al.,* J. Heat Transfer, *vol. 86, p. 537. Copyright © 1964, ASME.)*

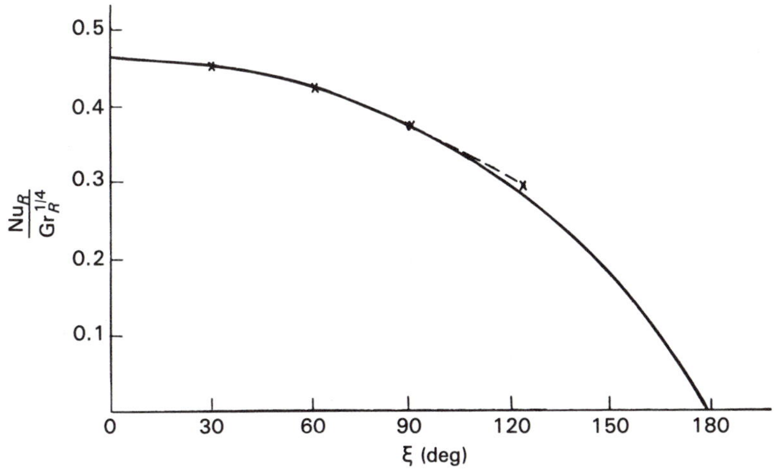

Figure 5.4.12 Local Nusselt number from an isothermal sphere, Pr = 0.7. (——) Potter and Riley (1980) and (– –) Chiang et al. (1964). *(Reprinted with permission from J. M. Potter and N. Riley,* J. Fluid Mech., *vol. 100, p. 769. Copyright © 1980, Cambridge University Press.)*

with the Grashof number taken as the expansion parameter. Both the temperature and the flow field were determined analytically. Based on a numerical integration, the following expression was given for the average Nusselt number for $0 < Gr_R < 1$:

$$Nu_D = \frac{hD}{k} = 2 + Gr_R + Gr_R^2(0.139 - 0.4519Pr + 1.1902Pr^2) \quad (5.4.60)$$

Fujii et al. (1981) presented numerical solutions for flow around an isothermal sphere for Pr = 0.7, 10, and 100 at small and moderate values of the Rayleigh number; Ra_D was 10^{-2}, 10^2, and 10^6. The curvature terms were retained in the x-momentum and energy equations. Based on the results, the following relation was proposed:

$$Nu_D = 2 + 0.507F(Pr)Ra_D^m \quad (5.4.61)$$

where

$$m = \frac{1}{4} + \frac{1}{4 + 7.5Ra_D^{0.08}} \quad (5.4.62)$$

and $F(Pr)$ may be evaluated from Table 3.4.1 or from the formulas of Churchill and Usagi (1972).

Other numerical solutions, by Geeola and Cornish (1981), used the complete set of the Navier-Stokes, continuity, and energy equations, which were solved by an extrapolated Gauss-Seidel method. The flow and temperature fields were presented for an isothermal sphere for Gr_D = 0.05, 1, 10, 25, and 50 for Pr = 0.72. The calculated variation of local Nusselt number $Nu_D(\xi) = h_\xi D/K$ with angle ξ

is shown in Fig. 5.4.13. As the Grashof number increases, the upstream region is thinned by convection, resulting in higher local heat transfer rates. The local values decrease downstream.

5.4.4 Other Two-Dimensional Planar or Axisymmetric Flows

In addition to the cylinder and sphere geometries discussed above, there is a body of information available for flows adjacent to several other geometries. Several of these that demonstrate procedures of most generality are considered next.

Raithby and Hollands (1975) developed an approximate method to obtain heat transfer from surfaces of different geometries and applied it to planar two-dimensional, axisymmetric, and enclosure flows. In a general application of the method, the natural convection laminar boundary layer adjacent to a surface is divided into two regions. The inner region extends from the wall to the position of the local velocity extremum. The outer region is the rest of the transport field. It is assumed that in the inner region inertial forces are not important. The energy transfer normal to the wall is by conduction only. The transport in this region is then quite similar to that in the liquid film in condensation problems. Using this analogy, an analysis is made that follows closely the standard Nusselt analysis for filmwise condensation. See Raithby and Hollands for details.

For two-dimensional and axisymmetric bodies, the general solution for the local Nusselt number given by Raithby and Hollands (1975) is

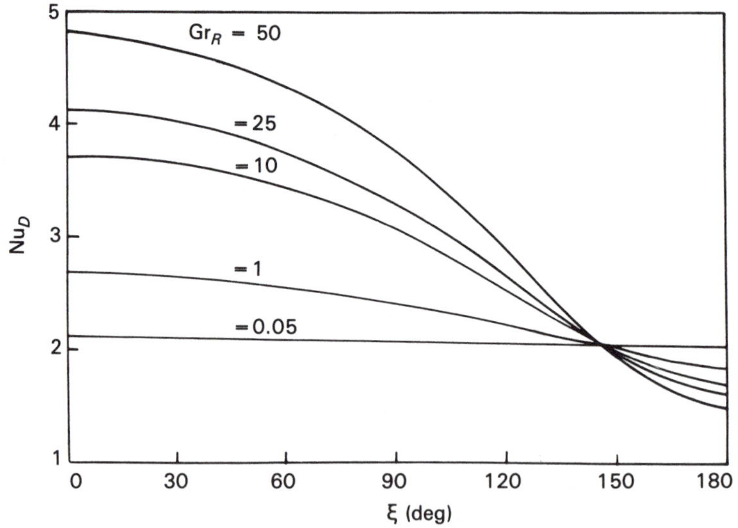

Figure 5.4.13 Variation of local Nusselt number with Grashof number for an isothermal sphere, Pr = 0.72. *(Reprinted with permission from* Int. J. Heat Mass Transfer, *vol. 24, p. 1369, F. Geoola and A. R. H. Cornish. Copyright © 1981, Pergamon Journals Ltd.)*

$$\mathrm{Nu}_x = C(\mathrm{Pr})\mathrm{Ra}_x^{1/4} \frac{\left\{(1/x)\int_0^x [r^{4i}(g_x/g)]^{1/3}\, dx\right\}^{-1/4}}{\left[(1/x)\int_0^x r^i\, dx\right]} \tag{5.4.63}$$

where g_x is the component of **g** acting along the surface and x the curvilinear distance along the surface from the stagnation point, as defined in Figs. 5.1.2 and 5.4.2 for particular geometries. In Eq. (5.4.63) $i = 0$ for two-dimensional bodies and $i = 1$ for axisymmetric bodies. As before, r is the horizontal distance from the surface element to the axis of symmetry and $C(\mathrm{Pr})$ is a "universal" function of Pr. Based on the proposal by Churchill and Usagi (1972), Raithby and Hollands (1976b) recommend $C(\mathrm{Pr}) = 0.503/[1 + (0.492/\mathrm{Pr})^{9/16}]^{4/9}$ for laminar flow. Equation (5.4.63) applies for boundary layer transport. This method has also been extended to incorporate both the curvature and nonboundary layer effects.

Axisymmetric bodies. For prolate and oblate spheroids, shown in Fig. 5.4.2, Raithby et al. (1976) also used this method. The following correlation was given for the average Nusselt number Nu_B based on dimension B:

$$\mathrm{Nu}_B = h\frac{B}{k} = \sqrt[4]{2}\,\frac{(f_2)^{3/4}}{f_1}\,\bar{C}(\mathrm{Pr})\left(\frac{m}{B}\mathrm{Ra}_B\right)^{1/4} \tag{5.4.64}$$

where $\bar{C}(\mathrm{Pr}) = \frac{4}{3}C(\mathrm{Pr})$, f_1 and f_2 are definite integrals listed in Table 5.4.2, $m = B$ for prolate spheroids, and $m = A$ for oblate spheroids.

Table 5.4.2 Constants f_1 and f_2

	Prolate spheroid		Oblate spheroid		Elliptic cylinder	
C/B	f_1	f_2	f_1	f_2	f_1	f_2
1.00	2.000	1.683	2.000	1.683	3.142	2.587
0.900	1.935	1.653	1.868	1.601	2.987	2.514
0.800	1.873	1.624	1.739	1.518	2.836	2.441
0.700	1.814	1.596	1.615	1.435		
0.600	1.759	1.570	1.494	1.352	2.553	2.300
0.500	1.709	1.545	1.380	1.269		
0.400	1.665	1.524	1.274	1.187	2.301	2.169
0.300	1.627	1.506	1.177	1.108		
0.200	1.598	1.491	1.094	1.033	2.101	2.059
0.100	1.578	1.481	1.030	0.969	2.032	2.019
0.050	1.573	1.479	1.009	0.945		
0.005	1.571	1.478	1.000	—		
0.001	$\approx\pi/2$	1.478	1.000	—		
0.000					2.000	2.000

Source: From Raithby et al. (1976) and Raithby and Hollands (1976b).

Also concerning spheroids, it is noted that, as the ratio of the minor to the major axis approaches zero, the prolate spheroid resembles a thin vertical needle tapered from the center toward both ends. The analysis for this special case of prolate spheroids has been treated by Raithby and Hollands (1976a). The related kind of transport is that from thin vertical needles tapered only toward the upstream end. Similarity solutions are possible for certain surface temperature conditions and were discussed in Section 4.3.

Two-dimensional planar bodies. Among such bodies, elliptic cylinders (Fig. 5.4.2) have also been studied in a number of investigations. Raithby and Hollands (1976b) gave the following expression for average heat transfer rate for an isothermal surface, using the approximate method of solution described above for spheroids:

$$\mathrm{Nu}_B = 2^{1/4}\left(\frac{f_2^{3/4}}{f_1}\right)\bar{C}(\mathrm{Pr})\mathrm{Ra}_B^{1/4} \tag{5.4.65}$$

where the numerical values of f_1 and f_2 are also given in Table 5.4.2 and $\bar{C}(\mathrm{Pr})$, as before, is the "universal" function of Pr given above.

Note that as $A/B \to 1$, the elliptic cylinder becomes circular. Then for an isothermal horizontal circular cylinder of diameter B, the result is

$$\mathrm{Nu}_B = 0.77\bar{C}(\mathrm{Pr})\mathrm{Ra}_B^{1/4} \tag{5.4.66}$$

When compared with the solution obtained by Saville and Churchill (1967), for $\mathrm{Pr} = 0.01, 0.7$, and 1000, good agreement was indicated. As $A/B \to 0$ the elliptic cylinder degenerates into an isothermal vertical plate of height B. Then,

$$\mathrm{Nu}_B = \bar{C}(\mathrm{Pr})\mathrm{Ra}_B^{1/4} \tag{5.4.67}$$

This relation is the same as that suggested by Churchill and Usagi (1972) and given in Eq. (3.9.4).

More detailed transport results for horizontal cylinders of elliptic cross section in both blunt (major axis horizontal) and slender (major axis vertical) orientations have also been obtained numerically by Merkin (1977). Boundary layer equations, Eqs. (5.4.20)–(5.4.23), for both isothermal and uniform flux surface conditions were solved. Lin and Chao (1978) obtained solutions for the same conditions by using a series expansion of the form proposed by Merk (1959) and indicated good agreement with Merkin's results. Harpole and Catton (1976) used a Blasius series expansion to determine heat transfer from downward-facing, blunt-nosed, two-dimensional or axisymmetric bodies to liquid metals. Both arbitrary surface heat flux and surface temperature conditions were considered. Universal functions were given that were said to be valid for generating solutions for any arbitrary blunt-nosed surface. However, Lin and Chao (1978) found that solutions obtained with these universal functions are quite accurate for circular cylinders, but when they are applied to blunt elliptic cylinders, large discrepancies occur.

Summary. Several approximate methods of solution are available to determine heat transfer rates from surfaces for which no similarity or other analytical solutions are available. These include the method proposed by Raithby and Hollands and the series method. The series used may be Blasius, Merk, or Görtler. Blasius-type series have been used by, for example, Chiang and Kaye (1962) for horizontal circular cylinders (Section 5.4.2) and Chiang et al. (1964) for spheres (Section 5.4.3). For the use of Görtler-type series for two-dimensional planar and axisymmetric boundary layer flows, see Saville and Churchill (1967), Wilks (1972), and Lin (1976). For a Merk-type series, see Lin and Chao (1974) and Chao and Lin (1975). A comparison of the predictive capabilities of these different series is reviewed by Lin and Chao (1978).

However, there are many uncertainties in some of these results. Also, some provide no information on local velocity or temperature profiles. Further, the agreement of such information with data for a few geometries and conditions is an inadequate basis for assuming universal applicability. For quite different body configurations or Prandtl numbers, such results may be oversimplified.

5.4.5 Three-Dimensional Flows

Such buoyancy-induced flows arise in many practical circumstances. A ready example is that adjacent to an inclined cylinder. The cross-sectional shape can be of a circular cylinder, as shown in Fig. 5.4.14, or a two-dimensional planar body, shown in Fig. 5.1.2a. There are three components of the buoyancy force. That is, $B = (B_t, B_n, B_a)$, where B_t and B_n are tangential and normal in a plane perpendicular to the cylinder axis and B_a is along the axis of the cylinder and generates an axial flow component. Such effects are much more complicated than two-dimensional flows over spheres and other symmetric bodies. However, the inclined cylinder has been studied and heat transfer results have been obtained through approximate analytical and numerical methods as well as through experiments. This configuration is described next, followed by some generalized solutions for other such orientations.

Transport from inclined cylinders. The earliest systematic study of this appears to be due to Farber and Rennat (1957). Their experiments were with a cylinder 1.829 m long with a 3.175-mm outside diameter at inclinations varying from horizontal to vertical. The cylinder was electrically heated to provide a constant heat flux surface condition. The heat transfer coefficient was found to decrease with an increase in inclination angle γ, measured from the horizontal. No general correlation of the results was given. Kato and Ito (1968) used similarity criteria to analyze this transport and obtained an equation for the average Nusselt number. Their experimental data indicated higher values of Nusselt number than predicted by the analysis.

Savage (1969) showed that for an infinitely long cylinder, that is, with no variation of flow quantities along the z direction, similarity solutions exist for the boundary layer equations for an isothermal surface condition. The cross-sectional shapes that permit similarity are those given in Fig. 5.1.2*a*, for which r is given by Eq. (5.4.4). In particular, velocity and temperature profiles were obtained for an inclined cylinder having a parabolic nose [$m = \infty$ in Eq. (5.4.4)] for $\Pr = 0.72$.

For an isothermal, infinite, inclined circular cylinder, numerical solutions were obtained by Peube and Blay (1978) for Pr = 0.72.

Oosthuizen (1976a) considered flow over an inclined circular cylinder of finite length, as shown in Fig. 5.4.14. It is assumed that both ends of the cylinder are insulated so that heat is transferred only from the cylindrical surface of length L. With the usual boundary layer approximations and neglecting B_n, the governing equations are

$$\frac{\partial u}{\partial x} + \frac{\partial v}{\partial y} + \frac{\partial w}{\partial z} = 0 \tag{5.4.68}$$

$$u\frac{\partial u}{\partial x} + v\frac{\partial u}{\partial y} + w\frac{\partial u}{\partial z} = \nu\frac{\partial^2 u}{\partial y^2} + g\beta(t - t_\infty)\cos\gamma\sin\xi \tag{5.4.69}$$

$$u\frac{\partial w}{\partial x} + v\frac{\partial w}{\partial y} + w\frac{\partial w}{\partial z} = \nu\frac{\partial^2 w}{\partial y^2} + g\beta(t - t_\infty)\sin\gamma \tag{5.4.70}$$

$$u\frac{\partial t}{\partial x} + v\frac{\partial t}{\partial y} + w\frac{\partial t}{\partial z} = \frac{\nu}{\Pr}\frac{\partial^2 t}{\partial y^2} \tag{5.4.71}$$

The boundary conditions for an isothermal surface condition are

$$\begin{aligned} z = 0&: \quad u = 0 = w \quad t = t_\infty \quad \text{for all } x \text{ and } y \\ y = 0&: \quad u = v = w = 0 \quad t = t_0 \quad \text{for all } x \text{ and } z \\ y \to \infty&: \quad w \to 0 \quad u \to 0 \quad t \to t_\infty \quad \text{for all } x \text{ and } z \end{aligned} \tag{5.4.72}$$

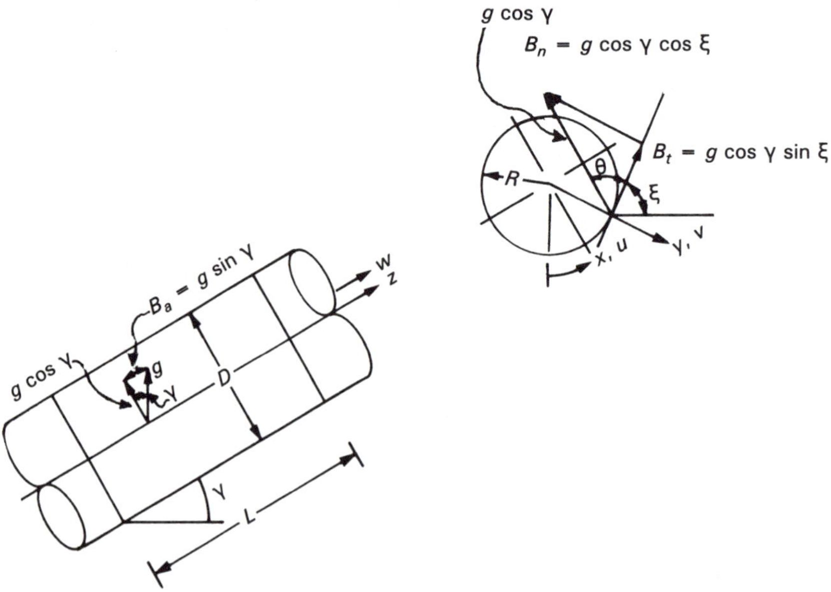

Figure 5.4.14 Coordinate system and nomenclature for an inclined cylinder.

In addition, boundary conditions at the symmetry location $x = 0$ are required. These are

$$x = 0: \qquad u = 0 = \frac{\partial w}{\partial x} = \frac{\partial t}{\partial x} \qquad \text{for all } y \text{ and } z \tag{5.4.73}$$

The Grashof number is defined as

$$\mathrm{Gr}_D = \frac{g\beta(t_0 - t_\infty)D^3 \cos\gamma}{\nu^2} \tag{5.4.74}$$

where $D = 2R$. Equations (5.4.68)–(5.4.73) were solved numerically by Oosthuizen (1976a). Mean heat transfer rates were calculated for different Prandtl numbers and various angles of inclination.

The results showed that the heat transfer coefficient decreases with increasing angle of inclination γ. For a given Prandtl number, an equation of the following form was found suitable for a heated axial length L:

$$\frac{\mathrm{Nu}_D}{(\mathrm{Gr}_D \cos\gamma)^{1/4}} = F(\mathrm{Pr})f\left(\frac{L}{D\tan\gamma}\right) = F(\mathrm{Pr})f(L^*) \tag{5.4.75}$$

where F is a function of Pr and f is some function of $L^* = L/D \tan\gamma$. It was found that for large values of L^* or z, the flow is essentially two dimensional in terms of u and v. The heat transfer rate, then, is essentially the same as that for two-dimensional flow over horizontal cylinders. For small L^* and near the bottom of any long cylinder, the heat transfer rate is effectively the same as if only the flow parallel to the z axis existed.

The associated experimental study of Oosthuizen (1976b) confirmed that the heat transfer rates may be correlated in terms of variables given in Eq. (5.4.75). Mean heat transfer rates were measured for inclined circular cylinders in air, from horizontal to vertical orientations, for $L/D = 8$, 10.67, and 16. For L^* greater than about 10, the heat transfer rate from the cylinder was found to be essentially the same as that over a horizontal cylinder, while for L^* less than 1, the heat transfer rate approached that for vertical cylinders. For intermediate values of L^* the following equation was seen to correlate well all of the data in air:

$$\frac{\mathrm{Nu}_D}{(\mathrm{Gr}_D \cos\gamma)^{0.25}} = 0.42\left(1 + \frac{8.673}{L^{*2}}\right)^{1/8} \tag{5.4.76}$$

This correlation applies for the data, which were in the limited range $4 \times 10^4 < \mathrm{Gr}_D < 9 \times 10^4$. It was suggested that this relation may be used for a broader laminar range of Grashof number, between 10^4 and 10^9.

Al-Arabi and Khamis (1982) obtained heat transfer data from isothermal inclined cylinders to air, at different inclinations, in both the laminar and turbulent range. In the laminar range, said to be $1.08 \times 10^4 < \mathrm{Gr}_D < 6.9 \times 10^5$, the data were about 20% higher than those of Oosthuizen (1976b). Al-Arabi and Salman (1980) had also measured heat transfer rates from a uniformly heated inclined cylinder (uniform heat flux condition). Correlations were given for local and average heat transfer rates in the laminar flow regime. Based on experimental data, Chand and Vir (1980) gave separate correlations for short and long cylinders when the inclination γ is less than $\tan^{-1}(L/D)$.

Other three-dimensional transport. Generalized solutions for transport from three-dimensional bodies of arbitrary shape have also been obtained. Raithby and Hollands (1978) extended the approximation method, discussed earlier in this section for two-dimensional planar and axisymmetric bodies, to include three-dimensional flows over isothermal surfaces. Solutions were given for heat transfer from an inclined cylinder and a horizontal prolate spheroid (with a vertical major axis). The predictions of the analysis for an inclined cylinder when curvature effects are considered show good agreement with the data of Oosthuizen (1976b) discussed above.

Suwono (1980) developed general series solutions for transport from three-dimensional bodies of arbitrary shape to a fluid of any Prandtl number. The method was applied to an inclined isothermal circular cylinder. Numerical solutions were obtained for stream function and temperature profiles for Pr = 0.72. These were compared with the experimental data of Deluche (1970) in air. Good agreement was indicated.

Stewart (1971) derived boundary layer solutions for large Prandtl number fluids for three-dimensional systems and gave the following expression for average Nusselt number Nu and vertical drag force F_D based on a characteristic dimension L:

$$\mathrm{Nu} = -\phi'(0)(|\mathrm{Ra}|)^{1/4}K_2 \tag{5.4.77}$$

$$F_D/\rho\nu^2\mathrm{Gr} = f''(0)(|\mathrm{Ra}|)^{1/4}K_1 \tag{5.4.78}$$

where $\phi(0)$ and f, as in Chapter 3, are the dimensionless temperature and reduced stream function, respectively. Grashof number Gr and Rayleigh number Ra are based on the characteristic length L. The absolute values in Eqs. (5.4.77) and (5.4.78) are necessary to take care of both heating and cooling. K_1 and K_2 are dimensionless constants whose values depend on the geometric configuration of the surface. The values of K_1 and K_2 for various surfaces are given in Table 5.4.3. Also included are some of the surfaces discussed earlier in this chapter. The computed values of $-\phi'(0)$ and $f''(0)$ are 0.5402 and 1.085, respectively. Comparison with the experimental data showed good agreement. Note again that these results apply only to fluids with a large Prandtl number and provide no information on local velocity or temperature profiles.

5.5 HEAT TRANSFER CORRELATIONS

Preceding sections indicate the transport information that has been obtained from analysis. For many geometries some of the flow conditions and physical effects have not yet been treated in a general and satisfactory way. However, additional information is available from experimental studies. This section contains a somewhat limited selection of these results. The contents have been selected for the usefulness of the results as well as the conceptual information conveyed. The following summary is by surface geometry.

Flat inclined surfaces. Based on the studies reviewed in Section 5.2, all the correlations in Section 3.9 for vertical surfaces may be used for inclined flat surfaces by substituting $g \cos \theta$ for g, for at least some values of θ, where θ is the angle of inclination from the vertical. For warm surfaces facing downward, $B_n < 0$, when this procedure is applied, correlations (3.9.3), (3.9.4), and (3.9.6) agree well with the data of Fussey and Warneford (1978), Moran and Lloyd (1975), Fujii and Imura (1972), and others for $0 \leq \theta \leq 85°$. See Churchill (1983). For warm surfaces facing upward, $B_n > 0$, these correlations apply until the flow separation and/or transition takes place. In the literature the terms separation and transition have sometimes been construed as meaning the same mechanisms for inclined flows, as discussed in Section 5.8.

Flows with B_n away from the surface are potentially less stable than those with B_n toward the surface. For the less stable case, Fujii and Imura (1972) determined experimentally the Rayleigh number value at which the separation took place, $\mathrm{Ra}_{x,c}$, for different inclinations. From these data, Churchill (1983) gave the following correlation for separation:

$$\mathrm{Ra}_{x,c} = 10^5 \times e^{11.5 \cos \theta} \tag{5.5.1}$$

where $\mathrm{Ra}_{x,c} = g\beta(t_0 - t_\infty)x_c^3/\nu\alpha$ and x_c is the downstream distance at which separation takes place. Fussey and Warneford (1978) measured heat transfer rates from an inclined uniform flux surface, with the heated surface facing downward, in water. The data indicated the following correlation for transition Rayleigh number $\mathrm{Ra}^*_{x,c}$, defined as the value of x at which the heat transfer rate deviated from the laminar trend:

$$\mathrm{Ra}^*_{x,c} = 6.31 \times 10^{12} e^{0.0705\theta} \qquad -70° < \theta < 0° \tag{5.5.2}$$

where $\mathrm{Ra}^*_{x,c} = \mathrm{Gr}^*_{x,c}\mathrm{Pr} = (g\beta x_c^4 q''/k\nu^2)\mathrm{Pr}$.

In the downstream turbulent regime, Vliet and Ross (1975) obtained the following heat transfer correlation for inclined uniform flux surfaces:

$$\mathrm{Nu}_x = 0.17(\mathrm{Gr}^*_x\mathrm{Pr})^{0.25} \tag{5.5.3}$$

This transport was independent of inclination θ for a warm surface facing upward. For the surface facing downward, the data correlated best when g was replaced by $g \cos^2 \theta$ in Gr^*_x above.

Horizontal surfaces, warm surface facing downward, $B_n < 0$. A variety of edge conditions used in different investigations has led to many correlations. Fishenden and Saunders (1950) proposed the following correlation for bare-edged square surfaces with L up to 61 cm in air:

$$\mathrm{Nu} = 0.31\mathrm{Ra}^{1/4} \qquad 10^5 \leq \mathrm{Ra} \leq 10^{10} \tag{5.5.4}$$

Birkebak and Abdulkadir (1970) presented results for a 19-cm bare-edged square surface in water:

Table 5.4.3 Factors K_1 and K_2 in Eqs. (5.4.77) and (5.4.78) for different geometric configurations

Geometry	Characteristic length	Other specifications	K_1	K_2
Sphere	Diameter, D		3.20	1.090
Rectangular plate (one side) 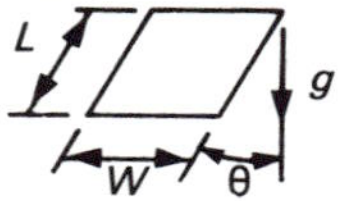	Slant height, L	$\theta \neq \pi/2$	$\frac{0.860(W/L)}{(\cos\theta)^{1/4}}$	$1.241(\cos\theta)^{1/4}$
Vertical cone, base insulated 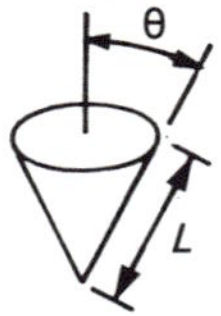	Slant height, L	Flow toward base, $\theta \neq \pi/2$	$\frac{2.43\sin\theta}{(\cos\theta)^{1/4}}$	$1.314(\cos\theta)^{1/4}$
		Flow toward vertex, $\theta \neq \pi/2$	$\frac{2.75\sin\theta}{(\cos\theta)^{1/4}}$	$1.314(\cos\theta)^{1/4}$
Inclined disk (one side) 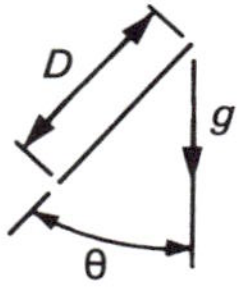	Diameter, D	$\theta \neq \pi/2$	$\frac{0.645}{(\cos\theta)^{1/4}}$	$1.305(\cos\theta)^{1/4}$

Inclined cylinder, ends insulated 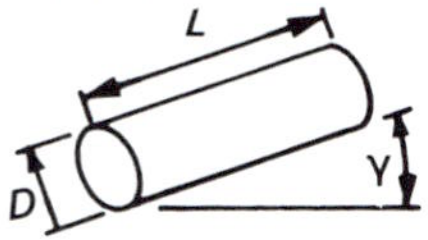	Diameter, D	$Z = (L/D) \cot \gamma \to 0$	$K_1(D/L)(\cos \gamma)^{1/4} = 2.70Z^{1/4}$	$K_2(\cos \gamma)^{-1/4} = 1.241\, Z^{-1/4}$
		$Z = 0.5$	2.26	1.486
		$Z = 1.0$	2.66	1.272
		$Z = 2.0$	3.04	1.128
		$Z = \infty$	3.61	0.958
Flat surfaces of isothermal disk fin on a horizontal tube 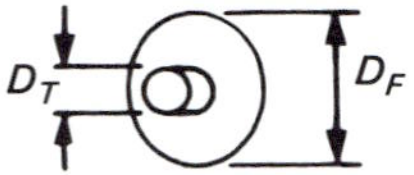	Fin diameter, D_F	$B = (D_T/D_F) \to 0$	$K_1 = 1.29$	$K_2 = 1.30$
		0.25	1.14	1.39
		0.50	0.83	1.54
		0.75	0.41	1.81
		1.00	$2.79(1 - B)^{5/4}$	$1.27(1 - B)^{-1/4}$
Flat surfaces of isothermal square fin on a horizontal tube 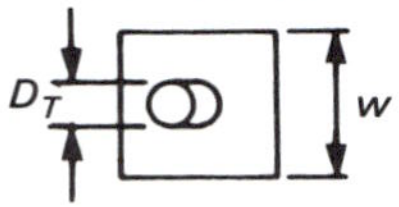	Fin width, W	$B = (D_T/W) \to 0$	$K_1 = 1.72$	$K_2 = 1.24$
		0.25	1.56	1.31
		0.50	1.25	1.40
		0.75	0.80	1.56
		1.00	0.25	1.94

Source: From Stewart (1971).

$$\mathrm{Nu} = 0.90\mathrm{Ra}^{1/5} \qquad 4 \times 10^8 \leq \mathrm{Ra} \leq 8 \times 10^8 \tag{5.5.5}$$

To determine the effect of edge conditions on Nusselt number, Restrepo and Glicksman (1974) measured heat transfer from a square horizontal heated 17.8 × 17.8 cm plate facing downward in air with three different edge conditions: (1) vertical surfaces extending up from the plate edge, cooled to ambient temperature t_∞, (2) vertical surfaces extending up from the edge and kept at the plate temperature t_0, and (3) adiabatic horizontal extensions outward from the outer edges of the plate. These edge conditions significantly affected the heat transfer rate, as indicated by the correlations given below:

Cooled edge:

$$\mathrm{Nu} = 0.587\mathrm{Ra}^{1/5} \tag{5.5.6}$$

Heated edge:

$$\mathrm{Nu} = 0.68\mathrm{Ra}^{1/5} \tag{5.5.7}$$

Horizontal extensions:

$$\mathrm{Nu} = 0.41\mathrm{Ra}^{1/5} \tag{5.5.8}$$

Hatfield and Edwards (1981) made measurements for square and rectangular plates in air, water, and a high Prandtl number oil. The warm surface faced downward. To investigate edge effects, data were also obtained with approximately adiabatic horizontal extensions added to the plates. The following correlation was proposed:

$$\mathrm{Nu} = C_1\left(1 + C_4\frac{L}{W}\right)[(1 + X)^{3m} - X^{3m}]\mathrm{Ra}_L^m \tag{5.5.9}$$

where

$$X = C_2\mathrm{Ra}_L^n + C_3\left(\frac{L_a}{L}\right)^p \tag{5.5.10}$$

and C_1, C_2, C_3, C_4, m, n, and p are empirical constants, W is the length of the long side of the plate, L the length of the short side, and L_a the length of the adiabatic extension. The coefficients were determined as $C_1 = 6.5$, $C_2 = 0.38$, $C_3 = 13.5$, $C_4 = 2.2$, $m = 0.13$, $n = -0.16$, and $p = 0.7$. This correlation was found to fit the data from various other studies to within 10% for aspect ratios from zero (i.e., an infinite strip) to unity (a square plate) for Prandtl numbers from 0.7 to 4800 with adiabatic extensions L_a/L up to 0.2. See also Goldstein and Lau (1983) for the effect of insulated vertical and horizontal extensions.

Horizontal surfaces, warm surface facing upward, $B_n > 0$. Fishenden and Saunders (1950), Wragg and Loomba (1970), Hassan and Mohamed (1970), Fujii and Imura (1972), and Goldstein et al. (1973) reported experiments for this configuration with edge extensions. Correlations were given. Lloyd and Moran (1974) measured mass transfer from horizontal surfaces of various planforms and showed that the data from this and other studies reduced to a single correlation for the laminar and turbulent regimes. The characteristic length is $L^* = A/P$, where A is the surface area and P the perimeter. This was first suggested by Goldstein et al. (1973). The laminar data were correlated as

$$\mathrm{Nu} = 0.54\mathrm{Ra}_{L^*}^{1/4} \qquad 2.2 \times 10^4 \le \mathrm{Ra}_{L^*} \le 8 \times 10^6 \tag{5.5.11}$$

For the turbulent regime

$$\mathrm{Nu} = 0.15\mathrm{Ra}_{L^*}^{1/3} \qquad 8 \times 10^6 \le \mathrm{Ra}_{L^*} \le 1.6 \times 10^9 \tag{5.5.12}$$

where Ra_{L^*} is the Rayleigh number based on characteristic length L^*.

Horizontal circular cylinders, $L/D >> 1$. Morgan (1975) made an extensive survey of the heat transfer data for long horizontal circular cylinders and reported the following correlations:

$$\mathrm{Nu}_D = 0.675\mathrm{Ra}_D^{0.058} \qquad 10^{-10} \le \mathrm{Ra}_D \le 10^{-2} \tag{5.5.13}$$

$$\begin{aligned} \mathrm{Nu}_D &= 1.02\mathrm{Ra}_D^{0.148} && 10^{-2} \le \mathrm{Ra}_D \le 10^2 \\ &= 0.850\mathrm{Ra}_D^{0.188} && 10^2 \le \mathrm{Ra}_D \le 10^4 \\ &= 0.480\mathrm{Ra}_D^{0.250} && 10^4 \le \mathrm{Ra}_D \le 10^7 \\ &= 0.125\mathrm{Ra}_D^{0.333} && 10^7 \le \mathrm{Ra}_D \le 10^{10} \end{aligned} \tag{5.5.14}$$

where all the properties in the calculation of Nu_D and Ra_D are evaluated at the film temperature $t_f = (t_0 + t_\infty)/2$. Equation (5.5.13) is the correlation given earlier by Collis and Williams (1954).

Churchill and Chu (1975) proposed the following correlation in the laminar range, $10^{-6} \le \mathrm{Ra}_D \le 10^9$, applicable for all Pr for an isothermal surface:

$$\mathrm{Nu}_D = 0.36 + 0.518\left\{\frac{\mathrm{Ra}_D}{[1 + (0.559/\mathrm{Pr})^{9/16}]^{16/9}}\right\}^{1/4} \tag{5.5.15}$$

This was modified to the following to cover both the laminar and turbulent regimes for $\mathrm{Ra}_D > 10^{-6}$:

$$\mathrm{Nu}_D = 0.60 + 0.387\left\{\frac{\mathrm{Ra}_D}{[1 + (0.559/\mathrm{Pr})^{9/16}]^{16/9}}\right\}^{1/6} \tag{5.5.16}$$

It was suggested that this correlation may be used as a good approximation for a uniform flux surface condition as well. Then Ra_D is based on the integrated local mean temperature difference. An alternative approach is to evaluate Nu_D and Ra_D based on the temperature difference at $\xi = \pi/2$. The constants in the above correlation must then be appropriately modified.

Fand et al. (1977) measured heat transfer from horizontal cylinders to air, water, and silicone oils in the experimental ranges $2.5 \times 10^2 \le \mathrm{Ra}_D \le 1.8 \times 10^7$

and $0.7 \leq \Pr \leq 3090$. The following correlation was given:

$$\mathrm{Nu}_D = 0.478\mathrm{Ra}_D^{0.25}\mathrm{Pr}^{0.05} \tag{5.5.17}$$

The fluid properties are to be evaluated at the reference temperature $t_r = t + 0.32(t_0 - t_\infty)$. Two other correlations, similar in algebraic form to the above equation, were also given. The difference is in the method of evaluation of fluid properties as a function of temperature. However, all three equations correlate the data extremely well—in fact, better than the correlations given in Eq. (5.5.13)–(5.5.16).

Fand and Brucker (1983) proposed a new correlation for heat transfer from horizontal circular cylinders that accounts for viscous dissipation effects:

$$\mathrm{Nu}_D = 0.400\mathrm{Pr}^{0.0432}\mathrm{Ra}_D^{0.25} + 0.503\mathrm{Pr}^{0.0334}\mathrm{Ra}_D^{0.0816} + \frac{0.958Ge^{0.122}}{\mathrm{Pr}^{0.0600}\mathrm{Ra}_D^{0.0511}} \qquad 10^{-8} < \mathrm{Ra}_D < 10^8, \quad 0.7 < \mathrm{Pr} < 4 \times 10^4 \tag{5.5.18}$$

where $Ge = g\beta D/c_p$ is the viscous dissipation parameter discussed in Section 3.6. This new result was shown to correlate the experimental data more accurately than some of the previous correlations, including Eqs. (5.5.13)–(5.5.16).

Short horizontal circular cylinders. As the aspect ratio decreases, the end effects become increasingly important and the above correlations for $L/D >> 1$ may not be accurate. These effects include heat conduction losses to the cylinder supports, flow effects arising from edge entrainment, and distortion of the flow and temperature fields by the supports. An important question is, what is the value of L/D above which these end effects are negligible? The experiments of Collis and Williams (1954), Gebhart et al. (1970), and Gebhart and Pera (1970) suggest that this value is very large, of the order of 10^4.

In some important applications, such as hot-wire anemometry, the wire probes used have values of L/D very much smaller than 10^4, typically around 200. Then the end effects are large and must be accounted for accurately if long-wire correlations are to be applied to interpret anemometer output. Accurate calculation of all these effects has not been done. Although the axial conduction loss to the wire supports may be calculated (Lowell, 1950; Gosse, 1956), the flow effects are very difficult to assess. For a discussion of these effects, see Gebhart and Pera (1970). On the other hand, the end effects have been determined experimentally by several investigators. Morgan (1975) proposed the following empirical equations to assess the apparent increase in heat transfer rate, $\delta = (\mathrm{Nu} - \mathrm{Nu}_\infty)/\mathrm{Nu}_\infty$, due to the finite-length effect, where Nu is the Nusselt number for a wire of finite aspect ratio and Nu_∞ applies for L/D very large:

$$\delta = \frac{7.5}{(L/D)^{1/2}} + \frac{3.5 \times 10^4}{(L/D)^2} \tag{5.5.19a}$$

or

$$\delta = \frac{13.3}{(L/D)^{0.566}} \tag{5.5.19b}$$

These relations apply in the range $300 \leq L/D \leq 10^5$.

The data of Gebhart and Pera (1970) for $10^3 \leq L/D \leq 1.6 \times 10^4$ are in good agreement with both these estimates. At a lower aspect ratio, $L/D = 250$, the data of Mahajan and Gebhart (1980b) indicated a better correlation with Eq. (5.5.19*b*) than with Eq. (5.5.19*a*).

Spheres. The correlation in Eq. (5.4.61) by Fujii et al. (1981) predicts heat transfer rates from isothermal spheres with reasonable accuracy for $10^{-2} \leq \text{Ra}_D \leq 10^6$ and all Pr. Based on the boundary layer solutions of Chiang et al. (1964) for $\text{Pr} = 0.7$, of Mori et al. (1976) for $\text{Sc} = 1$, 100, 500, and 650, and of Stewart (1971) for $\text{Sc} \to \infty$, Churchill (1983) gave the following expression:

$$\text{Nu}_D = \frac{0.589\text{Ra}_D^{1/4}}{[1 + (0.469/\text{Pr})^{9/16}]^{4/9}} \qquad 10^5 \leq \text{Ra}_D \leq 10^9 \tag{5.5.20}$$

It was indicated that this equation correlates well most of the laminar boundary layer data. However, some self-consistent data sets fall as much as 20% below or 50% above. As pointed out in Section 5.4, the Nusselt number for a sphere approaches 2.0, from pure conduction, as $\text{Ra} \to 0$. With this limiting value added to the right side of Eq. (5.5.20), Churchill suggested that the expression may also be used to determine heat transfer rates for small Gr_D. The correlation given earlier by Yuge (1960) includes this contribution, as seen below:

$$\text{Nu}_D = 2 + 0.45\text{Gr}_D^{1/4}\text{Pr}^{1/3} \qquad 10^0 \leq \text{Gr}_D \leq 10^5 \tag{5.5.21}$$

This is expected to apply for Prandtl numbers in the vicinity of 1.0. Properties at the film temperature are to be used. For weak flows, Eq. (5.4.60) given by Hossain and Gebhart (1970) may be used to calculate heat transfer rates.

Other surfaces. In addition to the correlation above and those discussed in Section 5.4, some generalized relations are available to estimate heat transfer from geometries for which more specific information is not available. In the laminar boundary layer regime, $10^4 \leq \text{Ra}_{L*} \leq 10^9$, the following correlation suggested by King (1932) has been used:

$$\text{Nu}_{L*} = 0.55(\text{Ra}_{L*})^{1/4} \tag{5.5.22}$$

where

$$\frac{1}{L^*} = \frac{1}{L_h} + \frac{1}{L_v} \tag{5.5.23}$$

where L^* is the characteristic length and L_h and L_v are the significant horizontal and vertical dimensions.

Churchill (1983) proposed the following general correlation for all geometries and both isothermal and uniformly heated surface conditions for the entire range of Ra_{L^*} and all Pr:

$$\mathrm{Nu}_{L^*}^{1/2} = \mathrm{Nu}_0^{1/2} + \left(\frac{\mathrm{Ra}_{L^*}/300}{[1 + (0.5/\mathrm{Pr})^{9/16}]^{16/9}} \right)^{1/6} \tag{5.5.24}$$

where the values of Nu_0 and the appropriate characteristic length L^* for different geometries are given in Table 5.5.1. Earlier, Churchill and Churchill (1975) had compared this correlation with experimental data for vertical surfaces, cylinders, and spheres. Good agreement was reported for $10^{-4} < \mathrm{Ra}_{L^*} < 10^{14}$.

A modified form of Eq. (5.5.22) with the coefficient 0.55 replaced by 0.60 has also been used; see Gebhart (1971), Kreith and Black (1980), and Holman (1981). Lienhard (1981) suggested the correlation

$$\mathrm{Nu} = 0.52\mathrm{Ra}_{L^*}^{1/4} \tag{5.5.25}$$

where L^* here equals the length of travel of the fluid in the boundary layer. It was suggested that with this definition of L^*, the above correlation should predict heat transfer from any submerged body within 10% if Pr is not much less than 1.0.

It should be kept in mind that these correlations cannot accurately determine transport from all kinds of different geometries, which is sometimes the impression gained from the literature. They do not take into consideration the specific fluid flow patterns associated with different configurations. The regions of flow separation generally are not included. Also, in a multidimensional body, heat transfer from a surface may be affected by the plume shed by another surface or by pre-

Table 5.5.1 Characteristic length and Nu_0 for Eq. (5.5.24), where s is the slant height

Geometry	L^*	Nu_0
Vertical surface	L	0.68
Inclined surface[a]	s	0.68
Inclined disk[a]	$9D/11$	0.56
Vertical cylinder	L	0.68
Horizontal cylinder	πD	0.36π
Sphere	$\pi D/2$	π
Vertical cone[a]	$4s/5$	0.54
Spherelike, area A and volume V	$3\pi V/A$	$A^3/36V^2$

[a]In Ra_{L^*}, g is to be replaced by $g \cos \theta$.
Source: From Churchill (1983).

heating of the fluid in passing over other surfaces of the body. Since such effects are excluded in the above correlations, estimates of heat transfer obtained by using the correlations may be quite inaccurate.

In this connection Sparrow and Ansari (1983) experimentally determined heat transfer from the top, bottom, and side of a heated vertical cylinder of equal height and diameter in air. Significant discrepancies were shown between the data and the estimates of heat transfer given by Eqs. (5.5.22) and (5.5.25). These were attributed to the flow effects mentioned above.

5.6 STRATIFIED AMBIENT MEDIA

The transport solutions and correlations discussed in the previous sections assumed an isothermal ambient body of fluid. However, ambient fluids often are appreciably thermally stratified. As for the vertical flows discussed in Section 3.11, stratification will, in general, modify transport. Assessment of the modifications, however, usually is more difficult with other configurations. Similarity solutions are not found and approximate methods are used. An exception is the classic solution given by Prandtl (1952) for flow adjacent to an infinite wall that is inclined at an angle θ from the vertical and has a surface temperature higher or lower by a constant difference $t_0 - t_\infty$ than a stably and linearly temperature-stratified ambient at t_∞. For these conditions one-dimensional parallel flow results, as discussed in Section 3.11. The solutions t and u are then given by Eqs. (3.11.1) and (3.11.2), respectively.

The heat transfer from spheres and horizontal circular cylinders immersed in a thermally stratified ambient fluid has been studied by Chen and Eichhorn (1979). The approximate method of Raithby et al. (1975) was used. The following general expression for the local heat transfer rate variation over a two-dimensional planar or an axisymmetric body, Figs. 5.1.2a and 5.1.2b, to a stratified fluid was obtained:

$$q'' = C(\text{Pr})\left(\frac{\beta}{\nu\alpha}\right)^{1/4} k\,(r^i)^{1/3}(g_x)^{1/3}(t_0 - t_\infty)^{5/3}\left(\frac{A}{B}\right)^{1/4} \tag{5.6.1}$$

where g_x is the local tangential component of g acting locally along the surface, the superscript $i = 0$ for two-dimensional planar and $i = 1$ for axisymmetric bodies, r is the horizontal distance locally from the surface element to the axis of symmetry, and x is the coordinate along the body surface. The constant $C(\text{Pr})$ is dependent on Prandtl number and, as pointed out in Section 5.4, may be taken as $C(\text{Pr}) = 0.503/[1 + (0.492/\text{Pr})^{9/16}]^{4/9}$. Also,

$$A = \exp\left[\frac{4}{3}(p+1)\int_0^x \frac{dt_\infty/dx}{t_0 - t_\infty}\,dx\right] \tag{5.6.2}$$

and

$$B = \int_0^x (r^i)^{4/3} (g_x)^{1/3} (t_0 - t_\infty)^{5/3} A \, dx \tag{5.6.3}$$

In Eq. (5.6.2) p is an undetermined constant that depends on the velocity and temperature distributions, which in the derivation of Eq. (5.6.1) are assumed to be functions of Prandtl number and ambient fluid stratification. An exact value of p is generally not known. However, an estimate can be made, as discussed below.

As in Section 3.11, the effect of stratification on heat transfer can be expressed in terms of a stratification parameter $S = LN_\infty/\Delta t_m$, where Δt_m is the temperature difference at midheight of the body and L is the height of the body. In terms of Figs. 5.1.2*a* and 5.1.2*b*, $N_\infty = dt_\infty/dz$, where z is the vertical coordinate in the ambient medium. The average Nusselt number with stratification, Nu, compared to that without, $\mathrm{Nu}_{\mathrm{iso}}$, is expressed as

$$\frac{\mathrm{Nu}}{\mathrm{Nu}_{\mathrm{iso}}} = F(S, p) \tag{5.6.4}$$

Equations (5.6.1)–(5.6.3) were applied to the cylinder and sphere geometries in a linearly stratified ambient fluid by Chen and Eichhorn (1979). First, the value of constant p was determined by comparing the approximate solution for the vertical plate in a linearly stratified ambient with the corresponding local nonsimilarity solutions of Chen and Eichhorn (1976). For $\mathrm{Pr} = 6.0$, $p = 2.5$ resulted in good agreement between the two results. The dependence of p on S was found to be weak.

Using this value of p, Eq. (5.6.1) was then solved for the Nusselt numbers for the cylinder and sphere geometries with various values of the stratification parameter S. Note that in Eqs. (5.6.1)–(5.6.3), $r^i = 1$, $g_x = g \sin(2x/D)$, and $z = (D/2)(1 - \cos 2x/D)$ for a horizontal circular cylinder of diameter D. For a sphere of diameter D, $r^i = (D/2)\sin(2x/D)$ and g_x and z are the same as for the cylinder. Curvature effects were also considered. The resulting values of $\mathrm{Nu}/\mathrm{Nu}_{\mathrm{iso}}$ are shown in Figs. 5.6.1 and 5.6.2. The experimental data of Eichhorn et al. (1974) for water are also plotted. Reasonable agreement is seen between the experiment and the approximate solution.

Other relevant work is due to Singh (1977). A perturbation solution was obtained for the flow around a sphere in a weakly stratified fluid for low values of Grashof number. Qualitative agreement between the theoretically obtained streamlines and the experimental ones reported by Eichhorn et al. (1974) was found.

5.7 INTERACTIVE FLOWS

In this and the previous chapters, natural convection flows about single bodies—a single plate, a single cylinder, or a single line heat source—have been dis-

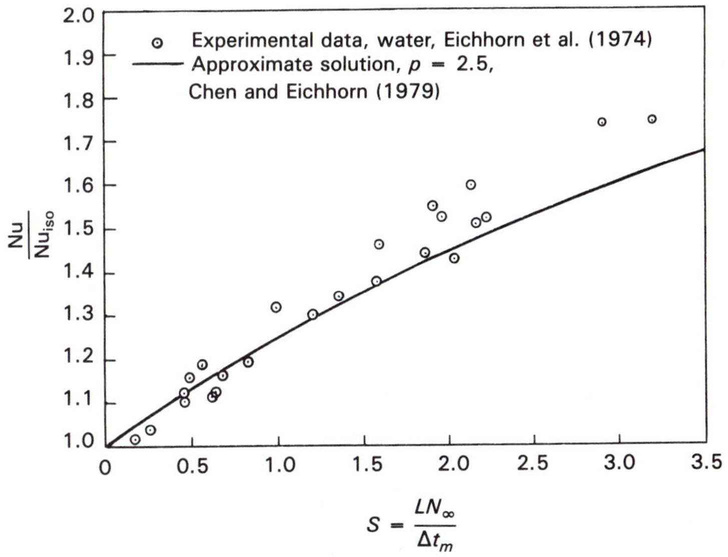

Figure 5.6.1 Influence of stratification on heat transfer from a horizontal cylinder. *(Reprinted with permission from C. C. Chen and R. Eichhorn,* J. Heat Transfer, *vol. 101, p. 566. Copyright © 1979, ASME.)*

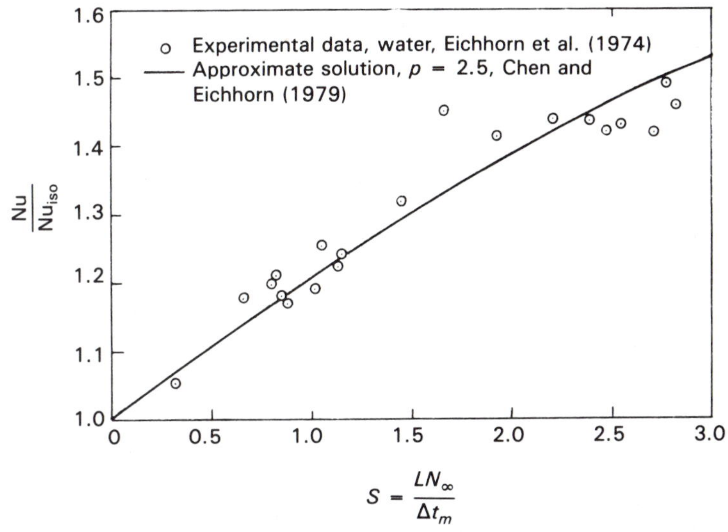

Figure 5.6.2 Influence of stratification on heat transfer from a sphere. *(Reprinted with permission from C. C. Chen and R. Eichhorn,* J. Heat Transfer, *vol. 101, p. 566. Copyright © 1979, ASME.)*

cussed. However, in many cases of practical interest multiple flows occur simultaneously or a flow interacts with nearby surfaces. These circumstances may arise in arrays of heat-dissipating electronic components, between elements of a processing plant, and with a group of heaters submerged in a fluid medium. The engineering aspect of such transport is the optimization of the effects of the interaction in terms of particular design objectives.

5.7.1 Arrays of Heated Cylinders

One of the earliest studies of such flows was the experimental investigation by Eckert and Soehngen (1948). Heat transfer rates from a vertical arrangement of warm horizontal tubes were measured. However, before presenting these results, it is instructive to understand the mechanisms that influence the heat transfer response of a second body that lies in the wake of an upstream buoyancy-induced flow.

The effect on transport from an upper body is a combination of two mechanisms. First, the flow generated by lower surfaces becomes a moving wake. Its velocity tends to increase the heat transfer from a surface lying in the wake above. Second, the hot wake reduces the effective temperature difference between this surface and the local adjacent fluid, compared to the difference with the distant environment at t_∞. This tends to decrease the heat transfer. Thus, there are opposite flow and temperature effects. Whether the downstream surface will have increased or decreased heat transfer will depend on the balance between these effects. Further, the magnitudes of these effects will depend on the streamwise and transverse distances separating the bodies.

The results of Eckert and Soehngen demonstrate these competing mechanisms. In an aligned vertical configuration of two tubes, 1 and 2, at the same temperature, Nu_2 was $0.87\mathrm{Nu}_1$. Thus, the temperature effects dominated, causing a decrease in heat transfer from the upper tube. When a third tube was added in line with the second tube, Nu_2 remained $0.87\mathrm{Nu}_1$ and Nu_3 was $0.65\mathrm{Nu}_1$, indicating again the domination of temperature effects. However, when the middle tube was offset horizontally by a distance $D/2$, Nu_2 was $1.03\mathrm{Nu}_1$. In this configuration tube 2 lay outside the wake of tube 1, but close enough to feel the effect of flow entrainment and other effects from the wake of the first tube. Heat transfer increased since there was no offsetting temperature effect.

A later study of the same and other effects by Lieberman and Gebhart (1969) used an array of horizontal wires in a plane. The wires were 0.127 mm in diameter and 0.184 m long. The Nusselt number was determined for each electrically heated parallel wire, separately, and flow was visualized for six different wire center spacings, from 37.5 to 225 diameters. For each spacing, experiments were carried out at four array angles: horizontal, vertical, and inclined at 60° and 30° from the vertical. The Grashof number based on the wire diameter, Gr_D, was of order 10^{-1}.

Interferograms of the downstream temperature field for the closest wire spacing, 37.5 diameters, for the four inclination angles are shown in Figs. 5.7.1–5.7.4. The temperature distributions and heat transfer data determined from such

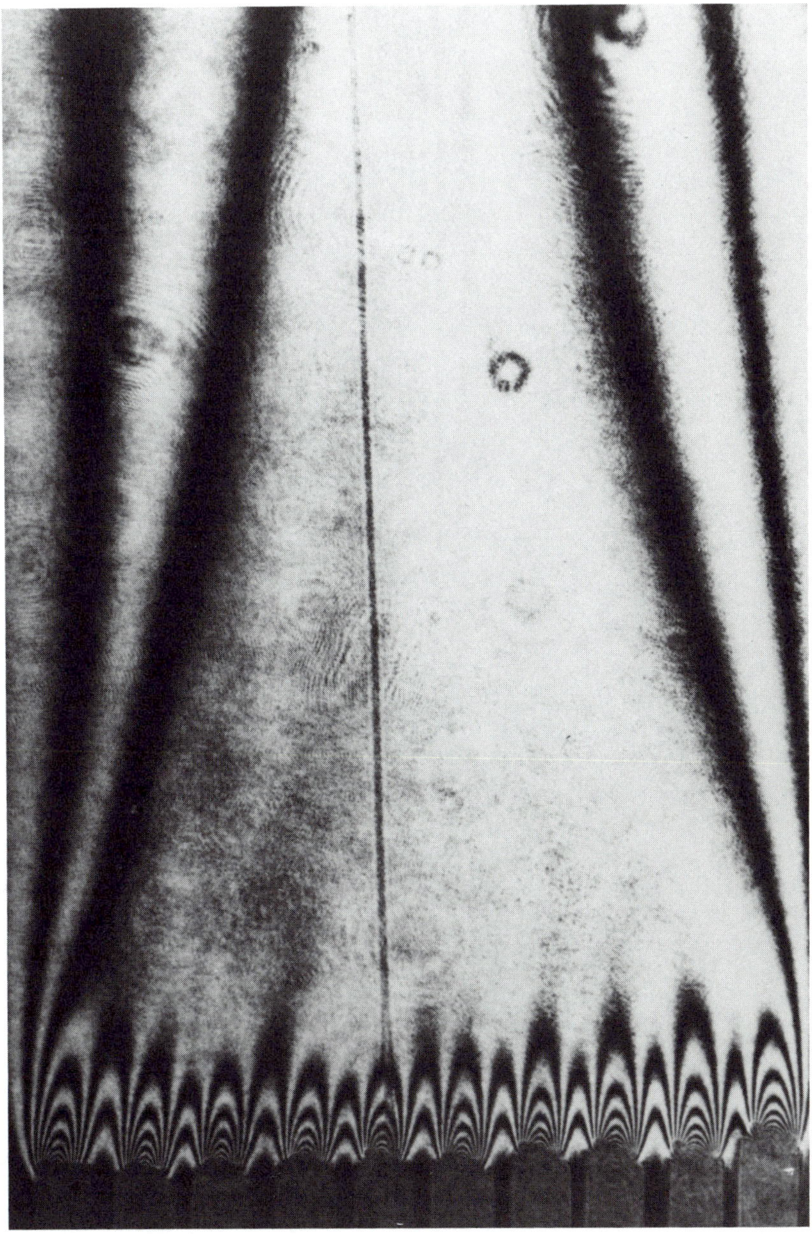

Figure 5.7.1 Interferograms of the flow adjacent to an array of electrically heated wires. Array angle, 90°; spacing, 0.47 mm. *(Reprinted with permission from* Int. J. Heat Mass Transfer, *vol. 12, p. 1385, J. Lieberman and B. Gebhart. Copyright © 1969, Pergamon Journals Ltd.)*

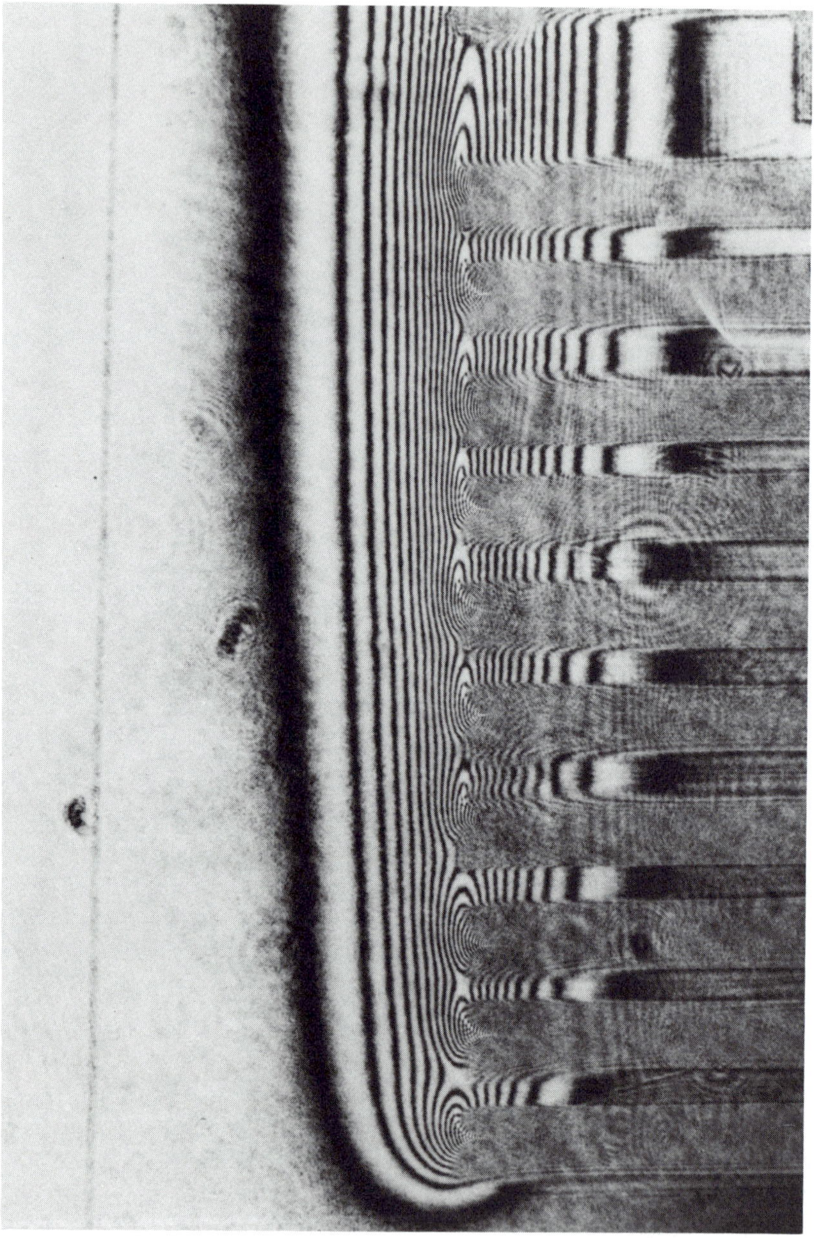

Figure 5.7.2 Interferogram for array angle 0°, spacing 0.47 mm. *(Reprinted with permission from* Int. J. Heat Mass Transfer, *vol. 12, p. 1385, J. Lieberman and B. Gebhart. Copyright © 1969, Pergamon Journals Ltd.)*

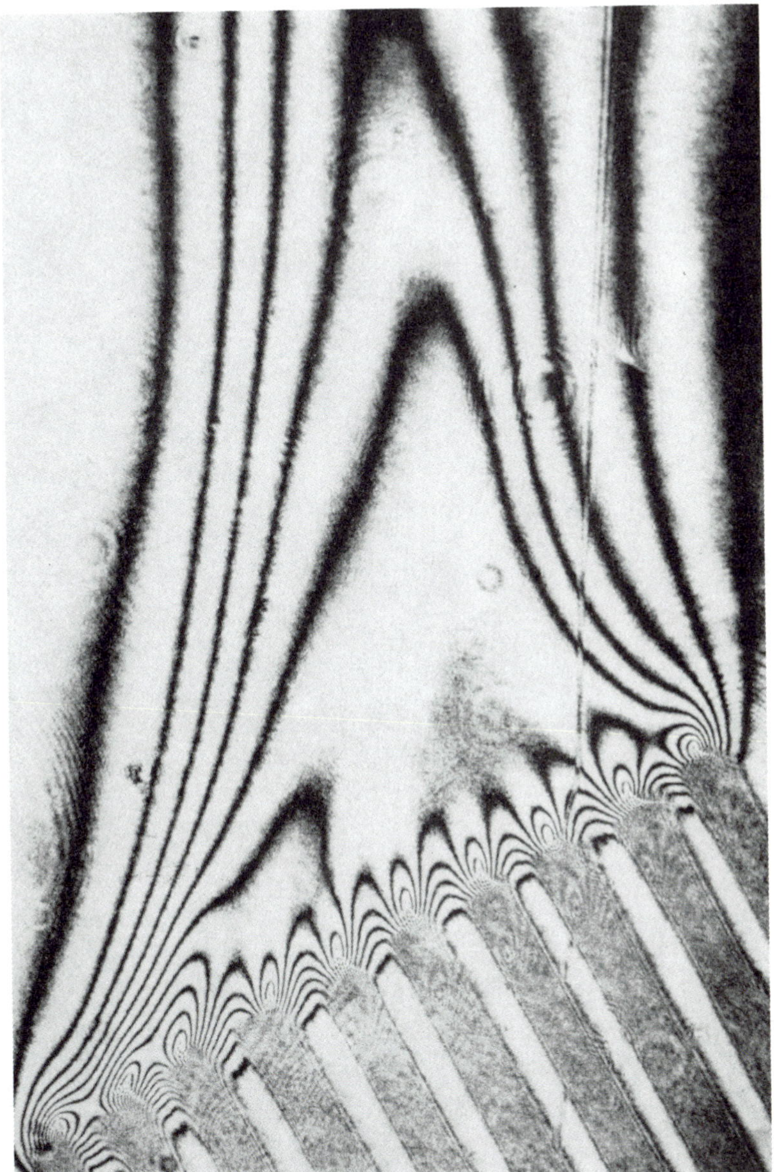

Figure 5.7.3 Interferogram for array angle 60°, spacing 0.47 mm. *(Reprinted with permission from* Int. J. Heat Mass Transfer, *vol. 12, p. 1385, J. Lieberman and B. Gebhart. Copyright © 1969, Pergamon Journals Ltd.)*

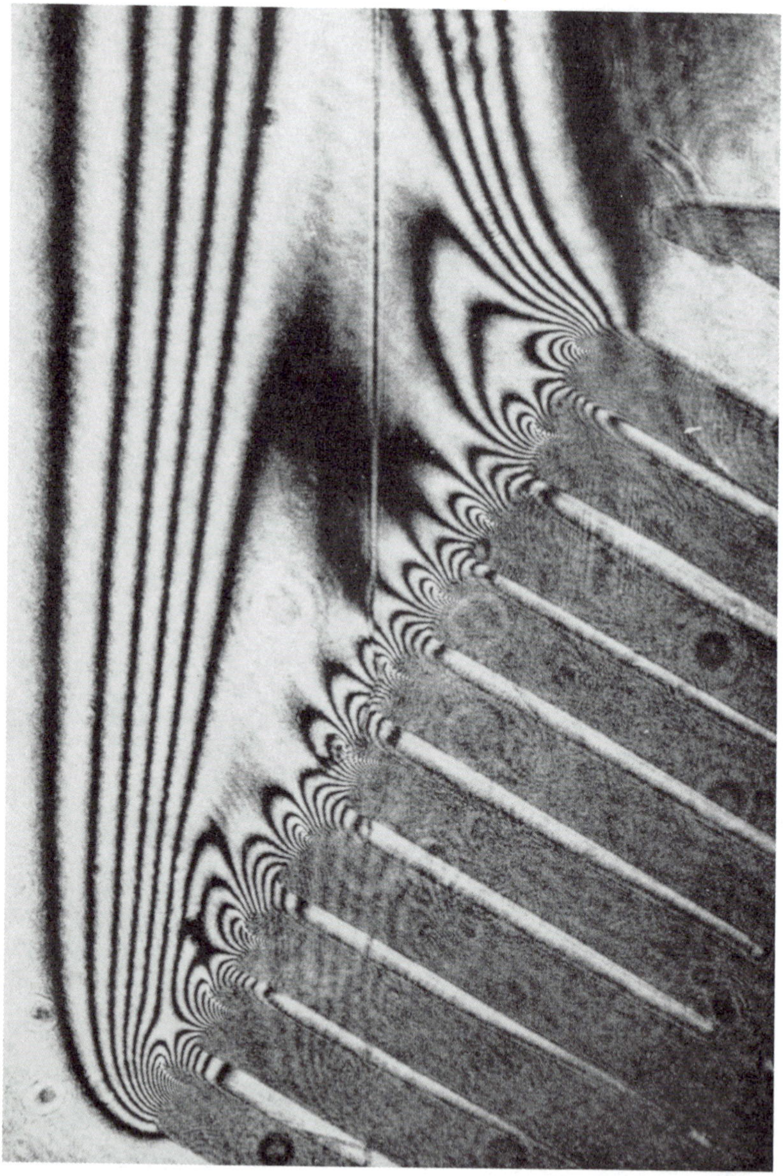

Figure 5.7.4 Interferogram for array angle 30°, spacing 0.47 mm. *(Reprinted with permission from* Int. J. Heat Mass Transfer, *vol. 12, p. 1385, J. Lieberman and B. Gebhart. Copyright © 1969, Pergamon Journals Ltd.)*

interferograms indicated that the plumes interacted. Both spacing and array orientation angle affected the transport.

For the vertical array it was found that, as spacing increases, the flow effects dominate the temperature effects of the thermal wake. Heat transfer increases downstream. For still larger spacing, the flow effect decreases. The heat transfer, therefore, decreases downstream.

Figure 5.7.5 is a plot of the average Nusselt number of the array, Nu_D, versus spacing for the four inclinations. For horizontal and inclined arrays, the wire heat transfer was higher in the center of the array than at the edges. This was attributed to the stronger upflow induced in midarray locations by the general entrainment inflows generated by the whole array. The average Nusselt number variation with spacing, however, follows the same pattern as for the vertical, although the maxima are attained at lower spacing. Clearly, there is an optimum spacing at each array angle for maximum Nusselt number.

Heat transfer from a vertical array of heated tubes in air has also been determined experimentally by Marsters (1972). The tubes were 6.35 mm in diameter and 58.5 cm long. Heat transfer results were obtained for arrays of three, five, and nine tubes. The spacing between the tube centers varied from 2 to 20 diameters for the three-tube array and 2 to 10 diameters for the five- and nine-tube arrays. Gr_D ranged from 750 to 2000.

Plots of Nu_i/Nu_0 vs. X/X_{max} for different array configurations are shown in Fig. 5.7.6, where Nu_i is the Nusselt number for a cylinder in the array, Nu_0 the Nusselt number predicted for a single heat cylinder at that Rayleigh number, X the distance upward in the array from the bottom cylinder, and X_{max} the spacing

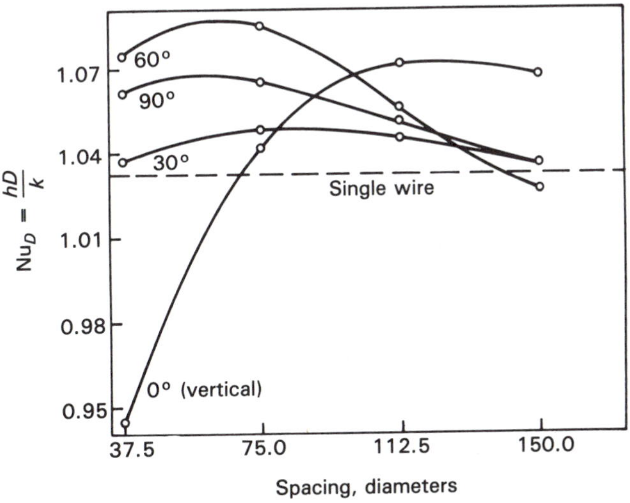

Figure 5.7.5 Average array Nusselt number for different inclinations and spacings; $Gr_D = 1.49 \times 10^{-1}$. *(Reprinted with permission from* Int. J. Heat Mass Transfer, *vol. 12, p. 1385, J. Lieberman and B. Gebhart. Copyright © 1969, Pergamon Journals Ltd.)*

between the top and bottom cylinders. It is clear that the bottom cylinder behaves almost as if the other cylinders were not present. The small enhancement in its heat transfer capability that is seen at large spacings is believed to be due to the flow induced by the entire array. The upper cylinders have reduced heat transfer at smaller spacings and enhanced heat transfer at larger spacings. As a result, the average array Nusselt number is smaller than for a single cylinder for small spacings and larger for larger spacings. This agrees with the trend in the results shown in Fig. 5.7.5.

Effect of confining walls. Marsters (1975) experimentally investigated the influence of unheated confining walls on the heat transfer from a single horizontal heated circular cylinder. The configuration is shown in Fig. 5.7.7. The working fluids were air, water, and Freon 113. Rayleigh number ranged from 10 to 5×10^5. The effect of wall spacing s and wall height H on heat transfer from the cylinder of diameter D was determined, where H/D varied from 5 to 128 and s/D from 2 to ∞. The unconfined case is $s/D \rightarrow \infty$, or $H/D \rightarrow 0$. The data indicated enhanced heat transfer from the cylinder in presence of the confining walls. This is likely due to entrainment effects as discussed in Section 5.7.3.

5.7.2 Colinear, Separated Plates

Two aligned isothermal surfaces separated by distance s are shown in Fig. 5.7.8. This configuration was considered by Sparrow and Faghri (1980). There are three different flow regions. Flow adjacent to the lower surface at t_0 is as discussed in Section 3.3. In the space between the surfaces, the wake tends toward a plume. The no-slip and isothermal conditions at $y = 0$ are replaced by symmetry conditions. For the upper surface the fluid arriving at the leading edge has velocity and temperature distributions. New boundary layers embedded within the arriving velocity and temperature layers begin to develop.

The boundary layer equations subject to the different boundary and initial conditions for the three flow regions were solved numerically for $\Pr = 0.7$ for different values of the spacing s, surface temperatures t_{01} and t_{02}, and lengths L_1 and L_2. In accordance with the wake velocity and temperature effects described earlier for cylinders, the heat transfer from the upper surface was again found to be subject to both flow and temperature effects. For $t_{01} = t_{02}$ the total heat transfer from the upper surface was less for small s and more for large s. For the upper surface maintained at a higher temperature than the lower one, the heat transfer rate was increased, of course. These results are summarized in Fig. 5.7.8, where the ratio of average heat transfer coefficient for the upper surface, h_2, to that for the lower surface, h_1, is shown for various relative lengths and relative plate temperature levels $\phi_{02} = (t_{02} - t_\infty)/(t_{01} - t_\infty)$.

5.7.3 Plume Interactions

Interaction between laminar two-dimensional thermal plumes has also been studied in detail; see Pera and Gebhart (1975) and Gebhart et al. (1976). Extensive

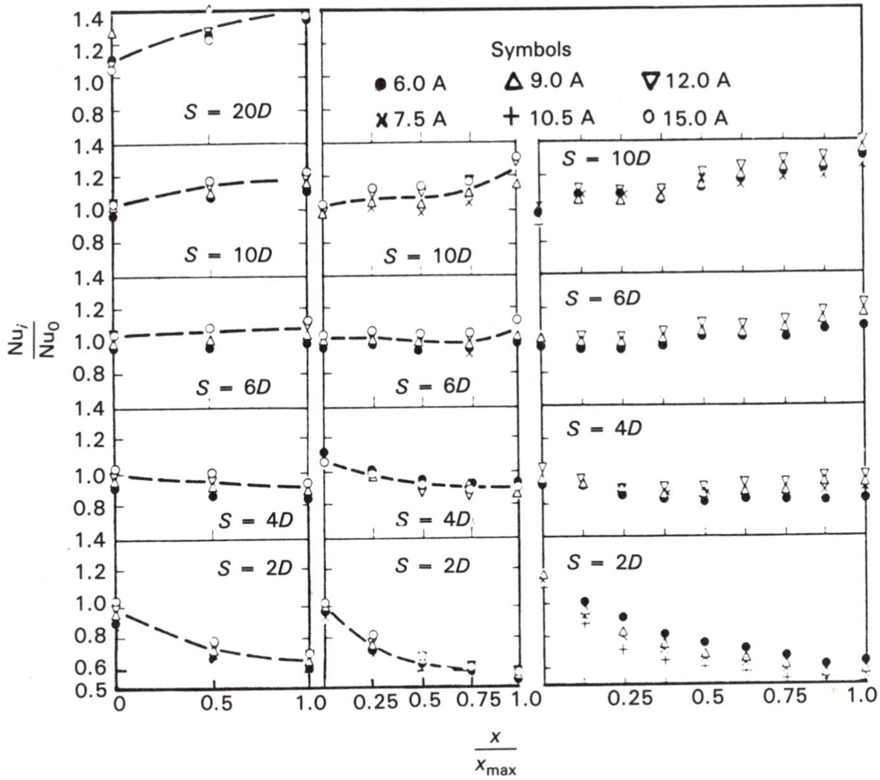

Figure 5.7.6 Nusselt number distribution for three-, five-, and nine-cylinder arrays at different spacings and heating rates given in amperes (A). The Nusselt number is normalized with respect to the Nusselt number predicted for a single cylinder at the Rayleigh number of that cylinder. *(Reprinted with permission from* Int. J. Heat Mass Transfer, *vol. 15, p. 921, G. F. Marsters. Copyright © 1972, Pergamon Journals Ltd.)*

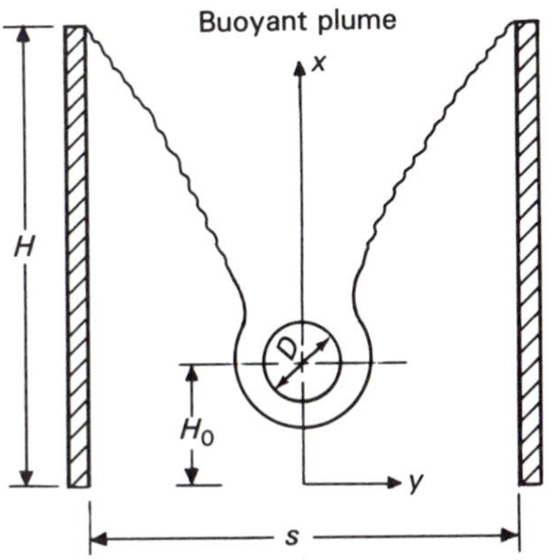

Figure 5.7.7 Interactive flow due to a heated horizontal cylinder in a vertical channel. *(Reprinted with permission from G. F. Marsters,* Can. J. Chem. Eng., *vol. 53, p. 144. Copyright © 1975, The Chemical Institute of Canada, Publisher.)*

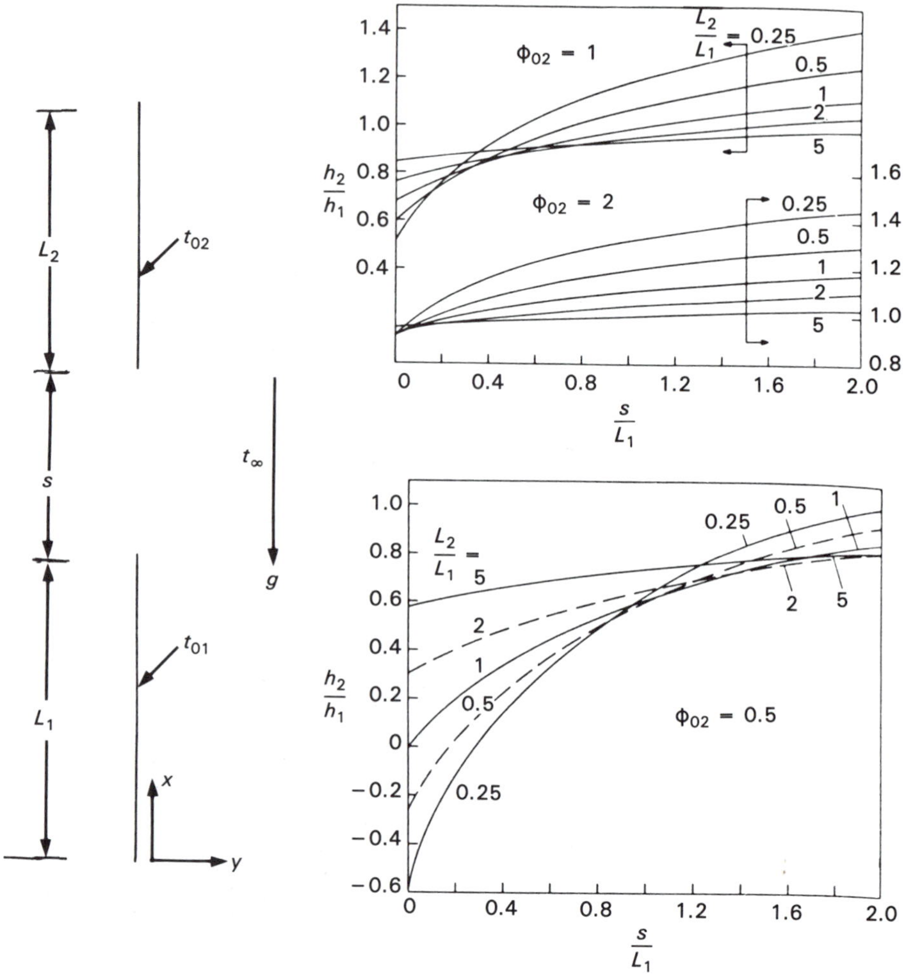

Figure 5.7.8 Comparison of average heat transfer rate for the upper plate with its single-plate counterpart for various values of ϕ_{02}. *(Reprinted with permission from E. M. Sparrow and M. Faghri,* J. Heat Transfer, *vol. 102, p. 623. Copyright © 1980, ASME.)*

experimental results on the effect of spacing and heat input on the interaction were obtained for plumes of both equal and unequal buoyancy. Interaction was analyzed. The effect of an adjacent vertical surface and semicylindrical surface on plane plume flow was also studied. Figures 5.7.9–5.7.12 show the nature of these kinds of flow interactions for plane plumes.

Figure 5.7.9 indicates that the two plumes of equal strength slant equally toward each other. For smaller initial gaps between the plumes, they incline more toward each other, indicating a strong interaction. For plumes of very unequal strength (Fig. 5.7.10) the weaker plume slants more toward the stronger, which

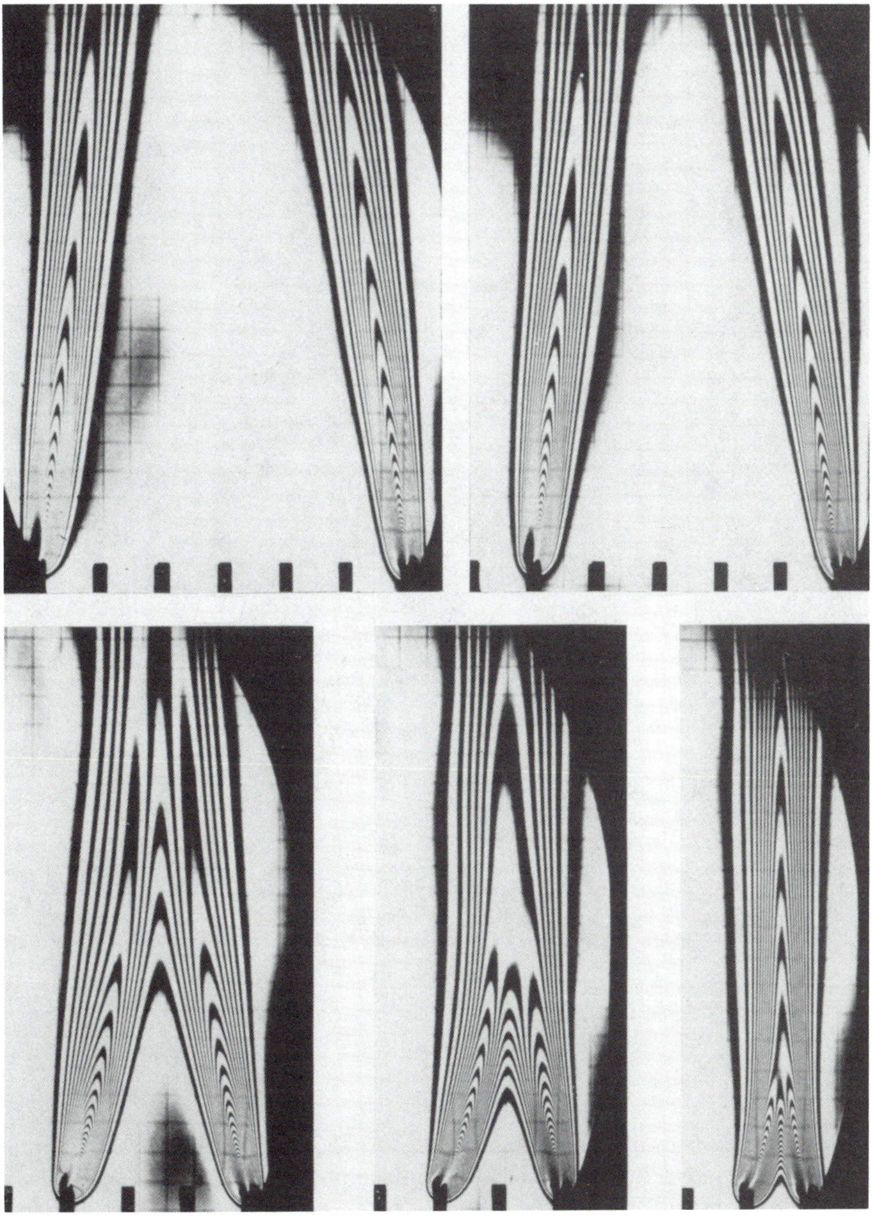

Figure 5.7.9 Interaction between adjacent plane plumes of equal strength in air. The energy input to each plume is 73 W/m of span. The span is 18 cm and the wire connecting posts seen are spaced at 1.4 cm. *(Reprinted with permission from L. Pera and B. Gebhart,* J. Fluid Mech., *vol. 68, p. 259. Copyright © 1975, Cambridge University Press.)*

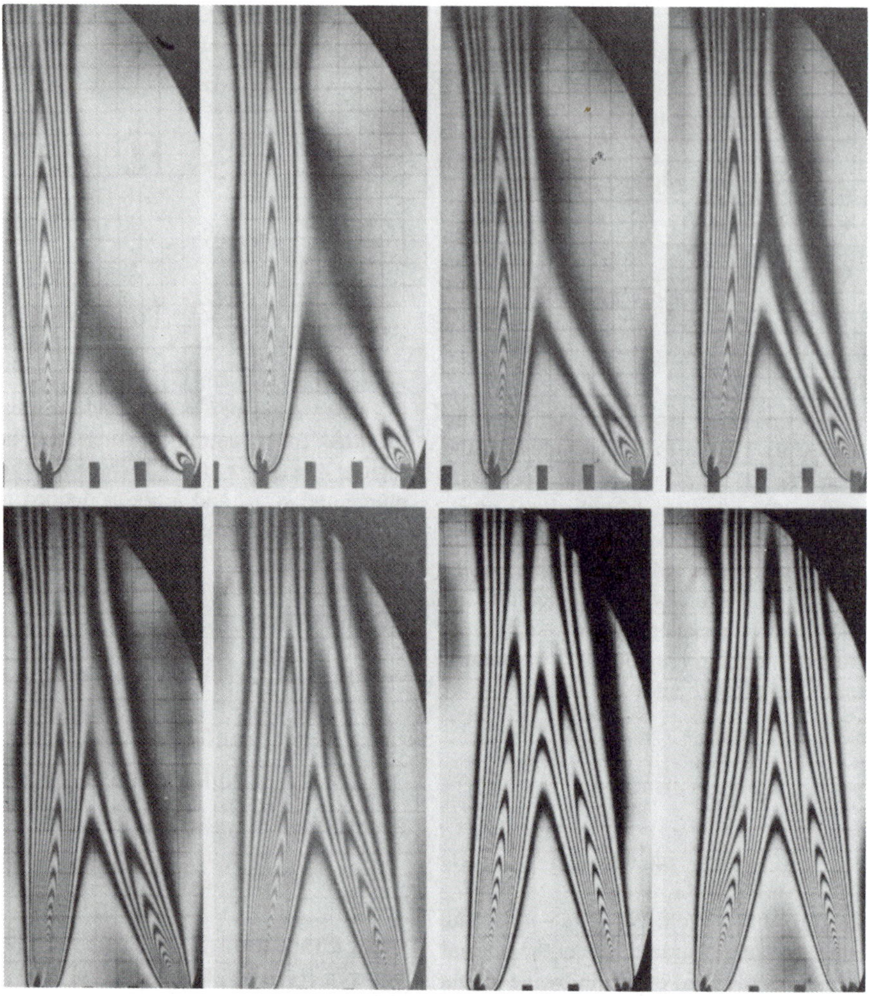

Figure 5.7.10 Interaction between unequal plumes; $Q_{\text{left}} = 73$ W/m and $Q_{\text{right}} = 1.3$, 3.0, 5.3, 12, 21, 32, 46, and 73 W/m. *(Reprinted with permission from* Int. J. Heat Mass Transfer, *vol. 19, p. 751, B. Gebhart et al. Copyright © 1976, Pergamon Journals Ltd.)*

is little affected until the merger takes place. As the strength of the weaker plume increases, the interaction is greater. For a plane plume adjacent to a vertical wall (Fig. 5.7.11) and a curved surface (Fig. 5.7.12), the plume attaches and follows the surface. The interaction of a two-dimensional thermal plume with a neighboring vertical surface has also been studied experimentally by Jaluria (1982a). More recently, Agarwal and Jaluria (1985) carried out a numerical and experimental study of this flow. The resulting pressure difference across the plume was also computed.

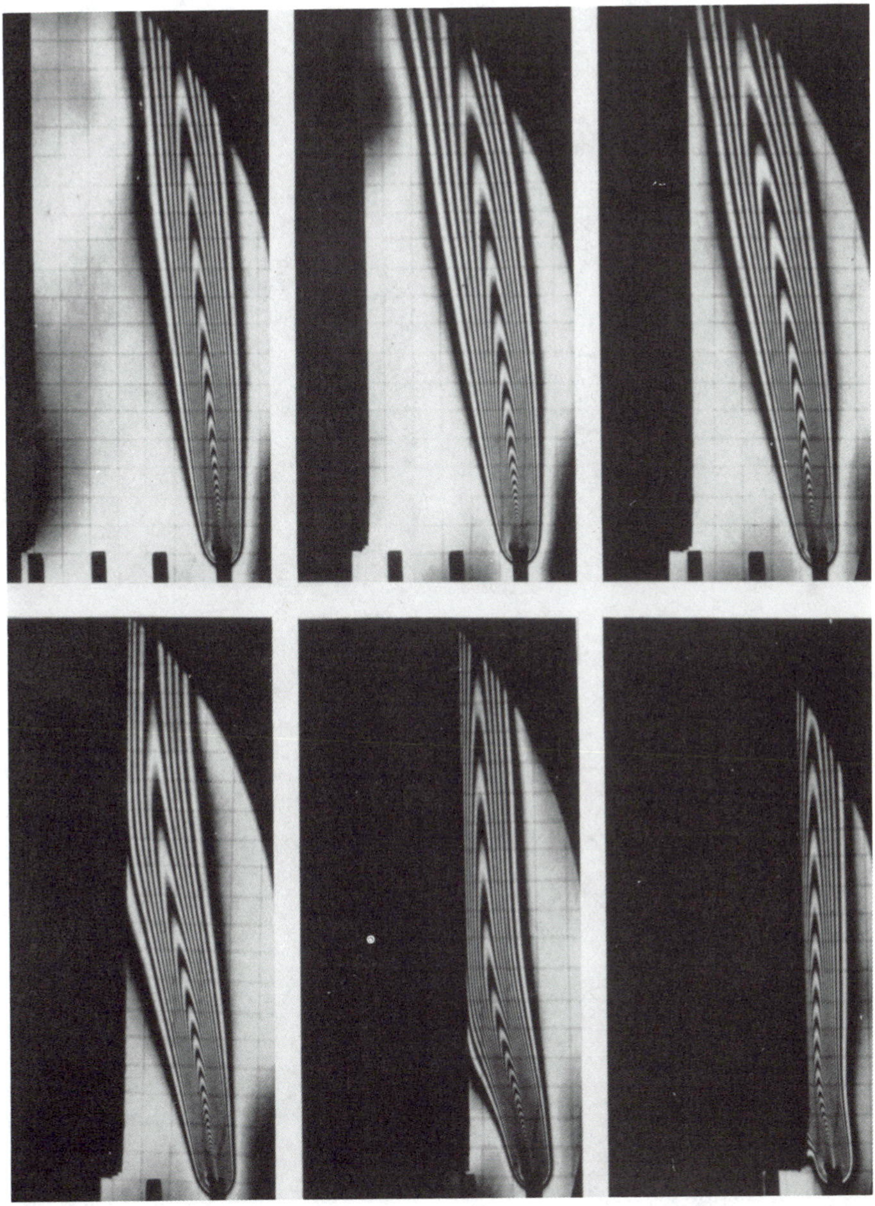

Figure 5.7.11 Effect of a vertical wall on a plane plume flow at various spacings, as in Fig. 5.7.9. *(Reprinted with permission from L. Pera and B. Gebhart,* J. Fluid Mech., *vol. 68, p. 259. Copyright © 1975, Cambridge University Press.)*

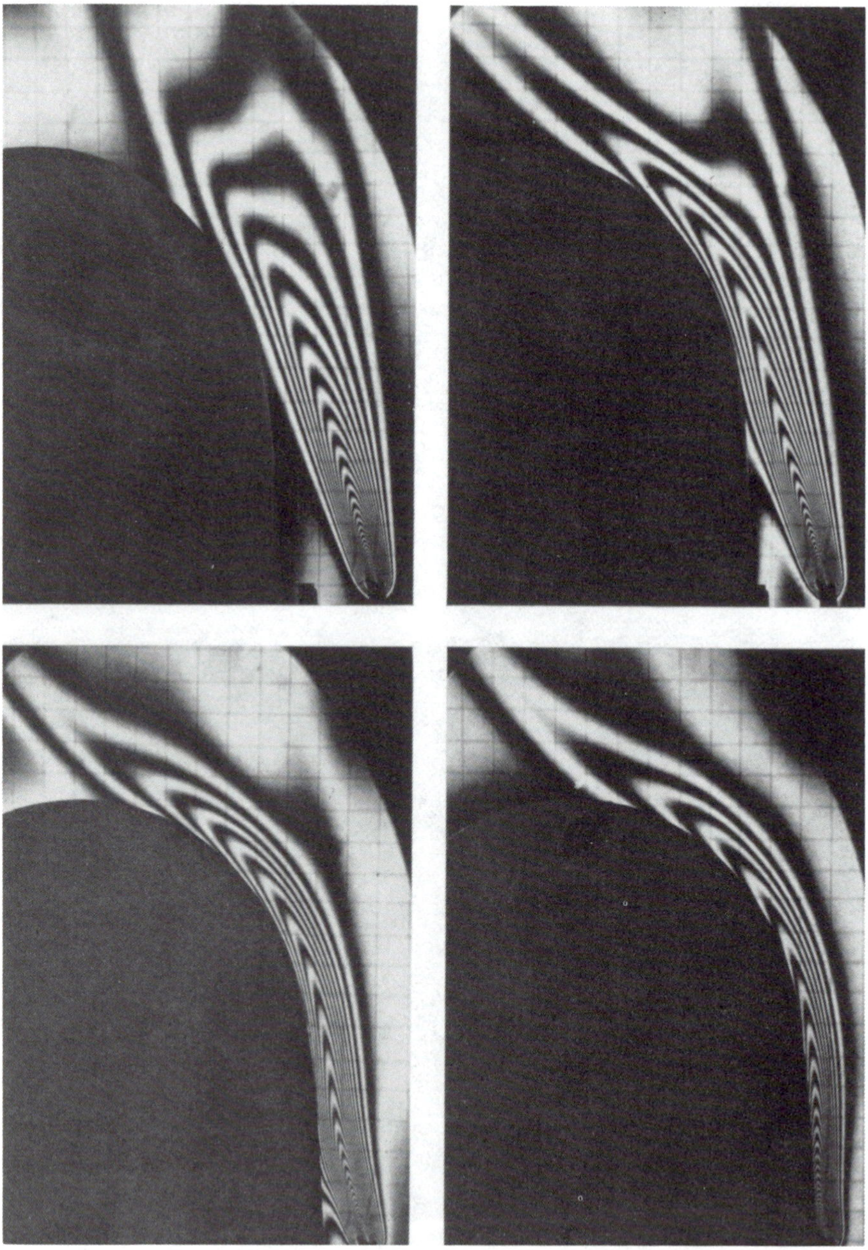

Figure 5.7.12 Interaction of a plane plume with a curved surface. *(Reprinted with permission from L. Pera and B. Gebhart,* J. Fluid Mech., *vol. 68, p. 259. Copyright © 1975, Cambridge University Press.)*

The underlying mechanism for the interactions observed in all these flows has been shown to be entrainment from the ambient, as analyzed by Pera and Gebhart (1975) and Gebhart et al. (1976). A schematic of the model for plume interaction is shown in Fig. 5.7.13. The force opposing the slanting of the plumes is the cross-plume buoyancy force component B_n. This is balanced by the motion pressure, arising in the ambient fluid, which drives in the entrainment flows. These augment the plumes downstream. The difference between the motion pressures $p_{m,o}$ and $p_{m,i}$ on the two sides of each plume balances B_n. The level of $p_{m,i}$ is less than that of $p_{m,o}$ because the entrainment fluid required on the inside must come upward through the restricted space between the plumes and will therefore be at a higher velocity U. Some fluid also enters through each end for plumes of finite length. Figures 5.7.9 and 5.7.11 indicate that the plumes remain essentially flat downstream. Also, B_n remains essentially constant downstream, as does $p_{m,o} - p_{m,i}$. Since the entrainment velocities $v(x, \infty)$ for a plume flow are not highly vari-

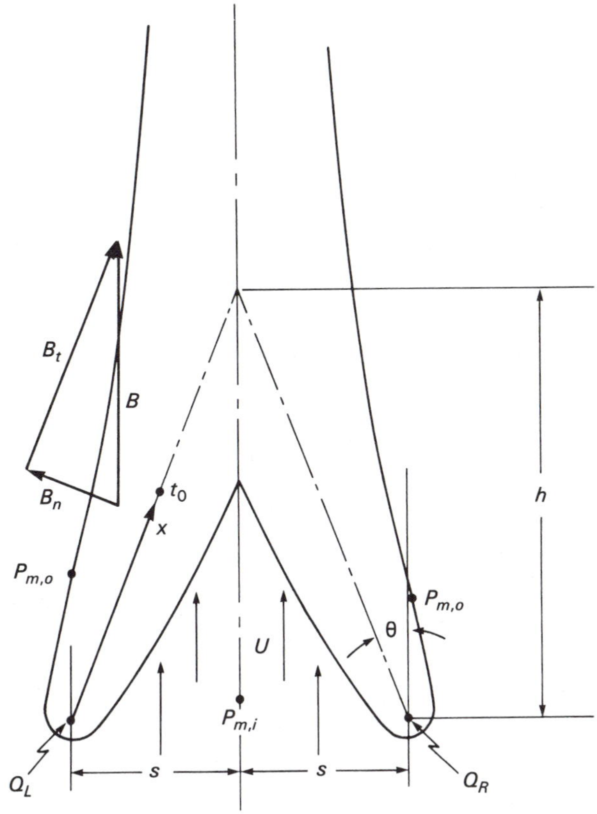

Figure 5.7.13 Model of the interaction between two adjacent plane plumes. *(Reprinted with permission from L. Pera and B. Gebhart,* J. Fluid Mech., *vol. 68, p. 259. Copyright © 1975, Cambridge University Press.)*

able downstream, U is essentially uniform over the whole extent of the inside region.

The plumes must merge when the volume flow rate into the inside region, approximately $U(2sL + sh)$ where L is the plume span, has been entrained by the two plumes along their inside surfaces. These entrainment rates may be calculated simply, even with inclination. Relations may then be written between U, entrainment rates, motion pressure, flow geometry, and B_n. This model predicts the interaction height h as a function of s, total heat input Q, and so on. The model was extended by Gebhart et al. (1976) to interpret the interaction between a pair of laminar plumes of unequal strength. The experimental results were in general agreement with the predictions of the model for both equal and unequal plumes.

Flow interaction between plumes has direct technological applications, particularly in turbulent flows, for instance, in the design of an array of cooling towers. The results discussed above suggest that a spacing that would enhance plume interaction might be desirable. Interactions generate a much stronger plume, which would be expected to rise much higher in the atmosphere and have greater resistance to side winds. In the arrangement of cooling towers, the considerations of optimum location could, therefore, be related to the interaction of buoyant plumes.

Pera and Gebhart (1975) also studied the interaction between axisymmetric plumes arising from diffusion flames above hypodermic tubing burners; see Fig. 5.7.14. The dominating mechanism of interaction again arises from the entrainment effects. However, the interaction is much weaker than that between plane plumes since the neighboring plumes offer less restriction to the supply of ambient fluid for downstream entrainment.

5.7.4 Interactive Flows from Thermal Sources on a Bounding Surface

Such flows are of particular relevance in the design of electronic components in circuit boards. Often, multiple heat-dissipating electronic components are stacked together; see Fig. 5.7.15*a*. An important design consideration is to ensure that the components and the boards do not reach excessively high temperatures. The heat removal is frequently by natural convection. Therefore, the natural convection flows generated on these devices determine the levels of energy input or the location of the electronic components for optimum heat removal and high reliability.

However, such flows are quite complex and difficult to analyze. Each dissipating component causes a natural convection effect. This interacts with the flow field of other components and with the circuit board. These interactive flows are further influenced in a complicated manner by the presence of multiple bounding surfaces.

The complexity of the flows in electronic cooling led Aung (1972) and Aung et al. (1972a, 1972b) to deal with the overall convective cooling of arrays of

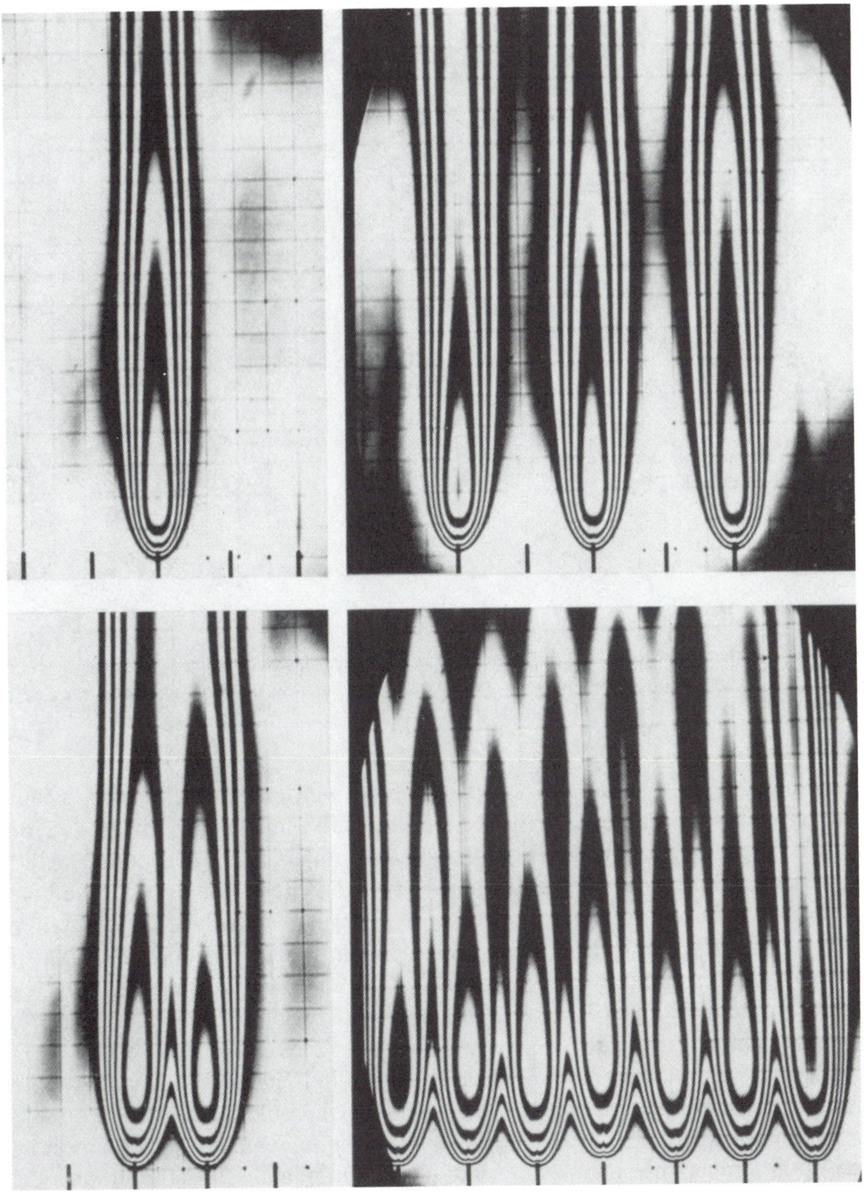

Figure 5.7.14 Plane arrays of axisymmetric plumes generated by combustion above 0.51-mm-diameter gas sources of equal Q. The spacing of the sources seen is about 1.4 cm. *(Reprinted with permission from L. Pera and B. Gebhart,* J. Fluid Mech., *vol. 68, p. 259. Copyright © 1975, Cambridge University Press.)*

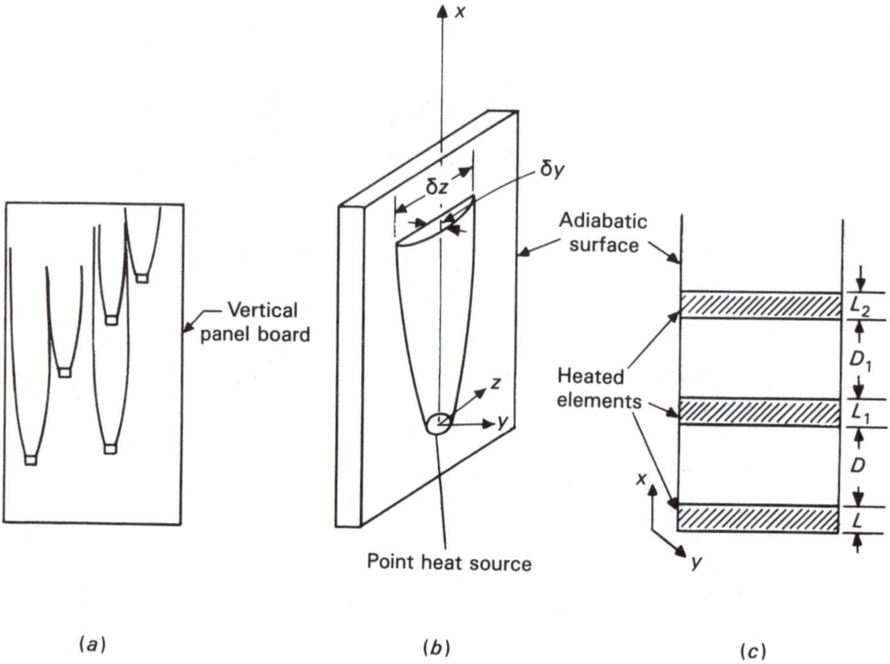

Figure 5.7.15 Examples of interactive flows. (*a*) Heat-dissipating electronic components on a circuit board; (*b*) concentrated heat source on a vertical adiabatic surface; (*c*) three uniform heat flux dissipating elements on a vertical adiabatic surface.

circuit boards. Aung et al. (1972a) considered a vertical parallel surface channel configuration and theoretically and experimentally studied the effect of circuit channel height on the maximum circuit board temperature. This study included the interaction of the convective flows from adjacent circuit boards, with the boards approximated as uniform heat flux surfaces. Aung (1972) solved the natural convection flow in a vertical parallel surface channel with asymmetric heating for both uniform temperature and uniform heat flux surfaces. Aung et al. (1972b) considered nondeveloped flow for the same kinds of conditions. For further details, see Chapter 14.

The flow and temperature fields arising from individual components were not considered. However, the thermal behavior of each component is often quite important in estimating its performance. Some of the relevant questions concern the maximum temperature attained by individual components and the effect of upstream flow and temperature fields on the performance of elements in their wake.

As a first step in determining such effects arising from an isolated thermal source on a bounding surface, Zimin and Lyakhov (1970) analyzed the flow downstream of a horizontal line thermal source on an otherwise adiabatic surface. Similarity solutions were calculated for Pr = 7.0. These calculations were extended by Jaluria and Gebhart (1977) to a wider range of Pr, from 0.01 to 100. They are discussed in detail in Section 3.7.

The three-dimensional analog of the wall plume is the flow arising due to a concentrated (or point) source of heat embedded on a vertical insulating surface; see Fig. 5.7.15*b*. This flow is much more complicated in analysis. It has, however, been investigated experimentally by Carey and Mollendorf (1977). Detailed temperature measurements of the laminar flow field in water were made and the spanwise and normal thermal boundary layers were visualized. The surface temperature above the source decayed as $x^{-0.77}$, faster than the two-dimensional wall plume, due to enhanced entrainment. However, this is slower than the decay for an axisymmetric plume in the absence of a bounding surface, where $t_0 - t_\infty$ decays as x^{-1}. The data also indicated that the thermal boundary layer thickness δ_y, in the direction normal to the surface, increases downstream linearly with x. In the spanwise direction, the thickness δ_x varies less rapidly, approximately as $x^{1/5}$.

Jaluria (1982b) analyzed the natural convection flow due to two or three isolated heated horizontal strips on a vertical adiabatic surface; see Fig. 5.7.15*c*. The elements are of finite width and are each assumed to have a uniform thermal energy surface flux q''. In a manner analogous to that of Sparrow and Faghri (1980), this transport was treated as a boundary region flow and the equations were solved numerically for Pr = 0.7. The surface temperature, velocity field, and heat transfer rate from the elements lying in the wake of the lower elements were determined for various spacings between the elements. Again, the flow and temperature effects on the heat transfer rate from the upper elements lying in the wake of lower element were found to be influenced by the element spacing. Heat transfer enhancement or degradation depended on whether the flow or temperature effect dominated. That is, for two similar elements with identical thermal input, the coefficient of heat transfer from the element was lowered by the upstream element for small spacings and increased for larger ones.

Heat transfer results for different ratios of heat inputs to the two or three elements, along with calculations of surface temperature and flow velocity, are also given. Some of these results are shown in Fig. 5.7.16, where X is the dimensionless distance x/L and ϕ the dimensionless temperature $(t_0 - t_\infty)k\text{Gr}^*/q''L$, where Gr* is based on the height L and heat flux input q'' of the lowermost element. Note that the temperature of the uppermost element is highest, resulting in the lowest heat transfer coefficient, for the spacings $D/L = D_1/L = 2$.

Figure 5.7.16*b* indicates that the maximum velocity increases with X. This is due to the buoyancy augmentation at the upper element from the heated fluid coming from below. The variation in velocity level is, however, very gradual compared to the abrupt change in temperature in the neighborhood of a heated element, as is expected. Buoyancy is a cumulative effect. The surface boundary conditions for the velocity field do not change, as they do abruptly for the temperature field. A larger velocity change would have occurred if the two heated elements were not located on a surface, as in Fig. 5.7.8. The flow discussed above has also been investigated by Jaluria (1985a), using the full equations. It was

shown that boundary layer assumptions are valid as long as Gr* is larger than about 10^3.

The natural convection cooling of electronic equipment has been reviewed by Jaluria (1985b). Many of the flow configurations discussed above, as well as other geometries that arise in electronic circuitry, are included.

The trend toward miniaturization in electronic equipment results in closer packaging of electronic components and in much greater heat dissipation levels than before. Cooling by natural convection alone is often insufficient to meet the critical temperature limitations of the components. Consequently, reliance on forced convection is increasing. Again, the flow passage diversities and irregularities make the analysis difficult. It is generally necessary to resort to experimental investigation of the different geometries and conditions that might arise in practice. The interested reader is referred to the experimental investigation by Sparrow et al. (1982), which deals with some of the questions involved in present-day cooling of electronic equipment.

5.8 FLOW SEPARATION

The discussion in Sections 5.1–5.4 was mainly concerned with the portions of the flows and transport that remain attached to the surface. Any separated flows that arise in subsequent downstream regions were not treated. Such flows are quite complex and difficult to analyze. Several experimental investigations, however, have increased our understanding of these flows.

From several of these studies, taken together, it appears that flow separation in buoyancy-induced flows may arise in two different ways: from direct interaction of several attached flows, such as the formation of a plume above a cylinder, or from a component of the buoyancy force normal to the surface, B_n. An example is the separation of flow at some distance downstream from the leading edge of a heated horizontal surface facing upward. These two different kinds of separation are discussed here. The emphasis is on the basic mechanisms that lead to and follow flow separation.

Separation arising from direct flow interactions. The first kind of separation, hereafter called "interaction-caused separation," was studied by Pera and Gebhart (1972) for flow above a cylindrical surface and by Jaluria and Gebhart (1975) for flow over a hemispherical surface, in water. A schematic of the flow-generating geometry used in the former study is shown in Fig. 5.8.1*a*. A vertical boundary layer flow develops adjacent to each of the two vertical side surfaces of the inverted *U* shape. The flows, following the curved sections, interact and join together to rise in a plume above the curved surface, as seen in Fig. 5.8.1*b*. The lengths of the streaks in Fig. 5.8.1*b*, caused by small illuminated Pliolite particles, indicate both the magnitude and the local direction of flow. There is clearly no appreciable vortical motion, backflow, or circulation, effects that are often associated with boundary layers in forced flows separating from downstream sur-

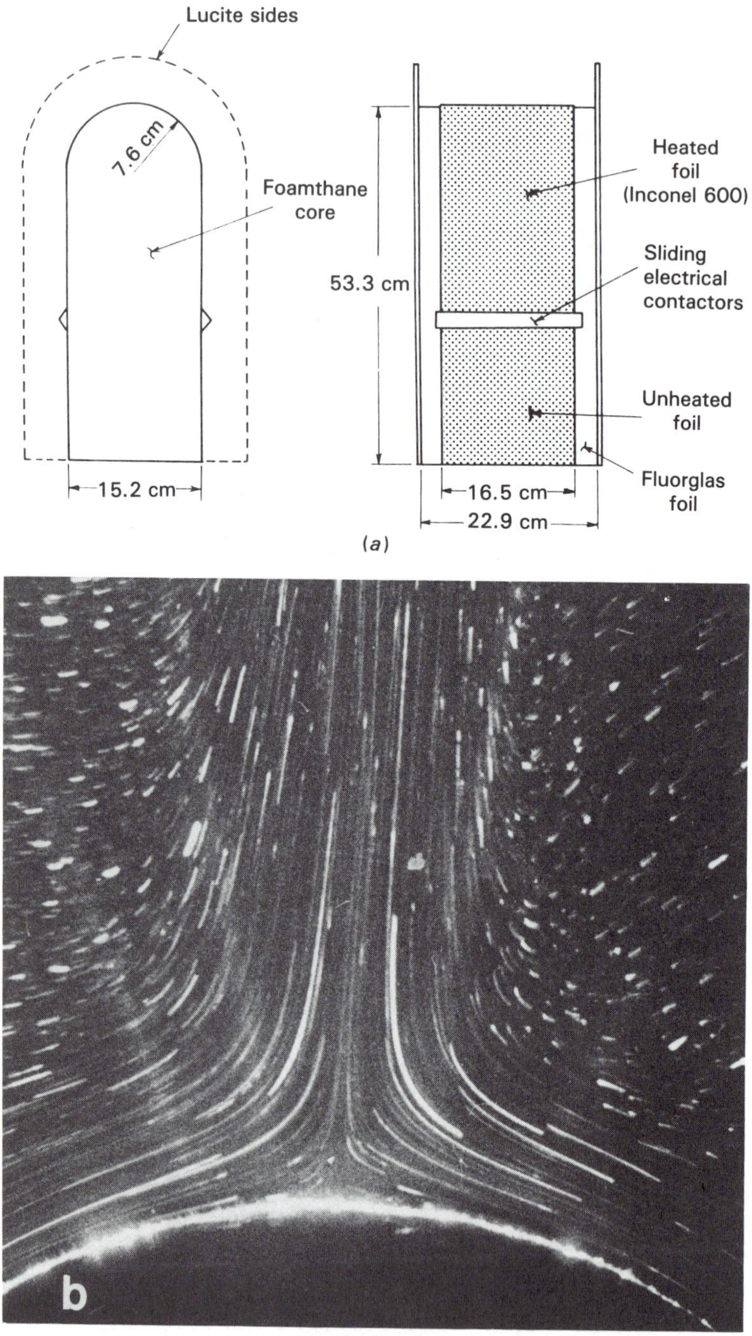

Figure 5.8.1 Steady natural convection in the wake of a horizontal heated cylindrical surface. (*a*) Flow-generating geometry; (*b*) pattern of flow at and above the top, $Gr_D = 0.25 \times 10^{10}$. *(Reprinted with permission from* Int. J. Heat Mass Transfer, *vol. 15, p. 175, L. Pera and B. Gebhart. Copyright © 1972, Pergamon Journals Ltd.)*

faces. The steady laminar flows simply join and separate smoothly. They must separate since they are flowing against each other.

To further investigate these aspects of flow separation and subsequent behavior, Jaluria and Gebhart (1975) took detailed measurements of temperature and velocity in the flow adjacent to and above heated isothermal hemispheres in water; see Fig. 5.8.2. These measurements indicated that three different flow regions arise downstream: the attached boundary layer region, which sometimes extends only to about $\xi = 45°$; the separating region, where the flow turns from tangential to upward; and region of shed flow at the top. This shed flow, during its ascent, develops toward axisymmetric plume behavior.

As for a cylindrical surface, these measurements again indicated that separation with backflow does not occur in such natural convection flows. The flow does separate from the surface as the boundary layer material from all sides meets at the top. However, this separation does not arise through a first-order interaction with an external pressure field, nor does it result in flow reversal downstream of separation. The heated fluid is, apparently, simply realigned upward.

This realignment begins well ahead of the topmost point, as indicated by an abrupt downstream change in the boundary layer thickness and the velocity level (Fig. 5.8.3). Measurement of the local heat transfer coefficient also shows this effect (Fig. 5.8.4). In the attached portion of the flow, the boundary layer thickens and the Nusselt number gradually decreases. The increased velocity level due to realignment of the flow causes an increase in heat transfer at large ξ, despite the increase in boundary layer thickness. Several of these aspects of separation and realignment were also considered by Jaluria (1976).

Separation arising with unstable thermal region stratification. This second

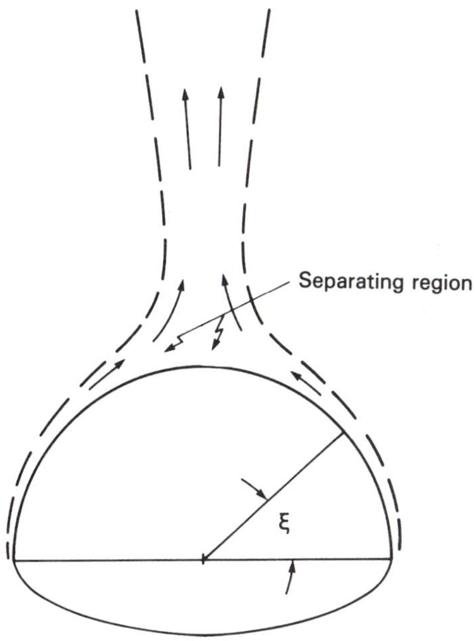

Figure 5.8.2 Heated hemispherical surface with the insulating bottom horizontal; an attached flow on the surface is shed into a developing axisymmetric plume.

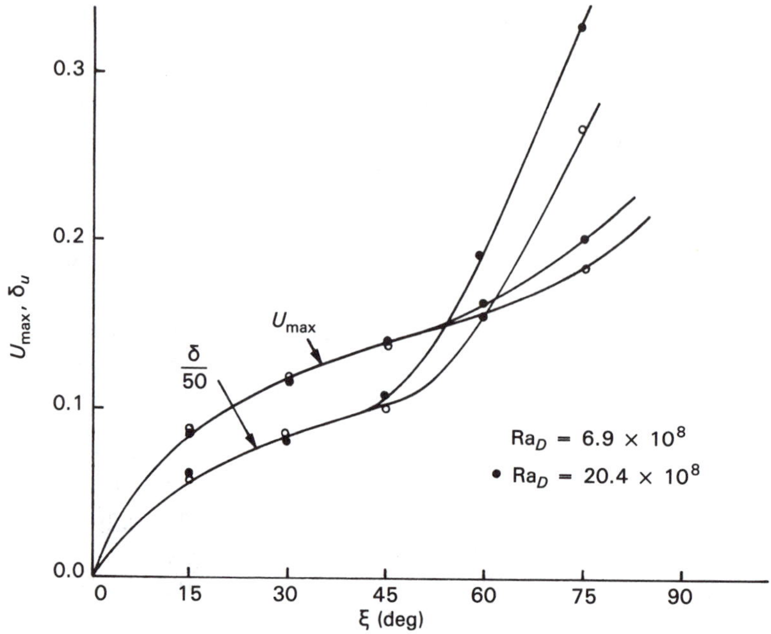

Figure 5.8.3 Variation of the maximum tangential velocity and the velocity boundary region thickness, δ_u, with ξ for an upright hemisphere representing the top portion of a sphere. *(Reprinted with permission from* Int. J. Heat Mass Transfer, *vol. 18, p. 415, Y. Jaluria and B. Gebhart. Copyright © 1975, Pergamon Journals Ltd.)*

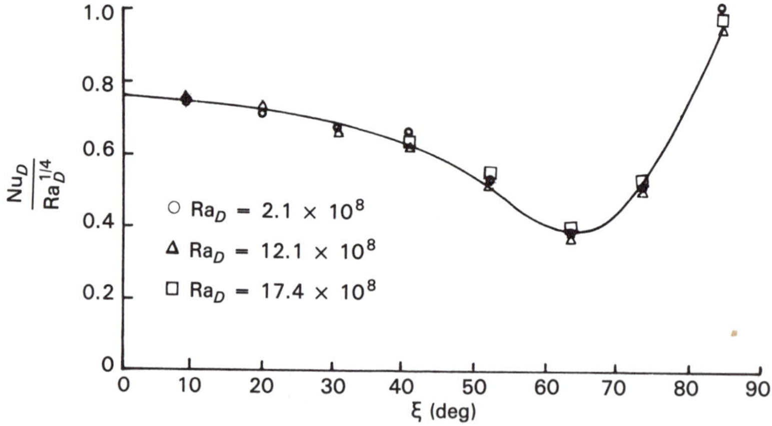

Figure 5.8.4 Variation of the local Nusselt number with ξ for the upright hemisphere. *(Reprinted with permission from* Int. J. Heat Mass Transfer, *vol. 18, p. 415, Y. Jaluria and B. Gebhart. Copyright © 1975, Pergamon Journals Ltd.)*

kind of flow separation is shown in Fig. 5.8.5 (Pera and Gebhart, 1973b). The upstream flow, here toward the left, is adjacent to a horizontal heated surface facing upward. Smoke filaments were introduced into the flow near the leading edge for visualization. In Fig. 5.8.5a, a controlled disturbance was introduced upstream. A rapidly amplifying longitudinal system of counterrotating vortices is seen to have developed a short distance from the leading edge. This motion in-

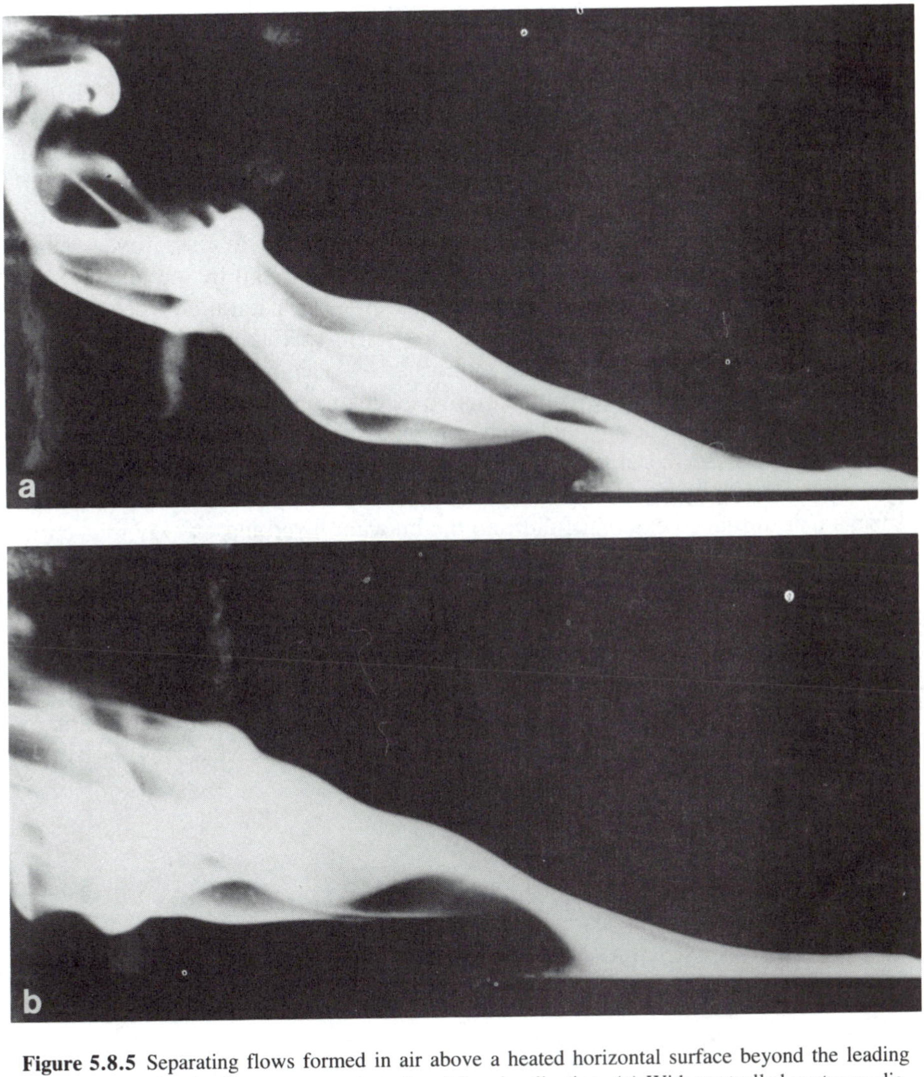

Figure 5.8.5 Separating flows formed in air above a heated horizontal surface beyond the leading edge for flow. Smoke filaments were introduced for visualization. (*a*) With controlled upstream disturbance at 1.7 Hz; (*b*) subject to naturally occurring disturbances. *(Reprinted with permission from* Int. J. Heat Mass Transfer, *vol. 16, p. 1147, L. Pera and B. Gebhart. Copyright © 1973, Pergamon Journals Ltd.)*

ducts a large amount of ambient fluid and results in very rapid thickening of the downstream region. In a flow subject only to naturally occurring disturbances, as in Fig. 5.8.5b, similar separation occurs, although the motive vortices are not as clearly defined. Similar characteristics for horizontal flows have also been observed by Hassan and Mohamed (1970), Fujii and Imura (1972), and Black and Norris (1975). However, in the last two studies, these effects were taken as events of transition from laminar to turbulent transport. Goel and Jaluria (1986) experimentally studied the transport from an isolated heat source on an inclined surface. The measured thermal field indicated an earlier downstream separation at larger inclinations with the vertical, as expected.

The impetus for this mode of separation is the upward normal component of the buoyancy force, B_n. It drives the flow but also amounts to an unfavorable density stratification in the thermal diffusion region. This results in thermal instability. It is believed that this instability, through early amplification of disturbances, gives rise to the observed secondary longitudinal vortex rolls; see Gebhart (1973, 1979). Apparently, similar mechanisms also arise in flows over heated surfaces at various inclinations. These are discussed in detail in Section 11.12.2.

Summary. The two kinds of separation observed in natural convection flows are quite different from those in forced flows. As in forced flows, thickening layers of irregular flow and viscous motion arise, accompanied by a change in the nature of transport. However, the motive mechanisms are different. For a flow generated in a quiescent ambient medium, large external pressure gradients are not present. Instead, separation is caused by the interaction of attached flows or by a cross-flow buoyancy force. Another fundamental difference lies in the mechanisms that introduce external fluid into the buoyant layer and quickly thicken it. In forced flows, a vortex system arises normal to the imposed flow direction. It continuously, or alternately as behind a cylinder, inducts fluid from the outer flow.

In interaction-caused separation, no vortical motion arises. The thickening of the boundary region accompanies the boundary layer realignment that occurs to generate a shed plume above the surface. In the second type of separation, a longitudinal secondary vortical motion is produced instead. That is, it is largely parallel to the flow in which it arises.

REFERENCES

Abdulkadir, A. (1968). M.S. Thesis, Univ. of Kentucky, Lexington.

Ackroyd, J. A. D. (1976). *Proc. R. Soc. London Ser. A 352,* 249.

Acrivos, A. (1960). *AIChE J. 6,* 584.

Agarwal, R., and Jaluria, Y. (1985). Paper 85-WA/HT-4, ASME Winter Annual Meeting, Florida.

Aihara, T., Yamada, Y., and Endo, S. (1972). *Int. J. Heat Mass Transfer 15,* 2535.

Akagi, S. (1965). *Trans. JSME 31,* 1327.

Alamgir, M. (1979). *J. Heat Transfer 101,* 174.

Al-Arabi, M., and El-Riedy, M. K. (1976). *Int. J. Heat Mass Transfer 19,* 1399.

Al-Arabi, M., and Khamis, M. (1982). *Int. J. Heat Mass Transfer 23,* 3.
Al-Arabi, M., and Salman, Y. K. (1980). *Int. J. Heat Mass Transfer 25,* 45.
Amato, W. S., and Tien, C. (1972). *Int. J. Heat Mass Transfer 15,* 327.
Aung, W. (1972). *Int. J. Heat Mass Transfer 15,* 1577.
Aung, W., Kessler, T. J., and Beitin, K. I. (1972a). Paper 72-WA/HT-40, ASME Winter Annual Meeting, New York.
Aung, W., Fletcher, L. S., and Sernas, V. (1972b). *Int. J. Heat Mass Transfer 15,* 2293.
Bandrowski, J., and Rybski, W. (1976). *Int. J. Heat Mass Transfer 19,* 827.
Birkebak, R. C., and Abdulkadir, A. (1970). *Proc. 4th Int. Heat Transfer Conf.,* Paris, Paper NC2.2.
Black, W. Z., and Norris, J. K. (1975). *Int. J. Heat Mass Transfer 18,* 43.
Blanc, P., and Gebhart, B. (1974). *Proc. 5th Int. Heat Transfer Conf.,* Tokyo, Paper NC1.5.
Braun, W. H., Ostrach, S., and Heighway, J. E. (1961). *Int. J. Heat Mass Transfer 2,* 121.
Carey, V. P., and Mollendorf, J. C. (1977). *Int. J. Heat Mass Transfer 20,* 1059.
Chand, J., and Vir, D. (1980). *Lett. Heat Mass Transfer 7,* 213.
Chao, B. T., and Lin, F. N. (1975). *J. Heat Transfer 97,* 294.
Chapman, D. R., and Rubesin, M. W. (1949). *J. Appl. Sci. 16,* 547.
Chen, C. C., and Eichhorn, R. (1976). *J. Heat Transfer 98,* 446.
Chen, C. C., and Eichhorn, R. (1979). *J. Heat Transfer 101,* 566.
Chiang, T., and Kaye, J. (1962). *Proc. 4th U.S. Natl. Cong. Appl. Mech.,* Berkeley, Calif., p. 1213.
Chiang, T., Ossin, A., and Tien, C. L. (1964). *J. Heat Transfer 86,* 537.
Churchill, S. W. (1983). In *Heat Exchanger Design Handbook,* pt. 2, E. Schlünder, Ed., Hemisphere, Washington, D.C.
Churchill, S. W., and Chu, H. H.-S. (1975). *Int. J. Heat Mass Transfer 18,* 1049.
Churchill, S. W., and Churchill, R. U. (1975). *AIChE J. 21,* 604.
Churchill, S. W., and Usagi, R. (1972). *AIChE J. 18,* 1121.
Clarke, J. F. (1973). *J. Fluid Mech. 57,* 45.
Clarke, J. F., and Riley, N. (1975). *Q. J. Mech. Appl. Math. 28(4),* 373.
Clifton, J. V. (1967). Ph.D. Thesis, Rice University, Houston, Texas.
Clifton, J. V., and Chapman, A. J. (1969). *Int. J. Heat Mass Transfer 12,* 1573.
Collis, D. C., and Williams, M. J. (1954). Aerodynamics Note 140, Aeronautical Research Laboratories, Melbourne, Australia.
Deluche, C. (1970). Thèse de troisième cycle, Faculté des Sciences de l'Universite de Poitier.
Eckert, E. R. G., and Soehngen, E. (1948). Tech. Rept. 5747, U.S. Air Force Air Material Command, Dayton, Ohio.
Eckert, E. R. G., Schneider, P. J., Hayday, A. A., and Larson, M. M. (1958). *Jet Propul. 28,* 34.
Eichhorn, R., Lienhard, J. H., and Chen, C. C. (1974). *Proc. 5th Int. Heat Transfer Conf.,* Tokyo, Paper NC1.3.
Elenbaas, W. (1942). *Physica (The Hague) 9,* 285.
Emery, A. F., Yang, A., and Wilson, J. R. (1976). ASME Paper 76-HT-46.
Fand, R. M., and Brucker, J. (1983). *Int. J. Heat Mass Transfer 26,* 709.
Fand, R. M., Morris, E. W., and Lum, M. (1977). *Int. J. Heat Mass Transfer 20,* 1173.
Farber, E. A., and Rennat, H. O. (1957). *Ind. Eng. Chem. 49,* 437.
Farouk, B., and Güceri, S. I. (1981). *J. Heat Transfer 103,* 522.
Fatt, L. F. (1976). *J. I.E.M. 20,* 24.
Fishenden, M., and Saunders, O. A. (1950, 1961). *An Introduction to Heat Transfer,* Oxford Univ. Press, London.
Fujii, T., and Imura, H. (1972). *Int. J. Heat Mass Transfer 15,* 755.
Fujii, T., Fujii, M., and Mastunaga, T. (1979). *Numer. Heat Transfer 2,* 329.
Fujii, T., Fujii, M., and Honda, T. (1981). *Numer. Heat Transfer 4,* 69.
Fussey, D. E., and Warneford, I. P. (1976). *Lett. Heat Mass Transfer 3,* 443.
Fussey, D. E., and Warneford, I. P. (1978). *Int. J. Heat Mass Transfer 21,* 119.
Gebhart, B. (1971). *Heat Transfer,* 2d ed., McGraw-Hill, New York.

Gebhart, B. (1973). *Adv. Heat Transfer 9,* 273.
Gebhart, B. (1979). *J. Fluids Eng. 101,* 5.
Gebhart, B., and Pera, L. (1970). *J. Fluid Mech. 45,* 49.
Gebhart, B., Audunson, T., and Pera, L. (1970). *Proc. 4th Int. Heat Transfer Conf.,* Paris, Paper NC3.2.
Gebhart, B., Shaukatullah, H., and Pera, L. (1976). *Int. J. Heat Mass Transfer 19,* 751.
Geeola, F., and Cornish, A. R. H. (1981). *Int. J. Heat Mass Transfer 24,* 1369.
Gill, W. N., Zeh, D. W., and Del Casal, E. (1965). *Z. Angew. Math. Phys. 16,* 539.
Goel, S., and Jaluria, Y. (1986). *Proc. 8th Int. Heat Transfer Conf.,* Hemisphere, Washington, D.C., vol. 3, p. 134, San Francisco.
Goldstein, R. J., and Lau, K. S. (1983). *J. Fluid Mech. 129,* 55.
Goldstein, R. J., Sparrow, E. M., and Jones, D. C. (1973). *Int. J. Heat Mass Transfer 16,* 1025.
Gosse, J. (1956). *Publ. Sci. Tech. Minist. Air (Fr.),* No. 322.
Gupta, A. S., and Pop, I. (1977). *Phys. Fluids 20,* 162.
Harpole, G. M., and Catton, I. (1976). *J. Heat Transfer 98,* 208.
Hasan, M. M. (1978). M.S. Thesis, Univ. of Kentucky, Lexington.
Hasan, M. M., and Eichhorn, R. (1979). *J. Heat Transfer 101,* 642.
Hassan, K. E., and Mohamed, S. A. (1970). *Int. J. Heat Mass Transfer 13,* 1873.
Hatfield, D. W., and Edwards, D. K. (1981). *Int. J. Heat Mass Transfer 24,* 1019.
Hering, R. G. (1965). *Int. J. Heat Mass Transfer 8,* 1333.
Hering, R. G., and Grosh, R. J. (1962). *Int. J. Heat Mass Transfer 5,* 1059.
Hermann, R. (1954). NACA Tech. Memo. 1366.
Holman, J. P. (1981). *Heat Transfer,* 5th ed., p. 281, McGraw-Hill, New York.
Hossain, M. A., and Gebhart, B. (1970). *Heat Transfer IV, NC1.6,* p. 1, Elsevier, Amsterdam.
Husar, R. B., and Sparrow, E. M. (1968). *Int. J. Heat Mass Transfer 11,* 1206.
Ingham, D. B. (1978). *Z. Angew. Math. Phys. 29,* 871.
Jaluria, Y. (1976). *Lett. Heat Mass Transfer 3,* 457.
Jaluria, Y. (1982a). *Lett. Heat Mass Transfer 9,* 107.
Jaluria, Y. (1982b). *J. Heat Transfer 104,* 223.
Jaluria, Y. (1985a). *J. Heat Transfer 107,* 883.
Jaluria, Y. (1985b). In *Natural Convection; Fundamentals and Applications,* W. Aung, S. Kakac, and R. Viskanta, Eds., Hemisphere, Washington, D.C.
Jaluria, Y., and Gebhart, B. (1975). *Int. J. Heat Mass Transfer 18,* 415.
Jaluria, Y., and Gebhart, B. (1977). *Int. J. Heat Mass Transfer 20,* 153.
Jodlbauer, K. (1933). *Forsch. Ver. Dtsch. Ing. 4,* 158.
Jones, D. R. (1973). *Q. J. Mech. Appl. Math. 26,* 77.
Kato, K., and Ito, H. (1968). *Trans. JSME 34,* 1430.
Kierkus, W. T. (1968). *Int. J. Heat Mass Transfer 11,* 241.
King, W. J. (1932). *Mech. Eng. 54,* 347.
Koh, J. C. Y. (1964). *Int. J. Heat Mass Transfer 7,* 811.
Koh, J. C. Y., and Price, J. F. (1965). *J. Heat Transfer 87,* 237.
Kranse, A. A., and Schenk, J. (1965). *Appl. Sci. Res. Sec. A15,* 397.
Kreith, F., and Black, W. Z. (1980). *Basic Heat Transfer,* p. 259, Harper & Row, New York.
Kuehn, T. H., and Goldstein, R. J. (1980). *Int. J. Heat Mass Transfer 23,* 971.
Kuiken, H. K. (1968). *Int. J. Heat Mass Transfer 11,* 1141.
Kyte, J. R., Madden, A. J., and Piret, E. L. (1953). *Chem. Eng. Prog. 49,* 653.
Lee, J. B., and Lock, G. S. H. (1972). *Trans. CSME 1,* 197.
Levy, S. (1955). *J. Appl. Mech. 22,* 515.
Lieberman, J., and Gebhart, B. (1969). *Int. J. Heat Mass Transfer 12,* 1385.
Lienhard, J. H. (1981). *A Heat Transfer Textbook,* p. 361, Prentice-Hall, Englewood Cliffs, N.J.
Lin, F. N. (1976). *Lett. Heat Mass Transfer 3,* 59.
Lin, F. N., and Chao, B. T. (1974). *J. Heat Transfer 96,* 435.
Lin, F. N., and Chao, B. T. (1978). *J. Heat Transfer 100,* 160.
Lin, F. N., and Chern, S. Y. (1981). *J. Heat Transfer 103,* 819.

Lloyd, J. R., and Moran, W. R. (1974). *J. Heat Transfer 96,* 443.
Lloyd, J. R., Sparrow, E. M., and Eckert, E. R. G. (1972). *Int. J. Heat Mass Transfer 15,* 457.
Lowell, H. H. (1950). NACA Tech. Note 2117.
Lugt, H. J., and Schwiderski, E. W. (1966). *J. Atmos. Sci. 23(1),* 54.
Mahajan, R. L., and Gebhart, B. (1980a). *J. Heat Transfer 102,* 368.
Mahajan, R. L., and Gebhart, B. (1980b). *J. Phys. E Sci. Instrum. 13,* 1110.
Mahony, J. J. (1956). *Proc. R. Soc. London Ser. A 238,* 412.
Marsters, G. F. (1972). *Int. J. Heat Mass Transfer 15,* 921.
Marsters, G. F. (1975). *Can. J. Chem. Eng. 53,* 144.
Merk, H. J. (1959). *J. Fluid Mech. 5,* 460.
Merk, H. J., and Prins, J. A. (1953, 1954). *Appl. Sci. Res. Sec. A 4,* part I, p. 11, part II, p. 195, part III, p. 207.
Merkin, J. H. (1975). *Int. J. Heat Mass Transfer 18,* 237.
Merkin, J. H. (1976). Presented at the ASME-AIChE Heat Transfer Conference, St. Louis, Mo.
Merkin, J. H. (1977). *J. Heat Transfer 99,* 453.
Minkowycz, W. J., and Cheng, P. (1976). *Int. J. Heat Mass Transfer 19,* 805.
Moran, W. R., and Lloyd, J. R. (1975). *J. Heat Transfer 97,* 472.
Morgan, V. T. (1975). *Adv. Heat Transfer 11,* 199.
Mori, S., Tanimoto, A., Yamashita, M., Aoki, M., and Kazuhiko, Y. (1976). *Kagaku Kugaku Ronbunshu 2,* 214.
Muntasser, M. A., and Mulligan, J. C. (1978). *J. Heat Transfer 100,* 165.
Nakai, S., and Okazaki, T. (1975). *Int. J. Heat Mass Transfer 18,* 387.
Oosthuizen, P. H. (1976a). *J. Heat Transfer 98,* 570.
Oosthuizen, P. H. (1976b). *J. Heat Transfer 98,* 672.
Oosthuizen, P. H., and Donaldson, E. (1972). *J. Heat Transfer 94,* 330.
Pera, L., and Gebhart, B. (1972). *Int. J. Heat Mass Transfer 15,* 175.
Pera, L., and Gebhart, B. (1973a). *Int. J. Heat Mass Transfer 16,* 1131.
Pera, L., and Gebhart, B. (1973b). *Int. J. Heat Mass Transfer 16,* 1147.
Pera, L., and Gebhart, B. (1975). *J. Fluid Mech. 68,* 259.
Peterka, J. A., and Richardson, P. D. (1969). *Int. J. Heat Mass Transfer 12,* 749.
Peube, J. L., and Blay, D. (1978). *Int. J. Heat Mass Transfer 21,* 1125.
Potter, J. M., and Riley, N. (1980). *J. Fluid Mech. 100,* 769.
Prandtl, L. (1952). *Essentials of Fluid Dynamics,* p. 422, Hafner, New York.
Raithby, G. D., and Hollands, K. G. T. (1975). *Adv. Heat Transfer 11,* 265.
Raithby, G. D., and Hollands, K. G. T. (1976a). *J. Heat Transfer 98,* 522.
Raithby, G. D., and Hollands, K. G. T. (1976b). *J. Heat Transfer 98,* 72.
Raithby, G. D., and Hollands, K. G. T. (1978). *Proc. 6th Int. Heat Transfer Conf.*, Toronto, NC187.
Raithby, G. D., Pollard, A., Hollands, K. G. T., and Yovanovich, M. M. (1976). *J. Heat Transfer 98,* 452.
Restrepo, F., and Glicksman, L. R. (1974). *Int. J. Heat Mass Transfer 17,* 135.
Rich, B. R. (1953). *Trans. ASME 75,* 489.
Riley, N. (1975). *Int. J. Heat Mass Transfer 18,* 991.
Rotem, Z. (1967). *Proc. 1st Can. Natl. Cong. Appl. Mech. 2-b,* 309.
Rotem, Z., and Claassen, L. (1969). *J. Fluid Mech. 39,* 173; see also *Can. J. Chem. Eng.* (1969), *47,* 461.
Rotem, Z., and Claassen, L. (1970). *Can. J. Chem. Eng. 47,* 460.
Savage, S. B. (1969). *AIAA J. 7,* 1628.
Saville, D. A., and Churchill, S. W. (1967). *J. Fluid Mech. 29,* 391.
Schlichting, H. (1968). *Boundary Layer Theory,* 6th ed., McGraw-Hill, New York.
Schmidt, E. (1932). *VDI Forschungsh. 3,* 181.
Schuh, H. (1948). Boundary Layers of Temperature, in *Boundary Layers,* W. Tollmien, Ed., Sec. B6, British Ministry of Supply, German Doc. Cent., Ref. 3220T.
Shaukatullah, H., and Gebhart, B. (1978). *Int. J. Heat Mass Transfer 21,* 1481.
Singh, S. N. (1977). *Int. J. Heat Mass Transfer 20,* 1155.

Singh, S. N., and Birkebak, R. C. (1969). *Z. Angew. Math. Phys. 20,* 454.
Singh, S. N., Birkebak, R. C., and Drake, R. M., Jr. (1969). *Prog. Heat Mass Transfer 2,* 87.
Smith, F. T., and Riley, D. S. (1979). *Int. J. Heat Mass Transfer 22,* 309.
Sparrow, E. M., and Ansari, M. A. (1983). *Int. J. Heat Mass Transfer 26,* 1357.
Sparrow, E. M., and Faghri, M. (1980). *J. Heat Transfer 102,* 623.
Sparrow, E. M., and Yu, H. S. (1971). *J. Heat Transfer 93,* 328.
Sparrow, E. M., Quack, H., and Boener, C. J. (1970). *AIAA J. 8,* 1936.
Sparrow, E. M., Niethammer, J. E., and Choboki, A. (1982). *Int. J. Heat Mass Transfer 25,* 961.
Stewart, W. E. (1971). *Int. J. Heat Mass Transfer 14,* 1013.
Stewartson, K. (1958). *Z. Angew. Math. Phys. 9,* 276.
Sugawara, S., and Michiyoshi, I. (1955). *Trans. JSME 21,* 651.
Suwono, A. (1980). *Int. J. Heat Mass Transfer 23,* 53.
Vliet, G. C. (1969). *J. Heat Transfer 91,* 511.
Vliet, G. C., and Ross, D. C. (1975). *J. Heat Transfer 97,* 549.
Wagner, C. (1956). *J. Appl. Mech. 23,* 320.
Weise, R. (1935). *VDI Forshungsh. 6,* 281.
Wilks, G. (1972). *Int. J. Heat Mass Transfer 15,* 351.
Wragg, A. A. (1968). *Electrochim. Acta 13,* 2159.
Wragg, A. A., and Loomba, R. P. (1970). *Int. J. Heat Mass Transfer 13,* 439.
Yamagata, K. (1943). *Trans. JSME 9,* 132.
Yousef, W. W., Tarasuk, J. D., and McKeen, W. J. (1982). *J. Heat Transfer 104,* 493.
Yuge, T. (1960). *J. Heat Transfer 82,* 214.
Zakerullah, M., and Ackroyd, J. A. D. (1979). *Z. Angew. Math. Phys. 30,* 427.
Zariffeh, E. K., and Daguenet, M. (1981). *Int. J. Heat Mass Transfer 24,* 1071.
Zimin, V. D., and Lyakhov, Y. N. (1970). *Zh. Prikl. Mekh. Tekh. Fiz. 11,* 159.

PROBLEMS

5.1 Consider the external natural convection flow induced on the upper side and near the leading edge of a horizontal half-infinite surface. A similarity solution results for $d(x) = Nx^n$. Determine the range of values of N and n for which boundary region transport results from the formulation.

5.2 A heated surface at 25°C, inclined at 50° with the vertical, is placed in water. Find the heat transfer by natural convection on the top and bottom sides if the water temperature is 15°C.

5.3 Consider the external natural convection flow induced on the upper side and near the leading edge of a horizontal half-infinite surface.

(a) For constant $t_0 > t_\infty$, write the relevant laminar boundary layer equations, using the Boussinesq approximation.

(b) Postulating the usual stream function ψ, eliminate the motion pressure between the two-force momentum balances and write the remaining differential equations in terms of ψ and $\phi = (t - t_\infty)/(t_0 - t_\infty)$.

5.4 Consider buoyancy-induced flow above an extensive horizontal surface [at $t_0(x)$] with a single straight leading edge. Flow up from below is prevented by setting this surface flush in an extensive horizontal insulated surface. The medium is unstratified and $t_0(x) > t_\infty$.

(a) Find an expression for the local rate of energy convection, assuming that a boundary region flow occurs.

(b) Find how t_0 varies for:

(1) An assigned uniform flux surface condition.

(2) A concentrated line source along the leading edge; the surface is otherwise adiabatic.

(c) Find how the boundary region thickness δ_H varies downstream compared with the thickness that would result with the surface vertical δ_V. Use an isothermal surface condition.

5.5 A heat exchanger pipe carries water at 60°C and is located in still air at 15°C. If its diameter is 3 cm and if the surface temperature may be taken as 45°C, find the total energy lost to the environment by natural convection.

5.6 Consider a sphere of diameter 10 cm, a horizontal cylinder of diameter 10 cm, and a plate of height 10 cm. Each geometry is to have the same surface area. If the surface temperature is maintained at 100°C above the ambient air temperature of 20°C, obtain the natural convection heat transfer from the three bodies. Also, find the heat transfer from the sphere in the absence of fluid flow by solving the conduction equation. Can you do the same for the other two geometries? Explain.

5.7 A thermocouple and a hot wire are located 1 mm away from a heated vertical surface at 400°C. If the ambient air is at 20°C, determine the velocity and temperature measured. The sensors are 1 m above the leading edge. Take the diameter of the hot wire and of the thermocouple junction as 0.001 cm and the junction as a black body.

5.8 A long pipe carries water at 80°C. If its diameter is 5 cm, determine the energy input by natural convection into the thermal plume. Also find the boundary layer thickness and the temperature at a height of 20 cm above the tube, assuming it to be a line source at its axis.

5.9 A sphere of diameter 0.5 cm dissipates 10 W in air. Determine the temperature, velocity, and boundary layer thickness in the plume 10 and 20 cm above the sphere, assuming it to be a point source at its center. Discuss the validity of this approximation.

5.10 For natural convection flow over a sphere at a surface temperature of 40°C in water at 15°C, find the maximum velocity that occurs in the boundary layer, before the occurrence of flow separation, and the downstream location where it occurs. Find the heat transfer coefficient at $\theta = 90°$ above the bottom.

5.11 The top of a furnace may be approximated as an isothermal horizontal surface at a temperature of 400°C. Determine the velocity and heat transfer coefficient 10 cm from the edge of the surface. Ambient air temperature is 30°C.

5.12 Develop an integral analysis for natural convection flow over an inclined plate and obtain the velocity and temperature distributions for $0 < \theta < 30°$ and for $\theta = 90°$, where Θ is the angle between the surface and the vertical. Consider only a heated surface facing upward.

5.13 A disk of diameter D at a temperature t_0 is placed in a stratified environment with $t_\infty - t_r = Mx$. Obtain an expression to determine the heat transfer from the disk to the air medium by natural convection.

5.14 Consider a two-dimensional plume in the neighborhood of a vertical surface. Determine the pressure difference that arises to tilt the plume toward the surface due to the difference in entrainment on either side. Consider laminar flow.

5.15 Derive the equivalent of Eqs. (5.3.14)–(5.3.17) when the normal buoyancy force B_n is directed toward the surface. Also, explain why it is incorrect to derive these equations by replacing ξ by $-\xi$.

CHAPTER

SIX

COMBINED MASS AND THERMAL TRANSPORT

6.1 INTRODUCTION

The preceding three chapters considered in detail the flows and transport in which the buoyancy force arises solely due to thermal diffusion. However, there is an important class of flows in which the driving force for the flow is provided by a combination of thermal and chemical species diffusion effects. Such circumstances arise, for example, during cleaning operations as residual fluid diffuses into the surrounding fluid at a different temperature, in the curing of plastics, and in the manufacture of pulp-insulated cables. Similar transport also arises in many other chemical processes in which concentration differences of dissimilar species exist. In atmospheric flows, thermal convection that results from heating of the earth by sunlight is affected by differences in water vapor concentration. Flow in a body of water is driven through comparable effects on density by temperature and the concentration of dissolved materials in the water. Often, interest lies mainly in the determination of the total energy and material transfer rates. Such processes are considered in this chapter, with the aim of estimating the transport quantities of interest from an understanding of the basic mechanisms of such flows.

6.1.1 Transport Equations

The general equations describing the simultaneous diffusion of thermal energy and chemical species were developed in Chapter 2. For a plane vertical two-dimensional steady laminar flow, assuming the Boussinesq approximations and constant transport properties μ, k, and D, the equations are (2.7.10) and (2.7.15)–(2.7.18), as reproduced below:

$$\frac{\partial u}{\partial x} + \frac{\partial v}{\partial y} = 0 \qquad (6.1.1)$$

$$u\frac{\partial u}{\partial x}+v\frac{\partial u}{\partial y}=\nu\left(\frac{\partial^2 u}{\partial x^2}+\frac{\partial^2 u}{\partial y^2}\right)-\frac{1}{\rho}\frac{\partial p_m}{\partial x}$$
$$+g\beta(t-t_\infty)+g\beta^*(C-C_\infty) \tag{6.1.2}$$

$$u\frac{\partial v}{\partial x}+v\frac{\partial v}{\partial y}=\nu\left(\frac{\partial^2 v}{\partial x^2}+\frac{\partial^2 v}{\partial y^2}\right)-\frac{1}{\rho}\frac{\partial p_m}{\partial y} \tag{6.1.3}$$

$$u\frac{\partial t}{\partial x}+v\frac{\partial t}{\partial y}=\alpha\left(\frac{\partial^2 t}{\partial x^2}+\frac{\partial^2 t}{\partial y^2}\right)+\frac{q'''}{\rho c_p} \tag{6.1.4}$$

$$u\frac{\partial C}{\partial x}+v\frac{\partial C}{\partial y}=D\left(\frac{\partial^2 C}{\partial x^2}+\frac{\partial^2 C}{\partial y^2}\right)+c''' \tag{6.1.5}$$

where c''' is the local rate of production of species C per unit volume. If there is simultaneous diffusion of several distinct chemical species, equations similar to (6.1.5) are written for each of these species and their buoyancy components are included in Eq. (6.1.2).

The usual boundary conditions on temperature and velocity fields, Eqs. (3.3.4) and (3.3.5), are supplemented by additional boundary conditions on the concentration field. That is,

$$\begin{aligned}&\text{at}\quad y=0: &&u=0 &&v=v_0(x)\approx 0 &&t=t_0(x) &&C=C_0(x)\\ &\text{as}\quad y\to\infty: &&u\to 0 &&t\to t_\infty &&C\to C_\infty\end{aligned} \tag{6.1.6}$$

The normal velocity component v_0, from diffusion of the species at the surface, is not zero in mass transfer processes.

6.1.2 Assumptions

It is instructive at this stage to recall the assumptions, particularly those related to the mass transfer, for which these equations result. The secondary effects of concentration gradient on thermal diffusion and of thermal diffusion on mass transfer have been neglected. In Eq. (6.1.5) it is assumed that mass diffusion is a function of concentration gradient only. It is known, however, that in addition to concentration gradients, other potential gradients, for instance, of temperature, pressure, and body forces, also often cause mass diffusion. The last two gradients generally have negligible effects. However, there are some applications in which temperature gradients cause appreciable mass diffusion. This mechanism is called the thermodiffusion, or Sorét, effect. Similar to the Sorét effect is the Dufour, or diffusion-thermal, effect, which is the contribution to the thermal energy flux due to concentration gradients. This effect usually is appreciable for large concentration differences. The above set of equations, therefore, applies for small concentration differences of the diffusing species.

In the estimation of the buoyancy force in Eq. (6.1.2) it has been assumed that density is a linear function of concentration and temperature; see Eqs. (2.5.2)–

(2.5.4). This is generally a good approximation for both liquids and gases for small Δt and ΔC. It is shown in Section 2.5 that use of the Boussinesq approximation, $\Delta\rho_t = \rho'\beta(t - t_\infty)$, gives a reasonable estimate of the temperature effect on density if $\beta(t_0 - t_\infty) << 1$. Proceeding in the same manner, it can be shown that $\Delta\rho_C = \rho'\beta^*(C_0 - C_\infty)$ represents quite accurately the concentration effect on density provided $\beta^*(C_0 - C_\infty) << 1$.

The assumption of small concentration difference allows significant simplifications in obtaining solutions to Eqs. (6.1.1)–(6.1.6), as shown later in this chapter. The normal velocity component v_0 is generally not zero in mass transfer. However, this effect is very small for low concentration differences and often may be approximated as zero. This allows the use of heat and mass transfer analogies in obtaining mass transfer results, as discussed in detail in Section 6.3.

Equation (6.1.5) assumes that the diffusion flux is described by Fick's law of diffusion, Eq. (2.1.6). In the absence of other potential gradients, this law is the relationship for the mass diffusion in a binary mixture. It also accurately describes the diffusion of any component in a multicomponent mixture provided the binary diffusion coefficients for each pair of components in the mixture are the same. In multicomponent mixtures where this condition is not met, the use of Fick's law introduces an approximation that depends on the actual differences in the binary diffusion coefficients for different pairs of components. In this chapter, only binary mixtures are considered for which Eq. (2.1.6) is a suitable relationship.

Binary diffusion coefficients. For a binary mixture, it may be shown that these coefficients for the two components are the same. To prove this, consider a binary mixture of species A and B. Because of the mass flow, the two species may counterdiffuse at different mean velocities. If v_A and v_B are the local mean velocities of components A and B with respect to stationary coordinate axes, the local mass average velocity v is calculated as

$$v = \frac{1}{\rho}(C_A v_A + C_B v_B) = m_A v_A + m_B v_B \tag{6.1.7}$$

where ρ is the density of the mixture and C_i the concentration (or mass density), defined as the mass of species i per unit volume of the mixture, where i means A or B. The mass fraction concentration is $m_i = C_i/\rho$. Note that ρv is the local rate at which mass passes through a unit cross section perpendicular to the velocity v. This is generally referred to as the absolute mass flux w. For the individual species A and B, the absolute fluxes w_A and w_B from Eq. (6.1.7) are

$$w_A = C_A v_A \qquad \text{and} \qquad w_B = C_B v_B \tag{6.1.8}$$

In flow systems one usually wishes to determine the mass fluxes of the species relative to surfaces moving with this bulk velocity v. This is termed the diffusion flux. It is obtained from the absolute flux by excluding the contribution due to the bulk velocity. For example, for species A the diffusion mass flux m''_A is given by

$$m''_A = C_A(v_A - v) = w_A - C_A v \tag{6.1.9}$$

where $v_A - v$ is the diffusion velocity of species A with respect to B.

Expressions similar to those above may also be developed in molar quantities. For example, the molar bulk velocity v^* is defined as

$$v^* = \frac{C_A^* v_A + C_B^* v_B}{\rho^*} \tag{6.1.10}$$

where ρ^* is the molar density of the mixture and C_A^* and C_B^* are the numbers of moles of species A and B per unit volume of the mixture, respectively. If M_A and M_B represent the molecular weights of species A and B,

$$C_A^* = \frac{C_A}{M_A} \qquad C_B^* = \frac{C_B}{M_B} \tag{6.1.11}$$

With this understanding of absolute and diffusion flux, it may now easily be shown that for a binary mixture, mass diffusion binary coefficients D_{AB} and D_{BA} are the same. From Eqs. (6.1.7)–(6.1.9) and (2.1.7),

$$w_A = C_A v + m''_A = m_A(w_A + w_B) - D_{AB}\,\nabla C_A = m_A(w_A + w_B) - \rho D_{AB}\,\nabla m_A$$

$$w_B = C_B v + m''_B = m_B(w_A + w_B) - D_{BA}\,\nabla C_B = m_B(w_A + w_B) - \rho D_{BA}\,\nabla m_B$$

Therefore,

$$w_A + w_B = (m_A + m_B)(w_A + w_B) - \rho(D_{AB}\,\nabla m_A + D_{BA}\,\nabla m_B)$$

Since

$$m_A + m_B = 1 \qquad \nabla m_A = -\nabla m_B$$

therefore,

$$\rho\,\nabla m_A(D_{AB} - D_{BA}) = 0$$

As this is applicable for all m_A, $D_{AB} = D_{BA}$. In a binary mixture, the indices on the mass diffusion coefficient are no longer necessary and may be omitted as in Eq. (6.1.5).

Summary of assumptions. Equations (6.1.1)–(6.1.6) are limited to processes that occur at low concentration differences, with the condition that there are negligible Sorét and Dufour effects and negligible interfacial velocities from diffusion of the species at the surface. Finally, they apply only to diffusing species that are chemically nonreactive.

Fortunately, these assumptions are not generally very restrictive. In most applications of interest, the Sorét and Dufour effects are negligible, especially for fluids in motion. The condition of low concentration is also often met in the atmosphere, in bodies of water, within a group of growing plants, and in many applications of technology. Therefore, unless otherwise specified, as in Sections 6.7 and 6.8, it is assumed throughout this chapter that these assumptions are valid and that Eqs. (6.1.1)–(6.1.6) are adequate.

Simplifications result when boundary region approximations are applied. The resulting equations again may be shown to admit similarity solutions for many conditions. This is discussed in the next three sections. The solutions for flows other than vertical are also discussed. In Section 6.5 the results for extreme values of the Schmidt and Prandtl numbers are described. This is followed by a summary of the correlations of transport based on the experimental results. The assumptions made in the earlier sections are relaxed in the next two sections. The Sorét and Dufour effects, along with the effects of comparable levels of concentration, are discussed in Section 6.7, while simultaneous heat and mass transport in the presence of chemically reactive diffusing species is described in Section 6.8. The last section considers the effect of ambient medium density stratification on the flows and transport characteristics of such flows.

6.2 BOUNDARY REGION APPROXIMATIONS FOR SPECIES DIFFUSION

The boundary region approximations for thermally buoyant flows were developed in Section 3.2. A similar treatment is possible for flows induced by mass diffusion alone or by combined thermal and mass diffusion. This is examined next, in detail for plane vertical flows and briefly for some of the other flow configurations discussed in Chapters 4 and 5.

6.2.1 Vertical Plane Flow Induced by Mass Diffusion

For such a two-dimensional steady laminar flow, the governing equations are Eqs. (6.1.1)–(6.1.3) and (6.1.5), without the thermal component of the buoyancy force in the momentum equation, Eq. (6.1.2). These, with the assumptions of constant transport properties μ, k, D and the Boussinesq approximations, are

$$\frac{\partial u}{\partial x} + \frac{\partial v}{\partial y} = 0 \tag{6.2.1}$$

$$u\frac{\partial u}{\partial x} + v\frac{\partial u}{\partial y} = \nu\left(\frac{\partial^2 u}{\partial x^2} + \frac{\partial^2 u}{\partial y^2}\right) - \frac{1}{\rho}\frac{\partial p_m}{\partial x} + g\beta^*(C - C_\infty) \tag{6.2.2}$$

$$u\frac{\partial v}{\partial x} + v\frac{\partial v}{\partial y} = \nu\left(\frac{\partial^2 v}{\partial x^2} + \frac{\partial^2 v}{\partial y^2}\right) - \frac{1}{\rho}\frac{\partial p_m}{\partial y} \tag{6.2.3}$$

$$u\frac{\partial C}{\partial x} + v\frac{\partial C}{\partial y} = D\left(\frac{\partial^2 C}{\partial x^2} + \frac{\partial^2 C}{\partial y^2}\right) + c''' \tag{6.2.4}$$

Following the approach in Section 3.2, the relative orders of magnitude of various terms in these equations are estimated by nondimensionalizing the variables as follows:

$$X,Y = \frac{x,y}{L} \qquad \Delta_C = \frac{\delta_C}{L} \qquad U = \frac{u}{U_c} \qquad V = \frac{v}{U_c} \qquad P = \frac{p}{\rho U_c^2} \tag{6.2.5}$$

$$\tilde{C} = \frac{C - C_\infty}{C_0 - C_\infty} \tag{6.2.6}$$

where $U_c = O(\sqrt{g\beta^* L(C_0 - C_\infty)})$ is the characteristic velocity and δ_C the species boundary layer thickness. Again, for a boundary layer analysis the flow is assumed vigorous. Then $\delta_C << 1$. Substituting the above transformations into Eqs. (6.2.1)–(6.2.4) and making suitable simplifications, the following nondimensional equations are obtained:

$$\frac{\partial U}{\partial X} + \frac{\partial V}{\partial Y} = 0 \tag{6.2.7}$$

$$U\frac{\partial U}{\partial X} + V\frac{\partial U}{\partial Y} = \frac{1}{\sqrt{\mathrm{Gr}_C}}\left(\frac{\partial^2 U}{\partial X^2} + \frac{\partial^2 U}{\partial Y^2}\right) - \frac{\partial P_m}{\partial X} + \tilde{C} \tag{6.2.8}$$

$$1 \quad 1 \qquad \Delta\ \Delta^{-1} \qquad\qquad \Delta \qquad \Delta^{-2} \qquad\qquad 1$$

$$U\frac{\partial V}{\partial X} + V\frac{\partial V}{\partial Y} = \frac{1}{\sqrt{\mathrm{Gr}_C}}\left(\frac{\partial^2 V}{\partial X^2} + \frac{\partial^2 V}{\partial Y^2}\right) - \frac{\partial P_m}{\partial Y} \tag{6.2.9}$$

$$1 \quad \Delta \qquad \Delta \quad 1 \qquad\qquad \Delta \qquad \Delta^{-1} \qquad\qquad \Delta$$

$$U\frac{\partial \tilde{C}}{\partial X} + V\frac{\partial \tilde{C}}{\partial Y} = \frac{1}{\mathrm{Sc}\sqrt{\mathrm{Gr}_C}}\left(\frac{\partial^2 \tilde{C}}{\partial X^2} + \frac{\partial^2 \tilde{C}}{\partial Y^2}\right) \tag{6.2.10}$$

$$1 \quad 1 \qquad \Delta\ \Delta_C^{-1} \qquad\qquad 1 \qquad \Delta_C^{-2}$$

where $\mathrm{Sc} = \nu/D$ is the Schmidt number and the concentration Grashof number is $\sqrt{\mathrm{Gr}_C} = U_c L/\nu = [g\beta^* L^3(C_0 - C_\infty)/\nu^2]^{1/2}$.

The orders of magnitude of various terms in the continuity and momentum equations, from Section 3.2, are written below the relevant terms. For species diffusion terms, since $\tilde{C}(X, 0) = 1$ and $\tilde{C}(X, \Delta_C) = 0$, then $\partial\tilde{C}/\partial Y = O(\Delta_C^{-1})$ and $\partial^2\tilde{C}/\partial Y^2 = O(\Delta_C^{-2})$. Since $\tilde{C} = 0$ for $X < 0$ and $\tilde{C} \approx 1$ downstream in the mass transfer region, $\partial\tilde{C}/\partial X = O(1)$ and $\partial^2\tilde{C}/\partial X^2 = O(1)$. Since both Δ and Δ_C are small compared to 1, it is seen that $\partial^2 U/\partial X^2$ and $\partial^2\tilde{C}/\partial X^2$ may be neglected in comparison to $\partial^2 U/\partial Y^2$ and $\partial^2\tilde{C}/\partial Y^2$. Also, the pressure term in Eqs. (6.2.9) is of order Δ and may be neglected. Thus, $\partial P_m/\partial X = O(\Delta^2)$ and the pressure field may again be omitted.

Species boundary layer thickness, δ_C. It is also possible to obtain estimates of Δ and Δ_C in terms of known quantities Gr_C and Sc. Since the largest of the viscous force terms must be of the same order as the convection and buoyancy force terms, it follows from Eq. (6.2.8) that

$$\frac{1}{\sqrt{\mathrm{Gr}_C}}\frac{\partial^2 U}{\partial Y^2} = O\left(\frac{1}{\sqrt{\mathrm{Gr}_C}}\frac{1}{\Delta^2}\right) = O(1) \tag{6.2.11}$$

or

$$\Delta = \frac{\delta}{L} = O\left(\frac{1}{\sqrt[4]{\mathrm{Gr}_C}}\right)$$

Similarly, from Eq. (6.2.10), comparison of the convection term with the largest diffusion term, for diffusion at moderate values of Sc, gives

$$\frac{1}{\mathrm{Sc}\sqrt{\mathrm{Gr}_C}} = O(\Delta_C^2) \qquad \text{or} \qquad \frac{\delta_C}{L} = O\left(\frac{1}{\sqrt[4]{\mathrm{Gr}_C}\sqrt{\mathrm{Sc}}}\right) \tag{6.2.12}$$

From Eqs. (6.2.11) and (6.2.12),

$$\frac{\delta_C}{\delta} = \frac{\Delta_C}{\Delta} = O\left(\frac{1}{\sqrt{\mathrm{Sc}}}\right) \tag{6.2.13}$$

Boundary region equations. With the permissible simplifications found above, boundary region equations for the vertical plane flow are

$$\frac{\partial u}{\partial x} + \frac{\partial v}{\partial y} = 0 \tag{6.2.14}$$

$$u\frac{\partial u}{\partial x} + v\frac{\partial u}{\partial y} = \nu\frac{\partial^2 u}{\partial y^2} + g\beta^*(C - C_\infty) \tag{6.2.15}$$

$$u\frac{\partial C}{\partial x} + v\frac{\partial C}{\partial y} = D\frac{\partial^2 C}{\partial y^2} + c''' \tag{6.2.16}$$

The boundary conditions for these equations, for flow adjacent to a vertical surface are

$$\text{at} \quad y = 0: \qquad u = 0 \qquad v = v_0 \approx 0 \qquad C = C_0 \tag{6.2.17}$$

$$\text{as} \quad y \to \infty: \qquad u \to 0 \qquad C \to C_\infty \tag{6.2.18}$$

where C_0 is determined from the physical condition imposed at $y = 0$ and C_∞ is the species concentration in the quiescent ambient medium. The boundary conditions on u follow from the usual no-slip condition at the surface and from the quiescent ambient medium. The normal velocity component v_0 generally is not zero in mass transfer. However, as shown later, this velocity effect is very small for low concentrations and may often be taken as zero.

6.2.2 Boundary Region Equations for Combined Mass and Heat Transport

With the boundary region approximations made above for the species diffusion effects and with those for thermal diffusion made in Section 3.2, the boundary region equations for simultaneous diffusion of heat and chemical species are as below. These are Eqs. (6.2.14)–(6.2.16), with the buoyancy force supplemented by the thermal effect $g\beta(t - t_\infty)$ and the complete boundary region form of the energy equation, Eq. (3.2.17).

$$\frac{\partial u}{\partial x} + \frac{\partial v}{\partial y} = 0 \tag{6.2.19}$$

$$u\frac{\partial u}{\partial x} + v\frac{\partial u}{\partial y} = \nu\frac{\partial^2 u}{\partial y^2} + g\beta(t - t_\infty) + g\beta^*(C - C_\infty) \tag{6.2.20}$$

$$u\frac{\partial t}{\partial x} + v\frac{\partial t}{\partial y} = \alpha\frac{\partial^2 t}{\partial y^2} + \frac{\beta T}{\rho c_p} u \frac{\partial p_h}{\partial x} + \frac{\mu}{\rho c_p}\left(\frac{\partial u}{\partial y}\right)^2 + \frac{q'''}{\rho c_p} \tag{6.2.21}$$

$$u\frac{\partial C}{\partial x} + v\frac{\partial C}{\partial y} = D\frac{\partial^2 C}{\partial y^2} + c''' \tag{6.2.22}$$

The boundary conditions are again Eqs. (6.2.17) and (6.2.18), with additional boundary conditions on the temperature field. That is,

$$\text{at} \quad y = 0: \qquad u = 0 \qquad v = v_0 \approx 0 \qquad t = t_0(x) \qquad C = C_0(x) \tag{6.2.23}$$

$$\text{as} \quad y \to \infty: \qquad u \to 0 \qquad t \to t_\infty \qquad C \to C_\infty \tag{6.2.24}$$

Similar boundary layer approximations often apply to the axisymmetric and inclined flows discussed in Chapters 4 and 5 when they are induced by either thermal or species diffusion buoyancy, or both together. Those equations simply are modified by including the added buoyancy component $g\beta^*(C_0 - C_\infty)$, along with the species equation, Eq. (6.2.22), and the additional appropriate boundary conditions. The equations and boundary conditions for some of these flows, with the usual assumptions and only the simplest effects, are given below.

Vertical axisymmetric flows:

$$\frac{\partial}{\partial x}(yu) + \frac{\partial}{\partial y}(yv) = 0 \tag{6.2.25}$$

$$u\frac{\partial u}{\partial x} + v\frac{\partial u}{\partial y} = \nu\frac{1}{y}\frac{\partial}{\partial y}\left(y\frac{\partial u}{\partial y}\right) + g\beta(t - t_\infty) + g\beta^*(C - C_\infty) \tag{6.2.26}$$

$$u\frac{\partial t}{\partial x} + v\frac{\partial t}{\partial y} = \alpha\frac{1}{y}\frac{\partial}{\partial y}\left(y\frac{\partial t}{\partial y}\right) \tag{6.2.27}$$

$$u\frac{\partial C}{\partial x} + v\frac{\partial C}{\partial y} = D\frac{1}{y}\frac{\partial}{\partial y}\left(y\frac{\partial C}{\partial y}\right) \tag{6.2.28}$$

For an axisymmetric free boundary or plume flow the boundary conditions are

$$\text{at} \quad y = 0: \qquad \frac{\partial u}{\partial y} = v = \frac{\partial t}{\partial y} = \frac{\partial C}{\partial y} = 0 = t - t_0 = C - C_0 \tag{6.2.29}$$

$$\text{as} \quad y \to \infty: \qquad u \to 0$$

Two additional boundary conditions, $t \to t_\infty$ and $C \to C_\infty$, in the ambient medium are also available. However, as for thermal axisymmetric plumes in Chapter 4, these two conditions result from the other boundary conditions and are not independent. For flow along a vertical cylinder (see Fig. 4.1.1*b*) the boundary conditions are

$$\begin{aligned} &\text{at} \quad y = R: \qquad u = v = 0 \qquad t = t_0 \qquad C = C_0 \\ &\text{as} \quad y \to \infty: \qquad u \to 0 \qquad t \to t_\infty \qquad C \to C_\infty \end{aligned} \tag{6.2.30}$$

Horizontal flows, B_n away from the surface. The equations below retain p_m, which arises from the buoyancy force in Eq. (6.2.33):

$$\frac{\partial u}{\partial x} + \frac{\partial v}{\partial y} = 0 \tag{6.2.31}$$

$$u \frac{\partial u}{\partial x} + v \frac{\partial u}{\partial y} = \nu \frac{\partial^2 u}{\partial y^2} - \frac{1}{\rho} \frac{\partial p_m}{\partial x} \tag{6.2.32}$$

$$0 = -\frac{1}{\rho} \frac{\partial p_m}{\partial y} + g\beta(t - t_\infty) + g\beta^*(C - C_\infty) \tag{6.2.33}$$

$$u \frac{\partial t}{\partial x} + v \frac{\partial t}{\partial y} = \alpha \frac{\partial^2 t}{\partial y^2} \tag{6.2.34}$$

$$u \frac{\partial C}{\partial x} + v \frac{\partial C}{\partial y} = D \frac{\partial^2 C}{\partial y^2} \tag{6.2.35}$$

$$\text{at} \quad y = 0: \qquad u = 0 \qquad v = v_0 \approx 0 \qquad t = t_0 \qquad C = C_0 \tag{6.2.36}$$

$$\text{as} \quad y \to \infty: \qquad u \to 0 \qquad t \to t_\infty \qquad C \to C_\infty \qquad p_m \to 0 \tag{6.2.37}$$

Inclined flow, B_n away from the surface. Only flows for which the pressure term in the x-momentum equation may be neglected are considered here. As given in Eq. (5.2.1), this requires that $\Delta \tan \theta << 1$. Then the governing equations and boundary conditions are

$$\frac{\partial u}{\partial x} + \frac{\partial v}{\partial y} = 0 \tag{6.2.38}$$

$$u \frac{\partial u}{\partial x} + v \frac{\partial u}{\partial y} = \nu \frac{\partial^2 u}{\partial y^2} + g\beta \cos\theta(t - t_\infty) + g\beta^* \cos\theta(C - C_\infty) \tag{6.2.39}$$

$$u \frac{\partial t}{\partial x} + v \frac{\partial t}{\partial y} = \alpha \frac{\partial^2 t}{\partial y^2} \quad (6.2.40)$$

$$u \frac{\partial C}{\partial x} + v \frac{\partial C}{\partial y} = D \frac{\partial^2 C}{\partial y^2} \quad (6.2.41)$$

$$\text{at} \quad y = 0: \quad u = 0 \quad v = v_0 \approx 0 \quad t = t_0 \quad C = C_0 \quad (6.2.42)$$

$$\text{as} \quad y \to \infty: \quad u \to 0 \quad t \to t_\infty \quad C \to C_\infty \quad (6.2.43)$$

A similarity analysis, as in Chapter 3, may be used to find conditions on C_0, t_0, C_∞, t_∞, and so forth for which similar solutions are possible. This is done in the following two sections.

6.3 SIMILARITY FORMULATION AND SOLUTIONS FOR PLANE VERTICAL FLOWS

For a similarity analysis for Eqs. (6.2.19)–(6.2.22), both C_0 and C_∞ are specified by some physical considerations to be functions of x as follows:

$$C_0 - C_\infty = e(x) \qquad C_\infty - C_r = r(x) \quad (6.3.1)$$

where $e(x)$ and $r(x)$ are general functions of x to be determined from requirements of similarity and C_r is some reference value of ambient concentration. The x dependences of t_0 and t_∞, $d(x)$ and $j(x)$, respectively, similarity variable η, nondimensionalized temperature ϕ, and generalized stream function $f(\eta)$ are those given in Eqs. (3.5.4)–(3.5.6). Using Eq. (6.3.1) above in conjunction with Eqs. (3.5.4)–(3.5.6) and

$$\tilde{C} = \frac{C - C_\infty}{C_0 - C_\infty} = \frac{C - C_\infty}{e(x)} \quad (6.3.2)$$

the complete Eqs. (6.2.19)–(6.2.22) and boundary conditions transform to

$$f''' + \frac{d}{cb^3} \frac{g\beta}{\nu^2} \phi + \frac{e}{cb^3} \frac{g\beta^*}{\nu^2} \tilde{C} + \frac{c_x}{b} ff'' - \left(\frac{c_x}{b} + \frac{cb_x}{b^2} \right) f'^2 = 0 \quad (6.3.3)$$

$$\frac{\phi''}{\mathrm{Pr}} + \frac{c_x}{b} f\phi' - \frac{c}{b}\frac{d_x}{d} f' \phi - \frac{cj_x}{bd} f' - \beta T \frac{c}{db}\frac{g}{c_p} f' + \frac{b^2 c^2 \nu^2}{dc_p} f''^2 + \frac{1}{db^2}\frac{q'''}{k\,\mathrm{Pr}} = 0 \quad (6.3.4)$$

$$\frac{\tilde{C}''}{\mathrm{Sc}} + \frac{c_x}{b} f \tilde{C}' - \frac{c}{b}\frac{e_x}{e} f' \tilde{C} - \frac{cr_x}{be} f' + \frac{1}{eb^2}\frac{c'''}{\nu} = 0 \quad (6.3.5)$$

$$f'(0) = f(0) = 1 - \phi(0) = 1 - \tilde{C}(0) = f'(\infty) = \phi(\infty) = \tilde{C}(\infty) = 0 \quad (6.3.6)$$

For similarity to result with combined thermal and species diffusion, the functions b, c, d, e, j, and r must be such that x disappears from these equations. The conditions on d and j and the associated values of b and c that meet these requirements for a purely thermal diffusion process were determined in Chapter 3. It remains to determine whether, in the presence of mass diffusion, there are some forms of $e(x)$ and $r(x)$ that would result in similarity, given the b, c, d, and j determined in Chapter 3. There is also the question of whether b and c, determined for thermal diffusion alone, are also the optimum formulation for combined thermal and mass diffusion.

6.3.1 Mass Transfer

The last question above is examined for a purely mass diffusion process. With no thermal effects, Eqs. (6.3.3) and (6.3.5) become identical to those for the pure thermally driven flow, that is, Eqs. (3.5.8) and (3.5.9), with the pressure and viscous terms ignored in Eq. (3.5.9) and with Pr = Sc and $c''' = q'''$. The boundary conditions are identical. Then, $\tilde{C}$ merely replaces ϕ. The x-dependent coefficients in these equations are of exactly the same form as C_1, C_2, C_3K_1, C_4, and C_5 in Section 3.5. The only difference is that $d(x)$ and $j(x)$ there now become $e(x)$ and $r(x)$. Therefore, for similarity in a purely mass diffusion process, $e(x)$ and $r(x)$ are of the same form as $d(x)$ and $j(x)$. Similarity is obtained for power law and exponential distributions of $e(x)$. Similarity is retained in the presence of ambient medium stratification, provided $r(x)$ varies as $e(x)$. The associated values of b and c are again those determined for thermal transport, with β replaced by β^* and t by C. Thus, for the power law distribution, for transport driven entirely by mass diffusion-caused buoyancy,

$$e(x) = N_C x^{n_C} \qquad r(x) = N_{C,\infty} x^{n_C} \tag{6.3.7}$$

$$b(x) = \frac{G_C}{4x} \qquad c(x) = G_C = 4\sqrt[4]{\frac{g\beta^* x^3(C_0 - C_\infty)}{4\nu^2}} = 4\left(\frac{\mathrm{Gr}_{x,C}}{4}\right)^{1/4} \tag{6.3.8}$$

The resulting equations in f and $\tilde{C}$ are identical to those for thermal transport. The parameter Pr becomes Sc, β is β^*, ϕ is $\tilde{C}$, and q''' is c'''. Therefore, all the similar solutions obtained for heat transfer apply for mass diffusion, with the appropriate substitutions. The analogy also extends to both experimental data and the resulting correlations on heat transfer. However, it must be kept in mind that this is all strictly true only for low concentration differences. It is also applicable only if viscous dissipation and pressure energy terms in the energy balance, Eq. (6.3.4), remain small in the mass transfer process.

The boundary condition $f(0) = 0$ in Eq. (6.3.6) with the mass transfer results for $v_0 = 0$. This is generally valid only for low concentration differences. The condition for this assumption to be valid, in terms of the similarity formulation, is developed below:

$$v_0(x) = -\psi_x)_{y=0} = -\nu(c_x f + cf'b)_{y=0}$$

Using Eqs. (6.3.7) and (6.3.8) with a power law variation of $C_0 - C_\infty$:

$$v_0 = -\frac{\nu}{x}(n+3)\sqrt[4]{\frac{\mathrm{Gr}_{x,C}}{4}}\, f(0) \tag{6.3.9a}$$

Therefore, the condition $f(0) \approx 0$ requires that

$$\frac{v_0 x}{\nu} << (n+3)\sqrt[4]{\frac{\mathrm{Gr}_{x,C}}{4}} \tag{6.3.9b}$$

Writing the interface mass flux from Fick's law, v_0 is evaluated:

$$m'' = \rho v_0 = -D\left.\frac{\partial C}{\partial y}\right)_{y=0}$$

$$= \frac{D}{x}\sqrt[4]{\frac{\mathrm{Gr}_{x,C}}{4}}\,[-\tilde{C}'(0)](C_0 - C_\infty) \tag{6.3.10}$$

Eliminating v_0 between Eqs. (6.3.9*a*) and (6.3.10) and rearranging gives

$$f(0) = -\frac{1}{n+3}\frac{1}{\mathrm{Sc}}[-\tilde{C}'(0)]\frac{C_0 - C_\infty}{\rho} \tag{6.3.11}$$

For Sc and $\tilde{C}'(0)$ of $O(1)$, this implies that $f(0) \approx 0$ if the mass concentration difference $(C_0 - C_\infty)$ is very small compared to the density of the fluid, ρ. For $\mathrm{Sc} >> 1$, the requirement on $(C_0 - C_\infty)$ is not so stringent.

6.3.2 Combined Heat and Mass Transport

Returning now to combined transport, with two buoyancy force terms, consider first Eqs. (6.3.3)–(6.3.5). The viscous dissipation, pressure, heat source, and stratification effects in the energy equation and stratification and species source term in the mass transfer equation are first omitted. For thermal buoyancy alone, it is known that similarity results for

$$d = Nx^n \qquad c = 4\sqrt[4]{\frac{g\beta x^3 d}{4\nu^2}} = 4\sqrt[4]{\frac{\mathrm{Gr}_x}{4}} = G \qquad b = \frac{G}{4x}$$

Considering the two buoyancy terms in Eq. (6.3.3), similarity arises when e varies as d, since $g\beta/\nu^2$ and $g\beta^*/\nu^2$ are constants. Thus, for the power law variation, $d = Nx^n$, $e = N_C x^n$ is appropriate. The values of c and b can be either those used for the thermal transport, in terms of βd, or that given in Eq. (6.3.8) for the mass diffusion process, $\beta^* e$. However, the optimum form of these functions is perhaps that given by Gebhart and Pera (1971), as follows. It includes both of the buoyancy effects.

$$c(x) = G = 4\sqrt[4]{\frac{P\mathrm{Gr}_x + Q\mathrm{Gr}_{x,C}}{4}} = (\mathrm{Gr}'_x)^{1/4} \qquad b = \frac{c}{4x} \tag{6.3.12}$$

where P and Q are constants and Gr'_x is the combined Grashof number. Any linear combination of Gr_x and $\mathrm{Gr}_{x,C}$ in the above equation gives similarity. In fact, each Gr_x and $\mathrm{Gr}_{x,C}$ may be the combination of many different effects, some aiding, others opposing. The constants P and Q are limited only by the condition that Gr'_x be positive, after taking x in the proper direction.

Similarity also results for exponential variation of $d(x)$ and $e(x)$ when d and e are of the same x dependence. That is, $d = Me^{mx}$ and $e = M_C e^{mx}$. The corresponding values of $c(x)$ and $b(x)$ are then

$$c(x) = G_m = 4\sqrt[4]{P\mathrm{Gr}_m + Q\mathrm{Gr}_{m,C}} \qquad b = \frac{mG_m}{4}$$

$$\mathrm{Gr}_m = \frac{g\beta d}{m^3\nu^2} \quad \mathrm{Gr}_{m,C} = \frac{g\beta^* e}{m^3\nu^2} \tag{6.3.13}$$

The similarity requirements for the inclusion of stratification, viscous dissipation, pressure term, and other additional effects follow very simply. The treatment is as in Chapter 3 and the conclusions are identical. For example, for the power law variation, similarity results in presence of thermal and/or species stratification of the ambient medium if $j(x)$ and $r(x)$ vary as $d(x)$ and $e(x)$. With the inclusion of pressure energy effects, similarity results only for $n = 1$. No similar solution includes the viscous dissipation term in the energy equation. For the rest of this and the following section, the power law variation is considered without the pressure and viscous dissipation effects. The transformed equations, Eqs. (6.3.3)–(6.3.6) become

$$f''' + (n + 3)ff'' - (2n + 2)f'^2 + \frac{\phi + \tilde{N}\tilde{C}}{P + Q\tilde{N}} = 0 \tag{6.3.14}$$

$$\phi'' + \mathrm{Pr}[(n + 3)f\phi' - 4\,nf'\phi] = 0 \tag{6.3.15}$$

$$\tilde{C}'' + \mathrm{Sc}[(n + 3)f\tilde{C}' - 4nf'\tilde{C}] = 0 \tag{6.3.16}$$

$$f'(0) = f(0) = f'(\infty) = 1 - \phi(0) = 1 - \tilde{C}(0) = \phi(\infty) = \tilde{C}(\infty) = 0 \tag{6.3.17}$$

Here the parameter $\tilde{N}$ is defined as

$$\tilde{N} = \frac{\mathrm{Gr}_{x,C}}{\mathrm{Gr}_x} = \left(\frac{G_C}{G}\right)^4 = \frac{\beta^*(C_0 - C_\infty)}{\beta(t_0 - t_\infty)} \tag{6.3.18}$$

It is a measure of the relative importance of chemical and thermal diffusion in causing the density difference that drives the flow. It is positive when both effects combine to drive the flow and negative when the effects are opposed. In many common flows of interest, the value of $\tilde{N}$ is of $O(1)$. For example, consider atmospheric flows at a temperature level of 20°C. With a temperature difference of 10°C and the difference in water vapor concentration that accompanies saturated conditions for these temperature levels, $\tilde{N} \approx 3$. Similarly, for a salinity difference of 100 ppm in seawater with an associated temperature difference of 1°C, $\tilde{N} \approx 0.2$.

If the flow is driven entirely by thermal diffusion, G in Eq. (6.3.12) involves only Gr_x and $\tilde{N} = 0$. Then for $P = 1$ the above equations reduce to Eqs. (3.5.26)–(3.5.28). Likewise, if the flow is driven entirely by species diffusion, G involves only $\mathrm{Gr}_{x,C}$, $\tilde{N} \to \infty$, and the term $(\phi + \tilde{N}\tilde{C})/(P + Q\tilde{N})$ becomes $\tilde{C}/Q$. For $Q = 1$ the equations then reduce to the mass transfer analog of Eqs. (3.5.26)–(3.5.28). For intermediate values of $\tilde{N}$, the combined flow is described by the equations given above. However, there are some limitations on the value of $\tilde{N}$. These result from the requirement that the net flow be zero at the leading edge. For $P = Q = 1$, this condition is met if $\mathrm{Gr}_x + \mathrm{Gr}_{x,C}$ is positive and the leading edge is at the bottom of the vertical surface. For $\mathrm{Pr} = \mathrm{Sc}$, this requires that $P = Q$ and $\tilde{N} > -1$. For $\mathrm{Pr} \neq \mathrm{Sc}$, P need not necessarily equal Q. This results in a less restrictive requirement on $\tilde{N}$. For example, $\mathrm{Gr}_x + \mathrm{Gr}_{x,C}$ may be negative and still result in a net positive buoyancy force and hence upward flow throughout all or most of the boundary region. This condition results for $\mathrm{Sc} > \mathrm{Pr}$ for $\mathrm{Gr}_{x,C} < 0$ and for $\mathrm{Pr} > \mathrm{Sc}$ for $\mathrm{Gr}_x < 0$. The difficulty of sign in Eq. (6.3.12) is overcome in the choice of P and Q in $P\mathrm{Gr}_x + Q\mathrm{Gr}_{x,C}$. This matter is discussed in detail when considering the numerical results below.

6.3.3 Similar Solutions

In the above formulation any value of n is algebraically permissible. However, as discussed in Chapter 3, conditions of physical realism indicate the range of n to be $-0.6 \leq n \leq 1$. The two flows considered here are those arising adjacent to a surface at uniform temperature and concentration, $n = 0$, and that in a freely rising plume, $n = -\frac{3}{5}$. The boundary conditions for the surface are given by Eq. (6.3.17). Those for the plume are

$$f''(0) = f(0) = f'(\infty) = \phi'(0) = \tilde{C}'(0) = 1 - \phi(0) = 1 - \tilde{C}(0) = 0 \qquad (6.3.19)$$

Numerical solutions for these two flow cases have been obtained by Gebhart and Pera (1971) for air ($\mathrm{Pr} = 0.7$) and water ($\mathrm{Pr} = 7.0$). The calculations were for Schmidt numbers of 0.1–10.0 for $\mathrm{Pr} = 0.7$ and of 1.0–500 for $\mathrm{Pr} = 7.0$. These values cover the range of diffusion species of common interest, in air and in water, as indicated by the values in Table 6.3.1. Some of the important features of these results are discussed next.

Transport equations and results for $n = 0$. The quantities of most interest are the local Nusselt number Nu_x, the local Sherwood number $\mathrm{Nu}_{x,C}$, and the tangential component of velocity u. These are

$$\mathrm{Nu}_x = h_x \frac{x}{k} = \left[\frac{-\phi'(0)}{\sqrt{2}}\right](P\mathrm{Gr}_x + Q\mathrm{Gr}_{x,C})^{1/4}$$

$$= \left[\frac{-\phi'(0)}{\sqrt{2}}\right](\mathrm{Gr}_x)^{1/4}(P + Q\tilde{N})^{1/4} \qquad (6.3.20)$$

$$\mathrm{Nu}_{x,C} = h_{x,C} \frac{x}{D} = \left[\frac{-\tilde{C}'(0)}{\sqrt{2}}\right](P\mathrm{Gr}_x + Q\mathrm{Gr}_{x,C})^{1/4}$$

$$= \left[\frac{-\tilde{C}'(0)}{\sqrt{2}} \right] \sqrt[4]{|\mathrm{Gr}_{x,C}|} \sqrt[4]{\left| \frac{P + Q\tilde{N}}{\tilde{N}} \right|} \tag{6.3.21}$$

$$u = \psi_y = \nu c f' b = \frac{\nu G^2}{4x} f' \tag{6.3.22a}$$

or

$$\frac{ux}{2\nu} = \sqrt{\mathrm{Gr}_x} \sqrt{P + Q\tilde{N}} f' \tag{6.3.22b}$$

The two absolute values in Eq. (6.3.21) are necessary since both $\mathrm{Gr}_{x,C}$ and $\tilde{N}$ may be negative. Clearly, Nu_x, $\mathrm{Nu}_{x,C}$, and u depend on Pr, Sc, and $\tilde{N}$.

Results for Pr = Sc. Equations (6.3.14)–(6.3.17), for $P = Q$, indicate that $\phi(\eta) = \tilde{C}(\eta)$. Thus only one of Eqs. (6.3.15) and (6.3.16) and Eq. (6.3.14) are necessary. The buoyancy force term in Eq. (6.3.14) becomes either ϕ or $\tilde{C}$. The solutions then are merely those discussed earlier for heat transfer. The combined effect is present in the interpretation of the results in that $b(x)$ and $c(x)$ depend on the sum of Gr_x and $\mathrm{Gr}_{x,C}$. The transport rates for Pr = Sc are determined from

Table 6.3.1 Schmidt number and buoyancy parameters for various species at low concentration in air and in water at approximately 25°C and 1 atm

Species	$Sc = \nu/D$	$Le = Sc/Pr$	$\rho\beta^* = -(\partial\rho/\partial C)_{T,p}$	$D \times 10^5$ (cm^2/s)
In air, Pr = 0.7				
Ammonia	0.78	1.11	+1.07	
Carbon dioxide	0.94	1.34	−0.34	
Hydrogen	0.22	0.314	+13.4	
Oxygen	0.75	1.07	−0.094	
Water	0.60	0.86	+0.61	
Benzene	1.76	2.51	−0.63	
Ether	1.66	2.37	−0.61	
Methanol	0.97	1.39	−0.095	
Ethyl alcohol	1.30	1.86	−0.37	
Ethylbenzene	2.01	2.87	−0.73	
In water, Pr = 7.0				
Ammonia	445	63.57	−0.5	2.0
Carbon dioxide	453	64.71		1.96
Hydrogen	152	21.71		5.85
Oxygen	356	50.86		2.5
Nitrogen	468	66.86		1.9
Chlorine	617	88.14		1.44
Sulfur dioxide	523	74.71		1.7
Calcium chloride	750	107.14	+0.8	1.188
Sodium chloride	580	82.86	+0.7	1.545
Methanol	556	79.43	−0.17	1.6
Sucrose	1,700	242.86		0.5226

Source: From Gebhart and Pera (1971).

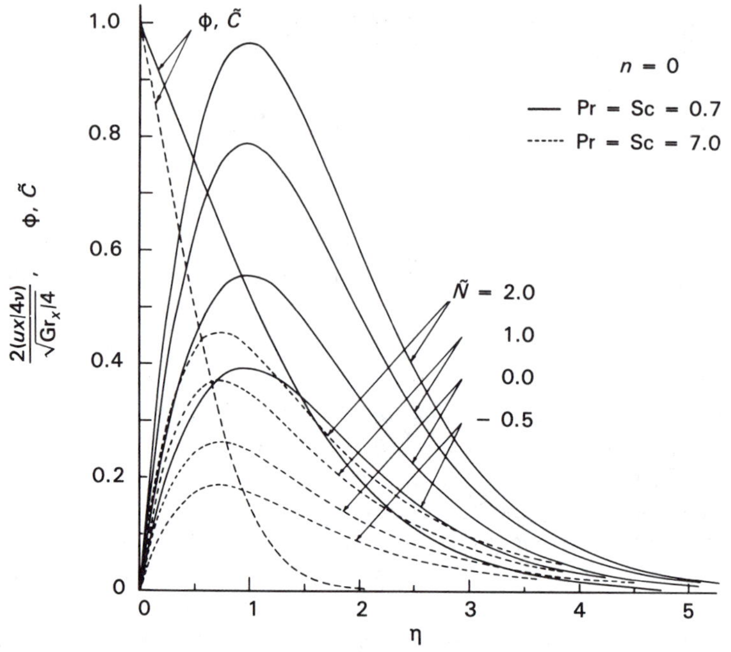

Figure 6.3.1 Variation of temperature, concentration, and velocity over the boundary region for Pr = Sc = 0.7 and 7.0, for flow adjacent to a vertical surface at various values of $\tilde{N}$. *(Reprinted with permission from* Int. J. Heat Mass Transfer, *vol. 14, p. 2025, B. Gebhart and L. Pera. Copyright © 1971, Pergamon Journals Ltd.)*

Eqs. (6.3.20) and (6.3.21) for $P = Q = 1$ from $[-\phi'(0)]$ and $\tilde{N}$, where the values of $-\phi'(0)$ are given in Table 3.4.1.

The effect of $\tilde{N}$ on the u and $\phi = \tilde{C}$ distributions is shown in Fig. 6.3.1. The distributions are plotted against η for Pr = 0.7 and 7.0 and $\tilde{N} = -0.5$, 0, 1, and 2. Since the results are plotted in terms of Gr_x, the curves for $\tilde{N} \neq 0$ show the added effect of mass diffusion. The value $\tilde{N} = 0$ is the pure thermal diffusion process, while negative and positive values of $\tilde{N}$ represent the opposing and aiding effects, respectively. From Fig. 6.3.1, the value of $\tilde{N}$ has no effect on the ϕ and $\tilde{C}$ distributions, but it does change the velocity distributions. The effect is to lower the velocity for opposing effects and vice versa. As expected, the effect of increasing Pr = Sc is thinning of the thermal and species diffusion layers.

Results for Pr ≠ Sc. The thermal and species diffusion regions are now inherently of different extents. Simplifications like those for Pr = Sc are not possible. All the equations, (6.3.14)–(6.3.17), must be solved simultaneously. Gebhart and Pera (1971) obtained these solutions for air and water at several values of Sc, for both aiding and opposing effects. The values of P and Q were taken as 1.0. The resulting values of various important transport parameters are listed in Table 6.3.2. These values may also be used in conjunction with Eqs. (6.3.20) and (6.3.21) to calculate Nu_x and $Nu_{x,C}$.

The distributions of temperature, concentration, and velocity based on these numerical solutions are shown in Figs. 6.3.2–6.3.4. Results are shown for two values of $\tilde{N}$, −0.5 and 1.0, to indicate the effect of opposing and aiding buoyancy mechanisms. As before, they are plotted in terms of Gr_x and therefore show the added effect of mass diffusion.

It is clear from these distributions that both aiding and opposing buoyancy effects alter the form of ϕ and $\tilde{C}$ distributions by an amount dependent on $\tilde{N}$. Opposed buoyancy effects thicken and aiding ones thin the regions, compared to flows caused by the purely thermal effect. The velocity distributions are also significantly affected in both magnitude and extent. An interesting result is the reversal of the flow in the outer part of the boundary layer for negative $\tilde{N}$ and

Table 6.3.2 Flow and transport quantities for flows adjacent to isothermal vertical surfaces

Pr	Sc	$\tilde{N}$	$f''(0)$	$-\phi'(0)$	$-\tilde{C}'(0)$
0.7	0.1	0.5	0.74221	0.54913	0.19561
		1	0.77213	0.56620	0.20656
		2	0.80148	0.58125	0.21554
	0.5	−0.5	0.64154	0.45891	0.37536
		0.5	0.69018	0.50768	0.42747
		1	0.69584	0.51125	0.43110
		2	0.70185	0.51559	0.43440
	0.7	All	0.67890	0.49950	0.49950
	0.94	−0.5	0.70824	0.51709	0.59582
		0.5	0.66880	0.49292	0.57019
		1	0.66378	0.48947	0.56659
		2	0.65873	0.48588	0.56286
	5.0	−0.5	0.85198	0.56603	1.2997
		0.5	0.61725	0.46880	1.11689
		1	0.58574	0.45026	1.08559
		2	0.55277	0.42829	1.04962
	10.0	−0.5	0.90099	0.57552	1.71754
		0.5	0.59885	0.46330	1.43962
		1	0.55728	0.44098	1.38965
		2	0.51428	0.41383	1.33242
7.0	7.0	All	0.45069	1.05411	1.05411
	1.0	0.5	0.51983	1.16002	0.48743
		1	0.55161	1.19963	0.51188
		2	0.58246	1.12348	0.53263
	100	−0.5	0.61831	1.21566	3.33176
		0.5	0.38933	0.97593	2.75894
		1	0.35748	0.92855	2.65410
		2	0.03244	0.87121	2.53325
	500	−0.3	0.54994	1.14494	5.58873
		0.5	0.36783	0.96154	4.79946
		1	0.32402	0.90298	4.56087
		2	0.27788	0.82889	4.27465

Source: From Gebhart and Pera (1971).

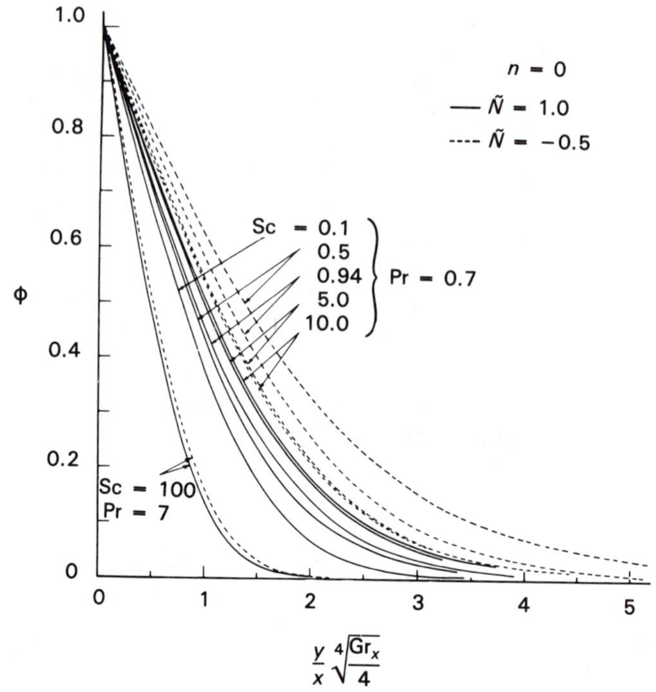

Figure 6.3.2 Temperature distributions for flow adjacent to a vertical surface, for varying Schmidt number at given Pr and $\tilde{N} = -0.5$ and 1. *(Reprinted with permission from* Int. J. Heat Mass Transfer, *vol. 14, p. 2025, B. Gebhart and L. Pera. Copyright © 1971, Pergamon Journals Ltd.)*

Sc << Pr, that is, for Le < 1. This is due to the incoming flow, at any location, experiencing the negative buoyancy effect of the thicker species diffusion layer. However, in these regions the boundary layer approximations are not applicable.

Another important result is the opposite effect of Sc on the temperature and velocity level for $\tilde{N} = -0.5$ and 1.0. For an aiding effect, decreasing Sc increases the velocity level and its extent, which in turn causes a thinning of the temperature region. This effect is reversed for opposed effects. In discussing the instability of these flows in Chapter 11, it will be shown that these large effects on the velocity distribution have a significant effect on the stability of such flows. In particular, the flow reversal observed for Sc < Pr for opposing effects reduces the stability of these flows.

The effect of Sc on Nu_x and $Nu_{x,C}$ is shown in Figs. 6.3.5 and 6.3.6 for Pr = 0.7 in terms of the parameter $\tilde{N}$. Around $\tilde{N} = 0$, largely a thermal diffusion process, heat transfer is not strongly affected by species diffusion, although the mass transfer becomes quite large. This is because the flow is induced almost entirely by thermal buoyancy, which causes a very effective species diffusion even at low concentration levels. For opposed effects, $\tilde{N} < 0$, both Nu_x and $Nu_{x,C}$ decrease sharply with decreasing $\tilde{N}$. For aiding effects, $\tilde{N} > 0$, the heat transfer is continuously enhanced by added buoyancy due to species diffusion. It becomes infinite as $\tilde{N} \rightarrow \infty$. The mass transfer, on the other hand, continuously decreases to the asymptotic values. These would be obtained for flows driven by species diffusion only. These characteristics were also found for other values of Pr.

Other studies. Bottemanne (1971) also obtained similarity solutions for the transport discussed above. The formulation is very similar to that used here and numerical results were presented for Pr = 0.71, Sc = 0.63, and $\tilde{N} = 1$. In a sequential experimental study, simultaneous heat transfer and evaporation of water vapor into still air were measured by Bottemanne (1972) adjacent to a vertical cylinder of large diameter. The results agreed well with the calculations.

Some aspects of the combined flows for aiding buoyancy forces have also been considered by Schenk et al. (1976). The approach and equations are identical to those presented above. In addition, the total flow rate was calculated to help explain the results of combined driving forces. The results were for Pr = 0.71 with Sc from 0.1 to 10. As before, heat and mass transfer were found to interact except for Pr = Sc. However, for $0.6 < \text{Sc} < 0.9$ the numerical results indicated that the two effects may be approximated as mutually independent, within 2%. These limits include the important example of water vapor, Sc = 0.63, and carbon

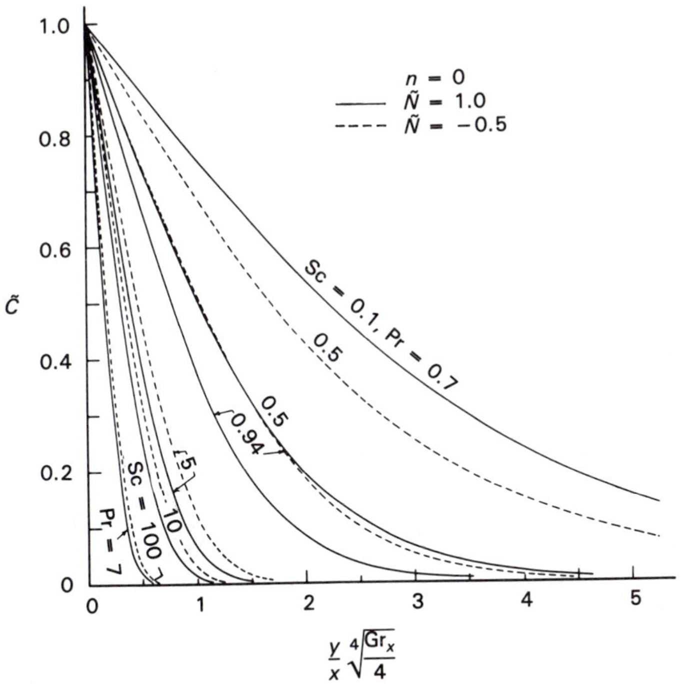

Figure 6.3.3 Species concentration distributions for flow adjacent to a vertical surface, for varying Schmidt number at given Pr and $\tilde{N} = -0.5$ and 1. *(Reprinted with permission from* Int. J. Heat Mass Transfer, *vol. 14, p. 2025, B. Gebhart and L. Pera. Copyright © 1971, Pergamon Journals Ltd.)*

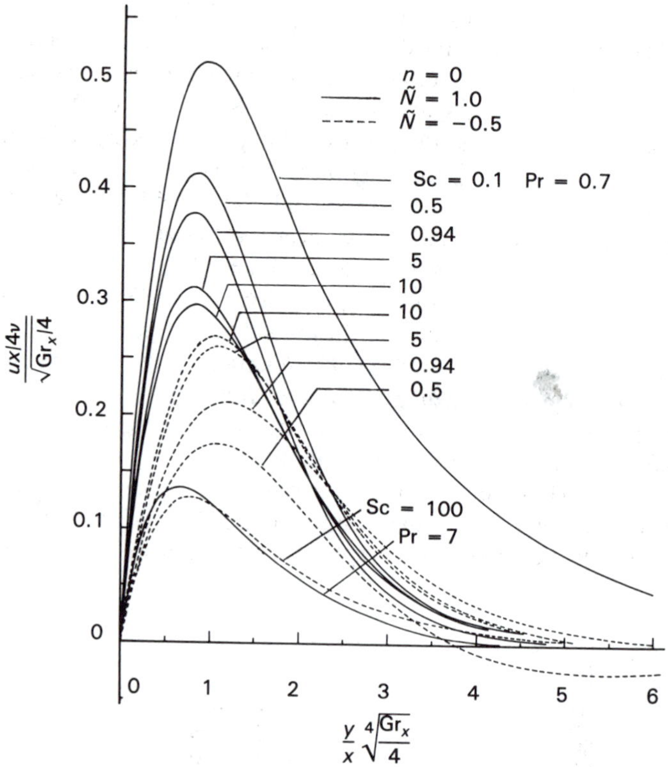

Figure 6.3.4 Flow velocity distributions for flow adjacent to a vertical surface, for varying Schmidt number at given Pr and $\tilde{N} = -0.5$ and 1. *(Reprinted with permission from* Int. J. Heat Mass Transfer, *vol. 14, p. 2025, B. Gebhart and L. Pera. Copyright © 1971, Pergamon Journals Ltd.)*

dioxide, Sc = 0.94. It was shown that the total flow rate and the temperature and concentration gradients tend to follow more closely the single effect that has the lowest Pr or Sc.

Taunton et al. (1970) also found a similarity solution for such transport and presented numerical results for $f''(0)$, $\tilde{C}'(0)$, and $\phi'(0)$ for Prandtl and Schmidt numbers ranging from 0.14 to 1000 and both opposing and aiding buoyancy effects. However, the nondimensionalization scheme used a reference velocity of $(\nu/L)\Lambda$, where Λ is the smaller of the Prandtl and Schmidt numbers. As a result, the equations are quite different from those obtained above. It is inconvenient to compare these results with other similarity solutions.

Additional studies related to these flows are due to Somers (1956), Wilcox (1961), Mathers et al. (1957), and Lightfoot (1968). In the first two studies, integral method analysis was used to obtain transport relations as a function of the various parameters involved. An important result was the expression of the species diffusion contribution as mediated by $\sqrt{\text{Le}}$, where Le = Sc/Pr = α/D. The results obtained, however, are approximate and are reasonably accurate only around

Le = 1.0, with one buoyancy effect being very small compared to the other. A comparison made by Gebhart and Pera (1971) of these results with the similar solutions presented above indicated some limitations and inaccuracies. See Figs. 6.3.5 and 6.3.6.

Mathers et al. (1957) formulated the same transport in terms of approximate boundary layer differential equations and obtained some numerical solutions. The inertia effects were omitted from the momentum equation. The solutions, therefore, are essentially valid for the asymptotic case Sc >> Pr, when Pr is large. The analysis of Lightfoot (1968) is also asymptotic, for Sc >> Pr and the species diffusion buoyancy effect very small compared to that caused by thermal diffusion. These solutions will be discussed in Section 6.5.

Transient natural convection along a vertical isothermal plane due to both temperature and concentration differences has been investigated by Callahan and Marner (1976), as discussed in Chapter 7. The stability characteristics of plane flows arising from combined heat and species diffusion have also been studied; see Boura and Gebhart (1976). These are discussed in Section 11.9. These and other studies related to combined mass transport have been reviewed by Ostrach (1980).

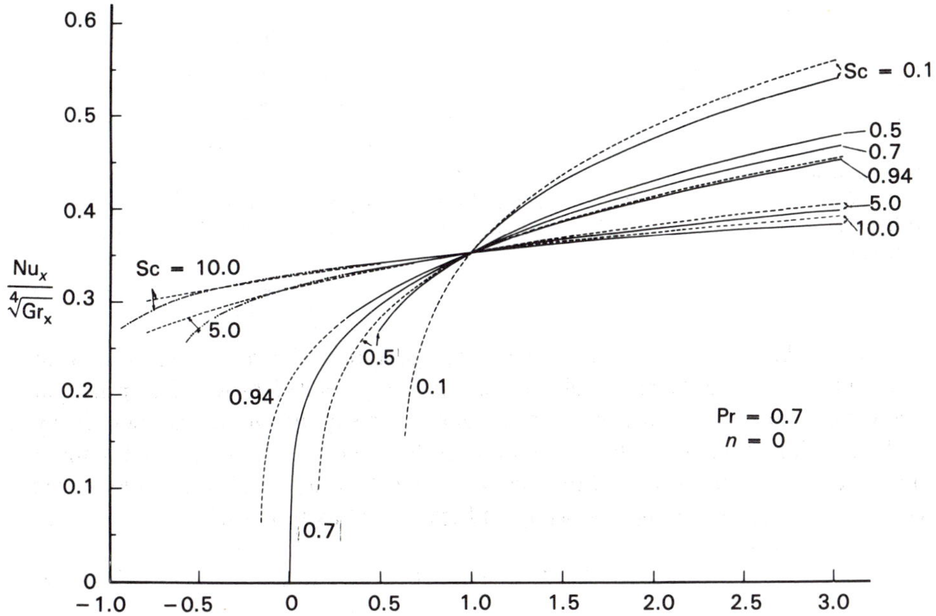

Figure 6.3.5 Values of the local Nusselt number for flow adjacent to a vertical surface, for various Schmidt numbers at Pr = 0.7 as a function of $\tilde{N}$. (——) Eq. (6.3.20) for $P = Q = 1$; (–··–) Eq. (6.3.20) for $P = \sqrt{Sc}$, $Q = \sqrt{Pr}$; (––––) integral method. *(Reprinted with permission from* Int. J. Heat Mass Transver, *vol. 14, p. 2025, B. Gebhart and L. Pera. Copyright © 1971, Pergamon Journals Ltd.)*

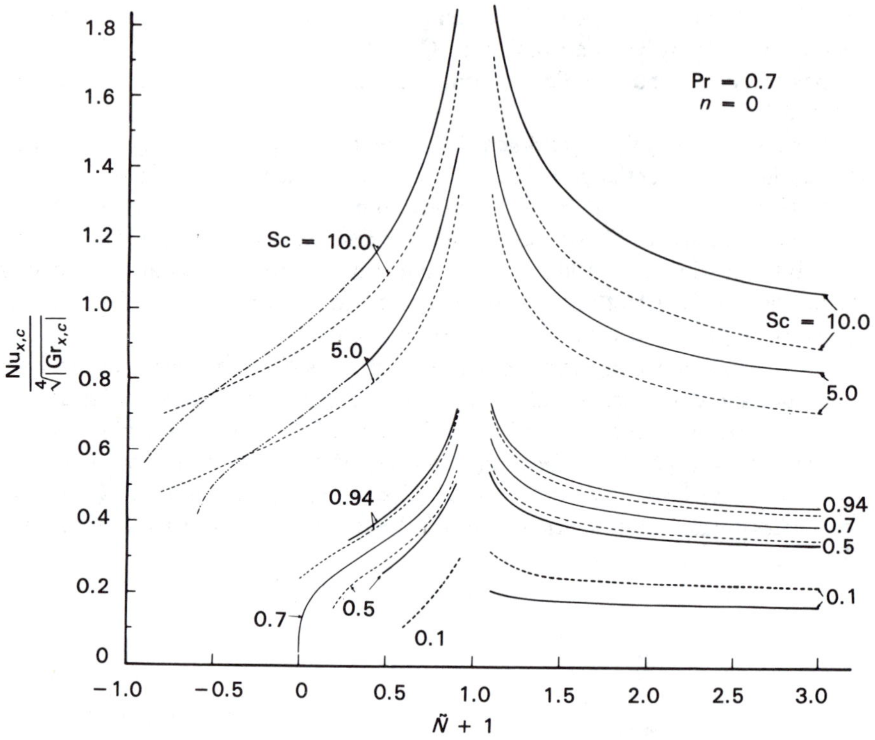

Figure 6.3.6 Values of the local Sherwood number for flow adjacent to a vertical surface, for various Schmidt numbers at Pr = 0.7, as a function of $\tilde{N}$. (——) Eq. (6.3.21) for $P = Q = 1$; (–··–) Eq. (6.3.20) for $P = \sqrt{\text{Sc}}$, $Q = \sqrt{\text{Pr}}$; (– – – –) integral method. *(Reprinted with permission from* Int. J. Heat Mass Transfer, *vol. 14, p. 2025, B. Gebhart and L. Pera. Copyright © 1971, Pergamon Journals Ltd.)*

Plume flow ($n = -\frac{3}{5}$). For plumes, the quantities of interest are somewhat different from those discussed above for surface-generated flows. The principal interest is in the temperature and concentration levels in the plume and their decay with elevation. Following the formulation in Section 3.7, the relations for these quantities are evaluated from functions f, ϕ, and $\tilde{C}$ in terms of $\tilde{N}$ and the plume convection rates of thermal energy Q and species M as follows:

$$t - t_\infty = Nx^{-3/5}\phi(n) \tag{6.3.23}$$

$$C - C_\infty = N_C x^{-3/5}\,\tilde{C}(\eta) \tag{6.3.24}$$

where

$$N = \left(\frac{Q^4}{4^3 g\beta\rho^2\mu^2 c_p^4 I^4}\right)^{1/5} (P + Q\tilde{N})^{-1/5} \tag{6.3.25}$$

$$N_C = \left(\frac{\tilde{N}M^4}{4^3 g\beta^* \nu^2 I_C^4}\right)^{1/5} (P + Q\tilde{N})^{-1/5} \tag{6.3.26}$$

$$I = \int_{-\infty}^{\infty} f'\phi \, d\eta \quad \text{and} \quad I_C = \int_{-\infty}^{\infty} f'\tilde{C} \, d\eta \tag{6.3.27}$$

The mass and momentum convection rate, m and M, in the plume are

$$m = 4^{3/5} J \left(\frac{g\beta\rho^2\mu^2 Q}{c_p I}\right)^{1/5} x^{3/5}(P + Q\tilde{N})^{1/5} \tag{6.3.28}$$

and

$$M = 8L\left(\frac{g\beta Q\rho^{2/3}\nu^{1/3}}{c_p I 4^{3/4}}\right) x^{4/5}(P + Q\tilde{N})^{3/5} \tag{6.3.29}$$

where

$$J = \int_{-\infty}^{\infty} f' \, d\eta \quad \text{and} \quad L = \int_{-\infty}^{\infty} f'^2 \, dn \tag{6.3.30}$$

The numerical results of Gebhart and Pera (1971) for quantities defined in Eqs. (6.3.27) and (6.3.30) along with $f'(0)$ are given in Table 6.3.3. These are for Pr = 0.7 and 7.0 at various values of Sc for both opposing and aiding effects.

The ϕ and $\tilde{C}$ distributions for Pr = Sc are plotted in Fig. 6.3.7. In general, these distributions indicate that the characteristics of plume flows are similar to those discussed above for $n = 0$. For Sc = Pr, the ϕ and $\tilde{C}$ distributions are again identical and velocities generalized with Gr_x dependent only on $\tilde{N}$.

For Pr $\neq$ Sc the temperature and velocity distributions are shown in Figs. 6.3.8 and 6.3.9, respectively. They indicate that aiding effects thin the thermal region while opposed effects greatly distort the velocity distribution. Large changes occur in the velocity field due to changes in Sc and in $\tilde{N}$. These should again have a significant effect on the laminar stability limits. Likewise, the Schmidt number effect on the temperature and velocity fields is reversed again for aiding and opposing buoyancy effects for the same reasons. This is also seen in Fig. 6.3.10, which is a plot of maximum midplane velocity level at various Schmidt numbers, at Pr = 0.7 and 7.0, as a function of $\tilde{N}$. The effect of increasing $\tilde{N}$ is simply to increase the velocity levels. Increasing Sc at a given value of Pr merely decreases velocity levels for aiding effects and increases them for opposing effects.

6.4 OTHER THAN PLANE VERTICAL FLOWS

Flows other than the plane vertical ones described in the previous section also admit similarity solutions. These are given here for vertical axisymmetric, horizontal, and inclined flows.

Table 6.3.3 Flow and transport quantities for plume flows

Pr	Sc	$\tilde{N}$	$f'(0)$	J	L	I	I_C
0.7	0.1	0.5	0.73587	3.251	1.339	1.398	2.463
		1	0.76508	3.628	1.564	1.458	2.658
		2	0.79178	3.881	1.769	1.508	2.809
	0.5	−0.5	0.62035	1.343	0.661	1.097	1.177
		0.5	0.67334	1.954	0.878	1.234	1.361
		1	0.67891	1.996	0.902	1.251	1.377
		2	0.68435	2.035	0.926	1.263	1.392
	0.7	All	0.66183	1.862	0.828	1.212	1.212
	0.94	−0.5	0.68855	1.996	0.929	1.264	1.149
		0.5	0.65243	1.810	0.793	1.192	1.089
		1	0.64763	1.782	0.775	1.182	1.081
		2	0.64273	1.751	0.757	1.171	1.072
	5.0	−0.5	0.79837	2.171	1.217	1.386	0.628
		0.5	0.60840	1.713	0.683	1.131	0.535
		1	0.57931	1.620	0.606	1.082	0.518
		2	0.54805	1.507	0.525	1.024	0.500
	10.0	−0.5	0.83104	2.187	1.271	1.405	0.460
		0.5	0.59386	1.701	0.661	1.119	0.383
		1	0.55628	1.600	0.572	1.062	0.369
		2	0.51532	1.474	0.477	0.993	0.353
7.0	7.0	All	0.44419	1.059	0.295	0.375	0.375
	100	0.5	0.57468	1.226	0.450	0.431	0.122
		0.5	0.39095	0.980	0.237	0.350	0.100
		1	0.36130	0.932	0.207	0.334	0.096
		2	0.32884	0.874	0.174	0.316	0.092
	500	−0.5	0.60360	1.248	0.476	0.440	0.056
		0.5	0.37628	0.966	0.226	0.344	0.044
		1	0.33798	0.936	0.192	0.325	0.042
		2	0.29200	0.802	0.147	0.299	0.039

Source: From Gebhart and Pera (1971).

Point source plume flows. These were analyzed by Mollendorf and Gebhart (1974). From Eqs. (4.2.1), (4.2.2), and (6.3.2), $f(\eta)$, ϕ, and $\tilde{C}$ are defined as follows:

$$\psi = \nu c f \qquad \eta = by \qquad \phi = \frac{t - t_\infty}{d} \qquad \tilde{C} = \frac{C - C_\infty}{e}$$

where

$$u = \frac{1}{y}\frac{\partial \psi}{\partial y} \qquad \text{and} \qquad v = -\frac{1}{y}\frac{\partial \psi}{\partial x}$$

Using these and the generalized similarity procedure, it is again shown that the governing equations, (6.2.25)–(6.2.29), admit similarity solutions for both power law and exponential variations of centerline temperature and concentration excess, at $y = 0$. That is, $d = Nx^n$ and Me^{mx} and $e = N_C x^n$ and $M_C e^{mx}$. The corresponding values of b and c and the governing equations are as follows.

Power law variation:

$$c = x \qquad b = \frac{1}{x}\sqrt[4]{\mathrm{Gr}_x} \tag{6.4.1}$$

$$\frac{f'''}{\eta} + \frac{f-1}{\eta}\left(\frac{f'}{\eta}\right)' - \frac{n+1}{2}\left(\frac{f'}{\eta}\right)^2 + \phi + \tilde{N}\tilde{C} = 0 \tag{6.4.2}$$

$$(\eta\phi')' + \mathrm{Pr}(f\phi' - nf'\phi) = 0 \tag{6.4.3}$$

$$(\eta\tilde{C}')' + \mathrm{Sc}(f\tilde{C}' - nf'\tilde{C}) = 0 \tag{6.4.4}$$

Exponential variation:

$$c = \frac{1}{m} \qquad b = m\left[\frac{g\beta(t_0 - t_\infty)}{m^3\nu^2}\right]^{1/4} = m\sqrt[4]{\mathrm{Gr}_m} \tag{6.4.5}$$

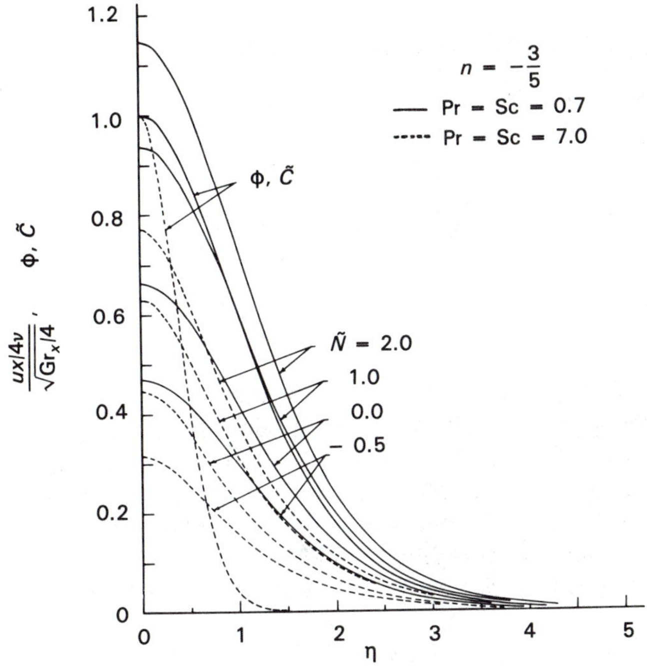

Figure 6.3.7 Distributions of temperature, concentration, and velocity for plume flows, for Pr = 0.7 and 7.0 and various values of $\tilde{N}$. *(Reprinted with permission from* Int. J. Heat Mass Transfer, *vol. 14, p. 2025, B. Gebhart and L. Pera. Copyright © 1971, Pergamon Journals Ltd.)*

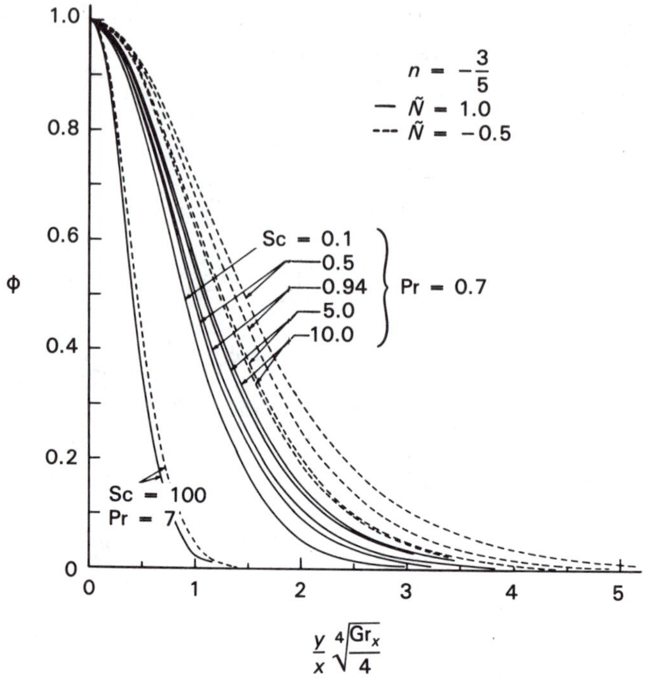

Figure 6.3.8 Temperature distributions for plane plume flows, for varying Schmidt number at given Pr, and $\tilde{N} = -0.5$ and 1. *(Reprinted with permission from* Int. J. Heat Mass Transfer, *vol. 14, p. 2025, B. Gebhart and L. Pera. Copyright © 1971, Pergamon Journals Ltd.)*

$$\frac{f'''}{\eta} - \frac{1}{\eta}\left(\frac{f'}{\eta}\right)' - \frac{1}{2}\left(\frac{f'}{\eta}\right)^2 + \phi + \tilde{N}\tilde{C} = 0 \tag{6.4.6}$$

$$(\eta\phi')' - \mathrm{Pr} f'\phi = 0 \tag{6.4.7}$$

$$(\eta\tilde{C}')' - \mathrm{Sc} f'\tilde{C} = 0 \tag{6.4.8}$$

The permissible values of n and m in the general formulation above are again to be determined from considerations of physical realism. For a point source of thermal and diffusing species concentration, that is, a point source plume, the condition is that the total energy Q and diffusing material M convected do not vary downstream. It was shown in Section 4.2 that this condition is not met for exponential variation. For the power law variation, this consideration requires that $n = -1$. The governing boundary layer equations for the point source then become

$$\frac{f'''}{\eta} + \frac{f-1}{\eta}\left(\frac{f'}{\eta}\right)' + \phi + \tilde{N}\tilde{C} = 0 \tag{6.4.9}$$

$$(\eta\phi')' + \mathrm{Pr}(f\phi' + f'\phi) = 0 \tag{6.4.10}$$

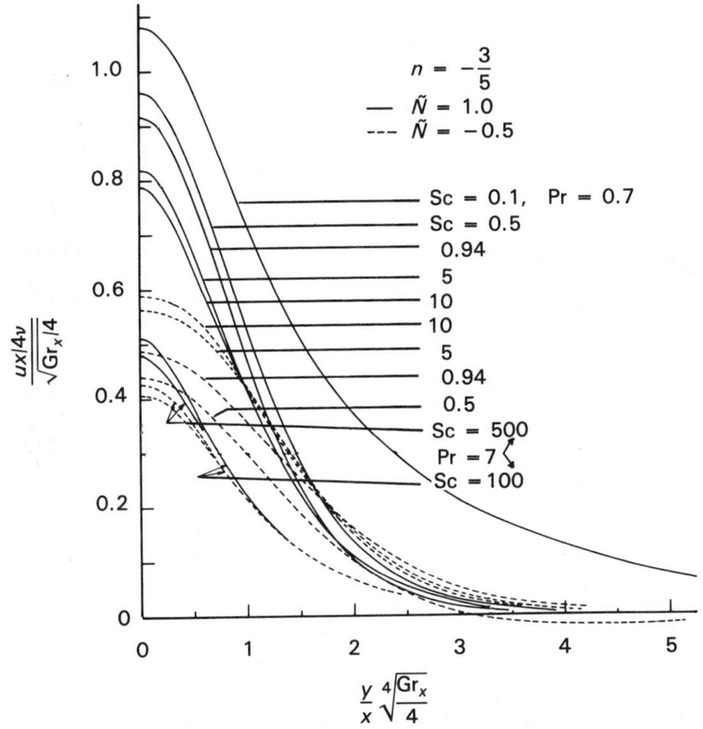

Figure 6.3.9 Flow velocity distributions for plane plume flows, for varying Schmidt number at given Pr, and $\tilde{N} = -0.5$ and 1. *(Reprinted with permission from* Int. J. Heat Mass Transfer, *vol. 14, p. 2025, B. Gebhart and L. Pera. Copyright © 1971, Pergamon Journals Ltd.)*

$$(\eta\tilde{C}')' + \text{Sc}(f\tilde{C}' + f'\tilde{C}) = 0 \tag{6.4.11}$$

The boundary conditions, Eq. (6.2.29), when transformed into similarity variables are

$$\text{at } \eta = 0: \qquad \left(\frac{f'}{\eta}\right)' = 0 = \frac{f}{\eta} - \frac{f'}{2} = \phi' = \tilde{C}' = 0 \qquad \phi = \tilde{C} = 1$$

$$\text{as } \eta \to \infty: \qquad \frac{f'}{\eta} \to 0$$

As shown in Section 4.4 for the point thermal source, these conditions reduce to

$$f(0) = f'(0) = \phi'(0) = \tilde{C}'(0) = \phi(0) - 1$$
$$= \tilde{C}(0) - 1 = 0 \qquad \text{and} \qquad f'(\infty) \text{ is bounded} \tag{6.4.12}$$

Note that Eqs. (6.4.10) and (6.4.11) are perfect differentials. This allows further simplifications. It can be shown from the integration of Eqs. (6.4.10) and (6.4.11) and the use of appropriate boundary conditions that

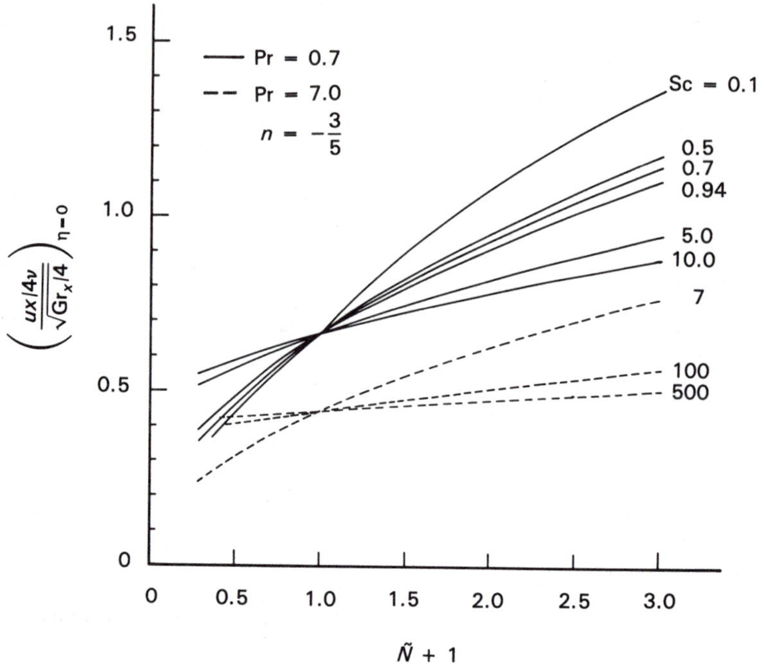

Figure 6.3.10 Values of the midplane flow velocity parameter for plane plume flows for various Schmidt numbers at Pr = 0.7 and 7.0 as a function of $\tilde{N}$. *(Reprinted with permission from* Int. J. Heat Mass Transfer, *vol. 14, p. 2025, B. Gebhart and L. Pera. Copyright © 1971, Pergamon Journals Ltd.)*

$$\tilde{C} = \phi^{\mathrm{Le}} \tag{6.4.13}$$

where Le = Sc/Pr (see Boura and Gebhart, 1976). The equations to be solved then are

$$\frac{f'''}{\eta} + \frac{f-1}{\eta}\left(\frac{f'}{\eta}\right)' + \phi + \tilde{N}\phi^{\mathrm{Le}} = 0 \tag{6.4.14}$$

$$\eta\phi' + \mathrm{Pr} f\phi = 0 \tag{6.4.15}$$

$$f'(0) = f(0) = 1 - \phi(0) = f'(\infty) = 0 \tag{6.4.16}$$

Mollendorf and Gebhart (1974) showed that closed-form solutions are possible for the above equations for Pr = Sc = 2. The solution is

$$f = \frac{4\gamma\eta^2}{1 + \gamma\eta^2} \tag{6.4.17}$$

$$\phi = \tilde{C} = (1 + \gamma\eta^2)^{-4} \tag{6.4.18}$$

where

$$\gamma = (\sqrt{1 + \tilde{N}})/8 \tag{6.4.19}$$

From Eq. (4.2.8), $u = (\nu/x)\sqrt{\text{Gr}_x}(f'/\eta)$ and $v = -(\nu/x)(\text{Gr}_x)^{1/4}(f/\eta - f'/2)$, then the values of u and v may readily be determined from Eq. (6.4.17).

For other values of Pr and Sc, Eqs. (6.4.14)–(6.4.16) were numerically integrated for Pr = 0.7 for Sc from 0.1 to 10 and for Pr = 7.0 for Sc from 1 to 700. Both aiding and opposed buoyancy effects were considered. The computed velocity, concentration, and temperature profiles are shown in Fig. 6.4.1 for Pr = 7 and Sc = 1. The results again indicate a strong effect of $\tilde{N}$ on the velocity distributions. Both the maximum and radial velocities increase with increasing $\tilde{N}$ this effect being more pronounced at lower Lewis numbers, as would be expected. The increase in velocity is provided by the additional component of the buoyancy force due to species diffusion. For a given Pr, as Sc decreases, the relative thickness of the concentration boundary layer increases and the contribution from mass diffusion-induced buoyancy becomes more significant. Many of the effects observed for plane plumes, discussed in the previous section, are also observed here.

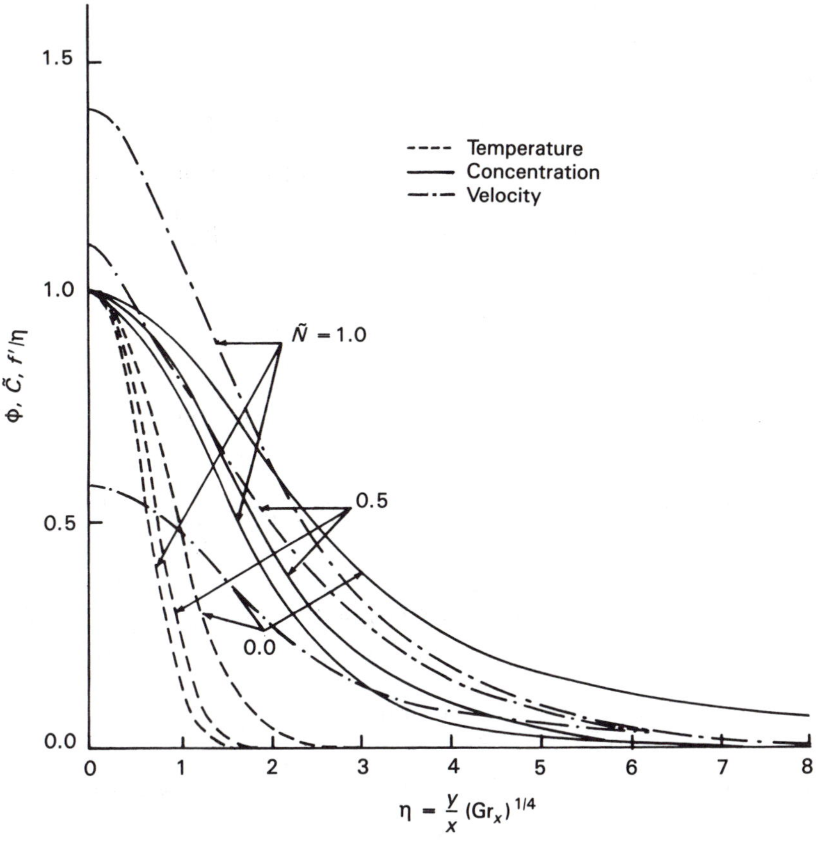

Figure 6.4.1 Temperature, concentration, and velocity distributions for Pr = 7 and Sc = 1 in an axisymmetric plume driven by combined buoyancy effects. *(From Mollendorf and Gebhart, 1974.)*

With increasing $\tilde{N}$, a thinning of the thermal region occurs for aiding effects and vice versa for opposing effects. However, unlike the case of plane plumes, regions of local flow reversals are not observed over the range of parameters investigated.

Flow adjacent to a vertical cylinder. Numerical solutions to the combined heat and mass transfer for this flow have been obtained by Chen and Yuh (1980) for the surface of the cylinder being either maintained at uniform temperature and concentration levels or subjected to a uniform heat and mass flux. For these conditions, Eqs. (6.2.25)–(6.2.28) and (6.2.30) do not admit a similarity solution. However, it will be shown later that a similar solution does again arise for a linear variation of both $t_0 - t_\infty$ and $C_0 - C_\infty$ with x.

For the uniform surface temperature and concentration condition, Eqs. (6.2.25)–(6.2.28) and (6.2.30) were transformed into a dimensionless form as follows:

$$\xi = 2\frac{x}{R}\left(\frac{\text{Gr}_x}{4}\right)^{-1/4} \qquad \eta = \frac{(y^2 - R^2)}{2Rx}\left(\frac{\text{Gr}_x}{4}\right)^{1/4}$$

$$\text{Gr}_x = \frac{g\beta(t_0 - t_\infty)x^3}{\nu^2} \qquad \psi(x, y) = 4\nu R\left(\frac{\text{Gr}_x}{4}\right)^{1/4} f(\eta, \xi) \tag{6.4.20}$$

$$\phi(\eta, \xi) = \frac{t - t_\infty}{t_0 - t_\infty} \qquad \tilde{C}(\eta, \xi) = \frac{C - C_\infty}{C_0 - C_\infty}$$

where ξ is the dimensionless axial coordinate, R the radius of the cylinder, η the pseudo-similarity variable, and f, ϕ, and $\tilde{C}$ are the dimensionless stream function, temperature, and concentration, respectively. The stream function $\psi(x, y)$, as before, is defined as follows:

$$u = \frac{1}{y}\frac{\partial\psi}{\partial y} \qquad v = -\frac{1}{y}\frac{\partial\psi}{\partial x} \tag{6.4.21}$$

Using Eqs. (6.4.20) and (6.4.21), Eqs. (6.2.25)–(6.2.28) and (6.2.30) transform to

$$(1 + \xi\eta)f''' + (\xi + 3f)f'' - 2f'^2 + \phi + \tilde{N}\tilde{C} = \xi\left(f'\frac{\partial f'}{\partial \xi} - f''\frac{\partial f}{\partial \xi}\right) \tag{6.4.22}$$

$$\frac{1}{\text{Pr}}(1 + \xi\eta)\phi'' + \left(\frac{\xi}{\text{Pr}} + 3f\right)\phi' = \xi\left(f'\frac{\partial \phi}{\partial \xi} - \phi'\frac{\partial f}{\partial \xi}\right) \tag{6.4.23}$$

$$\frac{1}{\text{Sc}}(1 + \xi\eta)\tilde{C}'' + \left(\frac{\xi}{\text{Sc}} + 3f\right)\tilde{C}' = \xi\left(f'\frac{\partial \tilde{C}}{\partial \xi} - \tilde{C}'\frac{\partial f}{\partial \xi}\right) \tag{6.4.24}$$

$$\begin{gathered} f'(\xi, 0) = 3f(\xi, 0) + \xi f_\xi(\xi, 0) = \phi(\xi, 0) - 1 = \tilde{C}(\xi, 0) - 1 = 0 \\ f'(\xi, \infty) = \phi(\xi, \infty) = \tilde{C}(\xi, \infty) = 0 \end{gathered} \tag{6.4.25}$$

In these equations the primes stand for partial derivatives with respect to η. The equations were solved numerically for a range of Sc, Pr, and $\tilde{N}$. The results for the local heat and mass transfer parameters are plotted in Figs. 6.4.2 and 6.4.3 in terms of axial coordinate ξ. The modification of the transport parameters for pure thermal (or mass) diffusion, to account for the added buoyancy force due to mass (or thermal) diffusion, is similar to that discussed earlier for flow over vertical surfaces. The additional consideration here is the effect of radius of curvature. Higher Nusselt and Sherwood numbers are seen for decreased radius of curvature R, that is, larger ξ, in Figs. 6.4.2 and 6.4.3.

The axial coordinate may also be written as $\xi = (y^2 - R^2)/\eta R^2$. This is a measure of the ratio of the boundary layer thickness to the cylinder radius R. When $\xi = 0$ the result reduces to that for a vertical flat surface. For small values of ξ the thickness of the boundary layer is much smaller than R and deviations from a flat surface are small. The reverse is true for larger values of ξ.

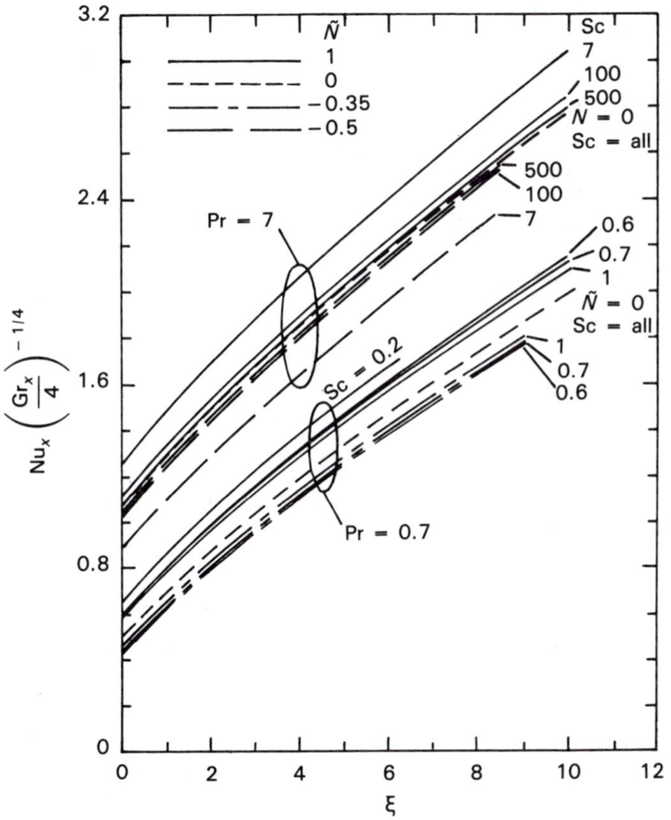

Figure 6.4.2 Local Nusselt number results for flow adjacent to a vertical cylinder maintained at uniform surface temperature/concentration for Pr = 0.7 and 7.0. *(Reprinted with permission from* Int. J. Heat Mass Transfer, *vol. 23, p. 451, T. S. Chen and C. F. Yuh. Copyright © 1980, Pergamon Journals Ltd.)*

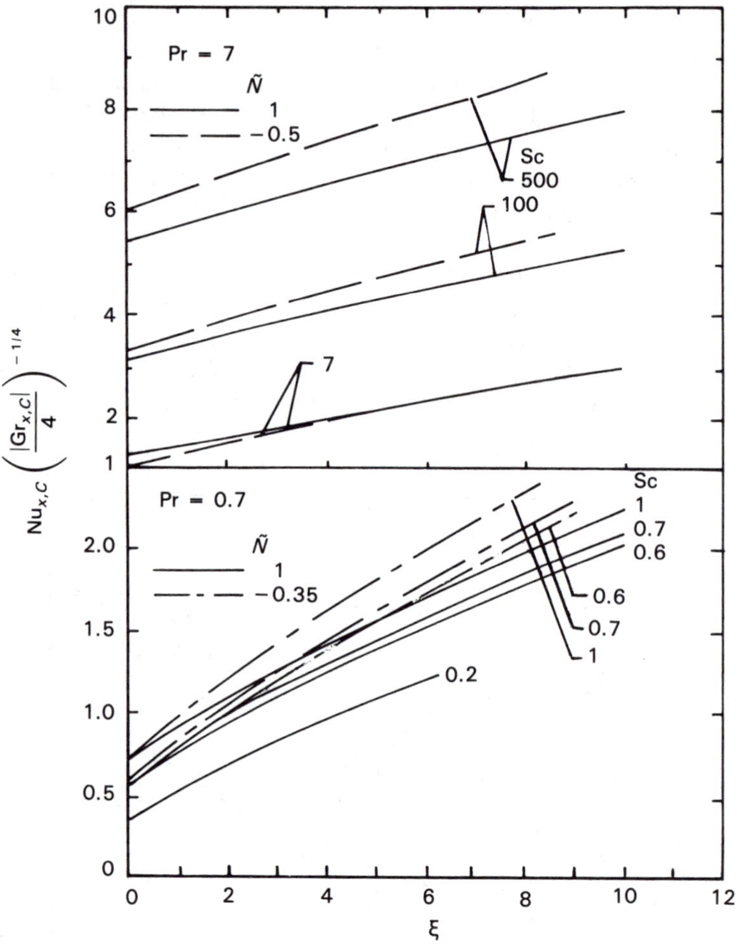

Figure 6.4.3 Local Sherwood number results for flow adjacent to a vertical cylinder maintained at uniform surface temperature/concentration for Pr = 0.7 and 7.0. *(Reprinted with permission from* Int. J. Heat Mass Transfer, *vol. 23, p. 451, T. S. Chen and C. F. Yuh. Copyright © 1980, Pergamon Journals Ltd.)*

For a linear variation of $t_0 - t_\infty$ and $C_0 - C_\infty$, that is, $t_0 - t_\infty = Nx$ and $C_0 - C_\infty = N_C x$, ξ and η from Eq. (6.4.20) are

$$\xi = \frac{2x}{R}\left(\frac{g\beta x^3 Nx}{4\nu^2}\right)^{-1/4} = \frac{2}{R}\left(\frac{g\beta N}{4\nu^2}\right)^{-1/4} \tag{6.4.26}$$

$$\eta = \frac{y^2 - R^2}{2xR}\left(\frac{g\beta x^3 Nx}{4\nu^2}\right)^{1/4} = \frac{y^2 - R^2}{2R}\left(\frac{g\beta N}{4\nu^2}\right)^{1/4} \tag{6.4.27}$$

That is, ξ is independent of x and η is a function of y only. The x-dependent terms in Eqs. (6.4.22)–(6.4.25) disappear. The remaining terms amount to ordinary differential equations in terms of the similarity variable η. In this connection, recall from Chapter 4 that, for the pure thermal diffusion process, similarity results for a linear variation of $t_0 - t_\infty$ with x.

Chen and Yuh (1980) also obtained numerical results for the surface maintained at uniform heat and mass flux. The local heat and mass transfer parameters for this case are plotted in Figs. 6.4.4 and 6.4.5, where $\xi^* = (2x/R)(\mathrm{Gr}_x^*/5)^{-1/5}$, $\mathrm{Gr}_{x,c}^* = g\beta^* m'' x^4/\rho D \nu^2$, $\mathrm{Nu}_{x,c} = m''x/\rho D(C_0 - C_\infty)$, and m'' is the mass flux of the diffusing species.

Horizontal flows, $B_n > 0$. Equations (6.2.31)–(6.2.37) for this flow may be analyzed for similarity by proceeding as for vertical plane flows. Using Eqs. (3.5.4)–(3.5.6) in conjunction with Eqs. (5.3.1), (6.3.1), and (6.3.2), similarly results

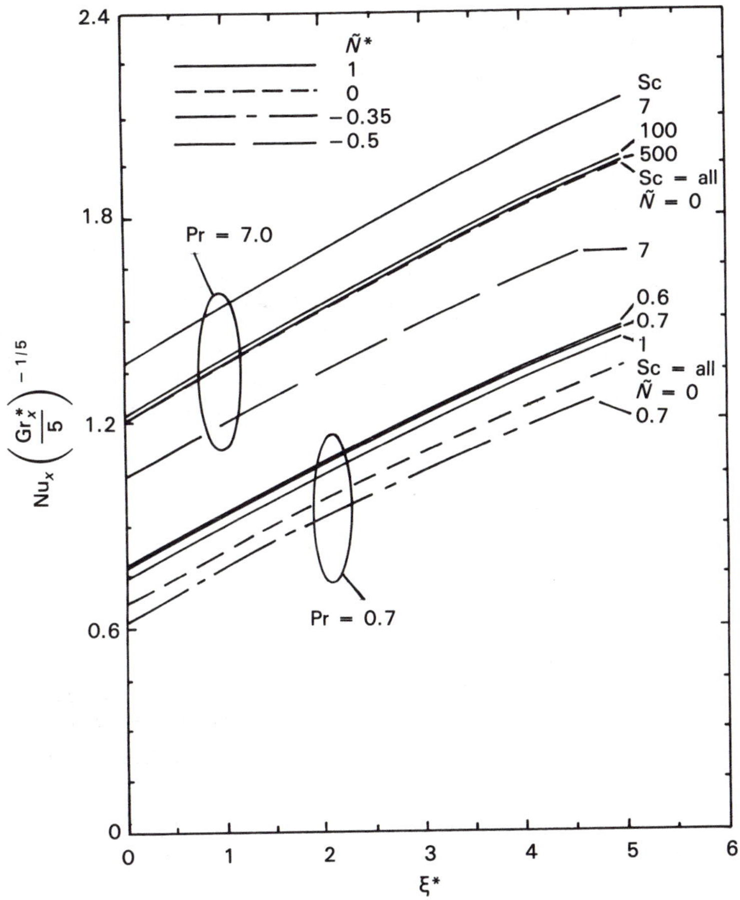

Figure 6.4.4 Local Nusselt number results for flow adjacent to a vertical cylinder maintained at uniform surface heat/mass flux for Pr = 0.7 and 7.0. *(Reprinted with permission from* Int. J. Heat Mass Transfer, *vol. 23, p. 451, T. S. Chen and C. F. Yuh. Copyright © 1980, Pergamon Journals Ltd.)*

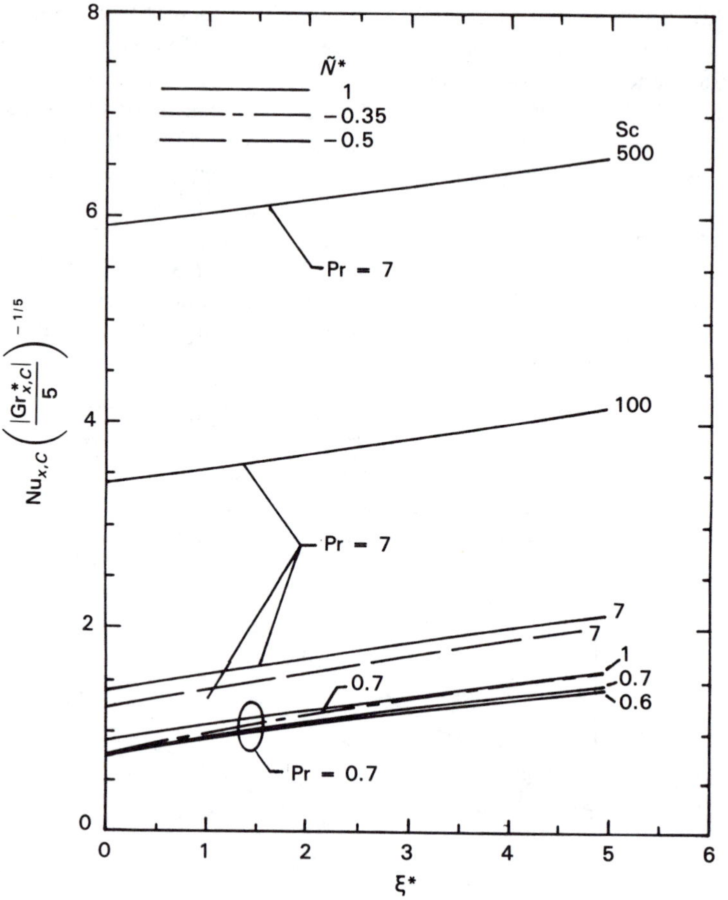

Figure 6.4.5 Local Sherwood number results for flow adjacent to a vertical cylinder maintained at uniform surface heat/mass flux for Pr = 0.7 and 7.0. *(Reprinted with permission from* Int. J. Heat Mass Transfer, *vol. 23, p. 451, T. S. Chen and C. F. Yuh. Copyright © 1980, Pergamon Journals Ltd.)*

again for power law surface temperature $d = Nx^n$ and concentration $e = N_C x^{n_C}$, provided $n = n_C$. Here, $P(\eta)$ in Eq. (5.3.1) will be denoted by $P_m(\eta)$ instead to avoid confusion with constant P. The most general forms of the associated values of $b(x)$ and $c(x)$ are found to be

$$c(x) = 5\sqrt[5]{\frac{P\mathrm{Gr}_x + Q\mathrm{Gr}_{x,C}}{5}} \qquad b(x) = \frac{c}{5x} \tag{6.4.28}$$

Using these values, the governing equations in similarity variables become

$$f''' + (n+3)ff'' - (2n+1)f'^2 = \frac{[(4n+2)/5]P_m + [(n-2)/5]P'_m\eta}{P + Q\tilde{N}} \tag{6.4.29}$$

$$P'_m = \phi + \tilde{N}\tilde{C} \tag{6.4.30}$$

$$\phi'' + \Pr[(n + 3)f\phi' - 5n\phi] = 0 \tag{6.4.31}$$

$$\tilde{C}'' + \mathrm{Sc}[(n + 3)f\tilde{C}' - 5n\tilde{C}] = 0 \tag{6.4.32}$$

$$\begin{gathered} f(0) = f'(0) = 1 - \phi(0) = 1 - \tilde{C}(0) = 0 \\ f'(\infty) = \phi(\infty) = \tilde{C}(\infty) = P_m(\infty) = 0 \end{gathered} \tag{6.4.33}$$

Here P and Q are again the scaling factors chosen so that $P + Q\tilde{N}$ is greater than zero. The limitations on n, as discussed in Chapter 5, are $-0.5 \le n \le 2$.

Numerical solutions to these equations for uniform surface temperature and concentration ($n = 0$) have been obtained by Pera and Gebhart (1972) for Pr = 0.7 and Schmidt numbers from 0.1 to 10. The computed values of $f''(0)$, $-P_m(0)$, $-\phi'(0)$, and $-\tilde{C}'(0)$ for this range are given in Table 6.4.1. The numerical values of integrals J, L, I, and I_C are also given, where

$$J = \int_0^\infty f'\, d\eta = f(\infty) \qquad L = \int_0^\infty f'^2\, d\eta$$

$$I = \int_0^\infty \phi f'\, d\eta \qquad I_C = \int_0^\infty \tilde{C} f'\, d\eta$$

These may be used to determine the total amount of heat, chemical species, momentum, and total mass convected in the boundary layer at a given location x.

Table 6.4.1 Flow and transport quantities for flow adjacent to isothermal horizontal surfaces, Pr = 0.7

Sc	$\tilde{N}$	$f''(0)$	$-P_m(0)$	$-\phi'(0)$	$-\tilde{C}'(0)$	J	L	I	I_C
0.1	0.5	0.7226	2.4619	0.5852	0.2259	1.3550	0.3656	0.2780	0.7155
	1.0	0.8153	3.6890	0.6158	0.2400	1.5156	0.4627	0.2926	0.7742
	2.0	0.9007	6.1295	0.6408	0.2415	1.6353	0.5479	0.3044	0.8178
0.5	−0.4	0.4814	0.7441	0.4367	0.3488	−0.0684	0.1267	0.2040	0.2195
	0.5	0.5426	1.9705	0.5016	0.4260	0.6987	0.1422	0.2383	0.2824
	1.0	0.5545	2.6759	0.5070	0.4309	0.7125	0.1494	0.2412	0.2861
	2.0	0.5660	4.0843	0.5123	0.4357	0.7308	0.1567	0.2438	0.2895
0.7	All	0.5191	1.2683	0.4890	0.4890	0.6457	0.1262	0.2326	0.2326
1.0	−0.5	0.5783	0.6988	0.5148	0.6056	0.7098	0.1553	0.2494	0.2018
	0.5	0.4982	1.8376	0.4788	0.5656	0.6183	0.1154	0.2277	0.1884
	1.0	0.4875	2.4070	0.4732	0.5595	0.6025	0.1097	0.2250	0.1864
	2.0	0.4765	3.5454	0.4673	0.5530	0.5857	0.1038	0.2221	0.1842
5.0	−0.5	0.7115	0.8857	0.5524	1.2403	0.7449	0.1880	0.2629	0.0826
	0.5	0.4433	1.6373	0.4577	1.0437	0.5938	0.1006	0.2176	0.0695
	1.0	0.4019	2.0031	0.4380	1.0046	0.5583	0.0859	0.2080	0.0700
	2.0	0.3572	2.7328	0.4136	0.9579	0.5132	0.0694	0.1961	0.0640
10.0	−0.5	0.7377	0.9338	0.5570	1.6174	0.7466	0.1904	0.2650	0.5557
	0.5	0.4312	1.5821	0.4545	1.3380	0.5522	0.0993	0.2160	0.0445
	1.0	0.3823	1.8896	0.4323	1.2799	0.5557	0.0838	0.2052	0.0426
	2.0	0.3286	2.4996	0.4040	1.2084	0.5073	0.0661	0.1914	0.0402

Source: From Pera and Gebhart (1972).

From the values of $-\phi'(0)$ and $-\tilde{C}'(0)$, Nu_x and $\mathrm{Nu}_{x,C}$ may be calculated as follows:

$$\mathrm{Nu}_x = \sqrt[5]{\frac{\mathrm{Gr}_x}{5}}\,[-\phi'(0)](P + Q\tilde{N})^{1/5} \tag{6.4.34a}$$

$$\mathrm{Nu}_{x,C} = \sqrt[5]{\frac{|\mathrm{Gr}_{x,C}|}{5}}\,[-\tilde{C}'(0)]\left(\left|\frac{P + Q\tilde{N}}{\tilde{N}}\right|\right)^{1/5} \tag{6.4.34b}$$

As indicated by the results in Table 6.4.1, the heat transfer is enhanced by aiding buoyancy effects with the increment depending strongly on Sc. Opposing buoyancy effects produce opposite results.

For Sc = Pr, simplifications similar to those discussed for vertical plane flows occur. The temperature and concentration distribution functions are equal, and only one of Eqs. (6.4.31) and (6.4.32) must be solved, in conjunction with Eqs. (6.4.29) and (6.4.30) and appropriate boundary conditions. The problem thus reduces to that for a single buoyancy effect. The transport rates, for example, are then determined from Eqs. (6.4.34*a*), for $P = Q = 1$, knowing only $-\phi'(0)$ and $\tilde{N}$, where $-\phi'(0)$ is given in Table 5.3.1. Again, the combined effect lies only in the definition of the Grashof number, which now becomes $\mathrm{Gr}_x + \mathrm{Gr}_{x,C}$.

For Sc $\neq$ Pr, the effects of added buoyancy due to species diffusion are significant and are quite different for aiding and opposing buoyancy effects. For Pr > Sc and for some negative values of $\tilde{N}$, a local flow reversal occurs in the outer part of the boundary layer. A similar effect appears in the inner region for Sc > Pr. These and other effects observed are very similar to those discussed in Section 6.3.

Flow adjacent to an inclined surface, $B_n > 0$. For small inclinations θ from the vertical, the boundary region equations for this flow, Eqs. (6.2.38)–(6.2.43), are identical to those for flow over a vertical surface except that g is now replaced by $g \cos \theta$. Therefore, all the similar solutions discussed in Section 6.3 for the vertical surface flow are applicable simply by replacing g in the Grashof number by $g \cos \theta$. In particular, the similarity variable η and dimensionless stream function f are now given as

$$\eta = by = \frac{y}{x}\sqrt[4]{\frac{(P\mathrm{Gr}_x + Q\mathrm{Gr}_{x,C})\cos\theta}{4}} \tag{6.4.35}$$

$$\psi = \nu c f(\eta) = 4\nu\sqrt[4]{\frac{(P\mathrm{Gr}_x + Q\mathrm{Gr}_{x,C})\cos\theta}{4}}\,f(\eta) \tag{6.4.36}$$

Chen and Yuh (1979) obtained similarity solutions for the surface either maintained at a uniform temperature and concentration or subject to uniform heat and mass fluxes. These surface conditions, as before, correspond to $n = 0$ and $n = \frac{1}{5}$, respectively. Chen and Yuh, however, used a somewhat different form of the variables. For the uniform temperature and concentration condition, the η and ψ used correspond to $P = 1$ and $Q = 0$ in Eqs. (6.4.35) and (6.4.36), and the governing equations are (6.3.14)–(6.3.17) with $P = 1$, $Q = 0$, and $n = 0$. For uniform heat and mass flux surface conditions, the formulation is in terms of flux Grashof numbers as follows:

$$\eta^* = \frac{y}{x}\left(\mathrm{Gr}_x^*\frac{\cos\theta}{5}\right)^{1/5} \tag{6.4.37}$$

$$\psi(x, y) = 5\nu\left(\mathrm{Gr}_x^*\frac{\cos\theta}{5}\right)^{1/5} f(\eta^*) \tag{6.4.38}$$

$$\phi(\eta^*) = (t - t_\infty)\left(\mathrm{Gr}_x^*\frac{\cos\theta}{5}\right)^{1/5} \frac{k}{q''x} \tag{6.4.39}$$

$$\tilde{C}(\eta^*) = \frac{(C - C_\infty)\rho D\left(\mathrm{Gr}_x^*\dfrac{\cos\theta}{5}\right)^{1/5}}{m''x} \tag{6.4.40}$$

$$\tilde{N}^* = \frac{\mathrm{Gr}_{x,C}^*}{\mathrm{Gr}_x^*} \tag{6.4.41}$$

$$\mathrm{Gr}_x^* = \frac{g\beta q''x^4}{k\nu^2} \quad \text{and} \quad \mathrm{Gr}_{x,C}^* = \frac{g\beta^* m''x^4}{\rho D\nu^2} \tag{6.4.42}$$

Using these, Eqs. (6.2.38)–(6.2.43) transform to

$$f''' + 4ff'' - 3f'^2 + \phi + \tilde{N}^*\tilde{C} = 0 \tag{6.4.43}$$

$$\frac{\phi''}{\mathrm{Pr}} + 4f\phi' - f'\phi = 0 \tag{6.4.44}$$

$$\frac{\tilde{C}''}{\mathrm{Sc}} + 4f\tilde{C}' - f'\tilde{C} = 0 \tag{6.4.45}$$

The appropriate boundary conditions are

$$f(0) = f'(0) = \phi'(0) + 1 = \tilde{C}'(0) + 1 = f'(\infty) = \phi(\infty) = \tilde{C}(\infty) = 0 \tag{6.4.46}$$

Note again that $f(0) = 0$ corresponds to the assumption that the interfacial velocity v_0 resulting from the mass diffusion at the wall is negligible. As for Eq. (6.3.9), it can be shown that

$$v_0(x) = -4\frac{\nu}{x}\left(\mathrm{Gr}_x^*\frac{\cos\theta}{5}\right)^{1/5} f(0) \tag{6.4.47a}$$

Therefore, $f(0) \approx 0$ requires that

$$\frac{v_0(x)x}{\nu} << 4\left(\mathrm{Gr}_x^*\frac{\cos\theta}{5}\right)^{1/5} \tag{6.4.47b}$$

Writing the interface mass flux $m''(0) = \rho v_0$ from Eq. (6.4.40) and eliminating v_0 between Eqs. (6.4.47*a*) and (6.4.40), we get

$$f(0) = \frac{1}{4\text{ Sc}} \frac{C_0 - C_\infty}{\tilde{C}(0)}$$

Therefore, for Sc $= O(1)$, $f(0) << 0$ requires that

$$\frac{1}{4\text{ Sc}} \frac{C_0 - C_\infty}{\tilde{C}(0)} << 1 \qquad (6.4.48)$$

Numerical results were obtained for the isothermal and uniform flux surface conditions for air for Sc from 0.2 to 10 and for water for Sc from 7.0 to 500, for both aiding and opposing buoyancy effects. The results for uniform surface conditions are the same as those given in Table 6.3.2 when a simple transformation is used to convert one set of results to the other. For the surface subject to uniform heat and mass flux, Nu_x and $\text{Nu}_{x,C}$ are shown in Figs. 6.4.6 and 6.4.7. The results are quite similar to those shown in Figs. 6.3.5 and 6.3.7. Considerations similar to those for vertical flows regarding the effect of aiding and opposed buoyancy flows on the transport parameters also apply here. Finally, the effect of angle of inclination θ on the local Nusselt number is shown in Figs. 6.4.8 and 6.4.9. For a given value of $\tilde{N}$ or $\tilde{N}^*$, Nu_x decreases with an increase in θ. A similar effect is observed for $\text{Nu}_{x,C}$.

Combined thermal and mass transfer convection in enclosures has also been investigated in some detail. See, for example, Chen et al. (1971), Wirtz et al. (1972), and Wirtz (1977). Much of the work is related to rectangular enclosures in which a stably stratified saline solution is heated laterally. Depending on the heating rate, convective layers are formed that are separated by diffusive interfaces containing large vertical gradients of temperature and concentration.

6.5 EXTREME SCHMIDT AND PRANDTL NUMBER RESULTS

The expressions for transport parameters from vertical surfaces in thermally driven natural convection flows for extreme values of the Prandtl number were derived in Section 3.8. A unique dependence on Pr at each end of the Pr range was indicated. The relations derived there are also applicable to extreme values of Sc in the analogous mass transfer processes, when the concentration difference and the Sorét and Dufour effects are small.

This section presents results for simultaneous heat and mass transfer at extreme values of both Prandtl and Schmidt numbers. Of the various possible combinations of extreme values of Sc and Pr, only those that correspond to most commonly occurring flow situations are considered. For gaseous mixtures Pr = Sc is a good approximation. Results are presented here for Pr = Sc $\rightarrow 0$ and Pr = Sc $\rightarrow \infty$. The asymptotic solutions are then presented for Sc >> Pr and Pr large, characteristic of many viscous liquids. Finally, the limiting condition of

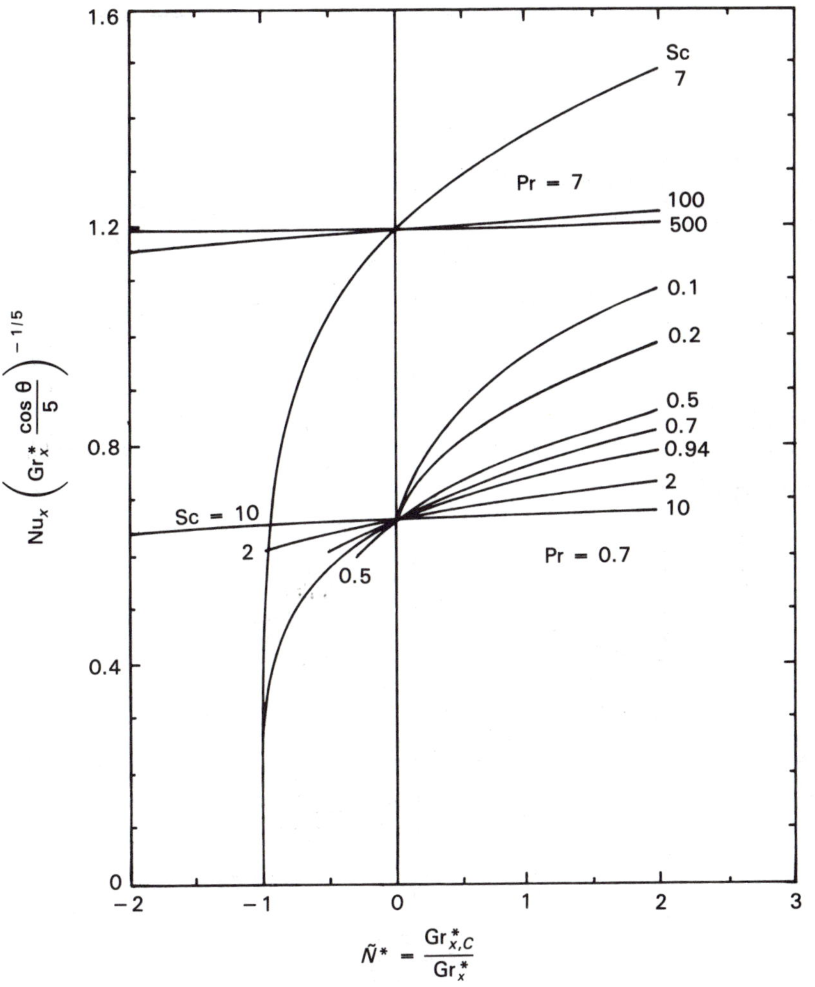

Figure 6.4.6 Local Nusselt number results for an inclined surface maintained at uniform heat/mass flux, for $B_n > 0$ and $Pr = 0.7$ and 7.0. *(Reprinted with permission from T. S. Chen and C. F. Yuh,* Numer. Heat Transfer, *vol. 2, p. 233. Copyright © 1979, Hemisphere Publishing Corporation.)*

$Pr \to 0$ and $Sc \to \infty$ is considered. This models the heat and mass transfer in liquid metals.

The asymptotic solutions for the combinations of Pr and Sc are obtained by using a perturbation analysis outlined in Section 3.8. It will be seen that the asymptotic expansions used there for flow and temperature fields also apply to the combined problem. Additional expansions, of course, are required for the concentration field. These perturbation and other solutions are described next in detail for the boundary layer flow adjacent to a vertical surface maintained at uniform temperature and concentration levels. Attention is restricted to flows for

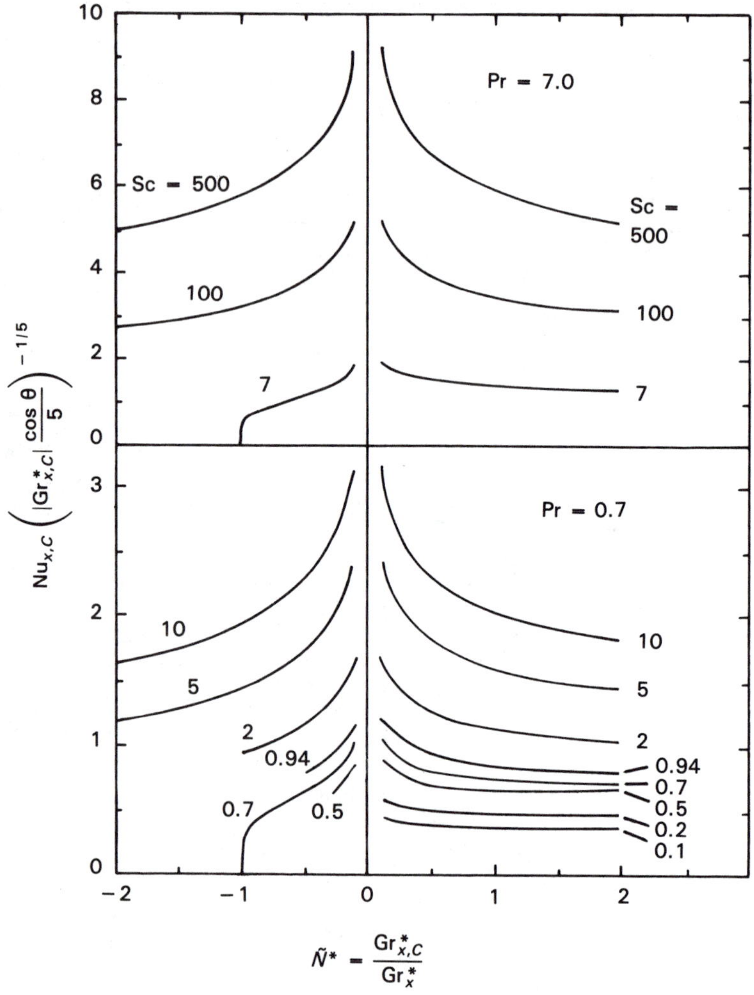

Figure 6.4.7 Local Sherwood number results for an inclined surface maintained at uniform surface heat/mass flux, for $B_n > 0$ and Pr = 0.7 and 7.0. *(Reprinted with permission from T. S. Chen and C. F. Yuh,* Numer. Heat Transfer, *vol. 2, p. 233. Copyright © 1979, Hemisphere Publishing Corporation.)*

which the buoyancy ratio parameter $\tilde{N}$ is positive. The problem posed for negative values of $\tilde{N}$, which results in local reversal of flow in the boundary layer, is quite complex and has not yet been considered.

Asymptotic solutions for Sc = Pr. In the discussion of similarity solutions for Pr = Sc in Section 6.3, it was shown that for $P = Q = 1$, $\phi = \tilde{C}$. Then the solutions for the simultaneous heat and mass transfer problem are merely those for a single buoyancy effect. The only change is in the replacement of Gr_x or $Gr_{x,C}$ by the combined Grashof number Gr'_x, which for $P = Q = 1$ in Eq. (6.3.12) is simply

$$\mathrm{Gr}'_x = \mathrm{Gr}_x + \mathrm{Gr}_{x,C} \tag{6.5.1}$$

This procedure is valid for both small and large values of Pr = Sc. The asymptotic solutions obtained in Section 3.8 for extreme values of Pr are therefore also the solutions for the combined problem, with Gr_x replaced by Gr'_x.

Solutions for Sc >> Pr → ∞. The asymptotic inner and outer expansions are obtained with the perturbation scheme used for large Pr in Section 3.8. As indicated there, the thicknesses of the inner and outer layer scales are $O(\mathrm{Pr}^{-1/4})$ and $O(\mathrm{Pr}^{1/4})$, respectively. The inner and outer variables remain as those in Eqs. (3.8.15) and (3.8.16), respectively, supplemented with the variables for the concentration field as follows.

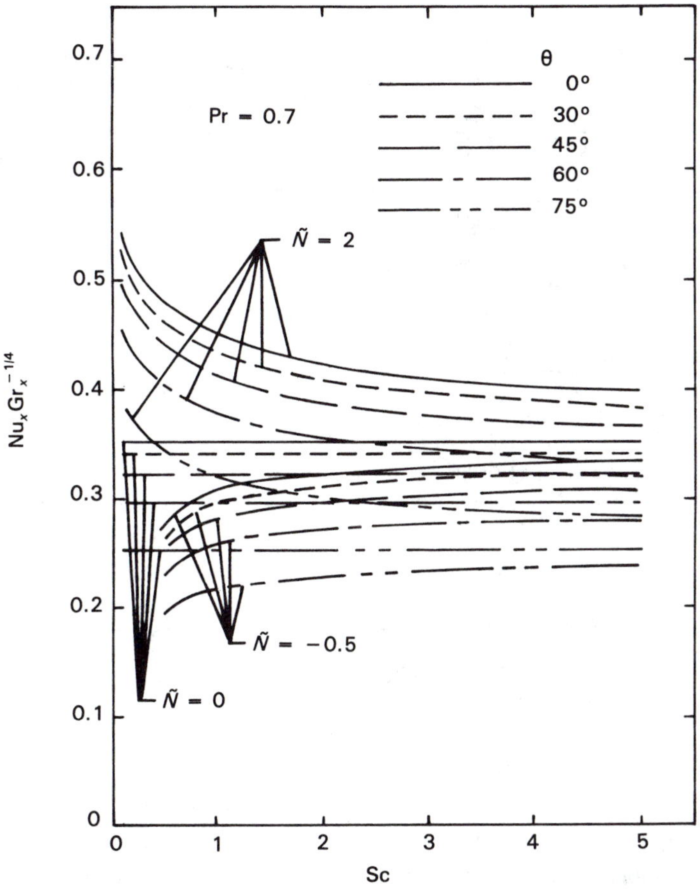

Figure 6.4.8 Effect of angle of inclination on the local Nusselt number for an inclined surface at uniform surface temperature/concentration, Pr = 0.7. *(Reprinted with permission from T. S. Chen and C. F. Yuh,* Numer. Heat Transfer, *vol. 2, p. 233. Copyright © 1979, Hemisphere Publishing Corporation.)*

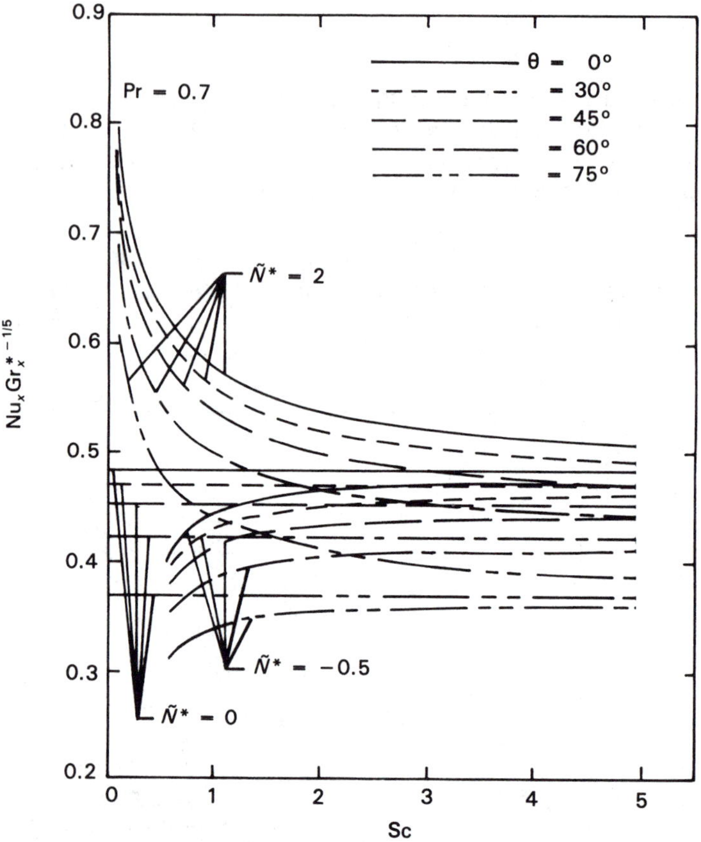

Figure 6.4.9 Effect of angle of inclination on the local Nusselt number for an inclined surface at uniform surface heat/mass flux surface condition, Pr = 0.7. *(From Chen and Yuh, 1979.)*

Inner layer:

$$\zeta = (3\mathrm{Pr})^{1/4}\eta \qquad \tilde{f}(\zeta) = (3\mathrm{Pr})^{3/4} f(\eta) \qquad H(\zeta) = \phi(\eta)$$
$$\bar{C}(\zeta) = \tilde{C}(\eta) \tag{6.5.2}$$

Outer layer:

$$\xi = A(3\mathrm{Pr})^{-1/4}\eta \qquad \tilde{G}(\xi) = (3\mathrm{Pr})^{1/4}\frac{f}{A} \tag{6.5.3}$$

where the temperature and concentration fields in the outer layer are zero and A, as before, is a constant to be determined from matching conditions. Using these transformations, Eqs. (6.3.14)–(6.3.16) result in the following governing equations for $\tilde{f}$, H, $\bar{C}$, and $\tilde{G}$:

$$\tilde{f}''' + \frac{1}{3\text{Pr}}(3\tilde{f}\tilde{f}'' - 2\tilde{f}'\tilde{f}') + H + \tilde{N}\bar{C} = 0 \tag{6.5.4}$$

$$H'' + \tilde{f}H' = 0 \tag{6.5.5}$$

$$\bar{C}'' + \frac{\text{Sc}}{\text{Pr}}\tilde{f}\bar{C}' = 0 \tag{6.5.6}$$

$$\tilde{G}''' + 3\tilde{G}\tilde{G}'' - 2\tilde{G}'\tilde{G}' = 0 \tag{6.5.7}$$

Expanding $\tilde{f}$, H, $\bar{C}$, and $\tilde{G}$ in a series expansion in ϵ, where ϵ is $(3\text{Pr})^{-1/2}$, the equations at various levels of approximation are obtained. The equations to the zeroth order, $O(\epsilon^0)$, in the inner region with the appropriate boundary conditions are

$$\tilde{f}_0''' + H_0 + \tilde{N}\bar{C}_0 = 0 \tag{6.5.8}$$

$$H_0'' + \tilde{f}_0 H_0' = 0 \tag{6.5.9}$$

$$\bar{C}_0'' + \frac{\text{Sc}}{\text{Pr}}\tilde{f}_0\bar{C}_0' = 0 \tag{6.5.10}$$

$$\tilde{f}_0(0) = \tilde{f}_0'(0) = \tilde{f}_0''(\infty) = H_0(\infty) = \bar{C}_0(\infty) = H_0(0) - 1$$
$$= \bar{C}_0(0) - 1 = 0 \tag{6.5.11}$$

Following a different approach, Mathers et al. (1957) obtained similar equations for the same circumstance. It was argued that for high Pr the inertia terms in the momentum equation may be dropped. That this is a reasonable procedure is seen by casting Eqs. (6.3.3)–(6.3.5) in a different form. Redefining $c(x)$ and $b(x)$ as

$$c(x) = 4\sqrt[4]{\frac{\text{Gr}_x}{4}}\,\text{Pr}^{-3/4} \qquad b(x) = \frac{c(x)}{4x}\,\text{Pr} \tag{6.5.12}$$

and using η, ϕ, and $\tilde{C}$ as before, Eqs (6.3.3)–(6.3.5) for the surface maintained at uniform temperature and concentration levels reduce to

$$f''' + \left(\frac{3ff'' - 2f'^2}{\text{Pr}}\right) + \phi + \tilde{N}\tilde{C} = 0 \tag{6.5.13}$$

$$\phi'' + 3f\phi' = 0 \tag{6.5.14}$$

$$\tilde{C}'' + 3\frac{\text{Sc}}{\text{Pr}}f\tilde{C}' = 0 \tag{6.5.15}$$

The terms in parentheses are the inertia terms. For $\text{Pr} \to \infty$, these may be neglected. The equations then reduce to Eqs. (6.5.8)–(6.5.10) except for the numerical constant 3 in Eqs. (6.5.14)–(6.5.15). However, using

$$\eta = 3^{-1/4}\zeta \qquad f(\eta) = 3^{-3/4}\tilde{f}_0(\zeta) \qquad \phi(\eta) = H_0(\zeta) \qquad \tilde{C}(\eta) = \bar{C}_0(\zeta) \tag{6.5.16}$$

one set of equations transforms into the other. The boundary conditions are the same as in Eq. (6.5.11).

The preceding equations were solved numerically by Mathers et al. (1957) for several values of Sc/Pr and $\tilde{N}$. The results are given below:

Sc/Pr	$\tilde{N}$	$-\phi'(0)$	$-\tilde{C}'(0)$
0.25	1.00	0.941	0.553
0.50	1.00	0.887	0.680
2.00	1.00	0.810	1.068
2.00	0.50	0.768	1.000
—	0	0.707	—

Based on these results, the following correlations are obtained for the transport parameters:

$$\mathrm{Nu}_x = -x\frac{(\partial t/\partial y)_{y=0}}{t_0 - t_\infty} = \left(\frac{\mathrm{Gr}_x\mathrm{Pr}}{4}\right)^{1/4}[-\phi'(0)] \tag{6.5.17a}$$

$$= 0.503(\mathrm{Gr}_x\mathrm{Pr})^{1/4}\left[1 + \left(\frac{\mathrm{Pr}}{\mathrm{Sc}}\right)^{1/2}\tilde{N}\right]^{1/4} \tag{6.5.17b}$$

$$\mathrm{Nu}_{x,C} = -x\frac{(\partial C/\partial y)_{y=0}}{C_0 - C_\infty} = \left(\frac{\mathrm{Gr}_x\mathrm{Pr}}{4}\right)^{1/4}[-\tilde{C}'(0)] \tag{6.5.18a}$$

$$= 0.503(\mathrm{Gr}_x\mathrm{Pr})^{1/4}\left[1 + \tilde{N}\left(\frac{\mathrm{Pr}}{\mathrm{Sc}}\right)^{1/2}\right]^{1/4}\left(\frac{\mathrm{Sc}}{\mathrm{Pr}}\right)^{3/8} \tag{6.5.18b}$$

$$= 0.503(\mathrm{Gr}_{x,C}\mathrm{Sc})^{1/4}\left[1 + \frac{(\mathrm{Sc}/\mathrm{Pr})^{1/2}}{\tilde{N}}\right]^{1/4} \tag{6.5.19}$$

A special case of the same result, with $\mathrm{Gr}_{x,C}/\mathrm{Gr}_x \to 0$, was considered by Lightfoot (1968). Using the mass transfer results of Acrivos (1962, 1966), the expression for $\mathrm{Nu}_{x,C}$ is

$$\mathrm{Nu}_{x,C} = 0.503(\mathrm{Gr}_x\mathrm{Pr})^{1/4}\left(\frac{\mathrm{Sc}}{\mathrm{Pr}}\right)^{1/3} \tag{6.5.20}$$

For $\mathrm{Gr}_{x,C} \to 0$, that is, $\tilde{N} \to 0$, Eq. (6.5.18*b*) reduces to Eq. (6.5.20), except for a slight difference in the exponent of Sc/Pr.

Asymptotic expansions for $Sc \to \infty$, $Pr \to 0$. The perturbation analysis is again carried out with the expansions in Section 3.8 for liquid metals ($\mathrm{Pr} \to 0$). Equations (3.8.31)–(3.8.37), describing the stream function and temperature field in the inner and outer regions, are supplemented by the following expansion for the concentration field in the inner region:

$$\tilde{C} = \tilde{C}_0 + \epsilon\tilde{C}_1 + \epsilon^2\tilde{C}_2 + \cdots \tag{6.5.21}$$

where the perturbation parameter ϵ, as before, is $\text{Pr}^{1/2}$. The concentration field in the outer region is taken to be zero. Using these inner and outer expansions with Eqs. (6.3.14)–(6.3.16) and collecting terms of equal powers of ϵ, equations of successive orders of approximation are obtained. The zeroth-order equations, along with the appropriate boundary conditions, are as follows.

Inner region:

$$f_0''' + 3f_0f_0'' - 2f_0'^2 + (1 + \tilde{N}\tilde{C}) = 0 \tag{6.5.22}$$

$$\tilde{C}_0'' + 3\text{Sc}f_0\tilde{C}_0' = 0 \tag{6.5.23}$$

$$f_0(0) = f_0'(0) = \tilde{C}_0(0) - 1 = \tilde{C}_0(\infty) = 0 \qquad f_0'(\infty) = \frac{1}{\sqrt{2}} \tag{6.5.24}$$

Outer region:

$$\tfrac{3}{2}F_0F_0'' - F_0'^2 + H_0 = 0 \tag{6.5.25}$$

$$H_0'' + F_0H_0' = 0 \tag{6.5.26}$$

$$F_0(0) = F_0'(\infty) = H_0(\infty) = H_0(0) - 1 = 0 \tag{6.5.27}$$

where the primes denote differentiation with respect to ξ defined in Eq. (3.8.35). These equations may be solved numerically for various values of $\tilde{N}$ to obtain the transport parameters of interest.

Other configurations. Saville and Churchill (1970) considered the four cases of Sc and Pr discussed above for flow adjacent to horizontal cylinders and vertical axisymmetric bodies with arbitrary body contours. The solution technique used is that developed by the same authors in 1967 for thermally driven natural convection flows. Briefly, the solutions for the stream function, temperature, and concentration are sought in the form of rapidly converging series that are universal with respect to the body contours within a specified class of body shapes. Using the first terms of the series, found to be sufficiently accurate for horizontal cylinders and vertical axisymmetric bodies, asymptotic expressions were developed for the shear stress and the Nusselt and Sherwood numbers. For equal Prandtl and Schmidt numbers, the effects due to temperature and concentration differences were again found to be simply additive. Therefore, the heat transfer results determined for these bodies in Chapter 5 are also applicable for the combined problem by replacing Gr_x by Gr_x', where $Gr_x' = \text{Gr}_x + \text{Gr}_{x,C}$. For the other two asymptotic conditions, Sc >> Pr and $\text{Sc} \to \infty$, $\text{Pr} \to 0$, the expressions are quite extensive. The reader is referred to Saville and Churchill (1970) for details.

6.6 CORRELATIONS

As discussed in Sections 3.9 and 5.6, there are several experimental and practical applications in which the idealized conditions assumed in obtaining similarity solutions are not fully realized. For example, the flow considered may not be laminar or the actual temperature and/or species conditions on the surface may not be those that admit similarity solutions. Correlations based on experimental data provide valuable information in indicating deviations from the theoretical solutions. A brief summary of the few relevant studies that are known is given here. Again, only those studies are discussed here in which the Sorét and Dufour effects are small and the concentration is low.

For Pr = Sc, the conclusion (Section 6.3) that heat transfer results may be used by simply replacing the thermal Grashof number Gr_x by a combined Grashof number $\mathrm{Gr}'_x = \mathrm{Gr}_x + \mathrm{Gr}_{x,C}$ has also been borne out by experimental results. For example, Bottemanne (1972) measured heat and mass transfer rates for a large-diameter vertical cylinder for Sc = 0.63 and Pr = 0.71. The equivalent measured values of $\phi'(0)$ and $\tilde{C}'(0)$ were in excellent agreement with the calculations. Pertinent to the discussion here, the experimental values of $-\phi'(0)$ or $-\tilde{C}'(0)$ for the simultaneous transfer, using Gr'_x, were very close to the measured value for pure heat transfer or mass transfer, using Gr_x or $\mathrm{Gr}_{x,C}$, respectively. It is therefore reasonable to postulate that for $\mathrm{Pr} \approx \mathrm{Sc}$ the correlations in the previous chapters for heat transfer may also be adopted as relations for simultaneous transport by replacing Gr_x by Gr'_x.

For $\mathrm{Sc} \neq \mathrm{Pr}$ there are no reliable correlations that cover the full ranges of Sc and Pr. The integral analysis results of Somers (1956) and Wilcox (1961) were directed at obtaining relations of the following forms:

$$\mathrm{Nu}_x = A(\mathrm{Gr}_x\mathrm{Pr})^{1/4}\left[1 + \tilde{N}\left(\frac{\mathrm{Pr}}{\mathrm{Sc}}\right)^{1/2}\right]^{1/4} \tag{6.6.1}$$

$$\mathrm{Nu}_{x,C} = A(\mathrm{Gr}_{x,C}\mathrm{Sc})^{1/4}\left[1 + \frac{1}{\tilde{N}}\left(\frac{\mathrm{Sc}}{\mathrm{Pr}}\right)^{1/2}\right]^{1/4} \tag{6.6.2}$$

where the factors $\mathrm{Pr}^{1/4}$ and $\mathrm{Sc}^{1/4}$ are those obtained in natural convection flows adjacent to vertical surfaces and A is a constant independent of Pr, Sc, and $\tilde{N}$. These equations are of the same form as those derived by Mathers et al. (1957) for the asymptotic condition $\mathrm{Sc} >> \mathrm{Pr}$ and Pr large; see Section 6.5. The value of the constant A in Eqs. (6.5.17*b*) and (6.5.19) is 0.503.

The few available experimental studies do not give unequivocal support to the above forms of correlation. Den Bouter et al. (1968) reported an experimental study of simultaneous heat and mass transfer between a vertical copper plate maintained at constant temperature and a copper sulfate-sulfuric acid solution. An electrochemical method was used for the mass transfer. Measurements were made with the two buoyancy effects aiding and opposed. For the aiding effects, the heat and mass transfer parameters agreed well with a single curve of the correlating equations, Eqs. (6.6.1) and (6.6.2), with the constant A taken as 0.503. For opposed effects, however, the disagreement between the data and the curve representing these relations is random and up to 30% in magnitude. For these experiments, with $\mathrm{Pr} \approx 10$ and $\mathrm{Sc} \approx 2000$, perhaps even that level of agreement for a relation for Pr and Sc near unity is surprising.

Adams and McFadden (1966) reported experimental measurements of heat and mass transfer parameters during the sublimation of *p*-dichlorobenzene ($\mathrm{Sc} = 2.23$) from a heated vertical surface into air. The two buoyancy effects were opposed. The data were 10–15% lower than the integral method result of Somers (1956). However, the data did correlate well when the factor $\sqrt{\mathrm{Sc}/\mathrm{Pr}}$ was omitted from those relations.

Mathers et al. (1957) obtained simultaneous heat and mass transfer data with naphthalene ($\mathrm{Sc} = 2.53$) and benzene ($\mathrm{Sc} = 1.79$) diffusing from a sphere into air. The effects were aiding. The data indicated that the dependence of the transport parameters on Pr/Sc and $\tilde{N}$, postulated in Eqs. (6.6.1) and (6.6.2), also applies for spheres. The following correlations were given:

$$\mathrm{Nu}_D = 2 + 0.282\hat{\mathrm{Gr}}_D^{0.37}\mathrm{Pr}^{0.37} \qquad \text{for } \hat{\mathrm{Gr}}_D\mathrm{Pr} < 10^2 \tag{6.6.3}$$

$$= 2 + 0.50\hat{\mathrm{Gr}}_D^{0.25}\mathrm{Pr}^{0.25} \qquad \text{for } 10^2 < \hat{\mathrm{Gr}}_D\mathrm{Pr} < 10^6 \tag{6.6.4}$$

$$\mathrm{Nu}_{D,C} = 2 + 0.282\hat{\mathrm{Gr}}_{D,C}^{0.37}\mathrm{Sc}^{0.37} \qquad \text{for } \hat{\mathrm{Gr}}_{D,C}\mathrm{Sc} < 10^2 \tag{6.6.5}$$

$$= 2 + 0.50\hat{\mathrm{Gr}}_{D,C}^{0.25}\mathrm{Sc}^{0.25} \qquad \text{for } 10^2 < \hat{\mathrm{Gr}}_{D,C}\mathrm{Sc} < 10^6 \tag{6.6.6}$$

where the characteristic length is sphere diameter D, $\hat{\mathrm{Gr}}_D = \mathrm{Gr}_D + \sqrt{\mathrm{Pr}/\mathrm{Sc}}\,\mathrm{Gr}_{D,C}$, and $\hat{\mathrm{Gr}}_{D,C} = \mathrm{Gr}_{D,C} + \sqrt{\mathrm{Sc}/\mathrm{Pr}}\,\mathrm{Gr}_D$.

From these results, it appears that the correlations of the form of Eqs. (6.6.1) and (6.6.2) may be used for aiding effects only. For opposing effects, these correlations are not adequate. There is also an uncertainty concerning the value of the constant A to be used in the correlations. It is also noted that Eqs. (6.6.1) and (6.6.2) indicate a lower limit of $\tilde{N} > -(\mathrm{Sc}/\mathrm{Pr})^{1/2}$. This limit has no known physical significance, and the boundary layer formulation in Section 6.2 does not suggest any such limit. In fact, the values of P and Q in Eqs. (6.3.20), (6.3.21), and (6.3.12) may be chosen as any pair of values that are consistent with physical realism.

All these considerations indicate the need for further and carefully designed experiments to cover a wide range of both Pr and Sc. The objective would be to obtain useful correlations, comparable to those known for thermally driven flows or for simultaneous heat and mass transfer for $\mathrm{Pr} = \mathrm{Sc}$. Opposed buoyancy effects should also be included.

6.7 FURTHER CONSIDERATIONS

Several assumptions were made in obtaining solutions in the previous sections. The normal component of fluid velocity at the surface was taken as zero, even with mass transfer. The fluid properties across the boundary region were assumed constant. The thermal-diffusion (Sorét) and diffusion-thermal (Dufour) effects were neglected. There are important applications in which these mechanisms may, singly or jointly, have large effects. For example, in transpiration cooling, where a cool gas is blown through the porous surface into the mainstream fluid, the blowing velocity v_0 is often appreciable. When the injected gas is very different from the mainstream gas, both the Sorét and Dufour effects may be significant under certain conditions. Finally, since physical properties depend on both temperature and concentration, a large variation in either will invalidate the assumptions of constant properties. This section concerns the effects of appreciable interfacial velocity and of the Sorét and Dufour mechanisms.

6.7.1 Effects of Appreciable Interfacial Velocity

Thermally driven flow in the presence of blowing or suction at the surface was considered in Sections 3.6 and 3.12. The solutions were obtained by assuming that the fluid added or removed is the same as the ambient fluid. Here, a binary system is considered instead. The material added may be different from the mainstream fluid.

For the analysis, a boundary region flow adjacent to a vertical surface is considered. Both the injected species A and the surrounding medium B are assumed to be perfect gases. It is further assumed that the molecular weight of component A is greater than that of component B. For these conditions, the experimental study of Sparrow et al. (1964a) indicates that the Sorét-Dufour effects may be neglected.

Governing equations. These, with the above assumptions, are Eqs. (6.2.19)–(6.2.24), except for modifications of the buoyancy term in the momentum equations and the boundary condition on v. Recall that in deriving the combined buoyancy force in Eq. (6.2.20), it was assumed that density varies linearly with both temperature and concentration. As pointed out in Chapter 2, this is often true for many actual fluids when both $\Delta t = t_0 - t_\infty$ and $\Delta C = C_0 - C_\infty$ are small. However, for large Δt and/or ΔC, the approximation may lead to large errors. A density equation must then be used. In general, such density relations are quite complicated. However, for a binary system of perfect gases, an exact relationship for the buoyancy force $\tilde{B} = g[(\rho_\infty - \rho)/\rho] = B/\rho$ may be developed.

Using the notation in Section 6.1, the mixture density ρ may be expressed as

$$\rho = C_{\mathrm{A}} + C_{\mathrm{B}} = M_{\mathrm{A}}C_{\mathrm{A}}^* + M_{\mathrm{B}}C_{\mathrm{B}}^* \tag{6.7.1}$$

If P, V, T, R, and C_T^* denote the total pressure, volume, absolute temperature, ideal gas constant, and total molar concentration of the mixture, respectively, then

$$C_T^* = C_{\mathrm{A}}^* + C_{\mathrm{B}}^* = \frac{P}{RT} \qquad C_{\mathrm{B}}^* = \frac{P}{RT} - C_{\mathrm{A}}^* \tag{6.7.2}$$

$$\begin{aligned} \rho &= C_{\mathrm{A}}^*(M_{\mathrm{A}} - M_{\mathrm{B}}) + \frac{P}{RT}M_{\mathrm{B}} \\ &= \frac{PM_{\mathrm{B}}}{RT}\left[1 + \frac{C_{\mathrm{A}}^*RT}{P}\left(\frac{M_{\mathrm{A}}}{M_{\mathrm{B}}} - 1\right)\right] \\ &= \frac{PM_{\mathrm{B}}}{RT}\left[1 + \frac{P_{\mathrm{A}}}{P}\left(\frac{M_{\mathrm{A}}}{M_{\mathrm{B}}} - 1\right)\right] \end{aligned} \tag{6.7.3}$$

Similarly,

$$\rho_\infty = \frac{P_\infty M_B}{RT_\infty}\left[1 + \frac{P_{A,\infty}}{P_\infty}\left(\frac{M_A}{M_B} - 1\right)\right] \tag{6.7.4}$$

For constant static pressure across the boundary layer,

$$\tilde{B} = g\left(\frac{\rho_\infty}{\rho} - 1\right) = g\left(\frac{1 + (P_{A,\infty}/P_\infty)[(M_A/M_B) - 1]}{\{1 + (P_A/P)[(M_A/M_B) - 1]\}T_\infty/T} - 1\right) \tag{6.7.5}$$

To express this in terms of mass fraction m_A, note that

$$m_A = \frac{m_A^* M_A}{m_A^* M_A + m_B^* M_B} \tag{6.7.6}$$

where m_A^* and m_B^* are the molar mass fractions of components A and B, respectively. Since $m_A^* + m_B^* = 1$ and $m_A^* = P_A/P = P_A/P_\infty$, this equation may be rewritten as

$$m_A = \frac{1}{1 + (M_B/M_A)[(P_\infty/P_A) - 1]} \tag{6.7.7}$$

Using this expression and assuming that the concentration of species A in the free stream is zero, that is, $P_{A,\infty} \rightarrow 0$, Eq. (6.7.5), as shown by Adams and Lowell (1968), reduces to

$$\tilde{B} = \frac{g(\rho_\infty - \rho)}{\rho} = g\left[\frac{1}{(T_\infty/T) + (1 - M_B/M_A)m_A} - 1\right] \tag{6.7.8}$$

When $M_A = M_B$ or $m_A = 0$ the body force terms reduce to the usual form $g\beta(T - T_\infty)$, where for a perfect gas $\beta = 1/T_\infty$.

Thus, the governing equations for the flow considered, neglecting the viscous dissipation, pressure, and source terms, are

$$\frac{\partial u}{\partial x} + \frac{\partial v}{\partial y} = 0 \tag{6.7.9}$$

$$u\frac{\partial u}{\partial x} + v\frac{\partial u}{\partial y} = \nu\frac{\partial^2 u}{\partial y^2} + g\left[\frac{1}{T_\infty/T + [1 - (M_B/M_A)]m_A} - 1\right] \tag{6.7.10}$$

$$u\frac{\partial t}{\partial x} + v\frac{\partial t}{\partial y} = \alpha\frac{\partial^2 t}{\partial y^2} \tag{6.7.11}$$

$$u\frac{\partial m_A}{\partial x} + v\frac{\partial m_A}{\partial y} = D\frac{\partial^2 m_A}{\partial y^2} \tag{6.7.12}$$

For convenience, the species conservation equation is written in terms of mass fraction rather than concentration.

Boundary conditions. To prescribe these, it is necessary to express the nonzero component of velocity v_0 in terms of m_A. From Eqs. (6.1.9) and (6.1.7),

$$w_A = m''_A + C_A v$$

$$= m''_A + C_A(m_A v_A + m_B v_B)$$

or

$$w_A = m''_A + m_A(w_A + w_B)$$

The surface is assumed to be impermeable to species B. That is, the mass flux of component B at the surface, w_B, is zero. This, for example, is the case for water evaporating from the surface of water into an adjoining airstream. Water vapor arises at the surface but essentially no air passes through it. The above equation then reduces to

$$w_A = \frac{m''_A}{1 - m_A} \tag{6.7.13}$$

Substituting this into Eq. (6.1.9) and noting that $m_A = C_A/\rho$, we get

$$\rho v = \frac{m''_A}{1 - m_A} \tag{6.7.14}$$

Using $m''_A = -D\,\partial C_A/\partial y)_{y=0}$ and noting that $v \to v_0$, $m_A \to m_{A,0}$, as $y \to 0$, the following value of v_0 is obtained:

$$v_0 = \frac{-(D/\rho)\,\partial C_A/\partial y)_{y=0}}{1 - m_{A,0}} = \frac{-D\,\partial m_A/\partial y)_{y=0}}{1 - m_{A,0}} \tag{6.7.15}$$

The complete boundary conditions, therefore, are

$$\text{at } y = 0: \quad u = 0 \quad v = \frac{-D\,\partial m_A/\partial y)_{y=0}}{1 - m_{A,0}} \quad T = T_0 \quad m_A = m_{A,0} \tag{6.7.16a}$$

$$\text{as } y \to \infty: \quad u \to 0 \quad m_A \to 0 \quad T \to T_\infty \tag{6.7.16b}$$

Similarity considerations. The question is whether Eqs. (6.7.9)–(6.7.12), subject to boundary conditions given by Eqs. (6.7.16*a*) and (6.7.16*b*), admit similarity solutions. In terms of the generalized similarity formulation used in Section 6.3, these equations reduce to

$$f''' + \frac{c_x}{b}ff'' - \left(\frac{c_x}{b} + \frac{cb_x}{b^2}\right)f'^2 + \frac{g}{\nu^2 cb^3}\left[\frac{1}{1/(1 + \beta d\phi) + [1 - (M_B/M_A)](\tilde{C}/\rho)e} - 1\right] = 0 \tag{6.7.17}$$

$$\frac{\phi''}{\mathrm{Pr}} + \frac{c_x}{b}f\phi' - \frac{c}{b}\frac{d_x}{d}f'\phi = 0 \tag{6.7.18}$$

$$\frac{\tilde{C}''}{\text{Sc}} + \frac{c_x}{b} f\tilde{C}' - \frac{c}{b}\frac{e_x}{e} f'\tilde{C} = 0 \tag{6.7.19}$$

where $\tilde{C} = (C_A - C_{A,\infty})/(C_{A,0} - C_{A,\infty}) = C_A/C_{A,0} = m_A/m_{A,0}$, since $C_{A,\infty}$ is assumed to be zero. It is clear from Eq. (6.7.17) that for similarity to exist, both $d(x)$ and $e(x)$ must be constant. That is, the surface must be isothermal and maintained at a uniform concentration. There must be no ambient stratification. For these conditions, the values of c and b that result in similarity are

$$c = 4\left(\frac{\text{Gr}_x}{4}\right)^{1/4} = G \qquad \text{and} \qquad b = \frac{G}{4x}$$

where

$$\text{Gr}_x = \frac{gx^3}{\nu^2}\left[\frac{1}{T_\infty/T_0 + [(M_A - M_B)/M_A]m_{A,0}} - 1\right] = \frac{x^3}{\nu^2}\tilde{B}_0 \tag{6.7.20}$$

where $\tilde{B}_0$ is the buoyancy force defined in Eq. (6.7.8), with m_A and T evaluated at the surface conditions.

The resulting ordinary differential equations are

$$f''' + 3ff'' - 2f'^2 + \frac{\tilde{B}}{\tilde{B}_0} = 0 \tag{6.7.21}$$

$$\phi'' + 3\text{Pr}f\phi' = 0 \tag{6.7.22}$$

$$\tilde{C}'' + 3\text{Sc}f\tilde{C}' = 0 \tag{6.7.23}$$

The transformed boundary conditions are

$$f'(0) = \phi(0) - 1 = \tilde{C}(0) - 1 = 0 \qquad f(0) = \frac{\tilde{C}'(0)}{3\text{Sc}}\frac{m_{A,0}}{1 - m_{A,0}} \tag{6.7.24}$$

$$f'(\infty) = \tilde{C}(\infty) = \phi(\infty) = 0 \tag{6.7.25}$$

The boundary condition on $f(0)$ above is derived from Eq. (6.7.15) and $v_0 = -\psi_x)_{y=0}$.

Equations (6.7.21)–(6.7.25) are the same as those given by Lowell and Adams (1967) and Adams and Lowell (1968) and similar to those analyzed by Gill et al. (1965). In the first study, a group theory method of analysis was used to show that similarity is permissible only for uniform surface temperature and concentration. In the second, the similarity variables were those used here and the equations above were solved numerically for transport parameters of interest, with different values of T_0/T_∞ and $P_\infty/P_{A,0}$, for three different organic compounds subliming from the vertical surface into the ambient air. The compounds chosen were p-dichlorobenzene ($p - C_6H_4Cl_2$, Sc = 2.23), naphthalene ($C_{10}H_8$, Sc = 2.57), and camphor ($C_{10}H_{16}O$, Sc = 2.60). The molecular weights of these compounds are much larger than that of air for which the Sorét and Dufour effects are small. The equations above, therefore, are applicable.

Results. The effect on the Nusselt or Sherwood number of changing $P_\infty/P_{A,0}$ was found to be different for $T_0/T_\infty < 1$ and $T_0/T_\infty > 1$. Note that varying

$P_\infty/P_{A,0}$ amounts to varying mass fraction at the surface, $m_{A,0}$. From Eq. (6.7.7), a decrease in $P_\infty/P_{A,0}$ results in an increase in $m_{A,0}$ and vice versa. For $T_0/T_\infty < 1$, the results indicated an increase in Nusselt number with increase in $m_{A,0}$. When the surface temperature is lower than the ambient, the natural convection flow induced by the temperature difference is downward. Sublimation of a heavier compound into the boundary layer tends to make the mixture heavier than air, which then generates additional downward flow. The net result is an increase in heat transfer. This behavior can easily be seen with the aid of Eq. (6.7.8). In the absence of component A, the buoyancy force at the surface $\tilde{B}_0$ is the conventional term $g(T_0 - T_\infty)/T_\infty$. For $M_B/M_A < 1$, the effect of injection of gas A into the boundary layer is to increase the buoyancy force and hence the heat transfer when $T_0 < T_\infty$.

For $T_0/T_\infty \gtrsim 1$, the thermal and subliming vapor concentration-induced buoyancy forces oppose each other. The net effect is to decrease the heat transfer. However, for the two effects nearly equal, no numerical solutions were obtained. The region of no solution was small for large $P_\infty/P_{A,0}$ values. It became larger as this ratio decreased. For example, for p-$C_6H_4Cl_2$, no solution was obtained for $P_\infty/P_{A,0} = 100$ and $0.998 < T_0/T_\infty < 1.05$. However, for $P_\infty/P_{A,0} = 10$, the range was $0.998 < T_0/T_\infty < 1.52$.

This behavior is to be expected. For $T_0 > T_\infty$ the flow induced by the temperature difference is upward and $\rho/\rho_\infty < 1$. The effect of the heavier vapor subliming into the boundary layer is that $\rho/\rho_\infty > 1$ near the surface. This induces a downflow that, combined with upward flow outside, is essentially unstable. The laminar boundary region formulation above may then not be applicable. For larger values of $P_\infty/P_{A,0}$ or smaller $m_{A,0}$, this situation occurs for small temperature differences. As the temperature difference is increased, the temperature-induced buoyancy overrides the concentration-induced buoyancy force and upward flow is generated. Similar results are observed at smaller $P_\infty/P_{A,0}$ or larger $m_{A,0}$, except that the upward flow is established at much larger temperature differences.

Processes involving mass transfer toward the surface have also been investigated. Oxidation of surface material or condensation through a noncondensing gas are some examples of such processes. Acrivos (1960) considered isothermal diffusion in the laminar boundary region flow of a binary mixture past a body of arbitrary geometry. Asymptotic results were given for the case of large interfacial velocity directed toward the surface. Both constant and variable physical properties were considered. Manganaro and Hanna (1970) extended the analysis to simultaneous heat and mass transfer and presented expressions for energy and mass fluxes.

6.7.2 Dufour and Sorét Effects

In the similarity analysis presented earlier in this section, it was assumed that the injected species is heavier than the ambient medium. This allowed the Sorét (thermal-diffusion) and Dufour (diffusion-thermal) effects to be neglected. However, when the species injected at the surface is lighter than the surrounding gas, these effects may be significant and must often be included in evaluating mass and heat fluxes. Following Hayday et al. (1962), the diffusive mass flow for component A in a binary mixture under the influence of both concentration and temperature gradients is given by

$$m''_A = -\rho D\left[\frac{\partial m_A}{\partial y} + k_T \frac{m_A(1 - m_A)}{T}\frac{\partial T}{\partial y}\right] \tag{6.7.26}$$

where k_T is a thermal diffusion ratio. Equivalent forms may be found in Baron's (1962) analysis based on the development by Chapman and Cowling (1952) as well as in other standard references such as Hirschfelder et al. (1954) and Grew and Ibbs (1952). The equation for thermal flux in a binary mixture, including the Dufour effect in addition to the heat conduction, is

$$q'' = -k\frac{\partial T}{\partial y} + k_T RT \frac{M^2}{M_A M_B} m''_A \tag{6.7.27}$$

where M is the molecular weight of the mixture. For the derivation of this equation, see Baron (1962).

One of the earlier studies dealing with such mechanisms in buoyancy-induced flows is that due to Tewfik and Yang (1963). In this experiment, helium was injected through a porous horizontal cylinder into ambient air. Data were obtained for a range of injection rates at various levels of surface temperature. The data indicated that the wall heat flux did not become zero for $T_0 = T_\infty$. For the adiabatic condition to be achieved, the wall temperature was found to be higher than T_∞ by as much as 31.7°C, depending on the injection rate. Similar results had been obtained earlier, for example, by Tewfik et al. (1962), in investigations of transpiration-cooled forced boundary layers in the helium-air binary system. Based on this similarity, it was concluded that diffusion-thermal or Dufour effects were also responsible for the experimental results in natural convection flows. Later, Sparrow et al. (1964b) analyzed these effects for flow in the neighborhood of the lower stagnation point of a horizontal cylinder for a helium-air system.

Governing equations. In terms of Fig. 5.4.2*a*, these are

$$\frac{\partial}{\partial x}(\rho u) + \frac{\partial}{\partial y}(\rho v) = 0 \tag{6.7.28}$$

$$\rho\left(u\frac{\partial u}{\partial x} + v\frac{\partial u}{\partial y}\right) = g(\rho_\infty - \rho)\frac{x}{R} + \frac{\partial}{\partial y}\left(\mu\frac{\partial u}{\partial y}\right) \tag{6.7.29}$$

$$\rho c_p v\frac{\partial T}{\partial y} + (c_{p,A} - c_{p,B})m''_A\frac{\partial T}{\partial y} = -\frac{\partial q''}{\partial y} \tag{6.7.30}$$

$$\rho v\frac{\partial m_A}{\partial y} = -\frac{\partial m''_A}{\partial y} \tag{6.7.31}$$

where m''_A and q'' are given by Eqs. (6.7.26) and (6.7.27). The energy equation contains two modifications due to the binary nature of the flow. These are an additional term representing convection due to diffusion on the left side and the diffusion-thermal term in q'' on the right side. Here $c_{p,\mathrm{A}}$ and $c_{p,\mathrm{B}}$ are the specific heats at constant pressure for A and B, respectively. The appropriate boundary conditions are

$$\text{at } y=0: \qquad u=0 \qquad v=v_0 \text{ (specified blowing velocity)}$$

$$m_\mathrm{A}=m_{\mathrm{A},0} \qquad T=T_0 \tag{6.7.32}$$

$$\text{as } y\to\infty: \qquad u=0=m_{\mathrm{A},\infty}=T-T_\infty \tag{6.7.33}$$

where the condition on $m_{\mathrm{A},\infty}$ follows from assuming that the ambient medium is pure air. These equations were then converted into ordinary differential equations by the following substitutions:

$$\eta=\sqrt[4]{\frac{g}{Rv_\infty^2}}\int_0^y \frac{1}{\Lambda_\mu}\,dy \qquad \psi=\sqrt[4]{\frac{g}{R}\,v_\infty^2}\,xf(\eta) \tag{6.7.34}$$

$$\phi(\eta)=\frac{T}{T_\infty} \qquad m_\mathrm{A}=m_\mathrm{A}(\eta) \tag{6.7.35}$$

where Λ denotes the ratio of a property locally to its value in the free stream. Thus $\Lambda_\mu=\mu/\mu_\infty$, $\Lambda_\rho=\rho/\rho_\infty$, and $\Lambda_{c_p}=c_p/c_{p,\infty}$.

Defining the stream function ψ by

$$u=\frac{1}{\Lambda_\rho}\psi_y=\sqrt{\frac{g}{R}\frac{x}{\Lambda_\rho\Lambda_\mu}}f' \qquad v=-\frac{1}{\Lambda_\rho}\psi_x=\frac{1}{\Lambda_\rho}\sqrt[4]{\frac{gv_\infty^2}{R}}f \tag{6.7.36}$$

and using Eqs. (6.7.34) and (6.7.35), the transformed equations are

$$\left(\frac{\Lambda_k}{\Lambda_\mu}\phi'\right)'+\mathrm{Pr}_\infty\left(\Lambda_{c_p}f+c_{p,\mathrm{AB}}\frac{m_\mathrm{A}'}{\mathrm{Sc}}\right)\phi'=-\mathrm{Pr}_\infty\left\{\frac{R\phi k_T M^2}{c_{p,\infty}\mathrm{Sc}M_\mathrm{A}M_\mathrm{B}}\left[m_\mathrm{A}'\right.\right.$$

$$\left.\left.+k_T m_\mathrm{A}(1-m_\mathrm{A})\frac{\phi'}{\phi}\right]\right\}'-\mathrm{Pr}_\infty c_{p,\mathrm{AB}}k_T m_\mathrm{A}(1-m_\mathrm{A})\frac{\phi'^2}{\phi\mathrm{Sc}} \tag{6.7.37}$$

$$\left(\frac{f'}{\Lambda_\mu\Lambda_\rho}\right)''+f\left(\frac{f'}{\Lambda_\mu\Lambda_\rho}\right)'=\frac{f'^2}{\Lambda_\mu\Lambda_\rho}-\Lambda_\mu(1-\Lambda_\rho) \tag{6.7.38}$$

$$\left(\frac{m_\mathrm{A}'}{\mathrm{Sc}}\right)'+f\,m_\mathrm{A}'=-\left[k_T m_\mathrm{A}(1-m_\mathrm{A})\frac{\phi'}{\phi\mathrm{Sc}}\right]' \tag{6.7.39}$$

where $c_{p,\mathrm{AB}}=(c_{p,\mathrm{A}}-c_{p,\mathrm{B}})/c_{p,\infty}$. The transformed boundary conditions are

$$f'(0)=0 \qquad f(0)=\frac{m_\mathrm{A}'(0)}{[1-m_\mathrm{A}(0)]\mathrm{Sc}}+\frac{k_T m_\mathrm{A}(0)\phi'(0)/\phi(0)}{\mathrm{Sc}}$$

$$m_\mathrm{A}(0)=m_{\mathrm{A},0} \qquad \phi(0)=\phi_0 \tag{6.7.40}$$

$$f'(\infty) = 0 = m_{A,\infty} = \phi(\infty) \tag{6.7.41}$$

The boundary condition on $f(0)$ is obtained, as earlier, by using Eqs. (6.7.14), (6.7.26), and (6.7.36).

Results. These equations were integrated numerically for a range of blowing parameter $f(0)$ and T_0/T_∞, where the blowing parameter $f(0)$, from Eq. (6.7.36), is defined as

$$f(0) = -\frac{\rho_0}{\rho_\infty}\sqrt[4]{\frac{R}{g\nu_\infty^2}}\,v_0 \tag{6.7.42}$$

The results for q''/q_0'' as a function of $f(0)$ are shown in Fig. 6.7.1 where q'' is the heat flux at the wall with blowing and q_0'' the heat flux without it. The two sets of curves are for $T_0/T_\infty = 1.1$ and 3. At each temperature ratio, results are presented for the equations solved without Dufour-Sorét effects, with the Dufour effect only, and with both Dufour and Sorét effects.

First consider the results for $T_0/T_\infty = 1.1$. Two conclusions are readily drawn. Of the Dufour and Sorét effects, the former is dominant and accounts for most

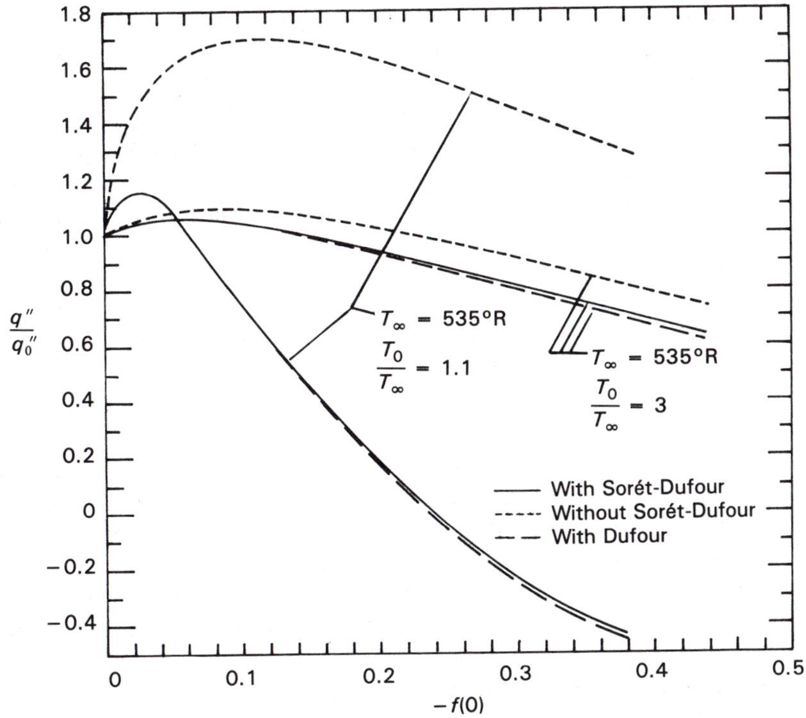

Figure 6.7.1 Heat transfer results for a buoyancy force around a horizontal cylinder with Sorét-Dufour effects. *(Reprinted with permission from E. M. Sparrow et al.,* J. Heat Transfer, *vol. 86C, p. 508. Copyright © 1964, ASME.)*

of the total diffusive effect. Second, the effect on the heat transfer is appreciable and increases with $f(0)$.

This behavior may be understood with the aid of Eq. (6.7.27). For $T_0 > T_\infty$ the conductive contribution to q'' is positive. Whether the terms due to the Dufour effect aid or oppose the conductive terms depends on the sign of product $k_T m''_A$. For gases lighter than air, such as helium, k_T is negative. For a gas injected at the surface, the concentration of the transpiring gas is highest at the surface. Then $\partial m_A/\partial y$ is negative. From Eq. (6.7.26), m''_A is then generally positive. The terms due to the Dufour effect thus contribute negatively to the surface heat flux. With an increase in $f(0)$, both m''_A and the contribution of the Dufour effect increase. At some value of $f(0)$, the conductive and diffusive contributions due to the Dufour effect cancel each other and the heat flux at the surface becomes zero. With further increase in blowing parameter, the Dufour effect completely dominates and heat is transferred from the fluid to the wall even when $T_0 > T_\infty$.

For the second set of curves, for $T_0/T_\infty = 3$, the trends observed are similar but the effects are much smaller. The conductive component of the heat transfer in Eq. (6.7.27) is large, and the impact of the Dufour effect is relatively small.

The wall temperature at which the net heat flux at the wall vanishes is denoted as adiabatic temperature, T_a. For example, in Fig. 6.7.1 this condition is achieved for $T_0/T_\infty = 1.1$ at $f(0) = -0.24$, that is, $T_a = 1.1T_\infty$. For $T_0/T_\infty = 3$, much higher blowing rates are required. In other words, T_a increases with an increase in injection rate. This trend can again be explained with the aid of Eq. (6.7.27). To satisfy the condition that q'' at the wall be zero, the increase in m''_A that accompanies a larger blowing rate must be balanced by an increase in $|\partial T/\partial y|$. A change in ambient temperature also affects T_a through a change in the thermodynamic and transport properties of the free stream, but only slightly. The adiabatic temperature decreases with increase in T_∞.

The conventional heat transfer relation $q'' = h(T_0 - T_\infty)$ is inadequate when diffusive effects are appreciable. The reversed heat flow for $T_0 > T_\infty$ would then require a negative value of h. However, the use of an adiabatic wall temperature instead of T_∞ avoids this difficulty:

$$q'' = h^*(T_0 - T_a) \tag{6.7.43}$$

Tewfik and Yang (1963) and Sparrow et al. (1964b) used this relation successfully to correlate their heat transfer results. The concept of adiabatic wall temperature was found earlier to be successful in applications involving aerodynamic heating; see, for example, Kays (1966) and Gebhart (1971).

The curves without the Sorét and Dufour effects in Fig. 6.7.1 also illustrate the effect of mass injection on the heat transfer rate when the injected species is lighter than the ambient medium. First, consider again the results for $T_0/T_\infty = 1.1$. For moderate mass blowing velocities, mass injection increases the heat transfer substantially. This is due to the increased buoyancy that the flow experiences due to injection of helium; see Eq. (6.7.8). For $M_B/M_A > 1$, the thermal buoyancy and the transpiration-induced buoyancy aid each other. The result is an increase in heat transfer. As the blowing rate increases, the thickening of the boundary layer that accompanies blowing becomes significant and diminishes the heat transfer rate. For $T_0/T_\infty = 3$, similar behavior is observed but again at a much reduced level. This follows from the relative ineffectiveness of the transpiration-induced buoyancy compared to the large thermal buoyancy that exists when the temperature difference is large.

Sparrow et al. (1964b) also gave solutions for $T_0/T_\infty < 1$ (not shown in Fig. 6.7.1). Then the transpiration and the temperature-induced buoyancy forces oppose each other; see Eq. (6.7.8). For moderate blowing rates, the two forces tend to cancel. As discussed earlier, for injection of heavier gases into air but for $T_0/T_\infty > 1$, this leads to unstable flow. Near the surface $\rho/\rho_\infty > 1$, but it is less than one in the outer part of the boundary region. No numerical solutions could be determined under these conditions. At larger blowing rates, however, the transpiration-induced buoyancy overrides the temperature-induced buoyancy and $\rho/\rho_\infty < 1$ across the entire boundary layer. The flow is upward everywhere and the solutions correspond to those for $T_0/T_\infty > 1$. The heat transfer first increases with increasing blowing rate but later decreases because of the increased boundary layer thickness. The Dufour and Sorét effects were again found to be quite significant.

Some of the analytical findings discussed above were verified experimentally by Sparrow et al. (1965). Hydrogen, helium, carbon dioxide, and Freon 12 were injected through a porous horizontal cylinder into an otherwise quiescent air environment. For the injected gases lighter than air (hydrogen and helium) the diffusion thermal effects were large and the adiabatic temperature exceeded the ambient temperature. For the injected gases heavier than air (carbon dioxide and Freon 12) the diffusion thermal effects were smaller and the adiabatic wall temperature was lower than the ambient.

Zeh and Gill (1967) extended the analysis of Sparrow et al. (1964b) to a vertical surface. The helium-air and hydrogen-air systems were considered, including the effects of variable fluid properties as well as Dufour and Sorét effects. Results very similar to those discussed above were obtained. Again, Sorét effects were found to have a significant effect on heat transfer, and the heat transfer results for these situations could be well correlated by using a driving force based on $T_0 - T_a$.

6.8 TRANSPORT IN FLAMES AND COMBUSTION

An important area involving combined thermal and mass transport is that of flames and combustion. Extensive work has been done on buoyancy-driven flows caused by or in the presence of combustion processes because of their importance in several practical problems, such as those related to furnaces, fires, chemical reactors, and engines. Much of the available information is presented in various books on flames and combustion, such as those by Williams (1965), Fristrom and Westenberg (1965), Afgan and Beer (1974), and Glassman (1977). Because of concern about fire safety and the need for efficient energy utilization in systems that involve combustion, considerable analytical and experimental efforts have been directed toward understanding the transport mechanisms that govern the buoyancy-induced flow in flames, fires, and related combustion processes. Since the literature is extensive, this section outlines some of the important considerations related to these flows and presents some typical results. Further information can be obtained from the books mentioned above and the Proceedings of the International Symposia on Combustion, published by the Combustion Institute, Pittsburgh, Pennsylvania.

Several studies have considered the natural convection flow arising adjacent to a flat vertical surface due to a fire burning along the surface. Such a fire may be divided into two regions. In the pyrolysis region, the wall material is gasified and partially burned in the gas phase adjacent to the surface. In the overfire region, the combustion process has been completed and the thermal plume generated by the fire rises freely or along a surface, if a noncombusting wall exists downstream. Its buoyancy decays downstream due to entrainment of ambient fluid and heat loss to the surface. Figure 6.8.1 shows a schematic of a steady, two-dimensional, laminar diffusion flame on a vertical pyrolyzing fuel slab. As the name suggests, the chemical reaction in a diffusion flame is governed by the diffusion of oxygen to the fuel.

In most studies of the laminar pyrolysis region of a wall fire, such as those by Kosdon et al. (1969) and Kim et al. (1971), a one-step reaction in an infinitely

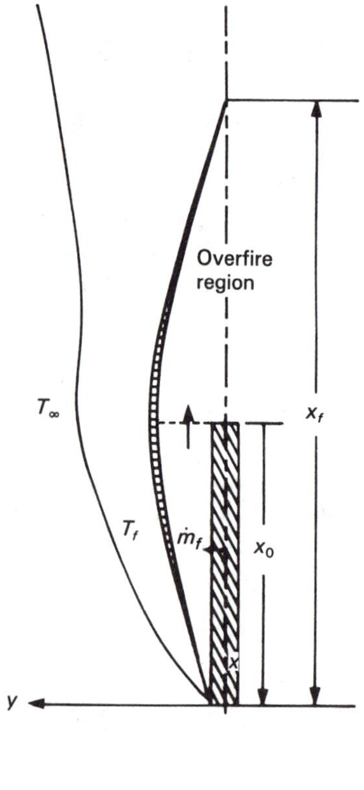

Figure 6.8.1 Sketch of a steady two-dimensional laminar diffusion flame on a vertical burning fuel slab of finite height. Also indicated is the overfire region.

thin diffusion flame sheet, with no oxygen present between the wall and the flame and no fuel present outside the flame, is assumed. The boundary between the pyrolysis and overfire regions is not precisely defined in natural fires, because the rate of pyrolysis varies continuously along the wall. However, pyrolysis rates vary strongly with temperature and a fairly short distance separates the two regions. We will consider first the buoyancy-driven flow in a diffusion flame and then the overfire region.

6.8.1 Laminar Flow in a Diffusion Flame

Consider a two-dimensional, steady-state, laminar, boundary layer flame adjacent to a pyrolyzing fuel slab, as shown in Fig. 6.8.2. The combustion zone is assumed to be thin, so the reactants do not coexist and are separated by the flame sheet. The governing boundary layer equations for this flow may be written as

$$\frac{\partial(\rho u)}{\partial x} + \frac{\partial(\rho v)}{\partial y} = 0 \tag{6.8.1}$$

$$\rho u \frac{\partial u}{\partial x} + \rho v \frac{\partial u}{\partial y} = \frac{\partial}{\partial y}\left(\mu \frac{\partial u}{\partial y}\right) + g(\rho_\infty - \rho) \tag{6.8.2}$$

$$\rho u \frac{\partial h}{\partial x} + \rho v \frac{\partial h}{\partial y} = \frac{\partial}{\partial y}\left(\frac{k}{c_p}\frac{\partial h}{\partial y}\right) + q''' - \frac{dq_r''}{dy} \tag{6.8.3}$$

$$\rho u \frac{\partial m_i}{\partial x} + \rho v \frac{\partial m_i}{\partial y} = \frac{\partial}{\partial y}\left(\rho D \frac{\partial m_i}{\partial y}\right) + m_i''' \tag{6.8.4}$$

where h, the specific enthalpy, and c_p, the specific heat of the gas mixture, may be defined as

$$h = c_p T \qquad \text{and} \qquad c_p = \sum_i m_i (c_p)_i \tag{6.8.5}$$

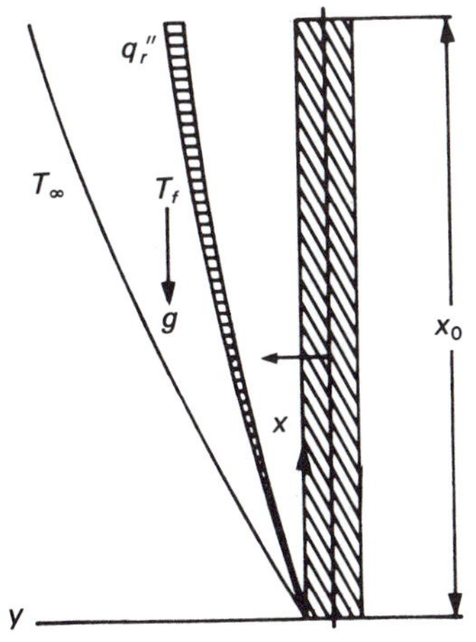

Figure 6.8.2 Coordinate system for a two-dimensional, steady-state, laminar boundary layer flame adjacent to a pyrolyzing vertical surface.

Also, q''' and m''' are the volumetric combustion heat and mass generation rates, respectively, q_r'' is the thermal radiative flux in the y direction, T the absolute temperature, and m_i the fractional mass concentration of species i. For boundary layer flow along a surface inclined at angle θ with the vertical, g is replaced by $g \cos \theta$ in the momentum equation, Eq. (6.8.2), as done by Kim et al. (1971).

In addition to these equations, an equation of state is needed for the fluid, along with a chemical reaction equation. For an ideal gas and no change in the average molecular weight,

$$\rho T = \rho_\infty T_\infty \tag{6.8.6}$$

The chemical reaction may be represented by

$$n_F' \text{ (fuel)} + n_o' \text{ (oxidant)} \rightarrow n_p'' \text{ (products)} + Q \text{ (heat of combustion)} \tag{6.8.7}$$

where the n's are the stoichiometric coefficients, single primes represent the reacting components and double primes the products. The subscripts F, o, and p denote fuel, oxygen, and the combustion products, respectively. In the combustion of gaseous methane, the gas mixture is composed of the fuel, the oxidizer, and the products of combustion. As discussed by Liu et al. (1981), the gas species methane, oxygen, water vapor, carbon dioxide, and nitrogen are present. Nitrogen does not participate in the reaction but is involved in the heat transfer and energy storage processes. The specific chemical reaction equation is

$$\begin{aligned} n_F'(CH_4) + n_o'(O_2) + n_p'(H_2O + CO_2) &\rightarrow n_F''(CH_4) \\ &+ n_o''(O_2) + n_p''(H_2O + CO_2) + Q \end{aligned} \tag{6.8.8}$$

From stoichiometry, the molecular coefficients in Eq. (6.8.8) are

$$\begin{gathered} n_F' = 1 \qquad n_o' = 2 \qquad n_p' = 0 \\ n_F'' = 0 \qquad n_o'' = 0 \qquad n_p'' = n_c'' + n_H'' = 1 + 2 = 3 \end{gathered} \tag{6.8.9}$$

where n_c'' and n_H'' represent the coefficients for CO_2 and H_2O, respectively. The species mass production rate is related to the energy generation by

$$\frac{m_F'''}{M_F(n_F'' - n_F')} = \frac{m_o'''}{M_o(n_o'' - n_o')} = \frac{m_p'''}{M_p(n_p'' - n_p')} = \frac{q'''}{Q} \tag{6.8.10}$$

where the M's denote the corresponding molecular weights.

The boundary conditions for Eqs. (6.8.1)–(6.8.4) are now considered. The distant ambient conditions yield

$$\text{as } y \rightarrow \infty: \qquad u = h = m_F = 0 \qquad m_o = m_{o,\infty} \tag{6.8.11}$$

where $m_{o,\infty}$ is the mass fraction of oxygen in the ambient medium. At the wall, the velocity component u parallel to the surface is zero and the diffusion flame approximation takes the oxygen concentration at the wall as zero. Therefore,

$$\text{at } y = 0: \qquad u = m_o = 0 \tag{6.8.12}$$

Pyrolysis at the wall provides the remaining conditions, as discussed by Kosdon et al. (1969) and Kim et al. (1971). These are given as

$$\text{at } y = 0: \qquad m'' = \rho v = \frac{k}{L_v c_p} \frac{\partial h}{\partial y}$$

$$h = h_w$$

$$\rho v(m_{FT} - m_F) = -\frac{k}{c_p} \frac{\partial m_F}{\partial y} \tag{6.8.13}$$

where m'' is the wall mass flux, L_v the effective heat of vaporization of the fuel, h_w the enthalpy at the wall, and m_{FT} the fuel mass fraction in the gas, due to the transfer of fuel into the gas phase. The heat flux to the wall is given by

$$\text{at } y = 0: \qquad q'' = \frac{k}{c_p} \frac{\partial h}{\partial y} \tag{6.8.14}$$

In addition to the above boundary conditions, the radiative heat transfer flux q_r'' is needed in the energy equation, Eq. (6.8.3). Various models for gas radiation have been used to obtain this term. Some of these considerations are discussed in greater detail in Section 17.6. Generally, one-dimensional gas radiation flux is assumed and $q_r''(y)$ is obtained by simplifying assumptions. Negrelli et al. (1977), Liu and Shih (1980), and Liu et al. (1981, 1982) discuss the determination of radiative transport based on the gray gas radiation model, the exponential wide-band gas radiation model, and other models.

The process of laminar natural convective burning of vertical fuel surfaces, as outlined above, is complicated by the presence of combined thermal and material transport, chemical reaction, and radiative heat transfer. It was first considered by Spalding (1954). After making several simplifying assumptions, a similarity solution similar to that of Pohlhausen (1921) was given for the boundary layer flow on a flat surface. Kosdon et al. (1969) and Kim et al. (1971) considered the processes in greater detail, leading to the formulation discussed above. A wide range of governing parameters was considered and numerical results were given for various fuels. Kim et al. (1971) also considered an approximate integral solution, using the procedure developed by Pohlhausen (1921). These and several other more recent studies, mentioned earlier, simplify the governing equations by eliminating the mass and energy generation terms through the use of the Schvab-Zeldovich variables, as discussed in detail by Zeldovich (1951) and Williams (1965). These variables are defined by Liu et al. (1981, 1982) as

$$W_i = \frac{m_i Q}{R_\infty M_i(n_i'' - n_i')} - \frac{(m_0 - m_{0,\infty})Q}{h_\infty M_0(n_0'' - n_0')} \tag{6.8.15}$$

and

$$\phi = \frac{h}{h_\infty} - \frac{(m_o - m_{o,\infty})Q}{h_\infty M_o(n_o'' - n_o')} \tag{6.8.16}$$

Similar transformations have been used by other investigators. The transformed equations are solved numerically to obtain the velocity, temperature, and concentration distributions.

A similarity analysis may be carried out for a semi-infinite vertical or inclined surface. Following the work of Kim et al. (1971), the similarity variable η and stream function $f(\eta)$ are defined as

$$\eta = \frac{\mathrm{Gr}_x^{1/4}}{x} \int_0^y \frac{\rho}{\rho_\infty}\, dy \tag{6.8.17}$$

$$\psi = 4\nu_\infty \mathrm{Gr}_x^{1/4} f(\eta) \tag{6.8.18}$$

where

$$\mathrm{Gr}_x = L_v g x^3 \cos\theta / 4 c_p T_\infty \nu_\infty^2 \tag{6.8.19}$$

The momentum equation is obtained as

$$f''' + 3ff'' - 2(f')^2 = -B + \frac{h_w}{L_v} - B\phi \qquad \text{for } \eta \le \eta_f$$

$$= -\left(B + \frac{h_w}{L_v}\right)\frac{\phi}{\phi_f} - B\phi \qquad \text{for } \eta > \eta_f \tag{6.8.20}$$

where η_f is the flame position. B, the mass transfer driving potential, is given by

$$B = \frac{m_{o,\infty} Q}{M_o n_o' L_v} - \frac{h_w}{L_v} \tag{6.8.21}$$

and ϕ is the normalized Schvab-Zeldovich variable, given by

$$\phi = -\frac{1}{B}\left(\frac{h}{L_v} + \frac{m_o - m_{o,\infty}}{M_o n_o' L_v}\right) \tag{6.8.22}$$

Here the subscript w refers to the wall and ϕ_f is the value of the variable at the flame, whose position is given by η_f. For Lewis number equal to 1.0, the energy and species equations are the same, as follows:

$$\phi'' + 3\mathrm{Pr} f \phi' = 0 \tag{6.8.23}$$

The corresponding boundary conditions are

$$\phi(0) = 1 \qquad \phi(\infty) = f'(0) = f'(\infty) = 0$$

and

$$f(0) = \frac{B}{3\mathrm{Pr}}\, \phi'(0) \tag{6.8.24}$$

These two ordinary differential equations were solved numerically by Kim et al. (1971). Figure 6.8.3 shows the computed temperature and velocity profiles for burning methanol. The maximum temperature occurs in the flame. In similarity flows the temperature peak then coincides with the position of chemical energy release. The maximum velocity is also seen to occur inside the flame. The position of the flame is indicated by $m_F = m_o = 0$ and may be determined from the concentration profiles shown in Fig. 6.8.4.

The mass transfer rate is indicated by the parameter $f(0)$. Kim et al. (1971) studied the variation of $f(0)$ with potential B for various important fuels. The streamlines for the flow were also obtained, as shown in Fig. 6.8.5, in terms of the physical coordinates. A demarcation, termed the converging line, separates the flow originating at the surface from that entrained from the ambient fluid. Therefore, convection aids the outward diffusion of fuel vapor between the wall and the converging line and opposes it beyond the converging line. The parameter B was found to be the dominant chemical parameter in determining burning rates. Good agreement between the theoretical and experimental results was obtained for lower molecular weight fuels. However, Lewis number effects were found to arise for higher molecular weight fuels.

Liu et al. (1981) carried out an analytical and experimental study of a methane diffusion flame in the region adjacent to a vertical flat plate burner of finite height,

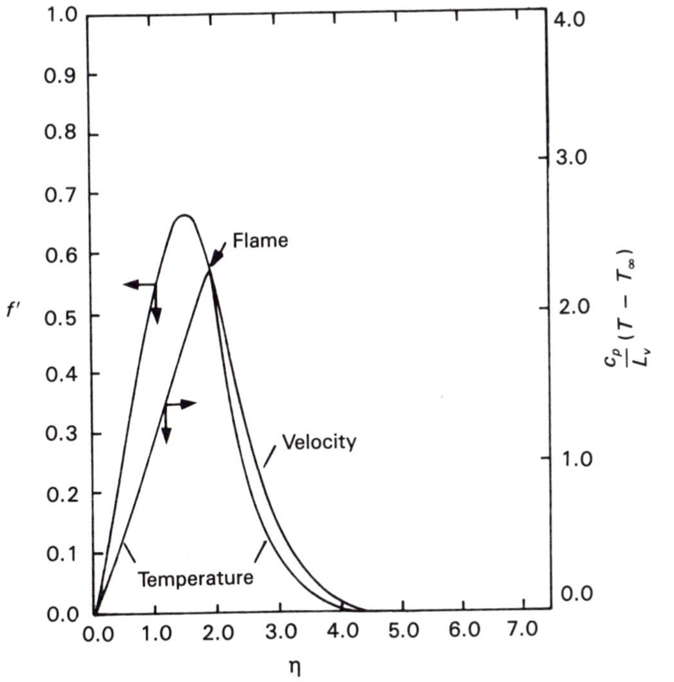

Figure 6.8.3 Temperature and velocity profiles for methanol burning along a vertical surface. *(Reprinted with permission from J. S. Kim et al.* 13th Symp. (Int.) Combust. *Copyright © 1971, Combustion Institute.)*

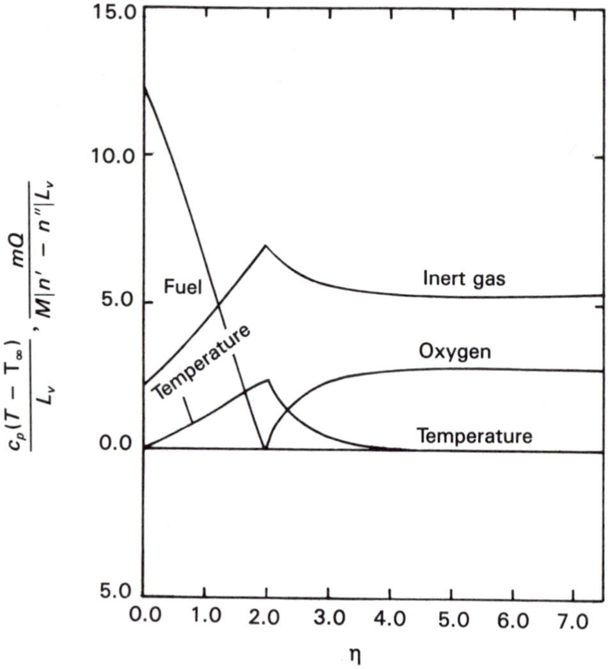

Figure 6.8.4 Concentration and temperature profiles for methanol burning along a vertical surface. *(Reprinted with permission from J. S. Kim et al.* 13th Symp. (Int.) Combust. *Copyright © 1971, Combustion Institute.)*

as shown in Fig. 6.8.6*a*. The local nonsimilarity method was applied in order to include the additional effect of thermal radiation. Gas species, laminar velocity, and temperature profiles inside the diffusion flame boundary layer were computed. The thermal radiation effect was found to reduce the peak flame temperature and to increase the boundary region thickness. Variable-property effects were also found to be important and the need to include them in the analysis was indicated.

Liu et al. (1982) made finite-difference calculations of the elliptic equations to include the non-boundary layer effects arising mainly from flow below the leading edge of the porous burner plate. Figure 6.8.6*b* shows the three computational regions considered. Local nonsimilarity results were used for region II. It was found that air preheating below the porous region does affect the thermal field, although this effect is largely limited to the lower half of that region.

The calculations are in better agreement with experimental data than the earlier local nonsimilarity results, as seen in Fig. 6.8.7. In the calculations, the measured temperature profile at $x = 0$ was used as the horizontal free boundary condition for region I. The leading edge of the burner was 1 cm above this location. Sibulkin et al. (1981, 1982) also studied this case, relaxing several of the earlier assumptions.

6.8.2 Transport in the Overfire Region

This flow has been studied by Pagni and Shih (1977), Groff and Faeth (1978), and Ahmad and Faeth (1978a). Upstream conditions for the overfire region follow from the pyrolysis zone. The flow in the overfire region is not similar. Ahmad and Faeth (1978a) used the numerical procedure developed by Hayday et al. (1967) (see Section 3.12). They considered a smooth flat wall at an angle θ with the vertical. A lower region of length x_0 undergoes pyrolysis. The flame stands away from the wall in this region. When fuel consumption is complete, the flame returns

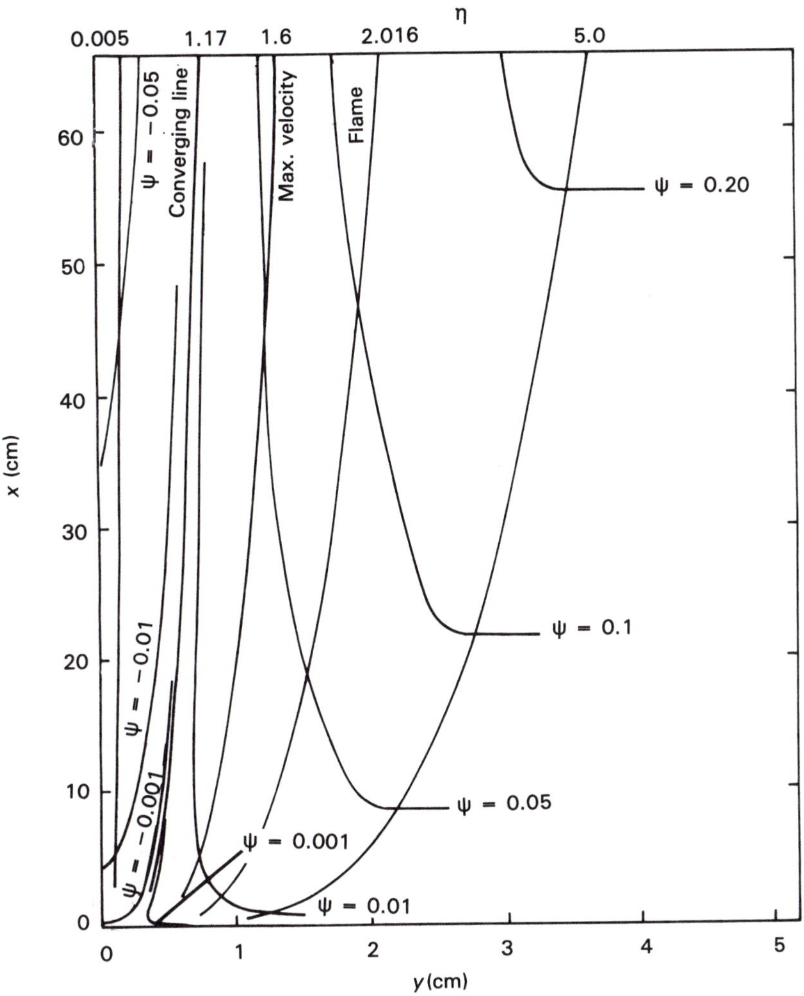

Figure 6.8.5 Streamline field for burning methanol. *(Reprinted with permission from J. S. Kim et al.* 13th Symp. (Int.) Combust. *Copyright © 1971, Combustion Institute.)*

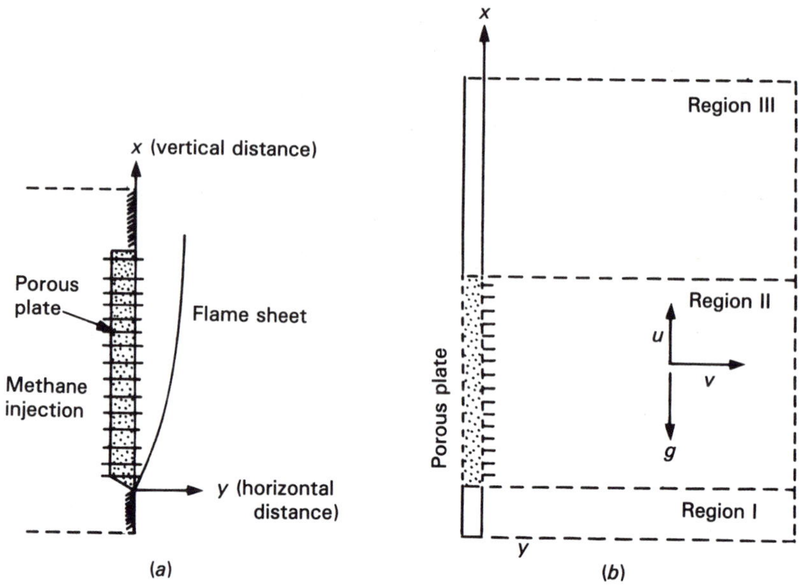

Figure 6.8.6 (*a*) Schematic drawing of methane flame adjacent to a vertical porous surface. *(Reprinted with permission from* Int. J. Heat Mass Transfer, *vol. 24, p. 1959, V. K. Liu et al. Copyright © 1981, Pergamon Journals Ltd.)* (*b*) Computational regions for the elliptic field calculation of a laminar diffusion flame. *(Reprinted with permission from* Int. J. Heat Mass Transfer, *vol. 25, p. 863, V. K. Liu, et al. Copyright © 1982, Pergamon Journals Ltd.)*

to the wall at $x_f > x_0$. The thermal plume with combustion extends from x_0 to x_f. Thereafter, the plume is without combustion. Figures 6.8.8–6.8.10 show the velocity, temperature, and concentration profiles, respectively, within a methanol-fueled wall fire in air, over the pyrolysis, combusting plume, and thermal plume regions. The position of the flame is given by $m_F = m_o = 0$ in Fig. 6.8.10. The flame moves toward the wall downstream, reaching it at $x/x_0 = 6.5$. This ends the combustion region. Above the combustion region, oxygen diffuses back to the wall. In the nonsimilar overfire region, the maximum temperature occurs outside the flame position. Beyond the flame tip the temperature profile continues to spread, similar to the velocity profile.

6.8.3 Turbulent Flames and Other Combustion Processes

Most studies of turbulent flames have considered flames along vertical or inclined plane surfaces and axisymmetric flames, all in stagnant environments. Much of this work is experimental, concentrating on measurements of burning rates, mean velocities, and temperatures. Examples are the work of deRis and Orloff (1974), Ahmad and Faeth (1978b), and Taminini and Ahmad (1979). Analysis by these investigators, using integral models, gave fairly good predictions of the burning rate and the wall heat flux in the plume region. The $k - \epsilon - g$ turbulent flow

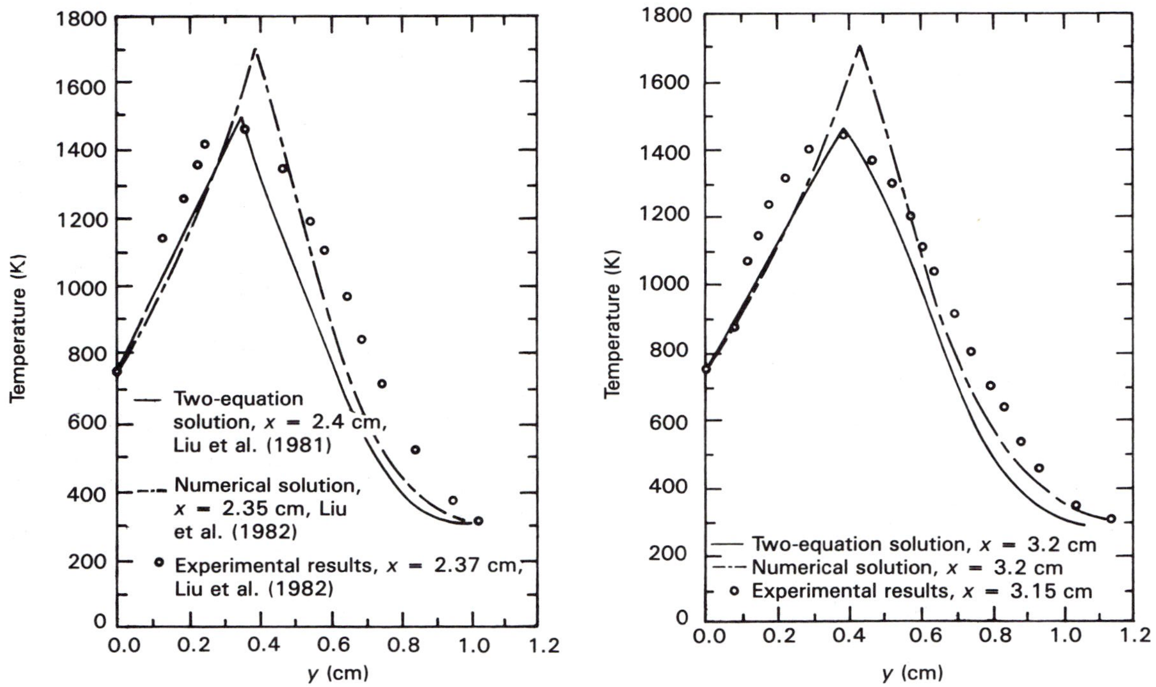

Figure 6.8.7 Comparison of calculated temperature profiles with experimental data for a methane diffusion flame. *(Reprinted with permission from* Int. J. Heat Mass Transfer, *vol. 25, p. 863, V. K. Liu et al. Copyright © 1982, Pergamon Journals Ltd.)*

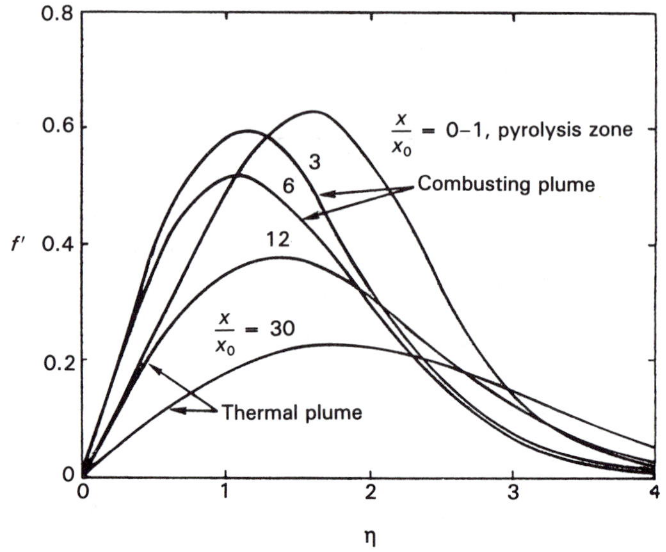

Figure 6.8.8 Velocity profiles in a methanol-fueled wall fire burning in air. *(Reprinted with permission from T. Ahmad and G. M. Faeth,* J. Heat Transfer, *vol. 100, p. 112. Copyright © 1978, ASME.)*

model (Chapter 11) has been used by Kennedy and Plumb (1976) and Taminini (1978). Some success was demonstrated by the former study in predicting flame structure and burning rates. However, uncertainties remain in the estimation of both the turbulence properties and thermal radiation.

Axisymmetric turbulent diffusion flames have been studied experimentally by Cox and Chitty (1980), You and Faeth (1982), and Becker and Liang (1982). Integral, mixing length, and $k - \epsilon - g$ turbulent flow models have also been used. See, for example, the papers mentioned above, along with those by Taminini (1977), Fishburne and Pergament (1978), and Jeng et al. (1982). Other geometries considered include horizontal plane disks, facing upward and downward, and cylinders. See the work of Orloff and deRis (1971), Tsuji and Yamaoka (1971), Negrelli et al. (1977), and Mao and Fernandez-Pello (1980).

Several other studies of combustion processes have considered the effect of thermal buoyancy. Fire spread over liquid fuels at subflash temperatures is known to be controlled mainly by flows induced in the liquid. These flows are driven by surface tension and buoyancy forces and have been analyzed by Sirignano and Glassman (1970), Sharma and Sirignano (1971), Torrance (1971), and Torrance and Mahajan (1974, 1975). Other studies of combustion processes include the work on flame spread and propagation by Hirano et al. (1974), Fernandez-Pello and Santoro (1978), Ray et al. (1980), and Arpaci and Tabaczynski (1982) and that on fires in buildings, as reviewed by Emmons (1978, 1980).

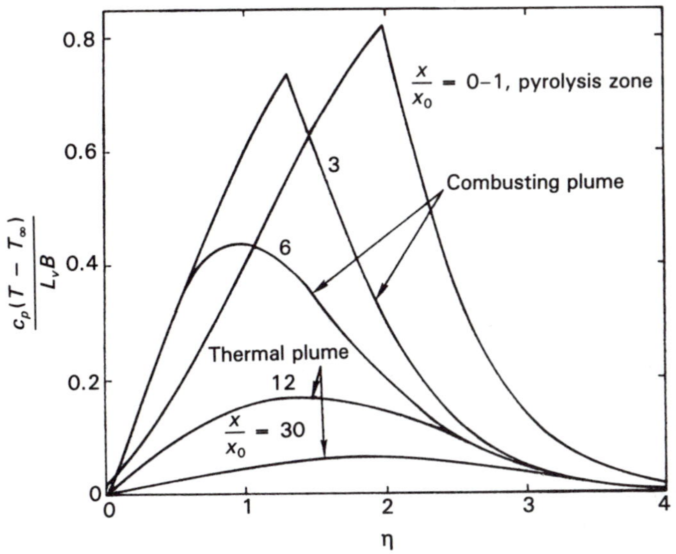

Figure 6.8.9 Temperature profiles in a methanol-fueled wall fire burning in air. *(Reprinted with permission from T. Ahmad and G. M. Faeth,* J. Heat Transfer, *vol. 100, p. 112. Copyright © 1978, ASME.)*

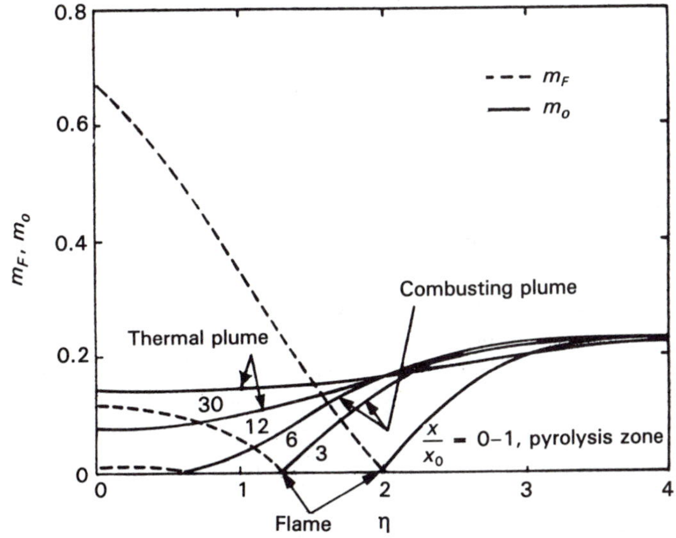

Figure 6.8.10 Concentration profiles in a methanol-fueled wall fire burning in air. *(Reprinted with permission from T. Ahmad and G. M. Faeth,* J. Heat Transfer, *vol. 100, p. 112. Copyright © 1978, ASME.)*

6.9 AMBIENT MEDIUM DENSITY STRATIFICATION

Stratification of the ambient medium that is due solely to a temperature variation, thermal stratification, was considered in Chapters 3 and 4. It was shown that a given stratification $d\rho_\infty/dx$ is stable if $d\rho_\infty/dx < (\partial\rho_\infty/\partial x)_S < 0$; see Eq. (3.6.7). For most fluids this condition is met if $dt_\infty/dx > 0$. Values of $dt_\infty/dx = j_x$, which permitted similarity, were considered. Similarity considerations require that $j(x) = t_\infty - t_r$ vary as $d(x) = t_0 - t_\infty$.

Stratification of the medium may also be caused by vertical concentration gradients. This occurs in many natural processes, such as transport processes in the sea. Stratification due to a concentration difference also arises in many chemical processes and systems. In a concentration-stratified medium, considerations similar to those for thermal stratification apply. The stratification $d\rho_\infty/dx$ is stable if $d\rho_\infty/dx < (\partial\rho_\infty/\partial x)_S < 0$. Again, for most fluids this condition would generally be met if $dC_\infty/dx < 0$ for more dense constituents.

For a purely mass diffusion process, recall that for small concentration differences, the governing equations are identical to those for pure thermally driven flow. The similarity requirements in terms of $e(x) = C_0 - C_\infty$ and $r(x) = C_\infty - C_r$ are identical to those for the thermal diffusion process in terms of $d(x)$ and $j(x)$. Similarly, all the solutions obtained in Chapters 3 and 4 for heat transfer in the presence of thermal stratification apply for mass diffusion, with the appropriate substitutions.

6.9.1 Convective Effects Due to Concentration Stratification

An interesting picture of the kind of mechanisms that arise in vertical buoyant flows in a stably stratified quiescent ambient medium was developed by Tenner and Gebhart (1971). The flow characteristics of upward low-momentum laminar buoyant axisymmetric jets of fresh and salt water in a linearly stratified saltwater environment were studied. Flow visualizations indicated several interesting features; see Fig. 6.9.1. Even in a stably stratified ambient, the buoyant jet induces the flow of a toroidal cell around itself, region 3 in Fig. 6.9.1*b*. This cell is drawn upward by the viscous shearing of the jet. As this more saline fluid moves up, the whole flow eventually becomes heavier than the surrounding stably stratified fluid. At some location the negative buoyancy exceeds the upward shear force, and the fluid turns and flows down to complete the cellular motion. Under these conditions, the descending flow forms a shroud around the jet. For such a shroud to form, stable stratification and low molecular diffusivities were found to provide favorable conditions.

Several studies dealing with concentration-driven buoyant flows in the presence of concentration stratification of the ambient medium are related to turbulent jets and are treated in Chapter 12. However, it is noted that vigorous and turbulent buoyant axisymmetric jets are not expected to produce a shroud.

The effects of ambient medium density stratification on flows that arise from

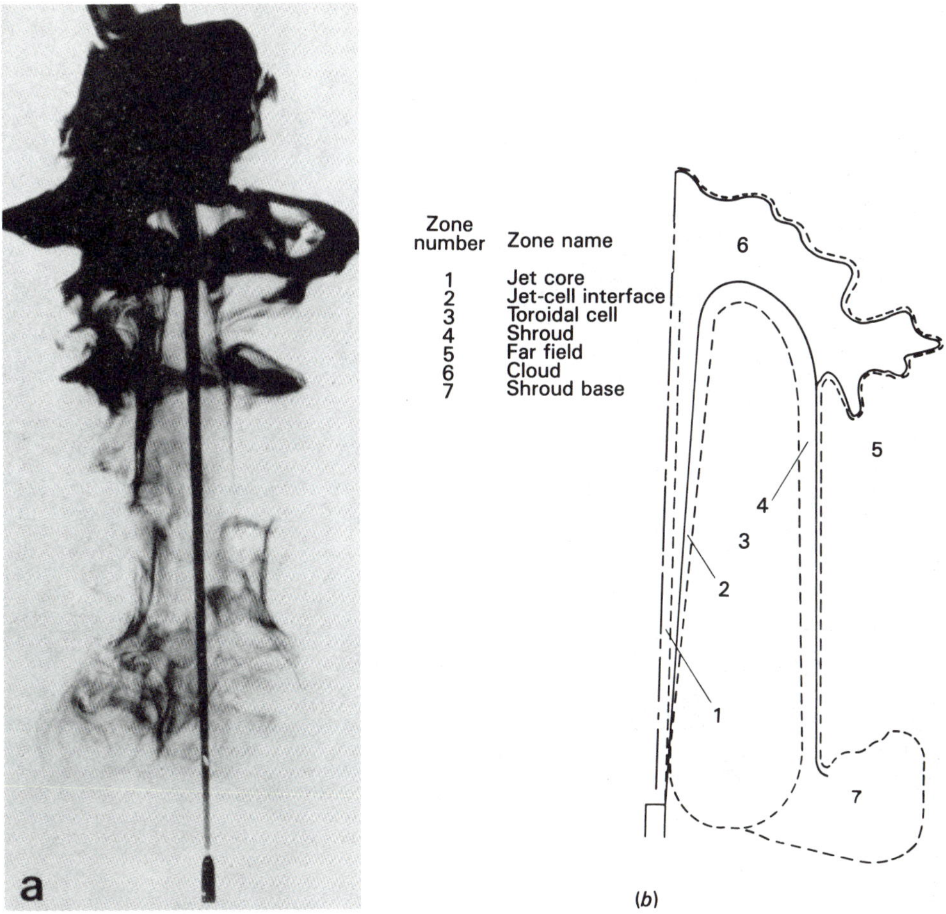

Figure 6.9.1 Laminar axisymmetric jet in a steady stratified environment. The flow in (*a*) was observed with Nigrosin dye. The various flow zones are shown in (*b*). *(Reprinted with permission from* Int. J. Heat Mass Transfer, *vol. 14, p. 2051, A. R. Tenner and B. Gebhart. Copyright © 1971, Pergamon Journals Ltd.)*

the combined buoyancy due to thermal and chemical species diffusion have also been investigated. Such flows arise, for example, when a heat-dissipating body is immersed in stably stratified salt water or when a saltwater buoyant jet is discharged in a thermally stratified pure water environment. The first kind of flow has been experimentally studied by Hubbell and Gebhart (1974). A horizontal copper cylinder was electrically heated to shed an upflow in a linearly and stably stratified saltwater solution in a tank. The stratification was achieved by filling the tank, layer by layer, with successively more concentrated saltwater solutions, introduced at the bottom. The measured concentration stratification after some time was found to be approximately linear. Several test runs were made for different values of the stratification parameter $\tilde{S}$, defined as follows:

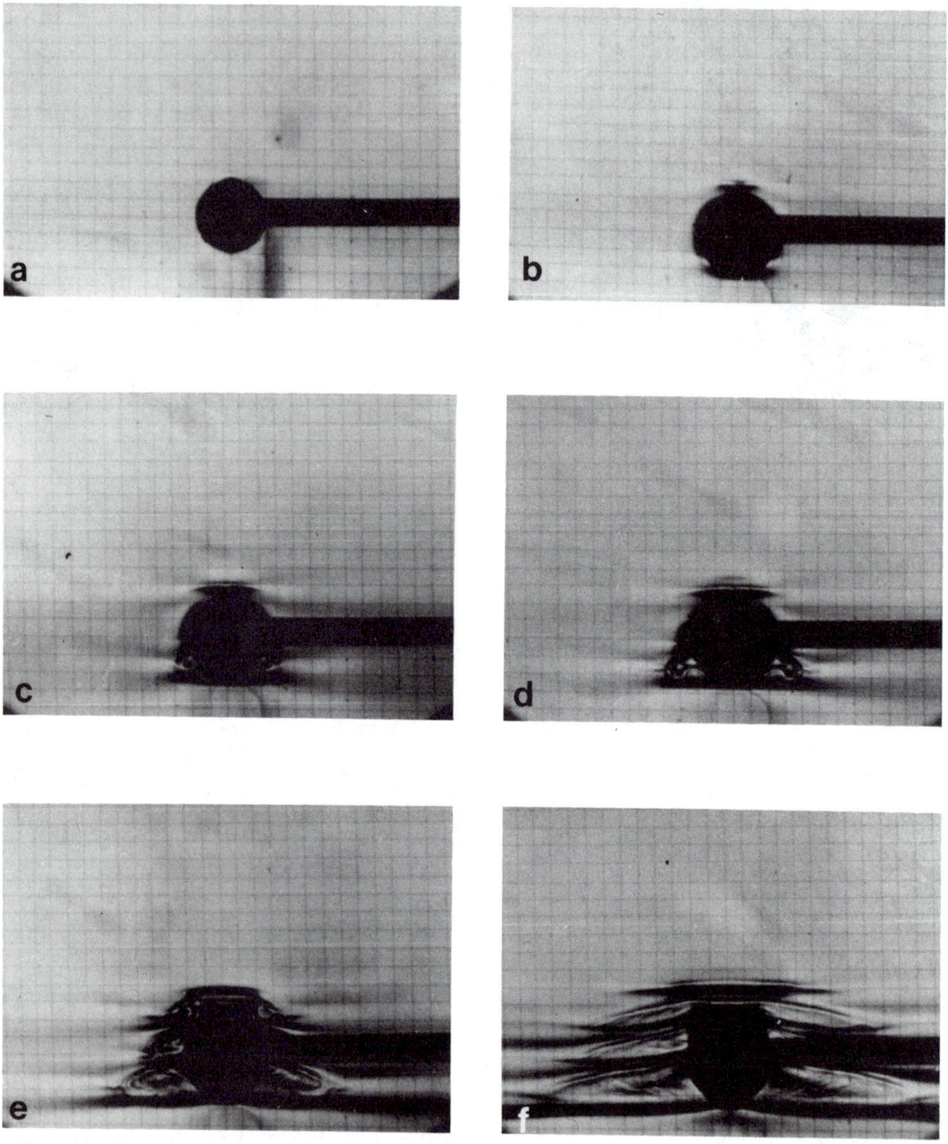

Figure 6.9.2 Growth of convection cells for a heated horizontal cylinder submerged in salt-stratified water. $\tilde{S} = 0.6$, and time given is in minutes: (*a*) 0; (*b*) 1.3; (*c*) 3; (*d*) 4; (*e*) 6, (*f*) 8. *(Reprinted with permission from R. H. Hubbell and B. Gebhart,* Proc. 24th Heat Transfer Fluid Mech. Inst. *Copyright © 1974, Stanford Univ. Press.)*

$$\tilde{S} = \frac{\partial \rho_t}{\partial \rho_C} = \frac{\rho_\infty \beta \, \Delta t}{-\partial \rho/\partial t)_C D} \tag{6.9.1}$$

where D is the cylinder diameter and $\tilde{S} = \infty$ corresponds to zero stratification. The numerator is the thermal effect $\Delta\rho$ and the denominator is the stratification density difference in vertical distance D. Note that $\tilde{S}$ is the reciprocal of the parameter $\tilde{N}$ defined in Eq. (6.3.18).

For salt in water the molecular diffusivity of salt, D, is much smaller than the molecular diffusivity of heat, α. The Lewis number D/α is about 100. This great difference between thermal and salt diffusion results in some decoupling of the thermal and salt transport and in trapping of the thermal discharge in a cell around and above the cylinder. This is seen in Figs. 6.9.2–6.9.4, which are Schli-

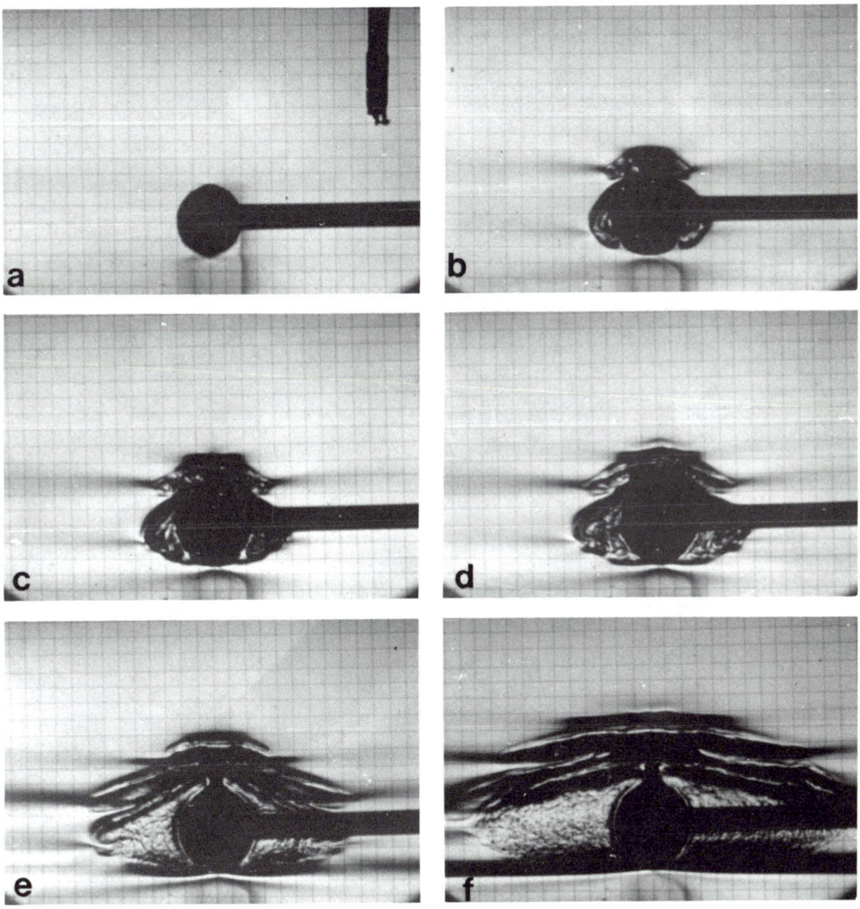

Figure 6.9.3 Growth of convection cells for a heated horizontal cylinder submerged in salt-stratified water. $\tilde{S} = 1.4$, and time given is in minutes: (*a*) 0; (*b*) 1.5; (*c*) 2; (*d*) 3; (*e*) 5; (*f*) 9. *(Reprinted with permission from R. H. Hubbell and B. Gebhart,* Proc. 24th Heat Transfer Fluid Mech. Inst. *Copyright © 1974, Stanford Univ. Press.)*

eren photographs of the developing transport region, through time, taken for the representative values $\tilde{S} = 0.6$, 1.4, and 2.2. The vertical extent of the discharge region is seen to depend strongly on $\tilde{S}$. It is larger for larger $\tilde{S}$, since then the relatively larger thermal buoyancy may push the fluid higher. This may also be seen in Fig. 6.9.5, where the cell volume is plotted as a function of time for the

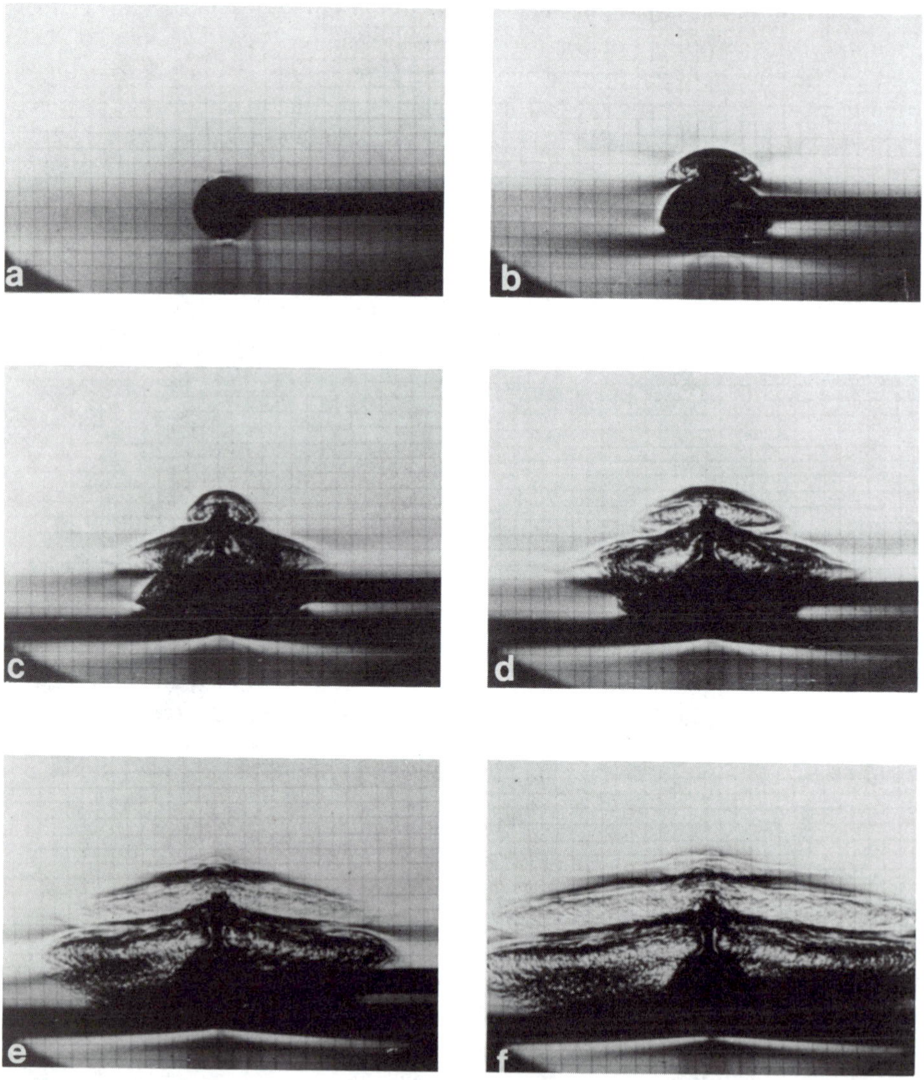

Figure 6.9.4 Growth of convection cells for a heated horizontal cylinder submerged in salt-stratified water. $\tilde{S} = 2.2$, and time given is in minutes: (*a*) 0; (*b*) 2; (*c*) 3; (*d*) 4; (*e*) 6; (*f*) 12. *(Reprinted with permission from R. H. Hubbell and B. Gebhart,* Proc. 24th Heat Transfer Fluid Mech. Inst. *Copyright © 1974, Stanford Univ. Press.)*

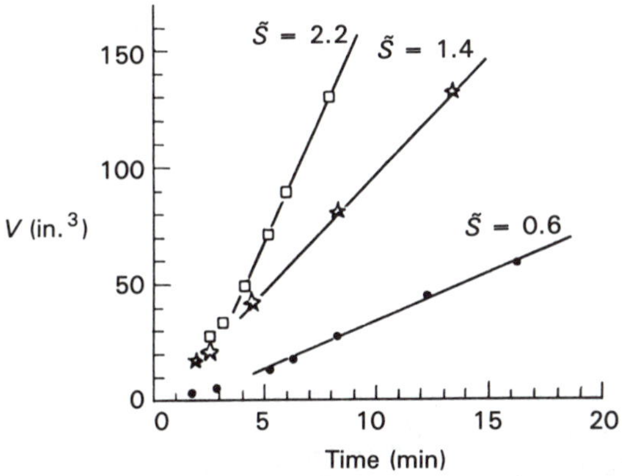

Figure 6.9.5 Measured cell volume as a function of time. (□) $\tilde{S} = 2.2$; (☆) $\tilde{S} = 1.4$; (●) $\tilde{S} = 0.6$. *(Reprinted with permission from R. H. Hubbell and B. Gebhart,* Proc. 24th Heat Transfer Fluid Mech. Inst. *Copyright © 1974, Stanford Univ. Press.)*

chosen values of $\tilde{S}$. At shorter times the cell volume is quite small and is larger for larger $\tilde{S}$. This concentration of the thermal energy in a relatively high-temperature small volume around the cylinder leads to decreased heat transfer compared to that in an unstratified medium. With time, the cell volume increases and the heat transfer increases, but not to the level reached without ambient stratification. There is always a heat transfer penalty to be paid as a result of the stratification trapping of the thermal wake. It is larger at smaller $\tilde{S}$.

The above characteristics may be expected in any concentration stratification of water, since the Lewis number is always much greater than 1.0. Stratification trapping of the thermal discharge would also be expected in a flowing stratified medium. Then the cell would be shed off downstream of the heated surface and would affect transfer characteristics only by its return. These results show the need to exercise caution in using results for an unstratified medium to predict transport in a concentration-stratified water environment. Neglecting stratification trapping may result in an overoptimistic prediction of transport rates.

6.9.2 Combined Stratification Effects

The different diffusion characteristics of salt and thermal energy discussed above result in some interesting convective flows in water with a combined vertical temperature gradient (dt/dx) and salinity gradient (dC/dx), where x upward is positive. Recall that $(\partial\rho/\partial s)_{t,p} > 0$, where s is the salinity. Depending on the sign of the two gradients, the following distinct cases are identified (see Turner and Stommel, 1964):

Case	dt/dx	dC/dx	$d\rho/dx$
1	+	−	−
2a	+	+	−
2b	+	+	+
3a	−	−	−
3b	−	−	+
4	−	+	+

Cases 2b, 3b, and 4 are gravitationally unstable and correspond to the thermal instability mechanism discussed in Chapter 13. Case 1 is gravitationally stable. The density decreases upward and no convection can occur. Cases 2a and 3a are also nominally gravitationally stable. However, because of the different diffusion characteristics of salt and thermal energy, it is possible to have convection, as shown by Stern (1960) and Turner and Stommel (1964). Recall that $Le = Sc/Pr \approx 100$ for saline water.

First consider case 2a. This occurs, for example, when cool fresh water is

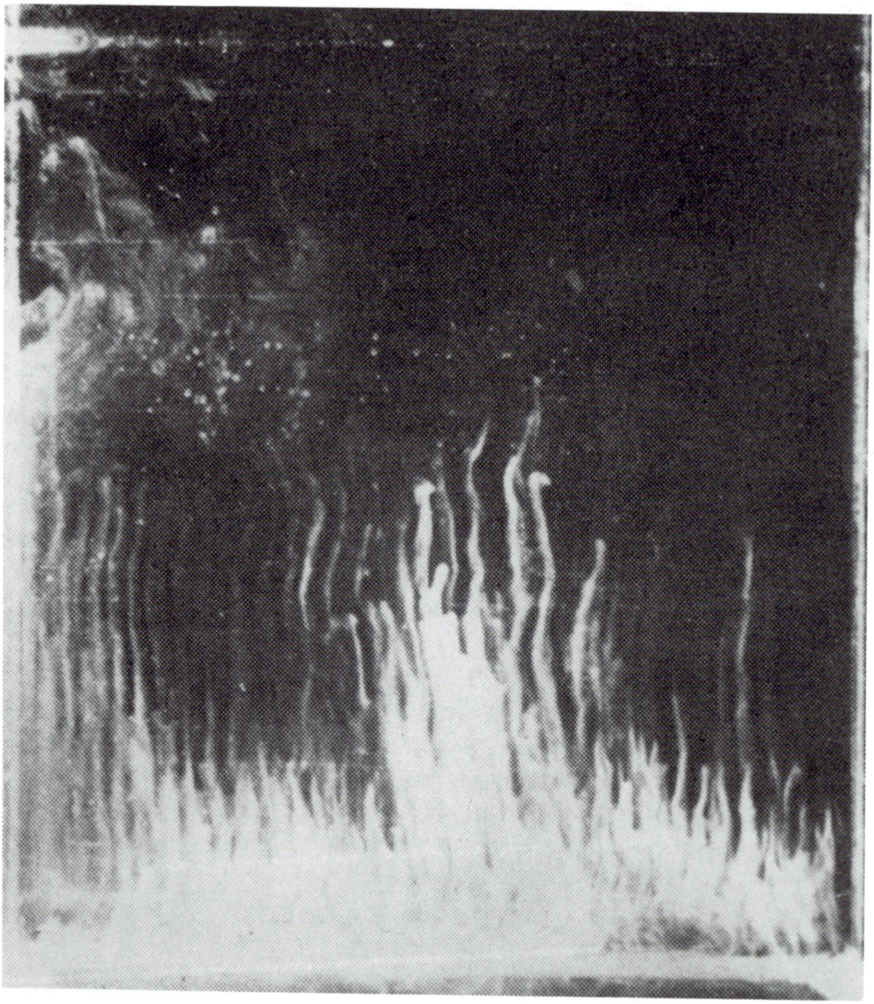

Figure 6.9.6 Vertical cross section of salt fingers, marked by fluorescein dye added to the upward-moving fingers. *(From Turner, 1974. Reproduced, with permission, from* Annu. Rev. Fluid Mech., *vol. 6, Copyright © 1974, Annual Reviews Inc.)*

overlaid with warm saline water. The resulting saline convection has been observed as a highly organized pattern of tall thin columns of fluid alternately ascending and descending. By conducting heat but not salinity laterally, since heat diffusion is much faster than salt diffusion, the fluid is able to overcome the stabilizing effect of the vertical temperature gradient and experiences a buoyancy force, due to its deficit in salt and density relative to the fluid in the ambient. This buoyancy force may be sufficient to drive convection even though the mean density field increases in the direction of gravity.

These experimentally observed patterns, called salt fingers, are shown in Fig. 6.9.6. The experiment was begun with a layer of cold water at the bottom and saline hot water at the top. The tank was square and 25 cm deep. Fluorescein dye injected at the bottom distinguishes the rising freshwater columns, or salt fingers.

The opposite situation, with warmer and saltier water under colder and fresher water, is an example of case 3a, since dt/dx, dC/dx, and $d\rho/dx$ are all negative. This system is gravitationally stable. However, convection may still arise. Stern (1960) was the first to note that for this circumstance the linear mode of instability is oscillatory, which allows the potential energy stored in the thermal field to be released. The subsequent behavior was studied later by Turner and Stommel (1964), who showed that horizontal layering occurs in this configuration. The consequences of heating from below an initially stably stratified salinity distribution at constant temperature are shown in Fig. 6.9.7. After an initial oscillatory instability, the layered structure is produced. As a result of heating, the elements near the bottom become buoyant with respect to the immediate surroundings. They rise and mix with the fluid around them until they reach a level where their temperature excess no longer compensates for the density excess due to their high salinity. Further growth of this layer ceases. Thereafter, additional convection layers are formed on top of it.

This behavior is again due to the large differences in diffusivity between salt and heat. Heat diffuses from the top of a layer more rapidly than salt and starts a new convective process above. The salt stays behind and preserves the stability of the interface. Such an interface, which separates warm salty water from colder fresher water above it, is also called a "diffusive interface" because heat and salt are transported through the interface solely by molecular diffusion.

Flow phenomena similar to those discussed above for a salt-heat diffusive system have also been observed in fluids in which two or more stratified chemical components have different molecular diffusivities. The components may make opposing contributions to the vertical density gradient. The term double-diffusive convection is used to describe all such flows.

Much of the work in double-diffusive convection has been stimulated by applications in oceanography. Existence of salt fingers and convection layers has been confirmed in oceanic measurements; see, for example, Tait and Howe (1968, 1971), Cooper and Stommel (1968), Neal et al. (1969), and Williams (1974). These measurements, along with laboratory experiments by Turner (1967, 1968), Stern and Turner (1969), Thorpe et al. (1969), Linden (1971, 1973), Chen et al.

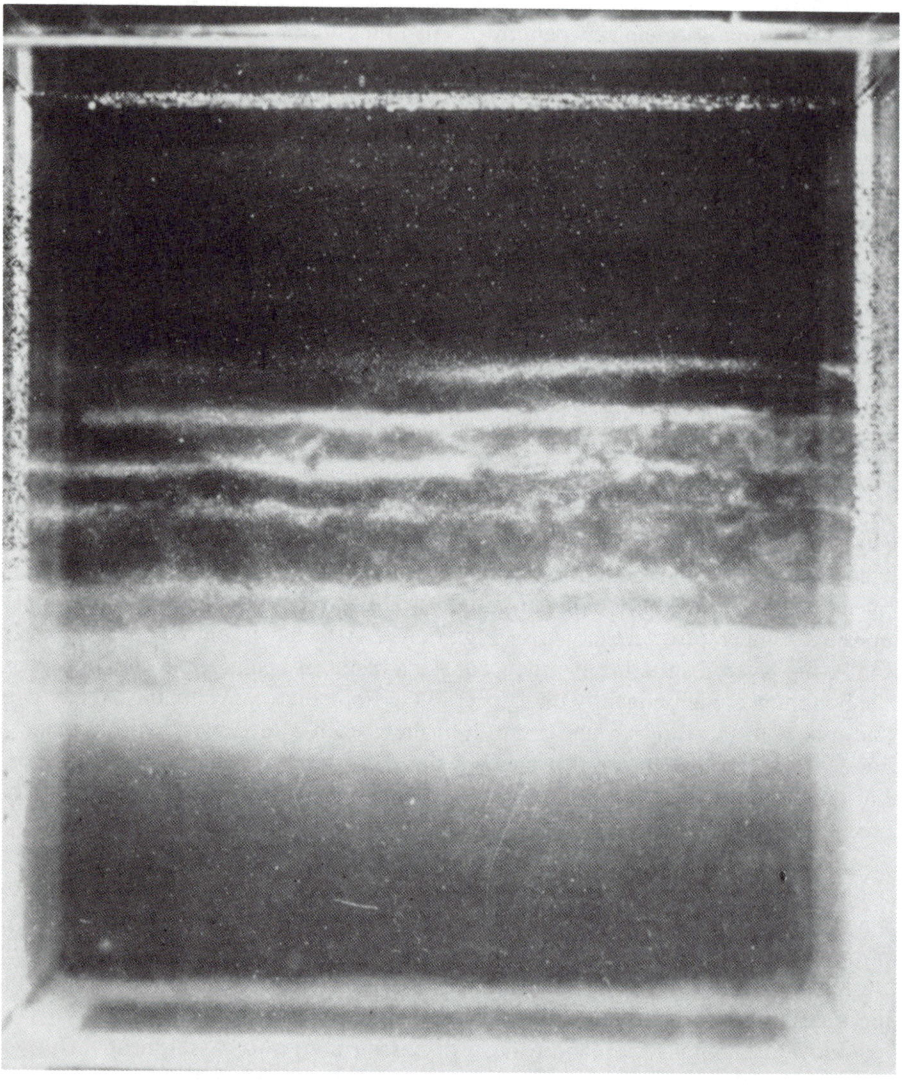

Figure 6.9.7 Horizontal layers formed in a laboratory tank of smoothly stratified salt solution by heating from below. The layers are marked with fluorescein dye and aluminum powder, and the tank is lit through a slit at the top. *(From Turner, 1974. Reproduced, with permission, from* Annu. Rev. Fluid Mech., *vol. 6, Copyright © 1974, Annual Reviews Inc.)*

(1971), Huppert and Linden (1979), Huppert and Turner (1978, 1980), and others, have contributed a great deal to a basic understanding of such oceanic mixing processes. For an excellent review of these and other double-diffusive processes, the reader is referred to Turner (1974, 1979) and Huppert and Turner (1981).

Applications to phenomena in astrophysics, geology, and many fields of engineering, such as solar ponds, storage and transport of liquid natural gas, and sewage disposal in the sea, are discussed. A brief description of the operation of a solar pond is given below as an example of applications of double-diffusive processes.

6.9.3 Solar Ponds

A solar pond is an artificially stratified shallow pool that depends on opposing density gradients due to salinity and temperature to maintain its stratification. Figure 6.9.8 shows a cross section of a solar pond with associated salinity and temperature profiles. The total depth may range from a fraction of a meter to several meters. Normally, there are three zones: a relatively thin mixed convection surface layer, a nonconvective zone having a stabilizing density gradient, and a mixed convective storage zone at the bottom. The gradient zone transfers heat by conduction only, since water is opaque to thermal radiation at these temperature levels. It thus acts as a selectively transparent window of radiation, of low thermal conductance. Some solar radiation penetrates into the bottom zone and heats it.

Typical salinity and temperature distributions are as shown. The essentially uniform temperature and salinity surface and storage zones are separated by the nonconvective gradient zone, in which there are gradients of salinity and temperature. As shown in Fig. 6.9.8, these gradients are generally not linear. With an increase in depth in the gradient zone, increasing temperature causes the fluid density to decrease. Offsetting this is the salinity gradient, whose effect is to increase the density.

Generally, the salinity gradient is sufficient to maintain a stabilizing density gradient. Successful operation of the pond depends on this. Otherwise, overall convection would result. The pond would be mixed and the stored heat lost at the surface. Further details on the operation of a solar pond are given in Elata and Levin (1965), Harris and Wittenberg (1979), Nielsen (1979), Tabor (1980), and Zangrando and Bryant (1977).

The energy extracted from the bottom of a solar pond may be used to operate a power plant. An interesting question is whether the surface layer of the pond, which is at a temperature close to the ambient, may be used for heat rejection from the plant. Such an arrangement would eliminate the need to provide an additional body of water for heat rejection. However, in this arrangement, the recirculating flow used for heat rejection would increase the temperature of the surface layer. It might also disturb the stabilizing gradient zone. This would adversely affect the performance of the solar pond. This issue has been addressed in an analytical and numerical study by Jaluria and Cha (1985). They found that the gradient zone is quite stable. It may be only slightly disturbed by the flow, if the inflow and outflow are located close to the surface. For other inflow-outflow locations and the inflow conditions, more care is needed to avoid significant disturbances of the gradient zone.

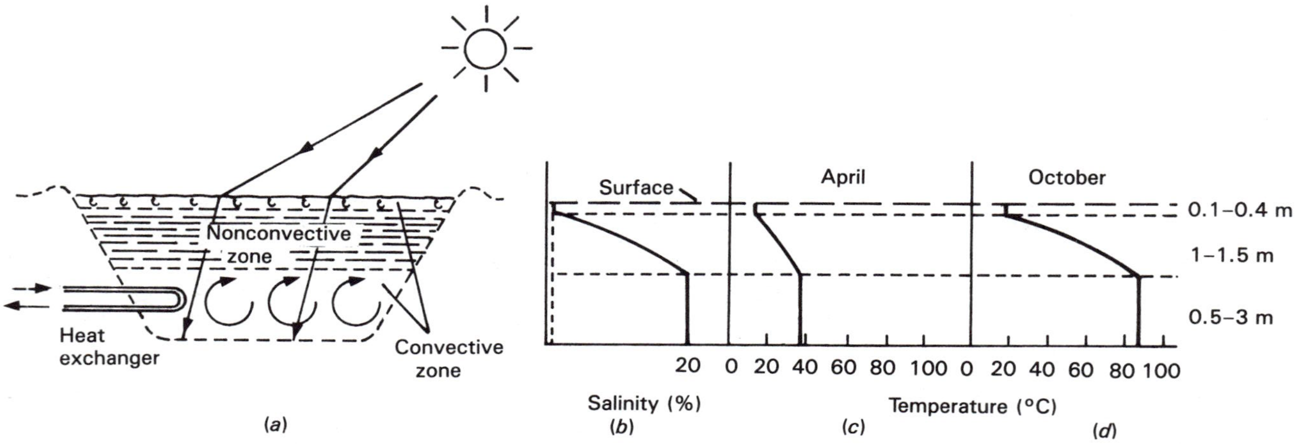

Figure 6.9.8 (*a*) Cross section of a salt gradient solar pond; (*b*) salinity profile, a possible stationary configuration; (*c* and *d*) temperature profiles, idealized, anticipated in space heating applications. *(From Nielsen, 1979.)*

REFERENCES

Acrivos, A. (1960). *AIChE J. 6,* 410.
Acrivos, A. (1962). *Chem. Eng. Sci. 17,* 457.
Acrivos, A. (1966). *Chem. Eng. Sci. 21,* 343.
Adams, J. A., and Lowell, R. L. (1968). *Int. J. Heat Mass Transfer 11,* 1215.
Adams, J. A., and McFadden, P. W. (1966). *AIChE J. 12,* 642.
Afgan, N. H., and Beer, J. M. (1974). *Heat Transfer in Flames,* Hemisphere, Washington, D.C.
Ahmad, T., and Faeth, G. M. (1978a). *J. Heat Transfer 100,* 112.
Ahmad, T., and Faeth, G. M. (1978b). *17th Symp. (Int.) Combust.,* Combustion Institute, 1149.
Arpaci, V. S., and Tabaczynski, R. J. (1982). *Combust. Flame 46,* 315.
Baron, J. R. (1962). *ARS J. 32,* 1053.
Becker, H., and Liang, D. (1982). *Combust. Flame 44,* 305.
Bottemanne, F. A. (1971). *Appl. Sci. Res. 25,* 137.
Bottemanne, F. A. (1972). *Appl. Sci. Res. 25,* 372.
Boura, A., and Gebhart, B. (1976). *AIChE 22,* 94.
Callahan, G. D., and Marner, H. W. (1976). *Int. J. Heat Mass Transfer 19,* 165.
Chapman, S., and Cowling, T. G. (1952). *The Mathematical Theory of Non-Uniform Gases,* Cambridge Univ. Press, London.
Chen, C. F., Briggs, D. G., and Wirtz, R. A. (1971). *Int. J. Heat Mass Transfer 14,* 57.
Chen, T. S., and Yuh, C. F. (1979). *Numer. Heat Transfer 2,* 233.
Chen, T. S., and Yuh, C. F. (1980). *Int. J. Heat Mass Transfer 23,* 451.
Cooper, J. W., and Stommel, H. (1968). *J. Geophys. Res. 85,* 2573.
Cox, G., and Chitty, R. (1980). *Combust. Flame 39,* 191.
Den Bouter, J. A., DeMunnik, B., and Heertjes, P. M. (1968). *Chem. Eng. Sci. 23,* 1185.
deRis, J., and Orloff, L. (1974). *15th Symp. (Int.) Combust.,* Combustion Institute, 175.
Elata, C., and Levin, O. (1965). Hydraulics of the solar pond. *Proc. 11th Cong. Int. Assoc. Hydraul. Res.,* Leningrad.
Emmons, H. W. (1978). *17th Symp. (Int.) Combust.,* Combustion Institute, 1101.
Emmons, H. W. (1980). *Annu. Rev. Fluid Mech. 12,* 223.
Fernandez-Pello, A. C., and Santoro, R. J. (1978). *17th Symp. (Int.) Combust.,* Combustion Institute, 1201.
Fishburne, E. S., and Pergament, H. S. (1978). *17th Symp. (Int.) Combust.,* Combustion Institute, 1063.
Fristrom, R. M., and Westenberg, A. A. (1965). *Flame Structure,* McGraw-Hill, New York.
Gebhart, B. (1971). *Heat Transfer,* 2nd ed., McGraw-Hill, New York.
Gebhart, B., and Pera, L. (1971). *Int. J. Heat Mass Transfer 14,* 2025.
Gill, W. N., Del Casal, E., Zeh, D. W. (1965). *Int. J. Heat Mass Transfer 8,* 1311.
Glassman, I. (1977). *Combustion,* Academic Press, New York.
Grew, K. E., and Ibbs, T. L. (1952). *Thermal Diffusion in Gases,* Cambridge Univ. Press, London.
Groff, E. G., and Faeth, G. M. (1978). *Combust. Flame 32,* 139.
Harris, M. J., and Wittenberg, L. J. (1979). Report 7P-245-430, in *Proceedings of the Solar Heating and Cooling Systems Conference—Operational Results,* Colorado Springs, November 27–30, 1979, SERI, Golden, Colo.
Hayday, A. A., Bowlus, D. A., and McGraw, R. A. (1967). *J. Heat Transfer 89,* 244.
Hayday, A. A., Eckert, E. R. G., and Minkowycz, W. J. (1962). Tech. Rep. 48, Univ. of Minnesota Heat Transfer Laboratory.
Hirano, T., Noreikis, S. E., and Waterman, T. E. (1974). *Combust. Flame 23,* 83.
Hirschfelder, J. O., Curtiss, C. F., and Bird, R. F. (1954). *Molecular Theory of Gases and Liquids,* Wiley, New York.
Hubbell, R. H., and Gebhart, B. (1974). *Proc. 24th Heat Transfer Fluid Mech. Inst.,* Corvallis, Ore.
Huppert, H. E., and Linden, P. F. (1979). *J. Fluid Mech. 95,* 431.

Huppert, H. E., and Turner, J. S. (1978). *Nature (London) 271,* 46.
Huppert, H. E., and Turner, J. S. (1980). *J. Fluid Mech. 100,* 367.
Huppert, H. E., and Turner, J. S. (1981). *J. Fluid Mech. 106,* 299.
Jaluria, Y., and Cha, C. K. (1985). *J. Heat Transfer 107,* 95.
Jeng, S. M., Chen, L. D., and Faeth, G. M. (1982). *19th Symp. (Int.) Combust.,* Combustion Institute, 349.
Kays, W. M. (1966). *Convection Heat and Mass Transfer,* McGraw-Hill, New York.
Kennedy, L. A., and Plumb, O. A. (1976). *16th Symp. (Int.) Combust.,* Combustion Institute, 1699.
Kim, J. S., deRis, J., and Kroesser, F. W. (1971). *13th Symp. (Int.) Combust.,* Combustion Institute, 949.
Kosdon, F. J., Williams, F. A., and Burman, C. (1969). *12th Symp. (Int.) Combust.,* Combustion Institute, 253.
Lightfoot, E. N. (1968). *Chem. Eng. Sci. 23,* 931.
Linden, P. F. (1971). *J. Fluid Mech. 49,* 611.
Linden, P. F. (1973). *Deep-Sea Res. 20,* 325.
Liu, V. K., Lloyd, J. R., and Yang, K. T. (1981). *Int. J. Heat Mass Transfer 24,* 1959.
Liu, V. K., and Shih, T. M. (1980). *J. Heat Transfer 102,* 724.
Liu, V. K., Yang, K. T., and Lloyd, J. R. (1982). *Int. J. Heat Mass Transfer 25,* 863.
Lowell, R. L., and Adams, J. A. (1967). *AIAA J. 5,* 1360.
Manganaro, R. L., and Hanna, O. T. (1970). *AIChE J. 16,* 204.
Mao, C. P., and Fernandez-Pello, A. C. (1980). Technical Meeting, Eastern Section, Combustion Institute, Pittsburgh, Pa.
Mathers, W. G., Madden, A. J., and Piret, E. L. (1957). *Ind. Eng. Chem. 49,* 961.
Mollendorf, J. C., and Gebhart, B. (1974). *Proc. 5th Int. Heat Transfer Conf.,* Tokyo, CT 1.3.
Neal, V. T., Neshyba, S., and Denner, W. (1969). *Science 166,* 373.
Negrelli, D. E., Lloyd, J. R., and Novotny, J. L. (1977). *J. Heat Transfer 99,* 212.
Nielsen, C. E. (1979). In *Solar Energy Handbook,* W. C. Dickinson and P. M. Chermisinoff, Eds., Dekker, New York.
Orloff, L., and deRis, J. (1971). *13th Symp. (Int.) Combust.,* Combustion Institute, 979.
Ostrach, S. (1980). *Physicochemical Hydrodynamics,* vol. 1, p. 233, Pergamon, Elmsford, N.Y.
Pagni, P. J., and Shih, T. M. (1977). *16th Symp. (Int.) Combust.,* Combustion Institute, 114.
Pera, L., and Gebhart, B. (1972). *Int. J. Heat Mass Transfer 15,* 269.
Pohlhausen, K. (1921). *Z. Angew. Math. Phys. 1,* 252.
Ray, S. R., Fernandez-Pello, A. C., and Glassman, I. (1980). *J. Heat Transfer 102,* 357.
Saville, D. A., and Churchill, S. W. (1967). *J. Fluid Mech. 29,* 391.
Saville, D. A., and Churchill, S. W. (1970). *AIChE J. 16,* 268.
Schenk, J., Altman, R., and DeWit, J. P. (1976). *Appl. Sci. Res. 32,* 599.
Sharma, O. P., and Sirignano, W. A. (1971). AIAA preprint 71-207.
Sibulkin, M., Kulkarni, A. K., and Annamalai, K. (1981). *18th Symp. (Int.) Combust.,* Combustion Institute, 611.
Sibulkin, M., Kulkarni, A. K., and Annamalai, K. (1982). *Combust. Flame 44,* 187.
Sirignano, W. A., and Glassman, I. (1970). *Combust. Sci. Technol. 1,* 307.
Somers, E. V. (1956). *J. Appl. Mech. 23,* 295.
Spalding, D. B. (1954). *Proc. R. Soc. London Ser. A 221,* 78.
Sparrow, E. M., Minkowycz, W. J., and Eckert, E. R. G. (1964a). *AIAA J. 2,* 652.
Sparrow, E. M., Minkowycz, W. J., and Eckert, E. R. G. (1964b). *J. Heat Transfer 86C,* 508.
Sparrow, E. M., Scott, C. J., Forstrom, R. J., and Ebert, W. A. (1965). *J. Heat Transfer 87C,* 321.
Stern, M. (1960). *Tellus 12,* 172.
Stern, M., and Turner, J. S. (1969). *Deep-Sea Res. 16,* 497.
Tabor, H. (1980). *Philos. Trans. R. Soc. London Ser. A 295,* 423.
Tait, R. I., and Howe, M. R. (1968). *Deep-Sea Res. 15,* 275.
Tait, R. I., and Howe, M. R. (1971). *Nature (London) 231,* 178.
Taminini, F. (1977). *Combust. Flame 30,* 85.

Taminini, F. (1978). *17th Symp. (Int.) Combust.*, Combustion Institute, 1075.
Taminini, F., and Ahmad, T. (1979). *7th Int. Colloq. Gas Dyn. Explosions React. Systems*, Gottingen.
Taunton, J. W., Lightfoot, E. N., and Stewart, W. E. (1970). *Chem. Eng. Sci. 25*, 1927.
Tenner, A. R., and Gebhart, B. (1971). *Int. J. Heat Mass Transfer 14*, 2051.
Tewfik, O. E., Eckert, E. R. G., and Shirtliffe, C. J. (1962). *Proc. 1962 Heat Transfer Fluid Mech. Inst.*, p. 42.
Tewfik, O. E., and Yang, J. W. (1963). *Int. J. Heat Mass Transfer 6*, 915.
Thorpe, S. A., Hutt, P. K., and Soulsby, R. (1969). *J. Fluid Mech. 38*, 375.
Torrance, K. E. (1971). *Combust. Sci. Technol. 3*, 133.
Torrance, K. E., and Mahajan, R. L. (1974). *15th Symp. (Int.) Combust.*, Combustion Institute, 281.
Torrance, K. E., and Mahajan, R. L. (1975). *Combust. Sci. Technol. 10*, 125.
Tsuji, H., and Yamaoka, I. (1971). *13th Symp. (Int.) Combust.*, Combustion Institute, 723.
Turner, J. S. (1967). *Deep-Sea Res. 14*, 599.
Turner, J. S. (1968). *J. Fluid Mech. 33*, 183.
Turner, J. S. (1974). *Annu. Rev. Fluid Mech. 6*, 37.
Turner, J. S. (1979). *Buoyancy Effects in Fluids*, Cambridge Univ. Press, London.
Turner, J. S., and Stommel, H. (1964). *Proc. Natl. Acad. Sci. U.S.A. 52*, 49.
Wilcox, W. R. (1961). *Chem. Eng. Sci. 13*, 113.
Williams, A. J. (1974). *Science 185*, 941.
Williams, F. A. (1965). *Combustion Theory*, Addison-Wesley, Reading, Mass.
Wirtz, R. A. (1977). *Int. J. Heat Mass Transfer 20*, 841.
Wirtz, R. A., Briggs, D. G., and Chen, C. F. (1972). *Geophys. Fluid Dyn. 3*, 265.
You, H. A., and Faeth, G. M. (1982). *Combust. Flame 44*, 261.
Zangrando, F., and Bryant, H. C. (1977). *Proc. Int. Conf. Alternative Energy Sources*, vol. 6, p. 2935, Miami Beach, Fla., December 5–7, Hemisphere, Washington, D.C.
Zeh, D. W., and Gill, W. N. (1967). *Int. J. Heat Mass Transfer 10*, 1159.
Zeldovich, Y. B. (1951). NACA Tech. Memo. 1296.

PROBLEMS

6.1 Consider a plane plume that arises under the combined (and aiding) buoyancy effects of both thermal and mass diffusion. Such a flow might occur, for example, over a long ridge heated by the sun and covered with vegetation releasing water vapor. Assume that the plume is laminar, that it comes from a line source, and that the environment is unstratified. Use the low-concentration approximation.

(a) Determine how both the midplane temperature and water vapor concentration vary with elevation in the boundary layer regime.

(b) How do plume thickness and maximum velocity vary with elevation and with other parameters of the problem?

(c) How would the answers to parts (a) and (b) change if buoyancy effects were opposed—for example, if the upward flow was driven by H_2O concentration but was opposed by a smaller thermal buoyancy force in the other direction, that is, $t_0 \leq t_\infty$?

6.2 A surface is placed vertically in quiescent dry ambient air at 70°F and 1 atm. The surface temperature is maintained at 90°F. The surface is porous and flooded to capacity with pure water, as it evaporates into the dry air. The concentration of water vapor in the air immediately adjacent to the surface will be saturated, at a partial pressure of 0.048 atm. The molecular weight of H_2O is 18, that of air is 28.97, and $\bar{R} = 1545$.

(a) Calculate the concentration of the water vapor at the surface.

(b) Find the ratio $\tilde{N}$ of the mass and thermal diffusion buoyancy force components adjacent to the surface.

(c) Do the two buoyancy effects aid or oppose each other?

(d) Calculate the local Grashof number 1 ft downstream from the leading edge.

6.3 For the combined buoyancy flow regime as in Problem 6.1 but adjacent to a vertical surface, find

(a) How the normal velocity component at the surface, $v_0(x)$, varies with x, the distance from the leading edge.

(b) Whether there is an imposed variation of v_0 that would result in similarity.

6.4 Find the mass transfer from a wetted vertical surface at 30°C to quiescent air at 30°C. The relative humidity of air is measured as 40%. Take the diffusion coefficient as 0.3 cm^2/s and the Schmidt number as 0.6. The density of air is 1 kg/m^3 and that of saturated water vapor at 30°C is 0.03 kg/m^3.

6.5 A 50-cm-high surface of porous wet material is placed vertically in completely dry air at 10°C for drying. The surface is radiantly heated by distant hot surfaces to a temperature of 30°C. The surface concentration of H_2O vapor at 30°C is 0.0304 kg/m^2. $\beta^* = 0.572$, ρ_0 and ρ_∞ are 1.148 and 1.248 kg/m^3.

(a) Determine the ratio of the magnitudes of the mass diffusion and thermal buoyancy force components and the direction of flow.

(b) Find the Nusselt and Sherwood numbers for the transport at the upper edge of the surface.

(c) Calculate the heat transfer rate and drying rate at the upper edge of the surface in kilograms per square meter per hour.

(d) Determine the resulting blowing parameter at the surface.

6.6 Using the method of asymptotic expansions,

(a) Derive the zeroth-order equations, Eqs. (6.5.22)–(6.5.27). Discuss how you obtain the perturbation parameter as $\mathrm{Pr}^{1/2}$.

(b) Derive the first-order, $O(\epsilon)$, equations with the appropriate boundary conditions for the combined heat and mass transfer problem when $\mathrm{Sc} >> \mathrm{Pr} \rightarrow \infty$.

6.7 Derive the governing boundary layer equations for horizontal flows when $B_n < 0$. Are similarity solutions possible for this flow circumstance?

6.8 Starting from the full boundary layer equations for inclined flows, $B_n > 0$, show that Eqs. (6.2.38)–(6.2.43) are justified if $\Delta \tan \theta << 1$.

CHAPTER

SEVEN

UNSTEADY EXTERNAL FLOWS

7.1 UNSTEADY-STATE PROCESSES

In many convection processes temporal variations in the temperature, velocity, and perhaps concentration of chemical species arise. Such flows are quite common in environmental processes and many technological and industrial applications. Many atmospheric, lake, and oceanic circulations are short- or long-term transients and usually are the result of a superimposition of multiple background processes. In technology, transients are common during start-up and shutdown. An important problem that arises is the possible occurrence of extreme conditions during such processes. A relevant example is the core of a nuclear reactor, where temperature overshoots during transients may be critical design considerations. The same considerations are important for electrical and electronic equipment cooled by natural convection processes.

Unsteady natural convection flows are subdivided into two general classes—internal and external. Internal flows are flows arising in a body of fluid partially or completely bounded by surfaces. These flows, both steady and transient, are discussed in Chapter 14. Steady external flows, arising in extensive media, were considered in preceding chapters. Some of these flows subject to temporal variations are considered here. Such flows are driven by changing imposed temperature or energy conditions or local inhomogeneities in density for concentration gradients of chemical species. External unsteady flows are further classified according to the geometry and the driving mechanism.

Flat vertical surfaces, vertical and horizontal cylinders, wires, and plumes are some of the configurations that have been investigated analytically, numerically, and experimentally. For all such configurations, a diverse number of flows may arise, depending on the driving mechanism for the flow, the ambient conditions, and the boundary conditions. Since the flat vertical surface is the most intensively investigated and best understood case, it is considered first. Several other geometries are treated later.

7.2 ANALYSIS AND MEASUREMENTS IN FLOWS ADJACENT TO VERTICAL SURFACES

When a flat vertical surface immersed in an extensive quiescent medium is suddenly heated at a thereafter constant rate, an unsteady transport response is initiated that continues until steady state is achieved. This transient process is often composed of distinctively different stages, depending on the heating process and the properties of the ambient fluid. The equations representing conservation of mass, momentum, and energy, after making the boundary layer and Boussinesq approximations, are

$$\frac{\partial u}{\partial x} + \frac{\partial v}{\partial y} = 0 \tag{7.2.1}$$

$$\frac{\partial u}{\partial \tau} + u\frac{\partial u}{\partial x} + v\frac{\partial u}{\partial y} = g\beta(t - t_\infty) + \nu\frac{\partial^2 u}{\partial y^2} \tag{7.2.2}$$

$$\frac{\partial t}{\partial \tau} + u\frac{\partial t}{\partial x} + v\frac{\partial t}{\partial y} = \alpha\frac{\partial^2 t}{\partial y^2} \tag{7.2.3}$$

For an initially quiescent condition everywhere, the following boundary and initial conditions apply:

$$\text{for } \tau \le 0: \qquad u = v = t - t_\infty = 0 \tag{7.2.4}$$

$$\text{at } y = 0: \qquad u = v = 0 \qquad \text{for all } \tau \tag{7.2.5}$$

$$\text{as } y \to \infty: \qquad u \to 0,\ t \to t_\infty \qquad \text{for all } \tau \tag{7.2.6}$$

Then for $\tau > 0$ a flux or temperature condition arises at the surface to cause buoyancy in the fluid.

Equation (7.2.4) specifies the initial quiescence and temperature uniformity of the ambient medium. The no-slip and nonpermeability of the surface lead to Eq. (7.2.5), while Eq. (7.2.6) states that the far ambient medium is at rest and not stratified. The driving mechanism for this process is determined by the manner in which the surface is heated. It may be a changing surface temperature as a function of time, $t_0 - t_\infty = F(\tau)$, an imposed heat flux at the surface, $q'' = G(\tau)$, or some other effect.

The distinctive feature of these equations, compared to steady-state transport, is that there are three independent variables, x, y, and τ. Similarity solutions are available for a few special surface temperature conditions, such as $1/(x + a\tau)$ and $(x + a)/(1 - b\tau)^2$, where a and b are constants; see Yang (1960, 1966). Other than numerically integrating the full equations in their time-dependent form, some simplifying assumptions must be made to obtain relatively simple solutions for this kind of transport, for bounding conditions of common practical importance.

During the early stages of the transient, heat is transferred from the surface to the fluid surrounding it in a one-dimensional conduction process, similar to heat conduction into a semi-infinite conducting medium. Then all streamwise derivatives are zero and, for that short period, solutions to Eqs. (7.2.2) and (7.2.3) are easily obtained. Although motion arises, there are no convective effects in this one-dimensional transport.

The earliest study of such transients appears to be that of Illingworth (1950) for locations along a heated surface above which the effects on transport due to the presence of a leading edge for the heated surface, called the leading edge effects, have not yet propagated. This phase is equivalent to transport adjacent to an infinite vertical surface, that is, one-dimensional transport. The ambient medium was considered to be quiescent and at uniform temperature. A step in wall temperature was assumed to start the transient. In effect, the wall temperature was assumed to be instantaneously and uniformly changed to some different value t_0, which was held constant thereafter.

Without a leading edge no x dependence arises, and the time-dependent boundary layer equations (7.2.1)–(7.2.3) were solved after omitting all derivatives with respect to x. This is a parallel flow approximation. Thus, heat is transferred solely by conduction. The resulting temperature response is the familiar one-dimensional distribution for a semi-infinite conducting medium whose surface temperature is suddenly changed, namely

$$\theta = t - t_\infty = (t_0 - t_\infty)\,\text{erfc}\left(\frac{y}{2\sqrt{\alpha\tau}}\right) \quad \text{or} \quad \frac{\theta}{\theta_0} = \phi = \text{erfc}\,\frac{y}{2\sqrt{\alpha\tau}} \tag{7.2.7}$$

where erfc is the conjugate error function, defined by

$$\text{erfc}(x) = 1 - \text{erf}\,(x) = 1 - \frac{2}{\sqrt{\pi}}\int_0^x \exp\,(-\xi^2)\,d\xi$$

This simple result is particularly valuable because at short times essentially all transient processes are one-dimensional, even in the presence of a leading edge. At any downstream location, the one-dimensional process may last until the leading edge effect arrives at that location.

Sugawara and Michiyoshi (1952) reported a numerical study of flow past a semi-infinite vertical flat surface subjected to a step change in temperature. The momentum, continuity, and energy equations were used without the usual boundary layer approximations. All convective terms were dropped and a short-time solution was obtained. This solution was then used to continue with the full equations including the convective terms. The approximate time at which steady state

Table 7.2.1 Summary of solutions for the one-dimensional conduction regime adjacent to a flat vertical surface, subject to various boundary conditions

Case	Wall boundary conditions	$\theta(y,\tau)$	$u(y,\tau)$: $\mathrm{Pr}=1$	$u(y,\tau)$: $\mathrm{Pr}\neq 1$
1	$\theta(0,\tau)=0 \quad \tau<0$ $=\theta_0 \quad \tau\geq 0$	$\theta_0 \operatorname{erfc}\dfrac{y}{\sqrt{4\alpha\tau}}$	$2(g\beta\theta_0\tau)\dfrac{y}{\sqrt{4\alpha\tau}}\, i\operatorname{erfc}\dfrac{y}{\sqrt{4\alpha\tau}}$	$\dfrac{4(g\beta\theta_0\tau)}{1-\mathrm{Pr}}\left[i^2\operatorname{erfc}\dfrac{y}{\sqrt{4\alpha\tau}} - i^2\operatorname{erfc}\dfrac{y}{\sqrt{(4\alpha\tau)\mathrm{Pr}}}\right]$
2	$\theta(0,\tau)=0 \quad \tau<0$ $=\theta_1 \quad 0<\tau<T$ $=\theta_2 \quad \tau>T$	$\theta_1 \operatorname{erfc}\dfrac{y}{\sqrt{4\alpha\tau}} + (\theta_1-\theta_2)\operatorname{erfc}\dfrac{y}{\sqrt{4\alpha(\tau-T)}}$	$2(g\beta\theta_1\tau)\dfrac{y}{\sqrt{4\alpha\tau}}\, i\operatorname{erfc}\dfrac{y}{\sqrt{4\alpha\tau}} + 2g\beta(\theta_2-\theta_1)(\tau-T) \times\left[\dfrac{y}{\sqrt{4\alpha(\tau-T)}}\, i\operatorname{erfc}\dfrac{y}{\sqrt{4\alpha(\tau-T)}}\right]$	
3	$\theta(0,\tau)=0 \quad \tau<0$ $=A\tau \quad \tau\geq 0$	$4(A\tau)i^2\operatorname{erfc}\dfrac{y}{\sqrt{4\alpha\tau}}$	$8[g\beta(A\tau)\tau]\dfrac{y}{\sqrt{4\alpha\tau}}\, i^2\operatorname{erfc}\dfrac{y}{\sqrt{4\alpha\tau}}$	$16[g\beta(A\tau)\tau] \times\left[i^4\operatorname{erfc}\dfrac{y}{\sqrt{4\alpha\tau}} - i^4\operatorname{erfc}\dfrac{y}{\sqrt{(4\alpha\tau)\mathrm{Pr}}}\right]$
4	$\theta(0,\tau)=0 \quad \tau<0$ $=\beta\tau^{1/2} \quad \tau\geq 0$	$\sqrt{\pi}\,(\beta\tau^{1/2})i\operatorname{erfc}\dfrac{y}{\sqrt{4\alpha\tau}}$	$2\sqrt{\pi}[g\beta(\beta\tau^{1/2})\tau]\dfrac{y}{\sqrt{4\alpha\tau}}\; i^a\operatorname{erfc}\dfrac{y}{\sqrt{4\alpha\tau}}$	$\dfrac{4\sqrt{\alpha}[g\beta(\beta\tau^{1/2})\tau]}{1-\mathrm{Pr}} \times\left[i^3\operatorname{erfc}\dfrac{y}{\sqrt{4\alpha\tau}} - i^3\operatorname{erfc}\dfrac{y}{\sqrt{(4\alpha\tau)\mathrm{Pr}}}\right]$

5	$\theta(0, \tau) = \theta_0 e^{i\omega\tau} \quad \tau >> 0$	$\theta_0 \exp\left(-\sqrt{\frac{\omega}{2\alpha}}\, y\right) \exp\left[i\left(\omega\tau - \sqrt{\frac{\omega}{2\alpha}}\, y\right)\right]$	$\frac{g\beta\theta_0 y}{\sqrt{8\omega\alpha}} \exp\left(-\sqrt{\frac{\omega}{2\alpha}}\, y\right) \exp\left[i\left(\omega\tau - \sqrt{\frac{\omega}{2\alpha}}\, y - \frac{\pi}{4}\right)\right]$	$\frac{g\beta\theta_0}{\omega(1 - \mathrm{Pr})} \exp\left(-\sqrt{\frac{\omega}{2\alpha}}\, y\right) \times \left\{\exp\left[i\left(\omega\tau - \sqrt{\frac{\omega}{2\alpha}}\, y + \frac{\pi}{2}\right)\right] - \exp\left[i\left(\omega\tau - \sqrt{\frac{\omega}{2\nu}}\, y + \frac{\pi}{2}\right)\right]\right\}$
6	$\theta(0, \tau) = 0 \quad \tau < 0$ $= \theta_0 \quad \tau \geq 0$ $g = g_0 e^{i\omega\tau}$	$\theta_0 \operatorname{erfc} \frac{y}{\sqrt{4\alpha\tau}}$	$\frac{g_0\beta\theta_0}{2\omega} e^{i(\omega\tau + \pi/2)} \left\{ e^{-y\sqrt{i\omega/\alpha}} \operatorname{erfc}\left(\frac{y}{\sqrt{4\alpha\tau}} - \sqrt{i\omega\tau}\right) + e^{y\sqrt{\omega i/\alpha}} \operatorname{erfc}\left(\frac{y}{\sqrt{4\alpha\tau}} + \sqrt{i\omega\tau}\right) - 2 \operatorname{erfc} \frac{y}{\sqrt{4\alpha\tau}} \right]$	
7	$\frac{\partial\theta}{\partial y}(0, \tau) = 0 \quad \tau < 0$ $= -\frac{q_0''}{k} \quad \tau \geq 0$	$\frac{2q_0''\sqrt{\alpha\tau}}{k}\, i \operatorname{erfc} \frac{y}{\sqrt{4\alpha\tau}}$	$2\frac{g\beta q_0'' y\tau}{k}\, i^2 \operatorname{erfc} \frac{y}{\sqrt{4\alpha\tau}}$	$\frac{8g\beta q_0''\sqrt{\alpha\tau}^{3/2}}{k(1 - \mathrm{Pr})}\left[i^3 \operatorname{erfc} \frac{y}{\sqrt{4\alpha\tau}} - i^3 \operatorname{erfc} \frac{y}{\sqrt{(4\alpha\tau)\mathrm{Pr}}}\right]$

Source: Reprinted with permission from J. A. Schetz and R. Eichhorn, *J. Heat Transfer,* vol. 4, p. 334. Copyright © 1962, ASME.

was reached was assumed to be the time at which the maximum velocity across the boundary region attained its steady-state value. It was concluded that in air the one-dimensional regime typically lasted for a very brief period.

Schetz and Eichhorn (1962) analyzed the one-dimensional conduction process in some detail. Again, the parallel flow approximation was made, reducing Eqs. (7.2.1)–(7.2.3) to

$$\frac{\partial v}{\partial y} = 0 \tag{7.2.8}$$

$$\frac{\partial u}{\partial \tau} = \nu \frac{\partial^2 u}{\partial y^2} + g\beta\theta \tag{7.2.9}$$

$$\frac{\partial \theta}{\partial \tau} = \alpha \frac{\partial^2 \theta}{\partial y^2} \tag{7.2.10}$$

These equations, subject to the boundary and initial conditions given by Eqs. (7.2.4)–(7.2.6) were then solved with Laplace transforms. The solutions were carried out for several different imposed surface temperature and flux conditions, $F(\tau)$ and $G(\tau)$, as summarized with the solutions in Table 7.2.1.

A similar analysis was reported by Menold and Yang (1962). Several different heating circumstances were considered, including a linear rise in temperature and a sinusoidal variation of the surface heat flux. Rao (1961, 1963) analyzed the same conditions with the inclusion of suction effects at the surface. Soundalgekar and Waver (1976) and Soundalgekar (1979) analyzed the transient buoyancy-induced and forced flow past an infinite vertical flat surface with variable suction, oscillating wall temperature, and mass transfer. Miyamoto (1976) studied the effects of variable fluid properties on the transient process, also for a one-dimensional conduction regime.

These one-dimensional solutions are useful, although they are valid only for a short period at the start of the transient. The duration of this process is often terminated by the arrival of the propagating leading edge effect. After the passage of this effect, the transient process becomes and remains two dimensional until the eventual steady-state condition is achieved. The following sections consider studies in which the whole transient response was investigated.

7.2.1 Integral Method Analysis for Transient Response

Siegel (1958) presented the results of an integral analysis for a vertical semi-infinite surface subject to step temperature and flux changes. The Karman-Pohlhausen method was used in the time-dependent integral momentum and energy boundary layer equations. Temperature and velocity profiles were assumed for each surface condition. For each set of profiles, the resulting simultaneous differential equations are hyperbolic, of the following form, where δ is the local boundary layer thickness and $\theta_0 = t_0 - t_\infty$ is the temperature difference:

$$\frac{\partial}{\partial \tau}\left(\frac{u\delta}{12}\right) + \frac{\partial}{\partial x}\left(\frac{u^2\delta}{105}\right) = \frac{g\beta\theta_0\delta}{3} - \frac{\nu u}{\delta} \quad (7.2.11)$$

$$\frac{\partial}{\partial \tau}\left(\frac{\delta}{3}\right) + \frac{\partial}{\partial x}\left(\frac{v\delta}{30}\right) = \frac{2\alpha}{\delta} \quad (7.2.12)$$

These equations were solved by the method of characteristics. At short times a one-dimensional conduction process is again found until the leading edge effect arrives. Then, after a transition period involving a convection transient, steady state is eventually reached. The time duration of the early one-dimensional conduction transient, τ_1, and the total time to steady state, τ_2, for an isothermal surface condition were found to be

$$\tau_1 = 1.8(1.5 + \text{Pr})^{1/2}(g\beta\theta_0)^{-1/2}x^{1/2} \quad (7.2.13)$$

$$\tau_2 = 5.24(0.952 + \text{Pr})^{1/2}(g\beta\theta_0)^{-1/2}x^{1/2} \quad (7.2.14)$$

The boundary layer thickness was predicted to reach a maximum value during the transient, before dropping to its steady-state value. Another interesting prediction was a minimum in the surface heat flux during the transient.

These predictions appeared to be in agreement with earlier experimental observations reported by Klei (1957). In that experimental study a flat vertical platinum foil 0.0005 in. (0.00127 cm) thick was heated by suddenly passing an electric current through it. The objective was to approximate a step rise in the surface heat flux. Surface temperature response was followed with 0.002-in. (0.00508-cm) copper-constantan thermocouples attached to the surface. Several different heat input levels were investigated. An important finding was the occurrence of a minimum in the heat transfer coefficient during the transient.

7.2.2 Other Changing Surface Conditions

In more extensive experiments, Goldstein (1959) and Goldstein and Eckert (1960) studied the transient response of a thin electrically heated metal foil in air and in water. A Mach-Zehnder interferometer was used to observe the transient development of the temperature field. The surface was a 0.001-in. (0.00254-cm) stainless steel foil about 4 in. (10.16 cm) wide and 6.5 in. (16.51 cm) high. An electric current was suddenly passed through the foil, generating heat in an approximately uniform fashion.

An initial one-dimensional conduction regime was observed, similar to conduction into a semi-infinite solid. During this period the temperature was uniform over most of the surface. The boundary region grew initially at a uniform spatial rate at all downstream locations. As predicted by earlier studies, the boundary region thickness reached a maximum during the transient and then decreased to the local steady-state value. The one-dimensional conduction process was observed to end at any location with the arrival of a leading edge effect. These observations are qualitatively in agreement with the theoretical predictions of Siegel (1958).

Table 7.2.2 $G(\mathrm{Pr})$ as a function of Prandtl number

Pr	$G(\mathrm{Pr})$
0.03	0.9875
0.72	1.530
1	3.983

Source: Reprinted with permission from E. M. Sparrow and J. L. Gregg, *J. Heat Transfer*, vol. 82, p. 258. Copyright © 1960, ASME.

Sparrow and Gregg (1960) calculated the transient response for a vertical surface whose temperature either experiences a step change or changes linearly with time. The stream function and temperature ratio (θ/θ_0) were expanded in a series about the quasi-static state and then substituted into the boundary layer equations. A quasi-static process is defined as a series of steady states, each consistent with the changing instantaneous bounding conditions. The ratio of the local instantaneous heat transfer coefficient h_i to the quasi-static one h_q was found to be

$$\frac{h_i}{h_q} = 1 + \frac{d\theta_0/d\tau}{\theta_0}\left(\frac{x}{g\beta\theta_0}\right)^{1/2} G(\mathrm{Pr}) \qquad (7.2.15)$$

where $G(\mathrm{Pr})$ is a function of Prandtl number as shown in Table 7.2.2. However, Goldstein (1959) had suggested that Eq. (7.2.15) was valid only close to the quasi-static conditions. A similar analysis was carried further by Chung and Anderson (1961), who used a slightly different perturbation expansion to apply this technique to larger temperature variations.

7.2.3 Response with More Realistic Input Conditions

Most calculations discussed thus far were based on idealized postulates concerning the transient behavior of either the surface temperature or the heat flux. Specified changes of surface temperature or surface heat flux, including steps, were assumed. These kinds of transient bounding conditions are neither found in practice nor usually attainable in the laboratory. The experimental arrangements in the studies of Klei (1957), Goldstein (1959), and Goldstein and Eckert (1960) did not provide a surface flux step, but rather provided a step in input to a thermal capacity element, the metal foil.

When a surface of finite thickness is suddenly heated, as by an electric current, it will dissipate heat to the fluid, but not at a constant rate through time. The surface absorbs some of the input energy as its temperature rises. The rate of heat storage in the surface element decreases as the transient proceeds and becomes zero at steady state. Thus, with negligible conductive resistance in the foil a more realistic boundary condition at the solid-fluid interface is

$$q'' = G(\tau) = c'' \frac{\partial t}{\partial \tau} - k \frac{\partial t}{\partial y} \tag{7.2.16}$$

where c'' is the surface thermal capacity per unit surface area. This condition results from an energy balance in the surface element. The total instantaneous input heat flux to the surface q'' is stored in the surface element and conducted into the fluid at the interface. The first term on the right-hand side is the rate of energy storage within the surface element; the second term is energy conducted away by the fluid.

The duration of any one-dimensional process in the fluid depends on the relative surface thermal capacity c''. If it is very small compared to the thermal capacity of the fluid, a one-dimensional process will persist for a major portion of the total transient response. Surfaces with large thermal capacities experience the one-dimensional conduction regime for a relatively smaller part of the total transient time, since more energy must be stored in the surface element, causing the transient to steady state to last longer. The leading edge effect then sweeps downstream earlier in the transient. The remaining transient response may then be quasi-static.

There is also a regime of true convection transients at intermediate levels of element thermal capacity. Then all terms in the equations, that is, all principal effects, are of comparable magnitude during most of the transient. Thus, there are three general kinds of transient response, depending on the relative thermal capacity.

Gebhart (1961) presented an integral method formulation of the surface temperature and buoyancy-induced flow response for a time-dependent heat flux $q''(\tau)$ generated internal to a vertical surface element of appreciable thermal capacity c'' and height L. This study accounts for the effect of the element thermal capacity. The downstream average surface temperature, boundary region thickness, and velocity responses were defined as follows, where the bars indicate instantaneous downstream averages:

$$\bar{\psi} = \frac{\bar{\theta}_m}{\bar{\theta}_{m,ss}} \tag{7.2.17a}$$

$$\bar{Y} = \frac{\bar{\Delta}_\theta}{\bar{\Delta}_{\theta,ss}} \tag{7.2.17b}$$

$$\bar{\chi} = \frac{u_c}{u_{c,ss}} \tag{7.2.17c}$$

Here $\theta_m = t_0 - t_\infty$ is the instantaneous local temperature difference, $\theta_{m,ss}$ applies at steady state, $\Delta_\theta = \delta_\theta/L$ is the normalized thermal boundary layer thickness, and u_c is a convection velocity. The quantities, $\bar{\psi}$, $\bar{Y}$, and $\bar{\chi}$ are zero initially and rise to one in the eventual steady state.

The resulting governing differential equations are then those of the energy balance in the fluid, the force-momentum balance, and the energy balance on the surface generating the flow:

$$\frac{\bar{\psi}}{\bar{Y}} - a(\bar{\psi}\bar{Y})' - \bar{\psi}\bar{Y}\bar{\chi} = 0 \tag{7.2.18}$$

$$S\bar{\psi}\bar{Y} - U\frac{\bar{\chi}}{\bar{Y}} - (\bar{\chi}\bar{Y})' - W\bar{Y}\bar{\chi}^2 = 0 \tag{7.2.19}$$

$$\frac{\bar{\psi}}{\bar{Y}} = \frac{q''}{q''_{ss}} - Q\bar{\psi}' \tag{7.2.20}$$

subject to $\bar{\psi} = \bar{\chi} = \bar{Y} = 0$ at $T = 0$. The single primes indicate derivatives with respect to nondimensional time T, which is defined in terms of physical time τ as

$$T = \frac{\alpha\tau}{L^2}(b\text{PrGr}^*)^{2/5} \tag{7.2.21}$$

Here Gr* is a modified Grashof number, defined as

$$\text{Gr}^* = \frac{g\beta L^4 q''_{ss}}{k\nu^2} \tag{7.2.22}$$

and q''_{ss} is the uniform steady-state value of the surface flux from the surface element to the fluid. The constants a, S, U, and W depend on Prandtl number and are collected in Table 7.2.3. These were evaluated from the steady-state temperature and velocity conditions adjacent to a flat vertical surface.

An important parameter that depends on the ratios of the thermal capacities also arises in the analysis. It characterizes the regimes of transient response. This parameter, Q, is obtained as

$$Q = \frac{c''}{\rho c_p LM}(b\text{PrGr}^*)^{1/5} \tag{7.2.23}$$

Table 7.2.3 Values of Prandtl number-dependent constants based on steady-state distributions for isothermal surface

Pr	a	$S(= U + W)$	U	W	M	$b \times 10^4$
0.01	0.1844	9.082	1.165	7.918	1.88	1.408
0.72	0.2000	16.13	9.242	6.886	1.79	40.25
1.0	0.1971	17.67	10.97	6.704	1.79	71.15
5[a]	0.1936	33.40	26.78	6.616	1.78	75.97
10	0.1894	41.69	35.29	6.398	1.77	87.43
100	0.1924	126.9	121.4	5.523	1.77	137.1
1000	0.1905	263.2	258.6	4.677	1.76	118.4

[a]Interpolated from other values.

Source: Reprinted with permission from B. Gebhart, *J. Heat Transfer,* vol. 83, p. 61. Copyright © 1961, ASME.

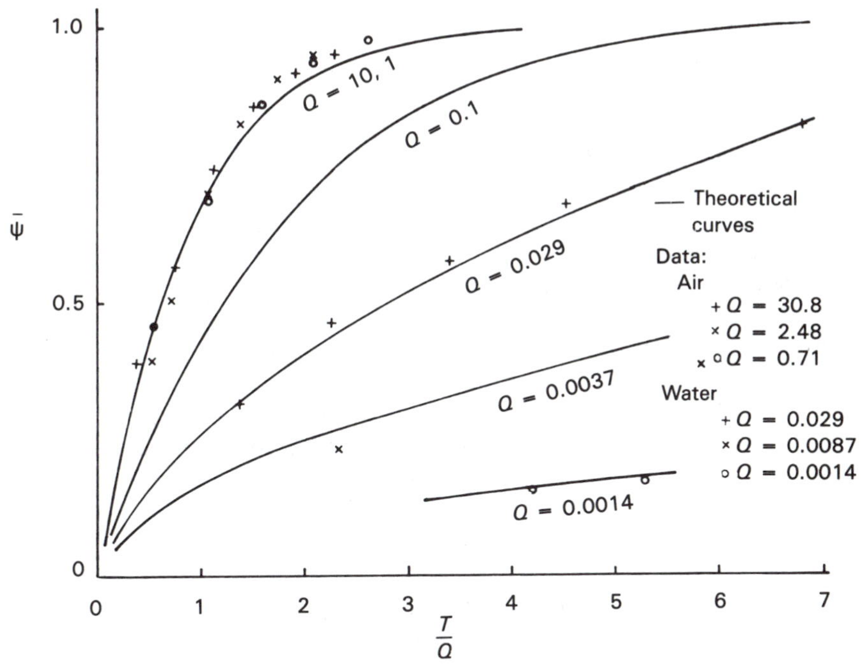

Figure 7.2.1 Transient temperature response for vertical surfaces, comparison of theory and experiment. *(Reprinted with permission from B. Gebhart and D. E. Adams,* J. Heat Transfer, *vol. 85, p. 25. Copyright © 1963, ASME.)*

Values of b and M are also given in Table 7.2.3. Large values of Q occur for elements with large c'' compared to ρc_p. This results in a quasi-static process. The large surface thermal capacity lengthens the transient time.

In related studies Gebhart (1963a, 1963b, 1963c, 1964) presented the results of the numerical solution of Eqs. (7.2.18)–(7.2.20) over a wide range of Q. One of the most important results of these calculations is the classification of the resulting transient regimes according to the value of Q. For an input flux step $\bar{q}''$ it was shown that for $Q \geq 1.0$ the process is quasi-static throughout. Then $\bar{\psi}$ is a function of T/Q alone. For $Q \leq 0.1$ the process is a one-dimensional conduction transient through most of the total transient process to steady state. True convection transients, where transient convection terms are important, occur for $0.1 \leq Q \leq 1.0$.

The results of these calculations, in terms of the transient temperature response of the surface element, $\bar{\Psi}$, were compared with experimental data taken in air, pressurized N_2, and water. Figure 7.2.1 shows the comparison for a wide range of Q values. The data were collected by three different experimental methods: interferometry, infrared detection, and resistance thermometry. The data are from Goldstein and Eckert (1960), Martin (1961), Lurie and Johnson (1962), and Gebhart and Adams (1963). This agreement in average response is very good.

7.2.4 Downstream Propagation of the Leading Edge Effect

Goldstein and Briggs (1964) presented the results of an analysis to predict the duration of a one-dimensional regime of velocity and heat transfer response at any downstream location before the arrival of the leading edge effect. The leading edge effect was considered to be like a wave, beginning at the leading edge and sweeping downstream. All locations downstream of the leading edge wave were assumed to be in the one-dimensional conduction regime and all locations upstream to be in a true convection transient, where the two-dimensional velocity and temperature fields are a function of x, y, and τ.

The leading edge effect propagation calculation was based on the transient flow response adjacent to a double infinite surface, that is, in a one-dimensional transient. The equations for conservation of mass, momentum, and energy are again Eqs. (7.2.8)–(7.2.10). These were solved for a step increase in the surface temperature and a step increase in the surface heat flux, for a surface of negligible thermal capacity as well as a uniform flux input condition, taking into account the surface element thermal capacity.

Table 7.2.4 Local heat transfer and percent overshoot for various Prandtl number values

(a) Step change in flux $q''(\tau) = \text{const.} \quad \tau > 0$

Pr	$Nu_x/(Gr_x^*)^{1/4}$ End of pure conduction	$Nu_x/(Gr_x^*)^{1/4}$ Steady state	Percent overshoot
0.1	0.237	0.2634	10.0
1.0	0.502	0.534	6.0
6.4	0.844	0.850	0.7
10.0	0.9473	0.9442	−0.3 (undershoot)
100.0	1.6409	1.5562	−5.4 (undershoot)

(b) Step change in surface temperature $\theta(0, \tau) = \theta_0 = \text{const.} \quad \tau > 0$

Pr	$Nu_x/(Gr_x)^{1/4}$ End of pure conduction	$Nu_x/(Gr_x)^{1/4}$ Steady state	Percent overshoot
0.01	0.0438	0.0574	23.7
1.0	0.315	0.401	21.4
10.0	0.697	0.827	15.7
100.0	1.386	1.549	10.5

Source: Reprinted with permission from R. J. Goldstein and D. G. Briggs, *J. Heat Transfer*, vol. 86, p. 490. Copyright © 1964, ASME.

It was assumed that the leading edge effect propagates downstream with the instantaneous velocity field. The instantaneous distance of penetration is thus

$$x_p = \int_0^\tau u(y,\tau)\, d\tau \tag{7.2.24}$$

where $u(y, \tau)$ is the velocity obtained from the one-dimensional conduction solution. The value of x_p is the maximum penetration distance for the leading edge effect at any time. Thus, the value $u(y, \tau)$ is taken as the maximum across the instantaneous velocity profile at each successive time.

Another important result of this analysis was the prediction of a minimum in the heat transfer coefficient during the transient. This was determined by comparing the values of Nu_x at the end of the pure conduction regime and at steady state. Table 7.2.4 lists some of those values for surfaces subjected to step heat flux and step temperature increases.

In related experimental studies, Gebhart and Dring (1967), Mollendorf and Gebhart (1970), and Mahajan and Gebhart (1978) determined the rate of propagation of the leading edge effect. For uniform energy generation in a surface element of finite thermal capacity it was shown that the predicted propagation rate of Goldstein and Briggs (1964) may be expressed as

$$x' = \left(\frac{ka^5}{q\beta q''\sqrt{\alpha}}\right) x_{p_{max}} = (t')^5 f(t') F(\mathrm{Pr}) \tag{7.2.25}$$

where $t' = a\sqrt{\tau}$, $a = (\rho c_p k)^{1/2}/c''$, $f(t')$ is a time-dependent function determined from the analysis of Goldstein and Briggs (1964), $x_{p_{max}}$ is the maximum value of the integral defined in Eq. (7.2.24) and $F(\mathrm{Pr})$ is the Prandtl number effect that arises from the integral in Eq. (7.2.24). For air $F(\mathrm{Pr}) = 1.17$.

Figure 7.2.2 shows interferometric measurements from Gebhart et al. (1967) and Mollendorf and Gebhart (1970), local sensor data from Mahajan and Gebhart (1978), and calculations from Goldstein and Briggs (1964) on an x' vs. t' plane. The local sensor data show the highest propagation rate for the leading edge effect, possibly because interferometers are insensitive to local disturbances. All the measurements, however, indicate a much higher propagation rate than does the analysis. This is probably due to effects not accounted for in the analysis, particularly since the characteristics of flow in the leading edge region and the development of the pressure field are not completely assessed.

Figure 7.2.3, from Gebhart and Dring (1967), illustrates several interesting phenomena. The first three frames show that at early times the temperature field is independent of x, except in the immediate vicinity of the leading edge. This substantiates the previous prediction of an early one-dimensional process. Frames four to seven show a temperature disturbance, the leading edge effect, traveling downstream. The temperature field in the wake of this disturbance is two dimensional. Downstream, however, the one-dimensional process still persists. Finally, in the last frame the whole temperature field is steady and two dimensional.

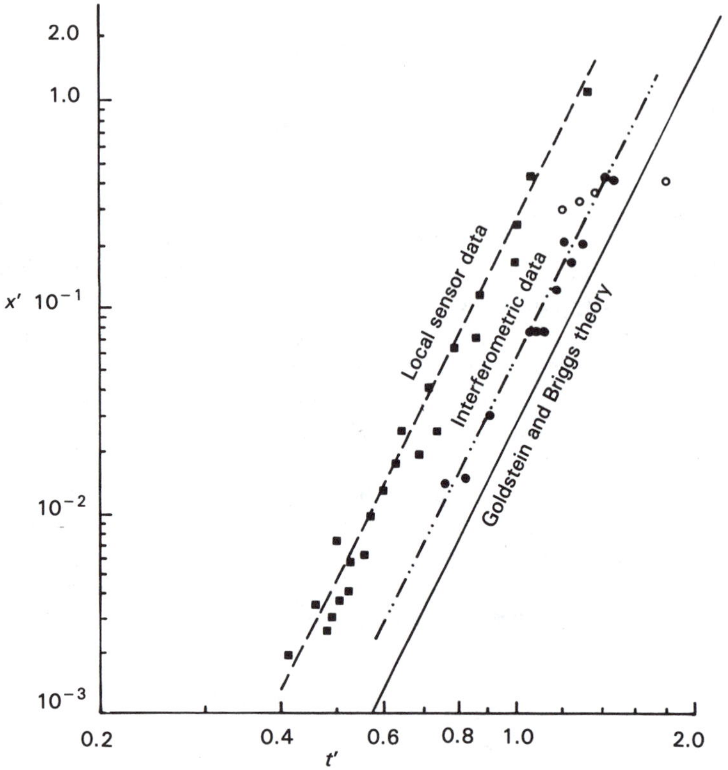

Figure 7.2.2 Comparison of first observed waves to the leading edge propagation rate theory and other data. (●) Gebhart and Dring (1967); (○) Mollendorf and Gebhart (1970); (■) Mahajan and Gebhart (1978). *(Reprinted with permission from R. L. Mahajan and B. Gebhart,* J. Heat Transfer, *vol. 100, p. 731. Copyright © 1978, ASME.)*

7.2.5 Numerically Determined Transient Responses

To obtain more detailed solutions the full partial differential equations (7.2.1)–(7.2.3) may be numerically integrated for specific conditions, as initially done by Sugawara and Michiyoshi (1952). Hellums and Churchill (1961) reported the results of a detailed numerical study for a flat vertical surface whose temperature experiences a step rise from t to t_0 to begin a transient. The equations in terms of u, v, x, y, t_0, and time τ were nondimensionalized as follows:

$$U = \frac{u}{[\nu g\beta(t_0 - t_\infty)]^{1/3}} \qquad V = \frac{v}{[\nu g\beta(t_0 - t_\infty)]^{1/3}} \tag{7.2.26}$$

$$X = x\left[\frac{g\beta(t_0 - t_\infty)}{\nu^2}\right]^{1/3} = (\mathrm{Gr}_x)^{1/3} \qquad Y = y\left[\frac{g\beta(t_0 - t_\infty)}{\nu^2}\right]^{1/3} \tag{7.2.27}$$

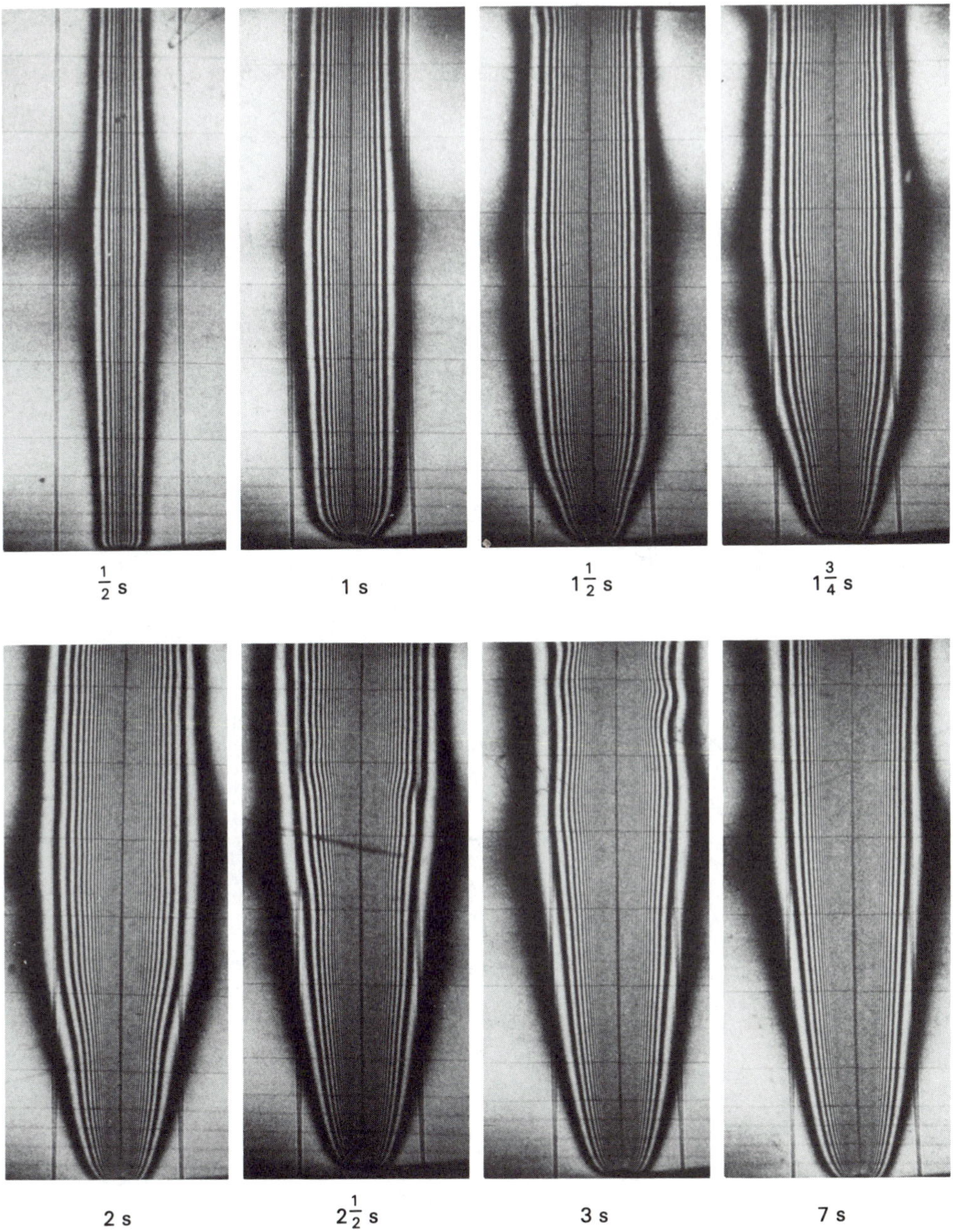

Figure 7.2.3 Interferograms at various times during the convection transient at Gr = 3.98×10^9, $P = 17.1$ atm. *(Reprinted with permission from B. Gebhart and R. P. Dring,* J. Heat Transfer, *vol. 89, p. 274. Copyright © 1967, ASME.)*

$$\bar{\tau} = \tau \frac{[g\beta(t_0 - t_\infty)]^{2/3}}{\nu^{1/3}} \qquad \phi = \frac{t - t_\infty}{t_0 - t_\infty} \tag{7.2.28}$$

The resulting generalized two-dimensional finite-difference transient equations are

$$\frac{U'_{i,j} - U'_{i-1,j}}{\Delta X} + \frac{V'_{i,j} - V'_{i,j-1}}{\Delta Y} = 0 \tag{7.2.29}$$

$$\frac{U'_{i,j} - U_{i,j}}{\Delta \bar{\tau}} + U_{i,j} \frac{U_{i,j} - U_{i-1,j}}{\Delta X} + V_{i,j} \frac{U_{i,j+1} - U_{i,j}}{\Delta Y}$$

$$= \phi'_{i,j} + \frac{U_{i,j+1} - 2U_{i,j} + U_{i,j-1}}{(\Delta Y)^2} \tag{7.2.30}$$

$$\frac{\phi'_{i,j} - \phi_{i,j}}{\Delta \bar{\tau}} + U_{i,j} \frac{\phi_{i,j} - \phi_{i-1,j}}{\Delta X} + V_{i,j} \frac{\phi_{i,j+1} - \phi_{i,j}}{\Delta Y}$$

$$= \frac{1}{\Pr} \frac{\phi_{i,j+1} - 2\phi_{i,j} + \phi_{i,j-1}}{(\Delta Y)^2} \tag{7.2.31}$$

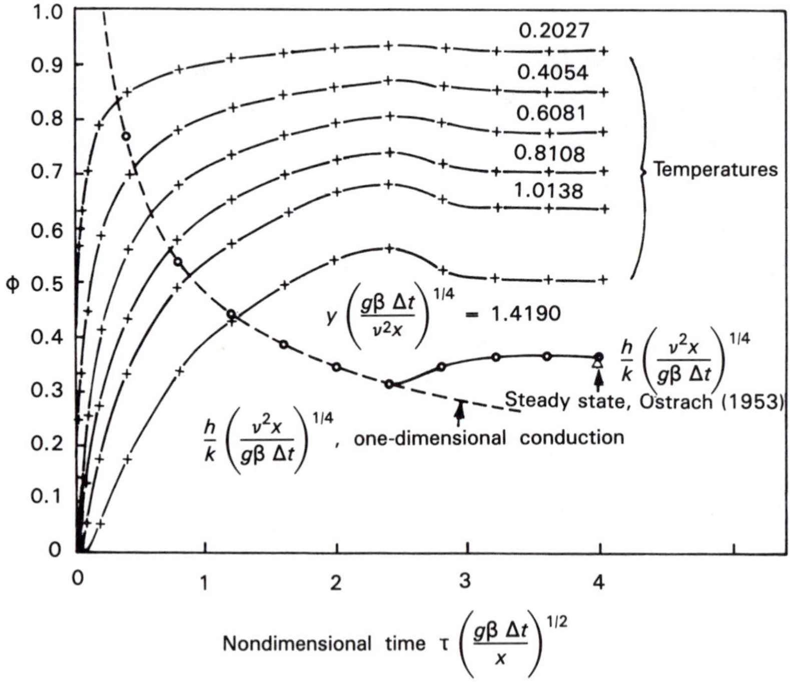

Figure 7.2.4 Transient temperatures and nondimensional heat transfer coefficient. The steady-state result is from Ostrach (1953). *(Reprinted with permission from J. D. Hellums and S. W. Churchill,* Proc. International Heat Transfer Conference, 1961. *Copyright © 1961, ASME.)*

subject to the following boundary and initial conditions:

$$X = 0: \quad U = \phi = 0 \tag{7.2.32}$$

$$Y = 0: \quad U = V = 0 \quad \phi = 1 \tag{7.2.33}$$

$$Y \to \infty: \quad U = V = \phi = 0 \tag{7.2.34}$$

$$\bar{\tau} = 0: \quad U = V = \phi = 0 \tag{7.2.35}$$

The primes in Eqs. (7.2.29)–(7.2.31) indicate values at the present time; all other values are from the previous time step. The subscripts i and j indicate location in the space grid for x and y, respectively. The equations were solved for Gr $= 10^6$ and Pr $= 0.733$, as for air at room temperature.

During the early transient heat is transferred mainly by conduction, as discussed earlier. Figure 7.2.4 shows the temperature response at various y locations as a function of time, as computed by Hellums and Churchill (1962). Also shown is the computed surface heat transfer coefficient in nondimensional form. At short

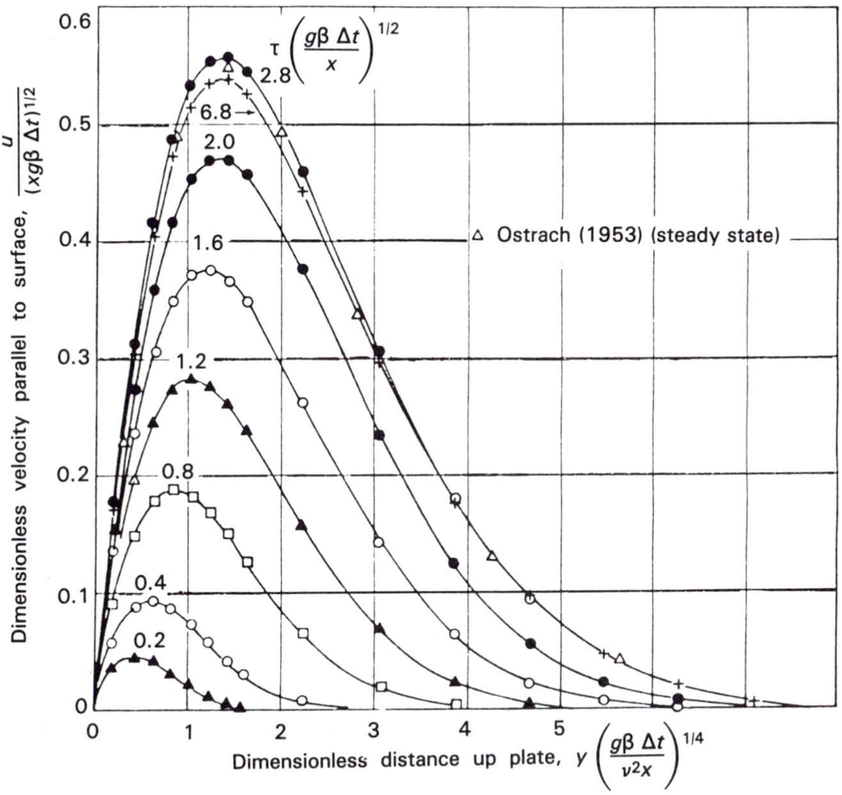

Figure 7.2.5 Transient velocity profiles. *(Reprinted with permission from J. D. Hellums and S. W. Churchill,* Proc. International Heat Transfer Conference, 1961. *Copyright © 1961, ASME.)*

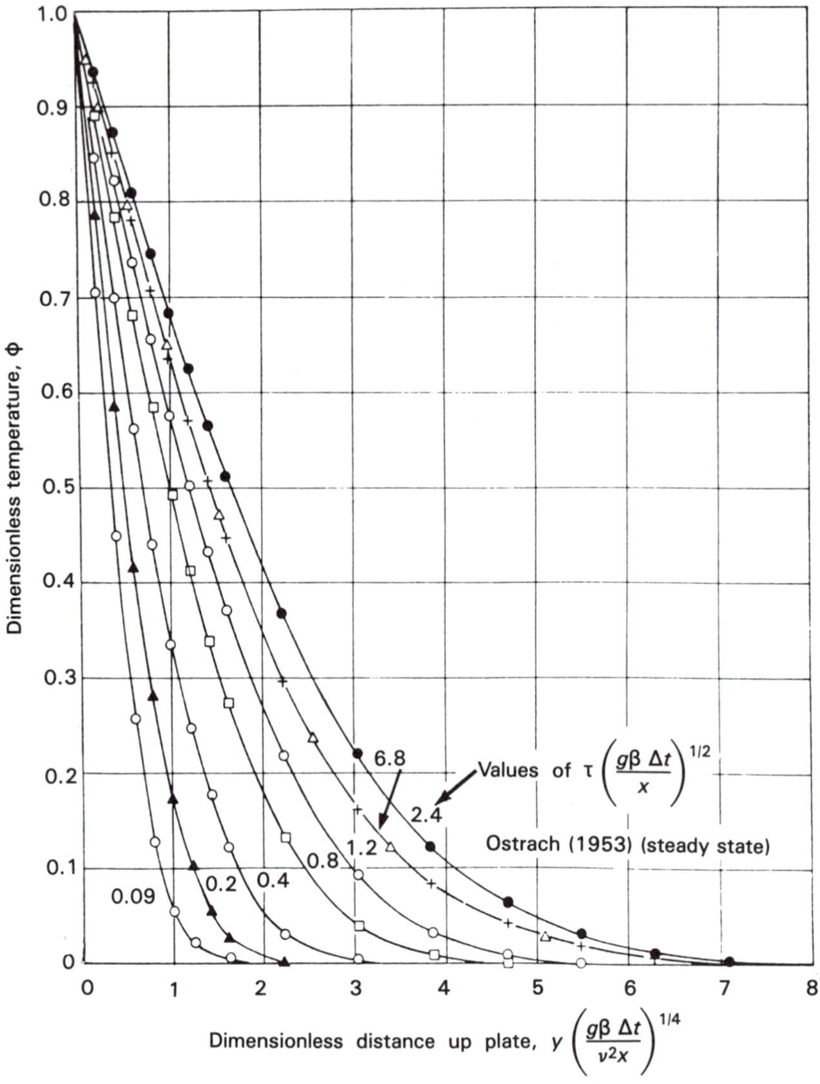

Figure 7.2.6 Transient temperature profiles. *(Reprinted with permission from J. D. Hellums and S. W. Churchill,* Proc. International Heat Transfer Conference, 1961. *Copyright © 1961, ASME.)*

times the computed heat transfer coefficient shows excellent agreement with the one-dimensional conduction solution. At later times the heat transfer coefficient reaches a minimum value before deviating from the one-dimensional conduction solution. Finally, at steady state good agreement is observed with the steady-state similarity solution.

Figures 7.2.5 and 7.2.6 show the transient temperature and velocity profiles, respectively, calculated by Hellums and Churchill (1962). Both the velocity and temperature levels locally exceed their eventual steady-state values during the transient. The overshoots occur near the end of the one-dimensional conduction regime. Note that in Figs. 7.2.4–7.2.6 a different normalization is used from that given in Eqs. (7.2.26)–(7.2.28) above.

7.2.6 Numerically Determined Response with Surface Thermal Capacity

Sammakia and Gebhart (1978, 1981) reported the results of a similar transient numerical analysis for a surface element with finite thermal capacity, suddenly subjected to a uniform heat flux q'', for air and water, respectively. Transient response was calculated after the energy supply to the surface was begun. The relevant conservation equations were nondimensionalized as follows:

$$U = \frac{u}{(\nu^2 g\beta q''/k)^{1/4}} = \frac{ux/\nu}{(\mathrm{Gr}_x^*)^{1/4}} \tag{7.2.36}$$

$$V = \frac{v}{(\nu^2 g\beta q''/k)^{1/4}} = \frac{vx/\nu}{(\mathrm{Gr}_x^*)^{1/4}} \tag{7.2.37}$$

$$X = \frac{x}{(\nu^2 k/g\beta q'')^{1/4}} = (\mathrm{Gr}_x^*)^{1/4} \tag{7.2.38}$$

$$Y = \frac{y}{(\nu^2 k/g\beta q'')^{1/4}} = \frac{y}{x}(\mathrm{Gr}_x^*)^{1/4} \tag{7.2.39}$$

$$\phi = \frac{t - t_\infty}{(\nu^2 q''^3/g\beta k^3)^{1/4}} = \frac{t - t_\infty}{q''x/k}(\mathrm{Gr}_x^*)^{1/4} \tag{7.2.40}$$

$$\bar{\tau} = \frac{\tau}{(k/g\beta q'')^{1/2}} = \frac{\alpha\tau}{x^2}\mathrm{Pr}(\mathrm{Gr}_x^*)^{1/2} \tag{7.2.41}$$

The resulting equations and relevant boundary conditions are

$$\frac{\partial U}{\partial X} + \frac{\partial V}{\partial Y} = 0 \tag{7.2.42}$$

$$\frac{\partial U}{\partial \bar{\tau}} + U\frac{\partial U}{\partial X} + V\frac{\partial U}{\partial Y} = \phi + \frac{\partial^2 U}{\partial Y^2} \tag{7.2.43}$$

$$\frac{\partial \phi}{\partial \bar{\tau}} + U\frac{\partial \phi}{\partial X} + V\frac{\partial \phi}{\partial Y} = \frac{1}{\mathrm{Pr}}\frac{\partial^2 \phi}{\partial Y^2} \tag{7.2.44}$$

$$\bar{\tau} = 0 \qquad U = V = \phi = 0 \tag{7.2.45}$$

$$X = 0 \qquad U = \phi = 0 \tag{7.2.46}$$

$$Y = 0 \qquad U = V = 0 \tag{7.2.47}$$

$$Y \to \infty \qquad U,T \to 0 \tag{7.2.48}$$

For $\bar{\tau} < \bar{\tau}_s$ and $Y = 0$, where $\bar{\tau}_s$ is the time at which the energy supply to the surface is switched off, the following boundary condition applies:

$$1 = Q^* \frac{\partial \phi}{\partial \bar{\tau}} - \frac{\partial \phi}{\partial Y} \tag{7.2.49}$$

For $\bar{\tau} \geq \bar{\tau}_s$ at $Y = 0$

$$Q^* \frac{\partial \phi}{\partial \bar{\tau}} = \frac{\partial \phi}{\partial Y} \tag{7.2.50}$$

The nondimensional thermal capacity parameter Q^* is given by

$$Q^* = c'' \left[\frac{\nu^2 g \beta q''}{k^5} \right]^{1/4} = \frac{c''}{\rho c_p x} \Pr (\mathrm{Gr}_x^*)^{1/4} \tag{7.2.51}$$

Equation (7.2.45) represents an initially quiescent ambient medium with a uniform temperature distribution, Eq. (7.2.46) specifies zero transport upstream of the leading edge, and Eq. (7.2.47) represents a no-slip, impervious boundary condition at the surface. Equation (7.2.49) results from an energy balance at the

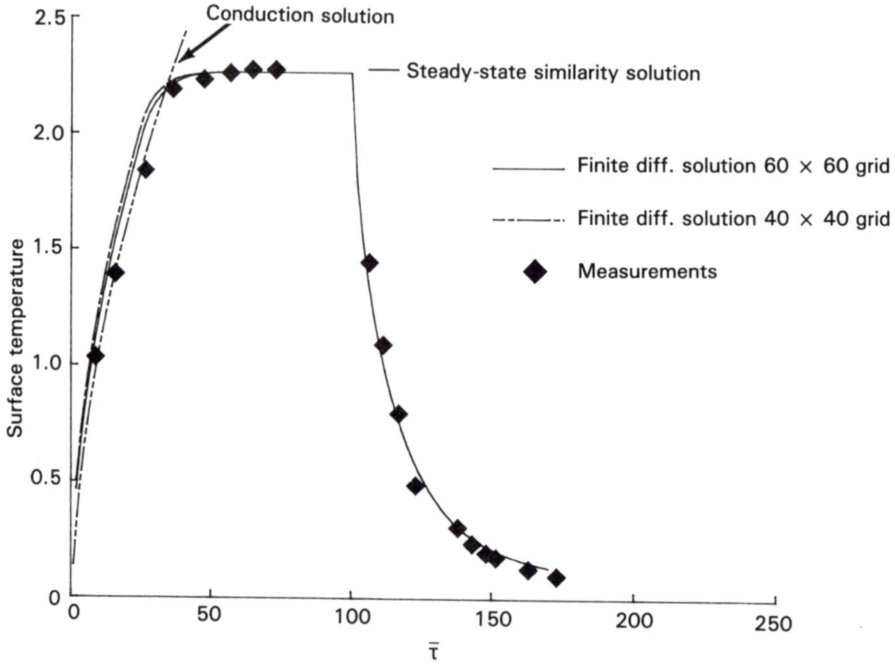

Figure 7.2.7 Surface temperature history for $X = 31.3$ and $Q^* = 2.0$. *(Reprinted with permission from B. Sammakia et al.,* J. Heat Transfer, *vol. 104, p. 644. Copyright © 1982, ASME.)*

solid-fluid interface during the heating process. The two terms on the right-hand side of Eq. (7.2.49) represent energy storage in the element of the surface and energy convected away by the fluid, respectively.

The energy storage rate within the surface has a maximum value of q'' at the beginning of the transient process. This storage rate decreases during the whole transient period and becomes zero as steady state is achieved. Meanwhile, energy transfer to the fluid increases from zero at the beginning of the transient to its final steady-state value q''. Equation (7.2.50) results from an energy balance at the solid-fluid interface during the cooling process and simply states that energy convected by the fluid comes from the energy previously stored in the surface. The cooling process continues until the temperature field reaches a uniform distribution at t_∞. The medium becomes quiescent once more.

Sammakia et al. (1980, 1982) reported numerical solutions to Eqs. (7.2.42)–(7.2.44) subject to boundary and initial conditions (7.2.45)–(7.2.50), for air and water, respectively. The equations were solved with an explicit finite-difference scheme. A uniform grid spacing was used in both the X and Y directions, and several different grid sizes were investigated. Experimental measurements were also reported.

Figures 7.2.7–7.2.9 show some of the computed results. In Fig. 7.2.7 the

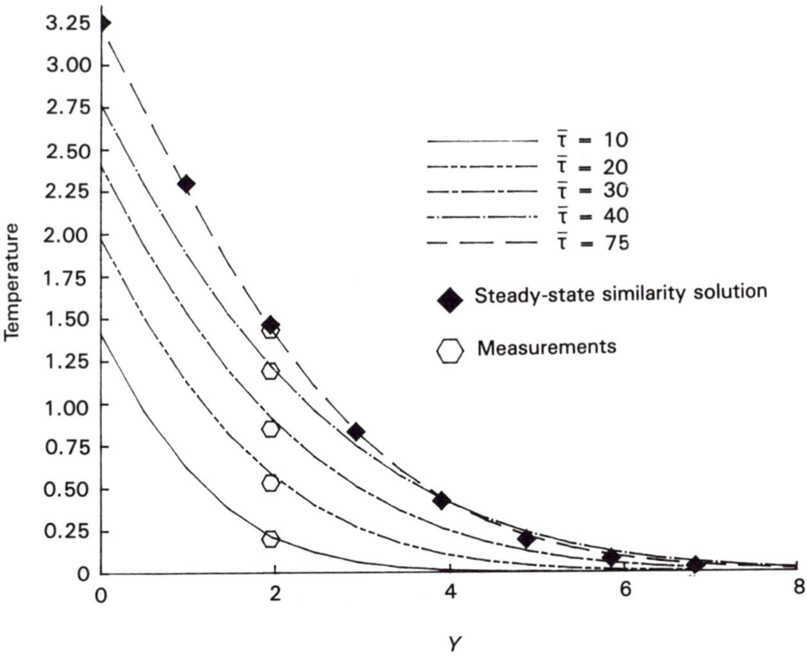

Figure 7.2.8 Transient temperature profiles during heating for $X = 187$ and $Q^* = 1.2$. *(Reprinted with permission from B. Sammakia et al.,* J. Heat Transfer, *vol. 104, p. 644. Copyright © 1982, ASME.)*

computed surface temperature response is shown for both the heating and cooling transients, for two grid sizes, with water. Also shown is the one-dimensional closed-form solution, which agrees with the numerical solution at early times. The steady-state similarity solution is also in good agreement with the eventual steady state predicted by the numerical scheme. Figures 7.2.8 and 7.2.9 show the computed transient temperature and velocity profiles in the fluid.

Sammakia et al. (1980, 1982) also reported the results of transient measurements in air and water during both the heating and cooling processes. Figure 7.2.7 compares the data with the computed surface temperature response for water. Good agreement was found during both parts of the transient. Temperature measurements at one location in the boundary layer are also shown in Figure 7.2.8 and compare well with the analysis.

Joshi and Gebhart (1987a) reported the results of extensive measurements of the transient response in water adjacent to a flat-vertical surface of finite thermal capacity. The surface dissipated a uniform and constant heat flux over a wide range. Temperature and velocity measurements were made in the boundary layer. The measurements compared well at short times with the one-dimensional conduction solution and at long times with the steady-state similarity solution, as shown in Fig. 7.2.10 and 7.2.11. Many different kinds of unstable and transition regimes were found.

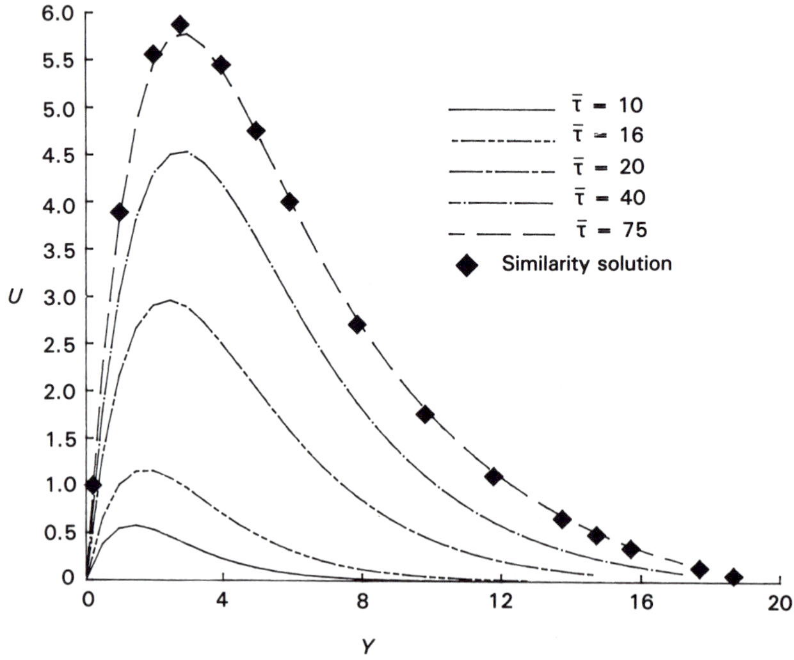

Figure 7.2.9 Transient velocity profiles during heating for $X = 187$ and $Q^* = 1.2$. *(Reprinted with permission from B. Sammakia et al.,* J. Heat Transfer, *vol. 104, p. 644. Copyright © 1982, ASME.)*

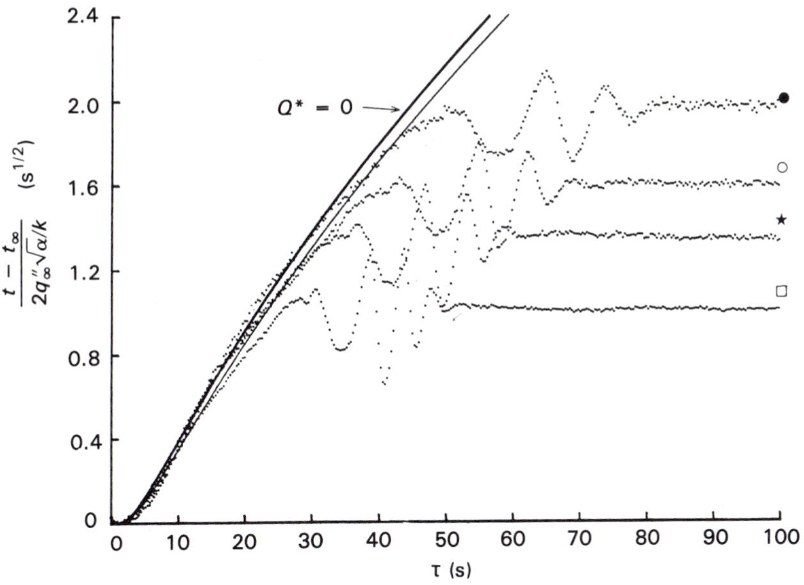

Figure 7.2.10 Transient temperature development for various q'' at x = 0.29 m and y = 1.68 mm. Short-time solutions with and without the effects of surface thermal capacity are shown as solid curves. Symbols at right are laminar boundary layer predictions at Pr = 6.2: (●) q'' = 587.6 W/m^2; (○) q'' = 842.7 W/m^2; (★) q'' = 1140.9 W/m^2; (□) q'' = 1873.6 W/m^2. *(From Joshi and Gebhart, 1987a.)*

Joshi and Gebhart (1987b) also measured the transient resulting from a sudden increase or decrease in surface heat flux, starting from an established boundary layer flow. The transient was found to initially be a superposition of a one-dimensional conduction regime on an established boundary layer flow. Table 7.2.5 summarizes all the experimental investigations discussed in this section.

7.2.7 Other Studies

Callahan and Marner (1976) reported a numerical analysis for transient natural convection with mass transfer adjacent to an isothermal surface. An explicit finite-difference scheme was used, similar to that of Hellums and Churchill (1961) and Sammakia and Gebhart (1978). Ingham (1978) pointed out that under some circumstances this finite-difference technique could yield inaccurate results at an intermediate time during the transient, after the early one-dimensional regime and before steady state is achieved.

Chung and Anderson (1961) perturbed both gravity and surface temperature around the steady-state solution to study the transient effects. Nanda and Sharma (1962) presented a possible transient similarity solution including the additional effect of suction at the surface. Singh (1964) also considered the effect of suction at the surface on transient response. Mizukami (1976) studied the propagation of

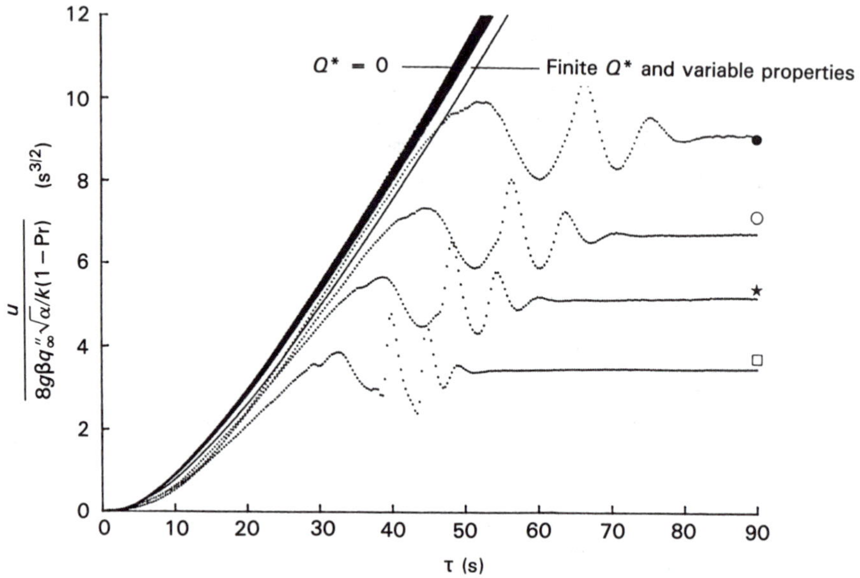

Figure 7.2.11 Transient velocity development for various q'' at $x = 0.29$ m and $y = 2.20$ mm. Short-time solutions with and without the effects of surface thermal capacity are contained within two separate bands. Symbols at right are laminar boundary layer predictions at Pr = 6.2: (●) $q'' = 587.6$ W/m^2; (○) $q'' = 842.7$ W/m^2; (★) $q'' = 1140.9$ W/m^2; (□) $q'' = 1873.6$ W/m^2. *(From Joshi and Gebhart, 1987a.)*

the leading edge effect for surfaces with time-dependent temperature or heat flux. Yang et al. (1974) used a finite-difference technique to study transients adjacent to surfaces with sinusoidal temperature variation.

Brown and Riley (1973) analyzed the transient response adjacent to an isothermal surface subjected to a sudden temperature change. They dealt with three different stages of the transient: the initial one-dimensional regime, the early departure from that regime, and an asymptotic solution that describes the manner in which the steady state is achieved.

7.3 TRANSIENT FLOW OVER CYLINDERS AND WIRES

Transient processes adjacent to cylinders and thin cylinders, or "wires," are diverse and interesting. Such processes may be classified according to several relevant parameters. Orientation of the cylinder or wire may play a major role in the transport process. Another consideration is the diameter-to-length ratio D/L.

Vertical cylinders. If D/L for a vertical cylinder is sufficiently large, the cylinder is well approximated by the vertical surface geometry of the previous section. The heating process, as for a flat vertical element, is another parameter that dictates the nature of the transient response. If the cylinder or wire has an

Table 7.2.5 Summary of experimental studies of natural convection transients adjacent to a vertical surface

Investigator(s)	Fluid(s)	Gr*	Gr	Q^*	ΔT (°C)	Height of surface (cm)	Experimental technique
Klei (1957)[a]	Air		$7.9–12 \times 10^4$	12.6–15.4	50.6–110.6	2.5	Thermocouple measurements
Goldstein (1959)	Water	$1.8–4.8 \times 10^8$	$3.4–7.8 \times 10^6$	0.6–0.08	0.3–0.6	16.5	Interferometry
	Air	$4.3–7.7 \times 10^8$	$1.3–1.9 \times 10^7$	20.1–23.2	5.9–21.2	20.3	
Martin (1961)	Water	$1.3–23 \times 10^7$	$4.3–34 \times 10^6$	0.4–0.8	6.7–52.8	2.5	Resistance thermometry
	Water	$2.8–21 \times 10^9$	$5.3–23 \times 10^6$	0.3–0.5	5.3–23.1	6.4	
	Water	$1.4–6.5 \times 10^9$	$1.9–5.9 \times 10^7$	0.3–0.5	6.9–21.4	8.9	
Lurie and Johnson (1962)	Water	1.8×10^{11}	9.2×10^8	1.2	63.3	7.6	Resistance thermometry
Gebhart and Adams (1963)	Air	7.7×10^6	5.0×10^5	282.3	129.4	3.8	Infrared detector
	Air	4.5×10^7	2.2×10^6	24.8	125.0	6.4	
	Air	8.7×10^7	4.6×10^6	7.3	110.0	7.6	
Gebhart, et al. (1967)	Pressurized nitrogen	$2.4–510 \times 10^9$	$5.4–400 \times 10^7$	2.3–12.8	2.7–41.9	18.4	Interferometry
Mollendorf and Gebhart (1970)	Pressurized nitrogen	$1.8–61 \times 10^{12}$	$9.1–120 \times 10^9$	2.7–5.1	5.6–27.3	50.2	Interferometry
Rajan and Picot (1971)	$Pr \sim 6–10^6$	$2.4–40 \times 10^{10}$			~3	25.4	Thermocouple measurements
Mahajan and Gebhart (1978)	Pressurized nitrogen	$2.5–150 \times 10^9$		2.9–32.3	2–80	38.5	Hot wire and thermocouple measurements
Sammakia et al. (1980)	Air	$3.7–69 \times 10^{11}$		284	~4.1–95.9	130.5	Thermocouple measurements
Sammakia et al. (1982)	Water	$6.7 \times 10^5–12 \times 10^8$		1.2–2.0		130.5	Thermocouple measurements
Joshi and Gebhart (1987a)	Water	$2.7 \times 10^8–10^{13}$		0.86–1.14	~2–8.7	124.0	Hot film and thermocouple measurements and flow visualization

[a]Based on data reported from Siegel (1958).

appreciable thermal capacity, the relative magnitude of that capacity will also shape the transient process, as seen for flat elements.

When a vertical cylinder in an infinite ambient medium is heated, the resulting transient process is similar to that for the flat vertical element. At first, conduction in the fluid dominates. Then the leading edge effect arrives. A true convection transient may prevail until steady state is achieved. A solution for the one-dimensional conduction transient is given in Carslaw and Jaeger (1959) for an infinite cylinder of radius R with a finite thermal capacity. Perfect thermal contact at the cylinder outer region interface and infinite thermal conductivity in the cylinder are postulated. The cylinder is assumed to be suddenly heated by internal generation at a constant rate per unit length, q'. Using Laplace transforms, two series solutions are found, Eqs. (7.3.1) and (7.3.2) below, for the cylinder temperature response $\theta = t(R, \tau) - t_\infty$ for short and long times, that is, for small and large Fourier number Fo, where $\mathrm{Fo} = \alpha\tau/R^2$.

$$\theta = \frac{q'c}{2\pi k}\left[\mathrm{Fo} - \frac{4c}{3\sqrt{\pi}}\mathrm{Fo}^{3/2} + O(\mathrm{Fo})^2\right] \tag{7.3.1}$$

$$\theta = \frac{q'}{4\pi k}\left[\ln\left(\frac{4\mathrm{Fo}}{e^\gamma}\right) + \frac{1}{2\mathrm{Fo}} - \frac{2-c}{2c\mathrm{Fo}}\ln\left(\frac{4\mathrm{Fo}}{e^\gamma}\right) + \cdots\right] \tag{7.3.2}$$

where k and α are the thermal conductivity and diffusivity of the fluid, c is twice the ratio of the material heat capacity of the fluid, ρc_p, to that of the cylinder, and $\gamma = 0.5772$ is Euler's constant.

Gebhart and Dring (1967) reported the results of an experimental investigation of the transient natural convection response adjacent to thin vertical cylinders. The D/L range for the experiments was 1.15×10^{-3} to 4.56×10^{-3}. Tests were run in air and in silicone oils. The early temperature response compares well with the conduction analysis discussed above, as shown in Fig. 7.3.1. However, the later response falls away, with convection, to the steady state.

The duration of the one-dimensional conduction regime adjacent to a vertical cylinder was calculated by Goldstein and Briggs (1964) in an analysis similar to that for the flat vertical surface discussed in Section 7.2. Three conditions were considered: a sudden step increase in cylinder temperature, beginning uniform heat flux generation from a cylinder with negligible thermal capacity, and uniform heat flux generation in a cylinder with appreciable thermal capacity but no internal conductive resistance. The solutions are in integral form, and either series expansion or numerical integration is required to obtain specific results.

Horizontal wires and cylinders. These transients are quite different. During the very early transient conduction stage, the orientation of the wire does not affect the resulting process. However, in contrast to vertical cylinders, there is no leading edge effect to terminate the one-dimensional regime along the axis. Instead, the conduction process lasts until buoyancy results in convective motion. Eventually steady state is reached as a plume is shed.

Temperature overshoot during early transient responses was observed by Os-

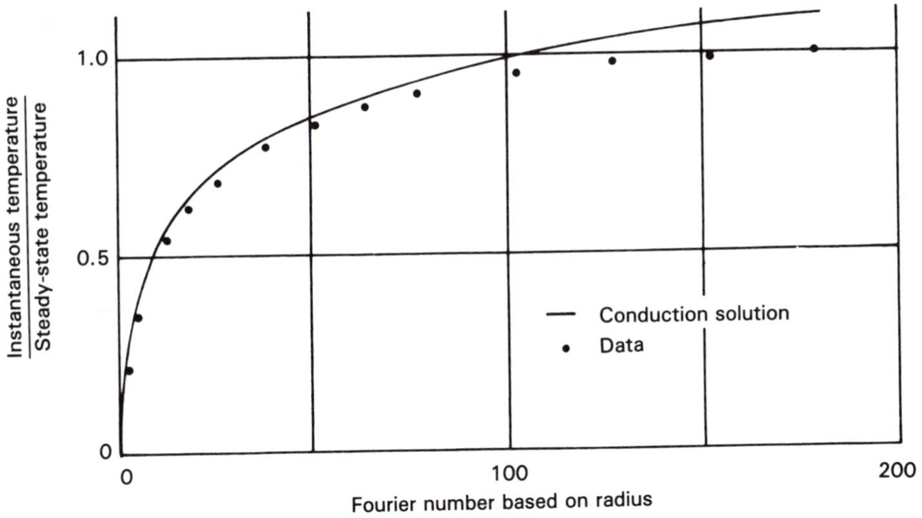

Figure 7.3.1 Conduction solution and data for a transient process adjacent to a vertical cylinder immersed in a silicone fluid. *(Reprinted with permission from B. Gebhart and R. P. Dring,* J. Heat Transfer, *vol. 89, p. 24. Copyright © 1967, ASME.)*

troumov (1956) under certain circumstances. In that experimental study, horizontal wires immersed in several different fluids were heated by passing an electric current through them. The total transient time was inversely proportional to the input heating rate. This is a trend similar to that predicted by Siegel (1958) for flat vertical surfaces. Soehngen (1968) observed similar temperature overshoots over horizontal wires and cylinders for high Prandtl number fluids.

The interferograms in Fig. 7.3.2 were taken during the early transient adjacent to a heated horizontal wire by Vest and Lawson (1972). One-dimensional conduction persists for about 8.0 s for the particular conditions shown, as indicated by the continuing symmetry of the isotherms in Figs. 7.3.2a and 7.3.2b. The duration of the conduction regime was correlated by the following equation, derived by Parsons and Mulligan (1978):

$$\tau_D = \frac{4\alpha\tau}{D^2} = 80.2(\mathrm{RaNu})^{-2/3} \tag{7.3.3}$$

where τ_D is the nondimensional time duration of the conduction process. The one-dimensional conduction regime was said to be terminated when the product RaNu reaches 1100. Convection effects were first observed at that value.

Measurements by Parsons and Mulligan (1978) showed agreement with the above correlation. Some of the latter results are shown in Fig. 7.3.3, along with measurements by Ostroumov (1956). The analysis by Sugawara and Michiyoshi (1952) was extended to include a cylindrical field of finite diameter by Parsons and Mulligan (1980).

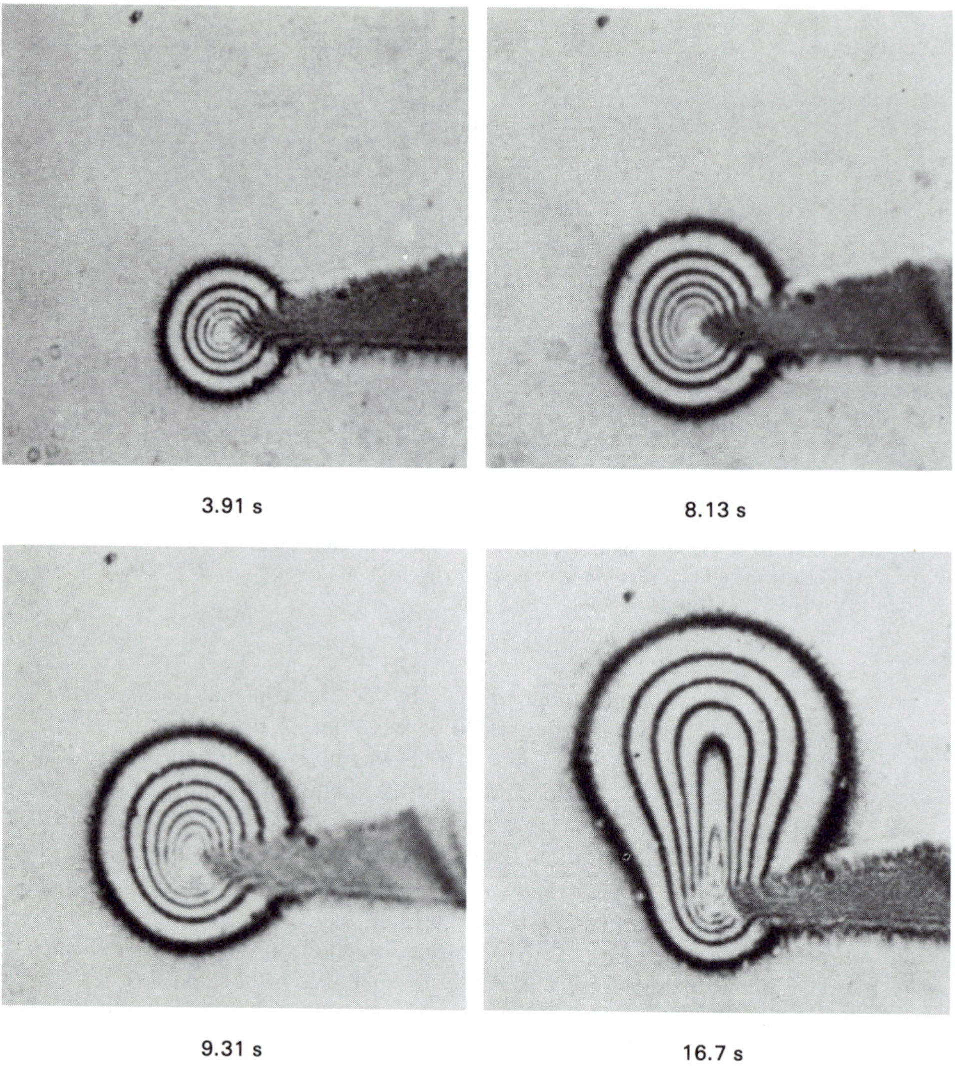

Figure 7.3.2 Developing temperature field around a horizontal wire in water for $q' = 11$ Btu/h ft. *(Reprinted with permission from* Int. J. Heat Mass Transfer, *vol. 15, p. 1281, C. M. Vest and M. L. Lawson. Copyright © 1972, Pergamon Journals Ltd.)*

In another study, Holster and Hale (1979) reported a numerical analysis for transient flow around an isothermal horizontal cylinder. A finite-element method was used. The calculations predict an early one-dimensional conduction regime during which a minimum in the Nusselt number occurs. This agrees with the observations of Vest and Lawson (1972) and Parsons and Mulligan (1978, 1980).

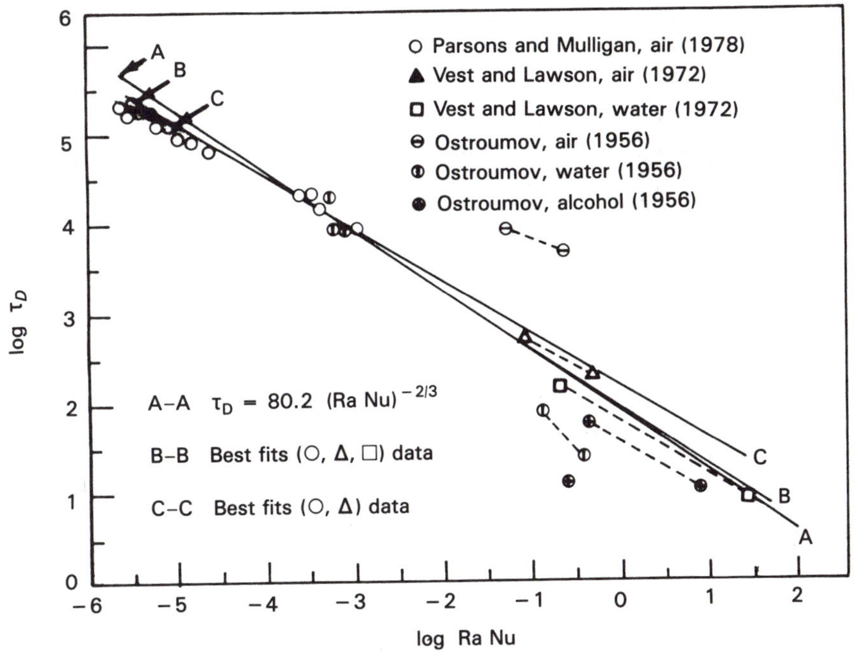

Figure 7.3.3 Correlation of the limit of pure conduction (convective delay) with Rayleigh number on the basis of thermal stability criteria. *(Reprinted with permission from J. R. Parsons and J. C. Mulligan,* J. Heat Transfer, *vol. 100, p. 423. Copyright © 1978, ASME.)*

7.4 OTHER TRANSIENTS

In addition to the transient flows discussed in the preceding sections, several other important classes of transients are discussed in later chapters, after the development of the appropriate formulations. Transients occurring due to ice melting are discussed in Chapter 9. Experimental and numerical investigations for vertical and horizontal surfaces are considered. Although all freezing and melting processes are transient, many may be treated as quasi-steady, provided the time scale is appropriate. Mixed convection transients adjacent to a vertical flat surface dissipating heat are discussed in Chapter 10. Several flow configurations are treated, representing different heating rates, velocity fields, initial conditions, and fluids. The transient transition to turbulent flow is discussed in Chapter 11, which examines the development in time of the several transition mechanisms. A survey of the different methods used for the linear stability analysis of time-dependent base flows is also presented. Several different formulations are compared and the suitability of each for different flow configurations is discussed.

Some inherently transient flows are also examined in Chapter 12, such as the starting plume that occurs during the early stages of development of thermal plumes.

Also considered are thermals, or isolated masses of fluid rising and expanding in an entraining ambient medium. Although the analysis for transient free-boundary turbulent flows is limited, some interesting experimental measurements and visualization studies are presented. Numerical results for transients occurring when water is cooled through its density extremum are examined in Chapter 14. Other transients in internal flows are also outlined in Chapter 14. Studies of transient effects in saturated porous media are discussed in Chapter 15 for several flow configurations.

REFERENCES

Brown, S. N., and Riley, N. (1973). *J. Fluid Mech. 2,* 225.

Callahan, G. D., and Marner, W. J. (1976). *Int. J. Heat Mass Transfer 19,* 165.

Carslaw, H. S., and Jaeger, J. C. (1959). *Conduction of Heat in Solids,* 2d ed., Oxford Univ. Press, New York.

Chung, P. M., and Anderson, A. D. (1961). *J. Heat Transfer 83,* 473.

Gebhart, B. (1961). *J. Heat Transfer 83,* 61.

Gebhart, B. (1963a). *J. Heat Transfer 85,* 184.

Gebhart, B. (1963b). *J. Heat Transfer 85,* 10.

Gebhart, B. (1963c). *Int. J. Heat Mass Transfer 6,* 951.

Gebhart, B. (1964). *Int. J. Heat Mass Transfer 7,* 479.

Gebhart, B., and Adams, D. E. (1963). *J. Heat Transfer 85,* 25.

Gebhart, B., and Dring, R. P. (1967). *J. Heat Transfer 89,* 274.

Gebhart, B., Dring, R. P., and Polymeropoulos, C. E. (1967). *J. Heat Transfer 89,* 53.

Goldstein, R. J. (1959). Ph.D. Thesis, Univ. of Minnesota.

Goldstein, R. J., and Briggs, D. G. (1964). *J. Heat Transfer 86,* 490.

Goldstein, R. J., and Eckert, E. R. G. (1960). *Int. J. Heat Mass Transfer 1,* 208.

Hellums, J. D., and Churchill, S. W. (1961). *Proc. International Heat Transfer Conference, 1961,* Boulder, Colo.

Hellums, J. D., and Churchill, S. W. (1962). *AIChE J. 8,* 690.

Holster, J. L., and Hale, L. A. (1979). *Trans. ASME,* Paper 79-WA/HT-27.

Illingworth, C. R. (1950). *Proc. Cambridge Philos. Soc. 46,* 603.

Ingham, D. B. (1978). *Int. J. Heat Mass Transfer 21,* 67.

Joshi, Y., and Gebhart, B. (1987a). *J. Fluid Mech. 179,* 407.

Joshi, Y., and Gebhart, B. (1987b). *Int. J. Heat Mass Transfer.* In press.

Klei, H. E. (1957). B.S. Thesis, Massachusetts Institute of Technology, Cambridge.

Lurie, H., and Johnson, H. A. (1962). *J. Heat Transfer 84,* 217.

Mahajan, R. L., and Gebhart, B. (1978). *J. Heat Transfer 100,* 731.

Martin, J. H. (1961). M.S. Thesis, Cornell Univ., Ithaca, N.Y.

Menold, E. R., and Yang, K. T. (1962). *J. Appl. Mech. 29,* 124.

Miyamoto, M. (1976). *Int. J. Heat Mass Transfer 20,* 1258.

Mizukami, K. (1976). *Int. J. Heat Mass Transfer 20,* 981.

Mollendorf, J. C., and Gebhart, B. (1970). *J. Heat Transfer 92,* 628.

Nanda, R. S., and Sharma, V. P. (1962). *AIAA J. 1,* 937.

Ostrach, S. (1953). NACA Rep. 1111.

Ostroumov, G. A. (1956). *Sov. Phys. Tech. Phys. 1,* 2627.

Parsons, J. R., and Mulligan, J. C. (1978). *J. Heat Transfer 100,* 423.

Parsons, J. R., and Mulligan, J. C. (1980). *J. Heat Transfer 102,* 636.

Rajan, V. S. V., and Picot, J. J. C. (1971). *Ind. Eng. Chem. Fundam. 10,* 132.

Rao, A. K. (1961). *Appl. Sci. Res. Sec. A 10,* 141.

Rao, A. K. (1963). *J. Phys. Soc. Jpn. 18,* 1829.
Sammakia, B., and Gebhart, B. (1978). *Numer. Heat Transfer 1,* 529.
Sammakia, B., and Gebhart, B. (1981). *Numer. Heat Transfer 4,* 331.
Sammakia, B., Gebhart, B., and Quershi, Z. H. (1980). *Int. J. Heat Mass Transfer 23,* 571.
Sammakia, B., Gebhart, B., and Quershi, Z. H. (1982). *J. Heat Transfer 104,* 644.
Schetz, J. A., and Eichhorn, R. (1962). *J. Heat Transfer 84,* 334.
Siegel, R. (1958). *Trans ASME 80,* 347.
Singh, D. (1964). *J. Phys. Soc. Jpn. 19,* 751.
Soehngen, E. (1968). *Proc. 3d All-Union Conf. Heat Mass Transfer,* Minsk, p. 125.
Soundalgekar, V. M. (1979). *Astrophys. Space Sci. 66,* 223.
Soundalgekar, V. M., and Waver, P. D. (1976). *Int. J. Heat Mass Transfer 20,* 1375.
Sparrow, E. M., and Gregg, J. L. (1960). *J. Heat Transfer 82,* 258.
Sugawara, S., and Michiyoshi, I. (1952). *Proc. 1st Jpn. Natl. Congr. Appl. Mech.,* p. 501.
Vest, C. M., and Lawson, M. L. (1972). *Int. J. Heat Mass Transfer 15,* 1281.
Yang, K. T. (1960). *J. Appl. Mech. 82,* 230.
Yang, K. T. (1966). *Int. J. Heat Mass Transfer 9,* 511.
Yang, K. T., Scaccia, C., and Goodman, J. (1974). *J. Heat Transfer 96,* 9.

PROBLEMS

7.1 Determine the time for which the one-dimensional conduction transient solution may be expected to be valid for a vertical surface 30 cm high, after a step change of 20°C in surface temperature, in water at 20°C.

7.2 For the conditions in Problem 7.1, consider a sudden 10°C increase in surface temperature.
(a) Plot the value of ϕ over the initial one-dimensional thermal layer in the water at times of 1, 10, and 20 s.
(b) Calculate the actual duration of the one-dimensional process at midheight of the surface.
(c) Calculate the time to steady state at the same location.

7.3 For a heated vertical surface in air at 20°C with a step change in the input flux, obtain the duration of the conduction regime and of the transient convection regime. The surface height is 30 cm, the thermal capacity per side 5×10^{-3} Btu/ft^2 °F, and the input flux 20 Btu/h ft^2 per side. Neglect radiation losses from the surface.

7.4 For the transient in Problem 7.3 consider the initial developing one-dimensional temperature field in the fluid. Model it as a thin highly conducting layer of thermal capacity c'' per side and very high relative thermal conductivity, in perfect thermal contact with the ambient air.
(a) Write the differential equation and all boundary and initial conditions for $t(y, \tau) - t_\infty = \theta(y, \tau)$, where y is the distance outward from the surface into the ambient.
(b) Find the solution, either from the literature or by analysis.
(c) What characteristic temperature difference parameter arises in the solution?
(d) Plot the ratio R of the storage and conduction (at $y = 0$) components of $q'' = 20$ Btu/h ft^2 against time until R decreases to 0.1.

7.5 Determine the transient temperature and velocity response for a step change in heat flux from 0 to 20 W/m^2 to a vertical surface of negligible thermal capacity in water. Determine these values at 1 and 2 s after the step change, at a distance of 0.1 cm out from the surface. Repeat the calculation for a step change of 50°C in the surface temperature.

7.6 For a step in total input heat flux of 100 W/m^2 into a vertical stainless steel surface 0.5 cm thick and 20 cm high, determine the regime of the transient. Determine the average final steady-state temperature of the plate and the time needed to achieve 95% of the total temperature rise. The fluid is water at 20°C.

7.7 A thin stainless steel foil is bonded to an insulating styrofoam sheet 1 cm thick. If the combination is held vertical in air at 20°C and a heat flux of 100 W/m^2 supplied electrically to the foil, obtain the natural convection transient time. The foil is 0.003 mm thick and 40 cm high. Neglect the thermal capacity of the insulation.

7.8 For the conditions in Problem 7.3, estimate the time required for the leading edge effect to pass over the whole surface.

7.9 Determine the transient regime for the conditions of Problem 7.6, had the surface been immersed in air at 20°C.

7.10 Determine the time duration of the purely conductive regime adjacent to a 0.1-cm-diameter stainless steel wire held horizontally in water. The wire dissipates 10 W/m. How long would the conductive regime last in air? In silicone oil?

CHAPTER

EIGHT

EFFECTS OF VARIABLE FLUID PROPERTIES

8.1 VARIABLE PROPERTY CONSIDERATIONS

The analyses of buoyancy-induced flows in previous chapters assume constant fluid properties, except for the density differences in the body force term in the momentum equation. These are essential for the generation of the buoyancy forces. However, the continuity equation used is still in its incompressible form. This approach is referred to as constant-property analysis. Most fluids, however, have temperature-dependent properties, and under circumstances where large temperature gradients exist across the fluid medium, fluid properties often vary significantly. Under many conditions, ignoring such variations may cause serious inaccuracies in estimating heat transfer rates. The thermophysical properties that appear in the governing equations include thermodynamic and transport properties. Thermodynamic properties define the equilibrium state of a system. Temperature, density, and specific heat are such properties. The transport properties include the diffusion rate coefficients such as the thermal conductivity and viscosity. Variation of these properties with temperature for various fluids of engineering interest has been extensively documented in the literature. A recommended reference is Touloukian and Ho (1970–1977).

A detailed study of approximations commonly used in natural convection analysis was reported by Gray and Giorgini (1976). It was concluded that constant-property analysis is accurate for rather small temperature differences. For example, in air at 15°C and 1 atm pressure, the difference should be less than 28.6°C. For larger temperature differences, property variations may cause considerable effects. For example, for gases ρ and β vary as T^{-1} while μ and k vary as $T^{0.65}$. On the other hand, for water, k and c_p show little variation with temperature, μ and Pr are highly variable, and the density has an anomalous behavior at near-freezing temperatures. The properties of various gases and liquids are given as a function of temperature in the property appendixes.

To account for variable properties in heat transfer correlations the two commonly used approaches are the *reference temperature method* and the *property ratio method*. In the reference temperature method a temperature is chosen at which all the properties are evaluated. These properties are then used in empirically or analytically derived constant-property results to approximate the variable-property behavior. The most commonly used reference temperature is the film temperature, $t_r = t_f = (t_0 + t_\infty)/2$, which is found to be adequate for moderate temperature differences. For large temperature variations a different value of t_r must be selected. Specific values and temperature ranges are discussed in later sections.

In the property ratio method, all properties are first evaluated at some reference temperature t_r, for which the film temperature t_f is commonly used. In this method, however, variable-property effects are also accounted for by multiplying the constant-property correlations by a function of the ratio of some pertinent property evaluated at the surface temperature to that property evaluated at the free-stream temperature. Studies by Clausing (1983) and Clausing and Kempka (1981) also indicate that multiplying the constant-property heat transfer correlations by a function of T_0/T_∞ results in good correlations for turbulent transport with variable-property effects.

Various analyses and experimental studies have shown that, in the laminar flow regime, the effects of property variations are small and may be adequately accounted for by using the reference temperature method. For the turbulent regime this method alone does not always correlate the data well. This regime requires the inclusion of temperature-dependent functions in the correlations. That is, the property ratio method is found to be more suitable for turbulent transport.

The following sections outline a general formulation for analyzing the effects of property variations in laminar flows. Since they are amenable to similarity techniques, analyses for vertical isothermal surfaces are given. Effects in gases for laminar flow are discussed on the basis of various analyses. For the turbulent regime, experimental findings are summarized. In general, fairly limited studies have been conducted to investigate variable-property effects on buoyancy-induced flows.

8.2 GENERAL FORMULATION

A general analysis is now presented for transport from a vertical isothermal surface allowing for the variation of all the fluid properties. The pressure and viscous dissipation terms are neglected. With the boundary layer approximations the governing equations for variable-property natural convection transport are

$$\frac{\partial}{\partial x}(\rho u) + \frac{\partial}{\partial y}(\rho v) = 0 \tag{8.2.1}$$

$$\rho\left(u\frac{\partial u}{\partial x} + v\frac{\partial u}{\partial y}\right) = g(\rho_\infty - \rho) + \frac{\partial}{\partial y}\left(\mu\frac{\partial u}{\partial y}\right) \tag{8.2.2}$$

$$\rho c_p\left(u\frac{\partial t}{\partial x}+v\frac{\partial t}{\partial y}\right)=\frac{\partial}{\partial y}\left(k\frac{\partial t}{\partial y}\right) \tag{8.2.3}$$

The necessary boundary conditions are

$$\begin{aligned} &\text{at } y=0: \quad && u=v=0 \quad && t=t_0 \\ &\text{as } y\to\infty: \quad && u\to 0 \quad && t\to t_\infty \end{aligned} \tag{8.2.4}$$

In the constant-property analysis the buoyancy force term in Eq. (8.2.2) is first replaced by a temperature difference, assuming that ρ, μ, c_p, and k are constant. Here instead the following properties are assumed to be temperature-dependent:

$$\rho=\rho(t) \qquad \mu=\mu(t) \qquad c_p=c_p(t) \qquad k=k(t) \tag{8.2.5}$$

Equation (8.2.1) is satisfied by defining the following stream function:

$$\frac{\rho}{\rho_r}u=\frac{\partial\psi}{\partial y} \qquad \frac{\rho}{\rho_r}v=-\frac{\partial\psi}{\partial x} \tag{8.2.6}$$

where the subscript r denotes properties evaluated at some reference temperature t_r. The momentum equation (8.2.2) and the energy equation (8.2.3) are transformed into ordinary differential equations by using the following Howarth (1948) transformations:

$$\eta=\frac{G}{4x}\int_0^y\frac{\rho}{\rho_r}\,dy \tag{8.2.7}$$

$$\phi(\eta)=\frac{t-t_\infty}{t_0-t_\infty} \tag{8.2.8}$$

$$\psi=\nu_r G f(\eta) \tag{8.2.9}$$

where

$$G=4\left(\frac{gx^3}{\nu_r^2}\frac{\rho_\infty-\rho}{\rho_0}\right)^{1/4} \tag{8.2.10}$$

and the subscript 0 denotes the quantities evaluated at the surface temperature t_0. With the above transformations the governing equations and the boundary conditions become

$$\left(\frac{\mu\rho}{\mu_r\rho_r}f''\right)'+3ff''-2(f')^2+\frac{(\rho_\infty/\rho)-1}{(\rho_\infty/\rho_0)-1}=0 \tag{8.2.11}$$

$$\left(\frac{k\rho}{k_r\rho_r}\phi'\right)'+3\mathrm{Pr}_r\frac{c_p}{c_{p_r}}f\phi'=0 \tag{8.2.12}$$

$$f(0)=f'(0)=\phi(0)-1=f'(\infty)=\phi(\infty)=0 \tag{8.2.13}$$

where primes denote differentiations with respect to η. Note that up to this point the analysis is valid generally, and Eqs. (8.2.11)–(8.2.13) embody only the approximation of properties that are solely a function of temperature. The "exact" solution to Eqs. (8.2.11)–(8.2.13) would thus require evaluating the properties locally as a function of temperature. Note also that the governing equations used in the constant-property analysis, Eqs. (3.3.19) and (3.3.20), may be recovered by setting all the property ratios equal to unity in Eqs. (8.2.11) and (8.2.12). Equations (8.2.11) and (8.2.12) may be solved by evaluating μ, ρ, c_p, and k from prescribed equations of state. Specific solutions for different gases and liquids are discussed in detail in later sections.

8.3 EFFECTS IN GASES

Several studies on variable-property effects of ideal and real gases have been reported. Analyses of laminar flows are considered first. For turbulent flows the experimental investigations will be discussed at the end of this section. For air, an early treatment is the perturbation analysis of Hara (1954). The solution is applicable for small values of the perturbation parameter $\epsilon = (T_0 - T_\infty)/T_\infty$, where T denotes the absolute temperature. Hara (1958) later extended the analysis to $\epsilon = 2$ and 4.

Sparrow and Gregg (1958) considered the variable fluid property effects in natural convection for a vertical isothermal surface. The objective was to find a suitable reference temperature t_r for evaluating properties of gases so that constant-property solutions could be used to calculate variable-property transport. Five property variation models were considered for gases, as described in Table 8.3.1. All obey the perfect gas law $p = \rho RT$. Gases A, B, and C represent simplified idealizations of real gas behavior. Power law variations for k and μ are commonly used approximations. Specific heat and Prandtl number were assumed constant because for real gases the variations of these properties are small compared to those of k, μ, and ρ. Gas D differs only in the thermal conductivity and viscosity variations. These are taken to be a more complicated function of temperature, as shown in Table 8.3.1. Gas E closely represents the properties of air. All the properties are taken to vary with temperature.

The results of the analysis are summarized below. Governing Eqs. (8.2.11)–(8.2.13) apply. Surface temperature t_0 was used as a reference temperature t_r in these equations by Sparrow and Gregg (1958). For a perfect gas the buoyancy force term in Eq. (8.2.11) is expressed in terms of a temperature difference as follows:

$$\rho_\infty - \rho = \frac{p}{RT_\infty} - \frac{p}{RT} = \rho\left(\frac{T}{T_\infty} - 1\right) = \frac{\rho}{T_\infty}(T - T_\infty) \tag{8.3.1}$$

For a perfect gas the coefficient of thermal expansion is

$$\beta = -\frac{1}{\rho}\left(\frac{\partial \rho}{\partial T}\right)_p = \frac{1}{T} \tag{8.3.2}$$

Table 8.3.1 Description of five different characterizations of gases used in modeling variable properties

Gas A	Gas B	Gas C	Gas D	Gas E
$p = \rho RT$	$p = \rho RT$	$p = \rho RT$	$p = \rho RT$	$p = \rho RT$
$k \sim T^{3/4}$	$k \sim T^{2/3}$	$\rho k = \text{const.}$	$k \sim \dfrac{T^{3/2}}{T + A_1}$	$k \sim \dfrac{T^{3/2}}{T + A_1}$
$\mu \sim T^{3/4}$	$\mu \sim T^{2/3}$	$\rho\mu = \text{const.}$	$\mu \sim \dfrac{T^{3/2}}{T + A_1}$	$\mu \sim \dfrac{T^{3/2}}{T + A_2} = b_0 + b_1 T$
$c_p = \text{const.}$	$c_p = \text{const.}$	$c_p = \text{const.}$	$c_p = \text{const.}$	$c_p = b_2 + b_3 T$
$\Pr = \text{const.}$	$\Pr = \text{const.}$	$\Pr = \text{const.}$	$\Pr = \text{const.}$	$\Pr = \text{variable}$

Source: Reprinted with permission from E. M. Sparrow and J. L. Gregg, *J. Heat Transfer,* vol. 80, p. 879. Copyright © 1958, ASME.

In constant-property analysis we write

$$\rho_\infty - \rho = \rho\beta(T - T_\infty) \tag{8.3.3}$$

Comparing the above equations, we conclude that the buoyancy term used in the constant-property analysis is correct for a perfect gas provided β is evaluated at T_∞. The buoyancy force term in Eq. (8.2.11) for an ideal gas is

$$\frac{(\rho_\infty/\rho) - 1}{(\rho_\infty/\rho_0) - 1} = \frac{T - T_\infty}{T_0 - T_\infty} = \frac{t - t_\infty}{t_0 - t_\infty} = \phi \tag{8.3.4}$$

Using t_0 as the reference temperature, the governing Eqs. (8.2.11) and (8.2.12) become

$$\left(\frac{\mu\rho}{\mu_0\rho_0} f''\right)' + 3ff'' - 2(f')^2 + \phi = 0 \tag{8.3.5}$$

$$\left(\frac{k\rho}{k_0\rho_0}\phi'\right)' + 3\Pr_0\left(\frac{c_p}{c_{p0}}\right)f\phi' = 0 \tag{8.3.6}$$

An interesting observation can be made from Eqs. (8.3.5) and (8.3.6) for a gas with the following property variations:

$$p = \rho RT \qquad \rho\mu = \text{const.} \qquad \rho k = \text{const.} \qquad c_p = \text{const.} \tag{8.3.7}$$

With these variations the resulting equations are identically the same as the constant-property Eqs. (3.3.19) and (3.3.20). This is gas C in Table 8.3.1. For such a gas all the solutions available for the constant-property equations apply precisely, provided $\beta = 1/T_\infty$.

In variable-property analysis a generalized Grashof number is defined as

$$\text{Gr}_x = \frac{gx^3(\rho_\infty - \rho_0)/\rho_0}{\nu^2} \tag{8.3.8}$$

For a perfect gas Eq. (8.3.8) reduces to

$$\mathrm{Gr}_x = \frac{gx^3(t_0 - t_\infty)}{T_\infty \nu^2} \tag{8.3.9}$$

Numerical heat transfer results, from Eqs. (8.3.5) and (8.3.6) with the boundary conditions of Eq. (8.2.13) for gas A in Table 8.3.1, are given in Table 8.3.2. For Pr = 0.7 a wide range of T_0/T_∞, from 4 to $\frac{1}{4}$, was investigated. The constant-property solution for Pr = 0.7 is

$$\frac{\mathrm{Nu}_x}{\mathrm{Gr}_x^{1/4}} = 0.353 = \frac{3}{4}\frac{\mathrm{Nu}}{\mathrm{Gr}^{1/4}} \tag{8.3.10}$$

Sparrow and Gregg (1958) suggested that if the properties in Eq. (8.3.10) are calculated at a suitable reference temperature, then the results of constant-property analysis, that is, Eq. (8.3.10), may be made to coincide with the variable-property results given in Table 8.3.2. By trial and error the following reference temperature was found for this gas model:

$$t_r = t_0 - 0.38(t_0 - t_\infty) \qquad \beta = \frac{1}{T_\infty} \tag{8.3.11}$$

The heat transfer rate predicted from the constant-property result, Eq. (8.3.10), by using this reference temperature was found to be in error of at most 0.6% over the entire range $\frac{1}{4} < T_0/T_\infty < 4$. It was further concluded that use of the constant-property result, with transport properties evaluated at the film temperature and

Table 8.3.2 Local and average Nusselt numbers for gas A[a]

$\frac{T_0}{T_\infty}$	$\frac{\mathrm{Nu}_x}{\mathrm{Gr}_x^{1/4}}$ or $\frac{\mathrm{Nu}}{\frac{4}{3}\mathrm{Gr}^{1/4}}$, Pr = 0.7	Pr = 1.0
4	0.371	—
3	0.368	0.418
5/2	0.366	—
2	0.363	—
3/4	0.348	—
1/2	0.339	—
1/3	0.330	0.375
1/4	0.323	—

[a] $\beta = 1/T_\infty$; other properties evaluated at t_0.

Source: Reprinted with permission from E. M. Sparrow and J. L. Gregg, *J. Heat Transfer*, vol. 80, p. 879. Copyright © 1958, ASME.

$\beta = 1/T_\infty$, is adequate for almost all engineering purposes. Similar but less extensive calculations for gases B through E (Table 8.3.1) supported the validity of the reference temperature given by Eq. (8.3.11).

Another gaseous substance of considerable practical importance is steam. Minkowycz and Sparrow (1966) reported an analysis, accounting for variable properties. All the relevant thermodynamic and transport properties of steam were considered to be both temperature and pressure dependent. The pressure was varied from 0.04 to 25 atm. Temperatures up to 954 K were considered for superheated steam. From extensive numerical calculations it was concluded that the following constant-property solution accurately predicts the laminar transport rates from a vertical isothermal surface in steam:

$$\mathrm{Nu}_x = (\mathrm{Gr}_x\mathrm{Pr})^{1/4}(0.3357 + 0.0885\mathrm{Pr} - 0.232\mathrm{Pr}^2) \tag{8.3.12}$$

where all properties except β are calculated at a reference temperature t_r given by

$$t_r = t_0 - 0.46(t_0 - t_\infty) \qquad \beta = \beta(T_\infty) \tag{8.3.13}$$

It was pointed out that, at temperatures near the saturation line, steam departs significantly from ideal gas behavior.

Variable-property effects with blowing at the surface were investigated experimentally by Gill et al. (1965). Various gases were injected at the surface, which was initially immersed in air. The variable-property effects were found to be most prominent for helium and hydrogen.

The analysis of variable-property effects for a vertical surface dissipating uniform heat flux is more difficult because of the downstream surface temperature variation. This makes it difficult to use similarity transformations. Analysis may be performed by either finite-difference or integral techniques. Ito et al. (1974) used an integral method and obtained solutions for carbon dioxide near the pseudocritical point.

The analyses mentioned above apply to laminar flows only when the property variations may be accounted for by calculating all the properties at a single suitable reference temperature, commonly the film temperature t_f. The coefficient of thermal expansion β is always calculated at T_∞.

Variable-property effects in turbulent flows. These have been examined both numerically and experimentally. As mentioned earlier, most analyses and measurements of turbulent transport indicate that the reference temperature method does not successfully correlate variable-property effects. There are, however, a few known exceptions. Pirovano et al. (1970) correlated their turbulent regime data in air, over the range $1.0 < T_0/T_\infty < 1.5$, with properties evaluated at a reference temperature $t_r = t_\infty + 0.2(t_0 - t_\infty)$. However, most studies indicate that the property ratio method is more suitable for turbulent transport correlations.

Clausing and Kempka (1981) found that their experimental results for turbulent natural convection from an isothermal vertical surface in nitrogen could not be correlated by the reference temperature method alone. They concluded that some suitable function of T_0/T_∞ may be used to modify constant-property results to account for property variations. The range $1 < T_0/T_\infty < 2.6$ was investigated.

High values of T_0/T_∞ were obtained experimentally by heating the surface isothermally in a cryogenic environment of gaseous nitrogen. The range $80 < T_0 < 320$ K resulted in Rayleigh numbers between 10^7 and 2×10^{10}. In the laminar regime the conclusions drawn from the data were consistent with the previous studies. The effect of variable properties was negligible if the properties were calculated at the film temperature t_f.

However, in the turbulent regime the Nusselt number was more strongly affected by property variations. It was suggested that the Bayley (1955) correlation, $\mathrm{Nu} = 0.1\mathrm{Ra}^{1/3}$, may be modified by multiplying the right-hand side by a function of T_0/T_∞. Two such functions were given for two ranges of T_0/T_∞. Clausing (1983) critically reviewed the modification of the Bayley correlation and found it unsuitable for correlating the data of Clausing and Kempka (1981) in the low T_0/T_∞ range.

Clausing (1983) also assessed the experimental data of Griffiths and Davis (1922), Saunders (1936), Warner and Arpaci (1968), and Vliet and Ross (1975) in terms of the effects of property variations. The following correlation was then given to account for property variations in the turbulent regime for isothermal vertical surfaces in gases:

$$\mathrm{Nu} = 0.082(\mathrm{GrPr})^{1/3}\left[-0.9 + 2.4\left(\frac{T_0}{T_\infty}\right) - 0.5\left(\frac{T_0}{T_\infty}\right)^2\right] \tag{8.3.14}$$

This correlation applies for $\mathrm{GrPr} > 1.6 \times 10^9$ and $1 < T_0/T_\infty < 2.6$. All the properties are calculated at the film temperature $t_f = (t_0 + t_\infty)/2$.

Another extensive experimental study of variable-property effects was reported by Siebers (1978) and Siebers et al. (1983). A 3.02-m-high surface placed in air was heated to surface temperatures in the range 60–520°C with $t_\infty = 20$°C. These conditions result in T_0/T_∞ ratios from 1.1 to 2.7. In the turbulent regime ($\mathrm{Gr} > 10^{10}$) the effect of property variations was accounted for by two approaches. The first was to evaluate all properties at t_∞ with the following correlation:

$$\mathrm{Nu}_x = 0.098(\mathrm{Gr}_x)^{1/3}\left(\frac{T_0}{T_\infty}\right)^{-0.14} \tag{8.3.15}$$

The other approach used reference temperature methods. Equation (8.3.15) applies without the term $(T_0/T_\infty)^{-0.14}$, provided properties are evaluated at

$$t_r = t_0 - 0.7(t_0 - t_\infty) \qquad \text{and} \qquad \beta = \frac{1}{T_\infty} \tag{8.3.16}$$

Transport in gases near the thermodynamic critical point. As the pressure and temperature of a gas approach the thermodynamic critical point, the thermophysical properties show large variations with small temperature and pressure changes. This, in turn, leads to considerable changes in the heat transfer rates. Fritsch and Grosh (1963) measured transport in water. The density and specific heat of the supercritical fluid were assumed to be temperature-dependent only. The resulting heat transfer rates were found to be strongly dependent on the surface and fluid temperatures, individually. Similar results were obtained by Hasegawa and Yoshioka (1966) when the properties were all assumed to be power functions of enthalpy.

In a later study Nishikawa and Ito (1969) carried out a numerical analysis of the flow near a vertical isothermal surface in fluids at supercritical pressures. By using similarity transformations of the same form as Eqs. (8.2.7)–(8.2.10), a set of governing equations similar to Eqs. (8.2.11)–(8.2.13) was obtained. These results compared well with the previous analyses of Fritsch and Grosh (1963) and Hasegawa and Yoshioka (1966). However, attempts to use the reference temperature method to correlate the heat transfer results with the constant-property analysis failed to produce good agreement. More recently, Ghajar and Parker (1981) reported the results of extensive calculations of laminar free flow adjacent to a vertical isothermal surface in carbon dioxide, Refrigerant 114, and water. All fluids were in their supercritical region. Heat transfer correlations were suggested for these fluids near their critical point.

Transport from heated horizontal wires immersed in fluids near their critical point was studied by Knapp and Sabersky (1966), Hahne (1965), Nishikawa et al. (1970), and Neumann and Hahne (1980). The heat transfer rates were found to increase as the critical point was approached. This was accompanied by oscillations of the wire if the temperature difference was about 70 K or more. No satisfactory explanation for this phenomenon has been published.

8.4 EFFECTS IN WATER

Variable-property effects on vertical laminar natural convection in water have been analyzed by Shaukatullah and Gebhart (1979). All the properties—density ρ, specific heat c_p, thermal conductivity k, and viscosity μ—were considered to be temperature dependent over the range 0–100°C. The analysis was complicated by the anomalous behavior of density in the low temperature range. This caused buoyancy force and flow reversals and even convective inversion (see Chapter 9). The normalized property variations for water are shown in Fig. 8.4.1. Properties computed at the wall temperature t_0 were used for normalization.

For an isothermal vertical surface the transformed boundary layer equations and boundary conditions in Eqs. (8.2.11)–(8.2.13) apply. Shaukatullah and Gebhart (1979) solved these equations numerically for all of the surface and ambient temperature conditions t_0 and t_∞ shown by the circles in Fig. 8.4.2. The resulting heat transfer and shear stress parameters were expressed as follows, where the properties in Nu_x and Gr_x are computed at t_∞:

$$\mathrm{Nu}_x = \frac{h_x x}{k} = [-\phi'(0)] \frac{k_0 \rho_0}{k_\infty \rho_\infty} (\mathrm{Gr}_x)^{1/4} \tag{8.4.1}$$

$$c_f = \frac{\tau_0}{\rho_\infty (\nu_\infty / x)^2} = \frac{\mu_0 \rho_0}{\mu_\infty \rho_\infty} f''(0)(\mathrm{Gr}_x)^{3/4} \tag{8.4.2}$$

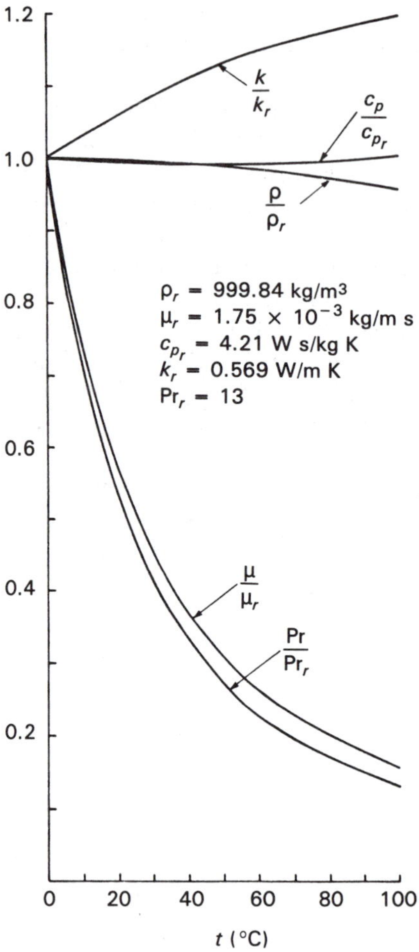

Figure 8.4.1 Properties of water as a function of temperature. *(Reprinted with permission from H. Shaukatullah and B. Gebhart,* Numer. Heat Transfer, *vol. 2, p. 215. Copyright © 1979, Hemisphere Publishing Corporation.)*

The temperature-dependent functions $(k_0\rho_0/k_\infty\rho_\infty)\phi'(0)$ and $(\mu_0\rho_0/\mu_\infty\rho_\infty)f''(0)$ in Eqs. (8.4.1) and (8.4.2) are given in Table 8.4.1. For the temperature conditions in the region shaded with vertical lines in Fig. 8.4.2, at higher temperatures, the following heat transfer correlation represents the calculated results with a maximum deviation of 3%:

$$\mathrm{Nu}_x = 0.416(\mathrm{Gr}_x)^{1/4}\frac{(\mathrm{Pr})^{0.422}}{(\mathrm{Pr}_0)^{0.122}} \tag{8.4.3}$$

All the properties in this correlation, except Pr_0, are evaluated at the ambient temperature t_∞. Note that for $t_0 \rightarrow t_\infty$, Eq. (8.4.3) reduces to the constant-property solution, with reasonable agreement for Pr between 1.5 and 13. The average Nusselt number is given by

$$\mathrm{Nu} = 0.555(\mathrm{Gr})^{1/4} \frac{(\mathrm{Pr})^{0.422}}{(\mathrm{Pr}_0)^{0.122}} \tag{8.4.4}$$

In the low-temperature region in Fig. 8.4.2 the calculated results are not correlated by the above equations because of the anomalous density behavior and its complicating effects on the transport. However, the transport may be determined from Eqs. (8.4.3) and (8.4.4) in conjunction with Table 8.4.1.

The results of an earlier study of the same circumstances by Piau (1974) compare very well with the above analysis except for those in the low-temperature region. The disagreement may be attributed to the inaccuracy in the assumption of linear dependence of μ and β on temperature in that work.

Padlog and Mollendorf (1983) extended the work of Shaukatullah and Gebhart (1979). The variable-property effects in water were analyzed around the density extremum in more detail. In particular, the temperature conditions under which convective inversion occurs were considered. The analysis also included the effect of blowing at the surface. Such a boundary condition approximates an ice surface melting in water. Blowing was seen to be the major effect, causing a maximum

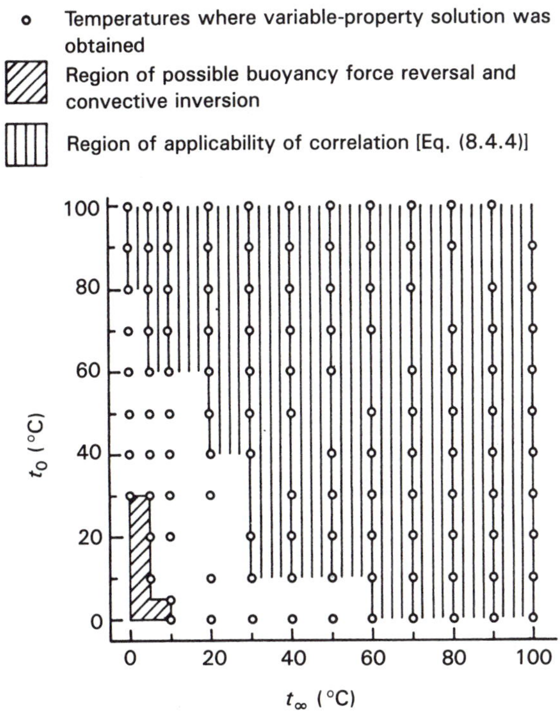

Figure 8.4.2 Region of investigation showing various surface and ambient temperature conditions. *(Reprinted with permission from H. Shaukatullah and B. Gebhart,* Numer. Heat Transfer, *vol. 2, p. 215, Copyright © 1979, Hemisphere Publishing Corporation.)*

Table 8.4.1 Values of transport quantities for natural convection from a vertical isothermal surface in water with variable properties

t_0 (°C)	t_∞ (°C)	$\frac{\rho_0\mu_0}{\rho_\infty\mu_\infty} f''(0)$	$\frac{\rho_0 k_0}{\rho_\infty k_\infty} \phi'(0)$	$f(\eta_e)$	t_∞ (°C)	$\frac{\rho_0\mu_0}{\rho_\infty\mu_\infty} f''(0)$	$\frac{\rho_0 k_0}{\rho_\infty k_\infty} \phi'(0)$	$f(\eta_e)$
0.0	0.0	0.561472	−0.890574	0.656722	50.0	1.170895	−0.537167	1.005583
10.0		—	—	—		1.017317	−0.547087	0.978385
20.0		—	—	—		0.914876	−0.561222	0.964645
30.0		0.326389	−0.822040	0.408303		0.838564	−0.576648	0.957422
40.0		0.323522	−0.905255	0.492327		0.778029	−0.592099	0.953560
50.0		0.315583	−0.964938	0.531844		0.728089	−0.606959	0.951417
60.0		0.306059	−1.013985	0.557340		0.685787	−0.620916	0.950067
70.0		0.296230	−1.056180	0.575675		0.649263	−0.633813	0.948963
80.0		0.286602	−0.093174	0.589479		0.617260	−0.645574	0.947775
90.0		0.277376	−1.125867	0.600049		0.588880	−0.656171	0.946302
100.0		0.268619	−1.154818	0.608137		0.563453	−0.665600	0.944422
0.0	5.0	—	—	—	60.0	1.241412	−0.494564	1.052331
10.0		0.474396	−0.792055	0.598606		1.095952	−0.505920	1.030391
20.0		0.430020	−0.819168	0.587299		0.993499	−0.519774	1.018811
30.0		0.404387	−0.859870	0.597237		0.914764	−0.534240	1.012466
40.0		0.383613	−0.900128	0.609505		0.851081	−0.548452	1.008881
50.0		0.365563	−0.937894	0.621122		0.797860	−0.561976	1.006711
60.0		0.349506	−0.972717	0.631346		0.752370	−0.574591	1.005178
70.0		0.335055	−1.004548	0.640013		0.712831	−0.586188	1.003805
80.0		0.321948	−1.033469	0.647152		0.678014	−0.596714	1.002307
90.0		0.309986	−1.059599	0.652855		0.647020	−0.606156	1.000502
100.0		0.299003	−1.083069	0.657237		0.619168	−0.614517	0.998278
0.0	10.0	1.369447	−1.049654	1.030929	70.0	1.309875	−0.460414	1.100139
10.0		0.601935	−0.809430	0.719381		1.170113	−0.472224	1.081779
20.0		0.514029	−0.812696	0.682030		1.067689	−0.485566	1.071766
30.0		0.468921	−0.837226	0.674426		0.987053	−0.499125	1.066093
40.0		0.437214	−0.866277	0.675536		0.920788	−0.512273	1.062734
50.0		0.412073	−0.895580	0.679570		0.864790	−0.524690	1.060554
60.0		0.390983	−0.923641	0.684352		0.816535	−0.536209	1.058881
70.0		0.372735	−0.949867	0.688949		0.774334	−0.546752	1.057301
80.0		0.356639	−0.974027	0.692940		0.736993	−0.556285	1.055554

90.0		0.342244	−0.996048	0.696143
100.0		0.329229	−1.015932	0.698494
0.0	20.0	0.988522	−0.765505	0.895563
10.0		0.739335	−0.735238	0.808954
20.0		0.638481	−0.743835	0.780078
30.0		0.577181	−0.762562	0.768999
40.0		0.533041	−0.784298	0.765223
50.0		0.498399	−0.806442	0.764689
60.0		0.469839	−0.827866	0.765519
70.0		0.445552	−0.848044	0.766758
80.0		0.424452	−0.866723	0.767898
90.0		0.405827	−0.883790	0.768666
100.0		0.389175	−0.899205	0.768916
0.0	30.0	1.032500	−0.662869	0.921066
10.0		0.842318	−0.660045	0.870246
20.0		0.740090	−0.672505	0.848725
30.0		0.671461	−0.689900	0.838924
40.0		0.619993	−0.708715	0.834583
50.0		0.578901	−0.727453	0.832885
60.0		0.544787	−0.745404	0.832393
70.0		0.515713	−0.762217	0.832322
80.0		0.490456	−0.777717	0.832228
90.0		0.468189	−0.791824	0.831853
100.0		0.448317	−0.804510	0.831051
0.0	40.0	1.099776	−0.591401	0.960968
10.0		0.933351	−0.597701	0.925386
20.0		0.830948	−0.611584	0.908545
30.0		0.757825	−0.627992	0.900174
40.0		0.701220	−0.644921	0.896013
50.0		0.655227	−0.661444	0.893971
60.0		0.616654	−0.677102	0.892915
70.0		0.583576	−0.691663	0.892196
80.0		0.554732	−0.705011	0.891434
90.0		0.529243	−0.717098	0.890401
100.0		0.506464	−0.727910	0.888963

	0.703624	−0.564801	1.053475
	0.673545	−0.572309	1.050962
80.0	1.375784	−0.432665	1.148437
	1.240347	−0.444459	1.132659
	1.138016	−0.457205	1.123826
	1.055870	−0.469929	1.118672
	0.987458	−0.482155	1.115488
	0.929085	−0.493634	1.113297
	0.878413	−0.504238	1.111507
	0.833846	−0.513909	1.109747
	0.794230	−0.522623	1.107783
	0.758696	−0.530379	1.105464
	0.726567	−0.537188	1.102694
90.0	1.439029	−0.409903	1.196886
	1.307080	−0.421455	1.183051
	1.204918	−0.433591	1.175132
	1.121571	−0.445559	1.170384
	1.051359	−0.456980	1.167335
	0.990939	−0.467655	1.165124
	0.938144	−0.477485	1.163225
	0.891463	−0.486421	1.161302
	0.849790	−0.494448	1.159139
	0.812277	−0.501568	1.156599
	0.778256	−0.507793	1.153589
100.0	1.499686	−0.391116	1.245270
	1.370684	−0.402323	1.232949
	1.268768	−0.413871	1.225756
	1.184470	−0.425158	1.221330
	1.112745	−0.435874	1.218380
	1.050552	−0.445856	1.216141
	0.995880	−0.455020	1.214135
	0.947305	−0.463329	1.212056
	0.903765	−0.470772	1.209706
	0.864438	−0.477353	1.206955
	0.828670	−0.483084	1.203717

Source: From Shaukatullah and Gebhart (1979).

decrease in the heat transfer rate of 7% at 20°C. Inclusion of property variations resulted in an additional decrease of only 1.7%.

8.5 EFFECTS IN OTHER LIQUIDS

Other liquids of common interest include different oils and liquid metals. Oils have high Prandtl numbers and liquid metals have very low ones. For oils, viscosity depends strongly on temperature. Ito et al. (1974) investigated the variable-property effect for spindle oil and Mobiltherm oil. For a vertical surface dissipating uniform heat flux q'', an integral method was employed. The results agreed well with both the experimental data and the following correlation of Fujii et al. (1970):

$$\text{Nu}_x = 0.62(\text{Gr}_x^*\text{Pr})^{1/5} \tag{8.5.1}$$

The effect of variable properties was taken into account as follows:

$$\beta = \frac{\rho_\infty - \rho_f}{\rho_f(t_f - t_\infty)} \qquad \text{where} \quad t_f = \frac{t_0 + t_\infty}{2}$$

All other properties were calculated at the reference temperature $t_r = t_0 - 0.25(t_0 - t_\infty)$.

Piau (1970) suggested that for high Prandtl number liquids, the variation of μ is often negligible and only the temperature dependence of viscosity is important. Carey and Mollendorf (1978) reported a detailed analysis of variable-property effects for liquids with temperature-dependent viscosity. It was pointed out that for many liquids, including petroleum oils, glycerine, glycols, silicone fluids, and some molten salts, the percentage variation of absolute viscosity with temperature is much higher than that of other properties. The boundary layer analysis for a vertical isothermal surface assumed μ as a general function of temperature:

$$\mu = \mu_f s(t) \tag{8.5.2}$$

where μ_f is the viscosity at the film temperature t_f, and therefore $s(t_f) = 1$. The governing Eqs. (8.2.1)–(8.2.3) apply with $g(\rho_\infty - \rho) = g\rho\beta(t - t_\infty)$. Defining

$$\eta = \frac{Gy}{4x} \qquad \psi = \nu_f G f(\eta)$$

$$\phi(\eta) = \frac{t - t_\infty}{t_0 - t_\infty} \qquad \text{where} \quad G = 4\left[\frac{g\beta x^3(t_0 - t_\infty)}{4\nu_f^2}\right]^{1/4} \tag{8.5.3}$$

Eqs. (8.2.1)–(8.2.3) reduce to

$$s(t)f''' + 3ff'' - 2(f')^2 + \phi = 0 \tag{8.5.4}$$

$$\phi'' + 3\text{Pr}_f f\phi' = 0 \tag{8.5.5}$$

From Eq. (8.5.4) it can be seen that similarity exists for an isothermal surface if $s(t)$ is a function of η only. This is true if $s(t)$ is a function of ϕ, that is, $s(t) = S(\phi)$. Then the film temperature corresponds to $\phi = \frac{1}{2}$ and similarity exists if the viscosity varies as

$$\mu = \mu_f S(\phi) \tag{8.5.6}$$

where $S(\frac{1}{2}) = 1$. Various functions of $S(\phi)$ satisfy this requirement—for instance, algebraic expressions, power series, and exponential forms. Carey and Mollendorf (1978) considered the following linear form of $S(\phi)$:

$$S(\phi) = 1 + \gamma_f \left(\phi - \frac{1}{2}\right) \tag{8.5.7}$$

where

$$\gamma_f = \frac{1}{\mu_f} \left(\frac{d\mu}{dt}\right)_f (t_0 - t_\infty) \tag{8.5.8}$$

The viscosity variation parameter γ_f signifies the rate of variation of viscosity with temperature. For a linear variation of viscosity with temperature, if μ_0 and μ_∞ are the values of viscosity evaluated at the surface and ambient temperatures, respectively,

$$\gamma_f = \frac{2(\mu_0 - \mu_\infty)}{\mu_0 + \mu_\infty} = \frac{2(\mu_0/\mu_\infty - 1)}{\mu_0/\mu_\infty + 1}$$

It is seen that $\gamma_f = -2$ for $\mu_0/\mu_\infty = 0$ and $\gamma_f = +2$ for $\mu_0/\mu_\infty \rightarrow \infty$. Thus γ_f varies from -2 to $+2$. However, the numerical solutions of Eqs. (8.5.4) and (8.5.5), along with boundary conditions (8.2.13), were obtained for γ_f between -1.6 and $+1.6$ and Prandtl numbers of 1, 10, 100, and 1000. The local heat transfer and shear stress are given by

$$\text{Nu}_x = \frac{hx}{k} = [-\phi'(0)]\left(\frac{\text{Gr}_x}{4}\right)^{1/4} \tag{8.5.9}$$

$$\tau_0(x) = \frac{\mu_f^2 G^3}{16\rho_f x^2}\left(1 + \frac{1}{2}\gamma_f\right) f''(0) \tag{8.5.10}$$

The functions $-\phi'(0)$, $(1 + \frac{1}{2}\gamma_f)f''(0)$, and other transport quantities are given in Table 8.5.1 for various values of γ_f and Prandtl number. The average Nusselt number is given by

$$\text{Nu} = \frac{hL}{k} = \frac{4}{3}[-\phi'(0)]\left(\frac{\text{Gr}}{4}\right)^{1/4} \tag{8.5.11}$$

Note that $\gamma_f = 0$ corresponds to the constant-property solution.

The effect of variable viscosity on shear stress and heat transfer is clearly shown by the parameters $(1 + \frac{1}{2}\gamma_f)f''(0)/f''(0)_{\gamma_f=0}$ and $-\phi'(0)/-\phi'(0)_{\gamma_f=0}$, respectively, in Table 8.5.1. Obviously, these parameters are equal to unity for

Table 8.5.1 Values of transport quantities for natural convection from an isothermal surface in liquids with temperature-dependent viscosity

Pr_f	γ_f	$f''(0)$	$(1 + \frac{1}{2}\gamma_f)f''(0)$	$\frac{(1 + \frac{1}{2}\gamma_f)f''(0)}{f''(0)_{\gamma f=0}}$	$-\phi'(0)$	$\frac{-\phi'(0)}{-\phi'(0)_{\gamma f=0}}$	$f(\infty)$	$\frac{f(\infty)}{f(\infty)_{\gamma f=0}}$
1.0	−1.6	2.0416	0.4083	0.6358	0.6514	1.1487	0.5981	1.1436
	−0.8	0.9256	0.5554	0.8648	0.5965	1.0518	0.5572	1.0654
	0.0	0.6422	0.6422	1.0000	0.5671	1.0000	0.5230	1.0000
	0.8	0.5050	0.7070	1.1009	0.5469	0.9644	0.4921	0.9409
	1.6	0.4233	0.7601	1.1836	0.5315	0.9372	0.4645	0.8881
10.0	−1.6	1.4981	0.2996	0.7147	1.4076	1.2038	0.3345	1.3423
	−0.8	0.6265	0.3759	0.8967	1.2476	1.0670	0.2883	1.1569
	0.0	0.4192	0.4192	1.0000	1.1693	1.0000	0.2492	1.0000
	0.8	0.3219	0.4507	1.0751	1.1190	0.9570	0.2083	0.8359
	1.6	0.2645	0.4761	1.1357	1.0843	0.9273	0.1591	0.6384
100.0	−1.6	0.9686	0.1937	0.7696	2.7168	1.2398	0.1961	1.4356
	−0.8	0.3848	0.2309	0.9174	2.3600	1.0769	0.1632	1.1947
	0.0	0.2517	0.2517	1.0000	2.1914	1.0000	0.1366	1.0000
	0.8	0.1904	0.2666	1.0592	2.0843	0.9511	0.1085	0.7943
	1.6	0.1545	0.2781	1.1049	2.0117	0.9180	0.0707	0.5176
1000.0	−1.6	0.5770	0.1154	0.7964	4.9827	1.2565	0.1125	1.4706
	−0.8	0.2238	0.1343	0.9268	4.2887	1.0815	0.0924	1.2078
	0.0	0.1449	0.1449	1.0000	3.9654	1.0000	0.0765	1.0000
	0.8	0.1089	0.1525	1.0524	3.7602	0.9483	0.0597	0.7804
	1.6	0.0878	0.1580	1.0904	3.6178	0.9123	0.0367	0.4797

Source: Reprinted with permission from V. P. Carey and J. C. Mollendorf, *Proc. 6th Int. Heat Transfer Conf.*, vol. 2, p. 211. Copyright © 1978, Hemisphere Publishing Corporation.

constant viscosity. Then the solution corresponds to that for constant properties. It was further concluded that temperature-dependent viscosity has a significant effect on the velocity and temperature distributions. At all values of the Prandtl number, for $\gamma_f = -1.6$, the velocity maximum shifted toward the surface and the momentum boundary layer thickness increased. For $\gamma_f = 1.6$ the opposite effects were found.

In addition to gases, Sparrow and Gregg (1958) analyzed the variable-property effects in mercury. They concluded that the results from constant-property analysis can be used accurately provided all the properties are calculated at

$$t_r = t_0 - 0.3(t_0 - t_\infty) \tag{8.5.12}$$

However, for most engineering calculations the film temperature for property evaluation is also a reasonable choice.

Brown (1975) used an integral approach to analyze the effect of temperature-dependent β on natural convection heat transfer. Heat transfer errors due to ignoring variable β were tabulated for water, ethyl alcohol, saturated Freon 12, benzene, and mercury. Constant-property results are in error by at most 5% if all the properties are evaluated at the film temperature.

REFERENCES

Bayley, F. J. (1955). *Proc. Inst. Mech. Eng. 169,* 361.
Brown, A. (1975). *J. Heat Transfer 97,* 133.
Carey, V. P., and Mollendorf, J. C. (1978). *Proc. 6th Int. Heat Transfer Conf.*, Toronto, vol. 2, p. 211.
Clausing, A. M. (1983). *J. Heat Transfer 105,* 138.
Clausing, A. M., and Kempka, S. N. (1981). *J. Heat Transfer 103,* 609.
Fritsch, C. A., and Grosh, R. J. (1963). *J. Heat Transfer 85,* 289.
Fujii, T., Takeuchi, M., Fujii, M., Suzaki, K., and Uehara, H. (1970). *Int. J. Heat Mass Transfer 12,* 753.
Ghajar, A. J., and Parker, J. D. (1981). *J. Heat Transfer 103,* 613.
Gill, W. N., Del Casal, E., and Zeh, D. W. (1965). *Int. J. Heat Mass Transfer 8,* 1135.
Gray, D. D., and Giorgini, A. (1976). *Int. J. Heat Mass Transfer 19,* 545.
Griffiths, E., and Davis, A. H. (1922). Food Investigation Special Rep. No. 9, London.
Hahne, E. (1965). *Int. J. Heat Mass Transfer 8,* 481.
Hara, T. (1954). *Trans. JSME 20,* 517.
Hara, T. (1958). *Bull. JSME 1,* 251.
Hasegawa, S., and Yoshioka, K. (1966). *Mem. Fac. Eng., Kyushu Univ. 26,* 1.
Howarth, L. (1948). *Proc. R. Soc. London Ser. A 194,* 1.
Ito, T., Yamashita, H., and Nishikawa, K. (1974). *Proc. 5th Int. Heat Transfer Conf.*, Tokyo, vol. 3, p. 49.
Knapp, K. K., and Sabersky, R. H. (1966). *Int. J. Heat Mass Transfer 9,* 41.
Minkowycz, W. J., and Sparrow, E. M. (1966). *Int. J. Heat Mass Transfer 9,* 1145.
Neumann, R. J., and Hahne, W. P. (1980). *Int. J. Heat Mass Transfer 23,* 1643.
Nishikawa, K., and Ito, T. (1969). *Int. J. Heat Mass Transfer 12,* 1449.
Nishikawa, K., Ito, T., and Yamashita, H. (1970). *Mem. Fac. Eng., Kyushu Univ. 30,* 2.
Padlog, R. D., and Mollendorf, J. C. (1983). *J. Heat Transfer 105,* 655.
Piau, J. M. (1970). *C. R. Acad. Sci. Ser. A 271,* 953.

Piau, J. M. (1974). *Int. J. Heat Mass Transfer 17,* 465.
Pirovano, A., Viannay, S., and Jannot, M. (1970). *Proc. 4th Int. Heat Transfer Conf.,* Paper NC-1.8.
Saunders, O. A. (1936). *Proc. R. Soc. London Ser. A 157,* 278.
Shaukatullah, H., and Gebhart, B. (1979). *Numer. Heat Transfer 2,* 215.
Siebers, D. L. (1978). SAND 78-8276, Sandia National Laboratories, Livermore, Calif.
Siebers, D. L., Schwind, R. G., and Moffat, R. J. (1983). SAND 83-8225, Sandia National Laboratories, Livermore, Calif.
Sparrow, E. M., and Gregg, J. L. (1958). *J. Heat Transfer 80,* 879.
Touloukian, Y. S., and Ho, C. Y., Eds. (1970–1977). *Thermophysical Properties of Matter,* Plenum, New York.
Vliet, G. C., and Ross, D. C. (1975). *J. Heat Transfer 97,* 549.
Warner, C. Y., and Arpaci, V. S. (1968). *Int. J. Heat Mass Transfer 11,* 397.

PROBLEMS

8.1 Consider a fluid for which the temperature dependence of density, $\rho(t)$, is best approximated by

$$\rho(t) = \rho(t_\infty)e^{-K(t-t_\infty)}$$

For a buoyancy-induced flow generated adjacent to a vertical isothermal surface at t_0 in a quiescent ambient medium:

(a) Calculate the buoyancy force $g(\rho_\infty - \rho)$.

(b) Find the coefficient of the buoyancy force term in the transformed force momentum balance, that is, in the form of $f''' + \text{etc.} = 0$.

(c) Considering only this buoyancy force term, would you expect a similarity solution?

8.2 For the above problem,

(a) Determine the proper Grashof number definition.

(b) Give one case in which a similarity solution results, in the absence of the pressure term, stratification, viscous dissipation, and a distributed energy source.

8.3 Consider external natural convection boundary layer flow.

(a) For an ideal gas having c_p constant and μ and k proportional to T, find the expressions for the local Nusselt number based on the local coordinate x and the average Nusselt number based on the average convection coefficient h, for a vertical isothermal plate, in terms of the solution of the differential equations.

(b) For a plate 1 in. high at 1000°R in a gas at 40°F and 1 atm, compare the heat transfer rates calculated assuming constant properties at 40°F and assuming the property behavior in part (a).

8.4 Starting with the variable-property boundary layer equations and conditions (8.2.1)–(8.2.4) and the transformation in Eqs. (8.2.6)–(8.2.10):

(a) Calculate the buoyancy force in Eq. (8.2.11).

(b) Show that the formulation (8.2.11)–(8.2.13) results.

8.5 A 20-cm-wide vertical surface, 10 cm high, at 10°C is in a quiescent water at 5°C:

(a) Calculate the heat transfer rate from the tabulated results of the numerical solution in Section 8.4.

(b) Repeat (a), using the correlation resulting from these calculations.

(c) Repeat (a), using the "film" temperature method.

(d) Comment on the reasons for any difference.

8.6 For the surface and temperature conditions in Problem 8.5, assume the ambient medium to be a light oil for which Pr = 1000:

(a) Determine the value of γ_f for some common oil.

(b) Calculate the heat transfer rate to the oil.

8.7 Consider laminar flow over a flat vertical surface 1 m high at 100°C in air at 20°C. Assuming constant fluid properties, compute the resulting heat transfer for properties evaluated at 20°C, at 100°C, and also at the film temperature. Compare these results and comment on the observed differences.

8.8 Repeat Problem 8.7 for a sphere of diameter 20 cm at 100°C in air at 20°C, using the results presented in Chapter 5.

8.9 Derive the governing equations for an axisymmetric plume, if property variations and non-Boussinesq effects are to be included, using the analysis for the constant-property case given in Chapter 4.

8.10 For a 1-m-high vertical plate at 60°C, in water, over the temperature range of 60 to 20°C, are the variable-property effects significant? If so, over what ambient temperature range can these effects be neglected? Assume laminar flow.

CHAPTER

NINE

TRANSPORT IN COLD PURE AND SALINE WATER

9.1 DENSITY BEHAVIOR OF COLD WATER

Transport processes arise in cold pure and saline water in many natural and technological processes. A common and important example is the formation and dissipation of river and oceanic ice. Flows occurring under such circumstances are often complicated by the presence of a density extremum in the water. This extremum results at decreasing temperatures from a balance of the competing density-controlling mechanisms of increasing hydrogen bonding and decreasing molecular thermal motion, as discussed by Pounder (1965). In pure water this extremum occurs at about 4°C. However, extrema also arise in saline water, up to a salinity level of about 26‰ (i.e., parts per thousand, ppt) and at pressures up to about 300 bars absolute, for systems remaining in local thermodynamic equilibrium. For nonequilibrium conditions, an extremum may be found well beyond these salinity and pressure limits.

Density correlation. There are many correlations representing the density of water as a function of temperature t, salinity s, and pressure p that account for the extremum effect. Such relations include those of Kell (1967), Bigg (1967), Fujii (1974), Brown and Lane (1976), Chen and Millero (1976), and Gebhart and Mollendorf (1977), among others. Most of these correlations are in close agreement. However, the correlation of Gebhart and Mollendorf (1977) is very accurate and the most amenable to analysis. Therefore it was used in many recent studies and will be adopted here. In that correlation the density is expressed as follows:

$$\rho(t, s, p) = \rho_m(s, p)\{1 - \alpha(s, p)[|t - t_m(s, p)|]^{q(s,p)}\} \tag{9.1.1}$$

$$\rho_m(s, p) = \rho_m(0, 1)[1 + f_1(p) + sg_1(p) + s^2h_1(p)] \tag{9.1.2}$$

Table 9.1.1 Parameters that arise in Eq. (9.1.1) for $n = 3$, $P = 48$

	j			
	0	1	2	3
f_{1j}	—	4.960998E − 05	−2.601973E − 09	7.842619E − 13
f_{2j}	—	1.377584E − 04	1.497648E − 06	2.903240E − 10
f_{3j}	—	−5.430000E − 03	7.720181E − 07	−7.038846E − 10
f_{4j}	—	−1.118758E − 04	−1.238393E − 07	5.857253E − 11
g_{1j}	7.992252E − 04	−5.194896E − 08	1.031185E − 10	−2.979653E − 14
g_{2j}	1.623355E − 02	1.129961E − 05	−8.053248E − 08	6.966452E − 12
g_{3j}	−5.265509E − 02	7.496781E − 05	−2.792053E − 07	1.411138E − 10
g_{4j}	−3.136530E − 03	2.983937E − 06	4.453557E − 09	−2.937601E − 12
h_{1j}	1.918334E − 07	1.347190E − 09	−2.203133E − 12	1.112440E − 15
h_{2j}	−4.565866E − 04	−4.352912E − 07	1.978675E − 09	−9.079379E − 13
h_{3j}	0.000000	−3.683650E − 06	7.694077E − 09	−4.561113E − 12
h_{4j}	7.599378E − 05	−8.718915E − 08	−4.166570E − 11	5.870105E − 14

Source: Reprinted with permission from B. Gebhart and J. C. Mollendorf, *J. Fluid Mech.*, vol. 89, p. 673. Copyright © 1978, Cambridge, University Press.

$$\alpha(s, p) = \alpha(0, 1)[1 + f_2(p) + sg_2(p) + s^2h_2(p)] \tag{9.1.3}$$

$$t_m(s, p) = t_m(0, 1)[1 + f_3(p) + sg_3(p) + s^2h_3(p)] \tag{9.1.4}$$

$$q(s, p) = q(0, 1)[1 + f_4(p) + sg_4(p) + s^2h_4(p)] \tag{9.1.5}$$

$$f_i(p) = \sum_{j=1}^{n} f_{ij}(p-1)^j \qquad g_i(p) = \sum_{j=0}^{n} g_{ij}(p-1)^j \qquad h_i(p) = \sum_{j=0}^{n} h_{ij}(p-1)^j \tag{9.1.6}$$

where $\rho_m(0, 1)$, $\alpha(0, 1)$, $t_m(0, 1)$, and $q(0, 1)$ are the values of those quantities for pure water at 1 bar absolute and f_i, g_i, and h_i are polynomials in $(p - 1)$.

Table 9.1.2 Parameters that arise in Eq. (9.1.1) for $n = 2$

	j		
	0	1	2
f_{1j}	—	4.955317E − 05	−1.950180E − 09
f_{2j}	—	5.181147E − 04	1.190039E − 06
f_{3j}	—	−5.430000E − 03	2.455177E − 07
f_{4j}	—	−1.898839E − 04	2.515528E − 08
g_{1j}	8.046157E − 04	−1.051410E − 09	3.304577E − 11
g_{2j}	−2.839092E − 03	−7.125734E − 06	−2.430584E − 09
g_{3j}	−5.265509E − 02	−6.824758E − 05	2.106695E − 09

Source: Reprinted with permission from B. Gebhart and J. C. Mollendorf, *J. Fluid Mech.*, vol. 89, p. 673. Copyright © 1978, Cambridge University Press.

Some polynomials are taken as zero in simpler, though less accurate, formulations. In Eq. (9.1.1), $\rho(t, s, p)$ varies with temperature as $|t - t_m|^q$. This form usually leads to extremely important and major simplifications of both the flow analysis and the parameter characterizations in particular convective circumstances. For example, gradients in salinity are often much more important than pressure gradients in wide ranges of applications.

The (0, 1) quantities and the pressure polynomials f_i, g_i, and h_i were deter-

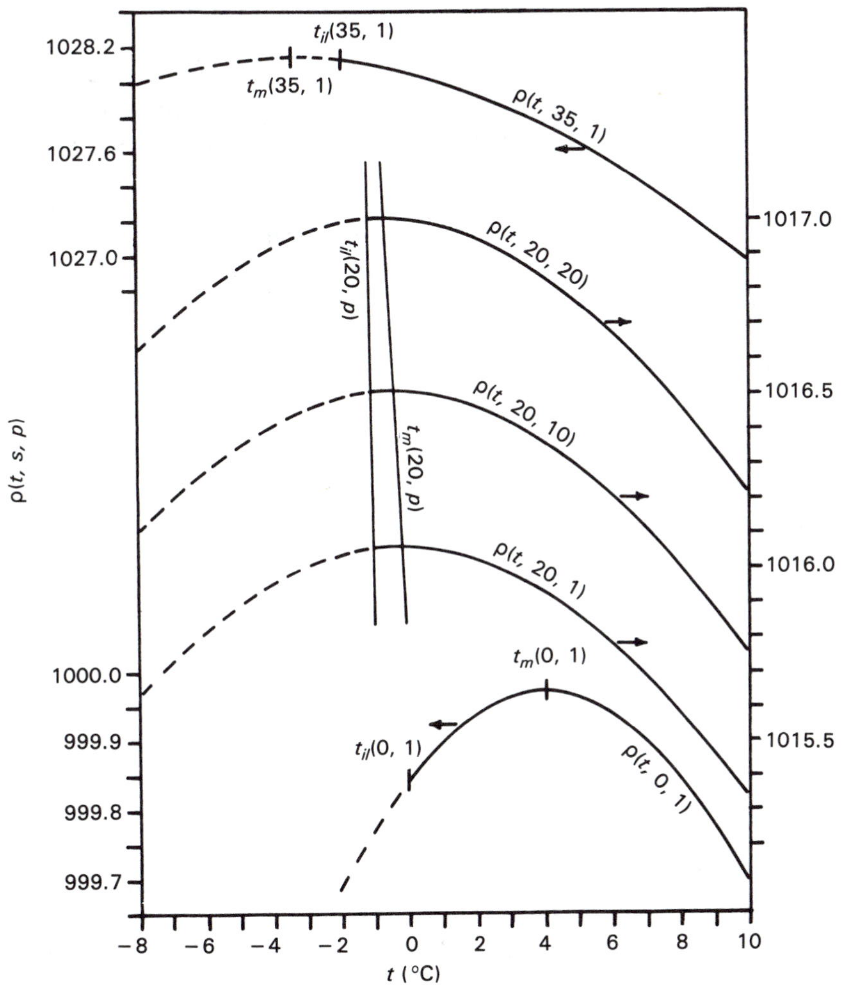

Figure 9.1.1 Density-temperature dependence at various s and p values as formulated by Gebhart and Mollendorf (1977). Units of $\rho(t, s, p)$ kilograms per cubic meter; t, degrees Celsius; s, parts per thousand; and p, bars absolute. Also shown is the equilibrium phase interface and the temperature of maximum density. The arrow associated with each curve indicates the vertical scale that applies. *(Reprinted with permission from B. Gebhart and J. C. Mollendorf,* J. Fluid. Mech., *vol. 89, p. 673. Copyright © 1978, Cambridge University Press.)*

mined by a nonlinear regression fit, with the smallest root mean square difference, to perhaps the best collection of information: the pure-water data of Fine and Millero (1973) and the saline-water density data of Chen and Millero (1976). The range of the regression was $t = 0–20°C$, $s = 0–40‰$, and $p = 1–1000$ bars absolute. This range of conditions includes most terrestrial surface water and many other conditions in technology.

The most accurate form of Eq. (9.1.1) obtained was with the third-order polynomials for f_i, g_i, and h_i, that is, $n = 3$. The resulting root mean square fit was within 3.5 parts per million (ppm) for pure water and within 10.4 ppm for the 309 saline-water data points of Chen and Millero (1976) that fell in the range of conditions given above. The resulting values of the coefficients are shown in Table 9.1.1. For $n = 2$, omitting all s^2 terms and assuming q independent of s, the root mean square differences are only 6.5 and 38.2 ppm over the same range. The resulting coefficients are shown in Table 9.1.2. The effects of salinity and pressure on ρ, ρ_m, and t_m are shown in Fig. 9.1.1. The variation of t_m with salinity and pressure from Eq. (9.1.1) is shown by the solid curves in Fig. 9.1.2.

Phase equilibrium. The temperature range of validity of any analysis of flows in cold water, for equilibrium phase changes, is bounded on the lower side by the equilibrium ice-melting temperature t_{il}. This was recently determined by Fujino et al. (1974) to be

$$t_{il}(s, p) = -0.02831 - 0.0499s - 0.000112s^2 - 0.00759p \qquad (9.1.7)$$

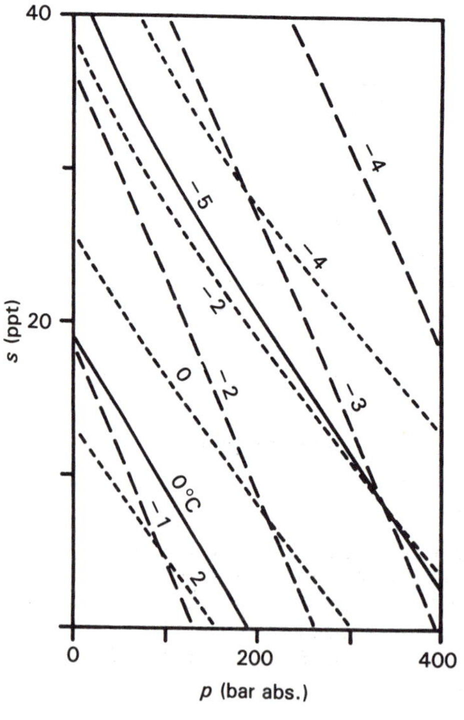

Figure 9.1.2 Variations of the temperature $t_m(s, p)$ at maximum density, the phase equilibrium temperature $t_{il}(s, p)$, and their difference $t_m(s, p) - t_{il}(s, p)$ over a range of salinities and pressures. (——) $t_m(s, p)$; (— — —) $t_{il}(s, p)$; (· · · · ·) $t_m - t_{il}$. Numbers on the curves are in degrees Celsius. *(Reprinted with permission from B. Gebhart and J. C. Mollendorf,* J. Fluid. Mech., *vol. 89, p. 673. Copyright © 1978, Cambridge University Press.)*

when corrected and converted to bars absolute. The data range was $17.1 < s < 35‰$ and $1 < p < 100$ atm. This result was corroborated by measurements by Doherty and Kester (1974). Contours of constant t_{il} vs. salinity and pressure are shown in Fig. 9.1.2 as dashed curves. Large depressions in the equilibrium melting temperatures are seen at high salinities and pressures. It can be seen in Fig. 9.1.1 that t_m decreases more rapidly than t_{il} with both increasing salinity and pressure.

Using Eq. (9.1.7) in conjunction with $t_m(s, p)$, Fig. 9.1.2 shows the variation of $t_m - t_{il}$ with salinity and pressure, also as dashed curves. The equilibrium limits for the occurrence of a density extremum are seen to be about $p < 300$ bars in pure water and about $s < 25.5‰$ at a pressure of 1 bar. However, recall that in freezing ice from pure water and possibly also from saline water, substantial temperature depressions below the equilibrium condition, Eq. (9.1.7), often occur and may persist for very long periods.

In the following sections, the general formulation of transport will be discussed first, followed by transport due to thermal buoyancy, in the absence of concentration gradients. External flows adjacent to flat vertical and horizontal surfaces are discussed in detail and both analytical and experimental results are presented. A discussion of transport due to combined thermal and saline effects follows, emphasizing the conditions of ice melting in saline water.

9.2 GENERAL FORMULATION OF TRANSPORT

The equations of steady laminar motion, with a Boussinesq approximation in Eq. (9.2.1) and with constant molecular diffusion properties μ, k, and D, are

$$\nabla \cdot \mathbf{V} = 0 \tag{9.2.1}$$

$$\rho_r(\mathbf{V} \cdot \nabla)\mathbf{V} = \mathbf{F} - \nabla p + \mu\nabla^2\mathbf{V} \tag{9.2.2}$$

$$\rho_r c_p(\mathbf{V} \cdot \nabla)t = k\nabla^2 t + \beta T(\mathbf{V} \cdot \nabla)p + \mu\Phi \tag{9.2.3}$$

$$(\mathbf{V} \cdot \nabla)s = D\nabla^2 s \tag{9.2.4}$$

where $\mathbf{F} = \mathbf{g}\rho$ is the body force per unit volume, p the local static pressure, and $\mathbf{V}$ the local fluid velocity. The constant-density approximation apparent in Eq. (9.2.1) is much more reliable for the conditions relevant to cold water than, in general, for liquids or gases. Note that the form of Eq. (9.2.1) leaves the specific value of ρ_r unspecified.

The salt concentration s is very commonly small compared with the density of the water. For seawater s is around 35‰. The formulation above neglects distributed energy and salinity sources, for instance, from chemical reactions. The Sorét effect is not included as it is a relatively small effect in the presence of appreciable convective motion. The Dufour effect is even smaller. The terms in the energy equation (9.2.3) corresponding to viscous dissipation and the pressure field will be ignored later, since they usually are very small in such flows. In addition, these terms do not admit similarity in some of the flows of greatest practical importance, as we will see later.

The x direction is first taken positive in the direction opposed to gravity—that is, $\mathbf{g} = -g\mathbf{i}$, where $\mathbf{i}$ is a unit vector in the x direction—for upward buoyancy. As presented in Chapter 2 for external flows, the local static pressure p is written as the sum of the local motion pressure p_m and the hydrostatic pressure p_h in the remote ambient medium, where $dp_h/dx = -g\rho_\infty$ and ρ_∞ is the local ambient density (at x). Therefore,

$$\mathbf{F} - \nabla p = -g\rho\mathbf{i} + g\rho_\infty\mathbf{i} - \nabla p_m = g(\rho_\infty - \rho)\mathbf{i} - \nabla p_m \tag{9.2.5}$$

The first term in Eq. (9.2.5) is the buoyancy force. In general, $\rho_\infty = \rho(t_\infty, s_\infty, p_\infty)$ and $\rho(t, s, p)$ is determined from Eq. (9.1.1). For internal flows, other reference pressure conditions may be used.

9.3 TRANSPORT DUE TO THERMAL BUOYANCY

Because of the anomalous density behavior of water, flows resulting under low-temperature conditions are often quite complex. This happens at about 4°C in pure water. Buoyancy force reversals, bidirectional velocity distributions, and total convective inversions are commonly encountered. Codegone (1939) was the first to demonstrate the existence of convective inversion due to the density extremum. Ede (1951, 1955) reported detailed heat transfer measurements in water under conditions of density inversion.

Merk (1953) was the first to analyze such flows. The density of water was modeled as a third-order polynomial in temperature and an integral method used to predict local heat transfer for ice melting in fresh water. Convective inversion was predicted to occur at about 5.3°C, and the heat transfer rate was found to undergo a minimum at that temperature. For ice melting in water at a temperature t_∞ below 5.3°C, the flow direction along the surface was found to be upward, while downward flow was predicted for $t_\infty > 5.3$°C. In a subsequent study, Schecter and Isbin (1958) also used an integral analysis to study the flow adjacent to a flat vertical surface in water at about 4°C.

Goren (1966) considered a vertical surface at a temperature t_0 in ambient water whose temperature t_∞ was equal to that for maximum density, that is, $t_\infty = t_m$. The usual equations of motion were used with the density difference $\Delta\rho$ taken as $\rho_m\alpha(t - t_m)^2$, where $\alpha = 8.0 \times 10^{-6}$ (°C)$^{-2}$. This is a conventional correlation said to give sufficient accuracy for $t_m \pm 4$°C. No additional parameters arise and an analog computer solution was given for Pr = 11.4. Vanier and Tien (1967) extended the study of Goren above its implied accuracy limit of $t_\infty = 8$°C by approximating the driving density difference by a sum of linear, square, and cubic terms in $t - t_m$, according to the density data of Perry (1963). This is similar to the treatment by Merk (1953). The penalty in the analysis is that there are two surface temperature-dependent parameters in two new terms in the differential equations. The formulation is still limited to $t_\infty = t_m$. Neglecting these terms, Vanier and Tien repeated Goren's calculations and obtained heat transfer values about 15% higher. Solutions were given for specific values in the range $0 \leq t_0 \leq 35$°C.

In a related study, Vanier and Tien (1968) discussed in detail the directional tendencies of flow across the boundary region formed adjacent to a vertical surface in ambient water at temperatures around its density extremum. The effects of the relation of t_0 and t_∞ to t_m were outlined. The density formulation of Merk was then used in the analysis. The equations were reduced to similarity form. Numerical results for $t_\infty = 0$ and $1 \leq t_0 \leq 14°C$ were compared with the predictions of Schecter and Isbin (1958). Considerable differences were found. Calculations were also made for other values of t_∞. The calculations were also compared with some of the data of Ede (1951) and Schechter and Isbin (1958) and showed fair agreement.

Other early studies include measurements by Oborin (1967) of the heat transfer from spheres and horizontal cylinders in cold water. Good agreement with Merk's prediction of convective inversion was reported. Schenk and Schenkels (1968) measured an ice sphere melting in cold water and observed fair agreement with the experimental results of Dumoré et al. (1953) and the analysis of Merk (1953).

Govindarajulu (1970) considered the boundary layer equations, again with $\Delta\rho \propto (\Delta t)^2$, for water at $t_\infty = t_m$ to investigate transport adjacent to both vertical and horizontal porous surfaces. Similarity was formulated for a power law downstream surface temperature variation $t_0(x)$, that is, $d(x) = t_0 - t_\infty = Nx^n$ in the notation that follows. The required x dependence of the blowing velocity for similarity was given. No solutions were obtained.

Roy (1972) repeated the calculations of Goren and obtained heat transfer results that were higher, as found by Vanier and Tien, by about 15%. Then a large Prandtl number approximation was made in a series solution in $\mathrm{Pr}^{-1/2}$. However, water is apparently the only prominent liquid of moderately high Prandtl number having a density extremum. Soundalgekar (1973) used the $\Delta\rho \propto (\Delta t)^2$ buoyancy formulation, again for $t_\infty = 4°C$, in an integral analysis to calculate the surface shear stress. Bendell and Gebhart (1976) determined the melting rates of vertical ice slabs in ambient water at temperatures from 2 to 20°C.

Since these studies there have been numerous experimental, analytical, and numerical investigations of various geometries and configurations. In the rest of this section external flows are analyzed for various effects and configurations.

9.3.1 Similarity Considerations for Thermally Driven Flows

Consider first a flat vertical heated or cooled surface immersed in pure or saline water, under circumstances where no mass diffusion effects arise. For steady, laminar, two-dimensional vertical plane flows, subject to boundary layer analysis, Eqs. (9.2.1)–(9.2.3) reduce to

$$\frac{\partial u}{\partial x} + \frac{\partial v}{\partial y} = 0 \tag{9.3.1}$$

$$\rho_r\left(u\frac{\partial u}{\partial x} + v\frac{\partial u}{\partial y}\right) = \mu\frac{\partial^2 u}{\partial y^2} + g(\rho_\infty - \rho) \tag{9.3.2}$$

$$\rho_r c_p\left(u\frac{\partial t}{\partial x} + v\frac{\partial t}{\partial y}\right) = k\frac{\partial^2 t}{\partial y^2} + \beta T u\frac{dp_h}{dx} + \mu\left(\frac{\partial u}{\partial y}\right)^2 \tag{9.3.3}$$

Gebhart and Mollendorf (1978) obtained general similarity solutions for a heated flat vertical surface immersed in a medium experiencing a density extremum, using the accurate density relation (9.1.1), to determine the buoyancy force. Following the notation of Gebhart (1971, 1973), and as in Chapter 3, a transformation is defined in terms of a similarity variable $\eta(x, y)$ and stream functions $\psi(x, y)$ and $f(\eta)$ as follows:

$$\eta = yb(x) \qquad \psi(x, y) = \nu c(x) f(\eta) \tag{9.3.4}$$

$$t_0 - t_\infty = d(x) \tag{9.3.5}$$

$$t_\infty - t_r = j(x) \tag{9.3.6}$$

$$\phi = \frac{t - t_\infty}{t_0 - t_\infty} \tag{9.3.7}$$

where t_r is a reference value and ν is taken as constant.

The local vigor of the flow is usually indicated by the local Grashof number, defined as $\mathrm{Gr}_x = g\beta x^3(t_0 - t_\infty)/\nu^2$. Around an extremum, however, linearizing the density temperature dependence as shown in Chapter 2 is inappropriate. Some other better measure of the density difference $\Delta\rho$ must be used, and Gr_x is written in terms of a reference density ρ_r as

$$\mathrm{Gr}_x = \frac{g x^3\, \Delta\rho}{\nu^2 \rho_r} \tag{9.3.8}$$

Introducing Eqs. (9.3.4)–(9.3.7) into Eqs. (9.3.1)–(9.3.3), the following equations are obtained with the appropriate boundary conditions:

$$f''' + \frac{c_x}{b} ff'' - \left(\frac{c_x}{b} + \frac{cb_x}{b^2}\right) f'^2 + \frac{g}{\nu^2 c b^3}\frac{\rho_\infty - \rho}{\rho_r} = 0 \tag{9.3.9}$$

$$\frac{\phi''}{\mathrm{Pr}} + \frac{c_x}{b} f\phi' - \frac{cd_x}{bd} f'\phi - \frac{cj_x}{bd} f' - \beta T\frac{c}{bd}\frac{g}{c_p} f' + \frac{b^2c^2}{d}\frac{\nu^2}{c_p} f''^2 = 0 \tag{9.3.10}$$

$$1 - \phi(0) = \phi(\infty) = f'(0) = f'(\infty) = 0 \tag{9.3.11}$$

The other boundary conditions result from additional considerations at $\eta = 0$. For a strictly impermeable surface

$$f(0) = 0 \tag{9.3.12}$$

However, this would admit no melting at an interface. Other surface conditions are considered later for melting and freezing ice surfaces.

Gebhart and Mollendorf (1978) showed that the effects of pressure on density are negligible for most flow circumstances, with the exception of regions of very large vertical extent. The pressure terms in Eq. (9.1.1) then pertain only to the ambient pressure level. The density difference is then

$$\begin{aligned}\rho_\infty - \rho &= \rho(t_\infty, s_\infty, p) - \rho(t, s_\infty, p) \\ &= \rho_m(s_\infty, p)\alpha(s_\infty, p)[|t - t_m|^q - |t_\infty - t_m|^q] \\ &= \rho_m\alpha[|t - t_m|^q - |t_\infty - t_m|^q]\end{aligned} \tag{9.3.13}$$

where $t_m = t_m(s_\infty, p)$. Both temperature terms are always taken positive, since ρ_m is a maximum. The density difference is next written in terms of a new parameter R, defined as

$$R = \frac{t_m(s_\infty, p) - t_\infty}{t_0 - t_\infty} \tag{9.3.14}$$

Therefore

$$B = g(\rho_\infty - \rho) = g\rho_m\alpha|t_0 - t_\infty|^q[|\phi - R|^q - |R|^q] = g\rho_m\alpha|t_0 - t_\infty|^q W \tag{9.3.15}$$

where $W(\phi, R)$ is also called the buoyancy force.

The parameter R indicates the relation between t_0, t_∞, and the extremum temperature t_m and, as a result, determines the distribution and direction of the buoyancy force W across the layer of temperature accommodation between t_0 and t_∞.

Table 9.3.1 Some transport conditions, values of R, and buoyancy directions

Condition	t_0	t_∞	R	Net buoyancy force
Heating water, $t_0 > t_\infty$	$\frac{1}{2}t_m$	0	2	Down
	t_m	0	1	Down
	$2t_m$	0	$\frac{1}{2}$	Down
	$3t_m$	0	$\frac{1}{3}$	Down
	$4t_m$	0	$\frac{1}{4}$	Up
	$2t_m$	t_m	0	Up
	$3t_m$	$2t_m$	-1	Up
	$4t_m$	$3t_m$	-2	Up
	$10t_m$	$9t_m$	-8	Up
Melting or freezing of ice at 1 bar, or cooling of water, $t_0 < t_\infty$	0	$\frac{1}{2}t_m$	-1	Up
	0	t_m	0	Up
	0	$\frac{2}{3}t_m$	$\frac{1}{2}$	Down
	0	$2t_m$	$\frac{1}{2}$	Down
	0	$4t_m$	$\frac{3}{4}$	Down
	$\frac{1}{2}t_m$	t_m	0	Up
	$\frac{1}{2}t_m$	$\frac{3}{4}t_m$	$\frac{1}{3}$	Down
	$\frac{1}{2}t_m$	$2t_m$	$-\frac{2}{3}$	Down
	t_m	$2t_m$	1	Down
	$8t_m$	$9t_m$	8	Down

Source: Reprinted with permission from B. Gebhart and J. C. Mollendorf, *J. Fluid Mech.*, vol. 89, p. 673. Copyright © 1978, Cambridge University Press.

For example, for $t_\infty = t_m$, that is, $R = 0$, the buoyancy force is always upward. However, for $t_0 = t_m$, that is, $R = 1$, it is always downward. There may also be buoyancy force reversals. For example, taking $t_0 = 0°\text{C}$ and $t_\infty = \frac{3}{2} t_m$ (°C), $R = \frac{1}{3}$, and the buoyancy force is upward very near the surface and downward otherwise. Large values of R result for both t_0 and t_∞ well away from t_m. Table 9.3.1 and Fig. 9.3.1 indicate some of the many possibilities. Note that the values of R shown in Fig. 9.3.1, as related to temperature, apply only for t_0 taken as 0°C.

Most of the earlier analyses of vertical flows were, in effect, for $q = 2$ and

$$R = \frac{t_m(0, 1) - t_\infty}{t_0 - t_\infty}, \quad R(t_0 = 0°\text{C}) = 1 - \frac{4}{t_\infty}$$

$$B = g(\rho_\infty - \rho) \propto |\phi - R|^q - |R|^q, \text{ where } \phi = \frac{t - t_\infty}{t_0 - t_\infty}$$

$$\text{Gr}_x = \frac{gx^3}{\nu^2} \alpha \, |t_0 - t_\infty|^q, \ \alpha \text{ and } q \text{ arise in } \rho(t, s, p)$$

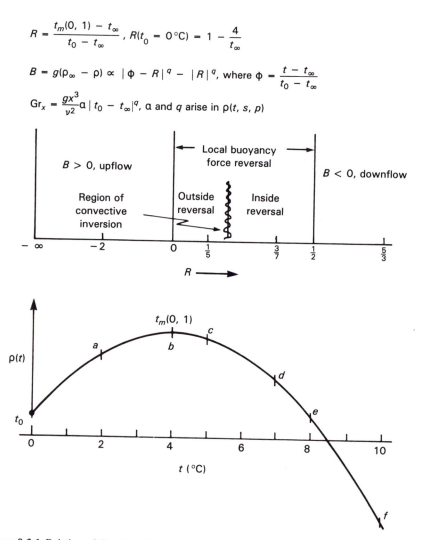

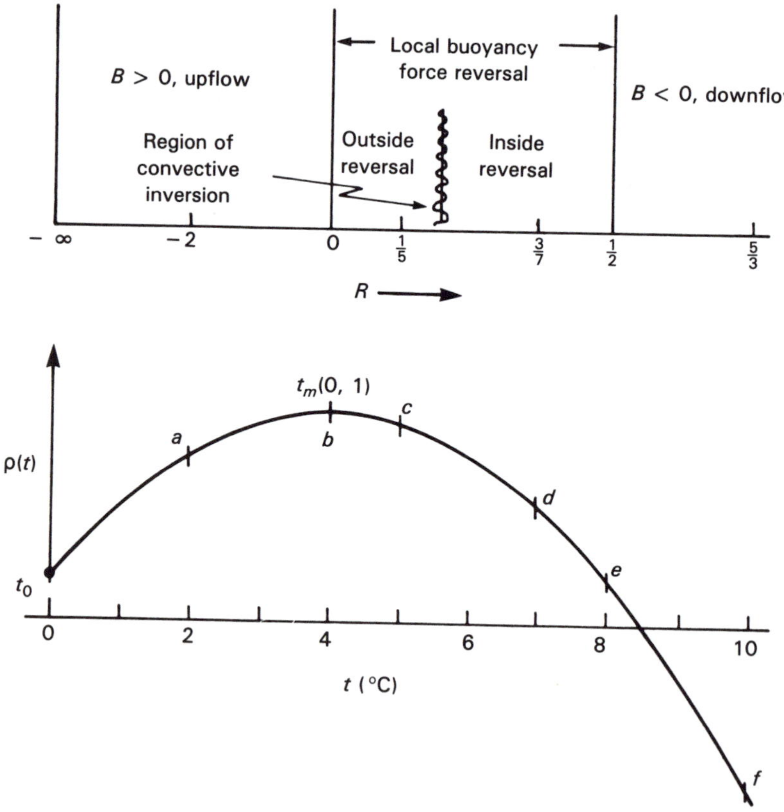

Figure 9.3.1 Relation of R to local buoyancy force and direction and reversal and to convective inversion. Various choices, a–e, for t_∞ are shown. For simplicity, t_0 is taken as the melting temperature of ice in pure water at 1 atm, 0°C.

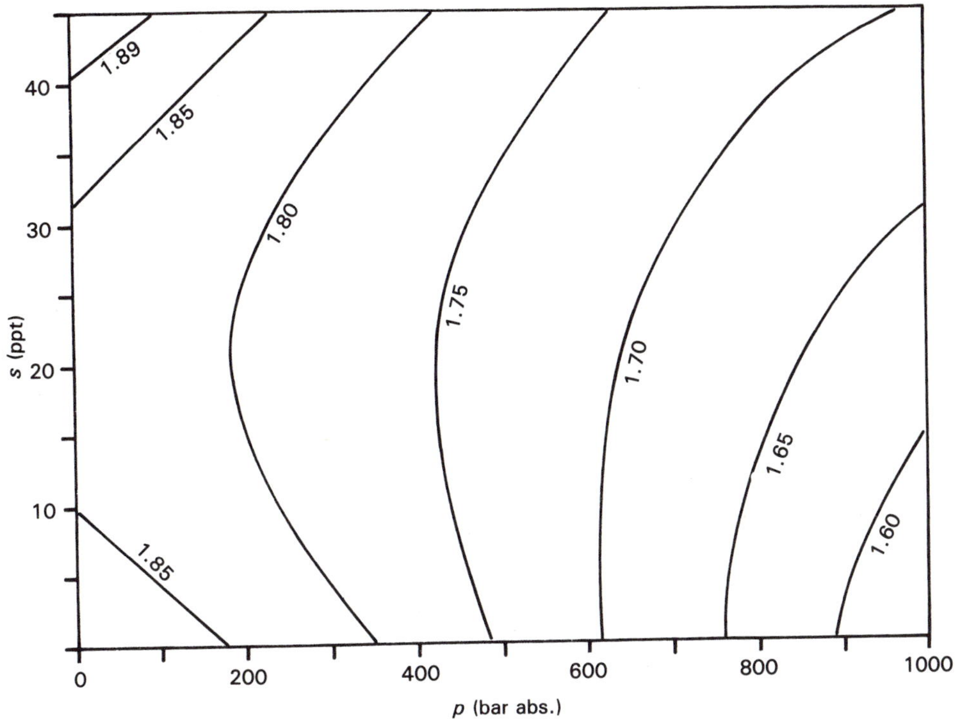

Figure 9.3.2 Variation of the exponent $q(s, p)$ with salinity and pressure from the correlation of Gebhart and Mollendorf (1977) for $n = 3$. *(Reprinted with permission from B. Gebhart and J. C. Mollendorf,* J. Fluid. Mech., *vol. 89, p. 673. Copyright © 1978, Cambridge University Press.)*

$R = 0$ in this formulation. This results in $W = \phi^2$ in Eq. (9.3.15). In Eq. (9.1.1) q varies between about 1.6 and 1.9 over the ranges of salinity and pressure. Contours of constant q have been determined to three significant digits from the roots of Eq. (9.1.5). The results are shown in Fig. 9.3.2 over a wide range of salinity and pressure.

Requirements for similarity. Sufficient conditions for similarity are found by examination of Eqs. (9.3.9), (9.3.10), and (9.3.15). Similarity is possible if the following quantities are independent of x:

$$C_1 = \frac{c_x}{b} \qquad C_2 = \frac{cb_x}{b^2} \qquad C_3 W = \frac{g\alpha\rho_m |t_0 - t_\infty|^q W}{\rho_r \nu^2 c b^3} \tag{9.3.16}$$

Clearly, the choice $\rho_r = \rho_m(s, p)$ is indicated.

It is seen from Eq. (9.3.15) that if $\phi = \phi(\eta)$, as is also necessary in Eq. (9.3.10), then W is independent of x only if R is. A sufficient condition for this is $R = 0$ or $t_\infty = t_m$, that is, the ambient medium is at the extremum density $\rho_m(s, p)$. This is the condition treated by Goren (1966), Govindarajulu (1970), Roy (1972), and Soundalgekar (1973), but with $s_\infty = 0$ and $q = 2$. Otherwise, from Eq. (9.3.14), the condition is that $t_0 - t_\infty = d(x) \propto t_m - t_\infty \propto j(x)$ must be satisfied. Then t_m is taken as the reference temperature, for example, t_r in Eq. (9.3.6). Setting temperature stratification, $j = j(x)$, aside for the moment, the requirement is that both t_0 and t_∞ be independent of x. Then R is a constant.

From the expression for C_1 and C_2 it may be shown that both b and c have either a power law or exponential dependence on x. Taking $C_1 = 3$, $C_3 = 1$, and $C_2 = -1$, for $n = 0$, $b(x)$ and $c(x)$ are

$$b(x) = \frac{1}{x}\left(\frac{g\alpha x^3 d^q}{4\nu^2}\right)^{1/4} = \frac{1}{x}\left(\frac{\mathrm{Gr}'_x}{4}\right)^{1/4} = \frac{G'}{4x} \qquad c(x) = G' \tag{9.3.17}$$

This is very similar to the Boussinesq approximation result, except that the Grashof number is now defined as follows:

$$\mathrm{Gr}'_x = \frac{gx^3}{\nu^2}[\alpha(s_\infty, p)|t_0 - t_\infty|^q] \tag{9.3.18}$$

However, as Gebhart and Mollendorf (1978) note, this quantity is always positive even though the buoyancy force $W(\eta)$ may be positive, negative, or of changing sign, depending on the values of t_0 and t_∞ relative to t_m. This deficiency may be removed by using instead a form of Gr_x dependent on some appropriate $\Delta\rho$, as in Eq. (9.3.8). Taking $\Delta\rho$ as the average value of the density difference $\rho_\infty - \rho$ across the convection layer amounts to integrating W over η [see Eq. (9.3.15)] as follows:

$$I_w = \int_0^\infty W\, d\eta = \int_0^\infty [|\phi - R|^q - |R|^q]\, d\eta \tag{9.3.19}$$

Then the Grashof number becomes

$$\mathrm{Gr}_x = \frac{gx^3}{\nu^2}[\alpha(s_\infty, p)|t_0 - t_\infty|^q]I_W \tag{9.3.20}$$

where $\alpha(s, p)$ is defined in Eq. (9.1.1). When this value of Gr_x is used in Eq. (9.3.17) instead of Eq. (9.3.18), the buoyancy force term in Eq. (9.3.9) becomes

$$\frac{g}{\nu^2 cb^3}\frac{\rho_\infty - \rho}{\rho_r} = F(\eta) = \frac{W}{I_W} \tag{9.3.21}$$

Thus, $I_W < 0$ will often signal a need in Eq. (9.3.20) to reinterpret x as positive in the direction of g. Then $F(\eta)$ would be, on the average, positive across the flow region. This question is that of "convective inversion" and will be clarified later. An additional consideration in the normalization of W by I_W is that the buoyancy force term becomes of order one over most of the region. On the other hand, the equations are now integrodifferential, since I_W must also be iterated in the numerical scheme.

Other conditions for similarity. Additional conditions under which similarity solutions exist will be set forth here before discussing numerical results. Recall that in addition to $R = 0$, that is, $t_\infty = t_m$, the condition $t_0 - t_\infty = d(x) \propto t_m - t_\infty = (t_m - t_r) - (t_\infty - t_r) = t_m - t_r - j(x)$ also results in $W = W(\eta, R)$ for $\phi = \phi(\eta)$. In addition, from Eq. (9.3.10) the parameter C_5, obtained as

$$C_5 = \frac{cd_x}{bd} \tag{9.3.22}$$

must be independent of x. This arises from the x dependence of $t_0 - t_\infty$ and, using Eq. (9.3.17), is satisfied by

$$d(x) = t_0 - t_\infty = Nx^n \qquad C_5 = 4n$$

With the above variation, Gr_x, b, and c are unchanged. However, the constants now become $C_1 = qn + 3$, $C_2 = qn - 1$, and $C_3 = 1$.. Admitting temperature stratification requires in addition that C_7, defined by

$$C_7 = \frac{cj_x}{bd} \tag{9.3.23}$$

be independent of x. This implies that $j(x) = t_\infty - t_r = (C_7N/4n)x^n$ and $C_7 = 4nN_\infty/N$. If $R = 0$, that is, $t_\infty = t_m$, we may not have stratification since

$$t_m = t_m(s_\infty, p) = \text{const.}$$

If $R \neq 0$, then

$$R = \frac{t_m - t_r - N_\infty x^n}{Nx^n}$$

is independent of x only for t_m chosen as t_r. Then $R = -N_\infty/N$.

The ambient medium will be quiescent only for stable stratification. The condition for this in the absence of mass diffusion is $j_x > -g(\partial T/\partial p)_s$, where s is the entropy. Since $(\partial t/\partial p)_s$ is positive and small for most states of liquids, the more conservative and convenient condition $j_x > 0$ is often taken; see Gebhart (1973). Since $(\partial T/\partial p)_s = \beta T/\rho c_p$, the exact condition is

$$j_x = N_\infty n x^{n-1} \geqq -\frac{g\beta T}{c_p} \qquad \beta = -\frac{1}{\rho}\left(\frac{\partial \rho}{\partial T}\right)_p \tag{9.3.24}$$

Noting that g, T, and c_p are positive, the sign of the limit is determined by that of β, which is negative below t_m. Thus, the stable limit allows decreasing t_∞ for $t_\infty > t_m$, but increasing t_∞ is required for $t_\infty < t_m$.

Now considering the viscous dissipation mode of thermal energy production in Eq. (9.3.10), we find

$$\frac{b^2c^2}{d}\frac{\nu^2}{c_p} = C_6 = \frac{4g\alpha N^{q-1}}{c_p} x^{n(q-1)+1} \tag{9.3.25}$$

This effect is of order $4g\alpha N \mathrm{Pr} L^{n(q-1)+1}/c_p$, compared with conduction, for example. Similarity results only for $n = -1/(q-1)$, which is an unrealistic circumstance, as shown below. This effect, dependent on $g\alpha$, is usually very small. Now the pressure term in Eq. (9.3.10) is rewritten as

$$C_{10} = \frac{gT}{c_p}\frac{c}{bd}\beta = \frac{gT}{c_p}\frac{c}{bd}\frac{\alpha q|\phi - R|^{q-1}|t_0 - t_\infty|^{q-1}}{(1 - \alpha|\phi - R|^q|t_0 - t_\infty|^q)}\frac{(\phi - R)(t_0 - t_\infty)}{|\phi - R||t_0 - t_\infty|} \tag{9.3.26}$$

where β is evaluated from Eq. (9.1.1). Neglecting the second term in the denominator compared with one, we have

$$C_{10} = \frac{4Tg\alpha q}{c_p}|\phi - R|^{q-2}(\phi - R)|N|^{q-2}x^{1-n(2-q)} \tag{9.3.27}$$

This is independent of x for $n = \frac{1}{2}(2 - q)$, quite a large value, if the variation of T in Eq. (9.3.24) across the flow field is neglected. This term also depends on $g\alpha$ and may be very small.

Neglecting viscous dissipation and pressure effects and taking $t_r = t_m$ if stratification is present, Eqs. (9.3.9) and (9.3.10) become

$$f''' + (3 + qn)ff'' - (2 + 2qn)f'^2 + F = 0 \tag{9.3.28}$$

$$\phi'' + \text{Pr}\left[(3 + qn)f\phi' - 4nf'\phi - \frac{4nN_\infty}{N}f'\right] = 0 \tag{9.3.29}$$

where Gr_x and buoyancy are defined as in Eqs. (9.3.20) and (9.3.21). If Gr_x is defined as in Eq. (9.3.18), then F in Eq. (9.3.28) is replaced by W. The boundary conditions in both formulations are

$$1 - \phi(0) = \phi(\infty) = f'(0) = f(0) = f'(\infty) = 0 \tag{9.3.30}$$

The choice of $\text{Gr}_x = (gx^3/\nu^2)\alpha(s_\infty, p)|t_0 - t_\infty|^q I_W$ and $F = W/I_W$ used by Gebhart and Mollendorf (1978) implies that x increases in the direction of net flow. This becomes ambiguous in the range $0 < R < 1/2$ since the net flow direction is not always dictated by the sign of I_W. An alternative formulation that overcomes this difficulty was presented by Carey et al. (1980) and will be discussed later.

The limits on the allowable values of n in $d(x) = Nx^n$ are determined from physical considerations, as for the Boussinesq approximation in Chapter 3. The local surface heat flux $q''(x)$, the energy convected locally by the flow, $Q(x)$, the local flow region thickness $\delta(x)$, and the local Nusselt number Nu_x are

$$q''(x) = -k\left(\frac{\partial t}{\partial y}\right)_0 = [-\phi'(0)]k\,db \propto x^{(1/4)[n(q+4)-1]} \tag{9.3.31}$$

$$Q(x) = \int_0^\infty \rho c_p(t - t_\infty)u\,dy = \rho c_p \nu c d\int_0^\infty \phi f'\,d\eta \propto x^{(1/4)[n(q+4)+3]} \tag{9.3.32}$$

$$\delta(x) = \frac{\eta_\delta}{b} \propto x^{(1/4)(1 - nq)} \tag{9.3.33}$$

$$\text{Nu}_x = \frac{h_x x}{k} = \frac{q''(x)}{d}\frac{x}{k} = [-\phi'(0)]\frac{G}{4} = \frac{[-\phi'(0)]}{\sqrt{2}}\text{Gr}_x^{1/4} \tag{9.3.34}$$

The requirement that the boundary layer thickness at the leading edge, $\delta(0)$, remain bounded or be zero results in $nq < 1$. That is, $n < 0.528$ at 1 bar in pure water, for which $q = 1.894816$. Also, with the positive x direction taken such that f' is essentially positive, $Q(x)$ must, for $N > 0$, be constant for a line source at $x = 0$ or increase with x with an energy input downstream, at, say, $y = 0$. Thus from Eq. (9.3.32)

$$n \geq \frac{-3}{q+4} = -0.509 \qquad \text{for } s = 0 \text{ and } p = 1 \text{ bar}$$

where $n = -0.509$ for the plume.

The limits are then $-0.509 < n < 0.528$. The comparable result for the Boussinesq approximation, $q = 1$ here, is $-0.6 < n < 1$, as presented in Chapter 3. The lower limit in both analyses is the plane plume or an adiabatic surface with a horizontal line source at the leading edge. The Nusselt number is the same as before, except for the definition of Gr_x. Also, the value of $\phi'(0)$ now depends on Pr, R, and the buoyancy formulation used in Eq. (9.3.21).

9.3.2 Thermally Driven Transport Adjacent to Flat Vertical Surfaces

Gebhart and Mollendorf (1978) presented solutions to Eqs. (9.3.28) and (9.3.29) subject to the boundary conditions given in Eq. (9.3.30) for isothermal surfaces immersed in unstratified quiescent media. The above formulation results in only three parameters, the Prandtl number, R, and q. Specifying $\mathrm{Pr} = 11.5$ and $q = 1.894816$, which represents fresh water at 1 bar at 4°C, leaves R as the only additional parameter.

In Table 9.3.1 and Fig. 9.3.1 the value of R is related to temperature conditions and to the corresponding direction of the buoyancy force. Although the role of R appears complicated at first, the buoyancy force, $W = |\phi - R|^q - |R|^q$, changes sign across the flow region only in the range $0 < R < \frac{1}{2}$. This is apparent in comparing the listed conditions in Table 9.3.1 with the density distributions in Fig. 9.1.1. This range of R is in accord with past observations of convective inversion with ice spheres in water. Dumoré et al. (1953) found convective inversion at $t_\infty = 4.8$°C, and Schenk and Schenkels (1968) estimated 5.3°C. These temperatures correspond to $R = 0.17$ and 0.25, respectively. These inversions were thought to be a complete flow reversal, which was accompanied by a drastic drop in transport, that is, in the ice melting rate. Bendell and Gebhart (1976) found that convective inversion occurred between 5.5 and 5.6°C for flow adjacent to a vertical ice surface. Taking $t_m(0, 1) = 4.03$°C, the resulting values of t_∞ and R at which convective inversion occurs are about 5.5°C and 0.27. Low melting rates extended over a much broader range of t_∞.

On the other hand, $R < 0$ invariably gives upflow and $R > \frac{1}{2}$ invariably gives downflow, independent of heating or cooling. Note that R is negative only if t_0 lies between t_∞ and t_m. Also note that a linear temperature approximation $\rho_\infty - \rho$ will become accurate at both large positive and large negative values of R, that is, for $|t_0 - t_\infty| << |t_m - t_\infty|$.

Table 9.3.2 Heat transfer and flow parameters for vertical flow without salinity gradients or local buoyancy force reversals[a]

R	Pr = 8.6	Pr = 9.6	Pr = 10.6	Pr = 11.6	Pr = 12.6	Pr = 13.6
−16.00	0.81163, 1.37935[a]	0.82000, 1.43954	0.82729, 1.49562	0.83370, 1.54822	0.83941, 1.59783	0.84453, 1.64484
	9.97224, 0.31966	9.54472, 0.31352	9.17826, 0.30813	8.85929, 0.30334	8.57825, 0.29903	8.32796, 0.29512
−14.00	0.81178, 1.37920	0.82015, 1.43938	0.82743, 1.49545	0.83384, 1.54804	0.83954, 1.59764	0.84466, 1.64465
	8.87115, 0.31957	8.49086, 0.31343	8.16489, 0.30804	7.88115, 0.30325	7.63116, 0.29894	7.40852, 0.29503
−12.00	0.81198, 1.37900	0.82034, 1.43917	0.82761, 1.49523	0.83402, 1.54781	0.83971, 1.59740	0.84483, 1.64439
	7.75370, 0.31944	7.42135, 0.31330	7.13646, 0.30792	6.88848, 0.30313	6.67000, 0.29882	6.47541, 0.29491
−10.00	0.81226, 1.37872	0.82061, 1.43887	0.82787, 1.49491	0.83427, 1.54748	0.83995, 1.59705	0.84507, 1.64403
	6.61691, 0.31927	6.33332, 0.31313	6.09022, 0.30775	5.87863, 0.30296	5.69220, 0.29865	5.52615, 0.29475
−8.00	0.81267, 1.37830	0.82100, 1.43843	0.82825, 1.49445	0.83463, 1.54699	0.84031, 1.59654	0.84541, 1.64350
	5.45652, 0.31900	5.22271, 0.31287	5.02228, 0.30749	4.84782, 0.30271	4.69411, 0.29841	4.55720, 0.29451
−4.00	0.81463, 1.37631	0.82289, 1.43631	0.83007, 1.49221	0.83640, 1.54465	0.84203, 1.59410	0.84708, 1.64095
	3.03451, 0.31774	2.90460, 0.31164	2.79323, 0.30628	2.69629, 0.30151	2.61086, 0.29723	2.53478, 0.29334
−3.00	0.81586, 1.37506	0.82408, 1.43498	0.83122, 1.49081	0.83751, 1.54317	0.84311, 1.59256	0.84813, 1.63935
	2.39704, 0.31694	2.29448, 0.31085	2.20656, 0.30551	2.13002, 0.30075	2.06257, 0.29648	2.00250, 0.29260
−2.00	0.81816, 1.37271	0.82629, 1.43249	0.83336, 1.48817	0.83959, 1.54041	0.84513, 1.58967	0.85010, 1.63635
	1.73858, 0.31544	1.66427, 0.30938	1.60056, 0.30406	1.54510, 0.29932	1.49623, 0.29507	1.45269, 0.29121
−1.00	0.82401, 1.36671	0.83193, 1.42611	0.83881, 1.48145	0.84487, 1.53335	0.85025, 1.58230	0.85508, 1.62868
	1.04801, 0.31154	1.00335, 0.30555	0.96504, 0.30030	0.93169, 0.29562	0.90230, 0.29142	0.87611, 0.28761
−0.50	0.83246, 1.35796	0.84007, 1.41681	0.84668, 1.47163	0.85249, 1.52305	0.85764, 1.57154	0.86227, 1.61749
	0.68233, 0.30570	0.65337, 0.29982	0.62853, 0.29467	0.60689, 0.29008	0.58782, 0.28596	0.57082, 0.28222

0.00	0.87166, 1.31584	0.87783, 1.37208	0.88316, 1.42445	0.88783, 1.47356	0.89195, 1.51986	0.89565, 1.56373
	0.28580, 0.27428	0.27393, 0.26910	0.26373, 0.26453	0.25483, 0.26046	0.24698, 0.25679	0.23998, 0.25346
0.50	0.71229, 1.47324	0.72425, 1.53934	0.73473, 1.60095	0.74401, 1.65875	0.75231, 1.71328	0.75978, 1.76494
	−0.16625, 0.37248	−0.15885, 0.36542	−0.15251, 0.35923	−0.14702, 0.35372	−0.14219, 0.34877	−0.13790, 0.34427
1.00	0.78491, 1.40612	0.79426, 1.46799	0.80241, 1.52564	0.80959, 1.57971	0.81600, 1.63072	0.82176, 1.67905
	−0.57804, 0.33574	−0.55296, 0.32930	−0.53149, 0.32366	−0.51281, 0.31864	−0.49637, 0.31412	−0.48173, 0.31003
2.00	0.80019, 1.39087	0.80898, 1.45178	0.81663, 1.50854	0.82338, 1.56177	0.82939, 1.61199	0.83478, 1.65957
	−1.30412, 0.32677	−1.24791, 0.32050	−1.19976, 0.31499	−1.15786, 0.31010	−1.12095, 0.30570	−1.08809, 0.30171
3.00	0.80405, 1.38700	0.81270, 1.44767	0.82022, 1.50420	0.82686, 1.55722	0.83276, 1.60723	0.83807, 1.65462
	−1.98107, 0.32442	−1.89585, 0.31819	−1.82281, 0.31272	−1.75926, 0.30786	−1.70327, 0.30348	−1.65342, 0.29952
4.00	0.80581, 1.38523	0.81440, 1.44579	0.82187, 1.50221	0.82845, 1.55514	0.83431, 1.60505	0.83957, 1.65235
	−2.63105, 0.32333	−2.51795, 0.31711	−2.42103, 0.31166	−2.33668, 0.30682	−2.26237, 0.30246	−2.19620, 0.29851
8.00	0.80828, 1.38274	0.81677, 1.44314	0.82416, 1.49942	0.83068, 1.55221	0.83647, 1.60200	0.84168, 1.64917
	−5.08154, 0.32179	−4.86335, 0.31560	−4.67634, 0.31018	−4.51359, 0.30535	−4.37021, 0.30102	−4.24252, 0.29708
10.00	0.80875, 1.38227	0.81723, 1.44264	0.82460, 1.49889	0.83110, 1.55165	0.83688, 1.60141	0.84208, 1.64857
	−6.25063, 0.32149	−5.98230, 0.31531	−5.75232, 0.30989	−5.55216, 0.30507	−5.37582, 0.30074	−5.21878, 0.29681
12.00	0.80906, 1.38195	0.81753, 1.44231	0.82489, 1.49854	0.83138, 1.55128	0.83715, 1.60103	0.84234, 1.64817
	−7.39439, 0.32130	−7.07701, 0.31512	−6.80498, 0.30970	−6.56822, 0.30489	−6.35963, 0.30056	−6.17388, 0.29663
14.00	0.80928, 1.38173	0.81774, 1.44207	0.82509, 1.49829	0.83158, 1.55102	0.83735, 1.60076	0.84253, 1.64789
	−8.51762, 0.32116	−8.15207, 0.31499	−7.83874, 0.30957	−7.56605, 0.30476	−7.32579, 0.30043	−7.11183, 0.29650
16.00	0.80944, 1.38157	0.81789, 1.44190	0.82525, 1.49811	0.83172, 1.55083	0.83749, 1.60056	0.84267, 1.64767
	−9.62365, 0.32105	−9.21065, 0.31489	−8.85667, 0.30947	−8.54858, 0.30466	−8.27715, 0.30033	−8.03542, 0.29641

[a]Values are shown for $f''(0)$, $-\phi'(0)$ on top line and I_w, $f(\infty)$ on bottom line, respectively, for each R and Pr value.

Source: Reprinted with permission from B. Gebhart and J. C. Mollendorf, *J. Fluid Mech.*, vol. 89, p. 673. Copyright © 1978, Cambridge University Press.

Table 9.3.3. Heat transfer and flow parameters for vertical flow without salinity gradients or local buoyancy reversals, all for Pr = 11.6[a]

R	q = 1.859663	q = 1.737147	q = 1.582950
−16.00	0.83366, 1.54827, 7.88200, 0.30336	0.83352, 1.54846, 5.05622, 0.30346	0.33336, 1.54867, 3.09809, 0.30356
−14.00	0.83379, 1.54810, 7.04406, 0.30328	0.83363, 1.54832, 4.59765, 0.30339	0.83345, 1.54855, 2.87072, 0.30351
−12.00	0.83396, 1.54787, 6.18946, 0.30316	0.83377, 1.54813, 4.12118, 0.30329	0.83357, 1.54840, 2.62965, 0.30343
−10.00	0.83420, 1.54756, 5.31507, 0.30300	0.83397, 1.54786, 3.62302, 0.30315	0.83373, 1.54819, 2.37158, 0.30332
−8.00	0.83456, 1.54709, 4.41639, 0.30276	0.83427, 1.54746, 3.09758, 0.30295	0.83397, 1.54787, 2.09160, 0.30316
−4.00	0.83626, 1.54484, 2.51351, 0.30161	0.83571, 1.54558, 1.92274, 0.30198	0.83511, 1.54637, 1.42694, 0.30238
−3.00	0.83732, 1.54343, 2.00404, 0.30088	0.83661, 1.54439, 1.58733, 0.30137	0.83583, 1.54543, 1.22355, 0.30180
−2.00	0.83932, 1.54078, 1.47204, 0.29951	0.83829, 1.54218, 1.22242, 0.30021	0.83718, 1.54368, 0.99214, 0.30098
−1.00	0.84438, 1.53403, 0.90515, 0.29596	0.84257, 1.53656, 0.80930, 0.29723	0.84060, 1.53926, 0.71216, 0.29861
−0.50	0.85171, 1.52419, 0.59914, 0.29065	0.84878, 1.52841, 0.56943, 0.29277	0.84559, 1.53288, 0.53607, 0.29505
0.00	0.88641, 1.47642, 0.25798, 0.26169	0.88070, 1.48733, 0.27076, 0.26655	0.87372, 1.49943, 0.28657, 0.27236
0.50	0.74437, 1.65907, −0.14946, 0.35349	0.74573, 1.66027, −0.15872, 0.35261	0.74731, 1.66167, −0.16870, 0.35161
1.00	0.81033, 1.57891, −0.50907, 0.31818	0.81323, 1.57569, −0.49427, 0.31638	0.81658, 1.57177, −0.47673, 0.31430
2.00	0.82373, 1.56133, −1.11550, 0.30987	0.82507, 1.55965, −0.96627, 0.30901	0.82655, 1.55776, −0.82127, 0.30805
3.00	0.82708, 1.55694, −1.66760, 0.30771	0.82794, 1.55584, −1.35859, 0.30715	0.82887, 1.55463, −1.07990, 0.30654
4.00	0.82861, 1.55492, −2.19052, 0.30671	0.82924, 1.55411, −1.71146, 0.30630	0.82993, 1.55322, −1.29973, 0.30585
8.00	0.83076, 1.55210, −4.12344, 0.30530	0.83106, 1.55171, −2.92285, 0.30510	0.83139, 1.55128, −1.99646, 0.30488
10.00	0.83116, 1.55157, −5.03117, 0.30503	0.83140, 1.55125, −3.45863, 0.30487	0.83167, 1.55091, −2.28491, 0.30470
12.00	0.83143, 1.55121, −5.91274, 0.30485	0.83163, 1.55095, −3.96478, 0.30472	0.83185, 1.55067, −2.54933, 0.30457
14.00	0.83162, 1.55096, −6.77326, 0.30473	0.83179, 1.55074, −4.44769, 0.30461	0.83198, 1.55050, −2.79541, 0.30449
16.00	0.83176, 1.55078, −7.61624, 0.30463	0.83191, 1.55058, − 4.91163, 0.30453	0.83207, 1.55037, −3.02686, 0.30443

[a] Values are shown for $f''(0)$, $-\phi'(0)$, I_w, and $f(\infty)$ at each q and R.

Source: Reprinted with permission from B. Gebhart and J. C. Mollendorf, *J. Fluid Mech.*, vol. 89, p. 673.

Tables 9.3.2 and 9.3.3 list the resulting transport parameters calculated by Gebhart and Mollendorf (1978) for a wide range of R and Pr. The Prandtl number range considered is $8.6 < \mathrm{Pr} < 13.6$. Table 9.3.2 shows the Prandtl number effect on transport and Table 9.3.3 shows the effect of pressure and salinity levels, in terms of $q(s, p)$.

For $\mathrm{Pr} = 11.6$ and $q(s, p) = q(0, 1)$, the vertical component of velocity adjacent to the surface is shown in Fig. 9.3.3 for various values of R. Recall that the flow is upward for $R < 0$ and downward for $R > \frac{1}{2}$. However, all are shown in the same direction in Fig. 9.3.3. The dashed curve represents the conventional Boussinesq results, $q = 1$ here. The magnitude of the maximum dimensionless velocity, $f'(\eta)$, is seen to increase by about 60% from $R = 0$ to $R = \frac{1}{2}$. Large calculated differences from conventional results are apparent.

A smaller effect on the calculated temperature distribution in the present coordinates is seen in Fig. 9.3.4. The curves of higher temperature gradients near the surface correspond to vertical velocity components of larger magnitude. The corresponding horizontal velocity component v is shown in Fig. 9.3.5. A 40% decrease in the entrainment velocity is seen between $R = 0.5$ and $R = 0$. These

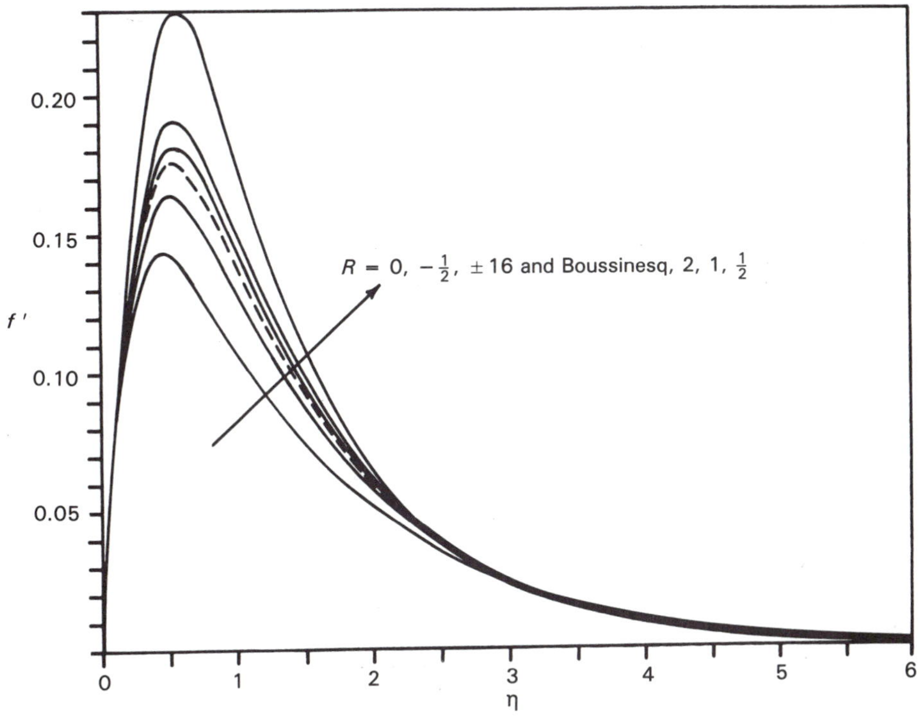

Figure 9.3.3 Calculated distribution of the velocity component parallel to the vertical surface for selected values of R; $\mathrm{Pr} = 11.6$ and $q(s, p) = q(0, 1) = 1.894816$. *(Reprinted with permission from B. Gebhart and J. C. Mollendorf,* J. Fluid. Mech., *vol. 89, p. 673. Copyright © 1978, Cambridge University Press.)*

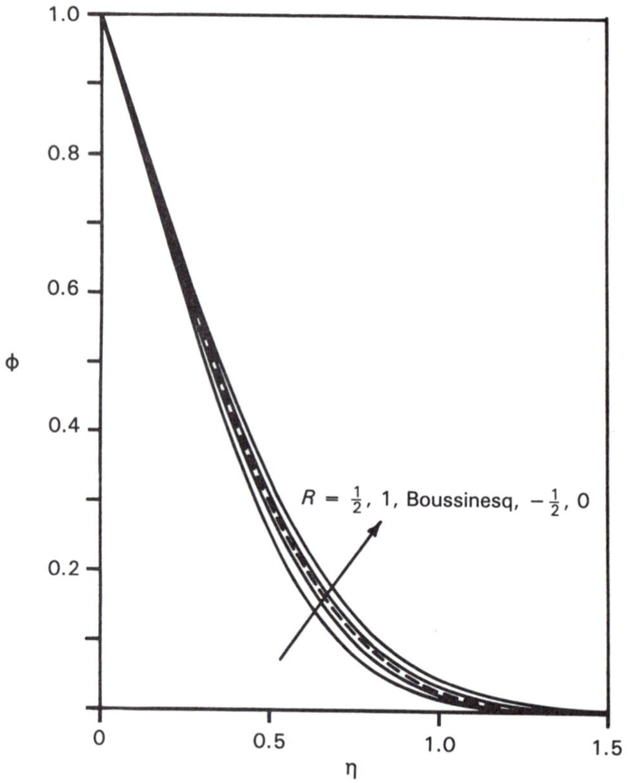

Figure 9.3.4 Calculated temperature distribution adjacent to a vertical surface for selected values of R; Pr = 11.6 and $q(s, p) = q(0, 1) = 1.894816$. *(Reprinted with permission from B. Gebhart and J. C. Mollendorf,* J. Fluid. Mech., *vol. 89, p. 673. Copyright © 1978, Cambridge University Press.)*

changes are related to those in the vertical velocity component seen in Fig. 9.3.3.

The distribution of the normalized local buoyancy force $W(\eta)/I_W$ is shown in Fig. 9.3.6. For $R = 0$, the upper bound for a uniformly upward buoyancy force, the buoyancy force is larger at the surface than for the other conditions shown. This buoyancy force distribution for $R = 0$ corresponds to a higher calculated average fluid temperature distribution and associated lower velocity levels, as seen in Figs. 9.3.3 and 9.3.4. For $R = \frac{1}{2}$, the lower bound for a uniformly downward buoyancy force, the buoyancy force is zero at the surface and has an extremum at $\eta = 0.3$. Recall that I_W is always negative for $R > \frac{1}{2}$ (downflow) and positive for $R < 0$ (upflow) and that

$$\int_0^\infty W(\eta)\, d\eta \equiv I_W$$

Figure 9.3.7 indicates the effects of the Prandtl number and R on the heat transfer parameter $\phi'(0)|I_W|^{1/4}$. As the Prandtl number decreases from 13.6 to 8.6, there is about a 14% decrease in heat transfer at all R. Also shown in Fig. 9.3.7 and Table 9.3.4 is a comparison of the measurements of Bendell and Gebhart (1976) and the calculations of Gebhart and Mollendorf (1978). The measurements agree with the analysis to within an average difference of about 6.5%. In the range $0 < R < \frac{1}{2}$ the data appear to lie on reasonable extrapolations of the computed results, and in general the Prandtl number trend in the data and the analytical results is the same.

Buoyancy force reversals. The calculations of Gebhart and Mollendorf (1978) were only for $R > \frac{1}{2}$ and $R < 0$. In the range $0 < R < \frac{1}{2}$ attempts at computations with regular two-point "shooting" numerical schemes were largely unsuccessful. In a subsequent study, however, Carey et al. (1980) were able to extend the calculations in on both sides of the range $0 < R < \frac{1}{2}$, leaving a residual gap in the approximate range $0.15 < R < 0.29$. These calculations used a different scheme, namely a shooting method augmented by the asymptotic behavior of f' and ϕ. El-Henawy et al. (1982) have since narrowed this gap and found multiple solutions near the edge of the gap on both sides.

The range $0 < R < 0.5$ includes all conditions of local buoyancy force re-

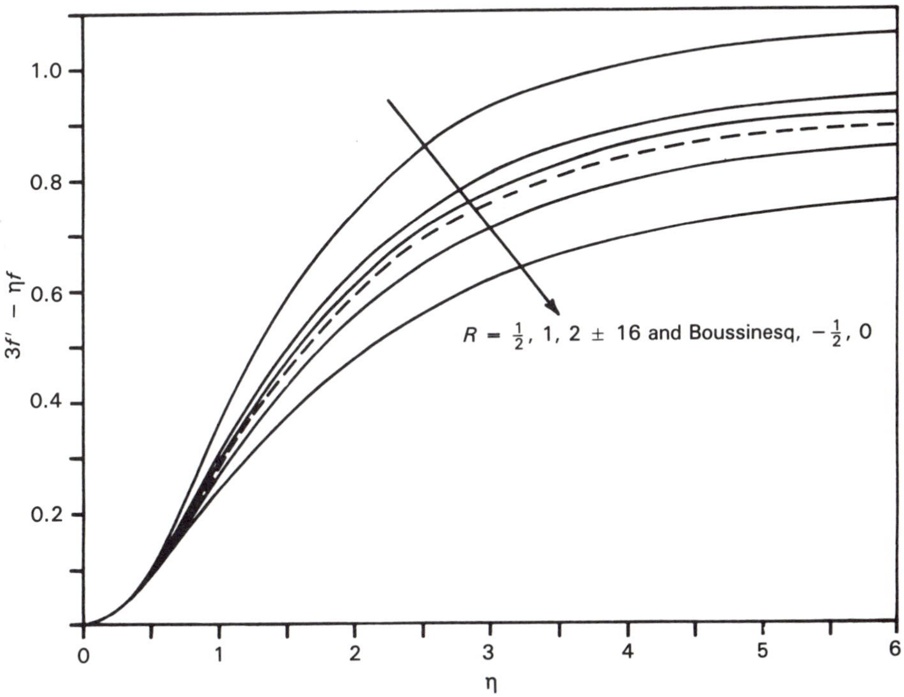

Figure 9.3.5 Calculated distribution of the velocity component normal to the vertical surface for selected values of R; Pr = 11.6 and $q(s, p) = q(0, 1) = 1.894816$. *(Reprinted with permission from B. Gebhart and J. C. Mollendorf,* J. Fluid. Mech., *vol. 89, p. 673. Copyright © 1978, Cambridge University Press.)*

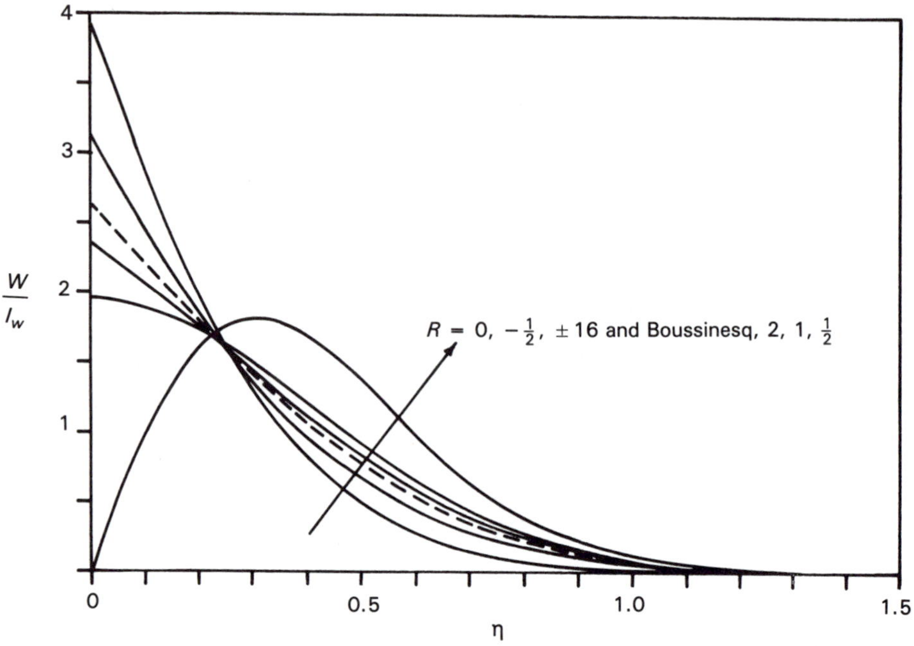

Figure 9.3.6 Calculated distribution of local buoyancy force for selected values of R; Pr = 11.6 and $q(s, p) = q(0, 1) = 1.894816$. *(Reprinted with permission from B. Gebhart and J. C. Mollendorf,* J. Fluid. Mech., *vol. 89, p. 673. Copyright © 1978, Cambridge University Press.)*

versal as well as of local flow reversal. The approach to convective inversion, coming in from both $R = 0$ and $R = \frac{1}{2}$, is characterized by a rapid decrease in flow vigor and surface heat transfer. For conditions very near the remaining gap, the buoyancy force is bidirectional. Reducing R from 0.5 leads to bidirectional flow, up near the surface and down in the outer part of the flow region.

Figure 9.3.8 shows the two groups of resulting tangential velocity distributions, $f'|I_W|^{1/2}$, for $q(0, 1)$ and Pr = 11.6. The upper group is for R increasing from zero, that is, below the gap. The other group is for R increasing above the gap to $R = 0.5$. For the downward flows the negative value of the nondimensional velocity f' is plotted. Thus the physical directions of the tangential velocity in each flow regime are properly oriented. Weakening upflow and downflow are seen toward the gap, from both $R = 0$ and $R = \frac{1}{2}$, as the buoyancy force reversals increase.

The simplest situation is the upper range of R, with inside reversal. Local upflow reversals are seen near the surface in Fig. 9.3.8 when R reaches 0.3 and 0.292. The relation of $W(\eta)$ to this effect is clear in Fig. 9.3.9, where W is seen to be negative over a large part of the boundary region. Even though it is large and positive over an appreciable region near the surface, only a small local flow reversal arises. This is due to the presence of the large downward viscous forces in the fluid near the surface, which resist the upward buoyancy there. For $R = 0.32$

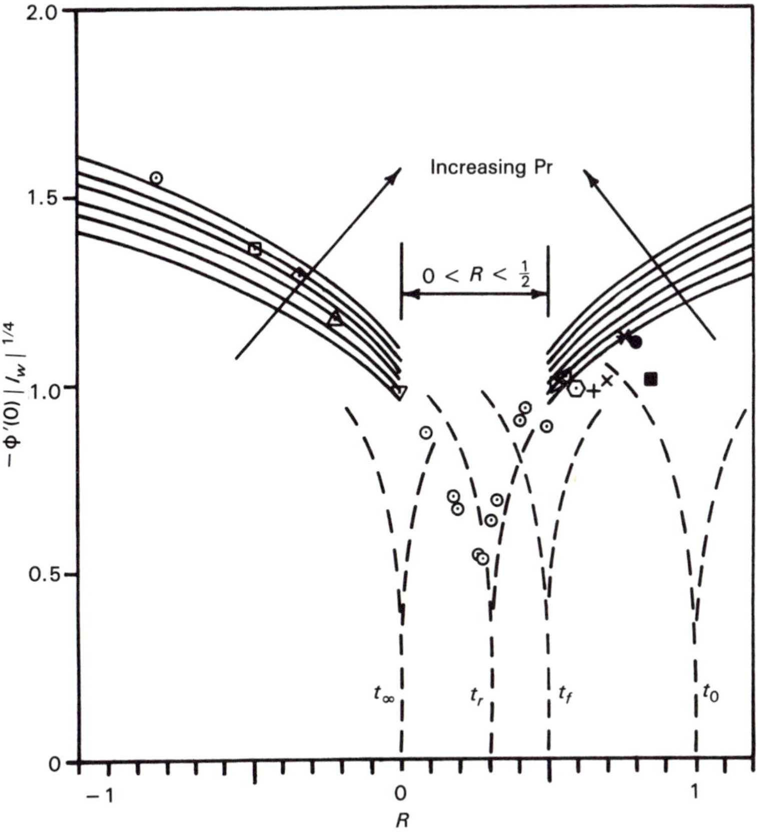

Figure 9.3.7 Calculated heat transfer variation with R near the region of net buoyancy force reversal compared with the measurements by Bendell and Gebhart (1976). Calculated results are compared with those obtained with the Boussinesq approximation with β evaluated at various reference temperatures t_r, where $t_r = t_0 - 0.69(t_0 - t_\infty)$. For symbols see Table 9.3.4. *(Reprinted with permission from B. Gebhart and J. C. Mollendorf,* J. Fluid. Mech., *vol. 89, p. 673. Copyright © 1978, Cambridge University Press.)*

and 0.40, appreciable local flow reversal does not occur in spite of wide regions of positive W.

For R increasing from zero, there is a range of R for which there is no local flow reversal even though W changes sign across the thermal region. These small buoyancy force reversals lie around $\eta = 1.5$, well inside the flow region. A large viscous force there, arising from high momentum, is able to overcome this effect.

Figure 9.3.10, for the buoyancy force reversal region $0 < R < 0.5$, indicates large decreases in heat transfer as convective inversion is approached. The data of Bendell and Gebhart (1976) and of Johnson (1978) are plotted for comparison. Both sets of data are systematically lower than the computed results. The difference, about 9%, is due mainly to neglecting interface motion or, equivalently,

Table 9.3.4 Comparison of the data of Bendell and Gebhart (1976) and the calculations of Gebhart and Mollendorf (1978)

t_∞	R	Pr at t_f	$Gr'_L = g\alpha L^3 \lvert t_0 - t_\infty \rvert^q / \nu^2$ at t_f	$\overline{Nu}_L$ (exp.)	$-\phi'(0)\lvert I_w \rvert^{1/4}$ (exp.)	Symbols in Fig. 9.3.7	$-\phi'(0)\lvert I_w \rvert^{1/4}$ (theor.)	Percent deviation
25.2	0.841	8.6	7.82×10^3	172.5	1.006	■	1.16	−15.3
19.9	0.798	9.3	4.33×10^3	163.4	1.104	●	1.17	−6.0
16.9	0.762	9.8	2.87×10^3	149.1	1.115	*	1.16	−4.0
13.3	0.697	10.3	1.67×10^3	116.9	1.001	×	1.14	−13.9
11.7	0.656	10.6	1.23×10^3	105.2	0.972	+	1.12	−15.2
10.1	0.601	10.9	9.02×10^7	98.4	0.982	⬡	1.09	−10.9
9.1	0.557	11.1	7.20×10^7	96.4	1.018	◁	1.07	−5.1
8.5	0.526	11.2	6.06×10^7	90.6	0.999	▷	1.04	−4.1
8.0	0.496	11.3	5.34×10^7	77.5	0.882	○	—	—
7.0	0.424	11.5	4.10×10^7	76.8	0.933	○	—	—
6.8	0.407	11.7	3.79×10^7	72.4	0.898	○	—	—
6.0	0.328	11.7	2.97×10^7	52.1	0.686	○	—	—
5.8	0.305	11.7	2.69×10^7	46.7	0.631	○	—	—
5.6	0.280	11.8	2.58×10^7	38.9	0.531	○	—	—
5.5	0.267	11.8	2.48×10^7	39.4	0.543	○	—	—
5.0	0.194	11.9	2.00×10^7	45.9	0.667	○	—	—
4.9	0.178	11.9	1.92×10^7	47.6	0.699	○	—	—
4.4	0.084	12.0	1.54×10^7	55.8	0.866	○	—	—
4.0	−0.007	12.1	1.30×10^7	60.0	0.972	▽	1.05	−8.0
3.3	−0.221	12.3	8.80×10^3	65.3	1.166	△	1.22	−4.6
3.0	−0.343	12.3	7.13×10^3	68.2	1.284	◇	1.29	−0.4
2.7	−0.492	12.4	5.91×10^3	68.8	1.357	□	1.37	−1.0
2.2	−0.832	12.5	3.87×10^3	70.5	1.546	○	1.49	−3.6
								−6.5% avg.
								8.6% r.m.s.

Source: Reprinted with permission from B. Gebhart and J. C. Mollendorf, *J. Fluid Mech.*, vol. 89, p. 673.

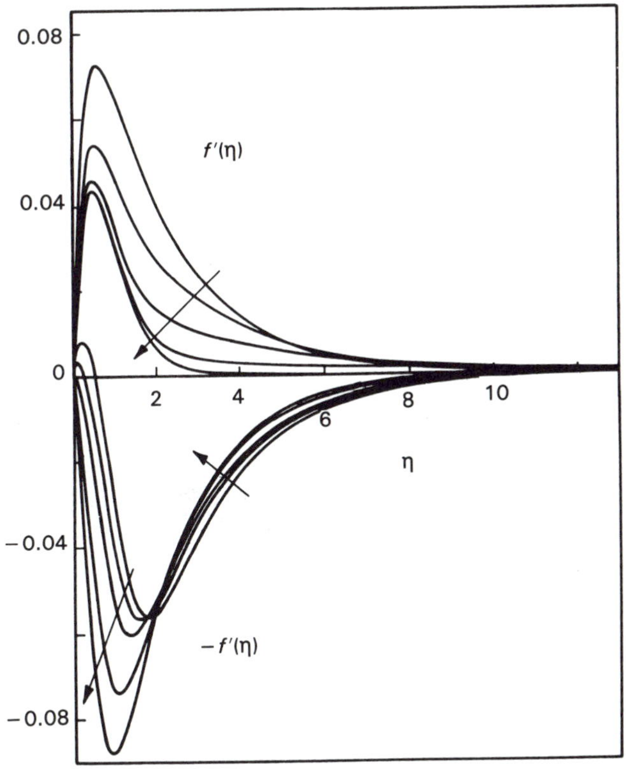

Figure 9.3.8 Distributions across the flow layer of the tangential component of velocity f' for upward flow and $-f'$ for downward flow, for $q(0, 1)$ and Pr = 11.6. For upflow the arrow indicates increasing R for $R = 0$, 0.10, 0.14, 0.15, and 0.15172. For downflow the arrow indicates increasing $R = 0.292$, 0.30, 0.32, 0.40, and 0.50. *(Reprinted with permission from V. P. Carey et al.,* J. Fluid. Mech., *vol. 97, p. 279. Copyright © 1980, Cambridge University Press.)*

interface blowing in the analysis. Over this range of R the greatest interface velocity occurs at $R = 0.5$ since the heat flux to the surface is greatest at this boundary.

To assess this effect, Carey et al. (1980) modified their analysis to include interface blowing. Calculations for $R = 0.5$ and Pr = 11.6 indicated that the surface heat transfer decreased by 5%. At yet lower values of R the blowing velocity was smaller, with a smaller effect on transport. The calculations of Carey et al. (1980) still left a gap, $0.15 < R < 0.29$, in which the numerical scheme failed to converge. Measurements by Wilson and Vyas (1979) and later by Carey and Gebhart (1981) showed that some flow regimes arising in this gap are not of the boundary layer type. Furthermore, many such flows were observed to be unsteady. Fluctuations in velocity and heat transfer were found.

Carey and Gebhart (1981) visualized the flow adjacent to a vertical ice surface melting in pure water, for ambient temperatures of 3.9, 4.05, 4.4, 4.7, 5.4, 5.9, and 8.4°C, by illuminating suspended particles in the flow field. The particles

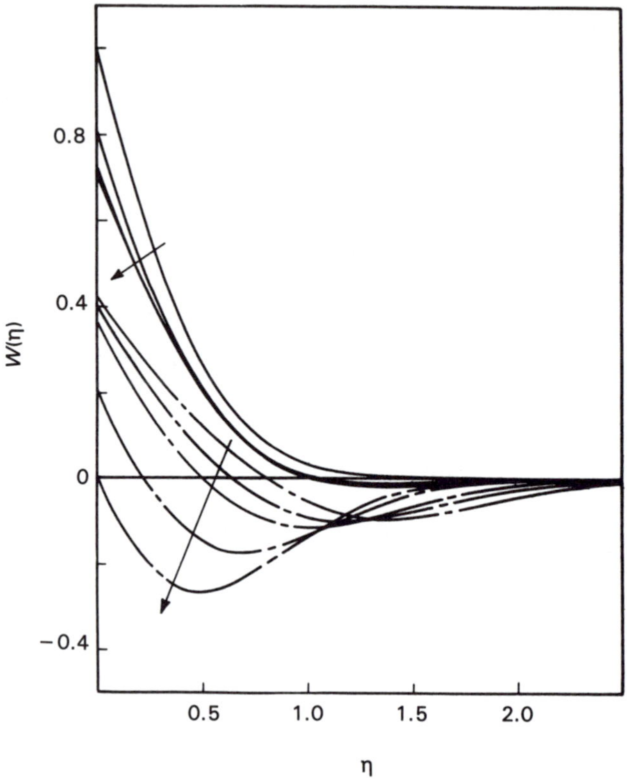

Figure 9.3.9 Distributions of local buoyancy force W across the thermal diffusion region for $q(0, 1)$ and Pr = 11.6. For net upward flow (——) the arrow indicates increasing R for $R = 0, 0.10, 0.14$, and 0.15172. For net downward flow (—‐‐—) the arrow indicates increasing R for $R = 0.292, 0.30$, 0.32, 0.40, and 0.50. *(Reprinted with permission from V. P. Carey et al.,* J. Fluid. Mech., *vol. 97, p. 279. Copyright © 1980, Cambridge University Press.)*

were illuminated by spreading a laser beam into a plane of light, then passing it perpendicular to the ice surface near the center of its horizontal span. Time exposure photographs of the flow field then revealed the flow pattern and velocity distributions. Figures 9.3.11 and 9.3.12 show the resulting flow regimes for the ambient temperature range $t_\infty = 3.9$–8.4°C, corresponding to $R = -0.033$ to 0.52. The sequence of photographs demonstrates the transition from total upflow at $R = -0.033$ to total downflow at $R = 0.52$. This confirms the analyses and numerical results presented to this point. Of more interest here, however, is the region around $0.15 \leqq R \leqq 0.29$. As seen in Fig. 9.3.11*d*, at $R = 0.143$ the flow is not of the boundary layer type. Furthermore, as seen in Fig. 9.3.13, the flow is also unsteady and fluctuates with time.

Such observations led to a renewed interest in calculations farther within the gap in R. El-Henawy et al. (1982) used COLSYS, an adaptive orthogonal collocation code, with BOUNDS, a multiple shooting code, to push into the gap from each side. The differential equations were reformulated to avoid reliance on unbounded quantities. The gap at Pr = 11.6 was narrowed to $0.15180 \le R \le 0.29181$. Of much more interest, however, was the fact that multiple solutions arose over a range of R on both sides. These are multiple steady-state solutions for given values of R and were found in subregions near 0.15 and 0.29. Figure 9.3.14 shows the variation of $\phi'(0)$ in each R region. This demonstrates the multiple steady-state nature of the heat transfer. In Fig. 9.3.15 the two calculated tangential velocity profiles arising at $R = 0.317$ are compared to the fluctuating velocity profiles observed by Wilson and Vyas (1979).

A uniform heat flux surface. Qureshi and Gebhart (1978) and Qureshi (1980) studied the flow and transport adjacent to a flat vertical surface dissipating a uniform heat flux in cold water. Similarity solutions were obtained for $t_\infty = t_m$, that is, $R = 0$. The range Pr = 8–13 was investigated. The governing equations were solved by integrating inward, starting with the asymptotic solutions at a large distance from the surface. The following heat transfer correlation was proposed:

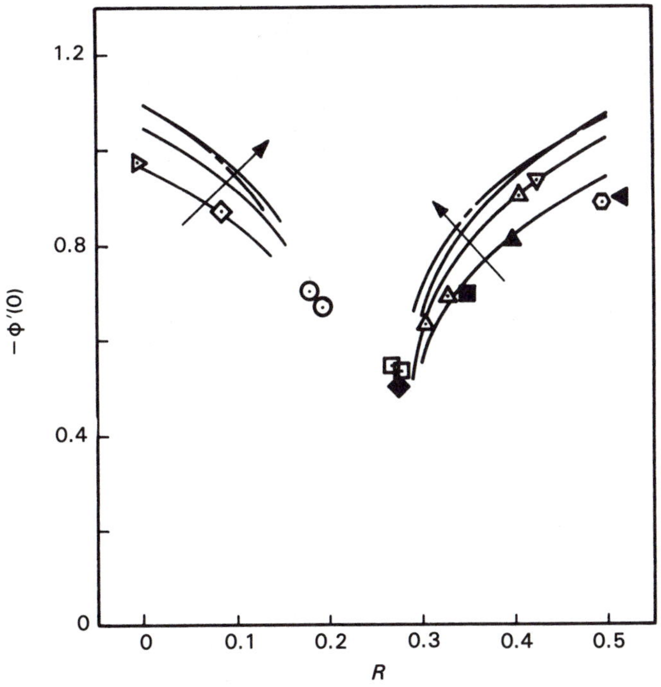

Figure 9.3.10 Variation of the calculated heat transfer parameter $-\phi'(0)$ with R. Dashed curves (— - —) are for $q = q(0, 1000)$ and Pr = 11.6. For the solid curves (——), $q = q(0, 1)$ and the arrows indicate increasing Pr for Pr = 8.6, 11.6, and 13.6. Also shown are the data of Bendell and Gebhart (1976) and Johnson and Mollendorf (1980), determined for ice melting in pure water. Bendell and Gebhart's (1976) film Prandtl numbers are: (⬡) 11.3; (▽) 11.5; (△) 11.7; (□) 11.8; (⊙) 11.9; (◇) 12.0; (▷) 12.1. Johnson and Mollendorf's (1980) film Prandtl numbers are: (◀) 11.6; (▲) 11.7; (■) 11.8; (◆) 12.0. *(Reprinted with permission from V. P. Carey et al.,* J. Fluid. Mech., *vol. 97, p. 279. Copyright © 1980, Cambridge University Press.)*

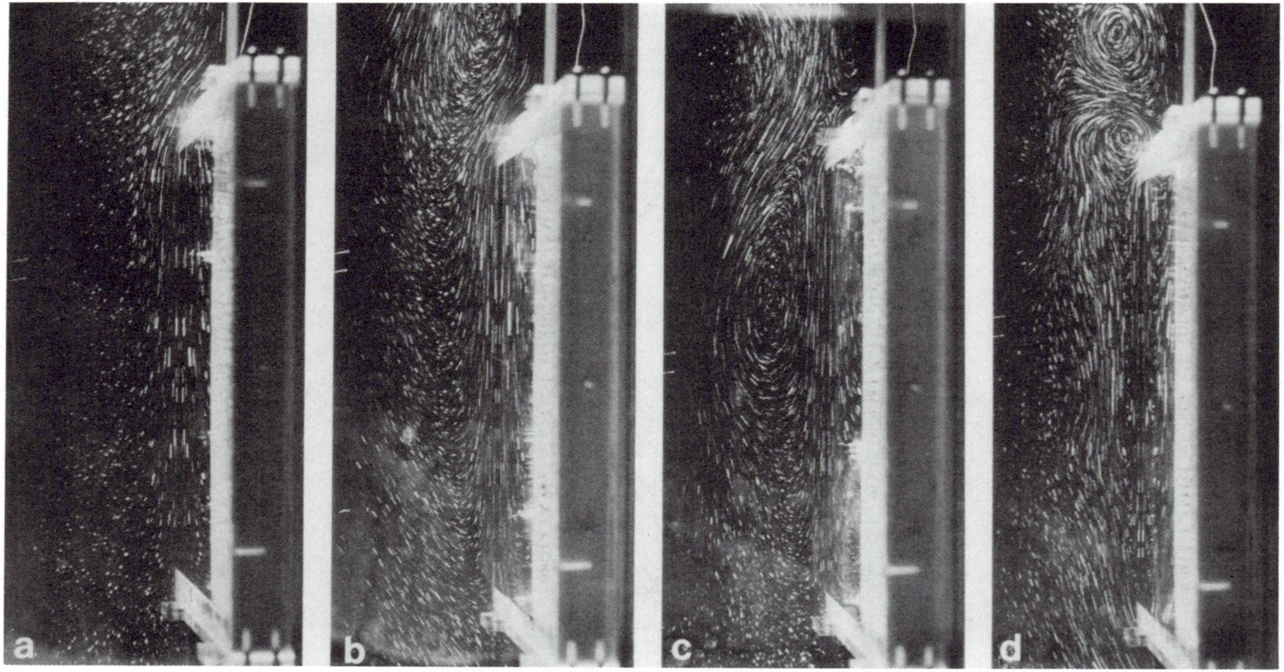

Figure 9.3.11 Flow adjacent to a vertical ice slab melting in pure water. The corresponding ambient temperatures, R values, and exposure times are (a) 3.90°C, $R = -0.033$, 6 s; (b) 4.05°C, $R = 0.005$, 10 s; (c) 4.40°C, $R = 0.084$, 10 s; (d) 4.70°C, $R = 0.143$, 10 s. The reference wires at the left edge of each photograph are 1 cm apart. *(Reprinted with permission from V. P. Carey and B. Gebhart,* J. Fluid. Mech., *vol. 107, p. 37. Copyright © 1981, Cambridge University Press.)*

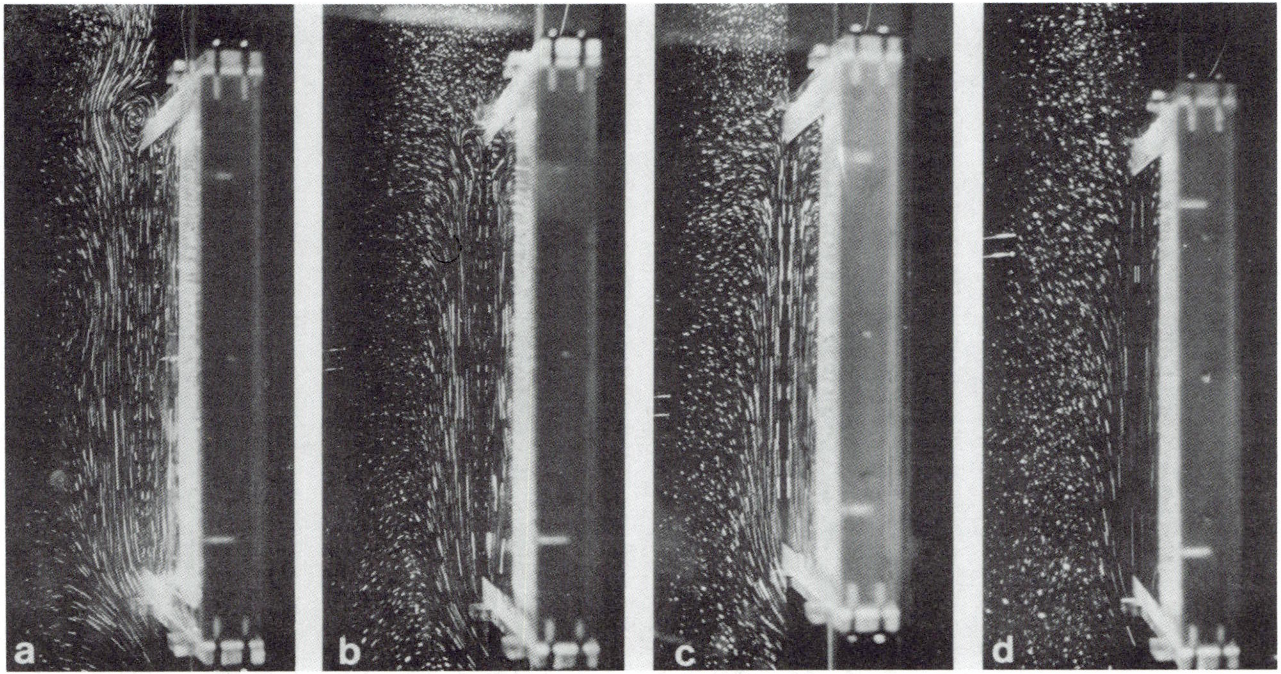

Figure 9.3.12 Flow adjacent to a vertical ice slab melting in pure water. The corresponding ambient temperatures, R values, and exposure times are: (a) 5.00°C, $R = 0.194$, 10 s; (b) 5.40°C, $R = 0.254$, 10 s; (c) 5.90°C, $R = 0.317$, 10 s; (d) 8.40°C, $R = 0.520$, 6 s. The reference wires shown at the left edge of each photograph are 1 cm apart. *(Reprinted with permission from V. P. Carey and B. Gebhart,* J. Fluid. Mech., *vol. 107, p. 37. Copyright © 1981, Cambridge University Press.)*

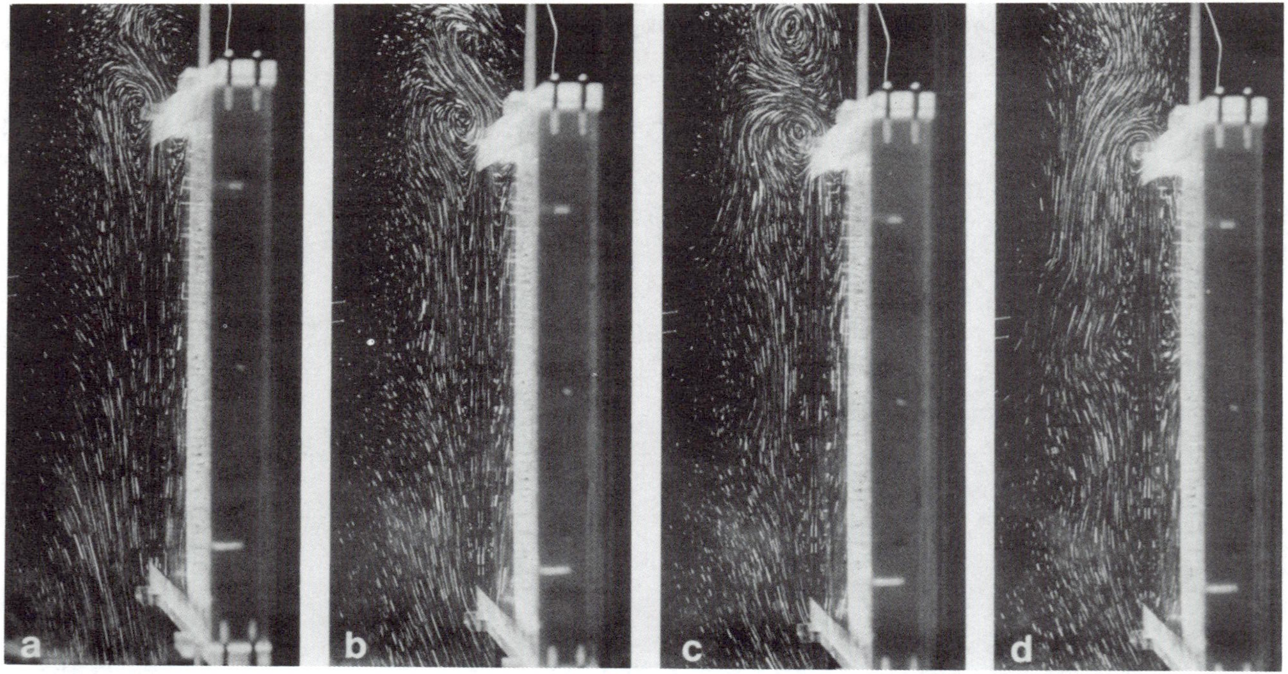

Figure 9.3.13 Flow adjacent to a vertical ice slab melting in pure water at 4.70°C ($R = 0.143$). The four photographs are 20-s time exposures taken in sequence (from left to right) with the shutter closed for 30 s between frames. The reference wires are 1 cm apart. *(Reprinted with permission from V. P. Carey and B. Gebhart,* J. Fluid. Mech., *vol. 107, p. 37. Copyright © 1981, Cambridge University Press.)*

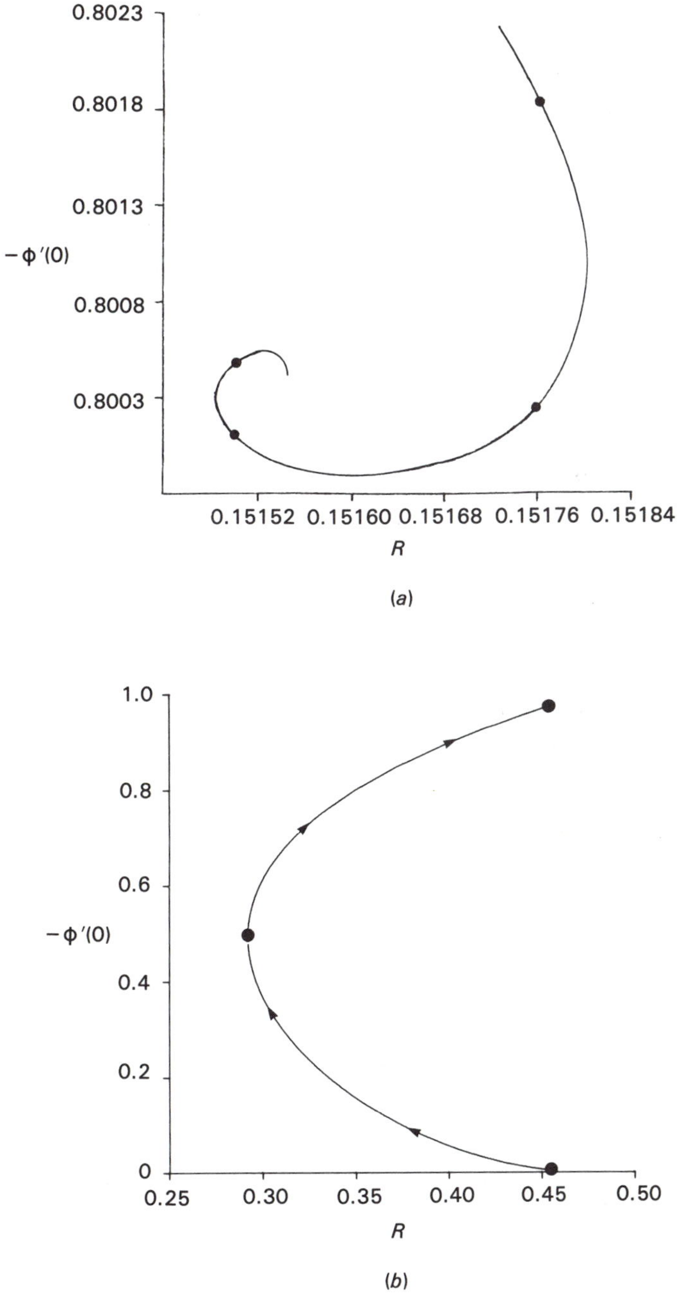

Figure 9.3.14 Details of the variation of heat transfer rate $-\phi'(0)$ with R in (*a*) the region of largely upflow and (*b*) the region of largely downflow. Arrows indicate increasing $-\phi'(0)$. *(Reprinted with permission from I. El-Henawy et al.,* J. Fluid. Mech., *vol. 122, p. 235. Copyright © 1982, Cambridge University Press.)*

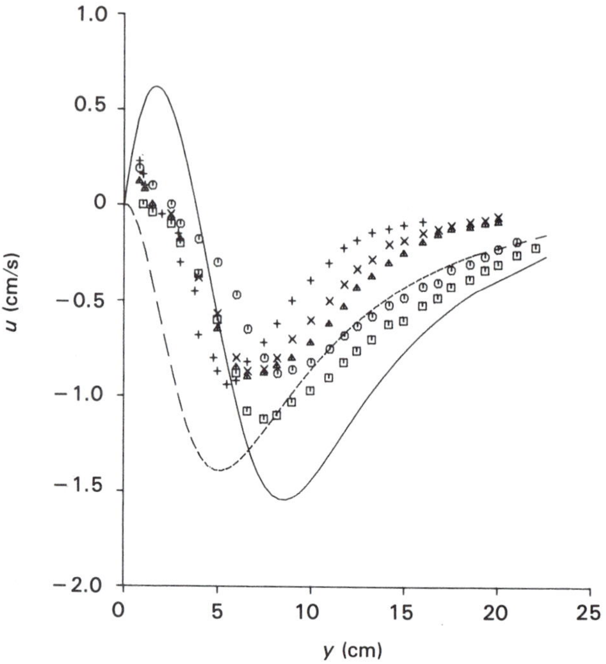

Figure 9.3.15 Experimental data from Wilson and Vyas (1979) for $R = 0.317$ showing unsteady behavior of the physical vertical velocity component u. Also shown are two curves corresponding to u as calculated from multiple steady-state solutions corresponding to (——) $R = 0.31889$, $-\phi'(0) = 0.26000$ and (– – –) $R = 0.31624$, $-\phi'(0) = 0.70000$. Symbols, in order of increasing time, are +, ×, ⊡, ▲, and ⊙. *(Reprinted with permission from I. El-Henawy et al.,* J. Fluid. Mech., *vol. 122, p. 235. Copyright © 1982, Cambridge University Press.)*

$$\mathrm{Nu}_x = 0.577(\mathrm{Ra}_x^*)^{1/(4+q)} \qquad 1.5829 \leqq q \leqq 1.8948 \tag{9.3.35}$$

where

$$\mathrm{Ra}_x^* = \frac{g\alpha x^3}{\nu^2}\left(\frac{q''x}{k}\right)^q \mathrm{Pr} \tag{9.3.36}$$

9.3.3 Flow Adjacent to Horizontal Surfaces

Such flows were studied by Gebhart et al. (1979), who considered boundary layer flows that arise when the net buoyancy force acts away from the surface. Two geometries were considered: plane two-dimensional onflows, where the fluid moves inward at the leading edge, and radial or disk outflows, with the fluid moving radially outward toward the edge of the surface. Solutions were obtained for isothermal and uniform flux surface conditions.

Plane flows. For plane flows the governing equations are

$$\frac{\partial u}{\partial x} + \frac{\partial v}{\partial y} = 0 \tag{9.3.37}$$

$$u\frac{\partial u}{\partial x} + v\frac{\partial u}{\partial y} = \nu\frac{\partial^2 u}{\partial y^2} - \frac{1}{\rho}\frac{\partial p_m}{\partial x} \tag{9.3.38}$$

$$0 = -\frac{1}{\rho}\frac{\partial p_m}{\partial y} + \frac{g(\rho_\infty - \rho)}{\rho} \tag{9.3.39}$$

$$u\frac{\partial t}{\partial x} + v\frac{\partial t}{\partial y} = \frac{k}{\rho c_p}\frac{\partial^2 t}{\partial y^2} \tag{9.3.40}$$

where Eq. (9.1.1) is used to calculate the buoyancy force. The following transformations are introduced:

$$\eta(x, y) = b(x)y \qquad \psi(x, y) = \nu c(x) f(\eta) \tag{9.3.41}$$

$$\phi(\eta) = \frac{t - t_\infty}{t_0 - t_\infty} = \frac{t - t_\infty}{d(x)} \tag{9.3.42}$$

$$p_m = a(x)P(\eta) \tag{9.3.43}$$

$$R = \frac{t_m(s, p) - t_\infty}{t_0 - t_\infty} \tag{9.3.44}$$

Introducing Eqs. (9.3.42) and (9.3.44) into Eq. (9.1.1), the buoyancy force is again

$$\rho_\infty - \rho = \alpha\rho_m|t_0 - t_\infty|^q(|\phi - R|^q - |R|^q) = \alpha\rho_m|t_0 - t_\infty|^q W \tag{9.3.45}$$

Equations (9.3.37)–(9.3.40) are thus reduced to

$$f''' + \frac{c_x}{b}ff'' - \left(\frac{c_x}{b} + \frac{cb_x}{b^2}\right)f'^2 = -\frac{1}{\rho\nu^2}\left(\frac{a_x}{cb^3}P + \frac{ab_x}{cb^4}\eta P'\right) \tag{9.3.46}$$

$$P' = -\frac{g\alpha\rho_m}{ab}|t_0 - t_\infty|^q(|\phi - R|^q - |R|^q) = -\frac{g\alpha\rho_m}{ab}|t_0 - t_\infty|^q W \tag{9.3.47}$$

$$\phi'' + \Pr\left|\frac{c_x}{b}f'\phi' - \frac{cd_x}{db}\phi f'\right| = 0 \tag{9.3.48}$$

For similarity, all the coefficients in the above equations must be constants or functions of η only. Similarity arises for an exponential and for a power law dependence in $d(x)$. Taking the power law variation

$$d(x) = t_0 - t_\infty = Nx^n \tag{9.3.49}$$

and again choosing the reference density as ρ_m, we find

$$c = G \tag{9.3.50}$$

$$b = \frac{G}{5x} \tag{9.3.51}$$

$$a = \frac{\nu^2\rho}{125x^2}(G)^4 \tag{9.3.52}$$

$$G = 5\left(\frac{\mathrm{Gr}_x}{5}\right)^{1/5} = 5\left(\frac{gx^3}{5\nu^2}\alpha|t_0 - t_\infty|^q\right)^{1/5} \tag{9.3.53}$$

In terms of these quantities the equations become

$$f''' + (nq+3)ff'' - (2nq+1)f'^2 - \frac{4nq+2}{5}P - \frac{nq-2}{5}\eta P' = 0 \tag{9.3.54}$$

$$P' = -\mathrm{sign}(g)(|\phi - R|^q - |R|^q) = -\mathrm{sign}(g)W \tag{9.3.55}$$

$$\phi'' + \mathrm{Pr}[(nq+3)f\phi' - 5n\phi f'] = 0 \tag{9.3.56}$$

For $q = 1$ and $R = 0$ these equations reduce to the Boussinesq approximation form presented earlier, Eqs. (5.3.3)–(5.3.5).

The limits of n are determined from the condition that the total local convected heat Q, for $t_0 > t_\infty$, must either increase with increasing x or remain constant. Together with other considerations, such as the value and sign of the pressure gradient and the boundary layer thickness behavior, this leads to the range $-1/(2q) < n < 2/q$. For an isothermal surface $n = 0$. For a uniform heat flux condition $n = 2/(q+5)$. In general, the local Nusselt number is

$$\mathrm{Nu}_x = [-\phi'(0)]\left(\frac{gx^3\alpha|t_0 - t_\infty|^q}{5\nu^2}\right)^{1/5} \tag{9.3.57}$$

For the isothermal surface, $n = 0$ in Eqs. (9.3.54)–(9.3.56), the boundary conditions are

$$f(0) = f'(0) = 1 - \phi(0) = \phi(\infty) = f'(\infty) = P(\infty) = 0 \tag{9.3.58}$$

Table 9.3.5 shows the resulting transport quantities for both an isothermal and a uniform flux surface over a range of Pr and R. Figure 9.3.16 shows the variation of $f'(0)$, $\phi'(0)$, I_w, and $f(\eta_\delta)$ for Pr = 11.6 and $q = 1.894816$ and 1.582950.

Disk flows. Gebhart et al. (1979) also analyzed axisymmetric outflows. The equations of motion representing conservation of momentum and energy are the same as Eqs. (9.3.38)–(9.3.40). Here x is the distance measured in the radial direction. The continuity equation is

$$\frac{\partial}{\partial x}(ux) + \frac{\partial}{\partial y}(vx) = 0 \tag{9.3.59}$$

This continuity equation is satisfied by use of the Stokes stream function ψ, defined as

$$u = \frac{1}{x}\frac{\partial \psi}{\partial y} \qquad v = -\frac{1}{x}\frac{\partial \psi}{\partial x} \tag{9.3.60}$$

Following the procedure outlined above for plane flow, it was shown that the similarity transformation for the above set of equations is

$$\eta = \frac{y}{5x} G \tag{9.3.61}$$

$$\psi = \nu x G f \tag{9.3.62}$$

$$p_m = \frac{\rho \nu^2}{125 x^2} G^4 P \tag{9.3.63}$$

$$\phi = \frac{t - t_\infty}{d(x)} = \frac{t - t_\infty}{N x^n} \tag{9.3.64}$$

where, as before,

$$G = 5\left(\frac{g\alpha|t_0 - t_\infty|^q x^3}{5\nu^2}\right)^{1/5} \tag{9.3.65}$$

$$R = \frac{t_m - t_\infty}{t_0 - t_\infty} \tag{9.3.66}$$

With this transformation and the density equation (9.1.1), Eqs. (9.3.37)–(9.3.40) become

$$f''' + (8 + nq)ff'' - (2nq + 1)f'^2 - \frac{nq - 2}{5}\eta P' - \frac{4nq + 2}{5} P = 0 \tag{9.3.67}$$

$$P = -\text{sign}(g)[|\phi - R|^q - |R|^q] = -\text{sign}(g)W \tag{9.3.68}$$

$$\phi'' + \text{Pr}[(nq + 8)\phi' f - 5nf'\phi] = 0 \tag{9.3.69}$$

The boundary conditions at the surface and in the distant medium in transformed variables are

$$f(0) = f'(0) = 1 - \phi(0) = \phi(\infty) = f'(\infty) = P(\infty) = 0 \tag{9.3.70}$$

The limits on n for a physically meaningful disk outflow are the same as those given above for horizontal plane flow.

The above equations were solved for an isothermal ($n = 0$) surface condition for Prandtl numbers 10.6, 11.6, and 12.6 with $q = 1.894816$ and 1.58295. The resulting buoyancy, drag, heat transfer, and entrainment velocity parameters are given in Table 9.3.6. For a uniform flux boundary condition $n = 2/(q + 5)$. Again, the requirement is $R = 0$. Results were obtained for Pr = 10.6, 11.6, and 12.5 and are also given in Table 9.3.6. The results of the conventional approximation, $q = 1$ when $R = 0$, are shown for comparison. The temperature, velocity, and

Table 9.3.5 Calculated values for isothermal and uniform flux horizontal plane flows

		(a) Uniform temperature, $n = 0$									
		$q = 1.894816$					$q = 1.582950$				
Pr	R	$f'(0)$	$P(0)$	$\phi'(0)$	$f(\eta_e)$	I_w	$f'(0)$	$P(0)$	$\phi'(0)$	$f(\eta_e)$	I_w
10.6	−2.000	0.42369	−1.88908	−1.24856	0.27868	1.88908	0.32702	−1.32297	−1.14721	0.25671	1.32297
	−1.000	0.30818	−1.26582	−1.11850	0.24812	1.26582	0.26552	−1.01752	−1.06763	0.23797	1.01752
	−0.500	0.23319	−0.90451	−1.01298	0.22243	0.94051	0.22087	−0.81421	−1.00026	0.22152	0.81421
	0.000	0.12525	−0.47260	−0.79006	0.15871	0.47260	0.14091	−0.51021	−0.83668	0.17314	0.51021
	0.500	0.13161	−0.27616	−0.88131	0.20750	−0.27616	0.14279	−0.30965	−0.90388	0.21120	−0.30965
	1.000	0.23728	−0.76695	−1.04749	0.23987	−0.76695	0.22317	−0.72663	−1.02238	0.23268	−0.72663
	2.000	0.37268	−1.48278	−1.20876	0.27398	−1.48278	0.30070	−1.12888	−1.12318	0.25387	−1.12888
11.6	−16.000	1.18707	−7.44726	−1.82398	0.38956	7.44726	0.63256	−3.21252	−1.47908	0.31600	3.21252
	−8.000	0.82458	−4.60059	−1.61438	0.34448	4.60059	0.49889	−2.34718	−1.36600	0.29167	2.34718
	−4.000	0.57705	−2.88114	−1.43159	0.30493	2.88114	0.39534	−1.73006	−1.26312	0.26939	1.73006
	−2.000	0.40948	−1.85009	−1.27408	0.27046	1.85009	0.31607	−1.29564	−1.17067	0.24913	1.29564
	−1.000	0.29781	−1.23976	−1.14134	0.24081	1.23976	0.25661	−0.99653	−1.08945	0.23095	0.99653
	−0.500	0.22532	−0.88595	−1.03364	0.21589	0.88595	0.21344	−0.79745	−1.02069	0.21499	0.79745
	0.000	0.12092	−0.46309	−0.80604	0.15414	0.46309	0.13608	−0.49986	−0.85365	0.16813	0.49986
	0.080	0.09542	−0.39077	−0.70082	0.10347	0.39077					
	0.301	0.05885	−0.00123	−0.74506	0.17815	−0.00123	0.07244	−0.02551	−0.79159	0.18760	−0.02551
	0.400	0.09739	−0.14638	−0.83771	0.19126	−0.14638	0.11027	−0.17709	−0.87169	0.19763	−0.17709
	0.500	0.12735	−0.27026	−0.89952	0.20134	−0.27026	0.13814	−0.30305	−0.92253	0.20494	−0.30305
	1.000	0.22943	−0.75093	−1.06900	0.23269	−0.75093	0.21577	−0.71149	−1.04336	0.22580	−0.71149
	2.000	0.36027	−1.45197	−1.23354	0.26588	−1.45197	0.20968	−1.10546	−1.14620	0.24637	−1.10546
	4.000	0.54142	−2.55477	−1.40869	0.30234	−2.55477	0.37924	−1.59956	−1.24991	0.26789	−1.59956
	8.000	0.79874	−4.33268	−1.60142	0.34301	−4.33268	0.48864	−2.25718	−1.35885	0.29086	−2.25716
	16.000	1.16833	−7.22727	−1.81665	0.38873	−7.22727	0.62604	−3.15037	−1.47520	0.31556	−3.15037

12.6	−2.000	0.39686	−1.81523	−1.29786	0.26318	1.81523	0.30633	−1.27120	−1.19253	0.24242	1.27120
	−1.000	0.28860	−1.21645	−1.16262	0.23433	1.21645	0.24869	−0.97776	−1.10978	0.22473	0.97776
	−0.500	0.21832	−0.86935	−1.05288	0.21009	0.86935	0.20684	−0.78246	−1.03972	0.20921	0.78246
	0.000	0.11708	−0.45456	−0.82092	0.15007	0.45456	0.13180	−0.49061	−0.86946	0.16368	0.49061
	0.500	0.12355	−0.26500	−0.91647	0.19589	−0.26500	0.13401	−0.29716	−0.93990	0.19940	−0.29716
	1.000	0.22244	−0.73661	−1.08904	0.22640	−0.73661	0.20918	−0.69797	−1.02689	0.21971	−0.69797
	2.000	0.34924	−1.42443	−1.25661	0.25871	−1.42443	0.28177	−1.08452	−1.16763	0.23972	−1.08452

(b) Uniform flux, $n = 2/(5 + q)$

		$q = 1.894816$, $n = 0.290073$					$q = 1.582950$, $n = 0.303815$				
Pr	R	$f'(0)$	$P(0)$	$\phi'(0)$	$f(\eta_e)$	I_w	$f'(0)$	$P(0)$	$\phi'(0)$	$f(\eta_e)$	I_w
10.6	0.0	0.13638	−0.40244	−0.98448	0.12764	0.40244	0.15035	−0.43733	−1.04569	0.14156	0.43733
11.6	0.0	0.13157	−0.39438	−1.00503	0.12394	0.39438	0.14509	−0.42852	−1.06648	0.13744	0.42852
12.6	0.0	0.12730	−0.38716	−1.02325	0.12065	0.38716	0.14043	−0.42063	−1.08584	0.13378	0.42063

(c) Conventional buoyancy force approximation

	Uniform temperature, $n = 0$					Uniform flux, $n = 1/3$				
Pr	$f'(0)$	$P(0)$	$\phi'(0)$	$f(\eta_e)$	I_w	$f'(0)$	$P(0)$	$\phi'(0)$	$f(\eta_e)$	I_w
10.6	0.18741	−0.61747	−0.95575	0.21484	0.61747	0.19198	−0.53937	−1.19988	0.18301	0.53937
11.6	0.18115	−0.60468	−0.97531	0.20850	0.60468	0.18542	−0.52829	−1.22388	0.17759	0.52829
12.6	0.17558	−0.59326	−0.99353	0.20288	0.59326	0.17960	−0.51837	−1.24623	0.17278	0.51837

Source: Reprinted with permission from *Int. J. Heat Mass Transfer,* vol. 22, p. 137, B. Gebhart et al. Copyright © 1979, Pergamon Journals Ltd.

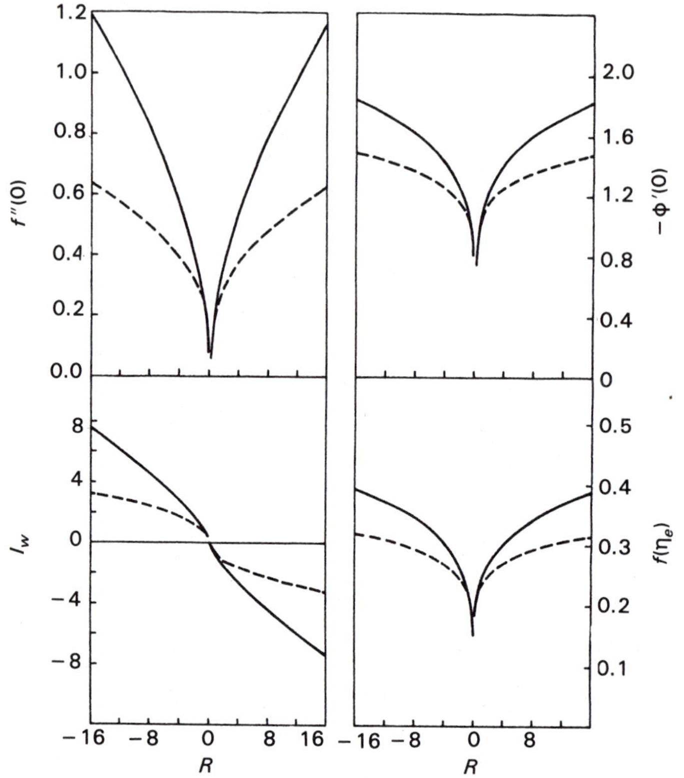

Figure 9.3.16 Heat transfer $\phi'(0)$, drag $f''(0)$, mass flow rate $f(\eta_e)$, and net buoyancy I_w over a range of R for an isothermal surface for Pr = 11.6; (——) $q = 1.894816$ and (– – – –) $q = 1.582950$. *(Reprinted with permission from* Int. J. Heat Mass Transfer, *vol. 22, p. 137, B. Gebhart et al. Copyright © 1979, Pergamon Journals Ltd.)*

pressure profiles in the boundary region are similar to those of the plane flow, as is the Prandtl number effect.

9.3.4 Plane and Axisymmetric Plumes

Mollendorf et al. (1981) presented solutions for plume flows generated in pure and saline water due to thermal diffusion arising from concentrated energy sources. Similarity plume solutions were obtained for $R = 0$, that is, $t_m = t_\infty$, for point and line sources as well as for plane wall plumes.

Taking x positive in the direction opposed to gravity and applying the boundary layer approximations, Eqs. (9.3.71)–(9.3.73) for plane flows and Eqs. (9.3.74)–(9.3.76) for axisymmetric flows are obtained as follows:

$$\frac{\partial u}{\partial x} + \frac{\partial v}{\partial y} = 0 \qquad (9.3.71)$$

$$\rho_r\left(u\frac{\partial u}{\partial x}+v\frac{\partial u}{\partial y}\right)=\mu\frac{\partial^2 u}{\partial y^2}+g(\rho_\infty-\rho) \tag{9.3.72}$$

$$\rho_r c_p\left(u\frac{\partial t}{\partial x}+v\frac{\partial t}{\partial y}\right)=k\frac{\partial^2 t}{\partial y^2}+\beta T u\frac{dp_h}{dx}+\mu\left(\frac{\partial u}{\partial y}\right)^2 \tag{9.3.73}$$

$$\frac{\partial}{\partial x}(yu)+\frac{\partial}{\partial y}(yv)=0 \tag{9.3.74}$$

$$\rho_r\left(u\frac{\partial u}{\partial x}+v\frac{\partial u}{\partial y}\right)=\mu\frac{1}{y}\frac{\partial}{\partial y}\left(y\frac{\partial u}{\partial y}\right)+g(\rho_\infty-\rho) \tag{9.3.75}$$

$$\rho_r c_p\left(u\frac{\partial t}{\partial x}+v\frac{\partial t}{\partial y}\right)=k\frac{1}{y}\frac{\partial}{\partial y}\left(y\frac{\partial t}{\partial y}\right)+\beta T u\frac{dp_h}{\partial x}+\mu\left(\frac{\partial u}{\partial y}\right)^2 \tag{9.3.76}$$

The boundary conditions for the plane plume, the wall plume, and the axisymmetric plume are, respectively,

$$\text{at } y=0:\quad v=\frac{\partial u}{\partial y}=\frac{\partial \phi}{\partial y}=0 \qquad \text{as } y\to\infty:\quad u=\phi=0 \tag{9.3.77}$$

$$\text{at } y=0:\quad v=u=\frac{\partial \phi}{\partial y}=0 \qquad \text{as } y\to\infty:\quad u=\phi=0 \tag{9.3.78}$$

$$\text{at } y=0:\quad v=\frac{\partial u}{\partial y}=\frac{\partial \phi}{\partial y}=0 \qquad \text{as } y\to\infty:\quad u=\phi=0 \tag{9.3.79}$$

For a power law temperature variation, $d = t_0 - t_\infty = Nx^n$, $n = -3/(q+4)$. Following the usual similarity transformations, the equations for the plane plume and the plane wall plume, along with the appropriate boundary conditions, become

$$f'''+\frac{12}{q+4}ff''+\frac{4(q-2)}{q+4}f'^2+\frac{\phi^q}{I_w}=0 \tag{9.3.80}$$

$$\phi''+\frac{12\mathrm{Pr}}{q+4}(f\phi)'=0 \tag{9.3.81}$$

or

$$\phi(\eta)=\exp\left[-\frac{12\mathrm{Pr}}{(q+4)}\int_0^\eta f\,d\eta\right] \tag{9.3.82}$$

Plume:

$$\text{at } \eta=0:\quad \phi-1=f''=f=0 \qquad \text{as } \eta\to\infty:\quad \phi=f'=0 \tag{9.3.83}$$

Wall plume:

$$\text{at } \eta=0:\quad \phi-1=f'=f=0 \qquad \text{as } \eta\to\infty:\quad \phi=f'=0 \tag{9.3.84}$$

Table 9.3.6 Calculated values for isothermal and uniform flux horizontal circular disk flows

(a) Uniform temperature, $n = 0$

		$q = 1.894816$					$q = 1.582950$				
Pr	R	$f'(0)$	$P(0)$	$\phi'(0)$	$f(\eta_e)$	I_w	$f'(0)$	$P(0)$	$\phi'(0)$	$f(\eta_e)$	I_w
10.6	−2.000	0.28836	−1.54214	−1.52797	0.13053	1.54214	0.22260	−1.07995	−1.40399	0.12024	1.07995
	−1.000	0.20968	−1.03347	−1.36871	0.11622	1.03347	0.18070	−0.83067	−1.30653	0.11146	0.83067
	−0.500	0.15859	−0.73861	−1.23943	0.10419	0.73861	0.15027	−0.66477	−1.22400	0.10376	0.66477
	0.000	0.08496	−0.38636	−0.96590	0.07440	0.38636	0.09567	−0.41694	−1.02317	0.08115	0.41694
	0.500	0.08991	−0.22511	−1.07938	0.09717	−0.22511	0.09752	−0.25244	−1.10690	0.09891	−0.25244
	1.000	0.16173	−0.62575	−1.28235	0.11230	−0.62575	0.15206	−0.59293	−1.25151	0.10898	−0.59293
	2.000	0.25386	−1.21007	−1.47957	0.12832	−1.21007	0.20480	−0.92132	−1.37477	0.11890	−0.92132
11.6	−16.000	0.80791	−6.08042	−2.23183	0.18250	6.08042	0.43053	−2.62289	−1.80981	0.14804	−2.62289
	−8.000	0.56118	−3.75629	−1.97534	0.16138	3.75629	0.33953	−1.91640	−1.67144	0.13644	1.91640
	−4.000	0.39268	−2.35248	−1.75163	0.14285	2.3248	0.26905	−1.41257	−1.54552	0.12620	1.41257
	−2.000	0.27861	−1.51072	−1.55885	0.12671	1.51072	0.21508	−1.05792	−1.43237	0.11671	1.05792
	−1.000	0.20257	−1.01245	−1.39635	0.11282	1.01245	0.17458	−0.81375	−1.33293	0.10820	0.81375
	−0.500	0.15319	−0.72364	−1.26443	0.10115	0.72364	0.14517	−0.65126	−1.24871	0.10072	0.65126
	0.000	0.08200	−0.37866	−0.98524	0.07227	0.37866	0.09237	−0.40858	−1.04371	0.07881	0.40858
	0.080	0.06455	−0.31993	−0.85579	0.04888	0.31993					
	0.301	0.04053	−0.00086	−0.91342	0.08344	−0.00086					
	0.400	0.06663	−0.11925	−1.02605	0.08958	−0.11925					
	0.500	0.08696	−0.22038	−1.10138	0.09431	−0.22038	0.09430	−0.24714	−1.12945	0.09600	−0.24714
	1.000	0.15632	−0.61285	−1.30836	0.10900	−0.61285	0.14696	−0.58074	−1.27687	0.10578	−0.58074
	2.000	0.24532	−1.18526	−1.50953	0.12455	−1.18526	0.19790	−0.90245	−1.40260	0.11541	−0.90245
	4.000	0.36858	−2.08567	−1.72378	0.14163	−2.08567	0.25816	−1.30589	−1.52946	0.12550	−1.30589
	8.000	0.54370	−3.53727	−1.95958	0.16069	−3.53727	0.33261	−1.84283	−1.66274	0.13626	−1.84283
	16.000	0.79524	−5.90058	−2.22292	0.18211	−5.90058	0.42611	−2.57208	−1.80510	0.14783	−2.57208

12.6	−2.000	0.26994	−1.48259	−1.58762	0.12332	1.48259	0.20839	−1.03821	−1.45881	0.11359	1.03821
	−1.000	0.19625	−0.99364	−1.42209	0.10980	0.99364	0.16915	−0.79861	−1.35753	0.10530	0.79861
	−0.500	0.14840	−0.71024	−1.28772	0.09845	0.71024	0.16915	−0.63916	−1.27173	0.09803	0.63916
	0.000	0.07939	−0.37176	−1.00325	0.07038	0.37176	0.08945	−0.40109	−1.06284	0.07674	0.40109
	0.500	0.08433	−0.21615	−1.12188	0.09178	−0.21615	0.09145	−0.24241	−1.15045	0.09342	−0.24241
	1.000	0.15151	−0.60132	−1.33259	0.10608	−0.60132	0.14243	−0.56984	−1.30050	0.10294	−0.56984
	2.000	0.23774	−1.16306	−1.53745	0.12122	−1.16306	0.19178	−0.88557	−1.42853	0.11232	−0.88557

(b) Uniform flux, $n = 2/(5 + q)$

		$q = 1.894816$, $n = 0.290073$					$q = 1.582950$, $n = 0.303815$				
Pr	R	$f'(0)$	$P(0)$	$\phi'(0)$	$f(\eta_e)$	I_w	$f'(0)$	$P(0)$	$\phi'(0)$	$f(\eta_e)$	I_w
10.6	0.0	0.10294	−0.35096	−1.09789	0.06868	0.35096	0.11312	−0.38081	−1.16168	0.07533	0.38081
11.6	0.0	0.09930	−0.34400	−1.11963	0.06670	0.34400	0.10915	−0.37321	−1.18473	0.07316	0.37321
12.6	0.0	0.09608	−0.33775	−1.13988	0.06495	0.33775	0.10564	−0.36640	−1.20620	0.07123	0.36640

(c) Conventional buoyancy force approximation

	Uniform temperature, $n = 0$					Uniform flux, $n = 1/3$				
Pr	$f'(0)$	$P(0)$	$\phi'(0)$	$f(\eta_e)$	I_w	$f'(0)$	$P(0)$	$\phi'(0)$	$f(\eta_e)$	I_w
10.6	0.12760	−0.50399	−1.16974	0.10063	0.50399	0.14301	−0.46703	−1.32412	0.09478	0.46703
11.6	0.12339	−0.49369	−1.19341	0.09768	0.49369	0.13810	−0.45753	−1.35054	0.09200	0.45753
12.6	0.11947	−0.48448	−1.21545	0.09506	0.48448	0.13374	−0.44903	−1.37517	0.08953	0.44903

Source: Reprinted with permission from *Int. J. Heat Mass Transfer*, vol. 22, p. 137, B. Gebhart et al. Copyright © 1979, Pergamon Journals Ltd.

For the axisymmetric plume, $n = -1$ and the result is

$$\frac{f'''}{\eta} + \frac{f-1}{\eta}\left(\frac{f'}{\eta}\right)' + \frac{q-1}{2}\left(\frac{f'}{\eta}\right)^2 + \frac{\phi^q}{I_w} = 0 \tag{9.3.85}$$

$$\eta\phi' + \Pr f\phi = 0 \tag{9.3.86}$$

or

$$\phi(\eta) = \exp\left[-\Pr\int_0^\eta\left(\frac{f}{\eta}\right)d\eta\right]$$

$$\text{at } \eta = 0: \quad f = f' = \phi - 1 = 0 \qquad \text{as } \eta \to \infty: \quad f' = 0 \tag{9.3.87}$$

In summary, the above sets of equations are derived by using $n = -3/(q+4)$ for the plane flows and $n = -1$ for the axisymmetric flows.

Table 9.3.7 shows the Pr and q effects on the integrated buoyancy force and the convected heat and momentum flux, that is, I_w, I_Q, and I_M, respectively. Figures 9.3.17–9.3.19 show the resulting velocity and temperature fields for these three plume flows. Calculations are made for Pr = 8.6 and 13.6 and $q = 1.894816$. Also shown are the Boussinesq results for comparison.

Gebhart et al. (1980) developed a perturbation analysis, perturbing t_∞ around t_m. This extended the range of the calculations beyond the similarity limit $R = 0$. A new perturbation parameter $R^* = (t_m - t_\infty)/(t_0 - t_\infty)_0$ was defined, where $t_m - t_\infty$ is the perturbation and $(t_0 - t_\infty)_0$ is the value that results in the similarity formulation when $t_m = t_\infty$, that is, for $R = 0$. Calculations were made for $R = 0$, for uniform flux surfaces, and for the isothermal condition. The results compared favorably with the similarity solutions for the range Pr = 8.6–13.6.

Joshi and Gebhart (1983) visualized line source plumes in pure cold water. Temperature measurements for $t_\infty = 4°\text{C}$ showed good agreement with the calculations of Mollendorf et al. (1981). Time exposure photographs of the flow field revealed that for $t_\infty < t_m$ the flow is not of boundary layer form. Very complicated circulations arose.

9.4 COMBINED THERMAL AND SALINE TRANSPORT

Inclusion of saline transport in the study of flows in water with a density extremum substantially changes the resulting flow regimes, heat transfer, and other transport quantities. This arises from the additional component in the buoyancy force due to mass diffusion. For example, when ice melts in saline water the resulting flow is driven by a buoyancy force having two components: thermal, due to temperature, and saline, due to melt. These two forces may be locally aiding or opposing, depending on ambient temperature and salinity levels.

Saline water also exhibits an extremum behavior, as shown in Fig. 9.1.1. The extremum occurs up to a salinity of about 26‰ at 1 bar. At higher salinities $t_{il} > t_m$ and no extremum arises in equilibrium. In circumstances where there are both

Table 9.3.7 Calculated flow and transport quantities for various flow configurations and values of Pr and q as indicated[a]

		(a) Plane plume					(b) Line source on adiabatic surface					(c) Axisymmetric plume				
Pr	q	$f'(0)$	I_w	$\int_0^\infty f'\phi\, d\eta$	$\int_0^\infty (f')^3\, d\eta$	$f(\infty)$	$f'(0)$	I_w	$\int_0^\infty f'\phi\, d\eta$	$\int_0^\infty (f')^3\, d\eta$	$f(\infty)$	$f'(0)$	I_w	$\int_0^\infty f'\phi\, d\eta$	$\int_0^\infty (f)^2 \frac{d\eta}{\eta}$	$f(\infty)$
8.6	1.894816	0.9446	0.2277	0.2686	0.4747	0.8280	0.8565	0.6078	0.1372	0.06813	0.4511	0.8261	0.5010	0.2004	0.5890	3.507
	1.859003	0.9369	0.2303	0.2065	0.4006	0.8212	0.8543	0.6112	0.1368	0.06798	0.4499	0.8189	0.5083	0.2003	0.5794	3.496
	1.727147	0.9042	0.2406	0.2590	0.4375	0.7964	0.8248	0.6248	0.1353	0.06749	0.4454	0.7950	0.5366	0.2001	0.5491	3.456
	1.582950	0.8709	0.2532	0.2512	0.4085	0.7707	0.8357	0.0416	0.1337	0.06709	0.4409	0.7737	0.5698	0.2001	0.5247	3.416
	1.000000	0.7444	0.3284	0.2212	0.3125	0.6787	0.7834	0.7449	0.1272	0.06770	0.4299	0.7168	0.7547	0.2029	0.4977	3.293
9.6	1.894816	0.9499	0.2147	0.2561	0.4747	0.8271	0.8020	0.5837	0.1286	0.06504	0.4429	0.8083	0.4787	0.1804	0.5623	3.504
	1.859663	0.9412	0.2171	0.2542	0.4666	0.8204	0.8609	0.5868	0.1283	0.00491	0.4417	0.8015	0.4856	0.1803	0.5530	3.493
	1.727147	0.9096	0.2208	0.2472	0.4375	0.7955	0.8526	0.5998	0.1269	0.06446	0.4372	0.7791	0.5123	0.1801	0.5236	3.452
	1.582950	0.8765	0.2386	0.2396	0.4085	0.7697	0.8430	0.6157	0.1253	0.06412	0.4328	0.7593	0.5435	0.1801	0.4999	3.411
	1.000000	0.7507	0.3090	0.2112	0.3125	0.6770	0.7924	0.7137	0.1194	0.06487	0.4219	0.7076	0.7174	0.1824	0.4722	3.283
10.6	1.894816	0.9544	0.2036	0.2454	0.4747	0.8265	0.8686	0.5628	0.1213	0.06237	0.4357	0.7925	0.4595	0.1639	0.5394	3.501
	1.859663	0.9458	0.2059	0.2435	0.4666	0.8197	0.8666	0.5658	0.1210	0.06224	0.4354	0.7860	0.4661	0.1639	0.5304	3.490
	1.727147	0.9143	0.2151	0.2308	0.4375	0.7948	0.8586	0.5782	0.1197	0.06184	0.4301	0.7649	0.4914	0.1637	0.5018	3.449
	1.582950	0.8813	0.2262	0.2297	0.4085	0.7688	0.8494	0.5934	0.1182	0.06153	0.4257	0.7463	0.5210	0.1637	0.4787	3.408
	1.000000	0.7563	0.2926	0.2024	0.3125	0.6757	0.8003	0.6869	0.1127	0.06239	0.4148	0.6990	0.6856	0.1657	0.4506	3.275
11.6	1.894816	0.9584	0.1941	0.2359	0.4747	0.8259	0.8736	0.5445	0.1150	0.06001	0.4292	0.7782	0.4429	0.1503	0.5194	3.499
	1.859663	0.9499	0.1963	0.2342	0.4606	0.8191	0.8716	0.5475	0.1147	0.05989	0.4280	0.7721	0.4491	0.1502	0.5107	3.488
	1.727147	0.9184	0.2049	0.2277	0.4375	0.7941	0.8639	0.5593	0.1134	0.05952	0.4237	0.7520	0.4733	0.1500	0.4829	3.447
	1.582950	0.8856	0.2155	0.2208	0.4085	0.7682	0.8549	0.6739	0.1121	0.05925	0.4194	0.7344	0.5015	0.1500	0.4604	3.405
	1.000000	0.7612	0.2785	0.1947	0.3125	0.6746	0.8074	0.6635	0.1069	0.06020	0.4085	0.6911	0.6581	0.1518	0.4320	3.208
12.6	1.894816	0.9620	0.1857	0.2275	0.4747	0.8254	0.8780	0.5284	0.1094	0.05791	0.4234	0.7652	0.4281	0.1387	0.5019	3.498
	1.859663	0.9534	0.1878	0.2258	0.4666	0.8186	0.8701	0.5312	0.1091	0.05781	0.4222	0.7593	0.4341	0.1386	0.4933	3.486
	1.727147	0.9221	0.1961	0.2196	0.4375	0.7936	0.8686	0.5426	0.1080	0.05747	0.4179	0.7402	0.4573	0.1385	0.4662	3.444
	1.582950	0.8894	0.2062	0.2130	0.4085	0.7676	0.8599	0.5566	0.1067	0.05722	0.4137	0.7236	0.4842	0.1385	0.4442	3.402
	1.000000	0.7655	0.2661	0.1879	0.3125	0.6737	0.8136	0.6429	0.1019	0.05824	0.4029	0.6836	0.6340	0.1401	0.4156	3.263
13.6	1.894816	0.9652	0.1784	0.2199	0.4747	0.8250	0.8819	0.5139	0.1045	0.08603	0.4180	0.7532	0.4150	0.1287	0.4862	3.496
	1.859663	0.9567	0.1803	0.2183	0.4666	0.8182	0.8801	0.5166	0.1042	0.05593	0.4169	0.7476	0.4207	0.1287	0.4779	3.485
	1.727147	0.9254	0.1883	0.2123	0.4375	0.7931	0.8729	0.5276	0.1031	0.05562	0.4127	0.7294	0.4430	0.1286	0.4514	3.443
	1.582850	0.8928	0.1979	0.2050	0.4085	0.7671	0.8644	0.5412	0.1020	0.05540	0.4085	0.7136	0.4690	0.1285	0.4299	3.399
	1.000000	0.7694	0.2553	0.1817	0.3125	0.6729	0.8193	0.6245	0.09738	0.05647	0.3977	0.6767	0.6126	0.1300	0.4012	3.258

[a]Note that rounding q from seven to four significant figures will result in an error of less than 0.1% in the calculated results.

Source: Reprinted with permission from J. C. Mollendorf et al., *J. Fluid Mech.*, vol. 113, p. 269. Copyright © 1981, Cambridge University Press.

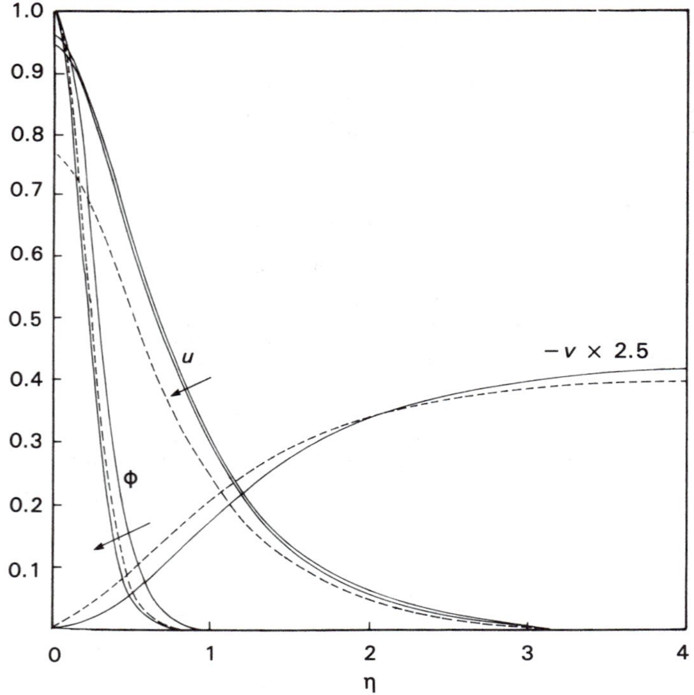

Figure 9.3.17 Calculated velocity and temperature fields for a freely rising plane plume for $q = q(0, 1) = 1.894816$. (– – –) Conventional Boussinesq results, i.e., $q = 1$ for Pr = 12.6. Arrows indicate increasing Pr: 8.6 and 12.6. The Prandtl number effect on $-v$ is indistinguishable on this figure. Rounding q from seven to four significant digits results in an error of less than 0.1% in the calculated results. *(Reprinted with permission from J. C. Mollendorf et al.,* J. Fluid. Mech., *vol. 113, p. 269. Copyright © 1981, Cambridge University Press.)*

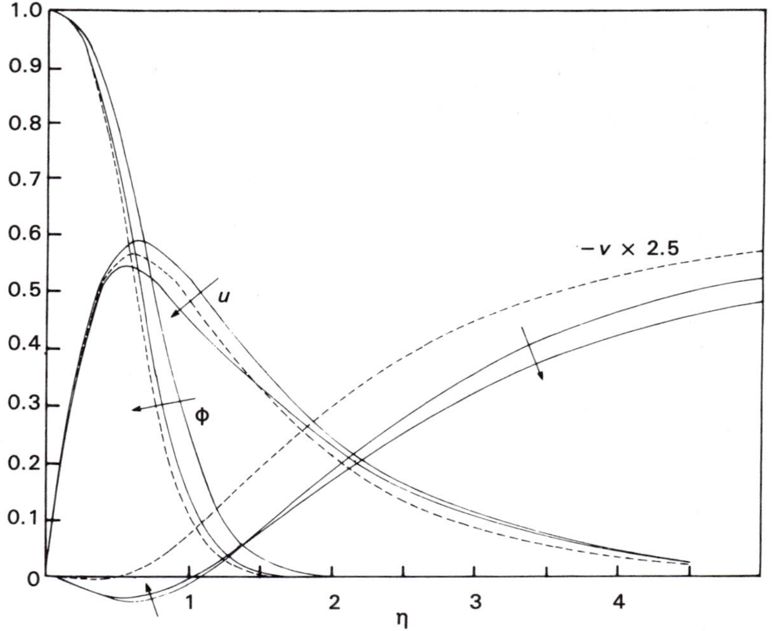

Figure 9.3.18 Calculated velocity and temperature fields for a thermal line source on an adiabatic surface for $q = q(0, 1) = 1.894816$. (– – –) Conventional Boussinesq results, i.e., $q = 1$ for Pr = 12.6. Arrows indicate increasing Pr: 8.6 and 12.6. *(Reprinted with permission from J. C. Mollendorf et al.,* J. Fluid. Mech., *vol. 113, p. 269. Copyright © 1981, Cambridge University Press.)*

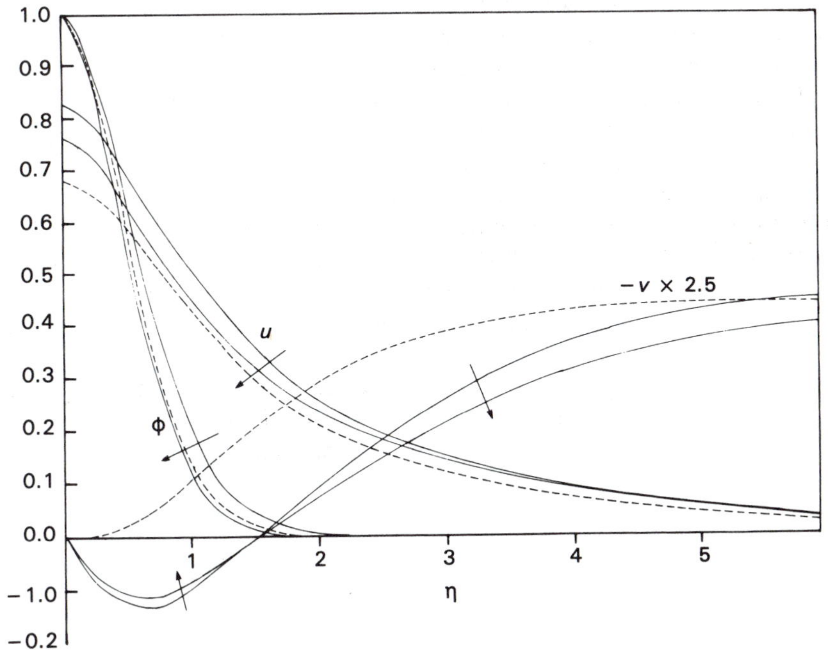

Figure 9.3.19 Calculated velocity and temperature fields for a freely rising axisymmetric plume for $q = q(0, 1) = 1.894816$. (– – –) Conventional Boussinesq results, i.e., $q = 1$ for Pr = 12.6. Arrows indicate increasing Pr: 8.6 and 12.6. *(Reprinted with permission from J. C. Mollendorf et al.,* J. Fluid. Mech., *vol. 113, p. 269. Copyright © 1981, Cambridge University Press.)*

temperature and salinity gradients, net local buoyancy force and flow reversals as well as total convective inversion may arise. The resulting flows are very interesting but commonly very complicated and not amenable to analysis. The general governing equations for these flows are Eqs. (9.2.1)–(9.2.5).

9.4.1 Transport Adjacent to Flat Vertical Melting Surfaces

A flat vertical ice surface immersed in an extensive quiescent saline water medium generates a buoyancy-driven flow as it melts. Under some circumstances the flow is of the boundary layer type and may be analyzed as such, as will be shown later. For constant transport properties the resulting steady-state continuity, momentum, and energy equations are Eqs. (9.3.1)–(9.3.3). The conservation of salinity equation is as given below, from Eq. (6.2.35):

$$u \frac{\partial s}{\partial x} + v \frac{\partial s}{\partial y} = D \frac{\partial^2 s}{\partial y^2} \tag{9.4.1}$$

Marschall (1977) solved the boundary layer equations, using a coordinate system fixed to the ice-water interface. An expression was derived for the resulting blowing velocity at the surface. Marschall also noted that the governing equations and boundary conditions admit a similarity solution. It is not clear, however, what density relation was used to formulate the buoyancy force. Gebhart and Mollendorf (1978) derived the similarity transformation for the boundary layer equations, including both thermal and saline effects. The density relation Eq. (9.1.1) was used. However, no solutions were presented.

Subsequent measurements and visualizations by Johnson (1978), Josberger and Martin (1981), Carey and Gebhart (1982b), Sammakia and Gebhart (1983), and Johnson and Mollendorf (1984) confirmed the existence of boundary layer flows at low salinities over a wide temperature range and at high salinities at quite low temperatures. Such observations motivated an analysis of the laminar boundary layer flows by Carey and Gebhart (1982a).

Because of the recession of the less dense ice during melting, the ambient fluid moves toward the surface, say at a velocity V_a. From the conservation of mass, this is related to the recession rate of the interface, $V_i(x)$, by

$$V_a(x) = V_i(x)\left[1 - \frac{\rho_i}{\rho_r(1 - 10^{-3}s_0)}\right] \tag{9.4.2}$$

Assuming an isothermal surface and a steady flow field, the interface and distant boundary conditions are

$$\text{at } y = -V_i\tau: \quad u = 0 \quad v = -V_a \quad t = t_0 \quad s = s_0$$

$$\text{as } y \to \infty: \quad u = 0 \quad t = t_\infty \quad s = s_\infty$$

where τ is time.

To simplify the numerical analysis, the equations and boundary conditions are transformed to a coordinate system with the $\bar{x}$ axis along the ice interface and moving with it. The following transformation is used:

$$\begin{gathered} x = \bar{x} \quad y = \bar{y} - V_i\bar{\tau} \quad \tau = \bar{\tau} \\ u = \bar{u} \quad v = \bar{v} - V_i \quad t = \bar{t} \quad s = \bar{s} \end{gathered} \tag{9.4.3}$$

where the overbars indicate variables in the moving coordinate system. The equations become

$$\frac{\partial \bar{u}}{\partial \bar{x}} + \frac{\partial \bar{v}}{\partial \bar{y}} = 0 \tag{9.4.4}$$

$$\rho_r\left(\bar{u}\frac{\partial \bar{u}}{\partial \bar{x}} + \bar{v}\frac{\partial \bar{u}}{\partial \bar{y}}\right) = \mu\frac{\partial^2 \bar{u}}{\partial \bar{y}^2} + g(\rho_\infty - \rho) \tag{9.4.5}$$

$$\rho_r c_p\left(\bar{u}\frac{\partial \bar{t}}{\partial \bar{x}} + \bar{v}\frac{\partial \bar{t}}{\partial \bar{y}}\right) = k\frac{\partial^2 \bar{t}}{\partial \bar{y}^2} \tag{9.4.6}$$

$$\bar{u}\frac{\partial \bar{s}}{\partial \bar{x}} + \bar{v}\frac{\partial \bar{s}}{\partial \bar{y}} = D\frac{\partial^2 \bar{s}}{\partial \bar{y}^2} \tag{9.4.7}$$

At the interface in the new moving coordinate system, fluid appears to be generated at the ice surface by the melting process with a velocity $V_0(x)$ normal to and outward from the surface. From conservation of mass and thermal energy at the interface, and including the effect of density change, V_0 is related to the temperature gradient at the surface as follows:

$$V_0 = V_i\left(\frac{\rho_i}{\rho_r(1 - 10^{-3}s_0)}\right) = \frac{k}{\rho_r(1 - 10^{-3}s_0)h_{il}}\left(\frac{\partial \bar{t}}{\partial \bar{y}}\right)_{\bar{y}=0} \tag{9.4.8}$$

where ρ_r is the fluid density at some reference condition and h_{il} is the latent heat of fusion for ice. After the transformation, the boundary conditions for Eqs. (9.4.4)–(9.4.7) become

$$\text{at } \bar{y} = 0: \qquad \bar{u} = 0 \qquad \bar{v} = V_0 \qquad \bar{t} = t_0 \qquad \bar{s} = s_0 \tag{9.4.9}$$

$$\text{as } \bar{y} \to \infty: \qquad \bar{u} = 0 \qquad \bar{t} = t_\infty \qquad \bar{s} = s_\infty \tag{9.4.10}$$

Conservation of salt and water as separate species at the interface also requires

$$\frac{ks_0}{h_{il}}\left(\frac{\partial \bar{t}}{\partial \bar{y}}\right)_{\bar{y}=0} - \frac{1}{1000}(1000 - s_0)D\rho_r\left(\frac{\partial \bar{s}}{\partial \bar{y}}\right)_{\bar{y}=0} = 0 \tag{9.4.11}$$

This equation relates the inward diffusion of salinity, the second term, to the amount of salinity required to raise the pure water melt to s_0. The first term represents the incoming heat flux, which determines the rate of production of the melt. As will be seen later, this condition is used to determine the unknown interface conditions t_0 and s_0.

Following Gebhart and Mollendorf (1978), a similarity variable $\eta(\bar{x}, \bar{y})$, stream functions $\psi(\bar{x}, \bar{y})$ and $f(\eta)$, and temperature and salinity functions are defined as

$$\eta = \bar{y}b(\bar{x}) \qquad \psi = \nu c(\bar{x})f(\eta) \tag{9.4.12}$$

$$\phi = \frac{\bar{t} - t_\infty}{t_0 - t_\infty} \qquad S = \frac{\bar{s} - s_\infty}{s_0 - s_\infty} \tag{9.4.13}$$

$$c(\bar{x}) = 4(\tfrac{1}{4}\mathrm{Gr}_{\bar{x}})^{1/4} = G \qquad b = \frac{G}{4\bar{x}} \tag{9.4.14}$$

Here t_0, t_∞, s_0, and s_∞ are taken as constant with t_0 and s_0 initially unknown, and $\mathrm{Gr}_{\bar{x}}$ is the Grashof number in the new coordinate system. Using the $n = 2$ form of Eq. (9.1.1), the density difference term in Eq. (9.4.5) becomes

$$\frac{\rho_\infty - \rho}{\rho_m(s_\infty, p)\alpha(s_\infty, p)|t_0 - t_\infty|^q} = (1 + AS)(1 + BS)|\phi - R - QS|^q - |R|^q - PS \qquad (9.4.15)$$

where

$$R = \frac{t_m(s_\infty, p) - t_\infty}{t_0 - t_\infty} \qquad (9.4.16)$$

$$A = \frac{g_1(p)(s_0 - s_\infty)\rho_m(0, 1)}{\rho_m(s_\infty, p)} \qquad (9.4.17)$$

$$B = \frac{g_2(p)(s_0 - s_\infty)\alpha(0, 1)}{\alpha(s_\infty, p)} \qquad (9.4.18)$$

$$Q = \frac{g_3(p)(s_0 - s_\infty)t_m(0, 1)}{t_0 - t_\infty} \qquad (9.4.19)$$

$$P = \frac{g_1(p)(s_0 - s_\infty)\rho_m(0, 1)}{\alpha(s_\infty, p)\rho_m(s_\infty, p)|t_0 - t_\infty|^q} \qquad (9.4.20)$$

For convenience, ρ_r is taken equal to $\rho_m(s, p)$. With t_0 and s_0 constant, similarity is obtained and the governing equations become

$$f''' + 3ff'' - 2f'^2 + W = 0 \qquad (9.4.21)$$

$$\phi'' + 3\mathrm{Pr}f\phi' = 0 \qquad (9.4.22)$$

$$S'' + 3\mathrm{Sc}fS' = 0 \qquad (9.4.23)$$

This formulation permits some flexibility in the definition of W and $\mathrm{Gr}_{\bar{x}}$. To conform to the nature of the numerical scheme used by Carey and Gebhart (1982a), two different definitions of $\mathrm{Gr}_{\bar{x}}$ and W were used. At low temperatures, below about 5°C, the flow is upward and dominated by the strong upward saline buoyancy force component. Therefore, $\bar{x}$ is taken positive in the upward direction and W and $\mathrm{Gr}_{\bar{x}}$ are defined as

$$W = -\frac{1}{P}[(1 + AS)(1 + BS)|\phi - R - QS|^q - |R|^q - PS] \qquad (9.4.24)$$

$$\mathrm{Gr}_{\bar{x}} = (-P)\frac{g\alpha(s_\infty, p)\bar{x}^3|t_0 - t_\infty|^q}{\nu^2} = \frac{\rho_m(0, 1)}{\rho_m(s_\infty, p)}\frac{gg_1(p)\bar{x}^3(s_\infty - s_0)}{\nu^2} \qquad (9.4.25)$$

At higher temperatures, above about 5°C, the flow is mostly downward and is dominated by the downward thermal buoyancy force component. Therefore, $\bar{x}$ is then taken to be positive downward and $\mathrm{Gr}_{\bar{x}}$ and W are defined as

$$W = PS + |R|^q - (1 + AS)(1 + BS)|\phi - R - QS|^q \qquad (9.4.26)$$

$$\mathrm{Gr}_{\bar{x}} = \frac{g\alpha(s_\infty, p)\bar{x}^3|t_0 - t_\infty|^q}{\nu^2} \qquad (9.4.27)$$

At intermediate temperatures, near about 5°C, it is not clear a priori which net flow direction results and which formulation, Eqs. (9.4.24) and (9.4.25) or Eqs. (9.4.26) and (9.4.27), applies. The range of ambient conditions that correspond to each formulation were determined by trial and error. The additional parameters that arise in W from saline diffusion effects are A, B, Q, and P. The magnitudes of A and B are usually small compared with one. They represent the effects of the local salinity level on the values of ρ_m and α. PS is the principal component of the contribution of the salinity gradient to the buoyancy force. From Eq. (9.4.20) it is seen that this is a very large term for $s_0 - s_\infty$ large and $t_0 - t_\infty$ small. This effect on $\rho_\infty - \rho$ is large, compared with that of temperature. However, since the Schmidt number Sc is large, the salinity diffusion layer is very thin. The other salinity contribution, Q, is the effect of local salinity on t_m. Although the term QS may be larger than ϕ for $s_0 - s_\infty$ large, its range of effect in η is small.

With the similarity transformation, the boundary conditions, Eqs. (9.4.9) and (9.4.10), for both transformations presented above become

$$f(0) = \frac{-\phi'(0)c_p(t_0 - t_\infty)}{3h_{il}\mathrm{Pr}(1 - 10^{-3}s_0)} \qquad f'(0) = 0 \qquad \phi(0) = 1 \qquad S(0) = 1 \qquad (9.4.28)$$

$$f'(\infty) = \phi(\infty) = S(\infty) = 0 \qquad (9.4.29)$$

The expression for conservation of species at the interface becomes

$$\frac{S'(0)}{\phi'(0)} - \frac{\mathrm{Sc}}{\mathrm{Pr}}\,\frac{c_p(t_0 - t_\infty)}{h_{il}}\,\frac{s_0}{(1 - 10^{-3}s_0)(s_0 - s_\infty)} = 0 \qquad (9.4.30)$$

In terms of similarity variables, the basic transport quantities as seen from a coordinate system fixed relative to the far ambient medium are

$$u = \nu cbf'(\eta) = \nu\frac{G^2}{4x}f'(\eta) \qquad (9.4.31)$$

$$-v = \frac{\nu G}{4x}3f - 3\frac{\rho_r}{\rho_i}(1 - 10^{-3}s_0)f(0) - \eta f' \qquad (9.4.32)$$

$$\tau(x) = \frac{\mu\nu G^3}{16x^2}f''(0) \qquad (9.4.33)$$

$$\mathrm{Nu}_x = \frac{h_x x}{k} = \frac{[-\phi'(0)]}{\sqrt{2}}\,\mathrm{Gr}_x^{1/4} \qquad (9.4.34)$$

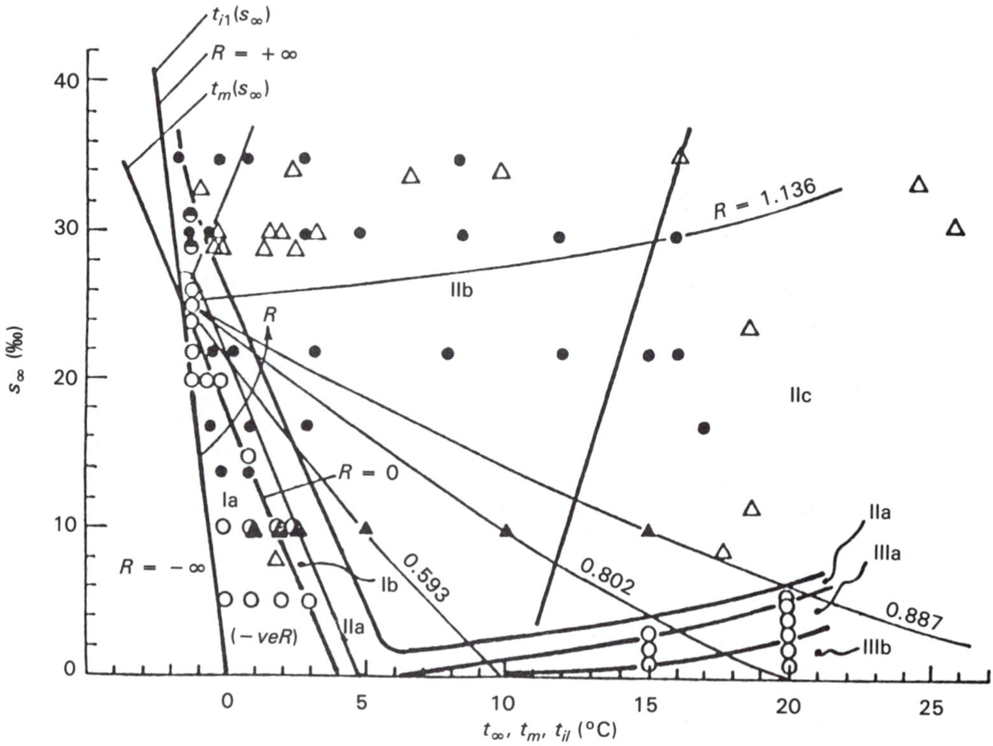

Figure 9.4.1 Flow regimes corresponding to ambient water salinity and temperature. Also shown are several lines of constant R values. Experimental points: (△) Josberger and Martin (1981); (▲) Carey and Gebhart (1982b); (●) Sammakia and Gebhart (1983). Conditions for the calculations: (○) Carey and Gebhart (1982a) (full equations); (⊖) Carey and Gebhart (1982a) (asymptotic solutions).

$$\mathrm{Sh}_x = \frac{h_{D,x}x}{\rho_r D} = \frac{[-S'(0)]}{\sqrt{2}}\,\mathrm{Gr}_x^{1/4} \tag{9.4.35}$$

where h_x and $h_{D,x}$ are the local heat and mass transfer coefficients, respectively, $\tau(x)$ is the local surface shear stress, Nu_x is the local Nusselt number, and Sh_x is the local Sherwood number. Here, η is evaluated at $\bar{y}$, $\eta = \bar{y}b(x)$, where $\bar{y} = y + V_i\tau$ is the instantaneous horizontal distance between the location of interest in the flow and the ice surface. Another quantity of interest is the integral of the buoyancy force

$$I_W = \int_0^\infty W(\eta)\, d\eta$$

where W is given by Eq. (9.4.24) or (9.4.26).

Figure 9.4.2 Some flow patterns corresponding to the regions of the (s_∞, t_∞) plane shown in Fig. 9.4.1. *(Reprinted with permission from* Int. J. Heat Mass Transfer, *vol. 26, p. 1439, B. Sammakia and B. Gebhart. Copyright © 1983, Pergamon Journals Ltd.)*

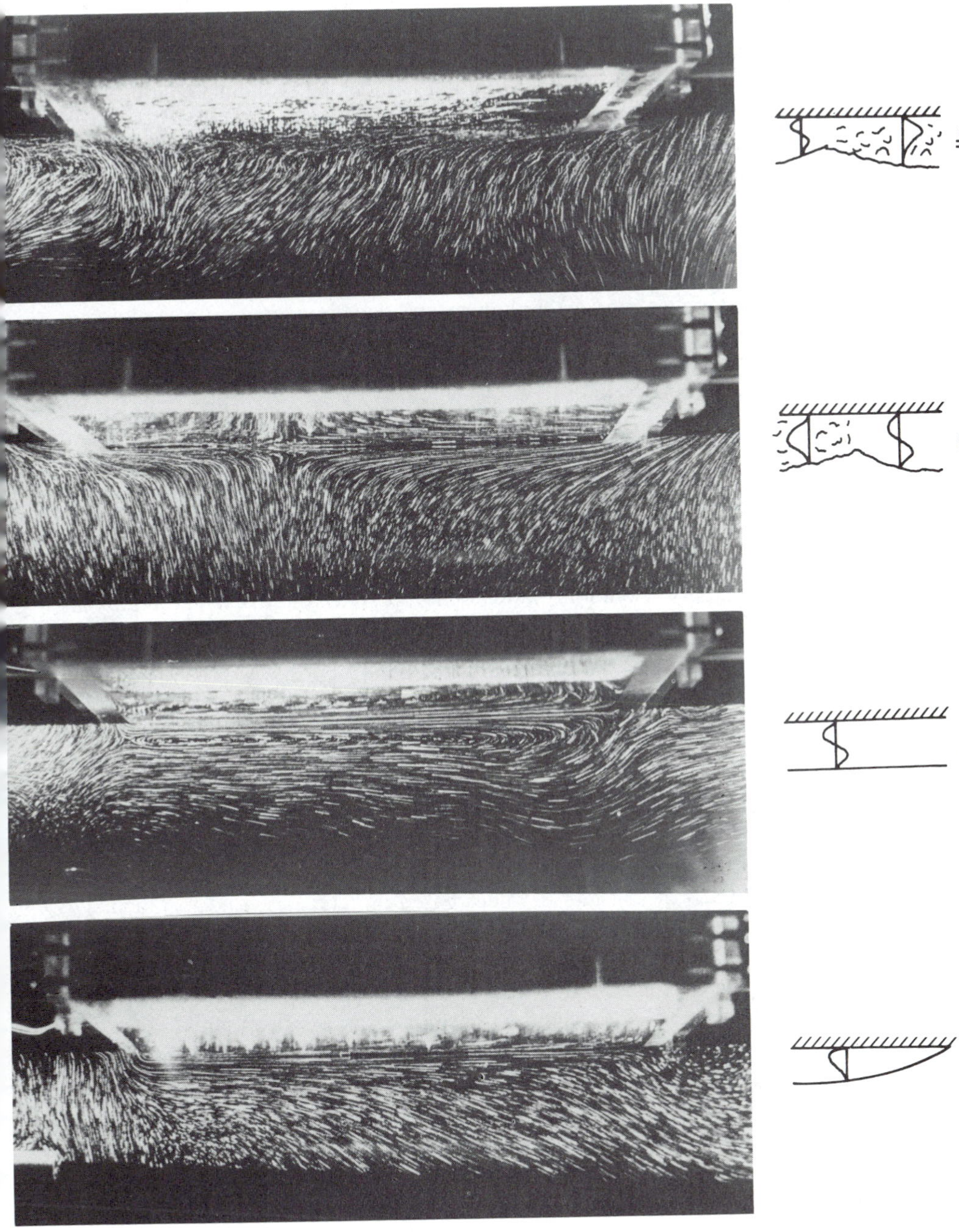

IIc

IIb

IIa

Ia

Note that from the condition given in Eq. (9.4.8) the x variation of V_0 is the same as that of the surface heat flux. For the isothermal surface condition assumed here, the heat flux and therefore V_0 are proportional to $x^{-1/4}$. Fortunately, this is the exact x dependence required to preserve similarity in the boundary condition for $f(0)$. Hence, the effect of interface motion is included without violating similarity.

If local thermodynamic equilibrium is assumed, t_0 will equal the equilibrium ice-melting temperature at the interface salinity s_0. These are related by the equilibrium melting temperature correlation determined by Fujino et al. (1974) as follows:

$$t_{il}(s, p) = -0.02831 - 0.0499s - 0.000112s^2 - 0.00759p \qquad (9.4.36)$$

This equation is in good agreement with measured data down to a salinity of about 4‰. However, for $p = 1$, it is in error very close to $s = 0$. Therefore, for $p = 1$, Eq. (9.4.36) was used down to $s_0 = 2$‰ and the following simple linear relation was used below 2‰:

$$t_{il}(s, 1) = -0.06807s \qquad (p = 1, s < 2) \qquad (9.4.37)$$

This form agrees with Eq. (9.3.36) at $s = 2$ and goes linearly to zero as s goes to zero. Since $t_0 = t_{il}(s_0)$, this, together with Eqs. (9.4.36) and (9.4.37), provides a relation between s_0 and the transport parameters $\phi'(0)$ and $s'(0)$.

The above equations were solved successfully in two ranges of ambient water temperature and salinity. Solutions were found possible for low temperatures at salinities up to 26‰. Solutions at temperatures up to 20°C were also obtained for low salinities, up to $s_\infty = 5.082$‰. Figure 9.4.1 is a schematic diagram of the whole s_∞, t_∞ plane of usual interest. The open circles indicate the conditions for which solutions to the full equations were found by Carey and Gebhart (1982a).

Figure 9.4.1 is divided into seven regions according to the kinds of flow regimes seen thus far over this range of s_∞ and t_∞. The regions were determined from the calculations of Carey and Gebhart (1982a) and the measurements and visualizations by Josberger and Martin (1981), Carey and Gebhart (1981), Sammakia and Gebhart (1983), and Johnson and Mollendorf (1984). Also shown in Fig. 9.4.1 are contours of constant R values based on $t_m(s_\infty, t_\infty)$, $t_0(s_\infty, t_\infty)$, and t_∞.

For freshwater ice melting in saline water, the saline buoyancy force always acts upward. In region Ia, where $t_\infty < t_m(s_\infty)$, a negative R, the thermal buoyancy force also acts upward. The resulting flow is upward everywhere, as seen in Fig. 9.4.2*a*. Tables 9.4.1 and 9.4.2 show the transport parameters computed in region Ia. In Fig. 9.4.3 the computed velocity profiles are given for $s_\infty = 10$‰ and $t_\infty = 1$, 2, and 2.5°C. This is in region Ia. Also shown in Fig. 9.4.3 are measurements from Carey and Gebhart (1982b).

Agreement is very good near the ice surface. For $t_\infty = 2$ and 2.5°C, however, the actual outside flow is seen to be downward. This is a consequence of the limited vertical extent of the ice surface and the downstream wake. The model assumes a semi-infinite surface, and any subsequent wake effects are not anticipated. For $t_\infty = 1$°C the flow is upward everywhere, and agreement is good across the extent of the boundary layer. Computed velocity, temperature, and salinity profiles for $t_\infty = -1$°C, $s = 20$, 22, 24, and 26‰, are shown in Fig. 9.4.4. How-

Table 9.4.1 Interface conditions and transport parameters computed with the full analysis for t_∞ and s_∞ in regions Ia and IIa of Fig. 9.4.1 and results computed using the asymptotic analysis (†)

t_∞	s_∞	t_0	s_0	$-A \times 100$	$B \times 100$	$-Q$	$-P$	R	Pr	Sc
0	5	−0.2677	4.598	0.03218821	0.1156753	0.3182985	426.5571	−11.08784	13.25	2938
1	5	−0.2095	3.452	0.1240212	0.4456976	0.2714653	94.36789	−1.627518	12.99	2804
2	5	−0.1679	2.629	0.1899807	0.6827370	0.2320075	47.84571	−0.4467510	12.74	2680
3	5	−0.1402	2.081	0.2339401	0.8407147	0.1972298	29.19547	0.01003031	12.50	2565
0	10	−0.4795	8.719	0.1022902	0.3744675	0.5670764	455.9298	−3.978765	13.31	2965
1	10	−0.3818	6.827	0.2532371	0.9270591	0.4871300	151.9019	−0.6568794	13.04	2825
2	10	−0.3134	5.492	0.3598003	1.317169	0.4134144	81.29250	0.03990745	12.79	2697
2.5	10	−0.2936	5.105	0.3907336	1.430410	0.3717828	61.75254	0.2120314	12.67	2637
0	15	−0.6786	12.53	0.1966207	0.7333833	0.7732722	460.5558	−1.248012	13.37	2992
1	15	−0.5566	10.20	0.3814605	1.422824	0.6539711	185.2861	0.09838049	13.09	2847
−1	20	−1.062	19.68	0.02506418	0.09527208	1.091950	5635.897	−12.78127	13.71	3183
−0.5	20	−0.9602	17.81	0.1733311	0.6588531	1.009118	860.1176	−0.6215387	13.56	3098
0	20	−0.8746	16.22	0.2994854	1.138381	0.9174134	440.1877	0.2446322	13.42	3018
−1	22	−1.145	21.21	0.06226374	0.2384827	1.153917	2780.294	−2.497995	13.73	3195
−1	24	−1.227	22.72	0.1010566	0.3900419	1.194283	1929.819	0.2753471	13.75	3207
−1	25	−1.269	23.48	0.1202221	0.4657989	1.201872	1674.628	1.021260	13.76	3213
−1	26	−1.311	24.24	0.1386698	0.5393464	1.199113	1469.709	1.564184	13.77	3219
−1†	20	−1.060	19.65	0.02733313	0.1038966	1.221710	6451.901	−13.11305	13.71	3183
−1†	29	−1.422	26.23	0.2175571	0.8560458	1.391099	1307.665	2.662165	13.80	3235
−1†	30	−1.462	26.95	0.2395699	0.9463308	1.400068	1216.257	2.890372	13.81	3241
−1†	31	−1.503	27.67	0.2610454	1.035185	1.403575	1133.509	3.079236	13.92	3247

Source: Reprinted with permission from V. P. Carey and B. Gebhart, *J. Fluid. Mech.*, vol. 117, p. 379. Copyright © 1982, Cambridge University Press.

Table 9.4.2 Interface conditions and transport quantities computed with the full analysis for t_∞ and s_∞ in regions Ia and IIa of Fig. 9.4.1 and results computed with the asymptotic analysis (†)

t_∞	s_∞	t_0	s_0	$f'(0)$	$-\phi'(0)$	$-S'(0)$	$-f(0) \times 10^3$	$f(\infty)$	I_w	$\dot{M}$[a] (cm/day)
0	5	−0.2677	4.598	0.13397	0.64328	5.4914	0.054434	0.10700	0.14149	0.463
1	5	−0.2095	3.452	0.14018	0.63407	4.6378	0.24714	0.10522	0.14741	2.91
2	5	−0.1679	2.629	0.14510	0.61577	3.9101	0.43848	0.10061	0.15155	5.65
3	5	−0.1402	2.081	0.14760	0.57815	3.3329	0.60763	0.088942	0.15231	8.17
0	10	−0.4795	8.719	0.12322	0.56080	5.1044	0.084349	0.087914	0.12748	0.957
1	10	−0.3818	6.827	0.12825	0.54612	4.4047	0.24151	0.084045	0.13206	3.40
2	10	−0.3134	5.492	0.13177	0.51427	3.8251	0.38802	0.073850	0.13455	5.88
2.5	10	−0.2936	5.105	0.13205	0.47708	3.6191	0.43871	0.059568	0.13380	6.73
0	15	−0.6786	12.53	0.11879	0.50879	4.8870	0.10749	0.74754	0.12151	1.45
1	15	−0.5566	10.20	0.12251	0.47777	4.2922	0.23636	0.064916	0.12444	3.71
−1	20	−1.062	19.68	0.11180	0.48303	5.3435	0.008999	0.068238	0.11389	0.0712
−0.5	20	−0.9602	17.81	0.11385	0.47183	5.0273	0.066478	0.064942	0.11570	0.877
0	20	−0.8746	16.22	0.11545	0.45081	4.7339	0.12193	0.058097	0.11688	1.83
−1	22	−1.145	21.21	0.11103	0.46606	5.2635	0.020388	0.063598	0.11276	0.221
−1	24	−1.227	22.72	0.11011	0.44335	5.1863	0.030377	0.057001	0.11143	0.361
−1	25	−1.269	23.48	0.10956	0.42838	5.1484	0.034680	0.052453	0.11065	0.427
−1	26	−1.311	24.24	0.10902	0.41091	5.1135	0.038436	0.046631	0.10986	0.488
−1†	20	−1.060	19.65	0.1123	0.5424	5.356	0.00985	—	0.1123	0.0823
−1†	29	−1.422	26.23	0.1088	0.4325	5.025	0.05461	—	0.1088	0.767
−1†	30	−1.462	26.95	0.1083	0.4209	4.995	0.05813	—	0.1083	0.837
−1†	31	−1.503	27.67	0.1079	0.4083	4.963	0.06126	—	0.1079	0.897

[a] Average melt rate computed for a surface 1 m long.

Source: Reprinted with permission from V. P. Carey and B. Gebhart, *J. Fluid. Mech.*, vol. 117, p. 379. Copyright © 1982, Cambridge University Press.

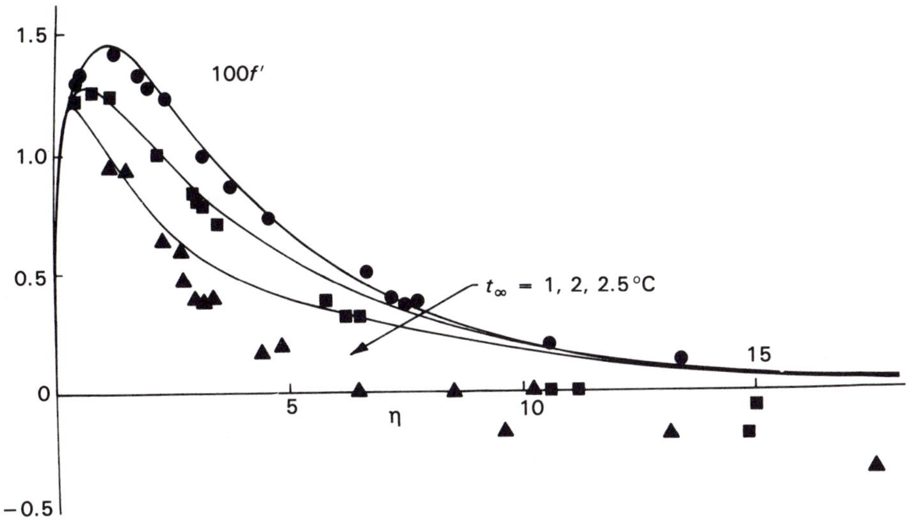

Figure 9.4.3 Measured velocity profiles for $s_\infty = 10‰$ and t_∞ = (●) 1°C, (■) 2°C, and (▲) 2.5°C. Also shown are the computed profiles from Carey and Gebhart (1982a). *(Reprinted with permission from V. P. Carey and B. Gebhart,* J. Fluid. Mech., *vol. 117, p. 403. Copyright © 1982, Cambridge University Press.)*

ever, as t_∞ is increased above $t_m(s_\infty)$, into region Ib, the thermal buoyancy force may reverse direction to downward. However, for a small increase in t_∞ in region Ib, this reversal is not enough to cause local flow reversal and pure upflow persists.

At the other extreme, high t_∞ and low s_∞, in region IIIb, the downward thermal buoyancy force overcomes the upward saline buoyancy force, and the resulting flow is purely down. In region IIIa, at a slightly higher salinity, the low-salinity melt causes a larger upward buoyancy force near the wall. However, the flow remains down everywhere. Tables 9.4.3 and 9.4.4 show the computed transport parameters in regions IIIa and IIIb. Figure 9.4.5 shows the resulting velocity, temperature, and salinity profiles for various s_∞, t_∞ conditions in these regions.

In regions II (a, b, and c) several different kinds of bidirectional flows arise. In IIa an upflow near the surface is driven by the saline buoyancy force. A thermally driven downflow is found on the outside (Fig. 9.4.2*b*). The flow is laminar and steady everywhere. The saline-driven layer is much thinner than the thermal layer, about 1:10, because of the high Schmidt number of about 10^3. This is a Lewis number Le of about 10^2.

For intermediate ambient water temperatures, region IIb, the flow is again different (Fig. 9.4.2*c*). It is turbulent adjacent to the upper part of the ice surface. On the other hand, the outer entraining flow splits at approximately three-fourths of the height of the slab into upward and downward flows. The flow adjacent to the lower portion of the ice slab is bidirectional and basically laminar. This observation is confirmed by the surface smoothness of the lower portion of the ice

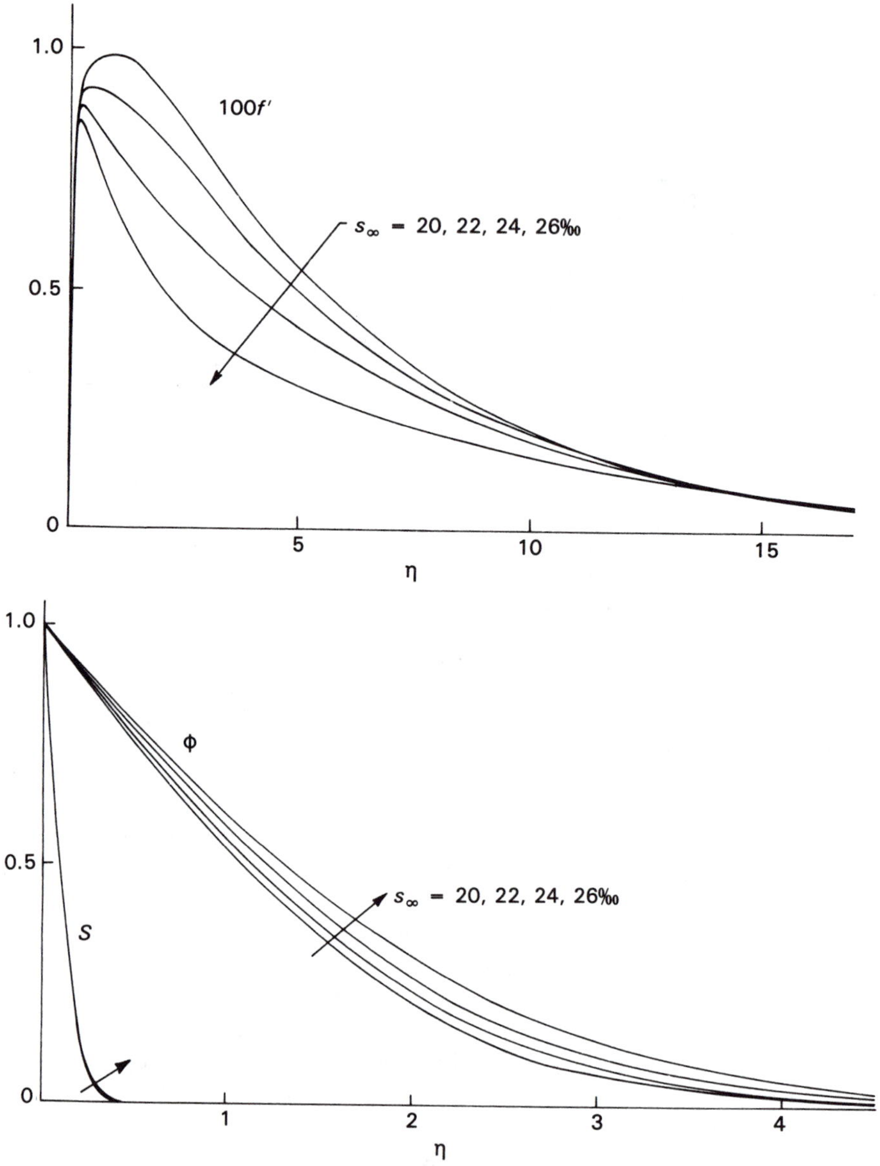

Figure 9.4.4 Velocity, temperature, and salinity profiles for $t_\infty = -1°\text{C}$, $s_\infty = 20, 22, 24, 26‰$. *(Reprinted with permission from V. P. Carey and B. Gebhart,* J. Fluid. Mech., *vol. 117, p. 403. Copyright © 1982, Cambridge University Press.)*

surface, which is seen even after considerable melting has occurred. The upper portion, however, has developed vertical longitudinal depressions, indicating the presence of streamwise vortices.

Near the ice surface, the saline buoyancy force still dominates the flow and

Table 9.4.3 Interface conditions and transport parameters for t_∞ and s_∞ in regions IIIa and IIIb of Fig. 9.4.1

t_∞	s_∞	t_0	s_0	$-A \times 100$	$B \times 100$	$-Q$	$-P$	R	Pr	Sc
10	1	−0.000410	0.00603	0.07991228	0.2830014	0.02108769	1.098114	0.6182586	10.99	1925
10	1.06	−0.000164	0.00241	0.08502277	0.3011657	0.02243791	1.168593	0.6195468	10.99	1925
15	1	−0.000161	0.00237	0.08020629	0.2840426	0.01411061	0.5112188	0.7455146	10.10	1607
15	2	−0.000203	0.00298	0.1604248	0.5702092	0.02824601	1.025430	0.7596566	10.10	1607
15	2.835	−0.000054	0.00079	0.2275255	0.8111861	0.04008769	0.457838	0.7714766	10.10	1607
20	1	−0.000036	0.00052	0.08035493	0.2845690	0.01060266	0.2969493	0.8091405	9.312	1361
20	2	−0.000057	0.00084	0.1605972	0.5708218	0.02120749	0.5951761	0.8197479	9.315	1361
20	3	−0.000064	0.00094	0.2407276	0.8587741	0.03181455	0.8946947	0.8303558	9.319	1361
20	4	−0.000054	0.00079	0.3207501	1.148454	0.04242436	1.195532	0.8409644	9.322	1361
20	5	−0.000018	0.00022	0.4006748	1.439912	0.05303829	1.497742	0.8515741	9.326	1361
20	5.082	−0.000011	0.00016	0.4072277	1.463903	0.05390928	1.522598	0.8524443	9.326	1361

Source: Reprinted with permission from V. P. Carey and B. Gebhart, J. *Fluid. Mech.*, vol. 117, p. 379. Copyright © 1982, Cambridge University Press.

Table 9.4.4 Interface conditions and transport parameters for t_∞ and s_∞ in regions IIIa and IIIb of Fig. 9.4.1 and melt rates M for a 1-m-long surface

t_∞	s_∞	t_0	s_0	$f'(0)$	$-\phi'(0)$	$-S'(0) \times 10^3$	$-f(0) \times 100$	$f(\infty)$	$-I_w$	M (cm/day)
10	1	−0.000410	0.00603	0.036213	0.92529	123.71	0.35289	0.23445	0.10249	32.9
10	1.06	−0.000164	0.00241	−0.031201	0.86089	43.211	0.32832	0.23191	0.036535	30.7
15	1	−0.000161	0.00237	0.26932	1.0303	73.393	0.64068	0.25879	0.73021	70.1
15	2	−0.000203	0.00298	0.16470	0.98841	44.218	0.61403	0.25742	0.26001	67.2
15	2.835	−0.000034	0.00079	−0.0046114	0.87711	7.3332	0.54488	0.25264	0.062605	59.6
20	1	−0.000036	0.00052	0.35328	1.0213	19.552	0.91747	0.27254	1.6020	112.0
20	2	−0.000057	0.00084	0.29658	1.0027	15.409	0.90048	0.27196	0.70210	109.0
20	3	−0.000064	0.00094	0.23272	0.97906	11.277	0.87786	0.27108	0.39393	107.0
20	4	−0.000054	0.00079	0.15581	0.94507	6.8503	0.84652	0.26965	0.22853	103.0
20	5	−0.000018	0.00022	0.031742	0.86757	1.7147	0.77614	0.26590	0.095956	94.9
20	5.082	−0.000011	0.00016	−0.00003166	0.83874	0.91413	0.75035	0.26442	0.072229	91.6

Source: Reprinted with permission from V. P. Carey and B. Gebhart, *J. Fluid. Mech.*, vol. 117, p. 379. Copyright © 1982, Cambridge University Press.

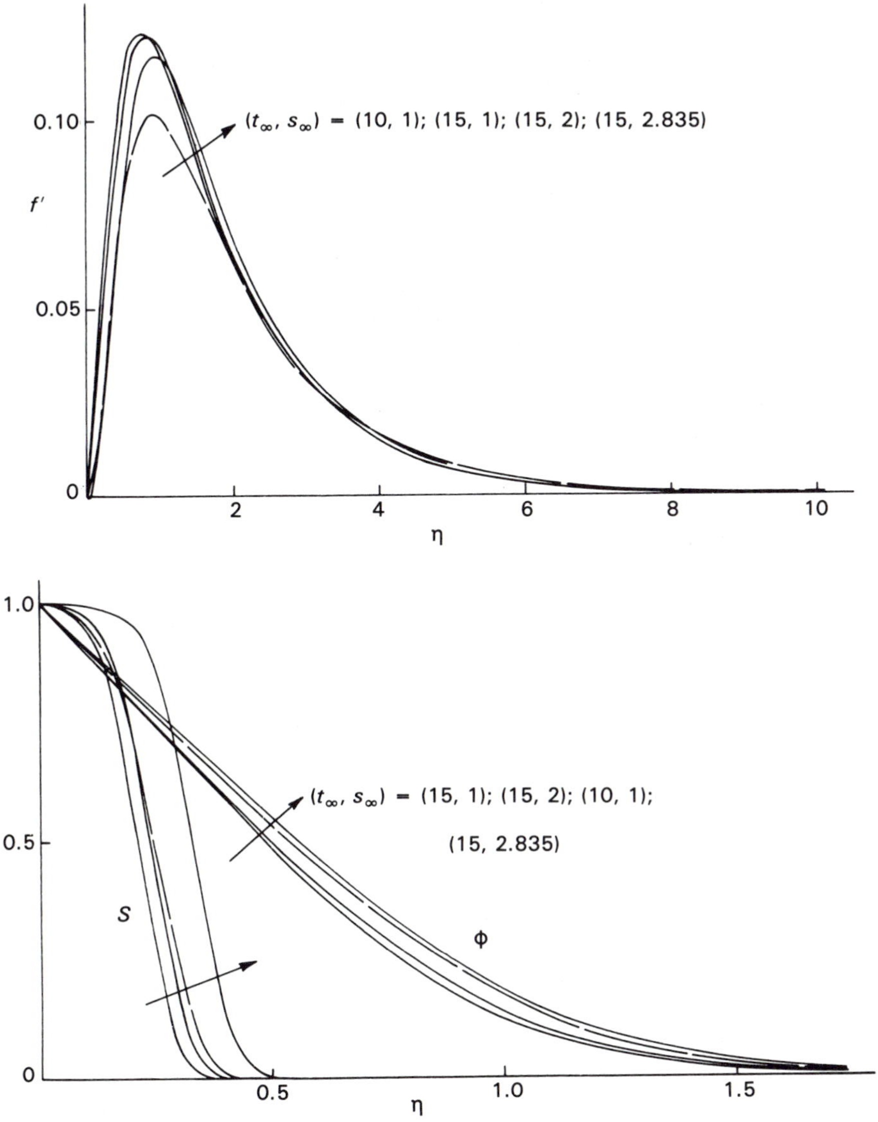

Figure 9.4.5 Velocity, temperature, and salinity profiles for $(t_\infty, s_\infty) = (10, 1)$, $(15, 1)$, $(15, 2)$, $(15, 2.84)$. *(Reprinted with permission from V. P. Carey and B. Gebhart,* J. Fluid. Mech., *vol. 117, p. 403. Copyright © 1982, Cambridge University Press.)*

causes the upward motion. At approximately midheight this upward-moving layer becomes unstable and undergoes a transition to turbulence. The downward thermal buoyancy force is overcome in the saline-driven inside region. However, the growing thermal effect eventually dominates and causes downward motion over the lower part of the slab.

As t_∞ is further increased, still within region IIb, the thermal buoyancy force becomes relatively stronger, as does the downflow. The inside upflow in the lower region disappears. The flow becomes split into a completely upflow region at the top and a downflow region at the bottom. With further increases of t_∞, the downflow becomes turbulent and covers a larger portion of the surface. Eventually most of the flow is downward and turbulent, in region IIc, as seen in Fig. 9.4.2*d*.

Figure 9.4.6 is a plot of the measured interface temperature $t_0 = t_{il}$ from various experiments at different ambient temperatures and salinities. Also plotted are the inferred interface salinities s_0 calculated from Eq. (9.4.36). It is seen that the interface temperature increases monotonically with increasing ambient water temperature at all given ambient salinity levels. Thus, s_0 decreases with increasing t_∞. This reflects the increased melt production at higher t_∞.

Figure 9.4.7 collects the measured heat transfer rates from Sammakia and Gebhart (1983), Josberger and Martin (1981), and Johnson and Mollendorf (1984). Increasing t_∞ is observed to increase Nu for the range of salinities and temperatures shown.

9.4.2 Other Geometries

Because of the frequent occurrence in nature of ice melting and freezing in saline water, there have been several other investigations of such processes. One of the

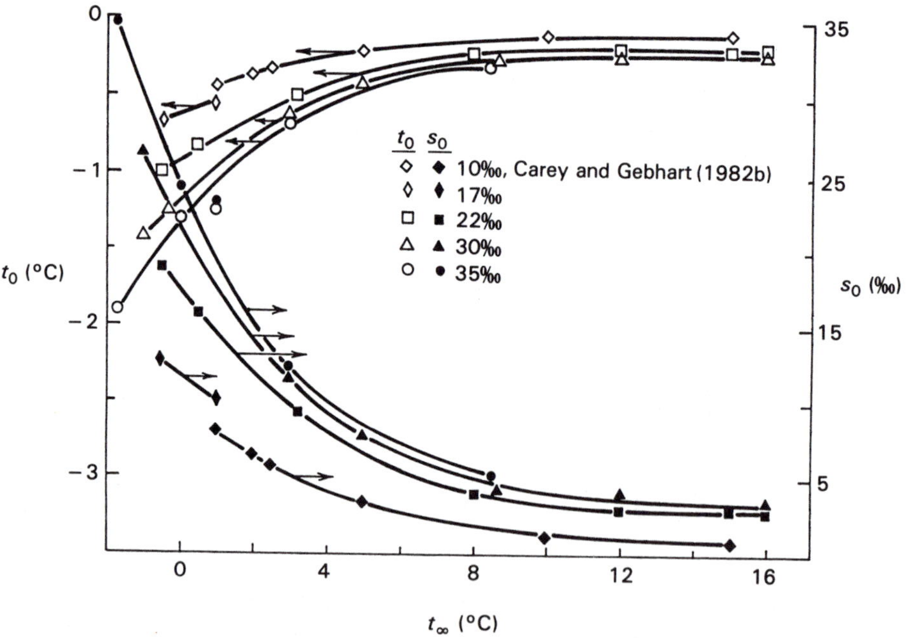

Figure 9.4.6 Measured interface temperatures and inferred salinities for different ambient water salinities and temperatures. *(Reprinted with permission from* Int. J. Heat Mass Transfer, *vol. 26, p. 1439, B. Sammakia and B. Gebhart. Copyright © 1983, Pergamon Journals Ltd.)*

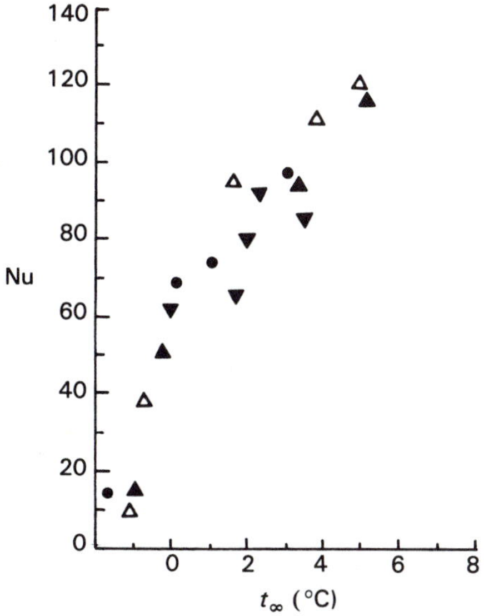

Figure 9.4.7 Measured average heat transfer as a function of s_∞ and t_∞; salinities are: (●) 35 and (▲) 30‰ from Sammakia and Gebhart (1983); (△) 35‰ from Johnson and Mollendorf (1984); and (▼) 35‰ from Josberger and Martin (1981). *(Reprinted with permission from* Int. J. Heat Mass Transfer, *vol. 26, p. 1439, B. Sammakia and B. Gebhart. Copyright © 1983, Pergamon Journals Ltd.)*

configurations most common in nature is a horizontal ice surface freezing or melting while floating on water of 35‰ salinity. Such transport is very complicated and variable because of the changing nature of the buoyancy force distribution, as seen in the preceding section. In addition, for a horizontal water-ice interface, any flow parallel to the surface is driven indirectly, through the pressure field, since the buoyancy force components both act only normal to the surface.

Although such flows will be very difficult to analyze, experimental measurements and flow visualizations have revealed much about the nature of the flow

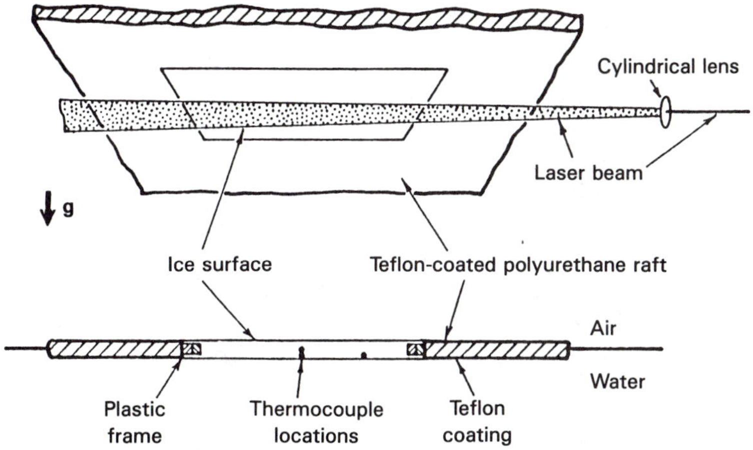

Figure 9.4.8 Schematic diagram showing the relative positions of the ice surface, laser beam, and thermocouples. *(Reprinted with permission from B. Gebhart et al.,* J. Geophys. Res., *vol. 88, p. 2935. Copyright © 1983, American Geophysical Union.)*

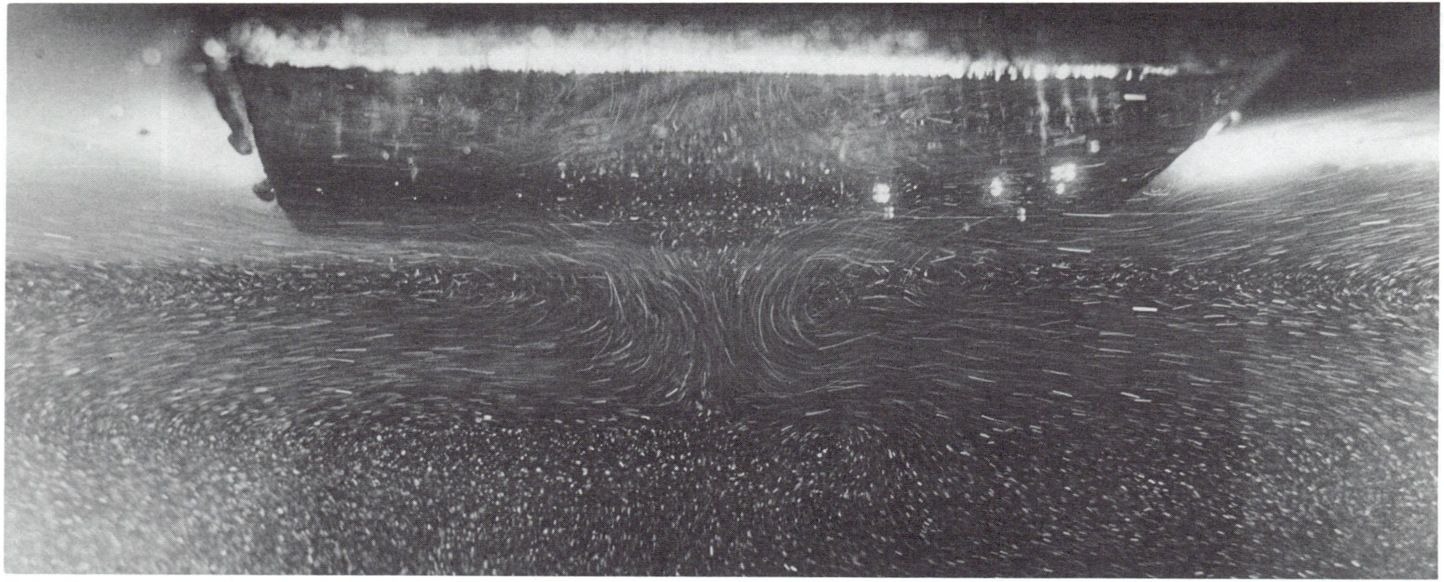

Figure 9.4.9 Ten-second time exposure of the flow field adjacent to a horizontal ice surface melting in water at $(s_\infty, t_\infty) = (35‰, 3°C)$. *(Reprinted with permission from B. Gebhart et al.,* J. Geophys. Res., *vol. 88, p. 2935. Copyright © 1983, American Geophysical Union.)*

and transport. Martin and Kauffman (1976) studied the flow generated under a melting ice surface floating on water at 37.6‰ salinity and at about 0°C. The ice surface was 0.45 × 0.45 × 0.1 m thick and fitted into a tank of the same cross-sectional area and 0.9 m deep. An array of thermocouples in the water revealed its temperature history. By removing fluid samples, salinity profiles were also obtained beneath the ice.

From the temperature and salinity profiles, it was surmised that convection occurs in three different regions. Just beneath the ice a narrow boundary region of about 6 cm thickness exists. Across this layer the salinity increases linearly from 18‰ at the surface to 37.6‰. Below this an unstable convective boundary layer was surmised to occur, due to double diffusion. Finally, deep below the ice surface pure thermal diffusion was inferred.

Gebhart et al. (1983) visualized the flow under a rectangular ice slab about 20 × 27 cm in cross section and 2.5 cm thick. The ice surface was mounted flush into a Teflon-coated polyurethane raft as shown in Fig. 9.4.8. The raft was floated on water at 35‰ salinity, and several ambient water temperatures ranging from −1.75 to 3°C were investigated.

Time exposure photographs revealed the resulting flow regimes. The principal flow regime seen was a plume downflow near the center of the ice surface. This downflow sinks into the ambient medium for only a short distance. It then splits and spreads out into a horizontal layer. The entrainment that produces this downflowing water arises from side induced inflows, just under the ice surface. Figure 9.4.9 shows the flow that arose at $t_\infty = 3°C$.

Foster (1969) visualized downward freezing in saline water at salinites ranging from 20 to 35‰. The tank was 25 × 25 × 25 cm, and freezing was initiated by a cooled chromium-plated copper surface across the top of the tank. Freezing rates up to 1.5×10^{-4} g/cm^2 s (about 8 cm/day) were achieved. A Schlieren system showed that convection was in the form of long vertical filaments with a horizontal spacing of 0.2–0.3 cm. The experiments were meant to simulate oceanic ice formation in the absence of waves. In an earlier study Foster (1968) had predicted that a secondary circulation with an average cell size of about 50 cm would develop. In the small tank in this experiment, however, such large circulations were not seen. One feature that arose was liquid supercooling before freezing actually began.

Tankin and Farhadieh (1971) and Farhadieh and Tankin (1972, 1975) used a Mach-Zehnder interferometer to investigate the freezing of seawater in a very small test section, 2.5 × 1.75 in., with very shallow depths, $\frac{1}{4}$, $\frac{3}{8}$, and $\frac{1}{2}$ in. Freezing was accomplished with thermoelectric cooling modules mounted on top of the test section. Salt plumes were observed and convection currents were found to be prominent for freezing from above during the early stages of the freezing process.

REFERENCES

Bendell, M. S., and Gebhart, B. (1976). *Int. J. Heat Mass Transfer 19,* 1081.
Bigg, P. H. (1967). *Br. J. Appl. Phys. 18,* 521.

Brown, I., and Lane, J. E. (1976). *Pure Appl. Chem. 45,* 1.
Carey, V. P., and Gebhart, B. (1981). *J. Fluid Mech. 107,* 37.
Carey, V. P., and Gebhart, B. (1982a). *J. Fluid Mech. 117,* 379.
Carey, V. P., and Gebhart, B. (1982b). *J. Fluid Mech. 117,* 403.
Carey, V. P., Gebhart, B., and Mollendorf, J. C. (1980). *J. Fluid Mech. 97,* 279.
Chen, C. T., and Millero, F. J. (1976). *Deep Sea Res. 23,* 595.
Codegone, C. (1939). *Acad. Sci. Torino 75,* 167.
Cormack, D. E., Leal, L. G., and Seinfeld, J. H. (1974). *J. Fluid Mech. 65,* 231.
Doherty, B. T., and Kester, D. R. (1974). *J. Mar. Res. 32,* 285.
Dumoré, J. M., Merk, H. J., and Prins, J. A. (1953). *Nature (London) 172,* 460.
Ede, A. J. (1951). *Proc. 8th Int. Congr. Refrigeration,* London, p. 260.
Ede, A. J. (1955). *Appl. Sci. Res. 5,* 458.
El-Henawy, I., Hassard, B., Kazarinoff, N., Gebhart, B., and Mollendorf, J. C. (1982). *J. Fluid Mech. 122,* 235.
Farhadieh, R., and Tankin, R. S. (1972). *J. Geophys. Res. 77,* 1647.
Farhadieh, R., and Tankin, R. S. (1975). *J. Fluid Mech. 71,* 293.
Fine, R. A., and Millero, F. J. (1973). *J. Chem. Phys. 59,* 5529.
Foster, T. D. (1968). *J. Geophys. Res. 73,* 1933.
Foster, T. D. (1969). *J. Geophys. Res. 74,* 6967.
Fujii, T. (1974). *Adv. Heat Transfer Eng. 3.*
Fujino, K., Lewis, E. L., and Perkin, R. G. (1974). *J. Geophys. Res. 79,* 1792.
Gebhart, B. (1971). *Heat Transfer,* 2d ed., McGraw-Hill, New York.
Gebhart, B. (1973). *Adv. Heat Transfer 9,* 273.
Gebhart, B., and Mollendorf, J. C. (1977). *Deep Sea Res. 24,* 831.
Gebhart, B., and Mollendorf, J. C. (1978). *J. Fluid Mech. 89,* 673.
Gebhart, B., Bendell, M. S., and Shaukatullah, H. (1979). *Int. J. Heat Mass Transfer 22,* 137.
Gebhart, B., Carey, V. P., and Mollendorf, J. C. (1980). *Chem. Eng. Commun. 3,* 555.
Gebhart, B., Sammakia, B., and Audunson, T. (1983). *J. Geophys. Res. 88,* 2935.
Goren, S. L. (1966). *Chem. Eng. Sci. 21,* 515.
Govindarajulu, T. (1970). *Chem. Eng. Sci. 25,* 1827.
Johnson, R. S. (1978). M.S. Thesis, State University of New York, Buffalo, N.Y.
Johnson, R. S., and Mollendorf, J. C. (1984). *Int. J. Heat Mass Transfer 27,* 1928.
Josberger, E. G., and Martin, S. (1981). *J. Fluid Mech. 111,* 439.
Joshi, Y., and Gebhart, B. (1983). *Trans. ASME, J. Heat Transfer 105,* 248.
Kell, G. S. (1967). *J. Chem. Eng. Data 12,* 66.
Marschall, E. (1977). *Lett. Heat Mass Transfer 4,* 381.
Martin, S., and Kauffman, P. (1976). *J. Fluid Mech. 64,* 507.
Merk, H. J. (1953). *Appl. Sci. Res. 4,* 435.
Mollendorf, J. C., Johnson, R. S., and Gebhart, B. (1981). *J. Fluid Mech. 113,* 269.
Newell, M. E., and Schmidt, F. W. (1970). *J. Heat Transfer 92,* 159.
Oborin, L. A. (1967). *J. Eng. Phys. 13,* 429.
Perry, J. H. (1963). *Chemical Engineering Handbook,* 4th ed., vol. 3, p. 70, McGraw-Hill, New York.
Pounder, E. R. (1965). *The Physics of Ice,* Pergamon, Elmsford, N.Y.
Qureshi, Z. H. (1980). Ph.D. Thesis, State University of New York.
Qureshi, Z. H., and Gebhart, B. (1978). *Proc. 6th Int. Heat Transfer Conf.,* Toronto, vol. 2, p. 217.
Roy, S. (1972). *Indian J. Phys. 46,* 245.
Sammakia, B., and Gebhart, B. (1983). *Int. J. Heat Mass Transfer 26,* 1439.
Schechter, R. S., and Isbin, H. S. (1958). *AIChE J. 4,* 81.
Schenk, J., and Schenkels, F. A. M. (1968). *Appl. Sci. Res. 19,* 465.
Soundalgekar, V. M. (1973). *Chem. Eng. Sci. 28,* 307.
Tankin, R. S., and Farhadieh, R. (1971). *Int. J. Heat Mass Transfer 14,* 953.
Vanier, C. R., and Tien, C. (1967). *Chem. Eng. Sci. 22,* 1747.

Vanier, C. R., and Tien, C. (1968). *Chem. Eng. Prog. Symp. Ser. 64,* 240.
Wilson, N. W., and Vyas, B. D. (1979). *J. Heat Transfer 101,* 313.

PROBLEMS

9.1 Concerning the formulation of vertical boundary layer transport and similarity.

(a) From the density equation and the definition of R, derive the buoyancy force W given by Eq. (9.3.15).

(b) Determine whether or not similarity arises for $d(x)$ of exponential form.

(c) Calculate the downstream variation of momentum to see if any other limits on n arise from considerations of reasonable behavior.

9.2 Calculate an expression for β from the density equation (9.1.1). For pure water at 1 bar, calculate the values of β at 5 and 10°C and compare them with the values given in the appendixes.

9.3 For water at ordinary temperatures, the general continuity relation, Eq. (2.1.1), is usually approximated as at constant density, that is, as Eq. (2.2.1), in two dimensions. Consider a 20-cm-high surface for which $t_0 - t_\infty = 2$°C. Compare the validity of this approximation at temperature levels $t_\infty = 20$ and 4°C. Make the comparison in terms of the value of the component $\partial\rho/\partial x$ in the full continuity equation (2.1.1). That is, find the ratio of the two values.

9.4 For the two conditions given in Problem 9.6 below, compare the maximum buoyancy force $g(\rho_\infty - \rho_0)$ across the thermal layer.

9.5 The limits for reasonable vertical boundary layer solutions are given in terms of n and q.

(a) Derive the expressions for $Q(x)$, $u(x, y)$, and $\delta(x)$ for cold water in terms of the similarity transformation.

(b) Show how these relations determine the limits given in Section 9.3.

9.6 A 10-cm-high, 10-cm-wide vertical surface at 8°C is in quiescent water. For ambient temperatures of 4 and of 3.5°C, calculate

(a) The two heat transfer rates, using the results of the similarity solution.

(b) The heat transfer rates, using the film temperature for property evaluation and the constant-property results in Chapter 3.

9.7 For the transport condition in Problem 9.6, for $t_\infty = 3.5$°C, determine

(a) The downstream variation of δ, δ_t, u_{max}, and $v(x, \infty)$.

(b) Calculate the specific values of these quantities at the trailing edge, $x = L$, in SI units.

(c) Compare the value obtained in (b) for u_{max} at $x = L$ with that from uniform property analysis in Chapter 3, using properties at the film temperature.

9.8 A gap in boundary layer solutions arises for vertical isothermal surfaces. For a surface temperature of 6°C, plot against t_∞ the variations of Nu_x and $v(x, \infty)$ as the gap is approached from each side, from $R = 0$ and $R = \frac{1}{2}$, respectively.

9.9 A vertical surface in pure water at $p = 1$ bar and $t_\infty = 4$°C dissipates energy uniformly at 200 W/m^2. Calculate the surface temperature t_0 and the variation of the maximum velocity downstream along the surface.

9.10 Consider the leading edge region of a horizontal surface at 0°C in water at 4°C.

(a) Sketch the configuration for which onflow would occur.

(b) Calculate the downstream variation of $\delta(x)$, h_x, and $q''(x)$.

(c) Calculate the heat transfer rate from $x = 0$ to L.

9.11 A plane plume rises above a long electrically heated horizontal wire, in water, dissipating electrical energy uniformly at 50 W/m.

(a) For $t_\infty = 30$°C calculate the centerline temperature and velocity downstream at $x = 10$ cm.

(b) For $t_\infty = 4$°C repeat the calculation in part (a).

(c) Discuss the flow likely to result for $t_\infty = 3$°C.

9.12 Consider a vertical wall plume for the same conditions as in Problem 9.11.

(a) For $t_\infty = 30°C$ calculate the surface temperature downstream at $x = 10$ cm.

(b) For $t_\infty = 4°C$ repeat the calculation in part (a).

9.13 Consider a vertical ice surface melting in water at $s_\infty = 10‰$, $t_\infty = 0°C$, and $p = 1$ bar.

(a) Determine the salinity level at the water-ice interface.

(b) Estimate the ice melting rate.

9.14 Repeat Problem 9.13 for ambient conditions of $s_\infty = 35‰$, $t_\infty = -1°C$, and $p = 1$ bar.

9.15 Consider a flat vertical ice surface melting in saline water. Assuming thermodynamic equilibrium, state which of the following conditions can be modeled with the usual boundary layer approximations: (s_∞, t_∞) = (2‰, 2°C), (2‰, 7°C), (2‰, 20°C), (20‰, 0.5°C), (20‰, 7°C). If the ice surface is 1 m high and 20 cm wide, estimate the melt rate M for these (s_∞, t_∞) values.

9.16 Consider a flat vertical ice surface melting in water at $s = 10‰$, $t_\infty = 1°C$. If the surface is 30 cm high and 15 cm wide:

(a) Determine the maximum fluid velocity in the upward-moving boundary layer.

(b) Determine the maximum thickness of the viscous, thermal, and saline boundary layers.

(c) Explain the difference between the measured and computed velocity profiles shown in Fig. 9.4.3, particularly the measured downward (negative) velocity values.

CHAPTER
TEN

MIXED CONVECTION

10.1 MECHANISMS OF TRANSPORT

The two distinct modes of convective transport are forced and natural convection. Generally, transport rates are estimated by considering one of these modes to be dominant. However, in any heat transfer circumstance, temperature differences exist in the boundary region near a heated or cooled surface. These differences cause density gradients in the ambient medium and, in the presence of a body force field such as gravity, natural convection effects arise. Thus, in any forced convection circumstance, natural convection effects are also likely to be present. From a practical standpoint, the important question is how large these buoyancy effects are and under what conditions they may be neglected compared to the forced convection effects. On the other hand, if the buoyancy effects are of greater relative magnitude, the question is when forced convection mechanisms may be neglected. In many practical situations these two effects are of comparable order. A convection situation in which both of these effects are significant is commonly referred to as mixed or combined convection. This mode arises, for instance, in the use of hot wire/film anemometry in low-velocity fields, in natural convection in the presence of ambient fluid circulations, in externally induced flow in heated channels, in the cooling of electronic circuitry by means of a fan, and in many other cases of practical interest.

Analysis indicates that the parameter that characterizes mixed convection is $\mathrm{Gr}/\mathrm{Re}^n$, where the Grashof number Gr and the Reynolds number Re represent the vigor of the natural convection and forced flow effects, respectively. The limiting values of $\mathrm{Gr}/\mathrm{Re}^n \rightarrow 0$ and $\mathrm{Gr}/\mathrm{Re}^n \rightarrow \infty$ correspond to the forced and natural convection limits, respectively. The exponent n depends on the geometry, the thermal boundary condition, the fluid, and so forth. Since the distinction between the pure natural and the pure forced convection regimes is gradual, approximate criteria are established to bound the three transport regimes: natural, mixed, and forced convection.

Analysis of a mixed convection flow usually requires an understanding of the two limiting regimes The complexity of the transport is largely due to the interaction of the buoyancy force with the externally induced flow field. If both of these effects are in the same direction, a higher transport rate will result. However, for some angles between the buoyancy force and the forced flow, the resultant transport may be less than that which would arise with either effect alone. Boundary layer approximations have been widely applied to analyze mixed convection. A boundary layer mechanism often arises, depending on the direction and magnitude of the two interacting motive effects. Even then, in a given mixed convection circumstance, $\mathrm{Gr}/\mathrm{Re}^n$ may be a local parameter and vary along the surface, causing additional complexity. A boundary layer flow may be attached to part of the surface, with separated or even reversed flow elsewhere. Furthermore, forced convection may dominate over part of the surface, with mixed convection elsewhere. Simplifying assumptions are often necessary to make the analysis amenable to the available techniques of analysis. Much of the past analytical work was based on one of two extreme conditions: a forced flow perturbed by small buoyancy effects and a buoyancy-dominated flow perturbed by small forced flow effects.

To illustrate the relative magnitudes of the two kinds of effects in thermal transport, consider an upward laminar flow of a uniform stream (U_∞, t_∞) over a flat vertical surface of height L. The surface temperature t_0 is taken as uniform and greater than t_∞. The buoyancy forces then aid the forced flow. The coordinate in the flow direction is taken as x. The x-direction component of the Navier-Stokes equation, for uniform properties except for buoyancy, is

$$u\frac{\partial u}{\partial x} + v\frac{\partial u}{\partial y} = -\frac{1}{\rho}\frac{\partial p}{\partial x} + \nu\nabla^2 u + g\beta(t - t_\infty) \tag{10.1.1}$$

This equation is normalized by introducing the following dimensionless quantities:

$$X = \frac{x}{L} \qquad Y = \frac{y}{L} \qquad \phi = \frac{t - t_\infty}{t_0 - t_\infty} \qquad P = \frac{p - p_\infty}{\rho U_\infty^2/2}$$
$$u' = \frac{u}{U_\infty} \qquad v' = \frac{v}{U_\infty} \qquad \nabla'^2 = L^2\nabla^2 \tag{10.1.2}$$

Then the normalized form of Eq. (10.1.1) is

$$u'\frac{\partial u'}{\partial X} + v'\frac{\partial u'}{\partial Y} = -\frac{1}{2}\frac{\partial P}{\partial X} + \frac{\nu}{U_\infty L}\nabla'^2 u' + \frac{g\beta L^3(t_0 - t_\infty)}{\nu^2}\frac{\nu^2}{U_\infty^2 L^2}\phi \tag{10.1.3}$$

The buoyancy effect will be comparable to the forced flow effects if the coefficient of ϕ is of order 1, as are the convection terms. This implies that

$$\frac{g\beta L^3(t_0 - t_\infty)}{\nu^2}\frac{\nu^2}{U_\infty^2 L^2} = \frac{\text{Gr}}{\text{Re}^2} = O(1)$$

$$\text{or} \qquad \text{Gr} = O(\text{Re}^2) \tag{10.1.4}$$

Physically, the magnitude of Gr/Re^2 indicates the relative effect of buoyancy on forced convection. High values of this parameter result in significant buoyancy effects.

For strongly buoyancy-driven transport, the velocities in Eq. (10.1.1) should be normalized with the buoyancy-induced velocity $U_c = O(\sqrt{gL\beta(t_0 - t_\infty)})$. Then the parameter $\text{Re}/\sqrt{\text{Gr}}$ arises in Eq. (10.1.3). Thus the relative magnitude of the forced convection effect on the transport mechanisms is characterized by $\text{Re}/\sqrt{\text{Gr}}$. The same kind of result is obtained when the effects are considered as comparable. Thus the full spectrum of the convection process may be described by the values of the parameter Gr/Re^n over the range from 0 to ∞. This will be seen to apply for both external and internal flows.

As mentioned above, mixed convective transport is of interest and importance in a wide variety of engineering applications. However, this area has received less attention than its practical importance warrants. Much of the effort in the past has been directed at delineating the transport regimes, mostly to determine when buoyancy effects are negligible in a forced flow circumstance and when the externally imposed flow may be neglected in a natural convection process. A wide variety of geometries, both internal and external, have been considered, by theoretical as well as experimental methods. Heat transfer coefficients in the various regimes, particularly for aiding forced and natural convection effects, have been determined. Some information has also been obtained on the velocity and temperature fields. However, opposing effects have not been considered in comparable detail, and there is little information on possible flow separation and reversal. Similarly, few three-dimensional and other complex flows of practical concern have been studied. Much of the research has concerned laminar flow in simple configurations and geometries. Transient and enclosure mixed convection flows have received little attention. In the following sections the past important analytical and experimental studies of mixed convection flows are outlined, along with the results obtained.

10.2 VERTICAL FLOWS

Mixed convection over a semi-infinite vertical surface has been analyzed by several investigators. Both isothermal and uniform heat flux surface conditions have been considered. The two important circumstances that arise are due to the buoyancy forces *aiding* or *opposing* the imposed forced flow. See Fig. 10.2.1. In the latter circumstance the buoyancy force causes an adverse pressure gradient.

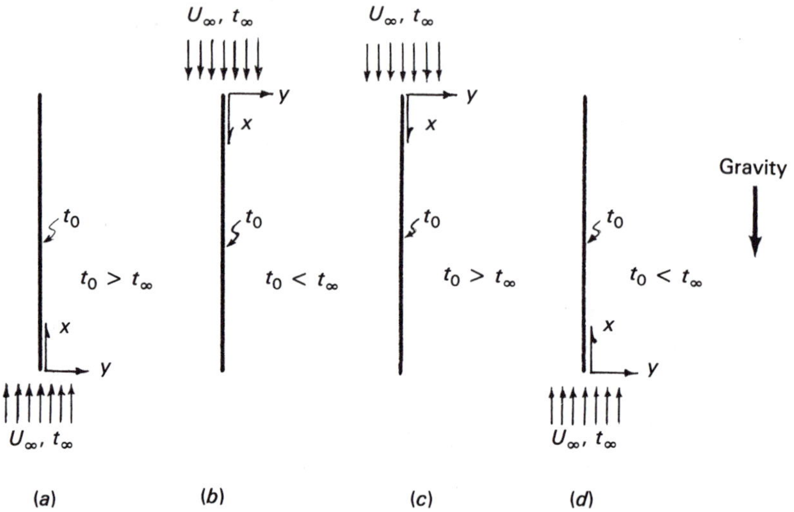

Figure 10.2.1 Various aiding (*a* and *b*) or opposing (*c* and *d*) flow circumstances in mixed convection.

Downstream of some location, the external flow separates and reverses direction. Since similarity solutions often are not possible for this and for many mixed convection flows in practical situations, other techniques such as perturbation, finite-difference, and local similarity methods are used.

10.2.1 Isothermal Surface

Among the earliest studies of mixed convection are the analyses of Sparrow and Gregg (1959), Szewczyk (1964), and Merkin (1969) of the flow over an isothermal vertical surface, by perturbation techniques. Of particular interest was the determination of the conditions under which the buoyancy effects may be neglected in forced flow. The analysis of Merkin (1969) is outlined here since it improves the earlier analyses and gives the proper physical interpretation of the results.

Consider a mixed flow along a vertical semi-infinite flat plate maintained at a uniform temperature t_0. A uniform laminar fluid stream at velocity U_∞ and temperature t_∞ flows parallel to the surface. Both $t_0 < t_\infty$ and $t_0 > t_\infty$ are considered in separate analyses. Following Merkin (1969), a Blasius flow dominates near the leading edge, where the buoyancy effects appear only as a perturbation of the imposed flow U_∞. Far downstream of the leading edge, for aiding effects, the flow is dominated by the buoyancy forces and the forced flow effect appears, instead, as a perturbation. In the intermediate zone both of these effects are of comparable order. For opposing buoyancy, separation and flow reversal arise downstream.

A boundary layer flow is assumed, this assumption being valid only up to any separation point for opposing effects. Transport properties are taken constant. Taking x as the downstream distance, the governing equations are

$$\frac{\partial u}{\partial x} + \frac{\partial v}{\partial y} = 0 \tag{10.2.1}$$

$$u\frac{\partial u}{\partial x} + v\frac{\partial u}{\partial y} = \pm g\beta(t - t_\infty) + \nu\frac{\partial^2 u}{\partial y^2} \tag{10.2.2}$$

$$u\frac{\partial t}{\partial x} + v\frac{\partial t}{\partial y} = \alpha\frac{\partial^2 t}{\partial y^2} \tag{10.2.3}$$

where the plus sign corresponds to an aiding and the minus sign to an opposing buoyancy force. The boundary conditions are

$$\begin{aligned} &\text{at } y = 0: \quad && u = v = 0 \quad && t = t_0 \\ &\text{as } y \to \infty: \quad && u \to U_\infty \quad && t \to t_\infty \end{aligned} \tag{10.2.4}$$

Various physical circumstances are shown in Fig. 10.2.1. The system of equations above is solved separately for the three downstream regions mentioned above. The formulations for Fig. 10.2.1*a* and *b* are the same as are those for *c* and *d*.

Near the leading edge. Forced flow dominates in this region since the buoyancy force has acted over a relatively small vertical distance. The stream function and the normalized temperature are then expanded around forced convection transport $f_0(\eta)$ and $\phi_0(\eta)$ as

$$\psi = (2\nu U_\infty x)^{1/2}[f_0(\eta) \pm \epsilon f_1(\eta) + \epsilon^2 f_2(\eta) \pm \cdots] \tag{10.2.5}$$

$$\phi = \frac{t - t_\infty}{t_0 - t_\infty} = [\phi_0(\eta) \pm \epsilon\phi_1(\eta) + \epsilon^2\phi_2(\eta) \pm \cdots] \tag{10.2.6}$$

where

$$\eta = y\left(\frac{U_\infty}{2\nu x}\right)^{1/2} = \frac{y}{x}\left(\frac{\text{Re}_x}{2}\right)^{1/2} \tag{10.2.7}$$

$$\epsilon = \frac{g\beta(t_0 - t_\infty)x}{U_\infty^2} = \frac{\text{Gr}_x}{\text{Re}_x^2} \tag{10.2.8}$$

The forced flow similarity variables are used. Nonsimilarity arises through the buoyancy force term in Eq. (10.2.2).

These expansions are substituted into Eqs. (10.2.2) and (10.2.3) and, on equating powers of ϵ, the following ordinary differential equations are obtained for the first three sets of functions of η, (f_0, ϕ_0), (f_1, ϕ_1), and (f_2, ϕ_2):

$$f_0''' + f_0 f_0'' = 0 \tag{10.2.9a}$$

$$\phi_0'' + \text{Pr} f_0 \phi_0' = 0 \tag{10.2.9b}$$

$$f_1''' + f_0 f_1'' - 2f_0' f_1' + 3f_0'' f_1 + 2\phi_0 = 0 \tag{10.2.10a}$$

$$\phi_1'' + \Pr(f_0\phi_1' - 2f_0'\phi_1 + 3f_1\phi_0') = 0 \tag{10.2.10b}$$

$$f_2''' + f_0f_2'' - 4f_0'f_2' + 5f_0''f_2 + 3f_1f_1'' - 2f_1'f_1' + 2\phi_1 = 0 \tag{10.2.11a}$$

$$\phi_2'' + \Pr(f_0\phi_2' - 4f_0'\phi_2 + 3f_1\phi_1' - 2f_1'\phi_1 + 5f_2\phi_0') = 0 \tag{10.2.11b}$$

The boundary conditions are

$$\begin{gathered} f_0(0) = f_1(0) = f_2(0) = 0 \\ f_0'(0) = f_1'(0) = f_2'(0) = 0 \\ \phi_0(0) = 1 \qquad \phi_1(0) = \phi_2(0) = 0 \\ f_0'(\infty) = 1 \qquad f_1'(\infty) = f_2'(\infty) = 0 \\ \phi_0(\infty) = \phi_1(\infty) = \phi_2(\infty) = 0 \end{gathered} \tag{10.2.12}$$

The resulting local shear stress and heat transfer at the surface are written as

$$\begin{aligned} \frac{\tau_0}{\rho U_\infty^2/2} &= \frac{1}{\rho U_\infty^2/2}\,\mu\left(\frac{\partial u}{\partial y}\right)_{y=0} \\ &= \sqrt{\frac{2}{\mathrm{Re}_x}}\,[f_0''(0) \pm \epsilon f_1''(0) + \epsilon^2 f_2''(0) \pm \cdots] \end{aligned} \tag{10.2.13}$$

$$\mathrm{Nu}_x = \frac{h_x x}{k} = -\sqrt{\frac{\mathrm{Re}_x}{2}}\,[\phi_0'(0) \pm \epsilon\phi_1'(0) + \epsilon^2\phi_2'(0) \pm \cdots] \tag{10.2.14}$$

Note that the zeroth-order equations, Eq. (10.2.9), and boundary conditions are for pure forced convection. For a given Prandtl number, Eqs. (10.2.9)–(10.2.11) along with the boundary conditions, Eq. (10.2.12), were solved numerically by Merkin (1969) for Pr = 1. These results, for the region near the leading edge, are

$$\frac{\tau_0(x)}{\rho U_\infty^2/2} = \sqrt{\frac{2}{\mathrm{Re}_x}}\left[0.4696 \pm 1.6216\frac{\mathrm{Gr}_x}{\mathrm{Re}_x^2} - 1.2699\left(\frac{\mathrm{Gr}_x}{\mathrm{Re}_x^2}\right)^2 \pm \cdots\right] \tag{10.2.15}$$

$$\mathrm{Nu}_x = \sqrt{\frac{\mathrm{Re}_x}{2}}\left[0.4696 \pm 0.3834\frac{\mathrm{Gr}_x}{\mathrm{Re}_x^2} - 0.6544\left(\frac{\mathrm{Gr}_x}{\mathrm{Re}_x^2}\right)^2 \pm \cdots\right] \tag{10.2.16}$$

Again, the plus and minus signs before the $\mathrm{Gr}_x/\mathrm{Re}_x^2$ term represent aiding and opposing flow circumstances, respectively.

A similar perturbation solution for the buoyancy effects in forced convection was obtained earlier by Szewczyk (1964) for Pr = 0.01, 0.72, 1.0, 5.0, and 10.0. Because of an error in the equations, incorrect f_2 and ϕ_2 distributions were computed. Sparrow and Gregg (1959) also perturbed the forced convection solution for Pr = 0.01, 1, and 10. The first perturbation, f_1 and ϕ_1, was computed.

Far from the leading edge. Far downstream from the leading edge, buoyancy forces dominate the flow. Only aiding effects produce boundary layer flow at large distances. The perturbation parameter is then $\mathrm{Re}_x/\mathrm{Gr}_x^{1/2}$. However, the same value of ϵ is again used and negative exponents arise. The following expansions are used, where η is now the similarity variable for the purely buoyancy-induced flow. The nonsimilar effect arises in the boundary condition $u \to U_\infty$ as $y \to \infty$.

$$\psi = \nu G[F_0(\eta) + \epsilon^{-1/2}F_1(\eta) + \epsilon^{-1}\log\epsilon\,\bar{F}_2(\eta) + \epsilon^{-1}F_2(\eta) + \cdots] \tag{10.2.17}$$

$$\phi = \frac{t - t_\infty}{t_0 - t_\infty} = H_0(\eta) + \epsilon^{-1/2}H_1(\eta) + \epsilon^{-1}\log\epsilon\,\bar{H}_2(\eta) + \epsilon^{-1}H_2(\eta) + \cdots \tag{10.2.18}$$

where

$$G = 4\left(\frac{\mathrm{Gr}_x}{4}\right)^{1/4} \qquad \mathrm{Gr}_x = \frac{g\beta x^3(t_0 - t_\infty)}{\nu^2} \tag{10.2.19}$$

$$\epsilon = \frac{\mathrm{Gr}_x}{\mathrm{Re}_x^2} \qquad \eta = \frac{yG}{4x} \tag{10.2.20}$$

The overbars indicate variables that are associated with the above transformation, which is applied far downstream, where buoyancy effects are dominant.

Merkin (1969) included terms up to $O(\epsilon^{-1})$ and showed that a logarithmic term must be included in the expansion to obtain a solution that becomes exponentially small as $\eta \to \infty$. Therefore, terms of $O[(\log\epsilon)/\epsilon]$ are included above. For a further discussion of this consideration, see Stewartson (1957).

Substitution of these expansions into Eqs. (10.2.2) and (10.2.3) result in the following ordinary differential equations:

$$F_0''' + 3F_0F_0'' - 2F_0'F_0' + H_0 = 0 \tag{10.2.21a}$$

$$H_0'' + 3\mathrm{Pr}F_0H_0' = 0 \tag{10.2.21b}$$

$$F_1''' + 3F_0F_1'' - 2F_0'F_1' + F_0''F_1 + H_1 = 0 \tag{10.2.22a}$$

$$H_1'' + \mathrm{Pr}(2F_0'H_1 + 3F_0H_1 + F_1H_0') = 0 \tag{10.2.22b}$$

$$\bar{F}_2''' + 3F_0\bar{F}_2'' - F_0''\bar{F}_2 + \bar{H}_2 = 0 \tag{10.2.23a}$$

$$\bar{H}_2'' + \mathrm{Pr}(3F_0\bar{H}_2' - \bar{F}_2H_0' + 4F_0'\bar{H}_2) = 0 \tag{10.2.23b}$$

$$F_2''' + 3F_0F_2'' - F_0''F_2 + F_1F_1'' + H_2 = 0 \tag{10.2.24a}$$

$$H_2'' + \mathrm{Pr}(F_1H_1' + 2F_1'H_1 - F_2H_0' + 3F_1H_2' + 4F_0'H_2) = 0 \tag{10.2.24b}$$

with the boundary conditions

$$\begin{gathered} F_0(0) = F_1(0) = \bar{F}_2(0) = F_2(0) = 0 \\ F_0'(0) = F_1'(0) = \bar{F}_2'(0) = F_2'(0) = 0 \\ H_0(0) = 1 \qquad H_1(0) = \bar{H}_2(0) = H_2(0) = 0 \\ F_0'(\infty) = 0 \qquad F_1'(\infty) = \frac{1}{2} \qquad \bar{F}_2'(\infty) = F_2'(\infty) = 0 \\ H_0(\infty) = H_1(\infty) = \bar{H}_2(\infty) = H_2(\infty) = 0 \end{gathered} \tag{10.2.25}$$

The second- and higher-order equations pose eigenvalue problems since the equations as well as the boundary conditions are homogeneous.

The local shear stress and heat transfer at the surface are obtained as

$$\tau_0 = \mu\left(\frac{\partial u}{\partial y}\right)_{y=0} = \frac{\mu\nu G^3}{16x^2}\,[F_0''(0) + \epsilon^{-1/2}F_1''(0)$$

$$+ \epsilon^{-1}\log\epsilon\,\bar{F}_2''(0) + \epsilon^{-1}F_2''(0) + \cdots] \tag{10.2.26}$$

$$\mathrm{Nu}_x = \frac{h_x x}{k} = -\frac{G}{4}\,[H_0'(0) + \epsilon^{-1/2}H_1'(0)$$

$$+ \epsilon^{-1}\log\epsilon\,\bar{H}_1'(0) + \epsilon^{-1}H_2'(0) + \cdots] \tag{10.2.27}$$

and for Pr = 1

$$\tau_0 = \frac{\mu\nu G^3}{16x^2}\,[0.6422 + 0.0830\epsilon^{-1/2} + 0.0105\epsilon^{-1}\log\epsilon$$

$$+ (0.0974 - 0.6422C)\epsilon^{-1} + \cdots] \tag{10.2.28}$$

$$\mathrm{Nu}_x = \frac{G}{4}\,[0.5671 + 0.0712\epsilon^{-1/2} - 0.0089\epsilon^{-1}\log\epsilon$$

$$+ 0.5671C\epsilon^{-1} + \cdots] \tag{10.2.29}$$

Here, C in the last term is a constant. It was determined by Merkin (1969) by comparing the temperature profiles obtained from the asymptotic series with those obtained from a numerical step-by-step computation. The value given is 0.03 ± 0.01.

The earlier calculations of Szewczyk (1964) included other values of Pr. However, the logarithmic terms above were not included in the expansions, which used a regular perturbation with ϵ as a parameter. As shown by Merkin (1969), the flow must be solved by coordinate perturbation since ϵ varies downstream. Treating ϵ as a constant is therefore physically improper.

For the intermediate region, Merkin (1969) solved the governing equations by a step-by-step numerical method. For an aiding buoyancy force, starting with the solution valid near the leading edge, numerical integration of the equations was continued in the increasing ϵ direction until it matched the solution obtained far downstream. For opposing buoyancy, numerical calculations proceeded until separation occurred. The results for the three regions are summarized below in terms of local shear stress and heat transfer.

$$\frac{\tau_0}{\rho U_\infty^2/2} = \frac{2}{\sqrt{\mathrm{Re}_x}}\left(\frac{\mathrm{Gr}_x}{\mathrm{Re}_x^2}\right)^{1/2} T(\epsilon) \tag{10.2.30}$$

$$\mathrm{Nu}_x = \sqrt{\mathrm{Re}_x}\left(\frac{\mathrm{Gr}_x}{\mathrm{Re}_x^2}\right)^{1/2} Q(\epsilon) \tag{10.2.31}$$

The functions $T(\epsilon)$ and $Q(\epsilon)$ are given for Pr = 1.0 in Table 10.2.1 for aiding

Table 10.2.1 Values of $T(\epsilon)$ and $Q(\epsilon)$ in Eqs. (10.2.30) and (10.2.31) for aiding effects; Pr = 1.0

ϵ	$T(\epsilon)$	$Q(\epsilon)$
0.00001	105.0	105.0
0.00448	5.038	5.978
0.04928	1.742	1.552
0.10048	1.388	1.122
0.20288	1.196	0.8135
0.30528	1.142	0.7066
0.46912	1.120	0.6011
0.55104	1.121	0.5672
0.67392	1.127	0.5284
0.79680	1.137	0.4989
0.87872	1.145	0.4827
1.00000	1.157	0.4624
1.2815	1.186	0.4266
1.5375	1.213	0.4028
2.1007	1.268	0.3659
2.5103	1.304	0.3468
4.0463	1.419	0.3018
5.2751	1.494	0.2799
10.190	1.714	0.2334
19.201	1.974	0.1970
29.032	2.171	0.1767
51.969	2.491	0.1512
78.184	2.747	0.137
150.27	3.218	0.1157
242.02	3.615	0.1025
346.88	3.949	0.09352
504.17	4.331	0.08507
102.8	5.615	0.07107
1343.0	5.519	0.06645
series	5.518	0.06647

Source: Reprinted with permission from J. H. Merkin, *J. Fluid Mech.*, vol. 35, p. 439. Copyright © 1969, Cambridge University Press.

Table 10.2.2 Values of $T(\epsilon)$ and $Q(\epsilon)$ in Eqs. (10.2.30) and (10.2.31) for opposing effects; Pr = 1.0

ϵ	$T(\epsilon)$	$Q(\epsilon)$
0.00249	7.284	7.324
0.01265	3.006	3.099
0.03174	1.654	1.813
0.08499	0.7764	1.045
0.11366	0.5492	0.8664
0.14234	0.3676	0.7321
0.16691	0.2224	0.6288
0.17920	0.1433	0.5271
0.18739	0.0782	0.5222
0.19098	0.0366	0.4852
0.19174	0.0289	0.4703
0.19200	0.0168	0.4628
0.19226	0.0081	0.4499
0.192337	0.0032	0.4399
0.192353	0.0012	0.4340
0.192355	0.0008	0.4320
0.192356	0.0002	0.4293
0.192357	0.0000	0.4281

Source: Reprinted with permission from J. H. Merkin, *J. Fluid Mech.*, vol. 35, p. 439. Copyright © 1969, Cambridge University Press.

effects and in Table 10.2.2 for opposing effects. For opposing effects separation occurs at $Gr_x/Re_x^2 = 0.1924$, as indicated by τ_0 approaching zero. The numerical solution is terminated at this point. The average heat transfer coefficient may be calculated by integrating Eq. (10.2.31) over a given length. $Q(\epsilon)$ is given in Tables 10.2.1 and 10.2.2. Thus, a numerical or graphical integration method may be used to obtain the average heat transfer coefficient.

Sparrow and Gregg (1959), in their earlier analytical study, determined the conditions under which the effects of the buoyancy force may be neglected in forced convection. They found that the effect of buoyancy on the local heat transfer coefficient h_x will be less than 5% if

$$|\mathrm{Gr}_x| \leq 0.075\mathrm{Re}_x^2 \tag{10.2.32}$$

Also, the effect of buoyancy on the average heat transfer coefficient is within 5% if

$$|\mathrm{Gr}_x| \leq 0.225\mathrm{Re}_x^2 \tag{10.2.33}$$

These limits hold for the range Pr = 0.01–10.0.

Lloyd and Sparrow (1970) analyzed mixed convection for small and moderate values of $\mathrm{Gr}_x/\mathrm{Re}_x^2$ by a local similarity technique, in which the nonsimilar terms at each given vertical location are neglected. For high values of $\mathrm{Gr}_x/\mathrm{Re}_x^2$, that is, near the pure natural convection limit, this technique is not applicable, as pointed out by Lloyd and Sparrow (1970). Heat transfer rates were obtained for Prandtl numbers ranging from 0.003 to 100, with aiding flow along an isothermal surface. The local heat transfer results shown in Table 10.2.3 compared very well with the experimental data of Kliegel (1959). The velocity and temperature profiles computed by Lloyd and Sparrow (1970) are shown in Fig. 10.2.2 over a wide range of the mixed convection parameter $\mathrm{Gr}_x/\mathrm{Re}_x^2$. A strong effect of buoyancy on both profiles is seen.

Oosthuizen and Hart (1973) presented a numerical solution for mixed convection along both an isothermal and a uniform heat flux surface. The isothermal results are in excellent agreement with those of Merkin (1969) and Lloyd and Sparrow (1970). The experimental results of Gryzagoridis (1975) were also in good agreement with the analysis of Lloyd and Sparrow (1970). Afzal and Banthiya (1977) investigated the effects of small and moderate values of the Reynolds number, for which the simplest boundary layer approximations are not applicable. For $\epsilon \to \infty$, in the pure natural convection regime, their results underestimated the

Table 10.2.3 Values of $\mathrm{Nu}_x/\sqrt{\mathrm{Re}_x}$ for an aiding mixed convection flow along a vertical isothermal surface

$\frac{\mathrm{Gr}_x}{\mathrm{Re}_x^2}$	Pr					
	0.003	0.01	0.03	0.72	10	100
0.0	0.02937	0.05159	0.08439	0.2956	0.7281	1.572
0.01	0.02966	0.05210	0.08524	0.2979	0.7313	1.575
0.04	0.03040	0.05346	0.08750	0.3044	0.7404	1.585
0.1	0.03160	0.05565	0.09118	0.3158	0.7574	1.605
0.4	0.03546	0.06264	0.1030	0.3561	0.8259	1.691
1.0	0.04000	0.07079	0.1168	0.4058	0.9212	1.826
2.0	0.04479	0.07936	0.1311	0.4584	1.029	1.994
4.0			0.1495	0.5258	1.173	2.232

Source: Reprinted with permission from *Int. J. Heat Mass Transfer*, vol. 13, p. 434, J. R. Lloyd and E. M. Sparrow. Copyright © 1970, Pergamon Journals Ltd.

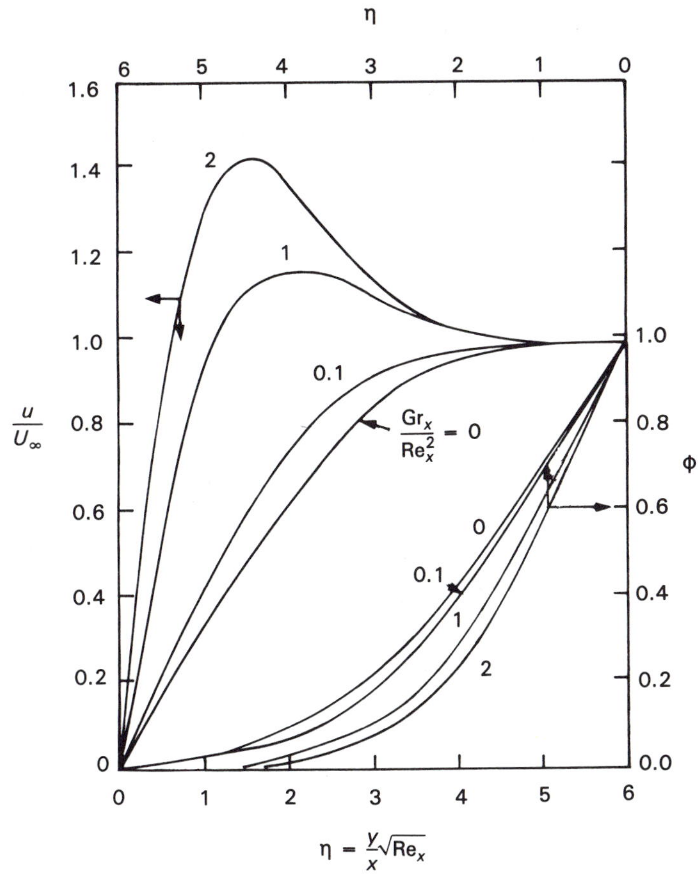

Figure 10.2.2 Computed velocity and temperature distributions for a mixed convection flow over a vertical surface at various values of the parameter Gr_x/Re_x^2; Pr = 0.72. *(Reprinted with permission from* Int. J. Heat Mass Transfer, *vol. 13, p. 434, J. R. Lloyd and E. M. Sparrow. Copyright © 1970, Pergamon Journals Ltd.)*

measured local heat transfer by 4% and overestimated the local wall shear stress by 22%. Mixed convection from an isothermal surface with blowing has been analyzed by Tsuruno and Iguchi (1980).

10.2.2 Uniform Heat Flux Surface

Surfaces dissipating a uniform heat flux are important for experiments, and this boundary condition also occurs frequently in heat exchangers. Wilks (1974) analyzed such a mixed convection flow along a vertical surface by an approach similar to that used by Merkin (1969) for an isothermal surface, considering only aiding effects with Pr = 1. The governing equations, Eqs. (10.2.1)–(10.2.3), still apply, with a positive sign on the buoyancy term in Eq. (10.2.2). The thermal boundary condition on the surface is $\partial t/\partial y = -q''/k$. Other boundary conditions are the same as those given in Eq. (10.2.4). The following expansions apply near the leading edge, where the forced flow is perturbed by small buoyancy effects:

$$\psi = (2\nu U_\infty x)^{1/2}[f_0(\eta) + \epsilon^{3/2}f_1(\eta) + \epsilon^3 f_2(\eta) + \cdots] \tag{10.2.34}$$

$$t - t_\infty = -\frac{q''}{k}\frac{2\nu x}{U_\infty}[\phi_0(\eta) + \epsilon^{3/2}\phi_1(\eta) + \epsilon^3\phi_2(\eta) + \cdots] \tag{10.2.35}$$

where

$$\eta = y\left(\frac{U_\infty}{2\nu x}\right)^{1/2} = \frac{y}{x}\left(\frac{\mathrm{Re}_x}{2}\right)^{1/2} \tag{10.2.36}$$

and

$$\epsilon = \left(\frac{2^3 g^2 \beta^2 q''^2 \nu}{5^2 k^2 U_\infty^5}\right)^{1/3} x = \left(\frac{2^3 \mathrm{Gr}_x^{*2}}{5^2 \mathrm{Re}_x^5}\right)^{1/3} \tag{10.2.37}$$

Far downstream, the flow is essentially natural convection perturbed by relatively small forced convection effects. On the basis of the study by Sparrow and Gregg (1956), the following expansions may be used for a uniform heat flux surface:

$$\psi = 2^{4/5}\nu G^*[F_0(\eta) + \epsilon^{-3/5}F_1(\eta) + \epsilon^{-6/5}F_2(\eta) + \cdots] \tag{10.2.38}$$

$$t - t_\infty = (2)^{1/5}\frac{q''x}{k}\frac{5}{G^*}[H_0(\eta) + \epsilon^{-3/5}H_1(\eta) + \epsilon^{-6/5}H_2(\eta) + \cdots] \tag{10.2.39}$$

where

$$\eta = \left(\frac{G^*}{5x}\right) y \tag{10.2.40}$$

and

$$G^* = 5\left(\frac{\mathrm{Gr}_x^*}{5}\right)^{1/5} \qquad \mathrm{Gr}_x^* = \frac{g\beta x^4 q''}{k\nu^2} \tag{10.2.41}$$

The perturbation parameter ϵ is given by Eq. (10.2.37).

Note that for the uniform heat flux boundary condition the logarithmic term does not arise in Eqs. (10.2.38) and (10.2.39). It can be shown that the contribution due to such a term is identically zero. This is discussed further by Wilks (1974). Both expansions are substituted into Eqs. (10.2.2) and (10.2.3) and the resulting sets of ordinary differential equations were solved numerically. For the intermediate region, the governing equations were again solved by a finite-difference method. The local wall shear stress and heat transfer results are summarized below for $\mathrm{Pr} = 1$.

$$\frac{\tau_0(x)}{\rho U_\infty^2/2} = \frac{\mu(\partial u/\partial y)_{y=0}}{\rho U_\infty^2/2} = 2(\mathrm{Re}_x)^{-1/2}\epsilon^{1/2}T^*(\epsilon) \tag{10.2.42}$$

$$\mathrm{Nu}_x = \sqrt{\mathrm{Re}_x}\,\sqrt{\epsilon}\,Q^*(\epsilon) \tag{10.2.43}$$

Here ϵ is defined by Eq. (10.2.37) over the whole range. The computed values of $T^*(\epsilon)$ and $Q^*(\epsilon)$ are listed in Table 10.2.4.

For other Prandtl numbers, a local similarity technique was also used by Wilks (1973) for a uniform heat flux vertical surface. As mentioned earlier, in Chapter 3, such a technique is inapplicable at large values of ϵ. For low and moderate values of ϵ the local shear stress and heat transfer are given by

Table 10.2.4 Values of $T^*(\epsilon)$ and $Q^*(\epsilon)$ in Eqs. (10.2.42) and (10.2.43) for aiding effects at Pr = 1 for a uniform heat flux surface

ϵ	$T^*(\epsilon)$	$Q^*(\epsilon)$
0.00001	105.0	145.0
0.00004	52.51	72.54
0.00016	26.26	36.27
0.00064	13.13	18.13
0.00256	6.573	9.069
0.01024	3.319	4.541
0.02	2.419	3.260
0.03	2.023	2.672
0.04	1.800	2.324
0.05	1.658	2.089
0.06	1.560	1.917
0.07	1.491	1.784
0.08	1.441	1.678
0.09	1.403	1.591
0.10	1.375	1.519
0.12	1.339	1.402
0.14	1.320	1.314
0.16	1.313	1.243
0.18	1.312	1.185
0.20	1.317	1.137
0.24	1.337	1.060
0.28	1.364	1.002
0.32	1.395	0.9551
0.36	1.428	0.9169
0.40	1.462	0.8847
0.48	1.531	0.8335
0.56	1.598	0.7939
0.64	1.664	0.7621
0.72	1.727	0.7358
0.80	1.788	0.7135
0.88	1.847	0.6946
0.96	1.903	0.6776
1.00	1.931	0.6699
1.2	2.107	0.6376
1.5	2.291	0.6014
2.0	2.559	0.5593

Table 10.2.4 Values of $T^*(\epsilon)$ and $Q^*(\epsilon)$ in Eqs. (10.2.42) and (10.2.43) for aiding effects at Pr = 1 for a uniform heat flux surface (*Continued*)

ϵ	$T^*(\epsilon)$	$Q^*(\epsilon)$
3.0	3.000	0.5070
4.0	3.364	0.4741
5.0	3.677	0.4506
10.0	4.857	0.3867
15.0	5.717	0.3545
20.0	6.418	0.3336
27.0	7.241	0.3133
36.0	8.128	0.2951
·48.0	9.123	0.2589
68.0	10.49	0.2780
100.0	12.25	0.2392
150.0	14.45	0.2203
250.0	17.68	0.1986
450.0	22.38	0.1764
800.0	28.18	0.1571
1500.0	36.24	0.1385

Source: Reprinted with permission from *Int. J. Heat Mass Transfer*, vol. 17, p. 743, G. Wilks. Copyright © 1974, Pergamon Journals Ltd.

$$\frac{\tau_0(x)}{\rho U_\infty^2/2} = \frac{\mu(\partial u/\partial y)_{y=0}}{\rho U_\infty^2/2} = 2(\mathrm{Re}_x)^{-1/2} f''(0) \tag{10.2.44}$$

$$\mathrm{Nu}_x = \sqrt{\mathrm{Re}_x}\,\frac{1}{[-\phi(0)]} \tag{10.2.45}$$

At various values of $[\mathrm{Gr}_x^{*2}/\mathrm{Re}_x^5]^{1/3}$, for Prandtl numbers in the range of 0.1–100, the resulting $f''(0)$ and $-\phi(0)$ are given in Table 10.2.5.

Carey and Gebhart (1982) analyzed the mixed convection flow along a uniform heat flux surface in the region far downstream from the leading edge. Higher-order corrections to the boundary layer analysis and the effect of nonzero free-stream velocity were considered simultaneously. A matched asymptotic expansion technique was used for the analysis. An error in the analysis of Wilks (1974), in the first-order correction equation for the expansions at large downstream distances, was found to have only a small effect on the resulting solution. The resulting local wall shear stress, surface tempertaure, and Nusselt number are given below.

For Pr = 0.733:

$$\frac{\tau_0(x)}{\rho U_\infty^2/2} = 1.1726\,\frac{\mathrm{Gr}_x^{*\,1/5}}{\mathrm{Re}_x}\quad (1 - 0.0274Re^{3/4} - 0.1033\epsilon$$

Table 10.2.5 Values of $(\partial^2 f/\partial\eta^2)_{\eta=0}$ and $-\phi(0)$ in Eqs. (10.2.44) and (10.2.45) at various Prandtl numbers for a uniform heat flux surface

	Pr = 0.1		Pr = 0.72		Pr = 1		Pr = 10		Pr = 100	
$\left(\frac{Gr_x^{*2}}{Re_x^5}\right)^{1/3}$	$\left(\frac{\partial^2 f}{\partial\eta^2}\right)_{\eta=0}$	$-\phi(0)$	$\left(\frac{\partial^2 f}{\partial\eta^2}\right)_{\eta=0}$	$-\phi(0)$	$\left(\frac{\partial^2 f}{\partial\eta^2}\right)_{\eta=0}$	$-\phi(0)$	$\left(\frac{\partial^2 f}{\partial\eta^2}\right)_{\eta=0}$	$-\phi(0)$	$\left(\frac{\partial^2 f}{\partial\eta^2}\right)_{\eta=0}$	$-\phi(0)$
0	0.33257	4.93984	0.3326	2.4375	0.3326	2.1175	0.3326	1.0016	0.3326	0.4644
0.1	0.58563	4.47398	0.4209	2.3258	0.4060	2.0916	0.3510	0.9903	0.3367	0.4632
1	3.53855	2.88811	1.7173	1.6375	1.5341	1.5015	0.7457	0.8270	0.4497	0.4340
2	6.56319	2.40940	3.0908	1.3791	2.7414	1.2678	1.2253	0.7216	0.6175	0.4014
4									0.9721	0.3552

Source: Reprinted with permission from *Int. J. Heat Transfer*, vol. 16, p. 1958, G. Wilks. Copyright © 1973, Pergamon Journals Ltd.

$$+ 0.0146R^2\epsilon^{3/2} + 0.1839R\epsilon^{7/4} + \cdots) \tag{10.2.46}$$

$$t_0 - t_\infty = 2.0417\frac{q''x}{k}(\mathrm{Gr}_x^*)^{-1/5}(1 - 0.0644R\epsilon^{3/4} - 0.2442\epsilon$$

$$- 0.0043R^2\epsilon^{3/2} + 0.4026R\epsilon^{7/4} + \cdots) \tag{10.2.47}$$

$$\mathrm{Nu}_x = 0.4898(\mathrm{Gr}_x^*)^{1/5}(1 + 0.0644R\epsilon^{3/4} + 0.2442\epsilon$$

$$+ 0.00844R^2\epsilon^{3/2} - 0.3712R\epsilon^{7/4} + \cdots) \tag{10.2.48}$$

For Pr = 6.7:

$$\frac{\tau_0(x)}{\rho U_\infty^2/2} = 0.51652\frac{(\mathrm{Gr}_x^*)^{1/5}}{\mathrm{Re}_x}(1 + 0.0173R\epsilon^{3/4} + 0.0188\epsilon$$

$$+ 0.0167R^2\epsilon^{3/2} + 0.0173R\epsilon^{7/4} + \cdots) \tag{10.2.49}$$

$$t_0 - t_\infty = 1.1613\frac{q''x}{k}(\mathrm{Gr}_x^*)^{-1/5}(1 - 0.0642R\epsilon^{3/4} - 0.0981\epsilon$$

$$- 0.0386R^2\epsilon^{3/2} + 0.4224R\epsilon^{7/4} + \cdots) \tag{10.2.50}$$

$$\mathrm{Nu}_x = 0.8611(\mathrm{Gr}_x^*)^{1/5}(1 + 0.0642R\epsilon^{3/4} - 0.0981\epsilon$$

$$- 0.0427R^2\epsilon^{3/2} - 0.4098R\epsilon^{7/4} + \cdots) \tag{10.2.51}$$

where

$$R = \mathrm{Re}_x\left(\frac{5}{\mathrm{Gr}_x^*}\right)^{1/4} \qquad \epsilon = \left(\frac{5}{\mathrm{Gr}_x^*}\right)^{1/5} \tag{10.2.52}$$

Throughout, Gr_x^* is the modified Grashof number defined in Eq. (10.2.41).

10.3 HORIZONTAL FLOWS

The preceding section concerned transport in which the externally induced flow is vertical and aligned with, or opposed to, the buoyancy force. However, there are many practical configurations in which the two are at an inclination with each other. An important example of this class of problems is the horizontal surface, where the buoyancy force is perpendicular to the flow direction; see Section 5.3. The effect of buoyancy on the flow is then much less than in vertical flows. Several analytical and experimental studies have considered this mechanism, as discussed below.

The effect of buoyancy in the boundary layer flow over a flat horizontal surface, facing upward and maintained at $t_0(x)$, has been calculated. For $t_0 > t_\infty$ the buoyancy force gives rise to a favorable pressure gradient and for $t_0 < t_\infty$ to an adverse pressure gradient; see Chapter 5. Since similarity solutions are not generally possible for mixed convection, other mathematical techniques are again used for the analysis.

10.3.1 Isothermal Surfaces

For laminar flow over a semi-infinite horizontal surface, x is taken as the distance along the plate measured from the leading edge and y as the distance out from the surface, as shown in Fig. 10.3.1 for $t_0 > t_\infty$. The buoyancy force B_n away from the surface causes a favorable pressure gradient. For constant properties, except for buoyancy, the governing equations are

$$\frac{\partial u}{\partial x} + \frac{\partial v}{\partial y} = 0 \tag{10.3.1}$$

$$u\frac{\partial u}{\partial x} + v\frac{\partial u}{\partial y} = -\frac{1}{\rho}\frac{\partial p_m}{\partial x} + \nu\frac{\partial^2 u}{\partial y^2} \tag{10.3.2}$$

$$\frac{\partial p_m}{\partial y} = g\beta\rho(t - t_\infty) \tag{10.3.3}$$

$$u\frac{\partial t}{\partial x} + v\frac{\partial t}{\partial y} = \alpha\frac{\partial^2 t}{\partial y^2} \tag{10.3.4}$$

The pressure term may be eliminated from Eq. (10.3.2) by use of Eq. (10.3.3). From Eq. (10.3.3) we obtain by integration, and after differentiating with respect to x,

$$\frac{\partial p_m}{\partial x} = -\rho\beta g\frac{\partial}{\partial x}\int_y^\infty (t - t_\infty)\, dy \tag{10.3.5}$$

With the above substitution, the horizontal momentum equation becomes

$$u\frac{\partial u}{\partial x} + v\frac{\partial u}{\partial y} = \rho\beta g\frac{\partial}{\partial x}\left[\int_y^\infty (t - t_\infty)\, dy\right] + \nu\frac{\partial^2 u}{\partial y^2} \tag{10.3.6}$$

The necessary boundary conditions are

$$\begin{aligned} &\text{at } y = 0: \quad && u = v = 0 \quad && t = t_0 \\ &\text{as } y \to \infty: \quad && u \to U_\infty \quad && t \to t_\infty \end{aligned} \tag{10.3.7}$$

As in vertical flows, in the leading edge region the effect of buoyancy forces is merely a perturbation on the forced flow, while farther downstream, far from the leading edge, the buoyancy forces dominate the flow. Perturbation analyses of the flow near the leading edge were given by Mori (1961), Sparrow and Min-

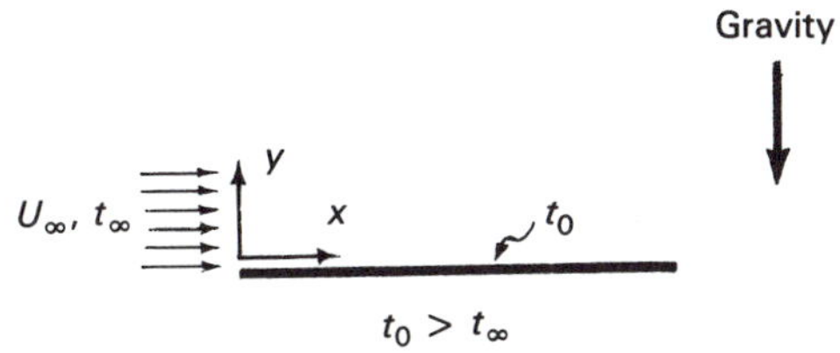

Figure 10.3.1 Coordinate system for mixed convection flow over a horizontal surface.

kowycz (1962), and Hieber (1973). The stream function and the normalized temperature distributions are expanded in a power series in terms of the parameter $Gr_x/Re_x^{5/2}$. Again, the zero-order terms represent the Blasius flow. In the far downstream region, the expansions are in terms of $Re_x/Gr_x^{2/5}$. Hieber (1973) calculated the first three terms of these expansions in the two regions and suggested a graphical interpolation for the intermediate region.

From a practical standpoint, it is important to know the conditions under which buoyancy effects become significant in a given flow situation. On the basis of their calculations, Sparrow and Minkowycz (1962) suggested that the change in the local Nusselt number, due to buoyancy effects, exceeds 5% when

$$Gr_x \geq C_1 Re_x^{5/2} \tag{10.3.8}$$

where $C_1 = 0.0085$, 0.0410, and 0.1590 for Pr = 0.01, 0.7, and 10.0, respectively.

Chen et al. (1977) analyzed the mixed convection flow over horizontal isothermal surfaces using both the local similarity and local nonsimilarity methods discussed in Section 3.13. Results were obtained for the buoyancy parameter $\epsilon = Gr_x/Re_x^{5/2}$ between -0.03 and 1.0 for Pr = 0.7. Local heat transfer and wall shear stress may be represented as

$$\frac{\tau_0}{\rho U_\infty^2/2} = \frac{2}{\sqrt{Re_x}}[f''(\epsilon, 0)] \tag{10.3.9}$$

$$Nu_x = \sqrt{Re_x}\,[-\phi'(\epsilon, 0)] \tag{10.3.10}$$

For Pr = 0.7, the values of $f''(\epsilon, 0)$ and $-\phi'(\epsilon, 0)$ from both the local similarity and local nonsimilarity methods are compared in Table 10.3.1. It was further shown that the wall shear stress is much more affected by buoyancy than the heat transfer.

Ramachandran et al. (1983) extended the analysis of Chen et al. (1977) to cover the entire range of mixed convection, from pure forced to pure natural convection. Furthermore, analyses were given for Pr = 0.01, 0.1, 0.7, and 7.0. The governing equations were solved numerically for $0 \leq Gr_x/Re_x^{5/2} \leq 10$ and a local nonsimilarity method was used for $Gr_x/Re_x^{5/2} > 8$. The two methods of solution compare very well in the overlapping region $8 \leq Gr_x/Re_x^{5/2} \leq 10$. The local heat transfer results are shown in Fig. 10.3.2, where $Nu_x/\sqrt{Re_x}$ is plotted against $Gr_x/Re_x^{5/2}$, for various values of the Prandtl number.

Therefore, the available calculations permit determination of the conditions under which buoyancy effects are important. The heat transfer rates across the whole mixed convection regime may also be computed from these results.

Work has also been done on continuous, moving horizontal flat surfaces. Chen and Strobel (1980) employed local similarity and local nonsimilarity methods to obtain the heat transfer from isothermal surfaces. Moutsoglou and Chen (1980) considered the same problem for inclined surfaces. Karwe and Jaluria (1986) solved the full equations for mixed convection on vertical moving sheets, including the

Table 10.3.1 Values of $f''(\epsilon, 0)$ and $-\phi'(\epsilon, 0)$ in Eqs. (10.3.9) and (10.3.10); Pr = 0.7

$\frac{Gr_x}{Re_x^{5/2}}$	Local similarity		Local nonsimilarity	
	$f''(\epsilon, 0)$	$-\phi'(\epsilon, 0)$	$f''(\epsilon, 0)$	$-\phi'(\epsilon, 0)$
−0.03	0.26576	0.28075	0.27851	0.28194
−0.02	0.28918	0.28510	0.29611	0.28515
−0.01	0.31119	0.28905	0.31553	0.28972
0	0.33206	0.29268	0.33206	0.29268
0.1	0.50358	0.31928	0.47673	0.31934
0.2	0.63974	0.33733	0.58915	0.33751
0.3	0.75769	0.35141	0.68915	0.35073
0.4	0.86387	0.36311	0.77849	0.36178
0.5	0.96162	0.37320	0.85920	0.37138
0.6	1.05290	0.38212	0.93694	0.37949
0.7	1.13900	0.39014	1.00878	0.38719
0.8	1.22084	0.39746	1.07592	0.39423
0.9	1.29909	0.40419	1.14158	0.40063
1.0	1.37421	0.41044	1.20469	0.40658

Source: Reprinted with permission from T. S. Chen et al., *J. Heat Transfer*, vol. 99, p. 66. Copyright © 1977, ASME.

effect of conduction into the plate. This flow is important in several processes of practical interest, such as continuous casting and hot rolling.

10.3.2 Uniform Heat Flux Surface Condition

The governing equations (10.3.1)–(10.3.6) still apply, with the boundary condition $\partial t/\partial y = -q''/k$. The other boundary conditions are the same as those in Eq. (10.3.7). The mixed convection parameter becomes $\epsilon = Gr_x^*/Re_x^3$, where $Gr_x^* = g\beta q'' x^4/k\nu^2$. For Pr = 0.7 and ϵ ranging from 0 to 1.0, Mucoglu and Chen (1978a) used the local similarity and local nonsimilarity methods of analysis. The resulting local wall shear stress and local heat transfer are given by

$$\frac{\tau_0}{\rho U_\infty^2/2} = \frac{2}{\sqrt{Re_x}}[f''(\epsilon, 0)] \tag{10.3.11}$$

$$Nu_x = \sqrt{Re_x}\frac{1}{[-\phi(\epsilon, 0)]} \tag{10.3.12}$$

The computed values of $F''(\epsilon, 0)$ and $H(\epsilon, 0)$ are given in Table 10.3.2. The buoyancy effects change the local forced convection Nusselt number by more than 5% if

$$\frac{Gr_x^*}{Re_x^3} \geq 0.025 \tag{10.3.13}$$

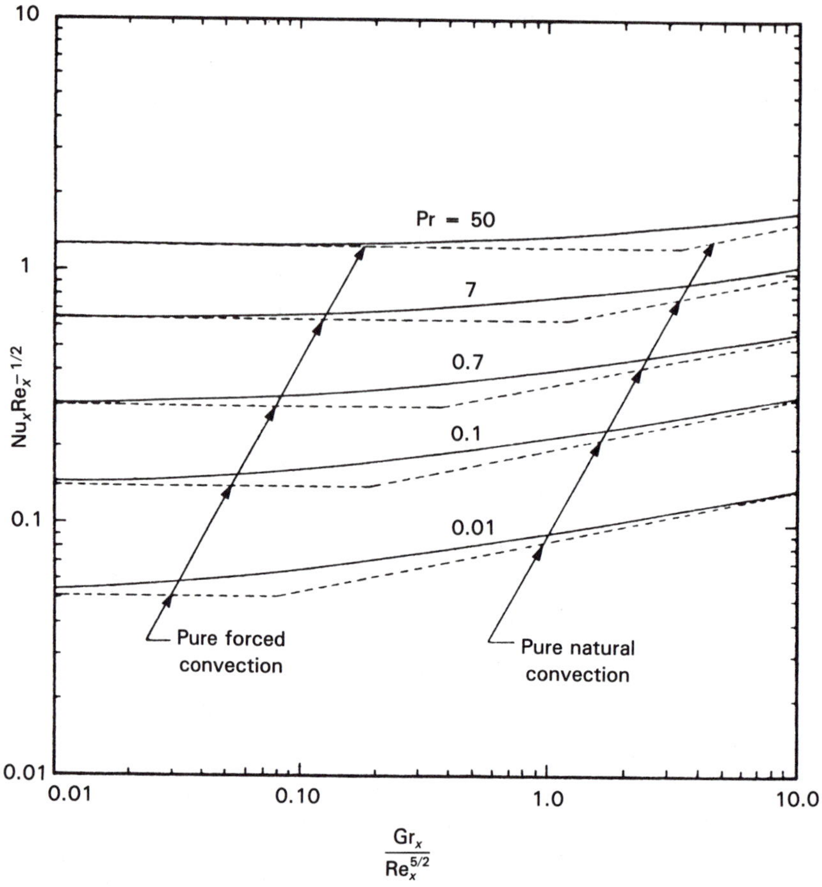

Figure 10.3.2 Variations of local Nusselt number in mixed convection over a horizontal isothermal surface. *(Reprinted with permission from N. Ramachandran et al.,* J. Heat Transfer, *vol. 105, p. 420. Copyright © 1983, ASME.)*

Wang (1982) reported an experimental investigation of mixed convection from a uniform heat flux horizontal plate in air. The effect of buoyancy on the forced convection is determined for two configurations, the upward-facing and the downward-facing surfaces. For the upward-facing surface, the following heat transfer correlations were suggested for air:

$$\mathrm{Nu}_x = 3 + 0.0253\mathrm{Re}_x^{0.8} \qquad \text{for } \mathrm{Gr}_x < 0.068\mathrm{Re}_x^{2.2} \tag{10.3.14}$$

$$\mathrm{Nu}_x = 2.7\left(\frac{\mathrm{Gr}_x}{\mathrm{Re}_x^{2.2}}\right)^{1/3} (3 + 0.0253\mathrm{Re}_x^{0.8})$$

$$\text{for } 0.068\mathrm{Re}_x^{2.2} \le \mathrm{Gr}_x \le 55.3\mathrm{Re}_x^{5/3} \tag{10.3.15}$$

$$\mathrm{Nu}_x = 0.14\mathrm{Gr}_x^{1/3} \qquad \text{for } \mathrm{Gr}_x > 55.3\mathrm{Re}_x^{5/3} \tag{10.3.16}$$

Table 10.3.2 Values of $f''(\epsilon, 0)$ and $-\phi(\epsilon, 0)$ in Eqs. (10.3.11) and (10.3.12); Pr = 0.7

$\frac{Gr_x^*}{Re_x^3}$	Local similarity		Local nonsimilarity	
	$f''(\epsilon, 0)$	$-\phi(\epsilon, 0)$	$f''(\epsilon, 0)$	$-\phi(\epsilon, 0)$
0	0.33206	2.46370	0.33206	2.46370
0.1	0.68661	2.09089	0.62908	2.14975
0.2	0.91133	1.94426	0.81651	2.01744
0.3	1.09090	1.85260	0.96718	1.93133
0.4	1.24521	1.78626	1.09691	1.86849
0.5	1.38175	1.73503	1.21203	1.81966
0.6	1.50754	1.69250	1.31449	1.77946
0.7	1.62324	1.65695	1.40520	1.74716
0.8	1.73116	1.62639	1.49632	1.71622
0.9	1.83247	1.59973	1.56952	1.69132
1.0	1.92846	1.57603	1.64799	1.66827

Source: Reprinted with permission from A. Mucoglu and T. S. Chen, *J. Heat Transfer*, vol. 100, p. 542. Copyright © 1978, ASME.

Fairly good agreement with earlier theoretical studies was obtained. Note that these correlations are given in terms of the Grashof number Gr_x. For surfaces dissipating a uniform heat flux, it is more convenient to use the modified Grashof number Gr_x^*. However, the correlations above may be expressed in terms of Gr_x^* by using

$$Gr_x^* = Nu_x Gr_x \tag{10.3.17}$$

10.4 MIXED CONVECTION TRANSPORT FROM A HORIZONTAL CYLINDER

Perhaps the most thoroughly investigated geometry for external mixed convection flows is the horizontal cylinder. This is due mainly to the use of hot wires or films for velocity measurements. In measuring low fluid velocities, buoyancy effects for warm sensors become increasingly important. In constant-temperature anemometers, the sensor wire or film is maintained at a certain temperature above that of the fluid stream whose velocity is to be measured. If the flow velocities are high, the buoyancy effects are negligible. This implies that the measurements are independent of the inclination of the flow with respect to gravity. However, for low velocities, buoyancy effects are significant and probe calibration and use become quite involved. From a practical standpoint, it is important to determine the conditions under which the buoyancy effects are insignificant. Most of the early investigations were experimental. However, in the past decade, many analyses have also been reported. Generally, aiding, opposing, or cross flows over horizontal cylinders have been investigated. Most of the studies have considered a large length-to-diameter ratio to avoid end effects, which give rise to three-dimensional aspects in the transport.

10.4.1 Experimental Results

The first quantitative information on mixed convection from long heated wires in air was from the experiments of Collis and Williams (1959). For horizontal air streams normal to horizontal wires, it was found that for $\mathrm{Re} \geq O(\mathrm{Gr}^{1/3})$ the natural convection effects are negligible. Furthermore, under certain other flow conditions, the resulting Nusselt number was less than that for pure natural convection. This strange phenomenon was apparently first observed by Ower and Johansen (1931). Generally, vectorial addition of the forced and natural convection Nusselt numbers, as suggested by Zijnen (1956), is an inaccurate approach.

Hatton et al. (1970) experimentally investigated mixed convection over horizontal cylinders in air for three forced flow directions: aiding, opposing, and cross flow. For different directions, different conditions for significant buoyancy effects were determined. Sharma and Sukhatme (1969) reported an experimental study of mixed convection from a horizontal tube in a horizontal air stream. Criteria were obtained for the change from pure natural to mixed convection, as indicated by significant forced flow effects, and from mixed to pure forced convection, as indicated by negligible buoyancy effects. The change from pure natural convection to mixed convection was found to occur when

$$\frac{\mathrm{Gr}}{\mathrm{Re}^{3.25}} = 0.185 \pm 0.01 \qquad (10.4.1a)$$

and the change from mixed convection to pure forced convection when

$$\frac{\mathrm{Gr}}{\mathrm{Re}^{1.8}} = 0.58 \pm 0.13 \qquad (10.4.1b)$$

The heat transfer data were correlated by plotting $\mathrm{Nu}/\mathrm{Gr}^{0.25}$ vs. $\mathrm{Gr}/\mathrm{Re}^{2.5}$. Considerable scatter resulted.

For aiding effects, Oosthuizen and Madan (1970) proposed the following heat transfer correlation:

$$\frac{\mathrm{Nu}}{\mathrm{Nu}_{\mathrm{forced}}} = 1 + 0.18\left(\frac{\mathrm{Gr}}{\mathrm{Re}^2}\right) - 0.011\left(\frac{\mathrm{Gr}}{\mathrm{Re}^2}\right)^2 \qquad \text{where } 0 < \frac{\mathrm{Gr}}{\mathrm{Re}^2} < 9 \qquad (10.4.2)$$

The Nusselt, Grashof, and Reynolds numbers are based on the cylinder diameter. The effect of buoyancy on the average heat transfer coefficient was found to be less than 5% for

$$\frac{\mathrm{Gr}}{\mathrm{Re}^2} < 0.28 \qquad (10.4.3)$$

Based on the experimental data of Oosthuizen and Madan (1970), Jackson and

Yen (1971) proposed the following simple correlation for aiding effects:

$$\frac{\mathrm{Nu}}{\mathrm{Nu}_{\mathrm{forced}}} = \left(1 + \frac{\mathrm{Gr}}{\mathrm{Re}^2}\right)^{0.25} \qquad \text{where } 0 < \frac{\mathrm{Gr}}{\mathrm{Re}^2} < 9 \tag{10.4.4}$$

The effects of fluid Prandtl number and length-to-diameter (L/D) ratio were investigated experimentally by Gebhart et al. (1970) and Gebhart and Pera (1970) for horizontal wires in a vertical fluid stream. The L/D necessary for effectively infinite length behavior, that is, negligible three-dimensional effects, in natural convection is apparently less than those in mixed and forced convection. However, the necessary L/D was found to be large, of order 10^4, in all regimes. It also increased with the Prandtl number.

Further experiments by Oosthuizen and Madan (1971) at cylinder inclinations of 0, 90, 135, and 180° assessed the mixed convection effects. The flow direction was shown to modify the transport rate drastically in mixed convection. Extensive experimental data for mixed convection in water have been reported by Fand and Keswani (1973). Aiding, opposing, and cross flows were considered. Four regimes were suggested, separated by values of $\mathrm{Gr}/\mathrm{Re}^2$, and different heat transfer correlations were suggested for the various regimes. A similar study by Nakai and Okazaki (1975), for air, considered aiding, opposing, and cross flow. Mixed convection was approached from the two limiting regimes, forced and natural convection. Experimental data compared with their analysis showed considerable scatter. It was suggested that $1/\mathrm{Nu}$ could be expressed as a function of $\mathrm{PrRe}^3/\mathrm{NuGr}$.

Morgan (1975) reviewed the literature on forced, natural, and mixed convection from circular cylinders and correlated the available experimental data for the three flow configurations—aiding, opposing, and cross flow. On the basis of the results presented above, Morgan considered the conditions under which the buoyancy effects on the forced convection Nusselt number are negligible, say less than 5%. The buoyancy effects may be neglected if

$$\mathrm{Re} \geq C_1(\mathrm{Gr})^m \tag{10.4.5}$$

For air, $\mathrm{Pr} = 0.7$, values of the empirical constants C_1 and m are shown in Table 10.4.1 for the aiding ($\hat{\theta} = 0°$), opposing ($\hat{\theta} = 180°$), and cross ($\hat{\theta} = 90°$) flows, where $\hat{\theta}$ is the angle between the direction of the external forced flow and the vertical buoyancy force. Similarly, the effects of forced convection on the natural convection heat transfer coefficient are negligible (less than 5%) if

$$\mathrm{Gr} \geq C_2(\mathrm{Re})^{1/m} \tag{10.4.6}$$

The empirical constants C_2 and m are given in Table 10.4.1 for air.

The average heat transfer from horizontal cylinders, in the various flow regimes and for various directions, is given by the following correlations.

Aiding flow ($\hat{\theta} = 0°$):

$$\frac{\mathrm{Nu}}{\mathrm{Nu}_f} = \left[1 + \frac{C_3(\mathrm{Gr})^m}{\mathrm{Re}}\right]^n \tag{10.4.7}$$

Table 10.4.1 Values of C_1, C_2, C_3, m, and n in Eqs. (10.4.5)–(10.4.9) for air; Pr = 0.7

GrPr		Re					$\hat{\theta} = 0°$		$\hat{\theta} = 90°$		$\hat{\theta} = 180°$	
From	To	From	To	C_3	m	n	C_2	C_1	C_2	C_1	C_2	C_1
10^{-10}	10^{-2}	10^{-4}	4×10^{-3}	102	0.648	0.0895	1.30×10^{-3}	141	4.69×10^{-4}	72.7	1.69×10^{-4}	37.5
		4×10^{-3}	9×10^{-2}	3.18	0.426	0.136	0.477	7.36	0.0629	3.10	8.28×10^{-3}	1.31
		9×10^{-2}	1	0.506	0.207	0.280	8.04×10^{4}	2.66	221	0.784	0.607	0.231
10^{-2}	10^{2}	4×10^{-3}	9×10^{-2}	52.2	1.09	0.136	0.0571	121	0.0258	51.0	0.0117	21.5
		9×10^{-2}	1	1.97	0.529	0.280	6.38	10.4	0.633	3.05	0.0628	0.900
		1	35	1.67	0.385	0.384	47.4	12.3	1.33	3.10	0.0370	0.781
10^{2}	10^{4}	9×10^{-2}	1	0.977	0.671	0.280	12.2	5.13	1.99	1.51	0.322	0.446
		1	35	1.00	0.490	0.384	59.4	7.38	3.55	1.86	0.212	0.468
		35	5×10^{3}	1.93	0.399	0.471	49.4	17.7	1.21	4.03	0.0296	0.916
10^{4}	10^{7}	1	35	0.213	0.651	0.384	232	1.57	27.9	0.396	3.35	0.0998
		35	5×10^{3}	0.548	0.531	0.471	202	5.02	12.4	1.14	0.762	0.260
		5×10^{3}	5×10^{4}	5.57	0.395	0.633	7.70	69.5	0.125	13.6	2.02×10^{-3}	2.68
10^{7}	10^{12}	35	5×10^{3}	0.0296	0.707	0.471	3330	0.271	411	0.0616	50.6	0.0140
		5×10^{3}	5×10^{4}	0.635	0.526	0.633	288	7.92	13.0	1.55	0.590	0.305
		5×10^{4}	2×10^{4}	7.82	0.409	0.814	5.91	127	0.0813	21.9	1.12×10^{-3}	3.79

Source: From Morgan (1975).

Opposing flow ($\hat{\theta} = 180°$):

$$\frac{\text{Nu}}{\text{Nu}_f} = \left[1 - \frac{C_3(\text{Gr})^m}{\text{Re}}\right]^n \tag{10.4.8}$$

Cross flow ($\hat{\theta} = 90°$):

$$\frac{\text{Nu}}{\text{Nu}_f} = \left[1 + \frac{C_3^2(\text{Gr})^{2m}}{\text{Re}^2}\right]^{n/2} \tag{10.4.9}$$

For air, the empirical constants C_3, m, and n are given in Table 10.4.1 for various ranges of the flow parameters Gr and Re and various flow directions. Here, Nu_f is the forced convection Nusselt number and is given by

$$\text{Nu}_f = C_4(\text{Re})^n \tag{10.4.10}$$

The constants C_4 and n are given in Table 10.4.2 over various ranges of the Reynolds number.

10.4.2 Theoretical Results

Calculations of mixed convection over a horizontal heated cylinder have been based on perturbation, local similarity, local nonsimilarity, and numerical methods. Only aiding or opposing effects have been considered, in unseparated boundary region flows. The analysis for isothermal and uniform heat flux boundary conditions will be summarized. The cylinder and the coordinate system are shown in Fig. 10.4.1. After applying the boundary layer approximations, the governing equations for constant fluid properties are

$$\frac{\partial u}{\partial x} + \frac{\partial v}{\partial y} = 0 \tag{10.4.11}$$

Table 10.4.2 Values of C_4 and n in Eq. (10.4.10) for air; Pr = 0.7

Re			
From	To	C_4	n
10^{-3}	4×10^{-3}	0.437	0.0895
4×10^{-3}	9×10^{-2}	0.565	0.136
9×10^{-2}	1	0.800	0.280
1	35	0.795	0.384
35	5×10^3	0.583	0.471
5×10^3	5×10^4	0.148	0.633
5×10^4	2×10^5	0.0208	0.814

Source: From Morgan (1975).

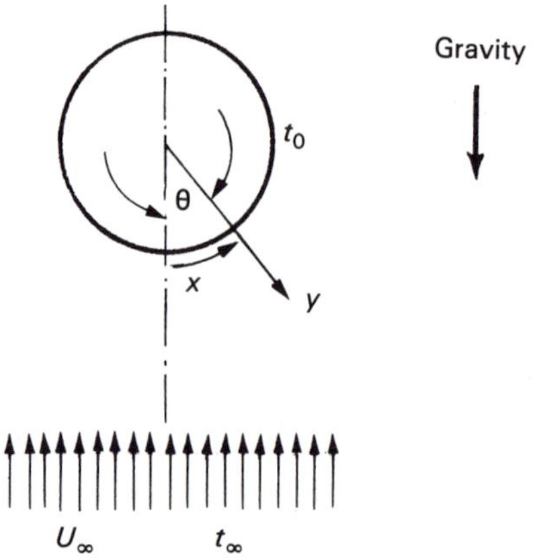

Figure 10.4.1 Orthogonal curvilinear coordinate system for mixed convection over a horizontal cylinder. Aiding effects for $t_0 > t_\infty$ and opposing effects for $t_0 < t_\infty$.

$$u \frac{\partial u}{\partial x} + v \frac{\partial u}{\partial y} = U \frac{dU}{dx} \pm g\beta(t - t_\infty) \sin \frac{x}{R} + \nu \frac{\partial^2 u}{\partial y^2} \tag{10.4.12}$$

$$u \frac{\partial t}{\partial x} + v \frac{\partial t}{\partial y} = \alpha \frac{\partial^2 t}{\partial y^2} \tag{10.4.13}$$

The y component of the momentum equation is neglected, taking B_n as zero. The forced flow pressure gradient is taken as that for potential flow for small buoyancy and boundary layer effects. However, as seen later, the equation for the y-direction momentum must also be included for inclined surfaces, near horizontal, in order to determine the inviscid flow and thus the imposed pressure gradient.

The boundary conditions for the above equations are

$$\begin{aligned} &\text{at } y = 0: \quad && u = v = 0 \quad && t = t_0 \\ &\text{as } y \to \infty: \quad && u \to U(x) \quad && t \to t_\infty \end{aligned} \tag{10.4.14}$$

For $t_0 > t_\infty$ the buoyancy forces aid the forced flow and for $t_0 < t_\infty$ they oppose it, as represented by the two signs for the buoyancy force term in Eq. (10.4.12). From the potential flow theory the local free-stream velocity $U(x)$ is

$$\frac{U(x)}{U_\infty} = 2 \sin \left(\frac{x}{R}\right) \tag{10.4.15}$$

where $R = D/2$ is the cylinder radius.

For high and moderate Reynolds numbers, flow separation occurs beyond a certain value of x. This distorts the streamlines and Eq. (10.4.15) becomes a poor approximation on the upstream side. The measured local free-stream velocities are then commonly approximated by

$$\frac{U(x)}{U_\infty} = A\left(\frac{x}{R}\right) + B\left(\frac{x}{R}\right)^3 + C\left(\frac{x}{R}\right)^5 + \cdots \tag{10.4.16}$$

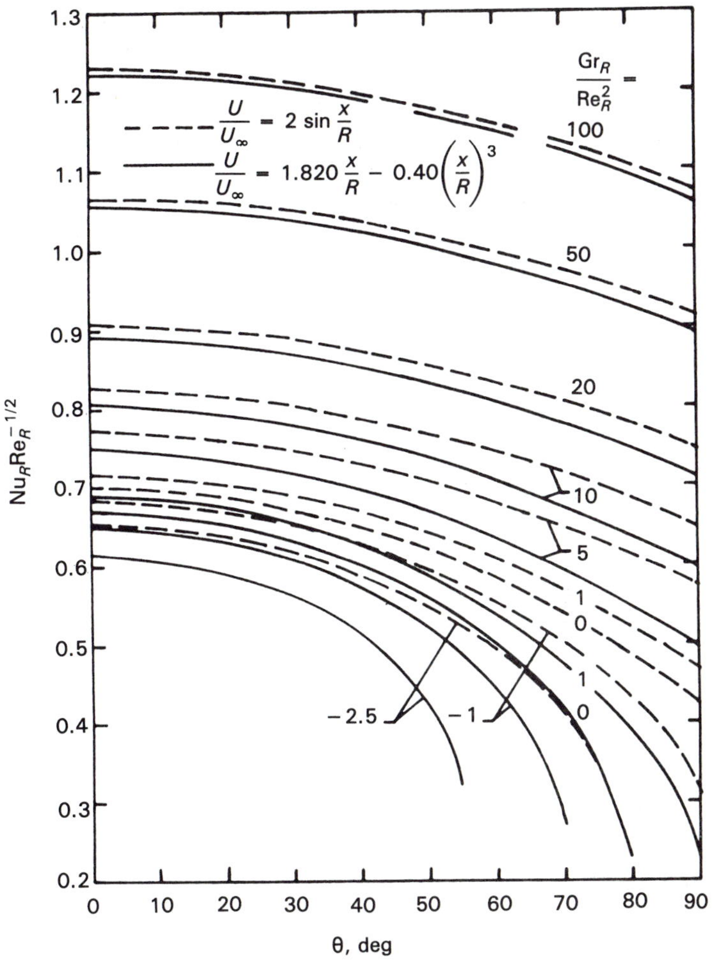

Figure 10.4.2 Angular distributions of the local Nusselt number in a mixed convection flow over a horizontal isothermal cylinder; Pr = 0.7. *(Reprinted with permission from A. Mucoglu and T. S. Chen,* Can. J. Chem. Eng., *vol. 55, p. 265. Copyright © 1977, The Chemical Institute of Canada, Publisher.)*

For pure forced flow over a cylinder, values of A, B, C, . . . have been given by several investigators. Sogin and Subramanian (1961) reported such values for high Reynolds numbers. No measurements are available for mixed convection flows.

Note that the potential flow solution, Eq. (10.4.15), may also be expanded in the same power series to obtain $A = 2$, $B = -\frac{1}{3}$, and $C = \frac{1}{60}$. Oosthuizen (1970) solved the governing equations by similarity methods in the region around the upstream stagnation point and by a finite-difference method farther downstream. Joshi and Sukhatme (1971) used a coordinate perturbation method for the solution.

Both of these studies used the potential flow solution, Eq. (10.4.15), for the local free-stream velocity. Sparrow and Lee (1976) reported an analysis using the Blasius series expansion along with measured local free-stream velocities.

Mucoglu and Chen (1977a) solved the governing equations for the upstream half of the cylinder, over the entire range of the mixed convection parameter Gr/Re_R^2, by a local nonsimilarity method. For the local free-stream velocities, they used both the potential solution given by Eq. (10.4.15) and the measured values from Sogin and Subramanian (1961), with $A = 1.82$ and $B = 0.4$ in Eq. (10.4.16). For an isothermal cylinder in air, Pr = 0.7, the resulting angular variation of the local Nusselt number is shown in Fig. 10.4.2. The various dimensionless groups shown in Fig. 10.4.2 are defined as

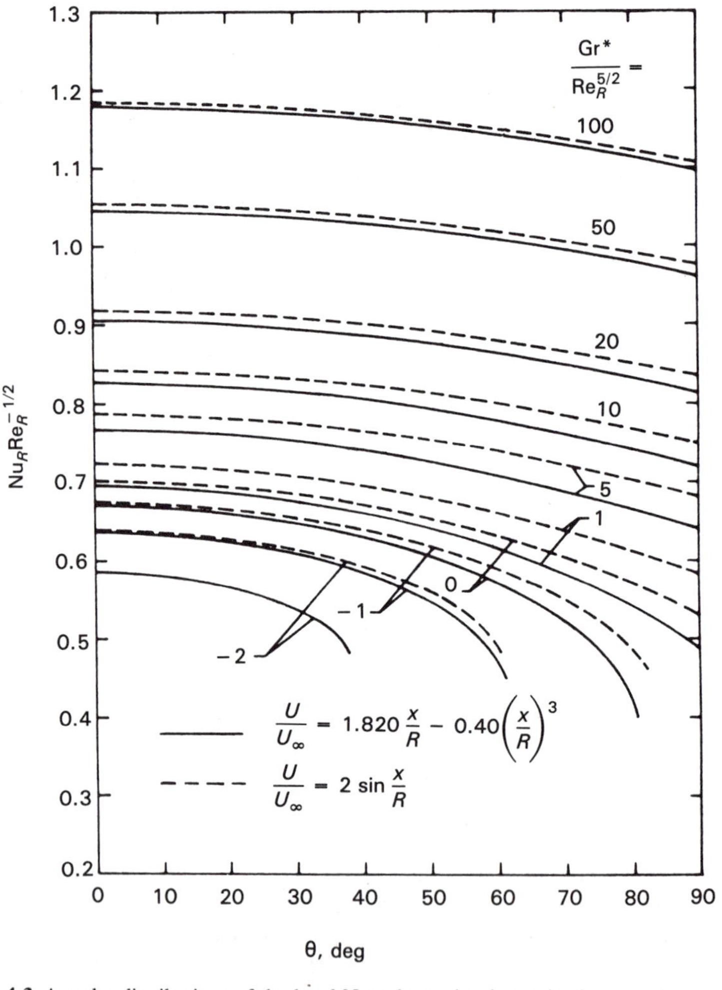

Figure 10.4.3 Angular distributions of the local Nusselt number in a mixed convection flow over a horizontal cylinder with a uniform heat flux boundary condition; Pr = 0.7. *(Reprinted with permission from A. Mucoglu and T. S. Chen,* J. Heat Transfer, *vol. 99, p. 679. Copyright © 1977, ASME.)*

$$\mathrm{Nu}_R = \frac{h_x R}{k} \qquad \mathrm{Gr}_R = \frac{gR^3\beta(t_0 - t_\infty)}{\nu^2} \qquad \mathrm{Re}_R = \frac{U_\infty R}{\nu} \tag{10.4.17}$$

The range of $\mathrm{Gr}_R/\mathrm{Re}_R^2$ is 0 to 100 for aiding effects and 0 to -2.5 for opposing effects. These results apply only up to $\theta = 90°$ from the lower stagnation point. Similar results for a uniform flux surface condition are shown in Fig. 10.4.3, from Mucoglu and Chen (1977b). For this boundary condition the modified Grashof number is defined as $\mathrm{Gr}_R^* = g\beta q'' R^4/k\nu^2$.

Merkin (1977) analyzed mixed convection over a horizontal isothermal cylinder for Pr = 1. Using the potential flow solution outside the boundary region, it was shown that the flow separation was completely suppressed for $\mathrm{Gr}_R/\mathrm{Re}_R^2 > 0.89$ when the buoyancy force aided the forced flow. For an opposing buoyancy force, the separation point moved toward the lower stagnation point. Under these circumstances, the boundary layer approximations become inapplicable.

10.5 MIXED CONVECTION IN OTHER EXTERNAL FLOWS

Various studies of external mixed convection flow over other geometries have been reported. These include flow over a wedge, an inclined surface, circular cylinders in a parallel flow, cones, and spheres. For each geometry, except spheres, the surface orientation with respect to gravity and to the direction of the forced flow influences the resulting transport rates. These geometries are considered below.

10.5.1 Mixed Convection from a Wedge

For pure forced convection onflow over a wedge, the similarity transformations for the momentum equations were given by Falkner and Skan (1931) and for the energy equation by Eckert (1942). The first analysis of mixed convection flow over a vertical wedge was given by Sparrow et al. (1959). By using boundary layer assumptions and neglecting the component of the buoyancy force normal to the wedge sides, that is, taking $B_n = 0$, similarity solutions were obtained for several conditions. The free-stream velocity $U(x)$ varies according to the power law

$$U(x) \propto x^m \tag{10.5.1}$$

where x is the distance from the leading edge and m is related to the wedge included angle $\pi\beta$ as $m = \beta/(2 - \beta)$; see Fig. 10.5.1. The surface temperature excess is assumed to vary as

$$t_0 - t_\infty \propto x^n \tag{10.5.2}$$

Sparrow et al. (1959) determined the following condition for similarity in the governing equations and the boundary conditions:

$$n = 2m - 1 \tag{10.5.3}$$

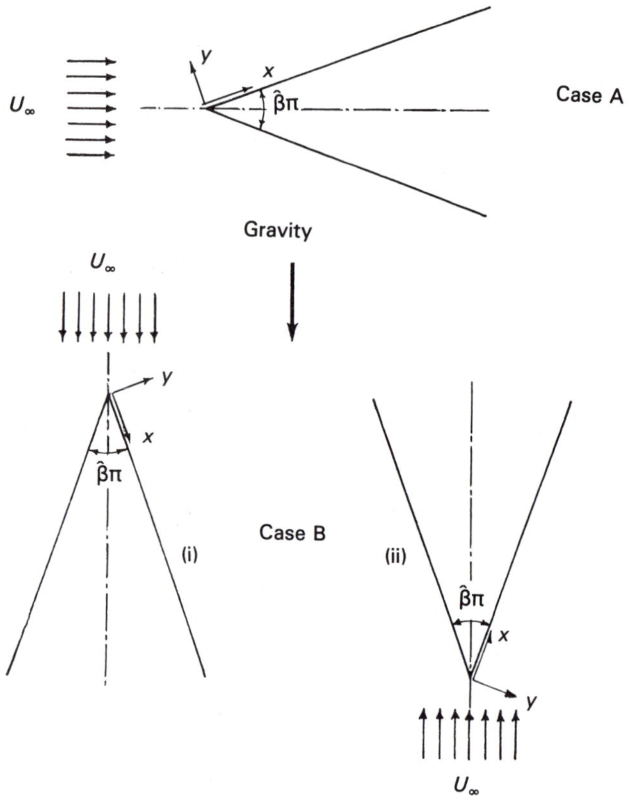

Figure 10.5.1 Various wedge configurations and the coordinate system.

For an isothermal wedge, $n = 0$. From Eq. (10.5.3), $m = \frac{1}{2}$ or $\hat{\beta} = \frac{2}{3}$. Thus, similarity results only for a wedge included angle of 120°. The calculated velocity profiles for this circumstance are shown in Fig. 10.5.2. For the uniform heat flux surface condition, $n = \frac{1}{5}$ and, therefore, $m = \frac{3}{5}$. This corresponds to a wedge included angle of 135°. For these two boundary conditions, $n = 0$ and $\frac{1}{5}$, and $\mathrm{Pr} = 0.7$, the local friction and heat transfer are

$$\frac{\tau_0}{\rho U^2/2} = \frac{1}{2}(\mathrm{Re}_x)^{-1/2} f''(0) \tag{10.5.4}$$

$$\mathrm{Nu}_x = \frac{1}{2}(\mathrm{Re}_x)^{1/2}[-\phi'(0)] \tag{10.5.5}$$

where $f''(0)$ and $[-\phi'(0)]$ are given in Table 10.5.1 for various values of $\mathrm{Gr}_x/\mathrm{Re}_x^2$ for both aiding and opposing buoyancy effects. The Reynolds and Grashof numbers are defined as

$$\mathrm{Gr}_x = \pm\frac{g\beta(t_0 - t_\infty)x^3}{\nu^2} \qquad \mathrm{Re}_x = \frac{Ux}{\nu} \tag{10.5.6}$$

Here the positive sign refers to the circumstance when g_x is positive and the negative to when g_x is negative, where g_x is the x component of the gravitational acceleration $\mathbf{g}$.

For the two thermal boundary conditions, uniform temperature and surface heat flux, the similarity analysis applies to only two wedge angles, even though B_n is neglected. Using perturbation methods, Gunness and Gebhart (1965) analyzed the effect of both components of the buoyancy force on forced convection over isothermal wedges of arbitrary angles. The two wedge orientations shown in Fig. 10.5.1—horizontal plane of symmetry (case A) and vertical plane of symmetry (case B)—were considered. For a zero wedge angle ($\pi\hat{\beta} = 0$) the first orientation is the same as a horizontal surface and the second represents a vertical surface. In general, both the streamwise and normal buoyancy force components are important. For the horizontal orientation, with a wedge included angle of more than 90°, and for the vertical orientation, with wedge angles up to 90°, the normal buoyancy component B_n is smaller in magnitude than the streamwise component.

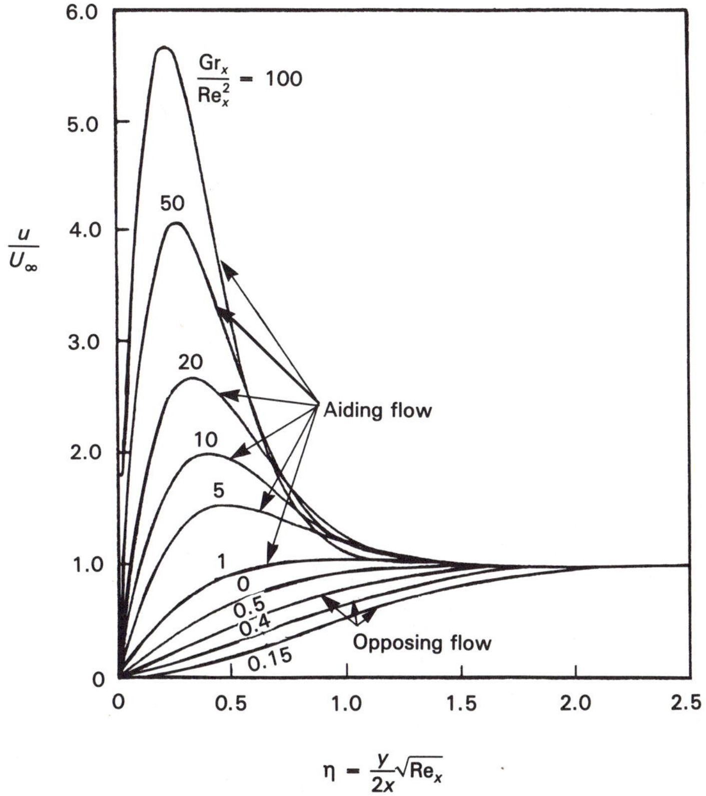

Figure 10.5.2 Velocity profiles from similarity analysis of mixed convection flow over a 120° isothermal wedge at various values of the parameter Gr_x/Re_x^2; Pr = 0.7. *(Reprinted with permission from E. M. Sparrow et al.,* Phys. Fluids, *vol. 2, p. 319. Copyright © 1959, American Institute of Physics.)*

Table 10.5.1 Values of $f''(0)$ and $-\phi'(0)$ in Eqs. (10.5.4) and (10.5.5) for uniform wall temperature (UWT) and uniform heat flux (UHF) boundary conditions; Pr = 0.7

$\frac{Gr_x}{Re_x^2}$	UWT		UHF	
	$f''(0)$	$-\phi'(0)$	$f''(0)$	$-\phi'(0)$
(a) Aiding flows				
100	122.59	2.2698	116.02	2.5882
50	73.449	1.9251	69.589	2.1952
20	37.733	1.5613	35.871	1.7816
10	23.187	1.3466	22.165	1.5389
5	14.652	1.1784	14.145	1.3504
25/9	10.290	1.0690		
2			8.5063	1.1761
1	6.2804	0.94252	6.3457	1.0927
0.8	5.7798	0.92401	5.8853	1.0732
0.5	5.0001	0.89353	5.1712	1.0412
0.25	4.3183	0.86492	4.5507	1.0117
0.05	3.7463	0.83927	4.0338	0.98574
0.02	3.6581	0.83516	3.9545	0.98162
0	3.5989	0.83238	3.9013	0.97883
(b) Opposing flows				
0	3.5989	0.83238	3.9013	0.97883
0.02	3.5394	0.82956	3.8479	0.97602
0.05	3.4495	0.82527	3.7673	0.97175
0.1	3.2980	0.81792	3.6318	0.96447
0.25			3.2152	0.94134
0.5	1.9852	0.74722	2.4795	0.89728
0.8	0.79060	0.66676	1.4957	0.83003
0.9	0.29531	0.62582		
0.94	0.065485	0.60449		
0.95	0.0033788	0.59841		
1.0			0.73003	0.76816
1.1			0.28082	0.72610
1.15			0.024373	0.69943
1.1544			0.00025963	0.69680

Source: Reprinted with permission from E. M. Sparrow et al., *Phys. Fluids*, vol. 2, p. 319. Copyright © 1959, American Institute of Physics.

The local shear stress and heat transfer results obtained by Gunness and Gebhart (1965) for Pr = 0.73 are now discussed.

The first question is whether the buoyancy forces aid or oppose the forced flow. This classification is shown in Table 10.5.2 for the various configurations and surface temperature conditions. The local transport rates are calculated from the equations given below, which are applicable for Gr_x/Re_x^2 small.

Case A—horizontal wedge orientation. The shear stress and the heat transfer rate are given in terms of perturbation functions F, G, H, and I, retaining only the first-order effects of buoyancy, as

$$\frac{\tau_0(x)}{\tau_f(x)} = 1 + \frac{\mathrm{Gr}_x}{\mathrm{Re}_x^2}\left[\left(\frac{F_1''}{f_0''}\right)_{\eta=0} \pm \frac{1}{\sqrt{\mathrm{Re}_x}}\left(\frac{G_1''}{f_0''}\right)_{\eta=0}\right] \tag{10.5.7}$$

$$\frac{\mathrm{Nu}_x}{(\mathrm{Nu}_x)_f} = 1 + \frac{\mathrm{Gr}_x}{\mathrm{Re}_x^2}\left[\left(\frac{H_1'}{\phi_0'}\right)_{\eta=0} \pm \frac{1}{\sqrt{\mathrm{Re}_x}}\left(\frac{I_1'}{\phi_0'}\right)_{\eta=0}\right] \tag{10.5.8}$$

where f_0 and ϕ_0 are the pure forced convection solution, obtained for flow over a wedge, and F_1, G_1, H_1, and I_1 are the first-order perturbation functions. Also, Gr_x and Re_x are as defined in Eq. (10.5.6). The plus sign in Eqs. (10.5.7) and (10.5.8) is used when g_x and g_y are of the same sign and the minus sign when these are of opposite sign, g_y being the y component of the gravitational acceleration; see Table 10.5.2. For details of the double perturbation scheme employed, see Gunness and Gebhart (1965). Here $\tau_f(x)$ and $(\mathrm{Nu}_x)_f$ denote the local shear stress and the heat transfer in the pure forced flow condition and are given by

$$\frac{\tau_f(x)}{\rho U^2/2} = \frac{1}{\sqrt{\mathrm{Re}_x}}(f_0'')_{\eta=0} \tag{10.5.9}$$

$$(\mathrm{Nu}_x)_f = \frac{1}{2}\sqrt{\mathrm{Re}_x}\,(-\phi_0')_{\eta=0} \tag{10.5.10}$$

The various functions at $\eta = 0$ in Eqs. (10.5.7)–(10.5.10) are given in Table 10.5.3.

Case B—vertical wedge orientation. This case is simpler than the above, for which a double perturbation scheme was needed. Here a single perturbation parameter arises and the resulting expressions are

$$\frac{\tau_0(x)}{\tau_f(x)} = 1 + \frac{\mathrm{Gr}}{\mathrm{Re}_x^2}\left(\frac{F_1''}{f_0''}\right)_{\eta=0} \tag{10.5.11}$$

Table 10.5.2 Aiding or opposing mixed convection flow circumstances for various wedge configurations and surface temperature conditions

Surface temperature condition	Horizontal wedge orientation, case A: Upper surface	Horizontal wedge orientation, case A: Lower surface	Vertical wedge orientation, case B: Case B(i) Fig. 10.5.1	Vertical wedge orientation, case B: Case B(ii) Fig. 10.5.1
$t_0 > t_\infty$	Aiding	Opposing	Opposing	Aiding
$t_0 < t_\infty$	Opposing	Aiding	Aiding	Opposing

Table 10.5.3 Various functions to be used in Eqs. (10.5.7)–(10.5.10) for a horizontal wedge; case A in Fig. 10.5.1, Pr = 0.73

$\hat{\beta}$	$(f''_0)_{\eta=0}$	$(\phi'_0)_{\eta=0}$	$-\left(\dfrac{F''_1}{f''_0}\right)_{\eta=0}$	$-\left(\dfrac{G''_1}{f''_0}\right)_{\eta=0}$	$-\left(\dfrac{H'_1}{\phi'_0}\right)_{\eta=0}$	$-\left(\dfrac{I'_1}{\phi'_0}\right)_{\eta=0}$
1.0	4.93035	−1.00836	0.4952		0.0000	
0.8	4.09064	−0.90448	0.6361		0.0678	
0.7	3.71805	−0.86006	0.6928		0.1359	
0.5	3.02979	−0.78080	0.7547		0.1737	
0.1	1.70353	−0.63561	0.4391	3.0849	0.1048	0.7487
0.0	1.32823	−0.59418	0.0000	5.0737	0.0000	1.1959
−0.1	0.88128	−0.54179	1.0520	10.5898	0.2196	2.2566
−0.15	0.59024	−0.50390	2.8887	21.1421	0.5300	3.9062
−0.175	0.39642	−0.47570	6.2807	42.1402	0.9772	6.4970
−0.180	0.34853	−0.46821	7.9394	52.8078	1.1624	7.6153
−0.185	0.29479	−0.45940	10.7212	70.9885	1.4407	9.3283
−0.190	0.23168	−0.44880	16.4464	109.195	1.9304	12.4069

Source: Reprinted with permission from *Int. J. Heat Mass Transfer,* vol. 8, p. 43, R. C. Gunness and B. Gebhart. Copyright © 1965, Pergamon Journals Ltd.

$$\frac{\mathrm{Nu}_x}{(\mathrm{Nu}_x)_f} = 1 + \frac{\mathrm{Gr}}{\mathrm{Re}_x^2}\left(\frac{H'_1}{\phi'_0}\right)_{\eta=0} \tag{10.5.12}$$

where $\tau_f(x)$ and $(\mathrm{Nu}_x)_f$ are given by Eqs. (10.5.9) and (10.5.10). The terms in the square brackets are given in Table 10.5.4. The plus and minus signs are used for aiding and opposing effects respectively; see Table 10.5.2.

Note that, with an opposing buoyancy force, flow separation may occur. It is indicated when the local shear stress becomes zero. This condition may be determined from Eq. (10.5.7) for case A and from Eq. (10.5.11) for case B. The total drag coefficient and average Nusselt number can be obtained by integrating $\tau_0(x)$ and Nu_x over the wedge surface.

10.5.2 Mixed Convection on Inclined Surfaces

Consider an inclined flat semi-infinite surface and aligned with the forced flow, at an angle θ from the vertical, as shown in Fig. 10.5.3. Both the isothermal and uniform heat flux surface conditions have been analyzed by Mucoglu and Chen (1979) and Chen et al. (1980). The external flow is taken as uniform and parallel to the surface. Note that the foregoing analysis for a wedge is not applicable here since the free-stream velocity is constant. The continuity and energy equations are Eqs. (10.2.1) and (10.2.3). Assuming a boundary layer flow, the x- and y-momentum equations are

$$u\frac{\partial u}{\partial x} + v\frac{\partial u}{\partial y} = -\frac{1}{\rho}\frac{\partial p_m}{\partial x} \pm g\beta(t - t_\infty)\cos\theta + \nu\frac{\partial^2 u}{\partial y^2} \tag{10.5.13}$$

Table 10.5.4 Various functions to be used in Eqs. (10.5.11) and (10.5.12) for a vertical wedge orientation; case B in Fig. 10.5.1, Pr = 0.73

$\hat{\beta}$	$-\left(\frac{F_1''}{f_0''}\right)_{\eta=0}$	$-\left(\frac{H_1'}{\phi_0'}\right)_{\eta=0}$
1.0	0.0000	0.0000
0.8	0.2067	0.0336
0.7	0.3530	0.0692
0.5	0.7547	0.1737
0.1	2.4973	0.5961
0.0	3.6526	0.8392
−0.1	6.6426	1.3867
−0.19	53.4595	6.2748

Source: Reprinted with permission from *Int. J. Heat Mass Transfer*, vol. 8, p. 43, R. C. Gunness and B. Gebhart. Copyright © 1965, Pergamon Journals Ltd.

$$0 = -\frac{1}{\rho}\frac{\partial p_m}{\partial y} + g\beta(t - t_\infty)\sin\theta \tag{10.5.14}$$

As discussed in Chapter 5, the y component of the momentum equations is needed to include the motion pressure effect.

The plus-minus sign in the buoyancy term Eq. (10.5.13) corresponds to upward (aiding buoyancy) and downward (opposing buoyancy) forced flow, respectively. In Eq. (10.5.14) positive and negative values of θ represent the flow above and below the inclined surface, as discussed in Chapter 5. For $\theta = 0$ the surface becomes vertical and Eq. (10.5.13) reduces to Eq. (10.2.2). For $\theta = 90°$

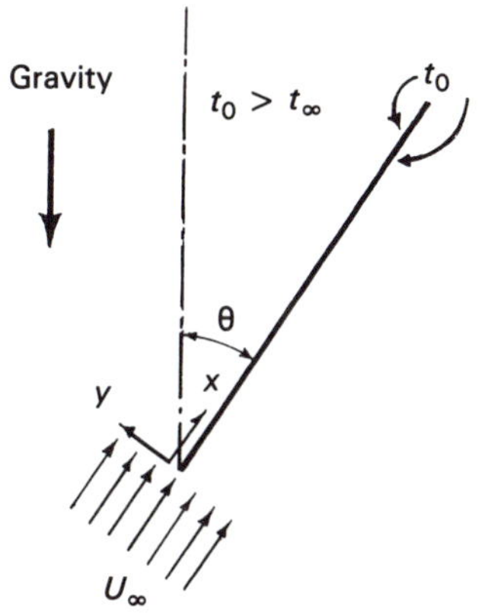

Figure 10.5.3 Coordinate system for a mixed convection flow over an inclined surface.

Eqs. (10.5.13) and (10.5.14) reduce to those for a horizontal surface, given in Section 10.3.

From an order of magnitude analysis, Mucoglu and Chen (1979) showed that, for $Re_x \simeq 10^3$, the pressure term in Eq. (10.5.13) is negligible if $\theta \leq 45°$. For $Re_x \simeq 10^5$ this approximation is valid if $\theta \leq 80°$. Within the range of this approximation, Eq. (10.5.14) is not used and Eq. (10.5.13) becomes

$$u \frac{\partial u}{\partial x} + v \frac{\partial u}{\partial y} = \pm g\beta(t - t_\infty) \cos \theta + \nu \frac{\partial^2 u}{\partial y^2} \qquad (10.5.15)$$

Mucoglu and Chen (1979) solved Eqs. (10.2.1), (10.2.3), and (10.5.15) by local similarity, local nonsimilarity, and finite-difference methods for an isothermal boundary condition over a wide range of the mixed convection parameter $Gr_x \cos \theta / Re_x^2$. The Grashof number for an isothermal boundary condition is defined as $g\beta(t_0 - t_\infty)x^3/\nu^2$. For Pr = 0.7 and 7.0, the local friction and heat transfer results are shown in Figs. 10.5.4 and 10.5.5, respectively. Here, $C_f = \tau_0/(\rho U_\infty^2/2)$, where τ_0 is the shear stress at the surface. For a uniform surface heat flux condition, the local friction and heat transfer results obtained by a finite-difference method are shown in Figs. 10.5.6 and 10.5.7, respectively. For this boundary condition the modified Grashof number is defined as $Gr_x^* = g\beta q''x^4/k\nu^2$ and the mixed convection parameter is $Gr_x^* \cos \theta / Re_x^{5/2}$.

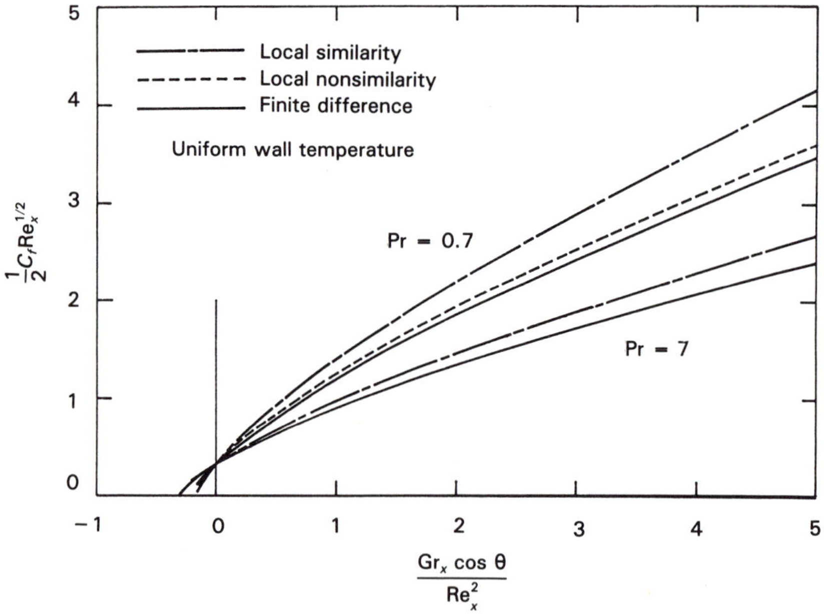

Figure 10.5.4 Local friction factor in a mixed convection flow over an inclined isothermal surface; Pr = 0.7. *(Reprinted with permission from A. Mucoglu and T. S. Chen,* J. Heat Transfer, *vol. 101, p. 422. Copyright © 1979, ASME.)*

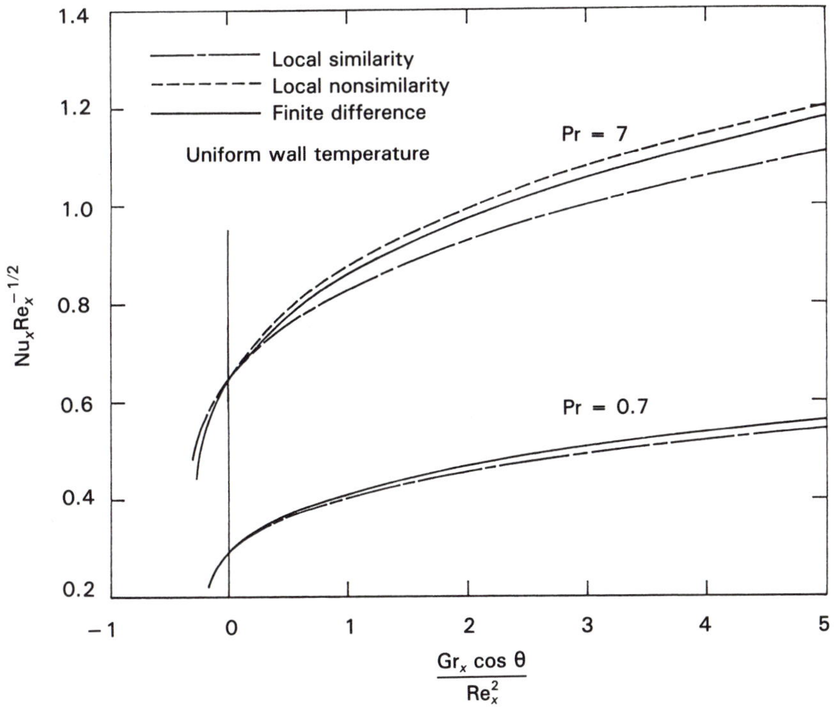

Figure 10.5.5 Local Nusselt number in a mixed convection flow over an inclined isothermal surface; Pr = 0.7. *(Reprinted with permission from A. Mucoglu and T. S. Chen,* J. Heat Transfer, *vol. 101, p. 422. Copyright © 1979, ASME.)*

10.5.3 Mixed Convection Transport from Spheres

A few analyses and experimental studies of mixed convection heat transfer from spheres have been reported. Three forced flow configurations—aiding, opposing, and cross flow—have been considered. Hieber and Gebhart (1969) analyzed the small Reynolds number mixed convection flow for Grashof numbers much less than the Reynolds number. A matched asymptotic expansion technique was used for the analysis. The Prandtl number was taken as unity. The drag coefficient is shown in Fig. 10.5.8. The positive and negative values of the Grashof numbers, based on diameter D, correspond to aiding and opposing effects, respectively. The levitational effects of buoyancy are quite clear.

For moderate and high values of the Reynolds and Grashof numbers, boundary layer approximations similar to those discussed earlier in Section 10.4.2 for a horizontal cylinder may be used. Chen and Mucoglu (1977) solved the boundary layer equations for aiding and opposing mixed convection flows over an isothermal sphere by finite-difference methods, assuming no downstream flow separation. For a Prandtl number of 0.7, the entire mixed convection regime was analyzed, starting with the two limiting cases of forced and natural convection.

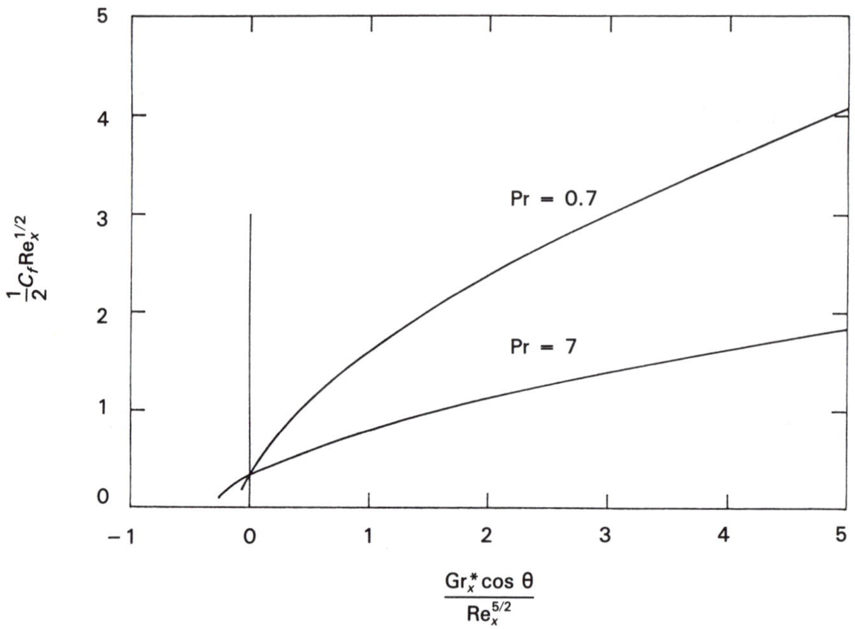

Figure 10.5.6 Local friction factor in a mixed convection flow over an inclined surface dissipating a uniform heat flux; Pr = 0.7. *(Reprinted with permission from A. Mucoglu and T. S. Chen,* J. Heat Transfer, *vol. 101, p. 422. Copyright © 1979, ASME.)*

Approaching from either side yielded the same results for the intermediate region. Two local free-stream velocity variations were considered, one from the potential flow solution around a sphere and the other from measured distributions. Angular distributions of the local friction factor and the Nusselt number were given. It was shown that the effect of buoyancy on forced convection heat transfer is significant if $\mathrm{Gr}_R/\mathrm{Re}_R^2 > 1.67$ or <-1.33, respectively, for aiding and opposing effects. Mucoglu and Chen (1978b) reported a similar analysis for a sphere with uniform heat flux surface condition.

Some experimental work on mixed convection over spheres has also been reported. One such study in air was reported for aiding, opposing, and crossed effects by Yuge (1960) for the Reynolds and Grashof number ranges 3.5 to 1.44×10^5 and 1 to 10^5, respectively. Klyachko (1963) critically reviewed the experimental data of Yuge (1960) and suggested heat transfer correlations for the mixed convection regime.

10.5.4 Mixed Convection from Cones

The effect of buoyancy forces on heat transfer from a cone rotating around its vertical axis was analyzed by Hering and Grosh (1963). A similarity solution arises for vertical forced flow when the surface temperature excess varies linearly

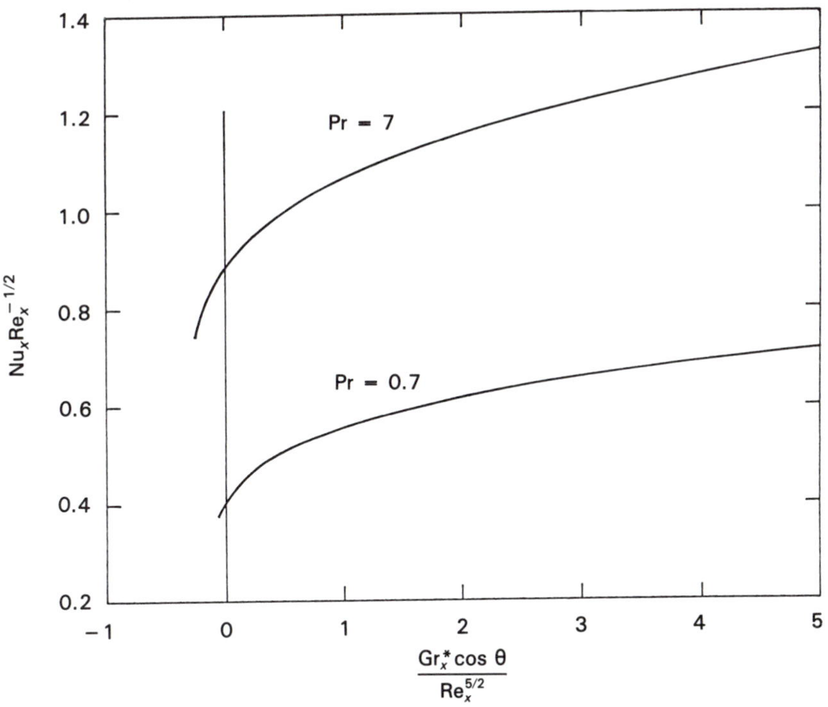

Figure 10.5.7 Local Nusselt number in a mixed convection flow over an inclined surface dissipating a uniform heat flux; Pr = 0.7. *(Reprinted with permission from A. Mucoglu and T. S. Chen,* J. Heat Transfer, *vol. 101, p. 422. Copyright © 1979, ASME.)*

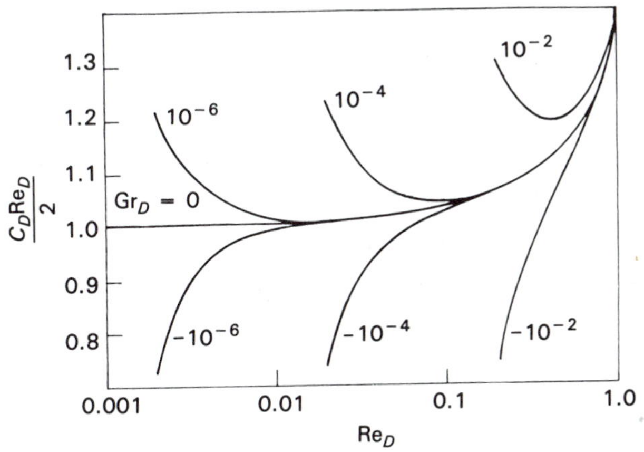

Figure 10.5.8 Drag coefficient for a sphere in a low Grashof number mixed convection flow. *(Reprinted with permission from C. A. Hieber and B. Gebhart,* J. Fluid Mech., *vol. 38, p. 137. Copyright © 1969, Cambridge University Press.)*

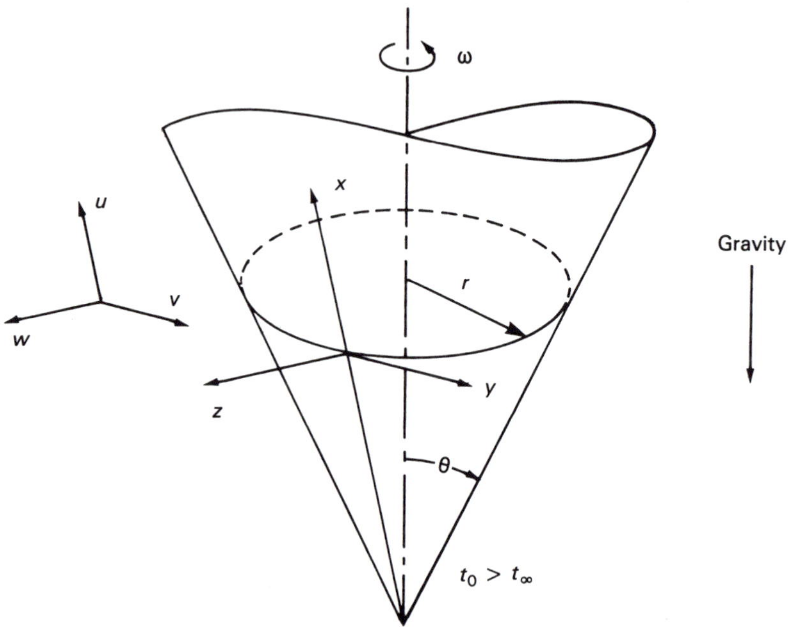

Figure 10.5.9 Coordinate system for mixed convection flow over a rotating cone. *(Reprinted with permission from R. G. Hering and R. J. Grosh,* J. Heat Transfer, *vol. 85, p. 29. Copyright © 1963, ASME.)*

with the distance from the cone apex and B_n is taken as zero. Only aiding effects were considered. The coordinate system is shown in Fig. 10.5.9. For a Prandtl number of 0.7, the local circumferential shear stress τ_y, axial shear stress τ_x, and heat transfer are given by

$$\frac{\tau_y}{\rho V^2/2} = \frac{\mu(\partial v/\partial z)_0}{\rho V^2/2} = 2(\mathrm{Re}_x)^{-1/2} G'(0) \tag{10.5.16}$$

$$\frac{\tau_x}{\rho V^2/2} = \frac{\mu(\partial u/\partial z)_0}{\rho V^2/2} = (\mathrm{Re}_x)^{-1/2} H''(0) \tag{10.5.17}$$

$$\mathrm{Nu}_x = (\mathrm{Re}_x)^{-1/2}[-\phi'(0)] \tag{10.5.18}$$

where the functions $G'(0)$, $H''(0)$, and $[-\phi'(0)]$ are given in Table 10.5.5 for various values of $\mathrm{Gr}_x/\mathrm{Re}_x^2$. The local Reynolds and the Grashof numbers are defined as

$$\mathrm{Re}_x = \frac{Vx}{\nu} \qquad \text{where } V = x\omega \sin\theta$$

$$\mathrm{Gr}_x = \frac{g\beta(t_0 - t_\infty)x\cos^2\theta}{\nu^2}$$

where ω is the angular velocity of the rotating cone and θ is the apex half-angle of the cone.

Using a perturbation method, Fox (1964) extended the above analysis, for $B_n = 0$, to a constant surface temperature condition. The average Nusselt number for Pr = 0.72 is given by

$$\text{Nu} = (\text{Re})^{1/2}\left[0.331 + 0.412\left(\frac{\text{Gr}}{\text{Re}^2}\right) + \cdots\right] \qquad (10.5.19)$$

where the slant height L of the cone is the characteristic length in Nu, Re, and Gr. The maximum circumferential velocity, which occurs at $x = L$, is used in the Reynolds number. For $\text{Gr}/\text{Re}^2 < 0.1$, Eq. (10.5.19) compares very well with experimental data.

The effect of buoyancy on the heat transfer from rotating vertical axisymmetric round-nosed bodies was analyzed by Suwono (1979). For a uniform surface temperature condition, the local shear stress and Nusselt number distributions were computed for Pr = 0.72 and 100 over a wide range of the parameter Gr/Re^2. An analytical study of mixed convection flow over a heated isothermal cone, with its axis horizontal, was reported by Yao and Catton (1978). Using regular perturbation expansions, the local heat transfer and the shear stress were determined for various values of the Prandtl number and the cone included angle. For a detailed discussion of the effect of rotation on the transport mechanisms, including Coriolis forces, see Chapter 17.

10.5.5 Mixed Convection from Isolated Thermal Sources

In recent years there has been growing interest in the cooling of electronic equipment. This has led to a considerable amount of study of mixed convection from isolated heat sources such as electronic components; see Jaluria (1985). Numerical methods have been used to determine the heat transfer rates for a wide variety of

Table 10.5.5 Values of $H''(0)$, $G'(0)$, and $\phi'(0)$ in Eqs. (10.5.16)–(10.5.18); Pr = 0.7

$\frac{\text{Gr}_x}{\text{Re}_x^2}$	$-H''(0)$	$-G'(0)$	$-\phi'(0)$
0	1.0205	0.61592	0.42852
0.1	1.13690	0.65489	0.46156
1.0	2.2078	0.85076	0.61202
10.0	8.5246	1.4037	1.0173
100.0	46.052	2.4738	1.7946
Free convection	1.4496	—	0.56699

Source: Reprinted with permission from R. G. Hering and R. J. Grosh, *J. Heat Transfer*, vol. 85, p. 29. Copyright © 1963, ASME.

boundary conditions and geometries. The transport process generally involves conjugate conduction heat transfer to the surfaces on which such sources are located. Experimental work has also been carried out on this mixed convection problem. For details of the various results obtained, reference is made to the work of Jaluria (1982, 1986), Moffat et al. (1985), and Oktay and Moffat (1985).

10.6 VERTICAL INTERNAL FLOWS

Mixed convection flows in vertical channels, tubes, and ducts have been extensively investigated because of their applications to nuclear reactors, heat exchangers, electronic equipment, and other cases of practical interest. Various boundary conditions have been considered in these studies. Perhaps the most important question is the effect of buoyancy on the forced convection transport rates. As discussed in earlier sections, the buoyancy forces may aid or oppose the forced flow, causing an increase or decrease in the heat transfer rates. An additional complexity in internal mixed convection flows is that a large variation in the pressure field may arise. Then a boundary layer formulation may not apply. Substantial experimental work has been done on these flows, along with some analysis. Most of the attention has been directed at determining the heat transfer rates, although some information has also been obtained on the velocity and temperature fields. Such work, for several different internal flow geometries, is reviewed here.

10.6.1 Flow between Two Vertical Surfaces

First considered is fully developed laminar mixed convection flow between two long parallel vertical surfaces at a distance $2a$ apart. The temperature of the two surfaces $t_0(x)$ is taken to vary linearly downstream. The buoyancy force is aligned with the forced flow. Figure 10.6.1 shows the coordinate system. For a fully developed laminar flow with constant properties except for the density variation, which is included with Boussinesq approximations, the governing equations are obtained by noting that the transverse velocity is zero. The normalized velocity and temperature distributions are taken as fully developed. Axial conduction is neglected. For this idealized circumstance, the governing equations become

$$\frac{\partial u}{\partial x} = 0 \tag{10.6.1}$$

$$-\nu \frac{\partial^2 u}{\partial y^2} = -\frac{1}{\rho}\frac{\partial p}{\partial x} + g\beta(t - t_0) \tag{10.6.2}$$

$$u \frac{\partial t}{\partial x} = \alpha \frac{\partial^2 t}{\partial y^2} \tag{10.6.3}$$

where the hydrostatic pressure is evaluated at t_0 and $\partial p/\partial x$ is the imposed pressure gradient driving the forced flow, neglecting motion pressure. The boundary conditions are

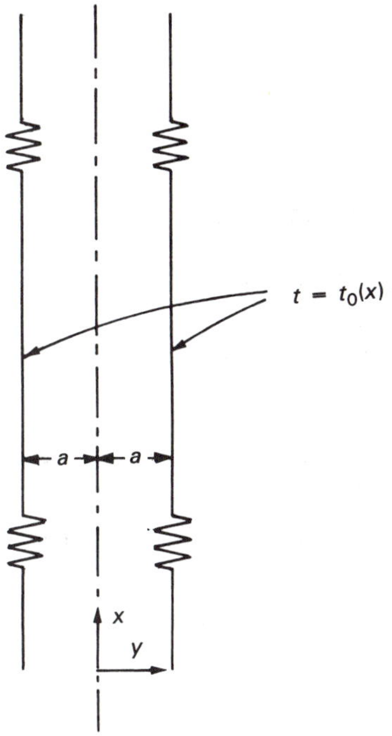

Figure 10.6.1 Coordinate system for flow between two vertical parallel surfaces.

$$u = 0 \qquad \text{at } y = \pm a$$
$$t = t_0(x) = C_1 x \qquad \text{at } y = \pm a \tag{10.6.4}$$

Since the surface temperature varies linearly and the form of the temperature distribution does not vary with x, the fluid temperature in the channel must also be of the form

$$t(x, y) - t_0 = t(x, y) - C_1 x = T(y) \tag{10.6.5}$$

The governing equations are now normalized by defining

$$X = \frac{x}{a} \qquad Y = \frac{y}{a} \qquad U = \frac{u}{u_m}$$

$$\phi = \frac{\alpha(t - t_0)}{C_1 u_m a^2} \qquad E = \frac{(\partial p/\partial x)a^2}{\mu u_m} \qquad \text{Ra} = \frac{g a^3 \beta a C_1}{\nu^2}\,\text{Pr} \tag{10.6.6}$$

where u_m is the mean axial velocity and C_1 the axial temperature gradient. With this normalization, the governing equations become

$$\frac{\partial^2 U}{\partial y^2} + Ra\phi = E \tag{10.6.7}$$

$$\frac{\partial^2\phi}{\partial Y^2} - U = 0 \tag{10.6.8}$$

The corresponding boundary conditions are

$$U = \phi = 0 \qquad \text{at } Y = \pm 1 \tag{10.6.9}$$

Equations (10.6.7) and (10.6.8) are combined by eliminating ϕ. The resulting fourth-order linear homogeneous equation is easily solved. Solutions obtained by different methods have been reported by Ostrach (1954), Tao (1960a, 1960b), and Agarwal (1962). Those obtained by Tao are given below.

The axial pressure gradient, given in terms of the parameter E, is obtained as

$$E = -(\text{Ra})^{1/2}K\left(\frac{\cosh 2K + \cos 2K}{\sinh 2K - \sin 2K}\right) \tag{10.6.10}$$

where $K = (\text{Ra})^{1/4}/\sqrt{2}$. The dimensionless axial velocity U and temperature ϕ distributions are given by

$$U = -(\text{Ra})^{-1/2}E\,\frac{\sinh K(1+Y)\sin K(1-Y) + \sin K(1+Y)\sinh K(1-Y)}{\cosh 2K + \cos 2K} \tag{10.6.11}$$

$$\phi = \frac{1}{\text{Ra}}E\left[1 - \frac{\cosh K(1+Y)\cos K(1-Y) + \cos K(1+Y)\cosh K(1-Y)}{\cosh 2K + \cos 2K}\right] \tag{10.6.12}$$

The mean values of ϕ and t across the channel are denoted by ϕ_m and t_m, where

$$\begin{aligned}\phi_m &\equiv \frac{\alpha(t_m - t_0)}{C_1 u_m a^2}\\ &= -\frac{E^2}{4(\text{Ra})^{3/2}}\left[\frac{3}{2K}\left(\frac{\sinh 2K - \sin 2K}{\cosh 2K + \cos 2K}\right)\right.\\ &\qquad \left. - \frac{\sinh 2K \sin 2K}{(\cosh 2K + \cos 2K)^2}\right]\end{aligned} \tag{10.6.13}$$

The Nusselt number is found to be

$$\text{Nu} = \frac{ha}{k} = \frac{2}{\phi_m} \tag{10.6.14}$$

Note that this situation amounts to a uniform heat flux surface condition.

The above analysis applies only to fully developed flow and temperature conditions. This simplifies the analysis, and an exact solution is obtained. In any actual circumstance, the surfaces are of finite length with an opening at each end.

With a forced flow at the inlet, transport develops downstream. Quintiere and Mueller (1973) developed an approximate method of analysis for such mixed convection flow. The method is based on a linearization of the governing equations, in terms of an average axial velocity. Both walls are taken as isothermal at t_0. The ambient fluid at t_∞ and U_0 enters at the bottom and is discharged at the top. The coordinate system is the same as that shown in Fig. 10.6.1 and $t_0 > t_\infty$. The channel length is L. The dimensionless temperature ϕ is found to be

$$\phi \equiv \frac{t - t_\infty}{t_0 - t_\infty} = 1 + 2\sum_{n=1}^{\infty} \frac{(-1)^n}{\beta_n} e^{-(\beta_n^2/\text{Pr})\xi} \cos\left(\beta_n \frac{y}{a}\right) \tag{10.6.15a}$$

where $\beta_n = [(2n - 1)/2]\pi$; $n = 1, 2, \ldots$; $\xi = x\nu/U_0a^2$ and $\text{Gr} = g\beta(t_0 - t_\infty)a^4/L\nu^2$. The velocity distribution is omitted here for brevity.

The Nusselt number (ha/k), averaged over the height L, is shown in Fig. 10.6.2 as a function of the Peclet (RePr), where $\text{Re} = U_0a^2/\nu L$, and Rayleigh numbers. The pressure variation in the channel is an important consideration. If a fan is to be selected for cooling purposes in the channel, then the required fan pressure level ΔP_{fan} is given by

$$\Delta P_{\text{fan}} = P_{\nu 0} + \frac{\rho_\infty U_0^2}{2} = \frac{\rho_\infty L^2\alpha^2}{a^4}\left[P'_{\nu 0} + \frac{(\text{RePr})^2}{2}\right] \tag{10.6.15b}$$

where $P_{\nu 0}$ is the pressure defect from stagnation, α is the fluid thermal diffusivity, and $\text{Re} = U_0a^2/\nu L$. For various values of the Rayleigh number (GrPr), a dimensionless pressure defect $P'_{\nu 0}$ was defined and computed as a function of the Peclet number, as shown in Fig. 10.6.3. For pure natural convection, the physical pressure defect $P_{\nu 0} = -\rho_\infty U_0^2/2$ from Bernoulli's equation and the fan pressure ΔP_{fan} is zero, as expected. In mixed convection, $P_{\nu 0}$ is larger than the corresponding

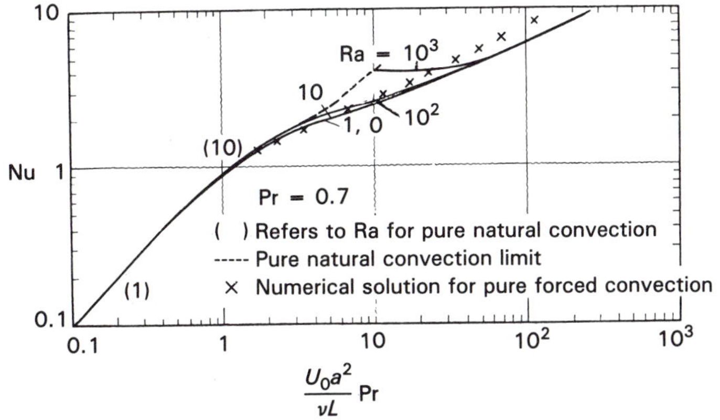

Figure 10.6.2 Nusselt number as a function of the Peclet number $(U_0a^2/\nu L\text{Pr})$ for developing mixed convection flow between two vertical surfaces. *(Reprinted with permission from J. Quintiere and W. K. Mueller,* J. Heat Transfer, *vol. 95, p. 53. Copyright © 1973, ASME.)*

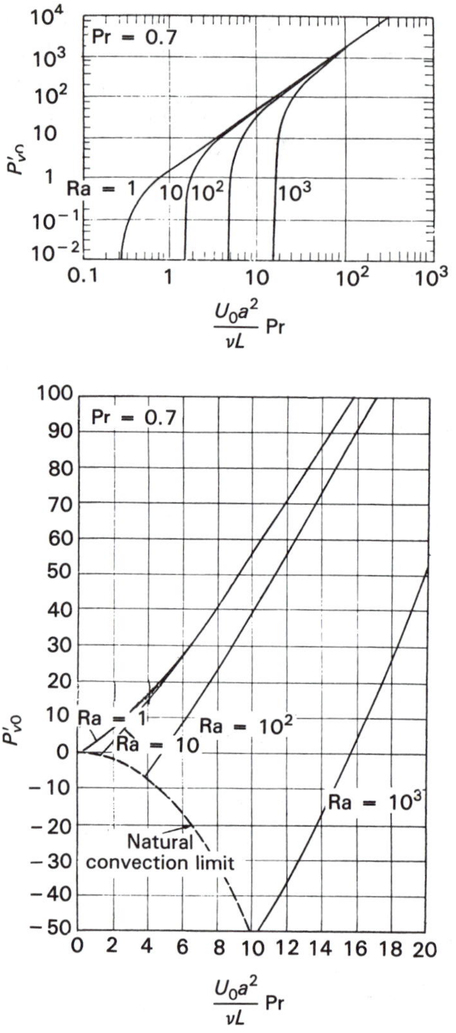

Figure 10.6.3 Dimensionless pressure defect P_{v0} as a function of the Peclet number $(U_0 a^2/\nu L\mathrm{Pr})$ for developing mixed convection flow between two vertical surfaces *(Reprinted with permission from J. Quintiere and W. K. Mueller,* J. Heat Transfer, *vol. 95, p. 53. Copyright © 1973, ASME.)*

natural convection value, and the required fan pressure may be calculated from Eq. (10.6.15*b*).

10.6.2 Mixed Convection in a Vertical Circular Tube

This is perhaps the most extensively investigated mixed convection problem because of its application in heat exchangers and in nuclear technology. The tube orientation substantially modifies the effect of buoyancy in the flow. A vertical flow is considered here. The buoyancy forces may aid or oppose the forced flow, depending on the direction of flow and the tube thermal boundary conditions. For a laminar flow, aiding buoyancy forces increase the heat transfer rates. However, in the turbulent flow regime the buoyancy effects are, surprisingly, found to have an opposite effect.

Developed flow. For laminar flow in a sufficiently long tube, the flow will develop completely. This assumption yields an exact solution for t_0 varying as $C_1 x$, as seen earlier for a parallel-surface channel. For constant properties, except for the buoyancy effect, the equations in axisymmetric fully developed flow are

$$\nu \frac{1}{r}\frac{\partial}{\partial r}\left(r \frac{\partial u}{\partial r}\right) + g\beta(t - t_0) = \frac{1}{\rho}\frac{\partial p}{\partial x} \tag{10.6.16}$$

$$\alpha \frac{1}{r}\frac{\partial}{\partial r}\left(r \frac{\partial t}{\partial r}\right) = u \frac{\partial t}{\partial x} \tag{10.6.17}$$

The analysis is for aiding effects, that is, for both buoyancy and the forced flow upward or downward. A uniform heat flux condition at the wall results for the linear wall temperature distribution considered. Solutions to Eqs. (10.6.16) and (10.6.17) have been reported by Hallman (1956), Hanratty et al. (1958), Morton (1960), and Tao (1960a, 1960b). Morton (1960) also considered an opposing buoyancy effect. These investigations also included the effect of volumetric heat generation in the ambient fluid, q'''. This is neglected in the following analysis.

The normalizing scales are the same as in Eqs. (10.6.6) except for $R = r/a$, where a is the tube radius. The governing equations and boundary conditions become

$$\frac{1}{R}\frac{\partial}{\partial R}\left(R \frac{\partial U}{\partial R}\right) + Ra\, \phi = E \tag{10.6.18}$$

$$\frac{1}{R}\frac{\partial}{\partial R}\left(R \frac{\partial \phi}{\partial R}\right) - U = 0 \tag{10.6.19}$$

$$\begin{aligned} &\text{at } R = 1: \quad U = \phi = 0 \\ &\text{at } R = 0: \quad \frac{\partial U}{\partial R} = \frac{\partial \phi}{\partial R} = 0 \end{aligned} \tag{10.6.20}$$

The solution to these equations is expressed in terms of Bessel and modified Bessel functions of zero order with complex arguments. It may also be written in terms of ber, bei, ker, and kei functions with real arguments. The pressure gradient parameter E, defined in Eq. (10.6.6), is given by

$$E = -\frac{\epsilon^2}{M} \tag{10.6.21}$$

where $\epsilon = (\text{Ra})^{1/4}$ and

$$M = \frac{\sqrt{2}}{\epsilon}\left[\frac{\text{ber}_0\epsilon(\text{ber}_1\epsilon + \text{bei}_1\epsilon) - \text{bei}_0\epsilon(\text{ber}_1\epsilon - \text{bei}_1\epsilon)}{\text{ber}_0^2(\epsilon) + \text{bei}_0^2(\epsilon)}\right]^2$$

The dimensionless axial velocity U and temperature ϕ are

$$U = \epsilon^{-2}E\left[\frac{\mathrm{ber}_0(\epsilon R)\mathrm{bei}_0(\epsilon) - \mathrm{bei}_0(\epsilon R)\mathrm{ber}_0(\epsilon)}{\mathrm{ber}_0^2(\epsilon) + \mathrm{bei}_0^2(\epsilon)}\right] \tag{10.6.22}$$

$$\phi = \epsilon^{-4}E\left[\frac{\mathrm{ber}_0(\epsilon R)\mathrm{ber}_0(\epsilon) + \mathrm{bei}_0(\epsilon R)\mathrm{ber}_0(\epsilon)}{\mathrm{ber}_0^2(\epsilon) + \mathrm{bei}_0^2(\epsilon)}\right] \tag{10.6.23}$$

To evaluate the Nusselt number, the local mean temperature, defined earlier, is determined. Therefore,

$$\begin{aligned}\phi_m &= \frac{k(t_m - t_0)}{\rho c_p C_1 u_m a^2}\\ &= -\frac{E^2}{2\epsilon^6}\left[M + \frac{(\mathrm{ber}_0\epsilon\ \mathrm{ber}_1\epsilon + \mathrm{bei}_0\epsilon\ \mathrm{bei}_1\epsilon)\ (\mathrm{ber}_0\epsilon\ \mathrm{bei}_1\epsilon - \mathrm{bei}_0\epsilon\ \mathrm{ber}_1\epsilon)}{\mathrm{ber}_0^2(\epsilon) + \mathrm{bei}_0^2(\epsilon)}\right]\end{aligned} \tag{10.6.24}$$

where M is defined in Eq. (10.6.21). The average Nusselt number based on the mixed mean temperature difference is

$$\mathrm{Nu} = \frac{ha}{k} = -\frac{1}{2\phi_m} \tag{10.6.25}$$

Developing flows. Again, in most practical applications, developing flows occur instead. Several analytical and experimental studies have been reported. Martinelli and Boelter (1942) analyzed mixed convection in vertical tubes and suggested a heat transfer correlation that included the effect of L/D ratio. Pigford (1955) and Rosen and Hanratty (1967) used an integral approach to analyze the flow in a vertical tube. Subsequent calculations used finite-difference methods. On the basis of an experimental study of mixed convection for air in a vertical tube, Jackson et al. (1967) found that the equation of Martinelli and Boelter did not correlate their data. The following correlation of the local Nusselt number for aiding effects in a vertical isothermal tube was suggested:

$$\mathrm{Nu}_x = \frac{h_x D}{k} = 1.128\left[\mathrm{RePr}\frac{D}{x} + \left(3.02\mathrm{GrPr}\frac{D}{L}\right)^{0.4}\right]^{0.5} \tag{10.6.26}$$

where x is the distance from the entrance to a heated portion in a tube of length L. Here, the Reynolds number $\mathrm{Re} = u_m D/\nu$.

Kemeny and Somers (1962) experimentally investigated aiding mixed convection flow in a vertical heated tube dissipating a uniform surface heat flux to the flowing fluid. They used water and oil. Most of the data were for laminar flow. Lawrence and Chato (1966) analyzed developing mixed convection flow in a vertical tube with uniform wall heat flux. Boundary layer approximations were used, and the equations were solved numerically. Experimental results for water in a tube with an L/D value of 203 were also reported. It was concluded that the velocity and temperature profiles did not become fully developed. The experimental data agreed well with the calculations if the density and viscosity variation with temperature was taken into account.

Marner and McMillan (1970) calculated a developing aiding mixed convection flow in a vertical isothermal tube by a finite-difference method. Typical developing dimensionless axial velocity profiles for Pr = 1 and Gr/Re = 120 are shown in Fig. 10.6.4, where both Gr and Re are based on the tube radius a. The fluid at $t_\infty < t_0$ enters the heated section at wall temperature t_0, with a fully developed parabolic velocity profile, at $x = 0$. As the fluid is heated near the wall, a buoyancy force arises. Fluid velocity near the wall increases and a decrease in the centerline velocity results. This deviation from the originally parabolic profile continues to (x/a)RePr = 0.102. At this x, the deviation is maximum and therefore the centerline velocity is minimum. Farther downstream, the velocity distribution again approaches a fully developed parabolic profile as the fluid temperature asymptotically approaches the wall temperature t_0. It was concluded that the decrease in the centerline velocity was even larger at higher values of Gr/Re.

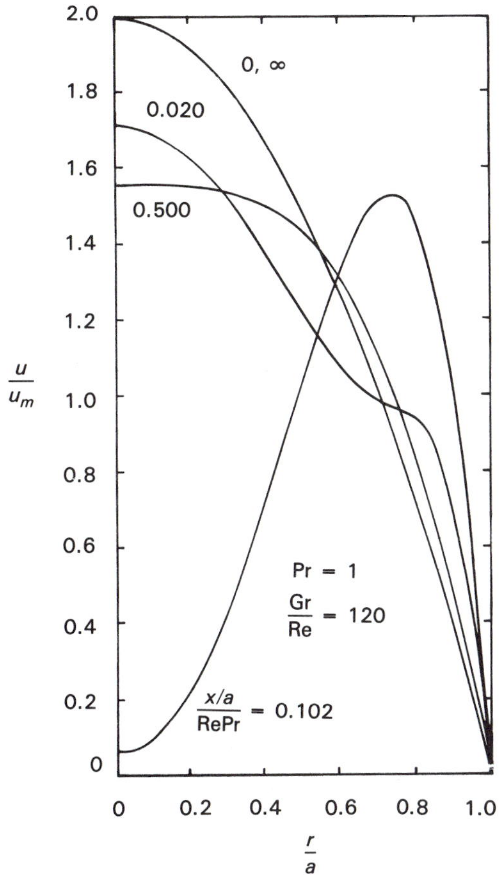

Figure 10.6.4 Typical developing dimensionless axial velocity profiles in the mixed convection flow in a heated vertical tube. *(Reprinted with permission from W. J. Marner and H. K. McMillan,* J. Heat Transfer, *vol. 92, p. 559. Copyright © 1970, ASME.)*

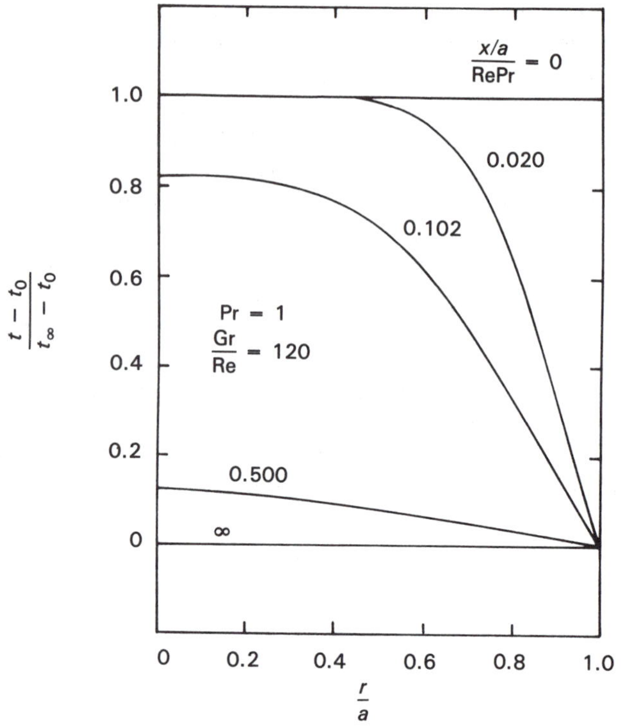

Figure 10.6.5 Typical developing dimensionless temperature profiles in the mixed convection flow in a heated vertical tube. *(Reprinted with permission from W. J. Marner and H. K. McMillan,* J. Heat Transfer, *vol. 92, p. 559. Copyright © 1970, ASME.)*

Typical developing dimensionless temperature profiles are shown in Fig. 10.6.5. At the entrance to the heated section, $\phi = 1$, where $\phi = (t - t_0)/(t_\infty - t_0)$. Then ϕ asymptotically approaches zero as the fluid temperature approaches the wall temperature. The variation of the local Nusselt number downstream is shown in Fig. 10.6.6 for Pr = 1 and Gr/Re = 120. It is interesting that the maximum corresponds to the downstream location of maximum velocity profile distortion.

Similar numerical calculations were made by Zeldin and Schmidt (1972) for two velocity distributions at the entrance to the heated length: a uniform velocity and a fully developed velocity profile. The latter condition arises with a long unheated upstream section of tube. Experimental measurements of the velocity and temperature profiles were also made with air for a uniform entrance velocity condition u_m. Two sets of data were taken: one in the forced convection-dominant regime, at Gr/Re = 0.735, and the other in the mixed convection regime, at Gr/Re = 33.25, where Re is based on the diameter $2a$, as defined below. It was concluded that natural convection effects may alter the velocity profile significantly. However, such effects were found to be negligible for

$$\frac{\text{Gr}}{\text{Re}} \leq 1 \tag{10.6.27}$$

where

$$\text{Gr} = \frac{ga^3\beta(t_0 - t_\infty)}{\nu^2} \quad \text{and} \quad \text{Re} = \frac{2au_m}{\nu}$$

Collins (1978) analyzed both aiding and opposing mixed convection flows in a vertical heated tube for both isothermal and uniform flux surface conditions. Computed heat transfer results for water were compared with the experimental data of Scheele et al. (1960) and Scheele and Hanratty (1962) with fairly good agreement.

Turbulent flows. The analyses discussed above concern only laminar flows for which the buoyancy force increases the heat transfer rate for aiding effects and decreases it for opposing ones. One is tempted to postulate similar behavior in turbulent mixed convection flows. Several experimental studies reveal just the

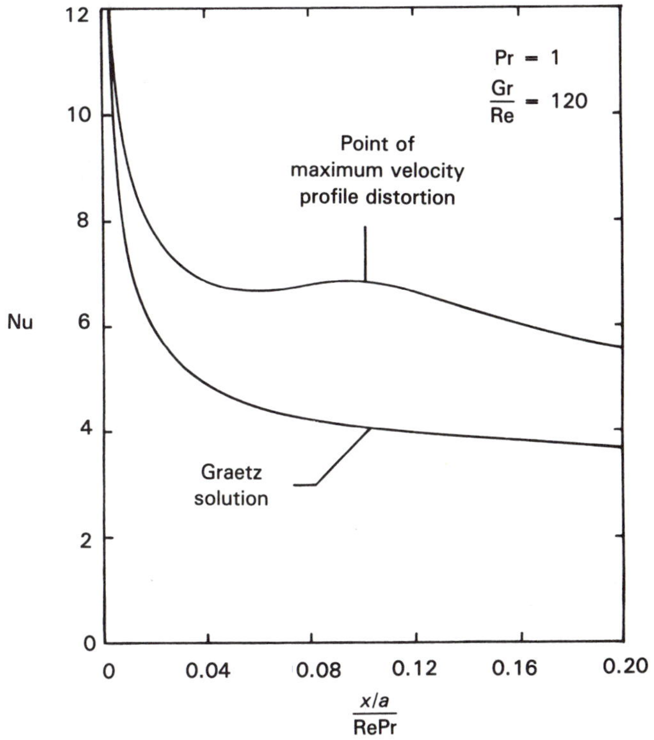

Figure 10.6.6 Variation of the local Nusselt number along the tube length in a developing mixed convection flow in a heated vertical tube. *(Reprinted with permission from W. J. Marner and H. K. McMillan,* J. Heat Transfer, *vol. 92, p. 559. Copyright © 1970, ASME.)*

opposite. For both isothermal and uniform flux surface conditions, aiding buoyancy forces tend to decrease local heat transfer rates while opposing buoyancy forces increase them.

Axcell and Hall (1978) proposed a mechanism that explains this phenomenon. For a given uniform surface heat flux, the temperature difference between the tube wall and the fluid flowing through it is determined by the convection velocities and the variation of thermal diffusivity in the fluid. Apart from variations with temperature, the effective thermal diffusivity is constant in a laminar flow and equal to the molecular diffusivity. However, it is usually orders of magnitude greater in a turbulent flow and varies sharply out from the wall. Figure 10.6.7 shows qualitatively the conjectured velocity and shear stress distributions in a turbulent mixed convection vertical flow. In laminar flows, the distortion of these distributions is the only effect, as borne out by experimental data. A similar shift in the distributions arises in turbulent flows. However, now the much higher turbulent diffusivity dominates.

The shear stress decreases across the flow cross section. Figure 10.6.7 suggests how this distribution changes with buoyancy. Away from the wall, the shear stress increases due to opposing buoyancy effects and decreases due to aiding effects. The effect is reversed at the wall. Since the turbulence production, which is proportional to the product of the turbulent shear stress and the velocity gradient, occurs at some distance away from the wall, opposing effects result in higher turbulence production and a larger heat transfer rate. Aiding effects result in the opposite tendency. For low surface heat flux, the effect of buoyancy on

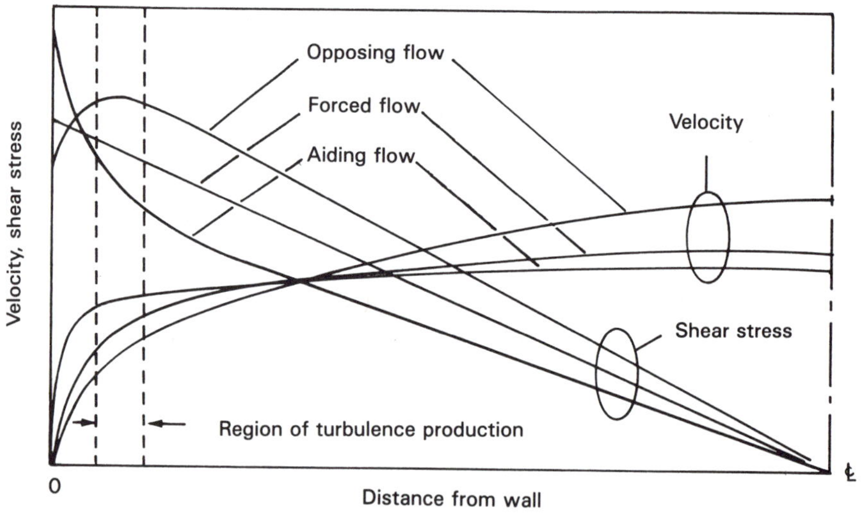

Figure 10.6.7 Qualitative description of velocity and shear stress distribution in a turbulent mixed convection flow in a heated vertical tube. *(Reprinted with permission from B. P. Axcell and W. B. Hall,* Proc. 6th Int. Heat Transfer Conf., *vol. 1, p. 37. Copyright © 1978, Hemisphere Publishing Corporation.)*

heat transfer is small. However, beyond a critical value of flux, it is large. Since the understanding of such flows is extremely important in nuclear reactor and heat exchanger applications, a large number of experimental and analytical studies have been performed.

Early experiments by Eckert and Diaguila (1954) determined local heat transfer rates in laminar and turbulent mixed convection in air. An isothermal vertical tube with a L/D ratio of 5 was used. For aiding effects at a given Reynolds number, the local Nusselt number was found to decrease as the Grashof number was increased and to increase slightly at large Gr. This behavior arose in the range $\mathrm{Re} = 10^5$ to 7×10^5, where $\mathrm{Re} = u_m x/\nu$ and x is again the distance from the tube entrance. However, for opposing effects the Nusselt number was found to be higher than those for either pure forced or natural convection flows.

Criteria were established for the limits between natural and mixed convection flow and between forced and mixed convection flow. Metais and Eckert (1964) reviewed all existing information and suggested boundaries between the natural, mixed, and forced convection regimes for both laminar and turbulent flow. Horizontal and vertical tubes were considered separately.

Ojalvo and Grosh (1962) reported finite-difference calculations concerning turbulent mixed convection. The results for pure forced convection did not agree with known experimental results, probably because of uncertainties regarding the closure model used. In a later paper Ojalvo et al. (1967) also included volume heat sources. A different model for eddy viscosity was used. Volume heat sources were found to have a negligible effect on the velocity profile. However, the temperature profiles were altered considerably. Brown and Gauvin (1965, 1966), in an experimental study of turbulent mixed convection, found that buoyancy forces cause large disturbances in the temperature field for opposing flow and thus enhance heat transfer. Tanaka et al. (1973) calculated the effects of buoyancy and acceleration due to the thermal expansion of fluid in a vertical tube. This acceleration is especially significant for fluids near their critical point. It was concluded that the effects of buoyancy and of acceleration on thermal transport are quite similar.

Carr et al. (1973) reported an analytical and experimental study of turbulent mixed convection flow of air with aiding effects in a vertical pipe with uniform surface heat flux. Measurements of various turbulent transport quantities were made for $L/D = 100$, $5000 < \mathrm{Re} < 14{,}000$, and the heat flux varying from about 15.8 to 63.1 $\mathrm{W/m^2}$. Buoyancy effects distorted the velocity profiles and caused low turbulent shear stress and heat transfer rates. Similar conclusions were reached by Buhr et al. (1974) for mercury. It was suggested that the assumption of negligible buoyancy effects is valid only if $\mathrm{Ra/Re} < 0.1$. The data indicated that as the surface flux increased the Nusselt number decreased to a minimum and then gradually increased. Petukhov (1976) reported wide-ranging experimental data and concluded that buoyancy has two distinct effects on turbulent transport: it influences the average velocity and temperature fields and strengthens or weakens the intensity of turbulent mixing. Depending on the conditions, both effects may be about the same or one may predominate.

An experimental study with opposing effects in the turbulent flow of nitrogen in a vertical isothermal pipe was reported by Easby (1978). Measurements were made for $2000 < \text{Re} < 10{,}000$ and $\text{Gr} < 10^6$. Compared with isothermal flow behavior, the measured mixed convection friction factors were reduced by up to 20%. Heat transfer increased by up to 40%. The increase in heat transfer was attributed to the distortion in the shear stress distribution, as discussed earlier. The heat transfer results are correlated in terms of the Stanton number, with the dimensionless parameters based on diameter, as

$$\frac{\text{St}}{\text{St}_f} = 1 + 8.9\frac{\text{Gr}}{\text{Re}^2} \tag{10.6.28}$$

This applies for opposing effects in the parameter ranges given above. Stanton number is defined as

$$\text{St} = \frac{h}{\rho u_m c_p} = \frac{\text{Nu}}{\text{RePr}} \tag{10.6.29}$$

St_f is the Stanton number for a pure forced flow, that is, as $\text{Gr} \to 0$. Measurements in turbulent mixed convection flow, reported by Connor and Carr (1978), confirm the heat transfer deterioration for aiding effects.

Abdelmeguid and Spalding (1979) used a finite-difference procedure to analyze the turbulent mixed convection flows of air in horizontal, inclined, and vertical pipes. A two-equation turbulent model proposed by Launder and Spalding (1974) was used. For a uniform heat flux boundary condition, both upward and downward flows were analyzed. For an upward flow the Nusselt number decreased first with an increase in Gr until a critical value of Gr was reached; then it began to increase monotonically. However, for a downward flow the Nusselt number increased monotonically with increasing Gr. These results were in good agreement with the experimental data of Buhr et al. (1974). There have been several experimental studies of turbulent mixed convection in vertical tube flow of supercritical fluids. Kakarala and Thomas (1980) have reviewed a large number of these studies.

10.6.3 Other Vertical Internal Flow Geometries

Tubes with noncircular cross section are common in heat exchangers, nuclear reactor cooling, magnetohydrodynamics (MHD), space heating and cooling ducts, and so forth. These may include rectangular, regular polygon, circular sector, annulus, and triangular cross sections. Vertical mixed convection flows in these noncircular geometries may also be classified as fully developed or developing flows. Most of the available information deals only with laminar flows. The heat transfer and pressure drop results for laminar forced convection in noncircular ducts have been reviewed by Shah and London (1978).

For fully developed laminar mixed convection flow in vertical rectangular ducts with a uniform heat flux boundary condition, exact solutions to the governing equations have been obtained by Han (1959), Tao (1960a, 1960b), and Agrawal (1962) by different methods. All the properties were considered constant except for the density effect, which gives rise to the buoyancy. These analyses included volumetric heat generation in the fluid. Also, for the uniform flux condition Iqbal et al. (1969) reported solutions for right triangle, isosceles triangle, and rhombic ducts. Aggarwala and Iqbal (1969) computed heat transfer rates and pressure drops for various triangular cross sections. Wall temperature was assumed to vary linearly in the flow direction and to be uniform around the periphery. For fully developed flow this corresponds to a uniform flux boundary condition. In the absence of volumetric heat generation, with aiding effects, the Nusselt number and the pressure drop parameter E for various values of the Rayleigh number are given in Table 10.6.1. These dimensionless groups are defined as

$$\mathrm{Nu} = \frac{hD_h}{k} \qquad E = \frac{-D_h^2(\partial p/\partial x)}{\mu u_m} \qquad \mathrm{Ra} = \frac{gD_h^3\beta a C_1}{\nu^2}\,\mathrm{Pr} \tag{10.6.30}$$

where D_h (= area/perimeter) is the hydraulic diameter of the duct, x the axial coordinate, u_m the mean axial velocity, and C_1 the specified axial temperature gradient. As shown in Table 10.6.1, the effect of buoyancy forces on an upward flow is to increase both the heat transfer and the pressure drop.

Table 10.6.1 Heat transfer and pressure drop characteristics in a developed mixed convection flow through vertical triangular ducts

Rayleigh number Ra	Nusselt number Nu			Pressure drop parameter E		
	60°–60°–60°	30°–60°–90°	45°–45°–90°	60°–60°–60°	30°–60°–90°	45°–45°–90°
0	3.1111	2.8875	2.9819	26.6668	26.0677	26.3063
10	3.1249	2.9067	2.9984	27.4686	26.9307	27.1424
10^2	3.2475	3.0730	3.1436	34.5306	34.4557	34.4690
3×10^2	3.4447	3.3280	3.3719	45.7385	46.1709	45.9797
5×10^2	3.7537	3.7017	3.7183	63.1045	63.9450	63.6066
8×10^2	4.0392	4.0250	4.0275	79.1441	80.1168	79.7410
10^3	4.3029	4.3091	4.3053	94.1270	95.1109	94.7386
2×10^3	5.1761	5.1903	5.1891	146.7824	147.5910	147.2683
3×10^3	5.8375	5.8314	5.8400	192.1201	192.8906	192.5375
5×10^3	6.7971	6.7632	6.7833	271.0402	272.0765	271.5378
8×10^3	7.6444	7.5975	7.6247	357.4286	358.9599	358.1352
10^4	8.2944	8.2406	8.2737	435.7798	437.8083	436.6856
2×10^4	10.0640	9.9778	10.0368	706.7318	710.7794	708.3187
3×10^4	11.2540	11.1304	11.2161	940.8602	947.2163	943.1565
5×10^4	12.9369	12.7397	12.8782	1353.0771	1364.9150	1357.0093
8×10^4	14.4346	14.1474	14.3506	1808.9477	1828.9385	1815.2493
10^5	15.5929	15.2170	15.4837	2224.8617	2254.2112	2233.8457

Source: Reprinted with permission from *Int. J. Heat Mass Transfer*, vol. 12, p. 737, D. Aggarwala and M. Iqbal. Copyright © 1969, Pergamon Journals Ltd.

Iqbal et al. (1970) considered a fully developed laminar mixed convection flow in vertical ducts with regular polygonal cross sections for aiding effects. Two thermal conditions were considered: uniform circumferential temperature and uniform circumferential heat flux. Values of the Nusselt number and the dimensionless pressure drop, for various values of the Rayleigh number and the number of sides of the polygon, are given in Tables 10.6.2 and 10.6.3. As the number of sides increases, the geometry approaches a circle. Results for a circular tube are also given in these tables. The results for $Ra = 1$ closely correspond to those for pure forced convection. Iqbal et al. (1972) also investigated the effect of peripheral wall conduction on mixed convection in a vertical rectangular duct. It was concluded that as the aspect ratio increases this effect becomes important.

Nayak and Cheng (1975) reported a finite-element analysis of fully developed laminar mixed convection flow through noncircular vertical ducts under the conditions of constant axial heat flux and uniform peripheral wall temperature. Numerical values of the Nusselt number at selected Rayleigh numbers were obtained for the special cases of square and triangular ducts. The numerical results were shown to approach the exact solution rapidly as the number of points used was increased.

Mixed convection in a vertical pipe with a cross section in the form of a sector of a circle was analyzed by Lu (1959). Steady, laminar, and fully developed flow was assumed. Heat transfer rates were computed for several such cross sections, including sectors of annuli.

Several studies of mixed convection in a vertical concentric annulus have been reported. Sherwin (1968) calculated mixed convection heat transfer for fully developed laminar flow at the inlet. Criteria for flow reversal were given. Velocity and temperature profiles and Nusselt number values for a radius ratio of 3 between the outer and the inner cylinders were also reported. Maitra and Raju (1975) extended the work of Sherwin to higher values of the Rayleigh number. In addition, experimental heat transfer data were reported. The inner wall was maintained at a constant heat flux while the outer wall was kept insulated. The analysis was in qualitative agreement with the data. Sherwin and Wallis (1970) provided solutions for developing flow in a vertical annulus. Ogunba and Barrow (1979) used a flow visualization technique to study the development of the velocity profile in the entrance region of a vertical annulus. El-Shaarawi and Sarhan (1980) analyzed laminar mixed convection annular flow with simultaneously developing hydrodynamic and thermal boundary layers. For $Pr = 0.7$ both aiding and opposing effects were considered with the thermal conditions of one wall being isothermal and the other adiabatic. Transport results were obtained for the values of Gr/Re between -700 and 1500, where both Gr and Re are based on the difference between the outer and inner diameters of the annulus as the characteristic length. Buoyancy forces were shown to have a significant effect on the hydrodynamic and heat transfer characteristics of a developing laminar annular flow. El-Shaarawi and Sarhan (1982) reported a finite-difference analysis of a similar geometry with a rotating inner cylinder.

Laminar mixed convection flow in vertical tubes with radial internal fins was

Table 10.6.2 Heat transfer characteristics in a developed mixed convection flow in vertical regular polygonal ducts

	Nusselt number															
	Uniform circumferential wall temperature								Uniform circumferential heat flux							
Ra	3[a]	4	5	6	7	8	12	Circle (exact solution)	3	4	5	6	7	8	12	Circle (exact solution)
1	3.11	3.61	3.87	4.01	4.10	4.16	4.27	4.36	1.90	3.23	3.65	3.88	4.02	4.13	4.28	4.36
100	3.25	3.70	3.95	4.08	4.17	4.23	4.34	4.43	2.53	3.32	3.72	3.96	4.09	4.17	4.34	4.43
500	3.75	4.26	4.26	4.32	4.45	4.51	4.60	4.69	3.35	3.69	4.02	4.24	4.36	4.44	4.61	4.69
1,000	4.30	4.47	4.63	4.72	4.78	4.85	4.92	4.99	3.61	4.14	4.37	4.57	4.69	4.77	4.93	4.99
2,000	5.18	5.20	5.30	5.35	5.38	5.46	5.50	5.56	4.69	4.95	5.00	5.18	5.28	5.34	5.52	5.56
5,000	6.79	6.80	6.88	6.89	6.90	6.91	6.91	6.94	6.01	6.34	6.47	6.61	6.70	6.73	6.93	6.94
10,000	8.27	8.40	8.43	8.46	8.47	8.48	8.48	8.49	7.84	8.10	8.17	8.17	8.26	8.31	8.48	8.49

[a]Number of sides of polygon.

Source: Reprinted with permission from M. Iqbal, *J. Heat Transfer*, vol. 92, p. 237. Copyright © 1970, ASME.

Table 10.6.3 Pressure drop characteristics in a developed mixed convection flow in vertical regular polygonal ducts

	Pressure drop parameter E											
	Uniform circumferential wall temperature						Uniform circumferential heat flux					
Ra	3[a]	4	6	8	16	Circle	3	4	6	8	16	Circle
1	26.75	28.50	30.04	30.17	31.98	32.06	27.03	28.56	30.06	30.17	32.00	32.06
100	34.53	35.26	36.16	36.86	37.16	37.69	53.39	41.90	38.07	36.87	37.32	37.69
500	63.10	61.03	59.81	59.81	59.72	59.63	129.02	94.67	68.73	62.94	60.15	59.63
1,000	94.13	90.35	87.30	86.67	85.40	85.45	202.39	158.35	104.99	93.58	87.01	85.45
2,000	146.80	142.01	136.97	135.57	133.36	132.81	347.16	278.30	171.04	149.73	136.29	132.81
5,000	271.27	266.35	259.92	257.95	253.31	252.56	810.05	570.88	341.22	292.42	261.21	252.56
10,000	437.32	429.78	423.25	421.70	414.98	413.91	1,310.54	853.63	575.40	487.34	430.58	413.91

[a]Number of sides of polygon.

Source: Reprinted with permission from M. Iqbal, *J. Heat Transfer*, vol. 92, p. 237.

calculated by Prakash and Patankar (1981). For a uniform heat input in the axial direction, the governing equations describing fully developed flow were solved by a finite-difference scheme. The effect of the buoyancy forces was larger with a smaller number of fins and with shorter fins.

10.7 HORIZONTAL INTERNAL FLOWS

In vertical internal mixed convection flows, discussed in Section 10.6, the buoyancy forces either aid or oppose the forced flow. Then the buoyancy forces and, hence, the resulting transport quantities are nominally symmetric about the vertical axis of a tube or about the symmetry plane for two parallel surfaces. However, the buoyancy effects in horizontal internal flows are different since the buoyancy forces are perpendicular to the forced flow. The otherwise symmetric forced flow is distorted increasingly as the buoyancy effects increase. For flow between two very wide parallel surfaces, the velocity maximum occurs above or below the center plane, depending on whether the fluid is being cooled or heated as it flows between the surfaces. This results in different viscous drag and heat transfer rates at the two walls. This shift in the velocity and temperature profiles is the main effect due to buoyancy in such plane two-dimensional horizontal flows.

However, in horizontal passages such as tubes, the interaction of the buoyancy and the pressure forces is much more complicated. When the fluid and wall temperatures are different, the buoyancy forces generate a secondary cross-flow pattern superimposed on the main flow. For a straight circular tube this motion is a pair of horizontal helical circulations. Circular symmetry is lost and analysis becomes very difficult. Furthermore, the vigor of the secondary flow usually increases downstream.

10.7.1 Flow in Horizontal Tubes

Consider forced flow in a horizontal tube maintained at a constant temperature t_0 higher than the fluid temperature at the inlet, t_i. Secondary motion begins as heat is transferred to the fluid near the tube wall. The buoyancy effects on heat transfer remain small near the tube entrance. They increase until a maximum in temperature is reached downstream as more fluid heats up. The effect then decays as the fluid bulk temperature approaches the tube wall temperature.

The great practical importance of horizontal tube flow and heat transfer has led to continuing investigation since the pioneering work of Graetz; see Shah and London (1978). The added effect of buoyancy was first examined quantitatively by Colburn (1933). A function of the Grashof number was suggested as a multiplier to the forced convection heat transfer rates. Later, Sieder and Tate (1936) gave an equation that accounted for variable fluid properties. Both these correlations were questioned later by Martinelli and Boelter (1942), who pointed out that they predict increasing importance of natural convection effects as the forced flow rate is increased, which is contrary to the data. The extensive data of Kern and Othmer (1943) showed that the equation of Sieder and Tate predicted too high heat transfer rates at high Reynolds numbers and too low values at small Re. Martinelli and Boelter (1942) suggested a correlation for mixed convection in a vertical tube, based on a theoretical model. A modification of this correlation for horizontal tubes was given by Eubank and Proctor (1951), based on data from the heating of oil in horizontal tubes. The following correlation was suggested for a constant tube wall temperature:

$$\mathrm{Nu}_{am}\left(\frac{\mu_w}{\mu_b}\right)^{0.14} = 1.75\left[\mathrm{Gz}_m + 12.6\left(\frac{\mathrm{Gr}_m\mathrm{Pr}_m D}{L}\right)^{0.4}\right]^{1/3} \qquad (10.7.1)$$

The Graetz number Gz is defined as Wc_p/kL, where W is the mass flow rate and L the tube length. The subscripts am, w, b, and m, respectively, denote quantities based on the arithmetic mean temperature difference between the fluid and the tube, the wall temperature, the fluid bulk (or mixed) temperature, and the film temperature. McAdams (1954) recommended a similar equation in which the empirical constants 12.6 and 0.4 in the above equation are replaced by 0.04 and 0.75, respectively. Jackson et al. (1967) reported data on developing laminar flow of air in a horizontal isothermal tube. In the range $60 < \mathrm{RePr} < 1300$, these data were well correlated by

$$\mathrm{Nu} = \frac{hD}{k} = 2.67[(\mathrm{Gz})^2 + (0.0087)^2(\mathrm{GrPr})^{1.5}]^{1/6} \qquad (10.7.2)$$

where the fluid properties in Gz and GrPr are determined at the bulk and tube wall temperatures, respectively. The Grashof number is based on the tube diameter and the logarithmic mean temperature difference (LMTD). The LMTD is defined as $(t_1 - t_i)/\log[(t_0 - t_i)/(t_0 - t_1)]$, where t_i is the inlet temperature of the fluid, t_1 its outlet temperature, and t_0 the wall temperature. Since the fluid outlet temperature is not known a priori, the Grashof number may not be explicitly calculated and an iterative procedure is required. Note that the ratio D/L does not appear in the natural convection term in Eq. (10.7.2). This was the first suggestion that the ratio was unnecessary in horizontal tube flow. This was later confirmed by Oliver (1962), based on laminar data obtained for liquids.

McComas and Eckert (1966) reported an experimental study of laminar mixed convection flow of air in a heated horizontal circular tube with isothermal walls. An unheated entrance length established the velocity field before the air entered the heated section. Data were obtained in the range $1 < \mathrm{Gr} < 1000$ and $100 < \mathrm{Re} < 900$. It was observed that the buoyancy-induced secondary flow assisted the main flow and resulted in higher heat transfer rates. At low Reynolds number values, a shorter thermal entrance length was observed. The complexity in correlating the heat transfer data was emphasized.

Mori et al. (1966) performed similar experiments to determine the buoyancy effects on heat transfer for air flow in a horizontal tube with constant heat flux input. It was concluded that for $\mathrm{ReRa} = 10^4$ the secondary flow was quite strong and that a pair of symmetric horizontal vortices formed. In laminar flow, appreciable buoyancy effects appeared for ReRa = 10^3. A critical Reynolds number, beyond which the flow was turbulent, was found to be a function of both the Rayleigh number and the turbulence level in the entering fluid. For high inlet turbulence levels, the critical Reynolds number was about 2000 without heating and increased with increasing Rayleigh number. This trend was explained in terms of the secondary flow suppressing the turbulence level. On the other hand, with low turbulence levels at the tube inlet, the critical Reynolds number was higher, about 7700. It decreased as the Rayleigh number increased. The increased secondary flow caused a transition to turbulence at smaller Re. The following equation correlates the critical Reynolds number data for low turbulence at the tube entrance:

$$\mathrm{Re}_{\mathrm{cr}} = \frac{\mathrm{Re}_{\mathrm{cr0}}}{1 + 0.14\mathrm{ReRa} \times 10^{-5}} \qquad \text{for ReRa} < 10^5 \tag{10.7.3}$$

where $\mathrm{Re}_{\mathrm{cr0}} = 7700$ is the critical Reynolds number with no heating at a low entrance turbulence level.

The following correlation was suggested for heat transfer in laminar flow:

$$\mathrm{Nu} = \frac{hD}{k} = 0.61(\mathrm{ReRa})^{1/5}\left[1 + \frac{1.8}{(\mathrm{ReRa})^{1/5}}\right] \tag{10.7.4}$$

where $\mathrm{Re} = u_m D/\nu$ and $\mathrm{Ra} = (g\beta/\alpha\nu)(dt_0/dx)D^4$. The term dt_0/dx denotes the axial temperature gradient along the wall. In turbulent flow the buoyancy effects on the Nusselt number were found to be small.

Mori and Futagami (1967) analyzed fully developed laminar mixed convection flow. The results were compared with the earlier experimental results. The following simple equation correlated the data very well:

$$\frac{\mathrm{Nu}}{\mathrm{Nu}_0} = 0.1634(\mathrm{ReRa})^{1/5} \qquad \text{for Pr} = 0.72 \tag{10.7.5}$$

Here Nu_0 is the Nusselt number for developed pure forced convection. For Poiseuille flow in a heated tube with a uniform heat flux input, $\mathrm{Nu}_0 = 48/11$.

Faris and Viskanta (1969) used a perturbation method to determine the buoyancy effects on a fully developed laminar flow in a horizontal tube under the condition of uniform wall heat flux. The average Nusselt number was substantially higher than that for forced convection alone. Note that the fully developed assumption implies a fully developed forced flow entering a heated section of the tube. A detailed finite-difference calculation of fully developed laminar mixed convection flow in a horizontal tube was reported by Newell and Bergles (1970). For a uniform heat flux surface condition, solutions for the heat transfer and the pressure drop were obtained for water, with two limiting wall conditions. A tube with a high thermal conductivity wall exhibited higher values of the Nusselt number and friction factors than one with a low conductivity wall. With the latter boundary condition, significant circumferential wall temperature variation was obtained. An experimental study by Bergles and Simonds (1971) followed the above calculations. Visual and quantitative observations were made with water flowing in a heated glass tube. A secondary flow was found to be established in a short distance in the presence of buoyancy effects.

Experimental data on the heat transfer in horizontal tubes by Ede (1961), Roy (1966), and Petukhov and co-workers (1967a, 1967b, 1969) were reviewed by Bergles and Simonds (1971). The resulting heat transfer correlation is plotted in Fig. 10.7.1. The dimensionless groups used in Fig. 10.7.1 are $\mathrm{Nu} = hD/k$, $\mathrm{Re} = u_m D/\nu$, and $\mathrm{Ra} = g\beta(t_0 - t_b)D^3/\alpha\nu$, where $t_0 - t_b$ is the difference between the wall temperature and the fluid bulk temperature. Properties are calculated at t_b. For an isothermal tube surface, Hieber and Sreenivasan (1974) reported an analytical study of laminar mixed convection flow of a large Prandtl number fluid. A good correlation with much of the available data was obtained.

For initially fully developed laminar mixed convection flow, variable-property and wall conduction effects on both the heat transfer and the pressure drop in horizontal tubes were measured by Morcos and Bergles (1975). Water and ethylene glycol were used in heated glass and stainless steel tubes. The following heat transfer correlation, including the effect of the wall thermal conductivity, was suggested:

$$\mathrm{Nu} = \left\{(4.36)^2 + \left[0.145\left(\frac{\mathrm{GrPr}^{1.35}}{P^{0.25}}\right)^{0.265}\right]^2\right\}^{1/2} \tag{10.7.6}$$

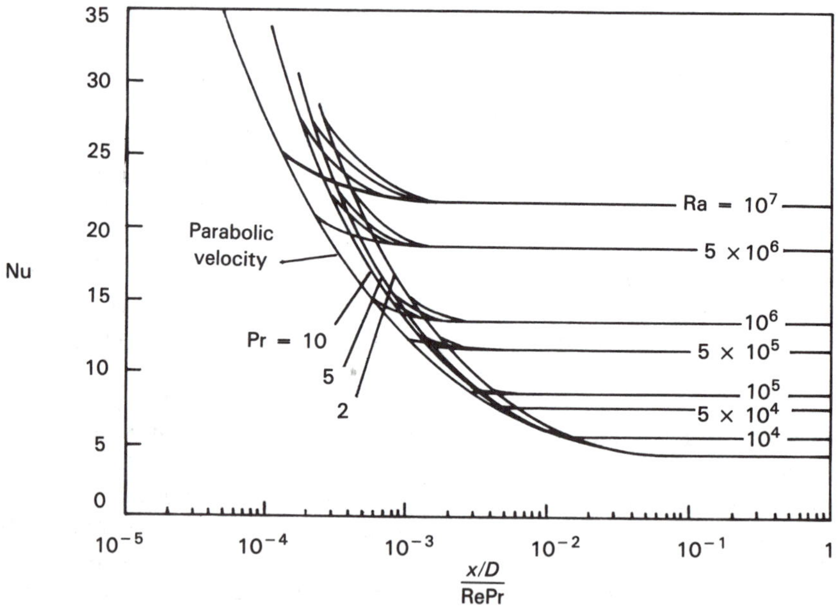

Figure 10.7.1 Heat transfer characteristics in a developing mixed convection flow in a horizontal circular tube with uniform surface heat flux. *(Reprinted with permission from* Int. J. Heat Mass Transfer, *vol. 14, p. 1989, A. E. Bergles and R. R. Simonds. Copyright © 1971, Pergamon Journals Ltd.)*

The tube wall parameter P is defined as $hD^2/k_0\delta$, where h is the heat transfer coefficient, k_0 the thermal conductivity of the wall material, and δ the tube wall thickness. The Grashof number is $g\beta D^3(t_0 - t_b)/\nu^2$. All fluid properties are evaluated at the bulk temperature. Equation (10.7.6) is recommended for $3 \times 10^4 < \text{Ra} < 10^6$, $4 < \text{Pr} < 175$, and $2 < P < 66$.

Large Prandtl number. Hong and Bergles (1976) reported a boundary layer analysis of fully developed laminar pipe flow of large Prandtl number fluids with temperature-dependent viscosity. A new viscosity parameter was included in the analysis. The Nusselt number for laminar convection in horizontal tubes with large Prandtl number fluids was found to be proportional to $(\text{Ra})^{1/4}$. Heat transfer correlations were suggested for different viscosity ranges.

Ou and Cheng (1977) calculated natural convection effects in initially developed laminar flow in horizontal isothermal tubes. For large Pr, numerical results were obtained for $\text{Ra} < 10^6$. The effect of circumferentially nonuniform heating on laminar mixed convection in a horizontal tube was studied by Patankar et al. (1978) with finite-difference techniques. Such boundary conditions arise in applications such as solar collector tubes. It was concluded that the buoyancy effects on laminar forced convection depend strongly on the circumferential distribution of the wall heat flux. When heat is added along the lower half of the tube, with the upper half insulated, a vigorous secondary flow enhances the heat transfer rates. When the upper half is heated and the lower half insulated, the induced secondary flow is much weaker. These trends were confirmed by the experiments of Schmidt and Sparrow (1978).

Other effects. Yao (1978) analyzed developing mixed convection flow and heat transfer in a horizontal tube. An asymptotic solution for the developing flow near the entrance is obtained by perturbing the developing flow in an unheated pipe. El-Hawary (1980) measured the effect of buoyancy on the flow stability in a horizontal tube. Turbulent flows were obtained in heated tubes at Reynolds number values much smaller than in unheated tubes. Yousef and Tarasuk (1981, 1982) measured the buoyancy effects on a developing flow of air in an isothermal tube. Heat transfer correlations were suggested for different regions along the entry length. A finite-difference analysis of the same physical circumstance is reported by Hishida et al. (1982). Developing velocity and temperature profiles were computed for $\text{Pr} = 0.71$ and for various values of Re and Gr for an isothermal tube. A secondary flow started near the tube entrance and was stronger in the upper region of the pipe. Its velocity over the whole cross section increased downstream to a maximum and then diminished gradually as the fluid bulk temperature approached the wall temperature, as seen in many other studies.

10.7.2 Mixed Convection in Noncircular Horizontal Channels

The basic transport mechanisms in mixed convection flow through noncircular passages remain quite similar to those in circular tubes. Temperature differences cause buoyancy forces and secondary flows are superimposed on the main flow. They start in the entrance region and become stronger downstream. As the fluid approaches the wall temperature, they decrease. In laminar flows these secondary motions enhance the transport rates. Cheng and Hwang (1969) reported the first analysis of initially fully developed laminar mixed convection flow in horizontal rectangular channels. Axially uniform wall heat flux and peripherally uniform temperature were assumed, as for large wall conductivity. Numerical solutions were obtained for aspect ratios γ of 0.2, 0.5, 1, 2, and 5 with $Pr = 0.73$. For square channels, calculations were also carried out for $Pr = 7.2$. Both the friction factor and the heat transfer rates increased with ReRa. At a given value of ReRa the friction factor and the heat transfer rates are maximum for a square channel. A plot of Nu/Nu_0 as a function of ReRa is shown in Fig. 10.7.2 for various aspect ratios. Here Nu_0 denotes the Nusselt number for pure forced convection. All the dimensionless groups are based on the hydraulic diameter D_h = area/perimeter. The Rayleigh number is given by $(gD_h^3\beta C_1 D_h/\nu^2)Pr$ where C_1 is the axial wall temperature gradient.

Cheng et al. (1972) extended the above analysis for a thermal entrance region to a large Prandtl number ($Pr > 10$) fluid. The local Nusselt number distribution was reported for aspect ratios of 0.2, 0.5, 1, 2, and 5. Because of entrance and secondary flow effects, a minimum in the Nusselt number occurred at some distance from the entrance. This distance depended on the value of the Rayleigh number. The effect of buoyancy was to decrease the thermal entrance length. Later, Ou et al. (1974) analyzed the same entrance region flow for an isothermal wall condition. For $0 < Ra < 5 \times 10^5$ the local heat transfer variation was determined for aspect ratios of 0.5, 1, and 2. Buoyancy effects were found to be significant only in some entrance length, which depended on the aspect ratio. Very near the entrance and in the thermally fully developed region, these effects were negligible. This behavior is different from that for a uniform wall heat flux boundary condition, where the fluid bulk temperature never reaches the wall temperature and the buoyancy effects persist throughout the length of the channel.

Buoyancy effects in forced flow through a horizontal annulus were studied by Hattori (1981), using a perturbation technique and a finite-difference method for low and high values of the Rayleigh number, respectively. A fully developed

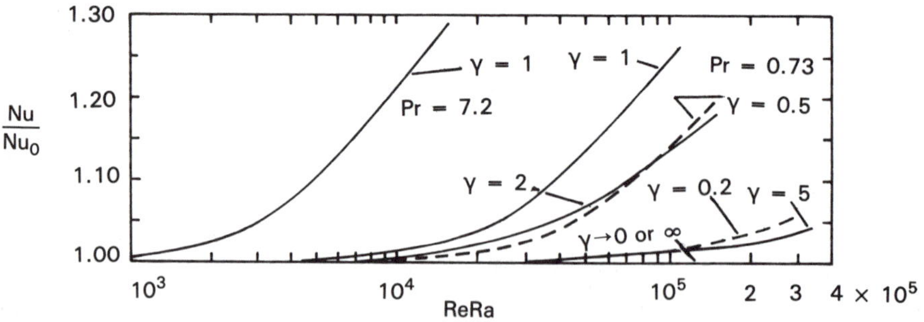

Figure 10.7.2 Nusselt number variation in developed mixed convection flow in horizontal rectangular ducts with uniform surface heat flux. *(Reprinted with permission from K. C. Cheng and G. Hwang,* J. Heat Transfer, *vol. 91, p. 59. Copyright © 1969, ASME.)*

laminar flow was analyzed for one wall uniformly heated and the other insulated and also for both walls heated uniformly. For a Prandtl number of unity and diameter ratios of 1.5, 2, 4, and 6, heat transfer characteristics and the nature of secondary flows were calculated. The secondary flow pattern changed markedly with increasing Rayleigh number and between different thermal boundary conditions. For only one wall heated, two secondary flow cells arose which were symmetric about the vertical. With both walls heated, the number of cells increased to four, two on either side of the vertical axis. The heat transfer results compared very favorably with experimental data reported by Hattori and Kotaki (1977).

10.8 OTHER INTERNAL FLOWS

Vertical and horizontal internal flows were discussed in Sections 10.6 and 10.7, respectively. However, many engineering applications, for example, solar collector tubes, involve inclined flow passages. The buoyancy force is again not parallel to the pressure forces, and a secondary flow arises. For laminar flow, the fluid moves along spiral paths and the flow is not axisymmetric.

Laminar mixed convection arising in a fully developed flow entering an inclined circular tube was first analyzed by Iqbal and Stachiewicz (1966). For a uniform heat flux boundary condition, the Poiseuille flow was perturbed for small values of the Rayleigh number, the perturbation parameter. For $Pr = 0.75$ and 5.0 the velocity and temperature distributions were determined for the case of the buoyancy force component in the flow direction aiding the flow. It was concluded that, for a given combination of the Rayleigh, Reynolds, and Prandtl numbers, there is an optimum value of tube inclination that gives the maximum value of the Nusselt number.

Cheng and Hong (1972) considered the same transport but used a numerical approach, a combination of a boundary vorticity method and a line iterative relaxation method. The accuracy of the method was checked by comparing the results for the limiting vertical and horizontal orientations with previous calculations. The results did not indicate the optimum value of the tube inclination, suggested by Iqbal and Stachiewicz (1966). However, for a given tube inclination, a reversal in the axial flow arises near the tube center when Ra exceeded a certain value. For $Pr = 0.75$ and $ReRa < 50{,}000$, this critical value of Ra is given below as a function of tube inclination ξ with the horizontal:

ξ (degrees)	0	10	20	30	45	60	75	90
Ra	∞	3800	1950	1350	930	780	690	670

Note that for a horizontal tube, $\xi = 0$, flow reversal does not occur. For a vertical tube, $\xi = 90°$, the critical value of the Rayleigh number is 670 and checks excellently with the results of Hallman (1958).

Since the Nusselt number is a function of ξ, Re, Ra, and Pr, it is very difficult to express the heat transfer results by a single correlation. Cheng and Hong (1972) reported the effect of various dimensionless groups on the heat transfer and the pressure loss. For Pr = 0.75 and ReRa = 4000, the heat transfer results are shown in terms of Nu/Nu_0 as a function of Ra sin ξ in Fig. 10.8.1. Here Nu_0 corresponds to the pure forced convection limit. The Nusselt number is defined as $2ha/k$. Also, $Ra = g\beta C_1 a^4/k\nu$ and $Re = Aa^3/4\rho\nu^2$, where a is the tube radius, C_1 the axial temperature gradient, and A the axial pressure gradient in the fluid.

For Ra sin ξ ≥ 70, all the heat transfer results for any value of tube inclination may be predicted by a single curve, as shown in Fig. 10.8.1. The effect of the parameter ReRa on heat transfer for ξ = 45° and Pr = 0.75 is shown in Fig. 10.8.2. It is seen that for Ra > 450 the ratio Nu/Nu_0 becomes independent of Re and is a function of Ra only. The effect of Pr on heat transfer is shown in Fig. 10.8.3 for ξ = 45° and ReRa = 4000. It is interesting that the effect of the Prandtl number is significant only for Ra < 450. It was further concluded that in the high Rayleigh number regime the effect of tube orientation on the flow and the heat transfer results is significant for angles near horizontal. Futagami and Abe (1972) and Sabbagh et al. (1976) reported measurements for mixed convection in inclined tubes. The experimental results of Sabbagh et al. (1976) are in qualitative agreement with the numerical results of Cheng and Hong (1972) discussed above.

Mixed convection in inclined rectangular channels was considered by Ou et al. (1976). Fully developed laminar aiding flow in a channel with a uniform wall heat flux was analyzed numerically for Pr = 5.0 and aspect ratios of 0.5, 1, and 2. The effect of channel inclination was accounted for by representing the heat transfer and friction results as functions of Re cot ξ and Ra sin ξ.

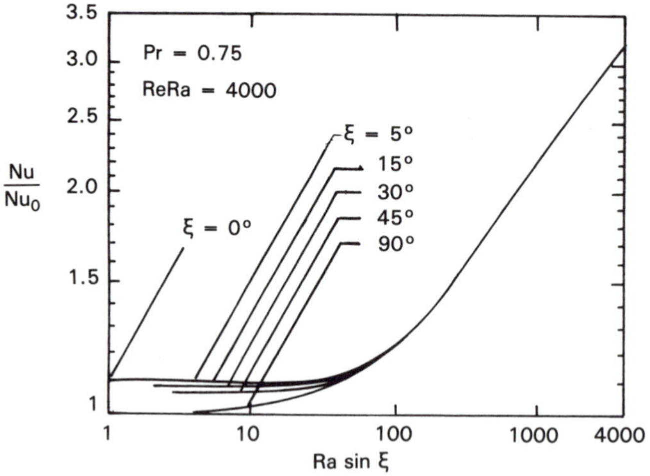

Figure 10.8.1 Nusselt number variation in developed laminar mixed convection flow in an inclined circular tube at an angle of inclination ξ from the horizontal, in degrees. *(Reprinted with permission from K. C. Cheng and S. W. Hong,* Appl. Sci. Res., *vol. 27, p. 19. Copyright © 1972, Martinus Nijhoff Publishers.)*

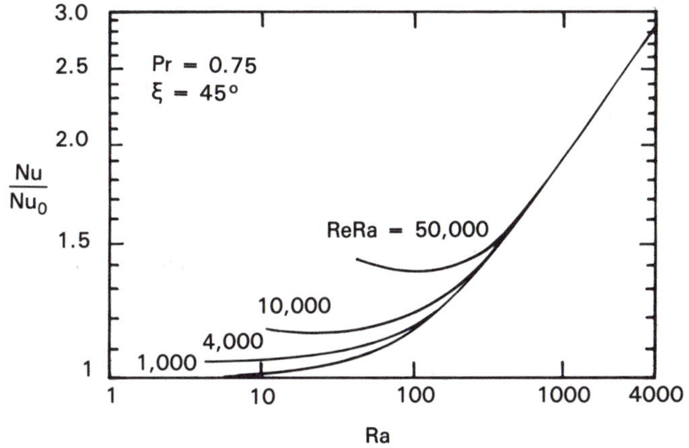

Figure 10.8.2 Effect of the parameter ReRa on the heat transfer in an inclined tube at an inclination with the horizontal of 45°. *(Reprinted with permission from K. C. Cheng and S. W. Hong,* Appl. Sci. Res., *vol. 27, p. 19. Copyright © 1972, Martinus Nijhoff Publishers.)*

Miyazaki (1971, 1973) reported analyses of mixed convection in curved circular and rectangular tubes rotating about the axis through the center of curvature. Such flow configurations are encountered in rotary machines such as turbines and motors. Chilurkuri and Humphrey (1981) analyzed the buoyancy effect on flow through a curved square duct.

Another important mixed convection flow is the one that arises in enclosed fluid regions due to the discharge of buoyant fluid into the region. This matter is of particular interest in heat rejection to water bodies and in energy extraction

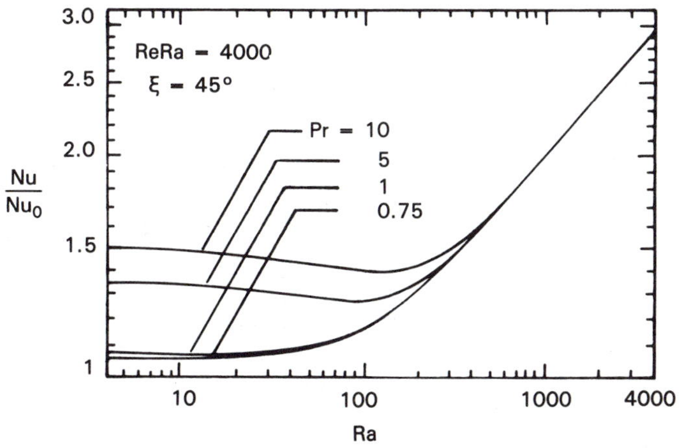

Figure 10.8.3 Effect of Prandtl number on heat transfer in an inclined tube at an inclination with the horizontal of 45°. *(Reprinted with permission from K. C. Cheng and S. W. Hong,* Appl. Sci. Res., *vol. 27, p. 19. Copyright © 1972, Martinus Nijhoff Publishers.)*

from energy storage systems such as solar ponds. An extensive amount of work has been done on such mixed connection flows; see, for instance, Oberkampf and Crow (1976), Cabelli (1977), Jaluria and Gupta (1983), and Cha and Jaluria (1984a, 1984b). The thermal and flow fields are found to be strongly affected by the presence of buoyancy effects, the relative magnitude of which is given by the parameter Gr/Re^2, where both Gr and Re are based on the inlet channel height or diameter, inlet temperature excess over ambient temperature, and inlet velocity.

10.9 OSCILLATORY EFFECTS

The preceding sections described the flow circumstances in which a steady forced flow, external or internal, is affected by the buoyancy forces arising from temperature differences in the fluid. Another circumstance in which buoyancy forces interact with imposed flow conditions arises when oscillations are imposed on the fluid or on the surface exchanging heat with the fluid. Such oscillations have been shown to increase heat transfer rates. However, under certain conditions a decrease in heat transfer rates has also been reported. Since a large number of variables are necessary to describe the interaction of imposed oscillations and buoyancy effects, a wide range of effects are expected.

Blankenship and Clark (1964) determined the effect of transverse ambient oscillations on natural convection from a vertical isothermal surface, both analytically and experimentally. It was shown that the transverse oscillation caused a slight reduction in the local Nusselt number from that of steady laminar natural convection. Another important effect observed experimentally was an earlier transition to turbulent flow. In the transition and turbulent regimes, the heat transfer rates increased rapidly, beyond the corresponding values in the absence of oscillations. The highest value of the improvement in heat transfer was about 60%, under fully turbulent conditions with intense oscillations.

The geometry most widely used to determine oscillatory effects has been a horizontal heated cylinder oscillating in a horizontal or vertical plane in a stationary ambient. Two flow regimes may be identified. At amplitudes A much larger than the cylinder diameter D, that is, $A/D >> 1$, the convection process near the cylinder surface may be assumed to be quasi-steady. The heat transfer may then be described by the steady convection correlations, neglecting conduction effects in the solid. Such a circumstance often arises for very small diameter cylinders. The experimental data on oscillating horizontal wires, reported by Deaver et al. (1962) and Penney and Jefferson (1966), fall in this regime. These results, summarized by Bergles (1973), are shown in Fig. 10.9.1. Depending on the amplitude of the vibration, three regimes arise: free convection dominated, mixed convection, and forced convection dominated. It is seen from Fig. 10.9.1 that in the last regime the forced convection correlation adequately describes the heat transfer rates.

For cylinders of large diameter vibrating at small amplitudes, $A/D << 1$, the transport mechanism is different. Then, the movement of the cylinder through the

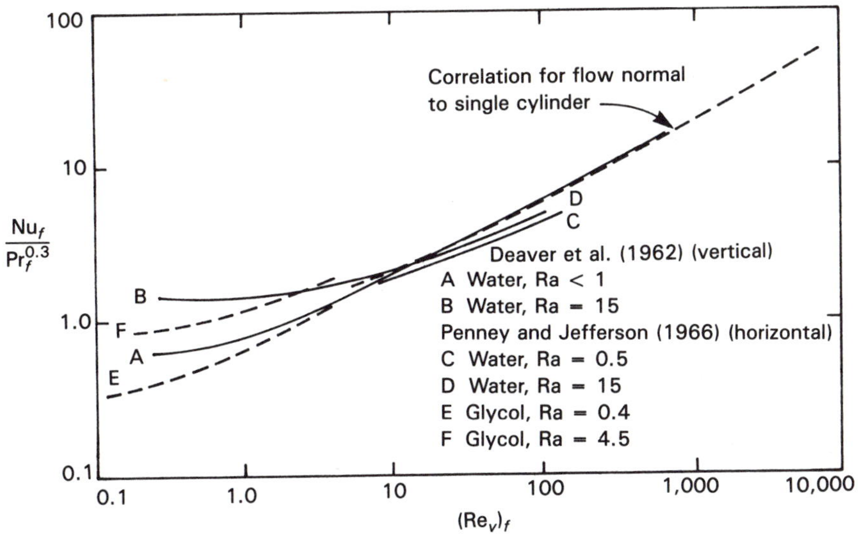

Figure 10.9.1 Effect of oscillations on the convection heat transfer from a cylinder. Amplitude/diameter >> 1 and $(\mathrm{Re}_v)_f = 2\pi ADF/\nu$, where F is the frequency and the subscript f refers to fluid properties evaluated at the film temperature. *(Reprinted with permission from A. E. Bergles, in* Handbook of Heat Transfer, *W. M. Rohsenow and J. P. Hartnett, Eds., Sec. 10. Copyright © 1973, McGraw-Hill Book Company.)*

fluid is not large enough to result in significant forced convection effects. Thus, natural convection effects dominate. However, as the vibration amplitude increases beyond a critical value, a secondary flow, commonly called acoustic or thermoacoustic streaming, develops. This causes an additional mechanism of heat transfer in the boundary layer. Experimental data in this transport regime have been reported by Martinelli and Boelter (1938), Mason and Boelter (1940), Lemlich (1955), Teleki et al. (1960), Fand and Peebles (1962), Shine (1962), Lemlich and Rao (1965), and Bergles (1969). The results, again summarized by Bergles (1973), are shown in Fig. 10.9.2. It is seen that the heat transfer coefficient remains at the natural convection value until a threshold value of the vibrational Reynolds number Re, based on the amplitude of vibration, is reached. Then it increases with increasing Reynolds number.

Another aspect of oscillatory effects in natural convection arises when there are fluctuations in the flow instead. The transport mechanisms are more complex than with surface oscillations. The effects of sonic vibrations on the heat transfer from horizontal cylinders to air have been investigated extensively. With planar sonic fields in a direction normal to the cylinder axis, improvements of 100 to 200% over the pure natural convection heat transfer coefficients were obtained by Holman and Mott-Smith (1959), Sprott et al. (1960), Fand and Kaye (1961), and Lee and Richardson (1965). The improvement in heat transfer occurs at a sound pressure level of about 134–140 dB and is associated with the formation of acoustically induced flow near the heated surface.

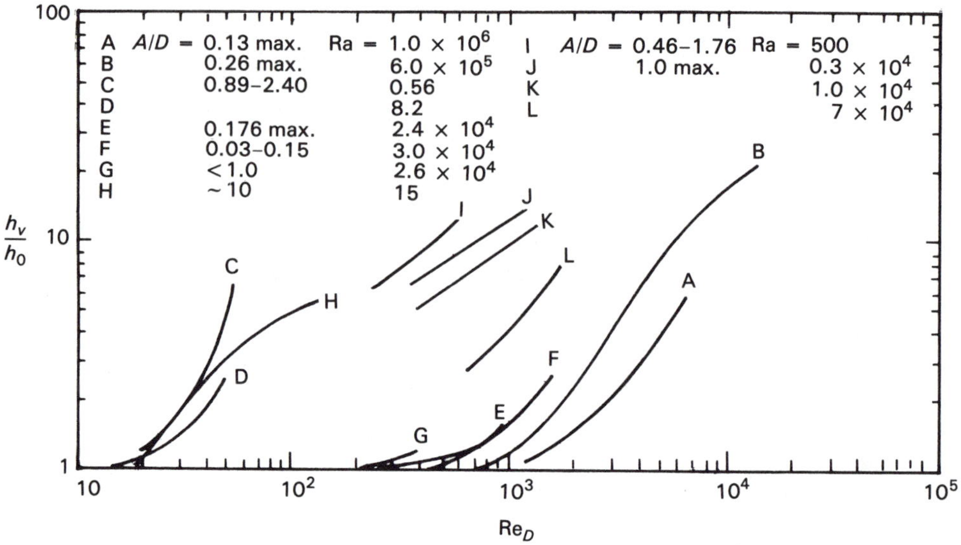

Figure 10.9.2 Effect of oscillations on the convective heat transfer from a horizontal cylinder. Here h_v denotes the convective heat transfer with vibration and h_0 that without vibration. Amplitude/diameter < 1. *(Reprinted with permission from A. E. Bergles, in* Handbook of Heat Transfer, *W. M. Rohsenow and J. P. Hartnett, Eds., Sec. 10. Copyright © 1973, McGraw-Hill Book Company.)*

Improvements in the heat transfer to liquids by using sonic and ultrasonic vibrations have been reported by Seely (1960), Zhukauskas et al. (1961), Larson and London (1962), Robinson et al. (1958), Fand (1965), Gibson and Houghton (1961), and Li and Parker (1967). The enhancement of heat transfer varies from 30 to 450%. In general, however, cavitation arises before these significant improvements in heat transfer are achieved.

10.10 TRANSIENTS

Transient mixed convection flows arise frequently in nature and in technology. However, very few studies of such mixed convection flows are available. If a surface is dissipating heat in a stream of fluid, then any variation in the surface heating rate or in the free-stream velocity will lead to a transient process. This will continue until a new steady state is achieved. In this section only external laminar aiding flows over flat vertical surfaces are considered. The surface is assumed to be semi-infinite and the flow to be two dimensional. Furthermore, constant fluid properties are assumed, and the density of the fluid is assumed constant except in calculating the buoyancy force term. The governing equations then are

$$\frac{\partial u}{\partial x} + \frac{\partial v}{\partial y} = 0 \qquad (10.10.1)$$

$$\frac{\partial u}{\partial \tau} + u\frac{\partial u}{\partial x} + v\frac{\partial u}{\partial y} = -\frac{1}{\rho}\frac{dp}{dx} + \nu\frac{\partial^2 u}{\partial y^2} + g\beta(t - t_\infty) \tag{10.10.2}$$

$$\frac{\partial t}{\partial \tau} + u\frac{\partial t}{\partial x} + v\frac{\partial t}{\partial y} = \alpha\frac{\partial^2 t}{\partial y^2} \tag{10.10.3}$$

For opposing effects, the last term in Eq. (10.10.2) is negative. For flows over flat surfaces the pressure term in Eq. (10.10.2) is retained when the local adjacent free-stream velocity U_∞ varies, as in wedge flows. Then

$$\frac{dp}{dx} = -\rho\frac{dU_\infty}{d\tau} \tag{10.10.4}$$

In addition to these equations, boundary and initial conditions are specified as follows:

$$y = 0: \qquad u = v = 0 \qquad t = t_0(x, \tau) \tag{10.10.5}$$

$$y \to \infty: \qquad u \to U_\infty(\tau) \qquad t \to t_\infty \tag{10.10.6}$$

Oosthuizen (1971) gave solutions of the above system of equations for three different conditions. The equations in their finite-difference algebraic form were solved by an implicit differencing technique. In all three flow configurations the surface temperature was assumed uniform. The first case was that of an impulsively heated flat surface in an impulsively moving parallel fluid stream. At time $\tau = 0$ the free-stream velocity is set equal to U_∞ and the surface temperature is increased to a constant value $t_0 > t_\infty$. The second case again concerns an impulsively started free stream. However, the surface remains unheated until some time τ_i later. Then, its temperature is raised to t_0 and kept at this value. The last case is that of an ambient stream with a periodically varying nonzero velocity flowing over an isothermal surface. All the solutions presented are for Pr = 0.7. Figure 10.10.1 shows the resulting heat transfer rate for the impulsively started flow with delayed heating. The nondimensional variables are

$$X = x\left(\frac{g\beta\,\Delta t}{U_\infty^2}\right) \qquad \bar{\tau} = \tau u_r\left(\frac{g\beta\,\Delta t}{U_\infty}\right) \tag{10.10.7}$$

$$\phi = \frac{t - t_\infty}{t_0 - t_\infty} \qquad Y = y\left(\frac{g\beta\,\Delta t}{\nu U_\infty}\right)^{1/2} \tag{10.10.8}$$

where $\Delta t = t_0 - t_\infty$. An undershoot in the heat transfer rate, proportional to $\partial\phi/\partial y$, is observed to occur during the transient for all the $\bar{X}$ values considered.

Sammakia et al. (1982) reported the results of a similar analysis. A vertical surface of finite thermal capacity was subjected to a step input of a uniform and thereafter constant heat flux. The ambient velocity U_∞ was constant. Solutions were considered for Pr = 0.7 and 7.6. By appropriately nondimensionalizing Eqs. (10.10.1)–(10.10.3) and the boundary conditions, the relevant parameters were found to be (see Chapter 7)

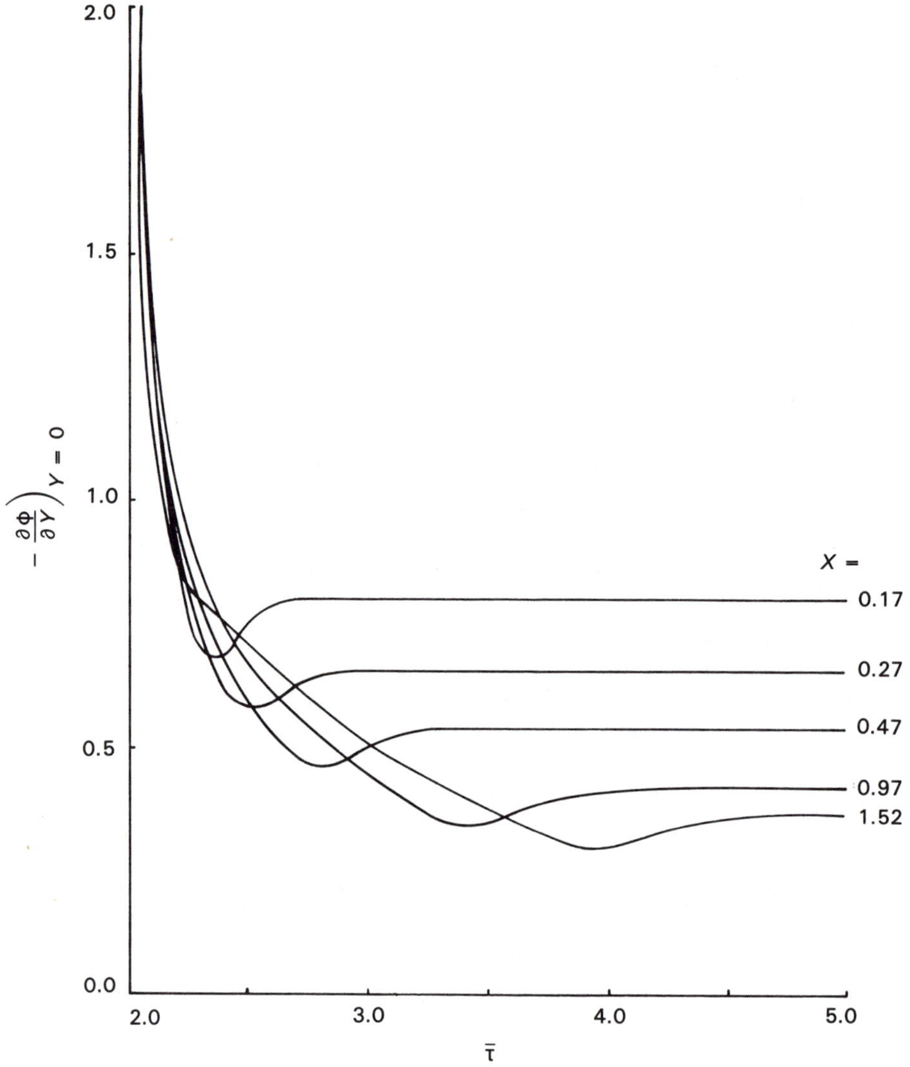

Figure 10.10.1 Variation of the dimensionless heat transfer rate with the dimensionless time for impulsively started flow with delayed heating. *(From Oosthuizen, 1971.)*

$$X = (\mathrm{Gr}_x^*)^{1/4} \qquad Y = \frac{y}{x}(\mathrm{Gr}_x^*)^{1/4} \tag{10.10.9}$$

$$U = \frac{ux/\nu}{(\mathrm{Gr}_x^*)^{1/4}} \qquad V = \frac{vx/\nu}{(\mathrm{Gr}_x^*)^{1/4}} \tag{10.10.10}$$

$$T = \frac{t - t_\infty}{q''x/k}(\mathrm{Gr}_x^*)^{1/4} \qquad \bar{\tau} = \frac{\alpha\tau}{x^2}\,\mathrm{Pr}(\mathrm{Gr}_x^*)^{1/2} \tag{10.10.11}$$

The boundary condition at the surface is

$$1 = Q^*\left(\frac{\partial T}{\partial \tau}\right) - \left(\frac{\partial T}{\partial Y}\right) \tag{10.10.12}$$

where

$$Q^* = \frac{c''}{\rho c_p x}\,\mathrm{Pr}(\mathrm{Gr}_x^*)^{1/4} \qquad \tau \text{ is the physical time}$$

and

$$\mathrm{Gr}_x^* = \frac{g\beta q'' x^4}{\nu^2 k}$$

Different values of Q^* are found to determine the resulting flow regime since Q^* represents the surface-fluid thermal capacity ratio. The equations of motion were solved by using an explicit finite-difference scheme. The initial flow condition was an established Blasius boundary layer, and the surface was then suddenly heated. Calculations were carried out in time to the steady state. The steady-state results were compared to the results of a steady-state perturbation analysis by Carey and Gebhart (1982), who used a matched asymptotic expansion.

Figures 10.10.2 and 10.10.3 show the resulting transient and steady-state temperature and velocity profiles, respectively, for Pr = 0.72 and $Q^* = 5.0$. Both

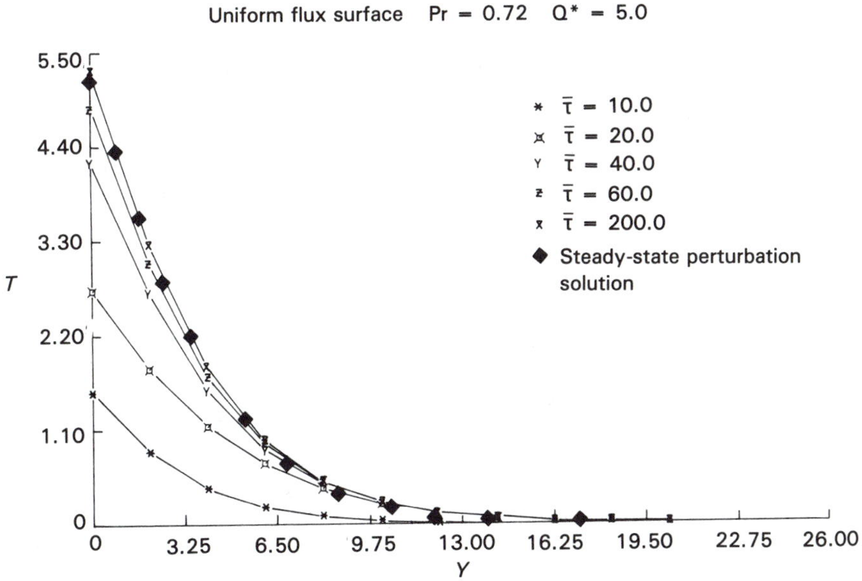

Figure 10.10.2 Transient temperature profiles adjacent to a uniform flux surface immersed in a uniform stream of air. *(Reprinted with permission from* Int. J. Heat Mass Transfer, *vol. 25, p. 835, B. Sammakia et al. Copyright © 1982, Pergamon Journals, Ltd.)*

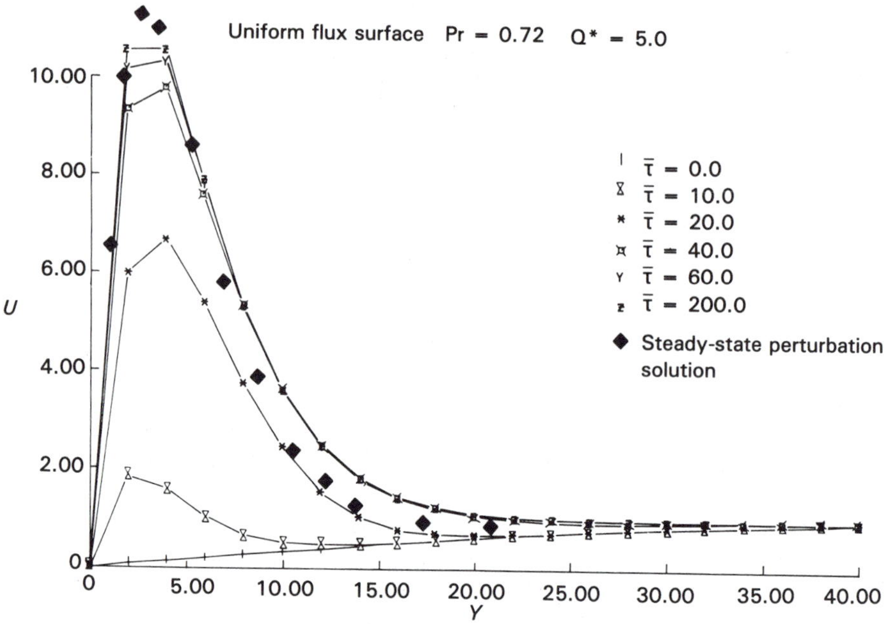

Figure 10.10.3 Transient velocity profiles adjacent to a uniform flux surface immersed in a uniform stream of air. *(Reprinted with permission from* Int. J. Heat Mass Transfer, *vol. 25, p. 835, B. Sammakia et al. Copyright © 1982, Pergamon Journals, Ltd.)*

temperature and velocity levels are observed to increase monotonically with time until steady state is reached. However, for smaller values of Q^*, temperature and velocity overshoots arise for both Pr = 0.72 and 7.6. Smaller values of Q^* correspond to surfaces with smaller heat capacity. Then more of the energy input to the surface element is transferred to the fluid. The result is an overshoot of both the temperature and the velocity during the transient. As entrainment develops, transport becomes fully developed.

REFERENCES

Abdelmeguid, A. M., and Spalding, D. B. (1979). *J. Fluid Mech. 94,* 383.
Afzal, N., and Banthiya, N. K. (1977). *J. Appl. Math. Phys. (ZAMP) 28,* 993.
Aggarwala, B. D., and Iqbal, M. (1969). *Int. J. Heat Mass Transfer 12,* 737.
Agrawal, H. C. (1962). *Proc. Int. J. Heat Mass Transfer 5,* 439.
Axcell, B. P., and Hall, W. B. (1978). *Proc. 6th Int. Heat Transfer Conf.* vol. 1, p. 37.
Bergles, A. E. (1969). *J. Heat Transfer 91,* 152.
Bergles, A. E. (1973). In *Handbook of Heat Transfer,* sec. 10, W. M. Rohsenow and J. P. Hartnett, Eds., McGraw-Hill, New York.
Bergles, A. E., and Simonds, R. R. (1971). *Int. J. Heat Mass Transfer 14,* 1989.
Blankenship, V. D., and Clark, J. A. (1964). *J. Heat Transfer 86,* 159.
Brown, C. K., and Gauvin, W. H. (1965). *Can. J. Chem. Eng. 43,* 306.
Brown, C. K., and Gauvin, W. H. (1966). *Chem. Eng. Sci. 21,* 961.

Buhr, H. O., Horsten, E. A., and Carr, A. D. (1974). *J. Heat Transfer 96,* 152.
Cabelli, A. (1977). *Sol. Energy 19,* 45.
Carey, V. P., and Gebhart, B. (1982). *Int. J. Heat Mass Transfer 25,* 255.
Carr, A. D., Connor, M. A., and Buhr, H. O. (1973). *J. Heat Transfer 95,* 445.
Cha, C. K., and Jaluria, Y. (1984a). *Int. J. Heat Mass Transfer 27,* 1801.
Cha, C. K., and Jaluria, Y. (1984b). *J. Sol. Energy Eng. 106,* 428.
Chen, T. S., and Mucoglu, A. (1977). *Int. J. Heat Mass Transfer 20,* 867.
Chen, T. S., and Strobel, F. A. (1980). *J. Heat Transfer 102,* 170.
Chen, T. S., Sparrow, E. M., and Mucoglu, A. (1977). *J. Heat Transfer 99,* 66.
Chen, T. S., Yuh, C. F., and Moutsoglou, A. (1980). *Int. J. Heat Mass Transfer 23,* 527.
Cheng, K. C., and Hong, S. W. (1972). *Appl. Sci. Res. 27,* 19.
Cheng, K. C., and Hwang, G. (1969). *J. Heat Transfer 91,* 59.
Cheng, K. C., Hong, S. W., and Hwang, G. (1972). *Int. J. Heat Mass Transfer 15,* 1819.
Chilurkuri, R., and Humphrey, J. A. C. (1981). *Int. J. Heat Mass Transfer 24,* 305.
Colburn, A. P. (1933). *Trans. AIChE 29,* 174.
Collins, M. W. (1978). *Proc. 6th Int. Heat Transfer Conf.,* Toronto, vol. 1, p. 25.
Collis, D. C., and Williams, M. J. (1959). *J. Fluid Mech. 6,* 357.
Connor, M. A., and Carr, A. D. (1978). *Proc. 6th Int. Heat Transfer Conf.,* Toronto, vol. 1, p. 43.
Deaver, F. K., Penney, W. R., and Jefferson, T. B. (1962). *J. Heat Transfer 84,* 251.
Easby, J. P. (1978). *Int. J. Heat Mass Transfer 21,* 791.
Eckert, E. R. G. (1942). *VDI-Forschungsh. 416.*
Eckert, E. R. G., and Diaguila, A. J. (1954). *Trans. ASME 76,* 497.
Ede, A. J. (1961). *Int. J. Heat Mass Transfer 4,* 105.
El-Hawary, M. A. (1980). *J. Heat Transfer 102,* 273.
El-Shaarawi, M. A. I., and Sarhan, A. (1980). *J. Heat Transfer 102,* 617.
El-Shaarawi, M. A. I., and Sarhan, A. (1982). *Int. J. Heat Mass Transfer 25,* 175.
Eubank, C. C., and Proctor, W. S. (1951). S.M. Thesis, Massachusetts Institute of Technology.
Falkner, V. M., and Skan, S. W. (1931). *Philos. Mag. 12,* 865.
Fand, R. M. (1965). *J. Heat Transfer 87,* 309.
Fand, R. M., and Kaye, J. (1961). *J. Heat Transfer 83,* 133.
Fand, R. M., and Keswani, K. K. (1973). *Int. J. Heat Mass Transfer 16,* 1175.
Fand, R. M., and Peebles, E. M. (1962). *J. Heat Transfer 84,* 268.
Faris, G. N., and Viskanta, R. (1969). *Int. J. Heat Mass Transfer 12,* 1295.
Fox, J. (1964). *J. Heat Transfer 86,* 560.
Futagami, K., and Abe, F. (1972). *Trans JSME 38,* 1799.
Gebhart, B., and Pera, L. (1970). *J. Fluid Mech. 45,* 49.
Gebhart, B., Audunson, T., and Pera, L. (1970). *Proc. 4th Int. Heat Transfer Conf.,* Paris, Paper NC3.2.
Gibson, J. H., and Houghton, G. (1961). *Chem. Eng. Sci. 15,* 146.
Gryzagoridis, J. (1975). *Int. J. Heat Mass Transfer 18,* 911.
Gunness, R. C., and Gebhart, B. (1965). *Int. J. Heat Mass Transfer 8,* 43.
Hallman, T. M. (1956). *Trans. ASME 78,* 1831.
Hallman, T. M. (1958). Ph.D. Thesis, Purdue University.
Han, L. S. (1959). *J. Heat Transfer 81,* 121.
Hanratty, T. J., Rosen, E. M., and Kabel, R. L. (1958). *Ind. Eng. Chem. 50,* 815.
Hatton, A. P., James, D. D., and Swire, H. W. (1970). *J. Fluid Mech. 42,* 17.
Hattori, N. (1981). *Heat Transfer Jpn. Res. 10,* 27.
Hattori, N., and Kotaki, S. (1977). *Trans. JSME 43,* 373.
Hering, R. G., and Grosh, R. J. (1963). *J. Heat Transfer 85,* 29.
Hieber, C. A. (1973). *Int. J. Heat Mass Transfer 16,* 769.
Hieber, C. A., and Gebhart, B. (1969). *J. Fluid Mech. 38,* 137.
Hieber, C. A., and Sreenivasan, S. K. (1974). *Int. J. Heat Mass Transfer 17,* 1337.
Hishida, M., Nagano, Y., and Montesclaros, M. S. (1982). *J. Heat Transfer 104,* 153.

Holman, J. P., and Mott-Smith, T. P. (1959). *J. Aerosp. Sci. 26,* 188.
Hong, S. W., and Bergles, A. E. (1976). *J. Heat Transfer 98,* 459.
Iqbal, M., and Stachiewicz, J. H. (1966). *J. Heat Transfer 88,* 109.
Iqbal, M., Aggarwala, B. D., and Fowler, A. G. (1969). *Int. J. Heat Mass Transfer 12,* 1123.
Iqbal, M., Ansari, S. A., and Aggarwala, B. D. (1970). *J. Heat Transfer 92,* 237.
Iqbal, M., Aggarwala, B. D., and Khatry, A. K. (1972). *J. Heat Transfer 94,* 52.
Jackson, T. W., and Yen, H. H. (1971). *J. Heat Transfer 93,* 247.
Jackson, T. W., Spurlock, J. M., and Prudy, K. R. (1967). *AIChE J. 7,* 38.
Jaluria, Y. (1982). *Comput. Fluids 10,* 95.
Jaluria, Y. (1985). Natural Convective Cooling of Electronic Equipment, in *Natural Convection: Fundamentals and Applications,* S. Kakac, W. Aung, and R. Viskanta, Eds., p. 91, Hemisphere, Washington, D.C.
Jaluria, Y. (1986). *Phys. Fluids. 29,* 934.
Jaluria, Y., and Gupta, S. K. (1983). *Int. J. Energy Res. 7,* 201.
Joshi, N. D., and Sukhatme, S. P. (1971). *J. Heat Transfer 93,* 441.
Kakarala, C. R., and Thomas, L. C. (1980). *Int. J. Heat Fluid Flow 2,* 115.
Karwe, M. V., and Jaluria, Y. (1986). ASME Paper 86-WA/HT-80.
Kemeny, G. A., and Somers, E. V. (1962). *J. Heat Transfer 84,* 339.
Kern, D. Q., and Othmer, D. F. (1943). *Trans. AIChE 39,* 517.
Kliegel, J. R. (1959). Ph.D. Thesis, University of California.
Klyachko, L. S. (1963). *J. Heat Transfer 85,* 355.
Larson, M. B., and London, A. L. (1962). ASME Paper 62-HT-44.
Launder, B. E., and Spalding, D. B. (1974). *Comput. Methods Appl. Mech. Eng. 3.*
Lawrence, W. T., and Chato, J. C. (1966). *J. Heat Transfer 88,* 214.
Lee, B. H., and Richardson, P. D. (1965). *J. Mech. Eng. Sci. 7,* 127.
Lemlich, R. (1955). *Ind. Eng. Chem. 47,* 1175.
Lemlich, R., and Rao, M. A. (1965). *Int. J. Heat Mass Transfer 8,* 27.
Li, K. W., and Parker, J. D. (1967). *J. Heat Transfer 89,* 277.
Lloyd, J. R., and Sparrow, E. M. (1970). *Int. J. Heat Mass Transfer 13,* 434.
Lu, P. C. (1959). ASME Paper 59-A-145.
Maitra, D., and Raju, K. S. (1975). *J. Heat Transfer 97,* 135.
Marner, W. J., and McMillan, H. K. (1970). *J. Heat Transfer 92,* 559.
Martinelli, R. C., and Boelter, L. M. K. (1938). *Proc. 5th Int. Congr. Appl. Mech.,* p. 578.
Martinelli, R. C., and Boelter, L. M. K. (1942). *Univ. Calif. Publ. Eng. 5,* 23.
Mason, W. E., and Boelter, L. M. K. (1940). *Power Plant Eng. 44,* 43.
McAdams, W. H. (1954). *Heat Transmission,* 3d ed., p. 235, McGraw-Hill, New York.
McComas, S. T., and Eckert, E. R. G. (1966). *J. Heat Transfer 88,* 147.
Merkin, J. H. (1969). *J. Fluid Mech. 35,* 439.
Merkin, J. H. (1977). *Int. J. Heat Mass Transfer 20,* 73.
Metais, B., and Eckert, E. R. G. (1964). *J. Heat Transfer 86,* 295.
Miyazaki, H. (1971). *Int. J. Heat Mass Transfer 14,* 1295.
Miyazaki, H. (1973). *J. Heat Transfer 95,* 64.
Moffat, R. J., Arvizu, D. E., and Ortega, A. (1985). Cooling Electronic Components: Forced Convection Experiments with an Air-Cooled Spray, in *Heat Transfer in Electronic Equipment—1985,* vol. 48, p. 17, ASME Heat Transfer Div.
Morcos, S. M., and Bergles, A. E. (1975). *J. Heat Transfer 97,* 212.
Mori, Y. (1961). *J. Heat Transfer 83,* 479.
Mori, Y., and Futagami, K. (1967). *Int. J. Heat Mass Transfer 10,* 1801.
Mori, Y., Futagami, K., Tokuda, S., and Nakamura, M. (1966). *Int. J. Heat Mass Transfer 9,* 453.
Morgan, V. T. (1975). *Adv. Heat Transfer 11,* 199.
Morton, B. R. (1960). *J. Fluid Mech. 8,* 227.
Moutsoglou, A., and Chen, T. S. (1980). *J. Heat Transfer 102,* 371.
Mucoglu, A., and Chen, T. S. (1977a). *Can. J. Chem. Eng. 55,* 265.
Mucoglu, A., and Chen, T. S. (1977b). *J. Heat Transfer 99,* 679.

Mucoglu, A., and Chen, T. S. (1978a). *Proc. 6th Int. Heat Transfer Conf.*,Toronto, vol. 1, p. 85.
Mucoglu, A., and Chen, T. S. (1978b). *J. Heat Transfer 100,* 542.
Mucoglu, A., and Chen, T. S. (1979). *J. Heat Transfer 101,* 422.
Nakai, S., and Okazaki, T. (1975). *Int. J. Heat Mass Transfer 18,* 397.
Nayak, A. L., and Cheng, P. (1975). *Int. J. Heat Mass Transfer 18,* 227.
Newell, P. H., and Bergles, A. E. (1970). *J. Heat Transfer 92,* 83.
Oberkampf, W. F., and Crow, L. I. (1976). *J. Heat Transfer 98,* 353.
Ogunba, V., and Barrow, H. (1979). *Int. J. Heat Fluid Flow 1,* 115.
Ojalvo, M. S., and Grosh, R. J. (1962). Report 12, Dept. of Mechanical Engineering, Purdue Research Foundation, West Lafayette, Ind.
Ojalvo, M. S., Anand, D. K., and Dunbar, R. P. (1967). *J. Heat Transfer 89,* 328.
Oktay, S., and Moffat, R. J., Eds. (1985). *Heat Transfer in Electronic Equipment—1985,* vol. 48, ASME Heat Transfer Div.
Oliver, D. R. (1962). *Chem. Eng. Sci. 17,* 335.
Oosthuizen, P. H. (1970). *Trans. Inst. Chem. Eng. 48,* T227.
Oosthuizen, P. H. (1971). Paper presented at the IUTAM Symposium, Quebec.
Oosthuizen, P. H., and Hart, R. (1973). *J. Heat Transfer 95,* 60.
Oosthuizen, P. H., and Madan, S. (1970). *J. Heat Transfer 92,* 194.
Oosthuizen, P. H., and Madan, S. (1971). *J. Heat Transfer 93,* 240.
Ostrach, S. (1954). NACA TN 3141.
Ou, J. W., and Cheng, K. C. (1977). *Int. J. Heat Mass Transfer 20,* 953.
Ou, J. W., Cheng, K. C., and Lin, R. C. (1974). *Int. J. Heat Mass Transfer 17,* 835.
Ou, J. W., Cheng, K. C., and Lin, R. C. (1976). *Int. J. Heat Mass Transfer 19,* 277.
Ower, E., and Johansen, F. C. (1931). R&M 1437, British ARC, Appendix IV.
Patankar, S. V., Ramadhyani, S., and Sparrow, E. M. (1978). *J. Heat Transfer 100,* 63.
Penney, W. R., and Jefferson, T. B. (1966). *J. Heat Transfer 88,* 359.
Petukhov, B. S. (1976). Seminar on Heat Transfer in Turbulent Free Convection, International Center for Heat Mass Transfer, Dubrovnik.
Petukhov, B. S., and Polyakov, A. F. (1967a). *Teplofiz. Vys. Temp. 5,* 87.
Petukhov, B. S., and Polyakov, A. F. (1967b). *Teplofiz. Vys. Temp. 5,* 384.
Petukhov, B. S., Polyakov, A. F., and Strigin, B. K. (1969). *Heat Transfer Sov. Res. 1,* 24.
Pigford, R. L. (1955). *Chem. Eng. Prog. Symp. Ser. 51,* 79.
Prakash, C., and Patankar, S. V. (1981). *J. Heat Transfer 103,* 566.
Quintiere, J., and Mueller, W. K. (1973). *J. Heat Transfer 95,* 53.
Ramachandran, N., Armaly, B. F., and Chen, T. S. (1983). *J. Heat Transfer 105,* 420.
Robinson, G. C., McClude, C. M., and Hendricks, R. (1958). *Am. Ceram. Soc. Bull. 37,* 399.
Rosen, E. M., and Hanratty, T. J. (1967). *AIChE J. 7,* 112.
Roy, D. N. (1966). Ph.D. Thesis, University of Calcutta.
Sabbagh, J. A., Aziz, A., El-Ariny, A. S., and Hamad, G. (1976). *J. Heat Transfer 98,* 322.
Sammakia, B., Gebhart, B., and Carey, V. P. (1982). *Int. J. Heat Mass Transfer 25,* 835.
Scheele, G. F., and Hanratty, T. J. (1962). *J. Fluid Mech. 14,* 244.
Scheele, G. F., Rosen, E. M., and Hanratty, T. J. (1960). *Can. J. Chem. Eng. 38,* 67.
Schmidt, R. R., and Sparrow, E. M. (1978). *J. Heat Transfer 100,* 403.
Seely, J. H. (1960). Master's Thesis, Syracuse University.
Shah, R. K., and London, A. L. (1978). *Advances in Heat Transfer,* Supl. 1, Academic Press, New York.
Sharma, G. K., and Sukhatme, S. P. (1969). *J. Heat Transfer 91,* 457.
Sherwin, K. (1968). *Br. Chem. Eng. 13,* 569.
Sherwin, K., and Wallis, J. D. (1970). *Proc. 4th Int. Heat Transfer Conf.,* Paris, vol. 4, NC3.9.
Shine, A. J. (1962). *J. Heat Transfer 84,* 225.
Sieder, E. N., and Tate, G. E. (1936). *Ind. Eng. Chem. 28,* 1429.
Sogin, H. H., and Subramanian, Y. S. (1961). *J. Heat Transfer 83,* 483.
Sparrow, E. M., and Gregg, J. L. (1956). *Trans. ASME 78,* 435.
Sparrow, E. M., and Gregg, J. L. (1959). *J. Appl. Mech. 81,* 133.

Sparrow, E. M., and Lee, L. (1976). *Int. J. Heat Mass Transfer 19,* 229.
Sparrow, E. M., and Minkowycz, W. J. (1962). *Int. J. Heat Mass Transfer 5,* 505.
Sparrow, E. M., Eichhorn, R., and Gregg, J. L. (1959). *Phys. Fluids 2,* 319.
Sprott, A. L., Holman, J. P., and Durand, F. L. (1960). ASME Paper 60-HT-19.
Stewartson, K. (1957). *J. Math. Phys. 336,* 173.
Suwono, A. (1979). *Int. J. Heat Mass Transfer 23,* 819.
Szewczyk, A. A. (1964). *J. Heat Transfer 86,* 501.
Tanaka, H., Tsuge, A., Hirata, M., and Nishiwaki, N. (1973). *Int. J. Heat Mass Transfer 16,* 1267.
Tao, L. N. (1960a). *J. Heat Transfer 82,* 233.
Tao, L. N. (1960b). *Appl. Sci. Res. 9,* 357.
Teleki, C., Fand, R. M., and Kaye, J. (1960). WADC TN 59-357.
Tsuruno, S., and Iguchi, I. (1980). *J. Heat Transfer 102,* 168.
Wang, S. A. (1982). *J. Heat Transfer 104,* 139.
Wilks, G. (1973). *Int. J. Heat Mass Transfer 16,* 1958.
Wilks, G. (1974). *Int. J. Heat Mass Transfer 17,* 743.
Yao, L. S. (1978). *J. Heat Transfer 100,* 212.
Yao, L. S., and Catton, I. (1978). *Proc. 5th Int. Heat Transfer Conf.,* Toronto, vol. 1, p. 13.
Yousef, W. W., and Tarasuk, J. D. (1981). *J. Heat Transfer 103,* 249.
Yousef, W. W., and Tarasuk, J. D. (1982). *J. Heat Transfer 104,* 145.
Yuge, T. (1960). *J. Heat Transfer 83,* 214.
Zeldin, B., and Schmidt, F. W. (1972). ASME Paper 71-HT-6.
Zhukauskas, A. A., Shlanch Yauskas, A. A., and Yanonees, Z. P. (1961). *J. Eng. Phys. 4,* 58.
Zijnen, Van der Hegge, B. G. (1956). *Appl. Sci. Res. Sec. A6,* 129.

PROBLEMS

10.1 Air at 60°F and 1 atm flows vertically over a flat surface 2 ft high. The surface is at 100°F. For what range of air velocity will forced convection effects be negligible over the region $0.5 \leq x < 2$ ft?

10.2 Air at 80°F and 1 atm flows vertically over a flat surface 2 ft high at a forced velocity of 5 ft/s. The surface is isothermal. For what range of plate temperature t_0 will natural convection effects be negligible over its whole extent?

10.3 A vertical plate of height 1 m is at a temperature of 200°C in air at 27°C. A downward uniform stream of air flows over the surface with velocity ranging from 0.01 to 1 m/s. Find the regime, forced, natural, or mixed, in which the flow lies for these two extreme cases. Develop a simple criterion for determining the separation point, assuming that the opposing mechanisms give rise to separation over the plate.

10.4 A vertical isothermal surface is at 60°C in air at 15°C. If a vertical external flow of 10 m/s is imposed parallel to the surface, determine whether natural convection effects are important. Does the conclusion change for a velocity of 1 m/s?

10.5 An electronic panel 10 cm high may be treated as a vertical surface at 50°C in air at 20°C. An externally imposed flow at 0.5 m/s occurs past the panel, in one case aiding the natural convection flow and in another case opposing it. Determine the resulting heat transfer in the two cases. Repeat the calculation if the surface temperature is 100°C.

10.6 A long, 0.0004-in.-diameter electrically heated wire at 90°F is in air at 70°F and 1 atm. The wire is horizontal.

(a) Find the upward velocity of the ambient air for which the buoyancy will not appreciably affect forced convection heat transfer.

(b) Would a natural convection effect increase or decrease the heat transfer rate?

10.7 A long wire of diameter 10^{-4} m is at 300°C and is placed in a cross flow of air at 1 m/s. Are buoyancy effects important if the air is at 20°C?

10.8 A long heated wire of diameter 1 mm is at a temperature of 200°C and is placed in a vertical air stream at 15°C and 10 cm/s. Determine the flow regime, natural or forced, in which the convection process lies. Also determine the heat transfer for the natural and forced convection problems separately and compare the results.

10.9 A hot-wire sensor of diameter 10^{-2} mm is at 320°C and is placed in air at 20°C. The cross-flow air velocity varies from 1 to 10 cm/s. For these two extreme values, find the flow regime (forced or natural) in which the convective process lies.

10.10 The sensor of a hot-wire anemometer is 0.001 cm in diameter and 0.2 cm long. It is used to measure natural convection velocities in air around 0.5 cm/s. Determine whether the natural convection effects in the sensor heat transfer are important. Would you expect the end conduction to be significant? The sensor and air temperatures are 100 and 20°C, respectively.

10.11 A sphere of diameter 10 cm is in an air stream at 20°C and 1 m/s. If the temperature of the sphere is 120°C, find the heat transfer from it. Would you expect natural convection effects to be significant?

10.12 Write down the full equations for mixed convection from a heated sphere, considering property variation due to temperature and pressure differences.

10.13 A horizontal heated surface is at 10°C above the ambient air at 25°C. A breeze flows past the surface at 0–5 m/s. Determine whether, at any velocity level, natural convection effects become insignificant.

10.14 For an axisymmetric plume in a uniform cross flow at velocity V_∞, write the governing equations and the boundary conditions. Discuss the various mechanisms that arise.

10.15 Consider an axisymmetric plume arising from a point heat source. If an upward velocity U_∞ exists outside the boundary layer flow, obtain the governing equations for this mixed convection flow. Formulate the solution of these equations by methods discussed in the text.

10.16 A vertical pipe of diameter 1 cm has an inflow of water at 1 cm/s at the bottom. At what height do you expect buoyancy effects to be important, if the wall temperature excess $t_s - t_a$ is ± 20°C?

(a) By physical reasoning or by using the governing equations, determine the nature of the velocity and temperature profiles.

(b) Determine the heat transfer for the first 1 m of the pipe for this internal flow.

10.17 Aerosol particles settle in the earth's atmosphere due to their own weight. Consider this process for particles assumed to be spherical, of diameter 100 μm and the density of limestone, in a solar intensity of 800 W/m^2. Take the absorptivity to be $\alpha_p = 0.9$.

(a) Calculate the settling rate, neglecting heat transfer and buoyancy effects.

(b) Calculate the particle temperature, neglecting buoyancy effects.

(c) Are mixed convection effects indicated?

(d) If mixed convection effects are indicated, calculate the temperature and settling rate, including such effects.

CHAPTER
ELEVEN

INSTABILITY, TRANSITION, AND TURBULENCE

11.1 INSTABILITY CHARACTERISTICS

The foregoing chapters considered many kinds of buoyancy-induced laminar boundary region flows. Many of the mechanisms that arise in downstream transport development have been set forth. In fact, many of the mechanisms that have been described occur in a similar fashion in the much larger-scale processes in which the transport is largely turbulent. The laminar flows considered to this point often become unstable to the disturbances that are always present, even at the small size scales common in technological and immediate environmental processes. The difference between the rates of laminar and turbulent transport is very large in most flows. Therefore, the question of how and when or where a flow becomes turbulent has a direct effect on the accuracy and reliability of estimates of transport.

Turbulence eventually follows the initial instability of a laminar flow to naturally occurring disturbances. A frequent source of such disturbances is external vibrations. Fluctuations in heat input to a heated surface also introduce disturbances in the flow. These disturbances may be fed into the flow anywhere at any time. Depending on the conditions, in terms of buoyancy, location, and so forth, they may grow in amplitude due to a balance of buoyancy, pressure, and viscous forces. However, the mechanisms are different for different flow configurations and bounding conditions. Also, between any initial laminar instability and eventual transition downstream to completely turbulent flow, the particular processes are often different. These aspects are considered in later sections of this chapter.

The first mode in this unstable progression for many flows is the initial growth of very small disturbances. This is studied by using a linear analysis. As seen in subsequent sections, considerable simplification arises in assuming that the velocity and temperature disturbances imposed by the environment on the flow are very small compared to the velocity and temperature differences in the developing laminar flow. Also, these disturbances are taken as periodic since a given disturbance may be represented by a Fourier series of periodic terms.

Consider first a vertical laminar flow developing downstream adjacent to a vertical surface. It eventually becomes unstable to some of the various components of the disturbance imposed on it by its environment. A particular component of the disturbance may then be isolated. It has been found that a frequent mechanism of disturbance growth in buoyancy-induced flows is a propagating downstream periodic wave. Early observations, such as those of Eckert and Soehngen (1951), indicated that the physical form is somewhat like the ocean surface wave motion, which approaches parallel to the shore and eventually becomes breaking surf. Such wave motion, in a forced boundary layer flow, is called a Tollmien-Schlichting wave.

These waves will grow in magnitude if conditions are such that the flow and buoyancy forces channel net flow energy into them. However, flows react differently to disturbances of different frequencies. They first become unstable to a given frequency at different downstream locations, in terms of the local flow vigor parameter Gr_x or G. Figure 11.1.1 shows the downstream progression of a disturbance that is first damped, then amplified.

A representative stability plane, such as would be calculated for small disturbance amplitude, is shown in Fig. 11.1.2. The region of disturbance damping is at low G. It is separated from the unstable and amplifying region by the neutral curve. The disturbance frequency is F or ω, where ω is the nondimensional frequency defined later in Eq. (11.2.29). A given condition on this plane, of ω and G or of frequency F and downstream location x, is either a stable, neutral, or unstable disturbance environment.

The question, however, really is how a given disturbance of frequency F is treated by the flow as it is convected and propagates downstream, to increasing G. In Fig. 11.1.2 the dashed curves show the paths of propagation of disturbances of three different given physical frequencies $F_1 > F_2 > F_3$. Such paths result, for example, in a flow generated adjacent to a vertical surface dissipating uniform heat flux. This is a frequently convenient experimental arrangement. For the path of F_2, the disturbance is seen first to damp and then to amplify at an increasing rate downstream, farther into the unstable region. The result, for this path and for others in its vicinity, is disturbance growth to large amplitude.

However, not all disturbance components behave this way, even for flow adjacent to a uniform heat flux surface. Note that F_1 and F_3 are damped. This example demonstrates the phenomenon of selective amplification observed by Dring and Gebhart (1968), which will be discussed quantitatively in the next section for this flow. Here, the general behavior of disturbances will be considered for different flow configurations. Paths of constant disturbance frequency propagation are shown in Fig. 11.1.3 for five flows.

The paths have strikingly different characteristics. For vertical surfaces, they

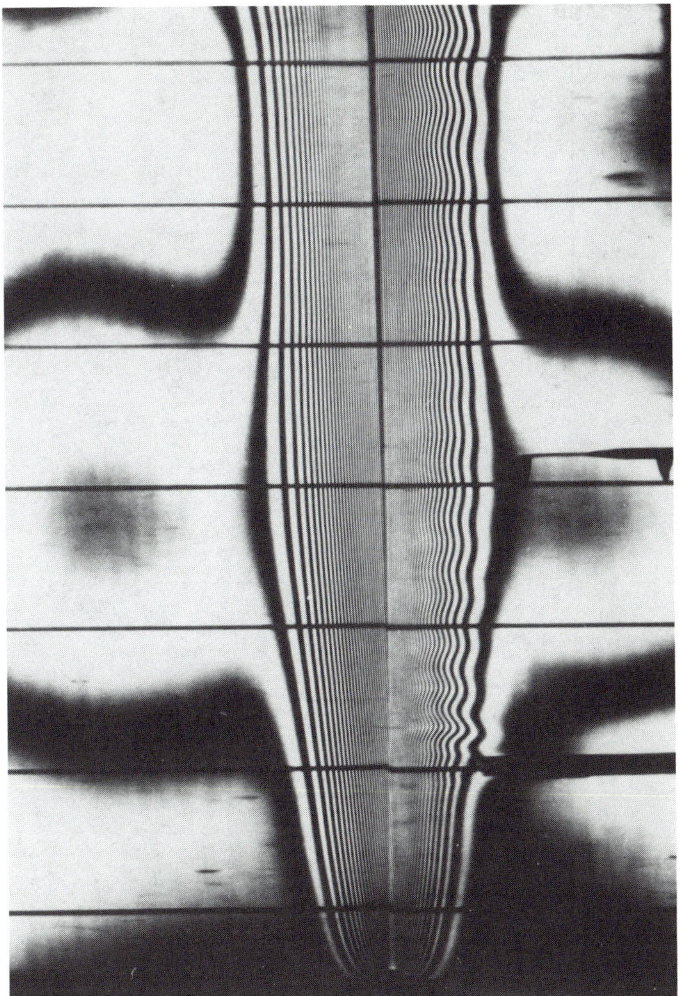

Figure 11.1.1 Response to a controlled disturbance introduced upstream in the stable region, crossing the neutral curve into the region of increasing downstream amplification, for flow next to an electrically heated foil in 0.65 centistokes silicone oil. *(Reprinted with permission from* Prog. Heat Mass Transfer, *vol. 2, p. 99, C. P. Knowles and B. Gebhart. Copyright © 1969, Pergamon Journals Ltd.)*

penetrate more deeply into the highly amplified region as they are convected downstream. Detailed behavior in the unstable region will later show, in Figs. 11.2.1 and 11.2.2, that only a very narrow band of frequencies is highly amplified. Thus such flows sharply filter a complicated disturbance for essentially a single frequency. However, disturbances in the flow over a horizontal surface, and in plume flows, cross the unstable region. The upper branch of the neutral curve is known to be bounded for many different flows. Therefore, any given

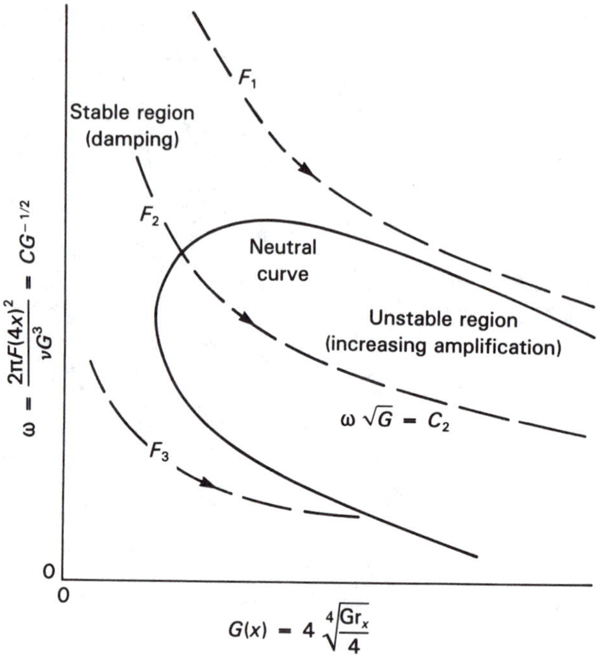

Figure 11.1.2 Typical stability plane for a buoyancy-induced flow generated adjacent to a vertical surface for downstream propagating disturbances of the Tollmien-Schlichting type.

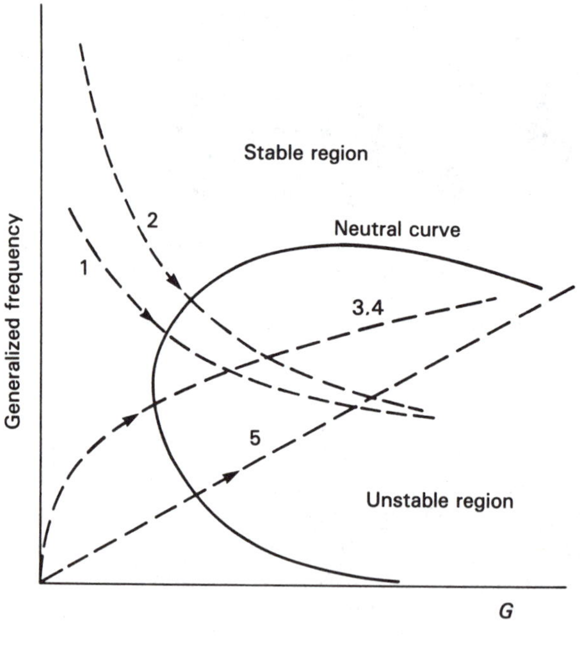

Figure 11.1.3 Typical stability plane for buoyancy-induced flows, showing downstream paths of propagation of a disturbance of frequency F in different kinds of flow. Numbers represent flow configurations: 1, isothermal vertical surface; 2, uniform heat flux vertical surface; 3, horizontal and slightly inclined surfaces; 4, plane plume; and 5, axisymmetric plume.

frequency over a very broad band is unstable only over a range of G, that is, of downstream distance x.

The implications of this are very interesting, since these kinds of disturbances are the origin of the later (in x) more complicated disturbances that disrupt a laminar flow. The results in Fig. 11.1.3 suggest that flows adjacent to vertical surfaces are inevitably unstable. However, the free-boundary flows 4 and 5, along with 3, in which the buoyancy force is normal to the flow direction, are eventually stable in the linear range of amplitude to all disturbances. But experimentally they are found to be much less stable than 1 and 2, because other mechanisms become important much more quickly (in x) for flows 3, 4, and 5. Disturbances grow quickly in free-boundary flows 4 and 5, presumably because of the absence of a surface to damp disturbances. The explanation for horizontal flow, 3, is likely to be the added thermal instability mechanism of unstable stratification, to be discussed in Chapter 13.

Detailed characteristics of instability to small disturbances are discussed for flow 3 by Pera and Gebhart (1973), for 4 by Pera and Gebhart (1971) and Haaland and Sparrow (1973a), and for 5 by Mollendorf and Gebhart (1973). Flows 1 and, in all particulars, 2 have been much more intensively studied—to, through, and even beyond transition into full turbulence. Therefore, 2, a vertical flow generated adjacent to a uniform heat flux surface, will be the example followed through in the following sections in an account of the detailed mechanisms to full turbulence.

In Sections 11.2–11.7 the vertical flows generated adjacent to surfaces are followed in detail through nonlinear disturbance growth and the transition process and on to a developed turbulence. Correlations for turbulent transport are given in Section 11.7. The measures of plane plume transition are given in Section 11.8. The stability of flows driven by the combined effects of thermal and mass transport and of mixed convection is considered in Sections 11.9 and 11.10. In Section 11.11 the improvement of linear stability analysis for vertical flows by consistent higher-order approximations is treated. Section 11.12 considers horizontal flows arising from a buoyancy force outward from a surface. Information concerning transient transition and turbulence is summarized in Section 11.13. Finally, the effects on stability of the anomalous density behavior of cold water, as set forth in Chapter 9, are considered in Section 11.14. Some of the mechanisms treated in this chapter have been reviewed in greater detail by Gebhart (1973a,b) and Gebhart and Mahajan (1982).

11.2 DOWNSTREAM DISTURBANCE GROWTH IN VERTICAL FLOWS

Analysis of the stability of laminar fluid motions began over a century ago. The modern linear formulation, which includes a viscous force mechanism of disturbance interaction, has been in use for forced flows for over a half-century. Most uses have been related to two-dimensional plane flows. The resulting stability equations, or Orr-Sommerfeld equations, are in linearized form in terms of disturbance quantities. These were generalized to include the effect of a buoyancy force by Plapp (1957) for flow generated adjacent to a vertical surface at a temperature t_0 in a quiescent ambient fluid medium at t_∞.

In this section the full equations governing transient flow will be set forth in terms of velocity components $u(x, y, \tau)$, $v(x, y, \tau)$, temperature $t(x, y, \tau)$, pressure $p(x, y, \tau)$, and density $\rho(x, y, \tau)$, where τ is time. Then disturbances $u'(x, y, \tau)$, $v'(x, y, \tau)$, and so forth will be postulated for a steady laminar base flow. This flow is steady in the sense that the average values of u, v, etc., that is, $\bar{u}(x, y)$, $\bar{v}(x, y)$, and so forth, are independent of the long time intervals over which the averages are taken.

The quantities $u = \bar{u} + u'$, $v = \bar{v} + v'$, and so on, are substituted into the full equations. A series of approximations are then made to reduce the formulation to a set of equations and boundary conditions for the conventional boundary region base flow and the disturbance quantities. After the disturbance forms are postulated, both the base flow and the disturbance formulations are converted to ordinary differential equations by the boundary layer similarity transformation. Section 11.11 then returns to improving these approximations in flow and stability analysis in a more consistent way.

11.2.1 Stability Equations for Two-Dimensional Plane Flows

The full equations in x and y are written out below, neglecting the pressure and viscous dissipation terms in the energy equation and assuming constant and uniform properties μ and k in the fluid. The positive x direction is again taken in the direction of gravity $\mathbf{g}$ but opposed to it. The motion pressure p_m is again defined as $p(x, y, \tau) = p_m(x, y, \tau) + p_h(x)$. That is, p_m is the difference between the actual static and the local external hydrostatic pressure. Using the Boussinesq approximations discussed in Chapter 2, pressure effects on fluid density are neglected and temperature effects are taken closely linear over the range from t_0 to t_∞. Then $g(\rho_\infty - \rho)$ may be taken as $g\rho\beta(t - t_\infty)$.

The resulting equations are Eqs. (2.7.10) and (2.7.15)–(2.7.17), with mass diffusion ignored. However, the transient terms in Eqs. (2.7.2) and (2.7.4) are retained since the disturbances have time-dependent behavior. The equations, along with boundary conditions for an impermeable vertical surface at $y = 0$ and $x \geq 0$, are then as follows. The ambient medium is assumed to be uniformly at t_∞.

$$\frac{\partial u}{\partial x} + \frac{\partial v}{\partial y} = 0 \tag{11.2.1}$$

$$\frac{\partial u}{\partial \tau} + u\frac{\partial u}{\partial x} + v\frac{\partial u}{\partial y} = \nu\nabla^2 u + g\frac{\rho_\infty - \rho}{\rho_r} - \frac{1}{\rho_r}\frac{\partial p_m}{\partial x} \tag{11.2.2}$$

$$\frac{\partial v}{\partial \tau} + u\frac{\partial v}{\partial x} + v\frac{\partial v}{\partial y} = \nu\nabla^2 v - \frac{1}{\rho_r}\frac{\partial p_m}{\partial y} \tag{11.2.3}$$

$$\frac{\partial t}{\partial \tau} + u\frac{\partial t}{\partial x} + v\frac{\partial t}{\partial y} = \alpha\nabla^2 t \tag{11.2.4}$$

$$u(x, 0, \tau) = v(x, 0, \tau) = t(x, 0, \tau) - t_0 = t(x, \infty, \tau) - t_\infty$$

$$= u(x, \infty, \tau) = 0 \tag{11.2.5}$$

where ∇^2 is $\partial^2/\partial x^2 + \partial^2/\partial y^2$ and ρ_r is a suitably chosen constant reference value of ρ.

The equations above are next written in terms of their mean and disturbance components. The relation between ρ' and t' is also given.

$$u(x, y, \tau) = \bar{u}(x, y) + u'(x, y, \tau) \qquad v(x, y, \tau) = \bar{v}(x, y) + v'(x, y, \tau) \tag{11.2.6}$$

$$t(x, y, \tau) = \bar{t}(x, y) + t'(x, y, \tau) \tag{11.2.7}$$

$$p_m(x, y, \tau) = \bar{p}_m(x, y) + p_m'(x, y, \tau) \tag{11.2.8}$$

$$\rho(x, y, \tau) = \bar{\rho}(x, y) + \rho'(x, y, \tau) = \bar{\rho}(x, y) - \rho_r\beta t' \tag{11.2.9}$$

These are introduced into Eqs. (11.2.1)–(11.2.5):

$$\frac{\partial \bar{u}}{\partial x} + \frac{\partial \bar{v}}{\partial y} + \frac{\partial u'}{\partial x} + \frac{\partial v'}{\partial y} = 0 \tag{11.2.10}$$

$$\bar{u}\frac{\partial \bar{u}}{\partial x} + \bar{v}\frac{\partial \bar{u}}{\partial y} + \frac{\partial u'}{\partial \tau} + \bar{u}\frac{\partial u'}{\partial x} + \bar{v}\frac{\partial u'}{\partial y} + u'\frac{\partial \bar{u}}{\partial x} + v'\frac{\partial \bar{u}}{\partial y} + u'\frac{\partial u'}{\partial x}$$

$$+ v'\frac{\partial u'}{\partial y} = \nu\nabla^2\bar{u} + \nu\nabla^2 u' + g\beta(\bar{t} - t_\infty) + g\beta t'$$

$$-\frac{1}{\rho_r}\frac{\partial \bar{p}_m}{\partial x} - \frac{1}{\rho_r}\frac{\partial p_m'}{\partial x} \tag{11.2.11}$$

$$\bar{u}\frac{\partial \bar{v}}{\partial x} + \bar{v}\frac{\partial \bar{v}}{\partial y} + \frac{\partial v'}{\partial \tau} + \bar{u}\frac{\partial v'}{\partial x} + \bar{v}\frac{\partial v'}{\partial y} + u'\frac{\partial \bar{v}}{\partial x} + v'\frac{\partial \bar{v}}{\partial y} + u'\frac{\partial v'}{\partial x}$$

$$+ v'\frac{\partial v'}{\partial y} = \nu\nabla^2\bar{v} + \nu\nabla^2 v' - \frac{1}{\rho_r}\frac{\partial \bar{p}_m}{\partial y} - \frac{1}{\rho_r}\frac{\partial p_m'}{\partial y} \tag{11.2.12}$$

$$\bar{u}\frac{\partial \bar{t}}{\partial x} + \bar{v}\frac{\partial \bar{t}}{\partial y} + \frac{\partial t'}{\partial \tau} + \bar{u}\frac{\partial t'}{\partial x} + \bar{v}\frac{\partial t'}{\partial y} + u'\frac{\partial \bar{t}}{\partial x} + v'\frac{\partial \bar{t}}{\partial y} + u'\frac{\partial t'}{\partial x}$$

$$+ v'\frac{\partial t'}{\partial y} = \alpha\nabla^2\bar{t} + \alpha\nabla^2 t' \tag{11.2.13}$$

These results are written in expanded form to make clear the nature and impact of the approximations that follow. These approximations will yield the Orr-Sommerfeld equations in terms of a conventional boundary layer solution. The collection of approximations may be classified as follows:

1. The base flow is taken as that resulting from the boundary layer theory formulation.
2. Only first-order, or linear, terms in the disturbance quantities u', v', t' are retained.
3. After approximations 1 and 2 are made, the base flow quantities $\bar{v}$ and the x derivatives of $\bar{u}$ and $\bar{t}$ are taken as zero. This is called the parallel flow approximation.
4. The postulated disturbances are assumed to be of the form so that the amplitude functions depend, as does the base flow, only on the similarity variable η. In addition, the disturbance amplification and wavelength are assumed to be independent of x. Implications of this approximation are discussed later.

Applying approximation 1 above, the following collection of terms in Eqs. (11.2.10)–(11.2.13) describe the base flow, $\bar{u}$, $\bar{v}$, and $\bar{t}$:

$$\frac{\partial \bar{u}}{\partial x} + \frac{\partial \bar{v}}{\partial y} = 0 \tag{11.2.14}$$

$$\bar{u}\frac{\partial \bar{u}}{\partial x} + \bar{v}\frac{\partial \bar{u}}{\partial y} = \nu\frac{\partial^2 \bar{u}}{\partial y^2} + g\beta(\bar{t} - t_\infty) \tag{11.2.15}$$

$$\bar{u}\frac{\partial \bar{t}}{\partial x} + \bar{v}\frac{\partial \bar{t}}{\partial y} = \alpha\frac{\partial^2 \bar{t}}{\partial y^2} \tag{11.2.16}$$

These equalities eliminate all these terms from Eqs. (11.2.10)–(11.2.13). Also eliminated there, as a result of conventional boundary layer approximations, are all other effects expressed purely in base flow quantities. These consist of $\bar{p}_m$, purely base flow terms in Eq. (11.2.12), and the streamwise second derivatives of $\bar{u}$, $\bar{v}$, and $\bar{t}$.

The second approximation listed above amounts to the small disturbance magnitude or linear approximation. This means, for example, that periodic disturbances will not generate either harmonics or mean flow components. From this assumption, all terms involving any product of u', v', and t' and/or their derivatives are eliminated.

The next approximation, 3, is an ad hoc one. A base flow and transport, $\bar{u}$, $\bar{v}$, and $\bar{t}$, were postulated in Eqs. (11.2.14)–(11.2.16). The value of $\bar{v}$ in the resulting solution is not zero. Nevertheless, both $\bar{v}$ and the x derivatives of $\bar{u}$ and $\bar{t}$ are next taken as zero in the disturbance equations that remain after approximations 1 and 2 are made. The principal result is that all remaining terms in $\bar{v}$ disappear. The residual equations then become as given below. The internal consistency of this collection of approximations is considered in the general discussion of such approximations in Section 11.11.

$$\frac{\partial u'}{\partial x} + \frac{\partial v'}{\partial y} = 0 \tag{11.2.17}$$

$$\frac{\partial u'}{\partial \tau} + \bar{u}\frac{\partial u'}{\partial x} + v'\frac{\partial \bar{u}}{\partial y} = \nu\nabla^2 u' + g\beta t' - \frac{1}{\rho_r}\frac{\partial p'_m}{\partial x} \tag{11.2.18}$$

$$\frac{\partial v'}{\partial \tau} + \bar{u}\frac{\partial v'}{\partial x} = \nu\nabla^2 v' - \frac{1}{\rho_r}\frac{\partial p'_m}{\partial y} \tag{11.2.19}$$

$$\frac{\partial t'}{\partial \tau} + \bar{u}\frac{\partial t'}{\partial x} + v'\frac{\partial \bar{t}}{\partial y} = \alpha\nabla^2 t' \tag{11.2.20}$$

Equations (11.2.14) and (11.2.17) suggest a stream function $\psi(x, y, \tau) = \bar{\psi}(x, y) + \psi'(x, y, \tau)$ such that

$$\bar{u} = \bar{\psi}_y \qquad -\bar{v} = \bar{\psi}_x \qquad u' = \psi'_y \qquad -v' = \psi'_x \tag{11.2.21}$$

The base flow similarity transformation in Section 3.5 is applied to Eqs. (11.2.14)–(11.2.16) for an impervious vertical surface beginning at $x = 0$ in an isothermal and quiescent ambient medium. The results are written below, along with boundary conditions, boundary layer thickness δ, and characteristic velocity U_c for a power law surface temperature variation.

$$f''' + (n + 3)ff'' - (2n + 2)(f')^2 + \phi = 0 \tag{11.2.22}$$

$$\phi'' + \Pr[(n + 3)f\phi' - 4nf'\phi] = 0 \tag{11.2.23}$$

$$f'(0) = f(0) = 1 - \phi(0) = f'(\infty) = \phi(\infty) = 0 \tag{11.2.24}$$

$$\delta(x) = \frac{4x}{G} \qquad \frac{U_c x}{\nu} = \frac{G^2}{4} \qquad G = 4\left(\frac{\mathrm{Gr}_x}{4}\right)^{1/4} \tag{11.2.25}$$

The form of the disturbance Eqs. (11.2.18) and (11.2.19) along with Eq. (11.2.21) suggests forming a disturbance vorticity equation in ψ' to eliminate p'_m. Then the remaining disturbance functions are only ψ' and t'. These are postulated for any given disturbance mode as follows:

$$\psi'(x, y, \tau) = \delta U_c\, \Phi(\eta)e^{i(\hat{\alpha}x - \hat{\omega}\tau)} \tag{11.2.26}$$

$$t'(x, y, \tau) = d(x)\, s(\eta)e^{i(\hat{\alpha}x - \hat{\omega}\tau)} \tag{11.2.27}$$

$$\alpha = \hat{\alpha}\delta = \hat{\alpha}_r\delta + i\hat{\alpha}_i\delta = \frac{2\pi\delta}{\lambda} + i\hat{\alpha}_i\delta = \alpha_r + i\alpha_i \tag{11.2.28}$$

$$\omega = \frac{\hat{\omega}\delta}{U_c} = \frac{\hat{\omega}_r\delta}{U_c} = \frac{2\pi F\delta}{U_c} \tag{11.2.29}$$

where $i = \sqrt{-1}$ and the dimensionless amplitude functions across the transport region, $\Phi(\eta)$ and $s(\eta)$, are in general complex. Note that α in the above equation is the complex wave number. The two disturbances, linked through their residence in mass, are taken to be of similar form. The frequency F appears in $\hat{\omega} = 2\pi F$, $\hat{\omega}$ being taken as real. Here $\hat{\alpha}$ is complex, where $\hat{\alpha}_r$ is the wave number, λ the wavelength, and $\hat{\alpha}_i$ the spatial (in x) exponential amplification rate. These quantities are generalized as in Eqs. (11.2.28) and (11.2.29). Neutral stability corresponds to $\alpha_i = 0$ and downstream amplification occurs for $\alpha_i < 0$.

At this point, the fourth approximation listed above is invoked to yield the form of Orr-Sommerfeld equations applicable here. These equations are in terms of the disturbance amplitude functions Φ and s. The first part of approximation 4 is that the amplitude functions depend only on η, that is, that $\Phi = \Phi(\eta)$ and $s = s(\eta)$. This is analogous to the conditions of the base flow, that ϕ and f depend only on η. The second part of approximation 4 is that the x derivatives of the disturbance spatial amplification rate $\hat{\alpha}_i$ and disturbance wavelength λ, or $\hat{\alpha}_r$, are taken as negligibly small.

The first three of these measures, concerning Φ, s, and $\hat{\alpha}_i$, are supported quite well by numerous accurate measurements in several different vertical buoyancy-induced flows. However, the same data, along with optical visualizations, as in Fig. 11.1.1, indicate that λ varies appreciably downstream. In fact it must, since frequency F is constant in an analysis by frequency modes, while the velocity of the base flow increases downstream.

These downstream effects may be much more severe in more complicated flows of different geometric and bounding conditions. The improvement of the analysis, in any of the four categories of approximation set forth above, must be done consistently. Any such improvement must be cognizant of the total body of approximations. It should not be merely an ad hoc or selective treatment of certain effects. The formulation of a consistent scheme of improved approximation is set forth in Section 11.11 for several vertical flows.

Substitution of Eqs. (11.2.26)–(11.2.29) into (11.2.17)–(11.2.20) yields the Orr-Sommerfeld equations in terms of disturbance amplitude functions $\Phi(\eta)$ and $s(\eta)$. Boundary conditions characteristic of a quiescent ambient medium are also given.

$$\left(f' - \frac{\omega}{\alpha}\right)(\Phi'' - \alpha^2\Phi) - f'''\Phi = \frac{\Phi'''' - 2\alpha^2\Phi'' + \alpha^4\Phi + s'}{i\alpha G} \qquad (11.2.30)$$

$$\left(f' - \frac{\omega}{\alpha}\right)s - \phi'\Phi = \frac{s'' - \alpha^2 s}{i\alpha \Pr G} \qquad (11.2.31)$$

$$\Phi(\infty) = \Phi'(\infty) = s(\infty) = 0 \qquad (11.2.32)$$

These coupled equations are seen to be sixth order. Three other boundary conditions, at $\eta = 0$, depend on the particular flow under consideration. The equations contain five independent parameters α_r, α_i, ω, Pr, and G. Also, $f(\eta)$ and $\phi(\eta)$ are dependent on Pr.

11.2.2 Stability Plane Calculations and Experimental Data

Solutions to the set of equations above amount to finding $G(\omega)$ for $\alpha_i = 0$, as the neutral stability condition or curve, in ω, G coordinates. Contours of constant amplification rate $-\alpha_i$ are then found in the unstable region. From Eq. (11.2.29), for constant physical frequency, $2\pi F = \hat{\omega} = \omega U_c/\delta = \nu G^3\omega/16x^2$ is constant. Since

$G \propto x^{(n+3)/4}$, this may be written in terms of either x or G. Using G, a relation is obtained between ω and G at constant physical disturbance frequency. In this way the constant-frequency paths in Fig. 11.1.3 were constructed for the different flows. Thus, we may follow a given disturbance in ω, G coordinates and determine the downstream change in its amplitude, in the manner of Dring and Gebhart (1968), as it is convected along to larger x (i.e., G). Note that from the relation $F = \nu G^3\omega/32\pi x^2$, a constant-frequency path for an isothermal surface condition, $n = 0$, is given by

$$\frac{2\pi F}{\nu}\left(\frac{\nu^2}{g\beta\,\Delta t}\right)^{2/3} = \omega G^{1/3} = \Omega \tag{11.2.33}$$

If a stability plane is plotted in terms of G and Ω, a more convenient interpretation of downstream amplification of disturbances at different frequencies results, because constant physical frequency paths are now horizontal lines in the Ω, G plane.

The cumulative downstream amplitude growth from the neutral condition at x_N or G_N, for a given disturbance frequency F, is calculated as follows. If A_N is the amplitude of the periodic disturbance as it reaches the location of neutral stability, x_N, and A_x is the amplitude farther downstream at x, then the amplitude ratio e^A and factor A are defined as follows:

$$\frac{A_x}{A_N} = e^A \qquad 4A = -\int_{x_N}^{x} \alpha_i \, dG \tag{11.2.34}$$

The neutral curve is thus $A = 0$.

Neutral curves and constant-amplification contours have been calculated for many buoyancy-generating processes since the first calculations of neutral stability by Plapp (1957), Szewcyzk (1962), Kurtz and Crandall (1962), and Nachtsheim (1963). The stability planes reproduced here are for a vertical surface dissipating a uniform heat flux q'', $n = \frac{1}{5}$ in Eq. (3.5.24). Figures 11.2.1 and 11.2.2 are stability planes for Pr = 0.733 and 6.7, respectively. However, the generalization in these figures is in terms of the flux Grashof number, defined in Chapter 3, as follows:

$$G^* = 5\sqrt[5]{\frac{\mathrm{Gr}_x^*}{5}} \qquad \mathrm{Gr}_x^* = \frac{g\beta x^4 q''}{k\nu^2} \tag{11.2.35}$$

$$\delta = \frac{5x}{G^*} \qquad \frac{U_c x}{\nu} = \frac{G^{*2}}{5} \tag{11.2.36}$$

The physical and nondimensional frequencies, F and ω, for the uniform flux condition are now related by $F = \nu G^{*3}\omega/50\pi x^2$. The constant physical frequency paths are now represented by

$$2\pi F\left(\frac{k}{g\beta q''}\right)^{1/2} = \omega G^{*1/2} = \Omega^* \tag{11.2.37}$$

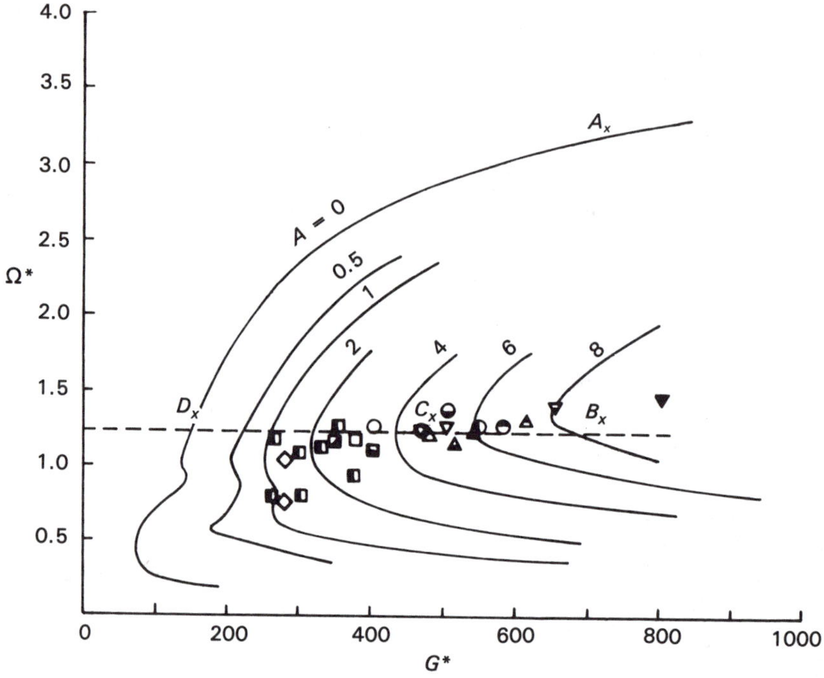

Figure 11.2.1 Stability plane for Pr = 0.733 for flow next to a uniform flux vertical surface, $s(0) = (0)$. Open symbols represent naturally occurring measured disturbance frequencies in the unstable laminar flow and partially or fully shaded symbols denote the locally laminar portion of the flow in the transition region. *(Reprinted with permission from R. L. Mahajan and B. Gebhart,* J. Fluid Mech., *vol. 91, p. 131. Copyright © 1979, Cambridge University Press.)*

Similar stability planes, for plane plumes and for combined buoyancy mode flows, are discussed in subsequent sections.

For the surface-generated flow on which Figs. 11.2.1 and 11.2.2 are based, the apparent additional boundary conditions are as given below, from Knowles and Gebhart (1968). The general surface condition on $s(0)$ takes into account thermal coupling between the surface and fluid temperature disturbances.

$$\Phi(0) = \Phi'(0) = 0 \qquad s(0) = \frac{i\, s'(0)}{\beta \hat{Q} (G^*)^{3/4}} \tag{11.2.38}$$

where $\hat{Q}$, the relative thermal capacity parameter, is

$$\hat{Q} = \left(\frac{\Pr c''}{\rho c_p}\right)\left(\frac{g\beta q''}{k\nu^2}\right)^{1/4} \tag{11.2.39}$$

Here c'' is the surface thermal capacity per unit surface area. Thus, $s(0)$ is not taken as zero unless the surface is massive, that is, $\hat{Q}$ is very large. Since that is usually the practical circumstance in air, Fig. 11.2.1 was calculated with $s(0) = 0$.

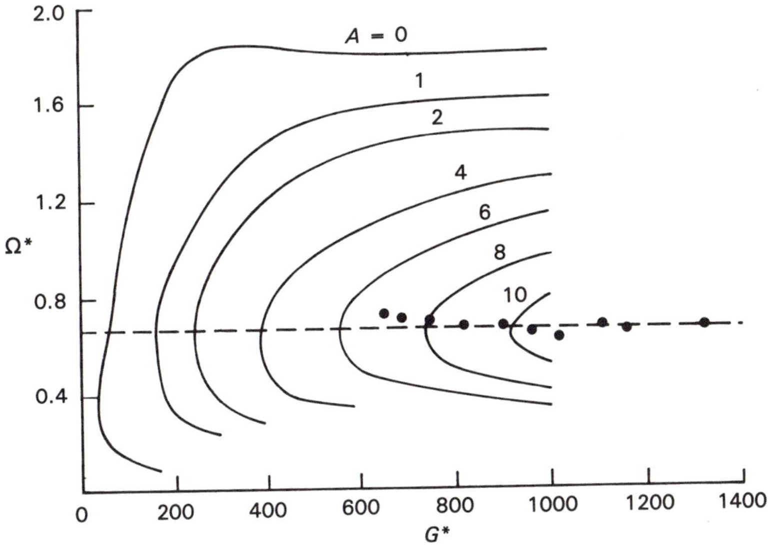

Figure 11.2.2 Stability plane for Pr = 6.7 for flow next to a uniform flux vertical surface, $s'(0) = 0$, showing amplitude curves in the unstable region. The dashed line represents the path of a rapidly amplifying frequency. The measured naturally occurring disturbance frequency data are from Qureshi and Gebhart (1978). *(Reprinted with permission from* Int. J. Heat Mass Transfer, *vol. 21, p. 1467, Z. M. Qureshi and B. Gebhart. Copyright © 1978, Pergamon Journals Ltd.)*

However, Fig. 11.2.2 is based on $s'(0) = 0$, that is, $\hat{Q} = 0$, which is more applicable to water.

It is seen in both figures, from the trajectories of the constant physical frequency paths in Ω^*, that these flows are sharply selective in their amplification characteristics. The disturbances are filtered for essentially a single frequency, or characteristic frequency, as they are convected downstream. The experimental data, like the points in Figs. 11.2.1 and 11.2.2, strongly substantiate this prediction.

Comparison of the A contours in Figs. 11.2.1 and 11.2.2 shows that the value of the most amplified Ω^* is Prandtl number dependent. The collection of available calculated and measured data shown in Fig. 11.2.3 was assembled by Gebhart and Mahajan (1975). The right ordinate, in Ω, corresponds to the isothermal surface condition; the left one, in Ω^*, is for uniform surface flux.

In Fig. 11.2.3 the asymptotic dependences shown at large and small Pr were inferred from the results of Hieber and Gebhart (1971a, 1971b). The crosses at intermediate values of Pr are derived from detailed stability planes for each specific Pr value. The data points shown in this range are seen to be in very good agreement with the calculations.

The kind of remarkable agreement seen above, between linear stability theory and experimental results, has been found to extend to other and much more subtle aspects of unstable flows and transport. Such success is not common across the

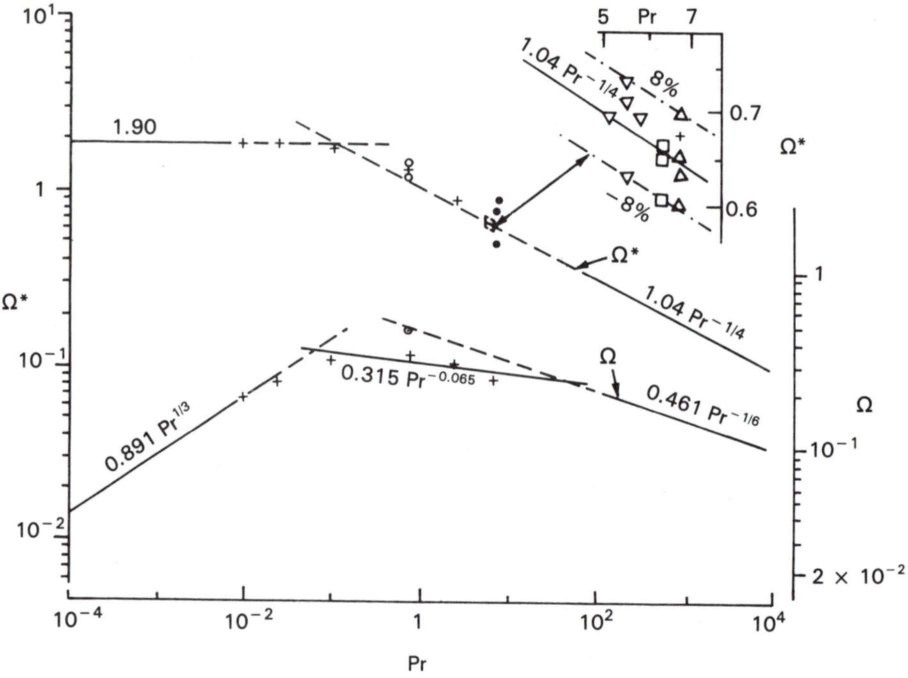

Figure 11.2.3 Characteristic frequency data for vertical natural convection flows. Large and small Prandtl number asymptotes and (+) data points are from Hieber and Gebhart (1971a, b). Other data: (○) Polymeropoulous and Gebhart (1967); (⊙) Eckert and Soehngen (1951); (●) Knowles and Gebhart (1969); (▽) Shaukatullah (1974); (□) Godaux and Gebhart (1974); (△) Jaluria and Gebhart (1974). *(Reprinted with permission from* Int. J. Heat Mass Transfer, *vol. 18, p. 1143, B. Gebhart and R. L. Mahajan. Copyright © 1975, Pergamon Journals Ltd.)*

broad range of such research in fluid flow. Perhaps the reason for such close agreement in these flows is that the disturbance amplification is so highly selective and thus not heavily dependent on particular aspects of each specific physical situation. Subsequent sections will indicate other very interesting downstream consequences of this initial characteristic.

The stability characteristics of a buoyancy-induced flow adjacent to a vertical surface with an imposed linear temperature variation, $n = 1$ in Eq. (3.5.24), have been analyzed in detail by Mahajan (1977). The effects of the downstream variation of flow quantities, stable stratification, and compression work were included. Stability planes were calculated for air, Pr = 0.733.

11.2.3 Spatial and Temporal Analyses

Spatial disturbance amplification was assumed in the above analysis. That is, the disturbances were assumed to vary sinusoidally in time and to be damped or amplified exponentially as they are convected downstream. This procedure is different from the frequent practice of assuming temporal amplification. Then a real wave number and complex frequency are postulated in an instantaneously changing base flow.

For nearly parallel flows, the spatial analysis has an obvious advantage. It faithfully models the actual mechanisms of instability downstream in a developing flow. Unstable disturbances grow as they are convected and propagate downstream. The predictions of the spatial analysis have been found to be very reliable compared to measured behavior, both in early instability and well beyond the linear disturbance growth range. This will be shown in detail in later sections.

In forced flows, spatial analysis has also been used instead of temporal modes. Landau and Lifshitz (1959) suggested the use of spatial modes on physical grounds. Gaster (1962) examined the relationship between the spatial and temporal formulations. It was shown that for small rates of amplification the spatial growth may be deduced from the temporal growth. Watson (1962) studied spatially growing finite disturbances in a plane Poiseuille flow. Gaster (1965) studied the development of spatially growing waves in a laminar boundary layer arising due to a continuously oscillating source, which is fixed in space and turned on at time $\tau = 0$. The asymptotic analysis of the disturbance at large times indicated that the disturbance is found only downstream of the source. Further, the solutions obtained are bounded at any instant of time since an unbounded disturbance magnitude, as $x \to \infty$, may arise only after first applying the limit $\tau \to \infty$ for fixed x.

However, spatial modes have sometimes been regarded as mathematically suspect; see, for example, Drazin and Reid (1981). The uncertainty arises because the spatial mode does not satisfy the far upstream and downstream boundary conditions at any instant. Also, the existence of an eigenvalue α with $\alpha_i < 0$ does not necessarily imply spatial instability. The solutions also include spatial modes that arise on both sides of a disturbance source. Therefore, if a given eigenvalue with negative α_i represented a mode for negative x, it would denote stability. It is argued that this distinction cannot be made without solution of the initial value problem; see Drazin (1977).

Several other criteria have been used to test the instability of spatial modes. These are $\alpha_i < 0$ and $\alpha_r > 0$ by Garg and Rouleau (1972) and $\alpha_i < 0$ and $\partial\omega/\partial\alpha_r > 0$, or $\alpha_i > 0$ and $\partial\omega/\partial\alpha_r < 0$, by Lessen and Singh (1973). However, as discussed recently by Garg (1987), all of the above criteria may not always lead to correct results. It was also pointed out that the plasma physicists had resolved this issue more than two decades ago. However, their work appears to have gone largely unnoticed in the research work in fluid mechanics. Criteria had already been proposed that unambiguously indicate whether the spatial mode is stable or unstable. Reference is made to early work by Sturrock (1958, 1961) and to a monograph by Briggs (1964).

First, a distinction is made between the two types of instabilities that might arise. Consider an imposed pulse disturbance, initially of finite spatial extent. It may grow in time without limit at every point in space. This is termed absolute instability. Also, the pulse may propagate and the disturbance may decay with time at a fixed point in space. This is called convective instability.

Sturrock (1958, 1961) concluded that a convective instability is basically of the same type as a spatially amplifying wave. Also, for a system that supports convective instability, solutions of the type given in Eq. (11.2.26) may exist only if ω and α are appropriately related. The equation that determines permissible solutions is called the dispersion relation $D(\omega, \alpha) = 0$. One or more roots of the dispersion equation must exist, corresponding to amplifying waves, where ω is real and α is complex. A criterion was proposed for determining whether the disturbances amplify or decay. However, this criterion is not easy to apply if the dispersion relation is not known explicitly.

Briggs (1964) proposed the following criterion for determining whether a mode with negative α_i is actually unstable for positive x. Consider a given wave with a complex $\alpha = \alpha_r + i\alpha_i$ for some real ω. Whether it is amplifying or decaying depends on whether α_i changes sign when the frequency has a large positive imaginary part. If it does, then the wave is amplifying. Otherwise, it is decaying. Using a different approach, Sudan (1965) arrived at the same conclusion.

Garg and Leibovich (1985) applied the above criterion to calculate the spatial stability of circular Poiseuille flow. They found that all the other criteria discussed earlier predicted that the flow would be unstable. The actual flow is known to be stable. However, correct interpretation of spatial modes, using Briggs's criterion, predicted that the flow would be stable.

Conclusions. The spatial theory has a sound mathematical basis. Succeeding sections also show, as does Fig. 11.2.2, that such predictions are well confirmed by experimental measurements of disturbance growth/decay mechanisms as disturbances are convected downstream. On the other hand, the temporal theory may correctly predict whether a given flow is stable, but it does not distinguish between convective and absolute instability.

This was recognized early in the studies of plasma instability; see Sturrock (1958). For a propagating system that admits a dispersion relation $D(\omega, \alpha) = 0$, where ω is complex and α is real, the temporal theory suggests that the system must disrupt some time after an arbitrary small disturbance is introduced. However, in systems exhibiting convective instability, such as a traveling-wave tube or a two-stream amplifier, a finite length of the system may persist in a quiescent state even in the presence of small random disturbances. The disturbances, although amplified, are carried away from the region in which they originate. Similar circumstances also arise in more ordinary fluid flows; see Garg (1987). The appropriate model for the study of convective instability is that provided by the spatial theory.

For the buoyancy-induced flows presented in the later sections, the criterion used for instability is $\alpha_i < 0$. The excellent agreement between the predictions of the spatial theory and data indicate that this criterion provides uncommonly accurate predictions. However, if this criterion results in disagreement between the predictions of the theory and the data for some flow configurations, the more general criterion of Briggs (1964) should be used for guidance.

11.3 NONLINEAR DISTURBANCE GROWTH

Sufficiently weak disturbances grow downstream as predicted by the linear stability theory. However, at increasing amplitude, nonlinear mechanisms arise and the growth mechanisms deviate from the predictions. Two possible growth mechanisms are the generation of higher harmonics and the generation of secondary mean flows. The latter are due to nonlinear interaction of two-dimensional disturbances, in x and y, and transverse disturbances, in z. However, there appears to be little likelihood that higher harmonics have any appreciable effect in the later stages preceding transition. The abundant experimental data for flows subject to both natural and controlled disturbances indicate that a simple sinusoidal form of the most highly amplified disturbances is retained to transition. Generation of a secondary mean flow, on the other hand, has been found to play an important role in the breakdown of the flow to turbulence. These effects have been both calculated and measured.

11.3.1 Calculations of Secondary Mean Motions

The theoretical treatment is by Audunson and Gebhart (1976), who postulated a two-dimensional disturbance modulated by a standing transverse disturbance. The relative amplitude of the two disturbance components was a parameter. However, their phase velocities and wavelengths were assumed equal. The latter assumption is similar to that used by Benny and Lin (1960) and Benny (1961) in the nonlinear analysis for a Blasius flow. Stuart (1965) has shown that two- and three-dimensional wavelengths may not be equal near the neutral curve in Blasius flow. Perhaps they are not equal in the amplified region downstream. Similar objections to the assumption of such two- and three-dimensional synchronization in forced flow have been raised by Hocking et al. (1972).

On the other hand, measurements by Jaluria and Gebhart (1973) of controlled nonlinear disturbance propagation in a buoyancy-induced flow indicate that the amplitudes of the disturbance velocities of the two waves are essentially the same; see Fig. 11.3.1. They remain out of phase by about one-quarter period. Nevertheless, the great simplification of the calculations following the assumption of equality of phases and the a posteriori good agreement between the calculations and the experimental results indicate that the assumptions are realistic.

Retaining nonlinear interactions, the disturbance response was calculated by a systematic perturbation of linear stability theory. The nonlinear interactions of solutions of the homogeneous Orr-Sommerfeld equations provide the driving functions for the first perturbation from linearized analysis. For details, see Audunson (1971) and Audunson and Gebhart (1976). Solutions were obtained for four characteristic downstream flow and disturbance conditions for a Prandtl number of 0.733. These conditions are shown on the stability plane in Fig. 11.2.1. Points B, C, and D lie close to the downstream path of the most amplified component; D is at the neutral condition and C and B are in the highly amplified region. Point A is at the same value of G^* as B, but it lies on the neutral curve.

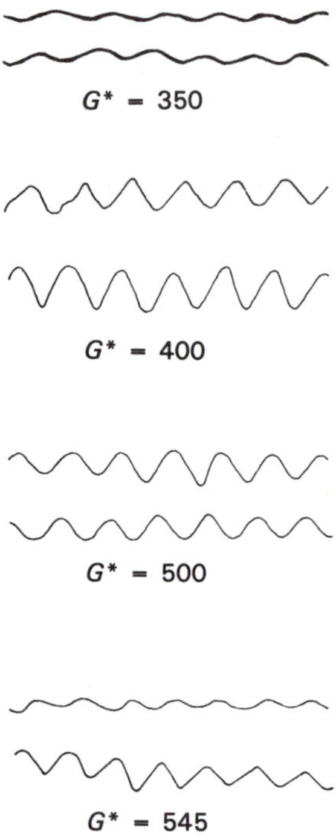

Figure 11.3.1 Two-dimensional and transverse velocity disturbances vs. time as measured at various downstream locations G^*, as indicated. Upper signal for each G^* is the transverse one. *(Reprinted with permission from* Int. J. Heat Mass Transfer, *vol. 19, p. 737, T. Audunson and B. Gebhart. Copyright © 1976, Pergamon Journals Ltd.)*

The results indicated a strong dependence of the resulting secondary flows on both G^* and frequency Ω^*, where Ω^* is defined in Eq. (11.2.37). Along the path of the most amplified disturbance, points B, C, and D, double longitudinal mean secondary vortex systems were found. For point B, at $G^* = 700$ in the highly unstable region, the calculated streamlines are shown in Fig. 11.3.2, where λ_1/λ_2 is a measure of the imposed initial relative strength of two-dimensional and transverse disturbances, $\tilde{\theta}$ is the transverse wave number, and z is the transverse coordinate. The conditions are from $\lambda_1/\lambda_2 = 100$, a highly two-dimensional disturbance, to $\lambda_1/\lambda_2 = 0$, a purely transverse oscillation. For $\lambda_1/\lambda_2 >> 1$, in Fig. 11.3.2*a*, the resulting cellular vortex motion has a spanwise periodicity of $2\pi/\tilde{\theta}$. With increasing three-dimensionality of the flow, that is, decreasing λ_1/λ_2, the center of the outer roll moves toward spanwise locations $\tilde{\theta}z = 2n\pi$ and the centers of the inner rolls are pushed toward spanwise locations $\tilde{\theta}z = (2n + 1)\pi$ in Figs. 11.3.2*b* and 11.3.2*c*. For the extreme case, a purely transverse primary oscillation, the spanwise period is $\pi/\tilde{\theta}$, in Fig. 11.3.2*d*.

These secondary mean motions imply a large transverse momentum transport in the boundary region. The result is a major modification of the undisturbed mean flow. Consider Fig. 11.3.2*a* further. These streamlines are for point B, in Fig. 11.2.1, which is closely associated with the observed beginning of transition. At $\tilde{\theta}z = (2n + 1)\pi$ the inner roll carries fluid of high downstream momentum from the inner part of the boundary layer to the outer, slower-moving region. On the other hand, the counterrotating outer vortex brings low-momentum fluid from the

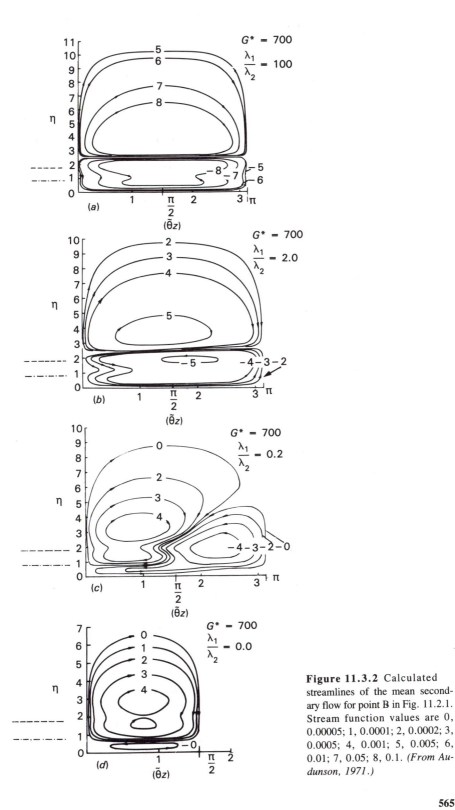

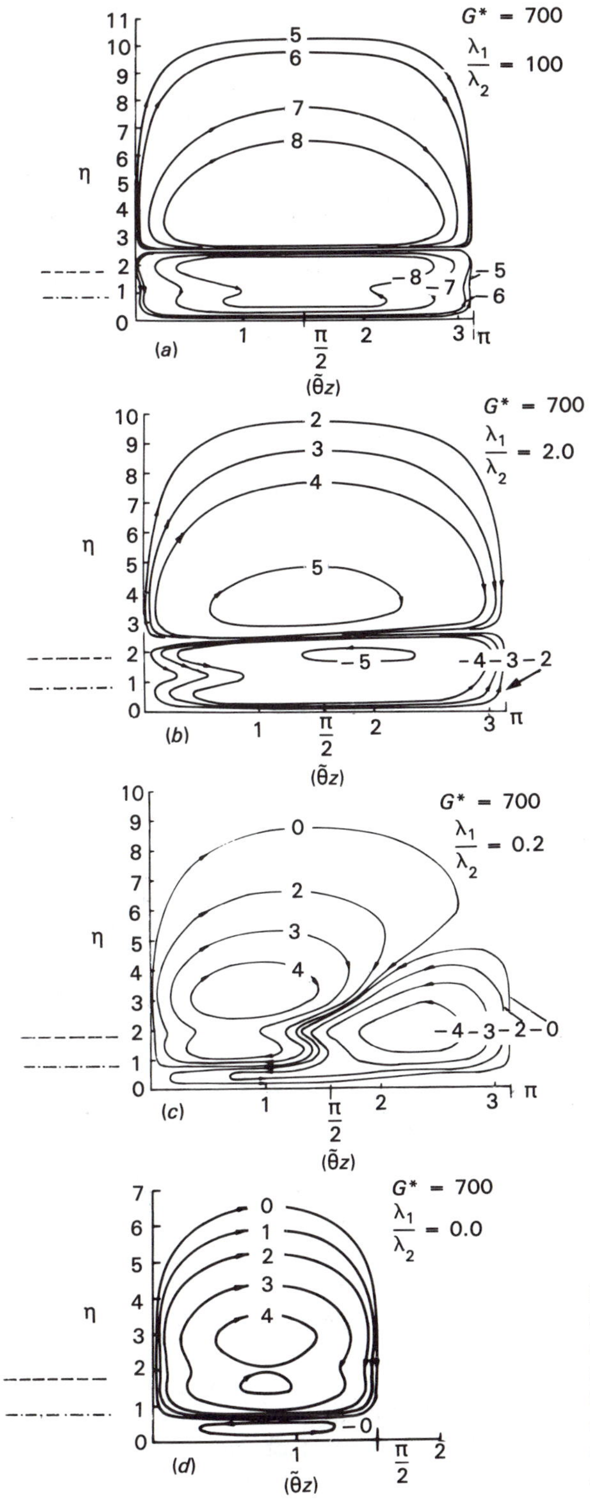

Figure 11.3.2 Calculated streamlines of the mean secondary flow for point B in Fig. 11.2.1. Stream function values are 0, 0.00005; 1, 0.0001; 2, 0.0002; 3, 0.0005; 4, 0.001; 5, 0.005; 6, 0.01; 7, 0.05; 8, 0.1. *(From Audunson, 1971.)*

far field into the boundary region at the same z location. These mean cross flows thus steepen the outer part of the mean velocity profile at locations $\tilde{\theta}z = (2n + 1)\pi$. It is flattened at locations $\tilde{\theta}z = 2n\pi$.

Since energy transfer to disturbances is at least approximately proportional to mean flow shear stress, disturbance growth rate is strongly augmented at $\tilde{\theta}z = (2n + 1)\pi$ locations and reduced at $\tilde{\theta}z = 2n\pi$ locations. The calculated locations of high shear are just the opposite of those found in analogous Blasius forced flow. The quiescent far field, rather than the region near the surface, is the source of the low momentum that causes the high-shear region. In forced flows only a single longitudinal vortex system was detected by Klebanoff et al. (1962). These vortices occupied only the inner half of the boundary region. However, in this buoyancy-induced flow the outer vortices reach across the boundary region, out in the quiescent fluid. Therefore, they may be expected to cause a great distortion in the tangential mean velocity profile. These secondary mean flows also alter the spanwise mean temperature distribution across the boundary region. At locations $\tilde{\theta}z = 2n\pi$ the local heat transfer is augmented, while at $\tilde{\theta}z = (2n + 1)\pi$ it is decreased.

The computed results for point C were the same as for point B. Recall that both points lie in the highly unstable region at the same physical frequency. For point D, at the same frequency but at the neutral curve, the outer vortex becomes even more dominant. The inner circulation is very weak. Nevertheless, the results for D, C, and B indicate that, along this path, a double longitudinal vortex system is produced at the earliest stages of instability. It grows into the downstream region of highly amplified disturbances.

Point A is at the same value of G^* as B but does not lie along the path of rapidly amplifying disturbances. The results were very different. Only a single longitudinal roll is observed. This also causes an alternating spanwise thinning and thickening of the boundary layer. However, it does not appear to produce significant steepening of the outer part of the tangential velocity profile. The profile merely shifts in and out from the surface while retaining its original form. Such changes would not be expected to greatly alter disturbance growth. For a more complete quantitative discussion see Audunson and Gebhart (1976) and Audunson (1971).

These calculations show that very similar double-vortex systems are consistent with conditions at points D, C, and B, as the most highly amplified disturbance is convected downstream. Therefore, such a secondary mean flow configuration need not appreciably change as it is convected downstream. It is merely amplified. This would occur simultaneously with the continued concentration of disturbance energy into the most amplified two-dimensional primary wave. Thus, the linear and nonlinear mechanisms appear to proceed together in a highly filtered way to concentrate disturbance energy and modify the tangential flow into high-shear regions. This significant result is very different from what is observed in forced flows. It would be interesting to calculate the integrated downstream effect of an initially three-dimensional disturbance along its path and compare the results with the experimental measurements of disturbance form and of transition.

11.3.2 Measurements of Nonlinear Growth in Controlled Experiments

Excellent corroboration of the above calculations has been provided by the detailed experimental results of Jaluria and Gebhart (1973). The measurements were in the flow generated adjacent to a vertical uniform flux surface in water. Controlled two-dimensional disturbances with a superimposed transverse variation were introduced in the flow by a vibrating ribbon, seen in Fig. 11.3.3c. The input disturbances were introduced at location $G^* = 140$, as discussed by Jaluria (1976), and the resulting downstream behavior of the disturbances and of the mean flow was measured in detail. Note that the actual local mean flow is the sum of the base flow $\bar{u}$ and $\bar{v}$ and any secondary mean flow that arises through interaction between the disturbances and this base flow. Denoting the local mean flow components as $\bar{U}$, $\bar{V}$, and $\bar{W}$, the components of the secondary mean flow are $\bar{U} - \bar{u}$,

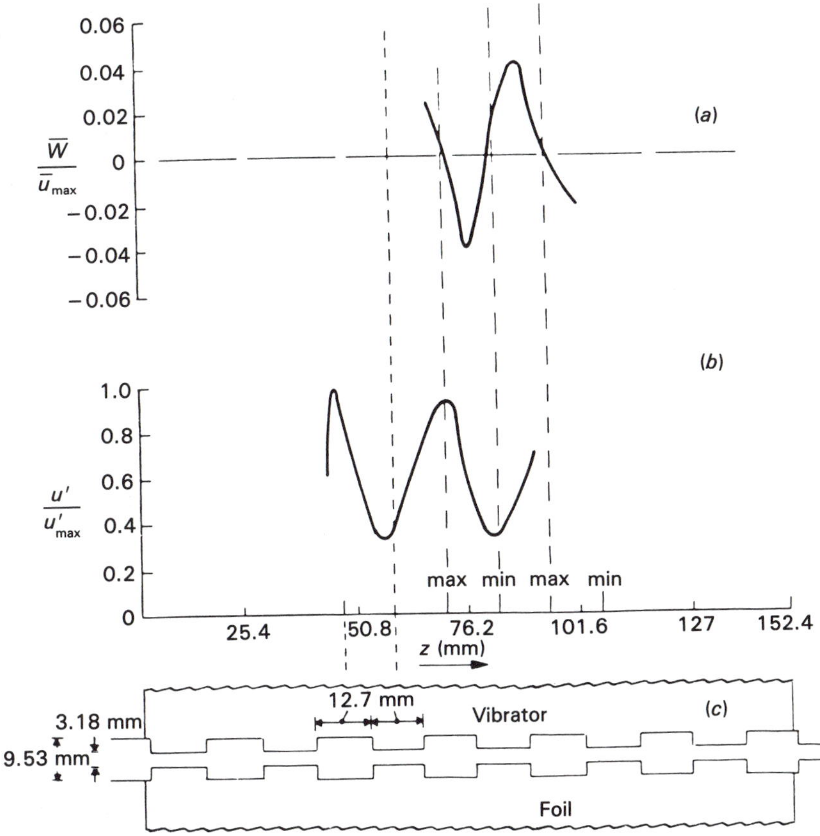

Figure 11.3.3 Configuration of vibrating ribbon and measured downstream spanwise distribution of u' and $\bar{W}$ in water at $G^* = 460$. The measurements from Jaluria and Gebhart (1973) are at a single value of each x and y.

$\bar{V} - \bar{v}$, and $\bar{W}$. The periodic parts of the velocity components, as before, are u', v', and w'.

For a range of input disturbance frequencies, downstream measurements of the velocity disturbance u' at different x and z locations indicated that the most rapidly amplifying frequency was almost exactly that predicted by linear stability theory. The conclusion was that the frequency filtering mechanism found in flows subject only to two-dimensional disturbances is not greatly affected by an additional spanwise variation in the input disturbance.

The spanwise measurements of the amplitude of u', normalized by the maximum value u'_{max}, are plotted in Fig. 11.3.3*b*. In these experiments the input disturbance was introduced at the most rapidly amplifying frequency. The initial z locations of maxima and minima in input disturbance, indicated in Fig. 11.3.3*c*, continue to be the respective locations for peaks and valleys in the u' vs. z distributions downstream. This indicates vertical propagation of the disturbance pattern. However, the peaks and valleys are sharper than the input disturbance, suggesting that these measurements were preceded by a region of amplifying transverse effects that, through linear and/or nonlinear interactions, accentuate the ribbon input spanwise variation of u'.

Associated with the spanwise variation of u' is the spanwise variation of the transverse mean velocity component $\bar{W}$, also shown in Fig. 11.3.3*a*. However, much more interesting are the measured distributions of $\bar{W}$ across the boundary region, as in Fig. 11.3.4, at $G^* = 400$ and 460 and at various spanwise locations.

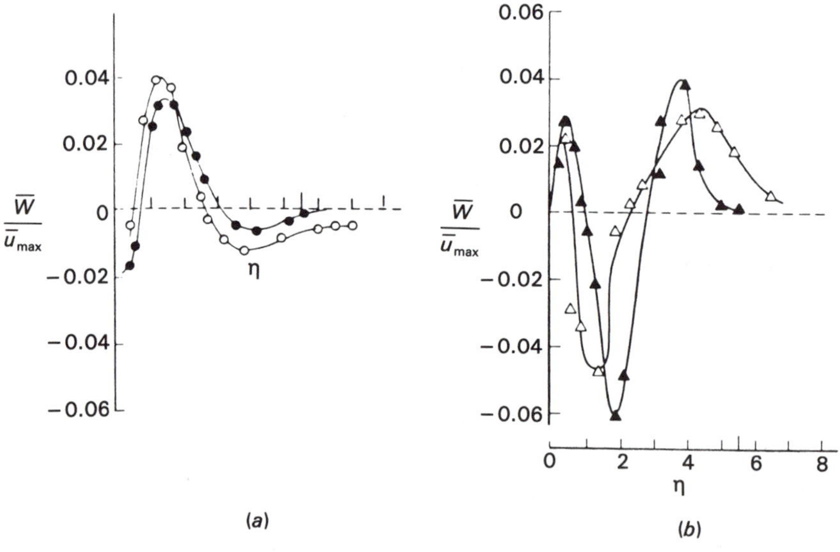

Figure 11.3.4 Distribution of the transverse component $\overline{W}$ across the boundary region. (*a*) ●, $G^* = 400$ at $z = 86.4$ mm; ○, $G^* = 460$ at $z = 86.4$ mm. (*b*) ▲, $G^* = 460$ at $z = 76.2$ mm; △, $G^* = 460$ at $z = 78.7$ mm. *(Reprinted with permission from Y. Jaluria and B. Gebhart,* J. Fluid Mech., *vol. 61, p. 337. Copyright © 1973, Cambridge University Press.)*

There, $\bar{u}_{max}$ is the measured maximum velocity in the base profile without disturbances. In each distribution $\bar{W}$ changes sign twice across the boundary region, indicating the presence of longitudinal rolls. The two sign reversal locations appear to be the centers of two rolls. On one side $\bar{W}$ is in one direction and on the other in the opposite direction. Assuming two rolls, their interface occurs where $\bar{W}$ attains the highest value, around $\eta = 1.7$. Therefore, the inner roll extends from $\eta = 0$ to 1.7 and the outer roll stretches from $\eta = 1.7$ to about 7. The extent of the inner roll for the two G^* values appears the same.

Also, the distribution in Fig. 11.3.4*b* at $z = 78.7$ mm is just the opposite of that in Fig. 11.3.4*a* at $z = 86.4$ mm. This implies that the double-vortex systems at these two locations rotate in opposite directions, with the minimum in the input disturbance, at $z = 83.3$ mm, being the boundary between these two adjacent systems. Thus, the maxima and minima in the input disturbance amplitude are locations of symmetry. Each longitudinal vortex system extends, in the transverse direction, from z positions of a maximum to a minimum in the input disturbance.

A schematic representation of the longitudinal vortex system that emerges from these measurements is shown in Fig. 11.3.5. It is in excellent agreement with the calculations of Audunson and Gebhart (1976). See, for example, Fig. 11.3.2*a*. Since the Prandtl number in these measurements, 6.7, is different from that of the analysis, 0.733, a quantitative comparison between the two results is not possible. However, the overall features are the same. At a higher Prandtl number one would expect the vortex system to be closer to the wall in η. These measurements in water, compared with the calculations for air, have that characteristic.

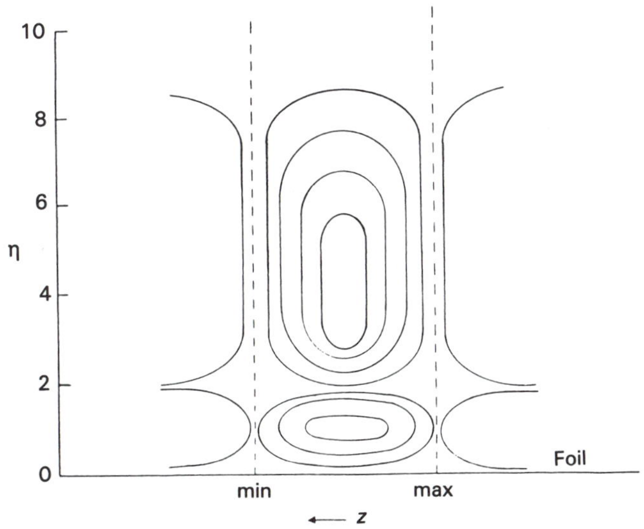

Figure 11.3.5 Sketch of the double mean longitudinal vortex system. *(Reprinted with permission from Y. Jaluria and B. Gebhart,* J. Fluid Mech., *vol. 61, p. 337. Copyright © 1973, Cambridge University Press.)*

11.3.3 Role of Mean Secondary Motions

A double longitudinal vortex system redistributes tangential momentum across the boundary layer. The outer vortex convects higher-velocity fluid outward at some transverse locations and inward, from the quiescent ambient, between the two locations. The inner vortex does just the opposite. The resulting mean flow downstream velocity is alternately steepened around the inflection point and flattened transversely. The measured profiles are seen in Fig. 11.3.6. There is an alternate spanwise steepening and flattening, accompanied by an alternate thinning and thickening of the boundary region. Local steepening augments disturbance growth. The measurements indicated that the spanwise distortion of the mean velocity profile increases with G^*, increasing disturbance growth. These results are again in good agreement with the analysis of Audunson and Gebhart (1976).

Prior to the analysis and the experimental results reported above, there were different views of the role and the form of any three-dimensional disturbances and of nonlinear mechanisms that may arise in this natural convection flow. Colak-Antic (1962) had suspended highly reflective aluminum particles in water and observed their behavior during transition. Two longitudinal vortices, similar to

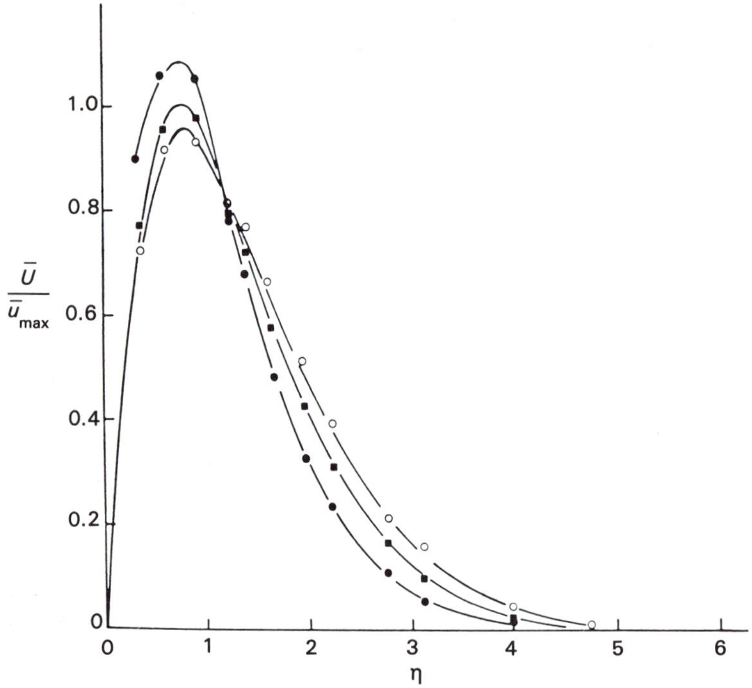

Figure 11.3.6 Measured longitudinal mean flow profiles compared with that for undisturbed flow. Data: (●) at spanwise location of primary disturbance minimum; (○) at spanwise location of primary disturbance maximum; (■) undisturbed flow. *(Reprinted with permission from Y. Jaluria and B. Gebhart,* J. Fluid Mech., *vol. 61, p. 337. Copyright © 1973, Cambridge University Press.)*

those discussed above, were seen. However, Szewcyzk (1962), using dye visualization in water, observed vortices whose axes were transverse. A vortex loop was postulated as a concentration of vorticity. However, there remained some ambiguity in relating dye injection to vorticity concentration. A similar difference in possible mechanisms had arisen for forced flows. Then Klebanoff et al. (1962) established, from detailed experiments with controlled three-dimensional disturbances, that secondary mean flow longitudinal vortices occur as a consequence of nonlinear and three-dimensional interactions. Several other mechanisms have also been offered, such as the generation of higher harmonics and the effects of the concave streamline curvature associated with the wave motion. However, it seems reasonable to conclude for these buoyancy-driven flows that such effects do not dominate the process. It appears that regions of high shear, along with other consequences of the longitudinal vortex system, are associated with the onset of transition. The evidence is given below.

11.3.4 Nonlinear Effects Resulting from Naturally Occurring Disturbances

The nonlinear growth mechanisms discussed above were initiated by artificially introduced disturbances. There is ample reason to believe that these effects also arise in flows subject only to naturally occurring disturbances. A frequent source of such disturbances is external vibrations. Tani (1969) followed such effects downstream in a forced flow boundary layer. A natural mode of disturbance origin is similar to that imposed by a vibrating ribbon. However, there are important differences. Artificial disturbances generally are introduced across the boundary region at one streamwise location x, in the unstable region. Natural disturbances, on the other hand, may be fed in anywhere at any time, over the entire boundary region. Further, they contribute a spectrum of frequencies, as indicated by hot-wire measurements of background disturbances by Jaluria and Gebhart (1977). No dominant frequency was found.

However, we have seen, for example in Figs. 11.2.1 and 11.2.2, that essentially one frequency component of any disturbance reaches the highest amplitude downstream. Therefore, the disturbance components that become large downstream are those that are impressed on the flow at a location corresponding to the early stages of initial instability and are in the narrow band of most rapid amplification. Thus, the end results for a vibrating ribbon and a spectrum of naturally occurring disturbances are expected to be similar; see also Jaluria (1985).

Experimental data confirm this. Jaluria and Gebhart (1974) investigated the transition mechanisms in a flow subject only to naturally occurring disturbances in water. The transverse secondary mean flow $\bar{W}$ was measured. Its small magnitude and the random noise level made complete measurements impossible. However, the measured variation of $\bar{W}$ over a portion of the boundary layer was in good agreement with the distribution shown in Fig. 11.3.4. This confirms the presence of longitudinal rolls. The double longitudinal vortex observed by Colak-Antic (1964) also arose in a flow subject only to naturally occurring disturbances, again confirming that the nonlinear interactions cause such mean flow modifications.

11.4 TRANSITION AND PROGRESSION TO DEVELOPED TURBULENCE

The previous sections discussed the sequence of events leading a laminar flow adjacent to a vertical surface toward eventual transition. Disturbances amplify. Later, nonlinear interactions arise. A double longitudinal mean vortex generates regions of higher shear, which result in rapid disturbance growth. These events are similar to those observed in forced convection flows. However, the analogy does not continue downstream. In forced flows, it is thought that the high-shear region, acting as a secondary instability, generates other rapidly oscillating effects intermittently in the boundary layer. Turbulent spots follow. However, the total transition process in natural convection proceeds differently and appears more complicated. The velocity and temperature fields are inevitably coupled together, and this causes significant additional effects on transition mechanisms. Since the interaction of the two fields and their influence on each other are also Prandtl number dependent, this then becomes an additional parameter.

Most studies of transition in natural convection flows have been in the two most common fluids in nature: water ($\Pr \simeq 6.7$) and air ($\Pr \simeq 0.7$). The overall transition mechanisms in these fluids appear quite similar. However, there are important differences. Some principal features of the transition processes are discussed in this section.

The most detailed investigations of transition mechanisms in water were by Godaux and Gebhart (1974), Jaluria and Gebhart (1974), and Qureshi and Gebhart (1978). Prior investigations, for instance by Vliet and Liu (1969) and Lock and Trotter (1968), dealt primarily with turbulent flows, although a few measurements during transition are also reported. An overall picture of transition that emerges from these studies is shown in Fig. 11.4.1 for a particular heating condition in water. Local turbulence first appears in the thicker velocity boundary layer and later in the thermal layer. As this grows downstream, the growth of the mean flow velocity lags behind the laminar trend. The mean velocity profile also deviates from its laminar form. The disturbances then become strong enough to diffuse the thermal layer material throughout the velocity layer, changing the mean temperature profile. Thermal transition has begun. The velocity and thermal boundary layers mix and thicken. The end of transition is marked simultaneously by the end of appreciable farther downstream changes in the distributions of local velocity and temperature intermittency factors I_v and I_t. These factors are defined as the fraction of the time the flow at any point is turbulent, in the velocity and temperature measures, respectively.

The end of transition in water (see Bill and Gebhart, 1979) is followed by a regime of spectral and transport development. The spectrum of velocity fluctuations broadens and temperature fluctuations decrease in magnitude. Also, the turbulent transport mechanisms become more effective, despite the leveling of the

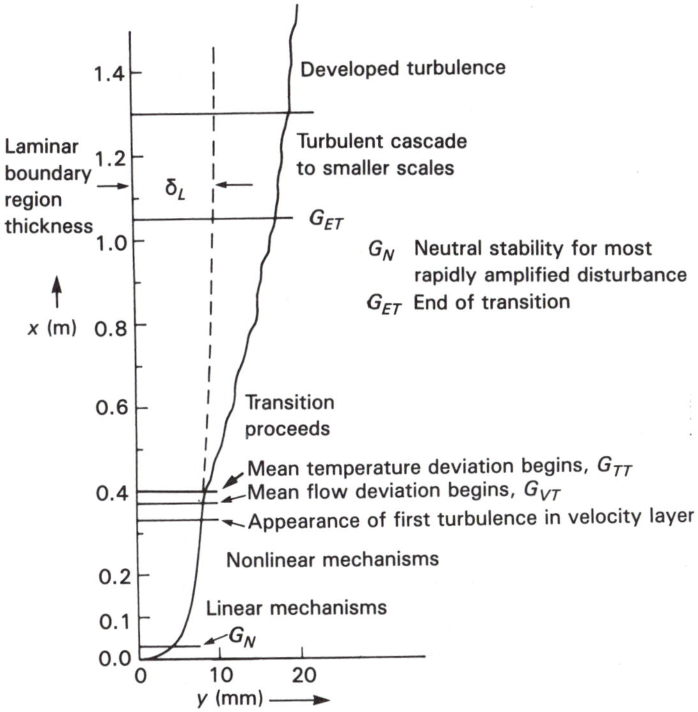

Figure 11.4.1 Sequence of downstream events during transition in water from a stable laminar flow to full turbulence. The spatial extent of each regime is shown to scale for a uniform flux surface condition of $q'' = 1000$ W/m^2. *(Reprinted with permission from B. Gebhart and R. L. Mahajan,* Advances in Applied Mechanics, *p. 231. Copyright © 1982, Academic Press.)*

growth of velocity disturbances. Development continues downstream until the distributions, scales, intensities, and other turbulent parameters adjust to the final trends of the turbulent flow.

Some data obtained during transition in gases are available from the experiments of Regnier and Kaplan (1963), Cheesewright (1968), Warner (1966), Warner and Arpaci (1968), and Smith (1972). A detailed determination of transition mechanisms in gases is provided by experiments of Mahajan and Gebhart (1979) on a vertical flow adjacent to a uniform heat flux surface in pressurized nitrogen. These measurements indicate that the overall mechanisms are similar to those in water. However, the temperature and velocity boundary regions are of comparable thickness, in contrast to water, and fluctuations in velocity more quickly affect the temperature field. Velocity and thermal transitions occur almost simultaneously.

The region of spectral development, beyond the end of the transition, has not been studied systematically in air. However, the experimental data reported by Smith (1972) do not indicate a well-defined region of further adjustment of turbulent transport parameters. This difference may be a Prandtl number effect. In water, the effects of the early lag in the development of temperature disturbance levels continue throughout the process.

Criteria for the beginning of transition. A detailed description of the stages of transition must begin with sharp definitions of the bounds of transition. The criterion given above for defining the end of transition appears to apply successfully to both liquids and gases. However, a multiplicity of criteria have been used in the past to mark the beginning of transition. Only recently have distinctions arisen between velocity and thermal transitions. Past criteria included the presence of significant temperature fluctuations, an increase in heat transfer effectiveness from the laminar trends, a decrease in the temperature difference across a uniform flux boundary layer from its laminar value, and a deviation from the laminar mean temperature profile. These have been used to indicate the beginning of what is now known as thermal transition.

It was found by Mahajan and Gebhart (1979) that the events on which the last three criteria are based occur almost simultaneously. Hereafter, the deviation of the mean temperature profile from the laminar trend is taken as the condition for the beginning of thermal transition. It is simple and reliable. It may be used for both isothermal and uniform flux surfaces.

An earlier criterion for the beginning of velocity transition, used by Jaluria and Gebhart (1974), was the appearance of a higher-frequency component superimposed on the single laminar filtered frequency. However, it was found by Mahajan and Gebhart (1979) that this criterion was sometimes ambiguous for gases. Instead, the deviation of the actual local maximum value of the tangential flow velocity across the boundary region, $\bar{U}_{max}$, from its laminar trend downstream, $\bar{u}_{max}$, was used. In the experiments of Jaluria and Gebhart (1974) in water, the latter event occurs downstream of their designated location for the beginning of velocity transition. Hereafter the criterion of Mahajan and Gebhart (1979) is always used.

Mean velocity and temperature distributions during transition. Figures 11.4.2*a* and 11.4.3 show the downstream progression of the velocity and temperature fields during transition in nitrogen gas at several heating rates q''. Also shown in Fig. 11.4.2*b* are the measured values of $\bar{U}_{max}$, normalized by the calculated laminar values, $\bar{u}_{max}$, at the same conditions. At $G^* = 434$, $\bar{U}_{max} = \bar{u}_{max}$. However, at $G^* = 470$ the mean profile has begun to deviate and the ratio is 0.96. It decreases at higher G^*. The flow continues to penetrate deeper into the ambient medium as a consequence of growing turbulence in the boundary region. The profile in the outer region is progressively flattened as the end of transition is approached, at around $G^* = 611$.

Measured mean temperature distributions are plotted as ϕ in Fig. 11.4.3. At $G^* = 470$ the flow is laminar. However, velocity transition had already begun. At $G^* = 503$ temperature profile deviation has begun. The thermal layer thickens. The profiles steepen on the inside and flatten on the outside. These deviations continue to follow those of the velocity field. The mean velocity and temperature distribution modifications during transition in water show similar trends; see Jaluria and Gebhart (1974).

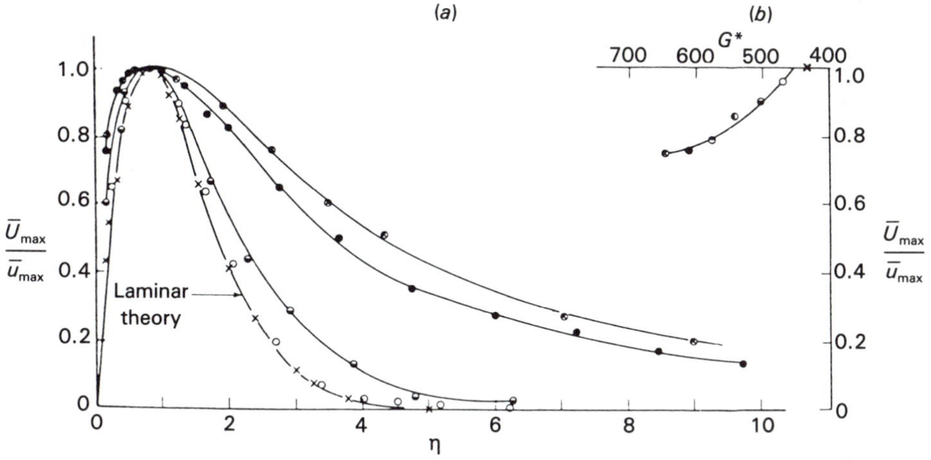

Figure 11.4.2 (*a*) Development of mean velocity profiles. (*b*) Variation of $\bar{U}_{max}/\bar{u}_{max}$ during transition in pressurized nitrogen for experimental conditions of $x = 22$ cm and $p = 8.36$ atm. Data: (×) $G^* = 434$; (○) $G^* = 470$; (◑) $G^* = 503$; (◐) $G^* = 543$; (◑) $G^* = 579$; (●) $G^* = 611$; (⊗) $G^* = 648$. *(Reprinted with permission from R. L. Mahajan and B. Gebhart,* J. Fluid Mech., *vol. 91, p. 131. Copyright © 1979, Cambridge University Press.)*

Figures 11.4.2*a* and 11.4.3 indicate that neither the mean velocity nor the temperature distributions may be correlated in terms of dimensionless distances out from the surface. The laminar similarity variable $\eta = y/\delta$ does not accomplish this in Fig. 11.4.2*a*, nor would it for the temperature distribution in Fig. 11.4.3. It is clear that no such simple scaling exists. Attempts have been made to correlate

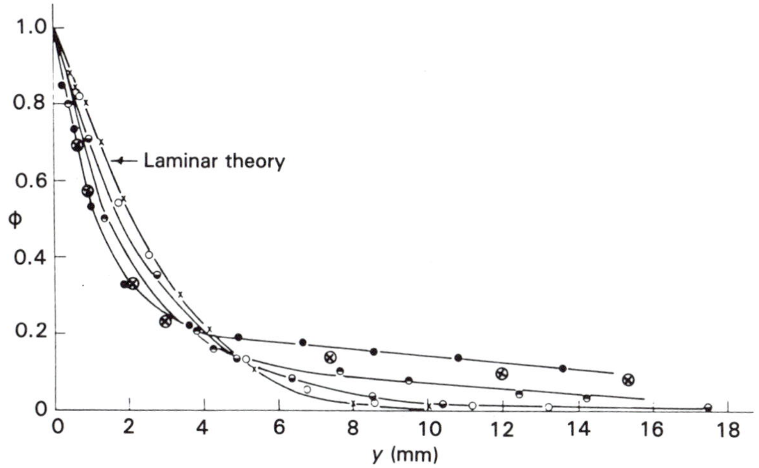

Figure 11.4.3 Development of mean temperature profiles for the same conditions as in Fig. 11.4.2. *(Reprinted with permission from B. Gebhart and R. L. Mahajan,* Advances in Applied Mechanics, *p. 231. Copyright © 1982, Academic Press.)*

fully turbulent flows by a single variable. The suggestion of $\eta = (y/x)\,\mathrm{Gr}_x^{0.1}$ by Cheesewright (1968) has not been successful. For fully turbulent flows, however, some order is suggested by George and Capp (1979) in terms of two different scaling parameters, one close to the wall and the other farther out. Very near the wall there is a laminar sublayer in which the mean temperature and velocity profiles are linear. In the next buoyant sublayer, the mean temperature and velocity profiles show a cube root and an inverse cube root dependence, respectively, on distance from the wall. These predictions have been corroborated by the experiments of Qureshi and Gebhart (1978). The data of Mahajan and Gebhart (1979) from the end of transition to early turbulence also support these measures. However, this scheme does not correlate the data during transition. The two layers, inner and outer, are not clearly differentiated and the temperature in the outer layer decays faster than the indicated $y^{-1/3}$.

Downstream growth of disturbances during transition. Measurements of downstream disturbance growth in air and in water are seen in Fig. 11.4.4. Normalized values of u' and t' indicate rapid growth immediately after the beginning

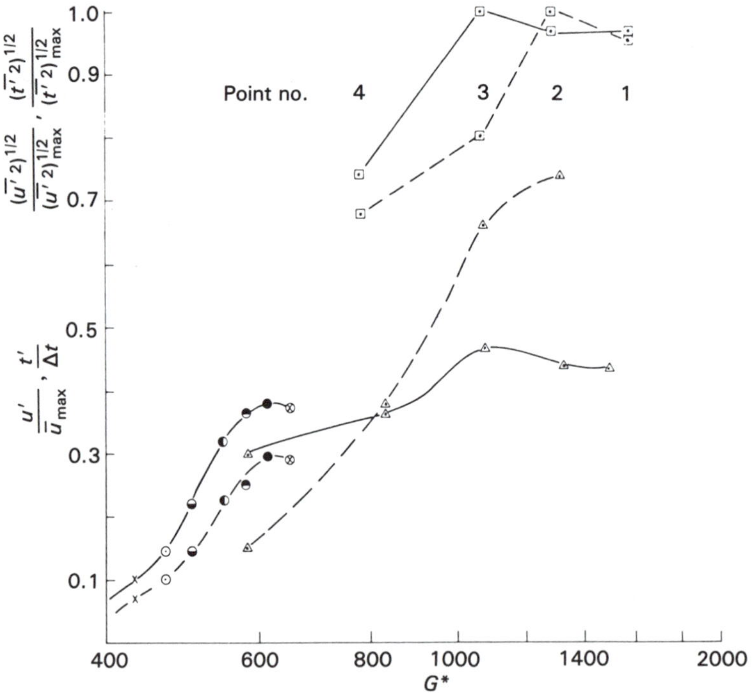

Figure 11.4.4 Downstream growth of disturbance level. (——) Velocity data; (– – –) temperature data. Data for water: (△) Jaluria and Gebhart (1974); (⊡) Bill and Gebhart (1979). Data for air are from Mahajan and Gebhart (1979); for key to the data see Fig. 11.4.2. *(Reprinted with permission from B. Gebhart and R. L. Mahajan,* Advances in Applied Mechanics, *p. 231. Copyright © 1982, Academic Press.)*

of transition. The growth slows downstream. Maximum values are reached and the ratios tend to decrease as the end of transition is approached. In air, the velocity and temperature fluctuation amplitudes reach their maximum values almost simultaneously. In water, the velocity fluctuation ratio reaches its maximum by the end of transition, while the temperature fluctuation ratio is still continuing to grow. This lag is a continuation of the lag in the earlier stages of transition. Note the large values of these ratios in both fluids.

Disturbance frequency during transition. The selective amplication, or filtering effect, predicted by linear instability was found to dominate both the earlier stages of instability and the region of nonlinear and three-dimensional disturbances, as discussed in Section 11.3. Measurements of Qureshi and Gebhart (1978) during transition in water indicate that, in locally laminar portions of the flow, the disturbance frequency also remains essentially unaltered to the end of transition. Data points in this region are those shown at large G^* in Fig. 11.2.2. The data of other investigators also show this result. See Fig. 11.2.1 for comparable data in air. The process of selective disturbance amplification extends far downstream beyond the range of linear amplification.

However, many other frequencies also arise later in transition. They come to involve a larger part of the flow field, at the expense of the characteristic frequency. Disturbance energy becomes distributed over a much broader frequency range. This is the beginning of the broad range of length scales, or eddy sizes, that characterizes the eventual completely turbulent flow.

An experimental determination of this frequency broadening during and after transition is provided by the spectrum analysis of data in air by Bill and Gebhart (1979). See Fig. 11.4.5. Cumulative spectra $\int_{F_1}^{F_2} \Phi_{u'}\, dF = \overline{u'^2}$ for the disturbance energy, $\overline{u'^2}$, are plotted for flows at the beginning of transition, at the end of transition, and in full turbulence farther downstream. The spectrum covers the frequency range from 0.006 to 10 Hz. Arrows in Fig. 11.4.5 correspond to the most amplified or preferred frequencies of the local flow. Near the beginning of transition, only about 5% of the disturbance energy lies above the most amplified frequency. About 45% of the energy is concentrated in a small range containing the characteristic frequency. By the end of transition, the turbulent energy extracted from the mean flow through nonlinear processes is more evenly distributed across the spectrum. Approximately 14% of the energy is now distributed in the frequency range above the filtered frequency.

Spectral broadening continues beyond the end of transition. The energy above the most amplified frequency increases. The spectrum continues to develop until a condition of local isotropic turbulence exists at the "inertial subrange," defined later. It is apparent from Fig. 11.4.5 that large-scale eddies break into smaller scales after transition begins. Taking disturbance frequency as a measure of turbulent length scales, the initially large scales result from the low laminar preferred frequencies. As transition progresses, energy is transferred to higher frequencies from a band of frequencies centered on the most amplified frequency. This indicates smaller-scale eddies.

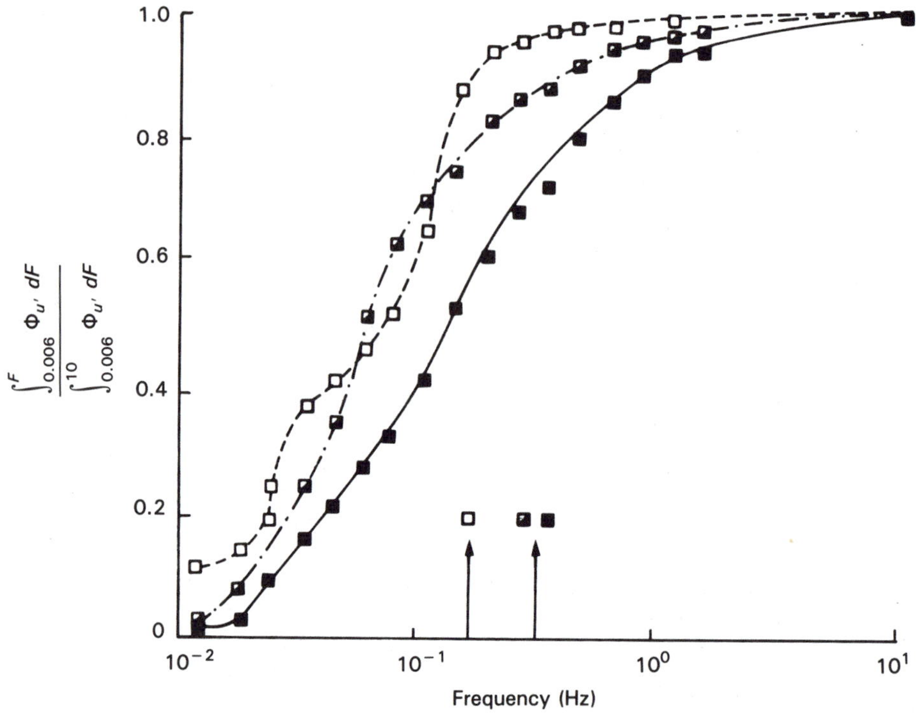

Figure 11.4.5 Cumulative energy distribution vs. frequency: (□) point 6, the beginning of transition; (◪) point 3, end of transition; (■) point 1, in turbulence. *(Reprinted with permission from* Int. J. Heat Mass Transfer, *vol. 22, p. 267, R. G. Bill, Jr., and B. Gebhart. Copyright © 1979, Pergamon Journals Ltd.)*

Thermal transport during transition. The most important practical aspect of transition is the improvement in the heat transfer mechanisms, compared to steady laminar flow. An example of the progression of local heat transfer from the laminar mechanism to developed turbulence is seen in Fig. 11.4.6. These data, from Qureshi and Gebhart (1978), were taken in flow induced adjacent to a vertical uniform flux surface in water. The increase seen in the local coefficient of heat transfer, h_x, accompanies a corresponding decrease in local surface temperature from its laminar value. The deviation from the laminar trend increases with the progression of transition for each of the five heat flux levels. Eventually a fully developed turbulent heat transfer trend is established farther downstream. The data points come into agreement with the correlation shown, from Vliet and Liu (1969).

Additional turbulent transport modes account for this much more effective transport. The modes are turbulent convection of heat downstream, $\rho c_p \overline{u't'}$, and increased transport of heat from the wall region to the outer boundary layer, $\rho c_p \overline{v't'}$. The results of Bill and Gebhart (1979) for water indicate that both increase during and after transition. This is discussed in detail in Section 11.6.

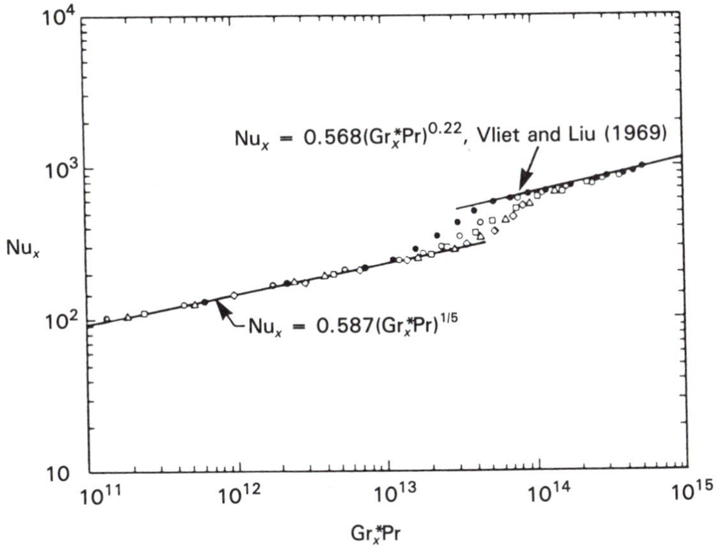

Figure 11.4.6 Variation of local heat transfer from laminar through transition to turbulence. (◇) $q'' = 583\ \mathrm{W/m^2}$; (△) $1323\ \mathrm{W/m^2}$; (□) $2326\ \mathrm{W/m^2}$; (○) $3714\ \mathrm{W/m^2}$; (●) $4488\ \mathrm{W/m^2}$. *(Reprinted with permission from* Int. J. Heat Mass Transfer, *vol. 21, p. 1467, Z. M. Qureshi and B. Gebhart. Copyright © 1978, Pergamon Journals Ltd.)*

11.5 PREDICTIVE PARAMETERS FOR TRANSITION

The most important aspect of instability and transition is the establishment of predictive parameters for the beginning and end of transition. Tradition, along with the success of linear stability theory in correctly predicting the dependence of the growth rate of two-dimensional disturbances solely on the basis of the local Grashof number, led to earlier assumptions that the Grashof number would correlate the beginning of transition. However, the accumulation of experimental data, as collected in Table 11.5.1, indicates that the beginning and end of transition each occur over a broad range of Grashof number. Some of the spread is clearly due to the use of different criteria. However, there is no compelling reason to believe that the occurrence of transition events should depend solely on a parameter whose primary importance first arises in laminar transport and in linear stability theory. Strong nonlinear interactions precede transition. There are many other important considerations.

11.5.1 Beginning of Transition

Vliet and Liu (1969) also concluded, from an experimental study of turbulent transport in water, that $\mathrm{Ra}_x^* = \mathrm{Gr}_x^*\mathrm{Pr}$ alone did not correlate the beginning of transition. In the first experimental study to investigate this matter, also in a flow

Table 11.5.1 Data for the beginning of transition

Reference	Fluid	Pr[a]	Transition G^*	Transition G	x (cm)	P (atm)	Instrument used to detect transition	E	$\frac{E}{E_p^b}$	Transition criterion
Godaux and Gebhart (1974)	Water	6.7	580–1030		36.2–100	1	Thermocouple	16.8–21.6	0.82–1.05	As proposed in Section 11.4
Qureshi and Gebhart (1978)	Water	6.25–6.81	749–950		43.3–78.6	1	Thermocouple	18.6–19.4	0.91–0.95	As proposed in Section 11.4
Mahajan and Gebhart (1979)	N_2	0.71	380–610		13.2–33	1–15.92	Hot wire	16.8–23	0.86–1.17	As proposed in Section 11.4
			400–645		13.2–33	1–15.92	Thermocouple	17.6–24	0.86–1.17	As proposed in Section 11.4
Warner and Arpaci (1968)	Air	0.71		580	91.5	1	Thermocouple	24	1.17	As proposed in Section 11.4
				466	65	1	Thermocouple	23	1.12	As proposed in Section 11.4
Jaluria and Gebhart (1974)	Water	6.7	504–802		38.1–121.9	1	Hot wire	13.6[c]	0.70	Jump in frequency of disturbance above laminar frequency
			563–802		38.1–121.9	1	Thermocouple	15.2	0.74	
Vliet and Liu (1969)	Water	5.05–6.4	855–960		28–79	1	Thermocouple	22.2[d]	1.08	Deviation of wall temperature from its maximum value
Colak-Antic (1964)	Air	0.71		572	84	1	Hot wire	25	1.28	
Cheesewright (1968)	Air	0.71		713	97	1	Thermocouple	30	1.46	During transition
Lock and Trotter (1968)	Water	11.0	485		24	1	Thermocouple	16.3	0.8	During transition
		11.4	665		27.4	1	Hot wire	20.5	1.03	During transition
Szewczyk (1962)	Water	6.7	665		60	1	Flow visualization with dye	17.2	0.88	Beginning of vortex formation
Eckert and Soehngen (1951)	Air	0.70		400	61	1	Interferometer	20	0.98	First appearance of Tollmien-Schlichting waves
Regnier and Kaplan (1963)	Air	0.71		622	92	1	Interferometer	26.4	1.29	First appearance of turbulent burst
	CO_2	0.77		460–547	12.5	4–9	Interferometer	23.8–24.9	1.16–1.21	First appearance of turbulent burst
				645–702	25	4–9	Interferometer	23.2–26.4	1.13–1.29	First appearance of turbulent burst
				541	17	5	Interferometer	24.4	1.19	First appearance of turbulent burst
				378	9	10	Interferometer	18.3	0.89	First appearance of turbulent burst
				605	25	11	Interferometer	19	0.93	First appearance of turbulent burst

[a]Prandtl number of ambient medium.
[b]Proposed values of E for beginning of transition: Velocity transition, 19.5; thermal transition, 20.5.
[c]Best fit value.
[d]Average value.
Source: From Gebhart and Mahajan (1982).

induced adjacent to a uniform flux vertical surface in water, Godaux and Gebhart (1974) used thermocouples to detect the downstream locations x of the beginning of thermal transition at different surface heat flux levels q''. The data proved unequivocally that the beginning of transition was not correlated by Gr_x^*, or G^*, alone. An additional dependence of q'' was indicated. Thermal transition, defined by the change in mean temperature profile from its laminar shape, began at an approximately constant value of $G^*/x^{3/5} \propto (q''x)^{1/5} \propto Q(x)^{1/5}$. That is, transition had begun when the total local thermal energy $Q(x)$ convected downstream in the boundary layer had reached a certain value.

Measurements of the mean and disturbance quantities during transition also confirmed the failure of G^* alone to correlate transition; see, for example, Fig. 11.5.1. The thermal intermittency factor I_t for each flow condition is given in the figure legend. At $x = 100$ cm, at $G^* = 948$, the mean temperature distribution has just deviated from the laminar. However, at $x = 36.2$ cm, it has already changed

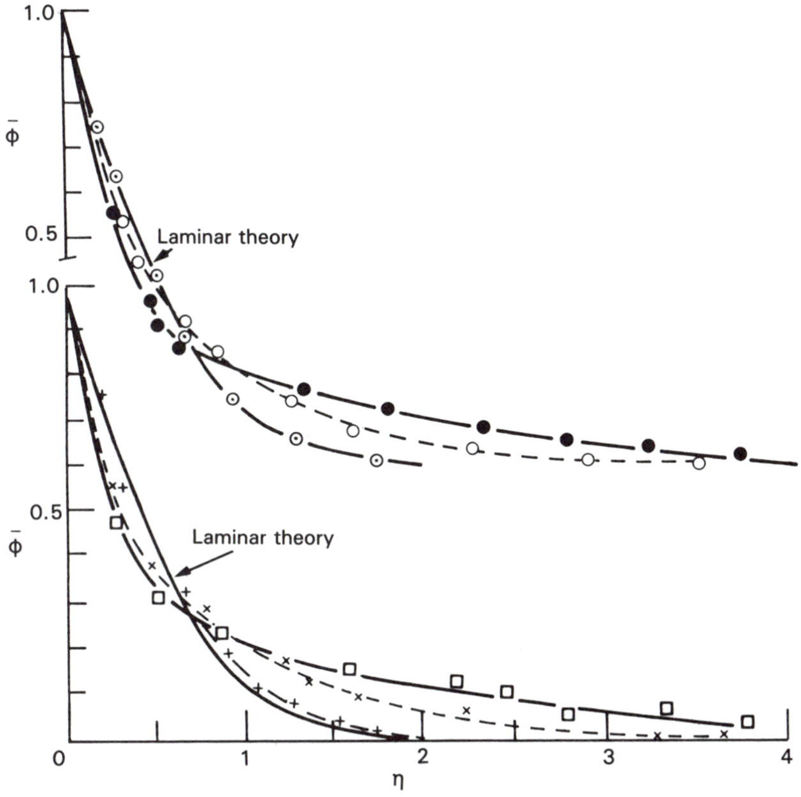

Figure 11.5.1 Dependence of mean temperature distributions on G^* and x. Values of x (cm), G^*, and I_t ($\eta = 2.5$), respectively, are: (⊙) 36.2, 485, 0; (○) 36.2, 608, 0.18; (●) 36.2, 625, 0.68; (+) 100, 948, 0.05; (×) 100, 1031, 0.6; (□) 100, 1131, 0.9. *(Reprinted with permission from* Int. J. Heat Mass Transfer, *vol. 17, p. 93, F. Godaux and B. Gebhart. Copyright © 1974, Pergamon Journals Ltd.)*

considerably at the much lower value $G^* = 625$. A similar conclusion follows from an examination of the disturbance data. An additional downstream parameter must arise.

Parameter $E = G^*/x^{2/5}$. The more detailed investigation, also in water, by Jaluria and Gebhart (1974) confirmed that neither the beginning nor the end of transition is a function only of the local Grashof number. It was found that velocity transition preceded thermal transition and that each began at a particular value of $G^*/x^{2/5}$. Mean flow and disturbance quantities confirmed the dependence. The end of transition was found to be approximately correlated by $G^*/x^{0.54}$.

The quantity $G^*/x^{2/5}$ also has a direct physical significance. It is proportional to the fifth root of the local boundary layer kinetic energy flux e.

$$e = \int_0^\infty \frac{\rho u^3}{2}\, dy = \frac{\rho \nu^3}{2}\frac{G^{*5}}{x^2}\int_0^\infty f'^3\, d\eta = \frac{\rho \nu^3}{x^2} G^{*5} F(\mathrm{Pr}) \tag{11.5.1}$$

or $e/\rho\nu^3 = (G^{*5}/x^2)F(\mathrm{Pr})$, where $F(\mathrm{Pr})$ is a nondimensional function that may be obtained from the laminar similarity calculations, as in Table 3.5.1. Nondimensionalization of $e/\rho\nu^3$ in terms of g and ν results in

$$e_0^{1/5} = \left[\frac{e}{\rho\nu^3}\left(\frac{\nu^2}{g}\right)^{2/3}\right]^{1/5} = \left(\frac{\nu^2}{gx^3}\right)^{2/15} G^* F(\mathrm{Pr})$$

or

$$\frac{e_0^{1/5}}{F(\mathrm{Pr})} = \left(\frac{\nu^2}{gx^3}\right)^{2/15} G^* = E \tag{11.5.2}$$

where gx^3/ν^2 is the unit Grashof number. Both velocity and thermal transitions, as judged from an increase in the disturbance frequencies above the laminar frequency, were found to begin at essentially fixed values of $E = 13.6$ and 15.2, respectively.

Later, Qureshi and Gebhart (1978), from experiments in water, established that this kinetic energy flux parameter E correlates well various other events commonly associated with the beginning of thermal transition. The appearance of significant temperature fluctuations, the deviation of the mean temperature profile from the laminar trend, and the attainment of maximum surface temperature for a uniform heat flux condition were found to be correlated approximately by unique values of $E = 17.5$, 19.2, and 22.7, respectively. The transition data in water obtained by several other investigators, when converted to E, fall close to one of these values of E, depending on the criterion used.

Parameter Q_{BT}. Despite this observed ability of E to predict the beginning of transition in water, it is still somewhat arbitrary in that the factor $(\nu^2/g)^{2/15}$ is introduced in the definition of E only to nondimensionalize x. This imposes a particular dependence of G^* on ν, at the beginning of transition, in Eq. (11.5.2). The validity of this dependence was tested by Mahajan and Gebhart (1979) in experiments on pressurized nitrogen. Since ν is inversely dependent on density in gases, ν was varied in experiments by changing the pressure level p of nitrogen from 1 to 15.92 atm. Measurements confirmed the dependence of the beginning of transition on $G^*/x^{2/5}$. However, the values of G^*_{VT} and G^*_{TT}, at which velocity and thermal transition began, showed a systematic additional dependence on pressure, both given closely by $G^* \propto p^{2/15}$, that is, by $G^* \propto \nu^{-2/15}$. This is different from the dependence in Eq. (11.5.2). Thus, the following new correlating parameter Q_{BT}, for the beginning of transition in gases, is indicated:

$$Q_{BT} \propto \frac{G^*}{x^{2/5}} \nu^{2/15} = \left(\frac{q''}{g\mu}\right)^{1/5} \left(\frac{gx^3}{\nu^2}\right)^{2/15} = q\left(\frac{gx^3}{\nu^2}\right)^{2/15} \tag{11.5.3}$$

Here q is the fifth root of the nondimensional local heat flux to the boundary region flow. Fixed values of $Q_{BT} = 290$ and 315 were found to characterize the beginning of the velocity and thermal transitions, respectively.

Prediction of the beginning of velocity and thermal transitions. The transition data of Mahajan and Gebhart (1979) for nitrogen gas (see Table 11.5.1), when calculated in terms of the parameter E, indicate an average value of $E = 20.5$ for the beginning of thermal transition. Noting that the criterion used was the deviation of the mean temperature profile from the laminar trend, this is very close to the value of $E = 19.2$ suggested by Qureshi and Gebhart (1978) for water, using the same criterion. The question then became whether a single value of E might, with reasonable accuracy, predict the beginning of transition in both liquids and gases.

From the data of various investigators for the beginning of transition in air and in water, values of E were calculated and are summarized in a review by Gebhart and Mahajan (1982). These are reproduced here in Table 11.5.1. The total body of data in Table 11.5.1 for the beginning of thermal transition, based on the deviation of the mean temperature profile, indicates an average value of $E = 20.5$, with a maximum difference of 17%, over a G^* range from 400 to 1030, an x range from 13.2 to 100 cm, for pressures from 1 to 15.92 atm. That is,

$$G^*_{TT} = 20.5\left(\frac{gx^3}{\nu^2}\right)^{2/15} \quad \text{or} \quad \mathrm{Gr}^*_{x,TT} = 5790\left(\frac{gx^3}{\nu^2}\right)^{2/3} \tag{11.5.4}$$

On the other hand, the beginning of thermal transition defined instead as the downstream location where the wall temperature begins to decrease from a maximum value, as done by Vliet and Liu (1969), occurred around an average value of $E \approx 22.2$. Interferometer determinations result in values of E that are substantially higher, typically of the order of 25%. This is due to the insensitivity of such measurements to small and/or concentrated turbulence. This effect was also seen in the study of transition in plane plumes by Bill and Gebhart (1975) and in supersonic boundary layers by Schubauer and Klebanoff (1956).

The beginning of velocity transition, marked by the deviation of $\bar{U}_{\max}$ from its laminar trend, appears from the data of Mahajan and Gebhart (1979) to center around the average value $E = 19.5$. This corresponds to a value of x upstream of that for the beginning of thermal transition. No such precise information based on this criterion is available for water. The presence of a higher-frequency component in the study of Jaluria and Gebhart (1974) was detected at a location G^* about 12% upstream of the location of the beginning of thermal transition. Since the deviation of $\bar{U}_{max}$ occurs between the locations of these two events, it is reasonable to assume that the beginning of velocity transition, according to the criterion used here, also occurs about 5% upstream of thermal transition. Using the value of $E_{TT} = 20.5$ then amounts to a value of $E_{VT} = 19.5$ and

$$G^*_{VT} = 19.5\left(\frac{gx^3}{\nu^2}\right)^{2/15} \qquad \text{or} \qquad \mathrm{Gr}^*_{x,VT} = 4510\left(\frac{gx^3}{\nu^2}\right)^{2/3} \tag{11.5.5}$$

11.5.2 End of Transition

As for the beginning of transition, the recent investigations have shown conclusively that the end of transition is also not correlated by G^* alone. Jaluria and Gebhart (1974) found that G^*/x^n, where $n \simeq 0.54$, approximately correlated the end of the transition. The more detailed study of Mahajan and Gebhart (1979) for pressurized nitrogen indicated $n \simeq 0.50$ and an additional dependence of G^* on ν as $G^* \propto \nu^{-1/3}$. The following correlating parameter was proposed:

$$\frac{G^*}{\sqrt{x}}\nu^{1/3} \propto G^*\left(\frac{\nu^2}{gx^3}\right)^{1/6} \propto G^*\mathrm{Pr}^{1/5}\left(\frac{\nu^2}{gx^3}\right)^{1/6}$$

$$= \left(5^4\mathrm{Ra}^*_x\right)^{1/5}\left(\frac{\nu^2}{gx^3}\right)^{1/6} = Q_{ET} \tag{11.5.6}$$

The value of Q_{ET} from these data for gases was found to be 11.4; see Table 11.5.2. The Prandtl number dependence in Eq. (11.5.6) correlates the effects of different fluids on Q_{ET}.

The data of other investigators for the end of transition in air and water, converted to Q_{ET}, are also included in Table 11.5.2. As for the beginning of transition, there is much scatter in the values of G^*_{ET} or G_{ET}. But, when converted to Q_{ET}, the data seem to collapse around the value $Q_{ET} = 11.4$. Based on this value of Q_{ET}, the following relation is suggested to determine the end of transition:

$$(\mathrm{Ra}^*_x)_{ET} = 308\left(\frac{gx^3}{\nu^2}\right)^{5/6} \tag{11.5.7}$$

11.5.3 Summary

Both the beginning and end of transition are not predicted by G^* alone. There are additional dependences on downstream location x and kinematic viscosity ν. Instead, two distinctive parameters arise at the two bounds of transition. Various events associated with the early stages of transition, such as the appearance of significant fluctuations or deviation of the mean profile from the laminar trend, seem to be well correlated by the kinetic energy flux parameter E defined in Eq. (11.5.2). Based on an extensive body of data, Eqs. (11.5.5) and (11.5.4) may be

Table 11.5.2 Data for the end of transition

Reference	Fluid	Pr[a]	Transition G^*	Transition G	x (cm)	P (atm)	Instrument used to detect transition	Q_{ET}	$\frac{Q_{ET}}{Q_{ET}}$[b]
Mahajan and Gebhart (1979)	N_2	0.71	495–980		13.2–33	4.18–15.92	Hot wire and thermocouple	11.4[c]	1.0
Jaluria and Gebhart (1974)	Water	6.7	870		61	1	Hot wire and thermocouple	10.94	0.96
			990		83.8			10.62	0.93
			1140		100.7			10.84	0.95
			1320		121.9			11.75	1.03
Cheesewright (1968)	Air	0.72		845	115	1	Thermocouple	11.89	1.04
Vliet and Liu (1969)	Water	6.2	1140		113	1	Thermocouple	10.18	0.89
		6.4	1195		80.7			12.67	1.11
		6.4	1385		60.5			16.96	1.49
		5.05	1615		53.6			18.73	1.64

[a]Prandtl number of ambient medium.
[b]Proposed value of $Q_{ET} = 11.4$ for end of transition.
[c]Best fit value.
Source: From Gebhart and Mahajan (1982).

used to provide a good estimate of G^* and Gr_x^* at which velocity and thermal transitions may begin. These transitions are marked by the first deviation of mean velocity and temperature profiles from their laminar trends, respectively. For the end of transition, indicated by no appreciable downstream changes in the distribution of velocity and temperature intermittency factors, the parameter is Q_{ET}, defined in Eq. (11.5.6). Equation (11.5.7) may then be used to provide a good estimate of Ra_x^* at which transition may end.

Correlations (11.5.4), (11.5.5), and (11.5.7) are simple to use. They are based on measurements with local sensors and should be applicable to at least gases and water at small temperature differences.

11.6 DEVELOPING TURBULENT TRANSPORT

Preceding sections have delineated the many steps in the transformation of a vertical laminar flow to turbulence. These advances in understanding have been achieved primarily through the direct measurement of velocities and temperatures, with little recourse to statistical interpretation. However, a complete picture of turbulence also concerns the way in which spectra develop downstream into an eventual pattern. The large scales that arise in transition disappear much farther downstream into smaller scales for fully developed turbulent flow.

For the fully developed turbulent regime, considerable statistical information is available for vertical flows. Kutateladze (1972) determined the profiles of the mean longitudinal velocity $\bar{u}$ and of the turbulent fluctuations $\overline{u'^2}$ adjacent to an isothermal surface in ethyl alcohol, using a method of stroboscopic flow visualization. The profiles of $\overline{u'^2}$ indicated a sharp maximum at the maximum of $\bar{u}$.

Papailou and Lykoudis (1974) measured mean temperature $\bar{t}$ and temperature fluctuation levels $\overline{t'^2}$ in mercury in the turbulent boundary layer formed along the isothermal wall of a cell. Values of $\overline{t'^2}$ first increased downstream, then decreased. The distribution of $\overline{t'^2}$ across the flow region indicated a fairly broad maximum near the wall. The spectra of these fluctuations in the outer portion of the boundary layer indicated a slight convective subrange; that is, the spectra decayed with wave number λ as $\lambda^{-5/3}$, as predicted by Corrsin (1951). This subrange was not clearly apparent near the wall. A substantial "-3" range, that is, with the decay proportional to λ^{-3}, was observed in all the spectra. It corresponds to the buoyancy range predicted by Shur (1962) and Lumley (1965) in which the cascade of energy in the inertial range is affected by the buoyancy production of turbulence.

More extensive measurements in turbulence have been reported by Smith (1972) adjacent to an isothermal surface in air. Profiles of mean longitudinal, $\bar{u}$, and normal, $\bar{v}$, velocities are given, along with mean temperature profiles $\bar{t}$. Profiles of turbulent fluctuations $\overline{u'^2}$, $\overline{v'^2}$, and $\overline{t'^2}$ are given along with covariances $\overline{u't'}$, $\overline{v't'}$, and $\overline{u'v'}$ and the associated correlation coefficients, spectra, and cospectra.

The transition results of Jaluria and Gebhart (1974), combined with the heat transfer results of Vliet and Liu (1969), for water indicate that an additional important adjustment stage occurs downstream of the transition regime considered in Sections 11.4 and 11.5. At the end of transition, both the thermal and velocity intermittency factors are one except in the outer region of the boundary layer, where entrainment occurs. However, the data of Vliet and Liu (1969) indicate that the local heat transfer coefficient h_x continues to rise sharply after the total disappearance of laminar flow. This trend then disappears and h_x decreases downstream at a rate similar to that of laminar flow.

The later spectral data of Bill and Gebhart (1979) indicate that these effects are the continuing consequences of the earlier concentration of disturbance energy in large-scale eddies. At the end of transition, a substantial fraction of the disturbance energy is distributed at and below the preferred or characteristic frequency. The flow beyond the end of transition is a subsequent regime of spectral development. The upward spectral broadening to higher frequency is eventually terminated at the upper frequencies associated with the smallest eddy sizes at which turbulence is dissipated by molecular viscosity.

Another feature of the delay in achieving developed turbulence, beyond the end of transition, is the continued lag in the development of temperature disturbance levels observed by Jaluria and Gebhart (1974). Bill and Gebhart (1979) found that temperature disturbance levels rose sharply beyond the end of transition, despite the leveling of the growth of velocity disturbances. The decrease of the growth rate of temperature disturbance levels, like the end of spectral development, corresponded well with the establishment of developed turbulence, as indicated by the heat transfer data of Vliet and Liu (1969).

An experimental investigation by Bill and Gebhart (1979) is discussed next. The flow was generated in water adjacent to an electrically heated vertical surface, 132 cm high and 41.5 cm wide, in a 2 × 1.7 × 3 m high insulated stainless steel tank. Water resistivity was always kept greater than 0.8 MΩ cm to enable use of bare hot wires.

The velocity and temperature were determined with a compact arrangement of three hot wires and a thermocouple. The hot-wire sensors were calibrated over the velocity range 0.06–3.3 cm/s. Outputs were digitalized and the operations were performed in real time, using a time series analysis to calculate variances, covariances, and spectral and cospectral densities. The variances and covariances are defined in terms of spectral and cospectral densities as

$$\overline{u'^2} = \int_{F_1}^{F_2} \Phi_{u'}(F)\, dF$$

$$\overline{v't'} = \int_{F_1}^{F_2} \Phi_{v't'}(F)\, dF$$

$$\overline{u't'} = \int_{F_1}^{F_2} \Phi_{u't'}(F)\, dF$$

where, for example, $\Phi_{u'}(F)$ is the contribution to the spectral density of $\overline{u'^2}$ in the frequency interval dF. The lower limit of integration F_1 is $1/T$, where T is the record length of the time series analyzed. The upper integration limit F_2, the Nyquist frequency, is at half the sampling rate.

A particular experiment amounted to traversing the boundary layer at given downstream distances x at a given surface heat flux q''. The experiments of Jaluria and Gebhart (1974) in water had established the approximate conditions for the beginning and end of transition, seen in Fig. 11.6.1. The points shown are the experimental conditions chosen for the investigation of further turbulence development, as given in Table 11.6.1. The slopes of the transition limits indicate that the regimes are dependent on both G^* and the heat flux q''. Note that conditions at any flux q'' correspond to increasing distances downstream along a uniform flux surface.

Distributions of velocity and temperature fluctuations. The growth of these fluctuation levels downstream is quite complicated. Jaluria and Gebhart (1974) reported that disturbances amplified at decreasing rates downstream. Temperature disturbances follow velocity disturbances after significant velocity fluctuations reach into the thermal layer.

Measured variations downstream of the maximum values of $\overline{t'^2}$ and $\overline{u'^2}$ are seen in Fig. 11.6.2, along with variations of $\overline{u't'}$ and $\overline{v't'}$. The variation from $x_4 = 48.3$ to $x_1 = 114.3$ cm is seen in Fig. 11.6.2. At all locations, x, the maximum in the velocity fluctuation u'_{max} occurred at approximately $\eta = 0.6$.

The maximum magnitude of $\overline{u'^2}$ occurs around point 3, approximately at the end of transition. However, the maximum in $\overline{t'^2}$ occurs later, a continuation of

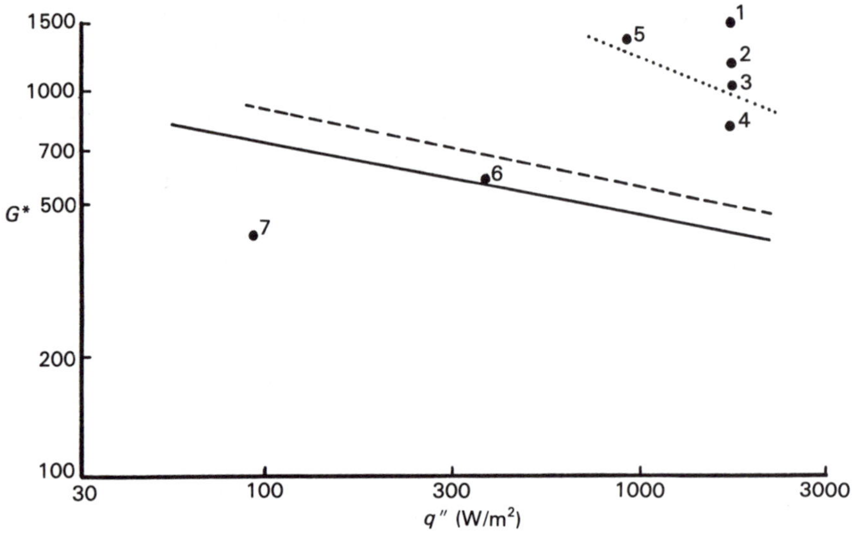

Figure 11.6.1 Location of (——) beginning of velocity transition, (– – –) thermal transition, and (· · ·) end of transition, from Jaluria and Gebhart (1974). Points 1–7, conditions of study of Bill and Gebhart (1979); see Table 11.6.1. *(Reprinted with permission from* Int. J. Heat Mass Transfer, *vol. 22, p. 267, R. G. Bill, Jr., and B. Gebhart. Copyright © 1979, Pergamon Journals Ltd.)*

the trend reported by Jaluria and Gebhart (1974) for the early stages of transition. Heat transfer results indicate that fully developed turbulent flow is achieved downstream of the locations along the upper boundary in Fig. 11.6.1. Thus, fully turbulent heat transfer is not achieved until after the maximum temperature fluctuation level is reached, despite the earlier full development of both the intermittency distributions and the velocity fluctuations. Increasing temperature fluctuation levels with increasing downstream location x, followed by gradually decreasing fluctuation levels, were also reported by Papailou and Lykoudis (1974) for mercury.

Table 11.6.1 Experimental conditions for the points in Fig. 11.6.1

Point	x (cm)	q'' (W/m^2)	G^*
1	114.3	1920	1574
2	88.9	1920	1284
3	71.1	1920	1077
4	48.3	1920	787
5	114.3	1160	1419
6	48.3	436	587
7	48.3	95.3	440

Source: From Bill and Gebhart (1979).

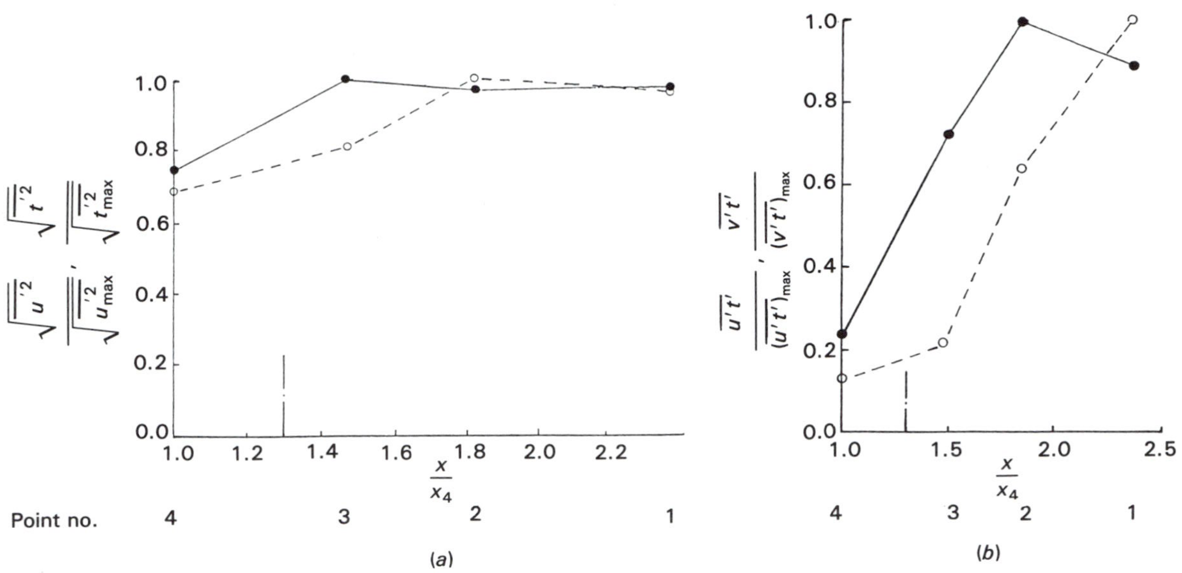

Figure 11.6.2 Downstream growth of turbulence parameters at $q'' = 1920\ \mathrm{W/m^2}$. (*a*) Downstream growth of turbulent fluctuation levels: (●) $(\overline{u'^2})^{1/2}/[(\overline{u'^2_{max}})^{1/2}]$; (○) $(\overline{t'^2})^{1/2}/[(\overline{t'^2_{max}})^{1/2}]$ vs. x/x_4 where $x_4 = 48.3$ cm. The vertical line is the end of transition determined by Jaluria and Gebhart (1974). (*b*) Downstream growth of turbulent heat transport at $\eta \approx 0.6$: (●) $\overline{u't'}/(\overline{u't'})_{max}$; (○) $\overline{v't'}/(\overline{v't'})_{max}$ vs. x/x_4 where $x_4 = 48.3$ cm. The vertical line is the end of transition from Jaluria and Gebhart (1974). *(Reprinted with permission from* Int. J. Heat Mass Transfer, *vol. 22, p. 267, R. G. Bill, Jr., and B. Gebhart. Copyright © 1979, Pergamon Journals Ltd.)*

Although Grashof numbers were not given, it may be inferred from the later results that these fluctuations were actually observed in transition.

The distributions of the normalized temperature and velocity fluctuation levels in Fig. 11.6.3*a* are for the two most evolved points, 1 and 2. No sharp peak is found, in contrast to the results of Kutateladze (1972). None is expected since turbulent diffusion smooths out gradients across the layer. These distributions are also quite different from those of Jaluria and Gebhart (1974), which were seen as fluctuation maxima in an analog record. Two peaks were reported, the highest at $\eta = 2.8$. It seems plausible that such multiple peaks result from the passage of particular bursts of turbulence. The observed random location shifts ensure that they would not be statistically significant in a root mean square representation.

The shape of $\overline{t'^2}$ distribution is similar to that reported by Papailou and Lykoudis (1974) and Smith (1972) for mercury and air. The velocity and thermal

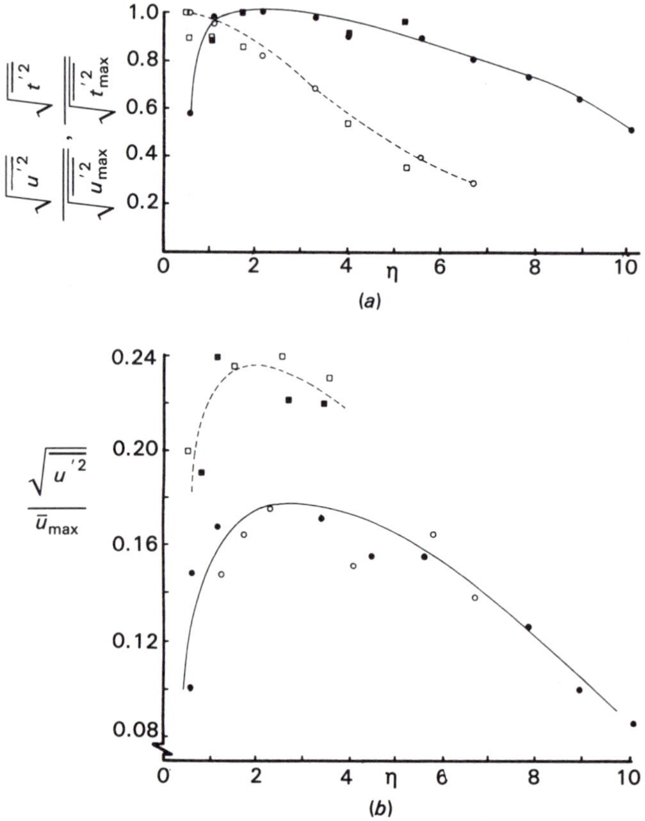

Figure 11.6.3 Turbulence parameter variations across the boundary region. (*a*) Turbulent fluctuation levels across the boundary layer. $(\overline{u'^2})^{1/2}/[(\overline{u'^2_{max}})^{1/2}]$; (●) point 1; (■) point 2. $(\overline{t'^2})^{1/2}/[(\overline{t'^2_{max}})^{1/2}]$ vs. η; (○) point 1; (□) point 2. (*b*) Distributions of turbulent intensity across the boundary layer. (●) Point 1; (○) point 2; (□) point 3; (■) point 5. *(Reprinted with permission from* Int. J. Heat Mass Transfer, *vol. 22, p. 267, R. G. Bill, Jr., and B. Gebhart. Copyright © 1979, Pergamon Journals Ltd.)*

layers in these fluids are of about the same thickness and the fluctuations overlap. On the other hand, in water the layer of large temperature fluctuation is seen to be approximately half the thickness of that in which velocity fluctuations arise.

Another primary parameter is turbulent velocity intensity $\overline{u'^2}/\bar{u}_{\max}$. It is plotted in Fig. 11.6.3*b* for points 1, 2, 3, and 5, where $\bar{u}_{\max}$ is the maximum local mean longitudinal velocity. For points 1 and 2, at the same q'', the turbulence intensity and the relative scale of turbulence are approximately the same. However, near the end of transition, points 3 and 5, much greater intensity is seen. The similarity in $\overline{u'^2}$ for 1 and 2 suggests that mechanisms for the creation of turbulent disturbances have been developed.

The intensities found are much greater than those observed in Blasius flow. The maximum, found near the plate surface, was 0.087; see Klebanoff (1954). In Fig. 11.6.3*b* the broad intensity maxima are about 0.18 and 0.24. As a further indication of the dependence of transition on both G^* and q'', the similarity of turbulent intensity at points 5 and 3 confirms the "end of transition" line drawn in Fig. 11.6.1.

Temperature and velocity spectra. The data of Bill and Gebhart (1979) indicate that, before transition, disturbance energy is concentrated in a small band of very low frequencies. In transition much of the disturbance energy also remains in a small band. Later, through nonlinear processes, energy is distributed over a much broader range. If frequency is a measure of the turbulent scale, these scales are large in transition. Subsequently, energy is transferred from these large scales into smaller scales.

A substitution of frequency for wave number frequently is justified by Taylor's hypothesis of frozen turbulence. The applicability of this hypothesis to shear flow has been discussed by Lumley (1965). The validity of the hypothesis is limited to small scales of turbulence $\overline{u'^2}/\bar{u}_{\max}$ and velocity gradients compared to the eddy sizes. Qualitatively, the frequency and wave number should be approximately proportional. In Fig. 11.4.5 cumulative spectra $\int_{F_1}^{F_2} \Phi_{u'}\, dF$ for the disturbance energy $\overline{u'^2}$ are seen for points 6, 3, and 1, respectively, in Fig. 11.6.1. Points 3 and 1 are downstream of the transition and 6 is at the beginning. As pointed out in Section 11.4, these spectra take into account only energy from 0.006 to 10 Hz. Above 10 Hz the spectral density approached the noise levels of the hot-wire anemometry. Arrows correspond to the single preferred frequency predicted by linear stability theory.

For point 6, approximately 45% of the disturbance energy is found in a small frequency range containing the predicted filtered frequency. Only 5% lies beyond. The unevenness of the distribution at lower frequencies is caused by changing patterns of flow, laminar to partially turbulent, occurring over relatively long periods of time, as well as by the development of large eddies. At the end of transition, point 3, energy has been more evenly distributed across the spectrum. Now about 14% lies above the filtered frequency. By point 1, the energy above the filtered frequency has increased, but to only 27% of the total disturbance energy. An increase indicates the development of small scales.

As the spectrum continues to develop, a condition may be reached in which regions of locally isotropic turbulence exist at some length, the familiar "inertial subrange." Within this subrange the spectral density of the kinetic energy of turbulence decays with wave number λ as $\lambda^{-5/3}$; see Obukhov (1941). The assumptions there concern the velocity spectra of incompressible flows. This development was extended by Corrsin (1951) to include the spectra of temperature fluctuations under conditions in which temperature could be considered a passive scalar in the flow field. This "convection subrange" also follows a $-\frac{5}{3}$ law.

Two of the major assumptions in the identification of an inertial subrange are that the spectral density is a function only of the spectral density flux and wave number and that the spectral density flux is constant. That is, within the inertial subrange, energy is not transferred into or out of the spectrum at particular wave numbers. With buoyancy in a turbulent flow it is possible that the spectral energy flux is not constant. The spectra of velocities measured in clear air turbulence (Shur, 1962) indicate a decay with wave number as λ^{-3} at low wave numbers. At higher wave numbers, the $-\frac{5}{3}$ subrange was observed. The assertion in Shur (1962) is that the inertial subrange decays as λ^{-3}, followed by a $-\frac{5}{3}$ range, if the transfer of energy due to work done by buoyancy forces against gravity is small compared to the local spectral density flux. For flows with buoyancy, Bolgiano (1962) assumed a wave number range in which the spectrum is a function only of the net rate of dissipation of "mean-square fluctuations of buoyancy forces." This leads to $-\frac{11}{5}$ and $-\frac{7}{5}$ forms for the velocity and temperature spectra.

Spectral densities $\Phi_{u'}(F)$ and $\Phi_{t'}(F)$ are seen in Figs. 11.6.4 and 11.6.5 for longitudinal velocity component and temperature fluctuations, respectively. These are at different locations across the boundary region for point 1 in Fig. 11.6.1. The temperature and velocity spectra are quite similar to each other and are very similar to those obtained at point 2. At low frequency there is a large scatter between the spectra at different y locations. By contrast, there is little scatter at higher frequencies. The low range or, correspondingly, the large length scales arise as energy is fed from the mean flow to the smaller turbulent eddies. The distribution at large scales results from the scale limitation imposed by the boundary layer thickness. This scatter indicates that the scale of large eddies varies across the boundary layer.

The similarity of the velocity spectra at different y locations at higher frequencies indicates the independence of small scales from the large-scale eddies. Above 0.5 Hz, the decay in the spectra closely follows the -3 law. The temperature spectra also follow this law. The spectrum corresponding to location $\eta = 6.72$ is considerably displaced to lower frequencies.

The temperature and longitudinal velocity spectra in air for an isothermal plate (Smith, 1972) were based on only a single point in the transition process. For this point, substantial regions of both spectra decay as λ^{-3}, as above and in the results of Papailou and Lykoudis (1974). This suggests the validity of the assumption by Lumley (1965). That is, the -3 region indicates that the spectral density may be considered constant and an inertial subrange exists across a wide range of wave numbers.

Turbulent transport. Heat transfer results, as in Vliet and Liu (1969) and

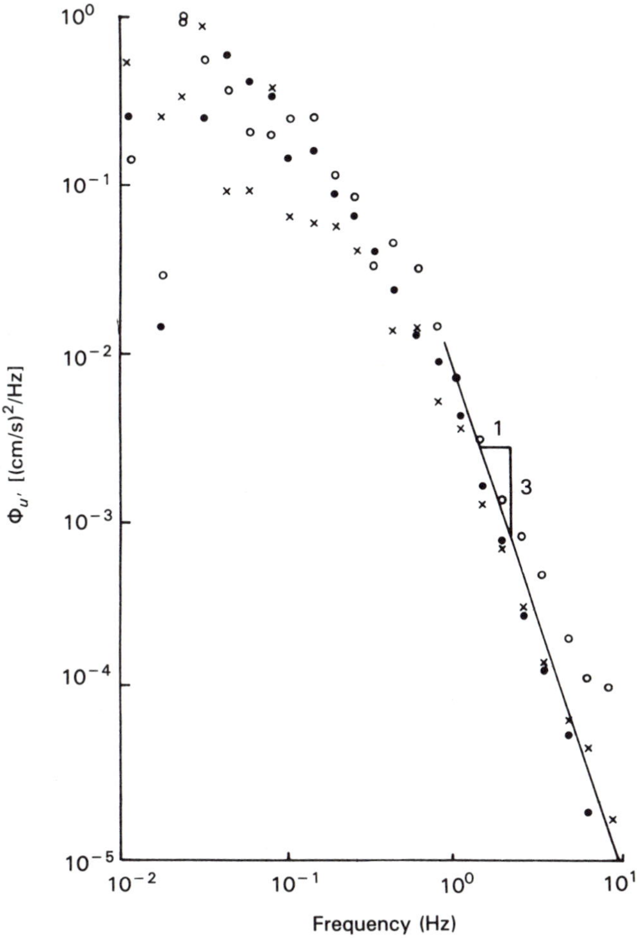

Figure 11.6.4 Spectral density of velocity fluctuations $\overline{u'^2}$ vs. frequency for point 1 in Fig. 11.6.1. (●) $\eta = 0.56$; (○) $\eta = 1.12$; (×) $\eta = 6.72$. *(Reprinted with permission from* Int. J. Heat Mass Transfer, *vol. 22, p. 267, R. G. Bill, Jr., and B. Gebhart. Copyright © 1979, Pergamon Journals Ltd.)*

Cheesewright (1968), indicate that turbulent transport modes strongly augment heat transfer mechanisms. This occurs either through turbulent convection downstream, $\rho c_p \overline{u't'}$, or through increased transport across the boundary layer region, $\rho c_p \overline{v't'}$. The downstream development of these quantities is shown in Fig. 11.6.2*b* for points 4, 3, 2, and 1 in Fig. 11.6.1. At each x the maximum of $\overline{u't'}$ or $\overline{v't'}$ across the boundary region was at approximately $\eta = 0.6$. Both modes are seen to increase into the completely turbulent regime. The maximum values occur beyond the end of transition, as designated by Jaluria and Gebhart (1974).

Downstream convection of heat is the sum of convection by the mean base flow and the turbulent downstream convection $\int_0^\infty \rho c_p \overline{u't'}\, dy$. During transition, the maximum of the mean longitudinal velocity falls from the laminar trend. For

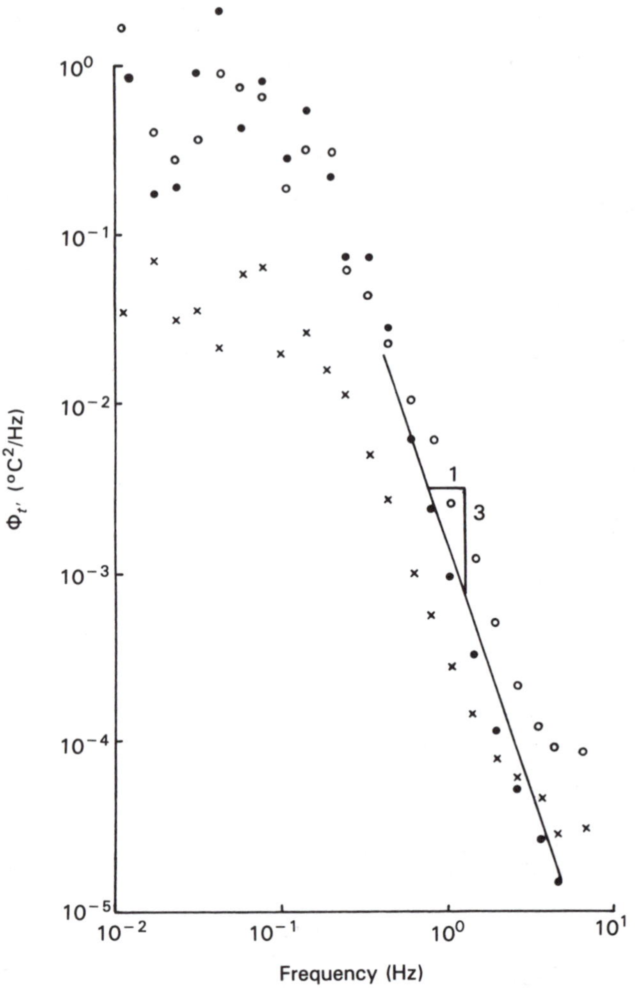

Figure 11.6.5 Spectral density of temperature fluctuations $\overline{t'^2}$ vs. frequency for point 1 in Fig. 11.6.1. (●) $\eta = 0.56$, (○) $\eta = 1.12$, (×) $\eta = 6.72$. *(From Bill and Gebhart, 1979.)*

all of the points 1, 2, 3, and 4 in Fig. 11.6.1, the measured velocity maxima are approximately 50% of the laminar value. Since the velocity decrease occurs in the region of steep temperature gradients, near the wall, mean thermal transport is reduced there. The turbulent transport mode $\rho c_p \overline{u't'}$ must account for this deficit, with an imposed uniform surface flux condition.

Turbulent thermal transport normal to the surface, $\overline{v't'}$, increases through and well beyond the end of transition. It lags the streamwise mode $\overline{u't'}$, which is coupled with the decrease in the maximum of the mean longitudinal velocity profile. It is also expected to be somewhat coupled to the growth of velocity fluctuations $\overline{u'^2}$.

A comparison of $\overline{v't'}$ and $\overline{u't'}$ with $\overline{u'^2}$ and $\overline{t'^2}$ in Fig. 11.6.2 suggests that the development of the normal heat transport mode is more closely coupled to the growth of high $\overline{t'^2}$ levels in the wall region. The sharp rise in both transport modes beyond the end of transition is expected since the heat transfer results in Vliet and Liu (1969) indicate that a fully developed heat transfer mechanism is not attained until beyond the end of transition designated in Jaluria and Gebhart (1974).

Table 11.6.2 shows transition events in terms of G^* for a given level of heat flux from the results of Jaluria and Gebhart (1974), Vliet and Liu (1969), and Bill and Gebhart (1979). The criterion of Jaluria and Gebhart (1974) is in terms of stabilization of the intermittency distributions. Figures 11.6.3*a* and 11.4.5 indicate that at the end of transition the level of velocity fluctuation $\overline{u'^2}$ has also stabilized. However, the lag in thermal transition in the early stages of transition is still in evidence in the growth lag of temperature fluctuation level $\overline{t'^2}$; see Fig. 11.6.2.

The local heat transfer coefficient h_x was measured for a constant-flux plate in water (Vliet and Liu, 1969) in the transition regime. The deviation from laminar theory abruptly increased and then decreased downstream. For conditions of point 2 in Table 11.6.2, h_x begins to decrease in developed turbulent flow.

Points 2 in Fig. 11.6.2 corresponds to the condition for which the maximum values of the temperature fluctuation level $\overline{t'^2}$ and the transport mechanism $\rho c_p \overline{u't'}$ were measured. As discussed above, development of these parameters occurs within a regime of spectral transition. Thus, the data of Vliet and Liu (1969) indicate that fully developed turbulent transport is not evolved until the energy spectrum and transport modes are fully developed.

It is thus apparent that the further development of turbulence late in and after transition amounts to the simultaneous development of several interdependent transport parameters. First, the intensity of turbulence grows as large eddies develop, disrupting the organized laminar flow. In water the velocity transition progresses first, driving turbulent temperature fluctuations. As the last of the laminar flow field vanishes, a regime of spectral and transport development begins. The disturbance field, initially dominated by selective amplification, expands to include a spectrum of turbulent frequencies. The turbulent eddies, over a wide range of wave numbers, are strongly affected by buoyant production of turbulence. Although the velocity disturbance magnitude has peaked at the end of the transition,

Table 11.6.2 Comparison of reported transition events as a function of G^*

Event	G^*	q'' (W/m^2)
End of transition	1000	1920
Developed turbulence	1245	1970
Point 2, region of spectral transition	1284	1920

Source: From Bill and Gebhart (1979).

temperature disturbance levels continue to grow. Turbulent transport quickly accounts for an increasing proportion of heat transport.

11.7 DEVELOPED TURBULENT TRANSPORT

Many buoyancy-induced flows of interest in nature and in technology are predominantly turbulent. A distinguishing feature of these flows is the presence of irregular fluctuations in the velocity, temperature, and pressure fields. The fluctuating or eddying motion leads to mixing, which gives rise to an additional transport process. An increase in shear stress and energy transfer occurs, and there is often interest in evaluating this additional transport.

Analyses of buoyant turbulent flows are generally cumbersome and unreliable. Some success has been achieved for flows adjacent to vertical surfaces, as reviewed in Section 11.7.1. For other flows and configurations, information is available in the form of empirical correlations based on experimental data and is described in Section 11.7.2.

First, however, a brief review of some of the fundamental concepts essential to understanding turbulent flows will be given. For a detailed treatment, the reader is referred to Schlichting(1968), Hinze (1975), Lumley and Panofsky (1964), Tennekes and Lumley (1972), and Arpaci and Larsen (1984).

Reynolds stress. As in Section 11.2, a steady turbulent flow may be described in terms of a mean motion with a fluctuating or eddy motion superimposed on it. That is, the instanteous velocity component u is written in terms of the time averaged velocity $\bar{u}$ and a fluctuating velocity component u' as

$$u = \bar{u} + u' \tag{11.7.1}$$

where

$$\bar{u} = \frac{1}{\tau} \int_{\tau_0}^{\tau_0+\tau} u \, d\tau \tag{11.7.2}$$

and

$$\overline{u'} = \frac{1}{\tau} \int_{\tau_0}^{\tau_0+\tau} u' \, d\tau = 0 \tag{11.7.3}$$

The time interval τ is taken long enough for the averages to be independent of time.

The governing equations for the instantaneous turbulent velocity, temperature, and pressure field for an incompressible fluid, neglecting viscous dissipation and pressure work in the energy equation, are given below in the indicial notation. The vector form of these equations is given in Chapter 2.

$$\frac{\partial v_i}{\partial x_i} = 0 \tag{11.7.4}$$

$$\frac{\partial v_i}{\partial \tau} + v_j \frac{\partial v_i}{\partial x_j} = g_i - \frac{1}{\rho}\frac{\partial p}{\partial x_i} + \nu \nabla^2 v_i \tag{11.7.5}$$

$$\frac{\partial t}{\partial \tau} + v_i \frac{\partial t}{\partial x_i} = \alpha \frac{\partial}{\partial x_i}\frac{\partial t}{\partial x_i} \tag{11.7.6}$$

where $v_i = \bar{v}_i + v_i'$. Equations for mean flow are obtained by averaging Eqs. (11.7.4)–(11.7.6). First consider the momentum equation. Using Eqs. (11.7.1)–(11.7.3) in conjunction with $p = \bar{p} + p'$, where $\overline{p'} = 0$, and noting that the following rule applies for obtaining the average of the product of two instantaneous velocity components:

$$\overline{uv} = \overline{(\bar{u} + u')(\bar{v} + v')} = \bar{u}\bar{v} + \overline{u'v'} \tag{11.7.7}$$

the average of Eq. (11.7.5) is written as

$$\bar{v}_j \frac{\partial \bar{v}_i}{\partial x_j} = g_i - \frac{1}{\rho}\frac{\partial \bar{p}}{\partial x_i} + \nu \nabla^2 \bar{v}_i + \frac{1}{\rho}\, \partial x_j(\tau_{ij}) \tag{11.7.8}$$

where $\tau_{ij} = -\rho\, \overline{v_i'v_j'}$ is the contribution of the turbulent motion to the mean stress and is called the Reynolds stress. In cartesian coordinates, components of this turbulent stress tensor and the system of equations are

$$\begin{pmatrix} \tau_{xx} & \tau_{xy} & \tau_{xz} \\ \tau_{xy} & \tau_{yy} & \tau_{yz} \\ \tau_{xz} & \tau_{yz} & \tau_{zz} \end{pmatrix}_t = -\begin{pmatrix} \rho\overline{u'^2} & \rho\overline{u'v'} & \rho\overline{u'w'} \\ \rho\overline{u'v'} & \rho\overline{v'^2} & \rho\overline{v'w'} \\ \rho\overline{u'w'} & \rho\overline{v'w'} & \rho\overline{w'^2} \end{pmatrix} \tag{11.7.9}$$

$$\rho\left(\bar{u}\frac{\partial \bar{u}}{\partial x} + \bar{v}\frac{\partial \bar{u}}{\partial y} + \bar{w}\frac{\partial \bar{u}}{\partial z}\right) = g - \frac{\partial \bar{p}}{\partial x} + \mu\nabla^2\bar{u} - \rho\left(\frac{\partial \overline{u'^2}}{\partial x} + \frac{\partial \overline{u'v'}}{\partial y} + \frac{\partial \overline{u'w'}}{\partial z}\right)$$

$$\rho\left(\bar{u}\frac{\partial \bar{v}}{\partial x} + \bar{v}\frac{\partial \bar{v}}{\partial y} + \bar{w}\frac{\partial \bar{v}}{\partial z}\right) = -\frac{\partial \bar{p}}{\partial y} + \mu\nabla^2\bar{v} - \rho\left(\frac{\partial \overline{u'v'}}{\partial x} + \frac{\partial \overline{v'^2}}{\partial y} + \frac{\partial \overline{v'w'}}{\partial z}\right) \tag{11.7.10}$$

$$\rho\left(\bar{u}\frac{\partial \bar{w}}{\partial x} + \bar{v}\frac{\partial \bar{w}}{\partial y} + \bar{w}\frac{\partial \bar{w}}{\partial z}\right) = -\frac{\partial \bar{p}}{\partial z} + \mu\nabla^2\bar{w} - \rho\left(\frac{\partial \overline{u'w'}}{\partial x} + \frac{\partial \overline{v'w'}}{\partial y} + \frac{\partial \overline{w'^2}}{\partial z}\right)$$

In a turbulent flow the total stresses are, then, the sum of the viscous stresses, $\mu\nabla^2\bar{v}_i$, and the additional Reynolds stresses, Eq. (11.7.9). Generally speaking, the Reynolds stresses far exceed the viscous components. In many flows the latter may even be neglected with a good degree of approximation.

Eddy viscosity. By analogy to the definition of laminar shear stress,

$$(\tau_{xy})_{\text{laminar}} = \tau_l = \rho\nu \frac{\partial \bar{u}}{\partial y} \tag{11.7.11}$$

it has been a conventional practice to express the Reynolds stress, considered in terms of some turbulent or eddy viscosity and mean flow quantities, as follows:

$$(\tau_{xy})_{\text{turbulent}} = \tau_t = \rho\epsilon_m \frac{\partial \bar{u}}{\partial y} \tag{11.7.12}$$

where ϵ_m is the eddy viscosity, also sometimes referred to as the "apparent" or "virtual" viscosity. Although the similarity of Eqs. (11.7.11) and (11.7.12) provides a familiar and conceptually easy formulation of the Reynolds stresses, ϵ_m is not a property of the fluid, like ν. It is variable and dependent on the nature of the flow. This is clear upon comparing the value of τ_{xy} in Eq. (11.7.12) with that in Eq. (11.7.9), which gives

$$\epsilon_m = -\frac{\overline{u'v'}}{\partial \bar{u}/\partial y} \tag{11.7.13}$$

From Eqs. (11.7.11) and (11.7.13), the total shear stress in a turbulent flow is expressed in terms of eddy viscosity as

$$\tau_{xy} = \tau = \rho(\mu + \epsilon_m) \frac{\partial \bar{u}}{\partial y} \tag{11.7.14}$$

Reynolds flux and thermal diffusivity. Turbulent mixing not only contributes to an exchange of momentum but also enhances the thermal energy transport. A treatment similar to that above for the Reynolds stresses is used to obtain an expression for the turbulent transport of thermal energy. The instantaneous value of temperature t is represented by the sum of a mean value $\bar{t}$ and a fluctuating component t'. That is,

$$t = \bar{t} + t' \tag{11.7.15}$$

where, as before,

$$\bar{t} = \frac{1}{\tau} \int_{\tau_0}^{\tau_0+\tau} t \, d\tau \tag{11.7.16}$$

and

$$\overline{t'} = 0 \tag{11.7.17}$$

Substituting Eqs. (11.7.1) and (11.7.15) in Eq. (11.7.6) and taking the average of all the terms yields

$$\bar{v}_i \frac{\partial \bar{t}}{\partial x_i} = \frac{\partial}{\partial x_i}\left(\alpha \frac{\partial \bar{t}}{\partial x_i} - \overline{v_i' t'}\right) \tag{11.7.18}$$

The total mean heat flux in a turbulent flow is then

$$q_i'' = \rho c_p\left(-\alpha \frac{\partial \bar{t}}{\partial x_i} + \overline{v_i' t'}\right) \tag{11.7.19}$$

That is, the turbulent heat flux consists of a laminar or molecular component and an eddy or mixing component. The latter is given by

$$(q_i'')_{\text{turbulent}} = \rho c_p \overline{v_i' t'} \tag{11.7.20}$$

This is called the Reynolds flux.

In a treatment similar to that for turbulent momentum transfer, the Reynolds flux may also be expressed after the formulation of laminar thermal energy transport. From the Fourier law of heat conduction, the heat flux in the y direction is given by

$$q''_{\text{laminar}} = q''_l = -\rho c_p \alpha \frac{\partial \bar{t}}{\partial y} \tag{11.7.21}$$

It is convenient to express the Reynolds flux component in the y direction as

$$q''_{\text{turbulent}} = q''_t = -\epsilon_t \rho c_p \frac{\partial \bar{t}}{\partial y} \tag{11.7.22}$$

where ϵ_t is the turbulent thermal diffusivity. From Eqs. (11.7.20) and (11.7.22)

$$\epsilon_t = -\frac{\overline{v_i' t'}}{\partial \bar{t}/\partial y} \tag{11.7.23}$$

The total heat flux in terms of ϵ_t is then

$$\frac{q''}{\rho c_p} = -(\alpha + \epsilon_t) \frac{\partial \bar{t}}{\partial y} \tag{11.7.24}$$

Scales in forced convection turbulent flows. Two different sets of scales are needed. They describe the mean and fluctuating components of turbulent flows, respectively, with the scales for the mean flow being the larger of the two.

Let $\hat{l}$, $\hat{u}$, and $\hat{\tau} = \hat{l}/\hat{u}$, respectively, represent the length, velocity, and time scales for the mean flow. Since the large eddies account for most of the transport and may be as big as the width of the flow, $\hat{l}$ is generally of the order of the width of the flow. For example, in a boundary layer flow $\hat{l}$ may be taken as the boundary layer thickness. The velocity scale $\hat{u}$ represents characteristic velocity fluctuations and is generally related to the mean square fluctuation $\overline{u'^2}$.

The length, velocity, and time scales for the fluctuating flow are taken as $\tilde{l}$, $\tilde{u}$, and $\tilde{\tau} = \tilde{l}/\tilde{u}$. They are associated with the small-scale turbulent motions. Note that the generation of these fluctuations is due to the nonlinear terms in the equations of motion, while the viscous terms determine the smallest eddy size. The viscous forces prevent generation of extremely small scales of motion by dissipating small-scale energy into heat. Thus, the small-scale structure of turbulence tends to be isotropic, that is, independent of any orientation effects introduced by mean shear. Therefore, all averages related to the small eddies do not change under rotation or reflections of the coordinate system.

It follows that small-scale motion should depend only on the rate at which it is supplied with energy by the large-scale motion and on kinematic viscosity. Thus, the parameters governing the small-scale motion include at least the dissipation rate per unit mass, ϵ (m^2 s^{-3}), and the kinematic viscosity ν. With ϵ and

ν, the $\tilde{l}$, $\tilde{u}$, and $\tilde{\tau}$ may be formed as follows:

$$\tilde{l} = \left(\frac{\nu^3}{\epsilon}\right)^{1/4} \qquad \tilde{u} = (\nu\epsilon)^{1/4} \qquad \tilde{\tau} = \left(\frac{\nu}{\epsilon}\right)^{1/2} \tag{11.7.25}$$

These scales are commonly referred to as the Kolmogorov or microscales. They are related to the mean flow scales as follows:

$$\Pi \sim \frac{\hat{u}^3}{\hat{l}} \sim \frac{\nu\tilde{u}^2}{\tilde{l}^2} \sim \epsilon \tag{11.7.26}$$

where Π is the production rate of turbulent energy by Reynolds stresses. Equation (11.7.26) states that the rate of turbulent production equals the rate of dissipation. This equality applies in steady, homogeneous, pure shear flows. However, production and dissipation do not balance in most shear flows, although they are nearly always of the same order of magnitude.

Scales in buoyancy-induced turbulent flows. For buoyancy-induced flows it is more appropriate to express the scales discussed above in terms of buoyancy. Let the length and velocity scales for the mean flow and the fluctuations be denoted by $\hat{l}_b$, $\hat{u}_b$ and $\tilde{l}_b$ and $\tilde{u}_b$. In addition, let $\hat{\vartheta}$ and $\tilde{\vartheta}$ represent the characteristic temperature scale for mean flow and the fluctuations, respectively, where $\hat{\vartheta}$ is typically the root mean square fluctuation $\sqrt{\overline{t'^2}}$.

To obtain relationships for $\tilde{l}_b$, $\tilde{u}_b$, and $\tilde{\vartheta}$, Eq. (11.7.26) for the kinetic equilibrium, after including the buoyancy production, may be rewritten as

$$\Pi_b \sim g\beta\hat{\vartheta}\hat{u}_b \sim \frac{\hat{u}_b^3}{\hat{l}_b} \sim \frac{\nu\tilde{u}_b^2}{\tilde{l}_b^2} \sim \epsilon_b \tag{11.7.27}$$

The thermal equilibrium for an arbitrary Prandtl number is

$$\Pi_b \sim \hat{\vartheta}\hat{u}_b\frac{\hat{\vartheta}}{\hat{l}_b} \sim \frac{\nu}{\Pr}\frac{\tilde{\vartheta}^2}{\tilde{l}_b^2} \sim \epsilon_b \tag{11.7.28}$$

For the eventual isotropic end of thermal equilibrium, $\hat{l}_b \to \tilde{l}_b$, $\hat{u}_b \to \tilde{u}_b$, and $\hat{\vartheta} \to \tilde{\vartheta}$. Then the balance between buoyant production, $g\beta\tilde{u}_b\tilde{\vartheta}$, and viscous dissipation, $\nu\tilde{u}_b^2/\tilde{l}_b^2$, in Eq. (11.7.27) gives

$$\tilde{u}_b \sim \frac{g\beta\tilde{\vartheta}\tilde{l}_b^2}{\nu} \tag{11.7.29}$$

From Eq. (11.7.28)

$$\tilde{\vartheta}\tilde{u}_b\frac{\tilde{\vartheta}}{\tilde{l}_b} \sim \frac{\nu}{\Pr}\frac{\tilde{\vartheta}^2}{\tilde{l}_b^2}$$

or

$$\tilde{u}_b \sim \frac{\nu}{\Pr}\frac{1}{\tilde{l}_b} \tag{11.7.30}$$

From Eqs. (11.7.29) and (11.7.30) we obtain

$$\tilde{l}_b \sim \left(\frac{\nu^2}{g\beta\tilde{\vartheta}\mathrm{Pr}}\right)^{1/3} \tag{11.7.31}$$

This relationship applies for fluids Pr ≈ 1 and Pr >> 1. For fluids with Pr ≈ 1

$$\tilde{l}_b \sim \left(\frac{\nu^2}{g\beta\tilde{\vartheta}}\right)^{1/3} \tag{11.7.32}$$

For fluids with Pr << 1 the appropriate length scale (see Arpaci and Larsen, 1984) is

$$\tilde{l}_b \sim \left(\frac{\nu^2}{g\beta\tilde{\vartheta}\mathrm{Pr}^2}\right)^{1/3} \tag{11.7.33}$$

A knowledge of these scales is useful in the development of turbulence models and in a conceptual understanding of heat transfer relations.

The above treatment of developed turbulent transport assumed isotropic turbulence, that is, uniformity of turbulent characteristics in all the directions. However, in recent years there has been a growing interest in large-scale coherent structures that are convected downstream into the flow. Studies have considered both naturally occurring large-scale structures as well as those induced through controlled excitation; see, for instance, the papers by Zaman and Hussain (1981, 1984).

11.7.1 Turbulent Boundary Layer Flow Adjacent to a Heated Vertical Surface

This flow has been analyzed by a number of investigators. The boundary layer form of the governing equations for mean momentum and energy transfer may be obtained from Eqs. (11.7.10) and (11.7.18). Using ϵ_m and ϵ_t from Eqs. (11.7.14) and (11.7.24), these are

$$\frac{\partial \bar{u}}{\partial x} + \frac{\partial \bar{v}}{\partial y} = 0 \tag{11.7.34}$$

$$\bar{u}\frac{\partial \bar{u}}{\partial x} + \bar{v}\frac{\partial \bar{u}}{\partial y} = g\beta(\bar{t} - t_\infty) + (\nu + \epsilon_m)\frac{\partial^2 \bar{u}}{\partial y^2} \tag{11.7.35}$$

$$\bar{u}\frac{\partial \bar{t}}{\partial x} + \bar{v}\frac{\partial \bar{t}}{\partial y} = (\alpha + \epsilon_t)\frac{\partial^2 \bar{t}}{\partial y^2} \tag{11.7.36}$$

The boundary conditions for an isothermal surface condition remain the same, Eqs. (3.3.4) and (3.3.5), with u, v, and t replaced by $\bar{u}$, $\bar{v}$, and $\bar{t}$, respectively.

Closure problem. The above formulation in terms of mean quantities is incomplete. There are three equations in three mean flow unknowns, $\bar{u}$, $\bar{v}$, and $\bar{t}$. There are also two other unknowns ϵ_m and ϵ_t or, equivalently, the Reynolds stress and Reynolds flux, respectively. Recall that the formulation for instantaneous turbulent transport, Eqs. (11.7.4)–(11.7.6) is complete since the number of unknowns matches the number of governing equations. Note, however, that these equations may not be solved since universal initial and boundary conditions are not known. The appearance of extra unknowns, as ϵ_m and ϵ_t or as the Reynolds stress and Reynolds flux, which results from the averaging of the instantaneous formulation, is the closure problem of turbulent transport calculations. A solution is possible if a formulation for the Reynolds stress and Reynolds flux is known.

Most of the analysis of turbulent transport is now based on solving the equations of the mean turbulence, subject to a set of closure hypotheses suitable for computation. This is discussed later. However, the earlier analytical work was done by integral methods and is discussed next.

Integral method solutions. The earliest such analysis for buoyancy-driven flows is due to Eckert and Jackson (1950). The following velocity and temperature profiles were assumed:

$$\frac{u}{u_1} = \left(\frac{y}{\delta}\right)^{1/7}\left(1 - \frac{y}{\delta}\right)^{1/4} \tag{11.7.37}$$

$$\frac{t - t_\infty}{t_0 - t_\infty} = 1 - \left(\frac{y}{\delta}\right)^{1/7} \tag{11.7.38}$$

where u_1 is a characteristic velocity given by $u_{\max} = 0.537u_1$. Using the frictional effect characteristics in forced convection flows, the mean Nusselt number valid for Prandtl numbers close to unity was obtained as

$$\mathrm{Nu}_x = 0.0246\mathrm{Gr}^{2/5}(\mathrm{Pr})^{7/5}[1 + 0.494(\mathrm{Pr})^{2/3}]^{-2/5} \tag{11.7.39}$$

Following a similar integral approach, Bayley (1955) extended the above analysis to liquid metals. For large Gr, 10^{10}–10^{15}, the following expression was given:

$$\mathrm{Nu} = 0.08\mathrm{Gr}^{1/4} \tag{11.7.40}$$

Both of these results agree well with the available heat transfer data. However, the assumed velocity and temperature profiles are quite approximate, particularly the $\frac{1}{7}$ power law in Eqs. (11.7.37) and (11.7.38). Instead of assuming the temperature and velocity profiles across the boundary layer, Kato et al. (1968) assumed shear stress, heat flux, and eddy diffusivity distributions. The equations were then solved for the velocity and temperature profiles and for the heat transfer coefficient.

Turbulence computational models. Most such work has followed the methods used in forced flows. A number of models have been developed and may be grouped into two categories, called first order and second order. In second-order models, balance equations for the Reynolds stresses are solved together with the equations of motion. These models are still under development. They are cumbersome to use and are not discussed further here.

In first-order models the procedure is to evaluate eddy diffusivity in terms of mean flow quantities. In one approach, eddy viscosity ϵ_m is evaluated in terms of the mixing length $\hat{l}$ and the mean flow velocity. That is,

$$\epsilon_m = \hat{l}^2 \left| \frac{\partial \bar{u}}{\partial y} \right| \tag{11.7.41}$$

where, as mentioned in Section 11.7.1, $\hat{l}$ is generally simply related to a characteristic length of the respective flow. Equation (11.7.41) is the Prandtl mixing length hypothesis. Because of the algebraic nature of Eq. (11.7.41), this is also called the algebraic model.

In another approach, called the one-equation model, the prescribed mixing length is supplemented by the solution of the balance equation for the kinetic energy of the turbulence, K. Then, there are two-equation models in which an estimate of ϵ_m is obtained through the solution of the balance equation for K and its dissipation ϵ. This is obtained together with the solution to the balance equations of the mean flow. These models are discussed by Arpaci and Larsen (1984) and Yang and Aung (1985).

Some of the calculations of buoyancy-induced turbulent flows adjacent to vertical surfaces with first-order models are those of Mason and Seban (1974), Cebeci and Khattab (1975), and Noto and Masumoto (1975). Simple eddy viscosity distributions were used, analogous to those used in forced convection. Nee and Yang (1970) and Plumb and Kennedy (1977) used the K-ϵ model to calculate the eddy viscosity from dynamical equations for turbulence. In the latter study, the approach of Jones and Launder (1972) for forced convection flows was followed. The length scale was taken to be the dissipation length scale. The conservation equations for K, ϵ, and $\overline{t'^2}$ were then solved numerically, along with the turbulent momentum and energy equations. Various transport quantities of interest were calculated and shown to be in good agreement with the available experimental data.

Solutions based on scaling laws. Still another approach in analyzing buoyancy-induced turbulent flows has been to use scaling laws in conjunction with the mean turbulence equations. See, for example, Priestly (1959), Elder (1965), Coutanceau (1969), Piau (1972), and George and Capp (1979). In the last study two distinct regions in the boundary layer were identified—an inner region, in which the mean convection terms are negligible, and an outer region consisting of most of the boundary layer, in which viscous and conduction terms are taken as negligible.

The governing equations for the inner region (near the wall), where the mean convection terms are neglected, reduce to

$$g\beta(t - t_\infty) + \frac{\partial}{\partial y}\left(\nu \frac{\partial \bar{u}}{\partial y} - \overline{u'v'}\right) = 0 \tag{11.7.42}$$

$$\frac{\partial}{\partial y}\left(\alpha \frac{\partial \bar{t}}{\partial y} - \overline{t'v'}\right) = 0 \tag{11.7.43}$$

Integrating Eq. (11.7.42) out to y yields

$$\int_0^y g\beta(t - t_\infty)\, dy + \nu \frac{\partial \bar{u}}{\partial y} - \overline{u'v'} = \frac{\tau_0}{\rho} \tag{11.7.44}$$

where τ_0 is the wall shear stress. Because of the presence of the buoyancy integral in Eq. (11.7.44), the inner region is not a constant-stress layer. Hence, for a turbulent natural convection boundary layer, the wall shear stress is not a fundamental parameter of the flow in the sense that it is imposed on both the inner and outer layers. Integration of Eq. (11.7.43) yields

$$\alpha \frac{\partial \bar{t}}{\partial y} - \overline{t'v'} = -\frac{q''_0}{\rho c_p} \tag{11.7.45}$$

where q''_0 is the wall heat flux. Clearly, the heat flux is constant across the inner layer. Thus, the heat flux is a fundamental parameter, not only for the inner layer but also for the outer layer.

The governing equations for the outer region, where the viscous and conduction terms are neglected, are

$$\bar{u} \frac{\partial \bar{u}}{\partial x} + \bar{v} \frac{\partial \bar{u}}{\partial y} = g\beta(\bar{t} - t_\infty) + \frac{\partial}{\partial y}(-\overline{u'v'}) \tag{11.7.46}$$

$$\bar{u} \frac{\partial \bar{t}}{\partial x} + \bar{v} \frac{\partial \bar{t}}{\partial y} = \frac{\partial}{\partial y}(-\overline{t'v'}) \tag{11.7.47}$$

The universal velocity and temperature profiles were derived in the two regions and then matched in the intermediate region, termed the buoyant sublayer by George and Capp (1979). In this region the velocity and temperature were found to be dependent on the cube root and the inverse cube root of distance from the wall, respectively. In the region next to the wall, linear velocity and temperature variations were found.

11.7.2 Experimental Data and Correlations

Analyses of turbulent flows generally have been confined to simple configurations such as the vertical surface discussed above. For the inclined flows discussed in Chapter 5, the analytical approach remains cumbersome and unreliable. One then relies on experiments to obtain the necessary heat transfer information. Over the years, some heat transfer data for various flow configurations have been obtained. Some of this information has already been presented in the various relevant sections. The results given next include other typical results for several important configurations.

Vertical surfaces. Among the earliest measurements were those by Griffiths and Davis (1922), who measured velocity profiles in the flow adjacent to a flat vertical surface. Saunders (1939) investigated natural convection in water and mercury and gave the following heat transfer correlation for $\mathrm{Ra} > 10^{10}$:

$$\text{Nu} = 0.17\text{Ra}^{1/3} \tag{11.7.48}$$

The results of turbulent heat transfer adjacent to an isothermal surface obtained by Warner and Arpaci (1968) for air show good agreement with Eq. (11.7.48). On the other hand, similar experiments by Cheesewright (1968) for air appear to support Eq. (11.7.39).

For uniform surface heat flux conditions, the results of Vliet and Liu (1969) for water, $3.6 < \text{Pr} < 10.5$, indicate the following correlation:

$$\text{Nu}_x = 0.568(\text{Gr}_x^*\text{Pr})^{0.22} \qquad \text{for } 10^{13} < \text{Gr}_x^*\text{Pr} < 10^{16} \tag{11.7.49}$$

The results of a recent experimental study, also in water, by Qureshi and Gebhart (1978) show good agreement with the above correlation. From data for air, Vliet and Ross (1975) gave the following relation:

$$\text{Nu}_x = 0.17(\text{Gr}_x^*\text{Pr})^{0.25} \tag{11.7.50}$$

For the whole laminar and turbulent range, Churchill and Chu (1975) recommended Eqs. (3.9.5) and (3.9.10) for isothermal and uniform heat flux surfaces, respectively. These two correlations are said to apply to all Pr.

Inclined and horizontal surfaces. For constant heat flux inclined surfaces, Vliet and Ross (1975) showed that for the turbulent regime Eq. (11.7.50) may be used for the upward-facing warm surface. For the downward-facing warm surface, Gr_x^* is replaced by $\text{Gr}_x^* \cos^2\theta$. The generalized equations, Eq. (3.9.10) for a uniform heat flux surface and Eq. (3.9.5) for an isothermal surface, may be modified in a similar manner to predict the turbulent heat transfer rates for flat inclined surfaces.

Several correlations exist for horizontal surfaces. For warm surfaces facing upward, Eq. (5.5.12) may be used to predict turbulent heat transfer rates with reasonable accuracy. That correlation is applicable for horizontal surfaces with different planforms if, as suggested by Goldstein et al. (1973), the characteristic length is taken as the area divided by the perimeter. For warm surfaces facing downward, there appear to be no correlations available for predicting turbulent heat transfer rates.

Cylinders and spheres. One of the earlier correlations for mean Nusselt number from horizontal circular cylinders, due to McAdams (1954), is as follows:

$$\text{Nu} = 0.13\text{Ra}^{1/3} \qquad 10^9 < \text{Ra} < 10^{12} \tag{11.7.51}$$

A more recent one by Churchill and Chu (1975), Eq. (5.5.16), correlates the experimental data quite well.

There are no reliable correlations available for predicting rates of turbulent heat transfer from spheres. Churchill (1983) indicated that Eq. (5.5.20), obtained by adding the conduction Nusselt number value of 2.0 on the right-hand side, correlated the mass transfer data of Schütz (1963) up to $\text{Gr}_{x,C}\text{Pr} = 1.5 \times 10^{10}$. However, the data of Kutateladze (1963) for $10^5 < \text{Ra} < 10^{13}$ deviated from the modified Eq. (5.5.10) at $\text{Ra} \approx 10^8$, and the following expression was proposed for a better fit:

$$\mathrm{Nu} - 2 = \frac{0.589\mathrm{Ra}^{1/4}}{[1 + (0.469/\mathrm{Pr})^{9/16}]^{4/9}} \times \left\{1 + \frac{7.44 \times 10^{-8}\mathrm{Ra}}{[1 + (0.469/\mathrm{Pr})^{9/16}]^{16/9}}\right\}^{1/12} \qquad (11.7.52)$$

11.8 PLUME INSTABILITY, TRANSITION, AND TURBULENCE

Free plume flows are very different from flows adjacent to a surface. A surface damps disturbances. Also, the image flow of a plane plume, across the plume midplane, may directly interact in disturbance mechanisms. Hence, such free-boundary flows are much less stable, in terms of G, than those adjacent to surfaces. Also, disturbance mechanisms that are asymmetric across the midplane are much less stable than those that are symmetric. A plane plume in air, subject to controlled disturbances of several frequencies, is seen in Fig. 11.8.1. The midplane temperature t_0 was shown in Chapter 3 to decrease downstream as $x^{-3/5}$. The velocity increases as $x^{1/5}$. The local flow parameter is again G, as for a vertical surface.

11.8.1 Instability Analysis

Two-dimensional disturbances were postulated by Pera and Gebhart (1971) as in Eqs. (11.2.26) and (11.2.27). The stability equations in terms of Φ and s are again Eqs. (11.2.30) and (11.2.31). The coupled base flow temperature and stream functions, ϕ and f, are found from Eqs. (11.2.22)–(11.2.24). The remote boundary conditions in ϕ and s are still the same, Eq. (11.2.32). The other three conditions admit the possibilities of disturbance motion at $\eta = 0$ and of disturbances being symmetric or nonsymmetric about $\eta = 0$. An extreme of nonsymmetry is complete asymmetry. This mode was found to be less stable, as seen in Fig. 11.8.1. The resulting neutral stability curve for air is seen in Fig. 11.8.2, where ω, as defined before in Eq. (11.2.29), is $2\pi F\delta/U_c$.

The first values of G for instability are predicted to be very low, an order of magnitude less than for flows adjacent to surfaces. The stability analysis was repeated by Haaland and Sparrow (1973a), retaining two of the several terms excluded in the conventional approximations as set forth in Section 11.2. A more consistent analysis followed by Hieber and Nash (1975). These results are compared in Fig. 11.11.1. The question of the consistency of higher-order approximations in stability analysis for vertical buoyancy-induced flows is considered in Section 11.11.

The paths that disturbances of a given frequency follow downstream are indicated in Fig. 11.8.2. The particular frequencies shown are in hertz, for experiments in air at a source strength per unit length, Q, of 56.3 W/m. These paths

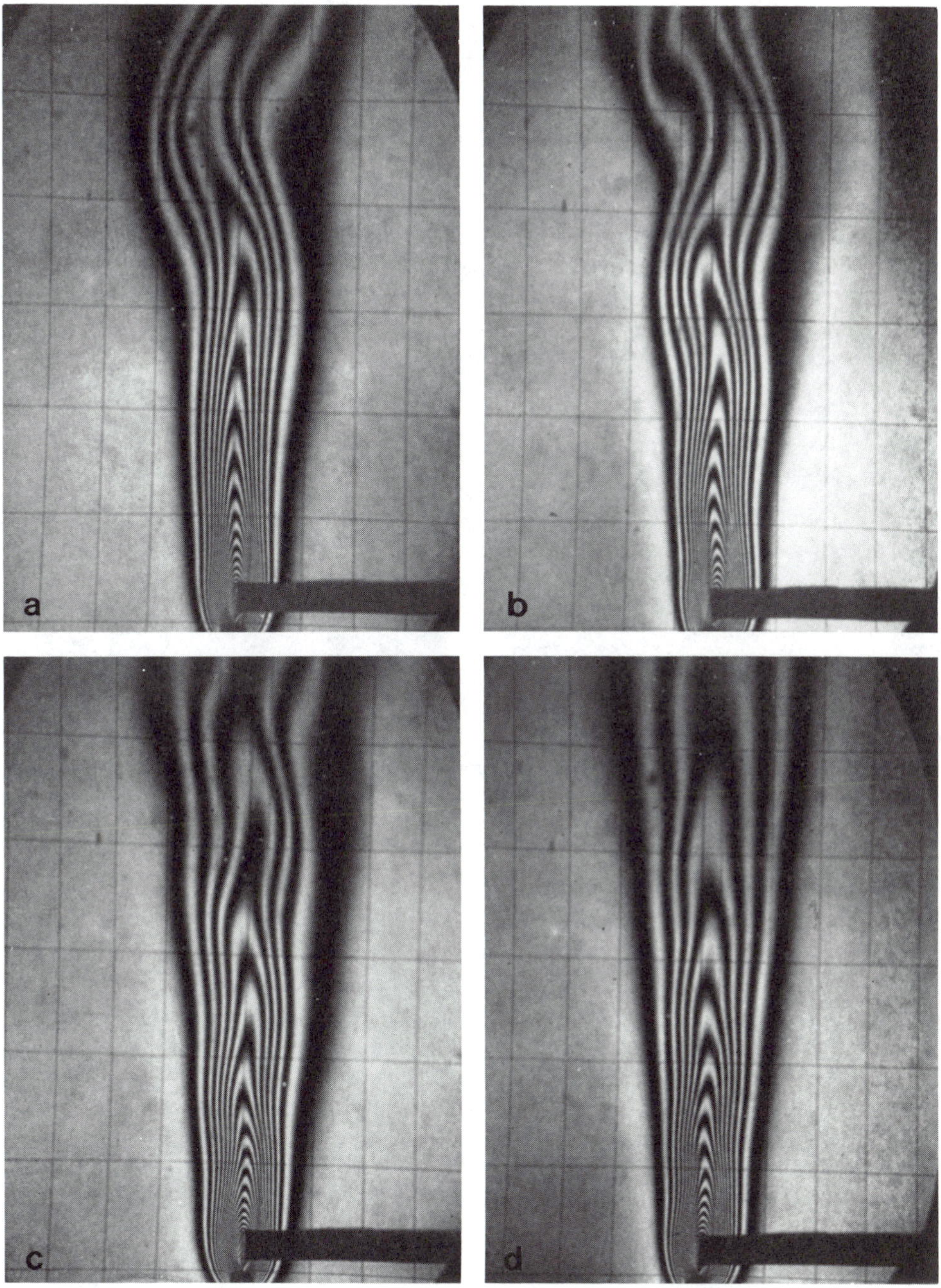

Figure 11.8.1 Plumes perturbed with sinusoidal disturbances at several frequencies for air at atmospheric pressure: (*a*) 2.4 Hz; (*b*) 3.6 Hz; (*c*) 5.1 Hz; (*d*) 7.0 Hz. $Q' = 56.3$ W/m; wire length, 15.3 cm; and wire diameter, 0.013 cm. *(Reprinted with permission from* Int. J. Heat Mass Transfer, *vol. 14, p. 975, L. Pera and B. Gebhart. Copyright © 1971, Pergamon Journals Ltd.)*

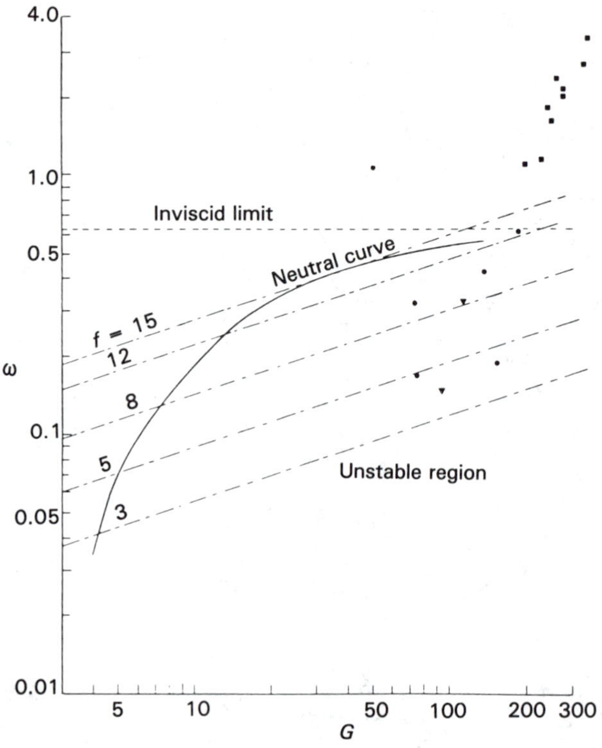

Figure 11.8.2 Computed neutral stability curve for a plane plume. Constant-frequency contours are shown for air for $Q = 56.3$ W/m. *(From Pera and Gebhart, 1971.)* Data of highest frequency velocity disturbances: (●) $Q = 65.9$ W/m; (■) $Q = 65.9$ W/m (turbulent flow); and (▼) $Q = 3.1$ W/m. *(Reprinted with permission from* Int. J. Heat Mass Transfer, *vol. 18, p. 513, R. G. Bill, Jr., and B. Gebhart. Copyright © 1975, Pergamon Journals Ltd.)*

are different from those for vertical flows adjacent to surfaces, recall Fig. 11.1.3. The base flow amplifies all frequencies below a certain limit, yet all are eventually stable. Of course, this does not happen in an actual plume. Other linear and nonlinear mechanisms intervene for some of the amplified frequencies.

Experiments by Pera and Gebhart (1971) tested these stability predictions. The interferograms of Fig. 11.8.1 show the extent of the thermal boundary region. Disturbances arise at the vibrator seen near the plume source, and those of lower frequency are very strongly amplified. This is in very good agreement with the predictions of Fig. 11.8.2.

Calculated spatial amplification rates $-\alpha_i$ are very large compared to those in flows adjacent to surfaces. The low levels of G for instability and the high amplification rates indicate that other and important nonlinear effects very quickly become important downstream. However, the calculated downstream range for growth is very short. As a result, even the relatively high values of $-\alpha_i$ result in maximum values of A of less than about 2.0 for the disturbance frequencies that disrupt such plumes. Thus, these plumes are relatively much less stable than vertical flows adjacent to surfaces.

11.8.2 Transition and Turbulence

There has been much study of plume transport beyond the initial laminar flow. Far-field turbulent free-boundary buoyant flows are treated separately in Chapter 12. However, plane plume transition is considered here. Forstrom and Sparrow (1967) generated flows sufficiently vigorous to disrupt the laminar patterns. The first appearance of turbulence, from thermocouple measurements, was taken as the beginning of transition. Characterizing the local flow vigor by a local Grashof number based on the heat input rate, this occurred at a flow Grashof number, $\mathrm{Gr}_{Q,x}$ of 5×10^8, where Q is the line source strength, per unit length:

$$\mathrm{Gr}_{Q,x} = \frac{gx^3}{\nu^2} \frac{\beta Q}{\rho c_p \nu} \tag{11.8.1}$$

Recall that in Chapter 3, Q was taken as half the total energy input to the plume. At the highest heating rate used, and far downstream, turbulence was often seen. A time-average temperature profile showed a thickening of the flow region with respect to laminar flow. It was concluded that full turbulence occurred at $\mathrm{Gr}_{Q,x} = 5 \times 10^9$. Only this single data point was measured in what was taken to be a turbulent region. No comparisons could be made with the downstream temperature decay for turbulent plumes as predicted by Zel'dovich (1937). Even though the flow was judged to be primarily turbulent, the maximum centerline temperature of the plume still closely followed the functional dependence on heat flux predicted by laminar theory. Temperature levels for laminar flow vary as $Q^{4/5}$, while in turbulent flow, either plane or axisymmetric, the variation is approximately as $Q^{2/3}$.

The turbulent data of Rouse et al. (1952) and Lee and Emmons (1961) are difficult to interpret as line source plumes since the plumes were generated by a row of gas burners and by burning liquid fuel in a channel, respectively. The sources were of appreciable size and introduced initial momentum flux, diffusing chemical species of different molecular weights, initial disturbances, and nonuniformity of energy production rate. Additional mechanisms may also have affected the observations of Miyabe and Katsuhara (1972) with spindle oil. Another mode of instability was seen and transverse sinusoidal oscillations were reported across the span of the plume.

In an experiment in atmospheric air, Bill and Gebhart (1975) used fine thermocouple and hot-wire anemometer probes with a 20-cm-aperture interferometer to study a plane plume flow subject only to naturally occurring disturbances. The measured disturbance frequencies were in accord with the predictions of linear stability theory. Increasing local Grashof numbers were obtained either by increasing the heat input or by moving the probes downstream. Velocity disturbance signals were decomposed and analyzed. Somewhat surprisingly, it was found that all resulting frequency components fell in the amplified region of the stability plane, even to the end of transition. This indicates that linear growth is important even in regions of large disturbance amplitude, as in flows generated adjacent to a vertical surface.

The visualizations indicated that the flow region, taken as turbulent bursts by Forstrom and Sparrow (1967), consisted of large-amplitude two-dimensional sinusoidal disturbances, with superimposed higher-frequency disturbances. These sometimes disappeared downstream, leaving a completely laminar boundary region. Farther downstream such disturbances became more frequent. Eventually the boundary layer broke down completely. This corresponds to the condition taken by Forstrom and Sparrow (1967) as complete turbulence. However, just upstream, the maximum instantaneous midplane temperature was still accurately predicted by the laminar theory.

Some of the characteristics of transition adjacent to a vertical surface, for example, as seen by Jaluria and Gebhart (1974), are paralleled by these plume data. Two-dimensional disturbances amplified selectively. Three-dimensional effects became apparent. However, a delayed thermal transition effect was not observed at $\mathrm{Pr} = 0.7$. After a period of intermittency the flow adjusted to turbulent parameters.

In the experiments of Bill and Gebhart (1975), plumes were generated inside a large enclosure from horizontal electrically heated wires of 25.4, 15.5, 5.1, and 2.5 cm length L with L/D values of 741, 445, 400, and 400, where D is the wire diameter. These different plume spans and levels of energy input resulted in downstream plume behavior varying from that of a plane plume toward that of an axisymmetric one.

Plume midplane temperature variation. A measure of the downstream laminar and turbulent transport in a plume is the nature of the decay of the temperature field, due to entrainment. This is expressed in terms of the actual midplane or axis fluid local temperature $t_0(x)$. For the plane or axisymmetric plume, P or A, either laminar or turbulent flow, L or T, may arise. The four downstream variations are written below in terms of a general temperature variable T, defined as follows, where $I = g(\mathrm{Pr})$ is the integral of $\phi f'$ across the boundary region:

$$T = \frac{\sqrt{32}\,(t_0 - t_\infty)(\mu c_p I)}{Q} \tag{11.8.2}$$

The centerline temperature dependence on the distance x downstream for the laminar and turbulent, plane and axisymmetric, plumes are given by

$$\text{Laminar plane, LP:} \qquad T = \frac{1}{\sqrt[4]{\mathrm{Gr}_x}} \tag{11.8.3}$$

$$\text{Turbulent plane, TP:} \qquad T \propto \frac{I}{\sqrt{\mathrm{Gr}_x}} \tag{11.8.4}$$

$$\text{Laminar axisymmetric, LA:} \qquad T \propto \sqrt{\frac{Q}{\mathrm{Gr}_x}} \tag{11.8.5}$$

Turbulent axisymmetric, TA: $$T \propto \frac{\sqrt{Q}}{\sqrt[4]{\mathrm{Gr}_x^5}} \tag{11.8.6}$$

The value of the Grashof number Gr_x in Eqs. (11.8.3) and (11.8.4) is based on $t_0 - t_\infty = d(x)$ resulting from Q, as for steady laminar boundary layer flow. A similar procedure is used in Eqs. (11.8.5) and (11.8.6), using Q to calculate $t_0 - t_\infty$ from laminar theory. Recall that Q is the energy input per unit span in a plane plume and is the total energy input for an axisymmetric plume. Comparisons of actual data with these trends were used to infer downstream plume transport. Typical examples of such transport are seen in Fig. 11.8.3 at two heat input conditions and different locations downstream.

The time variations of downstream plume midplane velocity and temperature were found to be strongly coupled, as would be surmised from the interferogram. Large fluctuations resulted from swaying and oscillations along the plume span. However, the plume remained primarily laminar. This would be expected from Fig. 11.8.3a, where the listed local value of G applies at the level of the arrow. In Fig. 11.8.3b the flow is apparently still laminar, despite the large spanwise distortion seen.

With increasing Grashof number, additional small-amplitude, higher-frequency disturbances appeared. Simultaneously, large unsteady and wavelike disturbances are visible in Fig. 11.8.3c immediately downstream of the arrow marking the local value $G = 186$. These are taken as the first sign of local turbulence, that is, the beginning of transition.

In the following transition region, the passage of disturbances was followed by relaminarization. Yet farther downstream, disturbance frequency and amplitude increased. The end of transition was taken as the location in x at which the mean flow boundary layer thickened and no relaminarization occurred. Beyond this, the sensor outputs were dominated by high-frequency components. The turbulent condition in Fig. 11.8.3d is a thickened boundary layer and a chaotic temperature field.

Since the flow was alternately laminar and turbulent early in transition, it was characterized in terms of the maximum measured instantaneous local temperature $t_0(x)$. These temperature trends downstream were then compared with Eqs. (11.8.2)–(11.8.6) to determine changing flow regimes.

Comparisons of measurements and calculations for laminar plane plumes in Section 3.7 indicated that the measured T was actually about 15% below Eq. (11.8.3). See Fig. 11.8.4. This deficiency, as discussed in Section 3.7, arises from flow generated upstream of the flow-generating electrically heated wire. Since such onflow is not included in the boundary layer formulation, which restricts the domain to $x \geq 0$ for $Q > 0$, this effect on T remains in downstream measurements.

In Fig. 11.8.4 it is seen that laminar transport continues often to penetrate, completely unchanged, into the transition region. The measured instantaneous temperature maxima t_0 continue to agree with the "corrected" LP (15% low) theory. Behavior farther downstream, in Fig. 11.8.5, for a plume with an initial span of only 25.4 cm is very different. The downstream LP, TP, TA, and LP (15%

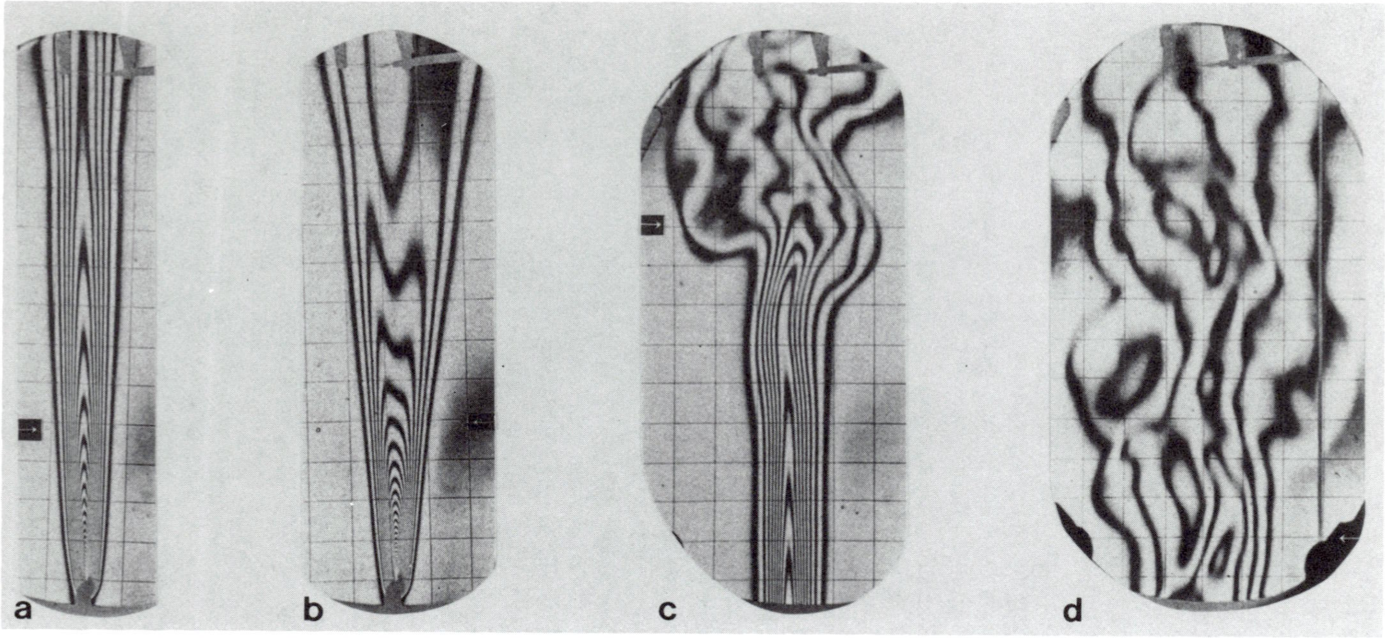

Figure 11.8.3 Interferograms of flow instability in two-dimensional plumes for $L = 25.4$ cm; $G = 68.8$, 68.8, 186.0, and 228.0 at arrow locations, respectively, which are $x = 5.1$, 5.1, 20.3, and 30.5 cm, respectively. The corresponding Q values are 50, 50, 98.1, and 98.1 W/m. *(Reprinted with permission from* Int. J. Heat Mass Transfer, *vol. 18, p. 513, R. G. Bill, Jr., and B. Gebhart. Copyright © 1975, Pergamon Journals Ltd.)*

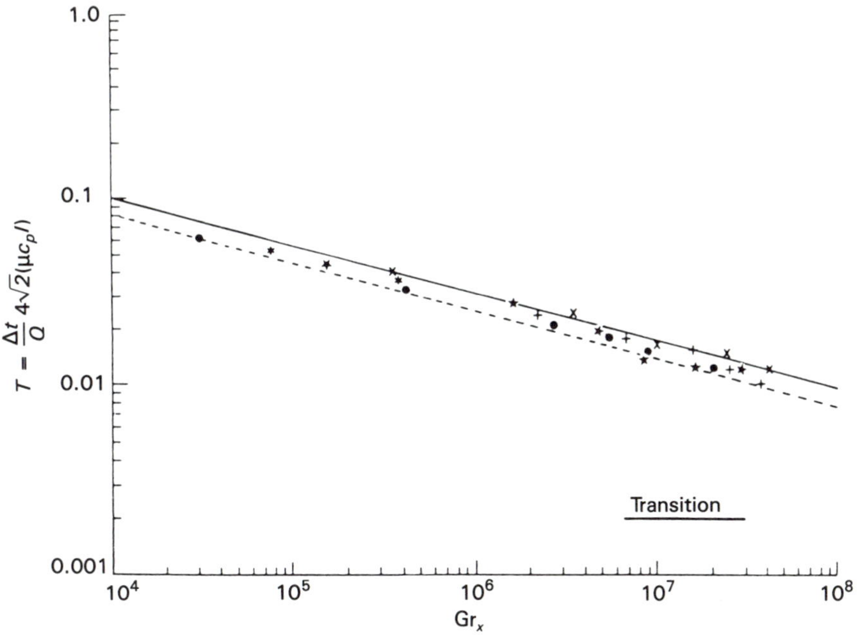

Figure 11.8.4 T vs. Gr_x data for laminar plane plumes in air; (——) LP; (– – –) experimental correction; (●) Forstrom and Sparrow (1967); (×) Brodwicz and Kierkus (1966); (+, ×) Bill and Gebhart (1975). *(Reprinted with permission from* Int. J. Heat Mass Transfer, *vol. 18, p. 513, R. G. Bill, Jr., and B. Gebhart. Copyright © 1975, Pergamon Journals Ltd.)*

low) trends, compared with the measurements, indicate that all but the first two data points are turbulent flow; that is, they diverge downward from the LP corrected trend. The variation of T is also shown for the turbulent plane and axisymmetric plumes, TP and TA. Since the constants of proportionality in Eqs. (11.8.4) and (11.8.6) are not known in general, these curves were placed to best agree with the data.

The data for the turbulent region do not correlate well with either theory. Immediately after the end of transition, at perhaps $Gr_x = 3 \times 10^7$, they conform most closely to TP plume behavior. Farther downstream the slope of the data decreases further to the TA plume trend. The theories for turbulent plumes had been derived for mean flow values. Neither their accuracy nor their applicability to temperature maxima is established.

Similar measurements for $L = 15.3$ cm indicated that transition is complete after $Gr_x = 7 \times 10^7$. For $L = 5.1$ cm the deviation, from the "corrected" LP theory, did not occur until about 2×10^8. Thereafter, the plane turbulent plume trend was followed. Surprisingly, for $L = 2.5$ cm the first deviation appeared to occur at $Gr_x = 1.5 \times 10^8$.

Disturbance growth and limits of transition. The other important aspect of such transitions is how the predictions of stability theory correspond to actual

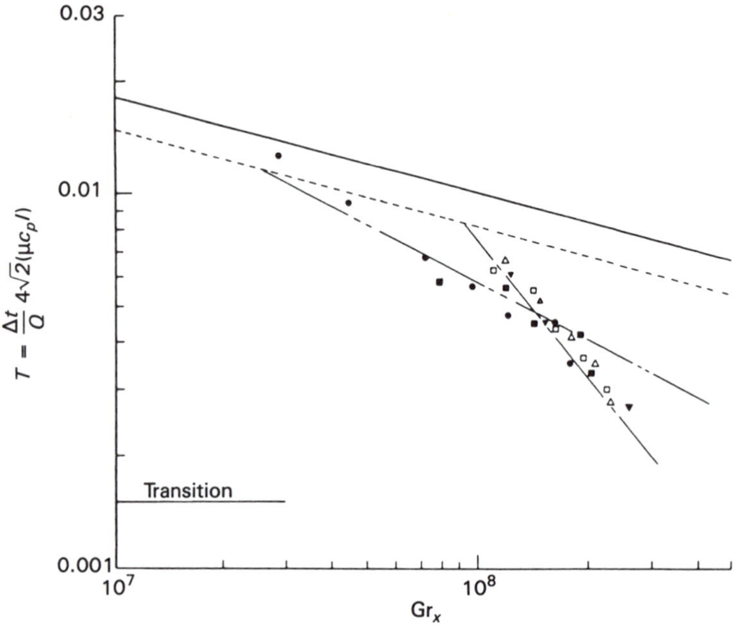

Figure 11.8.5 T vs. Gr_x data for turbulent plumes from Bill and Gebhart (1975). (——) Laminar theory, plane plume; (————) turbulent theory, plane plume; (———) turbulent theory, axisymmetric plume; (–––) experimental correction. (●) $Q = 50$ W/m, (■) $Q = 63.3$ W/m; (△) $Q = 75$ W/m; (□) $Q = 84.6$ W/m; (▼) $Q = 98.1$ W/m. *(Reprinted with permission from* Int. J. Heat Mass Transfer, *vol. 18, p. 513, R. G. Bill, Jr., and B. Gebhart. Copyright © 1975, Pergamon Journals Ltd.)*

disturbance growth and transition. Disturbance spectra, down to 2.5 Hz at different Gr_x values for a given plume, indicated which components had been amplified. The resulting points, in ω and G, are shown in Fig. 11.8.2. Clearly, disturbance energy is found at increasingly higher frequencies downstream. Also, with one exception, all disturbances up to $G = 194$ appear to have traversed the amplified region. The end of transition was at about $G = 208$. Thereafter, energy is at higher frequencies, indicating the kind of nonlinear mechanisms found by Jaluria and Gebhart (1973) in flows adjacent to a vertical surface. The spread of frequencies for $G < 208$ does not indicate a selected band but a one-sided process. The much higher frequencies indicate a conversion of disturbance energy.

Finally, the estimates of transition limits of Bill and Gebhart (1975) and Forstrom and Sparrow (1967) are compared. For the beginning of transition, the values are $Gr_{Q,x} = 11.2 \times 10^8$ from the interferometric measurements of Bill and Gebhart and 5.0×10^8 from the thermocouple measurements of Forstrom and Sparrow. This discrepancy may be due in part to the insensitivity of the integrated interferometer output to small local disturbances. For the end of transition, the value of Bill and Gebhart (1975) is $Gr_{Q,x} = 7.9 \times 10^9$, compared to a single data point at 5×10^9 of Forstrom and Sparrow. The criterion for the latter value was a thickened temperature profile, which does not define a precise point for the completion of transition. The latter data indicate that such a local flow was probably still within the transition region.

In summary, the measurements of Bill and Gebhart indicated that the bounds of the transition regime for the sources with $L = 25.4$ and 15.3 cm, converted to Gr_x, are $\mathrm{Gr}_x = 6.4 \times 10^7$ and 2.95×10^8, respectively. The comparison of centerline temperature with laminar theory provides a strong standard for the determination of the end of transition. After the complete disruption of the boundary layer, the flow begins to adjust to turbulent parameters and the laminar centerline temperature is no longer achieved. Nonlinear effects have spread disturbance energy to higher frequencies. Turbulence intensity and scale then begin to decrease.

11.9 INSTABILITY OF COMBINED BUOYANCY MODE VERTICAL FLOWS

Mass transfer also gives rise to buoyancy, and simultaneous transport of both thermal energy and one or more chemical constitutents frequently occurs, as discussed in Chapter 6. The additional buoyancy effect may cause major alterations of the stability characteristics of the resulting flow. The very complex interactions that arise between disturbances in velocity, temperature, and concentration follow from two important factors. First, there may be opposing or aiding buoyancy force components. Second, the chemical species and thermal transport layers are of different spatial extents if the Lewis number, $\mathrm{Le} = \mathrm{Sc}/\mathrm{Pr} = \alpha/D$, is different from 1.0. Recall Figs. 6.3.2 and 6.3.3. This effect is similar to that of the Prandtl number in expressing the relative extents of the velocity and thermal effects.

With small concentration differences the buoyancy force contribution is again calculated as $\Delta\rho_C = \rho\beta^*(C - C_\infty)$, analogous to the Boussinesq approximation for $\Delta\rho = \rho\beta(t - t_\infty)$, in that the density is assumed to vary linearly with concentration. Then the total buoyancy force $B = g(\Delta\rho_t + \Delta\rho_C) = B_t + B_C$. Recall that the ratio of these two effects, as used in Chapter 6, was

$$\tilde{N} = \frac{\mathrm{Gr}_{x,C}}{\mathrm{Gr}_x} = \frac{\beta^*(C_0 - C_\infty)}{\beta(t_0 - t_\infty)} \tag{11.9.1}$$

where $\tilde{N}$ may be positive, zero, negative, or infinity.

The governing equations for convection and Fickian diffusion of a species of local concentration $C(x, y, \tau)$, along with a general boundary condition, are written below as in Chapter 6:

$$\frac{\partial C}{\partial \tau} + u\frac{\partial C}{\partial x} + v\frac{\partial C}{\partial y} = D\nabla^2 C \tag{11.9.2}$$

$$C(x, \infty, \tau) - C_\infty = 0 \tag{11.9.3}$$

As before, the base flow and disturbance levels are formulated as $C(x, y, \tau) = \bar{C}(x, y) + C'(x, y, \tau)$. The additional buoyancy force associated with

C' and the boundary region base flow and disturbance equations are written, with the three approximations set forth in Section 11.2, as follows:

$$B_C = \bar{B}_C + B'_C \qquad B'_C = \rho_r \beta^* C' \tag{11.9.4}$$

$$\bar{u}\frac{\partial \bar{C}}{\partial x} + \bar{v}\frac{\partial \bar{C}}{\partial y} = D\frac{\partial^2 \bar{C}}{\partial y^2} \tag{11.9.5}$$

$$\frac{\partial C'}{\partial \tau} + \bar{u}\frac{\partial C'}{\partial x} + v'\frac{\partial \bar{C}}{\partial y} = D\nabla^2 C' \tag{11.9.6}$$

The base flow transformations in Section 3.5 may again be used to cast $\bar{C}$ into boundary layer similarity form, Eq. (6.3.16), where $\tilde{C} = (\bar{C} - C_\infty)/(C_0 - C_\infty)$.

The disturbance stream and temperature functions, ψ' and t', remain as defined in Eqs. (11.2.26) and (11.2.27). The disturbance C' is similarly defined in terms of an amplitude function $a(\eta)$:

$$C'(x, y, \tau) = (C_0 - C_\infty)a(\eta)\exp[i(\hat{\alpha}x - \hat{\omega}\tau)] \tag{11.9.7}$$

The equation for $a(\eta)$, from Eq. (11.9.6), becomes identical in form to that for $s(\eta)$, Eq. (11.2.31), namely

$$\left(f' - \frac{\omega}{\alpha}\right)a - \tilde{C}'\Phi = \frac{a'' - \alpha^2 a}{i\alpha \mathrm{Sc} G} \tag{11.9.8}$$

The previous disturbance force-momentum balance, Eq. (11.2.18), must now be augmented with the added buoyancy force component due to C' in Eq. (11.9.4). The result is that the disturbance equation (11.2.30) is the same, except that s' there is replaced by $s' + \tilde{N}a'$. The additional boundary conditions on a are analogous to those on s for an assigned surface condition at $\eta = 0$:

$$a(\infty) = 0 = a(0) \tag{11.9.9}$$

A more general form of the last condition, similar to that given for thermal disturbances in Eq. (11.2.38), is $a(0) = K_c a'(0)$, where K_c depends on the relative chemical species capacity at the surface compared with the adjacent fluid. The condition on $a(0)$ in Eq. (11.9.9) is for a surface of large relative chemical species capacity.

The above formulation for the amplitude functions Φ, s, and a, in terms of parameters α_r, α_i, ω, Pr, Sc, $\tilde{N}$, and G, is of eighth order. This, combined with the much greater complexity that may arise in the base flow with opposed effects when Pr $\neq$ Sc, makes calculations much more difficult. However, some results have been obtained by Boura and Gebhart (1976) for the surface conditions $t(x, 0, \tau) = t_0 =$ const. and $C(x, 0, \tau) = C_0 =$ const., that is, for $n = 0$. The full equations and boundary conditions are given below. Here G is as defined in Eq. (6.3.12) but with $P = 1$ and $Q = 0$ in Eq. (11.9.14) below:

$$\left(f' - \frac{\omega}{\alpha}\right)(\Phi'' - \alpha^2\Phi) - f'''\Phi = \frac{\Phi'''' - 2\alpha^2\Phi'' + \alpha^4\Phi + s' + \tilde{N}a'}{i\alpha G} \tag{11.9.10}$$

$$\left(f' - \frac{\omega}{\alpha}\right)s - \phi'\Phi = \frac{s'' - \alpha^2 s}{i\alpha \mathrm{Pr} G} \tag{11.9.11}$$

$$\left(f' - \frac{\omega}{\alpha}\right)a - \tilde{C}'\Phi = \frac{a'' - \alpha^2 a}{i\alpha \mathrm{Sc} G} \tag{11.9.12}$$

$$\Phi(0) = \Phi'(0) = s(0) = a(0) = \Phi'(\infty) = \Phi(\infty) = s(\infty) = a(\infty) = 0 \tag{11.9.13}$$

$$G = 4\left(\frac{P\mathrm{Gr}_x + Q\mathrm{Gr}_{x,C}}{4}\right)^{1/4} = 4\left(\frac{\mathrm{Gr}_x}{4}\right)^{1/4} \qquad \omega = 2\pi F\left(\frac{16x^2}{\nu G^3}\right) \tag{11.9.14}$$

where F, as before, is the physical frequency of the disturbance.

It is apparent from these relations that $a = s$ if Pr = Sc. Then the only effect of mass diffusion is through the coefficient of the buoyancy force term $(1 + \tilde{N})s$ in Eq. (11.9.10). Opposing buoyancy effects mean only that $\tilde{N}$ is negative and $1 + \tilde{N}$ is reduced. Should $1 + \tilde{N}$ be found to be negative, the assumed positive direction of x is reversed. In any event, the proper stability plane is that which applies for thermally caused buoyancy. However, the interpretation is different. The coordinates previously generalized in terms of Gr_x must now be interpreted in terms of $\mathrm{Gr}_x(1 + \tilde{N})$.

The interesting matters are the additional effects that arise when Pr $\neq$ Sc, that is, for Le $\neq$ 1. A buoyancy force reversal may even arise for $\tilde{N} < 0$, resulting in large effects on the velocity distribution. Stability calculations have been made by Boura and Gebhart (1976) for Pr = 0.7 and Sc = 0.2, 0.94, and 2.0. The remaining parameter is $\tilde{N}$ in Eq. (11.9.1). Figure 11.9.1 shows the effect of $\tilde{N}$ on neutral stability for Sc = 0.94 in terms of ω and G as defined in Eq. (11.9.14), with $P = 1$ and $Q = 0$. Curves for $\tilde{N}$ from -0.8 to $+0.5$ are seen. Increasing mass transfer buoyancy upward, that is, $\tilde{N} > 0$, appears to strongly destabilize the flow. This is because G and ω in Eq. (11.9.14) are not an accurate measure of the actual total buoyancy force. However, each of these neutral curves strongly suggests the very sharp selective disturbance amplification found in a purely thermally driven flow, that is, the amplification of only certain frequency components.

A more realistic plot is seen in Fig. 11.9.2 in the coordinates ω_1 and G_1. These are based on $P = Q = 1$ in Eq. (11.9.14) as follows:

$$G_1 = 4\sqrt[4]{\frac{\mathrm{Gr}_x + \mathrm{Gr}_{x,C}}{4}} = G(1 + \tilde{N})^{1/4}$$

$$\omega_1 = \omega(1 + \tilde{N})^{-3/4}$$

These are much more appropriate. The small effect on stability in the $\tilde{N}$ range from -0.2 to $+0.2$ agrees with the conclusions of Gebhart and Pera (1971). A single neutral curve results for Pr = Sc for all $\tilde{N}$ when G_1 is used. The small difference in Pr and Sc is first felt strongly for $\tilde{N} = -0.5$ and very strongly for $\tilde{N} = -0.8$. The effect for $\tilde{N} = -0.8$ is formally due to the singularity of the transformation of ω into ω_1 at $\tilde{N} = -1.0$. This singularity does not actually occur.

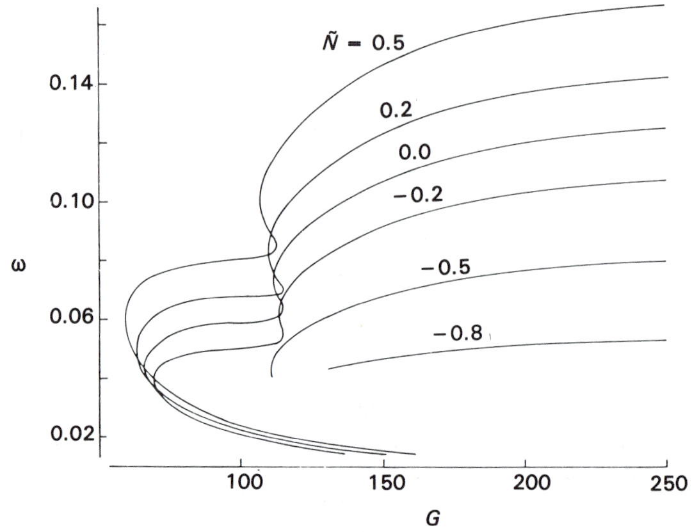

Figure 11.9.1 Neutral curves for flow next to a vertical surface, for Pr = 0.7 and Sc = 0.94 (carbon dioxide in air) in terms of the thermal Grashof number. *(Reproduced by permission of the American Institute of Chemical Engineers, A. Boura and B. Gebhart,* AIChE J., *vol. 22, p. 94. Copyright © 1976, AIChE.)*

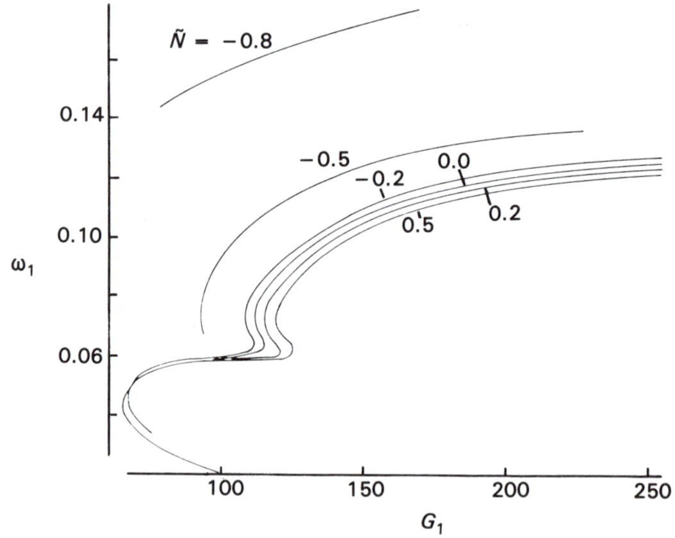

Figure 11.9.2 Neutral curves for flow next to a vertical surface, for Pr = 0.7 and Sc = 0.94 (carbon dioxide in air) in terms of the combined Grashof number. *(Reproduced by permission of the American Institute of Chemical Engineers, A. Boura and B. Gebhart,* AIChE J., *vol. 22, p. 94. Copyright © 1976, AIChE.)*

Taking $P = Q$ for Le $\neq 1.0$ is not completely representative, since the two transport processes have different spatial extents. Their simple sum does not then represent the actual buoyancy, their actual combined effect. Recall that for Le $= 1.0$ there is no flow for $\tilde{N} = -1.0$, no matter what values are assigned to P and Q.

The downstream, in $G(x)$, amplification rates are given by $-\alpha_i$. The amplitude ratio is again calculated as in Eq. (11.2.34), where $4A$ for the flux condition is replaced by $3A$ for the uniform temperature condition, $n = 0$. This amplitude growth calculation is again approximate to the extent that the form of the disturbance amplitude distributions across the boundary region changes downstream with G and is also subject to all other approximations made in preceding sections. The integration is performed in the ω_1, G_1 plane along paths of constant physical frequency F. These paths are $\omega_1 G_1^{1/3}$ = const. The A contours have been calculated for Pr $= 0.7$, for each Sc $= 0.94$, 2.0, and 0.2 or Le $= 1.34$, 2.86, and 0.286, for several values of $\tilde{N}$. The contours have been determined across the frequency band that experiences most rapid amplification in each case.

The results for Sc $= 0.94$, Le $= 1.34$, and $\tilde{N} = 0.5$, 0, and -0.5, seen in Fig. 11.9.3, again show the sharp downstream frequency filtering found for a purely thermally driven flow in Section 11.2 and abundantly corroborated by experiment. With combined buoyancy modes, the disturbances are amplified less rapidly for increasing $\tilde{N}$ in terms of G_1. The values $\tilde{N} = +0.5$ and $\tilde{N} = -0.5$ appear to cause opposite effects of comparable amounts, consistent with a value of Le $= 1.34$. The ω_1 location of the filtered band seems largely independent of

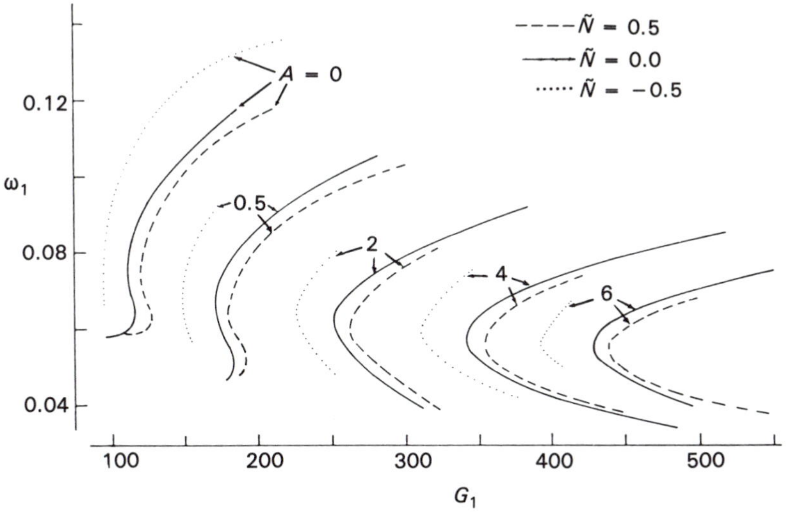

Figure 11.9.3 Downstream disturbance amplification for flow adjacent to a vertical surface, for Sc = 0.94 and Le = 1.34 in terms of ω_1 and G_1. *(Reproduced by permission of the American Institute of Chemical Engineers, A. Boura and B. Gebhart,* AIChE J., *vol. 22, p. 94. Copyright © 1976, AIChE.)*

$\tilde{N}$. Since Le = 1.34, these results do not amount to a demanding test of the effects of combined buoyancy modes on stability and growth mechanisms.

For Le = 2.86, the concentration boundary layer is relatively thin. In the plane in Fig. 11.9.4 the effects of $\tilde{N}$ on stability are much greater. An opposing buoyancy effect destabilizes the flow and an aiding one stabilizes. These large effects for negative $\tilde{N}$ are consistent with those found for Sc = 0.94. Recall that for values of Le > 1 the concentration boundary layer is thinner than the thermal one. Less extensive results for Le = 0.286 are seen in Fig. 11.9.5. The curves indicate that a positive $\tilde{N}$ again stabilizes.

These results indicate stabilization with increasing $\tilde{N}$ for all three values of the Schmidt number. This is rather surprising since this range of Sc spans the condition Le = 1. That is, the concentration gradient layer is both thinner and thicker than the thermal layer over the range. The explanation for this is not clear. There have been no experiments for comparison, even to the extent of measuring favored frequencies. Nevertheless, these results might be expected to be realistic estimates of the stability characteristics of actual flows because of the detailed past successes of linear stability theory compared to many experiments.

11.10 INSTABILITY OF MIXED CONVECTION FLOWS

As shown earlier, linear stability theory predicts that, for natural convection flows adjacent to a flat vertical heated surface, a narrow band of frequencies will be

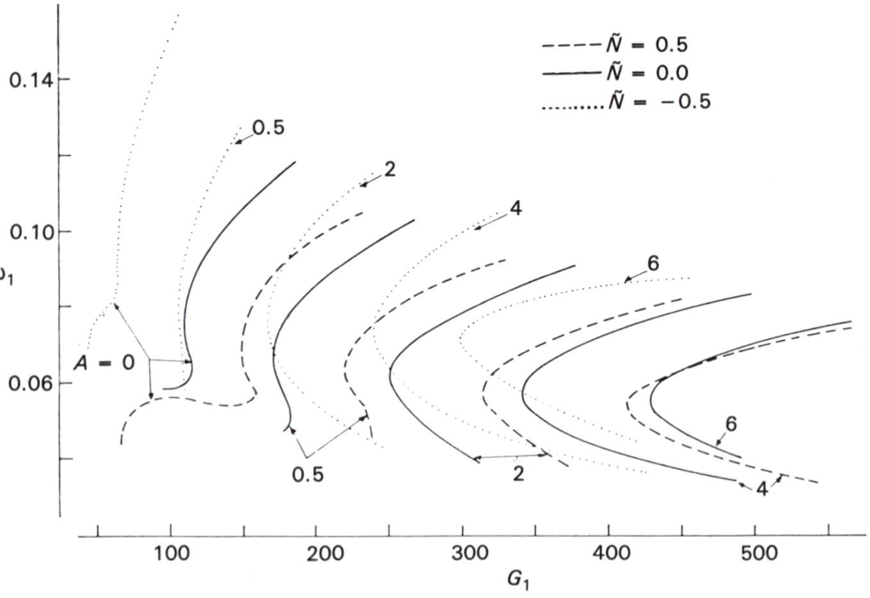

Figure 11.9.4 Downstream disturbance amplification for flow next to a vertical surface, for Sc = 2.0 and Le = 2.86 in terms of ω_1 and G_1. *(Reproduced by permission of the American Institute of Chemical Engineers, A. Boura and B. Gebhart,* AIChE J., *vol. 22, p. 94. Copyright © 1976, AIChE.)*

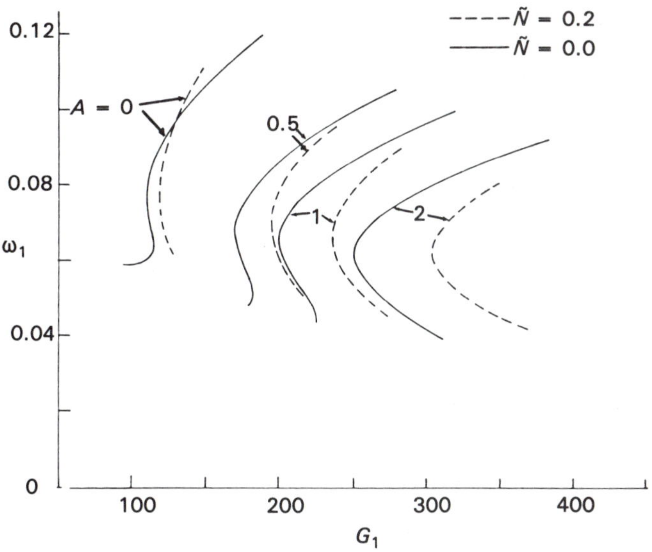

Figure 11.9.5 Downstream disturbance amplification for flow next to a vertical surface, for Sc = 0.2 and Le = 0.286 in terms of ω_1 and G_1. *(Reproduced by permission of the American Institute of Chemical Engineers, A. Boura and B. Gebhart,* AIChE J., *vol. 22, p. 94. Copyright © 1976, AIChE.)*

highly amplified as the disturbances are convected downstream. Measurements by Jaluria and Gebhart (1974) and Mahajan and Gebhart (1979) confirmed the existence of favored frequencies up to the early stages of transition. On the other hand, in forced convection flows, for example, Blasius flow, sharply favored frequencies are not predicted by linear theory. Instead, all frequencies below a certain level are predicted to first amplify and then decay, with the lower frequencies amplifying farther downstream, as discussed by Jordinson (1970).

Because of these fundamental differences in the amplification process, the study of the instability of mixed convection flow is of added interest and complexity. Since in both natural and forced convection flows small disturbances amplify initially in two-dimensional sinusoidal form, the same form of initial amplification may be expected for similar mixed convection flows.

Mucoglu and Chen (1978) studied the stability of mixed convection flows adjacent to a vertical isothermal surface. Both aiding and opposing effects were treated for fluid Prandtl numbers of 0.7 and 7.0. A forced flow was assumed to be perturbed by buoyancy effects. Increased buoyancy in aiding flows was found to stabilize the flow, and in opposed flows to destabilize it. Later Chen and Mucoglu (1979) and Chen and Moutsoglou (1979) analyzed the wave instability of mixed convection flows over horizontal and inclined surfaces.

Carey and Gebhart (1983) analyzed combined forced and natural, aiding and opposing flows adjacent to a flat vertical uniform flux surface. They considered relatively small forced convection effects, that is, a dominant natural convection effect, typical of far downstream locations. The usual boundary layer and Boussinesq approximations are applied to the base flow equations, as in Section 10.2, leading to Eqs. (10.2.1)–(10.2.3). The boundary conditions for uniform surface heat flux q'' are as follows:

$$y = 0: \qquad u = v = 0 \qquad \frac{\partial t}{\partial y} = -\frac{q''}{k}$$
$$y \to \infty: \qquad u = U_\infty \qquad t = t_\infty \tag{11.10.1}$$

The characteristic velocity, length, and temperature difference are defined as

$$U_C = \frac{\nu G^{*2}}{5x} \qquad \delta = \frac{5x}{G^*} \qquad t_0 - t_\infty = \frac{q''\delta}{k} = \frac{5xq''}{kG^*} \tag{11.10.2}$$

Because of the nonsimilarity of the base flow, the nondimensional stream and temperature functions are defined as

$$\psi = U_C \delta F(x, \eta) = \nu G^*[F_0(\eta) + \epsilon F_1(\eta) + \cdots]$$
$$\phi = \frac{t - t_\infty}{t_0 - t_\infty} = \phi_0(\eta) + \epsilon\phi_1(\eta) + \cdots \tag{11.10.3}$$

where

$$\eta = \frac{yG^*}{5x} \qquad u = \psi_y \qquad v = -\psi_x \qquad \epsilon = \frac{\mathrm{Re}_x}{(\mathrm{Gr}_x^*/5)^{2/5}} \tag{11.10.4}$$

Note that ϵ is proportional to U_∞. The forced convection effect enters the analysis through the functions F and ϕ, from the ambient boundary conditions on u. For the stability calculations, ϵ is rewritten as

$$\epsilon = \frac{R^*}{(G^*)^{3/4}} \qquad \text{and} \qquad R^* = \frac{5\mathrm{Re}_x}{(\mathrm{Gr}_x^*)^{1/4}}, \quad \mathrm{Gr}_x^* = \frac{g\beta q'' x^4}{k\nu^2} \tag{11.10.5}$$

Thus, R^* is independent of x for any given uniform external flow velocity U_∞. The particular value of R^* also depends on other fixed quantities for any given flow. Therefore, the downstream or x dependence of ϵ resides entirely in G^*. This is the parameter in the following stability analysis.

Substituting the above equations into the base flow equation yields the following equations to the first order:

$$F_0''' + 4F_0F_0'' - 3F_0'^2 + \phi_0 = 0$$
$$\phi_0'' + \mathrm{Pr}(4F_0\phi_0' - F_0'\phi_0) = 0$$
$$F_0(0) = F_0'(0) = \phi_0'(0) + 1 = F_0'(\infty) = \phi_0(\infty) = 0$$
$$F_1''' + F_1F_0'' + 4F_0F_1'' - 3F_0'F_1' + \phi_1 = 0 \tag{11.10.6}$$
$$\phi_1'' + \mathrm{Pr}(F_1\phi_0' + 4F_0\phi_1' + 2\phi_1F_0' - F_1'\phi_0) = 0$$
$$F_1(0) = F_1'(0) = \phi_1'(0) = F_1'(\infty) - \frac{1}{5} = \phi_1(\infty) = 0$$

Again, the velocity, temperature, and pressure are each assumed to be the sum of mean and fluctuating components. The disturbances are again taken two dimensional as

$$\psi' = U_C \delta \Phi(\eta) e^{i(\hat{\alpha}x - \hat{\omega}\tau)}$$
$$t' = (t_0 - t_\infty) s(\eta) e^{i(\hat{\alpha}x - \hat{\omega}\tau)} \tag{11.10.7}$$

The values of u, v, and t are substituted into the complete two-dimensional time-dependent governing equations. The base flow quantities are subtracted and the boundary layer and parallel flow assumptions are applied. The following equations for the disturbance amplitude functions $\Phi(\eta)$ and $s(\eta)$ and boundary conditions on Φ and s arise:

$$(\Phi'' - \alpha^2\Phi)\left(F' - \frac{\omega}{\alpha}\right) - \Phi F''' = \frac{1}{i\alpha G^*}(\Phi'''' - 2\alpha^2\Phi'' + \alpha^4\Phi + s') \tag{11.10.8}$$

$$s\left(F' - \frac{\omega}{\alpha}\right) - \Phi\phi' = \frac{1}{i\alpha G^* \mathrm{Pr}}(s'' - \alpha^2 s) \tag{11.10.9}$$

$$\Phi(0) = \Phi'(0) = s(0) = \Phi(\infty) = \Phi'(\infty) = s(\infty) = 0 \tag{11.10.10}$$

where, as before, $\alpha = \hat{\alpha}\delta$ and $\omega = \hat{\omega}\delta/U_C$.

Equations (11.10.8) and (11.10.9) are the usual Orr-Sommerfeld equations for buoyancy flow over a uniform flux surface. The difference here is that F and Φ are also functions of G^* through $\epsilon = \epsilon(R^*, G^*)$. Each physical flow circumstance with given U_∞ and q'' is indicated by a value of R^*. With q'' and R^* given, each downstream location has a corresponding G^*, which affects the stability of the flow as it appears in Eqs. (11.10.8) and (11.10.9).

Numerical solutions were obtained for Pr = 0.733 and 6.7, using the methods of Hieber and Gebhart (1971a) for a range of values of R^*. Again, for convenience, computations were done for a specific value of Ω^* rather than $\omega = 2\pi F$, where $\Omega^* = \omega G^{*1/2}$ is defined in Eq. (11.2.37).

The ratio of the downstream disturbance amplitude at any location L to that at the neutral curve N is

$$\frac{A_L}{A_N} = e^A \qquad A = -\frac{1}{4}\int_{G_N^*}^{G_L^*} \alpha_i \, dG^* \tag{11.10.11}$$

where the integration is at a constant Ω^*. At the neutral curve $A = 0$.

Figure 11.10.1 shows the amplification rate contours resulting for air, Pr = 0.733. Experimental data are also shown. The computations are for $R^* = 0$ corresponding to purely buoyant convection, and for $R^* = 20$, for aiding mixed convection flow, with $\mathrm{Re}_x = \frac{4}{5}G^{*-4/5}$. Comparison of the curves indicates that for aiding flows ($R^* > 0$) the amplification rates are lower than for purely buoyant convection flows. On the other hand, computations for opposing flows, $R^* = -10$, not shown in the figure, indicated higher amplification rates for $R^* = -10$ than for $R^* = 0$. Thus, as expected from physical considerations, an aiding free-

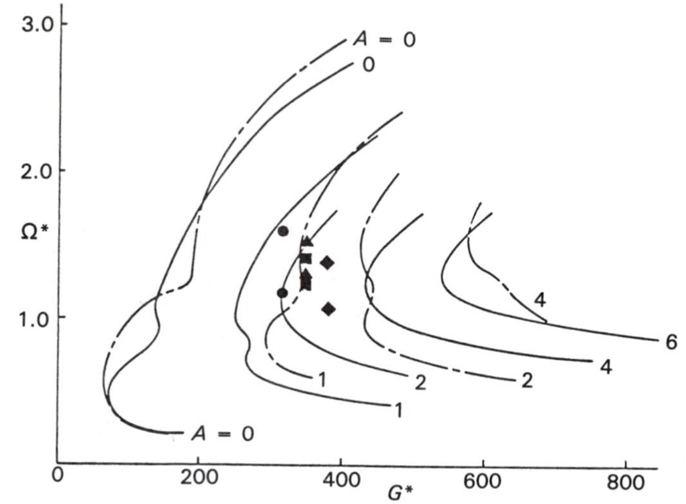

Figure 11.10.1 Characteristic frequencies observed in mixed convection flow along a vertical uniform heat flux surface in air, Pr = 0.733. Experimental conditions (R^*, G^*, ϵ_M) are: (●) 15, 319, 0.20; (■) 22, 350, 0.27; (▲) 20, 351, 0.24; and (◆) 16, 380, 0.19. Also shown are the calculated amplification rate contours for (——) $R^* = 0$ and (– – –) $R^* = 20$. *(Reprinted with permission from V. P. Carey and B. Gebhart,* J. Fluid Mech., *vol. 127, p. 185. Copyright © 1983, Cambridge University Press.)*

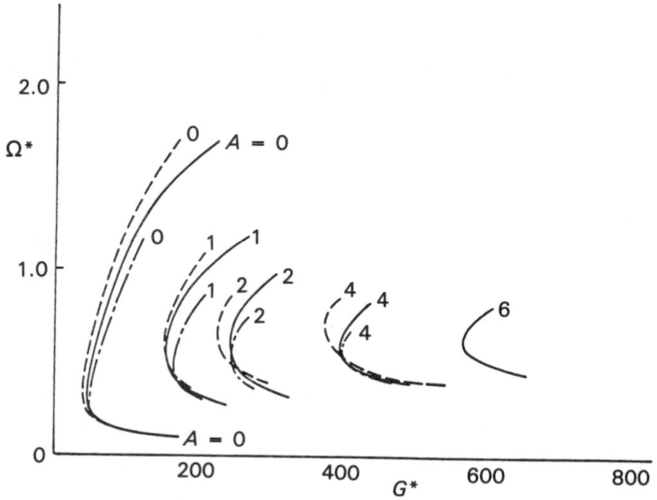

Figure 11.10.2 Amplification rate contours corresponding to (— - —) $R^* = -4$; (——) $R^* = 0$; and (– – –) $R^* = 8$ for Pr = 6.7. *(Reprinted with permission from V. P. Carey and B. Gebhart,* J. Fluid Mech., *vol. 127, p. 185. Copyright © 1983, Cambridge University Press.)*

stream velocity appears to stabilize the flow while an opposing one destabilizes it.

In addition, there are some differences in selective amplification. It is less sharp for $R^* = 20$ than for $R^* = 0$, for example, as seen from curves at $G^* = 350$ and farther downstream at $G^* = 600$. The band of most rapidly amplified frequencies appears to be higher for $R^* = 20$ than for $R^* = 0$.

For $\Pr = 6.7$ the amplification rate contours are shown in Fig. 11.10.2 for $R^* = -4$, 0, and 8. The free-stream velocity is seen to have very little effect on the disturbance growth characteristics. The neutral curve and amplification curves show very little change. Flows in air apparently are much more sensitive to the presence of a free-stream velocity than flows in water.

11.11 HIGHER-ORDER EFFECTS AND OTHER STABILITY ANALYSES

The instability characteristics of the different types of buoyancy-induced flows discussed in Sections 11.2, 11.8–11.10, and 11.12 are based on parallel flow and conventional boundary layer approximations. Using these approximations, Orr-Sommerfeld equations, (11.2.30) and (11.2.31), were obtained from a complete set of stability equations, (11.2.11)–(11.2.13), in effect omitting some terms of $O(G^{-1})$ from the latter equations. Terms of the same order are also neglected in the base flow. These terms may be shown to involve lower-order derivatives than appear on the right-hand side of Eqs. (11.2.30) and (11.2.31) and therefore may be neglected.

For many flow circumstances of interest, this procedure is justified. For example, for a flow adjacent to a vertical uniform flux surface, the location of first instability occurs around $G^* \sim 100$; see Figs. 11.2.1 and 11.2.2. The effect of $O(G^{*-1})$ terms would be small. Farther downstream, at higher G^*, it is even smaller. However, for a plane plume, unstable conditions arise at very low values of G; see the results of Pera and Gebhart (1971) in Section 11.8. The neutral curve extends down to about $G = 3$. Then the effects arising from $O(G^{-1})$ terms may not be negligible. An improved stability analysis of such flows at lower values of G must simultaneously consider both "nonparallel" and other associated approximations. These questions, however, have not been extensively investigated. Only plane plume flows have been analyzed, as discussed next.

11.11.1 Higher-Order Effects in Linear Stability

Some results for free plane plume flows were obtained by Haaland and Sparrow (1973a). They considered only the Orr-Sommerfeld equations, extended to include effects of the streamwise dependence of base flow quantities. These terms are shown underlined in Eqs. (11.11.1) and (11.11.2) below. Equation (11.11.1) is the vorticity equation obtained from Eqs. (11.2.10)–(11.2.12) by linearizing the disturbance quantities and following the standard procedure for obtaining the vorticity equation from the continuity and momentum equations. Here the base flow vorticity is $\bar{\zeta} = (\partial\bar{v}/\partial x) - (\partial\bar{u}/\partial y)$ and the disturbance vorticity is $\zeta' =$

$(\partial v'/\partial x) - (\partial u'/\partial y)$. Equation (11.11.2) is the energy equation derived from Eq. (11.2.13), linearized in disturbance quantities.

$$\frac{\partial \zeta'}{\partial \tau} + \bar{u}\frac{\partial \zeta'}{\partial x} + \underline{u'\frac{\partial \bar{\zeta}}{\partial x} + \bar{v}\frac{\partial \zeta'}{\partial y}} + v'\frac{\partial \bar{\zeta}}{\partial y} = \frac{\partial^2 \zeta'}{\partial x^2} + \frac{\partial^2 \zeta'}{\partial y^2} + g\beta\frac{\partial t'}{\partial y} \qquad (11.11.1)$$

$$\frac{\partial t'}{\partial \tau} + \bar{u}\frac{\partial t'}{\partial x} + \underline{u'\frac{\partial \bar{t}}{\partial x} + \bar{v}\frac{\partial t'}{\partial y}} + v'\frac{\partial \bar{t}}{\partial x} = \alpha\left(\frac{\partial^2 t'}{\partial x^2} + \frac{\partial^2 t'}{\partial y^2}\right) \qquad (11.11.2)$$

The underlined terms are $O(G^{-1})$ or smaller.

The analysis of Haaland and Sparrow (1973a) resulted in a neutral curve that exhibited both a lower branch and a critical Grashof number. See Fig. 11.11.1. However, as first pointed out by Hieber and Nash (1975), this recalculation neglected other terms of the same order as those underlined above. The largest terms neglected in boundary layer theory are also $O(G^{-1})$. These, through interaction with the disturbance quantities, also result in terms of the same order in the stability equations. Further, the terms arising from the x dependence in ζ' and from the x dependence of wave number $\hat{\alpha}$ for fixed physical frequency are also of the same order. An internally consistent higher-order stability analysis includes all terms, uniformly to the level of the approximation.

One such analysis for the plane plume was given by Hieber and Nash (1975). Using asymptotic matching, higher-order boundary layer effects were first calculated in terms of the perturbation parameter $\epsilon = (\mathrm{Gr}_x)^{-1/4}$. The leading, zero-order, terms satisfy the following governing equations:

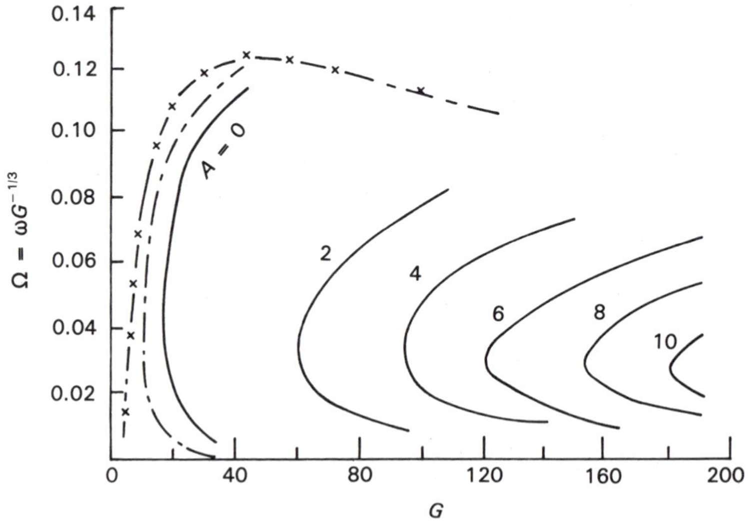

Figure 11.11.1 Constant amplitude ratio curves for a free plane plume flow, Pr = 0.7. (——) Hieber and Nash (1975); (– – –) Haaland and Sparrow (1973a); (– × – ×) Pera and Gebhart (1971). *(Reprinted with permission from B. Gebhart and R. L. Mahajan,* Advances in Applied Mechanics, *p. 231. Copyright © 1982, Academic Press.)*

$$f_0''' + \frac{3}{5} f_0 f_0'' - \frac{1}{5} f_0' f_0' + \phi_0 = 0 \tag{11.11.3}$$

$$\phi_0'' + \frac{3}{5} \Pr(f_0 \phi_0' + f_0' \phi_0) = 0 \tag{11.11.4}$$

$$f_0''(0) = f_0(0) = 1 - \phi_0(0) = f_0'(\infty) = \phi_0(\infty) = 0 \tag{11.11.5}$$

These are the same as Eqs. (11.2.22)–(11.2.24) with $n = -\frac{3}{5}$, except for some of the coefficients and the replacement of $f'(0) = 0$ for a vertical surface with $f''(0) = 0$ for a plane plume. This difference is due to definitions of similarity variables used by Hieber and Nash (1975). There η was taken as $\eta = (y/x)(\mathrm{Gr}_x)^{1/4}$.

The first-order equations for f_1 and ϕ_1 solved by Hieber and Nash (1975) are

$$f_1''' + \frac{3}{5} f_0 f_1'' + \frac{1}{5} f_0' f_1' + \phi_1 = 0 \tag{11.11.6}$$

$$\phi_1'' + \frac{3}{5} \Pr(f_0 \phi_1' + 2 f_0' \phi_1) + \frac{3}{5} \Pr f_1' \phi_0 = 0 \tag{11.11.7}$$

$$f_1(0) = 0 = f_1''(0) = \phi_1'(0) = \phi_1(\infty) \qquad f_1'(\infty) = \frac{3}{5} \cot \frac{2\pi}{5} f_0(\infty) \tag{11.11.8}$$

To assess the higher-order stability effects, a more general form of disturbance equations (11.2.26) and (11.2.27) was assumed:

$$\psi'(x, y, \tau) = \delta U_c \Phi(\eta, x) \exp \{i[\Lambda(x) - \hat{\omega}\tau]\} \tag{11.11.9}$$

$$t'(x, y, \tau) = d(x) s(\eta, x) \exp \{i[\Lambda(x) - \hat{\omega}\tau]\} \tag{11.11.10}$$

where $\Lambda = \int_0^x \hat{\alpha}(x)\, dx$.

As for the base flow, the disturbance quantities were expanded in terms of ϵ as

$$\Phi = \Phi_0(\eta) + \epsilon\Phi_1(\eta) + \cdots \tag{11.11.11}$$

$$s = s_0(\eta) + \epsilon s_1(\eta) + \cdots \tag{11.11.12}$$

$$\alpha = \hat{\alpha}\delta = \alpha_0 + \epsilon\alpha_1 + \cdots \tag{11.11.13}$$

$$c = \frac{\omega}{\alpha} = c_0 + \epsilon c_1 + \cdots \tag{11.11.14}$$

where $\omega = \hat{\omega}\delta/U_c$.

The linearized vorticity disturbance equation (11.11.1) was evaluated in terms of the higher-order base flow and disturbance quantities. Ordering resulted in equations for the disturbance functions. Calculating the values of α_0 and α_1 for a range of ω, the neutral stability and amplification curves were obtained. The series in α was truncated to two terms:

$$-\alpha_i = -(\alpha_{0i} + \epsilon\alpha_{1i}) \tag{11.11.15}$$

The neutral stability curve, $\alpha_i \equiv 0$, is therefore given by

$$\frac{1}{\epsilon} = (\mathrm{Gr}_x)^{1/4} = \frac{\mathrm{G}}{2\sqrt{2}} = -\frac{\alpha_{1i}}{\alpha_{0i}} \tag{11.11.16}$$

As in Section 11.2, it is shown from $\omega = 2\pi F\delta/U_c$, where $\delta = 4x/G$, $U_c = \nu G^2/4x$, $G = 4(g\beta N x^n x^3/4\nu^2)^{1/4}$, and $n = -\frac{3}{5}$, that

$$\Omega = \omega G^{-1/3} = \frac{\pi F}{\nu}\left(\frac{\nu^2}{g\beta N}\right)^{5/6} \tag{11.11.17}$$

The stability results are compared in Fig. 11.11.1, in terms of Ω vs. G, with contours of A. The results of Hieber and Nash (1975) have been converted to these coordinates. Horizontal lines are again constant physical frequency disturbance trajectories. The neutral curve from the parallel flow analysis of Pera and Gebhart (1971) is also shown. Clearly, the higher-order effects have a large influence on initial instability. A minimum Grashof number for instability and a lower branch in the neutral curve have arisen. Also shown in Fig. 11.11.1 is the neutral stability curve obtained from the calculations of Haaland and Sparrow (1973a). The results show the effect of the inconsistent approximations.

Higher-order corrections with the WKB method. Although the analysis of Hieber and Nash (1975) is an improvement over previous analyses, it is to be noted that the lowest-order equations, (11.11.3)–(11.11.5), are the standard inviscid Orr-Sommerfeld equations. The viscous and temperature coupling terms, as well as some of the nonparallel effects, appear in the next order equations. A more desirable approach is, perhaps, to apply the method of multiple scales, also known as the Wentzel-Kramers-Brillouin (WKB) method. This approach was used by Bouthier (1972, 1973), Gaster (1974), and Saric and Nayfeh (1975) to study the linear stability of the Blasius boundary layer flow.

Wakitani (1985), following the approach of Gaster (1974), has extended the use of this method to study nonparallel flow stability of a plane buoyant plume. As a first step, an initial disturbance is assumed to exist at a location x_0 and solutions are sought for $x > x_0$. New coordinates (ξ, η) are introduced:

$$\xi = \frac{x}{x_0} \qquad \eta = G_Q^* \frac{y}{x}$$

where

$$G_Q^* = \left(\frac{g\beta x^3 Q}{k\nu^2}\right)^{1/5} \tag{11.11.18}$$

Note that $G_Q^* = (\mathrm{Gr}_{x,Q}\mathrm{Pr})^{1/5}$, where $\mathrm{Gr}_{x,Q}$, as defined in Eq. (11.8.1), is the local Grashof number based on Q, the strength of the heat source per unit length. The disturbance quantities ψ' and t' are now taken to be of the generalized form

$$\psi' = \nu G_Q^* \Phi(\xi, \eta) \exp\left\{i\left[G_Q \int_1^{\xi} \xi^{-2/5}\alpha(\xi)\, d\xi - \omega\tau\right]\right\} \tag{11.11.19}$$

$$t' = \left(\frac{Q}{k}\right)(G_Q^*)^{-1}s(\xi, \eta) \exp\left\{i\left[G_Q \int_1^{\xi} \xi^{-2/5}\alpha(\xi)\, d\xi - \omega\tau\right]\right\} \quad (11.11.20)$$

where $G_Q = G_Q^* \xi^{-3/5}$ and Φ and s are given by

$$\Phi = A(\xi)\Phi_0(\xi, \eta) + \epsilon_1\Phi_1(\xi, \eta) + \cdots \quad (11.11.21)$$

$$s = A(\xi)s_0(\xi, \eta) + \epsilon_1 s_1(\xi, \eta) + \cdots \quad (11.11.22)$$

The leading terms Φ_0 and s_0 satisfy the Orr-Sommerfeld equations. $A(\xi)$ is a weak function of ξ and is included to take some account of the streamwise variations of the wave number and eigenfunctions. The terms $\epsilon_1\Phi_1$ and $\epsilon_1 s_1$ are the correction terms, where ϵ_1 is determined as $(G_Q^*)^{-1}$.

For the first-order corrections to be consistent, the base flow stream function and temperature are again expanded to include higher-order terms:

$$\psi = \nu[G_Q^* f_0(\eta) + f_1(\eta) + O(G_Q^{-2/3}) + \cdots] \quad (11.11.23)$$

$$t = t_\infty + \frac{Q}{k}(G_Q^*)^{-1}[\phi_0(\eta) + (G_Q^*)^{-1}\phi_1(\eta) + O(G_Q^{-5/3}) + \cdots] \quad (11.11.24)$$

Equations (11.11.19)–(11.11.22), along with Eqs. (11.11.23) and (11.11.24), are then substituted into Eqs. (11.11.1) and (11.11.2) to obtain governing equations in ascending order of ϵ_1, with the equations in ϵ_1^0 being those for standard parallel flow analysis. The equations obtained by collecting terms of $O(\epsilon_1)$ are inhomogeneous, with the inhomogeneous part involving terms containing A and $dA/d\xi$. These equations are solved to determine $A(\xi)$, which provides the first correction to the parallel flow solution. See Wakitani (1985) for details.

It is noted that in truly parallel flows the eigenfunctions are independent of the streamwise location ξ. For constant frequency, the exponential part of the disturbance stream function and temperature uniquely defines the wavelength and the amplification rate. In nonparallel flows, on the other hand, the eigenfunctions vary slowly with the streamwise location, and estimates of wavelength and amplification rate therefore depend on how these quantities are defined. For example, growth rate may be defined in terms of relative change of disturbance velocity u' and disturbance temperature t' as follows:

$$\frac{1}{u'}\frac{\partial u'}{\partial \xi} = G_Q^{-2/5}\xi(-\alpha) + \left(\frac{1}{A}\frac{dA}{d\xi} + \frac{1}{\Phi_0'}\frac{\partial \Phi_0'}{\partial \xi} + \frac{1}{5\xi} + \cdots\right) \quad (11.11.25)$$

$$\frac{1}{t'}\frac{\partial t'}{\partial \xi} = G_Q^{-2/5}\xi(-\alpha) + \left(\frac{1}{A}\frac{dA}{d\xi} + \frac{1}{s_0}\frac{\partial s_0}{\partial \xi} - \frac{3}{5\xi} + \cdots\right) \quad (11.11.26)$$

The leading term is the wave number, which arises from the Orr-Sommerfeld approximation; the other terms arise, in order of importance, from the amplitude function, the eigenvalue modification with Grashof number, and the coordinate system, respectively. The imaginary parts of Eqs. (11.11.25) and (11.11.26) give the wave numbers for u' and t' and the real parts give the amplification rate.

The neutral curves with the above definitions for amplification rate of u' and t' are then defined by

$$\xi^{-2/5}\alpha_i - G_Q^{-1}\left[\left(\frac{1}{A}\frac{dA}{d\xi}\right)_r + \left(\frac{1}{\Phi_0'}\frac{\partial \Phi_0'}{\partial \xi}\right)_r - \frac{1}{5\xi}\right] = 0 \qquad (11.11.27)$$

$$\xi^{-2/5}\alpha_i - G_Q^{-1}\left[\left(\frac{1}{A}\frac{dA}{d\xi}\right)_r + \left(\frac{1}{s_0}\frac{\partial s_0}{\partial \xi}\right)_r - \frac{3}{5\xi}\right] = 0 \qquad (11.11.28)$$

Note that when α_i is set equal to zero, that is, at the neutral point determined by the parallel flow theory, there is still growth or decay due to the nonparallel flow effects. Further, since the eigenvalue modification terms in Eqs. (11.11.27) and (11.11.28) depend on η, the values of these terms must be evaluated at some position, for instance, where $|u'|$ or $|t'|$ is a maximum, to determine the amplification rate.

Relations similar to those above may be obtained for other quantities such as the local disturbance kinetic energy $\overline{u'^2} + \overline{v'^2}$ or $\overline{u'^2}$ or $\overline{t'^2}$ or the relative amplitudes of u' and t' with respect to the base flow; see Wakitani (1985). From such relations, neutral curves may be constructed for a range of G_Q^* and disturbance frequencies.

The neutral curves based on the various disturbance quantities mentioned above differ from one another. However, all such curves show shifts from the parallel theory results. The curves obtained with the disturbance quantities relative to the base flow are in closer agreement with the parallel flow results. However, the amplification rates within an unstable region show substantial deviation from the parallel flow results. The nonparallel correction to the wave number was also found to be significant.

The experimental data of Dring and Gebhart (1969) and Wakitani and Yosinobu (1984) at low G_Q^*, for disturbance temperature amplification rate, showed better agreement with the nonparallel theoretical predictions than with the parallel theory results.

This treatment of the nonparallel flow effects seems reasonable. However, more experiments are needed to further verify the theoretical findings. The integrated disturbance amplitude curves for various values of A, defined in Eq. (11.2.34), should be obtained. The effect of nonparallel corrections on the phenomenon of frequency filtering should be examined. The theory should also be extended to the flows adjacent to a vertical surface and results compared with the extensive experimental data available for such flows.

11.11.2 Another Approach to the Mechanisms of Instability and Transition

An entirely different formulation of the onset of instability in plume flows has been suggested by Kimura and Bejan (1983). In the very early observations of Rayleigh (1880), it was shown that the minimum wavelength for instability, λ_{min}, for an inviscid jet of triangular profile, or of free shear flow profile, scales with the transverse dimension of the flow, D; see Fig. 11.11.2*a*. For an inviscid jet the flow was shown to be unstable to disturbances whose wavelengths exceeded $\lambda_{min} = 1.714D$. For the free shear flow, in Fig. 11.11.2*b*, λ_{min} was $4.914D$. Also shown in the figure (11.11.2*c*) is a plane of velocity discontinuity. This flow is unstable to any wavelength; see Lamb (1945). That is, the minimum wavelength is zero and is thus of the same order as the shear stress.

Based on these and similar results of Batchelor and Gill (1962) for radially symmetric flows and of Lopez and Kurzweg (1977) for annular shear layers, Kimura and Bejan (1983) argued that $\lambda_{min} \sim D$ scaling is an intrinsic property of inviscid flows and is responsible for instability (transition).

The particular flow considered to test the validity of the above approach was a point heat source plume. This flow, as pointed out by Rayleigh (1880), is unstable to all wavelengths $\lambda > \lambda_{min}$. Therefore, it may fluctuate with an infinity of periods τ as

$$\tau \geq \tau_{min} \sim \frac{\lambda_{min}}{U/2} \tag{11.11.29}$$

where $U/2$ is the plume mean velocity and τ_{min} is the minimum plume fluctuation time. Since $\lambda_{min} \sim D$, it follows that $\tau_{min} \sim D$. This relationship is shown as a straight line in Fig. 11.11.3. Fluctuations with a period shorter than τ_{min} are stable. This is the region to the left of the τ_{min} line. The region to the right of the line is that of inviscid instability.

To determine whether the plume flow will stay laminar, the minimum fluctuation time τ_{min} is compared with the time of viscous penetration normal to the plume from the plume ambient interface to the plume centerline, τ_v. Using the classical solution to Stokes' first problem (see Schlichting, 1968), τ_v is calculated as

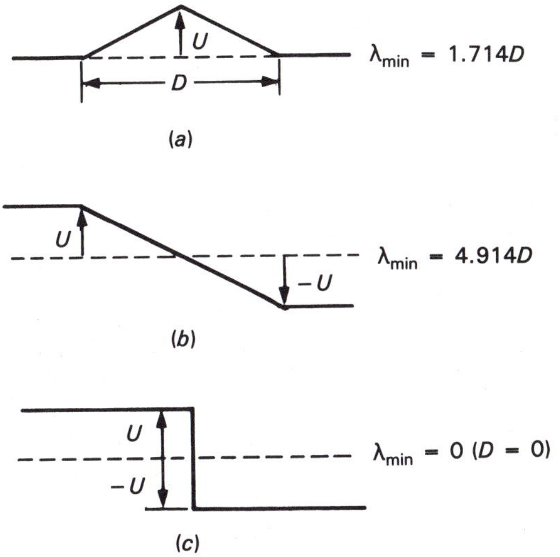

Figure 11.11.2 Minimum wavelength for instability in inviscid flow. (*a*) Free jet; (*b*) shear layer; (*c*) velocity discontinuity. *(Reprinted with permission from* Int. J. Heat Mass Transfer, *vol. 26, p. 1515, S. Kimura and A. Bejan. Copyright © 1983, Pergamon Journals Ltd.)*

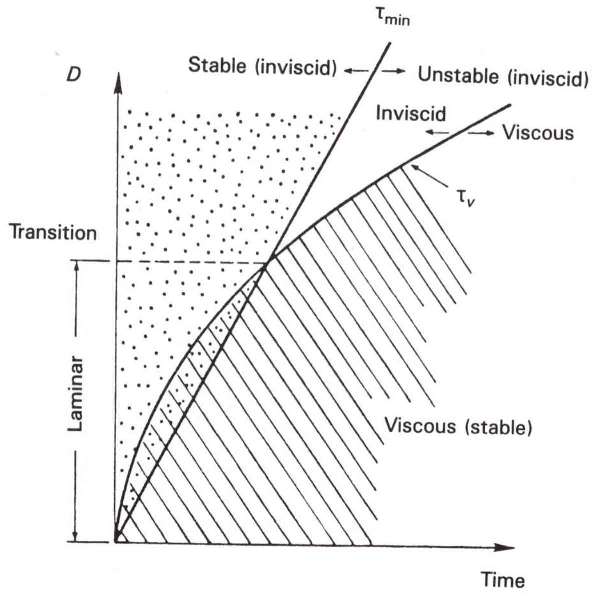

Figure 11.11.3 Schematic showing the role played by the minimum period for inviscid instability, τ_{min}, and the viscous penetration time, τ_v, in determining laminar-to-turbulence transition in a two-dimensional plane. *(Reprinted with permission from* Int. J. Heat Mass Transfer, *vol. 26, p. 1515, S. Kimura and A. Bejan. Copyright © 1983, Pergamon Journals Ltd.)*

$$\tau_v \sim \frac{D^2}{16\nu} \tag{11.11.30}$$

If τ_{min} exceeds τ_v, the plume will remain laminar since the ambient medium will restrain it successfully through viscous effects. Then, transition from laminar to turbulent flow conditions takes place when $\tau_v \sim \tau_{min}$.

The experimental part of the study visualized the transition of an axisymmetric air plume above a point heat source. Cigarette smoke was used for visualization. The experiment indicated that at transition, the plume assumes a meandering shape of characteristic wavelength that scales with the local plume diameter. The phenomenon of frequency filtering discussed earlier was also observed. The disturbances were generated by a loudspeaker suspended next to the plume at a certain height above the point source. Of the disturbances at several frequencies imposed on a plume, it was noted that the plume selectively filtered out the disturbance of a wavelength that scales with the plume diameter at the downstream location x where transition took place. The Grashof number for the onset of transition was found to be of $O(10^8)$.

This approach to the study of instability and transition is an interesting one. Although the experimental results reported above agree qualitatively with the theory proposed, much more detailed experiments for different flow circumstances would provide a fuller test of the theory.

11.12 INSTABILITY OF HORIZONTAL AND INCLINED FLOWS

Such flows were considered in Chapter 5. A similarity solution is given in Section 5.3 for a horizontal laminar flow away from a leading edge. This solution is then perturbed for the effect of small inclination, in terms of an inclination parameter $\epsilon(x, \xi)$. Given these solutions, a stability and disturbance growth analysis and measurements were reported by Pera and Gebhart (1973). This is described next, followed by a discussion of the instability mechanisms of other inclined flows.

11.12.1 Horizontal and Slightly Inclined Flows

The complete time-dependent two-dimensional base flow equations are obtained by retaining the terms $\partial u/\partial \tau$, $\partial v/\partial \tau$, and $\partial t/\partial \tau$ in Eqs. (5.1.1), (5.1.2), and (5.1.8), respectively. These are perturbed by disturbances u', v', t', p'_m, and ρ' as in Section 11.2 for vertical flows. The same four additional approximations are made. The disturbance forms of ψ' and t' are postulated as in Eqs. (11.2.26) and (11.2.27) except that the generalizations are instead those that arose in Section 5.3, as follows:

$$\mathrm{Gr}_{x,\xi} = \frac{gx^3 \cos \xi}{\nu^2} \beta(t_0 - t_\infty) \qquad G^+ = 5\left(\frac{\mathrm{Gr}_{x,\xi}}{5}\right)^{1/5} \tag{11.12.1}$$

$$\eta = \left(\frac{y}{5x}\right)G^+ \qquad \delta(x) = \frac{5x}{G^+} \qquad U_c = \frac{\nu G^{+2}}{5x} \qquad \epsilon(x, \xi) = \frac{G^+ \tan \xi}{5} \tag{11.12.2}$$

where ξ is the angle of inclination from the horizontal and $\epsilon(x, \xi)$ the perturbation parameter. The resulting Orr-Sommerfeld equations and boundary conditions are

$$\left(f' - \frac{\omega}{\alpha}\right)(\Phi'' - \alpha^2\Phi) - f'''\Phi + \frac{s}{5} = \frac{\Phi'''' - 2\alpha^2\Phi'' + \alpha^4\Phi - s'\epsilon}{i\alpha G^+} \tag{11.12.3}$$

$$\left(f' - \frac{\omega}{\alpha}\right)s - \phi\Phi' = \frac{s'' - \alpha^2 s}{i\alpha \mathrm{Pr} G^+} \tag{11.12.4}$$

$$\Phi(\infty) = \Phi'(\infty) = s(\infty) = \Phi(0) = \Phi'(0) = s(0) = 0 \tag{11.12.5}$$

where f and ϕ are the base flow solutions for $d(x) = t_0 - t_\infty = $ const. Instability characteristics were determined from these equations for $\mathrm{Pr} = 0.7$ and $\epsilon = 0$, 0.25, 0.75, and 1.0, out to a value of $G^+ = 200$, the region of most practical interest.

The stability plane for horizontal flow is seen in Fig. 11.12.1 in terms of $\omega = 2\pi F\delta/U_c$ and G^+; the amplitude ratio contours shown are for various values of A_x/A_N, defined as

$$\frac{A_x}{A_N} = \exp\left(-\frac{1}{5}\int_{G_N^+}^{G^+} \alpha_i \, dG^+\right) = e^A \tag{11.12.6}$$

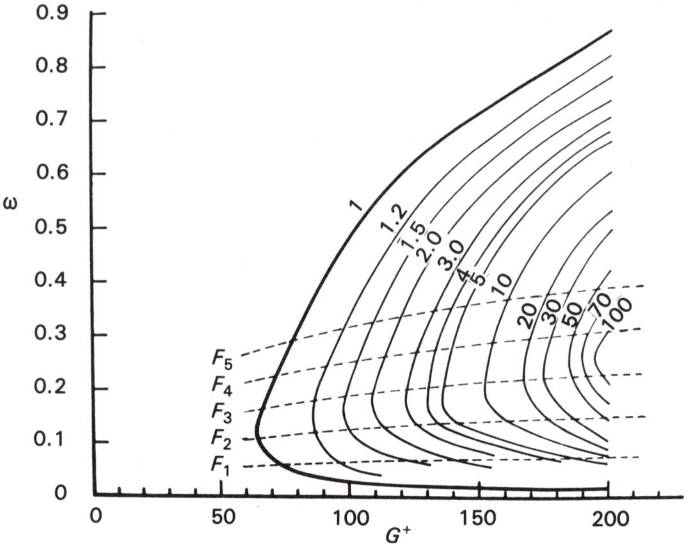

Figure 11.12.1 Neutral stability curve and amplitude ratio contours for the natural convection flow adjacent to a uniform temperature horizontal surface for Pr = 0.7. Also shown as dashed lines are several paths of constant disturbance frequency. *(Reprinted with permission from* Int. J. Heat Mass Transfer, *vol. 16, p. 1147, L. Pera and B. Gebhart. Copyright © 1973, Pergamon Journals Ltd.)*

where G_N^+ is again the neutral condition for any given frequency F. Recall that A was used to describe growth in the preceding sections of this chapter for vertical flows. For example, the contour $A_x/A_N = 100$ results in $A = 4.61$.

The paths of constant physical frequency in these coordinates are $\omega/\sqrt[3]{G^+} =$ const. The several such paths shown as dashed lines in Fig. 11.12.1 again indicate the selective amplification seen in Figs. 11.2.1 and 11.2.2 for vertical flows. However, the selectivity is not as strong. In fact, there is a continuing shift to higher favored frequencies downstream.

Figure 11.12.2 shows the effect of inclination on neutral stability for $\epsilon =$ 0.25, 0.5, 0.75, and 1.0. Increasing inclination, for example, at a given value of G^+, highly stabilizes the flow. Also, as ϵ increases the nose of the neutral curve becomes flatter, indicating that the width of the band of favored frequencies may become wider.

Experimental results. Pera and Gebhart (1973) also gave experimental results for an isothermal aluminum plate in air at various inclinations in the range $\xi =$ 0 to 6°. Sidewalls were used to prevent any edge flows. A vibrating ribbon in the boundary region, near the leading edge, provided a controlled sinusoidal disturbance. Downstream disturbance behavior was observed with a Mach-Zehnder interferometer.

Figure 11.12.3 is a composite interferogram of the results. Four inclinations are shown, each at four different disturbance frequencies. The most unstable conditions are found along a descending diagonal across the figure. At large ξ, higher

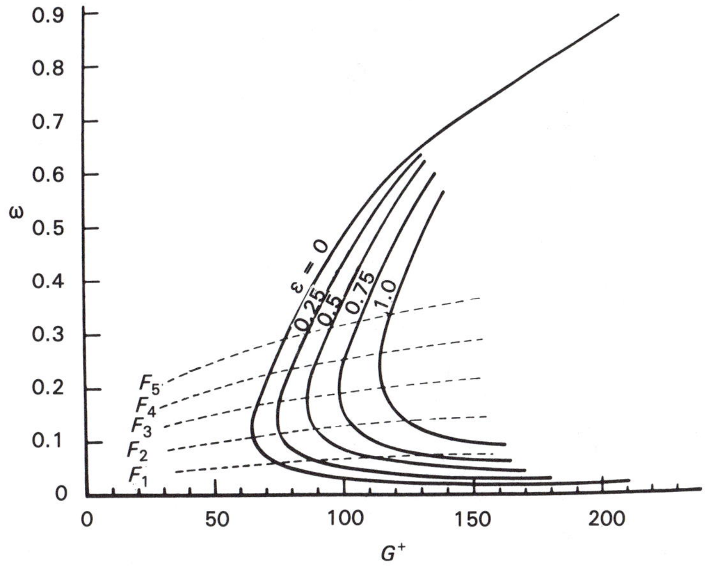

Figure 11.12.2 Neutral stability curves for flow adjacent to a horizontal and near-horizontal surface for Pr = 0.7; ε is the inclination parameter. *(Reprinted with permission from* Int. J. Heat Mass Transfer, *vol. 16, p. 1147, L. Pera and B. Gebhart. Copyright © 1973, Pergamon Journals Ltd.)*

frequencies are seen to be the most unstable. This is in agreement with the predictions in Fig. 11.12.2.

However, the frequencies calculated and measured to be the most unstable are somewhat different. For example, for $\xi = 0$, the value between F_2 and F_3 at the nose in Fig. 11.12.2 is 2.8 Hz. The most unstable frequency determined from these experiments is about 1.7 Hz. In addition, disturbance amplification was found to begin before the calculated neutral curve was reached, in G^+. Thus, the agreement here is not very good. One possible reason is that the somewhat larger flow-region thicknesses in this flow, at the same local value of the Grashof number, make the boundary layer and parallel flow approximations less reasonable. It is also impossible to define accurately a leading edge in experiments in this flow with gases, which have very low thermal conductivity.

There is another fundamental complication in this flow. It remains of simple two-dimensional boundary layer form for only a short downstream distance. Subsequent downstream flow mechanisms may greatly affect the upstream regions in such a weak and largely pressure-driven flow. For example, much-increased thickening of the convection region downstream would decrease the pressure through the increased buoyancy effect and increase the negative gradient $\partial p_m/\partial x$ that drives the flow. The base flow measurements consistently showed a slightly thinner convection region upstream than predicted by theory.

Although these calculations and measurements are in reasonable agreement, the discrepancies, along with the other observations, indicate that yet other effects

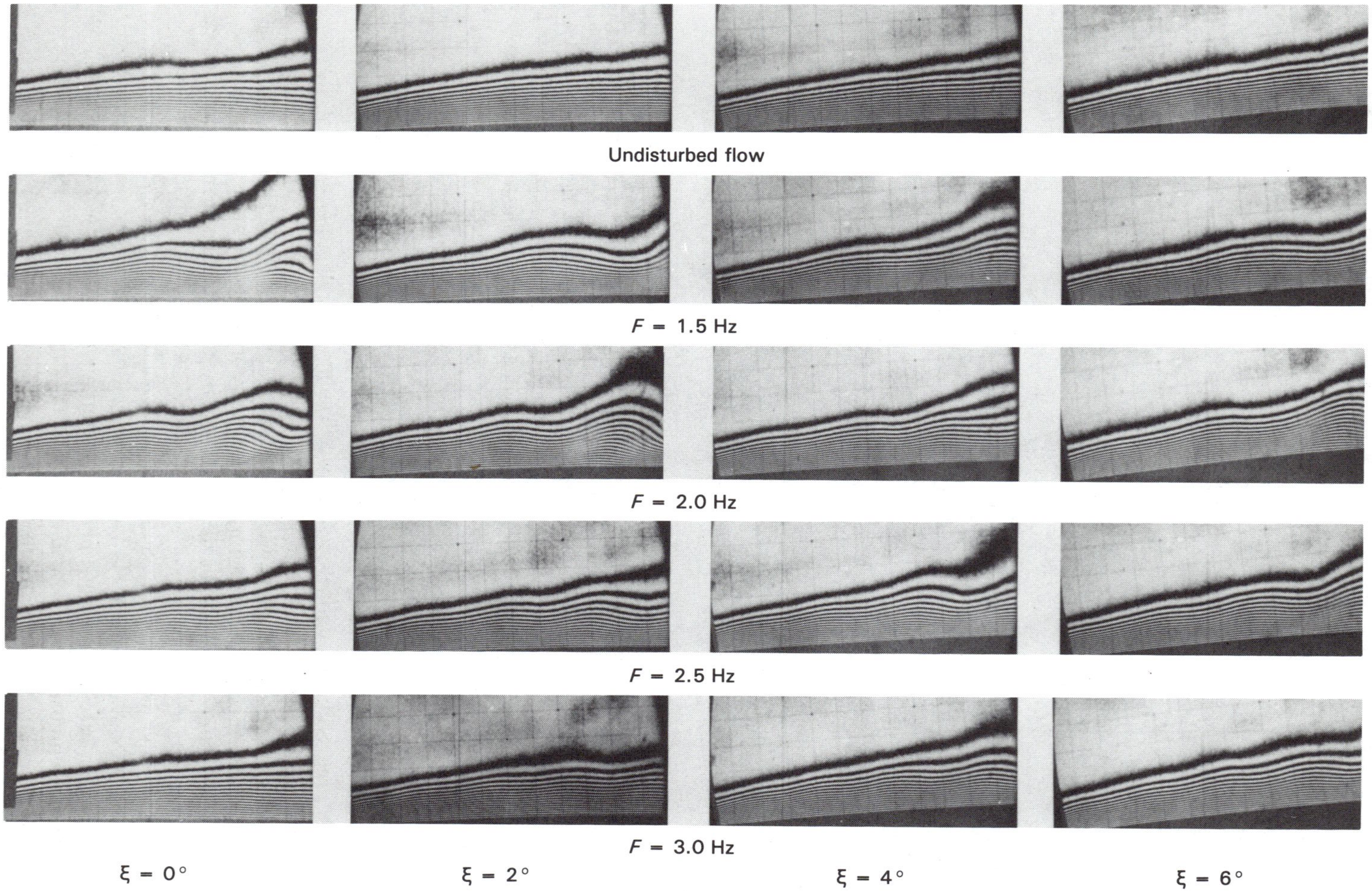

Figure 11.12.3 Disturbance amplification in air as a function of frequency and angle of inclination ξ with the horizontal. The leading edge is to the left. *(Reprinted with permission from* Int. J. Heat Mass Transfer, *vol. 16, p. 1147, L. Pera and B. Gebhart. Copyright © 1973, Pergamon Journals Ltd.)*

may also arise and may dominate downstream mechanisms. It may be that these mechanisms have a first-order interaction under some conditions.

Also, the assumption of two-dimensional disturbances, as in Eqs. (11.2.26) and (11.2.27), places a heavy and exclusive restriction on included mechanisms. These flows obviously have a possible component of thermal instability due to their unstable stratification. Omitting any spanwise dependence in z excludes several configurations of instability modes that may arise due to this unfavorable stratification. Including only disturbances expressible in x, y, and τ is not a generally reliable procedure. There is some experimental evidence that transverse effects and longitudinal rolls are important very early in the instability of inclined flows.

Approximate estimates indicate that the flow in Fig. 11.12.3 are not thermally unstable at early times. Visualizations of three-dimensional and secondary flow effects farther downstream by Pera and Gebhart (1973) indicated that such effects do arise later, perhaps in conjunction with amplified two-dimensional disturbances. Other modes of instability due to three-dimensional effects are implicated with the eventual flow separation in this geometry. For a discussion of the mechanisms of flow separation, the reader is referred to Section 5.8.

The calculations and experiments by Ishiguro et al. (1978) are related to the apparently similar, yet very different, circumstance of a long narrow heated strip embedded in an extensive horizontal insulating surface. The resulting flow, in from each side, gathers in a plume. Any instability involves additional and very different effects.

11.12.2 Other Inclined Flows

The instability of inclined flows other than those described in the preceding subsection has also been studied both experimentally and analytically. Lock et al. (1967) investigated the instability in a laminar inclined flow for inclinations from $\theta = 60°$ for an upward-facing heated surface to $\theta = -80°$ for a downward-facing heated surface, where θ is the angle from the vertical. A thermocouple probe and Schlieren visualizations detected wavelike disturbances as in vertical flows. Sparrow and Husar (1969), on the other hand, found longitudinal vortices with a visualization technique for $\theta \geq 15°$. Further studies by Lloyd and Sparrow (1970) indicated that the instability mode is wavelike for $\theta < 14°$ and of vortex form above $\theta = 17°$. In the range between $\theta = 14$ and $17°$, the two modes coexist.

A detailed experimental investigation of both flow mechanisms and heat transfer was carried out by Shaukatullah and Gebhart (1978). The surface condition was a uniform heat flux. Local measurements were made with a thermocouple and inverted-vee hot-film probes in water to inclinations of $\theta = 30°$. At about 10° incination, both disturbance forms appear amplified at about the same downstream location. Also, disturbance mechanisms below 10° are much like those in vertical flows. Beyond 10° the vortex mode prevails. The spanwise vortex wavelength is dependent on the inclination of the flow but not on the heat flux level.

Downstream, at higher inclinations, the local Nusselt number varied period-

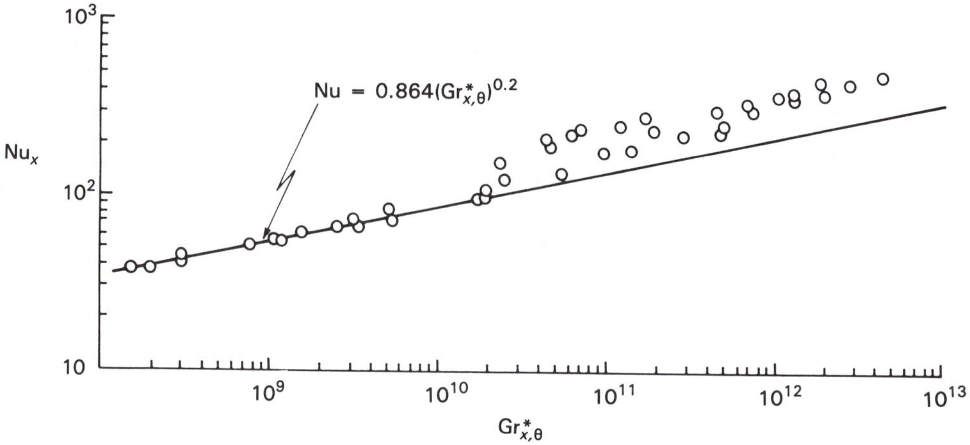

Figure 11.12.4 Heat transfer as a function of downstream location at an inclination of 29°, for different heat flux levels; (——) in the laminar boundary layer solution for vertical flow at Pr = 6.0. *(Reprinted with permission from* Int. J. Heat Mass Transfer, *vol. 21, p. 1481, H. Shaukatullah and B. Gebhart. Copyright © 1978, Pergamon Journals Ltd.)*

ically in the spanwise direction, a direct result of the vortices. The general downstream trend of the average transport was a substantial augmentation of Nu_x. An example of such behavior is seen in Fig. 11.12.4 in a plot of Nu_x vs. $Gr^*_{x,\theta}$ $(=g\beta x^4 q'' \cos\theta/k\nu^2)$. This effect was found to be a direct consequence of vortical motion and of its large-scale entrainment of cooler ambient fluid, as seen earlier in Fig. 5.8.5.

A similar experimental study by Miller and Gebhart (1978) measured transport in air from two inclined isothermal flat surfaces joined into a common edge to form a ridge. This geometry is characteristic of many common configurations. The transport, for symmetric included ridge angles from 120 to 160°, was found to be very similar to that from a single inclined surface. Above the ridge the separated flows from each inclined surface met and rose into a single plume, in a manner similar to that discussed in Chapter 5 for a half-cylinder and hemisphere. An interferogram of the shed plume for an included ridge angle of 120° is shown in Fig. 11.12.5. The shed flow is contained in a relatively well-defined plume. For flatter ridges the plume was highly disturbed. This is related to the more disturbed nature of the upstream flows at lower inclinations.

There have been a number of different kinds of instability analyses for such flows. In the analysis of Lee and Lock (1972), both wave and vortex modes were treated. Then the wavelike mechanism alone was considered by Haaland and Sparrow (1973b), as was the mean flow vortex by Hwang and Cheng (1973) and Haaland and Sparrow (1973c). In the latter two studies it was found that the susceptibility of the flow to the vortex mode of instability increases with increasing angle of inclination from the vertical. Calculations by Iyer and Kelly (1974) predicted that the vortex mode would be the first unstable effect for inclinations

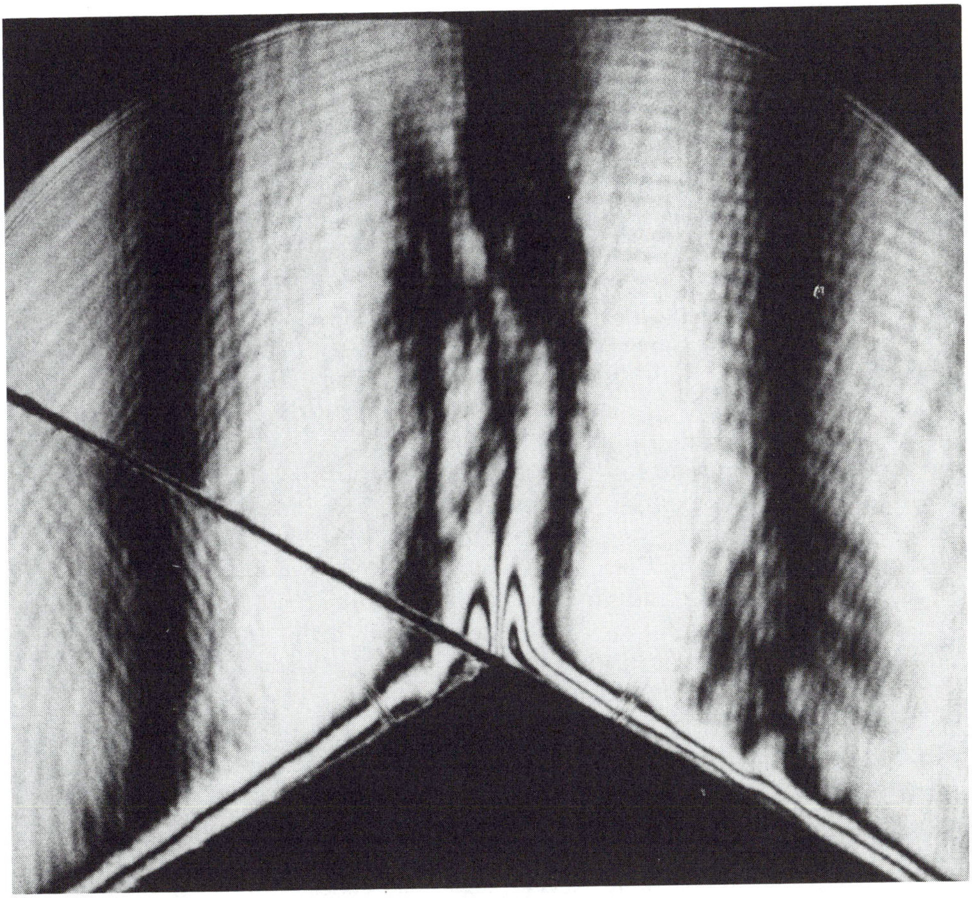

Figure 11.12.5 Interferogram of the flow shed from the ridge. Included ridge angle = 120°, Δt = 27.8°C. *(From Miller and Gebhart, 1978.)*

greater than 4°. Postulating three-dimensional disturbances, the wave mode was found by Kahawita and Meroney (1974) to be supplanted by the vortex mode at about 17°.

The base flow used in the analyses referred to above neglected the normal component of the buoyancy force, B_n. As described in detail in Chapter 5, this is not likely to be a valid assumption when angles of inclination θ from the vertical are large. Close to the horizontal, B_n causes longitudinal vortex rolls. Neglecting B_n in the base flow in instability calculations is generally not justified.

Recognizing these inadequacies of past analyses, Chen and Tzuoo (1982) analyzed instability of flow over warm upward-facing surfaces, for the vortex mode, from near-vertical to horizontal flows. Both the normal and streamwise components of buoyancy force were retained in the base flow. Numerically computed values of critical Grashof numbers and wave numbers for an isothermal surface condition were reported for Pr = 0.7 and 7.0. The results compared well with the previous approximate base flow vortex mode stability calculations, for θ up to 40° for Pr = 0.7 and to $\theta = 60°$ for Pr = 7. However, for larger values of θ, larger deviations occured. The deviations were largest for $\theta = 90°$.

Later, Tzuoo et al. (1985) reported a purely wave instability analysis for upward-facing heated surfaces for $0 \leq \theta \leq 90°$. The effects of the nonparallelism of the base flow were taken into account. The critical Grashof numbers were computed for Pr = 0.7 and 7.0. The results were in good agreement with those of Haaland and Sparrow (1973b) in the range $0 \leq \theta \leq 45°$. However, when compared with the wave instability calculations of Pera and Gebhart (1973) for $\theta = 90°$, a large discrepancy occurred. For Pr = 0.7, that study indicated a critical Grashof number $Gr_x = g\beta(t_0 - t_\infty)x^3/\nu^2$ of 1.8×10^6. The value of Tzuoo et al. was only 510. The discrepancy was attributed to the different modes of eigenvalues assumed in the two solutions. It was indicated that $Gr_x = 510$ corresponds to the least stable mode of solution while the calculations of Pera and Gebhart represent a mode of solution other than the least stable.

A comparison of the results of wave mode analysis with the earlier results of vortex mode analysis by Chen and Tzuoo (1982) indicated that the mode of instability shifts from wavelike to vortexlike at increasing θ. This shifting was predicted to occur at $\theta = 32°$ for Pr = 0.7 and $\theta = 50°$ for Pr = 7.0. These results are, however, only in qualitative agreement with previous experimental observations.

The above collection of experimental results and calculations points toward some similar conclusions. The wavelike mode, as in vertical flows, arises for small angles of inclination from the vertical. At higher inclinations, instability goes over toward a mean flow vortex mode, like that in horizontal flow. However, there are still considerable differences among the measurements and predictions taken together. These arise in terms of angular dependences, exact mechanisms, and the downstream appearance and growth of various effects. There is some uncertainty about internal consistency of some of the analyses. For example, as pointed out in Section 11.11.1, improved analyses must also take into account the streamwise dependence of the amplitude function and wave number. More experiments and analyses, therefore, are necessary to resolve the many remaining discrepancies between the experimental observations and predictions.

11.13 TRANSIENT TRANSITION AND TURBULENCE

Very little is known about transition during unsteady flows. For a suddenly heated flat vertical surface, the initial transient regime is a laminar one-dimensional conduction process at all downstream locations. In achieving steady flow, the transport may have changed from laminar to turbulent downstream. The occurrence of and changes in the transition in time have not been studied in detail. Also, the one-dimensional flow may undergo transition. On the other hand, instabilities occur after the arrival of the leading edge effect, in the establishment of a two-dimensional developing flow downstream.

Mollendorf and Gebhart (1970) observed transient transition adjacent to a flat vertical uniform flux surface. The temperature field was visualized with interferometry. Complex transient transition patterns were observed. At lower heat flux levels the flow was seen to relaminarize after becoming turbulent during the transient. At higher heat fluxes no return to laminar flow arose. This study appears to indicate that, for certain intermediate heat flux values, the transition region may shift with time before reaching its final steady-state location. The observations of Joshi and Gebhart (1987), in water, indicated that many regimes arise. Several kinds of disturbance growth and transition were seen.

Stability analysis of time-dependent flows is much more complex than steady-state analysis since both the base flow and the disturbances are time dependent. The first difficulty encountered in such studies is the establishment of a criterion for instability. Since the base flow is growing in time, mere growth of the disturbance quantities may not be sufficient evidence of flow instability.

Gunness and Gebhart (1969) analyzed the stability of the unsteady flow between two vertical flat surfaces. A perturbation analysis was performed for several values of Grashof and wave numbers and for a wide range of rate of acceleration and deceleration to determine the unsteady base flow stability characteristics. Accelerating flows were found to be more stable than steady flows. Decelerating flows were correspondingly less stable.

Shen (1961) suggested that if the energy of the disturbances is growing faster than the energy of the base flow, then the flow may be considered momentarily unstable; that is, the flow is unstable if

$$\frac{1}{2e}\frac{de}{d\tau} > 0 \tag{11.13.1}$$

where e is the ratio of the energy of the two-dimensional disturbances to the kinetic energy of the base flow:

$$e(\tau) = \frac{\int_0^\delta \int_0^\lambda (u'^2 + v'^2)\, dx\, dy}{\lambda \int_0^\delta \bar{u}^2\, dy} \tag{11.13.2}$$

where λ and δ are the disturbance wavelength and boundary layer thickness, respectively.

There have been several different approaches to the linear stability analysis of transient natural and forced convection flows. If the disturbances are growing much faster than the base flow, then the base flow is treated as quasi-steady, as by Schlichting (1932), Von Karman and Lin (1955), Morton (1957), Lick (1965), and Currie (1967). Under this postulate, no initial conditions are needed to carry out the analysis. However, this type of formulation is not applicable to impulsively started flows.

Another approach is to treat transient disturbance growth as an initial value problem. The method consists of initially distributing small random disturbances of given wavelength throughout the transient flow field. The linearized equations are then solved numerically. Foster (1965, 1968), Mahler et al. (1968), Mahler and Schecter (1970), Chen and Kirchner (1971), and Chen and Sanford (1977) applied this technique and reported good agreement between predicted and measured values for the fastest growing disturbance wavelengths.

Jhaveri and Homsy (1980, 1982) analyzed the onset of Rayleigh-Bénard convection from random disturbances arising within a fluid. The formulation is reduced to a random initial value problem for the Fourier modes. This is then solved by a Monte Carlo technique and by moment methods for various Rayleigh numbers. The disturbances are assumed to exist at the start as well as through the transient. The results calculated for Rayleigh-Bénard instability agree very well with past data.

11.14 INSTABILITY OF VERTICAL FLOWS IN COLD WATER

The entire consideration of instability, transition, and turbulence in buoyancy-induced flows presented in previous sections assumes a linear temperature dependence of the fluid density. Most gases and liquids closely approximate this behavior, at least over small ranges of temperature. Even the density of water well above the freezing condition varies linearly with temperature. However, for lower temperatures, the density variation with temperature gradually becomes nonlinear. For pure water at 1 atm, a density extremum occurs at about 4°C. A density extremum also occurs in saline water up to a salinity level of about 26‰ (parts per thousand) and in pure water at pressures up to about 300 atm, in local thermodynamic equilibrium. See Fig. 9.1.1 for examples of the temperature variation of the density of water at low temperatures.

Since buoyancy arises from very small density gradients in the transport region, the fluid density must be very accurately known as a function of temperature, salinity, and pressure. A simple, yet very accurate, density equation, by Gebhart and Mollendorf (1977), is given in Chapter 9. This equation will also be the basis for analysis in this section.

The linear stability analysis in earlier sections determines conditions under which small two-dimensional disturbances amplify selectively downstream toward the onset of transition. The energy for disturbance growth is extracted from the base flow, which is driven by buoyancy. For flows arising adjacent to heated surfaces in water at higher temperatures the base flow is unidirectional across the boundary region. The stability and disturbance growth characteristics of such flows, as predicted by a linear stability theory, are very well substantiated by the abundant experimental data.

However, flows arising in cold pure or saline water may be markedly different in direction and in velocity levels, as discussed in Chapter 9. These more complicated flows lead to different instabilty characteristics and additional mechanisms. The simplest such flows are those that have similarity solutions. The basic parameter R, which characterizes the buoyancy force and resulting flow direction, is defined by Eq. (9.3.14) as

$$R = \frac{t_m(s, p) - t_\infty}{t_0 - t_\infty} \tag{9.3.14}$$

where t_m is the temperature at which the density extremum occurs and t_0 and t_∞ are the surface and ambient temperatures, respectively. Table 9.3.1 and Fig. 9.3.1 indicate that for different values of R the flow may be downward, upward, or bidirectional in the transport region for various choices of t_0 and t_∞. For $R \le 0$ the flow is upward across the boundary region, while for $R > 0.5$ it is downward. For $0 < R \le 0.5$ the buoyancy force is bidirectional. Thus, additional limits arise for the usual boundary layer formulation.

This more complicated buoyancy effect also restricts similarity analysis to a narrower range of conditions, as seen in Chapter 9. Since R arises in the buoyancy force, it may not depend on downstream distance x. Therefore, for an isothermal vertical surface, $n = 0$, in a unstratified ambient, similarity results. For a uniform surface flux condition q'', R must be zero, that is, $t_\infty = t_m$. Very general base flow solutions have been calculated for both conditions. For $R = 0$ the buoyancy force is always upward and relatively simple solutions result. However, for t_0 constant, great complexity arises in the region $0 < R < \frac{1}{2}$.

There have been only a few studies of stability and disturbance growth in cold-water flows. They apply to vertical uniform flux and isothermal surface conditions. Qureshi and Gebhart (1986) considered a uniform flux condition, for pure and saline water, with similarity (i.e., $R = 0$) for several values of the exponent $q(s, p)$ in the density equation, Eq. (9.1.1). Results were also given for an isothermal condition for $R = 0$. Higgins and Gebhart (1983) determined neutral stability conditions for an isothermal surface for $R = -\frac{1}{2}$, 1, ± 2, and ± 4. Both studies used the linear stability formulation in Sections 11.1. and 11.2.

The $R = 0$ results indicate considerable effects of the cold-water behavior on instability. For the flux condition, later and less strongly selective amplification is found downstream. Another effect is initial stabilization of the flow. However, farther downstream the disturbance amplification rate is higher. For the isothermal condition at other values of R, neutral stability predictions are also different from those with the Boussinesq approximation. The accurate analysis predicts later instability for $R < 0$ and earlier for $R > \frac{1}{2}$.

A general formulation of instability and disturbance growth analysis is given below. It applies to both conditions. The effects of various bounding thermal conditions t_0, t_∞, and t_m are represented by R. The temperature difference is taken as

$$t_0 - t_\infty = Nx^n \tag{11.14.1}$$

where $n = 0$ is the isothermal condition and $n = 1/(4 + q)$ is the condition resulting from uniform flux. The density difference for cold water may be accurately represented by Eq. (9.3.15), rewritten as

$$\rho_\infty - \rho = \rho_m \alpha(s, p)|t_0 - t_\infty|^q(|\phi - R|^q - |R|^q) \tag{11.14.2}$$

The boundary layer formulation in Eqs. (9.3.28)–(9.3.30) is

$$f''' + (3 + qn)ff'' - (2 + 2qn)f'^2 \pm (|\phi - R|^q - |R|^q) = 0 \qquad (11.14.3)$$

$$\phi'' + \mathrm{Pr}[(3 + qn)f\phi' - 4nf'\phi] = 0 \qquad (11.14.4)$$

$$f(0) = f'(0) = 1 - \phi(0) = f'(\infty) = \phi(\infty) = 0 \qquad (11.14.5)$$

where

$$\eta = \frac{y}{\delta} \qquad \psi = \delta U_c f(\eta) \qquad \phi = \frac{t - t_\infty}{t_0 - t_\infty} \qquad (11.14.6)$$

with $\delta = 4x/G$ and $U_c = \nu G^2/4x$, and

$$G = 4\left(\frac{\mathrm{Gr}_x}{4}\right)^{1/4} \qquad \text{where } \mathrm{Gr}_x = \left(\frac{gx^3}{\nu^2}\right)\alpha(s, p)|t_0 - t_\infty|^q \qquad (11.14.7)$$

The disturbances in stream function and temperature are again postulated as in Eqs. (11.2.26)–(11.2.29), with δ and U_c defined in Eq. (11.14.6) above. The resulting disturbance equations and boundary conditions in similarity form in terms of disturbance amplitude functions $\phi(\eta)$ and $s(\eta)$ are

$$\left(f' - \frac{\omega}{\alpha}\right)(\Phi'' - \alpha^2\Phi) - f'''\Phi$$

$$= \frac{\Phi'''' - 2\alpha^2\Phi'' + \alpha^4\Phi \pm \left[q\dfrac{\phi - R}{|\phi - R|}|\phi - R|^{q-1}s' + q(q-1)|\phi - R|^{q-2}s\phi'\right]}{i\alpha G}$$

$$(11.14.8)$$

$$\left(f' - \frac{\omega}{\alpha}\right)s - \phi'\Phi = \frac{s'' - \alpha^2 s}{i\alpha \mathrm{Pr} G} \qquad (11.14.9)$$

$$\Phi(0) = \Phi'(0) = s(0) = \Phi(\infty) = \Phi'(\infty) = s(\infty) = 0 \qquad (11.14.10)$$

Comparing Eqs. (11.14.8) and (11.14.9) with Eqs. (11.2.30) and (11.2.31), which apply with the Boussinesq approximation, indicates that cold-water behavior effects arise only in the base flow and in the disturbance buoyancy force. The new parameters are q and R.

Stability calculations for uniform flux surface condition; $R = 0$. These have been done by Qureshi and Gebhart (1986) for Pr = 11.6 and q = 1.0, 1.5829, 1.8364, 1.8632, and 1.8948. These values of q correspond to the Boussinesq approximation, $q = 1$, and to the characteristic salinity and pressure levels shown in Table 11.14.1. For $R = 0$, $t_\infty = t_m$, and the flow is upward. Then the above equations are simpler. A plus sign applies to the buoyancy terms in Eqs. (11.14.3) and (11.14.8). The neutral stability curves for two values of q are seen in Fig. 11.14.1. The neutral curves for the Boussinesq approximation at Pr = 6.7 and 11.6 are also shown.

The flow appears to be stabilized by cold-water effects. However, the Grashof

Table 11.14.1 Values of $q(s,p)$ at different pressures and salinity levels

$q(s, p)$	Pressure p (bars)	Salinity s (‰)	$\alpha(s, p)$
1.0[a]			
1.5829	1000	0	27.164×10^{-6}
1.8364	1	25.25	10.417×10^{-6}
1.8632	1	35.0	9.381×10^{-6}
1.8948	1	0	9.297×10^{-6}

[a]$q = 1$ represents the linear density dependence approximation. Also for $q = 1$, $\alpha = \beta$ (coefficient of thermal expansion).

Source: From Qureshi and Gebhart (1986).

number, which contains q, appears in the coordinates ω and G. Therefore, a realistic comparison must be in terms of the physical quantities, such as the distance from the leading edge, x_N, where a disturbance becomes neutrally stable. A comparison is made for a vertical surface dissipating a flux of $q'' = 1000\ \text{W/m}^2$ in water at ambient temperatures of 20 and 4°C. For water at 20°C, $x_N = 2$ cm and for water at 4°C, and $x_N = 4.75$ cm. Thus, the flow is, in fact, stabilized.

The effect of ambient pressure and salinity levels is estimated by similarly

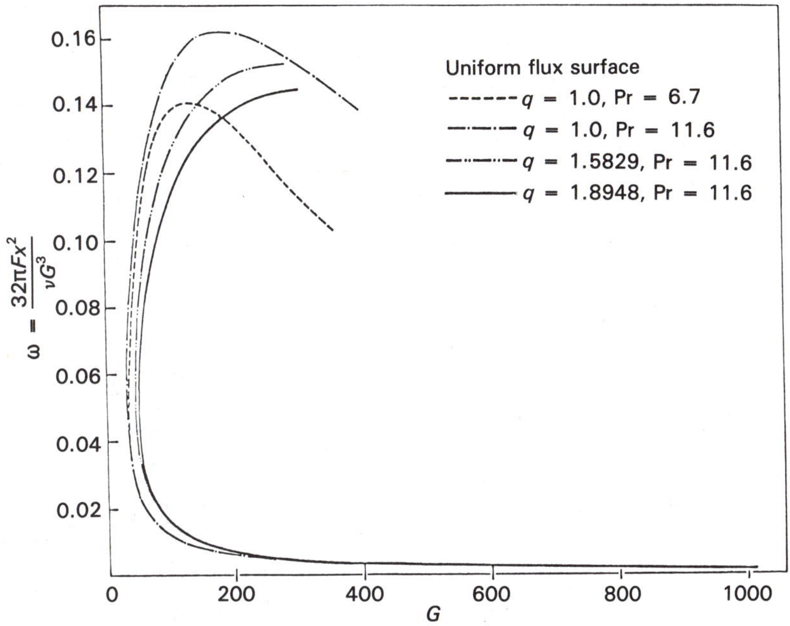

Figure 11.14.1 Neutral stability curves for a uniform flux vertical surface in cold water. *(Reprinted with permission from* Int. J. Heat Mass Transfer, *vol. 29, p. 1383, Z. Qureshi and B. Gebhart. Copyright © 1986, Pergamon Journals Ltd.)*

comparing the results for other values of $q(s, p)$. At any given salinity level, the value of $q(s, p)$ decreases as the ambient pressure level increases. The value of x_N is found to increase slightly with pressure. At a given pressure level, the variation of $q(s, p)$ with salinity is not monotonic. The initial effect of ambient salinity increase is to stabilize the flow. Note that $t_\infty = t_m(s, p)$. Hence, the ambient temperature that applies for the above conclusions varies according to the ambient salinity and pressure levels.

The disturbance amplitude contours for $q = 1.8948$ for Pr = 11.6 and $q = 1.0$ for Pr = 6.7 are compared in Fig. 11.14.2. As G increases the amplification rate becomes much greater. In the experiments of Qureshi and Gebhart (1981) the first downstream appearance of appreciable temperature and velocity disturbances was at the values of G at which $A = 6$–10. For a flux level of 1000 W/m^2 the values of x for $A = 6$ are actually 44 and 39 cm for $q = 1.8948$ and 1.0, respectively, that is, farther downstream in cold water.

As discussed in Section 11.2, the transition process begins with the selective amplification of two-dimensional disturbances. A narrow band of frequencies amplifies at a much higher rate than the others. Transition in cold water follows the same mechanism of downstream selective amplification. However, the band of frequencies is wider. The paths of three different frequencies, b, c, and d, are also shown in Fig. 11.14.2. It is seen that as G increases downstream the value

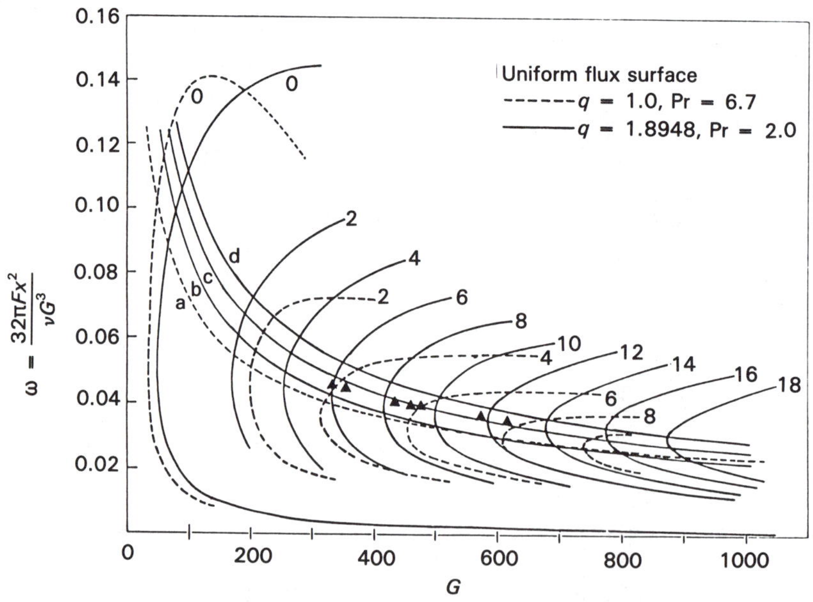

Figure 11.14.2 Amplification rate contours for a uniform flux surface in cold water. Curve a is for $q = 1$ and $\Omega = 0.73$; curves b, c, and d represent $\Omega = 1.32$, 1.51, and 1.69, respectively; (▲) characteristic frequencies in transition measured by Qureshi and Gebhart (1981). *(Reprinted with permission from* Int. J. Heat Mass Transfer, *vol. 29, p. 1383, Z. Qureshi and B. Gebhart. Copyright © 1986, Pergamon Journals Ltd.)*

of the most amplified frequency increases. This frequency is the one that passes through the minimum G values of the A contours. The suggested value of the most amplified frequency Ω for cold water is about 1.50, where Ω is related to the physical frequency F by

$$\Omega = \omega G^{(q+1)/(q+3)} = \frac{32\pi F}{\nu}\left[\frac{4^3 g\alpha(s,p)N^q}{\nu^2}\right]^{-(q+4)/2(q+3)} \tag{11.14.11}$$

where

$$N = \left\{\frac{\sqrt{2}\, q''\, \nu^{1/2}}{k[g\alpha(s,p)]^{1/4}\,[-\phi'(0)]}\right\}^{4/(4+q)}$$

The value $\Omega = 1.50$ closely corresponds to the experimental data of Qureshi and Gebhart (1981), as also shown in Fig. 11.14.2. For water at room temperature, $\text{Pr} = 6{,}7$, and $q = 1.0$, the characteristic frequency is more closely single valued, at $\Omega = 0.731$. For $q'' = 2000\ \text{W/m}^2$ the most amplified frequency in cold water, from Eq. (11.14.11), is 0.22 Hz. For water at 20°C it is 0.28 Hz.

Calculations for an isothermal surface condition. Computed neutral curves and disturbance amplification contours with $t_\infty = t_m(s, p)$, that is, $R = 0$, were reported by Qureshi and Gebhart (1986). Again, the density effect is to stabilize the flow. The ambient salinity and pressure effects on stability were also found to be the same. However, the value of the most amplified frequency Ω was found to be more nearly a single value. For $q = 1.8948$, $\Omega = 0.36$, where for an isothermal condition Ω is defined as

$$\Omega = \omega G^{1/3} = \frac{2\pi F}{\nu}\left[\frac{g\alpha(s,p)(t_0 - t_\infty)^q}{\nu^2}\right]^{-2/3} \tag{11.14.12}$$

Stability results for an isothermal surface for other values of R were given by Higgins and Gebhart (1983). Equations (11.14.3)–(11.14.10) were solved with $n = 0$. Neutral curves were determined for $R = \pm 4.0, \pm 2.0$, 1.0, and -0.5. Recall that for $R = -4$, -2, and -0.5 the buoyancy force and flow are always upward. For $R = 4$, 2, and 1.0 they are downward. These results are compared with those based on the Boussinesq approximations, for $\text{Pr} = 12.6$, in Figs. 11.14.3 and 11.14.4. The coordinates G and Ω are defined in Eqs. (11.14.7) and (11.14.12), respectively. The Boussinesq result consistently overpredicts the buoyancy force in upflow circumstances. Neutral stability conditions are predicted farther upstream. However, for conditions that result in downflow, $R > 0.5$, the predictions lie downstream of the cold-water results.

These different predictions are explained qualitatively in Fig. 11.14.5. An upflow case is shown in Fig. 11.14.5*a*, with temperatures $t_0 < t_\infty$ both less than t_m, $R < 0$. A line representing the linear approximation of $\rho_\infty - \rho$ is drawn tangent to the film temperature t_f. It is seen that at any location in the boundary region at a temperature t, the density difference $\rho(t) - \rho_\infty$ is overpredicted by the Boussinesq approximation. The resulting larger value of the buoyancy force gives higher flow vigor and causes earlier destabilization. For the downflow in Fig. 11.14.5*b*, again with $t_0 < t_\infty$ but both greater than t_m, $R > 0$. The density difference predicted

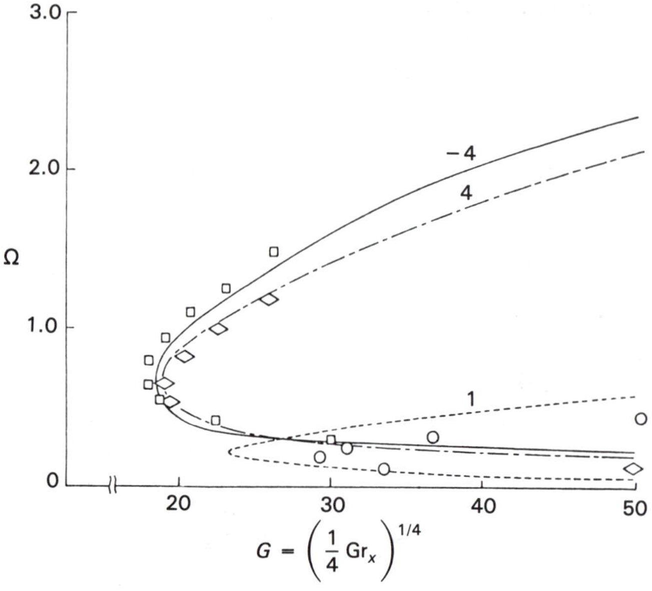

Figure 11.14.3 Stability plane for flow adjacent to a vertical surface in cold water for various values of R, Pr = 12.6; non-Boussinesq results: (——) $R = -4.0$; (– · –) $R = 4.0$, (– – –) $R = 1.0$. Boussinesq results: (□) $R = -4.0$; (◇) $R = 4.0$; (○) $R = 1.0$. *(Reprinted with permission from J. M. Higgins and B. Gebhart,* J. Heat Transfer, *vol. 105, p. 767. Copyright © 1983, ASME.)*

by the linear approximation is less than the actual one. The flow vigor is underestimated and calculated neutral conditions lie downstream of the correct values.

Transition limits. As discussed in Section 11.5, an important goal related to instability and transition is the establishment of predictive parameters for both the beginning and the end of the transition region. The parameters discussed in Section 11.5 are based on experimental data for gases or water at temperatures at which the density variation with temperature is almost linear. Two experimental studies concerning the transition of a vertical flow in cold water have been reported.

Qureshi and Gebhart (1981) made measurements in a flow adjacent to a vertical surface dissipating uniform heat flux in pure water at 4°C. Several experimental runs at different surface flux levels covered the laminar, transition, and turbulent regimes. The beginning of the velocity and thermal transition was determined as the downstream location where the maximum velocity and the surface temperature excess deviated from the laminar trends. Data indicated that the velocity and thermal transitions began simultaneously. The resulting predictive parameter was found to be of the form $G^*x^{1/5}$, nondimensionalized as $G^*(gx^3/\nu^2)^{1/15}$, where G^* is defined as

$$G^* = (4 + q)\left(\frac{\mathrm{Gr}_x^*}{4 + q}\right)^{1/(4+q)} \tag{11.14.13}$$

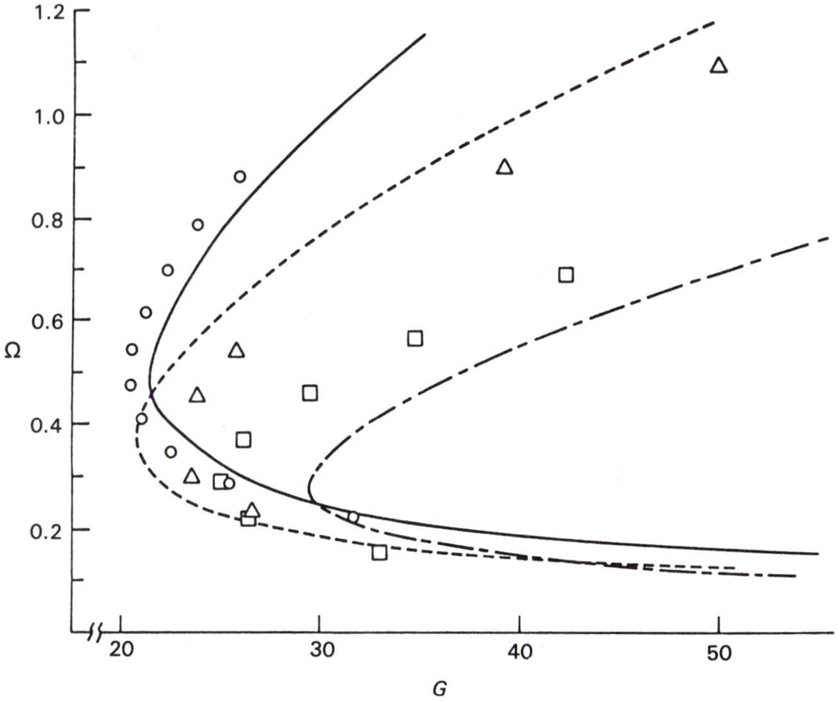

Figure 11.14.4 Stability plane for flow adjacent to a vertical surface in cold water for non-Boussinesq results: (——) $R = -2.0$; (– – –) $R = 2.0$; (– · –) $R = -\frac{1}{2}$. Boussinesq results: (○) $R = -2.0$; (△) $R = 2.0$; (□) $R = -\frac{1}{2}$. *(Reprinted with permission from J. M. Higgins and B. Gebhart,* J. Heat Transfer, *vol. 105, p. 767. Copyright © 1983, ASME.)*

and

$$\mathrm{Gr}_x^* = \frac{gx^3}{\nu^2}\,\alpha(s,p)\left(\frac{q''x}{k}\right)^q$$

a value of $G^*(gx^3/\nu^2)^{1/15}$ between 5500 and 5600 was found for the beginning of transition in all the experiments.

An experimental investigation of transition events in a flow adjacent to an isothermal vertical surface in cold pure water was reported by Higgins and Gebhart (1982). Instability and transition data were obtained with a hot-film anemometer and a fine thermocouple probe for $R = 0$, 0.1, and 0.4. The temperature difference $t_0 - t_\infty$ was limited to approximately 5°C. For $R = 0$ the flow is upward. For $R = 0.1$, that is, $t_0 \approx 9$°C and $t_\infty \approx 3.3$°C, buoyancy force reversal occurs in the outer part of the thermal layer. However, the overall flow is still upward. At $R = 0.4$, $t_0 \simeq 6.6$°C, and $t_\infty \simeq 2.3$°C the flow is downward, with an inside buoyancy force reversal. In all these experiments the thermal and velocity transitions were also found to occur simultaneously. The data suggest that the

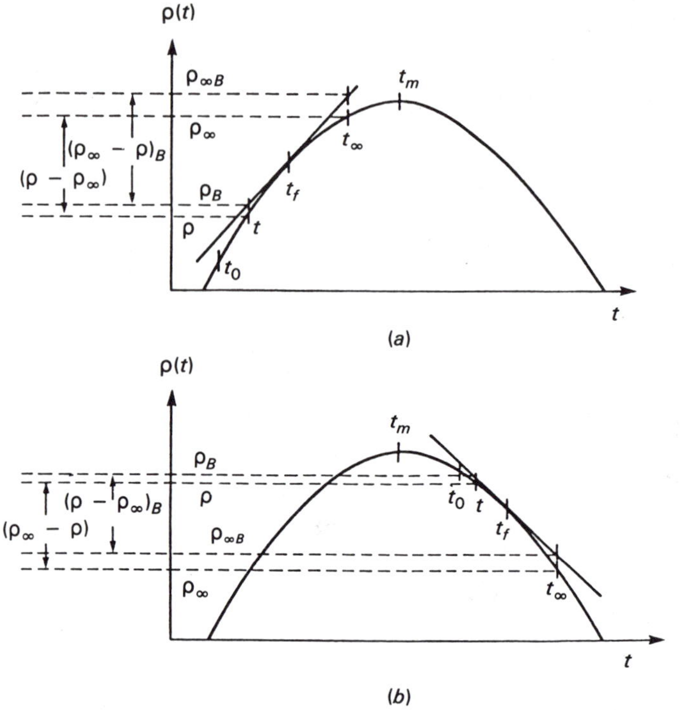

Figure 11.14.5 Illustration of the disparity between Boussinesq and true buoyancy forces near t_m; (*a*) upflow, $(\rho_\infty - \rho)_B > (\rho_\infty - \rho)$; (*b*) downflow, $(\rho_\infty - \rho)_B < (\rho_\infty - \rho)$. *(Reprinted with permission from J. M. Higgins and B. Gebhart,* J. Heat Transfer, *vol. 105, p. 767. Copyright © 1983, ASME.)*

predictive parameters E and Q_{BT} discussed in Section 11.5 closely predict these transition events also.

At $R = -0.01$, $t_0 = 9.07°C$, and $t_\infty = 4.08°C$ the flow is upward. The values of E and Q_{BT} corresponding to the beginning of transition were found to be 12.2 and 316.6, respectively, whereas for $R = 0.12$, with a buoyancy force reversal near the outer edge, the respective values of E and Q_{BT} were found to be 14.1 and 352.8. Note that for cold water the parameter G^* that appears in E is defined by Eq. (11.14.13). For an isothermal surface G is converted to G^*, using $q'' = k(t_0 - t_\infty)G/(4x[-\phi'(0)])$, where G is defined in Eq. (11.14.7).

REFERENCES

Arpaci, V. S., and Larsen, P. S. (1984). *Convection Heat Transfer,* Prentice-Hall, Englewood Cliffs, N.J.

Audunson, T. (1971). Ph.D. Thesis, Part II, Cornell Univ., Ithaca, N.Y.

Audunson, T., and Gebhart, B. (1976). *Int. J. Heat Mass Transfer 19,* 737.

Batchelor, G. K., and Gill, A. E. (1962). *J. Fluid Mech. 14,* 529.

Bayley, F. J. (1955). *Proc. Inst. Mech. Eng. 169*(20), 361.

Benny, J. D. (1961). *J. Fluid Mech. 10,* 209.
Benny, J. D., and Lin, C. C. (1960). *Phys. Fluids 3,* 656.
Bill, R. G., Jr., and Gebhart, B. (1975). *Int. J. Heat Mass Transfer 18,* 513.
Bill, R. G., Jr., and Gebhart B, (1979). *Int. J. Heat Mass Transfer 22,* 267.
Bolgiano, R. (1962). *J. Geophys. Res. 64,* 2226.
Boura, A., and Gebhart, B. (1976). *AIChE J. 22,* 94.
Bouthier, M. (1972). *J. Mec. 11,* 599.
Bouthier, M. (1973). *J. Mec. 12,* 75.
Briggs, R. J. (1964). *Electron Stream Interaction with Plasmas,* Research Monography No. 29, MIT Press, Cambridge, Mass.
Brodowicz, K., and Kierkus, W. T. (1966). *Int. J. Heat Mass Transfer 9,* 81.
Carey, V. P., and Gebhart, B. (1983). *J. Fluid Mech. 127,* 185.
Cebeci, T., and Khattab, A. (1975). *J. Heat Transfer 97,* 469.
Chessewright, R. (1968). *J. Heat Transfer 90,* 1.
Chen, C. F., and Kirchner, R. P. (1971). *J. Fluid Mech. 48,* 365.
Chen, C. F., and Sandford, R. D. (1977). *J. Fluid Mech. 83,* 83.
Chen, T. S., and Moutsoglou, A. (1979). *Numer. Heat Transfer 2,* 497.
Chen, T. S., and Mucoglu, A. (1979). *Int. J. Heat Mass Transfer 22,* 185.
Chen, T. S., and Tzuoo, K. L. (1982). *J. Heat Transfer 104,* 637.
Churchill, S. W. (1983). In *Heat Exchanger Design Handbook,* E. Schlünder, Ed., Pt. 2, Hemisphere, Washington, D.C.
Churchill, S. W., and Chu, H. H. S. (1975). *Int. J. Heat Mass Transfer 18,* 1049 and 1323.
Colak-Antic, P. (1962). *Sitzungsber. Heidelb. Akad. Wiss., Math. Naturwiss. Kl.,* 315.
Colak-Antic, P. (1964). *Jahrb.. WGLR,* 172.
Corrsin, S. (1951). *J. Appl. Phys. 22,* 469.
Coutanceau, J. (1969). *Int. J. Heat Mass Transfer 12,* 753.
Currie, J. G. (1967). *J. Fluid Mech. 29,* 337.
Drazin, P. G. (1977). *J. Mech. Appl. Math. 30*(1), 91.
Drazin, P. G., and Reid, W. H. (1981). *Hydrodynamic Stability,* Cambridge Univ. Press, London.
Dring, R. P., and Gebhart, B. (1968). *J. Fluid Mech. 34,* 551.
Dring, R. P., and Gebhart, B. (1969). *J. Fluid Mech. 36,* 447.
Eckert, E. R. G., and Jackson, T. W. (1950). NACA Tech. Note 2207.
Eckert, E. R. G., and Soehngen, E. (1951). *Proc. Gen. Discuss. Heat Transfer,* London, p. 321.
Elder, J. W. (1965). *J. Fluid Mech. 23*(1), 99.
Forstrom, R. J., and Sparrow, E. M. (1967). *Int. J. Heat Mass Transfer 10,* 321.
Foster, T. D. (1965). *Phys. Fluids 8,* 1249.
Foster, T. D. (1968). *Phys. Fluids 11,* 1257.
Garg, A. K. (1987). Ph.D. Thesis, Cornell Univ., Ithaca, N.Y.
Garg, A. K., and Leibovich, S. (1985). Presented at the 38th Annual Meeting of the Division of Fluid Dynamics, Tucson, Nov. 24–27, 1985.
Garg, V. K., and Rouleau, W. T. (1972). *J. Fluid Mech. 54*(1), 113.
Gaster, M. (1962). *J. Fluid Mech. 14,* 222.
Gaster, M. (1965). *Prog. Aerosp. Sci. 6,* 251.
Gaster, M. (1974). *J. Fluid Mech. 66*(3), 465.
Gebhart, B. (1973a). *Adv. Heat Transfer 9,* 273.
Gebhart, B. (1973b). *Annu. Rev. Fluid Mech. 5,* 213.
Gebhart, B., and Mahajan, R. L. (1975). *Int. J. Heat Mass Transfer 18,* 1143.
Gebhart, B., and Mahajan, R. L. (1982). *Advances in Applied Mechanics,* p. 231, Academic Press, New York.
Gebhart, B., and Mollendorf, J. C. (1977). *Deep Sea Res. 24,* 831.
Gebhart, B., and Pera, L. (1971). *Int. J. Heat Mass Transfer 14,* 2025.
George, W. K., Jr., and Capp, S. (1979). *Int. J. Heat Mass Transfer 22,* 813.
Godaux, F., and Gebhart, B. (1974). *Int. J. Heat Mass Transfer 17,* 93.
Goldstein, R. J., Sparrow, E. M., and Jones, D. C. (1973). *Int. J. Heat Mass Transfer 16,* 1025.

Griffiths, E., and Davis, A. H. (1922). The Transmission of Heat by Radiation and Convection, DSIR Special Rept. No. 9, British Food Investigation Board, London.
Gunness, R. C., and Gebhart, B. (1969). *Phys. Fluids 12,* 1968.
Haaland, S. E., and Sparrow, E. M. (1973a). *J. Heat Transfer 95,* 295.
Haaland, S. E., and Sparrow, E. M. (1973b). *J. Heat Transfer 95,* 405.
Haaland, S. E., and Sparrow, E. M. (1973c). *Int. J. Heat Mass Transfer 16,* 2355.
Hieber, C. A., and Gebhart, B. (1971a). *J. Fluid Mech. 48,* 625.
Hieber, C. A., and Gebhart, B. (1971b). *J. Fluid Mech. 49*(3), 577.
Hieber, C. A., and Nash, E. J. (1975). *Int. J. Heat Mass Transfer 18,* 1473.
Higgins, J. M., and Gebhart, B. (1982). *Int. J. Heat Mass Transfer 25,* 1397.
Higgins, J. M., and Gebhart, B. (1983). *J. Heat Transfer 105,* 767.
Hinze, J. O. (1975). *Turbulence, an Introduction to Its Mechanisms and Theory,* McGraw-Hill, New York.
Hocking, L. M., Stewartson, K., Stuart, J. T., and Brown, S. M. (1972). *J. Fluid Mech. 51,* 705.
Hwang, G. J., and Cheng, K. C. (1973). *Can. J. Chem. Eng. 51,* 659.
Ishiguro, R., Abe, T., Nagase, H., and Nakanishi, S. (1978). *Proc. 6th Int. Heat Transfer Conf.,* Toronto, Paper NC8.
Iyer, P.A., and Kelly, R. E. (1974). *Int. J. Heat Mass Transfer 17,* 517.
Jaluria, Y. (1976). *Int. J. Heat Mass Transfer 19,* 1057.
Jaluria, Y. (1985). In *Stability of Convective Flows,* W. S. Saric and A. A. Szewczyk, Eds., vol. 54, p. 1, ASME Heat Transfer Div.
Jaluria, Y., and Gebhart, B. (1973). *J. Fluid Mech. 61,* 337.
Jaluria, Y., and Gebhart, B. (1974). *J. Fluid Mech. 66,* 309.
Jaluria, Y., and Gebhart, B. (1977). *Int. J. Heat Mass Transfer 20,* 434.
Jhaveri, B., and Homsy, G. M. (1980). *J. Fluid Mech. 98,* 329.
Jhaveri, B., and Homsy, G. M. (1982). *J Fluid Mech. 114,* 251.
Jones, W. P., and Launder, B. E. (1972). *Int. J. Heat Mass Transfer 15,* 193.
Jordinson, R. (1970). *J. Fluid Mech. 43,* 801.
Joshi, Y., and Gebhart, B. (1987). *Int. J. Heat Mass Transfer.* In press.
Kahawita, R. A., and Meroney, R. N. (1974). *Int. J. Heat Mass Transfer 17,* 541.
Kato, H., Nishiwaki, N., and Hirata, M. (1968). *Int. J. Heat Mass Transfer 11,* 1117.
Kimura, S., and Bejan, A. (1983). *Int. J. Heat Mass Transfer 26,* 1515.
Klebanoff, P. S. (1954). NACA Tech. Note TN-3178.
Klebanoff, P. S., Tidstrom, K. D., and Sargent, L. M. (1962). *J. Fluid Mech. 12,* 1.
Knowles, C. P., and Gebhart, B. (1968). *J. Fluid Mech. 34,* 657.
Knowles, C. P., and Gebhart, B. (1969). *Prog. Heat Mass Transfer 2,* 99.
Kurtz, E. F., and Crandall, S. H. (1962). *J. Math. Phys. 41,* 264.
Kutateladze, S. S. (1963). *Fundamentals of Heat Transfer,* p. 293, Edward Arnold, London.
Kutateladze, S. S. (1972). *Int. J. Heat Mass Transfer 15,* 193.
Lamb, H. (1945). *Hydrodynamics,* Dover, New York.
Landau, L. D., and Lifshitz, E. M. (1959). *Fluid Mechanics,* vol. 6 of *Course of Theoretical Physics,* Pergamon, London.
Lee, J. B., and Lock, G. S. H. (1972). *Trans. Can. Soc. Mech. Eng. 1,* 197.
Lee, S. L., and Emmons, H. W. (1961). *J. Fluid Mech. 11,* 353.
Lessen, M., and Singh, P. J. (1973). *J. Fluid Mech. 60*(3), 443.
Lick, W. (1965). *J. Fluid Mech. 21,* 565.
Lloyd, J. R., and Sparrow, E. M. (1970). *J. Fluid Mech. 42,* 465.
Lock, G. S. H., and Trotter, F. J. de B. (1968). *Int. J. Heat Mass Transfer 11,* 1225.
Lock, G. S. H., Gort, C., and Pond, G. R. (1967). *Appl. Sci. Res. 18,* 171.
Lopez, J. L., and Kurzweg, V. H. (1977). *Phys. Fluids 20,* 860.
Lumley, J. L. (1965). *J. Atmos. Sci. 21,* 99.
Lumley, J. L., and Panofsky, H. A. (1964). *The Structure of Atmospheric Turbulence,* Interscience, New York.

Mahajan, R. L. (1977). Ph.D. Dissertation, Cornell Univ., Ithaca, N.Y.
Mahajan, R. L., and Gebhart, B. (1979). *J. Fluid Mech. 91,* 131.
Mahler, E. G., and Schechter, R. S.(1970). *Chem. Eng. Sci. 25,* 955.
Mahler, E. G., Schechter, R. S., and Wissler, E. H. (1968). *Phys. Fluids 11,* 1901.
Mason, H. B., and Seban, R. A. (1974). *Int. J. Heat Mass Transfer 17,* 1329.
McAdams, W. H. (1954). *Heat Transmission,* 3d ed., McGraw-Hill, New York.
Miller, R. M., and Gebhart, B. (1978). *Int. J. Heat Mass Transfer 21,* 1229.
Miyabe, K., and Katsuhara, T. (1972). *Mem. Kyushu Inst. Technol. Eng. 2,* 9.
Mollendorf, J. C., and Gebhart, B. (1970). *J. Heat Transfer 92,* 628.
Mollendorf, J. C., and Gebhart, B. (1973). *J. Fluid Mech. 61*(2), 367.
Morton, B. R. (1957). *Q. J. Mech. Appl. Math. 10,* 433.
Mucoglu, A., and Chen, T. S. (1978). *Numer. Heat Transfer 1,* 267.
Nachtsheim, P. R. (1963). NASA Tech. Note D-2089.
Nee, V. W., and Yang, K. T. (1970). *Heat Transfer,* vol. IV, *Proc. 4th Int. Heat Transfer Conf.,* NC1.12, Versailles, Elsevier, Amsterdam.
Noto, K., and Matsumoto, R. (1975). *J. Heat Transfer 97,* 621.
Obukhov, A. M. (1941). *Izv. Akad. Nauk SSSR Ser. Geogr. Geofiz. 4–5,* 453.
Papailou, D. D., and Lykoudis, P. S. (1974). *Int. J. Heat Mass Transfer 17,* 161.
Pera, L., and Gebhart, B. (1971). *Int. J. Heat Mass Transfer 14,* 975.
Pera, L., and Gebhart, B. (1973). *Int. J. Heat Mass Transfer 16,* 1147.
Piau, J. M. (1972). *C. R. Acad. Sci. Ser. A. 294,* 420.
Plapp, J. E. (1957). *J. Aeronaut. Sci. 24,* 318.
Plumb, O. A., and Kennedy, L. A. (1977). *J. Heat Transfer 99,* 79.
Polymeropoulos, C. E., and Gebhart, B. (1967). *J. Fluid Mech. 30,* 225.
Priestly, C. B. H. (1959). *Turbulent Transfer in the Lower Atmospheres,* Univ. of Chicago Press, Chicago.
Qureshi, Z. M., and Gebhart, B. (1978). *Int. J. Heat Mass Transfer 21,* 1967.
Qureshi, Z. M., and Gebhart, B. (1981). *Int. J. Heat Mass Transfer 24,* 1503.
Qureshi, Z. M., and Gebhart, B. (1986). *Int. J. Heat Mass Transfer 29,* 1383; see also Qureshi, Z. M. (1980). Ph.D. Dissertation, State Univ. of New York, Buffalo.
Rayleigh, L. (1880). *Proc. London Math. Soc. 11,* 57.
Regnier, G. M., and Kaplan, C. (1963). *Proc. Heat Transfer Fluid Mech. Inst.,* p. 94, Stanford Univ. Press, Stanford, Calif.
Rouse, H., Yih, C. S., and Humphreys, H. W. (1952). *Tellus 4,* 21.
Saric, W. S., and Nayfeh, A. H. (1975). *Phys. Fluids 18,* 945.
Saunders, O. A. (1939). *Proc. R. Soc. London Ser. A172,* 55.
Schlichting, H. (1932). *Nachr. Ges. Wiss. Goettingen Math. Phys. J.,* 196.
Schlichting, H. (1968). *Boundary Layer Theory,* McGraw-Hill, New York.
Schubauer, G. B., and Klebanoff, P. A. (1956). NACA Rept. No. 1289.
Shaukatullah, H. (1974). Graduate School of Aerospace and Mechanical Engineering, Cornell University, personal communications.
Shaukatullah, H., and Gebhart, B. (1978). *Int. J. Heat Mass Transfer 21,* 1481.
Shen, S. F. (1961). *J. Aerosp. Sci. 28,* 397.
Schütz, G. (1963). *Int. J. Heat Mass Transfer 6,* 873.
Shur, G. N. (1962). *Trudy 43,* 79; translated as AID Rept. T-63-55, Aerospace Information Division, Library of Congress, Washington, D.C.
Smith, R. R. (1972). Ph.D. Thesis, Univ. of London, Queen Mary College.
Sparrow, E. M., and Husar, R. B. (1969). *J. Fluid Mech. 37,* 251.
Stuart, J. T. (1965). *Appl. Mech. Rev. 18,* 523.
Sturrock, P. (1958). *Phys. Rev. 112,* 1488.
Sturrock, P. (1961). In *Plasma Physics,* J. E. Drummond, Ed., McGraw-Hill, New York.
Sudan, R. N. (1965). *Phys. Fluids 8,* 1899.
Szewczyk, A. A. (1962). *Int. J. Heat Mass Transfer 5,* 903.

Tani, I. (1969). *Rev. Fluid Mech. 1,* 169.
Tennekes, H., and Lumley, J. L. (1972). *A First Course in Turbulence,* MIT Press, Cambridge, Mass.
Tzuoo, K. L., Chen, T. S., and Armaly, B. F. (1985). *J. Heat Transfer 107,* 107.
Vliet, G. C., and Liu, C. K. (1969). *J. Heat Transfer 91,* 517.
Vliet, G. C., and Ross, D. C. (1975). *J. Heat Transer 97,* 549.
Von Karman, T., and Lin, C. C. (1955). *J. Franklin Inst. 259,* 517.
Wakitani, S., (1985). *J. Fluid Mech. 159,* 241.
Wakitani, S., and Yosinobu, H. (1984). *J. Phys. Soc. Jpn. 53,* 1291.
Warner, C. Y., (1966), Ph.D. Thesis, Univ. of Michigan, Ann Arbor.
Warner, C. Y., and Arpaci, V. S., (1968). *Int. J. Heat Mass Transfer 11,* 397.
Watson, J. (1962). *J. Fluid Mech. 14,* 211.
Yang, R. J., and Aung, W. (1985). In *Natural Convection Fundamentals and Applications,* S. Kakac, W. Aung, and R. Viskanta, Eds., Hemisphere, Washington, D.C.
Zaman, K. B. M. Q., and Hussain, A. K. M. F. (1981). *J. Fluid Mech. 112,* 379.
Zaman, K. B. M. Q., and Hussain, A. K. M. F. (1984). *J. Fluid Mech. 138,* 325.
Zel'dovich, Y. B. (1937). *Zh. Eksp. Teor. Fiz. 7,* 1463.

PROBLEMS

11.1 Reconsider the geometry and temperature conditions in Problem 6.2, but with a dry surface, that is, no evaporation. Find the distance downstream from the leading edge at which fluid mechanical transition would begin for a uniform surface heat flux condition, instead of an isothermal one, for $q'' = 12$ Btu/h ft^2.

11.2 A tall vertical flat surface in water at 20°C dissipates a steady heat flux of $q'' = 500$ W/m^2 from one side. The other side is insulated.

(a) Calculate the temperature distribution t_0 upward along the surface and the average value of t_0 between $x = 0$ and 1 m.

(b) Find the vertical location $x = x_N$ at which disturbances first amplify and the frequency of the first unstable disturbance component.

(c) For an initial disturbance amplitude of $10^{-5}u_{\max}$ at x_N, find the downstream location where the amplitude of the selected frequency has become $10^{-2}u_{\max}$.

(d) Estimate the downstream location $x = x_t$ where transition will begin.

11.3 For a vertical uniform flux flat plate in N_2 at n atm, observations are to be made on laminar instability. The flux level is to be sufficient that $t_0 - t_\infty$ at $x = 4$ in. is 30°F. Estimate necessary values of $\phi'(0)$ and η_δ from numerical results in the text.

(a) For a maximum plate height of 8 in., find the required pressure level so that naturally occurring disturbances will be seen to be amplified in the boundary layer.

(b) For $n = 16$ and $L = 8$ in. and the same q'' as in (a), find the location at which a disturbance would first be amplified and calculate its wavelength.

(c) For the conditions of (b) calculate $\Delta\delta/\delta$ for one wavelength of the disturbance and assess the assumption of locally one-dimensional flow.

11.4 The turbulent stress system is usually reduced to velocity variables. For constant properties show the equivalence of

$$\rho \frac{D\bar{u}}{D\tau} = X - \frac{\partial \bar{p}}{\partial x} + \mu\nabla^2\bar{u} + \frac{\partial \sigma_x)_T}{\partial x} + \frac{\partial \tau_{xz})_T}{\partial y} + \frac{\partial \tau_{xy})_T}{\partial z}$$

and

$$\frac{D\bar{u}}{D\tau} = \frac{X}{\rho} - \frac{1}{\rho}\frac{\partial \bar{p}}{\partial x} + \nu\nabla^2\bar{u} - \overline{u'\frac{\partial u'}{\partial x}} - \overline{v'\frac{\partial u'}{\partial x}} - \overline{w'\frac{\partial u'}{\partial z}}$$

where

$$\sigma_x)_T = -\rho\overline{u'^2} \qquad \tau_{xy})_T = -\rho\overline{u'v'} \qquad \tau_{xz})_T = -\rho\overline{u'w'}$$

11.5 A vertical surface in water at 20°C dissipates heat at a rate of 100 W/m^2.

(a) Calculate the downstream locations of the first unstable mode and the later most highly amplified mode.

(b) At what downstream location will the amplitude of the latter increase by 3000 times?

(c) Calculate the downstream locations of the beginning and end of transition.

11.6 For a plane plume in air at 20°C, rising from a long horizontal wire causing a thermal input of 10 W/m, consider the downstream instability characteristics. Determine the downstream distance at which a disturbance of 5 Hz first becomes unstable, both from the simplest and the higher-order stability results.

11.7 For the porous surface drying process specified in Problem 6.5, consider the stability of the resulting flow generated by the combination of the two buoyancy effects.

(a) Calculate the relative magnitude of the two buoyancy components, that is, $\tilde{N}$.

(b) The value of Le = α/D is about 0.86. Assuming instead Le = 1, calculate the downstream location at which neutral stability first arises.

(c) What is the frequency of the first unstable disturbance component?

11.8 A porous vertical surface at 30°C causes a surface concentration of 0.02 kg/m^3 of CO_2. The ambient is at 1 atm and 10°C, and the concentration of CO_2 is very small.

(a) Determine the ratio of the magnitudes of the two buoyancy force components, $\tilde{N}$.

(b) Find the downstream location at which neutral stability is reached for any disturbance.

(c) What is the frequency of that disturbance mode?

(d) At what distance downstream will the amplitude of the most favored frequency component have increased by a factor of 400?

11.9 A vertical surface in air at 20°C flowing vertically at 3 cm/s dissipates energy uniformly at 20 W/m^2.

(a) Calculate the mixed convection regime that arises.

(b) Estimate the downstream location of first disturbance instability, according to the stability plane given.

(c) At what location will the favored frequency mode have grown by a factor of 50 from its magnitude at the neutral location?

11.10 A vertical surface in water at 20°C, flowing vertically at 3 cm/s, dissipates energy uniformly at 100 W/m^2.

(a) Calculate the mixed convection regime that arises for a surface height of 50 cm.

(b) For the most highly amplified disturbance mode, determine the downstream locations of first instability and of a growth of 50 in magnitude.

11.11 For the conditions in cold water given in Problem 9.6, consider downstream instability and disturbance growth.

(a) At what downstream location does instability first arise, for what frequency mode?

(b) Considering amplification ratios to e^{12}, which is the characteristic frequency?

(c) At what downstream location was it first unstable?

CHAPTER

TWELVE

TURBULENT FREE–BOUNDARY BUOYANT FLOWS

12.1 FREE–BOUNDARY FLOWS AND TRANSPORT EQUATIONS

This chapter considers turbulent flows arising from input of both momentum and buoyancy directly into an extensive ambient body of fluid. Throughout, the buoyant fluid is the same fluid as the ambient. In the absence of interacting solid boundaries, the effects of buoyancy on the resulting flow regimes are more direct in terms of preferred flow directions, shapes, and internal circulations. In these free-boundary flows, either steady or transient, the driving buoyancy force and any initial momentum form and control the downstream flow. Earlier chapters considered several largely laminar free-boundary flows. However, most free-boundary flows of appreciable size are turbulent, even quite close to the source.

The different transport mechanisms that are discussed here are plumes, starting plumes, jets, thermals, and buoyant jets, as sketched in Fig. 12.1.1. Plumes result when energy is supplied continuously at some location in a fluid. The buoyant region causes flow, which continues downstream. Such flows may be planar, resulting from a "line" source, or axisymmetric, as from a "point" source. Thermals, on the other hand, result from a sudden discontinuous release or burst of buoyancy. These flows remain confined downstream to a limited, but increasing, fluid volume, as in Fig. 12.1.1. A starting plume is the advancing front of a buoyant plume as it is being established in its surroundings. Such flows resemble thermals near their advancing front, while upstream near the source they resemble steady plumes. Nonbuoyant jets, or simply jets, occur when material is discharged into surroundings at the same density. In buoyant jets or forced plumes the dis-

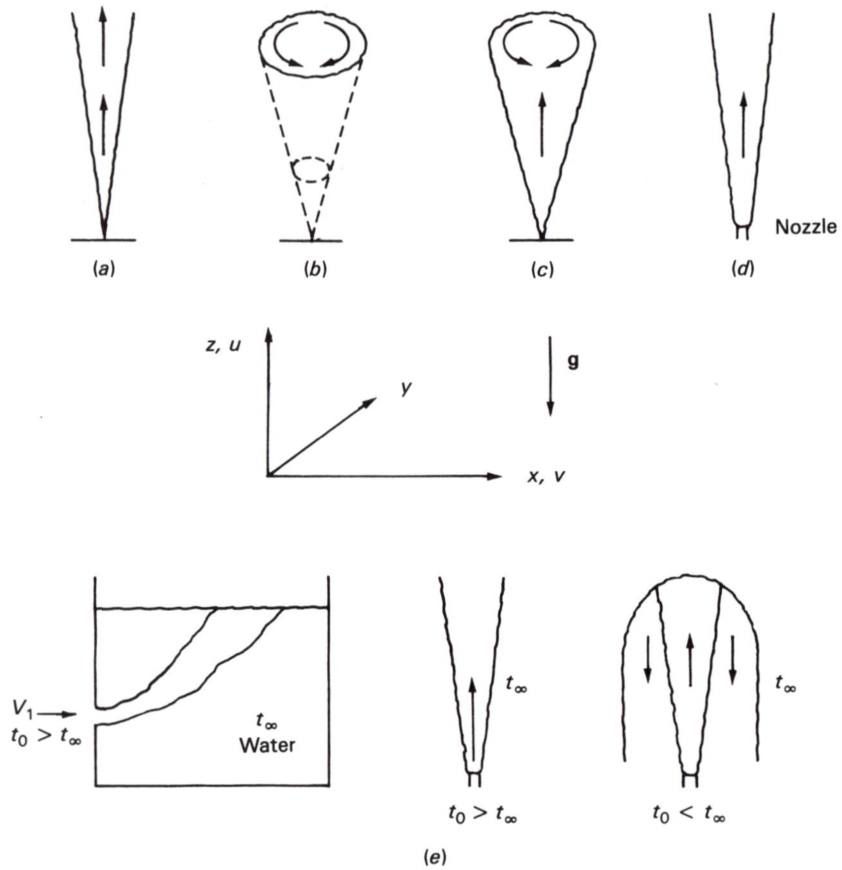

Figure 12.1.1 Plumes, thermals, starting plumes, jets, buoyant jets, and the coordinate system. (*a*) Plume; (*b*) thermal; (*c*) starting plume; (*d*) jet; (*e*) examples of buoyant jets.

charged fluid is at a density different from that of the ambient. Buoyant jets quickly become plumes when the initial momentum is small. On the other hand, they resemble jets when the input buoyancy is small.

The general governing equations of motion, representing conservation of mass, momentum, energy, and species, for turbulent free-boundary flows, are written below as in Chapter 2:

$$\frac{D\rho}{D\tau} = -\rho\nabla\cdot\mathbf{V} \tag{12.1.1}$$

$$\rho\frac{D\mathbf{V}}{D\tau} = \rho\mathbf{g} - \nabla p + \mu\nabla^2\mathbf{V} + \frac{1}{3}\mu\nabla(\nabla\cdot\mathbf{V}) \tag{12.1.2}$$

$$\rho c_p\frac{Dt}{D\tau} = \nabla\cdot k\nabla t + \beta T\frac{Dp}{D\tau} + \mu\Phi + q''' \tag{12.1.3}$$

$$\frac{DC}{D\tau} = \nabla \cdot D\nabla C + c''' \tag{12.1.4}$$

These are in terms of local instantaneous values of **V**, ρ, *p*, *t*, and *C*. The terms on the left side are the transient and convective effects. Those on the right side include the molecular diffusion and other terms. These equations represent the flow throughout the discharge field. An equation of state is also needed to relate the local density to the temperature and concentration and, perhaps, even to the pressure.

The mechanisms of turbulent free-boundary flows usually are very complicated. Most past modeling of such flows has therefore been based on greatly simplifying, though often reasonable, assumptions and approximations. For example, continuing flows are taken to be everywhere largely turbulent and of either planar or axisymmetric form. Molecular transport rates are assumed negligible. The density is assumed constant except in the buoyancy force term, unless ambient stratification is present. There are two common approaches to such calculations, within the limits of the above assumptions.

The first kind of modeling is in terms of the actual structure of the flow, by determining a relation between the Reynolds stress tensor and the deformation tensor of the fluid. Then the dependent variables are expressed as a sum of mean and fluctuating values, the equations are time-averaged, and an order-of-magnitude analysis is carried out. However, more unknowns arise than equations, necessitating further assumptions. Early investigators such as Tollmien (1926) and Taylor (1932) assumed self-preservation for the mean and fluctuating quantities. More elaborate approaches include those of Saffman (1970), Lumley and Khajeh-Nouri (1974), and Cormak (1975). Details of such techniques are given in Schlichting (1968), Tennekes and Lumley (1973), and Hinze (1975), among others.

The second approach is to quantify only the overall behavior of the flow, rather than the detailed structure. Integral methods are applied. Temperature and velocity profiles are assumed. The magnitude and functional dependence of the downstream entrainment of the ambient fluid are also assumed. Then the equations of motion are reduced to ordinary differential equations, which may be integrated downstream. This approach is discussed in some detail in the next sections for plumes, jets, buoyant jets, eventually including the effects of ambient stratification, and thermals. The objective, in all events, is to determine the trajectory of the entraining discharge, along with the decay of the buoyancy-producing effects of thermal and/or chemical species concentration. The practical questions usually are: where does the polluting discharge go and how fast does its concentration diminish?

12.2 STEADY AND STARTING PLUMES

Steady plumes are flows generated when buoyancy is supplied continuously in a medium free of solid boundaries, thereby creating a continuous buoyant downstream region. The resulting flow regime depends on the strength of the source. It also depends on the state of the environment: whether it is quiescent or moving, uniform or stratified. Most investigations of plume flows are concerned with steady conditions, which arise after the termination of the initial transient regime, referred to as the starting plume.

Steady plumes. The turbulent characteristics are taken to be steady; other usual principal assumptions are

1. Fully turbulent flow; molecular diffusion is neglected compared to turbulent transport.
2. Streamwise or downstream turbulent transport is negligible compared to convective transport.
3. The variation of density throughout the flow field is small compared to the density level. The density variation is included only in buoyancy, with the Boussinesq approximation.
4. Other fluid properties are taken constant.
5. Pressure is hydrostatic throughout the fluid field.
6. The flow remains axisymmetric or planar throughout the region of analysis.

Equations (12.1.1)–(12.1.4) for a steady vertical plume in a quiescent ambient medium, using the coordinate system shown in Fig. 12.1.1, reduce to

$$\frac{\partial}{\partial z}(\rho \bar{u} x^j) + \frac{\partial}{\partial x}(\rho \bar{v} x^j) = 0 \qquad (12.2.1)$$

$$\frac{\partial}{\partial z}(\rho \bar{u}^2 x^j) + \frac{\partial}{\partial x}(\rho \bar{u}\bar{v} x^j) = -g(\rho - \rho_\infty)x^j - \frac{\partial}{\partial x}(x^j \rho \overline{u'v'}) \qquad (12.2.2)$$

$$\frac{\partial}{\partial z}(\rho \bar{u}\bar{t} x^j) + \frac{\partial}{\partial x}(\rho \bar{v}\bar{t} x^j) = -\frac{\partial}{\partial x}(x^j \rho \overline{vt'}) \qquad (12.2.3)$$

$$\frac{\partial}{\partial z}(\rho \bar{u}\bar{C} x^j) + \frac{\partial}{\partial x}(\rho \bar{v}\bar{C} x^j) = -\frac{\partial}{\partial x}(x^j \rho \overline{v'C'}) \qquad (12.2.4)$$

where the superscript j equals one for axisymmetric plumes and zero for line plumes, $\bar{u}$ and $\bar{v}$ are the time-averaged velocities in the z and x directions, respectively, u' and v' are the fluctuating values, $\bar{t}$ and $\bar{C}$ are the time-averaged temperature and species concentration values, and t' and C' are the fluctuations in temperature and species concentration.

These partial differential equations are not a closed system; see Chapter 11. One way to avoid this difficulty is to use an integral method. Velocity, temperature, and concentration profiles are assumed, usually Gaussian, based on experimental observations. The equations are then integrated across the flow cross section to yield ordinary differential equations. Finally, the ambient fluid entrainment rate is assumed, usually as a function of the local centerline velocity.

Although the integral approach yields satisfactory results for many circumstances, it is difficult to extend to more complex turbulent flows such as inclined buoyant jets discharged into an arbitrarily stratified ambient medium, into a turbulent cross flow, or into a shallow medium. For such flows the downstream velocity and temperature profiles are not known, even approximately, and different formulations must be used. A common approach is to solve the partial differential equations numerically downstream.

Here and in Section 12.4 the integral method analysis is used. The ordinary differential equations that result from integrating Eqs. (12.2.1)–(12.2.4) over the whole jet cross section are given below:

$$\frac{d}{dz}\int_0^W \rho\bar{u}x^j\,dx = E \tag{12.2.5}$$

$$\frac{d}{dz}\int_0^W \rho\bar{u}^2x^j\,dx = g\int_0^W (\rho_\infty - \rho)x^j\,dx \tag{12.2.6}$$

$$\frac{d}{dz}\int_0^W \rho\bar{u}(\bar{t} - t_\infty)x^j\,dx = -\frac{dt_\infty}{dz}\int_0^W \rho\bar{u}x^j\,dx \tag{12.2.7}$$

$$\frac{d}{dz}\int_0^W \rho\bar{u}(\bar{C} - \bar{C}_\infty)x^j\,dx = -\frac{dC_\infty}{dz}\int_0^W \rho\bar{u}x^j\,dx \tag{12.2.8}$$

where E is the mass of fluid entrained from the ambient, per unit plume length downstream, and W is the plume width measured from the plane or axis of symmetry. In general, E is taken as the perimeter of the interface with the ambient, P, multiplied by the local velocity level $\bar{u}$ as follows:

$$E = P\bar{u}\alpha$$

Here α is a constant of proportionality called the entrainment coefficient.

Measurements by Rouse et al. (1952) and others indicate that the time-averaged profiles downstream are approximately similar and Gaussian, as shown in Figs. 12.2.1 and 12.2.2 for axisymmetric and planar plumes, respectively. For the axisymmetric plume, these are

$$\bar{u} = k_1\left(\frac{B_0}{z}\right)^{1/3}\exp\left(-96\frac{r^2}{z^2}\right) \tag{12.2.9}$$

$$g' = g\left(\frac{\rho - \rho_\infty}{\rho_\infty}\right) = 11(B_0)^{2/3}z^{-5/3}\exp\left(-71\frac{r^2}{z^2}\right) \tag{12.2.10}$$

where $B_0 = 2\pi\int_0^\infty (\bar{u}g(\rho - \rho_0)/\rho_0)r\,dr$. Here, the density at the source is ρ_0 and r is the radial distance. Thus, $\rho_0 B_0$ is the total weight deficiency produced per unit time at the source. Equations (12.2.9) and (12.2.10) apply for an axisymmetric plume, where $k_1 = 4.7$ was reported by Rouse et al. (1952). George et al. (1977), Nakagome and Hirata (1976), and Beuther (1980) reported values for k_1 in the range 3.4–3.9. For the planar plume the profiles are

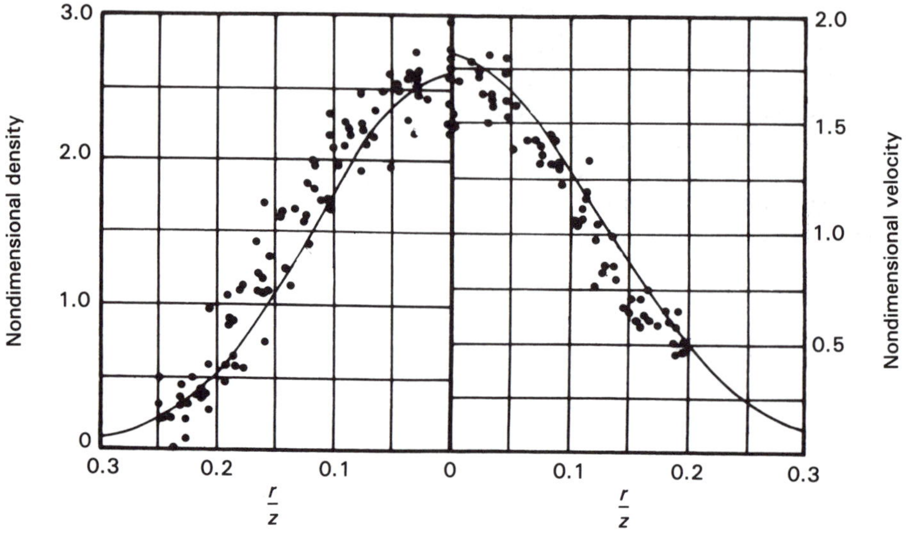

Figure 12.2.1 Nondimensional velocity and density profiles for an axially symmetric plume, from Eqs. (12.2.9) and (12.2.10). *(Reprinted with permission from H. Rouse et al.,* Tellus, *vol. 4, p. 201. Copyright © 1952, Swedish Geophysical Society.)*

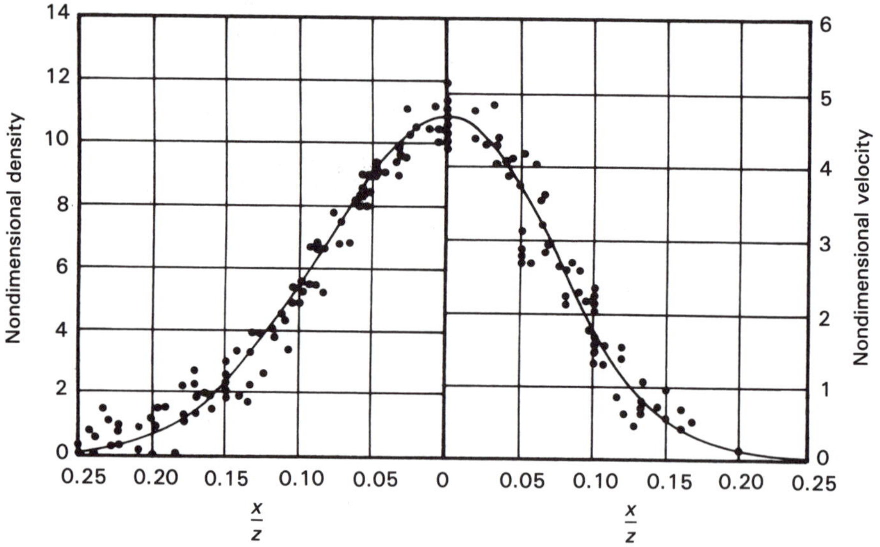

Figure 12.2.2 Nondimensional velocity and density profiles for a plane plume, from Eqs. (12.2.11) and (12.2.12). *(Reprinted with permission from H. Rouse et al.,* Tellus, *vol. 4, p. 201. Copyright © 1952, Swedish Geophysical Society.)*

$$\bar{u} = k_2 B_0^{1/3} \exp\left(-32\frac{x^2}{z^2}\right) \tag{12.2.11}$$

$$g' = 2.6 B_0^{2/3} z^{-1} \exp\left(-41\frac{x^2}{z^2}\right) \tag{12.2.12}$$

where $k_2 = 1.8$ was reported by Rouse et al. (1952). Kotsovinos (1975, 1977) reported experimental results for planar plumes. Laser Doppler anemometry was used to measure the axial velocity; the results are shown in Fig. 12.2.3 and indicate $k_2 = 1.66$. It should be noted that the z dependence in Eqs. (12.2.9) and (12.2.10) can be obtained directly from dimensional similarity considerations. This is demonstrated in detail by Chen and Rodi (1980) for several plume and jet flows. More detailed discussions of measurements in axisymmetric and turbulent plumes are given by Chen and Rodi (1980) and List (1982). Figures 12.2.4 and 12.2.5 from Rouse et al. show the mean isotherms and the streamlines for axisymmetric and plane plumes, respectively.

Integral analysis. The form of the entrainment velocity has not been specified up to this point since it was not needed for the dimensional similarity analysis discussed above. Taylor pointed out that the linear increase in plume thickness implies that the entrainment velocity must be proportional to the mean local upward velocity; see Morton et al. (1956). This entrainment hypothesis was invoked to find solutions for similar flows, where similarity here implies flows that maintain velocity and concentration profiles that have the same shape at different downstream locations. It was assumed that the velocity and buoyancy force are constant across the plume and zero outside it. These are called "top hat" profiles. The volume, momentum, and density deficiency conservation equations for an axisymmetric plume are then obtained by substituting the assumed top hat profiles into Eqs. (12.2.5)–(12.2.7) as follows:

$$\frac{d}{dz}(\pi b_T^2 \bar{u}_T) = P\bar{u}\alpha = 2\pi b_T \alpha_T \bar{u}_T = E \tag{12.2.13}$$

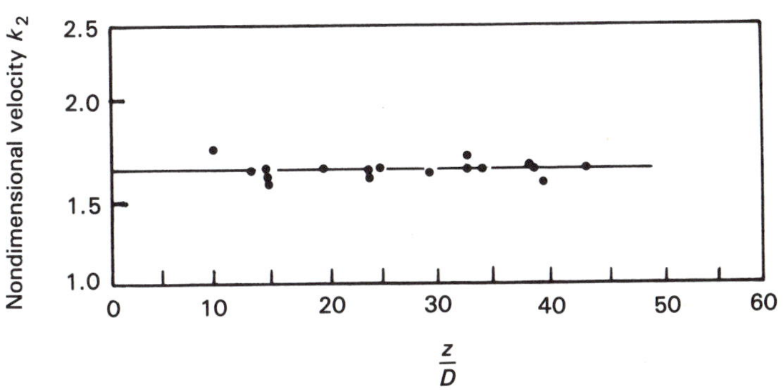

Figure 12.2.3 Nondimensional velocity on the axis of a plane plume, where D is the initial jet width. *(Adapted from List, 1982.)*

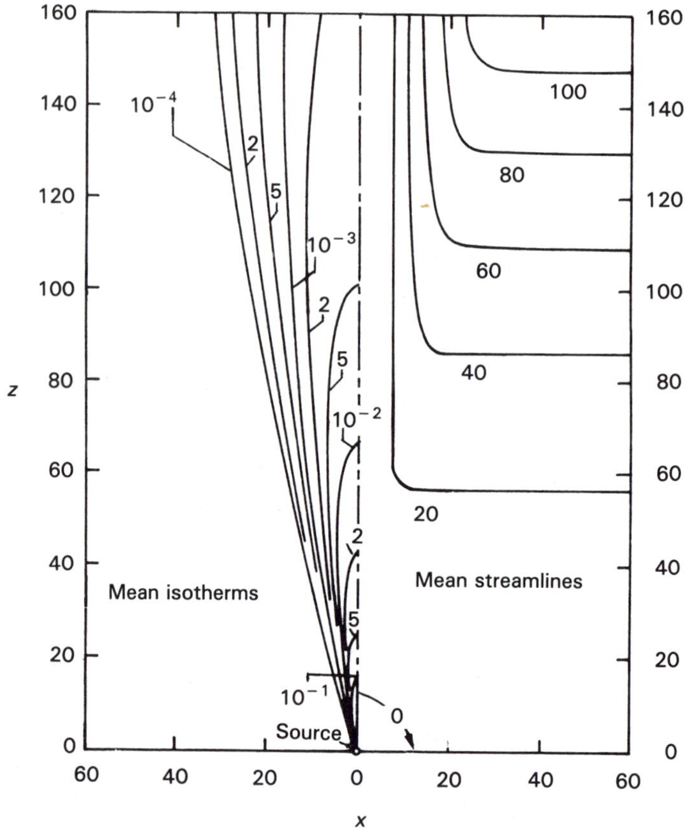

Figure 12.2.4 Convection pattern over a point source. *(Reprinted with permission from H. Rouse et al.,* Tellus, *vol. 4, p. 201. Copyright © 1952, Swedish Geophysical Society.)*

$$\frac{d}{dz}(\pi b_T^2 \bar{u}_T^2 \rho) = \pi b_T^2 g(\rho - \rho_\infty) \tag{12.2.14}$$

$$\frac{d}{dz}[\pi b_T^2 \bar{u}_T(\rho_0 - \rho)] = 2\pi b_T \alpha_T \bar{u}_T(\rho_0 - \rho_\infty) \tag{12.2.15}$$

where the subscript T refers to top hat profile, α_T is the proportionality constant in the entrainment hypothesis, and $P = 2\pi b_T$ is the plume perimeter. It should be noted that if the plume velocity distribution is taken as Gaussian, the entrainment constant becomes $\alpha = \alpha_T/\sqrt{2}$ and width $b = b_T/\sqrt{2}$. For small density variations the system of equations becomes

$$\frac{d}{dz}(b_T^2 \bar{u}_T) = 2\alpha_T b_T \bar{u}_T \tag{12.2.16}$$

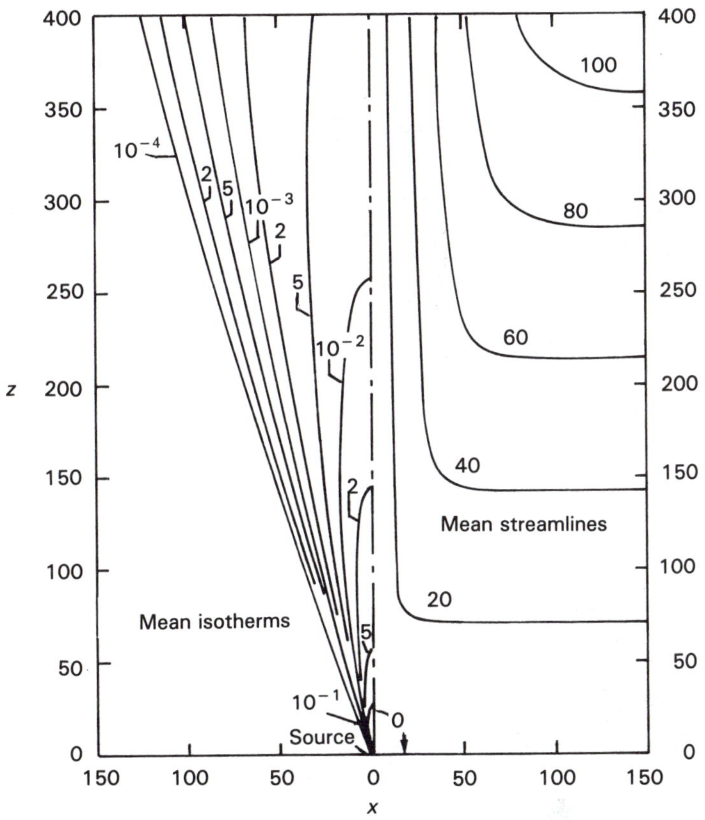

Figure 12.2.5 Convection pattern over a line source. *(Reprinted with permission from H. Rouse et al.,* Tellus, *vol. 4, p. 201. Copyright © 1952, Swedish Geophysical Society.)*

$$\frac{d}{dz}(b_T^2\bar{u}_T^2) = b_T^2 g\,\frac{\rho_\infty - \rho}{\rho_0} \tag{12.2.17}$$

$$\frac{d}{dz}\left[b_T^2\bar{u}_T g\left(\frac{\rho_\infty - \rho}{\rho_0}\right)\right] = b_T^2\bar{u}_T\,\frac{g}{\rho_0}\,\frac{d\rho_\infty}{dz} \tag{12.2.18}$$

For ρ_∞ constant, the last term in Eq. (12.2.18) is zero. Integration of this equation then gives $b_T^2\bar{u}_T g(\rho_\infty - \rho)/\rho_0 = \text{const.}$, indicating that the buoyancy flux is conserved. Integrating the above equations for constant ρ yields

$$\bar{u}_T = \frac{5}{6\alpha_T}\left(\frac{9}{10}\,\alpha_T B_0\right)^{-1/3} z^{-1/3} \tag{12.2.19}$$

$$g\left(\frac{\rho_0 - \rho}{\rho_0}\right) = \frac{5B_0}{6\alpha_T}\left(\frac{9}{10}\,\alpha_T B_0\right)^{-1/3} z^{-5/3} \tag{12.2.20}$$

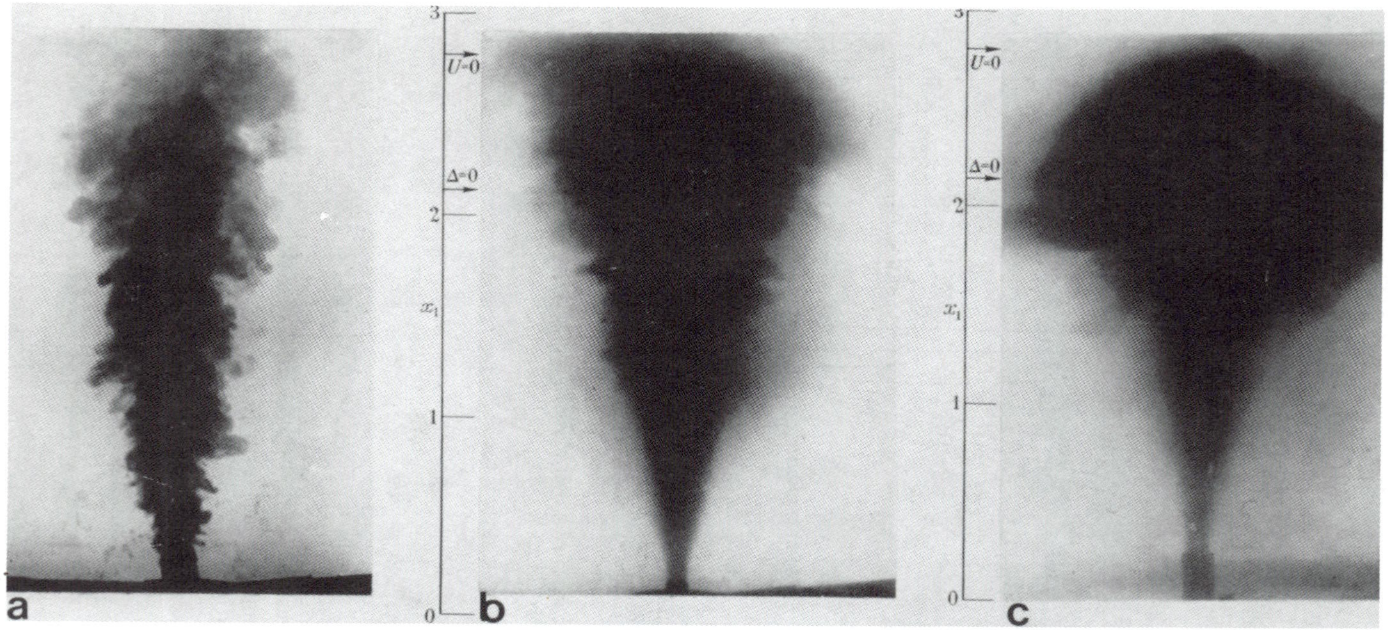

Figure 12.2.6 Steady turbulent plumes of dyed buoyant fluid (*a*) in a uniform environment, with a short exposure that shows the large eddy structure; (*b*) in a stably stratified environment, time exposure during an early stage of release; (*c*) in a stably stratified fluid, time exposure at a later stage when a layer is spreading out sideways at the top. *(Reprinted with permission from B. R. Morton et al.*, Proc. R. Soc. London Ser. A, *vol. 234, p. 1. Copyright © 1956, The Royal Society.)*

$$b_T = \frac{6}{5}\alpha_T z \qquad (12.2.21)$$

This is the same z dependence as predicted by dimensional similarity, Eqs. (12.2.9) and (12.2.10). The value of α_T must be determined experimentally. Morton (1959) adopted a value of 0.116 for top hat profiles in axisymmetric flows. George et al. (1977) reported measuring $\alpha_T = 0.153$.

Morton et al. (1956) gave solutions for the case of a stably stratified environment as well as experimental measurements for both uniform and stratified environments. It was found that as the buoyancy flux decreased with height, the

Figure 12.2.7 Streak picture of a "starting" plume of salt solution in fresh water, showing the cap (resembling a thermal) followed by a steady turbulent plume. *(Reprinted with permission from J. S. Turner,* J. Fluid Mech., *vol. 13, p. 356. Copyright © 1962, Cambridge University Press.)*

plume spread sideways more rapidly. Plumes thus terminate at a certain maximum height in stratified surroundings; see Fig. 12.2.6. Morton (1959) extended the above analysis to buoyant jets or forced plumes. These results were related to plume flows by finding a "virtual source" of buoyancy alone that would produce an equivalent plume flow downstream.

Starting plumes. A plume arising from a suddenly started source of buoyancy is composed of two sections. The advancing front behaves like a thermal, forming a cap. The upstream portion behaves like the initial portion of a steady plume; see Fig. 12.2.7. Turner (1962) pointed out that, although the solutions for these two configurations cannot be matched directly, a similarity solution for the flow may be obtained if it is assumed that the front has a lower velocity than the steady plume. A starting plume cap is found to spread at a smaller angle than an ordinary or pure thermal. Similarity of velocity and concentration profiles is preserved at different times if the cap velocity has a mean value about 61% of that at the center of an established plume at the same height.

Delichatsios (1979) gave a similarity solution of the flow equations for starting plumes with a time-varying source strength. Similarity was found for sources that produce a buoyancy flux which varies as a power function of time. The velocity of the plume cap was again found to be smaller than that of the trailing plume. For a steady source strength Delichatsios' results were in good agreement with those of Turner (1962). Measurements of the transient axial temperature in the buoyant plume generated by a fast-growing fire were also in good agreement with the similarity results of Delichatsios (1979).

12.3 NONBUOYANT JETS

Such jets occur when fluid is discharged into an ambient medium of the same material and density. Although no buoyancy effects arise, nonbuoyant jets are discussed here briefly because an understanding of jet behavior is essential to studying buoyant jets. Figure 12.3.1 shows an axisymmetric jet of air at a very moderate Reynolds number Re of 10,000 discharged into quiescent air. Near the nozzle exit a laminar shear flow region is seen. Farther downstream, as the jet entrains more fluid from the surroundings, axisymmetric vortex rings develop. These rings grow spatially until the flow abruptly becomes turbulent. Another interesting general observation is that at any instant in time a rather sharp boundary exists between the inlet jet region and its surroundings.

Many of the earlier experimental studies of jets focused on the fully developed turbulent region. It was found, however, that measurements of statistical quantities, such as the Reynolds stresses, or spatial correlations of the velocity fluctuations do not lead to a better understanding of the complicated flow dynamics. This led to increased interest in the development stages of the flow, particularly the behavior of the vortex ring structures and their effect on the subsequent stages of the flow.

The near-jet region is composed of a laminar unstable shear layer caused by

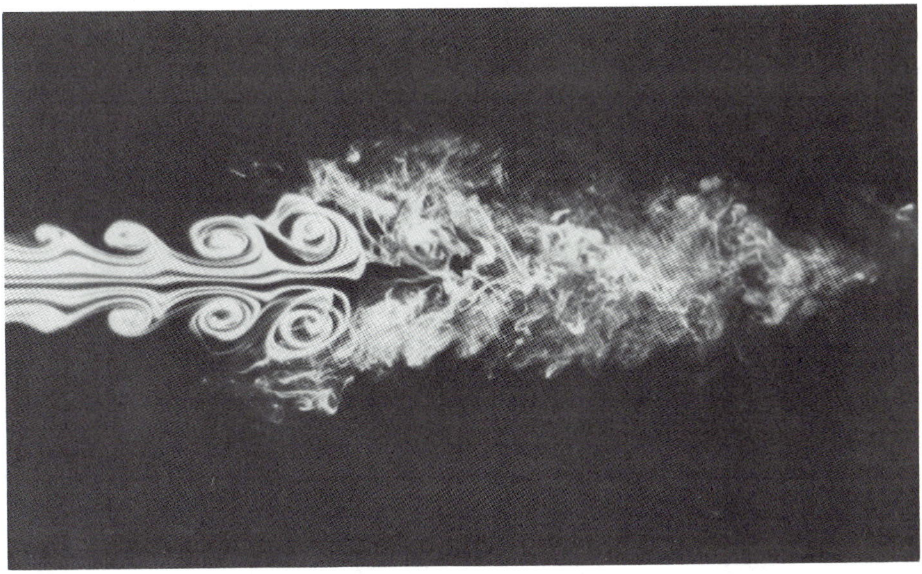

Figure 12.3.1 Instability of an axisymmetric jet. A laminar stream of air flows from a circular tube at Reynolds number 10,000 and is made visible by a smoke wire. The edge of the jet develops axisymmetric oscillations, rolls up into vortex rings, and then abruptly becomes turbulent. *(From Van Dyke, 1982.)*

the jet. This shear layer grows, rapidly entraining fluid from the ambient medium and thereby developing ring vortex-like structures. The ambient fluid entrained by these large-scale vortices is later broken down and digested by small-scale turbulence. On the other hand, for plane jets, the vortices may develop and grow in one of two modes. They may arise on alternate sides separately, as seen by Brown (1935), or on both sides simultaneously, as seen by Beavers and Wilson (1970) and Rockwell and Niccolls (1972).

As the vortex rings move downstream, they often coalesce with neighboring rings. This accounts for the observed increase in scale and separation of the rings with increasing distance from the nozzle. This phenomenon of vortex pairing was studied by Whremann and Wille (1957), Freymuth (1966), Yule (1978), and Schneider (1980). Several locations of vortex coalescence exist in the transition region, each producing vortex rings with more severely deformed cores than the preceding location. This process continues until a core deformation larger than some critical size occurs. Then turbulence arises. A detailed literature review concerning the jet exit region is given in List (1982).

Beyond a transition region, the turbulent flow becomes fully developed. This occurs about 10 jet diameters downstream of the nozzle exit. This is evident in numerous measurements of the turbulence intensity at different downstream locations, as shown in Fig. 12.3.2. After transition a steady decay of the intensity of turbulent velocity fluctuations begins in the flow, as measured over a wide range of Reynolds numbers. In the fully developed turbulent region, time-aver-

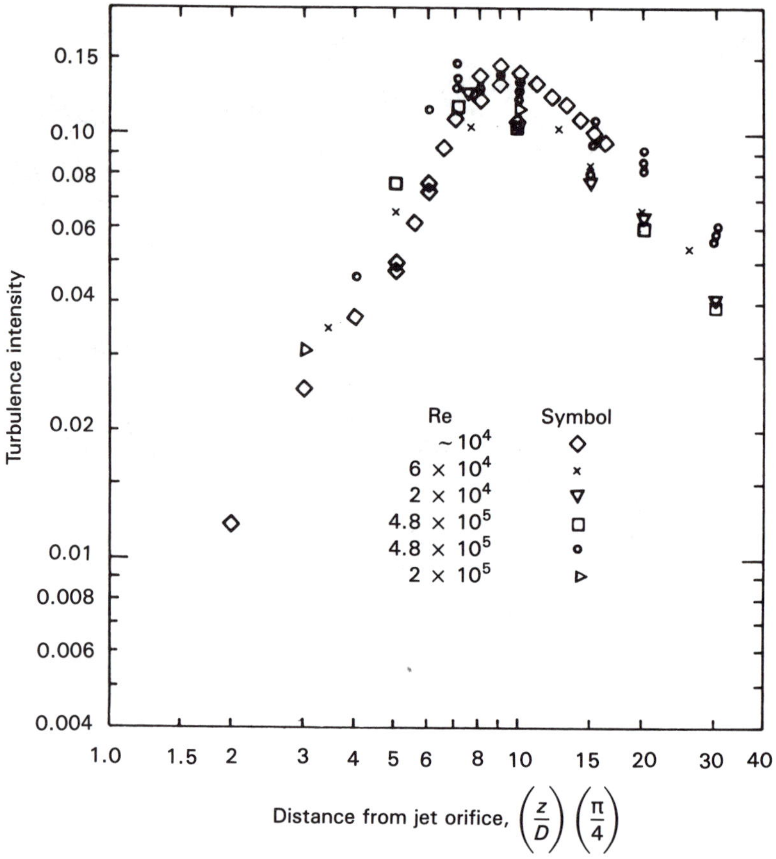

Figure 12.3.2 Intensity of turbulent velocity fluctuations on the axis of a round turbulent jet. *(Reprinted with permission from H. B. Fischer et al.,* Mixing in Inland and Coastal Waters. *Copyright © 1979, Academic Press, Inc.)*

aged measurements of the velocity and tracer concentration profiles in jets again indicate Gaussian distributions, similar to those found for plumes. However, the entrainment rate for jets is lower than that for plumes; $\alpha = 0.057$ was measured by Ricou and Spalding (1961) and others. This indicates a lower dilution rate for jets than for plumes with the same local momentum flux.

12.4 BUOYANT JETS

Buoyant jets arise when fluid is discharged continuously with initial momentum into a medium at a different density. Smokestacks discharging hot gases into the atmosphere and effluents from cooling towers of nuclear and fossil-fueled power plants are familiar examples of buoyant jets. Sewage and cooling water discharges are common flows that occur in water. Increasing concern over environmental pollution has led to numerous analytical, numerical, and experimental studies of such flows. These investigations deal with questions such as the velocity, trajectory, and decay of pollutants discharged in buoyant jets. The effect of ambient fluid stratification and motion on buoyant jet trajectory have also been investigated extensively.

The need for prediction of buoyant jet behavior, however, is not limited to environmental discharges. Rapid advances in the ability to detect small temperature and concentration differences and other anomalies may make it increasingly easy to detect many physical effects, changes, and motions. The potential for increasing knowledge of environmental, geophysical, and technological processes is great.

Given the wide range of applications, the range of possible jet and/or ambient characteristics that may be of interest is equally broad. The variables include initial jet geometry, discharge momentum, thermal and concentration loading, turbulence characteristics, ambient flow conditions, turbulence, and stratification. An extremely large number of appreciably different combinations arise. General jet-ambient interaction mechanisms are classified according to the following characteristics:

1. Source
   a. Circular
   b. Line or slit
2. Jet buoyancy
   a. Neutrally buoyant (jet)
   b. Buoyant (positively or negatively)
3. Orientation of discharge
   a. Horizontal (perpendicular to the gravity field)
   b. Inclined
4. Ambient motion
   a. Quiescent
   b. Flowing in the discharge plane or with crossflow
5. Ambient stratification
   a. Unstratified
   b. Linearly stratified
   c. Other stratifications
6. Other effects
   a. Multiple discharges
   b. Two-phase interfaces

Buoyant jets are driven by both buoyancy and momentum. Thus, purely buoyant plumes and pure momentum jets are limiting conditions of buoyant jets. A discharge in which the initial momentum is everywhere negligible, relative to the eventual momentum produced by buoyancy, is essentially a plume. On the other hand, if buoyancy effects are everywhere very small compared to discharge momentum, the resulting flow is a pure momentum jet.

A principal quantitative measure of relative jet momentum, with initial velocity u_0, buoyancy $(\rho_\infty - \rho_0)/\rho_0$, and diameter D, is the densimetric Froude number Fr, given by

$$\mathrm{Fr} = \frac{u_0}{\{gD[(\rho_\infty - \rho_0)/\rho_0]\}^{1/2}} \tag{12.4.1}$$

The momentum level, in the numerator, is given by the discharge velocity u_0. The buoyancy effect, in the denominator, in terms of the units of buoyancy $(\rho_\infty - \rho_0)/\rho_0$, is a measure of the velocity level generated by the buoyancy force. Thus, the densimetric Froude number ranges from near zero for plumes toward infinity for jets of small buoyancy. Hereafter, the Froude number means the densimetric form in Eq. (12.4.1).

Chen and Rodi (1980) combined a dimensional similarity analysis and experimental measurements to evaluate the flow regimes that occur downstream in a buoyant jet in a quiescent ambient medium. It was found that, in general, three distinctive regions exist, as shown in Figs. 12.4.1*a* and *b* for plane, slit-generated jets and in Fig. 12.4.2 for axisymmetric jets. In the first region, near the jet exit, the flow conforms to the trend of jets. However, in the third region, far from the source, plumelike behavior is observed. The intermediate region is one of transition from jetlike to plumelike behavior.

The exponents for the decay relations shown in Table 12.4.1 and Figs. 12.4.1*a*, 12.4.1*b*, and 12.4.2 were determined from dimensional similarity considerations for regions one and three, the jet and plume regions, respectively. For the intermediate region the decay relations were determined empirically. Although these results are valuable in providing insight into the physical behavior of buoyant jets in general, a somewhat different classification of the downstream flow regions is also appropriate for the purpose of modeling, as discussed below.

12.4.1 Downstream Flow Regimes

Independent of the buoyant jet-ambient mechanisms, all buoyant jets pass through several flow regions along their trajectory. These regions are shown for an inclined submerged buoyant jet in Fig. 12.4.3. The regimes are described below.

Zone of flow establishment. Flow characteristics are still dominated by the discharge conditions. The initial profiles of velocity and of the scalar quantities, temperature, salinity, and so forth, undergo transition from the initial uniform discharge configurations. The turbulent shear layer formed around the jet periphery grows inward. The core of undisturbed profiles becomes smaller. This zone of flow establishment ends where the turbulent mixing reaches the jet centerline. Jet behavior in this region is strongly influenced by initial discharge conditions. Ambient conditions have relatively little influence in this zone.

Zone of established flow. In this following region the motion of the buoyant jet and the mixing and entrainment are governed by the jet's initial and acquired momentum and its buoyancy, as well as by ambient stratification and flow conditions. Initial discharge conditions play a progressively smaller role downstream. The flow progresses from jetlike toward plumelike behavior.

Far field. In this region the effects of initial jet momentum and buoyancy are both negligible. The material is passively convected by any ambient flow and circulations. The fluid may be mixed and diffused further by ambient turbulence. The buoyant jet is gradually dispersed and ceases to be a separate entity.

12.4.2 Modeling Schemes

There are several different approaches to modeling buoyant jets. The simplest are algebraic models and algebraic equations, which are based on either empirical data or simplifications of a differential model. These usually predict only downstream trajectory and jet width. Some also predict velocity, concentration, and temperature excess over the ambient. Data-based algebraic models tend to become unreliable when the basic conditions on which they were based, such as the temperature and salinity range of the jet and ambient, are quite different from the actual application. Reviews of such models are given by Briggs (1975) and Gebhart et al. (1984).

Integral analysis and entrainment modeling. In Section 12.2 an integral analysis was given for a vertical fully developed turbulent plume in an unstratified ambient medium. The entrainment coefficient α must be determined from experimental measurements. For plumes, $\alpha = 0.082$ is reported by several investigators, while $\alpha = 0.057$ is commonly used for jets. Thus, in modeling buoyant jets, these different values indicate that some formulation is needed to represent the two limiting behaviors, the nonbuoyant and the plume regions. It must also represent the entrainment in the intermediate region.

The method of analysis in integral modeling is largely independent of the entrainment function used. First, the governing equations are integrated over the jet cross section. The results are in differential form, in terms of the trajectory coordinate S, as shown in Fig. 12.4.3. The equations are then nondimensionalized with respect to reference quantities. Initial conditions are specified. Then the resulting equations are numerically integrated downstream over the desired range of the streamwide path length S. For any given initial jet condition, different downstream trajectories and decay are found due to differences in the way any of the following aspects are modeled: (1) entrainment rate, (2) starting length processes, (3) initial conditions specified for the beginning of the zone of established flow, (4) equation of state used for the changing density of the fluid, and (5) computational technique.

12.4.3 Entrainment Models in a Quiescent Ambient

Morton et al. (1956) proposed an entrainment model for the whole range as

$$\alpha = 0.057 + \frac{a_2}{\mathrm{Fr}_L} \tag{12.4.2}$$

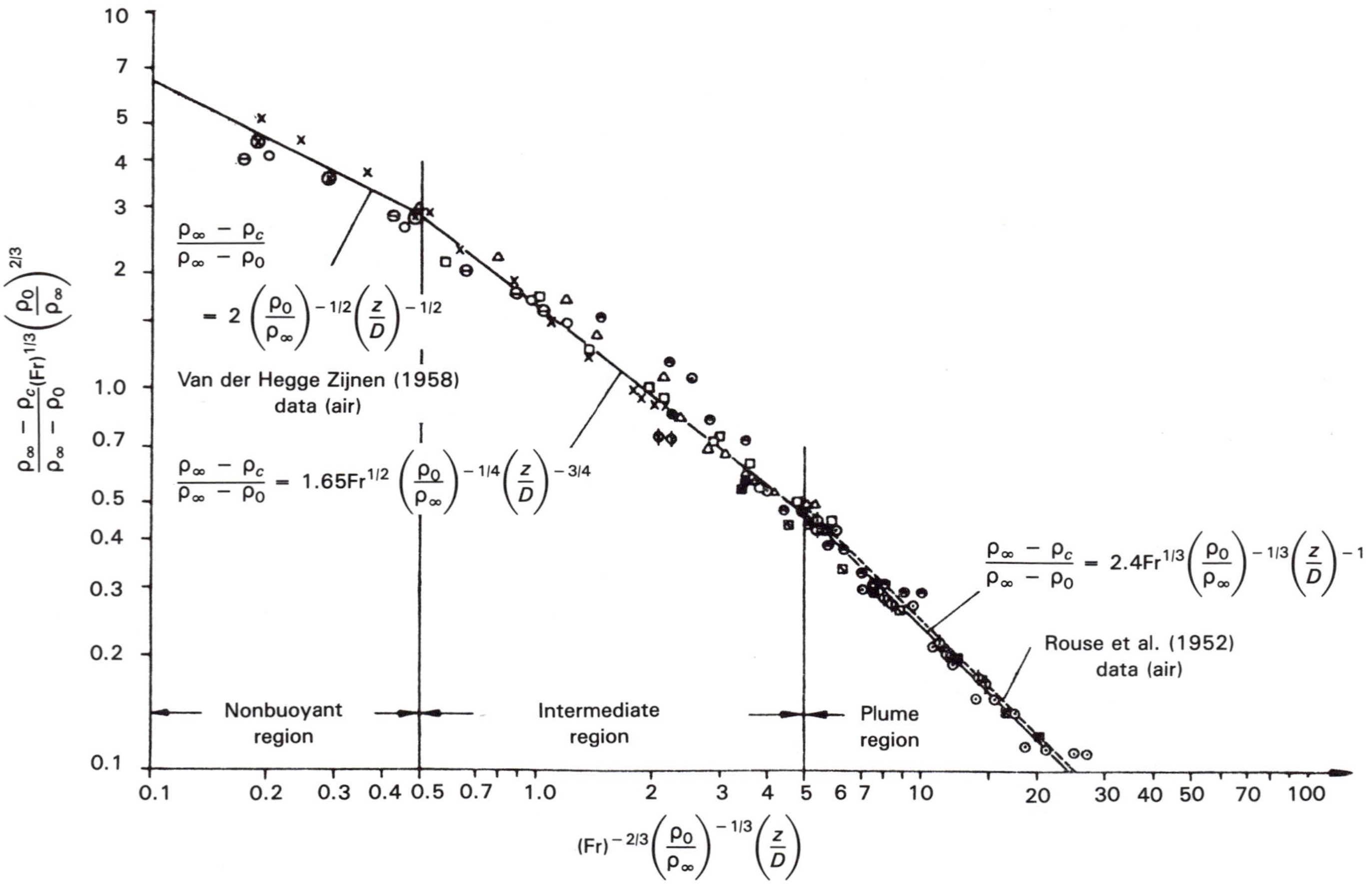

10
7
5
4
3
2
1.0
0.7
0.5
0.4
0.3
0.2
0.1
$\frac{\rho_\infty - \rho_c}{\rho_\infty - \rho_0}(Fr)^{1/3}\left(\frac{\rho_0}{\rho_\infty}\right)^{2/3}$
$\frac{\rho_\infty - \rho_c}{\rho_\infty - \rho_0} = 2\left(\frac{\rho_0}{\rho_\infty}\right)^{-1/2}\left(\frac{z}{D}\right)^{-1/2}$
Van der Hegge Zijnen (1958)
data (air)
$\frac{\rho_\infty - \rho_c}{\rho_\infty - \rho_0} = 1.65Fr^{1/2}\left(\frac{\rho_0}{\rho_\infty}\right)^{-1/4}\left(\frac{z}{D}\right)^{-3/4}$
$\frac{\rho_\infty - \rho_c}{\rho_\infty - \rho_0} = 2.4Fr^{1/3}\left(\frac{\rho_0}{\rho_\infty}\right)^{-1/3}\left(\frac{z}{D}\right)^{-1}$
Rouse et al. (1952)
data (air)
Nonbuoyant
region
Intermediate
region
Plume
region
0.1
0.2
0.3
0.4
0.5
0.7
1.0
2
3
4
5
6
7
10
20
30
40
50
70
100
$(Fr)^{-2/3}\left(\frac{\rho_0}{\rho_\infty}\right)^{-1/3}\left(\frac{z}{D}\right)$

(*a*)

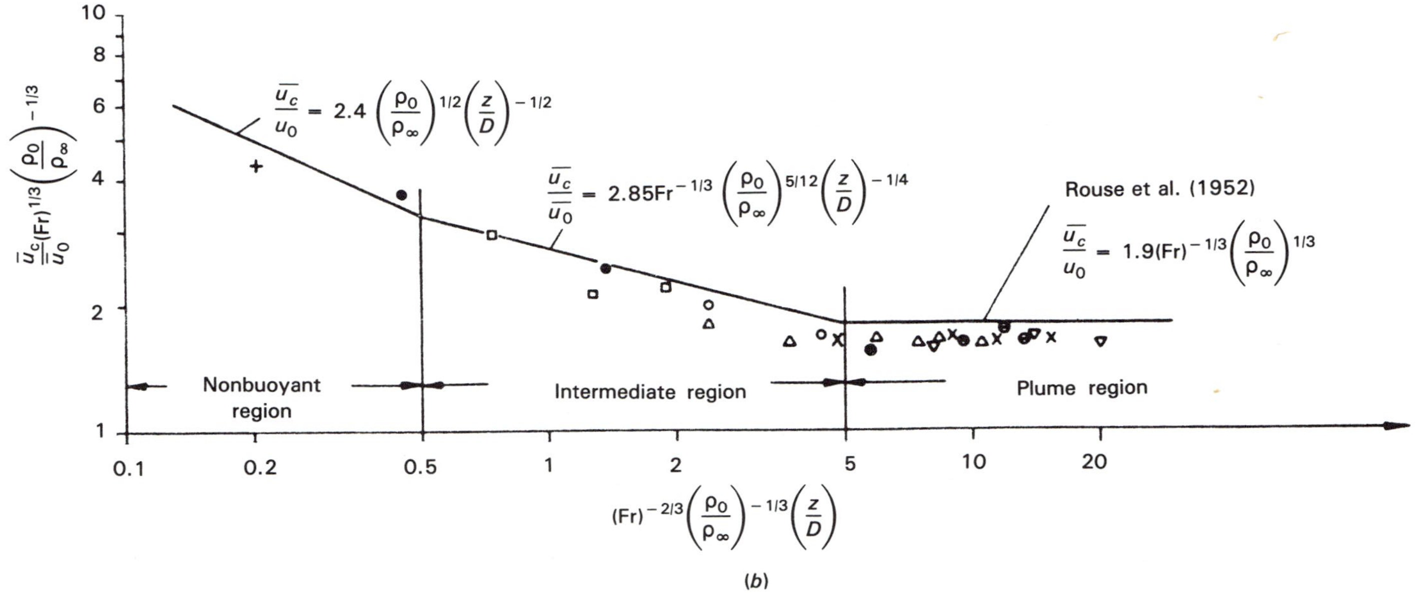

(*b*)

Figure 12.4.1 Decay of (*a*) centerline density ρ_c and (*b*) centerline velocity $\bar{u}_c$ in plane buoyant jets. *(Reprinted with permission from C. J. Chen and W. Rodi,* Vertical Turbulent Buoyant Jets: A Review of Experimental Data. *Copyright © 1980, Pergamon Books Ltd.)*

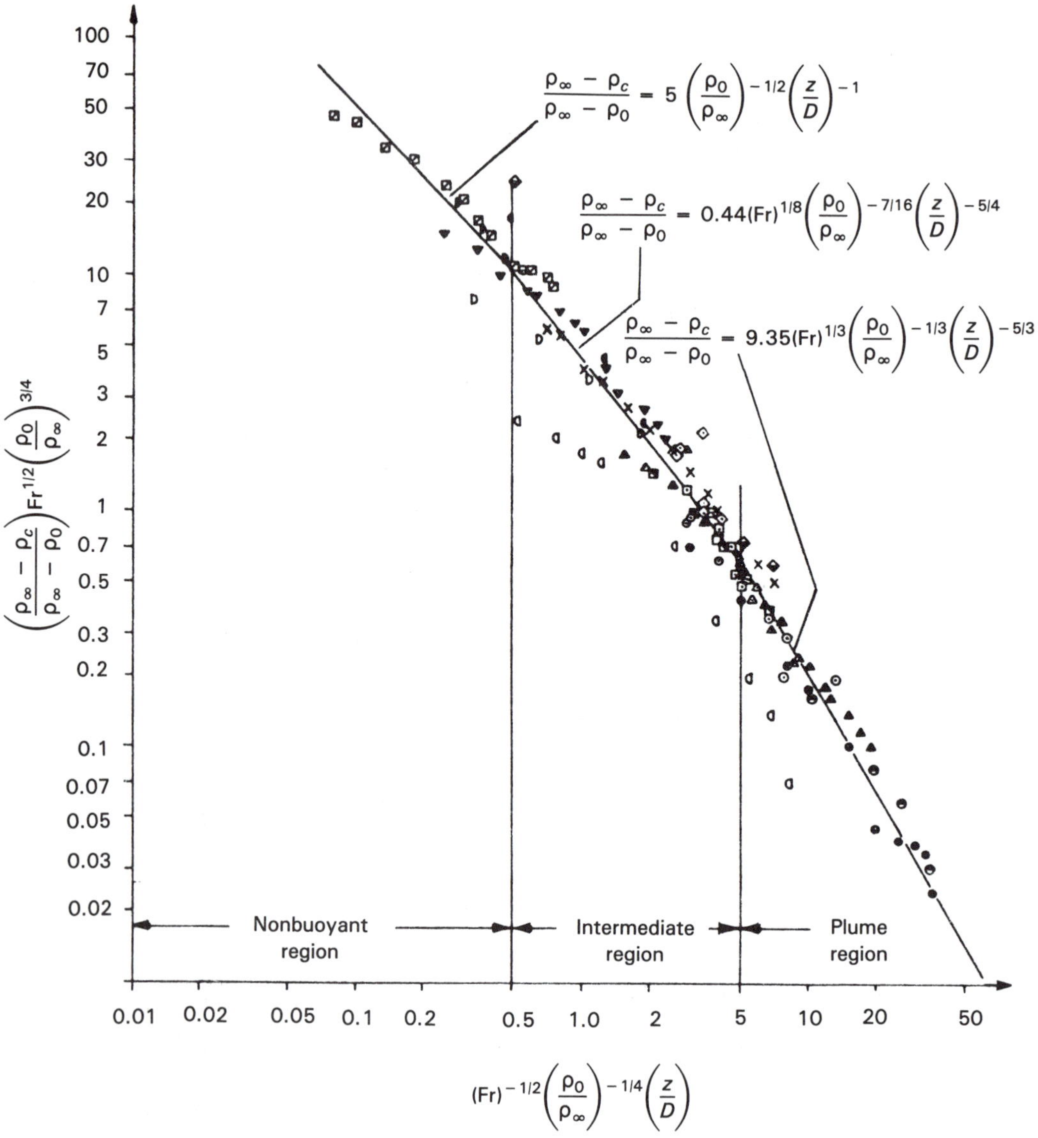

Figure 12.4.2 Decay of centerline density in axisymmetric buoyant jets. *(Reprinted with permission from C. J. Chen and W. Rodi,* Vertical Turbulent Buoyant Jets: A Review of Experimental Data. *Copyright © 1980, Pergamon Books Ltd.)*

where a_2 is an empirically determined constant and $\mathrm{Fr}_L = \bar{u}_c/[gD(\rho_\infty - \rho_0)/\rho_0]^{1/2}$ is a local Froude number. That is, Fr in Eq. (12.4.1) is instead based on the local centerline velocity $\bar{u}_c$. The same general form was given by Fox (1970) for a vertically discharged buoyant jet. Hirst (1971a) postulated that, for a buoyant discharge into a quiescent ambient, the entrainment function should depend on (1) local mean flow conditions in the jet, that is, $\bar{u}_c$, (2) local buoyancy within the

Table 12.4.1 Decay relations for centerline velocity and density in axisymmetric and plane buoyant plumes[a]

	Axisymmetric buoyant plumes	Plane plumes
Nonbuoyant region exists for	$(\mathrm{Fr})^{-1/2}\left(\frac{\rho_0}{\rho_\infty}\right)^{-1/4}\left(\frac{z}{D}\right)^{-1} < 0.5$	$(\mathrm{Fr})^{-2/3}\left(\frac{\rho_0}{\rho_\infty}\right)^{-1/3}\left(\frac{z}{D}\right) < 0.5$
Decay relations	$\frac{\bar{u}_c}{\bar{u}_0} = 6.2\left(\frac{\rho_0}{\rho_\infty}\right)^{-1/2}\left(\frac{z}{D}\right)^{-1}$	$\frac{\bar{u}_c}{\bar{u}_0} = 2.4\left(\frac{\rho_0}{\rho_\infty}\right)^{-1/2}\left(\frac{z}{D}\right)^{-1/2}$
	$\frac{\rho_\infty - \rho_c}{\rho_\infty - \rho_0} = 5\left(\frac{\rho_0}{\rho_\infty}\right)^{-1/2}\left(\frac{z}{D}\right)^{-1}$	$\frac{\rho_\infty - \rho_c}{\rho_\infty - \rho_0} = 2\left(\frac{\rho_0}{\rho_\infty}\right)^{-1/2}\left(\frac{z}{D}\right)^{-1/2}$
Intermediate region exists for	$0.5 \le (\mathrm{Fr})^{-1/2}\left(\frac{\rho_0}{\rho_\infty}\right)^{-1/4}\left(\frac{z}{D}\right) \le 5$	$0.5 \le (\mathrm{Fr})^{-2/3}\left(\frac{\rho_0}{\rho_\infty}\right)^{-1/3}\left(\frac{z}{D}\right) \le 5$
Decay relations	$\frac{\bar{u}_c}{\bar{u}_0} = 7.26(\mathrm{Fr})^{-0.1}\left(\frac{\rho_0}{\rho_\infty}\right)^{9/20}\left(\frac{z}{D}\right)$	$\frac{\bar{u}_c}{\bar{u}_0} = 2.85(\mathrm{Fr})^{-1/3}\left(\frac{\rho_0}{\rho_\infty}\right)^{5/12}\left(\frac{z}{D}\right)^{-1/4}$
	$\frac{\rho_\infty - \rho_c}{\rho_\infty - \rho_0} = 0.44(\mathrm{Fr})^{1/8}\left(\frac{\rho_0}{\rho_\infty}\right)^{-7/16}\left(\frac{z}{D}\right)^{-5/4}$	$\frac{\rho_\infty - \rho_c}{\rho_\infty - \rho_0} = 1.65(\mathrm{Fr})^{1/2}\left(\frac{\rho_0}{\rho_\infty}\right)^{-1/4}\left(\frac{z}{D}\right)^{-3/4}$
Plume region exists for	$(\mathrm{Fr})^{-1/2}\left(\frac{\rho_0}{\rho_\infty}\right)^{-1/4}\left(\frac{z}{D}\right) > 5$	$(\mathrm{Fr})^{-2/3}\left(\frac{\rho_0}{\rho_\infty}\right)^{-1/3}\left(\frac{z}{D}\right) > 5$
Decay relations	$\frac{\bar{u}_c}{\bar{u}_0} = 3.5(\mathrm{Fr})^{-1/3}\left(\frac{\rho_0}{\rho_\infty}\right)^{1/3}\left(\frac{z}{D}\right)^{-1/5}$	$\frac{\bar{u}_c}{\bar{u}_0} = 1.9(\mathrm{Fr})^{-1/3}\left(\frac{\rho_0}{\rho_\infty}\right)^{1/3}$
	$\frac{\rho_\infty - \rho_c}{\rho_\infty - \rho_0} = 9.35(\mathrm{Fr})^{1/3}\left(\frac{\rho_0}{\rho_\infty}\right)^{-1/3}\left(\frac{z}{D}\right)^{-5/3}$	$\frac{\rho_\infty - \rho_c}{\rho_\infty - \rho_0} = 2.4(\mathrm{Fr})^{1/3}\left(\frac{\rho_0}{\rho_\infty}\right)^{-1/3}\left(\frac{z}{D}\right)^{-1}$

[a]The subscript c indicates local centerline condition.
Source: Adapted from Chen and Rodi (1980).

jet, as indicated by Fr_L, and (3) initial jet orientation θ_0; see Fig. 12.4.3. The following form was proposed by Hirst (1971a):

$$\alpha = 0.057 + \frac{0.097}{\mathrm{Fr}_L} \sin \theta_0 \tag{12.4.3}$$

This is the general form also suggested by Morton et al. (1956). However, the constant in the above equation was obtained by fitting this function to the known discharge and end-point entrainment rates in such flows.

Another entrainment function, for initially horizontal buoyant momentum jets, is the jet-plume fit proposed by Riester et al. (1980):

$$\alpha = [(0.057 \cos \theta)^2 + (0.082 \sin \theta)^2]^{1/2} \tag{12.4.4}$$

From data on a buoyant jet discharged vertically downward into a quiescent ambient, Davis et al. (1978) proposed:

$$\alpha = 0.058 + \frac{0.083}{(\mathrm{Fr})^{0.3}} \tag{12.4.5}$$

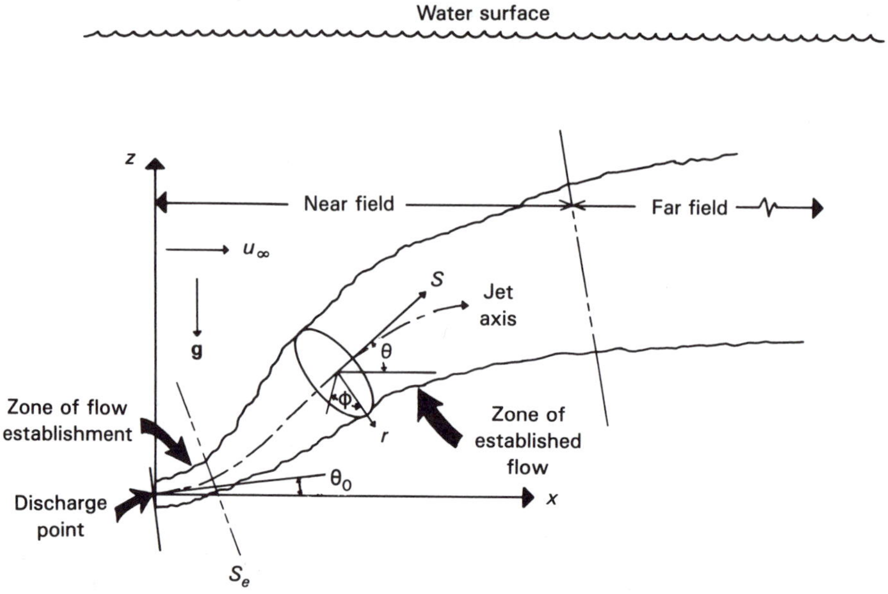

Figure 12.4.3 Flow regimes of a buoyant momentum jet and the coordinate system and dimensions in which its trajectory and growth are described. *(Reprinted with permission from B. Gebhart et al.,* Adv. Heat Transfer, *vol. 16, p. 1. Copyright © 1984, Academic Press, Inc.)*

A summary of these and other entrainment functions for discharges into quiescent ambient media is given in Table 12.4.2.

Flowing ambient media. Hirst (1971b) proposed an entrainment function applicable to general buoyant jet behavior in a horizontally flowing ambient. This function, given below, was proposed to apply to both horizontal and inclined jets:

$$E = P\bar{u}_c\alpha = 2\pi b\left(a_1 + \frac{a_2}{\mathrm{Fr}_L}\sin\theta\right)(|\bar{u}_c - \bar{u}_\infty\cos\theta| + a_3\bar{u}_\infty\sin\theta) \quad (12.4.6)$$

where $P = 2\pi b$ is a measure of the plume perimeter, $\bar{u}_c$ and $\bar{u}_\infty$ are the average centerline and ambient velocities, respectively, and θ is the local angle of inclination. The term $|\bar{u}_\infty - \bar{u}_c\cos\theta|$ represents the local relative velocity of the jet with respect to the ambient in the direction of initial jet flow. It is a pure "coflow" entrainment term. The last term, $a_3\bar{u}_\infty\sin\theta$, is the entrainment contribution arising from cross flow of ambient fluid as the jet turns upward due to buoyancy. The relative velocity is about $\bar{u}_\infty\sin\theta$ at large path length S.

The values of a_1 and a_2 were specified so that the flowing ambient entrainment function reduced to that for a quiescent ambient for $\bar{u}_\infty = 0$. A value of 9.0 was taken for a_3, based on a best fit of limited data. The resulting entrainment function is

$$E = 2\pi b\left(0.057 + \frac{0.97}{\mathrm{Fr}_L}\sin\theta\right)(|\bar{u}_c - \bar{u}_\infty\cos\theta| + 9.0\bar{u}_\infty\sin\theta) \quad (12.4.7)$$

Table 12.4.2 Entrainment functions for use with quiescent ambients[a]

No.[b]	α	$\theta = 0$	$\theta = \frac{\pi}{2}$	Fr or $Fr_L \to 0$	Fr or $Fr_L \to \infty$	Source
1	0.057 (jets)	0.057	0.057		0.057	Albertson et al. (1950)
2	0.082 (plumes)	0.082	0.082	0.082		List and Imberger (1973)
3	$0.057 + \frac{0.97}{Fr_L} \sin\theta$	0.057	$f(Fr_L)$	∞ ($\theta \neq 0$)	0.057	Hirst (1971a)
4	$0.057 + \frac{0.083}{Fr^{0.3}}$	$f(Fr)$	$f(Fr)$	∞	0.057	Davis et al. (1978)
5	$[(0.057 \cos\theta)^2 + (0.082 \sin\theta)^2]^{1/2}$	0.057	0.082	$f(\theta)$	$f(\theta)$	Riester et al. (1980)
6	0.085	0.085	0.085	0.085		Abraham (1965)
7	$0.057 + \frac{a_2}{Fr_L}$	$f(Fr_L)$	$f(Fr_L)$		0.057	General form of Morton et al. (1956)

[a]Form: $E = 2\pi\alpha\bar{u}_c B$.

[b]No. 1, applicable to simple momentum jet; 2, applicable to simple buoyant plume; 3, applicable to buoyant jet discharged at varying angles to a quiescent ambient; 4, empirical fit for a buoyant jet discharged vertically downward to a quiescent ambient; 5, applicable to buoyant jet discharged horizontally to a quiescent ambient; 6, applicable to simple buoyant plume; 7, empirically determined coefficient a_2, vertical discharge.

Source: From Gebhart et al. (1984).

Ginsberg and Ades (1975) performed a least-square analysis on a large set of cross-flow laboratory trajectory data. Using Hirst's model with a_1 and a_2 as specified, it was found that a large variation in the value of a_3 was necessary for the results of calculations to fit the data. The resulting correlation for a_3 as a function of Fr and the coflow ratio R is

$$a_3 = 25.81(Fr^{0.19464}R^{0.35155}) - 10.825 \tag{12.4.8}$$

where the coflow ratio R is defined as

$$R = \frac{\bar{u}_\infty}{u_0} \tag{12.4.9}$$

and u_0 is the jet exit velocity.

Other specific entrainment functions for flowing ambients have appeared. Schatzmann (1979) proposed an entrainment function similar in form to that of Hirst, but for use in a set of governing equations in which the Boussinesq approximation was not invoked. Fan (1967) proposed an entrainment function for coflowing ambients that included a drag term as well as the usual proportionality of entrainment to centerline velocity and jet width. It was found that the value of the drag coefficient and the entrainment constant, $\alpha = 0.082$, had to be readjusted to make the prediction of the model conform with the data with each change in discharge or ambient conditions. The entrainment functions of Hirst, Ginsberg and Ades, and Schatzmann are collected in Table 12.4.3.

12.4.4 Comparison of Predictions of Trajectory and Decay

Gebhart et al. (1984) compared the entrainment models for a wide range of initial conditions in both quiescent and flowing ambients. The governing equations in the zone of established flow are shown in Table 12.4.4. They were solved with entrainment models 1, 2, 3, and 5 in Table 12.4.2 for a quiescent ambient and models 8 and 9 in Table 12.4.3 for flowing ambients.

Calculations were also given for buoyant water jets in quiescent stratified ambients. The effects of temperature stratification were considered first. Results were then given for the kinds of stratification common in the oceans. The specific ambients were three characteristic oceanic conditions: the North Pacific, the tropical Atlantic, and the Arctic Ocean. Results and comparisons are given here for the quiescent and for stratified ambients.

Quiescent ambients. Resulting downstream trajectories or penetrations are seen in Figs. 12.4.4 and 12.4.5 for Fr = 1–8 and 10–200, respectively. These results

Table 12.4.3 Entrainment functions for buoyant jets issuing into flowing ambients[a,b]

No.[c]	Entrainment function	Source
8	$E = 2\pi B\left(0.057 + \dfrac{0.97}{\mathrm{Fr}_L}\sin\theta\right)(\lvert U^*\rvert + 9.0\bar{u}_\infty \sin\theta)$	Hirst (1971b)
9	$E = 2\pi B\left(0.057 + \dfrac{0.97}{\mathrm{Fr}_L}\sin\theta\right) \times \left(\lvert U^*\rvert + \left\{25.81\left[\mathrm{Fr}^{0.195}\left(\dfrac{\bar{u}_\infty}{\bar{u}_0}\right)\right]^{0.352} - 10.83\right\}\bar{u}_\infty \sin\theta\right)$	Ginsberg and Ades (1975)
10	$E = 2\pi B\left(0.057 - \dfrac{0.67}{\mathrm{Fr}_L}\sin\theta\right) \times \dfrac{\lvert U^*\rvert + 2\bar{u}_\infty \sin\theta}{1 + (5\bar{u}_\infty \cos\theta)/\lvert\bar{u}_c - \bar{u}_\infty\rvert}$	Schatzmann (1979)

[a]With coflowing ambients: complete volumetric entrainment.

[b]$\lvert U^*\rvert = \lvert\bar{u}_c - \bar{u}_\infty \cos\theta\rvert$.

[c]No. 8, Applicable to buoyant jet discharged at varying angles to a flowing ambient, cross-flow terms omitted; 9, applicable to buoyant jet discharged at varying angles to a flowing ambient; 10, same as (9), empirically determined coefficients.

Source: From Gebhart et al. (1984).

Table 12.4.4 General equations in dimensional form for entrainment modeling of buoyant momentum jets

Equation	Form
Continuity	$\frac{d}{dS}\left\{\int_0^{2\pi}\int_0^{\infty} \bar{u}r\,dr\,d\phi\right\} = 2\pi\alpha\bar{u}_c B = E$
Horizontal momentum	$\frac{d}{dS}\left\{\int_0^{2\pi}\int_0^{\infty} \bar{u}^2 \cos\theta\, r\,dr\,d\phi\right\} = \bar{u}_\infty E$
Vertical momentum	$\frac{d}{dS}\left\{\int_0^{2\pi}\int_0^{\infty} \rho\bar{u}^2 \sin\theta\, r\,dr\,d\phi\right\} = \int_0^{2\pi}\int_0^{\infty} (\rho_\infty - \rho)gr\,dr\,d\phi$
Energy	$\frac{d}{dS}\left\{\int_0^{2\pi}\int_0^{\infty} \bar{u}(t - t_\infty)r\,dr\,d\phi\right\} = \frac{dt_\infty}{dS}\int_0^{2\pi}\int_0^{\infty} \bar{u}r\,dr\,d\phi$
Concentration (or scalar species)	$\frac{d}{dS}\left\{\int_0^{2\pi}\int_0^{\infty} \bar{u}(c - c_\infty)r\,dr\,d\phi\right\} = \frac{dc_\infty}{dS}\int_0^{2\pi}\int_0^{\infty} \bar{u}r\,dr\,d\phi$
Horizontal component of trajectory	$dx = dS\cos\theta$
Vertical component of trajectory	$dz = dS\sin\theta$

Source: From Gebhart et al. (1984).

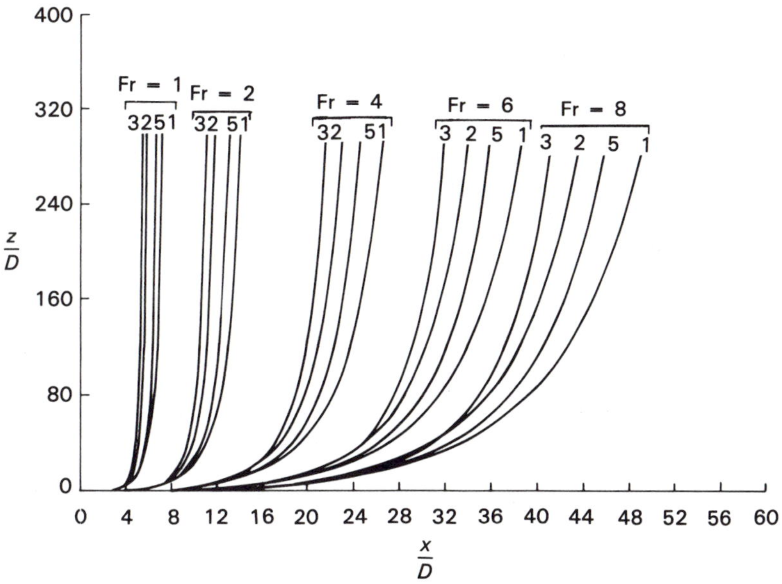

Figure 12.4.4 Comparison of four entrainment models in a quiescent ambient medium at low Fr values. The numbered labels on the curves indicate the entrainment model in Table 12.4.2. *(Reprinted with permission from B. Gebhart et al.,* Adv. Heat Transfer, *vol. 16, p. 1. Copyright © 1984, Academic Press, Inc.)*

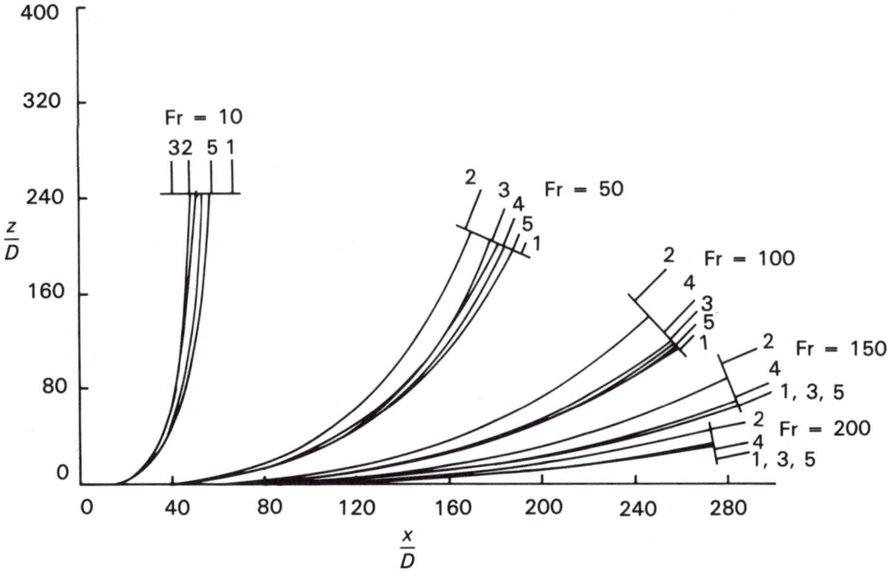

Figure 12.4.5 Comparison of five entrainment models in a quiescent ambient medium at high Fr values. The numbered labels on the curves indicate the entrainment model in Table 12.4.2, except that model 4 is omitted for Fr = 10. *(Reprinted with permission from B. Gebhart et al.,* Adv. Heat Transfer, *vol. 16, p. 1. Copyright © 1984, Academic Press, Inc.)*

show the great range of behavior, from Fr = 1, with approximately equal initial momentum and buoyancy, to Fr = 200, for which buoyancy is relatively very small. Note the different horizontal, x/D, scales on Figs. 12.4.4 and 12.4.5, where D is the initial jet diameter. For Fr = 1 entrainment quickly dilutes the horizontal penetrative power while constant buoyancy continues to generate essentially upward motion. For Fr = 200 penetration is little influenced by buoyancy, as this small buoyant effect is quickly diluted into a large mass by entrainment.

The entrainment functions that specifically relate to the nonbuoyant jet, 1, and to the plume, 2, often show wide downstream divergence from the mean of the other models, 3, 4, and 5. This is particularly true for the buoyant plume entrainment function, $\alpha = 0.082$, at large Fr. It consistently predicts higher trajectory at all but the lowest Froude numbers. The Albertson et al. (1950) momentum jet value, $\alpha = 0.057$, consistently predicts the highest entrainment and therefore the flattest trajectory at all values of Fr. The end point correlation of Riester et al. (1980) also predicts lower entrainment than the average for all values of Fr.

Recall that the accepted values of entrainment functions for the extremes of the Froude number are $\alpha = 0.082$ for Fr $\rightarrow 0$ and $\alpha = 0.057$ for Fr $\rightarrow \infty$. The model that most closely reflects the Froude number/entrainment relation over the range of Fr considered here is that of Hirst, model 3. This model predicts the highest entrainment for the lowest Froude numbers in Fig. 12.4.5.

On the other hand, all models predict strikingly similar behavior in the decays of centerline velocity and temperature level at high Fr, as seen in Figs. 12.4.6 and 12.4.7, respectively. The decays, in terms of path length S, are not strongly dependent on either Fr or the entrainment function used except for Fr = 10 for the velocity decay. Figure 12.4.6 shows the centerline velocity decay $\bar{u}_c/u_0$. The temperature decays in Fig. 12.4.7 are very similar. For all conditions and models, an extremely rapid initial velocity decrease occurs immediately after discharge. For all jets, $\bar{u}_c/u_0 < 0.4$ by about 10 diameters downstream. At 30 diameters along the trajectory, $\bar{u}_c/u_0 < 0.2$. Still farther downstream, the residual velocities decrease much less rapidly. Figure 12.4.6 also shows that higher velocities persist downstream at smaller values of Fr, that is, for relatively more buoyant or less vigorous jets. Such conditions are well outside the range of intended use.

Flowing ambients. However, similar comparisons by Gebhart et al. (1984) of the different entrainment model results for buoyant jets in flowing ambient media showed considerable disagreement in predictions of jet trajectory. Clearly, the models are very sensitive to differences in accounting for the cross-flow entrainment effect. For jets having both low Froude number and $R = \bar{u}_\infty/u_0$, the calculated difference in vertical rise is usually many diameters in magnitude. This disagreement is found at all discharge angles θ_0. At higher values of both Fr and R, the differences in predicted vertical rise become smaller. However, relative to the total rise of the jet, the difference may be many times greater than that for

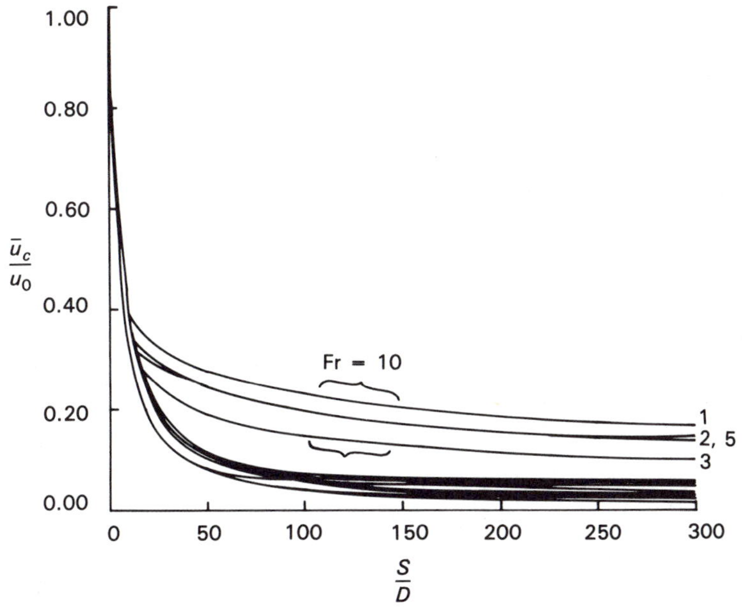

Figure 12.4.6 Velocity decay for 12 jets, Fr = 10, 50, and 200, for entrainment models 1, 2, 3, and 5. *(Reprinted with permission from B. Gebhart et al.,* Adv. Heat Transfer, *vol. 16, p. 1. Copyright © 1984, Academic Press, Inc.)*

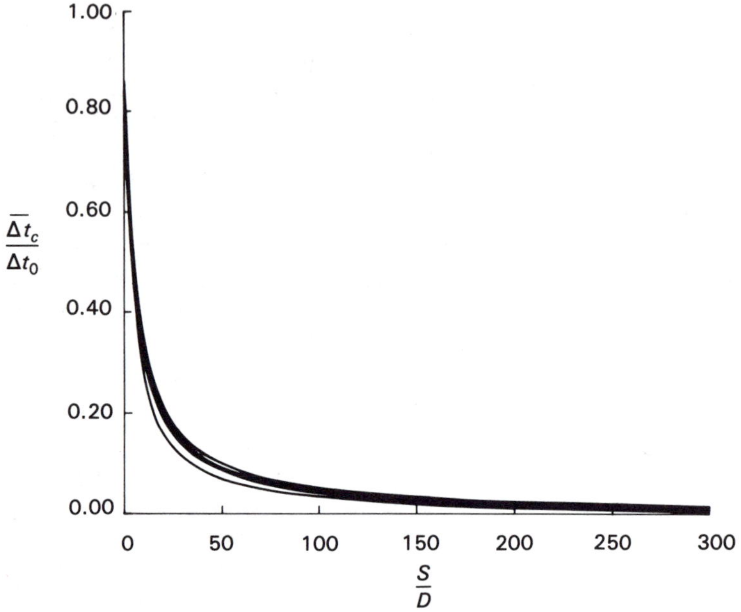

Figure 12.4.7 Temperature difference decay for a group of 15 jets, Fr = 10, 50, and 200, for entrainment model 1 through 5; model 4 is omitted for Fr = 10. *(Reprinted with permission from B. Gebhart et al.,* Adv. Heat Transfer, *vol. 16, p. 1. Copyright © 1984, Academic Press, Inc.)*

lower values of Fr and R. For a more detailed discussion of these differences see Gebhart et al. (1984).

12.4.5 Effects of Variable Properties and of Ambient Stratification

Determination of the discharge Froude number and the buoyancy force requires evaluation of both the temperature, t, and concentration, C, effects on density. Often the density, $\rho(t, C, p)$, is a sufficiently linear function of both t and C over the range of temperature and concentration difference between the jet and the ambient. Then density differences may be accurately estimated in terms of the two volumetric coefficients of expansion, as in Chapter 6:

$$\beta = -\frac{1}{\rho_r}\left(\frac{\partial \rho}{\partial t}\right)_{C,p} \quad \text{and} \quad \beta^* = -\frac{1}{\rho_r}\left(\frac{\partial \rho}{\partial C}\right)_{t,p} \tag{12.4.10}$$

where ρ_r is some reference value of ρ, say $\rho_r = \rho_0 = \rho(t_0, C_0, p_0)$. Then the overall density differences are

$$(\rho_\infty - \rho_0)_t = \beta\rho_r(t_0 - t_\infty)$$

and

$$(\rho_\infty - \rho_0)_C = \beta^* \rho_r (C_0 - C_\infty) \tag{12.4.11}$$

for the independent t and C effects on density. During integration, t and C are local values. The local density ρ is then approximated, neglecting the effect of pressure, as

$$\rho(t_0, C_0, p_0) - \rho(t, C, p_0) = \rho(t_0, C_0, p_0)[\beta(t - t_0) + \beta^*(C - C_0)] \tag{12.4.12}$$

or

$$\rho = \rho_0[1 - \beta(t - t_0) - \beta^*(C - C_0)] \tag{12.4.13}$$

The initial density difference between the local ambient and the initial jet, which appears in the jet Froude number, is written below, where $\rho_{\infty,0}$ is the ambient density at jet discharge:

$$\frac{\rho_{\infty,0} - \rho_0}{\rho_0} = \beta(t_0 - t_\infty) + \beta^*(C_0 - C_\infty) \tag{12.4.14}$$

Ambient medium stratification. Both temperature and concentration stratifications often arise. Then an equation of state is required for the solution of the governing equations. First, however, the simpler case of $\rho = \rho(t)$ only will be considered. The density stratification is then defined solely by a temperature stratification parameter

$$\frac{(dt_\infty/dz)D}{t_{\infty,0} - t_0}$$

Since β is taken as constant the local downstream Froude number changes only with changes in velocity and temperature. For such flows

$$\frac{\rho_\infty - \rho_c}{\rho_{\infty,0} - \rho_0} = \frac{t_c - t_\infty}{t_0 - t_{\infty,0}} \tag{12.4.15}$$

pertains throughout, where c indicates local centerline values. In addition, if $\rho \neq \rho(C)$, an assumption that might be made in a jet with a tracer dye only, the concentration and vertical momentum equations are not coupled. Concentration computations can proceed independently, even to the point of specifying a concentration stratification if necessary. Parallel reasoning holds true if the density variation is due to concentration stratification alone in a uniform-temperature medium. Then a density dependence on the concentration stratification parameter would couple the concentration and vertical momentum equations.

These simple examples of the overall modeling problem are relevent because they represent conditions under which measurements are often taken. These conditions are then specified to compare data with the analytical models. The underlying assumptions and limitations of such formulation are: (1) with no equation of state incorporated in the computational process, only unstratified ambients may be accommodated if $\rho = \rho(t, C)$, and then only if β and β^* are assumed constant; (2) if no equation of state is included in the model, density stratification in the ambient may be specified as either

$$\frac{(dt_\infty/dz)D}{t_{\infty,0} - t_0} \qquad \text{for } \rho = \rho(t)$$

$$\frac{(dC_\infty/dz)D}{C_{\infty,0} - C_0} \qquad \text{for } \rho = \rho(C)$$

or

$$\frac{(d\rho_\infty/dz)D}{\rho_{\infty,0} - \rho_0} \qquad \text{directly}$$

Ambients with more than one kind of density stratification cannot be accommodated, since temperature, concentration, and density differences would be independent of each other and downstream profiles could be dissimilar.

Temperature stratification. A stably stratified quiescent ambient in which density increases with increasing depth has the general effect of reducing the vertical motion and penetration of a buoyant jet. This results from the combined effects of stratification and jet buoyancy dilution due to entrainment. This change of trajectory takes place regardless of the direction of buoyancy, ambient stratification, or the magnitude of the stratification.

Figure 12.4.8 illustrates several of the effects of temperature stratification and of its strength on the trajectory for a thermally buoyant horizontal discharge in water for Fr = 50, 100, and 200, based on entrainment model 3 in Table 12.4.2. Stratification is assumed to be entirely in the form of a constant vertical temperature gradient, as expressed in the following parameter, for curves A, B, and C, respectively:

$$D\frac{dt_\infty}{dz} = 4 \times 10^{-4},\ 4 \times 10^{-3},\ \text{and}\ 4 \times 10^{-2}\,^\circ\text{C}$$

The effects of even small stratification are seen to be large because of the rapid dilution arising from entrainment. For the least stratification, A, the jets follow a trajectory similar to that in the unstratified ambient, with only small reduction in vertical penetration, for all three values of Fr. For B, a noticeable change in trend has already occurred at Fr = 50 near the end of the trajectory. A rebending of the path has occurred. The point of inflection in this curve is very significant. In the calculation, this was the vertical level at which the jet had become neutrally buoyant, due to entrainment on the one hand and the decreasing density of the surrounding fluid on the other. The further upward rise beyond this point is due solely to the vertical momentum remaining in the jet. This momentum is gradually decreased by the downward force of negative buoyancy.

Stratification C, at Fr = 50, shows a further and much greater effect. The jet is quickly becoming vertically "trapped" in this stronger stratification. The trajectory has undergone complete rebending to horizontal flow with eventual entrainment to zero buoyancy. Usually, in entrainment modeling, computations are stopped at the point of maximum rise. It is assumed that all but the most vigorous jets will have acquired "far-field" characteristics by this time. The assumption of

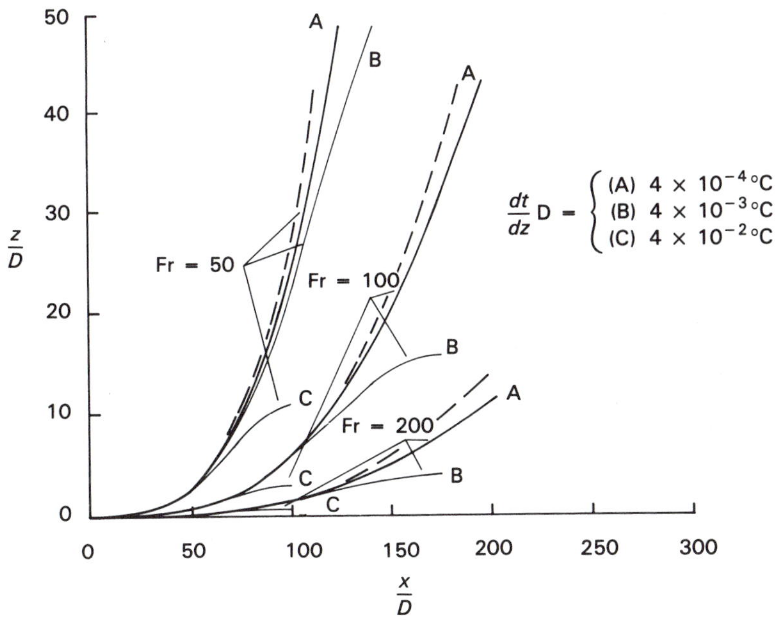

Figure 12.4.8 Buoyant jet behavior in a quiescent stratified ambient for horizontal discharges at Fr = 50, 100, and 200 at three temperature stratification levels, model 3. The dashed curve at each value of Fr is the unstratified trajectory. *(Reprinted with permission from B. Gebhart et al.,* Adv. Heat Transfer, *vol. 16, p. 1. Copyright © 1984, Academic Press, Inc.)*

an axisymmetric jet shape, already uncertain in a strongly stratified ambient, is even more suspect beyond this point. Also, the initial momentum and buoyancy have probably become largely irrelevant to further diffusion, compared to effects in most actual environments.

The trajectories for the same stratifications at Fr = 100 and 200 are similar to those at Fr = 50, except that the vertical range of rise is drastically reduced. This is because the higher momentum, for larger Fr, results in much higher entrainment and decrease of positive buoyancy, per unit of jet mass, much sooner along the trajectory.

Combined temperature and concentration stratification. The more general occurence of an ambient medium with both temperature and concentration stratification requires an equation of state for the solution of the governing equations. Then the equation of density excess, or deficiency, is of the form

$$\frac{d}{dS}\int_0^{2\pi}\int_0^{\infty} \bar{u}(\rho_\infty - \rho) r\,dr\,d\phi = -\frac{d\rho_\infty}{dS}\int_0^{2\pi}\int_0^{\infty} \bar{u} r\,dr\,d\phi \qquad (12.4.16)$$

The same Gaussian profile is assumed for density as for two of its constituent properties, temperature and concentration. This additional equation uncouples the temperature and concentration equations from the vertical momentum equation.

Hirst (1971b) specified particular constant values of β and β* throughout the downstream integration of the governing equations. The equation of state was taken as

$$\frac{\rho_\infty - \rho}{\rho_0} = \beta(t - t_{\infty,0}) + \beta^*(C - C_{\infty,0})$$

In reality, of course, the density of water is a complex and, at low temperatures, a nonlinear function of temperature. In many modeling situations the ranges of t, s, and p do not permit the use of constant values of β or β* or omission of the pressure dependence of density. Table 12.4.5 shows the variation of β, at $p = 1$ bar, encountered over a small but commonly occurring temperature range, for both saline water at 35%, typical of seawater, and fresh water.

The model of Riester et al. (1980) accounted for the effect of variable β by expresing it as a polynomial:

$$\beta = a_1 + a_2 t + a_3 t^2 \tag{12.4.17}$$

where the coefficients a_1, a_2, and a_3 were evaluated at a temperature pertinent to a given circumstance. The discharge Froude number was redefined as

$$\mathrm{Fr} = \frac{u_0}{[gD\beta(t_0 - t_\infty)]^{1/2}} \tag{12.4.18}$$

The need for an accurate equation of state was stressed for conditions in which $t_0 - t_\infty$ is greater than 3°C. This is a very significant matter when small-scale laboratory experiments are made at high temperature differences to simulate large-scale flows at low temperature differences. Also, saltwater jets cannot in general simulate freshwater jets operating over large temperature differences since the saline density effect does not generally duplicate the temperature-density characteristics of fresh water. This last observation indicates that, although the model proposed above attempts to compensate for the variable temperature effects on density, no dependence on salinity was included, except for any that may have been implicit in the choice of the three coefficients in Eq. (12.4.17).

Table 12.4.5 Variation of β over a limited temperature range for pure and saline water at 1 atm pressure[a]

	Temperature (°C)									
Salinity (‰)	−1	0	1	2	3	4	5	6	7	8
35	3.70	5.01	6.27	7.49	8.67	9.83	10.97	12.09	13.19	14.29
0				−3.31	−1.77	0.05	1.68	3.22	4.66	6.05

[a] $\beta = -(1/\rho)(\partial\rho/\partial t)_p \times 10^5\ (°\mathrm{C})^{-1}$.
Source: From Gebhart et al. (1984).

Gebhart et al. (1984) used the Gebhart-Mollendorf (1977) water density relation, Eq. (9.1.1), to calculate actual oceanic ambient water and jet densities and to compute buoyant jet trajectories and decays. Typical oceanic conditions were investigated for jets having various values of Fr.

Figure 12.4.9 shows an example of the results. The oceanic density variation is shown on the left. The trajectories shown are for five horizontal buoyant momentum discharges into that quiescent ambient, at that depth. The jets are assumed to have an initial salinity equal to that of the ocean at that depth, $s_{0,\infty} = 33.42‰ = s_0$. The temperature is higher than $t_{0,\infty} = 3.08°C$. Five values of t_0 were chosen to result in the values Fr = 10, 20, 30, 40, and 50.

As the jet rises, the net buoyancy is from a combination of temperature and salinity differences, in relation to the stratification. The buoyancy force along the trajectory is calculated from the density relation, Eq. (9.1.1), based on the local values of jet and ambient temperature, salinity, and pressure.

All five of these jets experience the progression from positive to negative buoyancy, as well as a level of maximum upward penetration. All have a rather smooth curvature and recurvature due to the relatively constant density gradient

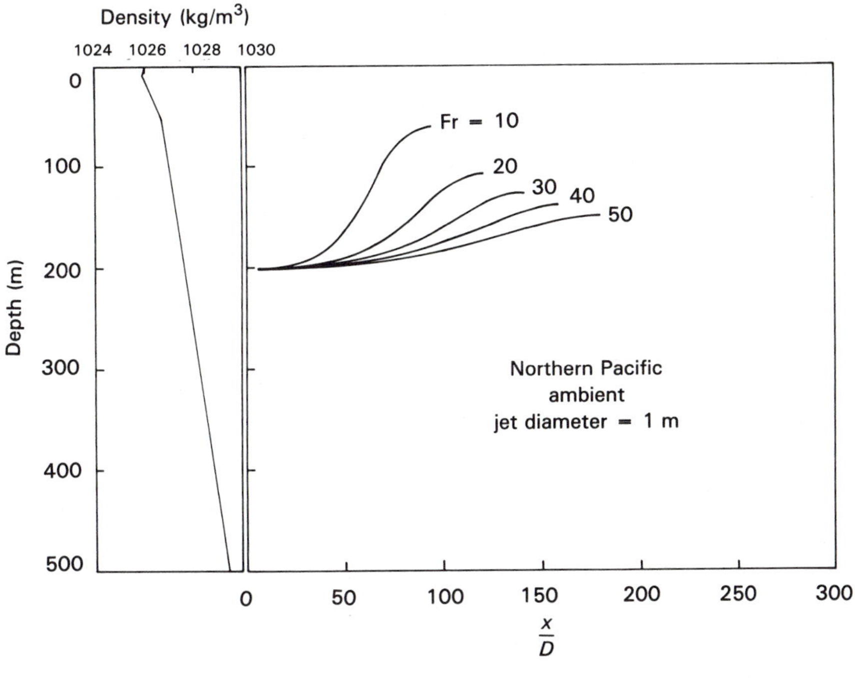

Figure 12.4.9 Trajectory at Fr = 10, 20, 30, 40, and 50 of 1-m-diameter jets discharged into the stratified ambient with the density distribution shown on the left side of the figure, model 3. *(Reprinted with permission from B. Gebhart et al.,* Adv. Heat Transfer, *vol. 16, p. 1. Copyright © 1984, Academic Press, Inc.)*

encountered in the vertical range of penetration. These results show that quite moderate environmental stratifications may have large effects on trajectory.

12.4.6 Other Studies of Buoyant Jets

Ero (1977) used an entrainment function that varies with the local densimetric Froude number to calculate the characteristics of buoyant plumes in an atmospheric inversion. In a series of papers, Madni and Pletcher (1975, 1977a, 1977b) and Hwang and Pletcher (1978), used a finite-difference method to analyze the behavior of turbulent buoyant jets in a variety of ambients, including quiescent, coflowing, and cross flowing, both with and without stratification. Savage and Chan (1970) studied the behavior of a hot vertical laminar jet flowing into a quiescent cold ambient, using a perturbation series expansion as well as integral methods, with hyperbolic secant distributions for the velocity and temperature distributions.

Yih and Wu (1981) analyzed the behavior of round buoyant plumes. For laminar flow a perturbation expansion about the exact solutions for Pr = 1 and 2 was used. An eddy viscosity formulation was applied for turbulent flow to obtain exact solutions for turbulent Prandtl numbers of 1 and 2. The approximate solutions were based on these exact solutions. Sforza and Mons (1978) used Reichardt's turbulence hypothesis (Schlichting, 1968) to analyze the mixing of various gas jets with quiescent ambient air. They also obtained measurements of the jet properties. Experiments with a downward-directed saltwater jet issuing into fresh water were performed by Wright (1977). In these experiments, the jet was also towed to investigate the effects of a cross flow.

Chu and Goldberg (1974) used a simple one-constant entrainment model to investigate the behavior of buoyant jets and plumes in a cross-flowing ambient. They also performed a series of experiments with very small jets (of the order of 1.8 mm in diameter) injected vertically downward. Chen and Nikitopoulos (1979) used a k-ϵ turbulence model to analyze the behavior of jets in a quiescent ambient.

In several circumstances of practical interest, the buoyancy force opposes the flow in the jet. Such flows are often referred to as negatively buoyant jets. The flow is retarded under the action of the opposing buoyancy force so that the flow reverses direction after reaching stagnation at a certain height, which is governed by the inlet momentum and buoyancy of the jet. Several experimental and analytical studies, based on the entrainment assumption, have been carried out. Turner (1966) and Seban et al. (1978) considered turbulent axisymmetric negatively buoyant jets. The penetration height and the temperature and velocity distributions were determined from the analytical model and compared with the experimental results, indicating fairly good agreement. A detailed experimental study of turbulent, negatively buoyant two-dimensional wall and free jets was carried out by Goldman and Jaluria (1986). The velocity and temperature distributions were measured. These were employed to compute the penetration distance of the jet. The results were shown to be well correlated in terms of the parameter Gr/Re^2, where both Gr and Re are based on the conditions at the jet inlet and the slot width.

12.5 THERMALS

A thermal is a discrete buoyant mass of fluid ascending or descending in an ambient medium due to a difference in density. Thermals arise from a sudden input of buoyancy in the fluid. For example, they are generated by explosions. They also form adjacent to a heated horizontal surface, due to the instability of the conduction layer adjacent to the surface. Masses of hot fluid then ascend through the ambient medium and, by entrainment, grow to several times their original size.

The earliest investigation of such flows was apparently an extensive study by Townsend (1959). Howard (1961) then formulated a theory to explain the generation of thermals. Sparrow et al. (1970) visualized the flow adjacent to a heated horizontal surface and observed thermal generation from the warm surface layer, as seen in Fig. 12.5.1. The thermals have a mushroomlike appearance, with a blunted hemispherical cap. They are produced at fixed, regularly spaced locations along the surface, as long as the heating rate is kept constant.

Their generation was observed to be periodic in time, and both the spatial frequency and the rate of production are found to increase with increased heating rate. When the surface is initially heated, a thermal boundary layer is formed in the fluid. This grows in thickness with increasing time according to

$$\delta = \sqrt{\pi \alpha \tau} \tag{12.5.1}$$

Sparrow et al. (1970) found that this process continues until some critical time τ_{cr} at which the Rayleigh number $R_\delta = g\beta(t_0 - t_\infty)\delta^3/\alpha\nu$ reaches a critical value, of order 1000. Then the thermal boundary layer breaks up and masses of hot fluid rise in the ambient medium as thermals. Then, the process starts over.

Howard (1964) proposed a model for evaluating the temperature profile above the heated surface. By assuming that heat is transferred only by conduction before the layer breaks up and rises in the form of thermals, it was possible to calculate the time-averaged conduction temperature profile above the surface. Figure 12.5.2 shows the computed results. These compare favorably with measurements by Townsend (1959) and Elder (1968).

Foster (1965, 1971) modeled thermal generation as a time-dependent stability problem and showed that, except for small heating rates, the flows are inherently unstable. After their initial generation, thermals ascend or descend into the ambient fluid and expand, growing to several times their original size. Turner (1973) pointed out that, although similarity solutions may be obtained to describe axisymmetric thermals during this process, they may not be strictly applicable to actual flows since they have insufficient time to adjust to any equilibrium process.

Scorer (1957) experimented with thermals by visualizing small quantities of heavy salt solutions sinking into fresh water. It was seen that their shape is slightly oblate spheroidal. Woodward (1959) and Saunders (1962) studied this type of flow experimentally and found that a vortexlike circulation is superimposed on the general vertical motion during the expansion period of the thermal. Figure 12.5.3 shows the streak lines of the flow. These are instantaneously the same as

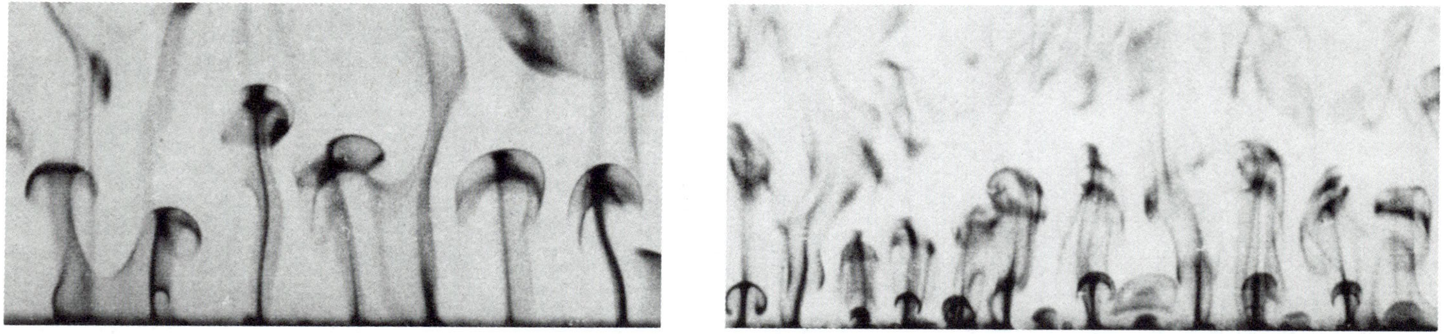

Figure 12.5.1 Buoyant thermals rising from a heated surface. Mushroom-shaped plumes rise periodically above a heated copper plate. They are made visible by an electrochemical technique using thymol blue. The heating rate is higher in the photograph at the right. *(Reprinted with permission from E. M. Sparrow et al.,* J. Fluid Mech., *vol. 41, p. 793. Copyright © 1970, Cambridge University Press.)*

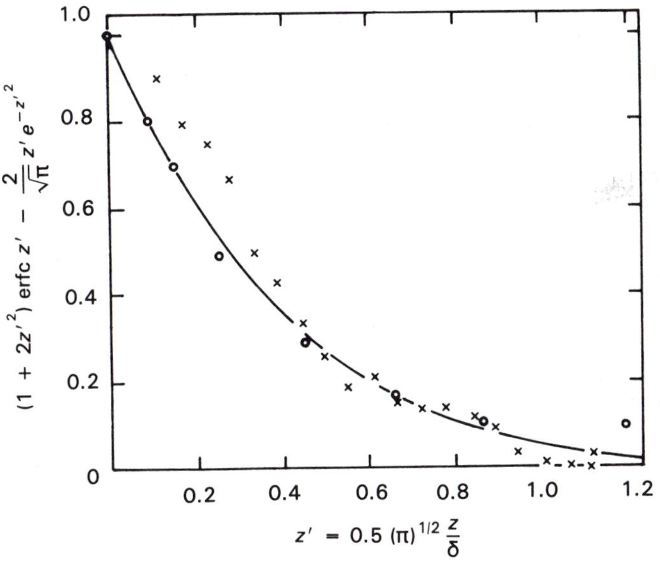

Figure 12.5.2 Comparison of the mean temperature profile with the experimental results of Elder (1968), Townsend (1959), and Howard (1964). Here $\delta = \sqrt{\pi\alpha\tau_{cr}}$. *(Reprinted with permission from J. S. Turner,* Buoyancy Effects in Fluids. *Copyright © 1973, Cambridge University Press.)*

those for a flattened spherical vortex, with an upflow near the core and a downflow on the outer edges. The resulting particle paths from that model are found to be in agreement with experimental observations, even though laboratory thermals are not spheres, but rather oblate spheriods.

A simple analytical model for a thermal may be developed on the basis of the entrainment hypothesis discussed earlier and as outlined by Turner (1973). Consider the rise of a thermal in an isothermal quiescent ambient medium. For a spherical thermal, the entrainment assumption, along with the continuity equation, gives

$$\frac{d(b^3)}{d\tau} = 3\alpha b^2 u \qquad (12.5.2)$$

where $b(z)$ is the horizontal radius of the thermal, τ is time, and the vertical velocity $u = dz/d\tau$. This gives $b = \alpha z$, which indicates a linear spread of the thermal with height z. Scorer (1957), who used thermals of heavy salt solution falling in lighter fresh water, summarized the experimental results as

$$u = C(gbB)^{1/2} \qquad (12.5.3)$$

where $B = \Delta\rho/\rho_\infty$ and C is a Froude number. Also, $\Delta\rho$ is the difference between the mean density ρ at a location z and that in the ambient medium, ρ_∞. Therefore,

$$C^2 = \frac{u^2}{gBb} \qquad \text{with } \Delta\rho = \rho_\infty - \rho \qquad (12.5.4)$$

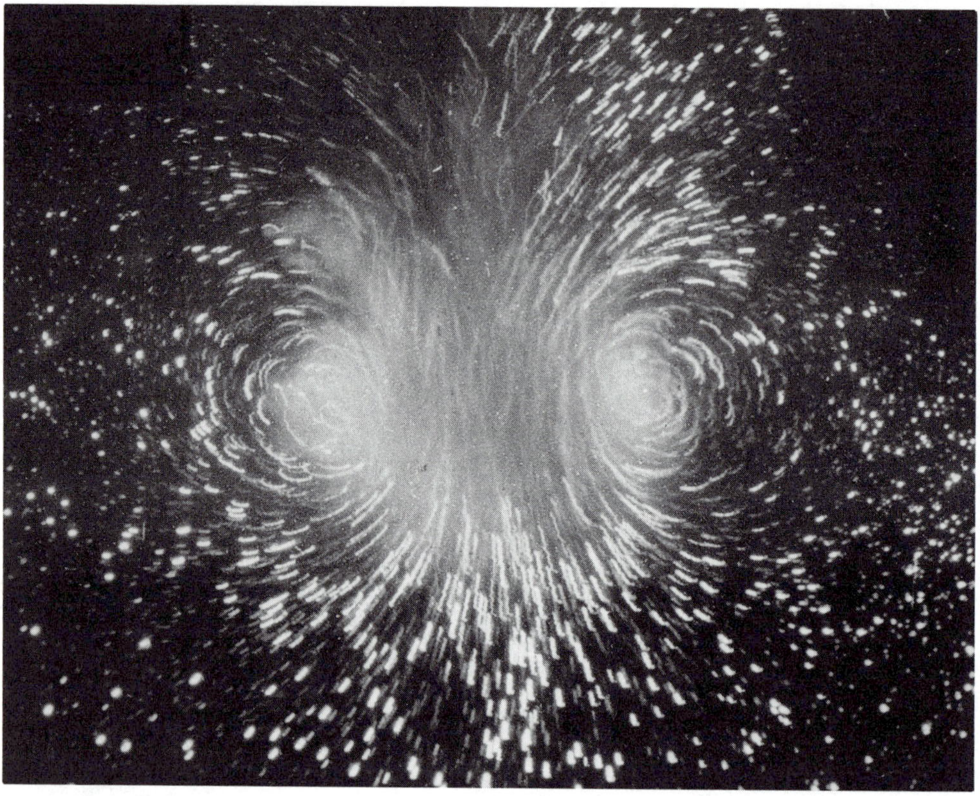

Figure 12.5.3 Streak picture of the flow in and around an isolated thermal. The exposure time (about 1.5 s) is short enough for this to give a good representation of the streamlines relative to axes at rest. The direction of motion in this experiment is downward. *(Reprinted with permission from J. S. Turner,* Buoyancy Effects in Fluids. *Copyright © 1973, Cambridge University Press.)*

This implies that C is the ratio of the inertia force to the buoyancy force. For clouds, C has a value of about 0.95. In laboratory experiments, C is about 1.2, with the mean value of α obtained as 0.25.

The volume of the thermal may be taken as mb^3 if it is an oblate spheroid. Scorer (1957) found the value of m to be 3.0. Therefore, buoyancy per unit volume is gB_0/mb^3, where B_0 is given as $(\rho_\infty - \rho_0)V_0/\rho_\infty$, the total buoyancy input. Here, ρ_0 and V_0 are density and volume of fluid at the source. Then

$$u = \frac{dz}{d\tau} = C\sqrt{gBb} = C\sqrt{\frac{gB_0}{mb^2}} = C\sqrt{\frac{gB_0}{m\alpha^2 z^2}} \tag{12.5.5}$$

or $z\,dz = C\sqrt{gB_0/m\alpha^2}\,d\tau$. Therefore,

$$z^2 = \frac{2C\tau}{\alpha}\sqrt{\frac{gB_0}{m}} \tag{12.5.6}$$

since z is zero at $\tau = 0$. This gives $z \propto \tau^{1/2}$, which implies

$$b \propto \tau^{1/2} \qquad \text{and} \qquad u \propto \tau^{-1/2} \tag{12.5.7}$$

The Reynolds number, given as ub/ν, is therefore a constant. The circulation Γ is $\oint \mathbf{V} \cdot \mathbf{ds}$, where $\mathbf{V}$ is the velocity vector and $\mathbf{ds}$ is along the circumference. Thus, it is proportional to ub and also remains constant downstream.

REFERENCES

Abraham, G. (1965). *J. Hydraul. Res. 3,* 1.

Albertson, M. L., Dai, Y. B., Jensen, R. A., and Rouse, H. (1950). *Trans. Am. Soc. Civ. Eng. 115,* 639.

Beavers, G. S., and Wilson, T. A. (1970). *J. Fluid Mech. 44,* 97.

Beuther, P. D. (1980). Experimental Investigation of the Turbulent Axisymmetric Plume, Ph.D. Thesis, State Univ. of New York, Buffalo.

Briggs, G. A. (1975). Plume Rise Predictions, in *Lectures on Air Pollution and Environmental Impact Analysis,* Chap. 3, American Meteorological Society.

Brown, G. B. (1935). *Proc. Phys. Soc. 47,* 703.

Chen, C. J., and Nikitopoulos, C. P. (1979). *Int. J. Heat Mass Transfer 22,* 245.

Chen, C. J., and Rodi, W. (1980). *Vertical Turbulent Buoyant Jets: A Review of Experimental Data,* Pergamon, New York.

Chu, V. H., and Goldberg, M. B. (1974). *Proc. Am. Soc. Civ. Eng., J. Hydraul. Div. 100,* 1203.

Cormack, D. E. (1975). Studies of a Phenomenological Turbulence Model, Ph.D. Thesis, California Inst. of Technology, Pasadena.

Davis, L. R., Shirazi, M. A., and Slegel, D. L. (1978). *Trans. ASME, J. Heat Transfer 100,* 442.

Delichatsios, M. A. (1979). *J. Fluid Mech. 93,* 241.

Elder, J. W. (1968). *J. Fluid Mech. 32,* 69.

Ero, M. I. O. (1977). *Trans. ASME, J. Heat Transfer 99,* 335.

Fan, L. H. (1967). W. M. Keck Lab. Rept. KH-R-15.

Fischer, H. B., List, E. J., Imberger, J., and Brooks, N. H. (1979). *Mixing in Inland and Coastal Waters,* Academic Press, New York.

Foster, T. D. (1965). *Phys. Fluids 8,* 1770.

Foster, T. D. (1971). *Geophys. Fluid Dyn. 2,* 201.

Fox, D. G. (1970). *J. Geophys. Res. 75,* 6818.

Freymuth, P. (1966). *J. Fluid Mech. 251,* 683.

Gebhart, B., and Mollendorf, J. C. (1977). *Deep Sea Res. 24,* 831.

Gebhart, B., Hilder, D. S., and Kelleher, M. (1984). *Adv. Heat Transfer 16,* 1.

George, W. K., Jr., Alpert, R. L., and Taminini, F. (1977). *Int. J. Heat Mass Transfer 20,* 1145.

Ginsberg, T., and Ades, M. (1975). *Trans. Am. Nucl. Soc. 21,* 87.

Goldman, D., and Jaluria, Y. (1986). *J. Fluid Mech. 166,* 41.

Hinze, J. O. (1975). *Turbulence: An Introduction to Its Mechanism and Theory,* McGraw-Hill, New York.

Hirst, E. A. (1971a). Oak Ridge Natl. Lab. Rept. ORNL-TM-3470.

Hirst, E. A. (1971b). Oak Ridge Natl. Lab. Rept. ORNL-4685.

Howard, L. N. (1961). *J. Fluid Mech. 10,* 509.

Howard, L. N. (1964). *Proceedings of the 11th Congress on Applied Mechanics,* Munich, Springer-Verlag, Berlin.

Hwang, S. S., and Pletcher, R. H. (1978). *Proc. 6th Int. Heat Transfer Conf., Toronto, General Papers 1,* 109.

Kotsovinos, N. E. (1975). A Study of the Entrainment and Turbulence in a Plane Buoyant Jet, Ph.D. Thesis, California Inst. of Technology, Pasadena.

Kotsovinos, N. E. (1977). *J. Fluid Mech. 81,* 45.
List, E. J. (1982). *Annu. Rev. Fluid Mech. 14,* 189.
List, E. J., and Imberger, J. (1973). *Proc. Am. Soc. Civ. Eng., J. Hydraul. Div. 99,* 1461.
Lumley, J. L., and Khajeh-Nouri, B. (1974). *Adv. Geophys. 18A,* 162.
Madni, I. K., and Pletcher, R. H. (1975). *Trans. ASME, J. Fluids Eng. 97,* 558.
Madni, I. K., and Pletcher, R. H. (1977a). *Trans. ASME, J. Heat Transfer 99,* 99.
Madni, I. K., and Pletcher, R. H. (1977b). *Trans. ASME, J. Heat Transfer 99,* 641.
Morton, B. R. (1959). *J. Fluid Mech. 5,* 151.
Morton, B. R., Taylor, G. I., and Turner, J. S. (1956). *Proc. R. Soc. London Ser. A 234,* 1.
Nakagome, H., and Hirata, M. (1976). *Proceedings of the ICHMT Seminar on Turbulent Buoyant Convection,* p. 361, Hemisphere, Washington, D.C.
Ricou, F. P., and Spalding, D. B. (1961). *J. Fluid Mech. 11,* 21.
Riester, J. B., Bajura, R. A., and Schwartz, S. H. (1980). *Trans. ASME, J. Heat Transfer 102,* 557.
Rockwell, D. O., and Niccolls, W. O. (1972). *Trans. ASME, J. Basic Eng. 94,* 720.
Rouse, H., Yih, C. S., and Humphreys, H. W. (1952). *Tellus 4,* 201.
Saffman, P. G. (1970). *Proc. R. Soc. London Ser. A 317,* 417.
Saunders, P. M. (1962). *J. Meteorol. 18,* 451.
Savage, S. B., and Chan, G. K. C. (1970). *Q. J. Mech. Appl. Math. 23*(3), 413.
Schatzmann, M. (1979). *Atmos. Environ. 13*(5), 721.
Schlichting, H. (1968). *Boundary Layer Theory,* 6th ed., McGraw-Hill, New York.
Schneider, P. E. M. (1980). *Z. Flugwiss. Weltraumforsch. 4*(5), 307.
Scorer, R. S. (1957). *J. Fluid Mech. 2,* 583.
Seban, R. A., Behnia, M. M., and Abreu, J. E. (1978). *Int. J. Heat Mass Transfer 21,* 1453.
Sforza, P. M., and Mons, R. F. (1978). *Int. J. Heat Mass Transfer 21*(4), 371.
Sparrow, E. M., Husar, R. B., and Goldstein, R. J. (1970). *J. Fluid Mech. 41,* 793.
Taylor, G. I. (1932). *Proc. R. Soc. London Ser. A 135,* 685.
Tennekes, H., and Lumley, J. L. (1973). *A First Course in Turbulence,* MIT Press, Cambridge, Mass.
Tollmien, W. (1926). *Z. Angew. Math. Mech. 6,* 468.
Townsend, A. A. (1959). *J. Fluid Mech. 4,* 361.
Turner, J. S. (1962). *J. Fluid Mech. 13,* 356.
Turner, J. S. (1966). *J. Fluid Mech. 26,* 779.
Turner, J. S. (1973). *Buoyancy Effects in Fluids,* Cambridge Univ. Press, Cambridge, England.
Van Dyke, M. (1982). *An Album of Fluid Motion,* Parabolic Press, Stanford, Calif.
Whremann, O., and Wille, R. (1957). *IUTAM Symp.,* H. Görtler, Ed., p. 387, Friburg.
Woodward, B. (1959). *Q. J. R. Meteorol. Soc. 85,* 144.
Wright, S. J. (1977). *Proc. Am. Soc. Civ. Eng., J. Hydraul. Div. 103*(5), 499.
Yih, C. S., and Wu, F. (1981). *Phys. Fluids 24*(5), 794.
Yule, A. J. (1978). *J. Fluid Mech. 89,* 413.
Zijnen, Van der Hegge, B. G. (1958). *Appl. Sci. Res. A7,* 256, 277.

PROBLEMS

12.1 A 5-cm-diameter jet of heated water at 40°C is discharged vertically into a tank of quiescent water at 25°C at a velocity of 10 cm/s. The jet is turbulent.

(a) Calculate the Froude number.

(b) Calculate the centerline temperature 4 m above the nozzle.

12.2 Plot the trajectory of the jet in problem 12.1 discharged horizontally. Also plot the average velocity and temperature decay against horizontal penetration. Use entrainment model 5, and compare the results with those obtained with the other models.

12.3 Water at 35°C is discharged horizontally into quiescent water at 20°C.

(a) Calculate the Froude number for discharge velocities of 10 and 100 cm/s for initial jet diameters of 5 cm and 1 m—that is, four values of the Froude number.

(b) For each condition, determine the downstream distance at which the jet rise has been to $z/D = 80$ for the most appropriate entrainment model.

(c) Estimate the length of the jet trajectory for each condition, at $z/D = 80$, and the jet diameter.

(d) For each condition, find the jet centerline temperature and velocity at $z/D = 80$.

12.4 Repeat the calculations for the conditions in Problems 12.1 and 12.2 if the ambient water is flowing horizontally at a velocity of 2 cm/s.

12.5 Cold air enters a storage chamber as a vertical jet, but negatively buoyant, at the floor of the room. Assuming turbulent flow and using the entrainment model of Taylor, obtain the conservative equations. Use top hat profiles. How will you solve this system of equations? Qualitatively plot the temperature and velocity distributions. Take inlet temperature as t_0 and ambient room air temperature as t_a, with $t_0 < t_a$.

12.6 For a turbulent axisymmetric plume generated by a small heated disk at 320°C in air at 20°C, calculate the velocity and temperature decay above the source. Also find the variation of the flow width or radius with height. Take $h = 15\ \text{W/m}^2\ \text{K}$ at the disk, of 2 cm diameter.

12.7 For the temperature and velocity conditions in Problem 12.1, consider a horizontal jet of 5 cm diameter flowing at a velocity of 1 m/s into a stably stratified quiescent ambient. The temperature gradient is 0.08°C/m.

(a) Find the maximum height of jet rise and the penetration at that location.

(b) Plot the ratio of actual rise to that in an unstratified ambient vs. horizontal distance.

12.8 Consider a seawater jet of 1 m diameter at 5.2°C discharged horizontally at 0.5 m/s at a depth of 200 m in the oceanic stratification shown in Fig. 12.4.11. The temperature at 200 m depth is 3.2°C.

(a) Calculate the Froude number, based only on the initial temperature difference, at 200 m.

(b) Estimate the distance to which the jet will rise and its diameter at that location.

CHAPTER

THIRTEEN

UNSTABLY STRATIFIED FLUID LAYERS

13.1 MECHANISMS OF INSTABILITY

Unstable stratification is a common cause of buoyancy-driven motions. It frequently occurs in bodies of fluid that would otherwise be quiescent. The impetus for such instability may be the vertical distribution of density. It was shown in Section 3.6 and in Fig. 3.6.1 that an existing vertical gradient of density in a medium, $d\rho/dx$, must satisfy the following condition for general stability:

$$\frac{d\rho}{dx} < \left(\frac{d\rho}{dx}\right)_S < 0 \qquad \text{for} \left(\frac{\partial\rho}{\partial p}\right)_S > 0 \tag{13.1.1}$$

where $(d\rho/dx)_S$ is the "adiabatic"—actually the reversible adiabatic—stratification. Since this quantity is negative, because pressure decreases upward, $d\rho/dx$ must decrease as fast or faster for complete stability.

The complexity of the above condition arises from the dependence of density on both temperature and pressure, as $\rho(t, p)$. However, if the density of the fluid being considered depends very weakly on pressure or if the vertical extent of the fluid medium is very small, as in a thin horizontal layer of fluid, then simpler measures of potential instability are a reasonable approximation. For example, in a very thin horizontal layer subject to different temperatures at its two bounding surfaces, the vertical difference in ρ is essentially due only to temperature differences. That is, the pressure effect is very small and $\rho = \rho(t)$ to a good approximation. For fluid states for which

$$\beta = -\frac{1}{\rho}\left(\frac{\partial\rho}{\partial t}\right)_p \geq 0 \tag{13.1.2}$$

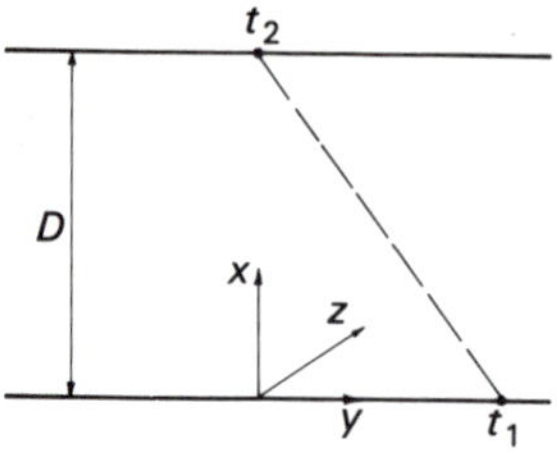

Figure 13.1.1 Temperature-stratified horizontal layer of quiescent fluid.

the relation in Eq. (13.1.1), for a tendency to instability, is replaced by Eq. (3.6.6) rewritten as

$$\frac{dt}{dx} \leq 0 \tag{13.1.3}$$

The average gradient of t in Fig. 13.1.1 is $(t_2 - t_1)/D$. This is negative and thus potentially unstable. If there is no motion in a fluid layer of constant thermal conductivity, the conduction solution, shown as the dashed line, applies as follows:

$$t(x) = t_1 - (t_1 - t_2)\frac{x}{D} \qquad \text{or} \qquad \phi = \frac{t - t_2}{t_1 - t_2} = 1 - \frac{x}{D} \tag{13.1.4}$$

Since $t_2 < t_1$, this variation may be unstable. However, actual instability is a fluid-mechanic question. An unstable case usually is considered to be one in which the ever-present disturbances will continue or amplify. Thus, actual instability and its consequences depend not only on unstable stratification but also on matters such as momentum production and viscous effects.

Instability mechanisms that arise primarily from thermal effects on density are called, variously, Bénard, Rayleigh, or thermal instability. The resulting thermal transport is often called thermal convection. Bénard (1900) observed hexagonal patterns in an unstably stratified horizontal fluid layer. Rayleigh (1916) analyzed an idealized model of such instability and found neutral stability limits in terms of $t_1 - t_2$, D, and the transport characteristic of the fluid layer.

Since such unstable stratification occurs very frequently in processes of common interest, there have been many subsequent investigations of such stability and of the complicated transport consequences of unstable conditions. The following sections will first consider stability and then the consequences of unstable circumstances in terms of the motions that arise and the resulting transport. Although thermal transport is largely considered here, similar effects arise for species diffusion as well and may often be treated in an analogous manner.

13.2 INITIAL INSTABILITY

Considering the geometry in Fig. 13.1.1, the disturbance forms are postulated in a manner similar to that followed in Chapter 11 for external flows generated by buoyancy. These disturbances are then superimposed on any base flow of interest, using the full equations of transport. Certain reasonable approximations are then made to simplify calculations. The first question that is asked is, under what conditions of D, μ, $t_1 - t_2$, and so forth will the postulated form of the disturbances be neutrally stable? Conditions on opposite sides of neutral stability will presumably damp and amplify, respectively, and a stability plane will result.

The "base" flow of Fig. 13.1.1 is zero velocity, a linear variation of $\bar{t}(x)$ between t_1 and t_2, and a hydrostatic pressure variation $d\bar{p}/dx = -g\bar{\rho}(\bar{t})$. The instantaneous or disturbed flow quantities are, therefore,

$$\mathbf{V}(x, y, z, \tau) = \mathbf{V}'(x, y, z, \tau) = (u', v', w') \tag{13.2.1}$$

$$t(x, y, z, \tau) = \bar{t}(x) + t'(x, y, z, \tau) \tag{13.2.2}$$

or

$$\frac{t - t_2}{t_1 - t_2} = \frac{\bar{t} - t_2}{t_1 - t_2} + \frac{t'}{t_1 - t_2} = \bar{\phi}(x) + \phi'(x, y, z, \tau) = 1 - \frac{x}{D} + \phi' \tag{13.2.3}$$

and

$$p(x, y, z, \tau) = \bar{p}(x) + p'(x, y, z, \tau) \tag{13.2.4}$$

$$\rho(x, y, z, \tau) = \bar{\rho}(\bar{t}) + \rho'(x, y, z, \tau) \tag{13.2.5}$$

where, from the Boussinesq approximation,

$$\rho' = \rho - \bar{\rho} = \bar{\rho}\beta(t - \bar{t}) = \bar{\rho}\beta t' = \bar{\rho}\beta(t_1 - t_2)\phi' \tag{13.2.6}$$

The body force in Eq. (2.1.2) is rewritten as

$$\begin{aligned} \mathbf{g}\rho - \nabla p &= -g\mathbf{i}(\bar{\rho} + \rho') - \nabla\bar{p} - \nabla p' \\ &= -g\mathbf{i}\rho' - \nabla p' = -g\mathbf{i}\bar{\rho}\beta(t_1 - t_2)\phi' - \nabla p'_m \end{aligned} \tag{13.2.7}$$

where for the undisturbed flow $g\mathbf{i}\bar{\rho} = -\nabla\bar{p}$ and the local reference density for the buoyancy force is taken as $\bar{\rho}$, as ρ_∞ was taken for external flows. The above evaluation of the disturbance buoyancy force results in the remaining disturbance motion pressure term, $\nabla p'_m$.

The general equations are again (2.1.1)–(2.1.3). Permissible approximations simplify them to Eqs. (2.7.19)–(2.7.21) for μ and k invariant. For $\mathbf{g}\rho - \nabla p$ evaluated in Eq. (13.2.7), when the density level in the equations is taken as some value of $\bar{\rho} = \rho$ and noting that $\mathbf{V} = \mathbf{V}'$, they become

$$\nabla \cdot \mathbf{V} = \nabla \cdot \mathbf{V}' = 0 \tag{13.2.8}$$

$$\frac{D\mathbf{V}}{D\tau} = g\mathbf{i}\beta(t_1 - t_2)\phi' - \frac{1}{\rho}\nabla p'_m + \nu\nabla^2\mathbf{V} \tag{13.2.9}$$

$$\frac{D\phi}{D\tau} = \alpha\nabla^2\phi \tag{13.2.10}$$

where the pressure, viscous dissipation, and distributed source effects are neglected in Eq. (13.2.10). The momentum convection term in Eq. (13.2.9) is rewritten for $\mathbf{V}'$ small as

$$\frac{D\mathbf{V}}{D\tau} = \frac{D\mathbf{V}'}{D\tau} = \frac{\partial \mathbf{V}'}{\partial \tau} + (\mathbf{V} \cdot \boldsymbol{\nabla})\mathbf{V}' \approx \frac{\partial \mathbf{V}'}{\partial \tau}$$

That is, the nonlinear terms in the disturbance velocity are neglected as relatively small in the present linear stability analysis. Now, from Eq. (13.2.10)

$$\frac{D\phi}{D\tau} = \frac{\partial \phi'}{\partial \tau} + u'\frac{d\bar{\phi}}{dx} + u'\frac{\partial \phi'}{\partial x} + v'\frac{\partial \phi'}{\partial y} + w'\frac{\partial \phi'}{\partial z} = \frac{\partial \phi'}{\partial \tau} + u'\frac{d\bar{\phi}}{dx} = \frac{\partial \phi'}{\partial \tau} - \frac{u'}{D}$$

where again nonlinear disturbance terms are neglected. The thermal conduction term is evaluated as

$$\nabla^2\phi = \nabla^2\bar{\phi} + \nabla^2\phi' = \frac{d^2\bar{\phi}}{dx^2} + \nabla^2\phi' = \nabla^2\phi'$$

Applying these results to Eqs. (13.2.8)–(13.2.10), the resulting equations governing the disturbances are

$$\boldsymbol{\nabla} \cdot \mathbf{V}' = 0 \tag{13.2.8}$$

$$\frac{\partial u'}{\partial \tau} = g\rho\beta(t_1 - t_2)\phi' - \frac{1}{\rho D}\frac{\partial p'_m}{\partial x} + \frac{\nu}{D^2}\nabla^2 u' \tag{13.2.11}$$

$$\frac{\partial v'}{\partial \tau} = -\frac{1}{\rho D}\frac{\partial p'_m}{\partial y} + \frac{\nu}{D^2}\nabla^2 v' \tag{13.2.12}$$

$$\frac{\partial w'}{\partial \tau} = -\frac{1}{\rho D}\frac{\partial p'_m}{\partial z} + \frac{\nu}{D^2}\nabla^2 w' \tag{13.2.13}$$

$$\frac{\partial \phi'}{\partial \tau} - \frac{u'}{D} = \frac{\alpha}{D^2}\nabla^2\phi' \tag{13.2.14}$$

where x, y, z are $(x, y, z)/D$ in Eqs. (13.2.8)–(13.2.10).

The five equations above are in u', v', w', p'_m, and ϕ', each of which may depend on x, y, z, and τ. Boundary conditions are to be applied at $x = 0$ and 1, the boundary locations. Many kinds of boundaries arise in applications. The simplest case in Fig. 13.1.1 might be the fluid layer bounded by solids above and below. Then $\mathbf{V}' = 0$ at $x = 0$ and 1. If the solids are of high thermal capacity and thermal conductivity, the temperature disturbance in the fluid will not penetrate them. Then $\phi' = 0$ at $x = 0$ and 1. The several interesting circumstances will be set forth after Eqs. (13.2.8) and (13.2.11)–(13.2.14) are simplified and disturbance forms are postulated.

Eliminating v', w', and p'_m from Eqs. (13.2.8) and (13.2.11)–(13.2.13), the following equation in u' and ϕ' results:

$$\left(\frac{\partial}{\partial \tau} - \frac{\nu}{D^2}\nabla^2\right)\nabla^2 u' = g\beta(t_1 - t_2)\left(\frac{\partial^2\phi'}{\partial y^2} + \frac{\partial^2\phi'}{\partial z^2}\right) \tag{13.2.15}$$

Thus, Eqs. (13.2.14) and (13.2.15) are only in terms of u' and ϕ'. The forms of the disturbance quantities are now postulated as

$$u'(x, y, z, \tau) = \frac{\alpha}{D} f(y, z) W(x) e^{\sigma\alpha\tau/D^2} = \frac{\alpha}{D} f W e^{\sigma t} \tag{13.2.16a}$$

$$\phi'(x, y, z, \tau) = f(y, z)\theta(x)e^{\sigma t} \tag{13.2.16b}$$

where the Fourier number $t = \alpha\tau/D^2$ and σ is taken real. Thus, $\sigma = 0$ is unchanging or neutral disturbance behavior, $\sigma < 0$ is temporal damping, and $\sigma > 0$ is temporal growth. The spatial form of the disturbances is assumed to be such that the y, z, and x dependencies may be separated into two functions, $f(y, z)$ and $W(x)$ or $\theta(x)$, as above. Disturbance amplitude is postulated to change with time as $e^{\sigma t}$.

Substituting Eqs. (13.2.16a) and (13.2.16b) into Eq. (13.2.14) yields the following, where $-a^2$ is the separation constant:

$$\frac{\nabla^2 f}{f} = \frac{f_{yy} + f_{zz}}{f} = \sigma - \frac{W}{\theta} - \frac{1}{\theta}\theta_{xx} = -a^2 \tag{13.2.17}$$

or

$$f_{yy} + f_{zz} + a^2 f = 0 \tag{13.2.18}$$

$$[\sigma - (d^2 - a^2)]\theta(x) = W(x) \tag{13.2.19}$$

where $d = d/dx$. A convenient consequence is apparent in Eq. (13.2.18). That is, as a result of the forms postulated in Eqs. (13.2.16) and (13.2.17), the y, z dependence $f(y, z)$ may be determined, independent of $W(x)$ and $\theta(x)$. A solution of Eq. (13.2.18) is

$$f(y, z) = a^2 \cos\frac{n_1\pi y}{L_1} \cos\frac{n_2\pi z}{L_2} \tag{13.2.20}$$

where L_1/n_1 and L_2/n_2, the wavelengths of periodic variations of u' and ϕ' in the y and z directions, may be any values constrained as follows:

$$\left(\frac{n_1}{L_1}\right)^2 + \left(\frac{n_2}{L_2}\right)^2 = \left(\frac{a}{\pi}\right)^2 \tag{13.2.21}$$

Another equation in $W(x)$ and $\theta(x)$ is obtained by substituting the disturbances, Eqs. (13.2.16), into Eq. (13.2.15). Then Eqs. (13.2.18) and (13.2.19) are used to obtain

$$\left[\frac{\sigma}{\text{Pr}} - (d^2 - a^2)\right](d^2 - a^2)W(x) = -a^2\text{Ra}\theta(x) \tag{13.2.22}$$

where Ra, the Rayleigh number, is

$$\text{Ra} = \frac{gD^3\beta(t_1 - t_2)}{\nu\alpha} = \text{GrPr} \tag{13.2.23}$$

Finally, Eqs. (13.2.19) and (13.2.22) are two equations in the two amplitude or eigenfunctions $W(x)$ and $\theta(x)$. The parameters or eigenvalues in the equations are σ, σ/Pr, a^2, and Ra, along with any that arise from boundary conditions. Thus, it might be expected that the following relations between the eigenvalues apply:

$$\mathrm{Ra} = F\left(a^2, \sigma, \frac{\sigma}{\mathrm{Pr}}\right) \tag{13.2.24}$$

$$= F(a^2, \sigma) \qquad \text{for any given fluid} \tag{13.2.25}$$

$$= F(a^2) \qquad \text{for neutral stability} \tag{13.2.26}$$

We will see later that practical applications of the above formulation usually limit interest to neutral stability predictions. Thus, Eq. (13.2.26) implies a single functional relation between Ra and a^2 for each different set of boundary conditions that may be imposed at $x = 0$ and 1.

Therefore, limiting interest here to neutral stability, the relevant equations for $W(x)$ and $\theta(x)$, Eqs. (13.2.19) and (13.2.22), become

$$(d^2 - a^2)\theta = -W \tag{13.2.27}$$

$$(d^2 - a^2)^2 W = a^2 \mathrm{Ra}\theta \tag{13.2.28}$$

or, eliminating W,

$$(d^2 - a^2)^3\theta = -a^2\mathrm{Ra}\theta \tag{13.2.29}$$

where $a^2 = F(\mathrm{Ra})$ will be determined from these in terms of boundary conditions.

Boundary conditions. Solutions of the above equations for $\mathrm{Ra} = F(a^2)$ have been determined for many kinds of bounding conditions. An interesting group of results is considered here. All have in common the assumption that the bounding regions above and below the horizontal layer of fluid in Fig. 13.1.1 are of very high thermal conductivity compared to the fluid. As a result, none of the spatial variations of the disturbed temperature field in the fluid, $f(y, z)$ and $\theta(x)$, penetrate the bounding regions. The resulting conditions on ϕ' and $\theta(x)$ in Eq. (13.2.17) are

$$\phi'(0, y, z, \tau) = \phi'(1, y, z, \tau) = 0 \tag{13.2.30}$$

or

$$\theta(0) = \theta(1) = 0 \tag{13.2.31}$$

The other conditions at the interfaces with the bounding regions concern the velocity. If these regions present solid surfaces to the fluid layer, then u', v', and w' are all zero at $x = 0$ and 1. For impermeability

$$u'(0, y, z, \tau) = u(1, y, z, \tau) = 0$$

or

$$W(0) = W(1) = 0 \tag{13.2.32}$$

Assuming no slip at the interface, v' and $w' = 0$, and therefore from continuity, Eq. (13.2.8), $\partial u'/\partial x = 0$ at $x = 0$ and 1. Thus, for $W(x)$

$$W'(0) = W'(1) = 0 \tag{13.2.33}$$

where the prime indicates differentiation with respect to x.

Another kind of bounding condition on velocity results in much simpler solutions for neutral stability. It was used by Rayleigh (1916) in the first analysis of this kind of transport. "Nonrigid" boundaries were involved. Specifically, it was assumed that no horizontal shear stress was imposed on the fluid layer by the bounding regions above and below, as though these regions were occupied by fluids of much lower viscosity, although still of much higher thermal conductivity. Regions of two different liquid metals might approximate these bounding conditions in a practical case. Then fluid in the layer must have an intermediate density and a much higher viscosity and be immiscible. It must also have a much lower conductivity to retain the general temperature conditions set forth in Eq. (13.2.31).

Nevertheless, assuming zero shear stress at $x = 0$ and 1 amounts to taking $\partial v'/\partial x = \partial w'/\partial x = 0$ there. Continuity, Eq. (13.2.8), yields

$$\frac{\partial}{\partial x}\,\nabla\cdot\mathbf{V}' = \frac{\partial^2 u'}{\partial x\partial x} + \frac{\partial^2 v'}{\partial x\partial y} + \frac{\partial^2 w'}{\partial x\partial z} = \frac{\partial^2 u'}{\partial x^2} = 0$$

or at $x = 0$ and 1

$$\frac{\partial^2 u'}{\partial x^2} \propto W''(0) = W''(1) = 0 \tag{13.2.34}$$

The above condition and $W(0) = W(1) = 0$, Eq. (13.2.32), still apply at $x = 0$ and 1 if the interface surfaces remain undeformed by motion that arises in the entrapped fluid layer.

Therefore, among the collected boundary conditions that are to be considered are invariably

$$\theta(0) = \theta(1) = 0 \tag{13.2.31}$$

Those on disturbance velocity and therefore on temperature, from Eq. (13.2.27), are

$$\text{at solid surfaces:} \qquad W = W' = 0 \tag{13.2.35a}$$

or

$$(d^2 - a^2)\theta = 0 \tag{13.2.35b}$$

and

$$d(d^2 - a^2)\theta = 0 \tag{13.2.35c}$$

$$\text{at zero-shear surfaces:} \qquad W = W'' = 0 \tag{13.2.36a}$$

or

$$(d^2 - a^2)\theta = 0 \tag{13.2.36b}$$

and

$$d^2(d^2 - a^2)\theta = 0 \qquad (13.2.36c)$$

Zero-shear surfaces. First consider the Rayleigh calculations following from Eqs. (13.2.29), (13.2.31), and (13.2.36), where Eq. (13.2.29) may be rewritten as

$$(d^2 - a^2)^3\theta = (d^6 - 3a^2d^4 + 3a^4d^2 - a^6)\theta = -a^2\text{Ra}\theta \qquad (13.2.37)$$

The following is tried as a solution:

$$\theta(x) = N \sin n\pi x \qquad n = 1, 2, 3, \ldots \qquad (13.2.38)$$

This satisfies Eqs. (13.2.36*b*) and (13.2.36*c*). Then from Eq. (13.2.37) the following relation is found:

$$(\pi^2n^2 + a^2)^3 = a^2\text{Ra} \qquad (13.2.39)$$

The earliest unstable condition for any wave number a is the smallest value of Ra that satisfies this relation. The minimum value for any value of a^2 occurs for $n = 1$. Therefore,

$$\text{Ra}_{\text{min}} = \frac{(\pi^2 + a^2)^3}{a^2} \qquad (13.2.40)$$

The minimum value is called the critical Rayleigh number Ra_c and occurs for $a_c^2 = \pi^2/2$. Therefore, Ra_c and the wave number are

$$\text{Ra}_c = \frac{27\pi^4}{4} = 657.5 \qquad \text{and} \qquad a_c = \frac{\pi}{\sqrt{2}} = 2.22 \qquad (13.2.41)$$

Measurements for a silicone oil by Goldstein and Graham (1969) yielded values within 10% of this prediction. This would be the minimum value of a neutral curve, which is Eq. (13.2.40), drawn as Ra vs. a. See Fig. 13.2.1. The amplitude functions $\theta(x)$ and $W(x)$ for all values of a are

$$\theta(x) = \sin \pi x \qquad (13.2.42)$$

$$W(x) = -(\pi^2 + a^2) \sin \pi x \qquad (13.2.43)$$

The additional modes of neutral stability, generated for $n = 2, 3, \ldots$, yield neutral curves at higher levels of Ra. The general relations for Ra_c and a_c are

$$\text{Ra}_{c,n} = \frac{27n^4\pi^4}{4} \qquad \text{and} \qquad a_{c,n} = \frac{n\pi}{\sqrt{2}} \qquad (13.2.44)$$

Exposed upper surface. Another similar circumstance of some interest arises in a layer of fluid on a horizontal heated surface and exposed on its upper surface. This is analogous to the actual experiment of Bénard (1900, 1901). The equations are as above with boundary conditions of Eq. (13.2.35) at $x = 0$ and Eq. (13.2.36) at $x = 1$. The principal results are $\text{Ra}_c = 1100.6$ and $a_c = 2.68$. The neutral curve, from Catton (1966), is b in Fig. 13.2.1.

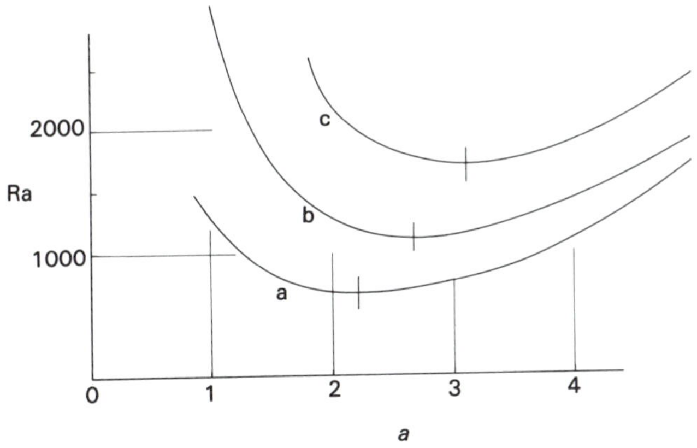

Figure 13.2.1 Neutral thermal stability curves for the following bounding surface conditions: (curve a) no shear stress at either boundary; (curve b) no shear stress at upper boundary, lower one rigid; (curve c) both boundaries rigid.

Rigid boundaries. The more interesting circumstance is a solid boundary at both $x = 0$ and 1. There are many practical applications of such results. The equations are again Eqs. (13.2.27) and (13.2.28) or (13.2.29) with boundary conditions of Eq. (13.2.35). Approximate estimates of the critical Rayleigh number for this solid-solid boundary problem were made by Jeffreys (1926). More complete results then appeared in Pellew and Southwell (1940).

Again, this configuration also has stability modes at increasingly high Ra levels, as in the Rayleigh result above, Eq. (13.2.39). Accurate calculations for the first two modes, $n = 1$ and 2, were given by Reid and Harris (1958, 1959), and by Catton (1966) for $n = 1$–10. The higher modes are higher wave numbers, that is, shorter wavelengths and disturbances of smaller scale. Values of $\mathrm{Ra}_{c,n}$ also again increase with n. The neutral curve for the first unstable mode, $n = 1$, appears as curve c in Fig. 13.2.1. The critical values are $\mathrm{Ra}_{c,1} = 1707.76$ and $a_{c,1} = 3.117$. The rigid boundaries are seen to stabilize with respect to the other two configurations considered above.

Other observations. There have been many experimental studies of neutral stability and/or initial instability of both the solid no-shear and two solid boundary configurations since the studies of Bénard. In fact, it appears that those observations of cellular motion in liquid spermaceti and paraffin, with a free upper surface, were subject to another quite different physical effect. For many substances, the surface tension at an interface, for instance, over a free surface, is also temperature dependent. Therefore, any temperature gradients along the surface would produce surface tension gradients. These would induce local surface motions, which would in turn induce, by shear below, liquid circulations such as those seen by Bénard. This thermocapillary effect was considered by Pearson (1958) and has come to be called Marangoni instability. A new parameter, called the Marangoni number, is introduced.

This aspect and the task of making visual observations through solid bounding surfaces of high thermal conductivity are special difficulties in arranging experiments that adequately simulate the model analyzed. As a result, much of the evidence gathered concerning neutral stability limits, which imply motion at slightly higher values of Ra, has come from consequences of this motion. In a stable circumstance, the Nusselt number is calculated as

$$q'' = -k\frac{dt}{dx} = k\frac{t_1 - t_2}{D}$$
$$\mathrm{Nu} = \frac{q''}{t_1 - t_2}\frac{D}{k} = \frac{hD}{k} = 1 \tag{13.2.45}$$

Motion causes an increase in heat transfer q''. Therefore, measuring q'' and $t_1 - t_2$ in experiments at increasing Ra would detect the beginning of motion and of convective transport as an increase of Nu above this lower limiting value, which applies for pure conduction. This would be expected to correspond at least approximately to the neutral stability limit. Such correspondence has been found by Silveston (1958), Rossby (1969), and Koschmieder (1974) and considered also by Thompson and Sogin (1966). Such data will appear in the next section, which considers the mechanisms following initial neutral instability.

13.3 RESULTING FLOWS

The linear stability theory discussed above allows determination of the critical Rayleigh number Ra_c for the onset of convection in a horizontal fluid layer heated from below. Although the geometric form of the most unstable disturbance is not determined uniquely, the growth or decay of the disturbance may be studied if the form is assumed. Extensive work has been done on the initial instability and on the form of the resulting disturbances, using linear theory for Ra values close to Ra_c. The state of knowledge is essentially complete, as discussed by Chandrasekhar (1961) and Turner (1973). Much of the recent effort has been on nonlinear finite-amplitude convection, as reviewed by Roberts (1966) and Joseph (1976). Of particular interest are the steady amplitude and preferred planform of the finite-amplitude motion established for $\mathrm{Ra} > \mathrm{Ra}_c$ and also the variation of the resulting heat transfer, in terms of the Nusselt number, with the Rayleigh number.

The earliest study of the convective flow arising in a fluid layer heated from below was that of Bénard (1900), who considered a layer with a rigid lower boundary and a free surface at the upper boundary. The hexagonal cellular flow seen is represented in Fig. 13.3.1. Warm fluid rises in the central cell region, spreads out over the upper surface, and flows downward at the perimeter with adjacent cells. Later reconsideration showed that surface tension effects along the free upper surface were the dominant mechanism in Bénard's experiments. However, even in the absence of surface tension effects, hexagonal cells and other cellular structures do arise. These result particularly for a significant variation of viscosity

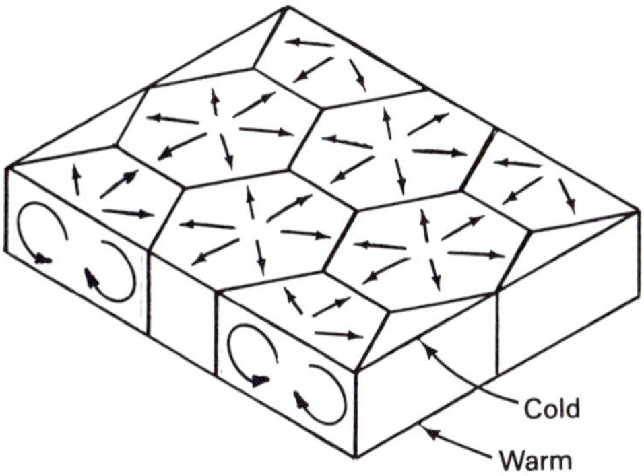

Figure 13.3.1 Bénard cells for natural convection in a horizontal fluid layer.

with temperature, due to subcritical and supercritical instabilities at Ra values close to Ra_c, as studied theoretically by Palm (1960), Segel and Stuart (1962), Busse (1962), and Krishnamurti (1968). The Rayleigh number range over which cellular motion arises was determined by Palm et al. (1967) for various boundary conditions. It was found to be strongly dependent on the physical depth of the fluid layer. For thinner layers, cellular motion was predicted up to larger Ra values, beyond Ra_c.

It was shown by the calculations of Segel (1965) and Schluter et al. (1965) and corroborated experimentally by Chen (1966) and Koschmieder (1966) that the cell structure should be primarily two-dimensional for negligible viscosity variation. This gives rise to a vertical symmetry. Busse (1967) calculated the stability of two-dimensional rolls for a fluid layer between two horizontal rigid boundaries, considering an infinite Prandtl number. Figure 13.3.2 shows the instability diagram obtained and indicates the region where two-dimensional rolls are stable. Experimental corroboration was provided by the work of Krishnamurti (1970). Solutions for steady two-dimensional flows for various wave numbers have been given over wide ranges of Ra by several workers, including Malkus and Veronis (1958), Fromm (1965), and Veronis (1966). These studies have received support from the experiments of Chen and Whitehead (1968), Rossby (1969), and Krishnamurti (1970). Figure 13.3.3 shows sketches of roll-shaped cells seen in fluid layers in rectangular and circular containers.

At yet higher Rayleigh numbers, several other flow patterns have been observed by Krishnamurti (1970), Willis and Deardorff (1970), and Busse and Whitehead (1971, 1974). Three-dimensional flow, which is periodic in space and steady in time, was found to replace the two-dimensional rolls. At high Prandtl numbers these changing patterns lead to steady "bimodal" convection. This term was used by Busse and Whitehead (1971) to describe the observed three-dimensional convection. It was found that the flow may be considered in terms of a superposition of two modes of roll-shaped convection patterns. A further increase in Rayleigh number results in time-dependent motion, also considered by Clever and Busse (1974). Ultimately, turbulent flow arises, as studied in detail by Goldstein and Chu (1969). The flow at large values of the Rayleigh number and the

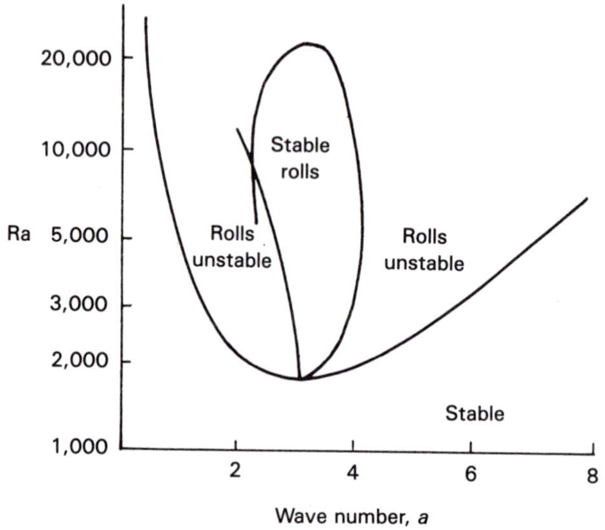

Figure 13.3.2 Stability diagram for convection in a fluid layer between two rigid horizontal surfaces for infinite Pr. *(Reprinted with permission from F. H. Busse,* J. Math. Phys., *vol. 46, p. 140. Copyright © 1967.)*

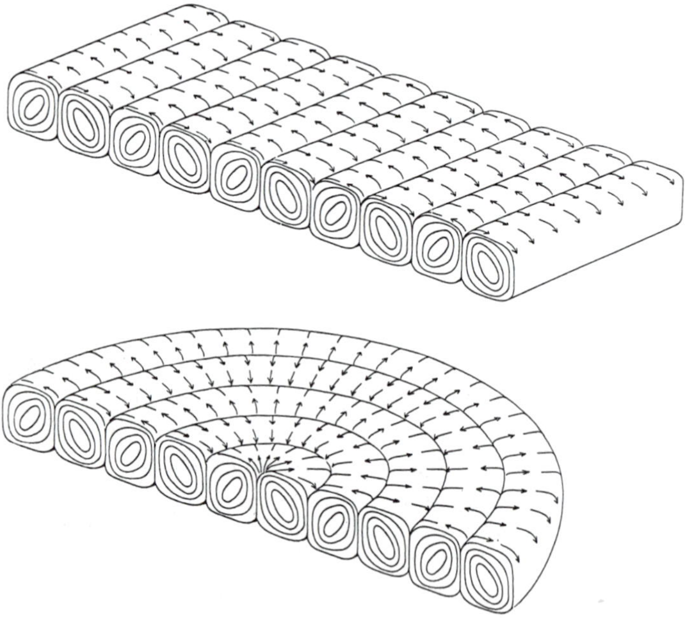

Figure 13.3.3 Roll-shaped cells in a rectangular or circular container.

corresponding heat transfer results are discussed in detail in Chapter 14 with respect to convection in horizontal rectangular cavities.

Several studies also considered the resulting flow in unstably stratified fluid layers under various other conditions. The stability of a thermally driven shear flow heated from below was studied by Weber (1978). Velocity measurements in transient Bénard convection, due to sudden cooling from above to a supercritical Rayleigh number, were made by Simpkins and Dudderar (1978) using laser speckle photography. Regular convective rolls were seen, oriented parallel to the short side of a Bénard cell. The number of rolls within the layer was approximately twice that seen under steady conditions. The onset of convection from random fluctuations arising within a fluid was analyzed by Jhaveri and Homsy (1980). The nonlinear convection in a layer with nearly insulating boundaries, with arbitrary three-dimensional disturbances, was analyzed by Busse and Riahi (1980). Turbulent natural convection in a horizontal layer heated from below has been studied experimentally and analytically by several workers, including Deardorff and Willis (1967), Chu and Goldstein (1973), Fitzjarrald (1976), and Tanaka and Miyata (1980). Several of these results are discussed in Chapter 14.

13.4 FLOW AND TRANSPORT REGIMES

As mentioned in the preceding section, there are several flow and transport regimes in an unstably stratified fluid layer. Among the most important are the conduction, laminar convection, and turbulent flow regimes. For Rayleigh numbers less than the critical value Ra_c, which is about 1708 for a fluid layer between two rigid horizontal surfaces, the fluid layer is stagnant and, for a vertically symmetric fluid condition, is stable to small disturbances. The Nusselt number Nu, based on the temperature difference and the depth of the layer, is unity. For vertical asymmetry, such as that produced by viscosity variation with temperature (see Richter, 1978) or by a nonlinear temperature profile, the conduction state is subcritically unstable to finite-amplitude disturbances. Subcritical instability arises for disturbances of small finite amplitude and increases the thermal energy transport beyond the conduction solution.

At Rayleigh numbers slightly greater than critical, steady laminar convection in the form of cells and rolls occurs. The stability of hexagonal cells vs. that of two-dimensional rolls has been studied for Ra less than and greater than Ra_c. Nonlinear thermal convection studies by Busse (1962), Segel and Stuart (1962), Segel (1965), Krishnamurti (1968), and Joseph and Sattinger (1972), as reviewed by Palm (1975), indicate that hexagonal cells are the only stable motion for $\mathrm{Ra} < \mathrm{Ra}_c$ and only two-dimensional rolls are stable at large Ra. Figure 13.4.1 shows a qualitative sketch of the variation of the amplitude with Ra.

Besides the transition from hexagonal cells to two-dimensional rolls in the neighborhood of Ra_c, a series of steady states arises in the flow as Ra is increased. An extensive experimental study on the transition from laminar to turbulent flow was carried out by Krishnamurti (1970) and precise transitions in the natural con-

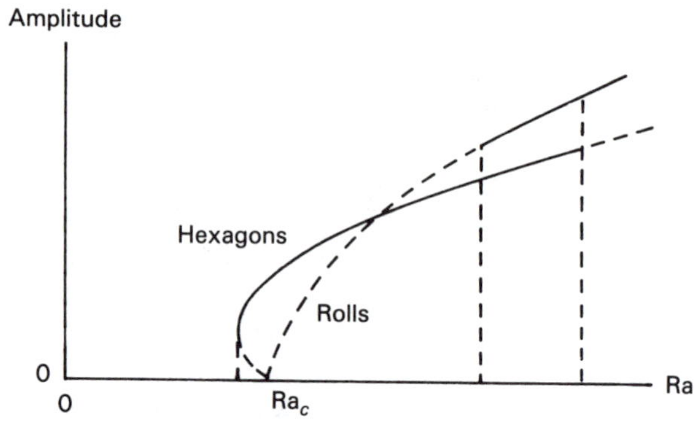

Figure 13.4.1 Qualitative sketch showing the dependence of the disturbance amplitude on the Rayleigh number. Solid lines refer to stable motion and dashed lines to unstable motion. *(From Palm, 1975. Reproduced, with permission, from* Annu. Rev. Fluid Mech., *vol. 7, Copyright © 1975, Annual Reviews, Inc.)*

vection flow and transport were determined. The dimensionless heat flux H, defined as $H = \mathrm{NuRa}$, is plotted against the Rayleigh number. The transitions from one flow pattern to another are determined by identifying discrete changes in the slope of the experimental data. Figures 13.4.2 and 13.4.3 together show three transitions in the flow at $\mathrm{Pr} = 100$. The first transition occurs at $\mathrm{Ra}_c \approx 1708$, with the onset of convection. The second transition occurs at Rayleigh number $\mathrm{Ra}_{\mathrm{II}}$, near $13\mathrm{Ra}_c$, with no definite Prandtl number dependence in the range of $10 < \mathrm{Pr} < 10^4$. This change in slope was identified with a change in planform

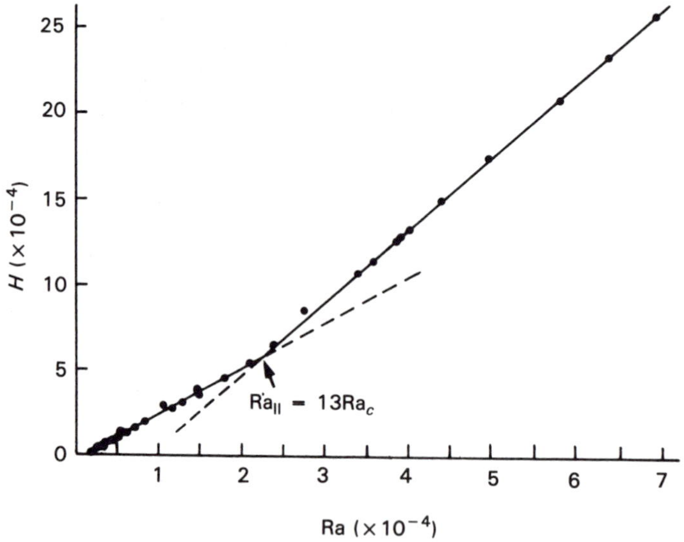

Figure 13.4.2 Heat flux $H = \mathrm{NuRa}$ as a function of Rayleigh number, showing the second transition for $\mathrm{Pr} = 100$. *(Reprinted with permission from R. Krishnamurti,* J. Fluid Mech., *vol. 42, pp. 295, 309. Copyright © 1970, Cambridge University Press.)*

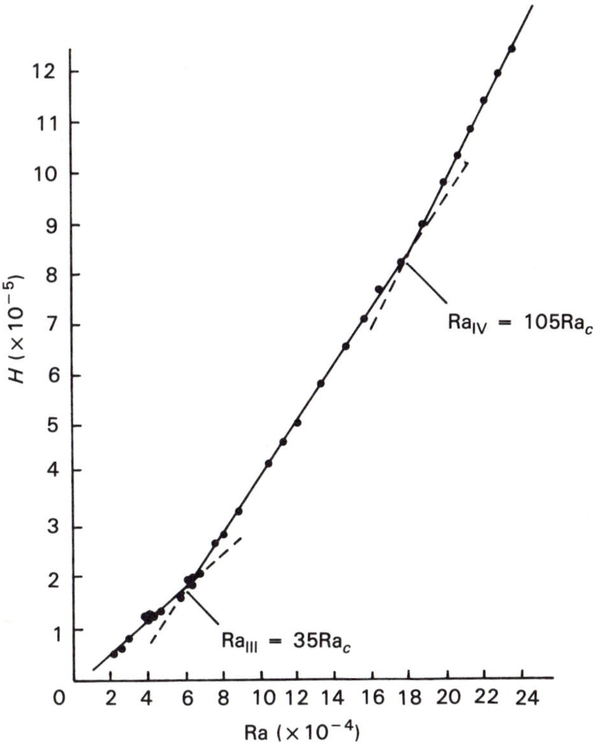

Figure 13.4.3 Heat flux H = NuRa as a function of Rayleigh number, showing the third and fourth transitions for Pr = 100. *(Reprinted with permission from R. Krishnamurti,* J. Fluid Mech., *vol. 42, pp. 295, 309. Copyright © 1970, Cambridge University Press.)*

from two-dimensional to three-dimensional circulations, caused by a finite-amplitude instability. This was indicated by a hysteresis in the heat flux and also in the flow pattern, as Ra is increased from smaller values or decreased from larger values. The third and fourth changes in slope, seen in Fig. 13.4.3, are related to transitions of the steady flow to time-dependent flows of several kinds.

These various regimes and the variation of the mean temperature distribution with Ra, particularly from thermals rising from the lower surface at large Ra and from turbulent flow, are again discussed for flow in horizontal cavities in Chapter 14. As Ra increases, the flow first becomes unsteady. Then the number and frequency of oscillations increase and eventually give way to turbulent flow. Busse and Whitehead (1974) carried out a detailed experimental study of the transitions from steady to time-dependent forms of convection, providing further information concerning the transition of the flow to turbulence.

13.5 INSTABILITY IN COLD WATER

Almost all of the foregoing analyses and experiments concerning thermal instability and convection are inapplicable to very common conditions arising in stratifications in cold water. The disturbance buoyancy force $g(\rho - \bar{\rho}) = g\rho'$ was calculated above by approximating $\rho(t)$ as a linear function of t. This amounts to taking ρ' as $(d\rho/dt)t'$, expressed in terms of β. The general support for this measure is that one may take the first term of a series, in approximating such a difference, if the Δt is sufficiently small. However, Fig. 13.5.1*a* shows a very commonly occurring counterexample: the low-temperature density behavior of pure and of saline water, $\rho(t, s, p)$, where s is salinity. For temperature differences around the temperature at which a density extremum occurs, $t_m(s, p)$, such a procedure is completely inapplicable; see Chapter 9.

Also, an additional diversity of instability and transport circumstances arises because differing bounding temperatures t_1 and t_2 in Fig. 13.1.1 result in basically different density stratifications and stability circumstances across the fluid layer. For some choices of $t_1 > t_2$ in Fig. 13.5.1, the stratification is potentially unstable, U, stable, S, and simultaneously both of these across the layer, SU. To assess the various consequences, consider thermal transport alone. In Fig. 13.5.1*a*, various choices of the pair $t_1 > t_2$, a through e, are located on a density variation that has an extremum. For both t_1 and t_2 on the same side and well away from t_m, choices a and b, the conventional analysis applies approximately when the proper algebraic value of β is used. Choice a is potentially unstable; b is stable.

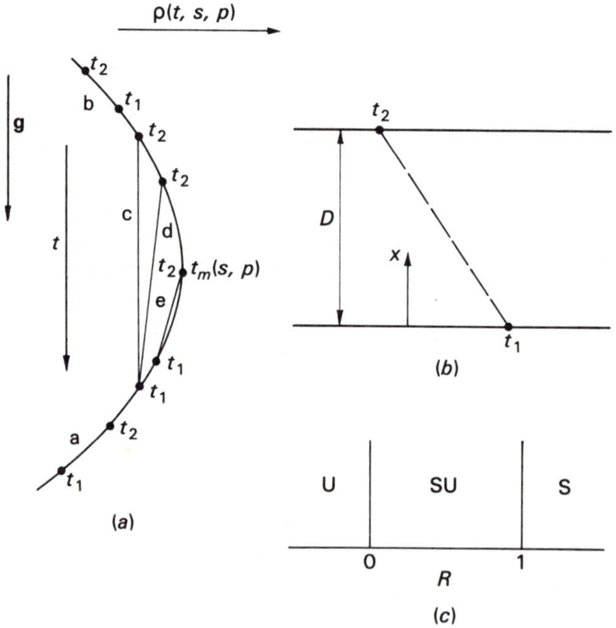

Figure 13.5.1 Temperature stratification characteristics for various choices of $t_1 > t_2$. A particular salinity level is indicated by s in $t_m(s, p)$. (*a*) Choices of t_1 and t_2; (*b*) fluid layer; (*c*) regimes in R. The stratifications corresponding to the choices in (*a*) are: a, unstable (R negative); b, stable (R positive); c, unstable-stable ($R = \frac{1}{2}$); d, unstable-stable ($R < \frac{1}{2}$); e, unstable ($R = 0$).

However, for both c and d the lower part of the layer is stable and the upper potentially unstable. Of course, similar effects may arise for $t_1 < t_2$.

The physical extent of this reversal condition and the nature of the stratification conditions in general depend on the relation of t_1 and t_2 to t_m. This relation has been expressed by Gebhart and Mollendorf (1978), as in vertical flows in Chapter 9, for pure and saline water as

$$R = \frac{t_m(s, p) - t_2}{t_1 - t_2} \tag{13.5.1}$$

Further and more careful consideration of the density variation in Fig. 13.5.1 indicates that for t_1, t_2, and t_m such that $R \geq 1$, the stratification is stable. On the other hand, for $R \leq 0$, it is potentially unstable; that is, $d\rho/dx \geq 0$ throughout the layer, from $x = 0$ to D. For $0 < R < 1$ an instability condition spans an increasingly large part of the fluid layer upward from $y = 0$, across this range downward toward $R = 0$. The conditions over this complete spectrum of R are shown in Fig. 13.5.1*c*.

There has been interest in this kind of circumstance because of both its unusual nature and its importance. There have been a number of measurements of transport. In some, one surface was ice. A number of calculations of instability and transport with such a density variation have appeared; see Veronis (1963), Debler (1966), Boger and Westwater (1967), Musman (1968), Sun et al. (1969), Tien et al. (1972), Moore and Weiss (1973), and Seki et al. (1974). It is very interesting to note that Debler (1966) was led to this mode of instability while considering instability with a uniform distributed energy source q''' in the fluid. The resulting conduction temperature distribution is parabolic, an approximation of the density-temperature dependence that arises in pure water around 4°C ($R = 0.5$). Several different kinds of boundary conditions have been considered. These and other related studies are reviewed by Mollendorf and Jahn (1983).

The foregoing studies, using one of several quite approximate equations for the density variation, are restricted to pure water at 1 atm. Pressure and salinity effects are unrepresented. The buoyancy force in water at low temperature should be calculated from a very accurate representation of density variation. For example, the change in density of pure water from 4 to 5°C at 1 atm is only about 9 ppm, compared to 205 ppm from 19 to 20°C. Most accurate density equations are far too complicated for any general calculation of instability and transport characteristics in saline, or even in pure water. Therefore, a more recent density equation, given in Chapter 9, is used here. It applies for pure and saline water, from 0 to 20°C, to 40‰ salinity, and to 1000 bars pressure, to very high accuracy.

This equation of state was used by Mollendorf and Jahn (1983) in the general disturbance formulation of Section 13.2, Eqs. (13.2.1)–(13.2.5). However, the body force and pressure term evaluation in Eqs. (13.2.6) and (13.2.7) must be carried out differently, without the Boussinesq approximations.

$$\mathbf{g}\rho - \nabla p = -g\mathbf{i}\rho' - \nabla p'_m \tag{13.5.2}$$

Then ρ' due to t' is to be evaluated from $\rho(t, s, p)$:

$$\rho(t, s, p) = \rho_m(s, p)[1 - \alpha(s, p)|t - t_m(s, p)|^{q(s,p)}] \tag{13.5.3}$$

or, at given s and p, from

$$\rho(t) = \rho_m(1 - \alpha_1|t - t_m|^q) \tag{13.5.4}$$

Here α_1 is a constant (see Section 9.1, where the symbol α has been used instead). Recalling that $\rho' = \rho - \bar{\rho}$ is a very small quantity, it is evaluated in a series expansion of ρ about $\bar{\rho}$, from Eq. (13.5.4), as follows:

$$\begin{aligned} \rho' = \rho - \bar{\rho} &= (t - \bar{t})\left(\frac{\partial \rho}{\partial t}\right)_{t=\bar{t}} + \cdots = t'\left(\frac{\partial \rho}{\partial t}\right)_{t=\bar{t}} + \cdots \\ &= \pm t'\rho_m\alpha_1 q|\bar{t} - t_m|^{q-1} + \cdots \end{aligned} \tag{13.5.5}$$

where $\pm$ apply for $\bar{t} < t_m$ and $\bar{t} > t_m$, respectively. Recall that the maximum density occurs at $t = t_m$; therefore, density is less for both lower and higher temperatures. Equation (13.5.5) is written in terms of ϕ, R, etc. as

$$\begin{aligned} \rho' &= \pm\rho_m\alpha_1 q(t_1 - t_2)|t_1 - t_2|^{q-1}|1 - x - R|^{q-1}\phi' \\ &= \rho_m KF(x, R)\phi' \end{aligned} \tag{13.5.6}$$

where $K = \alpha_1 q(t_1 - t_2)|t_1 - t_2|^{q-1}|1 - R|^{q-1}$, $F(x, R)$ is defined as

$$F(x, R) = \pm\left|\frac{1 - x - R}{1 - R}\right|^{q-1} \tag{13.5.7}$$

and $\pm$ apply for $1 - x < R$ and $1 - x > R$, respectively. In Eq. (13.5.7), x is x/D. Also, $\rho_m = \rho_m(s, p)$ will be taken as the reference density level ρ. Note that $F(x, R)$ in Eq. (13.5.7) becomes ± 1 for $q = 1$. This also occurs for very large values of R, at which the Boussinesq approximation becomes accurate for any value of q. Recall that the conditions for $\pm$ given above relate to the changing sign of β across t_m.

The disturbance buoyancy force from $\mathbf{g}\rho - \nabla p$ is seen in Eq. (13.2.7) to depend on ρ' as

$$-\mathbf{g}\mathbf{i}\rho' = -\mathbf{g}\mathbf{i}\rho KF(x, R)\phi' \tag{13.5.8}$$

Thus, the basic equations are the same as Eqs. (13.2.8)–(13.2.10), except that the buoyancy force is Eq. (13.5.8). Therefore, Eqs. (13.2.8) and (13.2.12)–(13.2.14) apply, with Eq. (13.2.11) becoming

$$\frac{\partial u'}{\partial \tau} = gKF(x, R)\phi' - \frac{1}{\rho D}\frac{\partial p'_m}{\partial x} + \frac{\nu}{D^2}\nabla^2 u' \tag{13.5.9}$$

Disturbances are again postulated exactly as in Eqs. (13.2.16) and (13.2.17). The resulting equations for the amplitude are also the same except that $F(x, R)$ appears:

$$f_{zz} + f_{yy} + a^2 f = 0 \tag{13.5.10}$$

$$[\sigma - (d^2 - a^2)]\theta = W \tag{13.5.11}$$

$$\left[\frac{\sigma}{\text{Pr}} - (d^2 - a^2)\right](d^2 - a^2)W = -a^2\text{Ra}F(x, R)\theta \tag{13.5.12}$$

However, the Rayleigh number is now different, since the buoyancy force in Eq. (13.5.9) is different from that in Eq. (13.2.11):

$$\text{Ra} = \frac{gKd^3}{\alpha\nu} = \frac{gd^3}{\alpha\nu}\alpha_1 q(t_1 - t_2)|1 - R|^{q-1} = \text{GrPr} \tag{13.5.13}$$

The boundary conditions are again as in Section 13.2:

$$W(0) = \theta(0) = W(1) = \theta(1) = 0$$

where, in addition,

$$W'(0) = W'(1) = 0 \qquad \text{rigid surface}$$

$$W''(0) = W''(1) = 0 \qquad \text{no-shear surface}$$

This formulation is different from that in which the Boussinesq approximation is permissible in that the Rayleigh number is defined differently; compare Eqs. (13.2.23) and (13.5.13). Also, $F(x, R) \neq 1$ arises in Eq. (13.5.12). This brings in R and $q = q(s, p)$ as additional parameters, beyond α, Ra, a, and Pr. Note that s and p are the salinity and pressure levels in the water. The above formulation does not include the effects of saline diffusion.

The effects of a density maximum and a consequent change in stratification arises entirely through $F(x, R)$ in Eqs. (13.5.7) and (13.5.12). For example, for R large, that is, both t_1 and t_2 far from t_m and on the same side, $R >> 1$ or $R << 1$ and $F \to \pm 1$. This becomes, in effect, the Boussinesq approximation, allowing for $(\partial\rho/\partial t)_p$ being both plus and minus. However, noting that $1 - x$ varies upward from 1 to 0, then $0 < R < 1$ results in $1 - x - R$ changing sign across the layer. This means that $\bar{t}(x)$ varies across t_m, reversing stratification.

The calculations by Mollendorf and Jahn (1983) for rigid bounding conditions considered the extreme values of $q(s, p)$ that arise over the whole range of salinity and pressure, to 40‰ and 1000 bars. The extreme values are $q(0, 1)$ and $q(0, 1000)$, for which a density extremum does not occur in thermodynamic equilibrium. Calculations were also made for $q = 2$, for comparison with the past results that used one of the less accurate density equations, a parabolic representation. The results for initial instability as a function of R are seen in Fig. 13.5.2, where Ra_c is plotted against R. The effect of q is seen to be relatively small across the whole range.

For all q, the occurrence of an extremum, or a tendency toward one, stabilizes. That is, $\text{Ra}_c \geq 1707.8$, although this limit is approached as $R \to -\infty$. Values of Ra_c to $R = -60$ have been calculated; see Table 13.5.1. Recall that $F \to \pm 1$ as $R \to -\infty$ and Ra in Eq. (13.5.13) becomes the conventional Rayleigh number. This enhanced stability results from reduced values of $(\partial\rho/\partial t)_p$ compared to behavior away from an extremum. All curves go toward a vertical asymptote at $R = 1$. Completely stable stratification is found when $R \geq 1$. These results are in

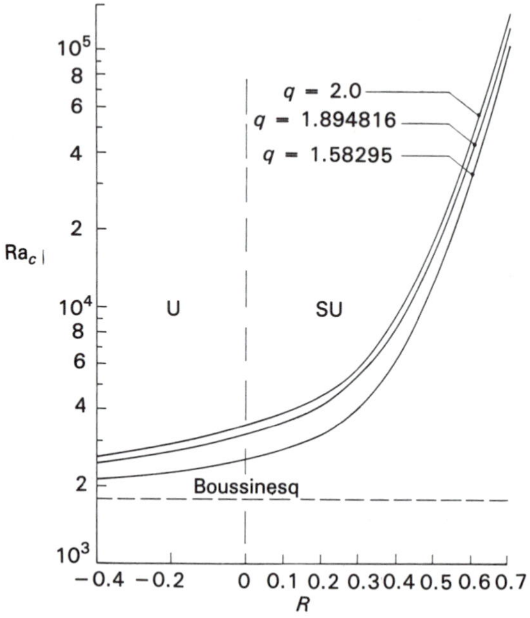

Figure 13.5.2 Variation of critical Rayleigh number with R for various values of $q(s, p)$, rigid bounding conditions.

some systematic disagreement with earlier calculations. An interesting study by Moore and Weiss (1973) considered the effects of finite-amplitude disturbances as well as the resulting transport, again for $q = 2$.

Hwang et al. (1984) investigated, analytically and experimentally, the effect of the density extremum and boundary conditions on the stability of a horizontally confined water layer. The density extremum was found to stabilize the layer.

Table 13.5.1 Variation of Ra_c with R and q^a

R	a $(2 \le q \le 1.58295)$	Ra_c $(q = 2)$	Ra_c $(q = 1.894816)$	Ra_c $(q = 1.58295)$
−60.0	3.120	1,722	1,721	1,716
−6.0	3.120	1,831	1,825	1,783
−3.0	3.120	1,951	1,924	1,846
0.0	3.126	3,390	3,172	2,587
0.1	3.130	3,798	3,524	2,794
0.2	3.140	4,461	4,105	3,163
0.3	3.175	5,715	5,229	3,945
0.4	3.325	8,778	8,030	6,076
0.5	3.995	18,663	17,083	12,955
0.6	5.095	46,110	42,210	32,018
0.7	6.775	145,510	133,176	100,952

[a] $q = 2$ corresponds to a parabolic form of the equation of state as presented by Debler (1966), Tien (1968), and Veronis (1963); $q = 1.894816$ corresponds to the equation of state as presented by Gebhart and Mollendorf (1977) for pure water at 1 bar; and $q = 1.58295$ represents pure water at 1000 bars.

Inclusion of the finite thermal capacity of the bounding surfaces resulted in destabilization. Different initial modes of instability were observed. The stability of the layer was studied over a wide range of R and increased stability was obtained as R was increased. Calculated and measured wavelengths of the observed distinct disturbance forms were found to agree quite well.

13.6 OTHER EFFECTS

Instability characteristics of a horizontal fluid layer have also been determined, by analysis and by experiment, for many other idealizations of circumstances found in applications. One of the most apparent additional ones arises when the imposed range of temperature also causes important variations of other fluid properties. Segel and Stuart (1962) found that subcritical instability may arise; that is, steady circulations arise for Rayleigh numbers below Ra_c. These circulations may be (see Busse, 1962) of the hexagonal form seen by Bénard.

There is also considerable interest in the effects on stability and transport in fluid layers subjected to a distributed thermal energy source q'''; see Sparrow et al. (1964), Debler (1966), Roberts (1967), and Thirlby (1970). The applications are in nuclear core meltdowns and in the quasi-fluid mantle of the earth. Various bounding thermal conditions have been considered, such as $t_1 = t_2$ and the bounding bottom surface adiabatic. An experimental study is reported by Kulacki and Emara (1977).

The discussion in Section 13.2 considered the realism of assuming no thermal disturbance coupling of the bounding surfaces with the fluid layer. This condition has been analyzed for arbitrary solid diffusivities. A related circumstance is that of two fluid layers separated by a slab of interacting solid material. Zero-shear bounding regions have been more carefully considered, for instance, by postulating a convective coupling with remote thermal conditions in the bounding regions. Related mechanisms arise when the imposed bounding thermal conditions are transient.

Another modification occurs often in applications when the horizontal layer is divided into small subregions by a perhaps regular structure of vertical partitions. This would arise in a flat-plate solar collector where the space between the glass cover and the actual collector plate is filled, for example, with vertical thin-walled tubes. The geometry of the dividers would be optimized to inhibit losses upward from the collector plate by convection; see Catton and Edwards (1967) and Edwards (1969). Complete stability might be sought. The conditions along these vertical boundaries would also depend on their contact with the bounding surfaces.

In many other circumstances the horizontal layer is a porous medium such as a layer of sand or pebbles saturated with water. Then, at small porosity, the viscous force $\nu\nabla^2\mathbf{V}$ may be replaced by the Darcy formulation. This simplifies the governing equations. Very good agreement for transport is found between calculations and measurements; see Horton and Rogers (1945), Lapwood (1948), and Schneider (1963).

The kind of combined buoyancy modes discussed in Chapter 6, in which the local buoyancy force arises from the simultaneous transport of thermal energy and chemical species, often occurs. An early study of instability was reported by Stern (1960). A prominent example is combined thermal and saline transport in seawater. An ice layer on the sea, either melting or freezing at the lower surface of the ice in contact with the water, results in such transport. Melting produces a freshwater melt, which is lighter and therefore may stabilize the layer, since saline effects on density are often greater than those of temperature. Downward-freezing ice excludes the components of salinity. The resulting high-concentration layer adds a powerful destabilizing effect to that of the temperature decrease upward. Both transport variations must be considered simultaneously to determine both instability conditions and any convective transport that may arise. Analysis is complicated because of the change in sign of the Sorét coefficient for salinity components at low temperature levels in water, according to Caldwell (1974, 1976) and Platten and Chavepeyer (1972). The Sorét effect is chemical species diffusion driven by temperature gradients.

Instability arising from other force fields such as electromagnetic or magnetohydrodynamic fields has also been studied. The results of Thompson (1951) and Chandrasekhar (1952, 1954) for the rigidly and freely bounded horizontal fluid layers are the earliest. The continuing concern with such transport is related to numerous applications in plasmas and perhaps also in planetary and celestial bodies. Many conditions and geometries have been considered. Rotating systems are subject to a Coriolis force. This effect is similar to that of a magnetic field in a conducting fluid. Both instability and convective motion have been considered.

Finally, thermal instability in several simple forced viscous flows has been assessed for an unstable stratification imposed by bounding temperature conditions. Examples are the classical developed parallel flows—Couette, Poiseuille, and a combination of the two effects, that is, of both applied shear and pressure gradient. The first question is whether the first mode of instability is hydrodynamic or thermal. For Couette flow, which is hydrodynamically stable for small disturbances, thermal instability was assessed by Deardorff (1965), Gallagher and Mercer (1965), and Ingersoll (1966). Poiseuille flow was found by Gage and Reid (1968) to be subject to thermal instability effects at sufficiently low Reynolds numbers. A number of developing flows, such as the Blasius boundary layer, have also been studied for thermal instability. Two-dimensional stagnation point boundary layers have been analyzed by Chen et al. (1983).

Clearly, a great diversity of physical effects has been considered, concerning both thermal instability and convection. Important aspects have been clarified for many and varied applications. A thorough, quantitative, and comprehensive treatment of such results is given by Gershuni and Zhukhovitskii (1976).

REFERENCES

Bénard, H. (1900). *Rev. Gen. Sci. Pures Appl. 11*, 1261.
Bénard, H. (1901). *Ann. Chim. Phys. 23*, 62.

Boger, D. V., and Westwater, J. W. (1967). *J. Heat Transfer 89*, 81.
Busse, F. H. (1962). Dissertation, Univ. of Munich; translation, The Rand Corp., 1966.
Busse, F. H. (1967). *J. Math. Phys. 46*, 140.
Busse, F. H., and Riahi, N. (1980). *J. Fluid Mech. 96*, 243.
Busse, F. H., and Whitehead, J. A. (1971). *J. Fluid Mech. 47*, 305.
Busse, F. H., and Whitehead, J. A. (1974). *J. Fluid Mech. 66*, 67.
Caldwell, D. R. (1974). *J. Fluid Mech. 64*, 347.
Caldwell, D. R. (1976). *J. Fluid Mech. 74*, 129.
Catton, I. (1966). *Phys. Fluids 9*, 2521.
Catton, I., and Edwards, D. K. (1967). *J. Heat Transfer 89*, 295.
Chandrasekhar, S. (1952). *Philos. Mag. 42*, 1417.
Chandrasekhar, S. (1954). *Philos. Mag. 45*, 1177.
Chandrasekhar, S. (1961). *Hydrodynamic and Hydromagnetic Stability*, Oxford Univ. Press, London.
Chen, K., Chen, M. M., and Sohn, C. W. (1983). *J. Fluid Mech. 132*, 49.
Chen, M. M. (1966). *Bull. Am. Phys. Soc. 11*, 617.
Chen, M. M., and Whitehead, J. A. (1968). *J. Fluid Mech. 31*, 1.
Chu, T. Y., and Goldstein, R. J. (1973). *J. Fluid Mech. 60*, 141.
Clever, R. M., and Busse, F. H. (1974). *J. Fluid Mech. 65*, 625.
Deardorff, J. W. (1965). *Phys. Fluids 8*, 1027.
Deardorff, J. W., and Willis, G. E. (1967). *J. Fluid Mech. 28*, 675.
Debler, W. R. (1966). *J. Fluid Mech. 24*, 165.
Edwards, D. K. (1969). *J. Heat Transfer 91*, 145.
Fitzjarrald, D. E. (1976). *J. Fluid Mech. 73*, 693.
Fromm, J. E. (1965). *Phys. Fluids 8*, 1757.
Gage, K. S., and Reid, W. H. (1968). *J. Fluid Mech. 33*, 21.
Gallagher, A. P., and Mercer, A. (1965). *Proc. R. Soc. London Ser. A286*, 117.
Gebhart, B., and Mollendorf, J. C. (1977). *Deep-Sea Res. 24*, 1.
Gebhart, B., and Mollendorf, J. C. (1978). *J. Fluid Mech. 89*, 673.
Gershuni, G. Z., and Zhukhovitskii, E. M. (1976). *Convective Instability of Incompressible Fluids*, Israel Program for Scientific Translations, Jerusalem.
Goldstein, R. J., and Chu, T. Y. (1969). *Prog. Heat Mass Transfer 2*, 55.
Goldstein, R. J., and Graham, D. J. (1969). *Phys. Fluids 12*, 1133.
Horton, C. W., and Rogers, F. T. (1945). *J. Appl. Phys. 16*, 367.
Hwang, L.-T., Lu, W.-F., and Mollendorf, J. C. (1984). *Int. J. Heat Mass Transfer 27*, 497.
Ingersoll, A. P. (1966). *Phys. Fluids 9*, 682.
Jeffreys, H. (1926). *Philos. Mag. 2*(7), 833.
Jhaveri, B., and Homsy, G. M. (1980). *J. Fluid Mech. 98*, 329.
Joseph, D. D. (1976). *Stability of Fluid Motions*, vol. 2, Springer, New York.
Joseph, D. D., and Sattinger, D. H. (1972). *Arch. Ration. Mech. Anal. 45*, 79.
Koschmieder, E. L. (1966). *Beitr. Phys. Atmos. 39*, 1.
Koschmieder, E. L. (1974). *Adv. Chem. Phys. 26*, 177.
Krishnamurti, R. (1968). *J. Fluid Mech. 33*, 445, 457.
Krishnamurti, R. (1970). *J. Fluid Mech. 42*, 295, 309.
Kulacki, F. A., and Emara, A. A. (1977). *J. Fluid Mech. 83*, 375.
Lapwood, E. R. (1948). *Proc. Cambridge Philos. Soc. 44*, 508.
Malkus, W. V. R., and Veronis, G. (1958). *J. Fluid Mech. 4*, 225.
Mollendorf, J. C., and Jahn, K. H. (1983). *J. Heat Transfer 105*, 460.
Moore, D. R., and Weiss, N. O. (1973). *J. Fluid Mech. 61*, 553.
Musman, S. (1968). *J. Fluid Mech. 31*, 343.
Palm, E. (1960). *J. Fluid Mech. 8*, 183.
Palm, E. (1975). *Annu. Rev. Fluid Mech. 7*, 39.
Palm, E., Ellingsen, T., and Gjevik, B. (1967). *J. Fluid Mech. 30*, 651.
Pearson, J. R. A. (1958). *J. Fluid Mech. 4*, 489.
Pellew, A., and Southwell, R. V. (1940). *Proc. R. Soc. London Ser. A176*, 312.

Platten, J. K., and Chavepeyer, G. (1972). *Phys. Fluids 15,* 1555.
Rayleigh, Lord, (1916). *Philos. Mag. 32,* 529.
Reid, W. K., and Harris, D. L. (1958). *Phys. Fluids 1,* 102.
Reid, W. K., and Harris, D. L. (1959). *Phys. Fluids 2,* 716.
Richter, F. M. (1978). *J. Fluid Mech. 89,* 553.
Roberts, P. H. (1966). In *Nonequilibrium Thermodynamics, Variational Techniques and Stability,* Univ. of Chicago Press, Chicago.
Roberts, P. H. (1967). *J. Fluid Mech. 30,* 23.
Rossby, H. T. (1969). *J. Fluid Mech. 36,* 309.
Schluter, A., Lortz, D., and Busse, F. (1965). *J. Fluid Mech. 23,* 129.
Schneider, K. J. (1963). *11th Int. Congr. Refrig.,* Munich, Paper 11-4.
Segel, L. A. (1965). *J. Fluid Mech. 21,* 359.
Segel, L. A., and Stuart, J. T. (1962). *J. Fluid Mech. 13,* 289.
Seki, N., Fukusako, S., and Sugawaro, M. (1974). Preprints JSME, 740, 91.
Silveston, P. L. (1958). *Forsch. Ingenieurwes. 24,* 29, 59.
Simpkins, P. G., and Dudderar, T. D. (1978). *J. Fluid Mech. 89,* 665.
Sparrow, E. M., Goldstein, R. J., and Jonsson, V. K. (1964). *J. Fluid Mech. 18,* 513.
Stern, M. E. (1960). *Tellus 12,* 172.
Sun, Z.-S., Tien, C., and Yen, Y.-C. (1969). *AIChE J. 15,* 910.
Tanaka, H., and Miyata, H. (1980). *Int. J. Heat Mass Transfer 23,* 1273.
Thirlby, R. J. (1970). *J. Fluid Mech. 44,* 673.
Thompson, H. A., and Sogin, H. H. (1966). *J. Fluid Mech. 24,* 451.
Thompson, W. B. (1951). *Philos. Mag. 42,* 1417.
Tien, C. (1968). *AIChE J. 14,* 652.
Tien, C., Yen, Y.-C., and Dotson, J. W. (1972). *AIChE Symp. Ser. 118,* 101.
Turner, J. S. (1973). *Buoyancy Effects in Fluids,* Cambridge Univ. Press, London.
Veronis, G. (1963). *Astrophys. J. 137,* 641.
Veronis, G. (1966). *J. Fluid Mech. 26,* 49.
Weber, J. E. (1978). *J. Fluid Mech. 87,* 65.
Willis, G. E., and Deardorff, J. W. (1970). *J. Fluid Mech. 44,* 661.

PROBLEMS

13.1 For air, water, and a light oil, calculate the least thickness of a horizontal layer bounded by solid surfaces on each side for incipient thermal instability for $t_1 = 27°C$ and $t_2 = 23°C$. Calculate the wavelengths of the neutral disturbances.

13.2 For the fluids and temperature conditions in Problem 13.1, calculate the least layer thicknesses and resulting wavelengths for incipient thermal instability with

(a) No-shear stress at either boundary.

(b) No-shear stress at the upper boundary.

13.3 For a horizontal layer of water between two surfaces at $t_1 = 27°C$ and $t_2 = 23°C$,

(a) Determine the layer thicknesses for incipient instability and for each of the subsequent transport regimes to arise.

(b) Calculate, for each of those conditions, the heat transfer rate and compare it to that by conduction alone.

13.4 A solar energy collection system may be approximated as a horizontal layer of water, 1 cm in height, bounded by solid surfaces. Compute the temperature difference $t_1 - t_2$ at which thermal instability first appears.

13.5 Repeat Problem 13.4 for air and compare the computed result with that for water. Comment on the physical implications of the difference between the two.

13.6 How would the results in Problems 13.4 and 13.5 change if the upper boundary were a zero-shear stress surface instead of a solid surface? Explain the result obtained on the basis of the underlying physical processes.

13.7 From the stability diagram in Fig. 13.3.2, determine the range of depth for which two-dimensional rolls are stable in a horizontal layer of air, with $t_1 = 30°C$ and $t_2 = 25°C$, as the depth of the layer is varied from 1 mm to 5 cm. Take wave number $a = 3.5$.

13.8 Determine the variation in heat transfer across a horizontal layer of water with depth of the layer, if $t_1 = 40°C$ and $t_2 = 30°C$, over the range 1 to 20 cm.

13.9 For a horizontal air layer 2 cm in height, compute the temperature differences at which the various transitions shown in Figs. 13.4.2 and 13.4.3 occur.

13.10 A horizontal layer of cold water is bounded by two surfaces at $t_1 - t_2 = 4°C$. Consider the four specific temperature conditions, $t_2 = 0$, 2, 4, and 6°C. For each condition

(a) Sketch the vertical temperature and density distributions across the layer, assuming it to be stable.

(b) Calculate the thickness that would result in incipient instability and the heat transfer rate.

CHAPTER

FOURTEEN

TRANSPORT IN ENCLOSURES AND PARTIAL ENCLOSURES

14.1 INTRODUCTION

Most of the discussion in the earlier chapters was related to external natural convection flow over surfaces and in free boundary flows. The ambient fluid was considered to be very extensive and unaffected by the buoyancy-driven flow under consideration. This leads to a considerable simplification in analysis since the conditions exterior to the flow region are then specified independent of the flow. However, in internal natural convection flows, as between parallel surfaces and in limited enclosures, the flows adjacent to the surfaces are inevitably coupled with the flow in the interior or core region enclosed by the bounding layers at the walls. As a consequence of this complexity, internal natural convection problems have received less attention than external flows. However, recently there has been a great increase in interest and research activity in natural convection flows in enclosures. Ostrach (1972) and Catton (1978) reviewed the earlier work on several internal flow problems.

Natural convection in enclosures and in partial enclosures occurs in many engineering applications. Building insulation often consists of air gaps in multilayered walls. The heat transfer then involves natural convection in enclosures filled with either an ordinary fluid or a fluid-saturated porous material. Reduction of heat losses in solar collectors involves a consideration of natural convection between the hot solar energy absorber and the transparent cover, and between the covers if more than one is used, for insulation. A honeycomb structure also may be used to reduce losses. Buchberg et al. (1976) reviewed the application of natural convection in enclosed spaces to solar energy collection. Energy storage as sensible heat in enclosed fluid regions, heat rejection to water bodies, flows arising in rooms due to thermal energy sources, and cooling of heat-generating components in the electrical and nuclear industries are all examples of practical systems where natural convection mechanisms in enclosures play a vital role. Similarly, detection of fire and prediction of its spread involve a study of the convection currents that arise in rooms, with or without ventilation. These flows are also usually complicated by the existence of significant radiation effects; see, for example, Torrance et al. (1969), Quintiere (1984), and Markatos et al. (1982).

Melting and solidification processes frequently occur in enclosed regions, and natural convection effects often have been found to be important. The effect of natural convection circulations in the liquid region on the quality of grown crystals has been an area of considerable interest, as reviewed by Pimputkar and Ostrach (1981). Mixed convection in enclosures is another area of practical importance, particularly in heat exchangers and power systems. These flows were considered in Chapter 10.

In this chapter several important flow configurations are considered. Rectangular (vertical, horizontal, and inclined) cavities have been studied extensively. The vertical cavity with two vertical walls at different temperatures, the remaining two surfaces being important mainly in completing the cavity, is probably the most studied configuration because of its relative simplicity and importance in many practical applications. In a horizontal or inclined cavity with heating from below, thermal instability may arise, as considered in Chapter 13. The flow between flat parallel surfaces is also of interest because many enclosure geometries may be approximated in this way and also because this configuration arises in practical problems such as those related to electronic equipment cooling and heat exchanger design. In this flow configuration the fluid layer is often taken as infinite and a condition of zero net mass flow at any cross section is invoked to simulate a completely enclosed fluid region.

Three-dimensional internal flows, thermal siphons, and internal flows in annuli and other geometries are also considered here. Partial enclosures are of particular interest in relation to convective flow in rooms. These have been studied extensively in fire research and in relation to the design of buildings, furnaces, energy storage and removal systems, and several other industrial systems.

This chapter discusses the physical processes underlying internal natural convection flows. The governing equations are obtained for the various flow geometries mentioned above. Experimental and analytical results are discussed in order to outline the dependence of the transport mechanisms on the important governing parameters. The nature of the flow that arises and the consequent heat transfer across the enclosed fluid region are discussed for several simple configurations, in most detail for two-dimensional flows in rectangular enclosures. Since, in practice, the ambient medium for external natural convection flows is also of finite extent, it is important to relate the results for internal flows to those obtained earlier for external ones. This point of view often allows the treatment of internal flows as external convection, particularly in the early stages of a starting transient. It also enables one to determine the effect of the neighboring surfaces on the flow and heat transfer mechanisms in experimental studies of external natural convection.

Before discussing the flow in various enclosure configurations, it is worthwhile to consider the formulation for the buoyancy term in an internal convection circumstance. Unlike the case of external natural convection, there are no distant ambient conditions that may be used for the hydrostatic pressure variation. However, the flow depends on the local buoyancy force, which, in turn, is determined by the density gradient. Therefore, in an enclosed region the temperature of the cold wall t_c is generally used as a reference to yield the buoyancy term, as $g\beta(t - t_c)$.

For flow between two infinite vertical parallel plates closed at the ends, the condition of zero net flow at any cross section is used to simulate a fully enclosed region. Therefore, for this flow a vertical pressure gradient term $\partial p_m/\partial x$ is retained in the equations, or t_c in the buoyancy term is replaced by a reference temperature t_r. This is determined by the condition of zero net flow in the vertical direction. These considerations are seen more clearly in terms of the governing equations obtained in the following sections.

A few comments may also be made regarding the numerical solution of natural convection in enclosures. Finite-difference methods have been used extensively. For two-dimensional flows, the vorticity equation, which is obtained by taking a curl of the momentum equation and thus eliminating the pressure term, is frequently employed. It is used in conjunction with a stream function, which satisfies the continuity equation, and the energy equation. This approach is often referred to as the vorticity-stream function formulation. For three-dimensional flows, the governing equations in terms of velocity, pressure, and temperature, frequently termed primitive variables, may generally be directly solved more easily than the corresponding vorticity-stream function formulation. For details of these methods, see Patankar (1980), Roache (1982), and Jaluria and Torrance (1986). Results obtained by both these approaches are discussed in this chapter.

14.2 FLOW BETWEEN FLAT PARALLEL SURFACES

Buoyancy-driven flow between two flat parallel surfaces at different temperatures may be considered in terms of two distinct configurations. The first consists of extensive flat surfaces closed at the ends, resulting in an enclosure, and the second of parallel surfaces with the ends open. Both of these flows have been studied, and we will discuss the enclosure first.

Convective flows in a fluid layer enclosed between two parallel plates are particular examples of flow in rectangular cavities. These result asymptotically as either the height H or the width d of a rectangular cavity is increased to a large value, keeping the other dimension unchanged, that is, H/d very large or small. Because of the resulting simplicity, infinite fluid layers have received a great deal of attention. The asymptotic condition $H/d << 1$, that is, a horizontal layer with heating from below, is the Bénard problem, which was considered in detail in

Chapter 13 as an often unstably stratified fluid layer. The layer thermal instability, the resulting flows, and the transport mechanisms were discussed.

Fully developed laminar flows between infinite vertical parallel plates and in tubes closed at the ends have also been analyzed because a one-dimensional simplification arises from the infinite height condition. The assumption of infinite height applies for a fairly wide range of geometries of practical interest. Ostrach (1964) reviewed the results of the early studies of such internal natural convection mechanisms. Much of this work was experimental and semiempirical; an example is the work of Elenbaas (1942), who considered flow between vertical parallel plates and in vertical tubes. We will first discuss the work done for flow between flat parallel surfaces, closed or open at the ends.

14.2.1 Developed Flow in a Vertical Cavity

Consider the flow between two isothermal vertical walls at different temperatures t_h and t_c. As $H/d \to \infty$, where H is the height and d the distance between the plates, the flow becomes fully developed and the analysis may be considerably simplified. As discussed by Batchelor (1954), the velocity and temperature are then functions only of the horizontal coordinate y, where y may be measured horizontally from the hot surface toward the cold one. The horizontal velocity is zero. This results in the convective transport terms becoming identically zero in the momentum and energy equations.

The governing equations for the dimensionless vertical velocity and temperature, U and ϕ, are obtained in terms of the dimensionless coordinate distance Y as

$$\frac{d^3U}{dY^3} = -\frac{g\beta(t_h - t_c)d^3}{\nu\alpha}\frac{d\phi}{dY} \tag{14.2.1}$$

$$\frac{d^2\phi}{dY^2} = 0 \tag{14.2.2}$$

where the pressure field has been eliminated between the two components of the momentum equation. The physical velocity u, coordinate distance y, and temperature t are nondimensionalized as

$$U = \frac{ud}{\alpha} \qquad Y = \frac{y}{d} \qquad \phi = \frac{t - t_c}{t_h - t_c} \tag{14.2.3}$$

where t_h is the temperature of the hot wall and t_c that of the cold one. From Eq. (14.2.2) the heat transfer across the fluid layer is by conduction alone. The temperature distribution is linear and independent of the flow field. Since the flow is driven by buoyancy, the velocity distribution depends on the temperature variation across the layer.

Since the two vertical surfaces are assumed to be infinite, the boundary conditions due to closed ends cannot be used. However, because it is a completely enclosed region, the condition of zero net vertical mass flow at any given cross section is used instead. This is employed in specifying the three boundary conditions needed for Eq. (14.2.1). The velocity distribution must be antisymmetric about the central axis, giving zero velocity there, as seen in Fig. 14.2.1. Thus, $U = 0$ at $Y = 0, 0.5$, and 1.0. Also, $\phi = 1.0$ at $Y = 0$ and $\phi = 0$ at $Y = 1.0$. Alternatively, the second-order momentum equation with the pressure term,

$$\nu \frac{d^2u}{dy^2} = \frac{dp_m}{dx} - g\beta(t - t_c)$$

may be used instead of Eq. (14.2.1). Then the pressure term is determined from the zero net vertical flow condition. The resulting temperature and velocity distributions as obtained from Eqs. (14.2.1) and (14.2.2) are

$$\phi = 1 - Y \tag{14.2.4}$$

$$U = \frac{\text{Ra}}{12} Y(1 - Y)(1 - 2Y) \tag{14.2.5}$$

where the only parameter that arises is the Rayleigh number Ra, defined as

$$\text{Ra} = \frac{g\beta(t_h - t_c)d^3}{\nu\alpha} \tag{14.2.6}$$

The U and ϕ distributions are shown in Fig. 14.2.1. For this purely conductive transport, the heat flux q'' across the fluid layer is obtained as $k(t_h - t_c)/d$. This gives the heat transfer coefficient h, based on the temperature difference $t_h - t_c$, as k/d. The Nusselt number, defined as $\text{Nu} = hd/k$, is 1.0. However, in this analysis no information is obtained about the flow and the heat transfer near the two ends of a tall vertical slot. The results apply only in the region far from the ends. The flow in an infinite vertical slot was also studied by Ostrach (1952), including the effect of viscous dissipation. This effect may be important and is similar to that of heat sources in the flow.

The fully developed flow in an infinite vertical channel, with the temperatures of the two vertical surfaces taken as functions of the vertical coordinate and differing by only a constant, was considered by Ostrach (1964). The general solutions obtained by an analytical iteration procedure for the constant and linearly varying wall temperature conditions were discussed. The circumstance that arises if the wall temperatures decrease in the vertical direction was considered by Ostrach (1955) and the stability characteristics were shown to be qualitatively similar to those obtained from thermal stability analyses of horizontal fluid layers. Maslen (1958) considered compressibility and property variation effects in these fully developed flows.

There have been several other studies of natural convection in an infinite vertical slot. The stability of the flow is relevant to the initial stages of transition to turbulence. Birikh (1966), Rudakov (1966), and Birikh et al. (1972) analyzed the stability of the above conduction regime and base flow between infinite iso-

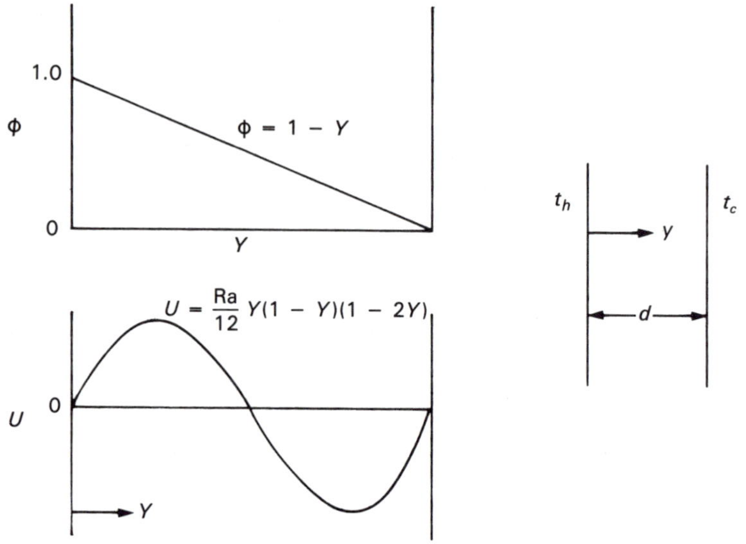

Figure 14.2.1 Temperature and velocity distributions in the fully developed natural convection flow between two infinite vertical isothermal flat plates. *(Reprinted with permission from G. K. Batchelor,* Q. Appl. Math., *vol. 12, p. 209. Copyright © 1954.)*

thermal vertical surfaces using linear analyses. Neutral stability curves were obtained for various Prandtl numbers, along with the disturbance eigenfunctions and frequency spectra.

14.2.2 Developing Flow between Vertical Surfaces

A boundary layer regime arises for the convection-dominated flow in a vertical cavity of limited height, if the aspect ratio H/d is sufficiently large for the base flow and the temperature field to be approximately one dimensional near the central portion of the cavity. A linear stability analysis of this quasi-one-dimensional flow in a vertical slot with isothermal sidewalls has been carried out by Vest and Arpaci (1969). A conduction regime in which the thermal transport is dominated by conduction mechanisms, and Nu = 1.0, arises for Ra $\lesssim$ 3000. At Ra $\gtrsim 8 \times 10^4$ thin boundary layers arise on the walls around a central core region. Both these regimes were studied by Vest and Arpaci and found to be unstable to stationary disturbances. Gill (1966) and Gill and Kirkham (1970) studied the boundary layer regime for two-dimensional flows. Bejan (1979) also considered the solution for this boundary layer regime. This flow is discussed in detail in the next section in the context of natural convection in a vertical rectangular enclosure.

Gill and Davey (1969) considered the flow over an infinite surface in a stably and linearly stratified medium with the surface at a constant temperature excess, $t_0 - t_\infty$, above the local ambient temperature. The exact solution for this flow was given in Section 3.11. It is an approximation to the solution in the boundary layers on the vertical walls of a rectangular cavity. Linear stability results were obtained by Gill and Davey (1969).

14.2.3 Transient Effects

Transient natural convection between infinite vertical parallel surfaces is also of interest in several engineering applications, such as the early stages of melting adjacent to a heated surface and in transient heating of insulating air gaps by heat input at the start-up of furnaces. The governing equations, in terms of the physical variables, are the following one-dimensional forms:

$$\frac{\partial u}{\partial \tau} = \nu \frac{\partial^2 u}{\partial y^2} + g\beta(t - t_r) \tag{14.2.7}$$

$$\frac{\partial t}{\partial \tau} = \alpha \frac{\partial^2 t}{\partial y^2} \tag{14.2.8}$$

The reference temperature t_r may be determined at a given time τ from the condition of zero net flow in the vertical direction. This condition is needed since the closed ends do not provide any boundary conditions for these equations; see Section 14.2.1. The temperature distribution is again independent of the flow and heat transfer is by conduction alone.

The buoyancy-driven flow that arises from the temperature field has been studied for solidification and melting processes by Szekeley and Stanek (1970) and Ramachandran et al. (1981). The liquid layer thickness $d = 2W$ changes with time. Figure 14.2.2 shows the time-dependent velocity distributions obtained numerically for two Prandtl number values. A decay of the flow occurs as the driving temperature difference in the fluid decreases with time. Although finite-sized enclosures are of greater interest and importance in such industrial applications, the infinite vertical slot frequently simulates the practical circumstance quite closely, especially at short times.

14.2.4 Developed Flow between Inclined Surfaces

The fully developed flow between inclined infinite parallel surfaces has also been analyzed. In this configuration, shown in Fig. 14.2.3, the axial velocity $u(y)$ and temperature $t(y)$ are simply those for a vertical flow when g is replaced by $g \cos \theta$, for $t_h > t_c$. Thus,

$$\phi = 1 - Y \tag{14.2.9}$$

$$U = \frac{\text{Ra}}{12} Y(1 - Y)(1 - 2Y) \cos \theta \tag{14.2.10}$$

where θ is the inclination of the parallel surfaces from the vertical. This assumes that no cross-flow effect arises due to instability or to edges. This result applies

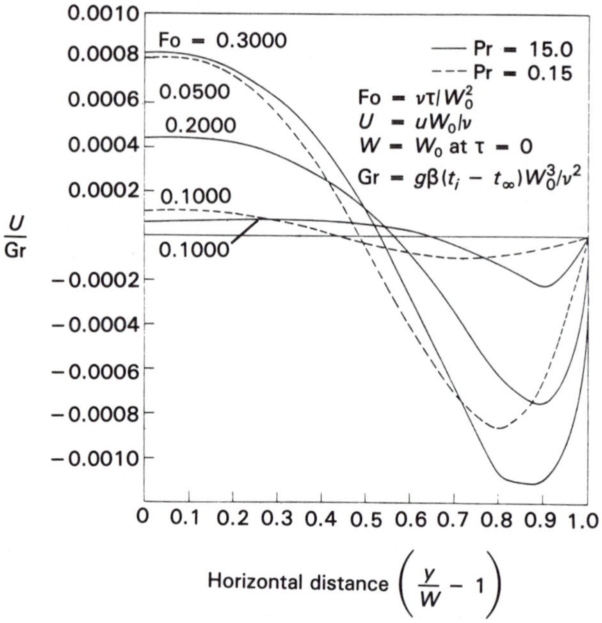

Figure 14.2.2 Time-dependent velocity distributions in the liquid region of instantaneous width $2W(\tau)$ for solidification on one side between two vertical infinite flat plates, at a separation of $2W_0$. The liquid is initially at a uniform temperature t_i and the outside ambient is maintained at temperature t_∞. *(Reprinted with permission from* Lett. Heat Mass Transfer, *vol. 8, p. 69, N. Ramachandran et al. Copyright © 1981, Pergamon Journals Ltd.)*

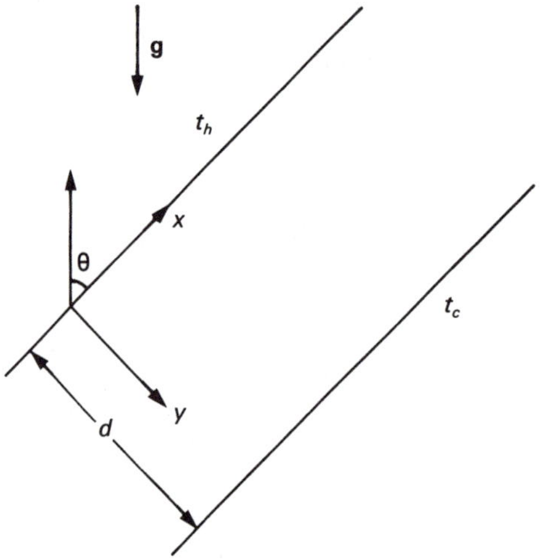

Figure 14.2.3 Coordinate system for flow between two infinite inclined flat plates.

if inclinations are small, the heating is from above, as shown, or the flow is otherwise stable. The heat transfer is again by conduction only and Nu = 1.0.

In a fluid layer between two horizontal surfaces, flow does not arise if the temperature of the upper wall is higher than that of the lower one, for $\beta > 0$. Stable thermal stratification arises, with heavier fluid lying below lighter fluid, and again Nu = 1.0. Luikov et al. (1969) indicated that convection could arise only if the upper wall temperature was either nonuniform or time dependent. The heating from below leads to unstably stratified fluid layers, in horizontal as well as in inclined layers. Chapter 13 discussed such thermal instability and the resulting transport.

14.2.5 Flow in Channels and Tubes with Open Ends

The discussion above concerned buoyancy-driven flow in the enclosed region between two extensive flat parallel surfaces, with the two ends closed. In this way the flow in a rectangular cavity with large height or length is approximated. A related problem that has been studied extensively is that of flow between two parallel surfaces, at t_0, with both ends open to an ambient at t_∞. This configuration exists in several practical systems such as electronic equipment, furnaces, and heat exchangers. For $t_0 > t_\infty$, flow enters at the bottom of the channel and rises due to buoyancy, as shown in Fig. 14.2.4*a*. The flow develops downstream and,

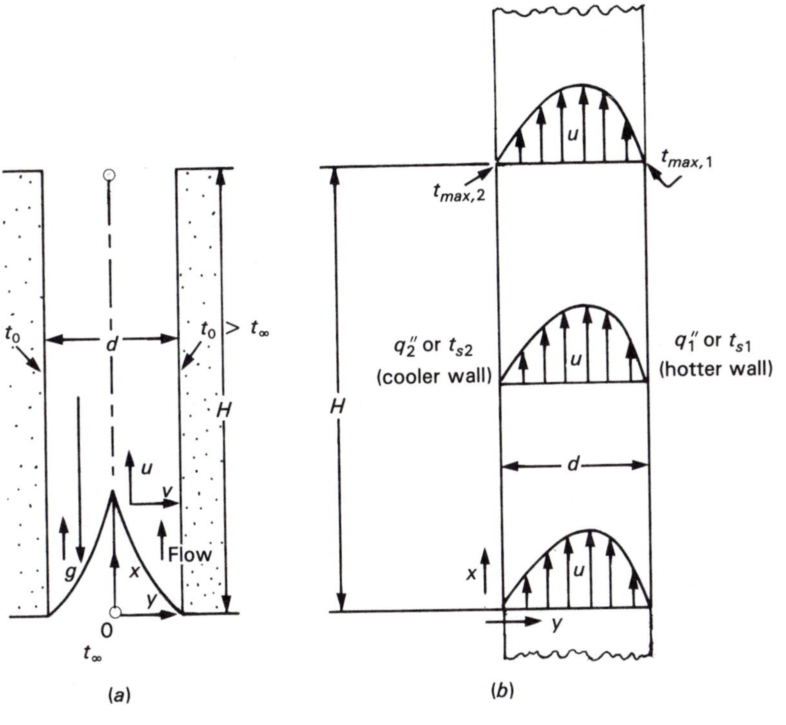

Figure 14.2.4 Flow between two parallel plates: (*a*) developing flow; (*b*) fully developed flow.

if the channel height is large compared to the spacing between the walls, a fully developed flow may arise far downstream. Often, the flow adjacent to the surfaces, near the entrance, may be analyzed as a boundary layer flow. Several of these aspects have been studied in detail experimentally and analytically, as outlined below.

The earliest such study was by Elenbaas (1942), who considered isothermal vertical tubes and square plates open to the ambient along all edges. Experimental results were obtained to provide Nusselt numbers, which were compared with semiempirical formulas also derived in that study. Badoia and Osterle (1962) analyzed the vertical channel flow for uniform wall temperature conditions and obtained numerical results for a limited range of Prandtl and Rayleigh numbers. Engel and Mueller (1967) used an integral technique to study the developing flow in a vertical channel of finite height and considered both uniform wall temperature and uniform surface heat flux boundary conditions. A wide range of Prandtl and Rayleigh numbers was investigated.

Aung (1972) considered the fully developed laminar natural convection flow between vertical surfaces supplying uniform heat flux inputs. Figure 14.2.4*b* shows the flow considered. Nusselt number values were obtained. Aung et al. (1972) carried out a numerical and experimental investigation of the developing flow in a vertical parallel-plate channel with asymmetric heating. The solution for this flow was shown to approach asymptotically the closed-form solution for fully developed flow, obtained by Aung (1972), as the flow proceeds downstream. Figure 14.2.5 shows the calculated velocity and temperature profiles at two values of the dimensionless flow rate parameter $\bar{M}$ for $r_H = 0$, where r_H is the ratio of the wall heat fluxes. In this formulation the motion pressure p_m is taken as zero at the two ends. The solution is obtained, at chosen values of $\bar{M}$, by computing the solution downstream from $x = 0$ until a height H is reached where $p_m = 0$ is attained. A better approach would be to solve for the flow at the two ends, using the full equations that apply in these regions. The relevance of this flow in the cooling of electronic equipment has been discussed by Jaluria (1985).

Sparrow and Bahrami (1980) carried out a detailed experimental study of this problem, considering the effect of blocking the flow across the lateral vertical edge openings. A mass transfer measurement technique was used to avoid extraneous heat losses and variable-property effects. Uncertainties due to large property variations were demonstrated in the results obtained by Elenbaas (1942). At low values of the parameter $(d/H)\mathrm{Ra}$, where d is the interplate spacing, H the cavity height, and Ra the Rayleigh number based on d, the effect of lateral inflow was found to be large, but was negligible for $(d/H)\mathrm{Ra} > 10$. Figure 14.2.6 shows the heat transfer results obtained for the lateral edge gaps open as well as blocked.

Several other studies have concerned similar transport. Sparrow and Tao (1982) considered two parallel adjacent vertical channels separated by an intervening wall across which energy is transferred from one channel to the other. Nakamura et al. (1982) made numerical calculations of the flow between two heated parallel plates, without using the boundary layer approximation. Reference may also be

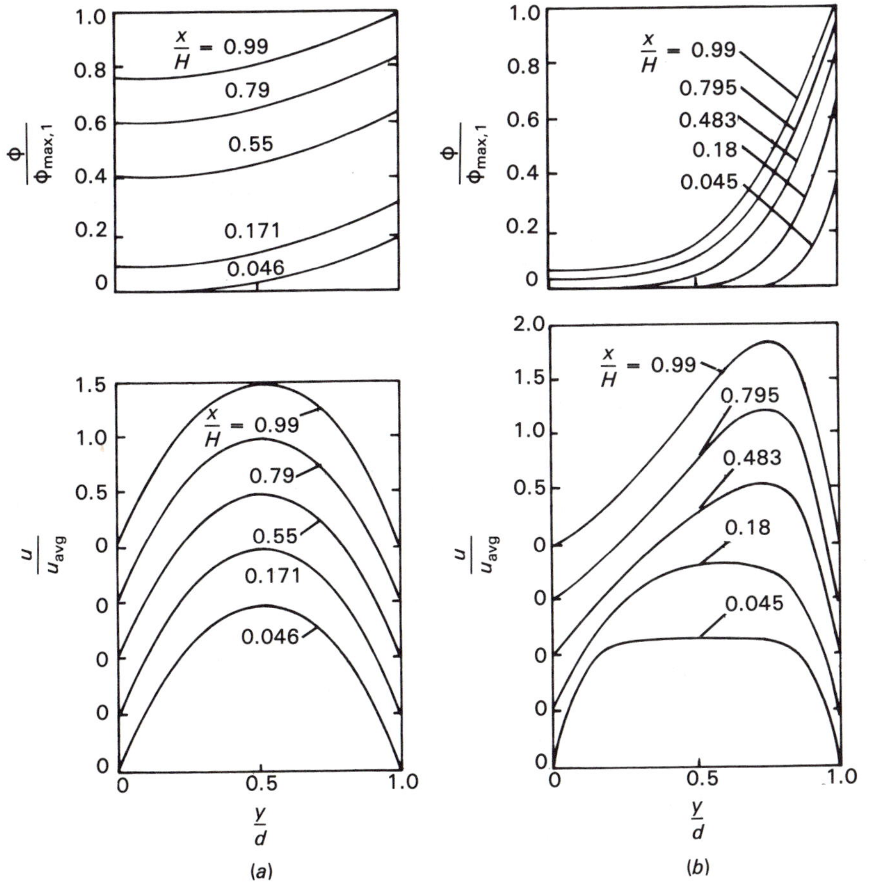

Figure 14.2.5 Calculated velocity and temperature profiles for the uniform heat flux condition, with asymmetric heating at the two vertical plates. (*a*) $r_H = 0$, $\bar{M} = 0.15$; (*b*) $r_H = 0$, $\bar{M} = 0.005$. *(Reprinted with permission from* Int. J. Heat Mass Transfer, *vol. 15, p. 2293, W. Aung et al. Copyright © 1972, Pergamon Journals Ltd.)*

made to the work of Kettleborough (1972), Miyatake and Fujii (1972), Miyatake et al. (1973), and Wirtz and Stutzman (1982).

Buoyancy-induced flow inside vertical tubes has also been studied. Ostroumov (1952) considered many of the aspects discussed above for this configuration. Hallman (1956) considered mixed and natural convection in vertical tubes, including the effect of heat generation in the fluid. Davis and Perona (1971) studied the developing flow in a heated vertical open tube. This configuration is of interest in heat exchangers, in chimney flows, in geothermal applications, and in several other problems of practical importance. The mixed convection flow in vertical tubes, particularly the developing flow in a semi-infinite tube, has been studied by several workers, as discussed in Chapter 10.

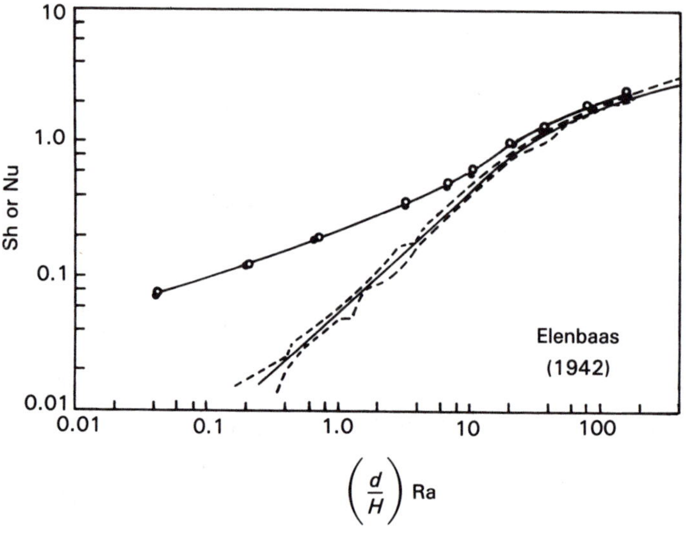

(a)

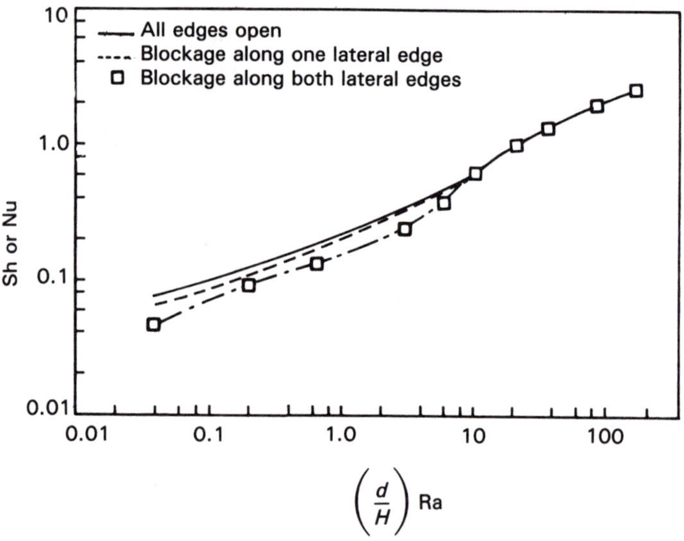

(b)

Figure 14.2.6 Sherwood and Nusselt numbers for square parallel plates. (*a*) All edge gaps open to ambient. (*b*) Flow blockage along the lateral edges. *(Reprinted with permission from E. M. Sparrow and P. A. Bahrami,* J. Heat Transfer, *vol. 102, p. 221. Copyright © 1980, ASME.)*

14.3 RECTANGULAR ENCLOSURES

Following consideration of the flow between extensive flat parallel surfaces, finite-sized cavities are discussed next; see Fig. 14.3.1. The two-dimensional flows in vertical, horizontal, and inclined rectangular cavities have been studied extensively because of the considerable simplification that arises compared to fully three-dimensional transport. Often, a cavity may be very long in one dimension. Then a two-dimensional representation may apply in the central region. Although the internal natural convection flows of interest in practical applications, such as crystal growth, solar collectors, and fires in enclosures, are in general three dimensional, two-dimensional results are often satisfactory. The effort and expense involved in extending the analysis from two to three dimensions are often quite substantial. Consequently, much of the information on internal flows has been obtained through two-dimensional analyses and comparable experimental studies. However, in recent years there has been a significant increase in three-dimensional calculations, as considered later in Section 14.5.

First considered here are vertical cavities in which a temperature difference is maintained between the two vertical surfaces. Flow in inclined and horizontal cavities, where the temperature difference is maintained between other than vertical surfaces, is discussed in Section 14.3.2.

14.3.1 Vertical Rectangular Enclosures

Basic considerations. A vertical rectangular cavity is defined conventionally as an enclosure bounded by two vertical surfaces held at different temperatures. The other two parallel surfaces are often taken as insulated or with their temperature varying linearly between those of the two vertical surfaces. Figure 14.3.1 shows the nomenclature. Of interest are the rate of heat transfer between the walls and

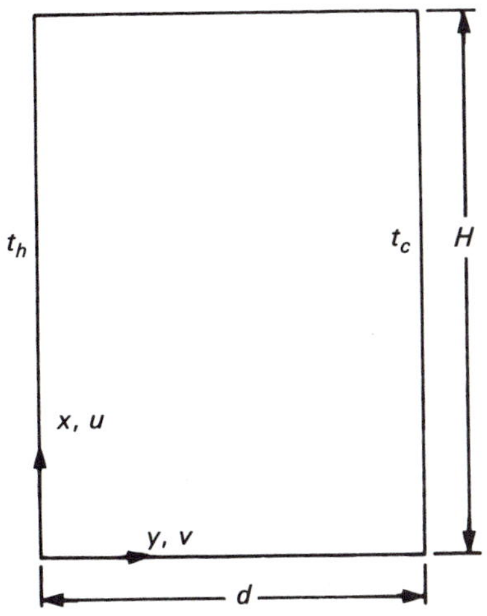

Figure 14.3.1 Coordinate system for a vertical rectangular enclosure.

the nature of the flow that arises, particularly the various flow regimes that occur, depending on the physical variables of transport. The governing equations are written below, assuming the motion to be steady and two dimensional, neglecting the pressure work and viscous dissipation effects, and using the Boussinesq approximations.

$$\frac{\partial u}{\partial x} + \frac{\partial v}{\partial y} = 0 \tag{14.3.1}$$

$$u\frac{\partial u}{\partial x} + v\frac{\partial u}{\partial y} = -\frac{1}{\rho}\frac{\partial p_m}{\partial x} + g\beta(t - t_c) + \nu\left(\frac{\partial^2 u}{\partial x^2} + \frac{\partial^2 u}{\partial y^2}\right) \tag{14.3.2}$$

$$u\frac{\partial v}{\partial x} + v\frac{\partial v}{\partial y} = -\frac{1}{\rho}\frac{\partial p_m}{\partial y} + \nu\left(\frac{\partial^2 v}{\partial x^2} + \frac{\partial^2 v}{\partial y^2}\right) \tag{14.3.3}$$

$$u\frac{\partial t}{\partial x} + v\frac{\partial t}{\partial y} = \alpha\left(\frac{\partial^2 t}{\partial x^2} + \frac{\partial^2 t}{\partial y^2}\right) \tag{14.3.4}$$

where p_m denotes the motion pressure arising due to the flow.

The boundary conditions are

$$y = 0 \text{ for } 0 \le x \le H: \quad u = v = 0 \quad t = t_h$$

$$y = d \text{ for } 0 \le x \le H: \quad u = v = 0 \quad t = t_c$$

$$x = 0, H \text{ for } 0 \le y \le d: \quad u = v = 0 \quad \text{and} \tag{14.3.5}$$

$$t = t_h - (t_h - t_c)\frac{y}{d} \quad \text{or} \quad \frac{\partial t}{\partial x} = 0$$

The two temperature conditions on the horizontal walls, $x = 0, H$, imply either a linearly varying surface temperature or an insulated condition.

The above equations are simplified by using a stream function ψ', defined below, and by eliminating the pressure between Eqs. (14.3.2) and (14.3.3). The dimensionless variables are

$$X = \frac{x}{d} \quad Y = \frac{y}{d} \quad U = \frac{ud}{\alpha} \quad V = \frac{vd}{\alpha}$$

$$\phi = \frac{t - t_c}{t_h - t_c} \quad \psi = \frac{\psi'}{\alpha} \tag{14.3.6}$$

where

$$u = \frac{\partial \psi'}{\partial y} \quad v = -\frac{\partial \psi'}{\partial x}$$

and ψ is the dimensionless stream function. The governing equations in nondimensional form become

$$\nabla^4\psi = \frac{1}{\Pr}\left[\frac{\partial\psi}{\partial Y}\frac{\partial(\nabla^2\psi)}{\partial X} - \frac{\partial\psi}{\partial X}\frac{\partial(\nabla^2\psi)}{\partial Y}\right] - \mathrm{Ra}\,\frac{\partial\phi}{\partial Y} \tag{14.3.7}$$

$$\nabla^2\phi = \frac{\partial\psi}{\partial Y}\frac{\partial\phi}{\partial X} - \frac{\partial\psi}{\partial X}\frac{\partial\phi}{\partial Y} \tag{14.3.8}$$

where $\mathrm{Ra} = g\beta(t_h - t_c)d^3/\nu\alpha$. The boundary conditions are

$$\begin{aligned}
&Y = 0 \text{ for } 0 \le X \le A: \quad \psi = \frac{\partial\psi}{\partial Y} = 0 \quad \phi = 1 \\
&Y = 1 \text{ for } 0 \le X \le A: \quad \psi = \frac{\partial\psi}{\partial Y} = 0 \quad \phi = 0 \\
&X = 0, A \text{ for } 0 \le Y \le 1: \quad \psi = \frac{\partial\psi}{\partial Y} = 0 \quad \text{and} \\
&\phi = 1 - Y \quad \text{or} \quad \frac{\partial\phi}{\partial X} = 0
\end{aligned} \tag{14.3.9}$$

Here $A = H/d$ is the aspect ratio.

The dimensionless parameters are seen to be the Rayleigh number Ra, Prandtl number Pr, and aspect ratio A. For insulated horizontal surfaces, heat transfer occurs only at the vertical surfaces. For other conditions at the horizontal surfaces, interest is mainly in the transport from or to the vertical surfaces. Effort has been directed largely in determining the corresponding heat transfer. However, heat transfer along and from the horizontal surfaces also can be determined from the computed temperature field if desired. The Nusselt number for the net heat transfer q'' between the two vertical walls of an enclosure, with the other surfaces adiabatic, is defined as

$$\mathrm{Nu} = \frac{hd}{k} \qquad \text{where } h = \frac{q''}{t_h - t_c} \tag{14.3.10}$$

The formulation above is complicated because of the nonlinear and partial differential nature of the equations. Approximate solutions for limiting values of the governing parameters were therefore considered by early investigators; see Jakob (1949). However, with the growth of computational capability, extensive numerical solutions have been obtained for wide ranges of the governing parameters. Nevertheless, much of the basic information for such transport has come from experimental work and analyses of various limiting regimes of flow.

Among the earliest investigations was the detailed analysis by Batchelor (1954), who considered H/d values ranging from about 5 to ∞. A perturbation scheme was used for small values of Ra, and it was concluded that there is little increase in the heat transfer over that due to conduction alone for $\mathrm{Ra} < 1000$ if H/d is large. Conduction was also shown to be the only mechanism of heat transfer for $H/d \to \infty$ at any Ra, as discussed in Section 14.2.1.

For large Ra, the core or interior region of the cavity, away from the boundary layers on the walls, was assumed to be isothermal and to rotate as a solid body, that is, with constant vorticity. This assumption had also been made earlier by Ostrach (1950) for natural convection flow in a horizontal circular cylinder and by Pillow (1952) for cellular flow between two parallel horizontal plates. Batchelor (1956) considered the interior or core region in a vertical cavity and showed that the vorticity there is constant if the core is not stagnant. This proof applies only to an isothermal core. Also, the boundary of the interior region must be a closed streamline that does not enter any part of the viscous boundary layer flow near the surfaces. Poots (1958) solved this problem numerically, using a method based on expansions in orthogonal polynomials. The results obtained by Batchelor (1954) and by Poots (1958) were compared by the latter with the experimental results of Mull and Reiher (1930). Fairly good agreement was obtained. Batchelor (1954) also considered turbulent flow in a vertical cavity.

Flow regimes. The experimental studies of Eckert and Carlson (1961) for air and of Elder (1965a) for silicone oil of Prandtl number around 1000 contributed significantly to understanding of the flow and heat transfer mechanisms in vertical rectangular cavities. In the first study, in air, the aspect ratio A was varied over the range 2.1–46.7 and the Rayleigh number was varied from 200 to 2×10^5. The temperature field was studied with a Mach-Zehnder interferometer. At low values of Ra, conduction was dominant and a linear temperature distribution was seen between the vertical walls in the region away from the ends. Convection effects were significant near the ends. At large Ra, boundary layers arose on the vertical surfaces and the core region was linearly and stably stratified.

Figure 14.3.2 shows the temperature profiles. The temperature scales shown are for the two end profiles, the rest being shifted vertically for clarity since all curves coincide at the isothermal vertical walls. The linear variation is seen in the central region of Fig. 14.3.2*a*, characterizing the "conduction regime." In the "boundary layer regime," Fig. 14.3.2*c*, thin boundary regions appear along vertical walls. Horizontal temperature uniformity is seen between the two layers. The core is also stably stratified since the temperature there increases with height. In the "transition regime" the two boundary regions are thicker and an isothermal interior region does not appear; see Fig. 14.3.2*b*.

Elder (1965a) used silicone oils of Prandtl number around 1000 and varied the aspect ratio from $H/d = 1$ to 60. Rayleigh numbers were up to about 10^8. Temperature and velocity distributions were determined. Aluminum powder was suspended in the fluid and the particle motions indicated the flow pattern. The upper end of the cavity was open to the atmosphere. At Ra less than about 1000, a weak, steady, unicellular circulation was observed, with fluid rising near the hot wall and descending adjacent to the cold one. This corresponded to the conduction regime of Eckert and Carlson (1961). For $10^3 < \text{Ra} < 10^5$, large temperature gradients near the walls and a uniform vertical gradient in the interior region were observed. Figure 14.3.3 shows the isotherms at $\text{Ra} = 4 \times 10^5$. The wall, the interior, and the end regions are clearly seen. The interior region is seen to have a small horizontal temperature variation and an almost linear vertical tem-

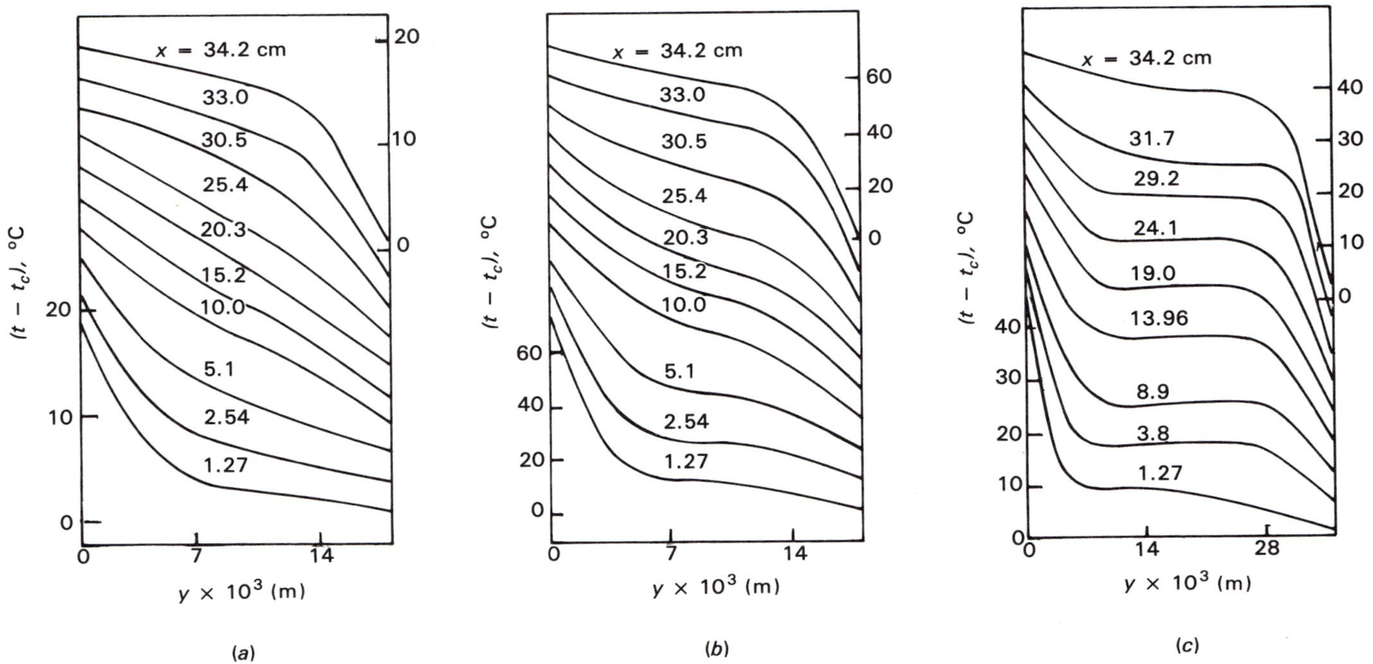

Figure 14.3.2 Measured temperature distributions in air in a vertical rectangular enclosure. (*a*) Conduction regime; (*b*) transition regime; (*c*) boundary layer regime. *(Reprinted with permission from* Int. J. Heat Mass Transfer, *vol. 2, p. 106, E. R. G. Eckert and W. O. Carlson. Copyright © 1961, Pergamon Journals Ltd.)*

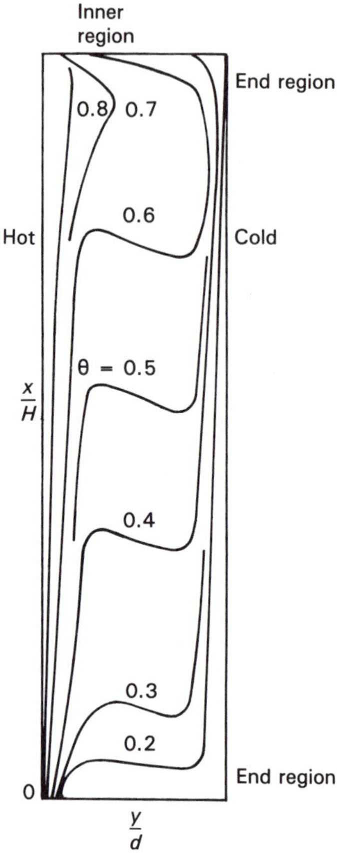

Figure 14.3.3 Experimentally determined isotherms in a rectangular enclosure at Ra = 4×10^5. *(Reprinted with permission from J. W. Elder,* J. Fluid Mech., *vol. 23, p. 77. Copyright © 1965, Cambridge University Press.)*

perature increase or stratification. Figure 14.3.4 shows the measured velocity profiles at $x = H/2$ for various Ra. At the largest values of Ra a relatively stagnant central region is seen to arise.

As Ra was increased further, Elder (1965a) observed secondary and tertiary flows. Secondary flows were found to arise for values around Ra = 3×10^5 and were very weak initially. At larger Ra, more cells appeared. For Ra $> 10^6$, tertiary flows arose in which further steady cellular motion was generated in the weak shear region between cells. Elder (1965b) also studied the transition from laminar to turbulent flow. Disturbances were observed to propagate up the hot surface and down the cold one. Turbulence arose around Ra = 10^9, at about half-height in the cavity, and then spread to the ends at higher Ra.

The experimental studies of Eckert and Carlson (1961) and of Elder (1965a), therefore, indicated that the core region was relatively stagnant and almost linearly stratified. This result contradicts the assumption of an isothermal interior region made by Ostrach (1950) and Batchelor (1954). Wilkes (1963) studied this flow numerically and also obtained an essentially stagnant core with an almost linear

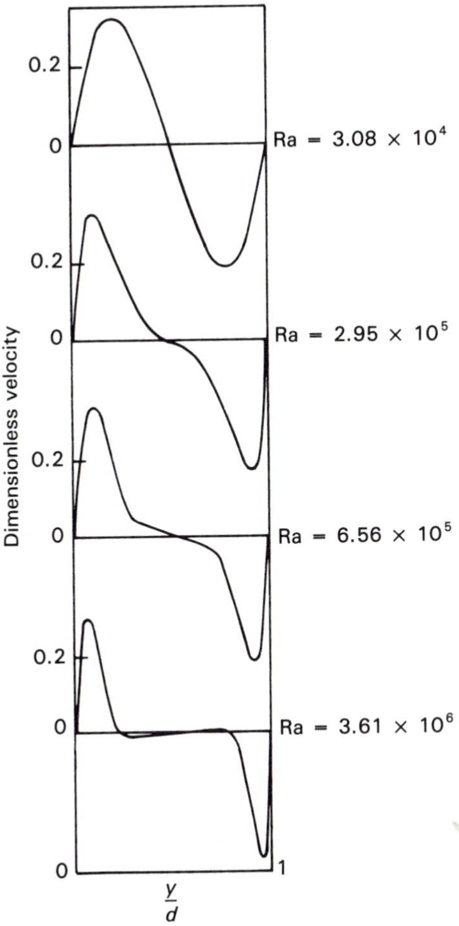

Figure 14.3.4 Measured velocity distributions at $x = H/2$ and various values of Ra. *(Reprinted with permission from J. W. Elder,* J. Fluid Mech., *vol. 23, p. 77. Copyright © 1965, Cambridge University Press.)*

temperature increase in the vertical direction. Gershuni et al. (1966) obtained a similar result in a numerical study. Various other numerical studies at about the same time also supported these findings. The work of Elder (1966) and of De Vahl Davis (1968) are mentioned in this regard.

Figure 14.3.5 shows the isotherms obtained by Elder (1966) from a numerical solution of the equations. It is seen that the slope of the isotherms in the core region is negative in the middle portion of the cavity. The calculations of De Vahl Davis (1968) also indicated horizontal isotherms and streamlines in the core, at lower values of Ra. As Ra increases, the slope of these isotherms becomes negative. Thus, the fluid temperature outside the boundary layer on the cold wall is greater than that at a corresponding point outside the layer on the hot wall. The vertical temperature gradient in the center of the cavity was found to be close to zero at small Rayleigh numbers and to approach a positive asymptotic value as Ra became very large. This value was also found to depend on the thermal bound-

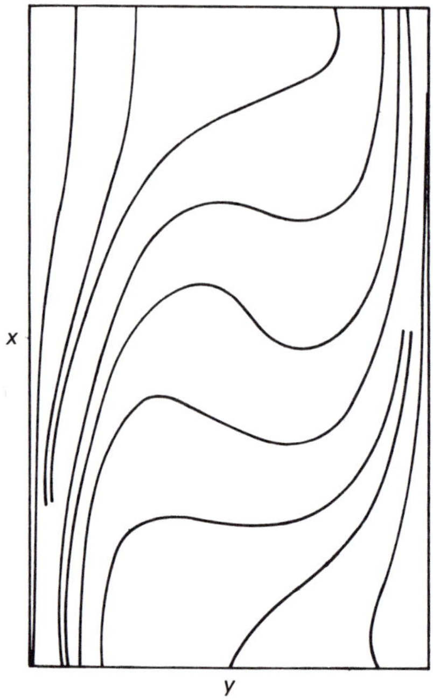

Figure 14.3.5 Calculated isotherms in a rectangular enclosure for Pr = 1.0 at Ra = 2×10^4. *(Reprinted with permission from J. W. Elder,* J. Fluid Mech., *vol. 24, p. 823. Copyright © 1966, Cambridge University Press.)*

ary conditions specified at the horizontal surfaces. These trends are shown in Fig. 14.3.6. Flow reversal near the center of the cavity at large Ra was also observed, indicating the onset of secondary flows, as seen in Fig. 14.3.7.

De Vahl Davis and Mallinson (1975) solved the equations numerically to study the secondary and tertiary flows observed by Elder (1965a), for aspect ratios up to 20, Rayleigh numbers up to 3×10^6, and a Prandtl number of 1000. The numerical model did generate the complex secondary and tertiary motions observed by Elder, without the assumptions generally employed in linear stability analysis, for example, the parallelism of the base flow and smallness of the disturbance amplitude. Figure 14.3.8 shows the flow field in terms of contours of the stream function ψ. A single roll is seen at Ra = 2.4×10^5. Secondary motion is most prominent at the center of the cavity for Ra = 5×10^5. As Ra increases, tertiary rolls become stronger.

Tall enclosures. The discussion above was concerned mainly with the effect of increasing Rayleigh number on the flow. Also, various values of the aspect ratio A and the Prandtl number were given. The corresponding heat transfer results are discussed later in this section. Let us now consider the two extreme cases, of tall and shallow enclosures, for which some analytical results exist.

For H/d large, Gill (1966) assumed a boundary layer flow in a rectangular cavity. That is, vertical fluid motion was assumed to be confined to layers near the two vertical surfaces. The fluid in the interior region was taken as stagnant and vertically stratified. This convection-dominated regime arises for sufficiently large Ra. Quon (1972) showed that the stream function calculated by Gill was

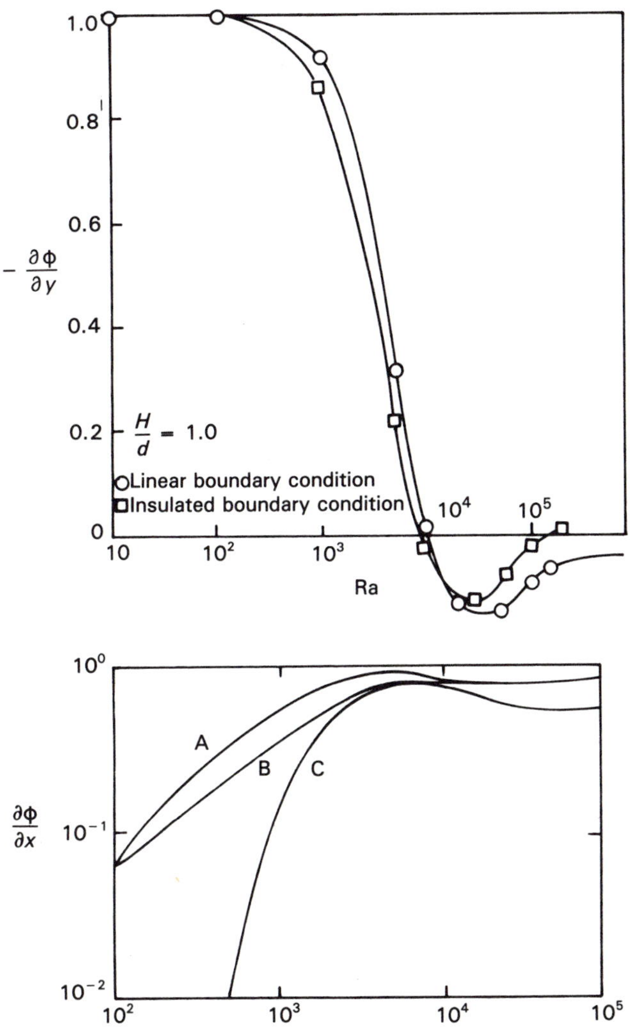

Figure 14.3.6 Computed dimensionless temperature gradients at cavity midheight, $y = H/2$, as functions of the Rayleigh number Ra. Curve A, $H/d = 1$, insulated; curve B, $H/d = 1$, linear; curve C, $H/d = 5$, insulated. *(Reprinted with permission from* Int. J. Heat Mass Transfer, *vol. 11, p. 1675, G. De Vahl Davis. Copyright © 1968, Pergamon Journals Ltd.)*

about 30% higher and the maximum vertical velocities about 25% higher than the corresponding values measured by Elder (1965a). Bejan (1979) proposed an extension of Gill's analysis by requiring that the net flow of energy vanish in the vicinity of the adiabatic top and bottom end walls, in order to determine an arbitrary constant that appears in the analysis. This leads to improved agreement between theory and experiment for tall cavities. Quon (1977) also proposed a

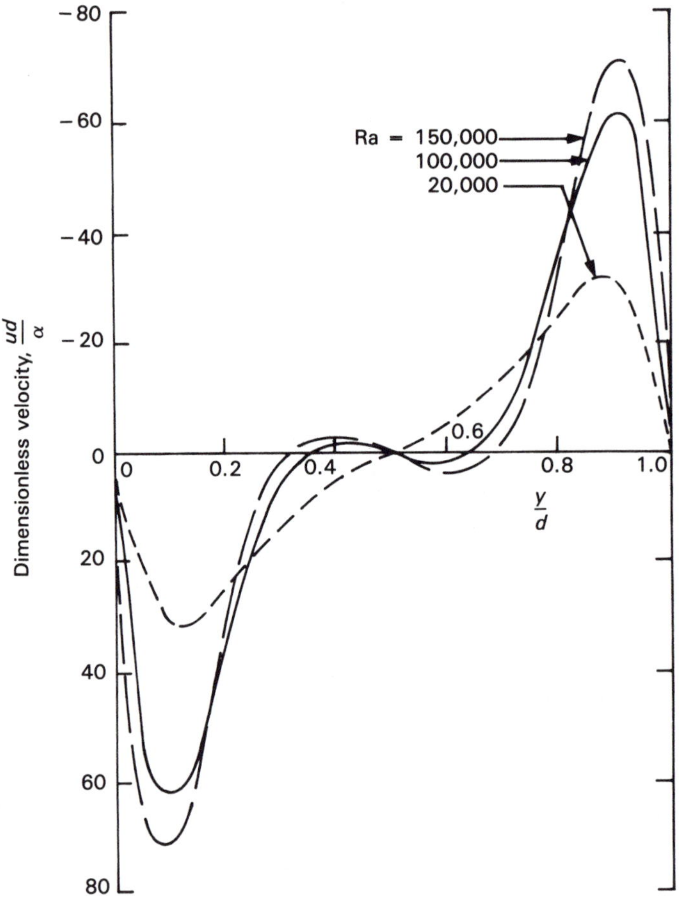

Figure 14.3.7 Distribution of the vertical velocity component at the midheight of the cavity for $H/d = 1.0$ and a linear temperature distribution boundary condition at the ends, with the right side wall heated and u taken positive downward. *(Reprinted with permission from* Int. J. Heat Mass Transfer, *vol. 11, p. 1675, G. De Vahl Davis. Copyright © 1968, Pergamon Journals Ltd.)*

method for determining the free constant in Gill's analysis from a numerical computation of the stream function.

Shallow enclosures. Some analytical work has been done on shallow cavities, that is, H/d small. Cormack et al. (1974a, 1974b) and Imberger (1974) made calculations with differentially heated end walls for application to convective heat transfer due to heat rejection in shallow water bodies. The horizontal walls were taken as adiabatic and the aspect ratio $A << 1$. The flow pattern found was composed of a core region, which contains counterflowing fluid with warmer fluid in the upper portion, and end regions where the flow is turned, as shown in Fig. 14.3.9. Cormack et al. (1974a) obtained an asymptotic solution for $A \rightarrow 0$ with Ra finite but fixed.

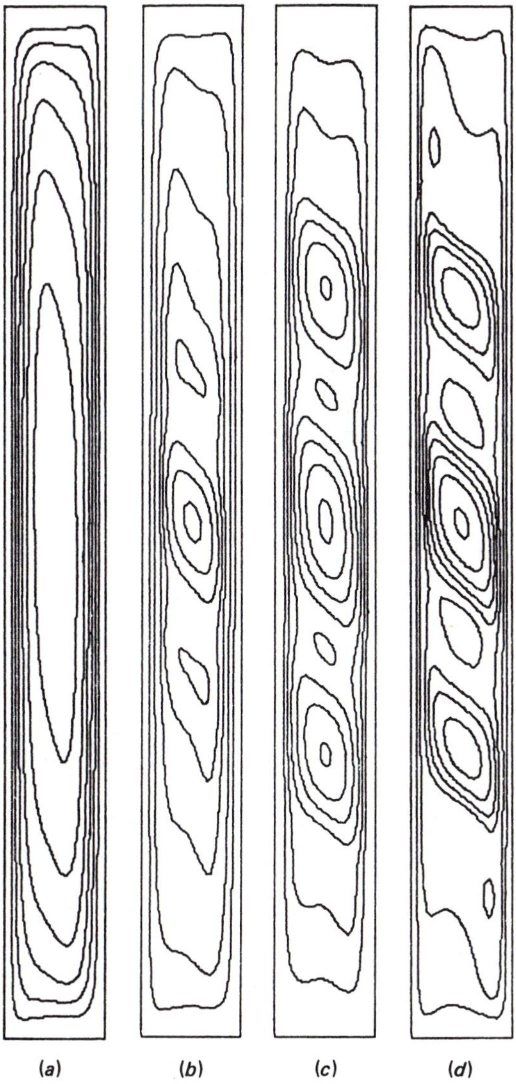

Figure 14.3.8 Streamlines in a vertical rectangular cavity for $H/d = 10$ and $\mathrm{Pr} = 10^3$. $\mathrm{Ra} =$ (*a*) 2.4×10^5; (*b*) 5.0×10^5; (*c*) 9.4×10^5; (*d*) 3.3×10^6. *(Reprinted with permission from G. De Vahl Davis and G. D. Mallinson,* J. Fluid Mech., *vol. 72, p. 87. Copyright © 1975, Cambridge University Press.)*

Bejan and Tien (1978) considered the flow in shallow cavities and coupled the asymptotic core solution of Cormack et al. (1974a) with integral solutions for the flow in the two end regions. The validity of the solution thus obtained is expected to extend well into the aspect ratio domain where the asymptotic solution does not apply. The three regimes, $\mathrm{Ra} \to 0$, intermediate Ra, and large Ra, were considered. Further work for this geometry has been done by Shiralkar and Tien (1981) and Shiralkar et al. (1981).

Other studies. There has been extensive additional numerical study of flows in vertical rectangular enclosures. With large computers and efficient numerical

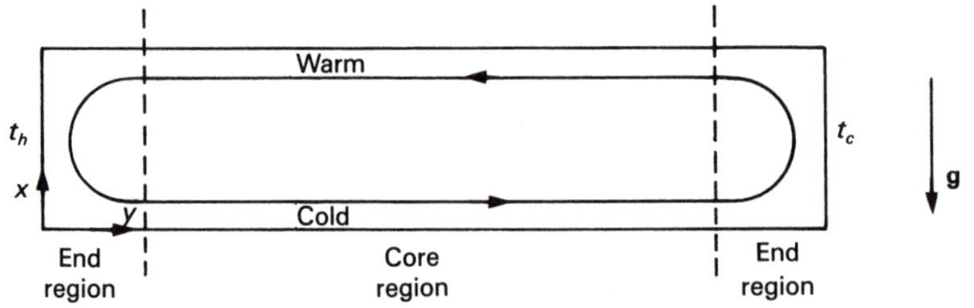

Figure 14.3.9 Schematic of natural convection flow in a horizontal cavity with different end temperatures.

techniques, solutions have been obtained over wide ranges of the governing parameters and for a variety of thermal boundary conditions. See, for instance, the symposium volume edited by Torrance and Catton (1980). Chu and Churchill (1977) developed a finite-difference method and considered the error associated with large grid size. Denny and Clever (1974) also emphasized the effect of grid size on the accuracy of the results. Although most calculations have been done with finite-difference methods, the Galerkin and finite-element techniques have also been used. Denny and Clever (1974) employed both of these methods for a high Prandtl number fluid in a square cavity with a linear temperature variation at the end walls. The Galerkin method was demonstrated to remain more accurate as the number of mesh points in the region was reduced.

There has also been considerable additional experimental work on vertical rectangular cavities. The effects of the Rayleigh number, aspect ratio, and Prandtl number on the flow and on the heat transfer have been determined. In general, these results compare quite well with both the numerical solutions and the few analyses available. Although heat transfer results were of prime concern in many of these studies, several also considered the nature of flow in the cavity. See, for example, the work of Simpkins and Dudderar (1981) and of Bejan et al. (1981). The work of Lykoudis (1974) on flows in mercury is also mentioned.

Heat transfer. The heat transfer results for these vertical cavities indicate that conduction mechanisms dominate in the core region for Ra ≤ 2000. Then the heat transfer across the middle height of the cavity, away from the horizontal surfaces, is given by

$$q'' = k\frac{t_h - t_c}{d} = h(t_h - t_c) \tag{14.3.11}$$

and therefore, as mentioned earlier,

$$\text{Nu} = \frac{hd}{k} = 1 \tag{14.3.12}$$

The heat transfer rate q'' across the upper and lower regions is different because of convection effects there. This is seen from the results of Eckert and

Carlson (1961) for air. The lower corner of the hot surface and the upper corner of the cold surface were termed starting corners, the other two being called departure corners. Local heat transfer rates q'' were found to be larger at the starting corners and smaller at the departure corners than at the midheight or core region of the cavity. If the end flow effects remain localized and do not affect the middle region, as seen from the results for the conduction regime, Fig. 14.3.2*a*, the heat transfer results near the ends should be independent of the height H of the cavity. The measurements of Eckert and Carlson (1961) near the starting corners, where the wall flows start, indicate

$$\mathrm{Nu}_x = \frac{h_x x}{k} = 0.256\mathrm{Gr}_x^{0.24} \tag{14.3.13}$$

where Nu_x is a local Nusselt number and x is distance measured along the vertical surfaces from the corners, being downward on the cold surface and upward on the hot surface. Also, h_x is based on the overall temperature difference $t_h - t_c$. This equation applies up to the value of x at which the heat transfer coefficient h_x becomes equal to the value of h for the central core, k/d. The average heat transfer coefficient may thus be obtained by integrating the expression for h_x up to this value of x. The result obtained is given as

$$\mathrm{Nu} = \frac{hd}{k} = 1.389 \tag{14.3.14}$$

Similarly, the heat transfer results for the departure corners were obtained in terms of the local and overall Nusselt numbers as

$$\mathrm{Nu}_x = 2.58\mathrm{Gr}_x^{0.4}\,\mathrm{Gr}^{-0.55} \tag{14.3.15}$$

and

$$\mathrm{Nu} = 0.835 \tag{14.3.16}$$

where Gr is the Grashof number based on the layer thickness d. From Eqs. (14.3.11)–(14.3.16), the average heat transfer from the hot to the cold surface may be obtained by combining the results for the corner and middle regions of the cavity, to yield the Nusselt number as

$$\mathrm{Nu} = \frac{hd}{k} = 1 + 0.00166\,\frac{d}{H}\,\mathrm{Gr}^{0.9} \tag{14.3.17}$$

Thus, it is seen that as the aspect ratio H/d becomes large, Nu approaches 1.

The heat transfer results for the transition and boundary layer flow regimes may also be obtained from the temperature profiles measured by Eckert and Carlson (1961). Correlations for the local heat transfer were attempted in terms of the difference between the wall and centerline temperatures, as well as the difference between the two wall temperatures. The overall temperature difference $t_h - t_c$ was used for average heat transfer results. The equation obtained is

$$\mathrm{Nu} = 0.119\mathrm{Gr}^{0.3}\left(\frac{H}{d}\right)^{-0.1} \tag{14.3.18}$$

This indicates the small effect of the aspect ratio H/d on the heat transfer.

Several other studies of transport in vertical rectangular enclosures have obtained expressions similar to the one above for the average convective transport; see Jakob (1949). The general expression is

$$\mathrm{Nu} = a\mathrm{Ra}^b A^c \tag{14.3.19}$$

where a, b, and c are constants. Table 14.3.1 lists the values for air obtained in several studies.

A heat transfer relationship presented by Berkovsky and Polevikov (1977) compares very well with the numerical results of Catton et al. (1974). It is expected to provide a reliable estimate of the effect due to the fluid Prandtl number over a wide range. The average Nusselt number is given as

$$\mathrm{Nu} = 0.22A^{-1/4}\left(\frac{\mathrm{RaPr}}{0.2 + \mathrm{Pr}}\right)^{0.28} \qquad \text{for } 2 < A < 10,\ \mathrm{Pr} < 10^5,\ \mathrm{Ra} < 10^{10}$$

and

$$\mathrm{Nu} = 0.18\left(\frac{\mathrm{RaPr}}{0.2 + \mathrm{Pr}}\right)^{0.29}$$

$$\text{for } 1 < A < 2,\ 10^{-3} < \mathrm{Pr} < 10^5,\ 10^3 < \frac{\mathrm{RaPr}}{0.2 + \mathrm{Pr}} \tag{14.3.20}$$

The above correlations cover a wide range of conditions. Although there are differences between the results obtained from the various studies, there is fair agreement between the data and calculations. Some low-Pr heat transfer results have also been given by Dropkin and Somerscales (1965).

Reference is also made here to some analytical results obtained for a few particular flow regimes. For the convection-dominated regime in a vertical cavity,

Table 14.3.1 Constants for the average Nusselt number relation of the form $\mathrm{Nu} = a(\mathrm{Gr})^b A^c$ for laminar flow in air

Reference	a	b	c	H/d
Newell and Schmidt (1970)	0.0547	0.397	—	1
Han (1967)	0.0782	0.3594	—	1
Elder (1966)	0.231	0.25	—	1
Newell and Schmidt (1970)	0.155	0.315	−0.265	2–20
Eckert and Carlson (1961)	0.119	0.3	−0.1	2–20
Jakob (1949)	0.18	0.25	−0.111	2–20
MacGregor and Emery (1969)	0.25	0.25	−0.25	2–20

Source: From Ostrach (1972).

studied by Gill (1966), the overall Nusselt number was calculated by Bejan (1979) as

$$\mathrm{Nu} = 0.364 \left(\frac{\mathrm{Ra}}{A}\right)^{1/4} \tag{14.3.21}$$

For shallow enclosures, Cormack et al. (1974a) obtained an asymptotic result for the overall Nusselt number, valid in the limit $A \to 0$, given as

$$\mathrm{Nu} = 1 + 2.86 \times 10^{-6} \mathrm{Ra}^2 A^8 \tag{14.3.22}$$

Bejan and Tien (1978) determined a theoretical Nusselt number correlation to cover all three laminar regimes, from the $\mathrm{Ra} \to 0$ limit to the laminar boundary layer limit. The expression is rewritten as

$$\mathrm{Nu} = A\left(1 + \left\{\left[\frac{(\mathrm{Ra}\,A)^2}{362{,}880}\right]^n + \left(0.623\,\frac{\mathrm{Ra}^{1/5}}{A}\right)^n\right\}^{1/n}\right) \tag{14.3.23}$$

where $n = -0.386$. Above $\mathrm{Ra} \approx 10^9$ the flow may become turbulent and this correlation would not apply. Further work has been done by Shiralkar and Tien (1981), who modified the above equation to

$$\mathrm{Nu} = A\left(1 + \left\{\left[\frac{(\mathrm{Ra}\,A)^2\gamma_1}{362{,}880}\right]^n + \left(\gamma_2\,\frac{\mathrm{Ra}^{1/5}}{A}\right)^n\right\}^{1/n}\right) \tag{14.3.24}$$

where $n = -0.386$ and

$$\gamma_1 = \left(0.811 - \frac{6.433 \times 10^{-3}}{A}\right)^{-2.5907}$$

$$\gamma_2 = \left(1.2425 + \frac{4.5 \times 10^{-4}}{A}\right)^{-2.5907}$$

For low Pr, the following correlation is proposed by Shiralkar and Tien (1981):

$$\mathrm{Nu} = 0.35(\mathrm{RaPr})^{0.25} \tag{14.3.25}$$

This expression is expected to apply for laminar flow in shallow and square enclosures and is obtained for the asymptotic condition $\mathrm{Ra} \to \infty$.

A synthesis of the analytical results for natural convection in vertical rectangular enclosures has been given by Bejan (1980). The heat transfer results obtained by various investigators are compared and the effect of the aspect ratio A is discussed. Figure 14.3.10 indicates the variation of the Nusselt number Nu with A at various values of the Rayleigh number. The limiting cases of shallow and tall enclosures, $A << 1$ and $A >> 1$, are shown, with conduction on the one hand and Gill's model on the other. The theories for these two extreme cases of A approach each other for a square enclosure, $A = 1$. At a constant Ra, the Nusselt number reaches a maximum at some critical aspect ratio.

Natural convection in vertical cavities has also been studied for fluid-saturated porous media; see Section 15.4. Because of its relevance to geophysical problems

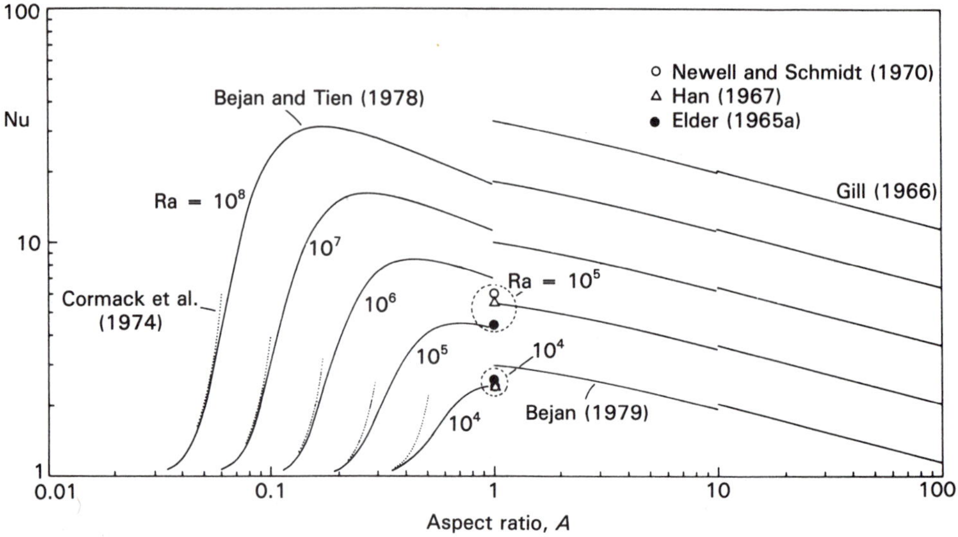

Figure 14.3.10 Heat transfer results from various studies for laminar natural convection in a vertical rectangular cavity. *(Reprinted with permission from* Int. J. Heat Mass Transfer, *vol. 23, p. 723, A. Bejan. Copyright © 1980, Pergamon Journals Ltd.)*

and to many other processes, heat transfer in porous media has begun to receive considerable attention. Cheng (1978) reviews the results. For the vertical enclosure, mention is made of the analytical work of Weber (1975), Walker and Homsy (1978), and Bejan and Tien (1978), the numerical work of Bankvall (1974) and Burns et al. (1977), and the experimental work of Schneider (1963) and Klarsfeld (1970). Trends similar to those discussed above are obtained. Bejan (1980) has also summarized the heat transfer results obtained in various investigations, including analytical as well as numerical and experimental work.

14.3.2 Horizontal and Inclined Rectangular Enclosures Heated from Below

Convection in horizontal and inclined cavities (see Fig. 14.3.11) heated from the lowest bounding surface has also received considerable attention in the past two decades. As discussed in detail in Chapter 13, this heating condition amounts to a potentially unstably stratified circumstance. This results in convective motion if the Rayleigh number Ra, based on height H, and the temperature difference is larger than a critical value Ra_c. The limiting case of infinite horizontal flat plates, $A \rightarrow 0$, is the Bénard configuration, considered in detail in Chapter 13. Of concern here is the two-dimensional transport arising in finite rectangular cavities with heating from below, at a high enough Rayleigh number for convective motion to arise. Both laminar and turbulent flow and heat transfer have been studied. Horizontal and inclined cavities are of interest in practical applications, such as solar ponds, solar collectors, thermal insulation by means of air gaps, and melting processes in manufacturing.

Information on the flow and heat transfer has been obtained from both measurements and calculations. The coupled system of equations for the two-dimensional steady flow in a horizontal cavity may be written in nondimensional form, employing the Boussinesq approximations, as

$$\left(u \frac{\partial}{\partial x} + v \frac{\partial}{\partial y} - \Pr \nabla^2\right)\nabla^2\psi = \Pr\mathrm{Ra} \frac{\partial \phi}{\partial y} \tag{14.3.26a}$$

$$\left(u \frac{\partial}{\partial x} + v \frac{\partial}{\partial y} - \nabla^2\right)\phi = 0 \tag{14.3.26b}$$

where

$$\mathrm{Ra} = \frac{g\beta(t_h - t_c)H^3}{\nu\alpha} \tag{14.3.26c}$$

The flow geometry is shown in Fig. 14.3.11*a* and the nondimensionalization is again as in Eq. (14.3.6), with d replaced by H. For an inclined cavity, shown in Fig. 14.3.11*b*, the inclination results in buoyancy components, $g\beta(t - t_c) \cos \theta$ and $g\beta(t - t_c) \sin \theta$, in the x and y momentum equations (14.3.2) and (14.3.3). The continuity and energy equations are unchanged and the momentum equations become

$$u \frac{\partial u}{\partial x} + v \frac{\partial u}{\partial y} = -\frac{1}{\rho} \frac{\partial p_m}{\partial x} + g\beta(t - t_c) \cos \theta + \nu\left(\frac{\partial^2 u}{\partial x^2} + \frac{\partial^2 u}{\partial y^2}\right) \tag{14.3.27a}$$

$$u \frac{\partial v}{\partial x} + v \frac{\partial v}{\partial y} = -\frac{1}{\rho} \frac{\partial p_m}{\partial y} + g\beta(t - t_c) \sin \theta + \nu\left(\frac{\partial^2 v}{\partial x^2} + \frac{\partial^2 v}{\partial y^2}\right) \tag{14.3.27b}$$

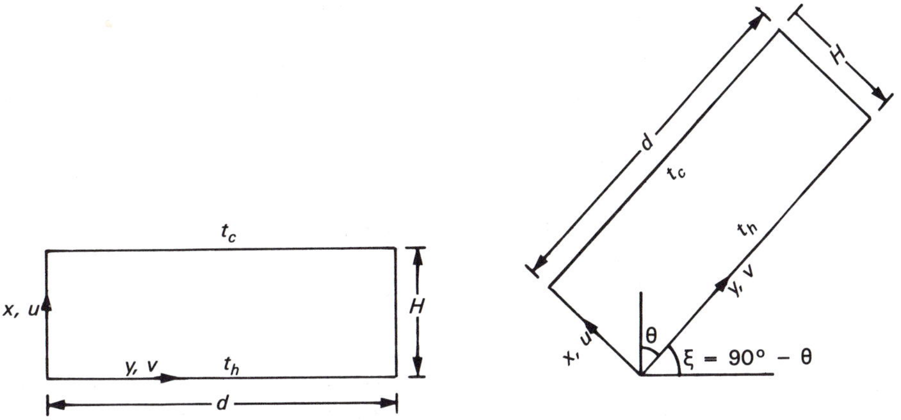

Figure 14.3.11 Coordinate system for horizontal and inclined rectangular enclosures.

The aspect ratio A will again arise in the boundary conditions. The Rayleigh number Ra and the Prandtl number Pr again appear. For an inclined cavity, the angle θ is an additional parameter. Experimental and numerical studies have considered wide ranges of all these parameters, and considerable information is available on both flow and heat transfer characteristics.

Horizontal enclosures. Among the earliest measurements of natural convection heat transfer in horizontal cavities were those of Soberman (1958), Silveston (1958), and Globe and Dropkin (1959). The last investigation covered a wide range of Pr, from 0.02 to 8750, and an Ra range from 3×10^5 to 7×10^9. The results suggested the following empirical correlation, which was also confirmed by Dropkin and Somerscales (1965):

$$\mathrm{Nu} = 0.069\mathrm{Ra}^{1/3}\mathrm{Pr}^{0.074} \qquad \text{with } \mathrm{Ra} = \frac{g\beta(t_h - t_c)H^3}{\nu\alpha} \tag{14.3.28}$$

Although the aspect ratio was varied, a negligible dependence of Nu on A was obtained. However, small Ra was obtained only for small Pr and large Ra only for large Pr, thus giving a coupled Pr and Ra dependence of the data. Silveston (1958) also gave another correlation over a similar range of parameters:

$$\mathrm{Nu} = 0.10\mathrm{Ra}^{0.31}\mathrm{Pr}^{0.05} \tag{14.3.29}$$

These results indicate that the Prandtl number dependence, beyond that in Ra, persists even at large Pr and Ra. However, as $\mathrm{Pr} \to \infty$, the dependence on Pr must disappear for Nu to remain finite. Kraichnan (1962) used similarity arguments to study this dependence on Pr and predicted the following expressions for the Nusselt number in the two limiting conditions:

$$\mathrm{Nu} = 0.089\mathrm{Ra}^{1/3} \qquad \text{at large Pr} \tag{14.3.30a}$$

$$\mathrm{Nu} = 0.17(\mathrm{PrRa})^{1/3} \qquad \text{at small Pr} \tag{14.3.30b}$$

The second result is said to be valid only for $(\mathrm{PrRa})^{1/3} >> 6$.

Rossby (1969) suggested that the transition to turbulent flow occurs at a Rayleigh number of about $14{,}000\mathrm{Pr}^b$, where $b \approx 0.6$ for $\mathrm{Pr} >> 1.0$. Thus, turbulence is predicted to arise at larger Ra when Pr is larger. At small Pr, the flow has been observed to be unsteady at Ra just above Ra_c (=1708) and to be always turbulent at large Ra. The various transport regimes observed in a horizontal fluid layer are shown in Fig. 14.3.12, from the experimental study of Krishnamurti (1970) for Pr from 1 to 10^4, Ra from 10^3 to 10^6, and $A << 1$; See Chapter 13.

O'Toole and Silveston (1961) suggested the following heat transfer correlation for turbulent transport:

$$\mathrm{Nu} = 0.104\mathrm{Ra}^{0.305}\mathrm{Pr}^{0.084} \tag{14.3.31}$$

for $\mathrm{Ra} > 10^5$. Rossby (1969) found the exponent of Ra to be 0.281 for silicone oil ($\mathrm{Pr} = 100$) and 0.257 for mercury. Goldstein and Chu (1969) gave the exponent as 0.294 for air and Chu and Goldstein (1973) gave 0.278 for water. Thus, the power law dependence on Ra is found to be less than $\frac{1}{3}$, which is obtained from similarity arguments for turbulent flow.

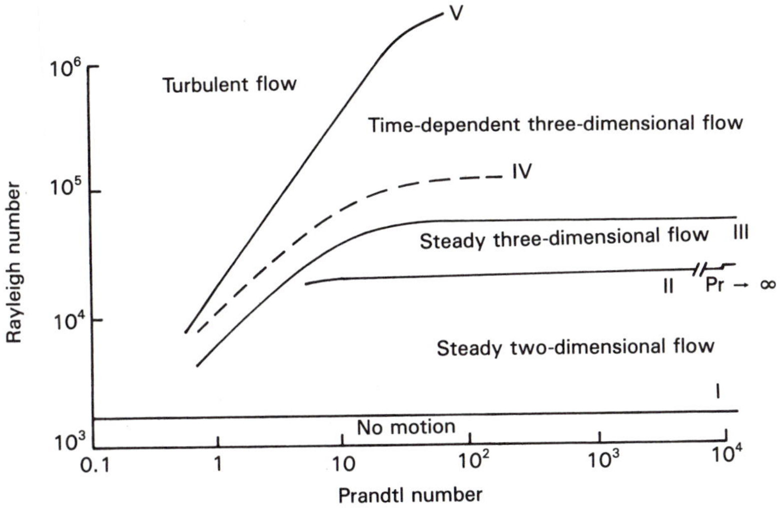

Figure 14.3.12 Various flow regimes in a horizontal fluid layer heated from below. *(Reprinted with permission from R. Krishnamurti,* J. Fluid Mech., *vol. 42, p. 295. Copyright © 1970, Cambridge University Press.)*

Mean temperature profiles were also measured in several of the studies mentioned above. The profiles obtained by Gille (1967) in air from measurements of the refractive index are shown in Fig. 14.3.13. A reversal of the gradient of the profiles occurs at high Ra/Ra_c. Chu and Goldstein (1973) also observed this effect in water and attributed it to thermals rising from the heated surface. Rossby (1969) and Sparrow et al. (1970) observed periodically rising thermals at the lower boundary.

Several other experimental measurements have been made, over wide ranges of the Rayleigh and Prandtl numbers. Hollands et al. (1975) correlated their own data and also those from other studies. For air and water the results are given for $Ra < 10^6$ as

$$\mathrm{Nu} = 1 + 1.44\left[1 - \frac{1708}{\mathrm{Ra}}\right] + \left[\left(\frac{\mathrm{Ra}}{5830}\right)^{1/3} - 1\right] \qquad \text{for air} \qquad (14.3.32a)$$

$$\mathrm{Nu} = 1 + 1.44\left[1 - \frac{1708}{\mathrm{Ra}}\right] + \left[\left(\frac{\mathrm{Ra}}{5830}\right)^{1/3} - 1\right] + 2\left(\frac{\mathrm{Ra}^{1/3}}{140}\right)\left(1 - \ln\frac{\mathrm{Ra}^{1/3}}{140}\right) \qquad \text{for water} \qquad (14.3.32b)$$

where the terms within square brackets are taken as zero if they have a negative value. The additional term for water is suggested by Hollands et al. (1975) to be due to the existence of thermals. For $Ra \leq 1708$, heat transfer across the fluid layer occurs by conduction and $Nu = 1$; see Chapter 13.

Effect of aspect ratio. This effect on the heat transfer has also been studied

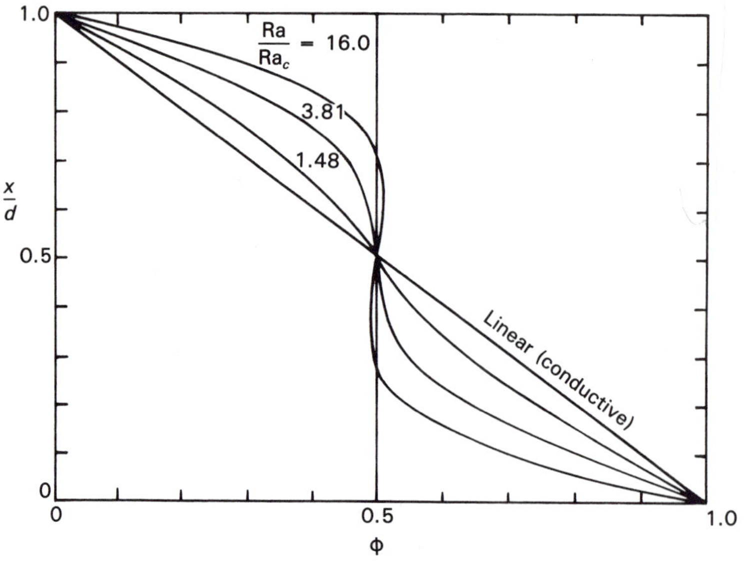

Figure 14.3.13 Mean temperature profiles in a horizontal fluid layer heated from below at various values of Ra/Ra_c. *(Reprinted with permission from J. Gille,* J. Fluid Mech., *vol. 30, p. 371. Copyright © 1967, Cambridge University Press.)*

experimentally. Catton and Edwards (1967) measured the effect of conducting and insulating sidewalls on natural convection between horizontal plates heated from below. The critical Rayleigh number was determined for the onset of convection. The resulting Nusselt number was also calculated. Figure 14.3.14 shows the regimes of transport, including the middle region, in which the sidewalls affect the heat transfer. The Nusselt number dependence on the aspect ratio and on the Rayleigh number was derived by Edwards and Catton (1969) by an integral method approach, as reviewed by Catton (1978).

Numerical results. Extensive numerical calculations have been made, largely using finite-difference methods to solve the governing time-dependent equations (14.3.26) and (14.3.27). Most results are in fairly good agreement with the experimental data, as outlined by Turner (1973). The numerical method allows consideration of wide ranges of the governing parameters and of boundary conditions. The availability of powerful computing techniques has resulted in a considerable amount of effort concerning enclosure transport. Only a few of the studies, related to horizontal cavities, are outlined here. Most of the numerical investigations solve for the time-dependent response, using various time-marching techniques, until steady state is obtained at large time. Interest is in the steady-state results. The initial condition is usually taken as zero flow and a step change is imposed in the boundary condition temperature at $\tau = 0$; see Jaluria and Torrance (1986). Some work has also been done specifically on the transient flow for various initial and boundary conditions, as discussed in Chapter 7 and also later in this chapter.

Deardorff (1964, 1965), Fromm (1965), and Cabelli and De Vahl Davis (1971)

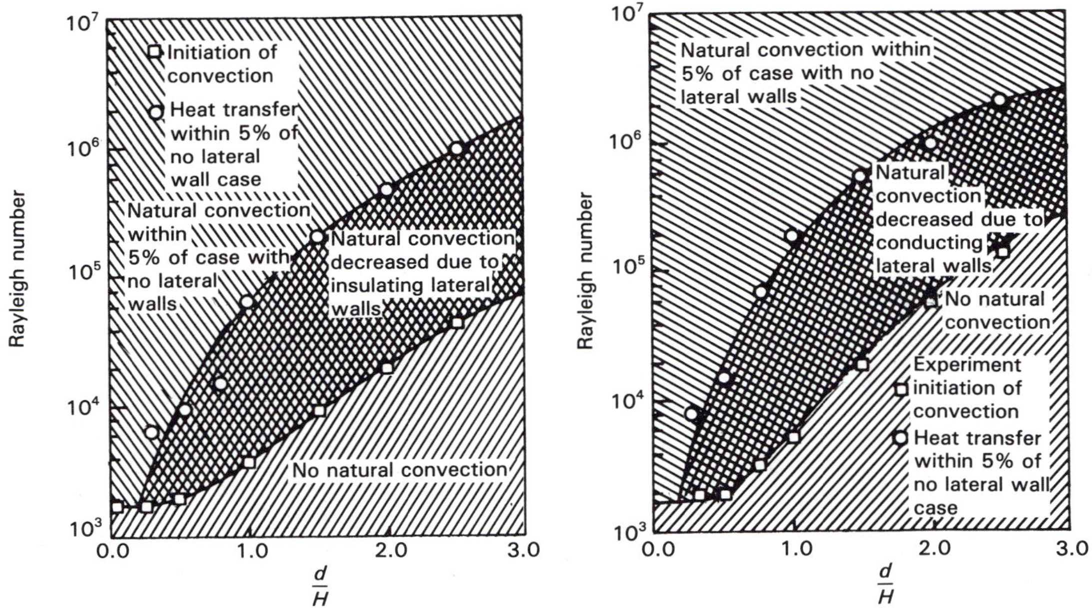

(a) (b)

Figure 14.3.14 Region of influence of the aspect ratio on heat transfer for (*a*) insulating lateral walls and (*b*) conducting lateral walls. *(Reprinted with permission from I. Catton and D. K. Edwards,* J. Heat Transfer, *vol. 89, p. 295. Copyright © 1967, ASME.)*

obtained numerical results for various boundary conditions and governing parameters. Figure 14.3.15 shows a steady-state cellular convection pattern, in terms of the streamlines and isotherms, at Pr = 0.71 and Ra = 6.75×10^5. The calculations of Deardorff (1965) were in close agreement with the heat flux measurements of Silveston (1958) and Globe and Dropkin (1959) for air. For large fluctuations in the flow, Herring (1963, 1964) carried out a numerical calculation of mean flow equations. The equations were obtained by averaging the governing equations in the horizontal direction and neglecting the nonlinear terms that describe the interaction between the fluctuating quantities. This study concerned the steady-state circumstance in the infinite Pr limit. At large Ra the heat transfer relationship obtained was $\mathrm{Nu} = 0.115\mathrm{Ra}^{1/3}$. Elder (1969) calculated the time development of the flow for Pr = 1, using Herring's model with a different finite-difference technique. The steady-state flow, as well as the temporal development of the flow at large Ra, was studied. A negative temperature gradient was found to arise in the central region, similar to the results of Gille (1967) discussed earlier.

Various other aspects have also been studied. Flow in a heat-generating fluid layer is of interest in geophysics, astrophysics, and nuclear power reactor safety and has been considered by several investigators, for example, Roberts (1967), Kulacki and Goldstein (1972), Kulacki and Emara (1977), and Emara and Kulacki (1980). Turbulent modeling of cavity flows has received some attention as outlined by Kaviany (1981). Three-dimensional flows have been considered in several studies, as discussed in Section 14.5.

Inclined enclosures heated at the bottom. Convective transport in inclined rectangular enclosures has also been studied experimentally and analytically, as reviewed in detail by Catton (1978). Much of the interest in such transport arises from the growing effort in both solar energy systems and thermal insulation for energy conservation. The inclination θ of the cavity with the vertical (see Fig. 14.3.11) is an additional parameter. For θ around 90°, the transport is expected to be similar to that in a horizontally heated cavity and for values around 0° to that in a vertical cavity. The Nusselt number in an inclined cavity can often be determined, to a reasonable approximation, from the Nusselt number relationships

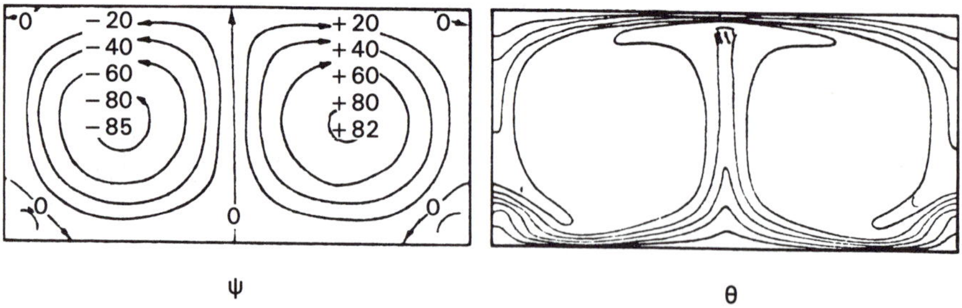

Figure 14.3.15 Calculated steady-state streamlines and isotherms for a horizontal layer of fluid at Pr = 0.71 and Ra = 6.75×10^5. *(Reprinted with permission from J. W. Deardorff,* J. Atmos. Sci., *vol. 21, p. 419. Copyright © 1964, American Meteorological Society.)*

for vertical and horizontal cavities by means of angular scaling laws, as described in detail by Raithby and Hollands (1985). In practice, the enclosure will have a finite width W. Then the horizontal aspect ratio A_H $(=W/d)$ is an additional parameter. It is of greater consequence for inclined than for vertical cavities, especially if the sides are open, because greater edge flows arise from the buoyancy component normal to the surfaces.

Hart (1971) carried out a detailed experimental study of flow and stability in a differentially heated inclined box. Ruth et al. (1980a, 1980b) also considered the flow instability in inclined air layers. Hollands and Konicek (1973) presented some heat transfer results. Several other workers have carried out heat transfer measurements for cavities with small aspect ratios, $A << 1$, at various inclinations. Reference may be made to the work of Ozoe et al. (1975), Hollands et al. (1976), Arnold et al. (1976), and Schinkel and Hoogendoorn (1978). Buchberg et al. (1976) reviewed much of the work on inclined enclosures and presented heat transfer correlations. Hollands et al. (1976), on the basis of several experimental studies, gave the following correlation for air in shallow cavities:

$$\mathrm{Nu} = 1 + 1.44\left[1 - \frac{1708}{\mathrm{Ra}\cos\xi}\right]\left\{1 - \frac{1708(\sin 1.8\xi)^{1.6}}{\mathrm{Ra}\cos\xi}\right\} + \left[\left(\frac{\mathrm{Ra}\cos\xi}{5830}\right)^{1/3} - 1\right] \quad \text{for } \frac{d}{H} \gtrsim 12 \quad \text{and} \quad 0 < \xi \leq \xi^* \qquad (14.3.33)$$

where $\mathrm{Ra} = g\beta(t_h - t_c)H^3/\nu\alpha$ and ξ is the inclination with the horizontal, that is, $\xi = 90° - \theta$. The quantities in square brackets are taken as zero if negative, and ξ^* is the limiting value for the validity of the correlation. This limit depends on the aspect ratio. For $d/H > 12$, ξ^* is given as 70°. For $d/H = 6$, Eq. (14.3.34), $\xi^* = 60°$. This correlation has been found to yield accurate heat transfer predictions.

For small values of d/H, Catton (1978) suggests the following correlation in terms of vertical and horizontal cavity Nusselt numbers, represented as Nu(90°) and Nu(0°), respectively:

$$\mathrm{Nu}(\xi) = \mathrm{Nu}(0°)\left[\frac{\mathrm{Nu}(90°)}{\mathrm{Nu}(0°)}\right]^{\xi/\xi^*} (\sin \xi^*)^{\xi/4\xi^*} \quad \text{for } \frac{d}{H} \lesssim 12 \quad \text{and} \quad 0 < \xi \leq \xi^* \qquad (14.3.34)$$

Beyond the critical angle of inclination ξ^*, Ayyaswamy and Catton (1973) and Arnold et al. (1975) recommend the following correlations, respectively:

$$\mathrm{Nu} = \mathrm{Nu}(90°)(\sin \xi)^{1/4} \qquad \text{for } \xi^* < \xi < 90° \qquad (14.3.35)$$

$$\mathrm{Nu} = 1 + [\mathrm{Nu}(90°) - 1]\sin\xi \qquad \text{for } \xi \geq 90° \qquad (14.3.36)$$

The various studies above have considered both the aspect ratio and the inclination. The effects of inclination are summarized in Fig. 14.3.16. It is seen that at given Ra the heat transfer increases with $\tilde{\theta}$, where $\tilde{\theta} = 180° - \xi = 90° + \theta$, until it reaches a maximum at $\tilde{\theta} = 90°$, that is, a vertical cavity. A local minimum then arises between $\tilde{\theta} = 90°$ and $180°$. This is the critical inclination ξ^*. For small $A \leq \frac{1}{12}$, this minimum is attained asymptotically in A at $\tilde{\theta} \approx 110°$. These trends agree qualitatively with the observations of Hart (1971) and Hollands et al. (1976).

The effect of a variation in the thermal conditions at the boundaries on natural convection in vertical and inclined air layers has been studied experimentally by

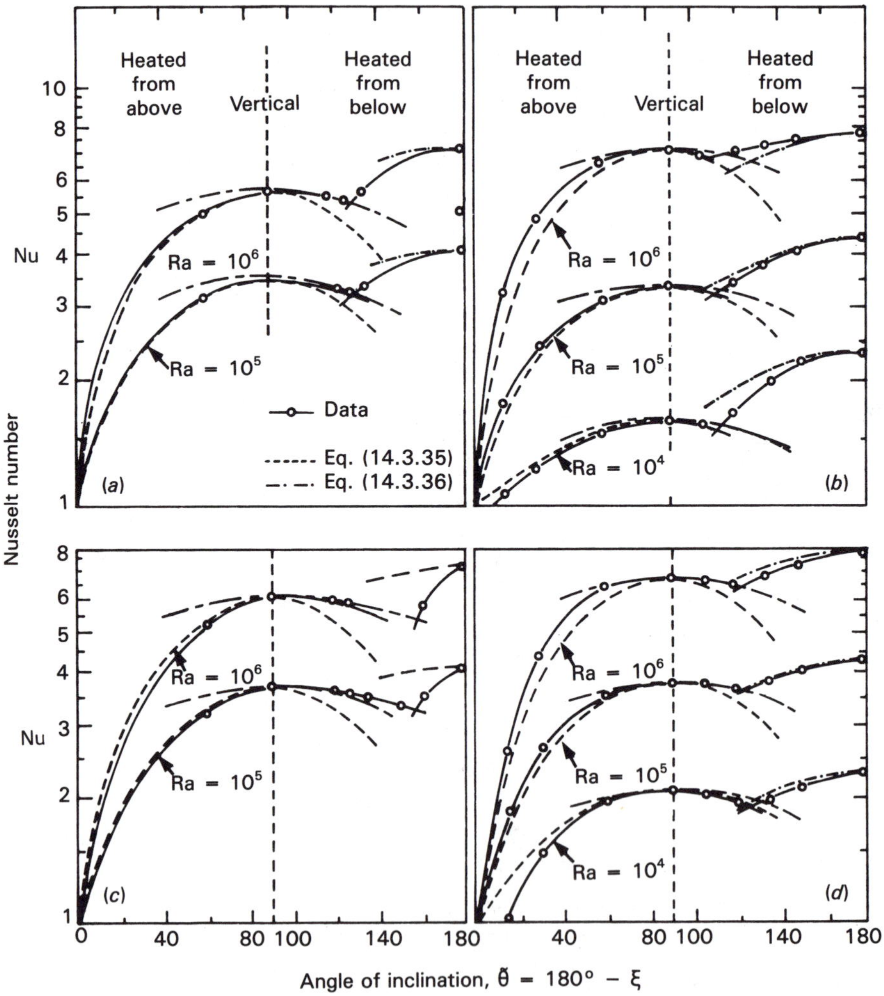

Figure 14.3.16 Effect of the angle of inclination on the Nusselt number for an inclined rectangular cavity. *(Reprinted with permission from J. N. Arnold et al.,* J. Heat Transfer, *vol. 98, p. 67. Copyright © 1976, ASME.)*

Elsherbiny et al. (1982). Catton (1978) has reviewed the work done on honeycomb structures heated from below, this geometry sometimes being of particular interest in solar collectors.

14.4 ANNULI AND OTHER GEOMETRIES

There has also been a considerable amount of study of transport in various other cavities, besides the two-dimensional rectangular enclosure and the parallel flat plate configurations considered in the preceding sections. Among these are cylindrical, spherical, and annular cavities. A few other geometries have also received some attention because of their importance in specific applications. Again, most studies concern two-dimensional flows, either plane or axisymmetric. The basic features of these flows and of the methods used for their study are similar to those pertaining to rectangular enclosures and may therefore be presented in terms of the earlier discussion.

14.4.1 Cylindrical Cavity

Transport in the fluid inside a horizontal circular cylinder with differential surface heating has been studied extensively, analytically as well as experimentally. In addition to the nature and angular location of the imposed wall conditions, there are only two governing parameters, Ra and Pr. There are no corner effects. Surface heating phase angles may specify the locations of various kinds of imposed surface temperature conditions. A review of such studies was given by Ostroumov (1952) and Ostrach (1972).

Lewis (1950) considered a cosine wall temperature variation for Rayleigh numbers Ra, based on the radius R of the cylinder and the maximum temperature difference Δt across the cylindrical region, of less than unity. A perturbation analysis indicated that conduction was the dominant mode of heat transfer. Ostrach (1950) considered large Rayleigh number transport for this configuration, for $\mathrm{Pr} = O(1)$, assuming an isothermal core of uniform vorticity throughout. Martini and Churchill (1960) made measurements in air in a horizontal cylinder whose two halves, on either side of a vertical plane, were maintained at different uniform temperatures. The Rayleigh number was varied from approximately 2×10^5 to 8×10^6. The interior region was found to be relatively stagnant and vertically stratified. The numerical results obtained by Hellums and Churchill (1962) were in good agreement with these experimental findings.

Consider the periodic surface temperature variation on the inside of a long cylindrical cavity as shown in Fig. 14.4.1. It is written as

$$t(\gamma) - t_0 = \Delta t \cos (\gamma + \gamma_0) \qquad \text{at } r = R \tag{14.4.1}$$

where R is the radius, γ the angular position, γ_0 the phase angle from horizontal at which the imposed temperature maximum appears, t_0 the average value of $t(\gamma)$, and Δt the amplitude of the variation. The governing equations in dimensionless form are

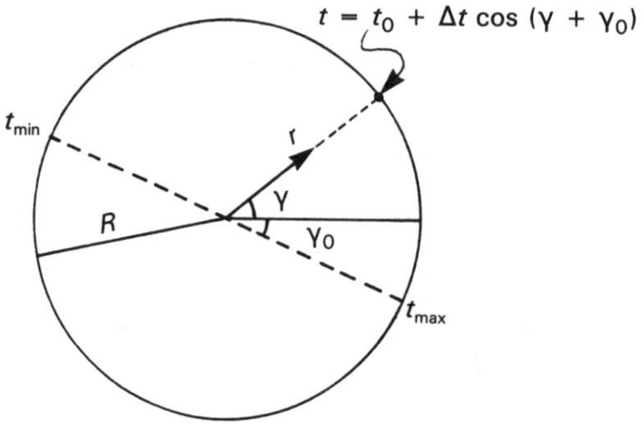

Figure 14.4.1 Coordinate system for natural convection flow inside a horizontal cylindrical enclosure with a periodic surface temperature variation.

$$\frac{1}{r}\left(\frac{\partial\psi}{\partial\gamma}\frac{\partial}{\partial r} - \frac{\partial\psi}{\partial r}\frac{\partial}{\partial\gamma}\right)\nabla^2\psi = \frac{1}{\mathrm{Gr}_R}\nabla^2\nabla^2\psi - \frac{1}{\mathrm{Gr}_R}\left(\cos\gamma\frac{\partial\phi}{\partial r} - \frac{\sin\gamma}{r}\frac{\partial\phi}{\partial\gamma}\right) \tag{14.4.2}$$

$$u\frac{\partial\phi}{\partial r} + \frac{v}{r}\frac{\partial\phi}{\partial\gamma} = \frac{1}{\mathrm{PrGr}_R}\nabla^2\phi \tag{14.4.3}$$

where the dimensionless stream function ψ, distance r, and temperature ϕ are given in terms of the respective physical quantities ψ', r', and t as

$$\psi = \frac{\psi'}{\nu\mathrm{Gr}_R} \qquad r = \frac{r'}{R} \qquad \phi = \frac{t - t_0}{\Delta t}$$

with

$$\mathrm{Gr}_R = \frac{g\beta\,\Delta t R^3}{\nu^2} \qquad u = \frac{1}{r}\frac{\partial\psi}{\partial\gamma} \qquad v = -\frac{\partial\psi}{\partial r} \tag{14.4.4}$$

The corresponding boundary conditions are

$$\text{at } r = 1: \quad \frac{\partial\psi}{\partial r} = \frac{\partial\psi}{\partial\gamma} = 0 \quad \text{and} \quad \phi = \cos(\gamma + \gamma_0) \tag{14.4.5}$$

This circumstance was analyzed by Weinbaum (1964) for large Ra in terms of large Gr and Pr $= O(1)$. By use of a linearization scheme known as the modified Oseen technique, as developed by Lewis and Carrier (1949), the governing equations were made linear and the energy equation was thereby decoupled. Assuming the core fluid region to be isothermal, results were obtained for heating on the bottom half, that is, $\gamma_0 = 90°$. Uniform vorticity was found in the core. For lateral heating instead, $\gamma_0 = 0°$, a stagnant core region arose and qualitative agreement was found with the experimental results of Martini and Churchill (1960).

Ostrach and Menold (1965) used a similar technique and considered large Prandtl number and unit-order Grashof number. Near the cylindrical surfaces the temperature gradients must be large, since the conduction terms in Eq. (14.4.3) are multiplied by a small quantity $1/\mathrm{PrGr}_R$ to keep the conduction and convection terms comparable. In the core region the conduction terms are expected to be negligible compared to the convection terms. For large Pr the nonlinear inertia terms become negligible and the equation of motion becomes linear.

Menold and Ostrach (1965) found that a rotating isothermal core was a good approximation if the heating phase angle γ_0 was not close to zero. At $\gamma_0 = 0°$, which amounts to side heating, the core was found to be relatively stagnant and stratified. Solutions for various γ_0 were obtained by using the modified Oseen linearization method. The stream function ψ and temperature ϕ are each taken as a combination of two variables, one for the core region and the other for the boundary layer region. Then ψ and ϕ are to satisfy the governing equations for the two regions. The total functions must satisfy the boundary conditions at the cylinder wall. Denoting the two regions by subscripts 0 and 1, the functions used are

$$\psi(r, \gamma) = \psi_0(r, \gamma) + \psi_1(r, \gamma) \tag{14.4.6}$$

$$\phi(r, \gamma) = \phi_0(r, \gamma) + \phi_1(r, \gamma) \tag{14.4.7}$$

Here ϕ_0 is a constant since the rotating core is taken as isothermal. This implies that the buoyancy terms in the core region are zero.

Considerable analytical and experimental work has been done for this configuration, as reviewed by Ostrach (1972). Figure 14.4.2 shows the streamline pattern for three heating phase angles, $\gamma_0 = 45$, 22.5, and 0°. The first result is typical of the general pattern obtained over the range from about 30 to 90°. The streamlines are closed and elliptic in shape with centers near the origin. With decreasing γ_0 the pattern shifts counterclockwise. By $\gamma_0 = 22.5°$ the pattern has split into two regions with vortices. The core region, where the isothermal assumption has been made on the basis of an argument similar to that of Batchelor (1956) and applied by Menold and Ostrach (1965), becomes smaller. For $\gamma_0 = 0°$, which corresponds to lateral heat input, the flow pattern is quite irregular. It was inferred that the velocity and temperature solutions are not compatible with each other at $\gamma_0 = 0°$ and that an approach different from that outlined above would be needed.

For phase angle γ_0 near zero, Hartman and Ostrach (1969) analytically studied the circumstance of a relatively stagnant, stratified core. The experimental results of Sabzevari and Ostrach (1966) and of Brooks and Ostrach (1970) agreed with the analysis for various phase angles. The temperature and velocity profiles were measured, along with the streamline pattern, to define the cellular motion inside the cylinder.

Although the calculated motion for bottom heating, $\gamma_0 = 90°$, is similar in form to that found for 45°, experimental results have indicated axial flow in a cylinder for $\gamma_0 > 60°$. Thus, a two-dimensional model is not valid for large phase angles. Ostrach and Pneuli (1963) considered the thermal instability for $\gamma_0 = 90°$. They showed that the flow always starts in the direction of the largest dimension in the body of fluid, the axial direction in this study. Therefore, three-dimensional analysis is needed.

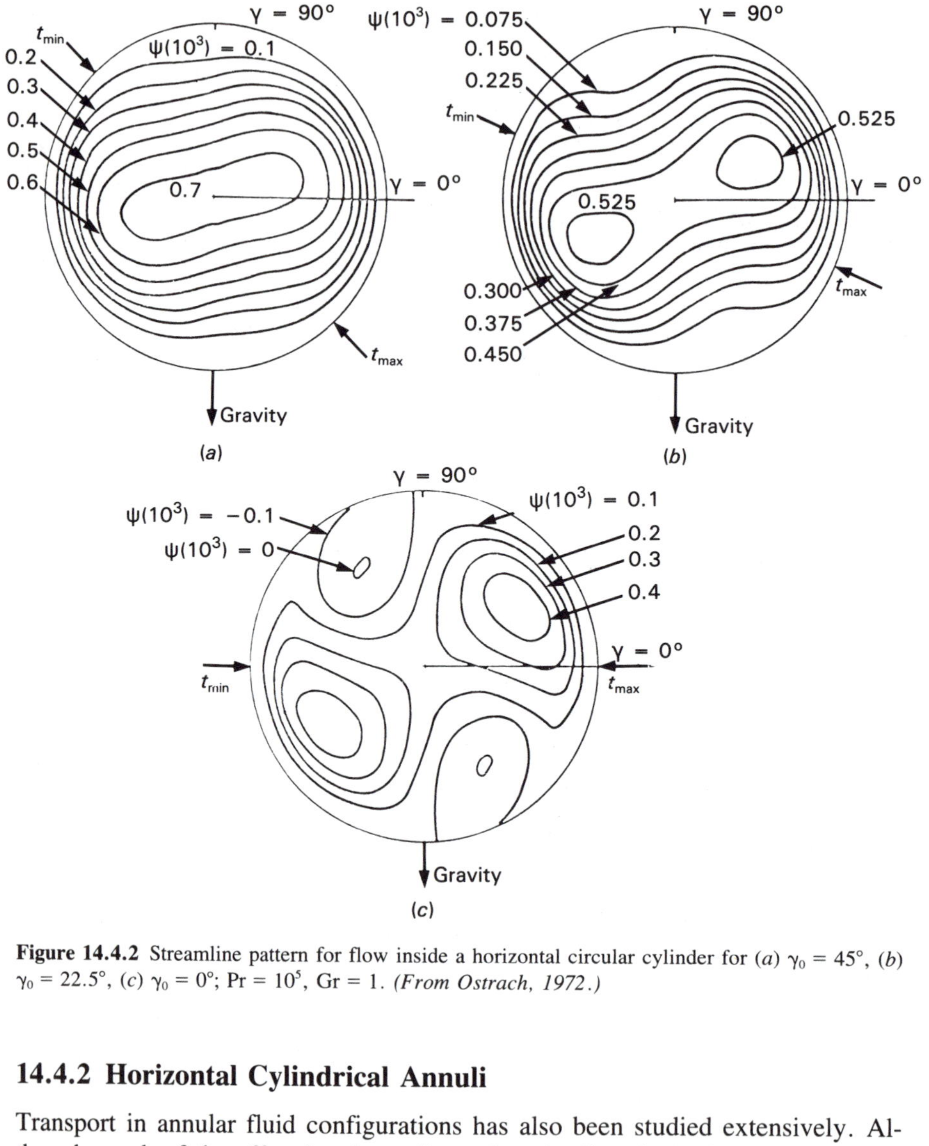

Figure 14.4.2 Streamline pattern for flow inside a horizontal circular cylinder for (*a*) $\gamma_0 = 45°$, (*b*) $\gamma_0 = 22.5°$, (*c*) $\gamma_0 = 0°$; Pr = 10^5, Gr = 1. *(From Ostrach, 1972.)*

14.4.2 Horizontal Cylindrical Annuli

Transport in annular fluid configurations has also been studied extensively. Although much of the effort has been directed at the flow in horizontal concentric cylindrical annuli, work has also been done on eccentric and vertical annuli as well as spherical annuli. Industrial systems in which these configurations are of interest include nuclear reactors, electronic equipment, transmission cables, thermal storage systems involving phase change around heating pipes, and concentrating solar collector receivers.

Transport between horizontal concentric isothermal cylindrical surfaces was first measured for various gases by Beckmann (1931). The overall heat transfer coefficients were determined. Kraussold (1934) and Liu et al. (1961) measured the effect of the Prandtl number on the overall heat transfer coefficient for various fluids. Eckert and Soehngen (1948) and Grigull and Hauf (1966) used interferometric techniques to obtain local heat transfer coefficients. Lis (1966) investigated the turbulent flow regime and measured overall heat transfer rates for Rayleigh numbers, based on gap thickness d, up to 10^{10}. Flow patterns in air have been observed by Grigull and Hauf (1966), Bishop and Carley (1966), Bishop et al. (1968), and Powe et al. (1969). The various flow regimes that arise, depending on the Grashof number and the diameter ratio, were delineated by Powe et al. (1969). Kuehn and Goldstein (1976a) carried out a detailed experimental and numerical study of annular transport. A Mach-Zehnder interferometer gave both temperature distributions and local heat transfer coefficients for air and water. The Rayleigh number, based on d, was varied from 2.11×10^4 to 9.76×10^5.

Analytical solutions valid at small Rayleigh numbers have been obtained by Mack and Bishop (1968), Hodnett (1973), and Huetz and Petit (1974). Raithby and Hollands (1975) considered the high Rayleigh number boundary layer limit, using a conductive layer model. Jischke and Farshchi (1980) used the boundary layer theory to analyze laminar flow. Several numerical studies of the flow have been carried out. Numerical solutions were obtained by finite-difference methods by Crawford and Lemlich (1962), Abbott (1964), Powe et al. (1971), Kuehn and Goldstein (1976a), and Ingham (1981). Turbulent transport in a horizontal cylindrical annulus has been modeled by Farouk and Guceri (1981), employing the K-ϵ turbulence model.

The governing equations for the laminar two-dimensional flow between horizontal concentric cylinders are similar to those for flow in a horizontal cylinder, Eqs. (14.4.2) and (14.4.3). However, a different nondimensionalization has been used in most studies. These equations, in terms of the stream function ψ and vorticity ω, after eliminating the pressure, and in temperature ϕ, are

$$\nabla^2\psi = -\omega \tag{14.4.8}$$

$$\nabla^2\omega = \frac{1}{\mathrm{Pr}}\left[u\frac{\partial\omega}{\partial r} + \frac{v}{r}\frac{\partial\omega}{\partial\theta}\right] + \mathrm{Ra}\left[\sin\theta\,\frac{\partial\phi}{\partial r} + \frac{1}{r}\cos\theta\,\frac{\partial\phi}{\partial\theta}\right] \tag{14.4.9}$$

$$\nabla^2\phi = u\frac{\partial\phi}{\partial r} + \frac{v}{r}\frac{\partial\phi}{\partial\theta} \tag{14.4.10}$$

where

$$\nabla^2 = \frac{\partial^2}{\partial r^2} + \frac{1}{r}\frac{\partial}{\partial r} + \frac{1}{r^2}\frac{\partial^2}{\partial\theta^2} \qquad u = \frac{1}{r}\frac{\partial\psi}{\partial\theta} \qquad v = -\frac{\partial\psi}{\partial r}$$

The nondimensionalization employed is given in terms of the physical variables, denoted by primes, as

$$\psi = \frac{\psi'}{\alpha} \qquad r = \frac{r'}{d} \qquad \phi = \frac{t - t_o}{t_i - t_o} \qquad u = \frac{u'd}{\alpha} \qquad v = \frac{v'd}{\alpha} \tag{14.4.11}$$

with $u' = (1/r')\partial\psi'/\partial\theta$, $v' = -\partial\psi'/\partial r'$, and $\mathrm{Ra} = g\beta(t_i - t_o)d^3/\nu\alpha$. Here d is the gap between the cylinders, t_o the outer cylinder temperature, t_i the inner cylinder temperature, r' the radial distance measured from the center of the system, θ the angle measured clockwise from the central vertical axis, with $\theta = 0°$ at the top, u' the radial velocity component, v' the circumferential velocity component, and Ra the Rayleigh number based on the gap thickness and the temperature difference between the two surfaces. The gap d is given by $d = R_o - R_i$, where R_o is the radius of the outer cylinder and R_i that of the inner cylinder. The corresponding boundary conditions are

$$\text{at } \theta = 0, \pi: \qquad \psi = \omega = \frac{\partial \phi}{\partial \theta} = 0$$

$$\text{at } r = \frac{R_i}{d} = H_i: \qquad \psi = 0 \qquad \omega = -\frac{\partial^2 \psi}{\partial r^2} \qquad \phi = 1 \tag{14.4.12}$$

$$\text{at } r = \frac{R_o}{d} = H_o: \qquad \psi = 0 \qquad \omega = -\frac{\partial^2 \psi}{\partial r^2} \qquad \phi = 0$$

The conditions at $\theta = 0, \pi$ arise from symmetry about the vertical plane through the center. Those on the surfaces are from the conditions at two impermeable isothermal walls at constant radii. The parameters are now Ra, Pr, H_i, and H_o.

Several flow regimes were found in the calculations of Kuehn and Goldstein (1976a). For Ra < 100 the flow was symmetric about the horizontal axis and the temperature distribution in the annulus was not affected by the flow. Thus, the isotherms were closely circular and heat transfer was mainly by conduction. As Ra and the flow increase, the isotherms resemble eccentric circles, as shown on the right in Fig. 14.4.3 for Ra = 10^3. Heat transfer remains essentially that due to conduction. At yet larger Ra, a separation of the inner and outer thermal boundary layers occurs. The center of flow rotation, as indicated by the maximum value of the stream function, moves up; compare Figs. 14.4.3 and 14.4.4. The vorticity

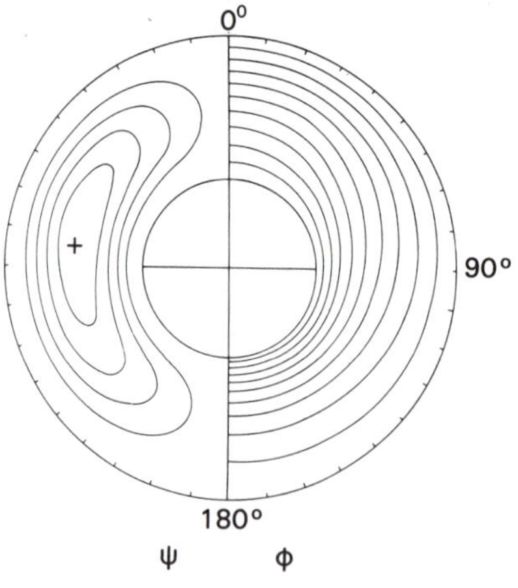

Figure 14.4.3 Streamlines and isotherms for flow within a concentric horizontal cylindrical annulus for Ra = 10^3, Pr = 0.7, $d/D_i = 0.8$, $\Delta\psi = 0.5$, and $\Delta\phi = 0.1$. *(Reprinted with permission from T. H. Kuehn and R. J. Goldstein,* J. Fluid Mech., *vol. 74, p. 695. Copyright © 1976, Cambridge University Press.)*

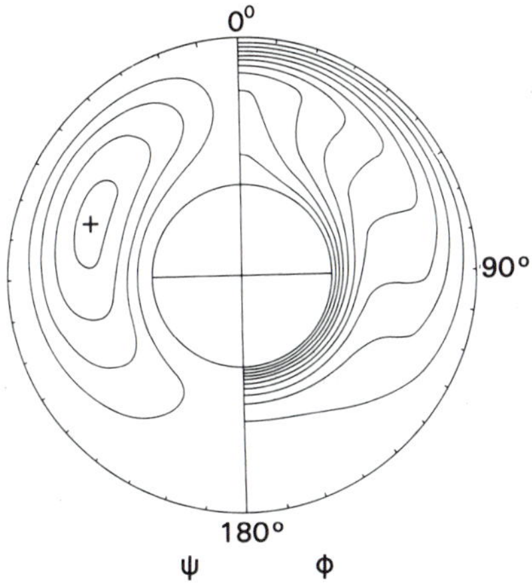

Figure 14.4.4 Streamlines and isotherms for flow within a concentric horizontal cylindrical annulus for $Ra = 10^4$, $Pr = 0.7$, $d/D_i = 0.8$, $\Delta\psi = 3.0$, and $\Delta\phi = 0.1$. *(Reprinted with permission from T. H. Kuehn and R. J. Goldstein,* J. Fluid Mech., *vol. 74, p. 695. Copyright © 1976, Cambridge University Press.)*

across the central core of each of these two circulations was found to be constant, as predicted by Batchelor (1954) for a vertical slot.

This transition regime, between the conduction and boundary layer regimes, was found experimentally by Kuehn and Goldstein (1976a) to span $10^2 < Ra < 3 \times 10^4$, for $Pr = 0.7$ and $d/D_i = 0.8$, where D_i is the diameter of the inner cylinder. A steady laminar boundary layer regime was found to arise in the range $3 \times 10^4 < Ra < 10^5$. Boundary layers arise on both cylinders. The lower portion of the annulus is essentially stagnant cold fluid, as shown in Fig. 14.4.5. At increasing Ra the vorticity approaches zero in the region away from the walls, indicating the beginning of a stationary core region, as discussed earlier for a vertical slot. As Ra is increased further, turbulence is inferred from temperature measurements to arise in the boundary layer on the upper cylinder while the inner boundary layer remains laminar. Eventually, with further increasing Ra, the inner boundary layer also becomes turbulent. Similar results were obtained in other

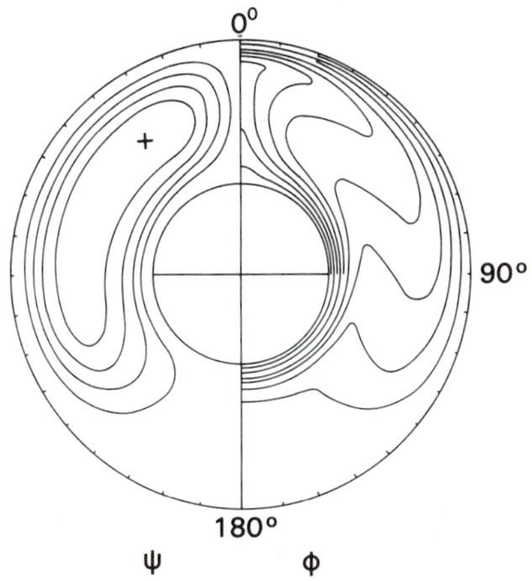

Figure 14.4.5 Streamlines and isotherms for flow within a concentric horizontal cylindrical annulus for $Ra = 5 \times 10^4$, $Pr = 0.7$, $d/D_i = 0.8$, $\Delta\psi = 5.0$, and $\Delta\phi = 0.1$. *(Reprinted with permission from T. H. Kuehn and R. J. Goldstein,* J. Fluid Mech., *vol. 74, p. 695. Copyright © 1976, Cambridge University Press.)*

calculations referred to above. The streamlines and isotherms calculated by Farouk and Guceri (1981) for turbulent flow at Ra = 10^7, using the K-ϵ model (refer to Section 11.7) in air, are shown in Fig. 14.4.6. The patterns are similar to those observed at lower Ra, shown in Fig. 14.4.5.

Eccentric annuli. Experimental and theoretical work has been done on transport in fluids in eccentric cylindrical annuli. Zagromov and Lyalikov (1966) measured overall heat transfer coefficients in air for various vertical and horizontal eccentricities of the inner cylinder. Kuehn and Goldstein (1978) made measurements of heat transfer in air and water in concentric and eccentric horizontal annuli. These results indicated substantial dependence of the local heat transfer rates on eccentricity for both surfaces. However, the change in the overall heat transfer coefficient from that for the comparable concentric arrangement is less than 10% for eccentricities $\epsilon/d \leq \frac{2}{3}$, where ϵ is the distance the inner cylinder is moved from a concentric position and $d = R_o = R_i$.

Analytical and numerical studies have been carried out by Ratzel et al. (1979), using a finite-element method to predict isotherms, flow patterns, and overall heat transfer up to Ra $\approx$ 22,000. Yao (1980) used perturbation techniques for small eccentricities. Cho et al. (1982) and Pepper and Cooper (1982) employed finite-difference methods. Projahn et al. (1981) obtained numerical solutions in the Ra

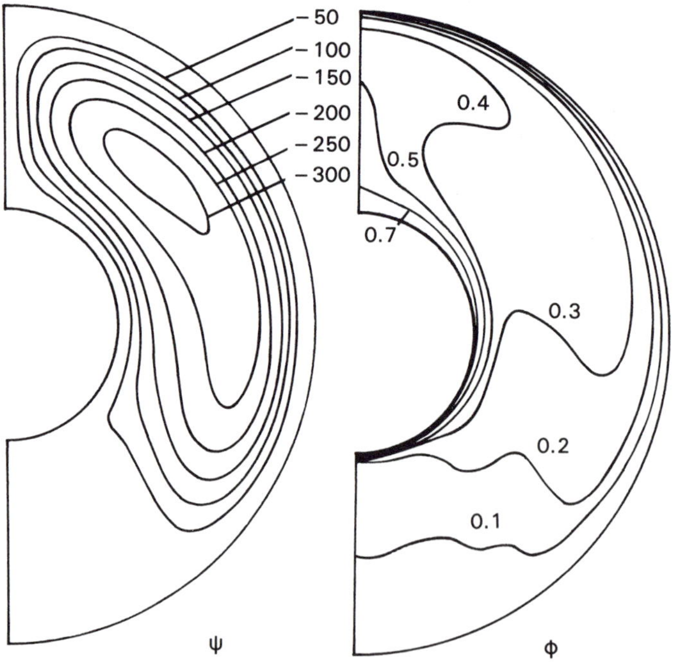

Figure 14.4.6 Streamlines and isotherms for turbulent natural convection flow between two concentric horizontal cylinders for Ra = 10^7, Pr = 0.721, and d/D_i = 0.8. *(Reprinted from B. Farouk and S. I. Guceri. Copyright © 1981, ASME.)*

range 10^2–10^5 for a Prandtl number of 0.7, considering positive and negative vertical and also horizontal eccentricities. Figure 14.4.7 shows typical results for a vertical eccentricity. The three regimes obtained earlier by Kuehn and Goldstein (1976a) were delineated. It was found that a downward vertical displacement of the inner cylinder leads to an increase of heat transfer. A similar configuration arises in melting around a heated horizontal cylinder, as considered by Yao and Chen (1980).

Heat transfer. Most of the studies mentioned above include heat transfer results, and correlations have been presented. Results are in the form of an equivalent conductivity k_{eq}. This is the ratio of the actual heat flow to that due to conduction alone across the region. It characterizes the effects of various parameters, such as Ra, d/D_i, and ϵ/d, on the heat transfer. For concentric cylinders, k_{eq} based on the respective inside and outside surface areas is obtained as

$$(k_{eq})_i = \frac{\text{Nu}_i}{\text{Nu}_{cond}} = \frac{h_i D_i}{2k} \ln\left(\frac{D_o}{D_i}\right) \tag{14.4.13}$$

$$(k_{eq})_o = \frac{\text{Nu}_o}{\text{Nu}_{cond}} = \frac{h_o D_o}{2k} \ln\left(\frac{D_o}{D_i}\right) \tag{14.4.14}$$

where

$$\text{Nu}_{cond} = \frac{2}{\ln(D_o/D_i)} \qquad d = \frac{D_o - D_i}{2} \tag{14.4.15}$$

The total energy lost by one cylinder equals that gained by the other. The subscript i refers to the inner cylinder and o to the outer one, and Nu_{cond} is the Nusselt number for pure conduction between concentric cylinders.

The experimentally obtained local heat transfer coefficients around the annulus for concentricity and for two vertical eccentricities at $\text{Ra} = 5 \times 10^4$ and $\text{Pr} = 0.706$ are compared in Fig. 14.4.8. The values at the top, $\theta = 0°$, deviate

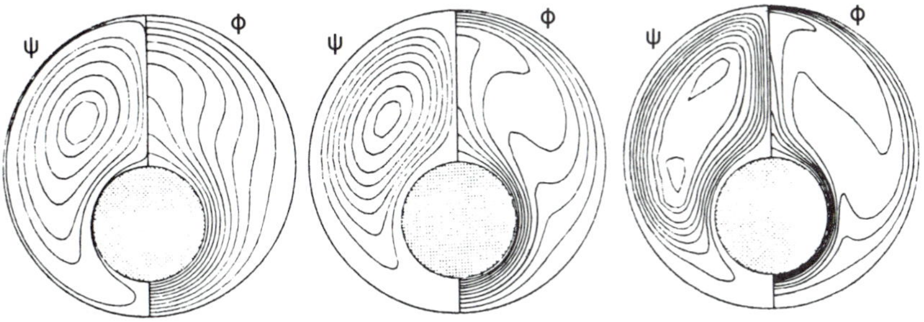

Figure 14.4.7 Streamlines and isotherms for flow in a vertically eccentric horizontal cylindrical annulus for $\text{Pr} = 0.7$, $d/D_i = 0.8$, $\epsilon/d = -0.625$, and Ra = (*a*) 10^3, (*b*) 10^4, (*c*) 10^5. *(Reprinted with permission from V. Projahn et al.,* Numer. Heat Transfer, *vol. 4, p. 131. Copyright © 1981, Hemisphere Publishing Corporation.)*

considerably from the concentric geometry when the inner cylinder is above the center. A similar effect arises at the bottom when it is moved down. When the cylinders are very close, due to eccentricity, the local conduction effect will be large and will increase the overall heat transfer considerably above that for concentric conduction. Zagromov and Lyalikov (1966) found this to occur when $\epsilon/d > 0.9$.

As seen in Fig. 14.4.8, the inner cylinder local heat transfer coefficient h_i varies with θ about as would be expected for an isolated cylinder in an extensive medium. The boundary layer develops from its bottom and is shed as a plume at the top. This plume hits the outer cylinder, causing the outer cylinder heat transfer coefficient h_o to be much higher than that for the inner one at the top. As the flow moves down along the outer cylinder surface, h_o decreases very rapidly.

Heat transfer correlations have been obtained by Kuehn and Goldstein (1976b, 1978). The correlation is expected to apply in the conduction, laminar, and turbulent flow regimes. The inner cylinder approaches the heat transfer correlation for a free horizontal cylinder as $D_o \to \infty$. The flow approaches that inside a horizontal cylinder as $D_i \to 0$. For Pr = 0.7 the equations are

$$\mathrm{Nu}_i = \frac{2}{\ln\{1 + 2/[(0.5\,\mathrm{Ra}_{D_i}^{1/4})^{15} + (0.12\,\mathrm{Ra}_{D_i}^{1/3})^{15}]^{1/15}\}}$$

$$\mathrm{Nu}_o = -\frac{2}{\ln\{1 - 2/[(\mathrm{Ra}_{D_o}^{1/4})^{15} + (0.12\,\mathrm{Ra}_{D_o}^{1/3})^{15}]^{1/15}\}}$$

$$\phi_b = \frac{\mathrm{Nu}_i}{\mathrm{Nu}_i + \mathrm{Nu}_o} \tag{14.4.16}$$

$$\mathrm{Nu}_{\mathrm{conv}} = \left(\frac{1}{\mathrm{Nu}_i} + \frac{1}{\mathrm{Nu}_o}\right)^{-1} \qquad \mathrm{Nu}_{\mathrm{cond}} = \frac{2}{\ln(D_o/D_i)}$$

$$\mathrm{Nu} = [(\mathrm{Nu}_{\mathrm{cond}})^{15} + (\mathrm{Nu}_{\mathrm{conv}})^{15}]^{1/15} \qquad k_{\mathrm{eq}} = \frac{\mathrm{Nu}}{\mathrm{Nu}_{\mathrm{cond}}}$$

where the Nusselt numbers are averaged values for the overall heat transfer, Ra_{D_i} is the Rayleigh number based on D_i and Ra_{D_o} that based on D_o, and ϕ_b is the spatial average dimensionless fluid temperature between the inner and outer cylinder boundary layers. This correlation was found to fit all the available overall data with a standard deviation of 4.5%. When only the data taken for Ra = $g\beta(t_i - t_o)d^3/\nu\alpha \geq 5 \times 10^3$ were considered, the standard deviation was only 1.7%.

Kuehn and Goldstein (1980) considered the effect of the Prandtl number and diameter ratio over the ranges $0.001 \leq \mathrm{Pr} \leq 1000$ and $1.0 \leq D_o/D_i \leq \infty$. The heat transfer coefficient was close to the conduction limit for $\mathrm{Pr} \to 0$. For $\mathrm{Pr} \geq 1.0$ the temperature profiles were almost independent of Pr. For low Pr, numerical results are given by Custer and Shaughnessy (1977). The flow in the annulus between horizontal elliptic cylinders was studied numerically by Lee and Lee (1981).

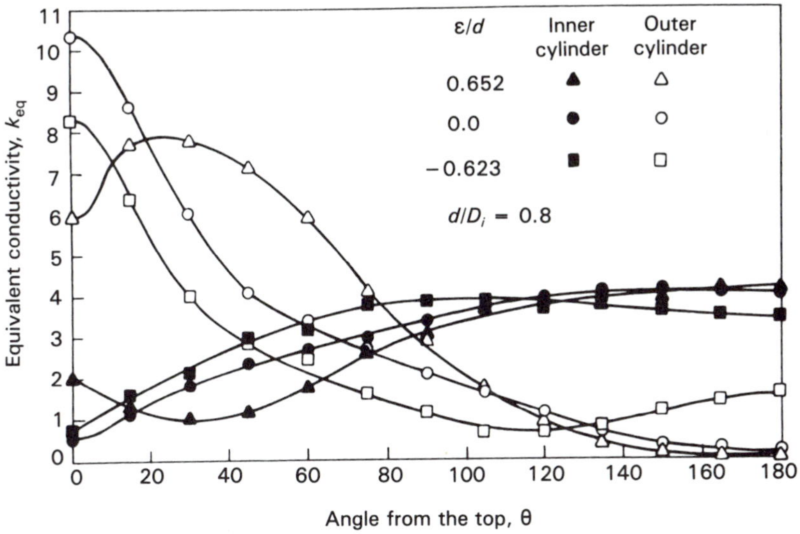

Figure 14.4.8 Comparison of measured heat transfer coefficients for concentric and vertically eccentric horizontal cylindrical annuli at Ra = 5×10^4, Pr = 0.706. *(Reprinted with permission from T. H. Kuehn and R. J. Goldstein,* J. Heat Transfer, *vol. 100, p. 635. Copyright © 1978, ASME.)*

Boyd (1981) obtained a unified theory to correlate steady laminar natural convection data for horizontal annuli.

14.4.3 Other Geometries

Work on several other enclosure geometries is mentioned here. Fluid transport in concentric spherical annuli has been studied by Bishop et al. (1964, 1966), Mack and Hardee (1968), Scanlan et al. (1970), Weber et al. (1973), Astill et al. (1980), and Ingham (1981). Experimental and numerical results for flow and heat transfer have been presented. Scanlan et al. (1970) gave the following correlation for heat transfer between concentric spheres:

$$\frac{k_{eq}}{k} = 0.202\text{Ra}^{0.228}\left(\frac{2d}{D_i}\right)^{0.252}\text{Pr}^{0.029} \qquad \text{with Ra} = \frac{g\beta(t_i - t_o)d^3}{\nu\alpha} \tag{14.4.17}$$

where d is the gap thickness, D_i the inner cylinder diameter, Ra the Rayleigh number, t_i the surface temperature of the inner sphere, and t_o the temperature of the outer sphere. The equivalent thermal conductivity k_{eq} is given in terms of the overall heat transfer rate Q as

$$k_{eq} = \frac{Qd}{2\pi(t_i - t_o)D_i D_o} \tag{14.4.18}$$

The general trends of transport are similar to those outlined above for horizontal cylindrical annuli.

Some work has also been done on vertical cylindrical annuli. De Vahl Davis and Thomas (1969) and Thomas and De Vahl Davis (1970) presented numerical results for isothermal surface conditions. Sheriff (1966) and Nagendra et al. (1970) reported experimental data. Schwab and DeWitt (1970) solved the problem numerically and obtained results qualitatively similar to those for vertical rectangular enclosures. For $\mathrm{Ra} > 5 \times 10^3$, where Ra is based on the gap thickness d, a fully developed boundary layer flow was found to exist in the cavity. Keyhani et al. (1983) presented heat transfer measurements in air and helium for $10^3 < \mathrm{Ra} < 2.3 \times 10^6$, with the inner wall at constant heat flux and the outer wall isothermal. Charmchi and Sparrow (1982) and Sparrow and Charmchi (1983) studied the transport in concentric and eccentric cylindrical annuli of different heights both experimentally and numerically. Choi and Korpela (1980) studied flow stability in a vertical cylindrical annulus, using the linear theory. The effect of Pr on the onset of instability was determined and the predicted mode of instability was verified experimentally for air.

Some work has been done on transient convection inside spheres. Drakhlin (1952), Shaidurov (1958), Pustovoit (1958), Whitley and Vachon (1972), and Chow and Akins (1975) presented experimental and numerical results for the flow and heat transfer. The last study gave the following heat transfer correlation for convection in water inside an isothermal sphere:

$$\mathrm{Nu}_D = \frac{hD}{k} = 0.8\mathrm{Ra}_D^{0.3} \qquad \text{for } 7.74 \times 10^4 < \mathrm{Ra}_D < 1.13 \times 10^7 \qquad (14.4.19)$$

where Ra_D is also based on the diameter of the sphere. In the experiment a pseudo-steady state was maintained by keeping the temperature difference across the sphere, between the outside and the center, constant.

Lyican et al. (1980a, 1980b) computed the flow pattern as well as the heat transfer in a trapezoidal enclosure. Flack et al. (1979) and Flack (1980) measured heat transfer rates in air contained in triangular enclosures for various thermal boundary conditions. Akinsete and Coleman (1982) also obtained finite-difference results for this flow, and Poulikakos and Bejan (1983) presented results for air and water. Van Doormal et al. (1981) considered nonrectangular enclosures, such as those bounded by vee-corrugated surfaces. A numerical technique was discussed for studying this configuration. Results were given for the flow and thermal fields, along with heat transfer predictions, for natural convection in air layers over vee-corrugated sheets.

14.5 THREE–DIMENSIONAL INTERNAL FLOWS

Most of the information available from calculations of buoyancy-driven transport in enclosures is related to two-dimensional plane and axisymmetric flows. These often only roughly approximate the three-dimensional fluid motions that arise in the physical world. Considerable mathematical simplification is obtained by neglecting the variation in the third direction. Experimental work has also largely been designed to simulate two-dimensional configurations. However, all studies other than those involving vertically axisymmetric geometries have inevitably been carried out for geometries in which three-dimensional effects may arise.

Many studies have concerned long cavities. Transport is studied far from the ends, where the effects of the end walls may not appear in the measurements. Flow visualization methods, on the other hand, often average out the variations across the third direction. The two-dimensional point of view has led to an understanding of many of the basic processes that govern internal flows. However, it is also often necessary to determine three-dimensional effects. In flows resulting from fires in rooms, for example, a two-dimensional approximation does not represent the important mechanisms.

There have been relatively few theoretical studies and experiments concerning three-dimensional transport. However, general numerical techniques for solving the three-dimensional Navier-Stokes equations have been developed. For internal buoyancy-driven flows, Aziz (1965), Aziz and Hellums (1967), Chorin (1968), Williams (1969), and Mallinson and De Vahl Davis (1973) have discussed suitable numerical methods. The Galerkin method was used by Davis (1967), Catton (1970, 1972), and others. As pointed out by Catton (1978), the Galerkin method is often particularly attractive for obtaining three-dimensional solutions. Numerical results for three-dimensional internal flows have also been presented by De Vahl Davis and Mallinson (1976), Ozoe et al. (1976), Mallinson and De Vahl Davis (1977), and Oertel (1980).

14.5.1 Basic Considerations

The governing equations for three-dimensional flow in an enclosure are obtained from the general equations given in vector form in Chapter 2. For two-dimensional flows these may be written in terms of the vorticity ω and the stream function ψ. Similarly, following the treatment of Mallinson and De Vahl Davis (1977) for three-dimensional flows, the governing equations may be written in terms of the vorticity vector $\boldsymbol{\omega}$ and the solenoidal vector potential $\boldsymbol{\psi}$. Considering the enclosure shown in Fig. 14.5.1, the flow is governed by

$$\frac{\partial \boldsymbol{\omega}}{\partial \tau} + \nabla \times (\boldsymbol{\omega} \times \mathbf{V}) = -\mathrm{RaPr}(\nabla \times \phi \hat{\mathbf{g}}) + \mathrm{Pr}\nabla^2 \boldsymbol{\omega} \tag{14.5.1}$$

$$\boldsymbol{\omega} = -\nabla^2 \boldsymbol{\psi} \tag{14.5.2}$$

$$\frac{\partial \phi}{\partial \tau} + \nabla \cdot (\mathbf{V}\phi) = \nabla^2 \phi \tag{14.5.3}$$

with

$$\mathbf{V} = \nabla \times \boldsymbol{\psi} \tag{14.5.4}$$

where the Rayleigh number Ra is based on d, the distance between the two opposite vertical walls maintained at temperatures t_h and t_c, and on the temperature difference $t_h - t_c$. A vector of magnitude unity in the direction of gravity is de-

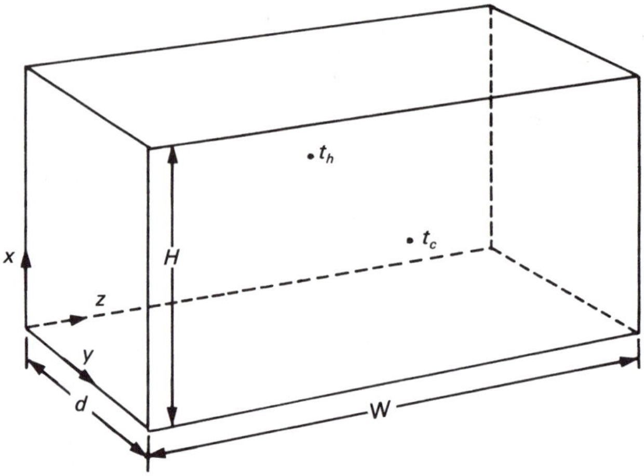

Figure 14.5.1 Coordinate system for three-dimensional flow in a box.

noted by $\hat{\mathbf{g}}$. The walls are all taken as rigid. Except for the two vertical isothermal walls, they are also taken as adiabatic.

The boundary conditions for $\boldsymbol{\psi}$ and $\boldsymbol{\omega}$ result from the velocity boundary conditions. At $x = 0$, for instance,

$$\frac{\partial \psi_1}{\partial x} = \psi_2 = \psi_3 = 0$$

$$\omega_1 = 0 \qquad \omega_2 = -\frac{\partial^2 \psi_2}{\partial x^2} \qquad \omega_3 = -\frac{\partial^2 \psi_3}{\partial x^2} \tag{14.5.5}$$

where ψ_1, ψ_2, ψ_3 and ω_1, ω_2, ω_3 are the components of $\boldsymbol{\psi}$ and $\boldsymbol{\omega}$, respectively. The governing equations can also be obtained for other flow configurations and boundary conditions.

Various numerical methods are available for solving these equations. The transient starting solutions are marched to steady state, within specified convergence criteria. The false transient method of Mallinson and De Vahl Davis (1973) adds a transient variation of $\boldsymbol{\psi}$ to Eq. (14.5.2). Thus, time marching may be used for this equation as well. The additional transient term approaches zero as steady state is attained so that Eq. (14.5.2) is satisfied. This method, therefore, does not give the true transient response. It is a numerical technique for reaching the steady-state results more quickly. See Guceri and Farouk (1985) and Jaluria and Torrance (1986) for further details on this and other numerical approaches for three-dimensional natural convection flows.

14.5.2 Vertical Cavities

This three-dimensional transport is governed by the parameters Ra, Pr, A_V, and A_H, where A_V and A_H are the two nondimensional aspect ratios, given by

$$A_V = \frac{H}{d} \qquad A_H = \frac{W}{d} \tag{14.5.6}$$

One of the dimensions, H or W, is taken as large to reach the two-dimensional approximation. Three-dimensional flow patterns are much harder to present than two-dimensional ones. Mallinson and De Vahl Davis (1977) employed steady-state streamlines through various points, as shown in Fig. 14.5.2, to study the flow effects that result from superposition of the axial flow on the single roll that arises in a two-dimensional enclosure. The axial flow is seen to be directed away from each end wall and the streamlines spiral toward the center of the box. The three-dimensional flow arises because of axial temperature gradients in the fluid and also from the interaction of the rotating roll with the stationary end walls. It is noted that a streamline through any given point returns to that point, as it must in steady enclosed flow.

The effect of the axial flow on the temperature field is seen in Fig. 14.5.3 at $\mathrm{Ra} = 10^4$. The convective effects are expected to be greatest near the center of the cavity. They will be reduced considerably due to shear stresses caused by the end walls. There the temperature pattern approaches that due to conduction alone. The effect of the end walls may also be seen in terms of the vertical average value of the Nusselt number Nu_v, defined as $\mathrm{Nu}_v(z) = (1/H)\int_0^H \mathrm{Nu}(x, z)\, dx$. Figure 14.5.4 shows its variation in the z direction and thus indicates the effect of the

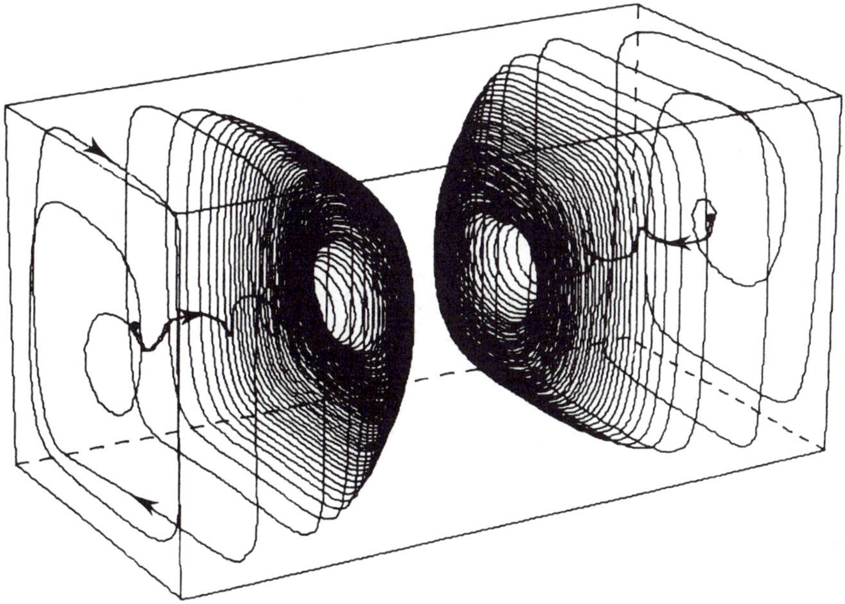

Figure 14.5.2 Streamlines through two points in the three-dimensional flow in a box at $\mathrm{Ra} = 10^4$, $\mathrm{Pr} = 1$, $A_V = 1.0$, and $A_H = 2.0$. *(Reprinted with permission from G. D. Mallinson and G. De Vahl Davis,* J. Fluid Mech., *vol. 83, p. 1. Copyright © 1977, Cambridge University Press.)*

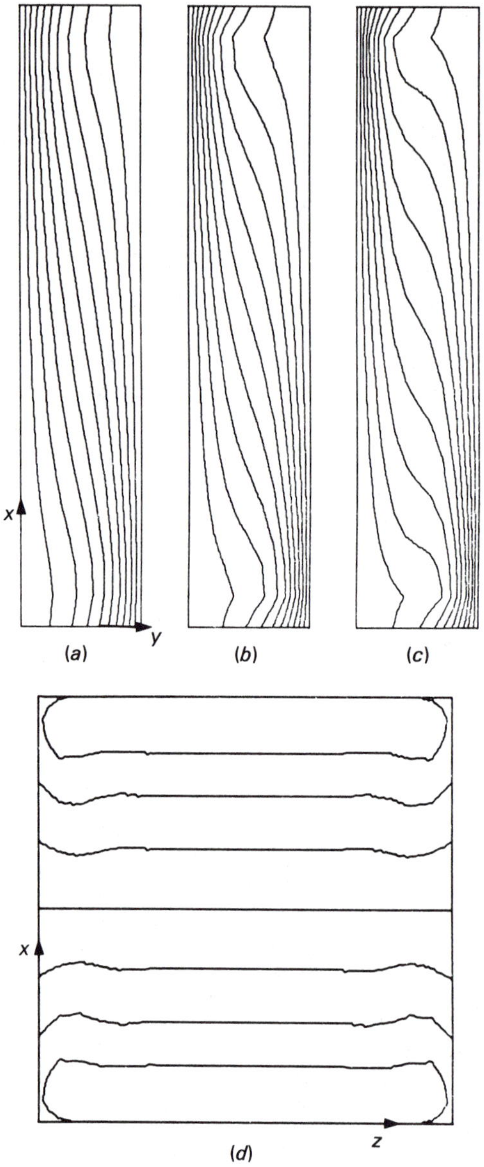

Figure 14.5.3 Temperature contours at various vertical cross sections for $W/D = H/d = 5$ at Ra = 10^4; see Fig. 14.5.1. (*a*) $z/d = 0$; (*b*) $z/d = 0.25$; (*c*) $z/d = 2.5$; (*d*) $y/d = 0.5$. *(Reprinted with permission from G. D. Mallinson and G. De Vahl Davis,* J. Comput. Phys., *vol. 12, p. 453. Copyright © 1973, Academic Press.)*

three-dimensional flow on this parameter. It is seen that Nu_v decreases as each end wall is approached. This indicates the extent of the end wall thermal boundary layer, where a reduced value of the convective transport is obtained. At larger Ra the thickness of this layer is seen to decrease slowly.

The overall average Nusselt number Nu_{av}, which is obtained by integrating $Nu(x, z)$ over x and z, was found to be always lower than the value obtained in the two-dimensional calculations. The longitudinal motion has the greatest effect

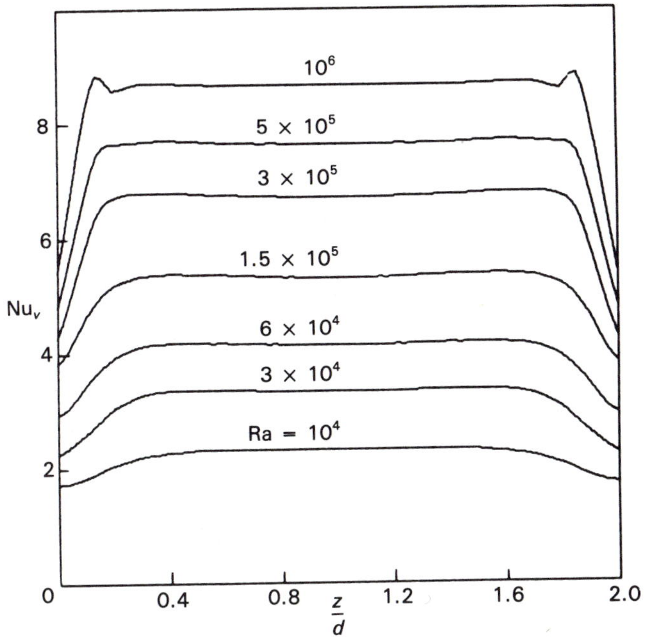

Figure 14.5.4 Variation of the vertically averaged Nusselt number Nu_v in the third direction, z, for $Pr = 0.71$, $A_V = 1.0$, and $A_H = 2.0$, at various values of Ra. *(Reprinted with permission from G. D. Mallinson and G. De Vahl Davis,* J. Fluid Mech., *vol. 83, p. 1 Copyright © 1977, Cambridge University Press.)*

for low Ra, small A_H, and low Pr. For A_H and Pr large at a given value of Ra, the effect was smallest. For high Ra and high Pr with $A_H > 1$, the two-dimensional model was found to be adequate. Multiple longitudinal flows developed at yet larger Ra and were strongly dependent on the governing parameters. For $Ra \approx 10^4$ an appreciable axial flow directed toward the center of the cavity arose. For $6 \times 10^4 \leq Ra \leq 10^6$ the end effects penetrated a distance less than 0.6 of the distance d between the isothermal walls. Therefore, for $A_H \geq 1.2$, there was a central region in which the flow was nearly two dimensional.

14.5.3 Heating at the Bottom Wall

Several studies have been carried out for heating from below. Computed fluid particle streak lines were used by Ozoe et al. (1976) to indicate the flow pattern. Aziz and Hellums (1967) presented a method for the numerical finite-difference solution of the governing equations in three dimensions and gave some results for the flow and temperature fields. Figure 14.5.5 shows the velocity components for a cubical enclosure heated from below, where u, v, and w correspond to x, y, and z in Fig. 14.5.1. The three-dimensional effects are seen in the decrease of these velocity levels away from the middle of the enclosure. The distribution of

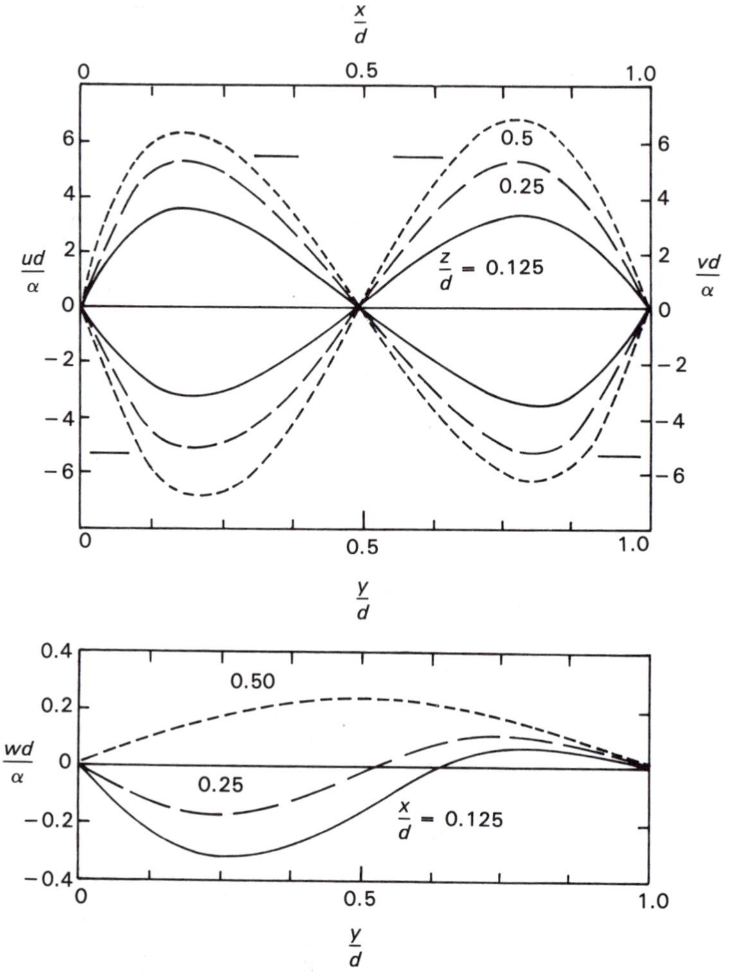

Figure 14.5.5 Distribution of velocity components for three-dimensional flow in a cubic enclosure heated from below at Ra = 4000 and Pr = 10. *(Reprinted with permission from H. Ozoe et al.,* J. Heat Transfer, *vol. 98, p. 202. Copyright © 1976, ASME.)*

w indicates the resulting flow in the third direction. Flows in a long channel with a square cross section and between infinite horizontal plates were also considered by Ozoe et al. (1976), the latter yielding a two-dimensional motion. For a cubic enclosure, the mode of circulation was found to depend on the initial conditions. Further experimental determination of these modes is indicated. Inclined enclosures were studied numerically by Chan et al. (1983) and Ozoe et al. (1983).

Clearly, relatively little work has been done on three-dimensional internal flow effects, despite their common occurrence. However, there is growing interest in such transport. Much numerical work is now based on the solution of the conservation equations in primitive form, that is, in terms of velocity, pressure, and temperature, rather than the solution of the derived equations in terms of $\boldsymbol{\omega}$ and $\boldsymbol{\psi}$. Chan and Banerjee (1979) discuss such a numerical technique for the solution of three-dimensional transient natural convection in rectangular enclosures. The method was applied to several enclosure flows and compared well with the earlier numerical and experimental studies.

14.6 THERMOSYPHONS

A thermosyphon is a fully or partially enclosed circulating fluid system driven by thermal buoyancy forces. Unconfined natural convection loops, where the fluid is heated from below and cooled from above, are often encountered in atmospheric and oceanic flows. Such flows have also long been attractive for several technological applications, including the cooling of gas turbines, electrical machinery, nuclear reactors, internal combustion engines, geothermal energy extraction, thermosyphon solar water heaters, and various applications in process industries. An extensive review of thermosyphon technology was given by Japikse (1973). Many possible configurations of thermosyphons were discussed and the results of various studies on flow and heat transfer were given.

A considerable amount of work has been done on thermosyphons, and this continues to be an area of significant research activity. In this section we will give a brief discussion of some of the important types of thermosyphons and the underlying physical processes. An important classification of thermosyphons is based on whether the system is open or closed to mass flow. Although various natural convection loops have been considered for circulating fluid, the basic characteristics of thermosyphons have been extensively studied by considering the simple two-dimensional configurations shown in Figs. 14.6.1 and 14.6.2.

14.6.1 Open Thermosyphon

Heating the wall of an open thermosyphon results in an upward flow along the walls, due to buoyancy, and a compensating return downward flow inside. A boundary layer regime may arise for large heat flux levels. A relatively weak flow is found at smaller values due to wall shear. A stagnant bottom region may arise at still lower heat flux input. The effect of geometry is expressed by the parameter L/a, where L and a are shown in Fig. 14.6.1.

For larger values of L/a, larger values of the Rayleigh number Ra, based on a, are needed to transport a given heat flux level. An increase in Pr results in an increase in the Nusselt number for the boundary layer regime, as expected, and in a decrease for the impeded flow condition. Lighthill (1953) analyzed this flow and obtained three laminar regimes, which have been very well confirmed experimentally. Results were also given for turbulent flow. However, these predictions differ significantly from the experimental results of Martin (1954, 1955), Hartnett and Welsh (1957), and Hasegawa et al. (1963), who made measurements

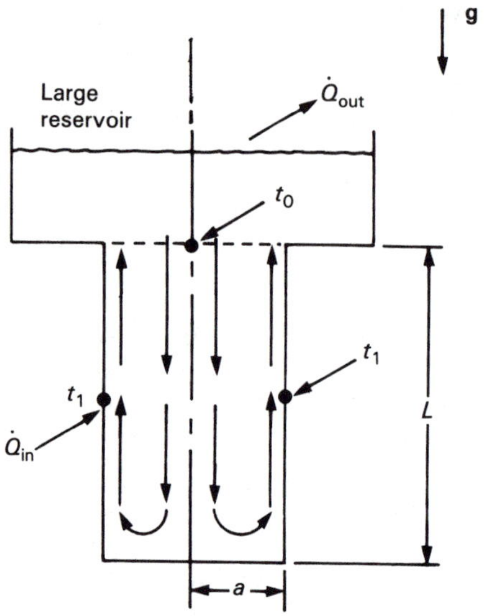

Figure 14.6.1 Open thermosyphon.

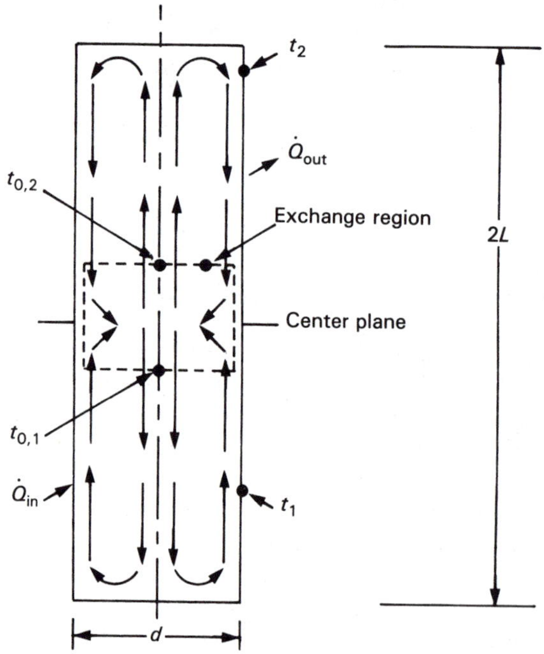

Figure 14.6.2 Closed thermosyphon.

in the open thermosyphon flow (Fig. 14.6.1) and reported heat transfer results. For laminar flow, Japikse (1973) has tabulated the results for the flow rate, temperature variation, and heat transfer, as given in Table 14.6.1 in terms of the dimensionless temperature difference across an open thermosyphon $T_{ot} = a\text{Ra}/L$, where $\text{Ra} = g\beta a^3(t_1 - t_0)/\nu\alpha$. Also presented were results for turbulent flow.

14.6.2 Closed Thermosyphon

The cylindrical closed thermosyphon, shown in Fig. 14.6.2, arises in applications that involve containment and pressurization, such as the cooling of turbine blades and the preservation of permafrost beneath structures in icebound regions. The walls of the lower half of the cylindrical cavity are maintained at a higher temperature t_1 than those of the upper half, at t_2. An unstable stratification is therefore created near the midheight. Fluid descends near the walls and ascends near the vertical axis in the upper half of the cylinder. In the lower half, it ascends near the walls and descends near the axis. If the motion is sufficiently strong, the

Table 14.6.1 Laminar open thermosyphon behavior[a]

$$\log_{10}(\text{Nu}_a T_{0t}) = M_1 \log_{10} T_{0t} + b_1$$

$$\log_{10}(\dot{m}) = M_2 \log_{10} T_{0t} + b_2$$

$$\log_{10}(T_{0t} - T_{mc}) = M_3 \log_{10} T_{0t} + b_3$$

$$M_1 = 1.256 \qquad M_2 = 0.272 \qquad M_3 = 0.984$$

Pr	b_1	b_2	b_3
∞	−0.243	0.172	−0.414
100	−0.246	0.161	−0.406
10	−0.265	0.0904	−0.354
3	−0.295	−0.00823	−0.285
1	−0.343	−0.137	−0.204

[a]These equations represent the results of the study by Japikse (1973) for the range $3.7 < \log T_{0t} < 6.75$, linearized on a log-log plot, using the values at $\log_{10} T_{0t} = 5.25$ and 6.75. They are related by $\dot{m}(T_{mc} - T_{0t}) = \text{Nu}_a T_{0t}$, which can be used as a check to show that the fits are indeed quite good. Here $\dot{M}$ and $\dot{m}$ are the physical and dimensionless mass flow rates, respectively, and t_{mc} is the mixing cup temperature.

$$T_{0t} = g\beta a^4(t_1 - t_0)/\nu\alpha L$$

$$T_{mc} = g\beta a^4(t_1 - t_{mc})/\nu\alpha L$$

$$\dot{m} = \dot{M}c_p/kL$$

Source: From Japikse (1973).

downward flow near the walls of the upper half feeds the flow near the axis of the lower half, while the flow near the walls of the lower half feeds the flow near the axis of the upper half of the cylinder. A complex three-dimensional exchange therefore arises in the midheight region of the tube, as indicated.

This closed thermosyphon may be treated as two simple open thermosyphons appropriately joined at the midheight exchange region. The modeling of the exchange region is very important since the temperatures $t_{0,1}$ and $t_{0,2}$ on the centerline must be known in order to apply open thermosyphon results to the rest of the enclosure. The three idealized exchange mechanisms proposed by Lock (1962) are shown in Fig. 14.6.3. In practice, the coupling mechanism may be expected to contain elements of more than one of these idealized motions.

Japikse (1969), Japikse and Winter (1970), and Japikse et al. (1971) carried out a detailed experimental study of circular closed thermosyphons. The results are presented in terms of the Prandtl number and a parameter T_{ct}, which is the dimensionless temperature difference across a closed thermosyphon and is given by

$$T_{ct} = \frac{g\beta d^4(t_1 - t_2)}{\nu\alpha L} = \text{Ra}\,\frac{d}{L} \tag{14.6.1}$$

where d is the inside tube diameter and L the half-length of the closed thermosyphon. The heat transfer coefficient h is based on the horizontal cross-sectional area $\pi d^2/4$. For $T_{ct} < 10^7$ the flow was found to be laminar and to consist of several fluid streams that cross at midheight of the tube. These streams increased in number with increasing T_{ct}. A stable steady state existed only for $\text{Pr} > 90$ and $T_{ct} \lesssim 5 \times 10^5$. Convection was the dominant mode of exchange for $T_{ct} \gtrsim 10^5$. For $T_{ct} \gtrsim 10^7$ the flow was turbulent and heat transfer occurred largely by an overall mixing process.

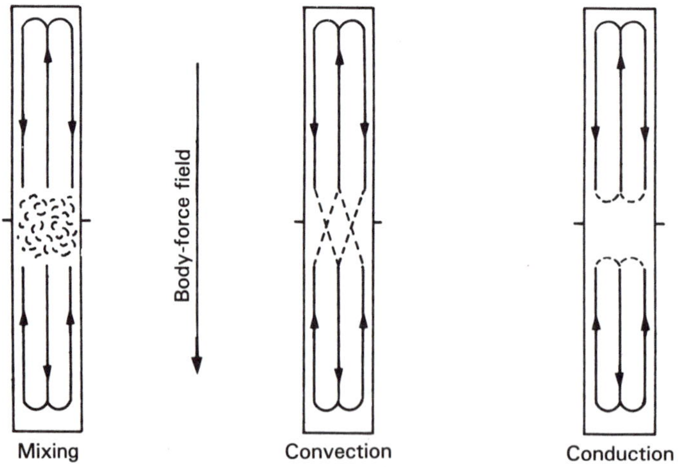

Figure 14.6.3 Ideal exchange mechanisms according to Lock (1962).

By using the condition that the heat flow into the closed thermosyphon $\dot{Q}_1$ must equal the heat flow out $\dot{Q}_2$, heat transfer relationships may be obtained from the open thermosyphon results. The center of the closed syphon is idealized as a reservoir for both ends. These results for a fully turbulent flow may be represented by

$$\mathrm{Nu} = \frac{ha}{k} = C_1\left(\frac{a}{L}\right)^{C_2} \mathrm{Ra}^{C_3} \qquad \text{where } \mathrm{Ra} = \frac{g\beta a^3(t_1 - t_0)}{\nu\alpha} \tag{14.6.2}$$

Here C_1 and C_2 are constants that depend on the radius a, C_3 depends on the heat transfer process (see Japikse, 1973), and a is $d/2$ in Fig. 14.6.1.

Mallinson et al. (1981) carried out numerical and experimental studies of the three-dimensional flow in a closed vertical rectangular thermosyphon like that in Fig. 14.6.2. For the experiments, the fluid Pr range was 5×10^3 to 10^5, for silicone oils. At $\mathrm{Ra} \lesssim 10^3$ conduction was found to be the dominant exchange mechanism for heat transfer in the midheight region of the thermosyphon. The complex flow patterns that arise at larger Ra were simulated numerically and verified by flow visualization.

Heat transfer results were given for $0 \le T_{ct} \le 4 \times 10^5$. For the convective heat transfer coefficient h, obtained by averaging the total heat flux over the area of the upper half of the thermosyphon as done by Japikse et al. (1971), the following correlations result for $\mathrm{Ra} \ge 2 \times 10^4$; h is based on the cross-sectional area of the thermosyphon.

$$\mathrm{Nu} = 0.035 T_{ct}^{0.4} = 0.092\left(\frac{2L}{d}\right)^{0.6} \mathrm{Ra}^{0.4} \tag{14.6.3}$$

Figure 14.6.4 compares the flow visualization and numerical results for $\mathrm{Ra} = 7 \times 10^4$. Such excellent agreement was also obtained for other values of Ra. Beyond $\mathrm{Ra} = 5 \times 10^5$ the flow was found to become unsteady for the large Pr values considered in the study, and turbulence developed near $\mathrm{Ra} = 5 \times 10^6$.

14.6.3 Fluid Loops

Fluid loops driven by buoyancy have also received considerable attention in recent years. Since they provide a simple means for circulating fluid without the use of pumps, they are of particular interest in solar heaters and emergency reactor core cooling. They are also of interest for understanding thermal springs, seawater circulations in the sea floor, and various other geological processes, as outlined by Torrance (1979a).

In a closed loop, fluid is recirculated continuously around a piping path, which forms the loop. The transient and stability characteristics of closed loops have been studied by Alstad et al. (1955), Creveling et al. (1975), and Greif et al. (1979). Steady-state performance and the stability of steady-state motion have been studied by Welander (1967) and Damerall and Schoenhals (1979). Oscillatory flows have been found to arise for both vertically symmetric and asymmetric

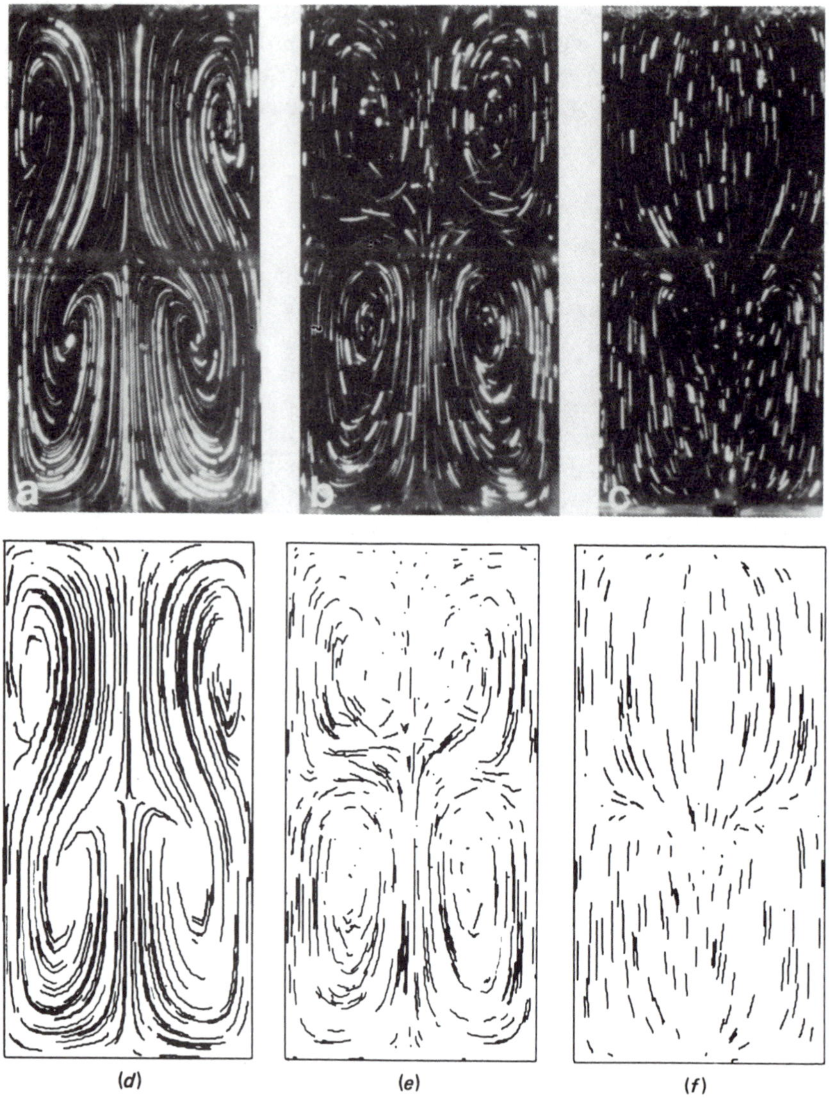

Figure 14.6.4 (*a–c*) Flow visualization and (*d–f*) numerical simulations at $Ra = 7 \times 10^4$. *(Reprinted with permission from G. D. Mallinson et al.,* J. Fluid Mech., *vol. 109, p. 259. Copyright © 1981, Cambridge University Press.)*

heating of the walls. These oscillations have been attributed to the phase lag between the heating process and the generation of the resulting buoyancy force.

Much less attention has been given to open loops, in which all or part of the circulating fluid may be exchanged with an external reservoir, whose temperature and pressure may be independent of the conditions in the loop. Lapin (1969),

Donaldson (1970), Torrance (1979a), Torrance and Chan (1981), and Bau and Torrance (1981a, 1981b) studied transient and steady behavior of such open loops. Figure 14.6.5 shows three types of loops, along with representative velocity transients. The third one, Fig. 14.6.5*c*, is considered in detail by Bau and Torrance (1981a, 1981b). Experimental and analytical results are presented for water and a water-saturated porous medium.

Figure 14.6.6 shows a model of an aquifer in which an open-loop thermosyphon arises. Solutions for the flow rate and the exit temperature were obtained by Torrance (1979a). Unique values were found for the maximum exit temperature and the critical Rayleigh number at the onset of convection. The maximum possible exit temperature excess over the surface temperature was found to be 68.8% of the temperature excess at the bottom of an elliptical thermosyphon.

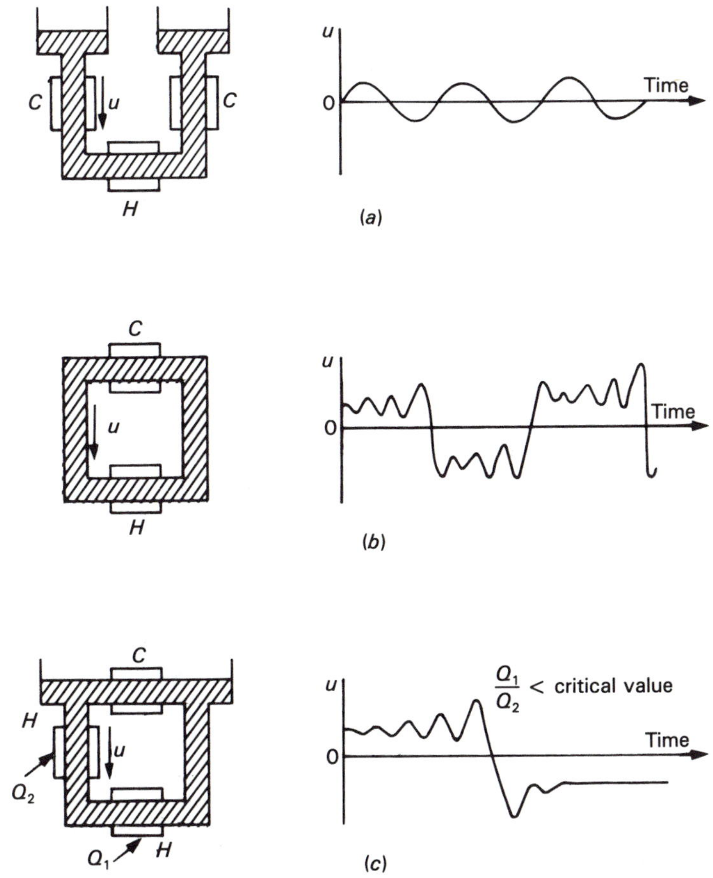

Figure 14.6.5 Three different free convection loops with heaters (H) and coolers (C). Representative velocity transients (u) are shown at the right. (*a*) Open loop; (*b*) closed loop; (*c*) open loop with a single isothermal reservoir at the top.

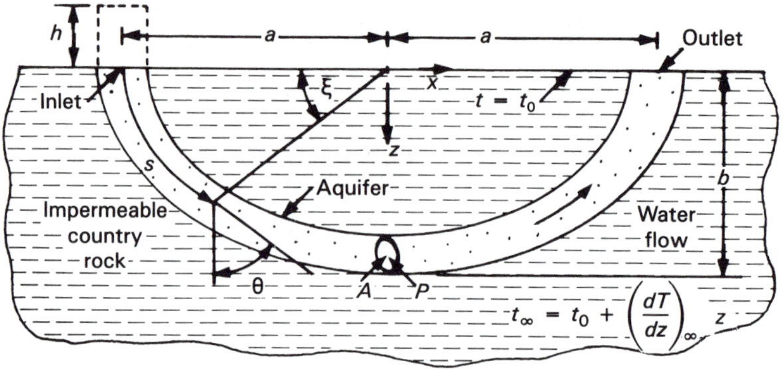

Figure 14.6.6 Schematic of an open-loop thermosyphon (the aquifer). *(Reprinted with permission from K. E. Torrance,* J. Heat Transfer, *vol. 101, p. 677. Copyright © 1979, ASME.)*

Higher exit temperatures would require longer residence times near the maximum depth of the thermosyphon. Zvirin (1979) has considered analytically the effect of viscous dissipation on open and closed natural convection loops. Mertol et al. (1982) have carried out a two-dimensional analysis of heat transfer and flow. Friction and heat transfer coefficients are presented. Their variation along the loop is also determined, with the Graetz number Gz, which may be related to Gr, Pr, and the diameter-to-length ratio (see Chapter 10), as the parameter. A detailed review of natural circulation loops has been presented by Mertol and Greif (1985).

14.7 LIMITED AMBIENT MEDIUM

The concept of purely external natural convection transport assumes that the flow occurs in an infinite quiescent ambient medium. This implies that the conditions far from the flow region remain unchanged during the period of interest. However, in practice, flows occur in finite regions and, with the passage of time, the presence of bounding walls far from the flow region will affect the flow. Therefore, experiments to determine external natural convection flow mechanisms, over surfaces and in free boundary flows, may actually be carried out in limited ambient media for only a limited period of time, before the effects of confinement interact with the flow.

The heat input into an enclosed ambient medium causes both recirculating flow and a change in the density distribution in the enclosure. The permissible time interval for simulating external natural convection conditions in the vicinity may be determined, at least approximately, by monitoring the conditions where measurements are being taken. Significant changes in local conditions indicate the limit. Ambient stratification, arising from a concentrated heat source, may also be calculated approximately in terms of the volume of the shed wake fluid.

Concentrated source. There has been some study of the natural convection flow arising from a relatively small energy source in a confined region. Besides the considerations mentioned above on simulating external natural convection, such configurations are also of interest in many practical applications. The flows due to fire in an enclosure, heaters in an oven or furnace, and heated electronic components in an enclosed region may initially be considered as purely external flow. The added effect of the confining walls gradually arises as time elapses.

Baines and Turner (1969) studied experimentally and analytically the turbulent convective flow induced by a buoyancy source in a confined region; see Fig. 14.7.1. The source gives rise to a thermal plume that rises and spreads over the ceiling of the enclosure, resulting in a stably stratified layer that increases in depth with time. The centerline temperature in the plume decreases with height above the source, as discussed in Chapter 12. The region below the upper stratified layer continues to be at the initial temperature in the enclosure before the onset of the flow. The temperature in the upper heated layer decreases downward from the ceiling to the interface between the upper and lower regions. Figure 14.7.1*b* shows the flow pattern, indicating side entrainment into the plume and the downward motion of the heated upper layer. Figure 14.7.1*a* gives a sketch of the variation with vertical distance x of the temperature, or the corresponding fluid density, in the plume and in the ambient environment away from the plume. The interface location and the temperature distribution are shown at two times, τ'_1 and τ'_2, after the heat input Q_0 is turned on.

The turbulent plume due to the source was analyzed by using the entrainment model of Morton et al. (1956); see Chapter 12. The induced flow was assumed to spread out horizontally over the ceiling of the enclosure. The downward movement of the interface, between the upper stratified layer and the lower region, and the density distribution were calculated. Experimentally, this flow circumstance

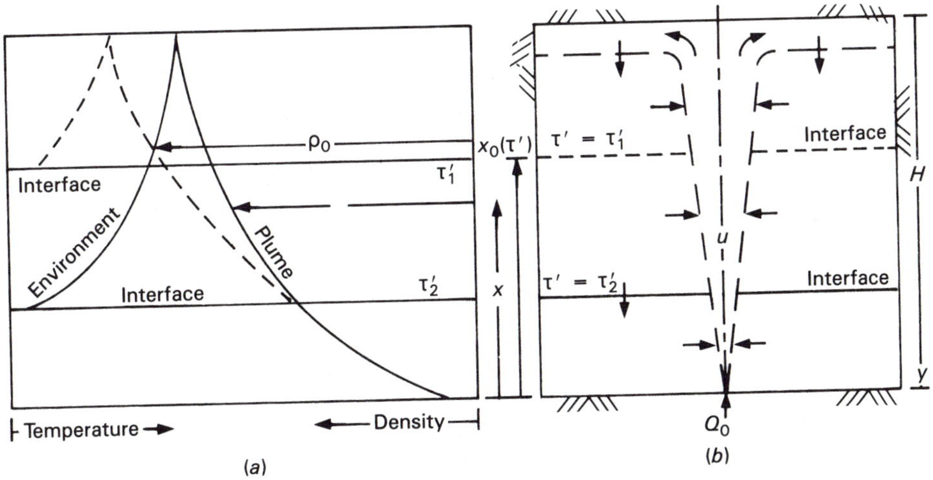

Figure 14.7.1 Sketch of (*a*) the development of stratified environment in an enclosure and (*b*) the flow pattern due to a heat source in the enclosure. *(Reprinted with permission from W. D. Baines and J. S. Turner,* J. Fluid Mech., *vol. 37, p. 51. Copyright © 1969, Cambridge University Press.)*

was simulated by saltwater jets descending in fresh water. The buoyancy effect was provided by the difference in density between the discharged saline water and the initial fresh water in the enclosure. The temperature was kept uniform during the experiment. Figure 14.7.2 is a comparison of the experimentally measured position of the interface, as a function of time, with the corresponding analytical results. A dimensionless time τ is employed, as defined in the figure. Good agreement between the calculations and the data is seen.

Similar models have been developed for fire in enclosed and partially enclosed regions, as discussed in greater detail in Section 14.8. Such "filling box" models, which determine the accumulation rate of buoyant fluid at the top of the enclosed region rather than the large-scale vertical circulation, are applicable when

$$\frac{\tau}{4\pi^{-1/3}\alpha_e^{4/3}} = \left(\frac{H}{R}\right)^2 \frac{Q_0^{1/3}\tau'}{H^{4/3}}$$

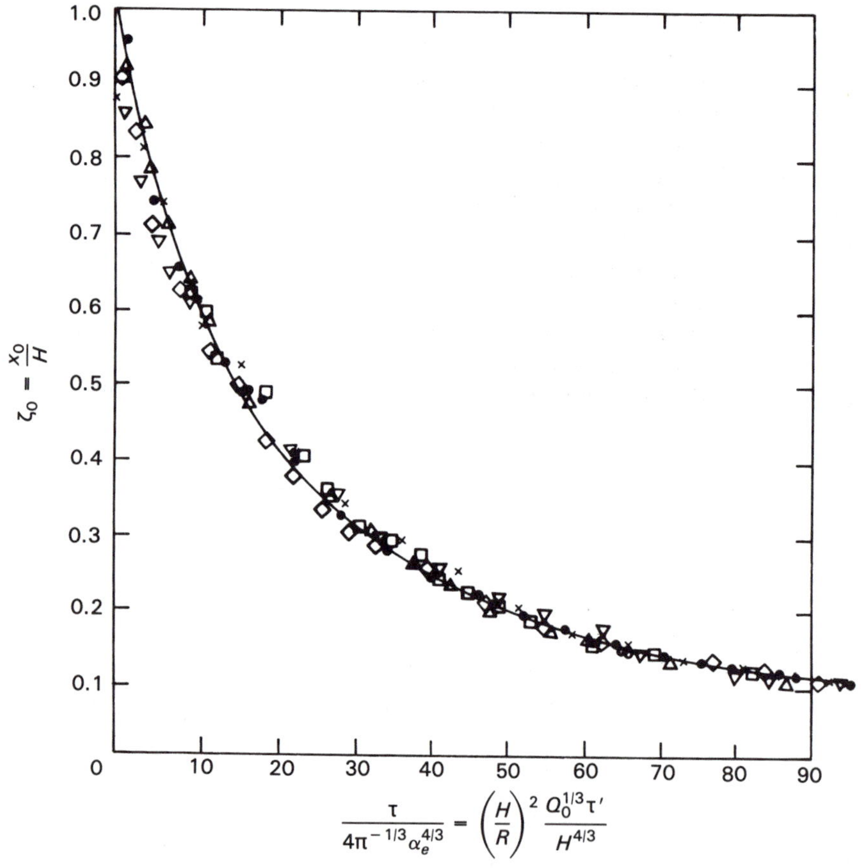

Figure 14.7.2 Dimensionless height ζ_0 of the interface between the lighter upper layer and the isothermal denser lower environment as a function of dimensionless time τ. The theoretical curve is obtained for the entrainment coefficient $\alpha_e = 0.1$. Here H is the height of the enclosed cylindrical region, R its radius, Q_0 the steady buoyancy input per unit time at the source, and τ' the physical time. *(Reprinted with permission from W. D. Baines and J. S. Turner, J. Fluid Mech., vol. 37, p. 51. Copyright © 1969, Cambridge University Press.)*

the buoyancy forces are large compared to the inertia of the plume as it reaches the constraining boundary of the enclosure. Otherwise, more vigorous momentum effects arise.

Depending on the dimensions of the enclosure and the buoyancy input at the source, strong recirculating flows may arise in the enclosure, rather than the growing nonconvecting stably stratified layer observed in the experiments of Baines and Turner (1969) and in studies of fires in rooms; see the review by Quintiere (1977). In actual flows the downward spread of the stable stratified layer, as the buoyant plume brings more fluid into it, is always accompanied by some internal flow circulation. This leads to enhanced convective mixing between the two regions. Both these effects were observed in the experimental study of Jaluria and Gebhart (1974). The study of the stability of stratified layers subjected to externally induced flow is relevant to the establishment and growth of stratification, as outlined by Turner (1973) and Jaluria (1980). The time-dependent stratification and circulation due to a heat source in a limited enclosure have also received some attention in the numerical study of Rehm and Baum (1978). Further detailed study is needed for both the laminar and turbulent flow regimes.

The steady recirculating flow due to small heat sources in a cylindrical region of air was studied by Torrance and Rockett (1969), Torrance et al. (1969), and Torrance (1979b). A steady laminar flow arose from a localized heat source, of width $0.2\,R$, at the center of the bottom surface of a vertical isothermal cylindrical enclosure. Both experimental and numerical results were given. Figure 14.7.3 shows the left side of the calculated symmetric steady-state streamlines at various values of the Grashof number Gr. The flow rises vertically above the source, flows along the ceiling, and turns back downward at the far corner. This hot fluid flowing downward on the cooler vertical wall is buoyant. Because of the buoyancy force acting against the flow, the fluid slows down, turns inward, and rises to a height determined by the flow field and local buoyancy. As Gr increases, the circulation becomes stronger and boundary regions develop. In such a limited ambient, this is more like enclosure transport, generated from a small wall source.

Calculations were also given for transient circulation following the start of the energy input into the source for an initially quiescent environment, as shown in Fig. 14.7.4. A small ring vortex of hot fluid forms near the source, which is located at the origin, and gradually spreads out to occupy the entire enclosed region. A good comparison with experimental results was obtained. A similar numerical study was carried out by Chu et al. (1976) to determine the effect of the size and location of a heater on the steady two-dimensional laminar natural convection flow in a rectangular enclosure. The heat transfer rates and circulation patterns were given.

Enclosed bodies. The heat transfer from heated bodies located inside relatively small enclosures has also been studied, to determine the effect of the enclosing surfaces on flow and heat transfer. McCoy et al. (1974), Powe (1974), and Powe et al. (1975, 1980) measured the transport arising from small cubic, spherical, and short cylindrical bodies at temperature t_1 located within spherical enclosures at surface temperature t_2. Steady-state heat transfer results were also

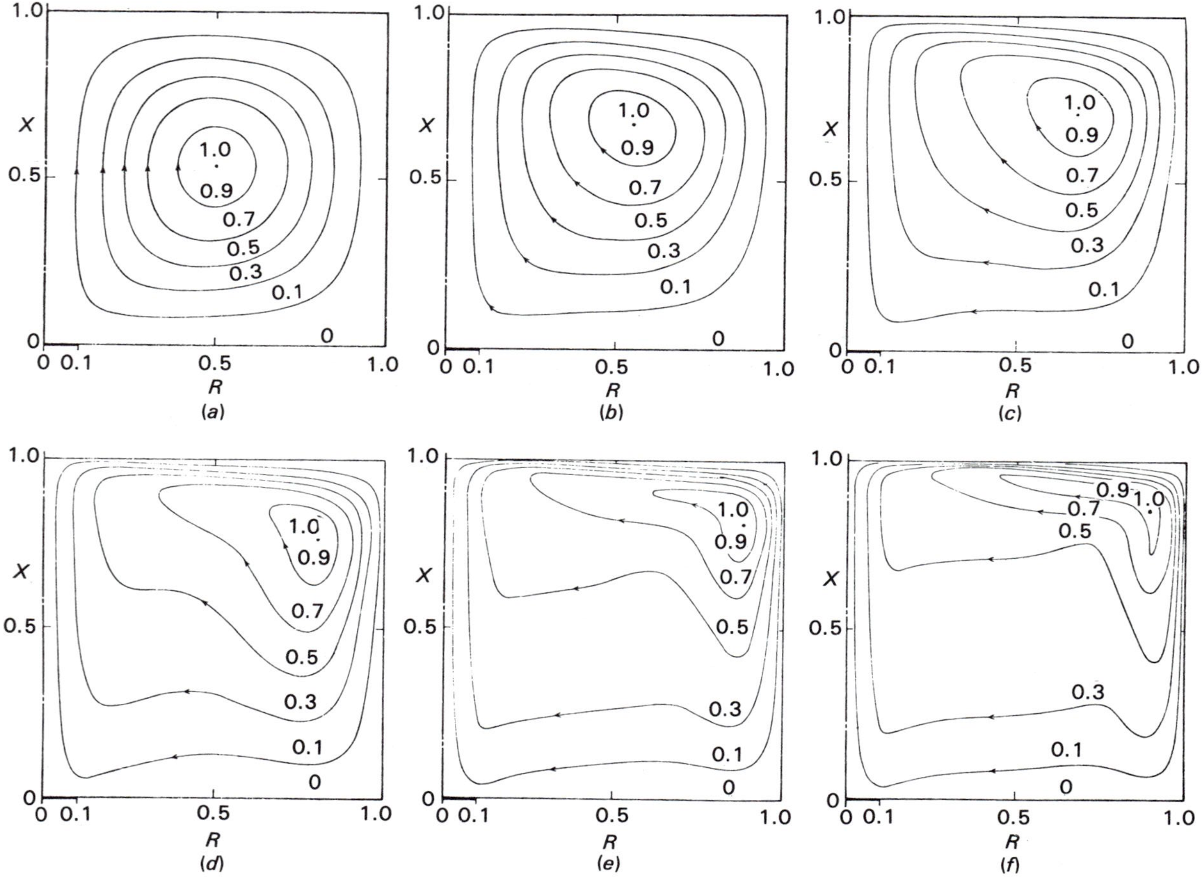

Figure 14.7.3 Steady-state streamline fields for various Grashof numbers Gr based on height H of the cylindrical enclosure and on the source temperature excess. The source is centered at $x = 0$, $R = 0$. (*a*) $\mathrm{Gr} = 4 \times 10^4$; (*b*) $\mathrm{Gr} = 4 \times 10^5$; (*c*) $\mathrm{Gr} = 4 \times 10^6$; (*d*) $\mathrm{Gr} = 4 \times 10^7$; (*e*) $\mathrm{Gr} = 4 \times 10^8$; (*f*) $\mathrm{Gr} = 4 \times 10^9$. *(Reprinted with permission from K. E. Torrance and J. A. Rockett,* J. Fluid Mech., *vol. 36, p. 33. Copyright © 1969, Cambridge University Press.)*

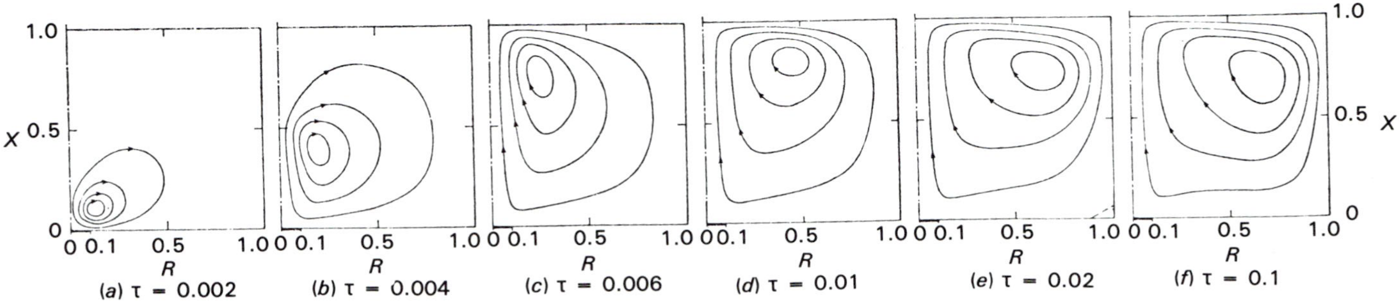

Figure 14.7.4 Transient streamlines for $\mathrm{Gr} = 4 \times 10^6$ at various times τ. (*a*) $\tau = 0.002$; (*b*) $\tau = 0.004$; (*c*) $\tau = 0.006$; (*d*) $\tau = 0.01$; (*e*) $\tau = 0.02$; (*f*) $\tau = 0.1$. Here dimensionless time $\tau = \alpha\tau'/H^2$, where τ' is the physical time. *(Reprinted with permission from K. E. Torrance and J. A. Rockett,* J. Fluid Mech., *vol. 36, p. 33. Copyright © 1969, Cambridge University Press.)*

obtained. Warrington and Powe (1981) studied similar inner body geometries located within a cubic enclosure. Various fluids, including air, water, and silicone oils, were used. Heat transfer correlations were given and compared with those for the spherical enclosure. The correlation that applies for both the spherical and the cubic enclosure data for Ra up to about 10^9 is given as

$$\mathrm{Nu} = 0.396\mathrm{Ra}^{0.234}\left(\frac{d}{R_i}\right)^{0.496} \mathrm{Pr}^{0.0162} \qquad 0.707 < \mathrm{Pr} < 1.4 \times 10^4 \quad (14.7.1)$$

where R_i is the radius of a sphere whose volume equals that of the inner body, d

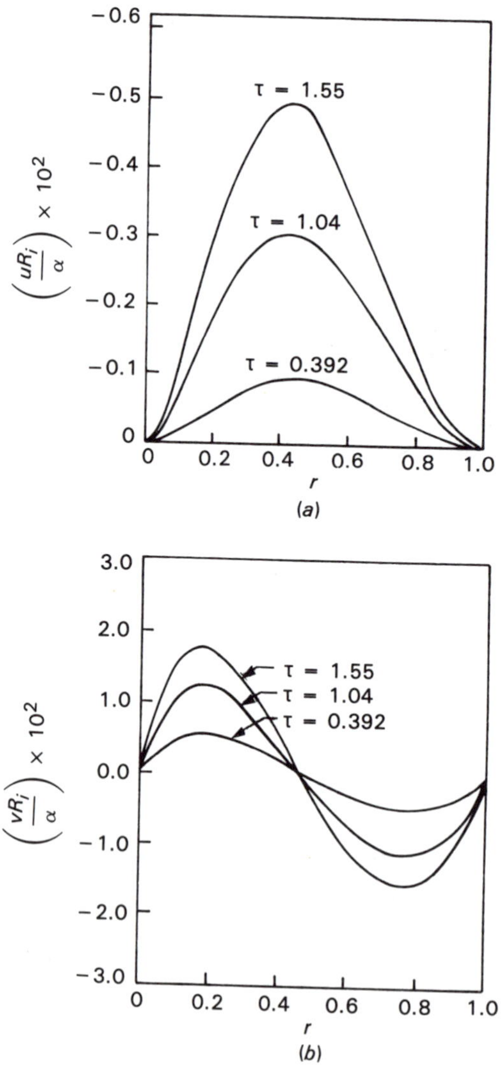

Figure 14.7.5 Distribution of (*a*) radial and (*b*) azimuthal velocity components as defined in Fig. 5.4.2 at Ra = 500 for various times. The former is a vertical downward traverse from the inner to the outer cylinder and the latter a horizontal traverse. Here r and τ are dimensionless radial coordinate distance and time, respectively. *(Reprinted with permission from L. S. Yao and F. F. Chan,* J. Heat Transfer, *vol. 102, p. 667. Copyright © 1980, ASME.)*

is the gap width $(R_o - R_i)$, R_o is the radius of a sphere whose volume equals that of the enclosure, and Nu and Ra are based on d.

It was found that the enclosure shape has a small effect on the temperature profiles and, thus, on the heat transfer. However, the enclosure dimensions have a significant effect on these results. The cubic enclosure resulted in a larger Nusselt number for given Ra and inner body size R_i. Powe and Warrington (1983) have obtained more data for these geometries.

Melting and solidification. Consideration of limited ambient media is also often important in melting and solidification. In melting, the enclosed fluid region increases in volume and the effect of the bounding surfaces decreases with time. In solidification, the reverse occurs. Even if the configuration initially corresponds to purely external natural convection, it eventually becomes a small enclosure.

Natural convection effects have been considered for a few important phase change configurations. As solidification proceeds, the effect of the decreasing fluid volume is seen in the numerical results of Ramachandran et al. (1981, 1982). Similarly, in melting, the flow change, as the distance between the receding surface and the boundaries increases, is seen in the calculations of Sparrow et al. (1977) and Yao and Chen (1980). The heat input is provided at a horizontal cylindrical surface. This gives rise to an approximately horizontal cylindrical annulus with an outward-growing boundary layer. Heat transfer is initially dominated by conduction across the melt layer since Ra, based on the gap thickness, is small. Natural convection effects increase as the gap thickness increases. The increase in velocity level with time is shown in Fig. 14.7.5. Such changing convection effects in melting and solidification are often very important in determining transport.

14.8 PARTIAL AND PARTITIONED ENCLOSURES

Buoyancy-induced flows frequently arise from surfaces in the presence of other surfaces, which collectively do not form a complete enclosure. Then the space is in contact with the outside environment through one or more openings. A common example is a fireplace enclosure, in which a bed of coals or burning material is partially surrounded by firebrick surfaces. There is inflow from the room and outflow up the chimney. The open thermosyphon in Fig. 14.5.1 is another example. Many complicated and important partial enclosures arise, for example, in heated spaces with open doors or windows, inlets, and outlets. Even in closed rooms, there are often small cracks and openings that allow infiltration. This interacts with any internal flow generated in the space.

Partial enclosure transport generally is quite different from that in similar complete enclosures. It is usually strongly dependent on the nature and size of the opening. A considerable amount of work has been done in fire research on transport in partial enclosures such as rooms, corridors, and buildings. Several solar energy collection systems also involve partial enclosures, with a wall removed for solar flux input and fluid flow.

Among earlier studies of partial enclosures was that of Lighthill (1953). Transport in a round tube with heated isothermal walls, closed at the bottom and open at the top to a region at a different temperature, was analyzed. This configuration is similar to the open thermosyphon considered in Section 14.6.1. It arose because of structure cooling problems in turbines. An integral method was used to analyze the flow for $\mathrm{Pr} \rightarrow \infty$. It was found that the flow is strongly dependent on the ratio of height H to radius R for given values of the Rayleigh and Prandtl numbers. For very small values of H/R, the effect of the confining walls is small and the internal flow is similar to the natural convection flow over a vertical surface. The boundary layer flow is modified by the entering downward core flow. On the other hand, for $H/R >> 1$, a fully developed pattern arises and the velocity and temperature distributions across the space do not vary, in form, with height. At intermediate values of H/R, the velocity and temperature profiles fill the tube completely but also vary along the height of the tube. These three regimes are shown in Fig. 14.8.1.

For the similarity regime, which arises at large H/R, Lighthill found that the flow arises over the entire height of the tube only for a discrete critical value of H/R. If H/R is larger than this value, a stagnant region arises near the closed bottom end. The flow above the stagnant region extends to a height corresponding to the critical value. The effective H/R thus becomes equal to the critical value.

For a linear temperature variation at the wall, Ostrach and Thornton (1958) considered the same configuration. Using an integral method, they found that for large H/R, flow arises in the entire tube over a range of H/R, rather than at a discrete value as found for the isothermal condition considered by Lighthill (1953). Experimental verification of the theoretical predictions of Lighthill was obtained by Martin (1955) for glycerin and rapeseed oil. Hartnett and Welsh (1957) performed similar experiments, with water as the fluid and a uniform heat flux condition along the vertical wall. The natural convection in a similar partially confined rectangular space was studied by Auxilien (1981) by both analysis and experiment. Similar trends and flow regimes were seen.

Room fires. Transport arising from a concentrated fire in a room with an

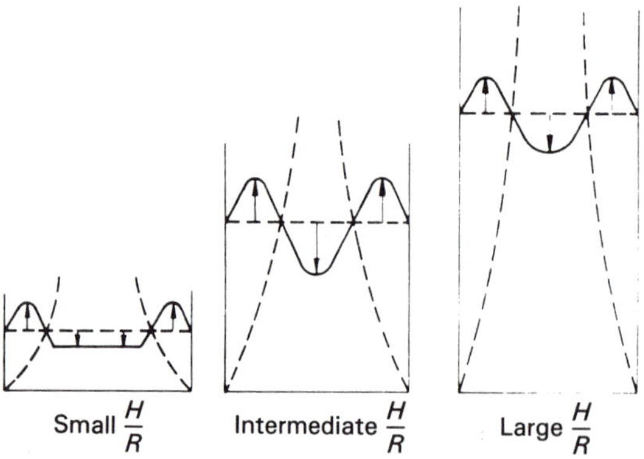

Figure 14.8.1 Flow regimes for the natural convection flow in a vertical tube closed at the bottom for three ranges of H/d. *(Reprinted with permission from* J. Mech. Appl. Math., *vol. 6, p. 398, M. J. Lighthill. Copyright © 1953, Pergamon Journals Ltd.)*

opening has been considered by several researchers, as reviewed by Quintiere (1977, 1981) and Emmons (1978, 1980). Interest lies in determining the resulting thermal field and flow rates. The flow configuration for one kind of single opening is shown in Fig. 14.8.2. The interaction between the internal flow and the outside environment is an exchange, due to the inflow of cold ambient air and the outflow of hot combustion products and entrained air. The opening keeps the pressure in the room close to atmospheric.

In practice, this circumstance is very complex. The flow is generally turbulent, there are large radiation effects, there are substantial fluid property variations, and the heat transfer processes are coupled with the uncertainties of the surface pyrolysis and resulting combustion mechanisms. The flow is usually three dimensional and the boundary conditions at the opening are not known explicitly, as at an impermeable surface. The outside ambient conditions do apply far out from the opening. As a consequence of these complexities in the analysis, fire research has depended heavily on relatively simple physical and analytical models and on experimentation to determine the validity of these models.

Two main approaches have been used in the study of flows due to fires in partial enclosures. The first involves an integral zone model and considers two distinct isothermal layers, a heated and stably stratified one, lying above a cooler layer. Mass and energy balances of the two layers are interpreted to determine both the movement of the interface between the layers and the changing temperature levels in the space. Extensive work has been done for room fires with such zone models, as discussed by Zukoski (1978), Quintiere et al. (1978), and Tanaka (1978).

The second approach employs the governing partial differential equations, along with a turbulent transport formulation, such as the K-ϵ model, to study the convective flow in the enclosure. Although some work has been done on three-dimensional flows, the complexity of the problem has restricted many studies to the two-dimensional approximation. Hasemi (1976), Satoh et al. (1980, 1981), and Markatos et al. (1982) followed this approach and modeled the buoyancy-induced flow of smoke and gases in enclosures with openings. A somewhat sim-

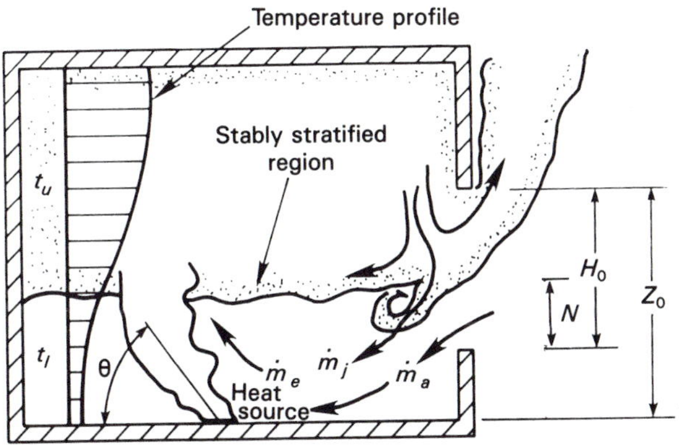

Figure 14.8.2 Schematic of the buoyancy-driven flow due to a thermal energy source in an enclosure with an opening.

ilar approach was applied by Rehm and Baum (1978) and Rehm et al. (1982) to rooms with an opening. An inviscid flow approximation was used in the equations.

These zone model studies have contributed substantially to the understanding of buoyancy-driven flow in partial enclosures and have received corroboration from experiments. The flow through the opening is determined in a manner similar to that used for flow through an orifice. The hydrostatic pressure difference due to the temperature difference across the opening is employed, along with an experimentally determined flow coefficient, to obtain the flow into the room at the bottom of the opening and the flow out at the top. This is discussed by Quintiere et al. (1978) and Zukoski and Kubota (1980). Figure 14.8.3 shows the typical resulting velocity distribution and Fig. 14.8.4 the temperature profiles inside, at a door, and outside.

McCaffrey and Quintiere (1977) measured velocity and temperature distributions for flows in corridors, considering the effects of fire size, location of

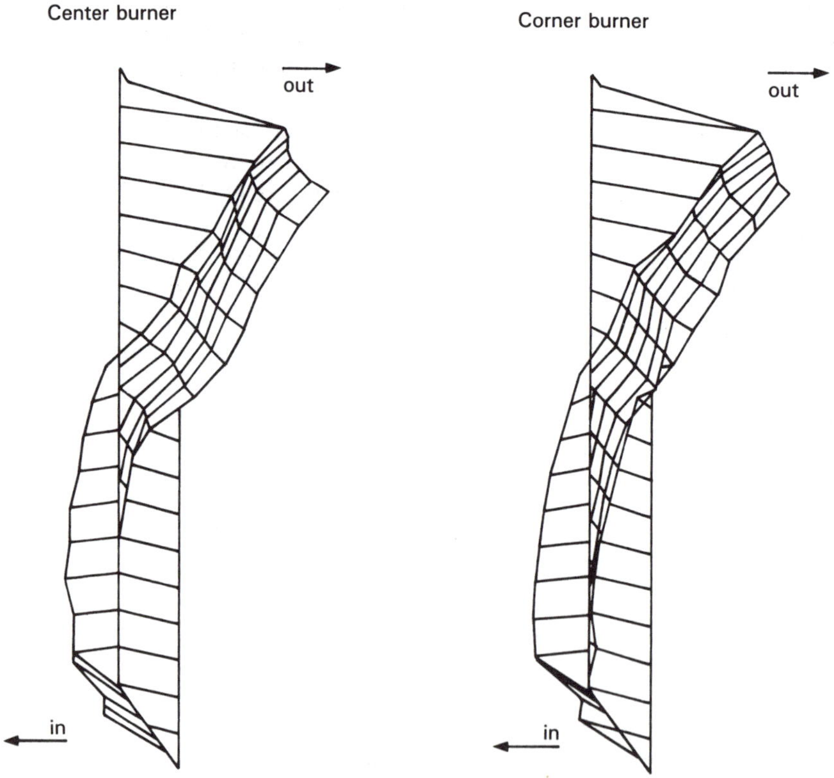

Figure 14.8.3 Experimentally determined velocity profiles at the doorway of a room with a gas burner at the center and then with one in a corner. The horizontal variation across the doorway is also indicated. *(Reprinted with permission from K. D. Steckler et al.,* 21st Symp. (Int.) Combust. *Copyright © 1982, Combustion Institute.)*

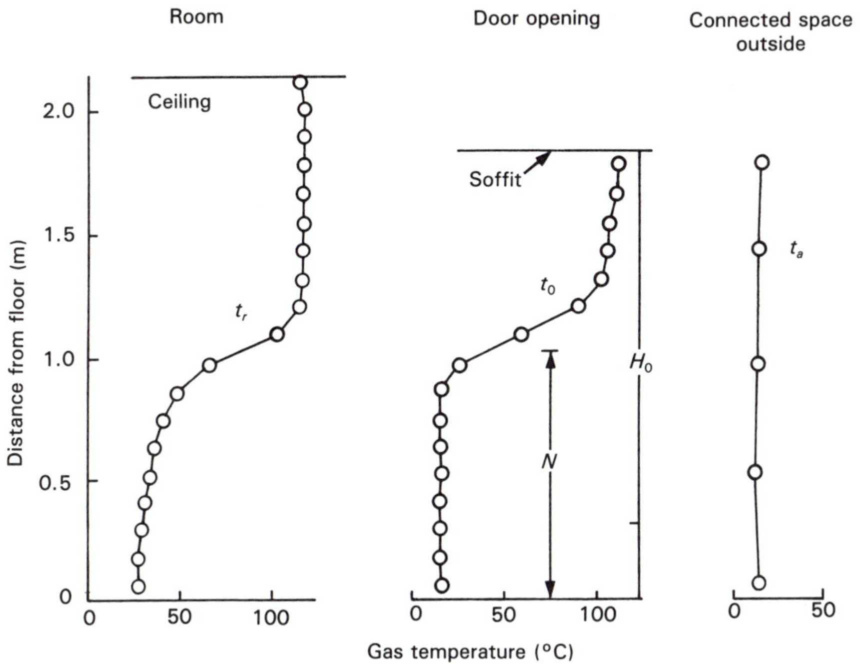

Figure 14.8.4 Measured vertical temperature profiles in the room, the doorway, and the space outside the opening, with a heat source just above the floor. *(Reprinted with permission from K. D. Steckler et al.,* 21st Symp. (Int.) Combust. *Copyright © 1982, Combustion Institute.)*

doorway openings, and space dimensions. Figure 14.8.5 shows the time-dependent variation of the interface height and of the upper layer fluid density as a function of the heat input. Zone models use turbulent plume analysis to determine the mass and energy transfer across the interface, considering mixing of fluid streams at the opening and the coupled radiation effects, to predict the growth of fire in rooms. Detailed work has therefore been done on entrainment mechanisms and combined convective and radiative transport, as outlined by Quintiere et al. (1981) and Quintiere (1981).

The major difficulty encountered in numerical solution of the governing equations is specification of the boundary conditions at openings. One approach is to use experimental or approximate analytical results to specify the pressure, velocity, and temperature distributions there. Another approach, outlined by Markatos et al. (1982), is to extend the flow domain to the free-boundary region outside the opening, as shown in Fig. 14.8.6. The dimensions needed for such an extension beyond the opening are determined numerically. This region is taken sufficiently large that a further increase does not significantly alter the results obtained for flow in the enclosure.

The resulting velocity and temperature distributions computed at the opening and in the room are shown in Fig. 14.8.7. Similar results have been obtained in

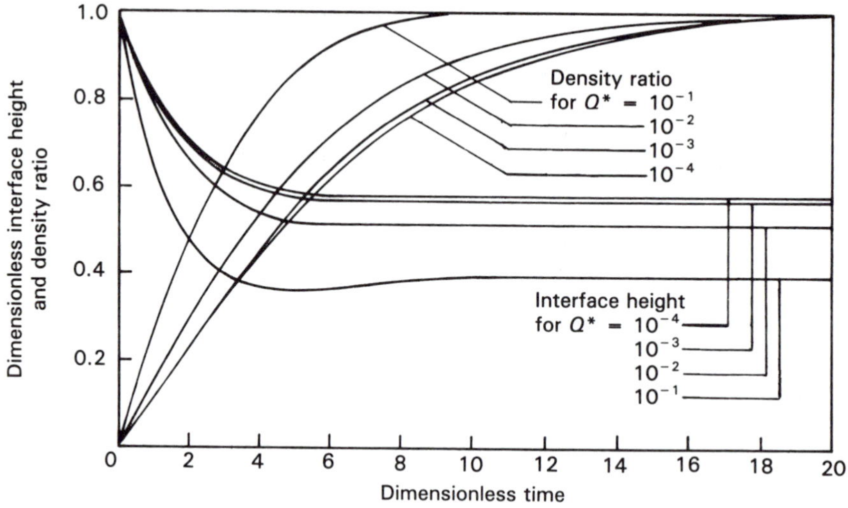

Figure 14.8.5 Computed dimensionless interface height and upper layer temperature as functions of dimensionless time for various values of the dimensionless heat input $Q^* = Q/\rho_\infty c_p T_\infty \sqrt{gH}\, H^2$, where Q is the heat release rate. *(Reprinted by permission of John Wiley & Sons, Ltd., E. E. Zukoski and T. Kubota,* Fire Mater., *vol. 4, p. 17. Copyright © 1980, John Wiley & Sons, Inc.)*

other numerical studies carried out by Ku et al. (1977), Yang and Chang (1977), and Liu and Yang (1978). The results of Yang and Chang are also shown in Fig. 14.8.7. Fairly good agreement between these results and those of Markatos et al. (1982) is seen. Under steady-state conditions the mass outflow equals the mass inflow at the opening, neglecting the fuel consumed. For very small openings a considerable amount of mixing occurs near the opening as the inflowing fluid interacts with the stratified upper layer. Thus, a gradual vertical temperature variation arises. A sharp change is found with large openings, which permits the use

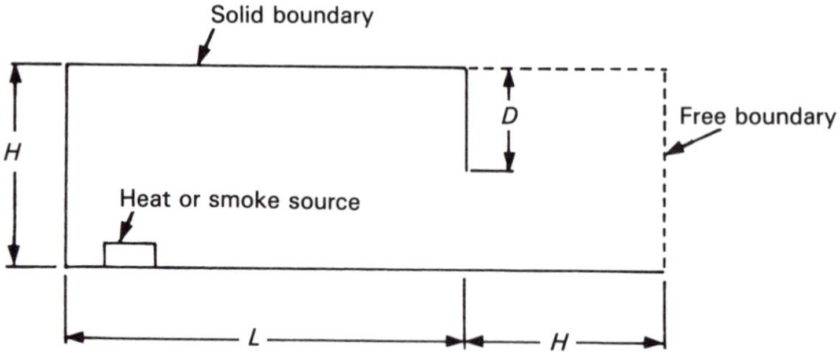

Figure 14.8.6 Computational domain for natural convection in an enclosure with an opening. *(Reprinted with permission from* Int. J. Heat Mass Transfer, *vol. 25, p. 63, N. C. Markatos et al. Copyright © 1982, Pergamon Journals Ltd.)*

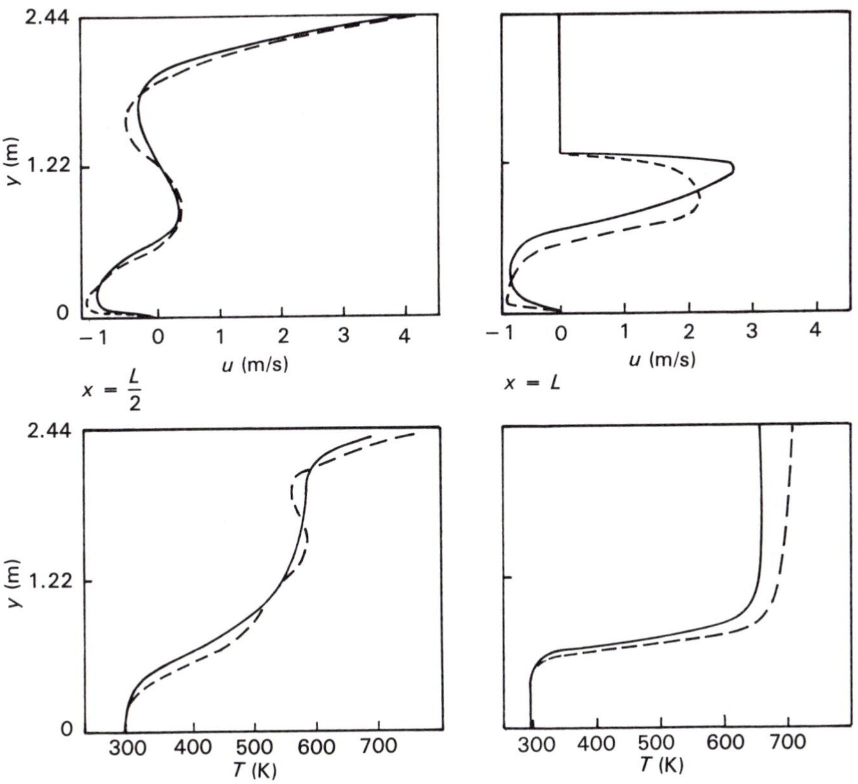

Figure 14.8.7 Horizontal velocity component u and temperature t fields at $x = L/2$ and $x = L$ in Fig. 14.8.6. Dashed curves are from Yang and Chang (1977). *(Reprinted with permission from* Int. J. Heat Mass Transfer, *vol. 25, p. 63, N. C. Markatos et al. Copyright © 1982, Pergamon Journals Ltd.)*

of a two-layer zone model. Extensive experimental results are presented for flow in a room with a single opening by Steckler et al. (1982).

Solar energy applications, particularly with respect to collector design and energy transfer within buildings, have also led to studies of natural convection flow in partial enclosures. Various orientations of a C-shaped enclosure have been considered experimentally and numerically to determine the transport that arises and the corresponding heat transfer from the solar-heated surfaces. As found experimentally by Sernas and Kyriakides (1982), the heat transfer from the vertical surface in this configuration is very close to that for external natural convection flow over a vertical surface in an extensive isothermal medium.

Several other studies have considered partitioned enclosures and other complex geometries that can be related to flow between rooms of a building and to the cooling of electronic components located in an enclosure with openings. See, for instance, the experimental work of Nansteel and Greif (1982) and the calculations of Chang et al. (1982a, 1982b).

14.9 COLD WATER EFFECTS

The anomalous density behavior of both pure and saline water has large effects in buoyancy-driven flows. Some of these are indicated in Chapter 9 for external flows. Characteristic density variations are seen in Fig. 9.1.1 for a symmetric density distribution around the temperature of the extremum, t_m. Buoyancy force and flow reversals arise. The various flow regimes were classified in Fig. 9.3.1 in terms of the parameter

$$R = \frac{t_m - t_\infty}{t_0 - t_\infty} \tag{9.3.14}$$

where t_0, t_∞, and t_m are the two imposed temperature conditions and the extremum temperature, respectively. Upflow and downflow result for $R \leq 0$ and $R \geq \frac{1}{2}$, respectively, and buoyancy force reversals are found only in the range $0 < R < \frac{1}{2}$.

In enclosed and partially enclosed bodies of cold water, similar considerations and regimes arise. Again R is the parameter. For enclosure bounding temperatures t_h and t_c, R is defined as

$$R = \frac{t_m - t_c}{t_h - t_c} \tag{14.9.1}$$

However, with multiple surfaces the significance of this quantity depends on the geometry. For example, consider the enclosure in Fig. 14.3.1, taking $t_c = 0$°C and the fluid to be pure water. If $t_h \leq 4$°C, that is, $R \geq 1$, there is upflow along the cold wall. A single cellular motion arises. However, for all $t_h > 4$°C, $0 \leq R < 1$, and maximum density fluid, at t_m, arises between the vertical boundaries. Thus, a buoyancy force reversal occurs across the region for $0 \leq R < 1$. A multicellular regime may arise, depending on the relative strength of the reversal compared to viscous forces. For $R = \frac{1}{2}$ the density extremum for the conduction temperature field lies halfway between the vertical boundaries. This occurs for all choices of t_h and t_c that are symmetric around t_m, that is, for $R = \frac{1}{2}$. Then symmetric counterrotating cells arise.

Increasing study has been given to such transport. Section 13.5 summarized results for thermal instability in a cold water layer between extensive parallel horizontal surfaces. Unstable stratification arises for $0 < R < 1$, as shown in Fig. 13.5.1. The critical Rayleigh numbers Ra_c are given across that range in Fig. 13.5.2. This section reviews the results for other enclosures, first for steady flow and then for transient flow.

14.9.1 Steady Flows in Enclosures

There have been a number of numerical calculations of such flows. Most are for the vertical rectangular enclosure shown in Fig. 14.3.1. The first calculation, by

Desai and Forbes (1971), considered aspect ratios $A = H/d$ of 1 and 3, with $d = 2$ cm and insulated horizontal boundaries. The two temperature conditions, t_h and t_c, were symmetrically placed about the extremum temperature t_m, taken as 4°C. Such conditions correspond to $R = \frac{1}{2}$ in Eq. (14.9.1). Calculations relied on sine and polynomial density relations $\rho(t)$.

A transient calculation procedure was used. The two steady-state cellular patterns found with the polynomial expression were almost symmetric. The asymmetry presumably results from the density relation used. It should be noted that the density changes in the range 0–8°C are extremely small. For example, the change from t_m to $t_m \pm 2$ and to $t_m \pm 4$°C are only 36 and 129 ppm, respectively, from Eq. (9.1.1). The uncertainty of the basic density data for pure water at 1 atm is variously estimated at about 3 ppm.

Nevertheless, a principal observation by Desai and Forbes was that heat transfer was mainly by conduction. Such small buoyancy effects produce only very weak flow in the small cavity considered. The value of Ra for these conditions was about 5×10^4.

Calculations were made for the same enclosure conditions and $A = 1$ by Watson (1972), using a polynomial equation for $\rho(t)$ and choosing $t_m = 3.98$°C. One surface was at $t_c = 0$°C. Patterns were given for t_h successively at 6, 7, 8, 9, and 10°C, for both uniform and variable viscosity, and Pr = 13.7. The values of R were thus about 0.67, 0.57, 0.50, 0.44, and 0.40. Therefore, a buoyancy force reversal arose under all conditions. However, a second cell, on the warm side, first became apparent at around $t_h = 7$°C. The cells were almost symmetric at $t_h = 8$°C. The cell on the cold side had almost disappeared at $t_h = 10$°C. The calculated effect of the anomalous density behavior on Nusselt number was very large over the range $t_h = 0$–16°C. A minimum value of Nu ≈ 1 was found at around $t_h = 8$°C or $R = 0.5$. The effect in external flows, seen in Fig. 9.3.7, is similar.

In an experimental and numerical investigation, Seki et al. (1978) considered $A = 1$–20 for $t_c = 0$°C and t_h from 1 to 12°C. Visualizations for $A = 1$ indicated that the flow patterns were similar to those calculated by Watson (1972). The calculated variation of Nu was almost the same. Measured temperature distributions are given for $A = 1$ and $t_h = 6$, 8, and 10°C, that is, $R \approx 0.67$, 0.5, and 0.4. Also, the large effect of enclosure width on the Nusselt number is shown over the range $A = 1$–5.

Lin and Nansteel (1987a) calculated the Rayleigh number effect over the range 10^3 to 10^6 in a square enclosure, L by L, that is, $A = 1$, for the same boundary conditions as above. The formulation of such enclosure transport is as given in Eqs. (14.3.1)–(14.3.4). However, the Boussinesq approximation appears in Eq. (14.3.2) as $g\beta(t - t_c)$. For cold water, using the density as Eq. (9.1.1), this term becomes

$$g\alpha(s, p)|t_h - t_c|^q[|\phi - R|^q - |R|^q]$$

This is analogous to Eq. (9.3.15) for external flows. The vertical force-momentum balance, Eq. (14.3.2), becomes

$$u\frac{\partial u}{\partial x} + v\frac{\partial u}{\partial y} = -\frac{1}{\rho}\frac{\partial p_m}{\partial x} + g\alpha(s,p)|t_h - t_c|^q(|\phi - R|^q - |R|^q) + \nu\left(\frac{\partial^2 u}{\partial x^2} + \frac{\partial^2 u}{\partial y^2}\right) \tag{14.9.2}$$

The equations were solved as a transient response, with the fluid initially at $(t_h + t_c)/2$. The four boundary conditions following Eq. (14.3.4) were suddenly applied, using the adiabatic condition at the top and bottom surfaces. For given values of the Prandtl and Rayleigh numbers, the only parameter that affects the eventual steady-state process is R. The definition of Ra is

$$\text{Ra} = \frac{g\alpha(s,p)L^3|t_h - t_c|^q \text{Pr}}{\nu^2} \tag{14.9.3}$$

Stream function distributions are given in Fig. 14.9.1 for Ra $= 10^3$ and $R = 0.67$, 0.55, 0.50, and 0.40. These may be interpreted as physical temperatures of $t_h = 6.0$, 7.3, 8.1, and 10.1°C, for t_c taken as 0°C. For all these conditions a downward buoyancy force occurs, between the warm boundary on the left and the cold one on the right. In Fig. 14.9.1*a* the small upward buoyancy effect at 6°C is effective in producing only two small clockwise patterns in the corners. In Fig. 14.9.1*c* the temperature field is symmetric and so is the flow. In Fig. 14.9.1*d* the more highly buoyant region on the left begins to dominate.

The pure Rayleigh number effect for $R = \frac{1}{2}$ is seen in Fig. 14.9.2. The flow pattern is about the same for Ra $= 10^4$ and 10^6. However, the higher velocities at 10^6 greatly deform the temperature field, thereby increasing the heat transfer rate. Analogous results are seen in Fig. 14.9.3, for $R = 0.55$. The buoyancy force on the left side is relatively much smaller. The temperature field is therefore even more deformed than at $R = \frac{1}{2}$.

These changing temperature fields indicate increasing heat transfer rates. A summary of such results is seen in Fig. 14.9.4. At large values of Ra the slopes of the curves are about 0.29, independent of R. The lowest values of Nu were found for $R = \frac{1}{2}$ because this is the symmetric condition around the extremum. This condition leads to the least buoyancy for any given value of the temperature difference $t_h - t_c$. Also, two equal cells arise, which isolate the two surfaces from each other. On the other hand, for $R \neq \frac{1}{2}$ a single cell may convect energy directly across the enclosure. This effect is seen more clearly in Fig. 14.9.5. For all values of Ra, Nu increases away from $R = \frac{1}{2}$. The pure conduction limit at low Rayleigh number is Nu $= 1$. At higher Ra the flow and heat transfer are much more vigorous. This general behavior is similar to that adjacent to vertical surfaces, as seen in Figs. 9.3.7 and 9.3.10.

The above formulation has also been used by Nansteel et al. (1987) in an asymptotic analysis, for small Ra, for enclosures of aspect ratios $A = \frac{1}{2}$, 1, and 2. The results are in close agreement with the numerical solutions above at small Ra. This confirms again that, with the small buoyancy forces around an extremum, heat transfer is essentially by conduction over an appreciable range of Ra.

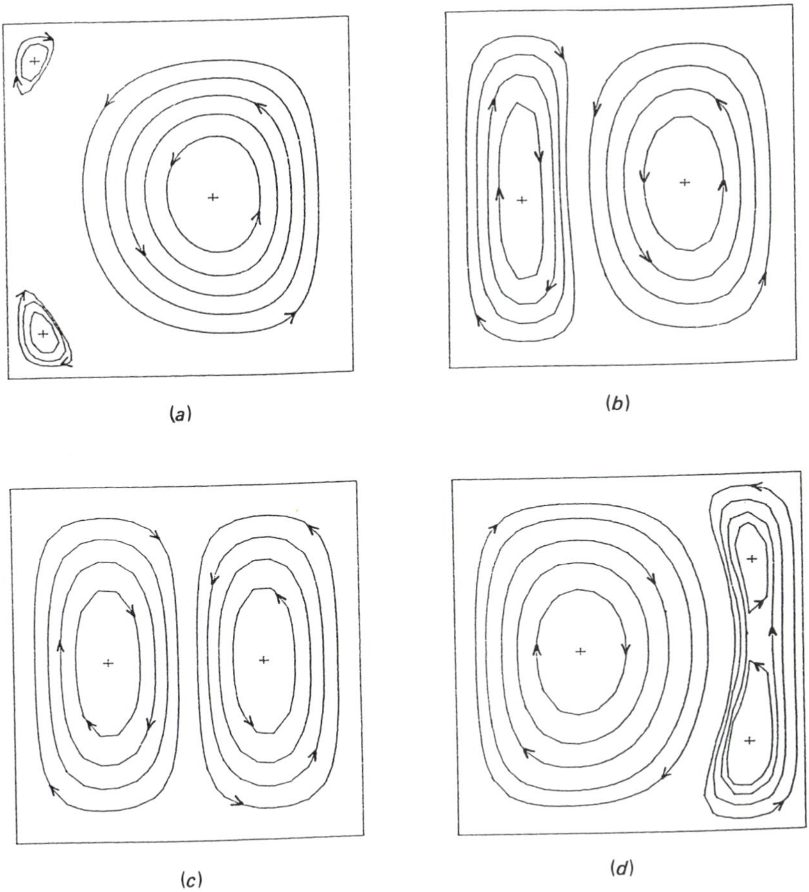

Figure 14.9.1 Convective motion of cold water in a square enclosure at Ra = 10^3 for Pr = 13 and for R = 0.67, 0.55, 0.50, and 0.4 in (*a*)–(*d*), respectively. *(From Lin and Nansteel, 1987a.)*

Inclined rectangular cavities. Calculations, visualizations, and measurements were reported by Inaba and Fukuda (1984a, 1984b) for a square water-filled enclosure. One wall was at 0°C and the other at various levels over the range 2–20°C. Transport at various angular orientations ξ was studied in 30° intervals of inclination angle from 0°, the warm surface at the bottom, to 180°, the warm surface at the top. Numerical calculations in general compared well with the observations. The flow and temperature patterns, and the variations of Nu, from 0 to 150°, were generally similar to those for vertical enclosures. The smallest values of Nu were typically for t_h = 8°C, which is $R = \frac{1}{2}$ in the calculation given above in detail.

Horizontal cylindrical annuli. Seki et al. (1975) reported visualizations and measurements for small water-filled horizontal concentric cylindrical annuli. The inside surface temperature t_i was 0°C and the outside temperature t_o was at dif-

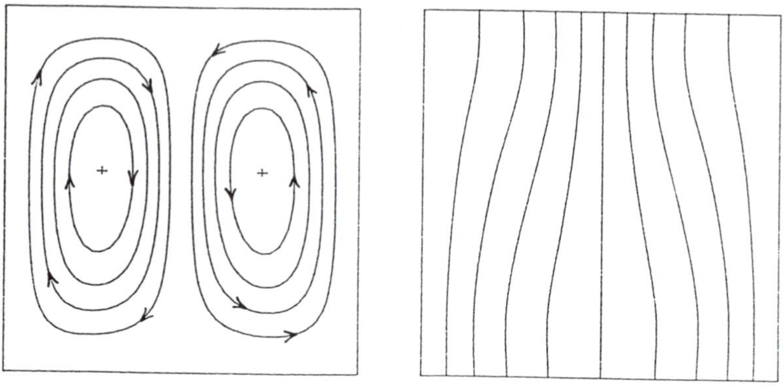

(*a*)

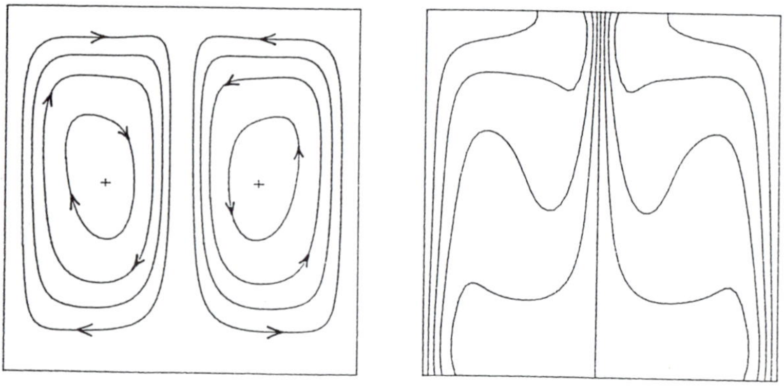

(*b*)

Figure 14.9.2 Stream function and temperature distributions for $R = \frac{1}{2}$ at Ra = (*a*) 10^4 and (*b*) 10^6. *(From Lin and Nansteel, 1987a.)*

ferent values over the range 1–15°C. The diameter ratio d_o/d_i was in the range 1.18–6.39, with L/d_i varying from 0.10–2.7. The Grashof number, based on annular length L, was between 3.2×10^1 and 2.7×10^5. The data indicate very large effects on the Nusselt number from the density inversion, which necessarily arises since $t_i = 0$°C. The visualizations showed very different and complicated patterns over this range of conditions. Minimum values of the Nusselt number resulted for values of t_o around 8°C, as for the rectangular enclosures.

Nguyen et al. (1982) followed the above measurements with a perturbation solution for the concentric horizontal cylindrical annulus. Expansions for nominally small Ra in terms of a parabolic equation of state were used. Calculations were made for bounding temperature conditions over a range of a parameter that is the same as R in Eq. (14.9.1), in terms of t_i and t_o. The results were in good agreement with the measurements of Seki et al. Transport was again much less in the range $0 < R < 1$.

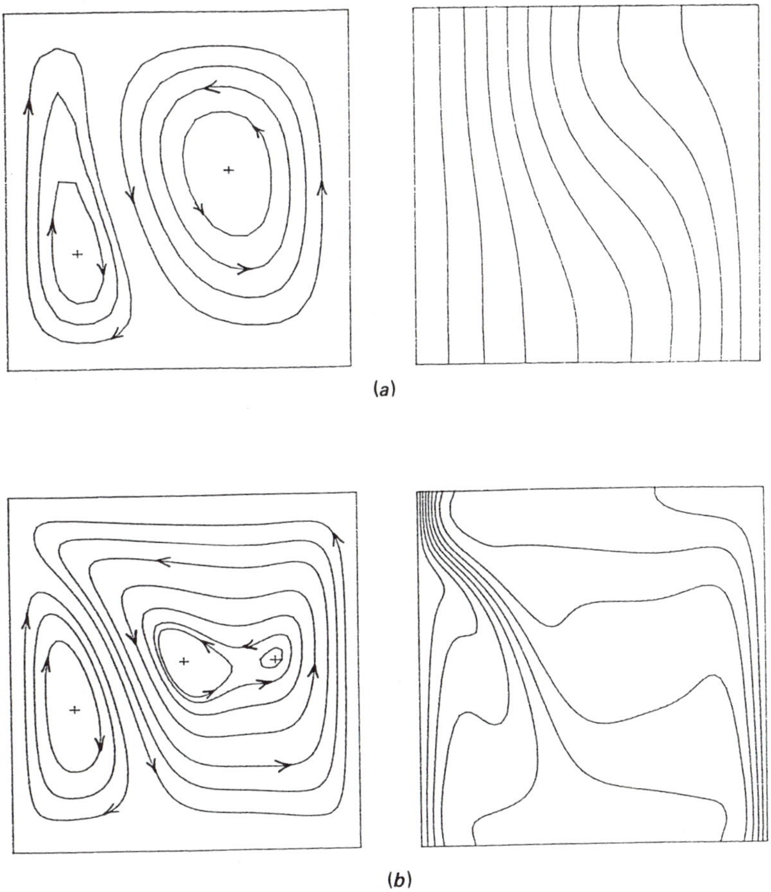

Figure 14.9.3 Stream function and temperature distributions for $R = 0.55$ at Ra = (*a*) 10^4 and (*b*) 10^6. *(From Lin and Nansteel, 1987a.)*

14.9.2 Transient Flows in Enclosures

Enclosure transport calculations often are carried out as a transient from an initial condition, with changed conditions at the boundaries. For example, the calculations of Lin and Nansteel (1987a,b), referred to above, were done in this way. On the other hand, the principal interest is sometimes in the transient development that follows changed surface conditions.

Rectangular enclosures. Robillard and Vasseur (1979) made such transient response calculations for a square enclosure filled with water initially uniformly at $t_i > 0$. The vertical surface temperatures were then changed to $t_c = 0°C$ and $t_h = 4$, 7, 8, 10, and 12°C in separate calculations. The steady-state distributions of stream function and temperature given are similar to those in other studies.

For the condition $t_h = 10°C$, the distributions are given at several time levels during the transient. A small cellular motion first arises in the corner at the lower

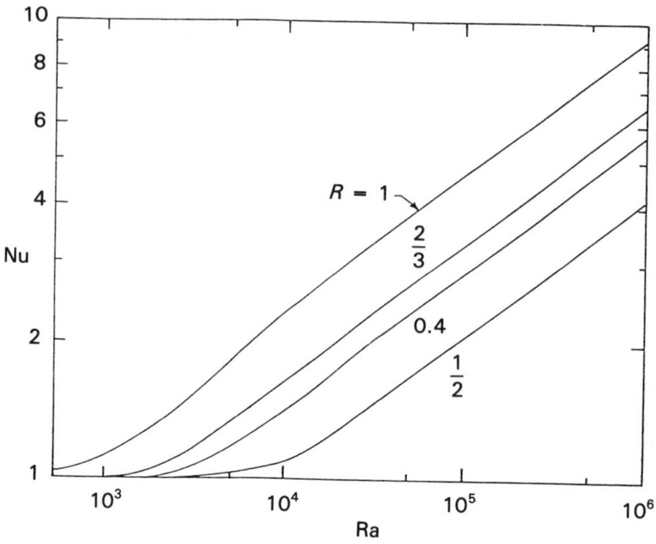

Figure 14.9.4 Nusselt number increase with Rayleigh number for representative values of R. *(From Lin and Nansteel, 1987a.)*

edge of the warm surface. It grows to eventually fill the whole cavity, except for a small counterrotating cell in the corner at the lower edge of the cold surface.

Transport response has also been assessed for three other kinds of boundary conditions, for the enclosure contents initially at t_i. Vasseur and Robillard (1980)

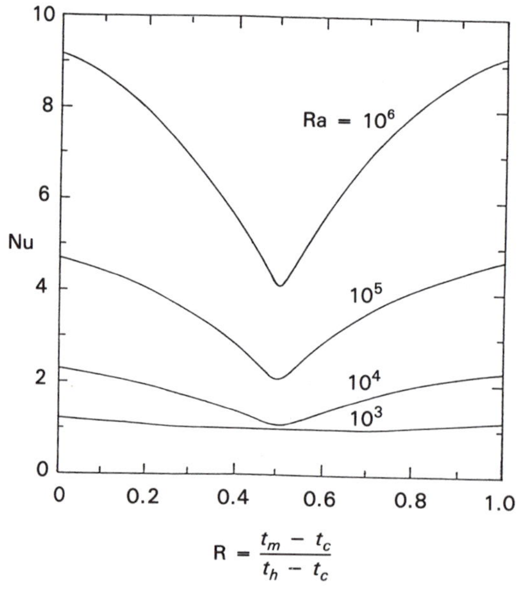

Figure 14.9.5 Increase in Nusselt number away from $R = \frac{1}{2}$ for Ra = 10^3–10^6. *(From Lin and Nansteel, 1987a.)*

considered t_i levels in the range 4–10°C, with the temperature of the four bounding surfaces decreased suddenly to 0°C. Examples of the evolution of flow and heat transfer are given. Robillard and Vasseur (1981, 1982) calculated the transient response for the four bounding walls convectively cooled by an outside environment at constant temperature and for the wall temperatures decreasing from t_i linearly with time. Transport distributions and temperature response information are given for representative conditions.

Horizontal cylindrical enclosures. Measurements and integral method boundary layer calculations were made by Gilpin (1975). Water in horizontal cylinders of $L/d \geq 3$ was cooled through 4°C. The water, initially usually at 20°C, was cooled by decreasing the wall temperature at a rate of 0.6–54 °C/h. Visualization was by dye streaks. Fixed thermocouples measured changing fluid temperature. Rayleigh numbers were in the range 10^5–10^8.

It was found that the transient internal circulation pattern soon became a quasi-static process. For $\text{Ra} > 10^6$ the internal flow amounted to a core, with boundary region flows adjacent to the surface. As the core temperature fell below 4°C, an inversion of the flow pattern was seen. The conduction field that arose in the core played an important role in this overall process. These results indicate very important and complicated mechanisms in this geometry, as the extremum effect is felt in the enclosure.

Cheng and Takeuchi (1976) numerically calculated the two-dimensional transient cooling of water in horizontal cylinders. The water was taken as initially at a temperature t_0 higher than t_m. Thereafter, the wall temperature was decreased from t_0 linearly with time. Results were given for $t_0 = 7$, 6, and 5°C. The unit Grashof number ga^3/ν^2, based on radius a, ranged from 10^7 to 2.5×10^8. The results supported the analysis of Gilpin (1975) and again indicated the severe effects of a density extremum on cooling.

Cheng et al. (1978) carried out additional numerical calculations for this geometry to assess supercooled conditions. The water was initially at $t_0 = 10$°C when a convection condition, h, to an ambient at $t_\infty = -10$°C was applied over the cylindrical surface. Here h is the total conductance through the wall and out to the exterior ambient. The Biot number ha/k was 2 and the unit Grashof number was 5×10^7. The numerical results were compared with photographs of dendritic ice growth. Additional numerical results were given for other values of t_0, t_∞, and unit Grashof number.

These studies indicate that transient cooling often departs from either a pure convection or a pure conduction mode. The net transient cooling process is often more complicated, in terms of the relation of the imposed temperatures to t_m. Predictions of water freezing in pipes must account for this, as well as for liquid supercooling and dendritic ice growth.

A partial enclosure. Forbes and Cooper (1975) considered the transient cooling of cold water in a rectangular enclosure, insulated across the bottom and up the sides but open across the top. The water was initially at $t_i = 4$ or 8°C. The transient response was calculated for two different cooling conditions at the upper surface of the water. In the first the surface temperature was taken as 0°C. In the second, convective processes with $h = 1$ and 50 Btu/h ft^2 °F were imposed between the water surface and ambient air at 0°C. Calculations were made for aspect ratios $A = 1$, 3, and 6, with a water layer 1 cm thick, using a quadratic density equation.

For $t_i = 4$°C no motion arises in the calculation since the fluid merely becomes stratified in an increasingly stable way. That is, the instantaneous R is 1.0. This stabilization arose for both kinds of surface conditions. Recall Fig. 13.5.1. However, for $t_i = 8$°C, R is initially $\frac{1}{2}$ and large convective circulations arise. They extend from the bottom to the top of the layer. Two layers then appear, approximately separated by a 4°C isotherm. Eddies remain in the upper unstable layer. These results emphasize the complexity that arises from vertical transient temperature fields across a fluid layer with extremum behavior.

REFERENCES

Abbott, M. R. (1964). *Q. J. Mech. Appl. Math. 17*, 471.

Akinsete, V. A., and Coleman, T. A. (1982). *Int. J. Heat Mass Transfer 25*, 991.

Alstad, C. D., Isbin, H. S., Amundson, N. R., and Silvers, J. P. (1955). *AIChE J. 1*, 417.

Arnold, J. N., Bonaparte, P. N., Catton, I., and Edwards, D. K. (1975). *Proc. 1974 Heat Transfer Fluid Mech. Inst.*, Stanford University.

Arnold, J. N., Catton, I., and Edwards, D. K. (1976). *J. Heat Transfer 98*, 67.

Astill, K. N., Leong, H., and Martorana, R. (1980). *ASME HTD Symp. 8*, 105.

Aung, W. (1972). *Int. J. Heat Mass Transfer 15*, 1577.

Aung, W., Fletcher, L. S., and Sernas, V. (1972). *Int. J. Heat Mass Transfer 15*, 2293.

Auxilien, J. M. (1981). *Lett. Heat Mass Transfer 8*, 303.

Ayyaswamy, P. S., and Catton, I. (1973). *J. Heat Transfer 95*, 543.

Aziz, K. (1965). Ph.D. Thesis, Rice Univ., Houston, Texas.

Aziz, K., and Hellums, J. D. (1967). *Phys. Fluids 10*, 314.

Badoia, J. R., and Osterle, J. F. (1962). *J. Heat Transfer 84*, 40.

Baines, W. D., and Turner, J. S. (1969). *J. Fluid Mech. 37*, 51.

Bankvall, C. G. (1974). *Waerme Stoffuebertrag. 7*, 22.

Batchelor, G. K. (1954). *Q. Appl. Math. 12*, 209.

Batchelor, G. K. (1956). *J. Fluid Mech. 1*, 177.

Bau, H. H., and Torrance, K. E. (1981a). *J. Fluid Mech. 106*, 417.

Bau, H. H., and Torrance, K. E. (1981b). *Int. J. Heat Mass Transfer 24*, 597.

Beckmann, W. (1931). *Forsch. Geb. Ingenieurwes.* 2(5), 165.

Bejan, A. (1979). *J. Fluid Mech. 90*, 561.

Bejan, A. (1980). *Int. J. Heat Mass Transfer 23*, 723.

Bejan, A., and Tien, C. L. (1978). *J. Heat Transfer 100*, 641.

Bejan, A., Al-Homoud, A. A., and Imberger, J. (1981). *J. Fluid Mech. 109*, 283.

Berkovsky, B. M., and Polevikov, V. K. (1977). In *Heat Transfer and Turbulent Buoyant Convection,* D. B. Spalding and H. Afghan, Eds., Hemisphere, Washington, D.C.

Birikh, R. V. (1966). *PMM 30*, 432.

Birikh, R. V., Gershuni, G. Z., Zhukhovitskii, E. M., and Rudakov, R. N. (1972). *PMM 36*, 745.

Bishop, E. H., and Carley, C. T. (1966). *Proc. 1966 Heat Transfer Fluid Mech. Inst.*, p. 63, Stanford University.

Bishop, E. H., Carley, C. T., and Powe, R. E. (1968). *Int. J. Heat Mass Transfer 11*, 1741.

Bishop, E. H., Kolflat, R. S., Mack, L. R., and Scanlan, J. A. (1964). *Proc. 1964 Heat Transfer Fluid Mech. Inst.*, p. 69, Stanford University.

Bishop, E. H., Mack, L. R., and Scanlan, J. A. (1966). *Int. J. Heat Mass Transfer 9*, 649.

Boyd, R. D. (1981). *Int. J. Heat Mass Transfer 24,* 1545.
Brooks, T., and Ostrach, S. (1970). *J. Fluid Mech. 44,* 545.
Buchberg, H., Catton, I., and Edwards, D. K. (1976). *J. Heat Transfer 98,* 193.
Burns, P. J., Chow, L. C., and Tien, C. L. (1977). *Int. J. Heat Mass Transfer 20,* 919.
Cabelli, A., and De Vahl Davis, G. (1971). *J. Fluid Mech. 45,* 805.
Catton, I. (1970). *J. Heat Transfer 92,* 186.
Catton, I. (1972). *Int. J. Heat Mass Transfer 15,* 665.
Catton, I. (1978). *Proc. 6th Int. Heat Transfer Conf.,* vol. 6, p. 13.
Catton, I., and Edwards, D. K. (1967). *J. Heat Transfer 89,* 295.
Catton, I., Ayyaswamy, P. S., and Clever, R. M. (1974). *Int. J. Heat Mass Transfer 17,* 173.
Chan, A. M. C., and Banerjee, S. (1979). *J. Heat Transfer 101,* 114.
Chan, P. K. B., Ozoe, H., Churchill, S. W., and Lior, N. (1983). *J. Heat Transfer 105,* 425.
Chang, L. C., Lloyd, J. R., and Yang, K. T. (1982a). *Proc. 7th Int. Heat Transfer Conf.,* Munich, Hemisphere, Washington, D.C.
Chang, L. C., Yang, K. T., and Lloyd, J. R. (1982b). Presented at the AIAA/ASME Conf., St. Louis, June 1982.
Charmchi, M., and Sparrow, E. M. (1982). *Numer. Heat Transfer 5,* 119.
Cheng, K. C., and Takeuchi, M. (1976). *J. Heat Transfer 98,* 581.
Cheng, K. C., Takeuchi, M., and Gilpin, R. R. (1978). *Numer. Heat Transfer 1,* 101.
Cheng, P. (1978). *Adv. Heat Transfer 14,* 1.
Cho, C. H., Chang, K. S., and Park, K. H. (1982). *J. Heat Transfer 104,* 624.
Choi, I. G., and Korpela, S. A. (1980). *J. Fluid Mech. 99,* 725.
Chorin, A. J. (1968). *Math. Comput. 22,* 745.
Chow, M. Y., and Akins, R. G. (1975). *J. Heat Transfer 97,* 54.
Chu, H. H. S., and Churchill, S. W. (1977). *Comput. Chem. Eng. 1,* 103.
Chu, H. H. S., Churchill, S. W., and Patterson, C. V. S. (1976). *J. Heat Transfer 98,* 194.
Chu, T. Y., and Goldstein, R. J. (1973). *J. Fluid Mech. 60,* 141.
Cormack, D. E., Leal, L. G., and Imberger, J. (1974a). *J. Fluid Mech. 65,* 209.
Cormack, D. E., Leal, L. G., and Seinfeld, J. H. (1974b). *J. Fluid Mech. 65,* 231.
Crawford, L., and Lemlich, R. (1962). *Ind. Eng. Chem. Fundam. 1,* 260.
Creveling, H. F., dePaz, J. F., Baladi, J. Y., and Schoenhals, R. J. (1975). *J. Fluid Mech. 67,* 65.
Custer, J. R., and Shaughnessy, E. J. (1977). *J. Heat Transfer 99,* 596.
Damerall, P. S., and Schoenhals, R. J. (1979). *J. Heat Transfer 101,* 672.
Davis, L. P., and Perona, J. J. (1971). *Int. J. Heat Mass Transfer 14,* 889.
Davis, S. H. (1967). *J. Fluid Mech. 30,* 465
Deardorff, J. W. (1964). *J. Atmos. Sci. 21,* 419.
Deardorff, J. W. (1965). *J. Atmos. Sci. 22,* 419.
Denny, V. E., and Clever, R. M. (1974). *J. Comput. Phys. 16,* 271.
Desai, V. S., and Forbes, R. E. (1971). *Environ. Geophys. Heat Transfer,* 41.
De Vahl Davis, G. (1968). *Int. J. Heat Mass Transfer 11,* 1675.
De Vahl Davis, G., and Mallinson, G. D. (1975). *J. Fluid Mech. 72,* 87.
De Vahl Davis, G., and Mallinson, G. D. (1976). *Comput. Fluids 4,* 29.
De Vahl Davis, G., and Thomas, R. W. (1969). *High Speed Comput. Fluid Dyn., Phys. Fluids, Suppl. II,* 198.
Donaldson, I. G. (1970). *U.N. Symp. Dev. Utili. Geothermal Resources,* Pisa, vol. 2, p. 649.
Drakhlin, E. (1952). *Zh. Tekh. Fiz. 22,* 829.
Dropkin, D., and Somerscales, E. (1965). *J. Heat Transfer 87,* 77.
Eckert, E. R. G., and Carlson, W. O. (1961). *Int. J. Heat Mass Transfer 2,* 106.
Eckert, E. R. G., and Soehngen, E. (1948). Tech. Rept. 5747, U.S. Air Force Air Material Command, Dayton, Ohio.
Edwards, D. K., and Catton, I. (1969). *Int. J. Heat Mass Transfer 12,* 23.
Elder, J. W. (1965a). *J. Fluid Mech. 23,* 77.
Elder, J. W. (1965b). *J. Fluid Mech. 23,* 99.
Elder, J. W. (1966). *J. Fluid Mech. 24,* 823.

Elder, J. W. (1969). *J. Fluid Mech. 35,* 417.
Elenbaas, W. (1942). *Physica 9,* 1.
Elsherbiny, S. M., Raithby, G. D., and Hollands, K. G. T. (1982). *J. Heat Transfer 104,* 96.
Emara, A. A., and Kulacki, F. A. (1980). *J. Heat Transfer 102,* 531.
Emmons, H. W. (1978). *17th Symp. (Int.) Combust.,* Combustion Institute, 1101.
Emmons, H. W. (1980). *Annu. Rev. Fluid Mech. 12,* 223.
Engel, R. K., and Mueller, W. K. (1967). ASME Paper 67-HT-16.
Farouk, B., and Guceri, S. I. (1981). In *Natural Convection, ASME HTD Symp. 16,* 143.
Flack, R. D. (1980). *J. Heat Transfer 102,* 770.
Flack, R. D., Konopnicki, T. T., and Rooke, J. H. (1979). *J. Heat Transfer 101,* 648.
Forbes, R. E., and Cooper, J. W. (1975). *J. Heat Transfer 97,* 47.
Fromm, J. E. (1965). *Phys. Fluids 8,* 1757.
Gershuni, G. Z., Zhukhovitskii, E. M., and Tarunin, E. L. (1966). *Mech. Liq. Gases Akad. Sci. USSR 5,* 56.
Gill, A. E. (1966). *J. Fluid Mech. 26,* 515.
Gill, A. E., and Davey, A. (1969). *J. Fluid Mech. 35,* 775.
Gill, A. E., and Kirkham, C. C. (1970). *J. Fluid Mech. 42,* 125.
Gille, J. (1967). *J. Fluid Mech. 30,* 371.
Gilpin, R. R. (1975). *Int. J. Heat Mass Transfer 18,* 1.
Globe, S., and Dropkin, D. (1959). *J. Heat Transfer 81,* 24.
Goldstein, R. J., and Chu, T. Y. (1969). *Prog. Heat Mass Transfer 2,* 55.
Greif, R., Zvirin, Y., and Mertol, A. (1979). *J. Heat Transfer 101,* 684.
Grigull, V., and Hauf, W. (1966). *Proc. 3rd Int. Heat Transfer Conf.,* vol. 2, p. 182, AIChE.
Guceri, S., and Farouk, B. (1985). In *Natural Convection Fundamentals and Applications,* S. Kakac, W. Aung, and R. Viskanta, Eds., Hemisphere, Washington, D.C.
Hallman, T. M. (1956). *Trans. ASME 78,* 1831.
Han, J. T. (1967). M.Sc. Thesis, Univ. of Toronto, Toronto, Ont., Canada.
Hart, J. E. (1971). *J. Fluid Mech. 47,* 547.
Hartman, R., and Ostrach, S. (1969). Rept. FTAS/TR-69-36, Case Western Reserve Univ., Cleveland, Ohio.
Hartnett, J. P., and Welsh, W. E. (1957). *Trans. ASME 79,* 1551.
Hasegawa, S., Nishikawa, K., and Yamagata, K. (1963). *Bull. JSME 6,* 930.
Hasemi, Y. (1976). Paper 66, Building Research Inst., Japan.
Hellums, J. D., and Churchill, S. W. (1962). *AIChE J. 8,* 692.
Herring, J. R. (1963). *J. Atmos. Sci. 20,* 325.
Herring, J. R. (1964). *J. Atmos. Sci. 21,* 277.
Hodnett, P. F. (1973). *J. Appl. Math. Phys. 24,* 507.
Hollands, K. G. T., and Konicek, L. (1973). *Int. J. Heat Mass Transfer 16,* 1467.
Hollands, K. G. T., Raithby, G. D., and Konicek, L. (1975). *Int. J. Heat Mass Transfer 18,* 879.
Hollands, K. G. T., Unny, T. E., Raithby, G. D., and Konicek, L. (1976). *J. Heat Transfer 98,* 189.
Huetz, J., and Petit, J. P. (1974). *Proc. 5th Int. Heat Transfer Conf.,* Tokyo, vol. 3, p. 169.
Imberger, J. (1974). *J. Fluid Mech. 65,* 247.
Inaba, H., and Fukuda, T. (1984a). *J. Heat Transfer 106,* 109.
Inaba, H., and Fukuda, T. (1984b). *J. Fluid Mech. 142,* 363.
Ingham, D. B. (1981). *Numer. Heat Transfer 4,* 53.
Jakob, M. (1949). *Heat Transfer,* Wiley, New York.
Jaluria, Y. (1980). *Natural Convection Heat and Mass Transfer,* Pergamon, Oxford, London.
Jaluria, Y. (1985). Natural Convective Cooling of Electronic Equipment, in *Natural Convection: Fundamentals and Applications,* S. Kakac, W. Aung, and R. Viskanta, Eds., Hemisphere, Washington, D.C.
Jaluria, Y., and Gebhart, B. (1974). *J. Fluid Mech. 66,* 593.
Jaluria, Y., and Torrance, K. E. (1986). *Computational Heat Transfer,* Hemisphere, Washington, D.C.

Japikse, D. (1969). *Adv. Heat Transfer 9,* 1.
Japikse, D. (1973). Heat Transfer in Open and Closed Thermosyphons, Ph.D. Thesis, Purdue Univ., West Lafayette, Ind.
Japikse, D., and Winter, E. R. F. (1970). *Heat Transfer 1970,* vol. 4, NC2.9, Elsevier, Amsterdam.
Japikse, D., Jallouk, P. A., and Winter, E. R. F. (1971). *Int. J. Heat Mass Transfer 14,* 896.
Jischke, M. C., and Farshchi, M. (1980). *J. Heat Transfer 102,* 228.
Kaviany, M. (1981). *ASME HTD Symp. 16,* 75.
Kettleborough, C. F. (1972). *Int. J. Heat Mass Transfer 15,* 883.
Keyhani, M., Kulacki, F. A., and Christensen, R. N. (1983). *J. Heat Transfer 105,* 454.
Klarsfeld, S. (1970). *Rev. Gen. Therm. 108,* 1403.
Kraichnan, R. H. (1962). *Phys. Fluids 5,* 1374.
Kraussold, H. (1934). *Forsch. Geb. Ingenieurwes. 5,* 186.
Krishnamurti, R. (1970). *J. Fluid Mech. 42,* 295.
Ku, A. C., Doria, M. L., and Lloyd, J. R. (1977). *16th Symp. (Int.) Combust.*, Combustion Institute, Pittsburgh, Pa., 1373.
Kuehn, T. H., and Goldstein, R. J. (1976a). *J. Fluid Mech. 74,* 695.
Kuehn, T. H., and Goldstein, R. J. (1976b). *Int. J. Heat Mass Transfer 19,* 1127.
Kuehn, T. H., and Goldstein, R. J. (1978). *J. Heat Transfer 100,* 635.
Kuehn, T. H., and Goldstein, R. J. (1980). *J. Heat Transfer 102,* 768.
Kulacki, F. A., and Emara, A. A. (1977). *J. Fluid Mech. 83,* 375.
Kulacki, F. A., and Goldstein, R. J. (1972). *J. Fluid Mech. 55,* 271.
Lapin, Y. D. (1969). *Therm. Eng. 16,* 94.
Lee, J. H., and Lee, T. S. (1981). *Int. J. Heat Mass Transfer 24,* 1739.
Lewis, J. A. (1950). Ph.D. Thesis, Brown Univ., Providence, R.I.
Lewis, J. A., and Carrier, G. F. (1949). *Q. Appl. Math. 7,* 228.
Lighthill, M. J. (1953). *J. Mech. Appl. Math. 6,* 398.
Lin, D. S., and Nansteel, M. W. (1987a). *Int. J. Heat Mass Transfer*. In press.
Lin, D. S., and Nansteel, M. W. (1987b). *J. Heat Transfer*. In press.
Lis, J. (1966). *Proc. 3rd Int. Heat Transfer Conf.,* AIChE, vol. 2, p. 196.
Liu, C. Y., Mueller, W. K., and Landis, F. (1961). *Int. Dev. Heat Transfer 4,* 976.
Liu, V. K., and Yang, K. T. (1978). Tech. Rept. TR-79002-78-3, Univ. of Notre Dame, Notre Dame, Ind.
Lock, G. S. H. (1962). Ph.D. Thesis, Univ of Durham, England.
Luikov, A. V., Berkovskii, and Fentman, V. E. (1969). *Prog. Heat Mass Transfer 2,* 77.
Lyican, L., Bayazitoglu, Y., and Witte, L. C. (1980a). *J. Heat Transfer 102,* 640.
Lyican, L., Witte, L. C., and Bayazitoglu, Y. (1980b). *J. Heat Transfer 102,* 648.
Lykoudis, P. S. (1974). *Int. J. Heat Mass Transfer 17,* 161.
MacGregor, R. K., and Emery, A. F. (1969). *J. Heat Transfer 91,* 391.
Mack, L. R., and Bishop, E. H. (1968). *Q. J. Mech. Appl. Math. 21,* 233.
Mack, L. R., and Hardee, H. C. (1968). *Int. J. Heat Mass Transfer 11,* 387.
Mallinson, G. D., and De Vahl Davis, G. (1973). *J. Comput. Phys. 12,* 435.
Mallinson, G. D., and De Vahl Davis, G. (1977). *J. Fluid Mech. 83,* 1.
Mallinson, G. D., Graham, A. D., and De Vahl Davis, G. (1981). *J. Fluid Mech. 109,* 259.
Markatos, N. C., Malin, M. R., and Cox, G. (1982). *Int. J. Heat Mass Transfer 25,* 63.
Martin, B. W. (1954). *Br. J. Appl. Phys. 5,* 91.
Martin, B. W. (1955). *Proc. R. Soc. London Ser. A 230,* 502.
Martini, W. D., and Churchill, S. W. (1960). *AIChE J. 6,* 251.
Maslen, S. H. (1958). NACA TN-4319.
McCaffrey, B. J., and Quintiere, J. G. (1977). In *Heat Transfer and Turbulent Buoyant Convection,* p. 457, Hemisphere, Washington, D.C.
McCoy, C. T., Powe, R. E., Bishop, E. H., Weber, N., and Scanlan, J. (1974). *Proc. 5th Int. Heat Transfer Conf.,* Tokyo.
Menold, E. R., and Ostrach, S. (1965). Rept. FTAS/TR-65-4, Case Western Reserve Univ., Cleveland, Ohio.

Mertol, A., and Greif, R. (1985). A Review of Natural Circulation Loops, in *Natural Convection: Fundamentals and Applications,* S. Kakac, W. Aung, and R. Viskanta, Eds., Hemisphere, Washington, D.C.
Mertol, A., Greif, R., and Zvirin, Y. (1982). *J. Heat Transfer 104,* 508.
Miyatake, O., and Fujii, T. (1972). *Heat Transfer Jpn. Res. 3,* 30.
Miyatake, O., Fujii, T., Fujii, M., and Tanaka, H. (1973). *Heat Transfer Jpn. Res. 4,* 25.
Morton, B. R., Taylor, G. I., and Turner, J. S. (1956). *Proc. R. Soc. London Ser. A 234,* 1.
Mull, W., and Reiher, H. (1930). *Beih. Z. Gesundh. Ing. 1,* 1.
Nagendra, H. R., Tirunarayanan, M. A., and Ramachandran, A. (1970). *Chem. Eng. Sci. 5,* 605.
Nakamura, H., Asako, Y., and Naitou, T. (1982). *Numer. Heat Transfer 5,* 95.
Nansteel, M. W., and Greif, R. (1982). *J. Heat Transfer 103,* 623.
Nansteel, M. W., Medjani, K., and Lin, D. S. (1987). *Phys. Fluids 30,* 312.
Newell, M. E., and Schmidt, F. W. (1970). *J. Heat Transfer 92,* 159.
Nguyen, T. H., Vasseur, P., and Robillard, L. (1982). *Int. J. Heat Mass Transfer 25,* 1559.
Oertel, H. H. (1980). *ASME HTD Symp. 8,* 11.
Ostrach, S. (1950). A Boundary Layer Problem in the Theory of Free Convection, Ph.D. Thesis, Brown Univ., Providence, R.I.
Ostrach, S. (1952). NACA TN-2863.
Ostrach, S. (1955). NACA TN-3458.
Ostrach, S. (1964). In *High Speed Aerodynamics and Jet Propulsion,* vol. 4, *Theory of Laminar Flows,* Chap. F, Princeton Univ. Press, Princeton, N.J.
Ostrach, S. (1972). *Adv. Heat Transfer 8,* 161.
Ostrach, S., and Menold, E. R. (1965). *Proc. All Union Conf. Heat Mass Transfer,* Minsk, U.S.S.R., vol. 1, p. 640.
Ostrach, S., and Pneuli, D. (1963). *Trans. ASME Ser. C 85,* 346.
Ostrach, S., and Thornton, P. R. (1958). *Trans. ASME 80.*
Ostroumov, G. A. (1952). Technio-Theoretical Literature, Moscow (also NACA TM-1407, 1958).
O'Toole, J. L., and Silveston, P. L. (1961). *AIChE Chem. Eng. Prog. Symp. Ser. 57*(32), 81.
Ozoe, H., Fujii, K., Lior, N., and Churchill, S. W. (1983). *Int. J. Heat Mass Transfer 26,* 1427.
Ozoe, H., Sayama, H., and Churchill, S. W. (1975). *Int. J. Heat Mass Transfer 18,* 1425.
Ozoe, H., Yamamoto, K., Churchill, S. W., and Sayama, H. (1976). *J. Heat Transfer 98,* 202.
Patankar, S. V. (1980). *Numerical Methods in Heat Transfer and Fluid Flow,* Hemisphere, Washington, D.C.
Pepper, D. W., and Cooper, R. E. (1982). AIAA Paper AIAA-82-0983.
Pillow, A. F. (1952). Aerospace Res. Lab. Rept. A79, Melbourne.
Pimputkar, S. M., and Ostrach, S. (1981). *J. Cryst. Growth 55,* 614.
Poots, G. (1958). *Q. J. Mech. Appl. Math. 11,* 257.
Poulikakos, D., and Bejan, A. (1983). *J. Heat Transfer 105,* 652.
Powe, R. E. (1974). *J. Heat Transfer 96,* 558.
Powe, R. E., and Warrington, R. O. (1983). *J. Heat Transfer 105,* 440.
Powe, R. E., Baughman, R. C., Scanlan, J. A., and Teng, J. T. (1975). *J. Heat Transfer 97,* 296.
Powe, R. E., Carley, C. T., and Bishop, E. H. (1969). *J. Heat Transfer 91,* 310.
Powe, R. E., Carley, C. T., and Carruth, S. L. (1971). *J. Heat Transfer 93,* 210.
Powe, R. E., Warrington, R. O., and Scanlan, J. A. (1980). *Int. J. Heat Mass Transfer 23,* 1337.
Projahn, U., Reiger, H., and Beer, H. (1981). *Numer. Heat Transfer 4,* 131.
Pustovoit, S. P. (1958). *P.M.M. 22,* 568.
Quintiere, J. G. (1977). ASTM Tech. Publ. 614.
Quintiere, J. G. (1984). *Combust. Sci. Technol. 39,* 11.
Quintiere, J. G., McCaffrey, B. J., and DenBraven, K. (1978). *17th Symp. (Int.) Combust.,* Combustion Institute, 1125.
Quintiere, J. G., Rinkinen, W. J., and Jones, W. W. (1981). *Combust. Sci. Technol. 26,* 193.
Quon, C. (1972). *Phys. Fluids 15,* 12.
Quon, C. (1977). *J. Heat Transfer 99,* 340.
Raithby, G. D., and Hollands, K. G. T. (1975). *Adv. Heat Transfer 11,* 265.

Raithby, G. D., and Hollands, K. G. T., (1985). In *Handbook of Heat Transfer,* W. M. Rohsenow, J. P. Hartnett, and E. N. Ganic, Eds., McGraw-Hill, New York.
Ramachandran, N., Gupta, J. P., and Jaluria, Y. (1982). *Int. J. Heat Mass Transfer 25,* 187.
Ramachandran, N., Jaluria, Y., and Gupta, J. P. (1981). *Lett. Heat Mass Transfer 8,* 69.
Ratzel, A. C., Hickox, C. E., and Gartling, D. K. (1979). *J. Heat Transfer 101,* 108.
Rehm, R. G., and Baum, H. R. (1978). *J. Res. Natl. Bur. Stand. 83,* 297.
Rehm, R. G., Baum, H. R., and Barnett, P. D. (1982). *J. Res. Natl. Bur. Stand. 87,* 165.
Roache, P. J. (1982). *Computational Fluid Dynamics,* Hermosa, Albuquerque, N.M.
Roberts, P. H. (1967). *J. Fluid Mech. 30,* 33.
Robillard, L., and Vasseur, P. (1979). *Can. J. Civ. Eng. 6,* 481.
Robillard, L., and Vasseur, P. (1981). *J. Heat Transfer 103,* 528.
Robillard, L., and Vasseur, P. (1982). *J. Fluid Mech. 118,* 123.
Rossby, H. T. (1969). *J. Fluid Mech. 36,* 309.
Rudakov, R. N. (1966). *P.M.M. 30,* 362.
Ruth, D. W., Hollands, K. G. T., and Raithby, G. D. (1980a). *J. Fluid Mech. 96,* 461.
Ruth, D. W., Raithby, G. D., and Hollands, K. G. T. (1980b). *J. Fluid Mech. 96,* 481.
Sabzevari, A., and Ostrach, S. (1966). Rept. FTAS/TR-66-8, Case Western Reserve Univ., Cleveland, Ohio.
Satoh, K., Lloyd, J. R., and Yang, K. T. (1980). *Proc. East. Sect. Combust. Inst.,* Paper 20.
Satoh, K., Lloyd, J. R., Yang, K. T., and Kanury, A. M. (1981). Paper presented at the 2d National Conference on Numerical Methods in Heat Transfer, Univ. of Maryland, College Park.
Scanlan, J. A., Bishop, E. H., and Powe, R. E. (1970). *Int. J. Heat Mass Transfer 13,* 1857.
Schinkel, W. M. M., and Hoogendoorn, C. J. (1978). *Proc. 6th Int. Heat Transfer Conf.,* Toronto.
Schneider, K. J. (1963). *Proc. Int. Inst. Refrig. 1963,* 247.
Schwab, T. H., and DeWitt, K. J. (1970). *AIChE J. 16,* 1005.
Seki, N., Fukusako, S., and Inaba, H. (1978). *Waerme Stoffuebertrag. 11,* 145.
Seki, N., Fukusako, S., and Nakaoka, M. (1975). *J. Heat Transfer 97,* 556.
Sernas, V., and Kyriakides, I. (1982). *Proc. 7th Int. Heat Transfer Conf.,* Munich, vol. 2, p. 275, Hemisphere, Washington, D.C.
Shaidurov, G. F. (1958). *Sov. Phys. Tech. Phys. 3,* 799.
Sheriff, N. (1966). *Proc. 3rd Int. Heat Transfer Conf.,* Chicago, vol. 2, p. 132,.
Shiralkar, G. S., and Tien, C. L. (1981). *J. Heat Transfer 103,* 226.
Shiralkar, G. S., Gadgil, A., and Tien, C. L. (1981). *Int. J. Heat Mass Transfer 24,* 1621.
Silveston, P. L. (1958). *Forsch. Geb. Ingenieurwes. 24,* 59.
Simpkins, P. G., and Dudderar, T. D. (1981). *J. Fluid Mech. 110,* 433.
Soberman, R. K. (1958). *J. Appl. Phys. 29,* 872.
Sparrow, E. M., and Bahrami, P. A. (1980). *J. Heat Transfer 102,* 221.
Sparrow, E. M., and Charmchi, M. (1983). *Int. J. Heat Mass Transfer 26,* 113.
Sparrow, E. M., and Tao, W. Q. (1982). *Numer. Heat Transfer 5,* 39.
Sparrow, E. M., Husar, R. B., and Goldstein, R. J. (1970). *J. Fluid Mech. 41,* 793.
Sparrow, E. M., Patankar, S. V., and Ramadhyani, S. (1977). *J. Heat Transfer 99,* 520.
Steckler, K. D., Quintiere, J. G., and Rinkinen, W. J. (1982). *21st Symp. (Int.) Combust.,* Combustion Institute.
Szekeley, J., and Stanek, V. (1970). *Metal. Trans. 1,* 2243.
Tanaka, T. (1978). BRI Rept. 79, Building Research Inst., Japan.
Thomas, R. W., and De Vahl Davis, G. (1970). *Proc. 4th Int. Heat Transfer Conf.,* Paper NC2.4, Elsevier, Amsterdam.
Torrance, K. E. (1979a). *J. Heat Transfer 101,* 677.
Torrance, K. E. (1979b). *J. Fluid Mech. 95,* 477.
Torrance, K. E., and Catton, I., Eds. (1980). *Natural Convection in Enclosures, ASME HTD Symp.,* vol. 8.
Torrance, K. E., and Chan, V. W. C. (1981). *Int. J. Heat Mass Transfer 23,* 1091.
Torrance, K. E., and Rockett, J. A. (1969). *J. Fluid Mech. 36,* 33.
Torrance, K. E., Orloff, L., and Rockett, J. A. (1969). *J. Fluid Mech. 36,* 21.

Turner, J. S. (1973). *Buoyancy Effects in Fluids*, Cambridge Univ. Press, London.
Van Doormal, J. P., Raithby, G. D., and Strong, A. B. (1981). *Numer. Heat Transfer 4*, 21.
Vasseur, P., and Robillard, L. (1980). *Int. J. Heat Mass Transfer 23*, 1195.
Vest, C. M., and Arpaci, V. S. (1969). *J. Fluid Mech. 36*, 1.
Walker, K. L., and Homsy, G. M. (1978). *J. Fluid Mech. 87*, 449.
Warrington, R. O., and Powe, R. E. (1981). In *Natural Convection, ASME HTD Symp.*, vol. 16, p. 111.
Watson, A. (1972). *Q. J. Mech. Appl. Math. 25*(4), 421.
Weber, J. E. (1975). *Int. J. Heat Mass Transfer 18*, 569.
Weber, N., Powe, R. E., Bishop, E. H., and Scanlan, J. A. (1973). *J. Heat Transfer 95*, 47.
Weinbaum, S. (1964). *J. Fluid Mech. 18*, 409.
Welander, P. (1967). *J. Fluid Mech. 29*, 17.
Whitley, H. G., and Vachon, R. I. (1972). *J. Heat Transfer 94*, 360.
Wilkes, J. O. (1963). The Finite Difference Computation of Natural Convection in an Enclosed Rectangular Cavity, Ph.D. Thesis, Univ. of Michigan, Ann Arbor.
Williams, G. P. (1969). *J. Fluid Mech. 37*, 727.
Wirtz, R. A., and Stutzman, R. J. (1982). *J. Heat Transfer 104*, 501.
Yang, K. T., and Chang, J. C. (1977). Tech. Rept. TR-79002-77-1, Univ. of Notre Dame, Notre Dame, Ind.
Yao, L. S. (1980). *J. Heat Transfer 102*, 279.
Yao, L. S., and Chen, F. F. (1980). *J. Heat Transfer 102*, 667.
Zagromov, Y. A., and Lyalikov, A. S. (1966). *Inzh. Fiz. Zh. 10*, 577.
Zukoski, E. E. (1978). *Fire Mater. 2*, 54.
Zukoski, E. E., and Kubota, T. (1980). *Fire Mater. 4*, 17.
Zvirin, Y. (1979). *Int. J. Heat Mass Transfer 22*, 1539.

PROBLEMS

14.1 Two vertical walls 10 cm high are separated by a distance D; their temperatures are 100 and 30°C. For air, compute the Rayleigh number if $D = 1.0$ and 10 cm. Determine the heat transfer across the air gap for all edges open and for both side edges closed.

14.2 Two vertical walls at temperatures 200 and 0°C are separated by a distance of 1 cm. For air, find the heat transfer across the air gap. What is the heat transfer if the gap is 2 cm wide? The bottom and top are closed with adiabatic surfaces.

14.3 Two horizontal plates 20 cm wide are 10 cm apart with water between them. One plate is at 80°C and the other at 0°C. Find the heat transfer when the hotter one is below the colder one and when the positions are reversed. All edges are closed with adiabatic surfaces.

14.4 Consider the air gap between two large flat vertical surfaces a distance d apart. The initial temperature is t_i for both surfaces, and suddenly one of them is raised to a temperature t_o. Assuming developed flow in the gap, obtain the governing equations and corresponding initial and boundary conditions. Give the final steady-state solution and qualitatively sketch the transient velocity and temperature variations.

14.5 A vertical rectangular channel 1 m high and 1 cm wide is filled with air. If the vertical walls are at 120 and 20°C, respectively, the horizontal walls being adiabatic, determine the heat transfer across the air gap, the Nusselt number, and the effective conductivity k_e. Find the heat transfer if the gap is made 10 cm wide.

14.6 A solar energy collector consists of a vertical enclosure 1 m high with the two vertical walls at a temperature difference of 100°C. The edges are closed with adiabatic walls. Determine the heat transfer across the air gap if the width of the enclosure is 1 cm and also if it is 10 cm. Would you expect an optimum width at which the heat transfer is

(a) Minimum?
(b) Maximum?

Determine these values.

14.7 Consider a vertical enclosure of height 5 cm filled with water. A temperature difference of 50°C is maintained between the two vertical walls. All edges are closed with adiabatic surfaces. If the distance d between the walls is varied, find the value of d for maximum and for minimum heat transfer across the water gap. Also give the Nusselt numbers at these two values. Use water properties at 40°C.

14.8 A small thermal energy source of strength Q_0 is located on the floor of a room that is cubic and has each side of dimension L. Estimate the nondimensional time required for the hot region at the top of the room to reach thickness l. Use simple modeling for the flow.

14.9 Repeat the calculation for the temperature conditions in Problem 14.7 if the fluid is air at 1 atm and the height is 20 cm.

14.10 A long rectangular enclosure of height H and width d has insulated horizontal walls. The vertical walls are at 65 and 55°F, respectively. Consider the enclosure filled with air and with water, in turn.

(a) For $H = 3$ in., find each Nusselt number for $d = 15$ in. and for $d = \frac{1}{2}$ in.

(b) For each fluid, over what range of H/d will the heat transfer rate be greater than 10 times that by pure conduction alone?

(c) For $H = d = 3$ ft, give a simple estimate of the Nusselt number for both fluids.

14.11 Consider two parallel wide plates of height 1 m separated by an air gap 1 cm thick. If the plates are maintained at temperatures 100 and 20°C, determine the heat transfer between the plates and the resulting flow rate. The top and the bottom are open to the ambient. Assume a fully developed flow between the plates.

14.12 In Problem 14.11 compute the approximate height of the region over which the flow is developing.

14.13 Show that the flow in a rectangular enclosure depends only on the local temperature gradients and is independent of the reference temperature t_r used in the buoyancy term.

14.14 Consider a rectangular enclosure of height 20 cm and width d, whose vertical walls are at 100 and 30°C and whose horizontal walls are adiabatic. If the fluid contained in the enclosure is air, determine the value of d when the flow changes from the conduction to the transition regime and also when it changes to the boundary layer regime.

14.15 In Problem 14.14 compute the heat transfer rates at the "starting" and "departure" corners. Compare the results with those for external natural convection transport from a flat vertical isothermal surface at 100°C in air at 30°C. Comment on the difference.

14.16 A rectangular enclosure containing water is 20 cm high. Its vertical walls are maintained at 80 and 20°C, while the horizontal walls are adiabatic. Calculate the heat transfer from the hot to the cold surface for $d = 1$, 5, 10, 100, and 500 cm, where d is the horizontal distance between the vertical walls. Compare the results for the last two values of d with those for the asymptotic case $A \to 0$, where A is the aspect ratio.

14.17 A shallow oven is 2 m long and 10 cm high. If two opposite vertical sides are at 400 and 50°C, respectively, compute the heat transfer across the enclosure by natural convection. Assume the top and bottom to be insulated and the fluid to be air.

14.18 Calculate the heat transfer in the above configuration if the vertical walls are insulated instead, with the bottom at 400°C and the top at 50°C. Also, find the resulting heat transfer if the enclosure is inclined at 30° with the horizontal. Explain the observed effect of inclination by physical reasoning.

14.19 A chemical reactor may be approximated as a horizontal cylindrical annulus of inner diameter 50 cm and outer diameter 1 m, with the two surfaces at 150 and 25°C, respectively. If air is the fluid contained, determine the heat transfer rate and compare it with that for the inner cylinder in an extensive environment at 25°C. Repeat this problem for inner and outer diameters of 10 and 20 cm, respectively, and compare the result obtained with the earlier one.

14.20 From the numerical results presented on three-dimensional natural convection flow in a box, estimate the effect of the three-dimensional flow on the average Nusselt number as a function of the Rayleigh number. Also, compare the results with those from earlier two-dimensional studies on convection in a rectangular enclosure.

14.21 Consider a point thermal source located at the center of the bottom of a vertical cylindrical enclosure. The fluid in the enclosure is initially at room temperature; then at time $\tau = 0$ a constant heat input Q_0 is supplied by the source. Assuming all the walls of the enclosure are adiabatic, write the governing equations for the flow and the relevant boundary conditions for this transient problem. Obtain the important dimensionless parameters. State the conditions for which the recirculating flow can be neglected in order to use a model similar to that of Baines and Turner (1969) to study the stratification in the enclosure.

14.22 For the partial enclosure shown in Fig. 14.8.6, write the governing equations for laminar flow, along with the boundary conditions that apply at the enclosure walls as well as those at the free boundaries in the extension.

14.23 For the open thermosyphon (Fig. 14.6.1), compute the heat transfer rate if $t_0 = 50°C$, $t_1 = 150°C$, $a = 5$ cm, and $L = 25$ cm. Also, determine the mass flow rate if water is the fluid driven by buoyancy.

14.24 Repeat the above calculation for a closed thermosyphon if $t_1 = 150°C$, $t_2 = 50°C$, $d = 10$ cm, and the half-length $L = 25$ cm. Compare the results obtained with those from Problem 14.23.

14.25 The problem of natural convective cooling of an electronic equipment may be approximated by a rectangular enclosure whose walls are at room temperature, except for finite-size thermal sources located at the bottom. Two small openings are provided near the top on both vertical walls. Sketch the expected flow pattern in the enclosure. Also, outline a method for solving the governing equations for laminar or turbulent flow to determine the temperatures at the sources, if they have constant imposed heat flux inputs, under steady-state conditions.

14.26 A horizontal square enclosure of $L = 10$ cm, containing water, has the left and right surfaces maintained at t_h and $t_c = 0°C$, respectively. The horizontal surfaces are insulated.

(a) For the values $t_h = t_m/2$, t_m, $2t_m$, and $5t_m/2$, calculate the values of R and Ra and indicate the expected general flow patterns.

(b) For $t_m = 4.029°C$, determine the heat transfer rate for each of the four values of t_h.

14.27 The curves in Fig. 14.9.4 at high Ra were shown to have a slope of about 0.29. That is, $\text{Nu} \approx F(R)\text{Ra}^{0.29}$. From the curves given, estimate the function $F(R)$.

CHAPTER
FIFTEEN

TRANSPORT IN SATURATED POROUS MEDIA

15.1 INTRODUCTION

Natural convection in fluid saturated porous media arises in a large number of fields of natural science as well as several branches of technology. These include geophysics, soil mechanics, rheology, metal casting, ceramic engineering, and the technologies of paper, textiles, and insulating materials. Because of these wide applications, there has been a considerable growth in interest and research during recent years. Most studies may be categorized according to criteria such as the scale of the process, extent of the surrounding medium, mechanism driving the flow, stability of the flow, imposed boundary conditions, and nature of the fluid.

Two pertinent properties associated with studies of flows in porous media are porosity and permeability. Porosity is defined here as the ratio of the pore volume to the total volume of a sample of material. As pointed out by Davis (1969), there are several difficulties in measuring porosity. Microscopically and submicroscopically there is a spectrum of size of voids in different materials. The fluid trapped within very small openings is almost stationary and is best considered part of the solid material in most transport circumstances. However, some of the methods used in measuring porosity involve the removal of such fluid and may yield a misleading value of the effective porosity.

Permeability is a measure of the ease with which fluids pass through a material. This is a property of the solid material and of the fluid passing through it. Permeability K is defined as

$$K = -\frac{Q}{A}\frac{\mu}{\rho g}\left(\frac{\Delta h}{\Delta x}\right)^{-1} \tag{15.1.1}$$

where Q is the volumetric flow rate, A is the cross-sectional area, ρ and μ are the fluid density and absolute viscosity, respectively, and $\Delta h/\Delta x$ is the gradient of head loss through the column. Thus, permeability, also called "intrinsic permeability," has units of L^2. A common unit for measuring permeability is the darcy and is equivalent to 0.987×10^{-12} m^2. Table 15.1.1 shows some typical values of porosity and permeability for various rocks and minerals, as given by Davis (1969).

In this chapter a general outline of the formulation of transport is given first, then several important flow configurations are discussed in detail. External flows or flows in extensive porous media adjacent to vertical, horizontal, and inclined flat surfaces are also considered. Natural as well as mixed convection flows have been studied, and the conditions under which similarity solutions exist have been established. Other flow configurations such as vertical cylinders and point sources are also treated. Then internal flows, in partial as well as complete enclosures, are discussed. Several different effects, such as angle of inclination, through-flow, constant and periodic boundary conditions, and insulated and conducting walls, are described.

Finally, other flow effects are discussed, including the effects of transport in cold water, of variable properties, of non-Darcian behavior, and of anisotropic media. Experimental results, where known, are given to supplement the analysis.

15.2 FORMULATION OF TRANSPORT

The first empirical law governing the laminar flow of homogeneous fluids in homogeneous porous media as formulated by Darcy (1856) is given below. The volumetric flow rate Q, or velocity u, of a fluid through a vertical column of a porous medium of cross-sectional area A is directly proportional to the head loss Δh and the cross-sectional area of the column and inversely proportional to the length of the flow path Δx; that is,

$$\frac{Q}{A} = u = -K_c\left(\frac{\Delta h}{\Delta x}\right) \quad \text{or, in the limit,} \quad u = -K_c\left(\frac{dh}{dx}\right) \tag{15.2.1}$$

where K_c is the hydraulic conductivity. Comparing Eqs. (15.1.1) and (15.2.1), $K_c = K\rho g/\mu$. It is a function of the properties of the fluid as well as the porous medium and has units of velocity.

Darcy's law avoids the often insurmountable difficulty of representing the detailed hydrodynamic microscopic picture by introducing an average velocity $u = Q/A$ through a given cross section of a porous medium. In a one-dimensional flow, this average velocity is known as the apparent velocity or the superficial velocity. It is expressed in terms of the total head h as

$$u = \frac{Q}{A} = -K_c\left(\frac{dh}{dx}\right) = -K_c\,\frac{d}{dx}\left(z + \frac{p}{\rho g}\right) \tag{15.2.2}$$

Table 15.1.1 Permeability and porosity of sedimentary rocks

		Permeability (darcys) (1 darcy = 0.987×10^{-12} m^2)		
Rock name	Porosity (%)	Vertical	Orientation not given	Horizontal
Arkose				
Coarse-grained	10.9	3.8×10^{-4}		5.5×10^{-4}
Fine-grained	14.4	1.6×10^{-4}		1.6×10^{-3}
Medium-grained	25.6	5.5×10^{-4}		1.1×10^{-3}
Chalk, Cretaceous	29.2[a]	—		—
Chert, Mississippian	3.8	—		—
Conglomerate				
Coarse-grained	17.3	3.8×10^{-4}		4.9×10^{-4}
Dolomite				
Mississippian	27.8		2.9×10^{-1}	
Ordovician	0.4[b]	—		—
	11.9		1.6×10^{-2}	
Limestone				
Cretaceous	4.6[c]	—		—
Oolitic	21.6		3.4×10^{-1}	
Pennsylvanian	6.3[d]	—		—
Permian	10.1		7.7×10^{-3}	
Sandstone				
Cambrian	11.2[e]	—		—
Cretaceous	—	4.8×10^{-2}		1.5×10^{-4}
Cromwell	16.6	1.7×10^{-1}		4.1×10^{-1}
Gilcrest	27.4	5.6×10^{-1}		8.0×10^{-1}
Pennsylvanian	17.4[f]	—		—
Prue	11.4	4.7×10^{-4}		3.4×10^{-3}
Wilcox	15.6	3.6×10^{-1}		7.6×10^{-2}
	12.4	8.5×10^{-2}		8.8×10^{-2}
Shale				
Cretaceous	—		4×10^{-6}	
Pennsylvanian	—		9×10^{-6}	
Oligocene and Miocene	21.1[g]	—		—
Silurian	5.2[h]	—		—
Siltstone	9.7	1.6×10^{-4}		1.2×10^{-4}
	—	1.5×10^{-6}		—

[a] Mean of 16 samples.
[b] Mean of 56 samples.
[c] Mean of 24 samples.
[d] Mean of 2109 samples.
[e] Mean of 24 samples.
[f] Mean of 587 samples.
[g] Mean of 9 samples.
[h] Mean of 5 samples.
Source: From Davis (1969).

where z is the local elevation and p is the local static pressure level. In terms of the permeability denoted by K, Eq. (15.2.2) becomes

$$u = -K_c \frac{dh}{dx} = -K_c \frac{d}{dx}\left(z + \frac{p}{\rho g}\right) = -\frac{K_c}{\rho g}\frac{d}{dx}(\rho g z + p)$$

$$= -\frac{K}{\mu}\frac{d}{dx}(\rho g z + p) \tag{15.2.3}$$

Thus, u is $K/\mu = K_c/\rho g$ times the gradient of the total head, where K has units of length squared. In general, for a three-dimensional flow field, Darcy's law is generalized as

$$\mathbf{V} = -\frac{K}{\mu}(\nabla p - \rho \mathbf{g}) \tag{15.2.4}$$

Several considerations arise in postulating Eq. (15.2.4). Inertial effects are ignored. The frictional loss is balanced by only the pressure and body forces. Thus, the equation is valid only for small fluid velocities. Also, since viscous effects are taken into account only through the proportionality constant (K/μ), the viscous continuum flow equations are not achieved as K becomes large, as for decreasing μ.

The governing equations, including thermal energy transport, are derived either by a macroscopic balance or by integration of the continuum equations, as done by Cheng (1978). The resulting equations for conservation of mass, momentum, and energy, along with a linear density variation, are reduced to

$$\nabla \cdot \mathbf{V} = 0 \tag{15.2.5}$$

$$\mathbf{V} = -\left(\frac{K}{\mu}\right)(\nabla p - \rho \mathbf{g}) \tag{15.2.6}$$

$$\hat{\sigma}\frac{\partial t}{\partial \tau} + \mathbf{V} \cdot \nabla t = \alpha \nabla^2 t \tag{15.2.7}$$

where

$$\rho = \rho_r[1 - \beta(t - t_r)] \tag{15.2.8}$$

$$\hat{\sigma} \equiv \frac{\lambda(\rho c_p)_f + (1 - \lambda)(\rho c_p)_s}{(\rho c_p)_f}$$

= ratio of the heat capacity of the saturated medium to that of the fluid (15.2.9)

$$\lambda = \text{porosity} = \frac{\text{volume of pores}}{\text{Total volume}}$$

$$\alpha = \text{effective thermal diffusivity} = \frac{\lambda k_f + (1 - \lambda)k_s}{(\rho c_p)_f}$$

and subscripts f and s designate fluid and solid, respectively.

Equations (15.2.5)–(15.2.8) result only after making a number of additional restrictive assumptions. The convective fluid is assumed to be a single phase. The porous material is assumed undeformable. Local thermodynamic equilibrium is assumed to exist everywhere between the fluid and solid. The physical properties of both are assumed to be uniform and to remain constant. Chemical reactions in the medium, viscous dissipation, and compression work are neglected. The density of the fluid is assumed to remain constant except in the body force term, to represent the buoyancy effect. Volumetric energy generation and radiation terms are neglected.

The usual "no-slip" boundary condition at solid-fluid interfaces may not be specified for the resulting solutions. This arises because Darcy's law neglects momentum effects. Hence, the resulting order of the force balance, Eq. (15.2.6), is one degree less than that of the Navier-Stokes equations. Thus, it is necessary to discard one boundary condition. It is important that the above equations are only weakly nonlinear, due only to the convection of energy in Eq. (15.2.7). They are therefore easier to solve than the corresponding viscous flow equations. Because of these simplifications, closed-form solutions are available for some simple configurations and finite-difference solutions are often much simpler.

15.3 FLOWS IN EXTENSIVE POROUS MEDIA

Natural convection flows adjacent to flat vertical surfaces immersed in extensive media are considered first. This is followed by mixed convection flows adjacent to vertical wedges. Natural convection and mixed convection adjacent to horizontal surfaces are then discussed. Finally, a stability analysis of flows in extensive porous media is presented.

15.3.1 Natural Convection Adjacent to Flat Vertical Surfaces

Because of the similarity of flows in porous media to viscous flows, it is expected that a thin vertical thermal boundary region may often arise adjacent to heated flat vertical surfaces. Various experimental and numerical studies confirm this surmise. This permits boundary layer approximations similar to those of classical boundary layer theory. This is generally true for flows with high Rayleigh number $\mathrm{Ra}'_x = g\rho_\infty xK\beta(t_0 - t_\infty)/\mu\alpha$. Wooding (1963) and McNabb (1965) investigated such flows and developed a boundary layer theory for vertical plane flows. The theory was substantiated by visualization experiments with a Hele-Shaw cell immersed in water. Some similarity solutions were also found for special flow configurations. Johnson and Cheng (1978) and Cheng (1978) obtained similarity solutions for boundary layers adjacent to flat vertical and inclined surfaces.

An order of magnitude analysis of Eqs. (15.2.5)–(15.2.8) leads to the following governing equations for plane boundary layers adjacent to a flat vertical surface:

$$\frac{\partial u}{\partial x} + \frac{\partial v}{\partial y} = 0 \tag{15.3.1}$$

$$u = \frac{K\rho_\infty g\beta(t - t_\infty)}{\mu} \tag{15.3.2}$$

$$p_m = 0 \tag{15.3.3}$$

$$u\frac{\partial t}{\partial x} + v\frac{\partial t}{\partial y} = \alpha\frac{\partial^2 t}{\partial y^2} \tag{15.3.4}$$

where p_m is the pressure difference between the actual local static pressure and the local hydrostatic pressure and is thus the motion pressure. By eliminating the pressure terms and using the stream function ψ so that $u = \psi_y$ and $v = -\psi_x$, Eqs. (15.3.1)–(15.3.4) are reduced to

$$\frac{\partial^2\psi}{\partial y^2} = \frac{\rho_\infty\beta g K}{\mu}\frac{\partial t}{\partial y} \tag{15.3.5}$$

$$\frac{\partial^2 t}{\partial y^2} = \frac{1}{\alpha}\left(\frac{\partial\psi}{\partial y}\frac{\partial t}{\partial x} - \frac{\partial\psi}{\partial x}\frac{\partial t}{\partial y}\right) \tag{15.3.6}$$

Similarity formulation. As demonstrated by Johnson and Cheng (1978), similarity solutions exist for the above equations when the temperature difference $t_0 - t_\infty$ varies as Nx^n, where x is the distance downstream from the leading edge. The complete boundary conditions for an impermeable surface are

$$y = 0 \qquad v = 0 \qquad t_0 - t_\infty = Nx^n \tag{15.3.7}$$

$$y \to \infty \qquad u = 0 \qquad t = t_\infty \tag{15.3.8}$$

By using the variables

$$\eta = (\mathrm{Ra}_x')^{1/2}\left(\frac{y}{x}\right) \tag{15.3.9}$$

$$\psi = \alpha(\mathrm{Ra}_x')^{1/2} f(\eta) \tag{15.3.10}$$

$$\phi(\eta) = \frac{t - t_\infty}{t_0 - t_\infty} \tag{15.3.11}$$

where

$$\mathrm{Ra}_x' = \frac{g\rho_\infty xK\beta(t_0 - t_\infty)}{\mu\alpha} = \frac{gxK\beta(t_0 - t_\infty)\mathrm{Pr}}{\nu^2} = \mathrm{Gr}_x'\mathrm{Pr} \tag{15.3.12}$$

is the local modified Rayleigh number in a porous medium, Eqs. (15.3.1)–(15.3.4) are transformed to

$$f'' - \phi' = 0 \tag{15.3.13}$$

$$\phi'' + \tfrac{1}{2}(1 + n)f\phi' - nf'\phi = 0 \tag{15.3.14}$$

The relevant boundary conditions are

$$f(0) = \phi(\infty) = f'(\infty) = 0 \qquad \phi(0) = 1$$

Darcy's velocities expressed in terms of the similarity variables are

$$\frac{u}{u_c} = f'(\eta) \tag{15.3.15}$$

$$v = \frac{1}{2}\left(\alpha\rho_\infty g\beta K \frac{t_0 - t_\infty}{\mu x}\right)^{1/2} [(1-\eta)\eta f' - (1+\eta)f] \tag{15.3.16}$$

where $u_c \equiv \rho_\infty g\beta(t_0 - t_\infty)K/\mu$ is the characteristic velocity. Integrating Eq. (15.3.13) and noting that $f'(\infty) = \phi(\infty) = 0$,

$$f' = \phi \tag{15.3.17}$$

Thus, the profiles for the vertical velocity u/u_c and the dimensionless temperature ϕ are the same. Also the Prandtl number does not appear. Therefore, there is only one parameter, n, in Eq. (15.3.14). The results of a numerical solution of these equations are shown in Fig. 15.3.1. The boundary layer thickness is found to be

$$\frac{\delta}{x} = \eta_T\,(\text{Ra}'_x)^{-1/2} \propto x^{-(1+n)/2} \tag{15.3.18}$$

where η_T is the location in the thermal boundary layer where $\phi = f'$ has a value

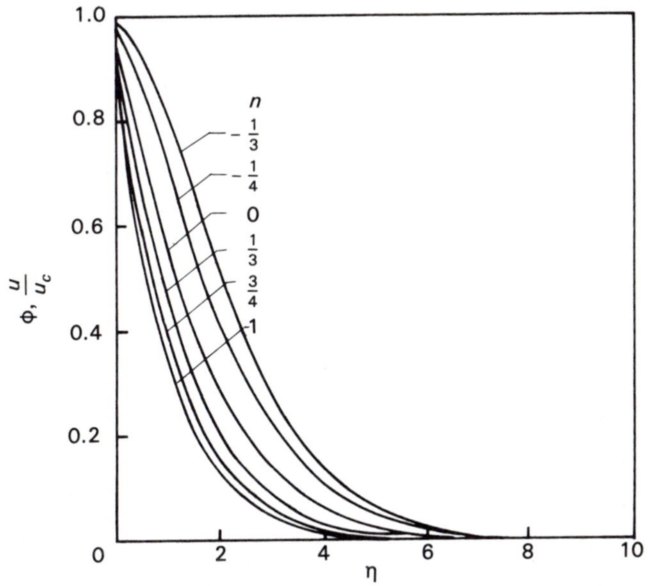

Figure 15.3.1 Dimensionless temperatures and vertical velocity vs. η for a free-convection porous layer adjacent to a vertical impermeable surface. *(Reprinted with permission from P. Cheng and W. J. Minkowycz,* J. Geophys. Res., *vol. 82, p. 2040. Copyright © 1977, American Geophysical Union.)*

of 0.01. Cheng and Minkoywcz (1977) found that for an isothermal wall $n = 0$, $\eta_T = 6.31$. The local surface heat flux at the heated surface is

$$q'' = -k\left(\frac{\partial t}{\partial y}\right)_0 = kN^{3/2}\left(\frac{\rho_\infty g \beta K}{\mu \alpha}\right)^{1/2} x^{(3n-1)/2}[-\phi'(0)] \qquad (15.3.19)$$

The local convection coefficient h_x is written in nondimensional form as

$$\frac{\mathrm{Nu}_x}{(\mathrm{Ra}'_x)^{1/2}} = -\phi'(0) \qquad (15.3.20)$$

where $-\phi'(0) = 0.444$ for $n = 0$.

The limits on the surface temperature variation as x^n are determined from physical considerations. In the range $0 \leq n \leq 1$, it is found that u, δ, and the total convected energy increase or at least remain constant downstream with increasing x. The value $n = \frac{1}{3}$ corresponds to a uniform heat flux at the surface. For an exponential temperature distribution at the surface, $t_0 - t_\infty = Me^{mx}$, where M and m are constants, Johnson and Cheng (1978) showed that a similarity solution also exists.

Cheng (1977a) investigated the effects of lateral mass flux, that is, blowing and suction, at the surface. This might arise in geothermal systems with the lateral flow directed into or out of the vertical surface. Similarity solutions are found for some special cases in which both the surface temperature and the blowing velocity have power law distributions. Merkin (1978) also derived a series solution valid near the leading edge of the surface and extended it by using a numerical solution of the full equations. At large downstream distances, asymptotic expansions are derived for both suction and blowing at the surface.

15.3.2 Mixed Convection Adjacent to Vertical Wedges

Cheng (1977b) also analyzed mixed convection flows adjacent to vertical wedges of included angle $\hat{\beta}\pi$, where $\hat{\beta} = 0$ corresponds to a vertical surface. Taking x as the direction along the surface, the local free-stream velocity is $U = Bx^m$ and $m = \hat{\beta}/(2 - \hat{\beta})$. It is shown that if the wall temperature $t_0 - t_\infty = \pm Nx^n$ and $m = n$, then a similarity solution is obtained. Equations (15.3.5) and (15.3.6) are then transformed, as in Chapter 10, into

$$f'' = \pm\left(\frac{\mathrm{Gr}'_x}{\mathrm{Re}_x}\right)\phi' \qquad (15.3.21)$$

$$\phi'' = n\phi f' - \tfrac{1}{2}(1 + n) f\phi' \qquad (15.3.22)$$

where

$$\mathrm{Gr}'_x = \frac{\mathrm{Ra}'_x}{\mathrm{Pr}} = \frac{gKx}{\nu^2}\beta(t_0 - t_\infty)$$

subject to boundary conditions

$$f(0) = 0 \qquad \phi(0) = 1 \qquad (15.3.23)$$

$$f'(\infty) = 1 \qquad \phi(\infty) = 0 \tag{15.3.24}$$

Here the similarity variables are

$$\eta = (\mathrm{Pe}_x)^{1/2} \frac{y}{x} \tag{15.3.25}$$

$$\psi(x, y) = \alpha(\mathrm{Pe}_x)^{1/2} f(\eta) \tag{15.3.26}$$

where

$$\mathrm{Pe}_x = \mathrm{Re}_x \mathrm{Pr} = \frac{U_\infty x}{\alpha} = \text{Peclet number} \tag{15.3.27}$$

The positive sign in Eq. (15.3.21) corresponds to aiding effects and the negative sign to opposing effects.

The quantity $\mathrm{Gr}'_x/\mathrm{Re}_x$ in Eq. (15.3.21) is the controlling parameter. It is seen to be independent of x for $n = 0$. The boundary layer thickness and the local Nusselt number are

$$\frac{\delta_T}{x} = \frac{\eta_T}{(\mathrm{Pe}_x)^{1/2}} \tag{15.3.28}$$

$$\frac{\mathrm{Nu}_x}{(\mathrm{Pe}_x)^{1/2}} = -\phi'(0) \tag{15.3.29}$$

Figures 15.3.2 and 15.3.3 show the boundary layer thickness η_T and the nondimen-

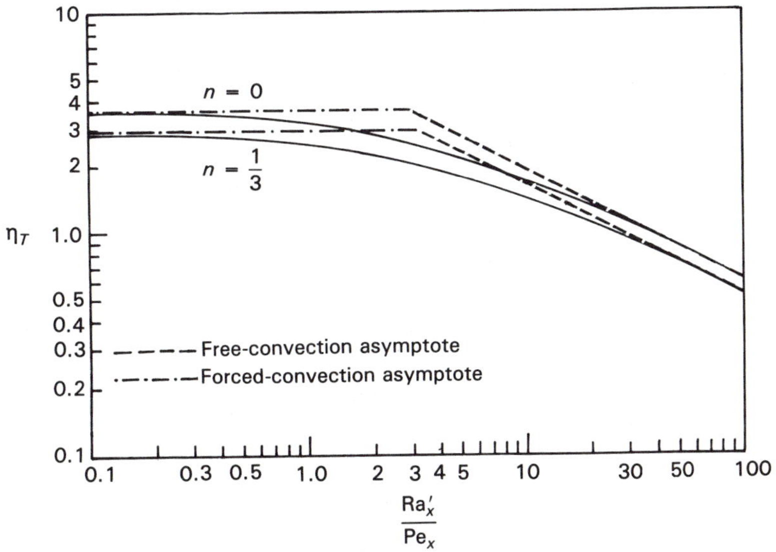

Figure 15.3.2 Dimensionless boundary layer thickness parameter for mixed convection porous layers adjacent to vertical ($n = 0$) and inclined ($n = \frac{1}{3}$) impermeable surfaces. *(Reprinted with permission from* Int. J. Heat Mass Transfer, *vol. 20, p. 807, P. Cheng. Copyright © 1977, Pergamon Journals Ltd.)*

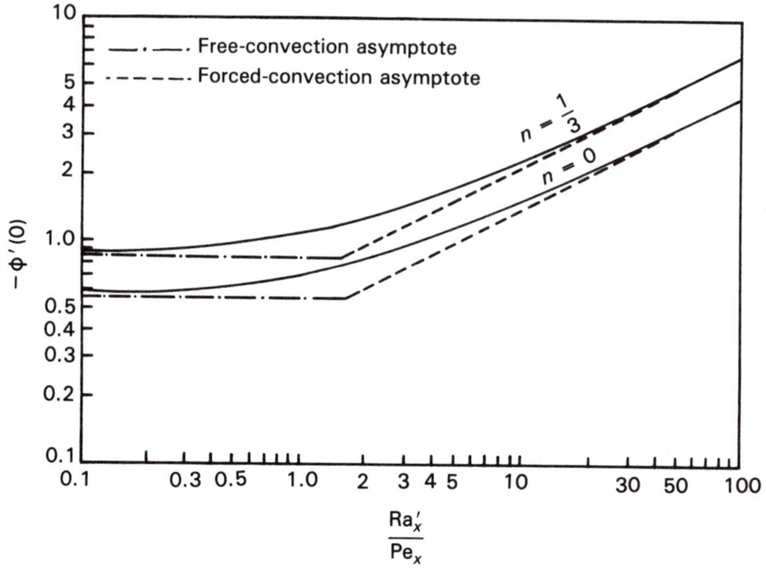

Figure 15.3.3 Heat transfer results for mixed convection porous layers adjacent to vertical ($n = 0$) and inclined ($n = \frac{1}{3}$) impermeable surfaces. *(Reprinted with permission from* Int. J. Heat Mass Transfer, *vol. 20, p. 807, P. Cheng. Copyright © 1977, Pergamon Journals Ltd.)*

sional heat transfer parameter $-\phi'(0)$ as functions of $\mathrm{Gr}'_x/\mathrm{Re}_x = gK\beta(t_0 - t_\infty)/\nu U_\infty$ for $n = 0$ and $\frac{1}{3}$. The corresponding wedge angles are $\hat{\beta} = 0$ and $\frac{1}{2}$; recall that $\hat{\beta} = 0$ corresponds to a flat vertical surface.

15.3.3 Natural Convection Adjacent to Horizontal Surfaces

Cheng and Chang (1976) gave numerical results for natural convection adjacent to a semi-infinite horizontal surface whose temperature is a power function of distance x from the leading edge. For horizontal surfaces, Darcy's law gives

$$u = -\frac{K}{\mu}\frac{\partial p_m}{\partial x} \tag{15.3.30}$$

$$0 = \frac{\partial p_m}{\partial y} - \rho_\infty g\beta(t - t_\infty) \tag{15.3.31}$$

Again, as seen in Chapter 5, the normal buoyancy force induces a pressure gradient that drives the flow through $\partial p_m/\partial x$. The governing conservation equations expressed in terms of a stream function ψ and temperature t are

$$\frac{\partial^2\psi}{\partial y^2} = -\frac{K\rho_\infty g\beta}{\mu}\frac{\partial t}{\partial x} \tag{15.3.32}$$

$$\frac{\partial^2 t}{\partial y^2} = \frac{1}{\alpha}\left(\frac{\partial\psi}{\partial y}\frac{\partial t}{\partial x} - \frac{\partial\psi}{\partial x}\frac{\partial t}{\partial y}\right) \tag{15.3.33}$$

Using the transformations

$$\eta = (\mathrm{Ra}_x')^{1/3}\frac{y}{x} \tag{15.3.34}$$

$$\psi = \alpha(\mathrm{Ra}_x')^{1/3}f(\eta) \tag{15.3.35}$$

the equations become

$$f'' + n\phi + \tfrac{1}{3}(n-2)\eta\phi' = 0 \tag{15.3.36}$$

$$\phi'' - n\phi f' + \tfrac{1}{3}(1+n)f\phi' = 0 \tag{15.3.37}$$

subject to

$$f(0) = 0 \qquad \phi(0) = 1 \qquad f'(\infty) = \phi(\infty) = 0$$

The boundary layer thickness δ_T and the local surface heat flux $q''(x)$ are

$$\frac{\delta_T}{x} = \frac{\eta_T}{(\mathrm{Ra}_x')^{1/3}} \tag{15.3.38}$$

$$q''(x) = kN^{4/3}\left(\frac{K\rho_\infty g\beta}{\mu\alpha}\right)^{1/3} x^{(4n-2)/3}[-\phi'(0)] \tag{15.3.39}$$

From Eqs. (15.3.38) and (15.3.39) it is seen that $\delta_T \propto x^{(2-n)/3}$ and the total convected energy $Q(x) \propto x^{(4n+1)/3}$. Therefore, for δ_T to remain finite at $x = 0$ and $Q(x)$ to remain constant or increase downstream, the limits on the power n are $-0.25 \le n \le 2$.

15.3.4 Mixed Convection Adjacent to Horizontal Surfaces

Cheng (1977c) considered two flows, an aiding flow parallel to the horizontal surface and the stagnation point flow. For the latter, the power law free-stream velocity is $U_\infty = Bx^m$. Then $m = 0$ corresponds to flow over a surface at zero angle of incidence and $m = 1$ represents a vertical stagnation point flow on a horizontal surface. Using the transformations

$$\eta = (\mathrm{Pe}_x)^{1/2}\frac{y}{x} \tag{15.3.40}$$

$$\psi = \alpha(\mathrm{Pe}_x)^{1/2}f(\eta) \tag{15.3.41}$$

the equations become

$$f'' = -\left[\frac{\mathrm{Ra}_x'}{2(\mathrm{Re}_x\mathrm{Pr})^{3/2}}\right][(3m+1)\phi + (m-1)\eta\phi'] \tag{15.3.42}$$

$$\phi'' = \frac{1}{2}[(3m+1)\phi f' - (m+1)f\phi'] \tag{15.3.43}$$

subject to

$$f(0) = \phi(\infty) = 0 \qquad \phi(0) = f'(\infty) = 1$$

For a power law surface temperature excess variation, $t_0 - t_\infty = Nx^n$, similarity arises when $n = (3m + 1)/2$. The governing parameter in Eq. (15.3.42) for the similarity solution is

$$\frac{\mathrm{Ra}_x'}{(\mathrm{Pe}_x)^{3/2}} = \frac{\rho_\infty g \beta K N}{\mu B}\left(\frac{\alpha}{B}\right)^{1/2}$$

which is independent of x. The boundary layer thickness and heat flux are given by

$$\frac{\delta_T}{x} = \frac{\eta_T}{(\mathrm{Pe}_x)^{3/2}}$$
$$q'' = -k\left(\frac{\partial t}{\partial y}\right)_0 = kNx^{2m}\left(\frac{B}{\alpha}\right)^{1/2}[-\phi'(0)] \qquad (15.3.44)$$

Figures 15.3.4 and 15.3.5 show the resulting thermal boundary layer thickness and heat transfer parameters as functions of $\mathrm{Ra}_x'/(\mathrm{Pe}_x)^{3/2}$. Also shown are the boundary layer thickness and resulting heat transfer for the free- and forced-convection asymptotes.

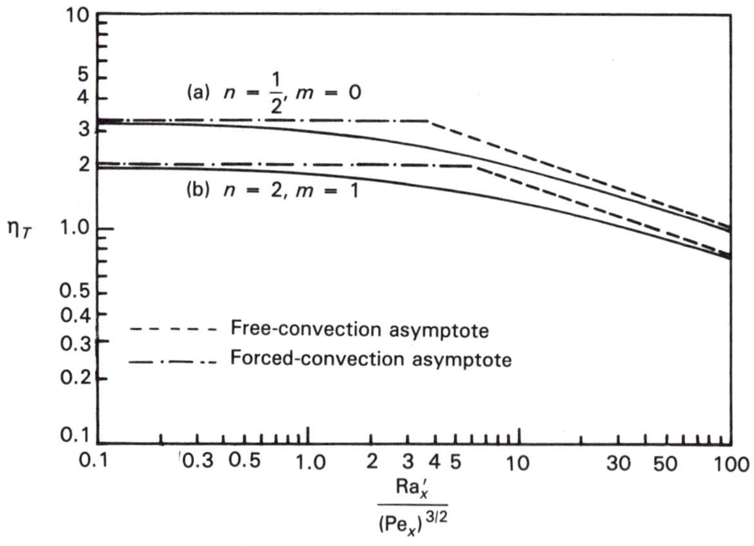

Figure 15.3.4 Dimensionless boundary layer thickness parameter for a mixed convection porous layer adjacent to a horizontal impermeable surface: (curves a) horizontal impermeable surface at zero angle of attack with constant heat flux ($m = 0$, $n = \frac{1}{2}$) and (curves b) stagnation point flow ($m = 1$ and $n = 2$). *(Reprinted with permission from* Int. J. Heat Mass Transfer, *vol. 20, p. 893, P. Cheng. Copyright © 1977, Pergamon Journals Ltd.)*

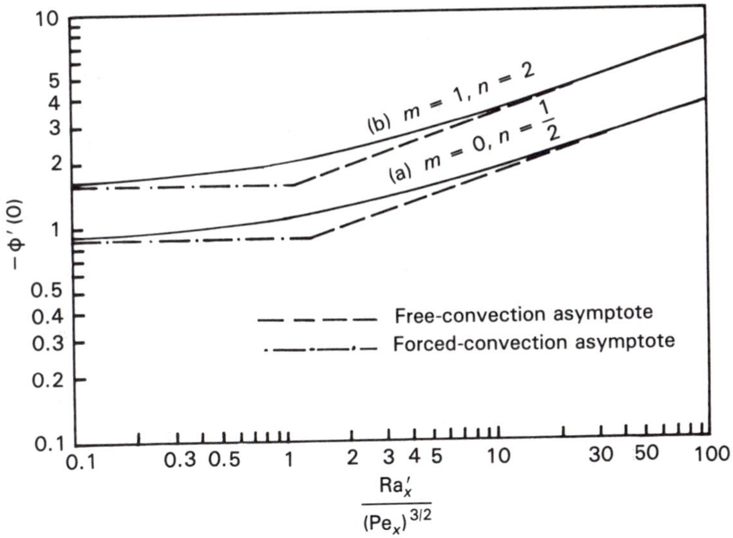

Figure 15.3.5 Heat transfer results for mixed convection porous layers with external flows: (curves a) horizontal impermeable surface at zero angle of attack with constant heat flux ($m = 0$ and $n = \frac{1}{2}$) and (curves b) stagnation point flow ($m = 1$ and $n = 2$). *(Reprinted with permission from* Int. J. Heat Mass Transfer, *vol. 20, p. 893, P. Cheng. Copyright © 1977, Pergamon Journals Ltd.)*

15.3.5 Other Flow Effects in Extensive Porous Media

Minkowycz and Cheng (1976) analyzed the flow about a vertical heated cylinder embedded in a saturated porous medium. The surface temperature difference $t_0 - t_\infty$ was considered proportional to a power of distance downstream from the leading edge. For a linear temperature distribution, $n = 1$, an exact solution is found for the boundary layer equations, as in Newtonian fluids. For other temperature distributions, approximate solutions based on local similarity and nonlocal similarity methods are presented.

Bejan (1978) considered the transient convection arising from a suddenly imposed point heat source in an infinite porous medium. Both the initial transient and the final steady state are investigated, including low Rayleigh number behavior. The solution is a perturbation analysis in Ra'_x. Hickox and Watts (1980) also found results for steady-state convection from a concentrated source for two configurations. Both are axisymmetric flows and apply for a wide range of Rayleigh numbers. The first configuration is a point source located on the lower boundary of a semi-infinite porous medium. The second is a point source located in an infinite medium. The governing equations are first transformed into nonlinear ordinary differential equations and then solved numerically.

Merkin (1979) analyzed the boundary region transport adjacent to axisymmetric and two-dimensional bodies of arbitrary shape. Both a horizontal cylinder and axisymmetric bodies oriented with their axes in the vertical direction were considered. For both cases, similarity transformations were given that reduced the equations to a form previously solved by Ackroyd (1967).

Stability of vertical flows in extensive porous media. Evans and Plumb (1978) report experiments in which instabilities were observed adjacent to a heated vertical impermeable surface in a porous medium. Hsu et al. (1978) and Cheng (1978) give the details of a linear stability analysis for boundary layer flows adjacent to horizontal heated surfaces. The base flow is assumed to be the two-dimensional steady-state boundary layer flow determined by Cheng and Chang (1976). The full three-dimensional disturbance equations are derived, taking into account the transverse velocity component of the base flow as well as the streamwise dependence of the base flow and temperature fields. These equations are then simplified by using the treatment proposed by Haaland and Sparrow (1973), as discussed in Section 11.8.1. That is, the disturbances at the onset of instability are contained within the boundary layer of the base flow.

For the simplified three-dimensional disturbance equations, the disturbances are assumed to be of the form

$$
\begin{aligned}
t' &= \hat{T}(x, y)e^{(i\hat{\alpha}z+\hat{\omega}\tau)} \\
\psi' &= \hat{\psi}(x, y)e^{(i\hat{\alpha}z+\hat{\omega}\tau)} \\
u' &= \hat{u}(x, y)e^{(i\hat{\alpha}z+\hat{\omega}\tau)}
\end{aligned}
\qquad (15.3.45)
$$

where $\hat{\alpha}$ is the spanwise periodic wave number, z the spanwise direction, and $\hat{\omega}$ the growth factor. Both $\hat{\alpha}$ and $\hat{\omega}$ are taken as real numbers. Substituting this expansion into the three-dimensional disturbance equations leads to

$$
\begin{aligned}
&(D^2 - \alpha^2)^2\Phi + \left(\frac{n+1}{3}\right) f_0 D(D^2 - \alpha^2)\Phi - n f_0'(D^2 - \alpha^2)\Phi \\
&\quad + (\mathrm{Ra}_x')^{2/3}\phi_0'\alpha^2\Phi + \frac{1}{3}\left[\frac{1}{3}(n-2)\eta\phi_0' + n\phi_0\right] \\
&\quad \times [(n-2)nD^2\Phi + (2n-1)D\Phi] = 0
\end{aligned}
\qquad (15.3.46)
$$

subject to

$$\Phi(0) = \mathrm{D}^2\Phi(0) = 0 \qquad (15.3.47)$$

$$\Phi(\infty) = D^2\Phi(\infty) = 0 \qquad (15.3.48)$$

where

$$\alpha = \frac{\hat{\alpha}x}{(\mathrm{Ra}_x')^{1/3}} \qquad \Phi(\eta) = \frac{\hat{\psi}}{i\alpha(\mathrm{Ra}_x')^{1/3}} \qquad D = \frac{d}{d\eta}$$

The functions f_0 and ϕ_0 are the base flow solution.

Integrating the equations numerically yields the critical Rayleigh numbers $(\mathrm{Ra}_x')_{\mathrm{cr}}$ and the associated wave numbers α_{cr}. These are plotted against n in Fig. 15.3.6. It is seen that $n = 0$, which represents a uniform excess wall temperature, is the

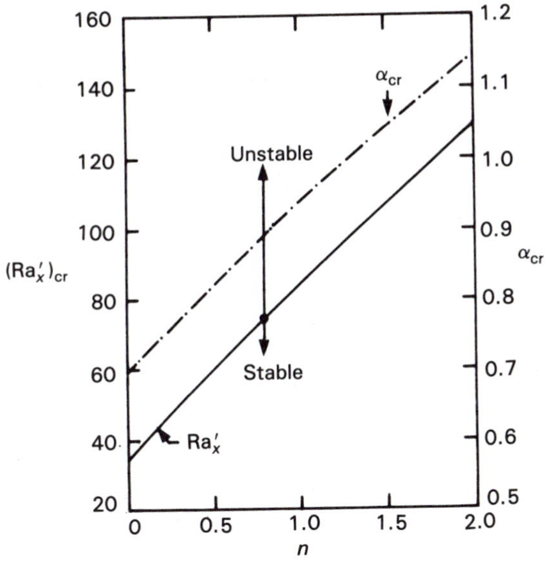

Figure 15.3.6 Critical Rayleigh numbers and associated wave numbers vs. n for free-convection boundary layers adjacent to a horizontal impermeable surface. *(Reprinted with permission from* Int. J. Heat Mass Transfer, *vol. 21, p. 1221, C. T. Hsu et al. Copyright © 1978, Pergamon Journals Ltd.)*

most unstable condition. Hsu and Cheng (1979) also carried out a stability analysis for a surface inclined from the vertical by an angle θ. Table 15.3.1 gives $(\mathrm{Ra}_x')_{cr} \tan^2\theta$ and α_{cr} values for several values of n.

15.4 ENCLOSURES

Heat transfer in enclosures and partial enclosures filled with saturated porous media arises under many natural and technological circumstances, as in geothermal

Table 15.3.1 Critical Rayleigh numbers and associated wave numbers for surfaces inclined from the vertical by angle θ

n	$(\mathrm{Ra}_x')_{cr} \tan^2\theta$	α_{cr}
−1/3	45.6	0.550
−1/4	64.7	0.570
0	120.7	0.636
1/4	176.5	0.698
1/3	195.1	0.717
1/2	232.2	0.758
3/4	287.9	0.814
1	343.7	0.865

Source: From Cheng (1978).

reservoirs and in building and other insulation. The effect of natural convection on the heat transfer under such circumstances is crucial and in many situations undesirable, since it could greatly increase heat flow. This led to many studies for the prediction of the onset of appreciable convection, its magnitude, and the resulting flow regimes. The primary parameters for convective heat transfer in enclosures filled with saturated porous media are the aspect ratio $A = H/d$ and the Darcy modified Rayleigh number

$$\mathrm{Ra}' = \frac{gKH\beta(t_h - t_c)}{\alpha\nu} \tag{15.4.1}$$

where H is the height of the enclosure, d is the width, and t_h and t_c are the hot and cold surface temperatures, respectively, as shown in Fig. 15.4.4.

15.4.1 Transport in a Medium Bounded by Two Flat Parallel Planes

The simplest idealized enclosure geometry is that between two infinite parallel flat surfaces. This is the rectangular cavity as either H or d is greatly increased. The orientation may be horizontal, vertical, or inclined. The horizontal enclosure, for $H/d \to 0$, is the Bénard configuration in a porous medium. To study the onset of convection, linear stability theory was applied by Lapwood (1948) and Homsy and Sherwood (1976) in a manner similar to that for flows in nonporous media, as described in Chapter 13. The variables are again taken as the sum of base and disturbance quantities. The surfaces are at t_1 and t_2, where $t_1 > t_2$, at $x = 0$ and H, respectively. Thus,

$$t = \bar{t} + t' \qquad \rho = \bar{\rho} + \rho' \qquad p = \bar{p} + p'$$
$$\mathbf{V}(x, y, z, \tau) = \mathbf{V}'(x, y, z, \tau) \tag{15.4.2}$$

where $\bar{t}$, $\bar{\rho}$, and $\bar{p}$ are the base flow quantities and $V = 0$ initially. The initial temperature, local static pressure, and density distributions are

$$\bar{t} = t_1 + (t_2 - t_1)\frac{x}{H}$$
$$\bar{p} = p_1 - \rho g x + \rho_1 g\beta \frac{t_2 - t_1}{H}\frac{x^2}{2} \tag{15.4.3}$$
$$\bar{\rho} = \rho_1\left[1 - \frac{\beta(t_2 - t_1)}{H}x\right]$$

where the subscript 1 denotes $x = 0$ and 2 denotes $x = H$.

The disturbances are assumed to be of the form

$$\phi(\hat{x}, \hat{y}, \hat{z}, \hat{\tau}) = \theta(\hat{x}) \exp\,[A\hat{\tau} + i(B\hat{z} + C\hat{y})] \tag{15.4.4a}$$

$$\hat{\omega}(\hat{x}, \hat{y}, \hat{z}, \hat{\tau}) = W(\hat{x}) \exp\,[A\hat{\tau} + i(B\hat{z} + C\hat{y})] \tag{15.4.4b}$$

where A is the disturbance temporal growth rate, B the wave number in the z direction, and C the wave number in the y direction. These are standing waves since B and C are real. Here

$$\phi = \frac{t'}{t_2 - t_1} \qquad \hat{\omega} = \frac{u'}{\alpha/H} \tag{15.4.5a}$$

$$\hat{\tau} = \frac{\alpha\tau}{\hat{\sigma}H^2} \qquad \hat{x} = \frac{x}{H} \qquad \hat{y} = \frac{y}{H} \qquad \hat{z} = \frac{z}{H} \tag{15.4.5b}$$

where $\hat{\sigma}$ is defined in Eq. (15.2.9).

Substituting the above into Eqs. (15.2.5)–(15.2.7), assuming small disturbances, and eliminating p', the following nondimensional equations are obtained:

$$\nabla^2\hat{\omega} = -\text{Ra}'\nabla_1^2\phi \tag{15.4.6a}$$

$$\frac{\partial\phi}{\partial\hat{\tau}} - \hat{\omega} = \nabla^2\phi \tag{15.4.6b}$$

where

$$\nabla_1^2 \equiv \frac{\partial^2}{\partial\hat{y}^2} + \frac{\partial^2}{\partial\hat{z}^2} \tag{15.4.6c}$$

The boundary conditions, as shown in Chapter 13, are

$$\phi = \hat{\omega} = 0 \qquad \text{at } \hat{x} = 0, 1 \tag{15.4.7}$$

Setting $A = 0$ for neutral stability results in

$$(D^2 - a^2)W(\hat{x}) = -a^2\text{Ra}'\theta(\hat{x}) \tag{15.4.8a}$$

$$(D^2 - a^2)\theta(\hat{x}) = -W(\hat{x}) \tag{15.4.8b}$$

or, combining these two equations,

$$(D^2 - a^2)^2\theta(\hat{x}) = a^2\text{Ra}'\theta(\hat{x}) \tag{15.4.9}$$

where

$$D \equiv \frac{d}{d\hat{x}} \qquad a^2 = B^2 + C^2 \tag{15.4.10}$$

Equation (15.4.9) is subject to the boundary conditions

$$\theta = D^2\theta = 0 \qquad \text{at } \hat{x} = 0, 1 \tag{15.4.11}$$

and admits solutions of the form

$$\theta(\hat{x}) = \sin(n\pi\hat{x}) \tag{15.4.12}$$

where n is an integer relating Ra′ and a as follows:

$$\text{Ra}' = \frac{[(n\pi)^2 + a^2]^2}{a^2} \tag{15.4.13}$$

To determine the minimum Ra′ that results in convective motion, $\partial \mathrm{Ra}'/\partial a$ is set equal to 0, giving $a = n\pi$. Thus, $n = 1$ determines the minimum Ra′, called the critical value. Therefore

$$\mathrm{Ra}'_{cr} = 4\pi^2 \tag{15.4.14}$$

This limit is confirmed by the experimental observations of many investigators, such as Schneider (1963), Elder (1967a), Combarnous and LeFur (1969), and Combarnous and Bories (1975).

A more general formulation. To investigate Bénard instability for media ranging from the nonporous-medium flow limit to the Darcy flow limit, Katto and Masuoka (1967) and Walker and Homsy (1977) used an alternative form of the momentum equation (15.2.6), as follows:

$$\nabla p - \rho \mathbf{g} = -\left(\frac{\mu V}{K}\right) + \mu \nabla^2 V \tag{15.4.15}$$

where the viscous term $\mu\nabla^2 V$ becomes larger as K increases. Eventually, for large K, results for nonporous media are obtained, as shown below. This form was originally suggested by Brinkman (1947) and Chan et al. (1970). From a derivation similar to that shown earlier, the equivalent forms of Eqs. (15.4.8) are

$$\left[(D^2 - a^2) - \frac{H^2}{K}\right](D^2 - a^2)W(\hat{x}) = a^2 \mathrm{Ra}\theta(\hat{x}) \tag{15.4.16a}$$

$$(D^2 - a^2)\theta(\hat{x}) = -W(\hat{x}) \tag{15.4.16b}$$

where the Rayleigh number Ra in Eq. (15.4.16*a*) is that for a nonporous-medium flow, $\mathrm{Ra} = g\rho H^3 \beta(t_1 - t_2)/\mu\alpha$, and is related to Ra′ by $\mathrm{Ra} = \mathrm{Ra}'(H^2/K)$. These equations reduce to Eqs. (15.4.8) as $K/H^2 \to 0$. They also reduce to the equations for nonporous-medium flow as $K/H^2 \to \infty$. The solution to Eqs. (15.4.16) for the

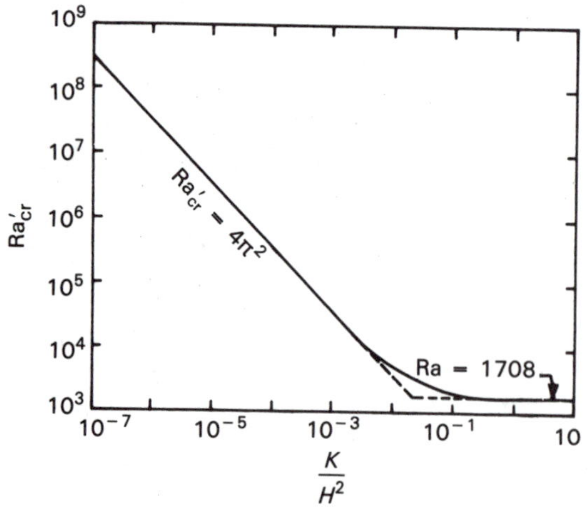

Figure 15.4.1 Critical Rayleigh number (for the Bénard problem) vs. dimensionless permeability from Darcy's limit to the viscous fluid limit. *(From Cheng, 1978.)*

critical Rayleigh number is plotted as a function of K/H^2 in Fig. 15.4.1. Both the Darcy limit and the viscous flow limit are seen to be reached for small and large K/H^2, respectively.

Thus, for porous-medium flows, when $\mathrm{Ra}' < \mathrm{Ra}'_{\mathrm{cr}} = 4\pi^2$, heat is transferred by conduction alone. For Ra′ slightly higher than $\mathrm{Ra}'_{\mathrm{cr}}$, convection effects arise in the form of adjacent polyhedral cells. These appear at fixed locations in porous media. Combarnous (1970) pointed out that the observed cells are not as uniform as those seen in Newtonian fluid layers.

Ordered cellular motion breaks down at higher Ra′ and a new convective regime appears. It is characterized by fluctuating temperature distributions within the medium, as reported by Combarnous and LeFur (1969). Because of the local heat exchange between the fluid and the porous solid, the temperature fluctuations are different from those in Newtonian fluids. Caltagirone et al. (1971) and Horne and O'Sullivan (1974) both interpreted the temperature fluctuations as the result of the continuous appearance and dissipation of the convective cells.

Other studies. Numerous other studies of the stability of horizontal porous layers subject to various boundary conditions have been carried out. Weber (1974) considered the effect on the critical Rayleigh number of specifying the temperatures of the top and bottom bounding surfaces to change linearly with horizontal distance x. The effect of superimposing horizontal flow through the medium was studied by Prats (1966). Homsy and Sherwood (1976) investigated the effect of a uniform vertical flow through the medium. It was concluded that the critical Rayleigh number increased with increasing values for the vertical through-flow. The direction of the through-flow with respect to gravity had no effect on $\mathrm{Ra}'_{\mathrm{cr}}$.

Other effects, such as temperature-dependent viscosity, were included by Kassoy and Zebib (1975) and Zebib and Kassoy (1978). Straus and Schubert (1977) also studied the effects of variable thermal expansion coefficient β and the density variation of water as a function of temperature on the onset of convection in a confined geothermal reservoir. Nield (1968) studied the onset of thermohaline convection in a horizontal porous medium. Beck (1972) considered the effect of lateral insulated walls on the stability. This amounts to the onset of convection in a three-dimensional rectangular cavity. It was found that as either of the horizontal dimensions is increased, the critical Rayleigh number approaches the $4\pi^2$ limit. Otherwise, $\mathrm{Ra}'_{\mathrm{cr}}$ is always greater than $4\pi^2$, indicating that surfaces further confining the medium stabilize the flow.

Heat transfer. Measurements for a horizontal porous medium heated from below have been reported by many investigators for a wide range of fluids, porous materials, and Rayleigh number values. Figure 15.4.2 is a collection of these data. As noted by Combarnous (1970), no single correlation between Nu and Ra′ successfully correlates all the different fluid-solid combinations. However, some quite specific correlations have been given. For a mixture of glass spheres ranging from 3–18 mm in diameter, Elder (1967a) reported

$$\mathrm{Nu} = \frac{hH}{k} = \frac{\mathrm{Ra}'}{40} \tag{15.4.17}$$

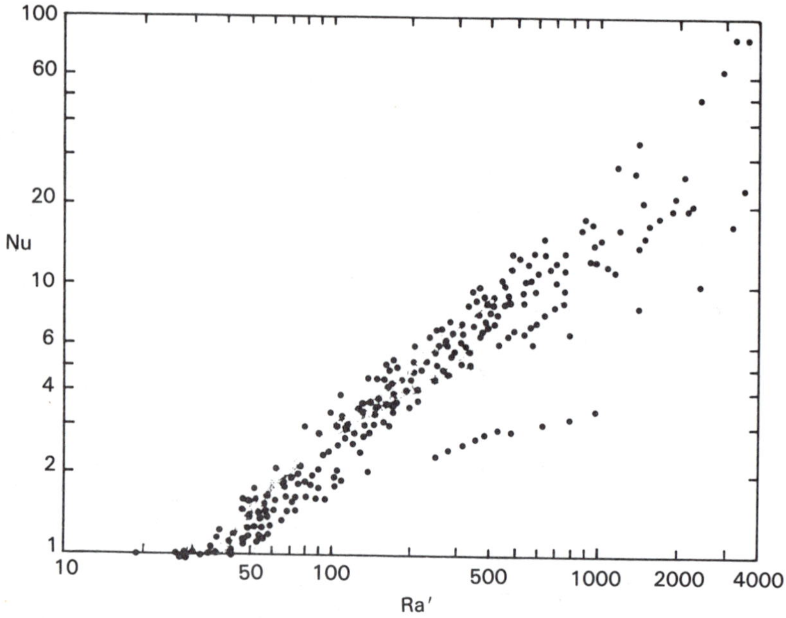

Figure 15.4.2 Compilation of experimental, analytical, and numerical results for Nusselt number vs. Rayleigh number for convection heat transfer in a horizontal porous layer heated from below. *(Reprinted with permission from P. Cheng,* Adv. Heat Transfer, *vol. 14, p. 1. Copyright © 1978, Academic Press, Inc.)*

Buretta and Berman (1976), for glass beads 3, 6, and 14 mm in diameter saturated with demineralized water, proposed the following correlation:

$$\log(\text{Nu} + 0.076) = 1.154 \log \text{Ra}' - 1.823 \qquad 40 < \text{Ra}' < 100 \qquad (15.4.18a)$$

$$\log(\text{Nu} + 0.35) = 0.835 \log \text{Ra}' - 1.214 \qquad 100 < \text{Ra}' < 1000 \qquad (15.4.18b)$$

For layers inclined at angles ξ from the horizontal, the onset of convection and the resulting flow regimes were observed by Bories and Combarnous (1973) to be substantially altered. Figure 15.4.3 summarizes the five regimes seen, in terms of Ra′ and ξ, for the range 0–90°. In contrast to horizontal layers, when the layer is inclined, a two-dimensional unicellular flow occurs (region A), even for $\text{Ra}' < 4\pi^2 = 39.5$. As Ra′ is further increased, this flow breaks down into polyhedral cells at low ξ (region B) or into stable adjoint coils (region C). A further increase in Ra′ leads to a fluctuating convective regime at small ξ (region D) and to large wavy coils at larger ξ (region E).

Similar measurements were made by Kaneko et al. (1974), who used heptane-sand and ethanol-sand systems and reported that their data correlated well with

$$\text{Nu} = 0.082(\text{Ra}' \cos \xi)^{0.76} \qquad 40 < \text{Ra}' \cos \xi \qquad (15.4.19)$$

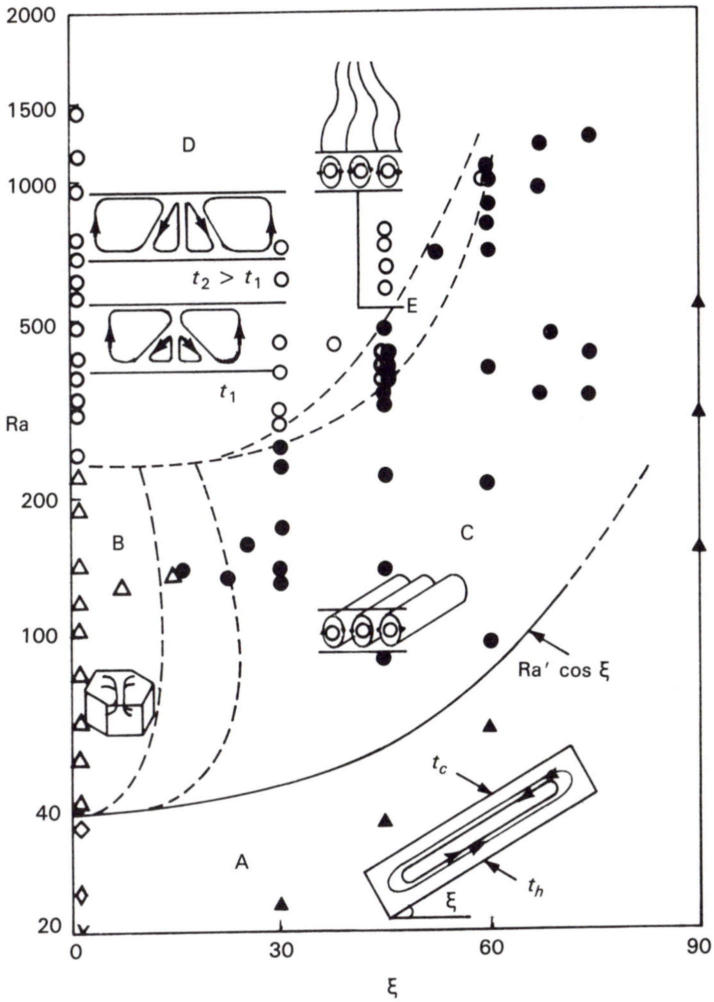

Figure 15.4.3 Experimental observation of different types of convective motions in a sloped porous layer: region A, unicellular flow; B, polyhedral cells; C, longitudinal stable coils; D, fluctuating regime; and E, oscillating longitudinal coils. *(Reprinted with permission from P. Cheng,* Adv. Heat Transfer, *vol. 14, p. 1. Copyright © 1978, Academic Press, Inc.)*

15.4.2 Transport in Rectangular Enclosures

The two-dimensional flow in this configuration approaches a vertical layer bounded by two flat surfaces as the height is increased or a horizontal layer if the width is increased. For two-dimensional flows the equations of motion (15.2.5)–(15.2.7) are expressed in terms of the stream function, after eliminating p_m, as

$$\frac{\partial^2\psi}{\partial X^2} + \frac{\partial^2\psi}{\partial Y^2} = -\text{Ra}' \frac{\partial\phi}{\partial X}$$
$$\frac{\partial^2\phi}{\partial X^2} + \frac{\partial^2\phi}{\partial Y^2} = \frac{\partial\phi}{\partial\tau} + \left(\frac{\partial\psi}{\partial Y}\frac{\partial\phi}{\partial X} - \frac{\partial\psi}{\partial X}\frac{\partial\phi}{\partial Y}\right) \qquad (15.4.20)$$

where

$$X = \frac{x}{H} \qquad Y = \frac{y}{H} \qquad \hat{\tau} = \frac{\alpha\tau}{\hat{\sigma}H^2}$$
$$u = \frac{\partial\psi}{\partial y}\alpha \qquad v = -\frac{\partial\psi}{\partial x}\alpha \qquad \phi = \frac{t - t_c}{t_h - t_c}$$

Here t_h is the temperature of the heated surface and t_c that of the colder one. The coordinate system is shown in Fig. 15.4.4.

Flows driven by a temperature difference between the side walls. Under some circumstances certain simplifications of the above equations arise. For shallow cavities the aspect ratio $A = H/d << 1$. For such cavities and an end-to-end temperature difference, with insulated top and bottom surfaces, Bejan and Tien (1978) and Walker and Homsy (1977) determined the heat transfer in terms of the Rayleigh number and the aspect ratio. The cavity flow was divided into end regions on each side and a core region. By assuming that the core region is sufficiently far removed from the end regions, the vertical component of velocity there, v, is set equal to zero. Then the horizontal velocity $u(y)$ and the temperature distribution in the core, denoted by subscript c, are found to be

$$u_c(y) = -\text{Ra}'K_1\left(y - \frac{1}{2}\right)$$
$$t_c(x, y) = K_1 x + K_2 + \text{Ra}'K_1^2\left(\frac{y^2}{4} - \frac{y^3}{6}\right) \qquad (15.4.21)$$

where $\text{Ra}' = g\rho HK\beta(t_h - t_c)/\mu\alpha$.

The constants of integration K_1 and K_2 were determined from the temperature conditions at the location where the core and end regions interact. By assuming

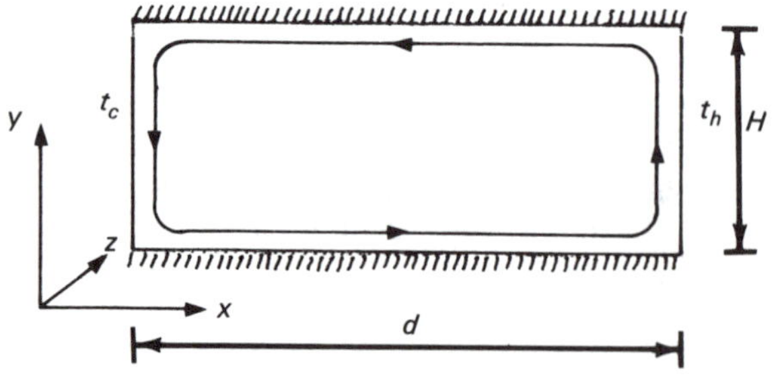

Figure 15.4.4 Cavity subjected to a horizontal temperature gradient.

velocity and temperature profiles in the end regions and applying an integral analysis, the Nusselt number was found to be

$$\text{Nu} = \frac{hd}{k} = 1 + \frac{1}{120}(\text{Ra}'_d)^2 A^4 \tag{15.4.22}$$

where $\text{Ra}'_d = \text{Ra}'\, d/H$.

At the other extreme, for tall, slender, rectangular enclosures filled with porous media, Weber (1975), emulating Gill's theory (1966) for nonporous-medium flow in such an enclosure, assumed boundary layer flow adjacent to the vertical walls. By postulating a stably stratified core, solutions were obtained and matched with those in the core region. The Nusselt number was found to be

$$\text{Nu} = \frac{hd}{k} = \frac{1}{\sqrt{3}}\left(\frac{\text{Ra}'}{\text{A}}\right)^{1/2} \tag{15.4.23}$$

Bejan (1979a, 1979b) pointed out that, with the boundary layer assumptions of Weber (1975), only two boundary conditions out of a possible four are satisfied at the bottom and top insulated boundaries. In satisfying the impermeability conditions at these boundaries, the calculation does not invoke the zero heat flux conditions. Bejan modified the analysis by calculating the net flow of energy in the vertical direction, which was set equal to zero at the top and bottom boundaries, thereby satisfying all four conditions. Figure 15.4.5 summarizes the calculated heat transfer results. Also shown are the numerical results for square cavities ($A = 1$) of Bankvall (1974) and Horne (1975).

Blythe et al. (1982) and Daniels et al. (1982) investigated the same flow configuration for high Ra′ flows. The region of calculation included the formation of horizontal boundary layers on the insulated surfaces. The structure of the flow field in the corner regions, where the boundary layers interact, was considered in detail. It was shown that the mass flux in the horizontal boundary layers along

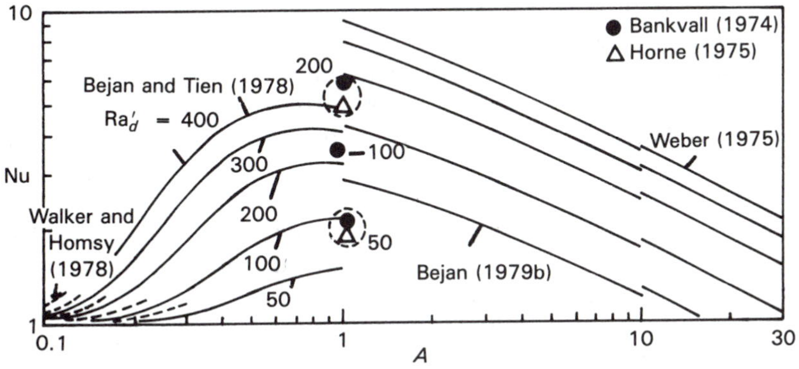

Figure 15.4.5 Summary of heat transfer calculations for a cavity filled with a fluid-saturated porous medium. *(Reprinted with permission from* Lett. Heat Mass Transfer, *vol. 6, p. 93, A. Bejan. Copyright © 1979, Pergamon Journals Ltd.)*

the horizontal insulated walls is small compared with the mass transported by the core flow.

End effects in rectangular enclosures. Earlier in this section, flows between parallel horizontal flat surfaces heated from below were discussed. Analytical and experimental results were reported. Measurements, however, are necessarily always conducted in a region of limited extent, and flows in such configurations are subject to end effects. These vary in severity depending on the particular flow configuration. For infinite horizontal layers, the resulting flow for $\mathrm{Ra}' > (\mathrm{Ra}')_{\mathrm{cr}}$ was found by Beck (1972) and others to be adjacent cells of similar structure and equal spacing. The deviations from this structure found in experiments are attributed to end effects.

Using a finite-difference scheme and a Hele-Shaw rectangular cell, which is commonly used to visualize flows in porous media (e.g., see Schlichting, 1978), Elder (1967a) studied the influence of end effects in steady flow. By numerically solving Eqs. (15.4.20), subject to the appropriate boundary conditions, the stream function and temperature distributions were obtained. By symmetrically varying the heated portion of the bottom surface, L, the effect of L on the flow was also studied. Figure 15.4.6 shows the steady-state isotherms and streamlines, on the left side, for four different values of $l = L/H$. H is again the depth of the cavity, at $\mathrm{Ra}' = 80$.

For $l = 2$ in Fig. 15.4.6*a* and *b*, two regions of equal and opposite circulation arise. These "cells" raise the heated fluid at the center of the cavity in a narrow vertical column. The fluid turns and diffuses, forming a weaker return flow. As the size of the heated portion l is increased, an even number of nearly square cells appear. The cells are "pushed" outward, as seen in Fig. 15.4.6*c–h*. Figure 15.4.7 shows the resulting variation in the Nusselt number with l. The breaks in the curve at $l = 2$, 4.3, and 5.4 correspond to the appearance of additional pairs of cells due to secondary flow.

Another important parameter affecting the influence of the end effects is Ra'. Figure 15.4.8 is a flow visualization in a Hele-Shaw cell for several values of Ra'. In Fig. 15.4.8a an even number of nearly square equally spaced cells are seen. At higher Ra' the end cells have grown and now "squeeze" the central cells in Fig. 15.4.8b–e to smaller size and number. Eventually only the end cells remain in Fig. 15.4.8f. As Ra' is decreased, in Fig. 15.4.8g, the square cells return, but the flow is clearly different from the one obtained by increasing Ra'. Thus, multiple steady-state flows have arisen, indicating that the steady-state flow depends on the path followed to its establishment.

Transients. Elder (1967b) also studied two kinds of transients in rectangular cavities. The first was the rise of a body of hot fluid, released at the base of a porous surface. The second was the flow that arises when a centered portion of the bottom surface is suddenly heated. Numerical and experimental results were presented. The numerical solution was based on a finite-difference representation of Eqs. (15.4.20) in their time-dependent form. The experiment was a flow visualization with a Hele-Shaw cell.

The resulting calculated streamlines and isotherms for the partially heated

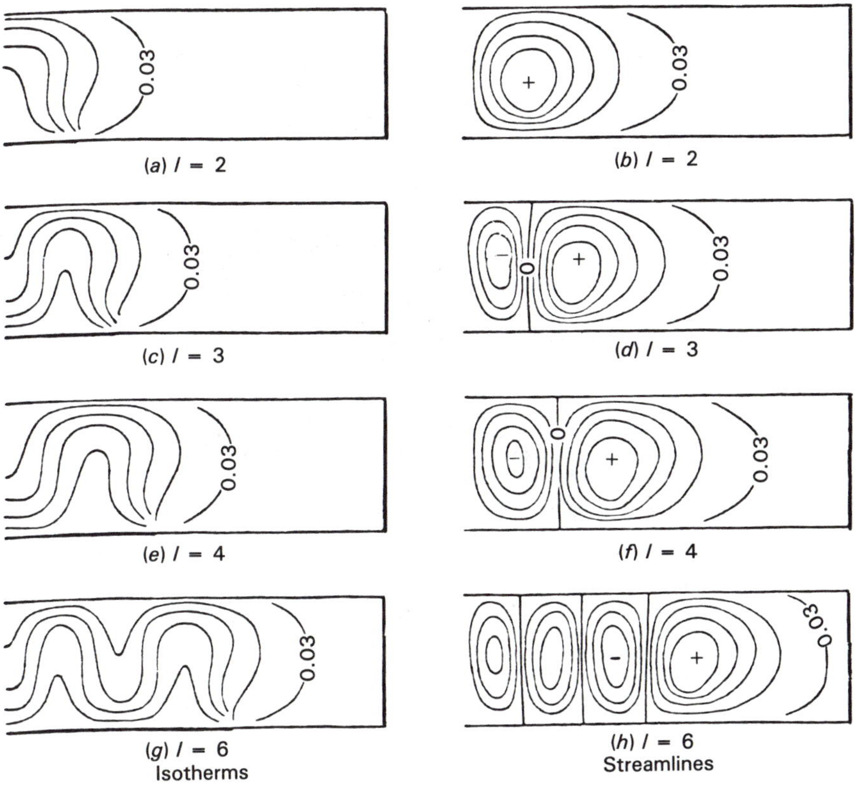

Figure 15.4.6 End effects. Isotherms and streamlines in a layer of horizontal extent $d/H = 10$ calculated at Rayleigh number $\mathrm{Ra}' = 80$ and various values of the heater length $l = L/H$, where d is the total horizontal length of the cavity, L is the heated portion of the bottom surface, and H is the depth of the cavity. *(Reprinted with permission from J. W. Elder,* J. Fluid Mech., *vol. 27, p. 29. Copyright © 1967, Cambridge University Press.)*

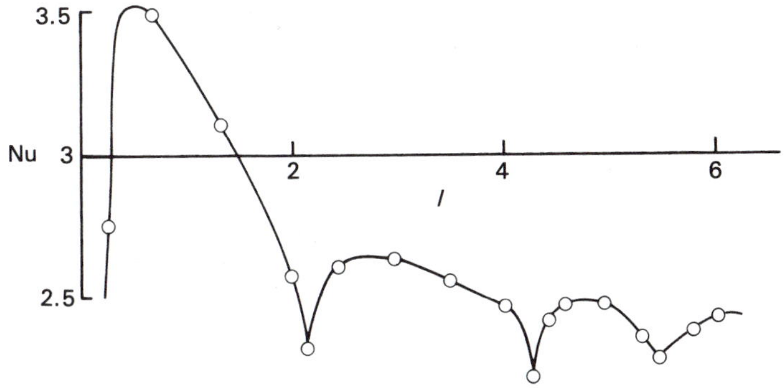

Figure 15.4.7 End effects. Effect of heater length l on the Nusselt number Nu. *(Reprinted with permission from J. W. Elder,* J. Fluid Mech., *vol. 27, p. 29. Copyright © 1967, Cambridge University Press.)*

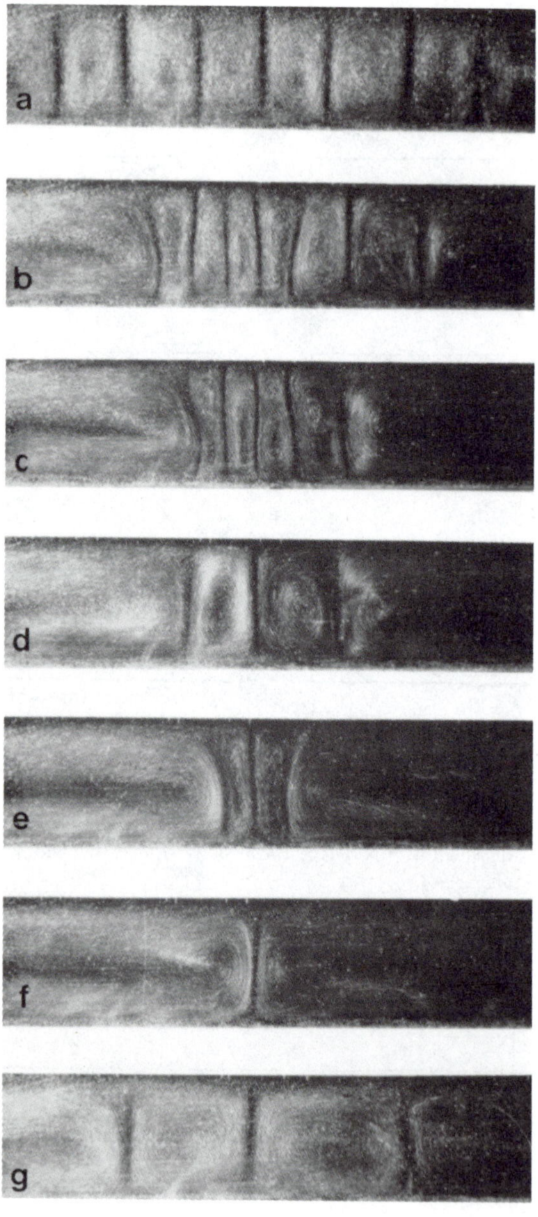

Figure 15.4.8 End effects in a thin cavity. Photographs of steady flows with medicinal paraffin in a Hele-Shaw cell of depth 2.0 cm, length 30.0 cm, and width 0.6 cm, with heated strip of length 10 cm. Values of 10^{-4} Ra: (*a*) 1.02, (*b*) 1.72, (*c*) 2.24, (*d*) 2.32, (*e*) 2.90, (*f*) 3.97, (*g*) 1.10. *(Reprinted with permission from J. W. Elder,* J. Fluid Mech., *vol. 27, p. 29. Copyright © 1967, Cambridge University Press.)*

lower surface condition are shown in Fig. 15.4.9 at various times. At short times the temperature field over most of the region is a one-dimensional conduction regime, in Fig. 15.4.9*a*, except near the corners. There the isotherms penetrate farther out into the fluid. Predictably, the fluid convective motion starts in those regions, in the form of growing eddies, convecting heat away from the ends of the heater. These eddies then increase in magnitude, and new adjacent pairs appear with increasing time, in Fig. 15.4.9*b* and *c*. The eddies eventually merge, in Fig. 15.4.9*d–f*, and grow to occupy most of the cavity, raising hot fluid in a vertical column near the center of the cavity. Figure 15.4.10 is the flow visualization for the same circumstances. It shows remarkable agreement of the flow patterns with the numerical results in Fig. 15.4.9.

The above transient heating condition was also studied by Horne and O'Sullivan (1974), using the more sophisticated numerical scheme introduced by Arakawa (1966) and Busbee et al. (1970). Two different flow regimes were found to occur at long times, depending on the extent of the heated portion of the bottom surface, L. For small L at sufficiently high Ra$'$, the boundary conditions force the solution into a steady periodic state. Otherwise, a stable steady multicellular flow was reached.

Other enclosures. Zebib (1978) analyzed the onset of convection in a vertical cylinder saturated with water. It was insulated on the vertical side wall, and the top and bottom horizontal rigid boundaries were taken at temperatures t_h and t_c. Using linear stability theory, the critical Ra$'$ was determined as a function of the aspect ratio of cylinder radius R to height H, $S = R/H$. The confining cylinder wall also stabilizes the fluid. As S is increased, the $4\pi^2$ limit for horizontal layers is approached, as shown in Fig. 15.4.11.

Bau and Torrance (1981) considered the stability of a similar geometry, a vertical annulus. Two types of boundary condition, permeable and impermeable, were considered at the top horizontal surface. The effect of permeability at the top was to destabilize the flow, as found previously for horizontal layers by Lapwood (1948). In a related study, Bau and Torrance (1982) experimentally investigated the onset of convection in a cylindrical cavity with a permeable upper boundary and reported good agreement with the earlier calculations.

Hickox and Gartling (1982) numerically determined the natural convection flow arising in a vertical annular space insulated at the top and bottom and subjected to inside and outside temperatures of t_i and t_o, respectively. Such a geometry is relevant to the insulation around a vertical cylindrical tank. The geometry and coordinate system are shown in Fig. 15.4.12. The two-dimensional equations of motion and energy, in radial coordinates r and z, are

$$\frac{1}{r}\frac{\partial}{\partial r}(ru) + \frac{\partial w}{\partial z} = 0 \tag{15.4.24}$$

$$\frac{\mu}{K_r}u = -\frac{\partial p_m}{\partial r} \tag{15.4.25}$$

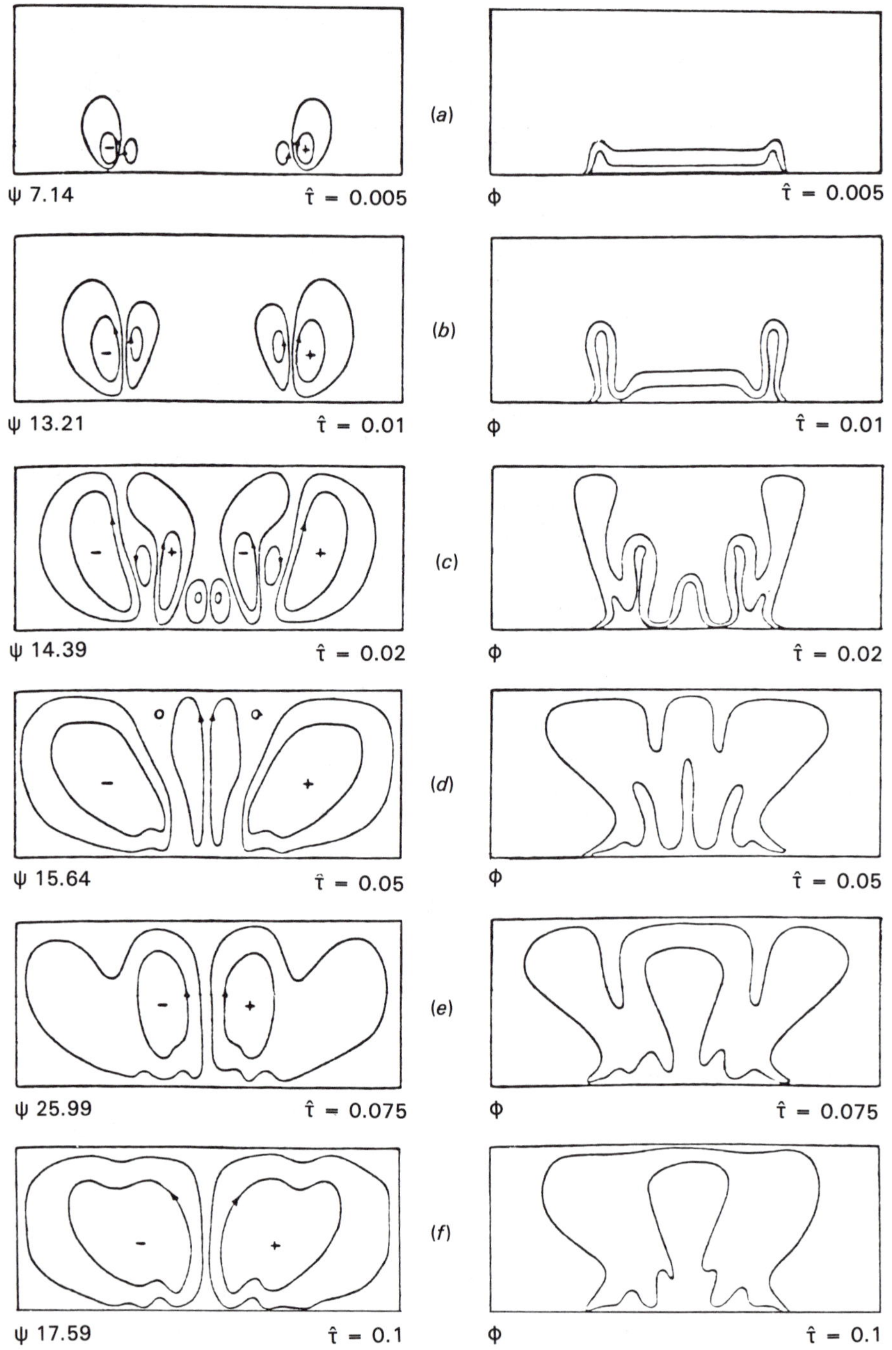

Figure 15.4.9 Stream function and temperature distributions for the short heater problem at various times; Ra = 400, $l = L/H = 2$. *(From Elder, 1967b.)*

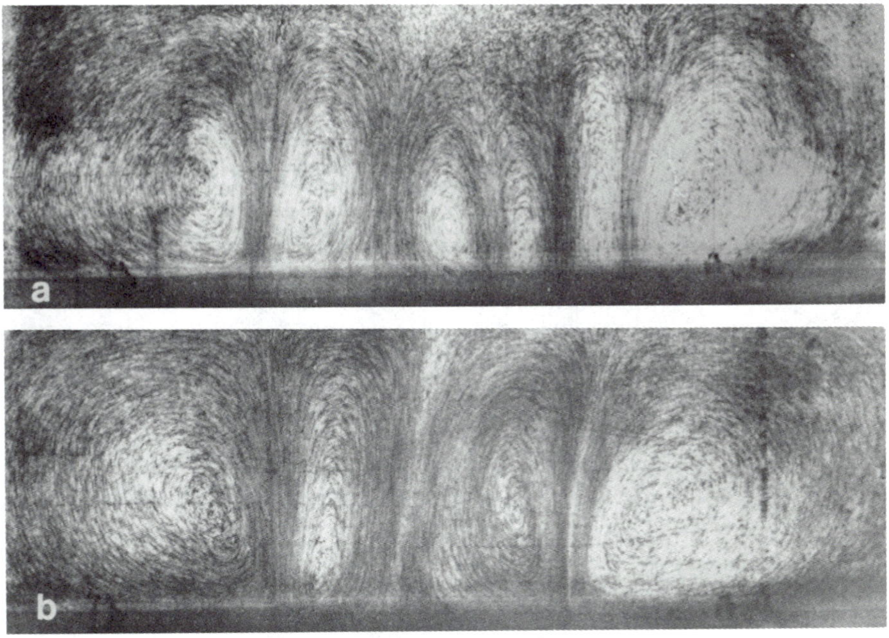

Figure 15.4.10 Visualizations of the streamlines in a Hele-Shaw cell for the same parameters as in Figure 15.4.9 at (*a*) $\hat{\tau} = 0.025$ and (*b*) $\hat{\tau} = 0.05$. *(Reprinted with permission from J. W. Elder,* J. Fluid Mech., *vol. 27, p. 609. Copyright © 1967, Cambridge University Press.)*

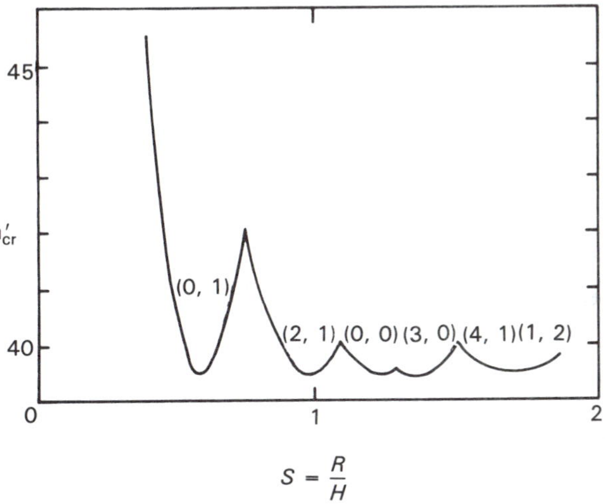

Figure 15.4.11 Critical Rayleigh number Ra'_{cr} as a function of $S = R/H$ for a cylinder filled with a saturated porous medium. The numbers on the plots refer to the modes of instability. *(Reprinted with permission from A. Zebib,* Phys. Fluids, *vol. 21, p. 699. Copyright © 1978, American Institute of Physics.)*

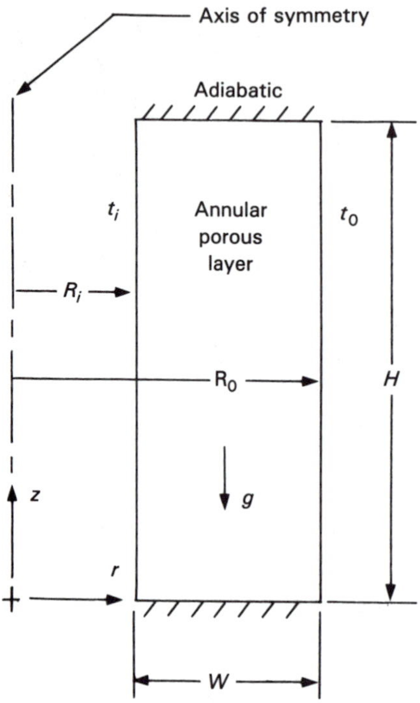

Figure 15.4.12 The vertical annular porous layer; one half is shown.

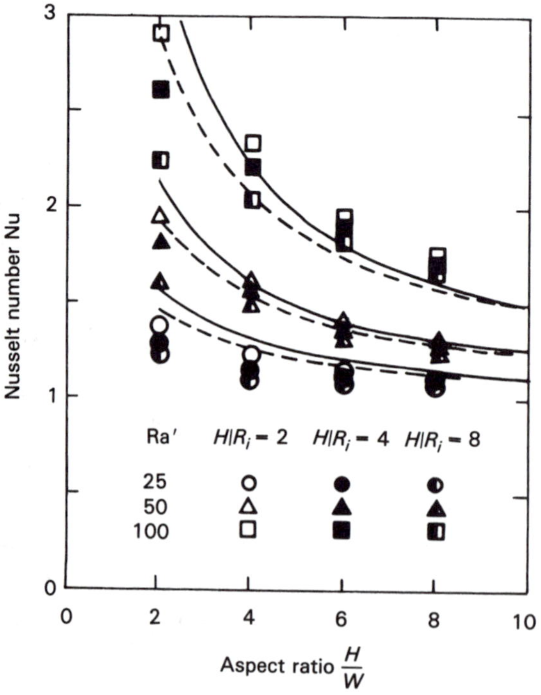

Figure 15.4.13 Predicted values of the Nusselt number. Symbols represent numerical results. Solid curves are predicted by an approximate analysis for $H/R_i = 2$; dashed curves are for $H/R_i = 8$. *(Reprinted with permission from C. E. Hickox and D. K. Gartling, ASME Paper 82-HT-68. Copyright © 1982, ASME.)*

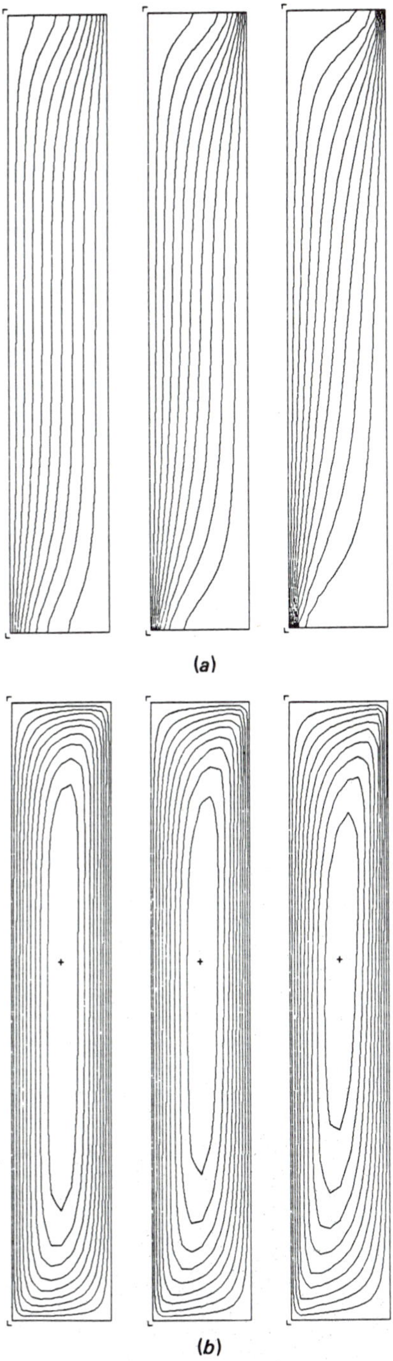

Figure 15.4.14 (*a*) Computed isotherms for $H/W = 6$ and $H/R_i = 4$. Rayleigh numbers are 25, 50, and 100, from left to right. (*b*) Computed steamlines for $H/W = 6$ and $H/R_i = 4$. Rayleigh numbers are 25, 50, and 100 from left to right. Maximum values of the nondimensional stream function are 4.047, 7.984, and 15.084 from left to right. *(Reprinted with permission from C. E. Hickox and D. K. Gartling, ASME Paper 82-HT-68. Copyright © 1982, ASME.)*

$$\frac{\mu}{K_z} w = -\frac{\partial p_m}{\partial z} + \rho_0 \beta g(t_i - t_o) \tag{15.4.26}$$

$$\rho_0 c_p \left(u \frac{\partial t}{\partial x} + w \frac{\partial t}{\partial z} \right) = k \left(\frac{\partial^2 t}{\partial r^2} + \frac{1}{r} \frac{\partial t}{\partial r} + \frac{\partial^2 t}{\partial z^2} \right) \tag{15.4.27}$$

where K_r and K_z are the possibly different permeabilities in the radial and vertical directions, respectively, k is the effective thermal conductivity, u and w are the radial and vertical velocity components, and ρ_0 is a reference fluid density.

A Galerkin form of the finite-element method was used for different values of the aspect ratio H/W, parameter H/R_i, and Ra$'$. The Nusselt number Nu $= hH/k = Q/[2\pi K_e H(t_i - t_o)/\ln(R_o/R_i)]$, where Q is the total heat transfer rate. Figure 15.4.13 shows the resulting Nu for different Ra$'$, aspect ratios, and H/R_i values. Increasing Ra$'$ or H/R_i is seen to enhance the heat transfer, while increasing H/W decreases it. Figure 15.4.14 shows the calculated isotherms and streamlines for Ra$' = 25$, 50, and 100 and $H/W = 6$.

15.5 OTHER EFFECTS OF POROUS MEDIA

Many flows in porous media occur under circumstances where additional effects are important. For example, many such flows are not well represented by Darcy's law. If the porous medium is anisotropic, additional effects arise. Flows in cold water near the density extremum must be formulated with an adequate density equation. The present section deals with these topics.

15.5.1 Non-Darcian Effects

Deviations from Darcy's law occur when the Reynolds number based on the mean pore diameter exceeds a value in the range 1–10. Inertial forces then become comparable to viscous forces. Several empirical formulations have been developed to account for such effects. One of the earliest is the addition of a square term by Forchheimer (1901):

$$Au + Bu^2 = J = -\frac{dp}{dx} \tag{15.5.1}$$

where J is the hydraulic gradient, u the apparent velocity, and A and B are determined experimentally. Thus, Bu^2 is the only difference from Eq. (15.2.2).

More recent formulations, such as those developed by Ergun (1952), Schneebeli (1955), Ward (1969), and others, account for the thermal properties of both fluid and porous material, to gain generality. The Ergun model is expressed as

$$u + \frac{\rho}{\mu} K' u^2 = -\frac{K}{\mu}\frac{dp}{dx}$$

$$K = \frac{D_p^2 \lambda^3}{150(1-\lambda)^2} \quad \text{and} \quad K' = \frac{1.75 D_p}{150(1-\lambda)} \tag{15.5.2}$$

where K and K' are the permeability and inertial coefficient in terms of the pore diameter D_p and porosity λ. Equation (15.5.2) approaches Darcy's law for K' very small. The major difference between the Ergun model and others is the explicit dependence of K and K' on the properties of the medium.

Plumb and Huenfeld (1981) applied Eq. (15.5.2) to study the boundary layer flow adjacent to a heated flat vertical surface immersed in a porous medium in which non-Darcian effects arise. The inertia effects were found to be significant when

$$\frac{g\beta KK'(t_0 - t_\infty)}{\nu^2} > 0.1 \tag{15.5.3}$$

where t_0 and t_∞ are the surface and ambient fluid temperatures, respectively. Cheng et al. (1981) experimentally studied non-Darcian flows and concluded that the Plumb and Huenfeld model overestimates the heat transfer rate.

15.5.2 Anisotropic Porous Media

Many porous media, such as sediments, rocks, and fibrous materials, are anisotropic. Epherre (1975, 1977) calculated stability limits in a medium in which the permeability and thermal conductivity are anisotropic in a relatively simple fashion. Along horizontal planes of constant properties, the permeability K_h and thermal conductivity k_h were taken to be the same in both directions. Perpendicular to that plane $K_v \neq K_h$ and $k_v \neq k_h$ were taken. The following criterion was obtained for the onset of convection:

$$(\mathrm{Ra}_h')_{\mathrm{cr}} = \pi^2\left[1 + \left(\frac{k_v K_v}{k_h K_h}\right)^{1/2}\right]^2 \qquad (\mathrm{Ra}_v')_{\mathrm{cr}} = \pi^2\left[1 + \left(\frac{k_h K_h}{k_h K_h}\right)^{1/2}\right]^2 \tag{15.5.4}$$

Figure 15.5.1 shows the values of $(\mathrm{Ra}_h)_{\mathrm{cr}}$ and $(\mathrm{Ra}_v')_{\mathrm{cr}}$ as a function of K_v/K_h and k_v/k_h. Castinel and Combarnous (1975, 1977), in an experimental investigation of flows in such anisotropic media, confirmed the criterion in Eq. (15.5.4).

Other studies of flows in anisotropic porous media include that of Tyvand (1977), which accounted for the effect on stability of hydrodynamic dispersion caused by a uniform base flow. Burns et al. (1977) analyzed convection in vertical slots filled with anisotropic porous media, while Kvernvold and Tyvand (1979) analyzed the flow in a homogeneous horizontal layer bounded by two flat planes at different temperatures.

15.5.3 Non-Boussinesq Fluid Behavior

In the previous sections of this chapter, the fluids considered were assumed to have a linear density variation with temperature. However, in many circumstances that is not a good approximation. For example, terrestrial water is commonly at temperature levels near its density extremum. This greatly complicates the buoyancy force behavior, causing possible buoyancy force reversals, local flow re-

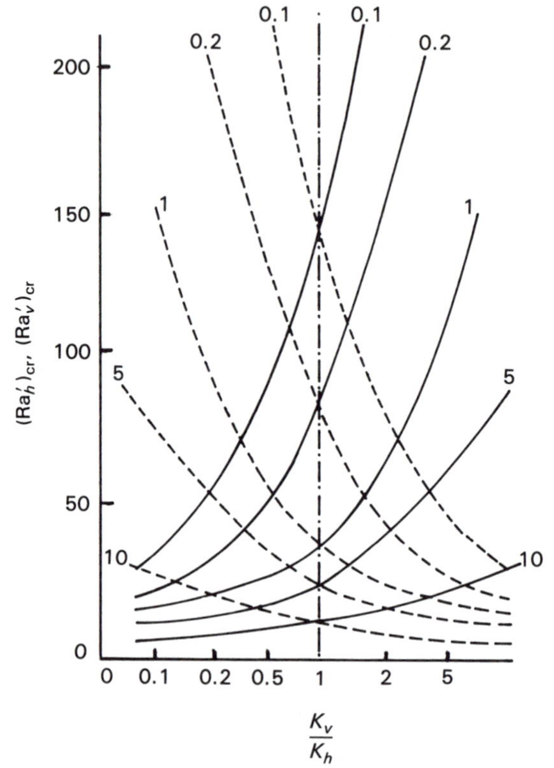

Figure 15.5.1 Criteria for the onset of convection: (– – –) $(\mathrm{Ra}'_h)_{cr}$ and (——) $(\mathrm{Ra}'_v)_{cr}$. Each curve of the two patterns corresponds to a given k_v^*/k_h^* value. *(Reproduced by permission of the American Institute of Chemical Engineers, G. Castinel and M. Combarnous,* Int. Chem. Eng., *vol. 17, p. 605. Copyright © 1977, AIChE.)*

versal, and convective inversion. Such effects were discussed in detail in Chapter 9 for flows in nonporous media.

Ramilison and Gebhart (1980) studied the flow adjacent to a heated flat vertical surface embedded in a porous medium saturated with cold water. The density equation used was that of Gebhart and Mollendorf (1977) (see Section 9.1), namely

$$\rho = \rho_m(s, p)[1 - \alpha(s, p)\,|t - t_m|^q] \tag{15.5.5}$$

This expression leads to the following modified form of Darcy's velocity in vertical plane external flows:

$$u(x, y) = \frac{K\alpha(s, p)\rho_m g}{\mu}\left(|t - t_m|^q - |t_\infty - t_m|^q\right) \tag{15.5.6}$$

The following similarity transformations are introduced:

$$\eta = yb(x) \qquad f(\eta) = \frac{\psi(x, y)}{\alpha c(x)} \qquad \phi(\eta) = \frac{t - t_\infty}{t_0 - t_\infty} \tag{15.5.7}$$

where b and c are to be determined from the similarity requirements and α is the effective thermal diffusivity, defined in Eq. (15.2.9).

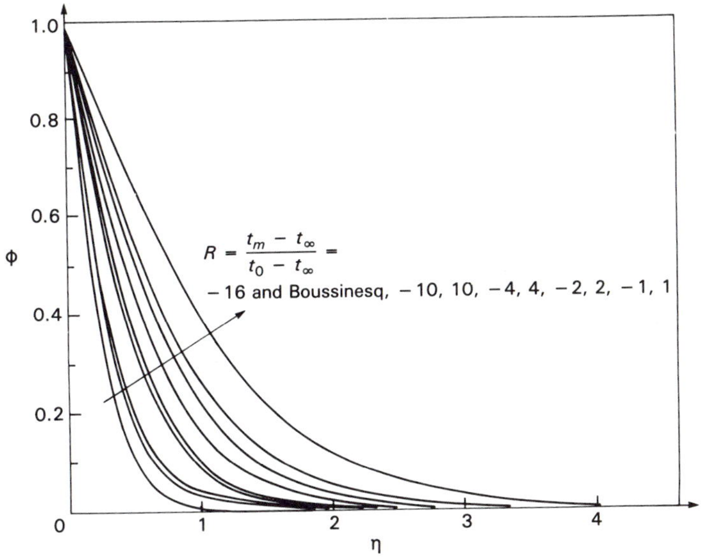

Figure 15.5.2 Distribution of the temperature variation $\phi(\eta)$ for selected values of R. The Boussinesq results are computed from Cheng and Minkowycz (1977). *(Reprinted with permission from* Int. J. Heat Mass Transfer, *vol. 23, p. 1521, J. M. Ramilson and B. Gebhart. Copyright © 1980, Pergamon Journals Ltd.)*

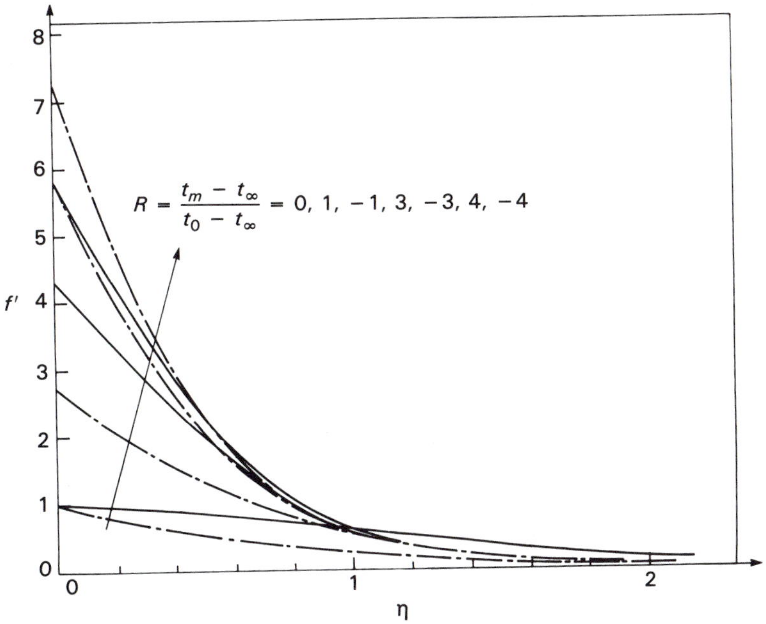

Figure 15.5.3 Distribution of the tangential component of velocity, f', for selected values of R outside the buoyancy reversal region. The solid lines are for downward flow, the dashed lines for upward flow. *(Reprinted with permission from* Int. J. Heat Mass Transfer, *vol. 23, p. 1521, J. M. Ramilson and B. Gebhart. Copyright © 1980, Pergamon Journals Ltd.)*

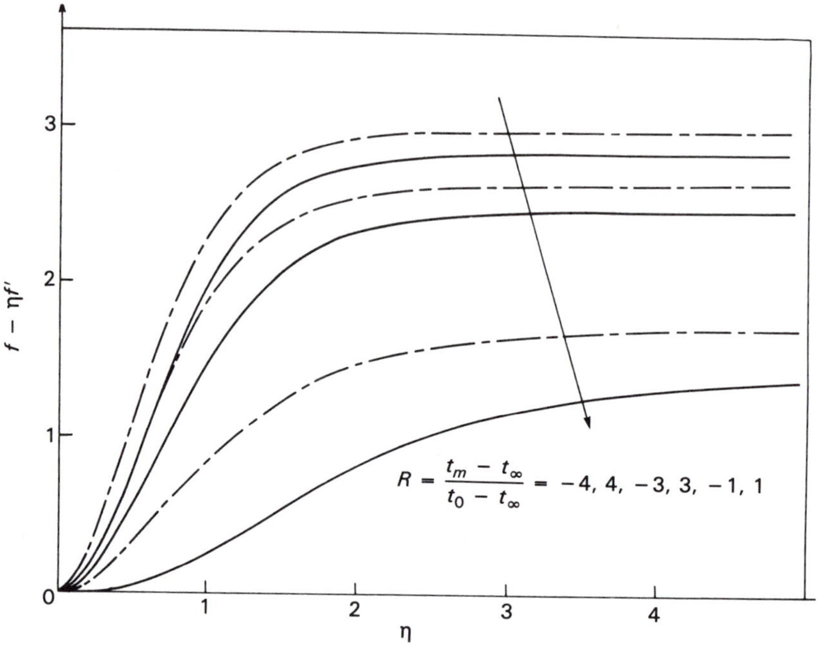

Figure 15.5.4 Distribution of the normal velocity component for selected R, outside the buoyancy reversal region. Solid lines represent downward flow and dashed lines upward flow. *(Reprinted with permission from* Int. J. Heat Mass Transfer, *vol. 23, p. 1521, J. M. Ramilison and B. Gebhart. Copyright © 1980, Pergamon Journals Ltd.)*

This formulation is substituted into the equations of motion. Following the usual procedure for finding similarity conditions, it is determined that similarity again arises for surface temperatures varying as a powerlaw, as x^n:

$$d(x) = t_0 - t_\infty = Nx^n \tag{15.5.8}$$

Figure 15.5.2 shows the resulting temperature profiles for an isothermal surface ($n = 0$) for various values of the only parameter that arises, $R = (t_m - t_\infty)/(t_0 - t_\infty)$. See Chapter 9 for its significance. No convergence was found over the range $0.195 < R < 0.4$. Small flow reversals were observed for $0.401 < R < 0.5$. The variation of the tangential component of the filtration velocity is seen in Fig. 15.5.3 for selected values of R, and the normal velocity component is shown in Fig. 15.5.4. See also the results of Gebhart et al. (1983) related to multiple solutions that arise at the edges of the region $0.401 < R < 0.5$.

REFERENCES

Ackroyd, J. A. D. (1967). *Proc. Cambridge Philos. Soc. 63*, 871.
Arakawa, A. (1966). *J. Comput. Phys. 1*, 119.
Bankvall, C. G. (1974). *Waerme Stoffuebertrag. 7*, 22.

Bau, H. H., and Torrance, K. E. (1981). *Phys. Fluids 24,* 382.
Bau, H. H., and Torrance, K. E. (1982). *J. Heat Transfer 104,* 166.
Beck J. L. (1972). *Phys. Fluids 15,* 1977.
Bejan, A. (1978). *J. Fluid Mech. 89,* 97.
Bejan, A. (1979a). *Int. J. Heat Mass Transfer 23,* 723.
Bejan, A. (1979b). *Lett. Heat Mass Transfer 6,* 93.
Bejan, A., and Tien, C. L. (1978). *J. Heat Transfer 100,* 191.
Blythe, P. A., Daniels, P. G., and Simpkins, P. G. (1982). *Proc. R. Soc. London Ser. A 380,* 119.
Bories, S. A., and Combarnous, M. (1973). *J. Fluid Mech. 57,* 63.
Brinkman, H. C. (1947). *Appl. Sci. Res. Sec. A 1,* 27.
Buretta, R. J., and Berman, A. S. (1976). *J. Appl. Mech. 98,* 249.
Burns, P. J., Chow, L. C., and Tien, C. L. (1977). *Int. J. Heat Mass Transfer 20,* 919.
Busbee, B. L., Golub, C. H., and Nielson, D. W. (1970). *SIAM J. Numer. Anal. 7,* 627.
Caltagirone, J. P., Cloupeau, M., and Combarnous, M. (1971). *C.R. Acad. Sci. Ser. B 273,* 833.
Castinel, G., and Combarnous, M. (1975). *Rev. Gen. Therm. 168,* 937.
Castinel, G., and Combarnous, M. (1977). *Int. Chem. Eng. 17,* 605.
Chan, B. K. C., Ivey, C. M., and Barry, J. M. (1970). *J. Heat Transfer 92,* 21.
Cheng, P. (1977a). *Int. J. Heat Mass Transfer 20,* 201.
Cheng, P. (1977b). *Int. J. Heat Mass Transfer 20,* 807.
Cheng, P. (1977c). *Int. J. Heat Mass Transfer 20,* 893.
Cheng, P. (1978). *Adv. Heat Transfer 14,* 1.
Cheng, P., and Chang, I. D. (1976). *Int. J. Heat Mass Transfer 19,* 1267.
Cheng, P., and Minkowycz, W. J. (1977). *J. Geophys. Res. 82,* 2040.
Cheng, P., Ali, C. L., and Verma, A. K. (1981). *Lett. Heat Mass Transfer 8,* 261.
Combarnous, M. (1970). Thesis, Univ. of Paris.
Combarnous, M., and Bories, S. A. (1975). *Adv. Hydrosci. 10,* 231.
Combarnous, M., and LeFur, B. (1969). *C.R. Acad. Sci. Ser. B 269,* 1009.
Daniels, P. G., Blythe, P. A., and Simpkins, P. G. (1982). *Proc. R. Soc. London Ser. A 382,* 135.
Darcy, H. (1856). *Les Fontaines Publique de la Ville de Dijon.* Victor Dalmont, Paris.
Davis, S. N. (1969). In *Flow through Porous Media,* J. M. DeWiest, Ed., Academic Press, New York.
Elder, J. W. (1967a). *J. Fluid Mech. 27,* 29.
Elder, J. W. (1967b). *J. Fluid Mech. 27,* 609.
Epherre, J. F. (1975). *Rev. Gen. Therm. 168,* 949.
Epherre, J. F. (1977). *Int. Chem. Eng. 17,* 615.
Ergun, S. (1952). *Chem. Eng. Prog. 48,* 89.
Evans, G. H., and Plumb, O. A. (1978). Presented at the AIAA/ASME Thermophysics and Heat Transfer meeting.
Forchheimer, P. (1901). *Forschtlft. Ver. D. Ing. 45,* 1782.
Gebhart, B. G., and Mollendorf, J. C. (1977), *Deep Sea Res. 24,* 831.
Gebhart, B., Hassard, B., Hastings, S. P., and Kazarinoff, N. D. (1983). *Numer. Heat Transfer 6,* 337.
Gill, A. E. (1966). *J. Fluid Mech. 26,* 515.
Haaland, S. E., and Sparrow, E. M. (1973). *Int. J. Heat Mass Transfer 16,* 2355.
Hickox, C. E., and Gartling, D. K. (1982). ASME Paper 82-HT-68.
Hickox, C. E., and Watts, H. A. (1980). *J. Heat Transfer 102,* 248.
Homsy, G. M., and Sherwood, A. E. (1976). *AIChE J. 22,* 168.
Horne, R. N. (1975). Transient Effects in Geothermal Convective Systems, Ph.D. Thesis, Univ. of Auckland, Auckland, New Zealand.
Horne, R. N., and O'Sullivan, M. J. (1974). *J. Fluid Mech. 66,* 339.
Hsu, C. T., and Cheng, P. (1979). *J. Heat Transfer 101,* 660.
Hsu, C. T., Cheng, P., and Homsy, G. M. (1978). *Int. J. Heat Mass Transfer 21,* 1221.
Johnson, C. H., and Cheng, P. (1978). *Int. J. Heat Mass Transfer 21,* 709.
Kaneko, T., Mohtadi, M. F., and Aziz, K. (1974). *Int. J. Heat Mass Transfer 17,* 485.

Kassoy, D. R., and Zebib, A. (1975). *Phys. Fluids 18,* 1649.
Katto, Y., and Masuoka, T. (1967). *Int. J. Heat Mass Transfer 10,* 297.
Kvernvold, O., and Tyvand, P. A. (1979). *J. Fluid Mech. 90,* 609.
Lapwood, E. R. (1948). *Proc. Cambridge Philos. Soc. 44,* 508.
McNabb, A. (1965). *Proc. 2d Australas. Conf. Hydraul. Fluid Mech.,* p. C161.
Merkin, J. H. (1978). *Int. J. Heat Mass Transfer 21,* 1499.
Merkin, J. H. (1979). *Int. J. Heat Mass Transfer 22,* 1461.
Minkowycz, W. J., and Cheng, P. (1976). *Int. J. Heat Mass Transfer 19,* 805.
Nield, D. A. (1968). *Water Resour. Res. 4,* 553.
Plumb, O. A., and Huenfeld, J. C. (1981). *Int. J. Heat Mass Transfer 24,* 765.
Prats, M. (1966). *J. Geophys. Res. 71,* 4835.
Ramilison, J. M., and Gebhart, B. G. (1980). *Int. J. Heat Mass Transfer 23,* 1521.
Schlichting, H. (1978). *Boundary Layer Theory,* 6th ed., McGraw-Hill, New York.
Schneebeli, G. (1955). *Huille Blanche No. 2 10,* 141.
Schneider, K. J. (1963). Die Warmeleitfahigkeit Korniger stoffe und ihre Beeinflussung durch freie Konvektion, Thesis, Univ. of Karlsruhe.
Straus, J. M., and Schubert, G. (1977). *J. Geophys. Res. 82,* 325.
Tyvand, P. A. (1977). *J. Hydrol. 34,* 335.
Walker, K., and Homsy, G. M. (1977). *J. Heat Transfer 99,* 338.
Walker, K. L., and Homsy, G. M. (1978). *J. Fluid Mech. 87,* 449.
Ward, J. C. (1969). *Proc. Soc. Civ. Eng. No. H45 90,* 1.
Weber, J. E. (1974). *Int. J. Heat Mass Transfer 17,* 241.
Weber, J. E. (1975). *Int. J. Heat Mass Transfer 18,* 569.
Wooding, R. A. (1963). *J. Fluid Mech. 15,* 526.
Zebib, A. (1978). *Phys. Fluids 21,* 699.
Zebib, A., and Kassoy, D. R. (1978). *Phys. Fluids 20,* 4.

PROBLEMS

15.1 For buoyancy-driven vertical flows in a quiescent ambient medium, determine the range of n, in $d = Nx^n$, for which reasonable solutions arise. Also, determine the value of n for both free-boundary and wall line source plumes.

15.2 For an extensive quiescent water-saturated porous sandstone at 18°C, consider an immersed vertical surface of height L at 22°C. Take the permeability to be 2×10^{-2} darcy.

(a) Calculate the boundary layer thickness as a function of downstream distance x, as δ/x.

(b) At $x = 10$ m, calculate δ, $u_{\max}$, and the surface heat flux.

15.3 For the conditions in Problem 15.2, consider that the ambient water is flowing upward at a velocity of U_∞ m/s. Find the range of U_∞ in which mixed convection effects are important. Calculate the total heat transfer rate at the upper limit of this range.

15.4 Consider a 20-m-thick horizontal layer of the sandstone medium in Problem 15.2, bounded by impervious surfaces on both sides.

(a) What temperature difference between the boundaries, Δt_c, would result in initial instability?

(b) Determine the heat flux.

(c) Determine the heat flux for $\Delta t = 2\,\Delta t_c$ and $10\,\Delta t_c$.

15.5 A horizontal rectangular region, H by d, of a porous medium saturated with water is maintained at 25 and 15°C on its right and left sides, respectively. The value of K is 2×10^{-3} darcy.

(a) For $H = 10$ m calculate the heat transfer rate across the enclosure for both $H/d << 1$ and $H/d >> 1$.

(b) Repeat (a) for $H/d = 1$ for the various predictions given in Section 15.4.

(c) Comment on any disagreements between the results in part (b).

15.6 Consider an isothermal vertical surface 10 m high, embedded in a porous medium saturated with water at 2°C. The permeability is 10^{-2} darcy.

(a) Calculate the surface temperatures at the two edges of the gap found in solutions of the boundary layer equations.

(b) Find the values of t_o that result in $R = -1$ and 1.

(c) Calculate the heat transfer rate and vertical velocity at 10 m for each of the conditions in part (b).

CHAPTER

SIXTEEN

NON–NEWTONIAN TRANSPORT

16.1 CLASSIFICATION AND CONSTITUTIVE EQUATIONS

The preceding chapters considered various aspects of buoyancy-induced flows and transport in Newtonian fluids. These fluids have a linear relationship between the shear stress and shear rate. However, many fluids of practical interest exhibit non-Newtonian behavior. Examples are solutions and melts of polymers, emulsions, blood, paints, nuclear slurries, and materials having both viscous and elastic properties. Since these fluids are increasingly encountered in a wide spectrum of the processing industry, it has become important to understand their transport characteristics. In the past two decades extensive research has been reported in the literature.

To differentiate between the flow behavior of a Newtonian fluid and that of a non-Newtonian fluid, consider a fluid between two parallel surfaces that are separated by a distance d as shown in Fig. 16.1.1. When a force F in the x direction is applied to the upper plate, a steady velocity U is generated. The fluid is merely pulled along because of its adherence to the upper plate. The resulting shear stress is given by $\tau_{yx} = F/A$, where A is the surface area of the upper plate.

When the fluid is Newtonian, the velocity distribution is linear, as shown. The velocity gradient, du/dy, or shear rate, is constant across the fluid layer. The slope of the shear stress variation with the velocity gradient is the coefficient of viscosity of the fluid, μ. It depends only on the temperature and pressure. It is independent of the shear rate.

For non-Newtonian fluids, the plot of τ_{yx} vs. du/dy is not a straight line through the origin. The apparent viscosity, that is, the ratio of the shear stress to the resulting rate of shear, may increase or decrease with the shear rate du/dy. In

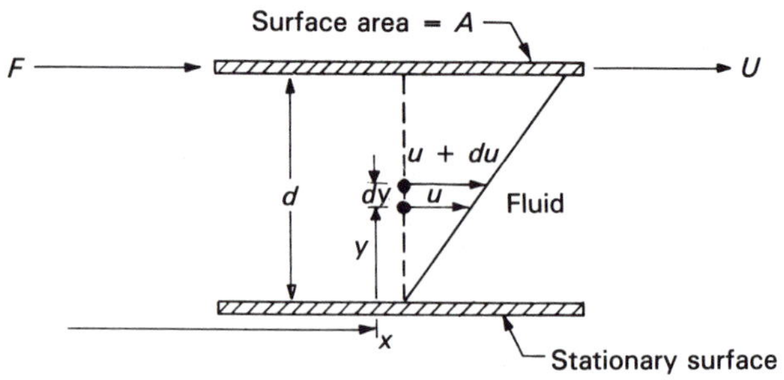

Figure 16.1.1 Steady laminar shearing motion of a Newtonian fluid between two parallel plates.

some fluids it may even depend on the duration of the shear. These anomalies in viscosity are discussed next in a description of non-Newtonian fluid behavior.

16.1.1 Classification

Non-Newtonian fluids are described by a broad classification system in terms of the complexities that arise in describing the flow properties. Metzner (1961, 1965) has given a comprehensive rheological classification. A simple classification given by Shenoy and Mashelkar (1982) is shown in Fig. 16.1.2.

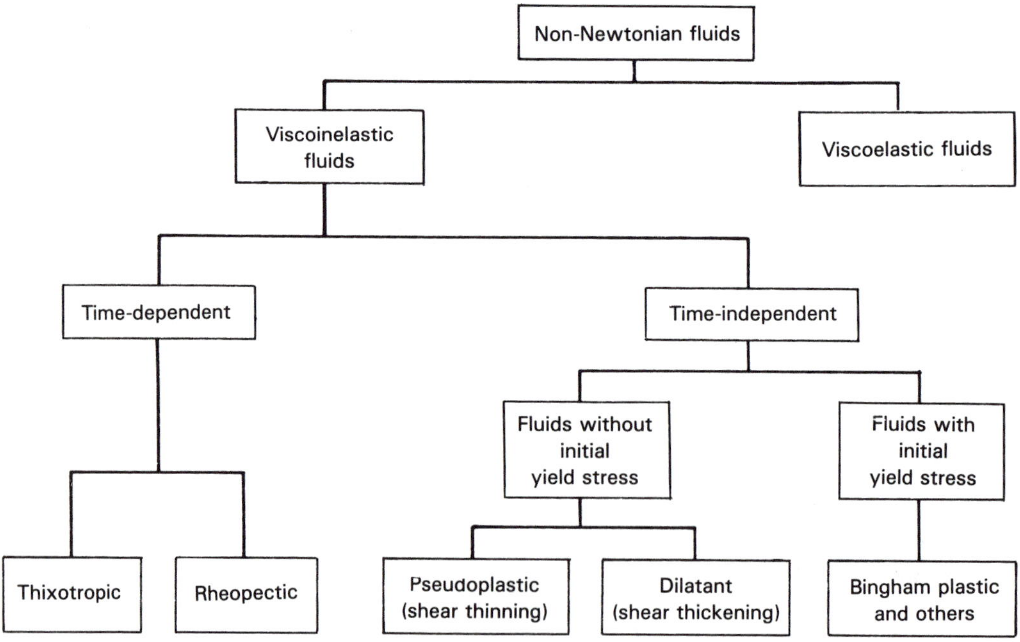

Figure 16.1.2 Rheological classification of non-Newtonian fluids. *(Reproduced by permission of the American Institute of Chemical Engineers, A. V. Shenoy and R. A. Mashelkar,* AIChE J., *vol. 24, p. 344. Copyright © 1978, AIChE.)*

Depending on whether the fluid has an elastic response to the applied stress, there are two general categories: viscoinelastic (or purely viscous) and viscoelastic. If there is no recovery of deformation on removal of the applied stress, that is, there is no elastic response, the fluid is called viscoinelastic. Recall that most solids exhibit some degree of elastic response, with complete recovery of deformation on removal of the applied stresses. The simplest such body is the Hookean elastic solid, for which the deformation is directly proportional to the applied stress. Viscoelastic fluids have properties of both viscous fluids and elastic solids. They show partial elastic recovery on removal of a deforming shear stress. The peculiarities of viscoinelastic and viscoelastic fluids are discussed briefly below.

Viscoinelastic (purely viscous) fluids. These fluids may be further subdivided as time-dependent and time-independent fluids. Time-independent fluids are those for which the duration of the shear has no effect on viscosity. The shear rate at a given point is solely dependent on the shear stress at that point. Plots of flow curves showing the shear stress τ_{xy} vs. shear rate du/dy for a group of such fluids are given in Fig. 16.1.3*a*.

As seen in Fig. 16.1.3*a*, the time-independent fluids may or may not have an initial yield stress τ_0. Fluids with no initial yield stress are called pseudoplastic if the apparent viscosity decreases with an increase in shear rate. They are called dilatant if the apparent viscosity increases with increasing shear rate. These behaviors are also known as shear thinning and shear thickening, respectively. Examples of shear-thinning fluids are dilute solutions of high polymers, most printing inks, paper pulp, and napalm. Examples of materials that have been found to show dilatancy are starch, potassium silicate, quicksand, wet beach sand, some cornstarch/sugar solutions, and some aqueous suspensions of titanium dioxide; see Calderbank and Moo-Young (1959), Metzner and Whitlock (1958), Wilkinson (1960), and Skelland (1967).

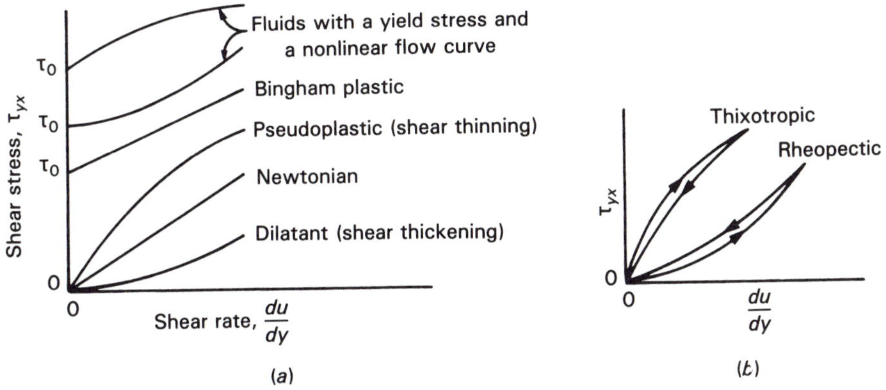

Figure 16.1.3 Plots of shear stress vs. shear rate for viscoinelastic non-Newtonian fluids. (*a*) Time-independent and (*b*) time-dependent fluids. *(Reprinted by permission of John Wiley & Sons, Inc., from A. H. P. Skelland,* Non-Newtonian Flow and Heat Transfer. *Copyright © 1967, John Wiley & Sons, Inc.)*

Many other important fluids are approximately pseudoplastic. Examples are rubber solutions, adhesives, polymer solutions or melts, greases, paints, dispersion media in certain pharmaceuticals, and biological fluids; see Skelland (1967). For such fluids, the apparent viscosity in the high-shear region, that is, large du/dy, is called "viscosity at infinite shear" and is denoted by μ_∞. Similarly, μ_0 denotes the "viscosity at zero-shear rate." As seen in Fig. 16.1.3*a*, the viscosity begins as the initial slope, μ_0, and gradually decreases toward μ_∞ as the shear rate increases.

For the time-independent fluids with a yield stress τ_0, this stress must be exceeded before flow starts. The apparent viscosity, as before, may increase or decrease with an increase in shear rate, as shown by the upper two curves in Fig. 16.1.3*a*. Or, in a Bingham plastic fluid, the viscosity is taken as independent of shear rate. However, all these fluids act as elastic solids below τ_0 and as viscous fluids above it. The explanation of this behavior is that the fluid at rest contains a three-dimensional structure of sufficient rigidity to resist any stress less than τ_0. Once this stress is exceeded, the internal structure breaks down and shearing motion begins. Certain plastic melts, oil well drilling muds, detergent slurries, thorium and uranium oxide slurries, paper pulp, toothpaste, margarine, and shortenings are examples of these fluids.

Time-dependent fluids are those for which the shear rate is a function of both the magnitude and the duration of shear. Such fluids are called thixotropic if they exhibit a reversible decrease in viscosity with time at a constant rate of shear and fixed temperature. On the other hand, if a constant shear rate causes a reversible increase in viscosity, the fluid is called rheopectic. The flow curves for these fluids are shown in Fig. 16.1.3*b*, where the arrows indicate the chronological progress of an experiment in which the shear rate is steadily increased from zero to a maximum value and then decreases toward zero.

Examples of thixotropic properties have been found in some solutions or melts of high polymers, oil drilling muds, many food materials, and paints. Rheopectic fluids are relatively rare, although bentonite clay suspensions, gypsum suspensions, and dilute suspensions of ammonium oleate exhibit rheopectic characteristics at moderate shear rates; see Metzner (1956), Wilkinson (1957), Calderbank and Moo-Young (1959), and Skelland (1967).

Viscoelastic fluids. These fluids show partial elastic recovery on removal of a deforming shear stress. They exhibit a combination of properties of both viscous fluids and elastic solids. In a purely Hookean elastic solid, the stress that produces a given strain is independent of time. However, for viscoelastic fluids the stress will gradually dissipate. In contrast to purely viscous liquids, on the other hand, viscoelastic fluids flow due to stress, but part of the deformation is gradually recovered when the stress is removed. Therefore, the time derivatives of both shear stress and shear rate must be included in the governing equations that describe the flow of such fluids. Stress relaxation, strain recovery, and die swell are some of the phenomena characteristic of such fluids. Solids and molten polymers and their solutions are usually strongly viscoelastic in nature. Some soap solutions, silicone putty, some condensed soups, the thick part of egg white, several shampoos, some types of condensed milk, and gelatin in water also exhibit viscoelastic effects.

16.1.2 Fluid Models

The major difficulty that arises in the analysis of non-Newtonian transport is the lack of any generally acceptable equation of state between the stress tensor and the shear rate. Several empirical models have been proposed to express the relation between the shear stress τ_{yx} and the shear rate du/dy for viscoinelastic fluids. Each of these models contains empirical positive parameters, which are determined numerically to fit experimental data on τ_{yx} vs. du/dy at constant temperature and pressure. The models described below are those that have been used frequently in analyzing buoyancy-driven transport. A detailed discussion of other models may be found in Skelland (1967), Middleman (1977), and Bird et al. (1977).

Time-independent viscoinelastic fluids without a yield stress. The model most commonly used to describe the relationship between the shear stress and the shear rate for these fluids has been the Ostwald-de Waele (1926) power law model given as:

Power law model:

$$\tau_{yx} = K \left| \frac{du}{dy} \right|^{n-1} \frac{du}{dy} \tag{16.1.1}$$

The parameters K and n are empirical constants and are generally referred to as the consistency index and the power law fluid index, respectively. Note that $n = 1$ represents a Newtonian fluid behavior with $K = \mu$. For $n < 1$ the behavior is pseudoplastic, whereas for $n > 1$ it is dilatant. Although this model has been widely used, it has certain limitations. It is an excellent approximation at intermediate shear rates but fails at the extremes of both very low and very high shear rates. It does not predict the behavior (usually Newtonian) observed at either of these extremes. Buoyancy-induced flows are more often low-shear flows. Under such conditions this model becomes less reliable. Two other models, the Ellis model (Reiner, 1960) and the Sutterby model (Sutterby, 1966), have been commonly used and are described below.

Ellis model:

$$\tau_{yx} = \frac{1}{A + B \left| \tau_{yx} \right|^{\alpha-1}} \frac{du}{dy} \tag{16.1.2}$$

This model contains three adjustable positive parameters, A, B, and α. Typically, A is taken to be $1/\mu_0$, where μ_0 is the zero shear rate viscosity. For $\alpha < 1$ the model approaches Newtonian behavior for small τ_{yx} and thus is better than the power law model. For $\alpha > 1$ the model again approaches Newtonian behavior at large τ_{yx}; however, the value of apparent viscosity in the high-shear region, μ_∞, is not correctly predicted.

Sutterby model:

$$\tau_{yx} = \mu_a \frac{du}{dy} \tag{16.1.3a}$$

where

$$\mu_a = \mu_0 \left[\frac{\text{arc sinh}\,(B\, du/dy)}{B\, du/dy} \right]^A \tag{16.1.3b}$$

The three adjustable positive parameters are A, B, and μ_0. In the low shear rate limit $du/dy \to 0$, $\mu_a \to 0$. Furthermore, at high shear rates, μ_a decreases. This describes a pseudoplastic behavior. However, μ_∞ again is not correctly predicted. This is not necessarily a defect, since for many systems the apparent viscosity approaches μ_∞ only at extremely high shear rates rarely encountered in practice. Like the Ellis model, this model is applicable for low and intermediate shear rates.

Time-independent viscoinelastic fluids with a yield stress. Of these fluids, the Bingham plastic fluid has been used most. The relevant model equations for this fluid are given below.

Bingham plastic fluid:

$$\tau_{yx} = \pm\tau_0 + \mu_B \frac{du}{dy} \qquad |\tau_{yx}| > \tau_0 \tag{16.1.4a}$$

$$\frac{du}{dy} = 0 \qquad |\tau_{yx}| < \tau_0 \tag{16.1.4b}$$

where in Eq. (16.1.4a) the plus sign applies for a positive velocity gradient and the minus sign for a negative velocity gradient. The parameter μ_B is the so-called plastic viscosity. This model is used primarily for slurries and pastes.

Although true Bingham plastic behavior is rarely encountered, departures from this model are sometimes small enough for solutions based on this model to be useful in design. For a description of fluid models for other time-independent fluids, with yield stress, for which the flow curves are nonlinear, the reader is referred to Skelland (1967).

Time-dependent viscoinelastic fluids. Constitutive equations for these fluids have been discussed by Hahn et al. (1959), Denney and Brodkey (1962), and Ritter (1962). These are quite complex and, to date, engineering problems of interest for such fluids have not been analyzed. However, under steady-state conditions these fluids may be assumed to behave as time-independent fluids. In the sections to follow in this chapter, discussion will therefore be mostly confined to time-independent viscoinelastic fluids. Fluid models described by Eqs. (16.1.1–16.1.4) will be used.

Viscoelastic fluids. The rheological properties of these fluids at any instant are functions of the recent history of the fluid and cannot be described by a relationship between shear stress and shear rate only. Choosing an appropriate constitutive equation and solving the thermal convection problem is then a difficult task. As a result, studies that accurately treat buoyancy-induced convection in viscoelastic fluids are few. A brief summary will be presented in the last section of the chapter.

16.2 VISCOINELASTIC FLUIDS: GOVERNING EQUATIONS

This section presents the transport equation for inelastic fluids. Recall that these fluids deform continuously in response to shear. In Chapter 2 the general equations (2.1.1)–(2.1.3) describe transport in a Newtonian fluid. For non-Newtonian flow behavior the continuity equation (2.1.1) still applies. However, the momentum and energy equations take a more general form, as given below:

$$\frac{D\rho}{D\tau} = -\rho\,\nabla\cdot\mathbf{V} \tag{16.2.1}$$

$$\rho\frac{D\mathbf{V}}{D\tau} = \rho\mathbf{g} - \nabla p + \nabla\cdot\boldsymbol{\tau}' \tag{16.2.2}$$

$$\rho c_p\frac{Dt}{D\tau} = \nabla\cdot k\nabla t + \beta T\frac{Dp}{D\tau} + (\boldsymbol{\tau}'\cdot\nabla\mathbf{V}) + q''' \tag{16.2.3}$$

where the extra stress tensor $\boldsymbol{\tau}'$ is now specified in terms of the rheological equation of state. Four specific rheological models were discussed in Section 16.1.

The physical significance of various terms in the above equations, except for non-Newtonian effects in $\nabla\cdot\boldsymbol{\tau}'$ [Eq. (16.2.2)] and in $(\boldsymbol{\tau}'\cdot\nabla\mathbf{V})$ [Eq. (16.2.3)], was discussed in Chapter 2. The term $\nabla\cdot\boldsymbol{\tau}'$ represents the viscous force on an element per unit volume. The viscous dissipation term $(\boldsymbol{\tau}'\cdot\nabla\mathbf{V})$ in Eq. (16.2.3) is evaluated as

$$\boldsymbol{\tau}'\cdot\nabla\mathbf{V} = \tau_{xx}\left(\frac{\partial u}{\partial x}\right) + \tau_{yy}\left(\frac{\partial v}{\partial y}\right) + \tau_{zz}\left(\frac{\partial w}{\partial z}\right) + \tau_{xy}\left(\frac{\partial u}{\partial y} + \frac{\partial v}{\partial x}\right) + \tau_{yz}\left(\frac{\partial v}{\partial z} + \frac{\partial w}{\partial y}\right) + \tau_{zx}\left(\frac{\partial w}{\partial x} + \frac{\partial u}{\partial z}\right) \tag{16.2.4}$$

For Newtonian fluids, the above equations result in Eqs. (2.1.1)–(2.1.4).

The fluid motion again is driven by the buoyancy forces arising from density differences due to temperature differences. The above equations remain coupled, requiring a simultaneous solution. Additional complexity arises if the properties are appreciably temperature dependent. In addition to ρ, k, and c_p, the parameters that describe the rheology may also be temperature dependent.

Solutions of the above equations are more difficult than before. However, for many flows, simplifying approximations may be invoked. For low temperature differences across the transport region, the property variations may be negligible except for the driving density differences. For steady flows the time derivatives are zero. Boundary layer approximations again result in large simplifications. These approximations are discussed in detail in Chapters 2–5 for many flows in Newtonian fluids. Similar simplifying arguments are also applicable to non-Newtonian transport.

Many important flows adjacent to heated or cooled surfaces are steady and two dimensional; an example is the flow adjacent to a vertical flat surface or a cylinder. Then the boundary layer approximations may be used for many thermal boundary conditions. For such flows external to a flat inclined or two-dimensional planar or curved surface at t_0 in an infinite quiescent medium at t_∞, the general governing equations, with viscous dissipation and pressure work terms neglected in the energy equation, then become

$$\frac{\partial u}{\partial x} + \frac{\partial v}{\partial y} = 0 \tag{16.2.5}$$

$$u\frac{\partial u}{\partial x} + v\frac{\partial u}{\partial y} = g\beta(t - t_\infty)\cos\theta + \frac{1}{\rho}\frac{\partial \tau_{xy}}{\partial y} - \frac{1}{\rho}\frac{\partial p_m}{\partial x} \tag{16.2.6}$$

$$0 = g\beta(t - t_\infty)\sin\theta - \frac{1}{\rho}\frac{\partial p_m}{\partial y} \tag{16.2.7}$$

$$u\frac{\partial t}{\partial x} + v\frac{\partial t}{\partial y} = \frac{k}{\rho c_p}\frac{\partial^2 t}{\partial y^2} \tag{16.2.8}$$

The boundary conditions are again

$$\text{at } y = 0: \qquad u = v = 0 \quad \text{and} \quad t = t_0 \tag{16.2.9a}$$

$$\text{as } y \to \infty: \qquad u \to 0 \quad \text{and} \quad t \to t_\infty \tag{16.2.9b}$$

where θ, as before, is the angle between the vertical and the local tangent to the surface, as shown in Figs. 5.1.1 and 5.1.2. For vertical or near-vertical surfaces, Eq. (16.2.7) may again be neglected. However, for horizontal or near-horizontal surfaces this equation must be retained to give the streamwise pressure gradient that drives the flow. A detailed discussion of these matters was given in Section 5.1 for Newtonian fluids.

The above equations are similar to those given in Section 5.1 for Newtonian fluids. Here, the shear stress terms in Eqs. (16.2.6) and (16.2.7) are determined by the rheological behavior of the fluid. They may be evaluated by using any of the models discussed in the previous section. For simple geometries, solutions to the above equations for external flows have been obtained. Some are discussed in the succeeding sections. Where possible, comparison with experimental data will be made.

16.3 VISCOINELASTIC (PURELY VISCOUS) FLUIDS: EXTERNAL FLOWS

Buoyancy-induced flows of such non-Newtonian fluids external to surfaces of various geometries frequently are encountered in the process industry. Qualitatively, the transport mechanisms in these fluids are similar to those in Newtonian fluids. However, the transport rates and velocity and thermal fields may differ drastically.

Several experimental studies and analyses have been reported for various geometries, such as a vertical surface, a horizontal cylinder, a sphere, a vertical cone, and an axisymmetric body. Two thermal boundary conditions, uniform surface temperature and uniform surface heat flux, are commonly considered. In general, similarity solutions in non-Newtonian transport are very few, and even those dictate very stringent conditions on surfaces, temperature variations, and geometric shape. This is primarily due to the complex nature of the relationship between the shear stress and the shearing rate. The only rheological model that admits any similarity solution is the power law model, and apparently only for fluids with high Prandtl numbers. For surfaces with arbitrary temperature variation, integral methods have been employed. Such methods have also been used with other rheological models.

Similarity considerations. For non-Newtonian fluids, these seem to have been first investigated by Acrivos (1960). The flow considered was for a power law fluid past a two-dimensional surface or an axisymmetric surface, shown in Figs. 5.1.1 and 5.1.2. For flow adjacent to a flat inclined surface (Fig. 5.1.1) or a symmetric two-dimensional planar surface (Fig. 5.1.2*a*), the governing equations are Eqs. (16.2.5)–(16.2.8). For vertical or near-vertical surfaces, Eq. (16.2.7) and the pressure term in Eq. (16.2.6) are negligible. Then, using the power law model, Eq. (16.1.1), the momentum equation (16.2.6) becomes

$$u\frac{\partial u}{\partial x} + v\frac{\partial u}{\partial y} = g\beta(t - t_\infty)\cos\theta + \frac{K}{\rho}\frac{\partial}{\partial y}\left(\frac{\partial u}{\partial y}\left|\frac{\partial u}{\partial y}\right|^{n-1}\right) \tag{16.3.1}$$

where the parameters K and n are the consistency index and the power law fluid index, respectively.

For a Newtonian fluid ($n = 1$ and $K = \mu$) a similarity transformation is possible if the surface geometry is of the form $\cos\theta = ax^m$, where a and m are arbitrary constants such that $a > 0$ and $m \geq 0$. Acrivos pointed out that for an isothermal surface in a power law fluid ($n \neq 1$) similarity exists if $\cos\theta = ax^{-1/3}$. For such a geometry, however, surface orientation cannot be defined for small values of x. It was thus concluded that no useful similarity solution exists for an isothermal surface exchanging heat with a power law fluid.

Allowing a nonisothermal surface condition, Na and Hansen (1966) found a similarity transformation for power law fluids if the temperature of a vertical surface varies as

$$t_0 - t_\infty \propto x^{-1/3} \tag{16.3.2}$$

As shown by Acrivos (1960), physically acceptable similarity solutions are permissible for power law fluids for flat isothermal inclined surfaces, shown in Figs. 5.1.1 and 5.1.2, provided the Prandtl number is large. Later, Chen and Wollersheim (1973) showed that similarity solutions are also permissible for high Prandtl numbers for flow adjacent to a vertical surface with uniform heat flux.

These similar solutions for steady laminar flow adjacent a vertical surface are discussed next in Section 16.3.1. Integral method solutions for the isothermal and uniform flux surface thermal conditions and for other than power law fluids are also described. Non-Newtonian transport from other surface geometries will also be discussed briefly.

Prandtl and Grashof numbers. Before describing these solutions, it is noted here that, unlike the case of Newtonian fluids, there does not exist a single definition of Prandtl number that characterizes all non-Newtonian fluids. Different definitions have been used for different surface conditions and different classes of non-Newtonian fluids. Some of the frequently used expressions for Prandtl number and Grashof number for power law fluids are given in Table 16.3.1. Expressions for these parameters for other than the power law model will be given as they appear in the text.

The definition of Prandtl number for non-Newtonian fluids differs from that for Newtonian fluids in that a reference length arises. This is rather unfortunate since the Prandtl number, which usually represents other fluid characteristics, then contains a length that depends on the experiment or application.

16.3.1 Flows Adjacent to a Vertical Isothermal Surface

Power law fluids. For such fluids, Acrivos (1960) reported solutions for the laminar equations, assuming a large Prandtl number, say $\mathrm{Pr} \geq 10$. Equations (16.2.5), (16.3.1), and (16.2.8) apply. In Eq. (16.3.1) the inertia terms are omitted, as small, and $\cos\theta = 1$. Boundary conditions are given by Eq. (16.2.9). The equations were then transformed by using the following similarity formulation:

$$u = \sqrt{g\beta L(t_0 - t_\infty)} \left(\frac{3n+1}{2n+1}\frac{x}{L}\right)^{n(n+1)/3n+1} \frac{f'(\eta)}{\hat{\mathrm{Pr}}^{(n+1)/(3n+1)}} \tag{16.3.3}$$

$$\phi(\eta) = \frac{t - t_\infty}{t_0 - t_\infty} \tag{16.3.4}$$

$$\eta(x, y) = \frac{(y/L)\hat{\mathrm{Gr}}^{1/(2n+2)}\hat{\mathrm{Pr}}^{n/(3n+1)}}{\{[(3n+1)/(2n+1)](x/L)\}^{n/(3n+1)}} \tag{16.3.5}$$

where $\hat{\mathrm{Gr}}$ and $\hat{\mathrm{Pr}}$ are as defined in Table 16.3.1. Using these transformations, the momentum equation (16.3.1) without inertia terms and with $\cos\theta = 1$, the energy equation (16.2.8), and the boundary conditions Eq. (16.2.9) become

$$\frac{d}{d\eta}[f'']^n + \phi = 0 \tag{16.3.6}$$

$$\phi'' + f\phi' = 0 \tag{16.3.7}$$

$$f(0) = f'(0) = \phi(0) - 1 = f'(\infty) = \phi(\infty) = 0 \tag{16.3.8}$$

Acrivos (1960) solved these equations numerically and reported the transport results. The local Nusselt number was defined as

Table 16.3.1 Definitions for the Prandtl number and Grashof number for power law fluids

Surface	Characteristic length	Prandtl number Symbol	Prandtl number Expression	Grashof number Symbol	Grashof number Expression	Reference
Vertical surface, isothermal	L	$\hat{\mathrm{Pr}}$	$\frac{\rho c_p}{k}\left(\frac{K}{\rho}\right)^{2/(1+n)} L^{(n-1)/(2n+2)}[g\beta(t_0 - t_\infty)]^{3(n-1)/2(n+1)}$	$\hat{\mathrm{Gr}}$	$\frac{\rho^2 L^{n+2}[g\beta(t_0 - t_\infty)]^{2-n}}{K^2}$	Acrivos (1960)
	x	$\hat{\mathrm{Pr}}_x$	$\hat{\mathrm{Pr}}\left(\frac{x}{L}\right)^{(n-1)/(2n+2)}$	$\hat{\mathrm{Gr}}_x$	$\hat{\mathrm{Gr}}_x\left(\frac{x}{L}\right)^{n+2}$	
	L	$\overline{\mathrm{Pr}}$	$\frac{\rho c_p}{k}\left(\frac{K}{\rho}\right)^{1/(2-n)} L^{(2n-2)/(n-2)}$	$\overline{\mathrm{Gr}}$	$\frac{g\beta(t_0 - t_\infty)L^{(n+2)/(2-n)}}{(K/\rho)^{2/(2-n)}}$	Reilly, et al. (1965)
	x	$\overline{\mathrm{Pr}}_x$	$\overline{\mathrm{Pr}}\left(\frac{x}{L}\right)^{(2n-2)/(n-2)}$	$\overline{\mathrm{Gr}}_x$	$\overline{\mathrm{Gr}}\left(\frac{x}{L}\right)^{(n+2)/(2-n)}$	
Vertical surface, uniform flux	L	$\hat{\mathrm{Pr}}^*$	$\frac{\rho c_p}{k}\left(\frac{K}{\rho}\right)^{5/(n+4)} L^{(2n-2)/(n+4)}\left(\frac{g\beta q''}{k}\right)^{(3n-3)/(n+4)}$	$\hat{\mathrm{Gr}}^*$	$\left(\frac{\rho}{K}\right)^2 L^4\left(\frac{g\beta q''}{k}\right)^{2-n}$	Chen and Wollersheim (1973)
	x	$\hat{\mathrm{Pr}}^*_x$	$\hat{\mathrm{Pr}}^*\left(\frac{x}{L}\right)^{(2n-2)/(n+4)}$	$\hat{\mathrm{Gr}}^*_x$	$\hat{\mathrm{Gr}}^*\left(\frac{x}{L}\right)^4$	
Cylinder, sphere, isothermal	R	$\hat{\mathrm{Pr}}_R$	$\hat{\mathrm{Pr}}\left(\frac{R}{L}\right)^{(n-1)/(2n+2)}$	$\hat{\mathrm{Gr}}_R$	$\hat{\mathrm{Gr}}\left(\frac{R}{L}\right)^{n+2}$	Acrivos (1960)
Cylinder, sphere, uniform flux	R	$\hat{\mathrm{Pr}}^*_R$	$\frac{\rho c_p}{k}\left(\frac{K}{\rho}\right)^{2/(n+1)} R^{(2n-2)/(n+1)}\left(\frac{g\beta q''}{k}\right)^{(3n-3)/(2n+2)}$	$\hat{\mathrm{Gr}}^*_R$	$\hat{\mathrm{Gr}}^*\left(\frac{R}{L}\right)^4$	Kim and Wollersheim (1976)

$$\mathrm{Nu}_L = \frac{h_x L}{k} = -L\left(\frac{\partial \phi}{\partial y}\right)_{y=0} \tag{16.3.9}$$

and is given by

$$\mathrm{Nu}_L = [-\phi'(0)]\left(\frac{2n+1}{3n+1}\right)^{n/(3n+1)} (\hat{\mathrm{Gr}})^{1/(2n+2)}(\hat{\mathrm{Pr}})^{n/(3n+1)}\left(\frac{x}{L}\right)^{-n/(3n+1)} \tag{16.3.10}$$

Note that Eq. (16.3.9) is different from the customary definition of local Nusselt number, given by

$$\mathrm{Nu}_x = -x\left(\frac{\partial \phi}{\partial y}\right)_{y=0}$$

In line with the above common practice, Eq. (16.3.10) is recast as follows:

$$\mathrm{Nu}_x = \mathrm{Nu}_L \frac{x}{L} = \frac{h_x x}{k}$$

$$= [-\phi'(0)]\left(\frac{2n+1}{3n+1}\right)^{n/(3n+1)} (\hat{\mathrm{Gr}})^{1/(2n+2)}(\hat{\mathrm{Pr}})^{n/(3n+1)}\left(\frac{x}{L}\right)^{(2n+1)/(3n+1)} \tag{16.3.11}$$

The average Nusselt number for a surface from $x = 0$ to $x = L$ is then easily determined as

$$\mathrm{Nu} = \frac{hL}{k} = \left(\frac{1}{L}\int_0^L h_x\, dx\right)\frac{L}{k} = C(\hat{\mathrm{Gr}})^{1/(2n+2)}(\hat{\mathrm{Pr}})^{n/(3n+1)} \tag{16.13.12a}$$

where

$$C = [-\phi'(0)]\left(\frac{3n+1}{2n+1}\right)^{(2n+1)/(3n+1)} \tag{16.13.12b}$$

Equations (16.3.11) and (16.3.12) apply to power law fluids having $\hat{\mathrm{Pr}} \geq 10$. Plots of the transport parameters $-\phi'(0)$ and $[f''(0)]^n$ as functions of the power law parameter n are shown in Fig. 16.3.1.

It is interesting to note that for a Newtonian fluid ($n = 1$), the correlation equation (16.3.12) reduces to $\mathrm{Nu} = C(\mathrm{GrPr})^{1/4}$, a familiar transport correlation for many natural convection flows for $\mathrm{Pr} \gtrsim 1$. Another observation may be made. The value of C in Eq. (16.3.12b) is a weak function of n in the range $0.1 \leq n \leq 1.5$, which includes many of the power law fluids of interest; see Table 16.3.2. The large Prandtl number assumption has been substantiated by several experiments, as discussed later in this section.

Integral methods. Procedures described in Section 3.13 have also been used to obtain solutions for laminar thermal convection in a power law fluid from an isothermal vertical surface. See Tien (1967) and Shenoy and Ulbrecht (1979). Results from both of these studies indicate the same dependence of Nusselt number on Grashof and Prandtl numbers as that obtained by the similarity solution. The magnitude of the coefficient C in Eq. (16.3.12b), however, differs slightly.

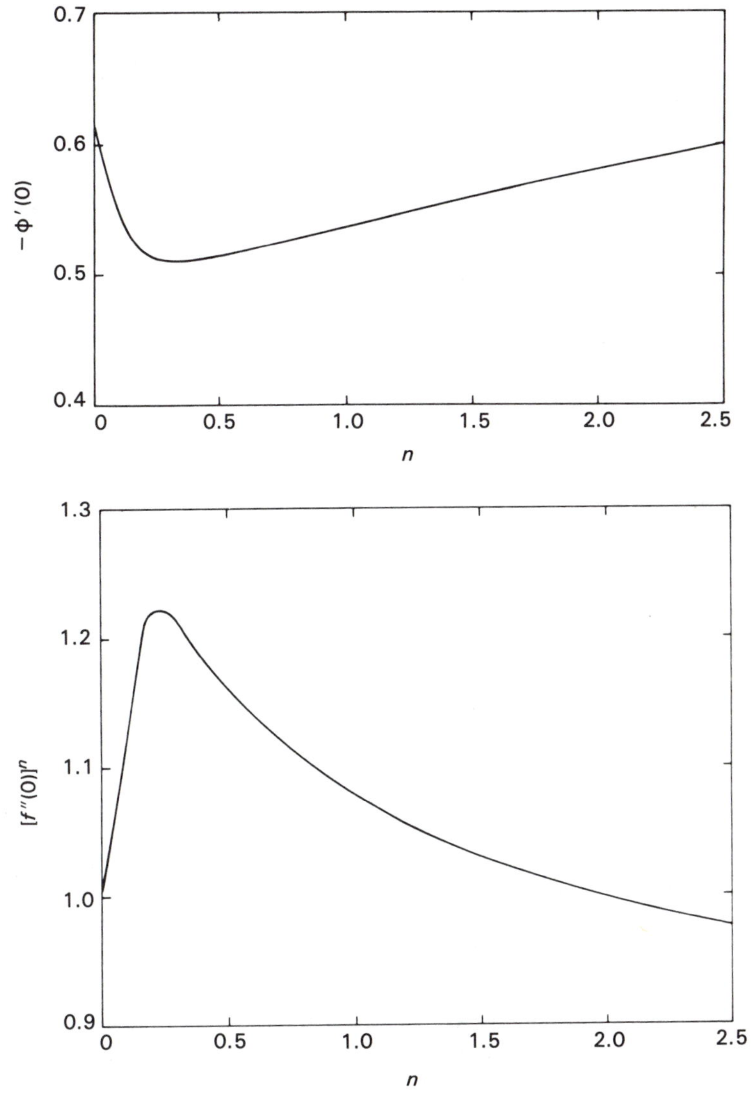

Figure 16.3.1 Dependence of $-\phi'(0)$ and $[f''(0)]^n$ on the power law index n. *(Reproduced by permission of the American Institute of Chemical Engineers, A. Acrivos,* AIChE J., *vol. 6, p. 584. Copyright © 1960, AIChE.)*

Table 16.3.3 shows a comparison of the integral method solutions and the similarity solution.

Experimental results. Some data have been obtained for non-Newtonian transport from an isothermal vertical surface. Reilly et al. (1965) reported results for an electrically heated copper plate immersed in 0.5 and 1% aqueous solutions of carboxypolymethylene (Carbopol), for which $0.7 < n < 1$. The results for the av-

Table 16.3.2 Values of C in Eq. (16.3.12) for various surface geometries and values of the power law index n

	$n = \frac{1}{10}$	$n = \frac{1}{2}$	$n = 1$	$n = \frac{3}{2}$
Flat surface	0.60	0.63	0.67	0.71
Horizontal cylinder (L = radius)	0.36	0.38	0.42	0.45
Sphere (L = radius)	0.44	0.45	0.49	0.52
Vertical cone	0.61	0.65	0.71	0.75
Stagnation region on a horizontal cylinder	0.36	0.45	0.54	0.60
Stagnation region on a sphere	0.45	0.55	0.64	0.70

Source: From Acrivos (1960).

erage Nusselt number were expressed as

$$\mathrm{Nu} = C(\overline{\mathrm{Gr}}\ \overline{\mathrm{Pr}}^n)^{1/(3n+1)} \tag{16.3.13}$$

where $\overline{\mathrm{Gr}}$ and $\overline{\mathrm{Pr}}$ are as defined in Table 16.3.1. Since the product $(\overline{\mathrm{Gr}}\ \overline{\mathrm{Pr}}^n)^{1/(3n+1)}$ may be shown to be equal to the product $\hat{\mathrm{Gr}}^{1/2n+2}\hat{\mathrm{Pr}}^{n/3n+1}$, the constant C in Eq. (16.3.13) is the same as that in Eq. (16.3.12*b*). The constant C determined from the data was found to be 5–10% lower than calculated by Acrivos (1960).

Later, Sharma and Adelman (1969a) extended the experimental work to a more non-Newtonian fluid, $0.2 < n < 0.69$, using 1.25, 1.5, and 1.75% aqueous solutions of Carbopol. The data could not be correlated adequately by using $\hat{\mathrm{Gr}}$ and $\hat{\mathrm{Pr}}$ defined in the solution by Acrivos. Instead, the data were reduced in terms of $\overline{\mathrm{Gr}}$ and $\overline{\mathrm{Pr}}$. For the ranges $6.24 < \overline{\mathrm{Gr}} < 3.17 \times 10^4$ and $1.95 \times 10^2 < \overline{\mathrm{Pr}} < 1.84 \times 10^4$, the following correlation for average Nusselt number was suggested:

$$\mathrm{Nu} = 0.511(\overline{\mathrm{Gr}}\ \overline{\mathrm{Pr}})^{0.263} \tag{16.3.14}$$

Table 16.3.3 Comparison of the average Nusselt numbers for natural convection from an isothermal vertical surface to a power law fluid

	$C = \dfrac{\mathrm{Nu}}{\hat{\mathrm{Gr}}^{1/(2n+2)}\hat{\mathrm{Pr}}^{n/(3n+1)}}$		
n	Acrivos (1960)	Tien (1967)	Shenoy and Ulbrecht (1979)
0.5	0.63	0.6098	0.5957
1.0	0.67	0.6838	0.6775
1.5	0.71	0.7229	0.7194

Source: From Shenoy and Mashelkar (1982).

Shenoy and Ulbrecht (1979) reported data on temperature profiles in the transport region. These compared very well with their own approximate integral solution.

Apparently, there are no velocity profiles or local heat transfer measurements. Such data would be useful in determining which of the two generalized definitions of the Grashof and Prandtl numbers ($\hat{\text{Gr}}$, $\hat{\text{Pr}}$ or $\overline{\text{Gr}}$, $\overline{\text{Pr}}$) better fit the data.

Other fluid models. The above results are for a power law fluid. However, as discussed in Section 16.1, this two-parameter model does not accurately represent the fluid behavior at the low shear rates characteristic of most buoyancy-induced flows. Several analyses with three-parameter models, which describe the limiting behavior in the range of low shear rates, have been reported.

Flow adjacent to an isothermal surface in a Sutterby fluid, Eq. (16.1.3), was studied by Fujii et al. (1972, 1973). The governing equations are Eqs. (16.2.5), (16.2.6) without the motion pressure term, and (16.2.8), subject to boundary conditions in Eq. (16.2.9). As mentioned before, no similarity solution exists for this flow circumstance. The equations were solved numerically. The following correlation for local Nusselt number, based on the computed results, was given:

$$\text{Nu}_x = 0.50(\text{Gr}_{x,S}\text{Pr}_S)^{0.25(1+m)} \tag{16.3.15}$$

where

$$m = 0.04\text{Pr}_S^{-0.23}A^{3.7\text{Pr}_S^{-0.34}}Z_S^{0.63A^{0.66}} \tag{16.3.16}$$

$$Z_S = \frac{\rho^2 g\beta(t_0 - t_\infty)(B\mu_0/\rho)^{3/2}}{\mu_0^2} \tag{16.3.17}$$

$$\text{Gr}_{x,S} = \frac{g\beta(t_0 - t_\infty)x^3}{(\mu_0/\rho)^2} \qquad \text{Pr}_S = \frac{\mu_0/\rho}{\alpha} \tag{16.3.18}$$

A, B, and μ_0 are the parameters in Eq. (16.1.3) that describe the rheology of a Sutterby fluid, where μ_0, the viscosity at zero-shear rate, is the slope of the shear stress τ_{yx} vs. du/dy curve in Fig. 16.1.3, in the immediate vicinity of the origin.

Equation (16.3.15) agrees with the computed local Nusselt number within an accuracy of $\pm 10\%$ in the range $A = 0$–1, $Z_S = 0$–10^3, $\text{Pr}_S = 100$–3000, and $\text{Gr}_{x,S}\text{Pr}_S = 10^6$–$10^{11}$. Fujii et al. (1973) also reported experimental results for 0.2 and 0.5% aqueous solutions of polyethylene oxide and a 2% aqueous solution of carboxymethylcellulose. The shear behavior for these fluids was fitted by the Sutterby model. The experimental results were found to be in good agreement with the numerical computations.

The Ellis model, Eq. (16.1.2), was used by Tien and Tsuei (1969) in an integral analysis for a vertical isothermal surface. Assuming large Prandtl number, heat transfer results were reported that were found to be higher than the experimental data of Reilly et al. (1965) by 10–40%.

Turbulent convection. The above analyses assume laminar flow. For surfaces of large vertical extent, systematic deviations of the heat transfer rates from the laminar trend have been observed at high Grashof numbers. These have been attributed to the appearance of turbulence in the flow at some downstream location. For Newtonian fluids, instability, transition, and turbulent transport have been investigated extensively, as described in Chapter 11. These flow regimes in non-Newtonian fluids are much less well understood. Shenoy and Mashelkar (1978a) used an integral approach to analyze fully developed turbulent transport adjacent to an isothermal surface. The procedure followed the integral analysis of Eckert and Jackson (1951) for a Newtonian fluid. For a power law pseudoplastic, shear-thinning fluid, the correlation obtained for the local Nusselt number was

$$\mathrm{Nu}_x = \frac{h_x x}{k} = C\mathrm{Gr}^a_{x,ps}\mathrm{Pr}^b_{x,ps} \tag{16.3.19}$$

where

$$\mathrm{Gr}_{x,ps} = \left(\frac{\rho}{\gamma}\right)^{8\beta_1} x^{4\beta_1(2+n)}[g\beta(t_0 - t_\infty)]^{4\beta_1(2-n)}$$

$$\mathrm{Pr}_{x,ps} = \frac{c_p}{k}\gamma^{4\beta_1}\rho^{1-4\beta_1}x^{[3-4\beta_1(2+n)]/2}[g\beta(t_0 - t_\infty)]^{[1-4\beta_1(2-n)]/2}$$

$$\gamma = 8^{n-1}K\left(\frac{3n+1}{4n}\right)^n$$

K and n are the power law parameters. Values of β_1, a, b, and C as a function of n, are given in Table 16.3.4.

Transient natural convection. Little work has been reported. For an isothermal surface, Kleppe and Marner (1972) numerically computed transient buoyancy-induced flow of a Bingham plastic fluid. Recall that this fluid has a linear relationship between shear stress and shear rate but has an initial yield stress τ_0. In regions where the shear stress exceeds τ_0, the material flows as a Newtonian fluid; otherwise, it flows without shear. Thus, the velocity profiles in a Bingham plastic typically consist of two regions: a viscous flow region where the shear

Table 16.3.4 Values of β_1, C, a, and b in Eq. (16.3.19) for various values of power law index n

n	β_1	C	a	b
1.0	0.250	0.0402	0.400	0.200
0.9	0.255	0.0428	0.405	0.199
0.8	0.263	0.0443	0.410	0.192
0.7	0.270	0.0450	0.416	0.187
0.6	0.281	0.0464	0.422	0.174
0.5	0.290	0.0477	0.429	0.165
0.4	0.307	0.0483	0.438	0.138
0.3	0.325	0.0497	0.448	0.106
0.2	0.349	0.0501	0.463	0.054

Source: From Shenoy and Mashelkar (1978a).

stress is greater than τ_0, and an outer plug flow region of uniform velocity where it is less than τ_0.

The governing parameters were shown to be the Prandtl number $\Pr_B = \mu_B c_p/k$ and a dimensionless group involving the Hedstrom and Grashof numbers, $\mathrm{He}/\mathrm{Gr}^{3/4}$, where μ_B is the Bingham plastic viscosity, $\mathrm{He} = \rho\tau_0 L^2/\mu_B^2$, and $\mathrm{Gr} = g\rho^2\beta(t_0 - t_\infty)L^3/\mu_B^2$. The Hedstrom number is a dimensionless yield stress that indicates the importance of Bingham plastic behavior. As He $\rightarrow$ 0, the plug flow region becomes smaller and Newtonian behavior is approached. A numerically computed Nusselt number results. Velocity and temperature profiles were presented for a range of $\Pr_B$ and $\mathrm{He}/\mathrm{Gr}^{3/4}$.

Flow in a Bingham plastic fluid does not start until the buoyancy forces become sufficiently large to cause shear stresses in the material that exceed the yield stress τ_0. Until that occurs, heat is transferred entirely by transient conduction. After flow starts, the computed temperature and velocity profiles reach a transient maximum and then decrease to their steady-state values. The time required to reach steady-state conditions is found to increase sharply with Pr. A temporal minimum in the average Nusselt number is noted. Similar trends for Newtonian fluids are discussed in Chapter 7. However, because of the longer period of pure conduction, as a consequence of the yield stress, the temporal minimum is more pronounced for Bingham plastic than for Newtonian fluids.

The average Nusselt number values for steady-state conditions were found to be up to 15% higher than for Newtonian fluids. A temporal maximum was noted in the mean friction coefficient since the maximum velocity gradient at the wall occurred before steady-state conditions were reached. Furthermore, both transient and steady-state surface friction coefficients were found to be significantly larger for Bingham plastics than for Newtonian fluids. However, this increase was attributed primarily to τ_0, with a relatively small contribution due to a steeper velocity gradient at the wall.

16.3.2 Vertical Flow Adjacent to a Uniform Flux Surface

Power law fluids. The governing equations are again Eqs. (16.2.5), (16.2.8), and (16.3.1). Boundary conditions (16.2.9) apply, except for the thermal boundary condition at $y = 0$. This is now replaced by $(\partial t/\partial y)_{y=0} = -q''/k$. Chen (1971) attempted to obtain similarity solutions, following the technique of Acrivos (1960) for an isothermal surface. This approach was not successful. The equations were solved numerically instead.

For large Prandtl numbers, however, similarity again results. Chen and Wollersheim (1973), after neglecting inertia terms in Eq. (16.3.1), transformed the governing equations with the following formulation:

$$\eta = \frac{(y/L)\hat{\mathrm{Gr}}^{*\,1/(n+4)}\hat{\mathrm{Pr}}^{*\,n/(3n+2)}}{[(3n+2)x/L]^{n/(3n+2)}} \tag{16.3.20}$$

$$u = \frac{[(3n+2)x/L]^{(n+2)/(3n+2)} f'(\eta)}{(k/g\beta q'' L^2)^{1/2}\hat{\mathrm{Gr}}^{*\,1/2(n+4)}\hat{\mathrm{Pr}}^{*\,(n+2)/(3n+2)}} \tag{16.3.21}$$

$$v = \frac{n}{[(3n+2)x/L]^{n/(3n+2)}} \frac{\eta f'(\eta) - [2(n+1)/n] f(\eta)}{(k/g\beta q''L^2)^{1/2} \hat{\mathrm{Gr}}^{*3/2(n+4)} \hat{\mathrm{Pr}}^{*2(n+1)/(3n+2)}} \tag{16.3.22}$$

$$\phi = \frac{(t - t_\infty) \hat{\mathrm{Gr}}^{*1/(n+4)} \hat{\mathrm{Pr}}^{*n/(3n+2)}}{(q''L/k)[(3n+2)x/L]^{n/(3n+2)}} \tag{16.3.23}$$

The resulting ordinary differential equations and boundary conditions were

$$\frac{d}{d\eta}\{[f'']^n\} + \phi = 0 \tag{16.3.24}$$

$$\phi'' + 2(n+1) f \phi' - n f' \phi = 0 \tag{16.3.25}$$

$$f'(0) = f(0) = \phi'(0) + 1 = f'(\infty) = \phi(\infty) = 0 \tag{16.3.26}$$

Numerical integration of these equations resulted in the following transport results. The surface temperature variation is given by

$$t_0(x) - t_\infty = [(3n+2)x/L]^{n/(3n+2)} \frac{q''L}{k} \frac{\phi(0)}{\hat{\mathrm{Gr}}^{*1/(n+4)} \hat{\mathrm{Pr}}^{*n/(3n+2)}} \tag{16.3.27}$$

Note that the surface temperature excess varies with the downstream distance as

$$t_0(x) - t_\infty \propto x^{n/(3n+2)} \tag{16.3.28}$$

For a Newtonian fluid, $n = 1$, Eq. (16.3.28) reduces to $t_0 - t \propto x^{1/5}$, as given in Section 3.5. Most non-Newtonian fluids are pseudoplastics, for which $n < 1$. Then the surface temperature excess becomes a weaker function of the downstream distance. The expressions for local and average convection coefficients are

$$\mathrm{Nu}_x = \frac{h_x x}{k} = \frac{\hat{\mathrm{Gr}}^{*1/(n+4)} \hat{\mathrm{Pr}}^{*n/(3n+2)} (x/L)^{2(n+1)/(3n+2)}}{(3n+2)^{n/(3n+2)} \phi(0)} \tag{16.3.29}$$

$$\mathrm{Nu} = \frac{hL}{k} = \left(\frac{2}{3n+2}\right)^{n/(3n+2)} \frac{\hat{\mathrm{Gr}}^{*1/(n+4)} \hat{\mathrm{Pr}}^{*n/(3n+2)}}{\phi(0)} \tag{16.3.30}$$

Values of the dimensionless local surface temperature excess [$\phi(0)$], as computed by Chen and Wollersheim (1973), are 1.3188, 1.3026, 1.1474, and 1.0286 for $n = 0.1$, 0.5, 1.0, and 1.5, respectively.

Also, for power law fluids, integral method solutions for a uniform surface flux condition were reported by Tien (1967) and Shenoy (1977), including the effects of Pr. Dale and Emery (1972) investigated this case both numerically and experimentally. In the numerical solutions the inertial terms in the momentum equation were retained in Eq. (16.3.1), whereas they had been omitted under the large Prandtl number assumption. Local heat transfer results were correlated by

$$\mathrm{Nu}_x = \frac{h_x x}{k} = C(\mathrm{Gr}^*_{xE} \overline{\mathrm{Pr}}^n_x)^m \tag{16.3.31}$$

where

$$\mathrm{Gr}^*_{xE} = \frac{g\beta q'' x^{4/(2-n)}}{k(K/\rho)^{2/(2-n)}}$$

and $m = 1/(3n + 2)$ as suggested by the integral analysis of Tien (1967).

The experimental data of Dale and Emery (1972) obtained with water, carboxymethylcellulose, and Carbopol, in the range $0.4 < n < 1$, were also correlated well by Eq. (16.3.31). The experimental and computed values of C and m in Eq. (16.3.31) are shown in Table 16.3.5 for different values of n. Good agreement may be seen for the values of n near 0.8. Computed temperature and velocity fields were also found to be in good agreement with the measurements.

Other fluid models. As for the isothermal surface condition, some work has been reported for a uniform flux surface condition, using a three-parameter model. Fujii et al. (1974) numerically solved the laminar transport in a Sutterby fluid and gave the following correlation for local heat transfer results:

$$\mathrm{Nu}_x = \frac{h_x x}{k} = 0.62(\mathrm{Gr}^*_{xS}\mathrm{Pr}_S)^{0.2(1+m^*)} \tag{16.3.32}$$

where

$$m^* = 0.06\mathrm{Pr}_S^{-0.28} A^{3.7\mathrm{Pr}_S^{-0.34}} Z_S^{*\,0.35A^{0.66}} \tag{16.3.33}$$

and

$$Z_S^* = \frac{g\beta q'' B^2}{k} \tag{16.3.34}$$

Table 16.3.5 Values of *C* and *m* in Eq. (16.3.31) for a uniform flux vertical surface

		Computed		Experiment	
Fluid	n	C	m	C	m
Water	1.0	0.615	0.200	0.607	0.200
0.05% carboxymethylcellulose	0.888	0.59	0.214	0.600± 0.0027	0.2101± 0.0003
0.06% Carbopol	0.828 (0.815–0.847)[a]	0.57	0.223	0.576	0.219
0.06% Carbopol	0.787 (0.778–0.795)	0.56	0.229	0.595	0.223
0.065% Carbopol	0.816 (0.811–0.821)	0.57	0.225	0.505	0.226
0.065% Carbopol	0.754 (0.711–0.766)	0.55	0.235	0.546	0.233
0.52% Carbopol	0.893 (0.881–0.918)	0.59	0.214	0.570	0.207
0.57% Carbopol	0.807 (0.787–0.828)	0.56	0.226	0.538	0.222
0.68% Carbopol	0.564 (0.535–0.611)	0.48	0.271	0.593	0.265
0.75% Carbopol	0.395 (0.383–0.416)	0.41	0.314	0.660	0.299

Source: From Dale and Emery (1972).

[a]Numbers in parentheses indicate the range of flow index n.

$$\mathrm{Gr}_{xS}^* = \frac{g\beta q'' x^4}{k(\mu_0/\rho)^2} \qquad \mathrm{Pr}_S = \frac{\mu_0/\rho}{\alpha} \tag{16.3.35}$$

A, B, and μ_0, as before, are the parameters of the Sutterby fluid as given in Eq. (16.1.3). Equation (16.3.32) correlates the computed results within an accuracy of ±5% in the ranges $A = 0\text{–}1$, $Z_S^* = 0\text{–}10^4$, $\mathrm{Pr}_S = 100\text{–}3000$, and $\mathrm{Gr}_{xS}^* \mathrm{Pr}_S = 10^9\text{–}10^{13}$. Experiments were also conducted by Fujii et al. (1974) to obtain local heat transfer rates with aqueous solutions of polyethylene oxide and carboxymethylcellulose. Good agreement was indicated between the data and the numerical results.

16.3.3 Laminar Natural Convection in Other External Flows

Other geometries for which non-Newtonian transport has been investigated include horizontal cylinders, spheres, slender vertical cones, and an axisymmetric body about a vertical axis.

Horizontal cylinder. Transport from a two-dimensional surface, shown in Fig. 5.1.2*a*, to a power law fluid was analyzed by Acrivos (1960). The governing equations were taken to be Eqs. (16.2.5), (16.2.8), and (16.3.1). Assuming a large Prandtl number, the following similarity transformations were used:

$$\phi = \frac{t - t_\infty}{t_0 - t_\infty} \tag{16.3.36}$$

$$u = [g\beta L(t_0 - t_\infty)]^{1/2} (\hat{\mathrm{Pr}})^{-(n+1)/(3n+1)} (\sin \xi)^{1/n} \times \left[\frac{3n+1}{2n+1}\left(\frac{1}{\sin \xi}\right)^{(3n+1)/n(2n+1)} \int_0^{x_1} (\sin \xi)^{1/(2n+1)}\, dx_1\right]^{n(n+1)/(3n+1)} f'(\eta) \tag{16.3.37}$$

$$\eta = \frac{y_1}{\left\{[(3n+1)/(2n+1)](1/\sin \xi)^{(3n+1)/n(2n+1)} \int_0^{x_1} (\sin \xi)^{1/(2n+1)}\, dx_1\right\}^{n/(3n+1)}} \tag{16.3.38}$$

where $x_1 = x/L$, $y_1 = (y/L)\hat{\mathrm{Gr}}^{1/(2n+2)} \hat{\mathrm{Pr}}^{n/(3n+1)}$, and $\xi = 90° - \theta$, as in Chapter 5, is the angle of inclination with the horizontal.

The resulting transformed equations are again Eqs. (16.3.6) and (16.3.7) subject to the boundary conditions in Eq. (16.3.8), as for a vertical surface. The resulting local and average Nusselt numbers for a horizontal isothermal cylinder are

$$\mathrm{Nu}_x = \frac{h_x x}{k} = \frac{x}{L}[-\phi'(0)]\left(\frac{2n+1}{3n+1}\right)^{n/(3n+1)} \times \hat{\mathrm{Gr}}^{1/2(n+1)} \hat{\mathrm{Pr}}^{n/(3n+1)} \frac{(\sin \xi)^{1/(2n+1)}}{\left[\int_0^x (\sin \xi)^{1/(2n+1)}\, dx\right]^{n/(3n+1)}} \tag{16.3.39}$$

$$\text{Nu} = \frac{hL}{k} = C\hat{\text{Gr}}^{1/(2n+2)}\hat{\text{Pr}}^{n/(3n+1)} \tag{16.3.40}$$

where C is calculated by integrating Eq. (16.3.39) over the surface. Values of C for various surface geometries were computed by Acrivos (1960) and are listed in Table 16.3.2. A plot of $\text{Nu}_R/\hat{\text{Gr}}_R^{1/2(n+1)}\hat{\text{Pr}}_R^{n/(3n+1)}$ for a horizontal isothermal cylinder is shown in Fig. 16.3.2 for $n = 0.5$, 1.0, and 1.5, where $\text{Nu}_R = hR/K$ and $\hat{\text{Gr}}_R$ and $\hat{\text{Pr}}_R$ are as defined in Table 16.3.1.

Integral solutions for isothermal and uniform flux surface conditions for a horizontal cylinder exchanging heat with a power law fluid were given by Gentry and Wollersheim (1974) and Kim and Wollersheim (1976), respectively. In addition, both of these studies provided experimental data for the two surface conditions. Good agreement was reported between the similarity and integral solutions and the data. The local Nusselt number variations with angle ξ are compared in Fig. 16.3.3 for an isothermal condition and in Fig. 16.3.4 for a uniform flux surface condition.

Axisymmetric surface. For such a surface, Fig. 5.1.2*b*, the governing equations are (16.2.5)–(16.2.8), except that the continuity equation is now

$$\frac{\partial}{\partial x}(ru) + \frac{\partial}{\partial y}(rv) = 0 \tag{16.3.41}$$

Neglecting the y-momentum equation, Eq. (16.2.7), and the motion pressure term in Eq. (16.2.6), Acrivos (1960) analyzed the heat transfer from an isothermal

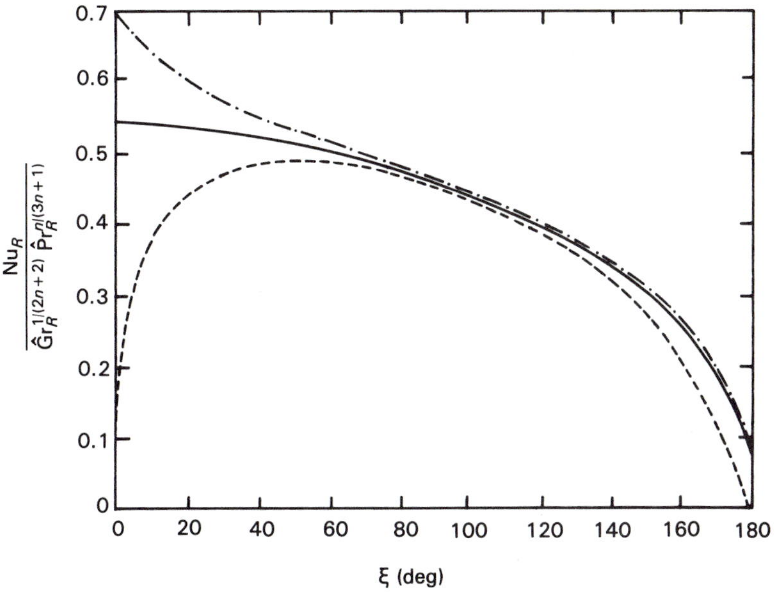

Figure 16.3.2 Variation of the local heat transfer rate along the surface of a horizontal isothermal cylinder. (– – –) $n = 0.5$; (——) $n = 1.0$; (–·–) $n = 1.5$. *(Reproduced by permission of the American Institute of Chemical Engineers, A. Acrivos,* AIChE J., *vol. 6, p. 584. Copyright © 1960, AIChE.)*

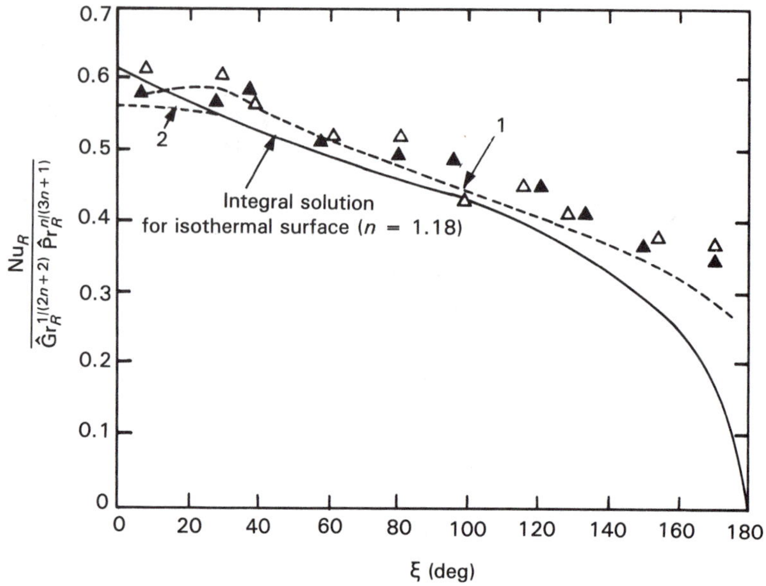

Figure 16.3.3 Variation of the local heat transfer along the isothermal surface of a horizontal cylinder for 38% cornstarch in aqueous sucrose solution. (▲) $\Delta t = 2.2°C$; (△) $\Delta t = 8.9°C$; curve 1, experimental results for Newtonian fluids; curve 2, integral solutions for Newtonian fluids. *(Reprinted with permission from C. B. Kim and D. E. Wollersheim,* J. Heat Transfer, *vol. 98, p. 144. Copyright © 1976, ASME.)*

surface, assuming a large Prandtl number. The following expression for the local Nusselt number was given.

$$\mathrm{Nu}_R = \frac{h_x R}{k} = -\phi'(0)\left(\frac{2n+1}{3n+1}\right)^{n/(3n+1)} \hat{\mathrm{Gr}}_R^{1/(2n+2)} \hat{\mathrm{Pr}}_R^{n/(3n+1)} \times \frac{[(r/L)^n \sin \xi]^{1/(2n+1)}}{\left[\int_0^{x_1} (r/L)^{(3n+1)/(2n+1)} (\sin \xi)^{1/(2n+1)}\, dx_1\right]^{n/(3n+1)}} \tag{16.3.42}$$

where x_1, as before, is x/L.

Amato and Tien (1976) reported experimental data on natural convection from an isothermal sphere to aqueous polymer solutions described by the power law model. The results were correlated by an expression of the form

$$\mathrm{Nu}_R = C \hat{\mathrm{Gr}}_R^{1/(2n+2)} \hat{\mathrm{Pr}}_R^{n/(3n+1)} = CZ$$

where $C = 0.489 \pm 0.005$ with a mean error of 7.6% for $10 < Z < 40$ and $5 < \mathrm{Nu}_R < 20$. For $Z < 10$ the following correlation was seen to apply:

$$\mathrm{Nu}_R = (0.996 \pm 0.120) Z^{(0.682 \pm 0.062)}$$

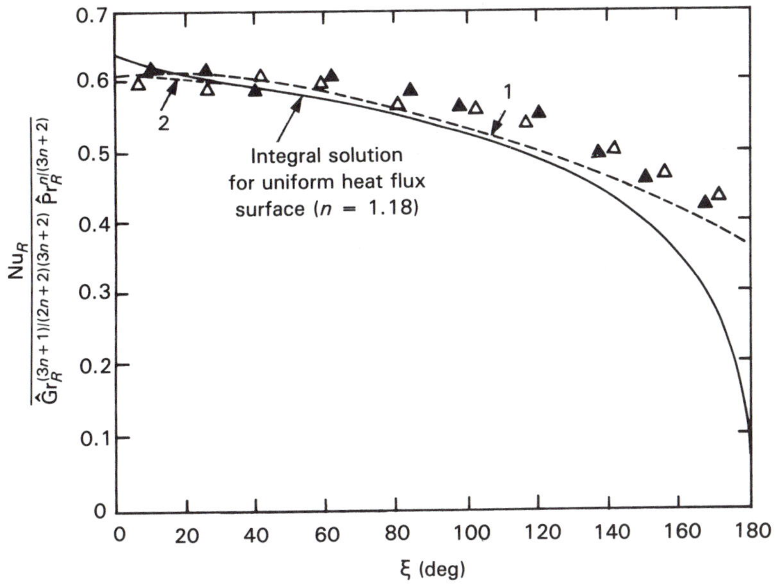

Figure 16.3.4 Variation of the local heat transfer rate along the uniform flux surface of a horizontal cylinder for 38% cornstarch in aqueous sucrose solution. (▲) q'' = 465 Btu/h ft^2; (△) q'' = 625 Btu/h ft^2; curve 1, experimental results for Newtonian fluids; curve 2, integral solution for Newtonian fluids. *(Reprinted with permission from C. B. Kim and D. E. Wollersheim,* J. Heat Transfer, *vol. 98, p. 144. Copyright © 1976, ASME.)*

with a mean error of 8% for $1.5 < \overline{Nu}_R \leq 5.0$. These correlations are shown in Fig. 16.3.5. The experimentally determined constant $C = 0.489$ for $\hat{Gr}_R^{1/(2n+2)} \hat{Pr}_R^{n/(3n+1)} > 10$ corresponds very closely with the value of 0.49 for $n = 1$ predicted by Acrivos; see Table 16.3.2. The local heat transfer variation around a sphere measured by Amato and Tien (1976) agreed well with that given by Eq. (16.3.42).

The analysis of Acrivos (1960) also included a vertical cone exchanging heat with a power law fluid. Shenoy and Mashelkar (1982) report results for heat transfer rates from slender nonisothermal vertical cones. The effect of curvature on the mean Nusselt number is also given.

Horizontal surfaces. Measurements of non-Newtonian transport from a horizontal isothermal surface of dimensions L_1 by L_2 to a power law fluid have been reported by Reilly (1964) and Sharma and Adelman (1969b). The latter study suggested the following correlations for the average Nusselt number. For a warm surface facing up, that is, B_n directed away from the surface,

$$\mathrm{Nu} = \frac{hL}{k} = 0.493(\overline{\mathrm{Gr}}\,\overline{\mathrm{Pr}})^{0.307} \qquad \text{for } 0.243 < n < 1.0$$

$$\text{and} \quad 1.76 \times 10^5 < \overline{\mathrm{Gr}}\,\overline{\mathrm{Pr}} < 10^9 \qquad (16.3.43)$$

For a warm surface facing down, that is, B_n directed toward the surface,

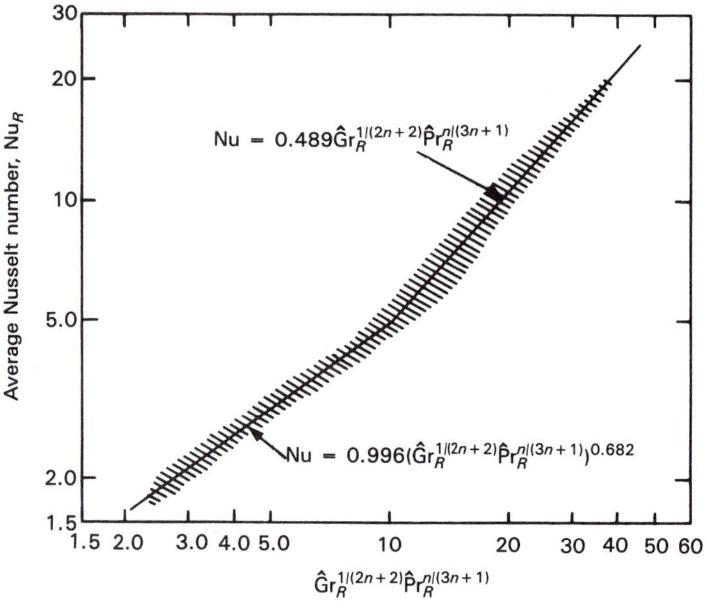

Figure 16.3.5 Correlation of thermal convection from isothermal spheres in polymer solutions as given by Amato and Tien (1976). The shaded area represents the experimental data. *(Reprinted with permission from* Int. J. Heat Mass Transfer, *vol. 19, p. 1257, W. S. Amato and C. Tien. Copyright © 1976, Pergamon Journals Ltd.)*

$$Nu = 0.161(\overline{Gr}\,\overline{Pr})^{0.308} \qquad \text{for } 0.19 < n < 1.0$$
$$\text{and} \quad 2.21 \times 10^5 < \overline{Gr}\,\overline{Pr} < 7.68 \times 10^8 \qquad (16.3.44)$$

The definitions of $\overline{Gr}$ and $\overline{Pr}$ are as given in Table 16.3.1 and the characteristic length L in Nu, $\overline{Gr}$, and $\overline{Pr}$ is

$$L = \sqrt{L_1 L_2} \qquad (16.3.45)$$

Mixed convection. Kubair and Pei (1968) reported a numerical analysis of combined free and forced convection heat transfer to power law fluids adjacent to an isothermal vertical surface. The following dimensionless parameter was found to characterize the mixed convection transport:

$$\epsilon = \frac{\overline{Gr}_x}{Re_x^{2/(2-n)}} \qquad (16.3.46)$$

where

$$Re_x = \frac{\rho u_\infty^{2-n} x^n}{K} \qquad (16.3.47)$$

It was concluded that the natural convection effects are negligible if $\epsilon < 0.1$. For $\epsilon > 5.0$ natural convection becomes the dominant transport mode.

Shenoy (1980) pointed out some of the discrepancies in the analysis of Kubair and Pei (1968). The dimensionless groups of Kubair and Pei do not satisfy the continuity equation. In the transformed dimensionless forms of the boundary layer equations given by Kubair and Pei, the parameters ϵ and Prandtl number are both functions of x and are constant only under certain quite restrictive conditions.

Following the approach of Churchill (1977) for Newtonian fluids and of Ruckenstein (1978) for power law fluids, Shenoy (1980) used a correlating equation of the form

$$\mathrm{Nu}_{x,M}^3 = \mathrm{Nu}_{x,F}^3 + \mathrm{Nu}_{x,N}^3 \tag{16.3.48}$$

where $\mathrm{Nu}_{x,M}$, $\mathrm{Nu}_{x,F}$, and $\mathrm{Nu}_{x,N}$ are the local Nusselt numbers for a surface transferring heat by mixed, forced, and natural convection, respectively. The following correlation for $\mathrm{Nu}_{x,M}$ resulted:

$$\begin{aligned}\frac{\mathrm{Nu}_{x,M}}{\mathrm{Re}_x^{1/(n+1)}} = \Bigg\{ &\left(\frac{1}{0.893}\right)^3 \frac{1}{18}\left(\frac{117}{560(n+1)}\right)^{1/(n+1)} \left(\frac{2n+1}{n+1}\right)\hat{\mathrm{Pr}}_{x,F} \\ &+ 8\left[\frac{1}{2}\left(\frac{3}{10}\right)^{1/n} f(n)\left(\frac{2n+1}{3n+1}\right)\hat{\mathrm{Pr}}_{x,F}\right]^{3n/(3n+1)} \\ &\times \left[\frac{\hat{\mathrm{Gr}}_x}{\mathrm{Re}_x^2}\right]^{3/(3n+1)(2-n)}\Bigg\}^{1/3}\end{aligned} \tag{16.3.49}$$

where

$$\hat{\mathrm{Pr}}_{x,F} = \frac{\rho c_p}{k}\left(\frac{K}{\rho}\right)^{2/(n+1)} (u_\infty)^{3(n-1)/(n+1)} x^{(1-n)/(n+1)} \tag{16.3.50}$$

$$f(n) = \frac{1}{15} - \frac{5}{126n} + \frac{1}{84n^2} - \frac{1}{486n^3} + \frac{1}{5103n^4} - \frac{1}{124{,}740n^5} \tag{16.3.51}$$

where $\hat{\mathrm{Gr}}_x$ and Re_x are as defined in Table 16.3.1 and Eq. (16.3.47), respectively. No experimental data are available to test the validity of this correlation.

16.4 VISCOINELASTIC (PURELY VISCOUS) FLUIDS: INTERNAL FLOWS

Process industry applications frequently encounter non-Newtonian transport in internal flows, such as flow of paints, polymers, suspensions, and plastics, in pipes and channels of various configurations. However, the majority of these flows, involving exchange of thermal energy, are essentially forced convection. Here, attention will be limited to pure natural and mixed convection internal flow transport.

The analysis of buoyancy-induced transport in an enclosed region is usually more complex than for external flows since the fluid motion near the walls is coupled with the core region. Then the pressure terms in the flow equations may not generally be neglected, as is commonly done for most external flows. Chapter 14 considers transport to Newtonian fluids in fully and partially enclosed spaces. For non-Newtonian fluids, very few results have been reported for internal flows solely due to buoyancy. However, relatively more information is available on the effect of buoyancy forces on forced convection or mixed convection.

16.4.1 Buoyancy-Driven Transport

Buoyancy-induced flow of a power law fluid between two vertical parallel surfaces at different thermal conditions has been investigated. For a finite height of the surfaces, H, end conditions must be included. Both ends or either end may be open or closed. Very different flow patterns result. For heated surfaces with both ends open, the fluid is drawn in at the bottom and discharged at the top. If both ends are closed and the two surfaces are differentially heated, an enclosure flow results. Although the flow patterns in non-Newtonian transport resemble qualitatively those in Newtonian fluids, the velocity levels may be markedly different.

For such an enclosure, of height H and width d, Emery et al. (1971) measured transport rates for several pseudoplastic power law fluids. One surface was heated electrically, at a constant heat flux condition, while the other was cooled to be isothermal. Aspect ratios H/d of 10, 20, and 40 were used. For the ranges $\overline{\mathrm{Pr}}_H = 10\text{–}500$ and $n = 0.7\text{–}1.0$, the Nusselt number based on average h was correlated by

$$\mathrm{Nu} = \frac{hH}{k} = 0.331(\overline{\mathrm{Gr}}_H \overline{\mathrm{Pr}}_H^n)^{1/(3n+1)} \tag{16.4.1}$$

where

$$\overline{\mathrm{Gr}}_H = \frac{g\beta(t_h - t_c)H^{(n+2)/(2-n)}}{(K/\rho)^{2/(2-n)}} \quad \text{and}$$

$$\overline{\mathrm{Pr}}_H = \frac{c_p}{k}\left(\frac{K}{\rho}\right)^{1/(2-n)} H^{(2n-2)/(n-2)} \tag{16.4.2}$$

where t_h and t_c are the hot and cold wall temperatures, respectively, and K and n are again the power law parameters. Equation (16.4.1) is a least-squares fit to the data in the range $6 \times 10^6 < \overline{\mathrm{Gr}}_H \overline{\mathrm{Pr}}_H^n < 10^{10}$.

Between vertical surfaces. The buoyancy-induced flow of a power law fluid between two vertical parallel isothermal surfaces each at t_0, open at both the top and bottom, was analyzed by Irvine et al. (1983), where t_0 is the temperature of the entering fluid. A finite-difference method was used. The dimensionless groups are given as

$$\hat{\mathrm{Gr}}_d = \frac{g\beta(t_0 - t_\infty)\, d^{2(1+n)/(2-n)}}{H^{n/(2-n)}\,(K/\rho)^{2/(2-n)}} \qquad \text{and}$$

$$\hat{\mathrm{Pr}}_d = \left(\frac{\rho c_p}{k}\right)\left(\frac{K}{\rho}\right)^{1/(2-n)} d^{3(1-n)/(2-n)} \tag{16.4.3}$$

where d is the plate spacing. The transport results were correlated in terms of a generalized Rayleigh number $\hat{\mathrm{Gr}}_d^{1/n}\hat{\mathrm{Pr}}_d$. For small values of this parameter, that is, small d, the flow approaches a fully developed condition. Transport rates are then given by

$$\mathrm{Nu} = \frac{hd}{k} = \frac{n}{2n+1}\left(\frac{1}{2}\right)^{(n+1)/n} \hat{\mathrm{Gr}}_d^{1/n}\hat{\mathrm{Pr}}_d \tag{16.4.4}$$

As the plate spacing and hence $\hat{\mathrm{Gr}}_d^{1/n}\hat{\mathrm{Pr}}_d$ increase, the transport approaches a single isolated plate correlation. Values of hd/k as a function of $\hat{\mathrm{Gr}}_d^{1/n}\hat{\mathrm{Pr}}_d$ are plotted in Fig. 16.4.1 for various values of power law index n. Also shown in Fig. 16.4.1 are plots of dimensionless induced inlet velocity u_0, defined as

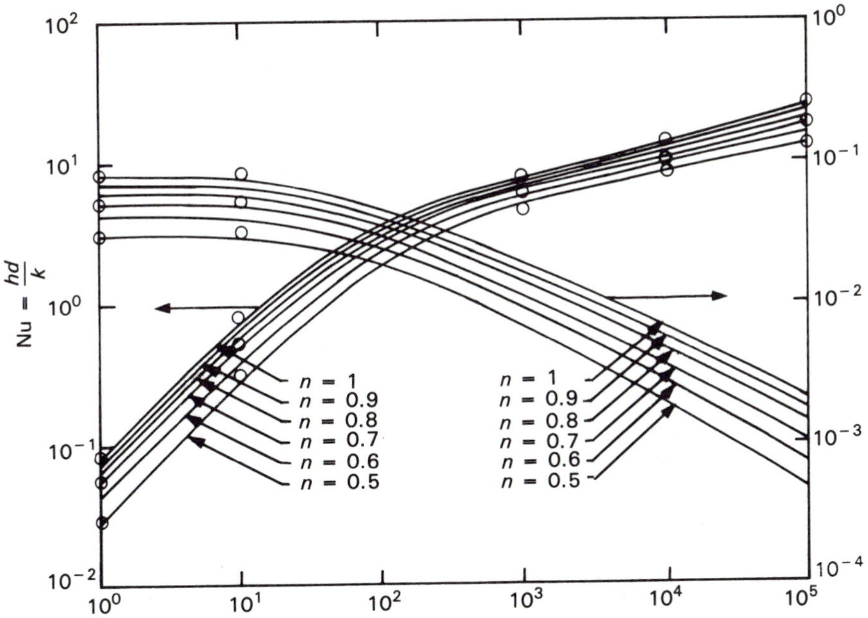

Figure 16.4.1 Dimensionless heat transfer and entrance velocity for vertical channel natural convection to a power law fluid. *(Reprinted with permission from T. F. Irvine et al., Brookhaven National Laboratory Rep. 31640. Copyright © 1983, Brookhaven National Laboratory.)*

$$u_0 = \frac{n}{2n+1}\left(\frac{1}{2}\right)^{(n+1)/n} \hat{\mathrm{Gr}}_d^{1/n} \tag{16.4.5}$$

The computed results given above are valid in the Prandtl number range $7 \leq \hat{\mathrm{Pr}}_d \leq 1000$. By use of a numerical example, it was shown that adding only 1000 ppm by weight of carboxymethylcellulose to water reduced the heat transfer by a factor of 3 and the average velocity by a factor of 11.7, with the plates kept at 25°C, for the entering fluid at 15°C. This suggests that the performance of a buoyancy-driven system with power law fluids may be varied quite significantly with small changes in the rheological properties.

Fully developed natural convection flow of a Bingham plastic between two infinite parallel vertical surfaces maintained at different temperatures was analyzed by Yang and Yeh (1965). Computed results for the flow field and the wall shear stress were reported.

16.4.2 Mixed Convection Effects

The effects of buoyancy forces on non-Newtonian internal forced convection flows often modify the transport rates considerably. These effects have been reported mostly for flow in horizontal, vertical, and circular tubes.

Horizontal tubes. Consider a horizontal tube in which the fluid bulk temperature is different from the tube wall temperature t_0. The buoyancy force then acts normal to the pressure gradient driving the main flow, and secondary motions arise that enhance the transport rates. For laminar mixed convection flow of power law fluids, Metzner and Gluck (1960) proposed the following correlation for a Nusselt number Nu_D based on tube diameter D:

$$\mathrm{Nu}_D = \frac{hD}{k} = 1.75\left(\frac{3n+1}{4n}\right)^{1/3}\left(\frac{K_\infty}{K_0}\right)^{0.14}\left[\mathrm{Gz} + 12.6\left(\frac{\mathrm{Gr}_0\mathrm{Pr}_0 D}{L}\right)^{0.4}\right]^{1/3} \tag{16.4.6}$$

where the subscript 0 denotes quantities calculated at t_0, K_∞ is the consistency index evaluated for the fluid bulk, h is the heat transfer coefficient based on an arithmetic mean local temperature difference, and $\mathrm{Gz} = \dot{m}c_p/kL$ is the Graetz number, where $\dot{m}$ is the mass flow rate. The correlation applies for $\mathrm{Gz} > 20$ and $n > 0.10$. The experimental data of Gluck (1959), Charm (1957), and Hirai (1956) agreed with Eq. (16.4.6) within 25%.

Oliver and Jenson (1964) measured mixed convection effects in pseudoplastic fluids and concluded that Eq. (16.4.6) did not fit their data. The aspect ratio L/D was not found to affect heat transfer rates. The following correlation was proposed:

$$\mathrm{Nu}_D = 1.75\left(\frac{K_\infty}{K_0}\right)^{0.14}[\mathrm{Gz} + 0.0083(\mathrm{Gr}_0\mathrm{Pr}_0)^{0.75}]^{1/3} \tag{16.4.7}$$

However, as pointed out by Shenoy and Mashelkar (1982), there are very few data available to support this correlation.

Vertical tubes. For mixed convection flow in vertical tubes, the forced flow may be upward or downward, as may the buoyancy force. Aiding effects usually cause an increase in heat transfer rates. DeYoung and Scheele (1970) and Marner and Rehfuss (1972) analyzed an upward fully developed flow of a power law fluid, with the tube wall dissipating uniform heat flux to the fluid inside. The heat transfer results for a fully developed flow are shown in Fig. 16.4.2. Nusselt number Nu_D is plotted against mixed convection parameter $\hat{\mathrm{Gr}}^*_{De}/\mathrm{Re}_R$ for various values of n, where

$$\mathrm{Nu}_D = \frac{hD}{k} \qquad \hat{\mathrm{Gr}}^*_{De} = \frac{R^4\rho_{\mathrm{av}}^2 g\beta q''}{k\mu_{\mathrm{eff}}^2} \qquad \mathrm{Re}_R = \frac{Rv_{\mathrm{av}}\rho_{\mathrm{av}}}{\mu_{\mathrm{eff}}} \tag{16.4.8}$$

$$\mu_{\mathrm{eff}} = K\left(\frac{v_{\mathrm{av}}}{R}\right)^{n-1} \tag{16.4.9}$$

Here v_{av} and ρ_{av} are average axial fluid velocity and radial average density, respectively. As pseudoplasticity increases (decreasing n), the heat transfer rates increase at a given value of $\hat{\mathrm{Gr}}^*_{De}/\mathrm{Re}_R$, whereas for dilatant fluids ($n > 1$), heat transfer rates decrease. The effect of increasing buoyancy is to increase heat transfer rates for all values of n.

Fully developed downward forced flow subject to opposed buoyancy, again with a uniform input heat flux condition, was also analyzed by DeYoung and

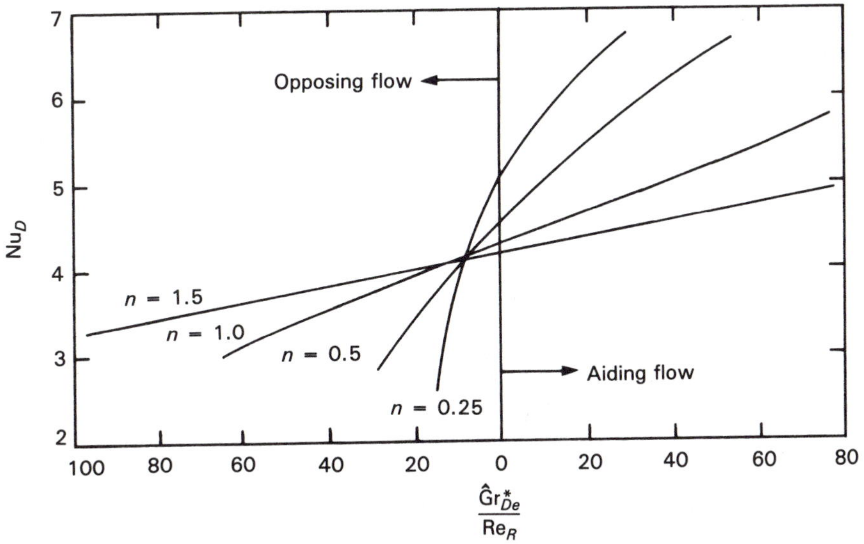

Figure 16.4.2 Variation of Nusselt number with $\hat{\mathrm{Gr}}^*_{De}/\mathrm{Re}_R$ in mixed convection flow of a power law fluid in a vertical tube with uniform heat flux for upflow and downflow. *(Reproduced by permission of the American Institute of Chemical Engineers, S. H. DeYoung and G. F. Scheele,* AIChE J., *vol. 16, p. 712. Copyright © 1970, AIChE.)*

Scheele (1970) for power law fluids. Lower heat transfer rates were found for opposing effects. The results are shown in Fig. 16.4.2.

For vertical tubes maintained at a constant temperature, the mixed convection heat transfer to a power law fluid with an initially fully developed velocity profile was studied both theoretically and experimentally by Marner and McMillan (1972). Only aiding flow was considered. Plots of theoretically predicted Nusselt numbers for $n = 0.2$, 0.5, 1.0, and 1.5 for different values of $\hat{\mathrm{Gr}}^*_{De}/\mathrm{Re}_R$ in the Prandtl number range 1–1000 were presented. The location and magnitude of the maximum increase in mean Nusselt number due to natural convection were investigated. In each case, the maximum was found to occur for $\hat{\mathrm{Pr}}_{De} = \mu_{\mathrm{eff}} c_p/k = 1000$. The maximum increase in the Nusselt number was lower at larger n. Experimental data on heat transfer to Carbopol solutions indicated agreement with the theoretical results to within $\pm 15\%$.

16.5 VISCOELASTIC FLUIDS

Compared to viscoinelastic fluids, little work has been reported on viscoelastic fluids. The governing equations are more complex, and choosing an appropriate constitutive equation is not an easy task. Some of the earlier work is due to Mishra (1966a, 1966b), Amato and Tien (1970), and Soundalgekar (1971, 1972). The review by Shenoy and Mashelkar (1978b) indicates some inadequacies of these solutions. It was also shown that the appropriate boundary layer equations for laminar flow past a two-dimensional curved surface, Fig. 5.1.2*a*, are

$$\frac{\partial u}{\partial x} + \frac{\partial v}{\partial y} = 0 \qquad (16.5.1)$$

$$u\frac{\partial u}{\partial x} + v\frac{\partial u}{\partial y} = \frac{1}{\rho}\frac{\partial \tau_{xy}}{\partial y} + \frac{1}{\rho}\frac{\partial(\tau_{xx} - \tau_{yy})}{\partial x} + g\cos\theta\,\beta(t - t_\infty) \qquad (16.5.2)$$

$$u\frac{\partial t}{\partial x} + v\frac{\partial t}{\partial y} = \alpha\left(\frac{\partial^2 t}{\partial y^2}\right) \qquad (16.5.3)$$

The second term on the right side in Eq. (16.5.2) is the elastic stress term, where τ_{xx} and τ_{yy} are the normal stresses in the x and y directions, respectively. For a Newtonian or a purely viscous (inelastic) fluid, $\tau_{xx} - \tau_{yy} = 0$, and these equations reduce to the usual boundary layer equations of natural convection flow over a two-dimensional surface, such as Eqs. (16.2.5), (16.2.6), without the motion pressure term, and (16.2.8).

The above equations, again subject to the boundary conditions in Eq. (16.2.9), were solved by Shenoy and Mashelkar (1978b). For a viscoelastic fluid, stress components τ_{xy} and $\tau_{xx} - \tau_{yy}$ were given by

$$\tau_{xy} = K\left(\frac{\partial u}{\partial y}\right)^n - m\left(\frac{\partial u}{\partial y}\right)^{s-2}\left(u\frac{\partial^2 u}{\partial x\partial y} + v\frac{\partial^2 u}{\partial y^2} + 2\frac{\partial u}{\partial x}\frac{\partial u}{\partial y}\right) \qquad (16.5.4)$$

$$\tau_{xx} - \tau_{yy} = 2m\left(\frac{\partial u}{\partial y}\right)^{s} \tag{16.5.5}$$

where m, n, s, and K are fluid parameters.

Neglecting inertia terms, with the high Prandtl number assumption, integral method solutions were obtained to these equations for different two-dimensional surfaces. It was shown that a similarity solution is possible only for a second-order fluid, for which $s = 2$ and $n = 1$, in the stagnation region of an isothermal horizontal cylinder. The results for local and average values of the Nusselt number were presented for this condition. Qualitative agreement was reported between the analytical results and the experimental data of Lyons et al. (1972) for natural convection with a horizontal cylinder to moderately elastic drag-reducing polyethylene oxide solutions.

Clearly, there is a need for further research on viscoelastic fluids. Numerical methods generally will be required to solve equations of transport with such complicated constitutive equations. Careful experiments should supplement these studies.

REFERENCES

Acrivos, A. (1960). *AIChE J. 6,* 584.

Amato, W. S., and Tien C. (1970). *Chem. Eng. Prog. Symp. Ser. No. 102 66,* 92.

Amato, W. S., and Tien, C. (1976). *Int. J. Heat Mass Transfer 19,* 1257.

Bird, R. B., Armstrong, R. C., and Hassager, O. (1977). *Dynamics of Polymer Liquids,* Wiley, New York.

Calderbank, P. H., and Moo-Young, M. B. (1959). *Trans. Inst. Chem. Eng. 37,* 26.

Charm, S. E. (1957). Sc.D. Thesis, Masssachusetts Institute of Technology, Cambridge.

Chen, T. Y. W. (1971). Ph.D. Thesis, Univ. of Missouri, Columbia.

Chen, T. Y. W., and Wollersheim, D. E. (1973). *J. Heat Transfer 95,* 123.

Churchill, S. W. (1977). *AIChE J. 23,* 10.

Dale, J. D., and Emery, A. F. (1972). *J. Heat Transfer 94,* 64.

Denney, D. A., and Brodkey, R. S. (1962). *J. Appl. Phys. 33,* 2269.

DeYoung, S. H., and Scheele, G. F. (1970). *AIChE J. 16,* 712.

Eckert, E. R. G., and Jackson, T. (1951). NACA Rept. 1015.

Emery, A., Chi, H. W., and Dale, J. D. (1971). *J. Heat Transfer 93,* 164.

Fujii, T., Miyatake, O., Fujii, M., and Tanaka, H., (1972). *Jpn. Soc. Mech. Eng. 38,* 2883.

Fujii, T., Miyatake, O., Fujii, M., Tanaka, H., and Murakami, K. (1973). *Int. J. Heat Mass Transfer 16,* 2177.

Fujii, T., Miyatake, O., Fujii, M., Tanaka, H., and Murakami, K. (1974). *Int. J. Heat Mass Transfer 17,* 149.

Gentry, C. C., and Wollersheim, D. E. (1974). *J. Heat Transfer 96,* 3.

Gluck, D. F. (1959). M.Ch.E. Thesis, Univ. of Delaware, Newark.

Hahn, S. J., Ree, T., and Eyring, H. (1959). *Ind. Eng. Chem. 51,* 856.

Hirai, E. (1956). *J. Chem. Eng. Jpn. 20,* 440.

Irvine, T. F., Wu, K. C., and Schneider, W. J. (1983). Brookhaven National Laboratory Rept. 31640.

Kim, C. B., and Wollersheim, D. E. (1976). *J. Heat Transfer 98,* 144.

Kleppe, J., and Marner, W. J. (1972). *J. Heat Transfer 94,* 371.

Kubair, V. G., and Pei, D. C. T. (1968). *Int. J. Heat Mass Transfer 11,* 855.

Lyons, D. W., White, J. W., and Hatcher, J. D. (1972). *Ind. Eng. Chem. Fundam. 11*, 586.
Marner, W. J., and McMillan, H. K. (1972). *Chem. Eng. Sci. 27*, 473.
Marner, W. J., and Rehfuss, R. A. (1972). *Chem. Eng. J. 3*, 294.
Metzner, A. B. (1956). *Advances in Chemical Engineering*, vol. 1, Academic Press, New York.
Metzner, A. B. (1961). In *Handbook of Fluid Dynamics*, V. L. Streeter, Ed., McGraw-Hill, New York.
Metzner, A. B. (1965). *Adv. Heat Transfer 2*, 357.
Metzner, A. B., and Gluck, D. F. (1960). *Chem. Eng. Sci. 12*, 185.
Metzner, A. B., and Whitlock, M. (1958). *Trans. Soc. Rheol. II*, 239.
Middleman, S. (1977). *Fundamentals of Polymer Processing*, McGraw-Hill, New York.
Mishra, S. P. (1966a). *Indian Chem. Eng. 8*, 26.
Mishra, S. P. (1966b). *Proc. Indian Acad. Sci. Sec. A 64*, 291.
Na, T. Y., and Hansen, A. G. (1966). *Int. J. Heat Mass Transfer 9*, 261.
Oliver, D. R., and Jenson, V. G. (1964). *Chem. Eng. Sci. 19*, 115.
Ostwald, W. (1926). *Kolloid Z. 38*, 261.
Reilly, T. G. (1964). M.A.Sc. Thesis, Univ. of Windsor, Windsor, Ont.
Reilly, T. G., Tien, C., and Adelman, M. (1965). *Can. J. Chem. Eng. 43*, 157.
Reiner, M. (1960). *Deformation, Strain and Flow*, p. 246, Interscience, New York.
Ritter, R. A. (1962). Ph.D. Thesis, Univ. of Alberta, Edmonton, Alta.
Ruckenstein, E. (1978). *AIChE J. 24*, 940.
Sharma, K. K., and Adelman, M. (1969a). *Can. J. Chem. Eng. 47*, 553.
Sharma, K. K., and Adelman, M. (1969b). *Can. J. Chem. Eng. 47*, 556.
Shenoy, A. V. (1977). Ph.D. Thesis, Univ. of Salford, U.K.
Shenoy, A. V. (1980). *AIChE J. 26*, 505.
Shenoy, A. V., and Mashelkar, R. A. (1978a). *AIChE J. 24*, 344.
Shenoy, A. V., and Mashelkar, R. A. (1978b). *Chem. Eng. Sci. 33*, 769.
Shenoy, A. V., and Mashelkar, R. A. (1982). *Adv. Heat Transfer 15*, 143.
Shenoy, A. V., and Ulbrecht, J. J. (1979). *Chem. Eng. Commun. 3*, 303.
Skelland, A. H. P. (1967). *Non-Newtonian Flow and Heat Transfer*, Wiley, New York.
Soundalgekar, V. M. (1971). *Chem. Eng. Sci. 26*, 2043.
Soundalgekar, V. M. (1972). *Int. J. Heat Mass Transfer 15*, 1253.
Sutterby, J. L. (1966). *AIChE J. 12*, 63.
Tien, C. (1967). *Appl. Sci. Res. Sec. A 17*, 233.
Tien, C., and Tsuei, H. S. (1969). *Appl. Sci. Res. 20*, 131.
Wilkinson, W. L. (1957). *Ind. Chem.*, 595.
Wilkinson, W. L. (1960). *Non-Newtonian Fluids*, Pergamon, London.
Yang, W. J., and Yeh, H. C. (1965). *J. Heat Transfer 87*, 319.

PROBLEMS

16.1 For a certain fluid, the following measurements of shear stress and shear rate were obtained.

Shear stress (lbf/ft^2)	Shear rate (s^{-1})
20	30
15	22
10	13
5	7
3	3

Plot the data on a log-log scale and determine whether the material is a power law fluid. If the answer is yes, determine the value of the power law index n and the consistency index K. How would the flow curve appear on this figure if the fluid were Newtonian?

16.2 To determine the relationship between shear stress and shear rate, a capillary tube viscometer is used. The following measurements of flow rate and pressure drop were obtained for the fluid flowing through a horizontal section of 0.1 in. (inner diameter) stainless steel tube. The fluid density is 50 lbm/ft^3. Determine a plot of shear stress vs. du/dy.

Flow rate W (lb/h)	Pressure drop ΔP (lbf/in.^{2})
7.152	445
1.032	167
3.435	333
0.642	125
2.430	277.5
0.344	83.3

Hint: for calculating the flow curve, shear stress = $D\ \Delta P/4L$ and shear strain = $(8V/D)[\frac{3}{4} + \frac{1}{4}\ d \ln (8V/D)/d \ln (D\ \Delta P/4L)]$, where V, D, d, and L are velocity, outer diameter, inner diameter, and length, respectively.

16.3 Compare the flow curve in Problem 16.2 with those given in Fig. 16.1.3. Can you identify the class of non-Newtonian fluid that this curve corresponds to? Also, determine μ_0 and μ_∞ for this fluid.

16.4 Consider natural convection from a vertical surface at 122°F in a power law fluid at 86°F. Properties of this fluid are $c_p = 0.5$ Btu/lbm °F, $k = 0.08$ Btu/ft °F h, $K = 0.2$ lbm s^{n-2}/ft, $\rho = 75$ lbm/ft^3, $\beta = 0.00035$ °R^{-1}, $n = 0.6$.

(a) Calculate the maximum velocity and the total heat transfer per side if the plate is 2.5 ft high. How do these values compare with those calculated for a Newtonian fluid in Problem 3.2?

(b) Calculate the boundary layer thickness and check whether the boundary layer solutions in (a) are applicable.

16.5 For flow adjacent to a vertical surface, in a power-law fluid, derive expressions for $q''(x)$, $\delta(x)$, and $Q(x)$. On physical grounds, do these quantities indicate any limits on the power law index n?

16.6 From Eqs. (16.3.3) and (16.3.5) for power law fluids, derive an expression for stream function ψ. Use $\eta = by$ and $\psi = \nu cf$.

(a) Derive an expression for normal velocity v.

(b) Evaluate $b(x)$ and $c(x)$ when $n = 1$ and $K = \mu$. How do these values compare with those given in Chapter 3 for an isothermal vertical surface?

16.7 Using the general similarity formulation of Chapter 3, show that similarity solutions for a heated vertical surface in a power law fluid are possible only for the temperature variation given in Eq. (16.3.2).

16.8 A plate 4 in. high is heated electrically to obtain a uniform heat flux of 31.71 Btu/ft^2 h in a power law fluid at 68°F. The properties of the fluid are $n = 0.50$, $c_p = 0.50$ Btu/lbm °F, $K = 0.20$ lbm s^{n-2}/ft, $k = 0.08$ Btu/ft °F h, $\rho = 60$ lbm/ft^3, $\beta = 0.00045$ °R^{-1}.

(a) Calculate the maximum and average surface temperature in natural convection.

(b) Is the average surface temperature different from the temperature calculated at the mid-height of the plate?

16.9 (a) Repeat Problem 5.5 when the ambient fluid is a power law fluid whose properties at 86°F are $\rho = 50$ lbm/ft^3, $c_p = 0.5$ Btu/lbm °F, $k = 0.4$ Btu/ft °F h, $K = 1.87$ lbm s^{n-2} ft^{-1}, $n = 0.6$, and $\beta = 0.0004$ °R^{-1}. Compare the result with that obtained for Problem 5.5.

(b) Now assume that the ambient fluid is Newtonian and has the physical properties given above (recall that then $n = 1$ and $K = \mu$). Calculate the total energy lost to the environment and compare it with the results in part (a). What conclusions may be drawn?

16.10 A heated horizontal surface at 88°F is placed in a 0.83% solution of ammonium alginate in water. Find the average rate of heat transfer by natural convection on the top and bottom sides if the fluid temperature is 77°F. The plate length is 2 ft and the fluid properties are $n = 0.78$, $c_p = 1.0$ Btu/lbm °F, $K = 0.06$ lbm s^{n-2}/ft, $k = 0.37$ Btu/ft °F h, $\rho = 60$ lbm/ft^3, and $\beta = 1.3 \times 10^{-4}$ °F^{-1}.

16.11 (a) In Problem 16.4, calculate the distance from the leading edge at which the flow will become turbulent. Assume that the limit for Grashof number or Rayleigh number at which the flow becomes turbulent is the same as that for Newtonian fluids, provided non-Newtonian Grashof and Prandtl numbers are used.

(b) In Problem 16.4, take the height of the plate to be 3 ft. Calculate the maximum value of the Grashof number, Prandtl number, and Nusselt number.

16.12 Two vertical walls 3 ft high are separated by a distance $d = 2$ in. and their temperatures are 122 and 86°F. For the power law fluid whose properties are given in Problem 16.4, calculate the Nusselt number based on the height of the walls. The bottom and top are closed with adiabatic surfaces.

CHAPTER

SEVENTEEN

SOME OTHER ASPECTS

17.1 OTHER MECHANISMS

The foregoing chapters cover in considerable detail the kinds of buoyancy-driven transport that most commonly arise and that have received the most attention in both experiment and calculation. The buoyancy agency throughout was terrestrial gravity interacting with a density variation in a body of fluid. The transport in such flows was treated, including many physical mechanisms and kinds of bounding conditions. The effects of highly and anomalously varying properties, of transition and turbulence, and of non-Newtonian behavior were discussed.

However, other transport-driving mechanisms also arise, as in rotating bodies of fluids with an internal density variation and in moving conductive and inductive fluids interacting with electric and magnetic fields. Also, many other effects resemble buoyancy-induced transport effects, as in noncontinuum conditions of gas-surface interaction and in random motions imposed on an otherwise quiescent fluid.

The foregoing discussion was also limited to the conductive and convective transport of thermal energy and to relatively simple characteristic boundary conditions. There are many important processes in which radiative thermal transport may be of comparable or greater magnitude, as in combustion and in the atmosphere. Also, in these and in other processes, multiple region convective processes and other transport processes may be coupled as conjugate effects. The effectiveness of transport may be assessed by several different measures. Recent calculations of entropy generation lead to interesting interpretations of different kinds of transport processes.

The following sections of this chapter consider these matters in some detail.

The next section treats rotational effects, followed by a section concerning other force field effects. Next, the transport arising from random inputs is discussed. Consideration of conjugate effects is then followed by sections concerning coupling radiation effects and entropy generation characteristics.

17.2 BUOYANCY WITH ROTATIONAL EFFECTS

Rotating solid-fluid boundaries impart rotation to the adjacent fluid through viscous forces generated by shear. Fluid rotation results in both centrifugal and Coriolis forces. The Coriolis force arises due to relative motions or flows generated in a rotating system, such as motions in the seas and atmosphere. The two forces interact with gravitational buoyancy forces, which in turn result from fluid density inhomogeneities. The density variation may be concentrated, as in a plume, or generalized in an ambient fluid, as in a density-stratified layer.

In either event, the motion of fluid along a curved path produces an additional tendency to motion that is density dependent. For example, the normal pressure gradient necessary to support such motion is $\partial p/\partial n = -\rho r\Omega^2$, where r and Ω are the local radius of streamline curvature and the angular velocity, respectively. Consider a vertical density stratification, for example, in the z direction. Then a pressure gradient component $\partial p/\partial z$ will also arise from stratification due to this density gradient. If the forces due to rotation overcome the stabilizing hydrostatic pressure gradient $-g\rho$, the fluid moves in the z direction. The coordinate system is shown in Fig. 17.2.1.

In a rotating horizontal stably stratified fluid in a container, the fluid near the bottom would move outward and upward and displace the lighter fluid inward

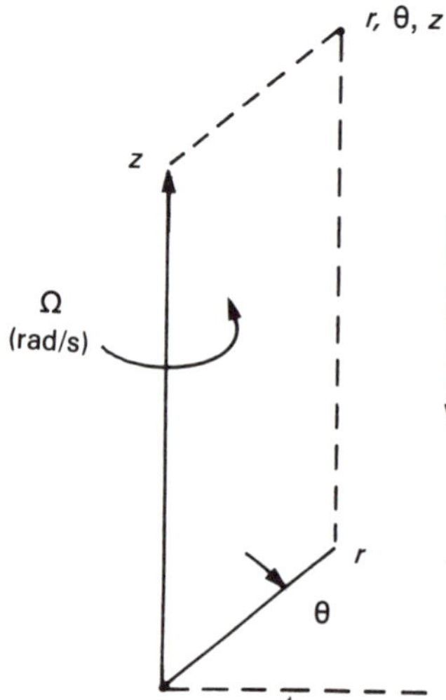

Figure 17.2.1 The stationary coordinate system r, θ, z in a buoyant body of fluid.

toward the vertical axis of rotation. The resulting motion then gives rise to an additional force due to mass motion at velocity **V** with respect to a rotating system, the Coriolis force. In general, this force is $\rho\mathbf{\Omega} \times \mathbf{V}$. These force fields then interact with buoyancy, momentum, and viscous effects, in conjunction with mass, energy, and any other conservation considerations, to determine the actual fluid motion. The resulting transport may be very complicated.

Combined rotational and buoyancy effects are very common in nature and in devices and process equipment. Most flows have regions of rotational motion and density differences. Heating a fluid in a pipe bend, separating density components in a centrifuge, and spinning objects and gas flows in turbines and compressors are important examples.

These matters have long been studied, both in establishing proper formulations of the transport and in determining actual transport rates. Much of the early research determined levels of rotational motion that would first produce instability and appreciable effects on transport. Later research has determined actual transport in various characteristic configurations: external flows, as in a boundary region on a rotating body; internal flows, as in rotating closed containers; and flows in curved or rotating passages.

This brief section does not cover any of these characteristic configurations in detail. For most, some physical aspects are discussed and some transport information given. References are then given to the most readily available literature. The following subsections briefly treat the formulation of transport and specific internal and external flow configurations that demonstrate the several results of increasing rotation.

17.2.1 Formulation of Transport

The additional centrifugal and Coriolis effects that arise are to be incorporated in the force-momentum balance. This relation and the other balances are then simplified for specific applications in terms of geometry and permissible additional approximations. The many applications that arise have resulted in numerous special forms. This brief summary mainly concerns only one general configuration. The rotation is about a vertical axis at an angular velocity of Ω (rad/s), with all bounding conditions axisymmetric. The coordinates are r, θ, and z. The only density variation accounted for is that which causes buoyancy, taken as the Boussinesq approximation $\Delta\rho = \rho\beta(t - t_r)$, where t_r is a reference temperature. Thus, centrifugally caused pressure difference effects on density are not included. This is a reasonable approximation for centrifugal forces that are small compared to gravitational acceleration, that is, for $A << 1$, where

$$A = \frac{\Omega^2 r}{g} \tag{17.2.1}$$

In terms of the velocity components u, v, and w in the r, θ, and z directions, the equations for axisymmetric transport, that is, no θ dependence, are

$$u \frac{\partial u}{\partial r} + w \frac{\partial u}{\partial z} - \frac{v^2}{r} - 2\Omega v = -\frac{1}{\rho}\frac{\partial p}{\partial r} + \nu\left(\nabla^2 - \frac{1}{r^2}\right)u \tag{17.2.2}$$

$$u \frac{\partial v}{\partial r} + w \frac{\partial v}{\partial z} - \frac{uv}{r} + 2\Omega u = \nu\left(\nabla^2 - \frac{1}{r^2}\right)v \tag{17.2.3}$$

$$u \frac{\partial w}{\partial r} + w \frac{\partial w}{\partial z} - g\beta(t - t_r) = -\frac{1}{\rho}\frac{\partial p}{\partial z} + \nu\nabla^2 w \tag{17.2.4}$$

$$\frac{1}{r}\frac{\partial(ur)}{\partial r} + \frac{\partial w}{\partial z} = 0 \tag{17.2.5}$$

$$u \frac{\partial t}{\partial r} + w \frac{\partial t}{\partial z} = \alpha\nabla^2 t \tag{17.2.6}$$

where

$$\nabla^2 = \frac{\partial^2}{\partial r^2} + \frac{1}{r}\frac{\partial^2}{\partial r^2} + \frac{\partial^2}{\partial z^2} \tag{17.2.7}$$

and p is the departure of local static pressure from the local hydrostatic value when the system is at t_r. The continuity equation (17.2.5) may be eliminated by defining the conventional Stokes stream function.

Comparing the above relations with those in Chapters 3 and 4, in the absence of rotation, indicates that the additional physical effects in Eqs. (17.2.2) and (17.2.3), for the meridional velocities u and v, are the terms in Ω. This gives rise to an additional dimensionless quantity—a gradient of the vertical convection velocity, over the height L, divided by an angular velocity, that is, $(U_c/L)/\Omega = U_c/\Omega L$. Taking $U_c = \sqrt{gL\beta\,\Delta t}$, this becomes

$$\frac{\sqrt{gL\beta\,\Delta t}}{\Omega L} = \sqrt{\mathrm{Gr}}\,\frac{\nu}{\Omega L^2} = 2\sqrt{\mathrm{Gr}}\,Ek = \frac{2\sqrt{\mathrm{Gr}}}{\sqrt{\mathrm{Ta}}} \tag{17.2.8}$$

where the Taylor and Ekman numbers, Ta and Ek, relate the centrifugal and viscous forces as follows:

$$\mathrm{Ta} = \frac{4\Omega^2 L^4}{\nu^2} = \mathrm{Ek}^{-2} \qquad \text{and} \qquad \mathrm{Ek} = \frac{\nu}{2\Omega L^2} \tag{17.2.9}$$

Very small values of Ta, that is, $\nu^2 >> \Omega^2 L^4$, high viscosity, and/or slow rotation indicate essentially solid-body rotation. Large values of Ta suggest boundary layer flows, with appreciable Coriolis effects.

Sometimes the Rayleigh number Ra = GrPr is used. In some applications some form of a Rossby number Ro is used instead. One version is

$$\mathrm{Ro} = \frac{U_c}{\Omega L} \tag{17.2.10}$$

The characteristic velocity U_c may be defined in several ways. In any case, Ro above is a kinematic quantity that relates U_c to the rotational velocity ΩL, as does the Taylor number.

17.2.2 Transport in a Rotating Vertical Cylinder

As an example of an internal flow, consider a fluid-filled right circular cylinder of height H and diameter D rotating about its axis at Ω rad/s. The bottom and top surfaces are taken at t_B and t_T and the cylindrical sidewall is taken as insulating. There are two possibilities. For heating from above, $t_T > t_B$, the conduction temperature field in the fluid would amount to stable stratification for $\beta > 0$. For heating from below, $t_T < t_B$, the stratification would be potentially unstable, as treated in Chapters 13 and 14, in the absence of rotation.

Heating from above. Rotation has many additional effects, even in the normally stable circumstance $t_T > t_B$. The outward centrifugal force is higher in the denser fluid in the bottom. Outflow results, with a return inflow in the upper region of the container. This circulation becomes increasingly vigorous at increasing values of $A = \Omega^2 r/g$. The kind of resulting flow is then indicated by the value of $\text{Ta} = 4\Omega^2 L^4/\nu^2$. For very small values, as for a relatively viscous liquid, the flow may be nearly solid-body rotation, perturbed by a small convective motion. For Ta very large, on the other hand, high velocities result and Ekman boundary region transport may be found adjacent to all surfaces inside the cylinder. Coriolis acceleration may dominate the core flow.

The transport calculations of Barcilon and Pedlosky (1967) and Homsy and Hudson (1969, 1971a, 1971b) for various levels of A indicate some of the properties of these circulations. Often of greatest interest, however, is the resulting heat transfer across the container. Hudson et al. (1978) measured the heat transfer with two different silicone oils, one with a viscosity of 0.65 centistokes, a water-like liquid, and the other with a viscosity of 350 centistokes. In the first fluid the Ekman layers form and Coriolis effects are very important. In the much thicker silicone, viscous effects are much more important. The experiments were in the parameter range

$$A = \frac{\Omega^2 D}{2g} \geq 40 \qquad \text{and} \qquad [\beta(t_T - t_B)]\text{Pr}\sqrt[4]{\text{Ta}} > 0.04$$

Here $\beta(t_T - t_B)$ is in units of buoyancy and $\text{Ta} = 4\Omega^2 H^4/\nu^2$, where H is the container height. The resulting correlation for Nusselt number Nu may be written as

$$\text{Nu} = \frac{hH}{k} = 0.334[\beta(t_T - t_B)]^{0.822}\left(\frac{D}{H}\right)^{0.173}\sqrt[4]{\text{Ta}} \tag{17.2.11}$$

Note that as Nu is defined, the value for pure conduction is 1.0.

Heating from below. On the other hand, $t_B > t_T$ results in convective motion even without rotation if $\text{Ra} = gH^3\beta(t_B - t_T)/\nu\alpha > 1708$, for D/H large, as discussed in Chapter 13. Also, the centrifugal effect now tends to move the denser upper fluid outward, in opposition to an inward motion in the bottom region. Some of the many studies of such transport are summarized by Bühler and Oertel (1982).

Early measurements of this configuration, with mercury, were reported by Dropkin and Globe (1959). Their results indicated that instability was deferred because of stabilization by rotation. A number of subsequent studies determined heat transfer over a wide range of Ra, Ta, and Pr; see, for example, Rossby (1969), Globe and Dropkin (1959), and Dropkin and Gelb (1964). Tang and Hudson (1983) used two silicones of viscosity 0.65 and 350 centistokes, Pr = 7.2 and 3100, at rotation rates A up to 115. For Pr = 7.2 the two acceleration effects were thought to be comparable.

However, for the fluid with Pr = 3100, rotation had a small effect on heat transfer for $A < 0.3$. Thereafter, heat transfer decreased with a further increase in A up to much larger values. Then heat transfer increased again due to centrifugally driven convection. For Pr = 7.2, a high Ta value stabilized the unstably stratified fluid.

For any such rotating cylindrical flow, the Nusselt number would depend on Pr, Ta, and $\text{Ra} = g\beta(t_B - t_T)H^3/\nu\alpha$, in general, in a very complicated way. An example of this dependence for water is seen in Fig. 17.2.2, from Rossby (1969). The large Ra effect at low Ta is much reduced at higher Ta as rotational effects come to dominate. The line C–C is the marginal stability curve of Chandrasekhar (1961), where Nu = 1 is the conduction solution, for no motion.

17.2.3 Buoyancy Effects External to Rotating Surfaces

The many applications in technology have inspired much analytical and experimental determination of heat transfer between rotating surfaces and surrounding fluids. Three distinct agencies of convection commonly arise: imposed motion in the ambient, rotational acceleration, and buoyancy in a gravity field. An early review of this field, as well as of internal flows, was given by Kreith (1968); see also Dorfman (1963).

Although there are three convective effects, ambient motion, rotation, and buoyancy, almost all studies have concerned cases in which only one or two of these arise. Most studies have concerned very high centrifugal effects, that is, $A >> 1$, where buoyancy is unimportant. There has been some study of mixed forced and centrifugal effects. There has been relatively little study of transport processes in which buoyancy and rotational effects are comparable.

However, buoyancy and rotational effects are often comparable in geophysical processes. Lugt and Schwiderski (1965, 1966) calculated convective transport for a rotating atmosphere above a locally heated surface. This is the circumstance that often gives rise to "dust devils" and other rising atmospheric circulations. In a similar study by Rotem and Claassen (1970), but one more related to technological applications, transport was calculated and visualized above a rotating horizontal disk of radius a, at t_0. The effect of rotation relative to buoyancy was in terms of

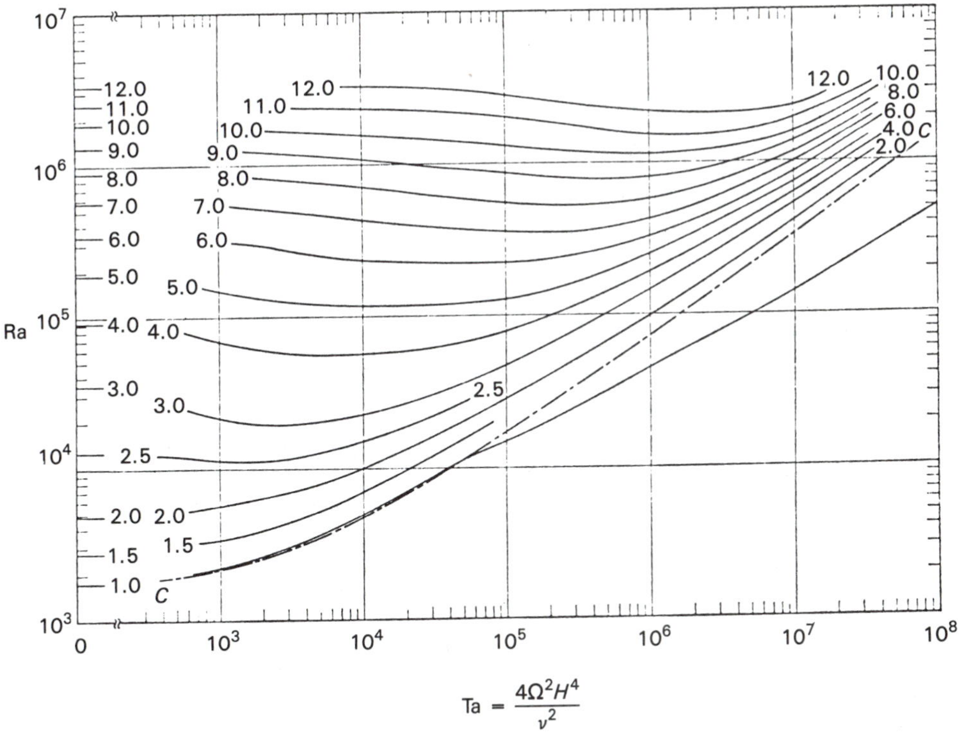

Figure 17.2.2 Lines of constant Nusselt number as a function of the Rayleigh and Taylor numbers in water with Prandtl number 6.8 and heating from below. Except for small values of the Nusselt number at large Taylor numbers, the uncertainty in the Nusselt numbers is less than about 2%. *(Reprinted with permission from H. T. Rossby,* J. Fluid Mech., *vol. 36, p. 309. Copyright © 1969, Cambridge University Press.)*

$$\frac{\sqrt{R^5}}{\mathrm{Gr}_a} = \frac{a^2\sqrt{\Omega^5}}{g\sqrt{\nu}\,\beta(t_0 - t_\infty)} \qquad \text{where } R = \frac{a^2\Omega}{\nu} \tag{17.2.12}$$

At small values of $\sqrt[2]{R^5}/\mathrm{Gr}$ transport is dominated by buoyancy and the flow is inward. The breakdown of laminar flow occurred for $\sqrt[3]{R} = O(100)$.

Popiel and Bosuslawski (1975) report heat transfer measurements for a rotating vertical isothermal disk in air. Results are reduced in terms of Nu and a combined parameter $\sqrt{\mathrm{Gr} + \mathrm{Re}} = N$, where $\mathrm{Re} = \Omega R^2/\nu = (2Ek)^{-1}$. Three regions clearly arise and are called laminar, for $N < 2 \times 10^5$, transition, for $2 \times 10^5 < N < 2.5 \times 10^5$, and turbulent, for $N > 2.5 \times 10^5$. Correlating equations are given for each range.

Kreith and Kneisel (1963) report heat transfer measurements for rotating heated downward-facing isothermal cones of vertex angle 30–80°, at t_0, in quiescent air at t_∞. Measurements indicated that buoyancy has an appreciable effect compared to rotation for

$$\beta(t_0 - t_\infty) > 0.05A \frac{\sin^2 \alpha}{\cos \alpha} \tag{17.2.13}$$

where α is half the vertex angle and A is as given in Eq. (17.2.1), where r becomes the cone slant height. This agrees with the calculations of Kreith (1968). Similar solutions for mixed convection in this geometry are given by Hering and Grosh (1963). Suwono (1980) reports a similar analysis for rotating round-nosed bodies with vertical axes. Numerical results are given for a sphere.

17.3 OTHER FORCE FIELDS

Often, force fields other than terrestrial gravity and rotation act on a body of fluid and affect the flow field and transport quantities. This occurs, for example, in electrically conducting fluids in electric and magnetic fields. The study of continuous electrically conducting fluids under the influence of electromagnetic fields is known as magnetohydrodynamics (MHD). Early interest in such phenomena arose in astrophysics, geophysics, and controlled nuclear fusion.

The forces that act on a conducting fluid at rest are summarized by Stratton (1941) and Romig (1964) as follows. Ponderomotive force is the force on a volume distribution of current in a magnetic field. Magnetostriction and electrostriction are elastic deformations of a fluid under the influence of electromagnetic fields. The electrostatic force is the body force on free-charge carriers in the fluid. For fluids of constant permeability and dielectric constant, neglecting displacement and polarization currents, the magnetostrictive, electrostrictive, and electrostatic forces are negligible. Only the ponderomotive force remains.

Most natural convection MHD investigations were conducted with the above restrictions. There are circumstances in which electrostrictive forces are important, as in heated, low-velocity flows of conducting fluids subjected to an electric field, as shown by Senftleben and Braun (1936), Schmidt and Leidenfrost (1953), and Lykoudis (1962). However, such effects are not commonly encountered, and the discussion here will be restricted to flows driven by only the ponderomotive force.

Two different physical effects will be discussed in this section. Several developed buoyancy-induced flows with MHD effects will be considered. These are flows adjacent to a flat vertical surface and between parallel vertical surfaces. Thereafter, the MHD effect on thermal instability of an electrically conducting fluid layer heated from below will be examined.

17.3.1 Natural Convection with MHD Effects

The conservation equations for a steady, two-dimensional, laminar, incompressible flow of a conducting fluid passing through a transverse constant, uniformly distributed field of magnetic induction B_0 are

$$\frac{\partial u}{\partial x} + \frac{\partial v}{\partial y} = 0 \tag{17.3.1}$$

$$u\frac{\partial u}{\partial x} + v\frac{\partial u}{\partial y} = g\beta(t - t_\infty) - \frac{\sigma_0 B_0^2}{\rho}u + \nu\frac{\partial^2 u}{\partial y^2} \tag{17.3.2}$$

$$u\frac{\partial t}{\partial x} + v\frac{\partial t}{\partial y} = \alpha\frac{\partial^2 t}{\partial y^2} \tag{17.3.3}$$

Here σ_0 is the fluid electrical conductivity taken as a scalar and $B_0 = \mu H$ is the magnetic induction, where H is the magnetic field strength. The second term on the right side of Eq. (17.3.2) is the magnetic ponderomotive force and is the only term appearing due to MHD effects. In deriving Eqs. (17.3.1)–(17.3.3), in addition to the boundary layer and Boussinesq approximations, the following effects are assumed negligible:

1. Slip between the charged and neutral particles
2. Imposed and induced electric fields
3. Viscous and Joulean dissipation
4. Electrical convection and displacements
5. Components of the magnetic field in the x and z directions
6. Induced magnetic fields

Equations (17.3.1)–(17.3.3) have been solved for various configurations and boundary conditions.

Vertical isothermal surface with constant B_0. No similarity solutions are available for constant B_0. Sparrow and Cess (1961) obtained a series expansion solution in the absence of an electric field, in terms of $(Mx/\text{Gr}^{1/4})^2$, and found that

$$\frac{\overline{\text{Nu}}}{\text{Gr}^{1/4}} = \frac{\overline{\text{Nu}}_0}{\text{Gr}^{1/4}} - 0.3145[-\phi'(0)]\frac{2M^2}{\text{Gr}^{1/2}} \tag{17.3.4}$$

where $\text{Gr} = g\beta L^3\,\Delta t/\nu^2$, $M = (\sigma_0 B_0^2 L^2/\mu)^{1/2}$ = ponderomotive force/viscous force is the Hartmann number, and $\overline{\text{Nu}}_0 = 0.404[-\phi'(0)]\,\text{Gr}^{1/4}$ is the mean Nusselt number when no magnetic field is present.

Solutions were obtained for Pr = 10, 0.73, and 0.02. Figure 17.3.1 shows Eq. (17.3.4) as the dashed curves for Pr = 0.73 and 0.02. The oustanding effect of the magnetic field is to inhibit convection, thereby reducing heat transfer.

Vertical isothermal surface with variable B. For $B(x) = B_0 x^m$, Lykoudis (1962) found a similarity solution for $m = -\frac{1}{4}$. The solutions are also plotted in Fig. 17.3.1 for Pr = 0.72, 0.3, 0.1, 0.03, 0.02, and 0.01. Again, for all values of Pr the effect of the magnetic field is to inhibit convection and heat transfer.

Vertical nonisothermal surface with constant B_0. For $t_0 - t_\infty = Nx^n$, where $n = 1$, Cramer (1962) found a similarity solution for B_0 constant. Calculations

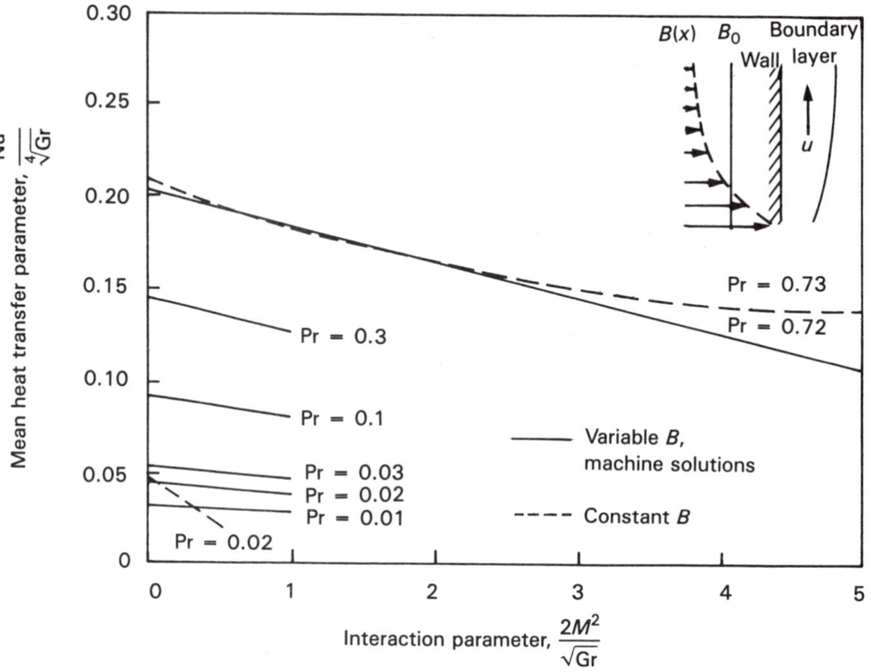

Figure 17.3.1 Variation of the mean heat transfer with uniform and variable magnetic fields. *(Reprinted with permission from M. F. Romig,* Adv. Heat Transfer, *vol. 1, p. 267. Copyright © 1964, Academic Press, Inc.)*

were made for Pr = 0.01 and 0.005 for various values of the magnetic field strength. Increasing the magnetic field or flux predictably inhibits convection, reduces heat transfer, and flattens the velocity and temperature profiles, as shown in Figs. 17.3.2 and 17.3.3.

Other studies. Riley (1964), Singh and Cowling (1963), and Kuiken (1970) analyzed the flow adjacent to a vertical isothermal surface with MHD effects. A matched asymptotic expansion was deemed necessary because near the leading edge the magnetic force, which is seen in Eq. (17.3.2) to be proportional to u, is very small. There the flow will then be essentially buoyancy driven. Downstream, as u grows, so will the magnetic force, until a balance eventually is reached between the buoyancy and magnetic forces. Thus, as pointed out by Kuiken (1970), a constant characteristic velocity is reached and thereafter maintained by the fluid.

Poots (1961) studied the flow between parallel vertical plates at different temperatures, including MHD effects, viscous and Joulean dissipation, and internal energy sources. The effect of a magnetic field was found to be a reduction in velocity and heat transfer, as found for flows generated adjacent to a flat vertical surface. Other similar studies of flow between vertical parallel surfaces were carried out by Cramer (1962) and Gershuni and Zhukhovitskii (1958).

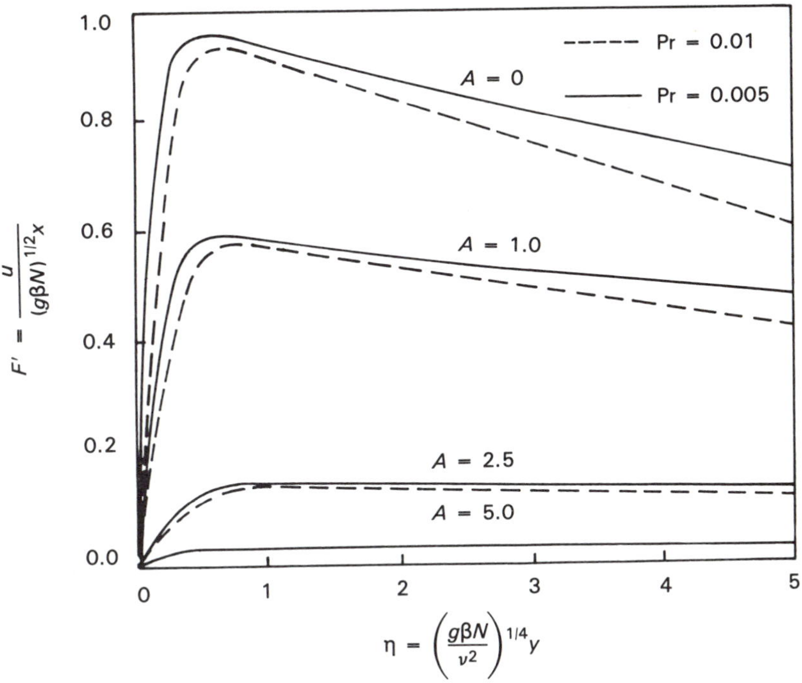

Figure 17.3.2 Flat surface velocity profiles. F' is the nondimensional velocity and η the nondimensional distance from the plate; $A = M/(\mathrm{Gr})^{1/4}$. *(Reprinted with permission from K. R. Cramer, ASME Paper 62-HT-22. Copyright © 1962, ASME).*

17.3.2 Thermal Instability with MHD Effects

When an electrically conducting fluid layer, heated from below, is subjected to a magnetic field, the field will inhibit instability and the onset of convection. As pointed out by Chandrasekhar (1961), the buoyancy force must balance not only the viscous effects but magnetic effects as well. Chandrasekhar found the critical Rayleigh number Ra_c to be

$$\mathrm{Ra}_c = \pi^2 M^4 \qquad \text{as } M \to \infty \tag{17.3.5}$$

where $M = B_0 d \cos\theta \sqrt{\sigma_0/\mu}$ is the Hartmann number and θ is the angle of incidence between the magnetic field and the liquid surface layer of thickness or depth d.

The above asymptotic limit applies for all three surface conditions: both surfaces free, one free and one rigid, and both rigid. In a series of experiments, Nakagawa (1955, 1957, 1960) and Nakagawa and Goroff (1961) verified Chandrasekhar's predictions, for rigid bounding surfaces, over a wide range of values of the Hartmann number, as shown in Fig. 17.3.4.

In another experiment, Lehnert and Little (1957) visualized the flow at the

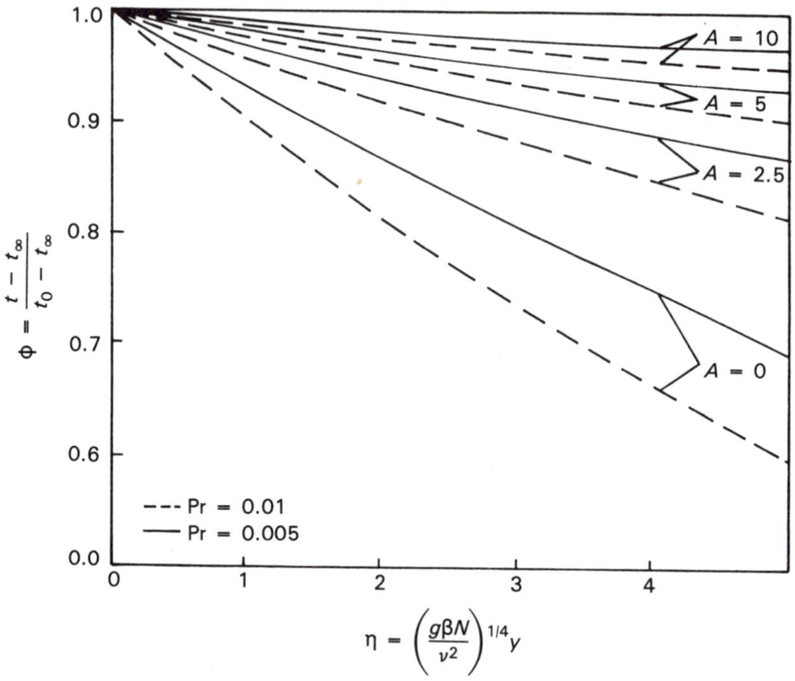

Figure 17.3.3 Flat surface temperature profiles. *(Reprinted with permission from K. R. Cramer, ASME Paper 62-HT-22. Copyright © 1962, ASME).*

top surface of a layer of mercury heated from below and subjected to magnetic fields of varying intensity and angle of inclination. Figure 17.3.5 shows the flow patterns (viewed from the top) that resulted when the bottom surface was placed on the edge of a magnetic pole piece. The left side of the picture shows liquid in the region subjected to a vertical magnetic field, while the right side shows liquid subjected to the weaker oblique field. The resulting flow field consisted of a pure conduction region to the left, adjacent to a cellular convection region to the right.

17.4 RANDOM CONVECTION

The bulk of all formulation, analysis, and experimentation concerning transport processes presumes that the component physical inputs and resulting effects are principally deterministic. For example, the stated geometry, bounding conditions, and characteristics of the fluid are assumed sufficient to specify any given transport result. In analysis, if the equations and boundary conditions are given, a solution is assumed to be formulated. Even with turbulence we are accustomed

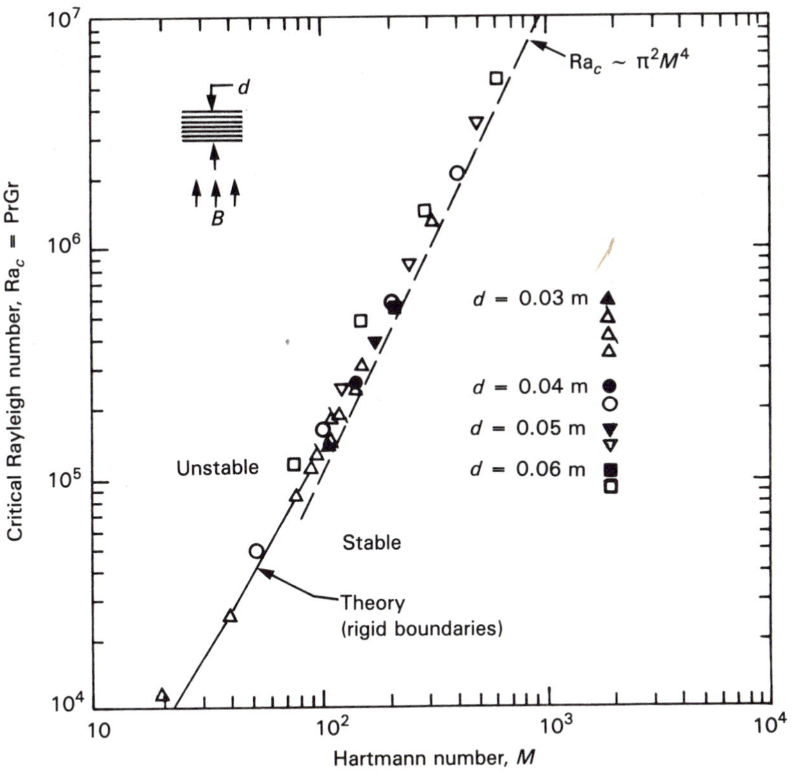

Figure 17.3.4 Variation of critical Rayleigh number with Hartmann number for a horizontal layer of liquid mercury heated from below. *(Reprinted with permission from M. F. Romig,* Adv. Heat Transfer, *vol. 1, p. 267. Copyright © 1964, Academic Press, Inc.)*

to often think of the added transport mechanisms as average effects, such as turbulent diffusivity.

However, other kinds of processes sometimes arise that must be or are best understood by postulating some random input, either to simulate or to account for appreciable effects that are not simply deterministic. An example of this is noncontinuum transport between a sphere of diameter D and a gas at low density. This arises when the gas mean free molecular path length λ is appreciable compared to D. Then the statistical properties of molecular motion must be taken into account. Other examples are random macroscopic inputs in nature or in devices, for example, where a bounding surface is moved in a random way.

Some attention has been given to such effects. Two examples of specific kinds of studies and methods of treatment are briefly summarized below. The first concerns random fluid motions or disturbances and their downstream or later effects on transport and on laminar stability. The second kind of effect considered below is the transport effect arising from the random motion of solid boundaries in contact with a fluid.

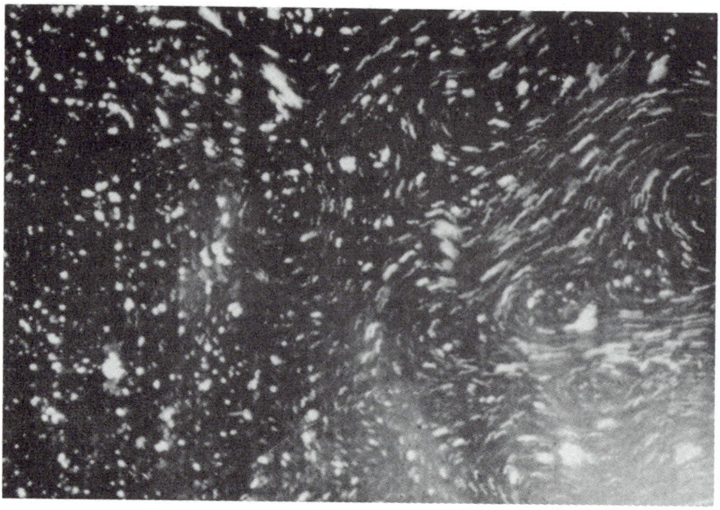

Figure 17.3.5 Magnetically inhibited convection. *(Reprinted with permission from B. Lehnert and N. C. Little,* Tellus, *vol. 9, p. 97. Copyright © 1957, Swedish Geophysical Society.)*

17.4.1 Random Motions Internal to a Body of Fluid

Such motions may result from imposed inputs, such as random local heating, or from radiation, electric currents, or a heater. They may also occur spontaneously. For example, in a gas, local thermal fluctuations are expected to arise from the statistical behavior of the molecular motion. There is a certain probability of small local volumes of warmer gas arising due to the random accumulation of subpopulations of molecules with higher velocities. These may interact, through viscosity and/or buoyancy effects, to affect the flow or its stability.

Indeed, small random effects are often thought to be the initial motions that trigger instability and lead to additional convection mechanisms, such as transition and turbulence. One method of stability analysis assumes the form of the disturbances, with a random magnitude, and determines how they grow downstream in a developing boundary region flow or in time in an unstably stratified fluid layer.

Another form of instability determination has been used in conjunction with the numerical calculation of transport; see Section 11.13. Random numerical transport inputs are contributed at the grid points. Their subsequent growth or decay indicates local flow stability. The energy stability criterion of Shen (1961) for transients is sometimes used, as also discussed in Section 11.13.

Random initial disturbances were applied by Chen and Kirchner (1971) to determine the stability of a transient circumferential flow in an annulus. Chen and Sandford (1977) reported similar calculations concerning stability in doubly diffusive transient flow in an inclined slot. The fluid was initially stratified. Random initial perturbations of vorticity were supplied. Such results have confirmed experimental observations.

Concerning unstably stratified horizontal fluid layers, considered in Chapter 13, both experiments and analyses indicate an important role for random disturbances in both initial and finite amplitude instability, as well as in transitional events. The limited success of some of the earlier conventional formulations is summarized in the review by Davis (1976). The data of Busse and Whitehead (1974) suggest a major role for random inputs in determining eventual transport. Newell et al. (1970) considered the convective response to an initial random disturbance spectrum. The onset of convection arising from random thermal inputs was analyzed by Jhaveri and Homsy (1980) as an initial-value problem in terms of the Fourier modes. Allowing for weak nonlinear effects, three stages of ordered convection evolve. These correspond to different time intervals during which fluctuations and nonlinear effects have different relative importance. The results also suggest that a large population of convection cells may arise in an extensive layer. A similar formulation of random effects was applied to transient stratification by Jhaveri and Homsey (1982).

17.4.2 Random Motion of Bounding Surfaces

Moving containers of fluids are commonplace. They may translate and rotate, either steadily, periodically, in a transient, or randomly. Section 17.2 considered buoyancy-driven transport affected by steadily rotating bounding surfaces. The present section concerns a uniform body of fluid in a container that, in turn, is subject to random perturbations of both its linear and angular velocities, that is, of its translation and rotation rates. The transmission of linear velocity fluctuations is largely by normal, or pressure, stresses at the solid-fluid interfaces. Equilibration of internal fluid motion by pressure waves is rapid in all but very large containers. However, changes in container angular motion are transmitted only by shear stresses and initially only at the interface. This effect then propagates inward to the contained fluid, relatively very slowly.

Early studies of such transport were related to the operation of fluid-filled devices on a nominally ballistic trajectory in space. The buoyancy-driven motions may be very small. Then the lower limit of diffusive heat transfer through the fluid would be pure conduction. However, such devices are subject to many velocity and rotational disturbances due to attitude control measures, engine burns, vibrations, relative motions of components of the spacecraft, motion of occupants, and particle impacts. The actual motion of the device is modeled as a zero average gravity trajectory, without rotation, perturbed by imposed random velocity and angular motion fluctuations. It might be modeled as a random distribution of both the time interval τ_c between instantaneous or abrupt fluctuations and the magnitude of the fluctuation. See Gebhart (1963).

In fact, on-board accelerometers in space devices have repeatedly documented such fluctuations, in terms of an "effective" gravity g_e, as related to velocity fluctuations. The characteristic magnitude has been $g_e = O(10^{-6}g)$. Such fluctuations are called "g-jitter." In a gas such effects may cause compressive temperature fluctuations adjacent to surfaces. These augment heat transfer. See, for example, Spradley (1974). A review of such effects and of some on-board heat transfer data from the Apollo spacecraft is given by Grodzka and Bannister (1972, 1974).

Resulting heat transfer. However, an enhancement of heat transfer, over pure conduction, also arises from angular perturbations. The mechanism may often be very different. A sequence of abrupt orientation changes are not rapidly propagated across an appreciable contained fluid volume. Think of a cup of coffee sitting on a horizontal surface as the fluid container. Quick small purely angular rotations of the cup leave the main body of the fluid almost motionless. The cup moves with respect to the bulk of the coffee.

As a model of the resulting heat transfer effects, consider a heated object of area A uniformly at t_0, well away from the wall, immersed in the fluid at t_∞. If the object is attached to the randomly moving enclosing container, it moves in a similar random way. In the absence of gravity and any container motion, this object would lose heat only by conduction. However, any random angular perturbations of the container would repeatedly move the object away from its conduction field into new fluid at t_∞. After each of the abrupt movements, spaced randomly at time interval τ_c apart, a new transient conduction field would begin to propagate outward from the object, into the adjacent fluid. This transient would, in turn, cease at the next movement of the container. Then a new transient would begin at a new location in the fluid.

The resulting average heat transfer rate is the average of the rates during these successive conduction transients. This average is calculated as follows. If $Q(\tau_c)$ is the total heat transferred in the time interval τ_c, the average heat transfer rate during this interval is $Q(\tau_c)/\tau_c$. Then the long-term average heat transfer from the surface, $\bar{Q}$, is written as the average of the value of $Q(\tau_c)/\tau_c$ in all of the intervals, as

$$\bar{Q} = \int_0^\infty \frac{Q(\tau_c)}{\tau_c} f(\tau_c)\, d\tau_c = \bar{q}''A \tag{17.4.1}$$

where $f(\tau_c)$ is the probability distribution of the time interval τ_c. The interval τ_c may be from 0 to ∞.

Now τ_c is written as a "disturbance" Fourier number, as $\text{Fo} = F = \alpha\tau_c/s^2$, where s is a characteristic length. Then Eq. (17.4.1), for average flux $\bar{q}'' = \bar{Q}/A$, becomes

$$\bar{q}'' = \frac{\bar{Q}}{A} = \frac{\alpha}{s^2}\int_0^\infty \frac{Q(F)f(F)\, dF}{AF} \tag{17.4.2}$$

The Nusselt number based on the average heat transfer rate is then

$$\text{Nu} = \frac{hs}{k} = \frac{\bar{q}''s}{(t_0 - t_\infty)k} = \frac{\alpha}{sk(t_0 - t_\infty)}\int_0^\infty \frac{Q(F)f(F)\, dF}{AF} \tag{17.4.3}$$

It is convenient to replace F by $P = F/F_m = \tau_c/\tau_m$, where τ_m and $F_m = \alpha\tau_m/s^2$ are the mean or most probable values of τ and F, distributed as the function $f(F)$. Then

$$\text{Nu} = \frac{\alpha}{skF_m(t_0 - t_\infty)} \int_0^\infty \frac{Q(P)f(P)\,dP}{AP} \tag{17.4.4}$$

The random events, now spaced in terms of the random time interval P as $f(P)$, may arise from one repeated effect or from a combination of many independent effects. If each effect follows an exponential distribution $f(P) = e^{-P}$, then the probability distribution of the sum of $n + 1$ such effects is the following gamma distribution f_n:

$$f_n\left(\frac{F}{F_m}\right) = f_n(P) = \frac{n+1}{n!} P^n e^{-nP} \tag{17.4.5}$$

The relation for Nu for any choice of n becomes

$$\text{Nu}_n = \frac{(n+1)/n!}{skF_m(t_0 - t_\infty)} \int_0^\infty \frac{Q(P)P^{n-1}e^{nP}\,dP}{A} \tag{17.4.6}$$

The value of Nu_n may be calculated for any kind of heating element. Geometry determines $Q(\tau_c) = Q(P)$ for any number of combined disturbance effects, $n + 1$.

Results were given in Gebhart (1963) for an extensive flat surface, a long cylinder, and a sphere. For the flat surface, resulting in a semi-infinite solid transient conduction response, the heat transfer in the first τ_c time interval of such a response lasting for time τ_c is

$$\frac{Q(P)}{A} = \frac{2k(t_0 - t_\infty)\sqrt{\tau_c}}{\sqrt{\pi\alpha}} = \frac{2sk(t_0 - t_\infty)}{\alpha}\sqrt{\frac{F}{\pi}} = \frac{2sk(t_0 - t_\infty)}{\alpha}\sqrt{\frac{F_m P}{\pi}} \tag{17.4.7}$$

Substituting this into Eq. (17.4.6), the effective Nusselt number and convection coefficient become

$$\text{Nu}_n = \frac{C_n}{F_m^{1/2}} \quad \text{and} \quad h_n = C_n\sqrt{\frac{\rho ck}{\tau_m}} \tag{17.4.8}$$

where τ_m is the mean or most probable value of τ_c. The result for a spherical surface is different from this only in the steady-state conduction contribution during τ_c, as follows:

$$\text{Nu}_n = 1 + \frac{C_n}{\sqrt{F_m}} \tag{17.4.9}$$

where the value of C_n in both relations above is

$$C_n = \frac{\sqrt{n}\,(2n-1)(2n-3)\cdots 3\cdot 1}{n!2^{n-1}} \tag{17.4.10}$$

The value of C_n across the range $n = 1$ to ∞ lies in the narrow range from 1 to 1.128. These solutions indicate that heat transfer, in a given circumstance, increases with increasingly frequent disturbing events, that is, small τ_c and τ_m.

17.5 CONJUGATE HEAT TRANSFER

In many natural convection transport processes, the conductive transport mechanisms in bounding walls are directly coupled with the convective processes. In analysis, boundary conditions arise at the interface. However, the two mechanisms generally are decoupled by expressing the boundary conditions in terms of a specified distribution of the temperature or the heat flux. Similarly, conduction in the solid region generally is analyzed by introducing an assumed heat transfer coefficient or surface temperature as one of the boundary conditions. The foregoing chapters present extensive analysis and results for flows adjacent to surfaces assumed to have either an isothermal, adiabatic, or uniform heat flux condition. Except for Chapter 7, where the coupling between the solid wall and the flow was considered, very little has been said about transport in which such uniform surface or bounding conditions may not be assumed. Then, the transport in the two regions jointly determines, or results in, the interface condition of temperature, flux, and so forth.

Despite the obvious importance of such heat transfer processes in practical problems, such as cooling of electronic circuitry and solidification in a mold (see Fig. 17.5.1), they have received very little attention. In many applications this is justified since the coupling effects are often small, as in flow over insulated or thin, highly conducting surfaces. Then the specification of a given temperature or heat flux distribution at the interface leads to analytical or numerical results of adequate accuracy.

Often, the coupling between two flows on either side of a thin wall is of considerable interest and importance. Such fluid-to-fluid conjugate heat transfer has been considered by Sparrow and Faghri (1980) and Sparrow and Prakash (1981), for natural convection on one side and forced convection on the other. Temperature and heat flux continuity requirements at the wall couple the two flows, leading to a significant increase in the complexity of the numerical solution obtained. This is a coupling of two convection processes in which the spatial distribution of convection coefficients differs on the two sides of the thin wall.

In coupled convection and conduction heat transfer, the thermal boundary condition at the interface also results from the interaction at the interface of the distributed processes in the two regions. This makes analysis fairly involved because of the elliptic nature of the conduction transport mechanisms. There has been some study of the interaction between forced convection in ducts and wall heat conduction; see the review by Shah and London (1978). Similarly, conjugate heat transfer involving external forced convection, that is, boundary layers on walls, has been studied, for instance, by Luikov (1974), Payvar (1977), and Sparrow and Chyu (1982). Very few studies have treated conjugate heat transfer with natural convection. In applications such as building insulation, electronic circuitry cooling, fin heat transfer, energy storage in enclosures, and furnace design, conjugate transport may be very important. This has stimulated a growing interest in recent years.

The natural convection boundary layer flow generated adjacent to a semi-

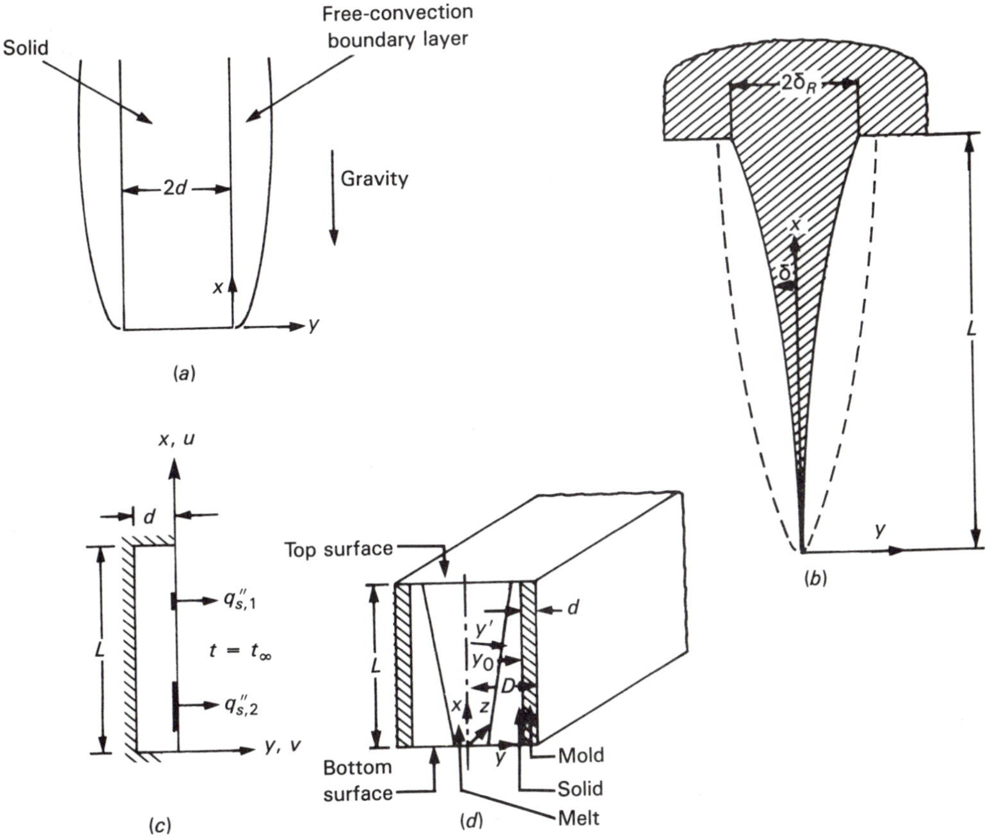

Figure 17.5.1 Geometry of some of the conjugate heat transfer flows that have been studied. *(a)* Kelleher and Yang (1967). *(Reprinted with permission from M. D. Kelleher and K.-T. Yang,* Appl. Sci. Res., *vol. 17, p. 249. Copyright © 1967, Martinus Nijhoff Publishers.) (b)* Lock and Gun (1968). *(Reprinted with permission from G. S. H. Lock and J. C. Gunn,* J. Heat Transfer, *vol. 90, p. 63. Copyright © 1968, ASME.) (c)* Zinnes (1970). *(Reprinted with permission from A. E. Zinnes,* J. Heat Transfer, *vol. 92, p. 528. Copyright © 1970, ASME.) (d)* Ramachandran et al. (1982). *(Reprinted with permission from* Int. J. Heat Mass Transfer, *vol. 25, p. 187, N. Ramachandran et al. Copyright © 1982, Pergamon Journals Ltd.)*

infinite vertical slab of finite thickness was considered by Kelleher and Yang (1967). The slab was assumed to contain an arbitrary distribution of heat sources, the energy generated being lost to the fluid by laminar natural convection under steady-state conditions. By using Fourier transforms for the conduction transport and a series solution procedure for the boundary layer equations, the temperature and heat flux distributions in the slab were determined. Lock and Gunn (1968) studied the laminar natural convection transport along a tapered, downward-projecting fin. One-dimensional conduction was assumed in the fin and the boundary layer approximations were made for the flow. Velocity and temperature profiles were obtained. Zinnes (1970) considered the flow along a vertical plate of finite thickness with specified uniform heat flux inputs at the surface and with the associated conductive transport in the plate. Figure 17.5.1*c* shows the geometry considered. The uniform flux condition results in a transverse temperature gradient at $y = 0$. This causes conduction internal to the plate.

The governing boundary layer equations for the natural convection fluid flow over the vertical surface shown in Fig. 17.5.1*c* were taken by Zinnes (1970) as

$$\frac{\partial u}{\partial x} + \frac{\partial v}{\partial y} = 0 \tag{17.5.1}$$

$$u\frac{\partial u}{\partial x} + v\frac{\partial u}{\partial y} = g\beta(t - t_\infty) + \nu\frac{\partial^2 u}{\partial y^2} \tag{17.5.2}$$

$$u\frac{\partial t}{\partial x} + v\frac{\partial t}{\partial y} = \alpha\left(\frac{\partial^2 t}{\partial x^2} + \frac{\partial^2 t}{\partial y^2}\right) \tag{17.5.3}$$

where the longitudinal conduction term is retained only in the energy equation. This includes the effect of large vertical temperature gradients in the fluid near the interface, due to the presence of localized heat sources. Boundary layer approximations were used for the momentum equation. The energy equation for steady conduction in the solid is

$$\frac{\partial^2 t}{\partial x^2} + \frac{\partial^2 t}{\partial y^2} = 0 \tag{17.5.4}$$

The plate was taken as insulated at the back, top, and bottom. The usual conditions for boundary layer flow arise, except for the interface thermal boundary condition at $y = 0$. This is now given by

$$\left(k\frac{\partial t}{\partial y}\right)_w - \left(k\frac{\partial t}{\partial y}\right)_f = q_s''(x) - q_r''(x) \tag{17.5.5}$$

Here the subscripts w and f refer to conduction in the plate and in the fluid. The imposed surface heat flux q_s'' is due to the heat input at the sources and q_r'' is the radiative loss out into the ambient environment.

Figure 17.5.2 shows the typical temperature variation with x and y obtained in the solid and in the fluid, taken as air. There is good agreement with the experimental results, which were obtained by using multiple isolated thermal sources located at $x/L = 0.25$, 0.5, and 0.75. In Fig. 17.5.2*b*, only two sources, at $x/L = 0.25$ and 0.75, are turned on, the third one being off. Peaks arise in the surface temperature at the heat sources and are found to be more pronounced for glass, which has a lower thermal conductivity than ceramic. The temperature profiles in the boundary region flow are also shown. For the case of a single source

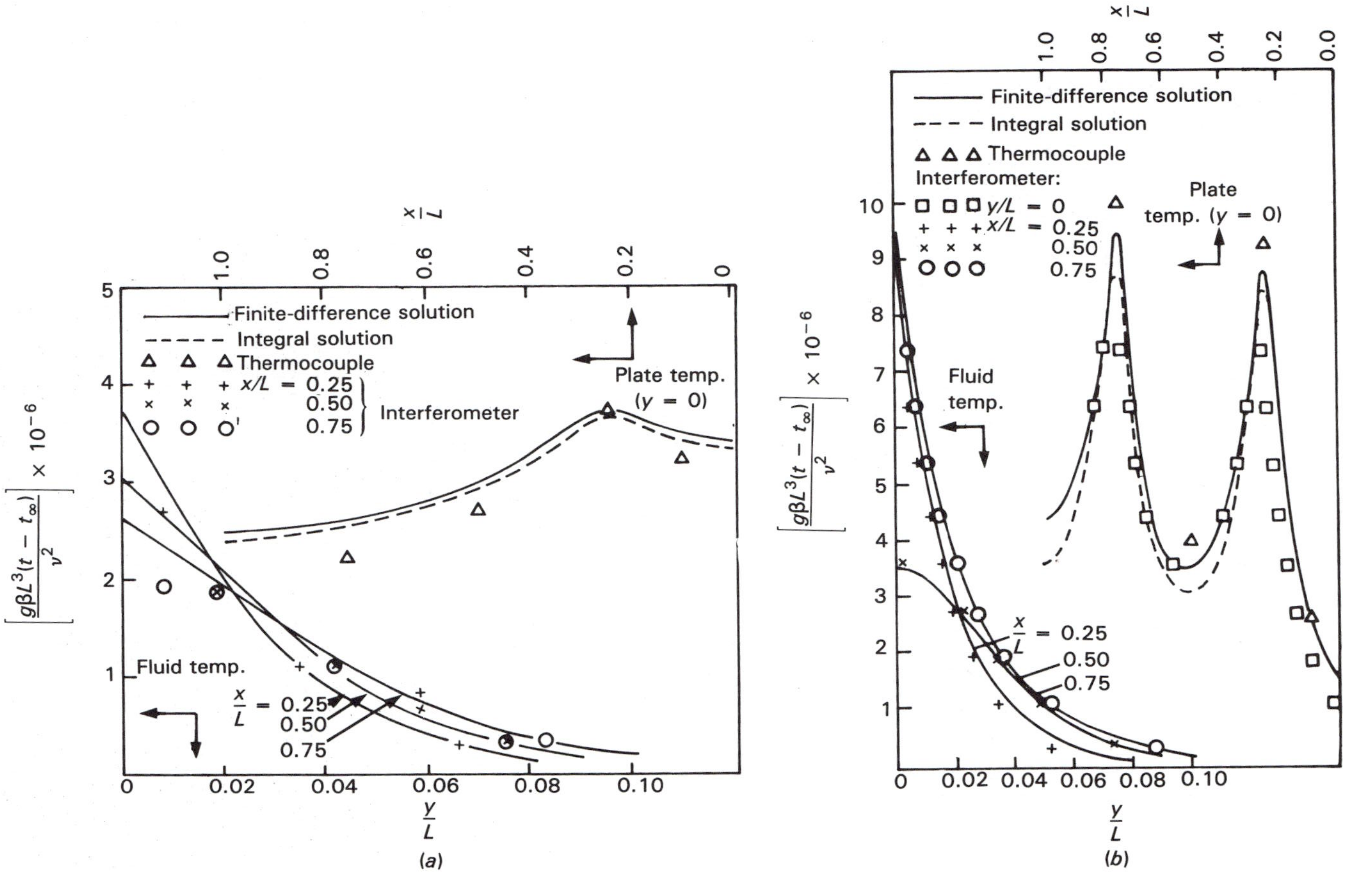

Figure 17.5.2 Plate and fluid temperature distributions: *(a)* ceramic plate with the bottom heater active; *(b)* glass plate with the upper and lower heaters active. *(Reprinted with permission from A. E. Zinnes,* J. Heat Transfer, *vol. 92, p. 528. Copyright © 1970, ASME.)*

at $x/L = 0.25$, seen in Fig. 17.5.2*a*, the temperature gradient in the fluid at the interface, $-(\partial t/\partial y)_f$, decreases with x, indicating the approach to adiabatic conditions far downstream. For multiple sources, this gradient increases in the vicinity of the downstream heaters. As expected, the boundary layer thickness increases downstream due to entrainment. At large x, in some cases, transfer of heat from the fluid to the plate occurs.

Additional work has been done on similar problems. Meyer et al. (1982) studied, numerically and experimentally, the effect of conduction in the end walls on the natural convection inside inclined rectangular cells, the other two enclosing surfaces being isothermal. It was found that the calculated wall convection heat transfer coefficients in cells with adiabatic end walls are substantially higher than those in cells with conducting end walls. Ramachandran et al. (1981, 1982) considered natural convection flow, with solidification, in a rectangular mold enclosure. The coupled conduction in the solid and in the mold was also considered. Numerical solutions were obtained for the time-dependent velocity and temperature distributions. The thermal characteristics of the solid and mold were found to have a strong effect on the flow for parametric ranges of practical interest.

Kwon et al. (1983) considered the conjugate problem for a short vertical plate fin attached along the bottom of a heated horizontal cylinder. There are many other circumstances in which conjugate heat transfer mechanisms are of particular interest. These include buoyancy-induced flows in room fires. Then heat transfer to the walls is an important consideration because of the transient nature of the problem, as outlined by Quintiere (1981). Heat conduction in the ground and sides of solar ponds has an important effect on the overall efficiency, requiring a solution of the conjugate problem; see the review by Nielsen (1979). Obviously, much remains to be done. There are very many applications of practical interest where conjugate transport is important.

17.6 COUPLED RADIATION EFFECTS

In the preceding chapters, the discussion of natural convection flows did not take into account other heat transfer modes or mechanisms that may arise simultaneously. The effects of the coupling of transport mechanisms in adjacent regions were discussed in the last section. Here we consider the simultaneous parallel effects of conductive-convective transport, coupled with radiation. In some conjugate transport circumstances such as those discussed in Section 17.5 (e.g., the boundary layer flow over a heated vertical surface), radiative heat transfer often becomes substantial even at relatively low temperatures, since the natural convection heat transfer rates are often small, especially in gases. Depending on the surface properties and the geometry, radiative transport is often comparable to or larger than the convective heat transfer in many practical situations. It is therefore important to determine its effect on the flow and the heat transfer.

An important class of transport concerns fluids such as air, inert gases, and nitrogen, which are largely nonabsorbing and nonemitting. Then the fluid convective and surface radiative transport mechanisms may be essentially independent of each other, although coupled through the boundary conditions. If the wall temperature is specified, the radiative transport may be computed independent of the flow, using available surface property data, temperatures, and relevant configuration factors. See, for example, Siegel and Howell (1981). Similarly, natural convection flow and transport may be determined for the given temperature condition and fluid properties.

However, if an interface heat flux q'' is imposed instead, the boundary condition couples the three transport mechanisms as

$$q'' = q''_w + q''_f + q''_r \tag{17.6.1}$$

where the subscripts w, f, and r refer to the transport inside the wall due to conduction, into the fluid due to convection, and into the surroundings due to radiation, respectively. For natural convection flow over a heated gray vertical surface in extensive surroundings at T_∞, this equation may be written as

$$q'' = \left(k\frac{\partial t}{\partial y}\right)_w - \left(k\frac{\partial t}{\partial y}\right)_f + \epsilon_w\sigma(T^4 - T_\infty^4) \tag{17.6.2}$$

where ϵ_w is the surface emissivity, σ the Stefan-Boltzmann constant, y the horizontal coordinate distance measured outward from the surface into the fluid, and T the absolute local surface temperature, which may vary over the surface. Similar expressions for the radiative heat transfer may be obtained for other applications of interest. See, for instance, the study by Jaluria (1982).

The nonlinear equation above couples the effects of the otherwise independent conductive, convective, and radiative transport processes. The surface temperature distribution $t_0(x)$ results from the interaction of all these modes and must be determined from a solution of the governing equations. Because of nonlinearity, iteration is usually needed to solve Eq. (17.6.2). If the conductive transport is small, as in insulation, it may be neglected. Often, the radiative heat transfer is not so simply formulated as above but must be obtained from integral or algebraic equations. Then the calculation is usually numerical for the velocity and temperature fields; see Jaluria and Torrance (1986).

A much more complicated and often more important transport arises when the convecting fluid absorbs and emits, as do ammonia, carbon dioxide, and water. These coupled radiation and natural convection transport processes arise in furnaces, natural water bodies, flames and fires, solar energy collection and storage, crystal growth and environmental heat transfer. Because of the importance of these processes, there has been considerable research effort in this area, as reviewed by Cess (1964), Viskanta (1966, 1982), Audunson and Gebhart (1972), and Buchberg et al. (1976). The extensive work on flames and combustion processes has also, inevitably, considered the interaction of gaseous thermal radiation with natural convection; see, for instance, the work of Negrelli et al. (1977), Ahmad and

Faeth (1978), Liu and Shih (1980), and Liu et al. (1981, 1982). Some of this work was outlined in Section 6.8.

Several of the natural convection flows considered in the preceding chapters have also been studied with coupled radiation effects. Cess (1966), Arpaci (1968), England and Emery (1969), Audunson and Gebhart (1972), and Bankston et al. (1977) have studied the natural convection boundary layer flow adjacent to a flat vertical surface in the presence of appreciable coupled radiation effects. With the usual boundary layer and Boussinesq approximations, the governing continuity and momentum equations remain the same as those for flow with negligible radiation effects, Eqs. (3.2.15) and (3.2.16). For plane flow and a one-dimensional radiation field q_r'', the energy equation becomes

$$\rho c_p\left(u\frac{\partial t}{\partial x}+v\frac{\partial t}{\partial y}\right)=k\frac{\partial^2 t}{\partial y^2}-\frac{\partial q_r''}{\partial y} \tag{17.6.3}$$

where q_r'' is the local outward radiation flux in the fluid region.

In analysis it is usually assumed that the surface radiates as a gray body and that the fluid is gray and absorbing but nonscattering. Since radiation is taken as a one-dimensional heat flux, relatively simple results may be obtained for the optically thick and thin approximations for the fluid. Cess (1966) used singular perturbation methods to study small radiation effects. Also, for the limiting case of an optically thin gray gas, England and Emery (1969) expressed local radiant absorption as

$$-\frac{\partial q_r''}{\partial y}=4a_f\phi(T_\infty^4-T^4) \tag{17.6.4}$$

where a_f is the absorption coefficient of the fluid, obtained from available charts for the given gas composition, pressure, and geometry. See Siegel and Howell (1981). Experimental results were also obtained for a uniform heat flux boundary condition in air and in carbon dioxide. Close agreement with the analytical results was obtained over the range of parameters considered. The local velocities and temperatures were found to increase slightly due to radiation effects and the temperature gradient at the wall decreased by less than 1%.

Arpaci (1968) used an integral analysis and considered absorption in the two limiting cases of the optically thin and the thick gas approximations for a vertical isothermal surface in a stagnant radiating gas. Thus, the gas domains from transparent to opaque were interpreted. For the optically thick gas model, q_r'' was approximated as

$$q_r''=-\frac{16\sigma}{3a_f}\left[1-\left(1-\frac{\epsilon_w}{2}\right)\exp\left(-\frac{3}{2}a_f y\right)\right]T^3\frac{\partial T}{\partial y} \tag{17.6.5}$$

where ϵ_w is the emissivity of the adjacent surface driving the convection. Figure 17.6.1 shows the effects on the local Nusselt number of the temperature and absorptivity of the gas, of the surface emissivity, and of $\lambda=(T_0-T_\infty)/T_\infty$, the ratio of the temperature difference between the wall and the ambient gas, to the ambient

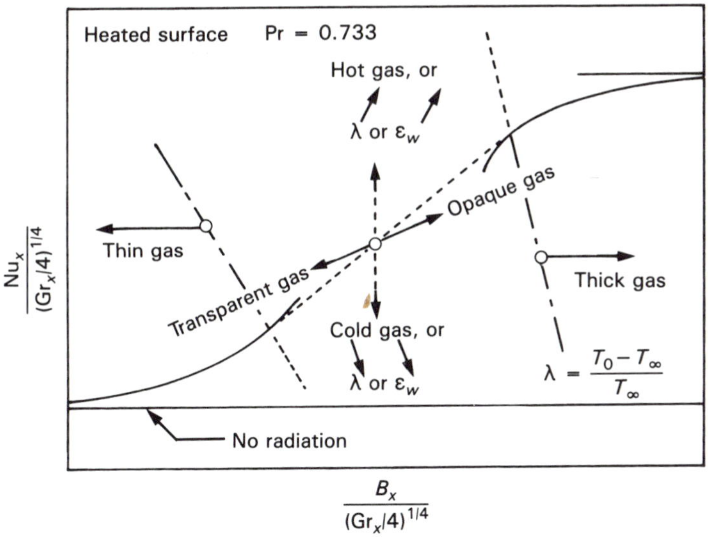

Figure 17.6.1 Variation of the local Nusselt number Nu with the Bouguer number B_x, where $B_x = a_f\delta$, δ being the boundary layer thickness. *(Reprinted with permission from* Int. J. Heat Mass Transfer, *vol. 11, p. 871, V. S. Arpaci. Copyright © 1968, Pergamon Journals Ltd.)*

gas temperature. The results are presented in terms of the variation of Nu_x with the Bouguer number B_x, given by $B_x = a_f\delta$, where a_f is the absorption coefficient of the fluid and $\delta(x)$ the local boundary layer thickness. With increasing B_x, the gas radiation characteristics move from transparent to opaque. Figure 17.6.1 shows that this results in an increase in the Nusselt number. The effects of the variables λ and ϵ_w are indicated by arrows, which show the change in the heat transfer rate. Three regimes arise in terms of B_x.

Audunson and Gebhart (1972) carried out an extensive experimental and analytical study of the natural convection boundary layer flow adjacent to a flat vertical surface with a uniform heat flux input. Both nonabsorbing and absorbing media—air, argon, and ammonia—were used in the experiments, which employed a Mach-Zehnder interferometer for the thermal field. A perturbation analysis yielded results in general agreement with the experimental results. The presence of a participating gas increased the convective heat transfer by as much as 40% for the experimental conditions studied. The effects of variations in the surface emissivity and in gaseous absorption and emission were also studied. A strong influence on the temperature field was found. The measured temperature distributions through the convective layer agreed closely with the theoretical predictions obtained by treating the convection and radiation processes as independent and superimposed.

Figure 17.6.2 shows the variation of the gas temperature gradient at the surface-gas interface, with the optical thickness τ_δ of the boundary layer, for various surface emissivity values ϵ_w. Here $\tau_\delta = \kappa_p\delta$, where κ_p is the absorption coefficient

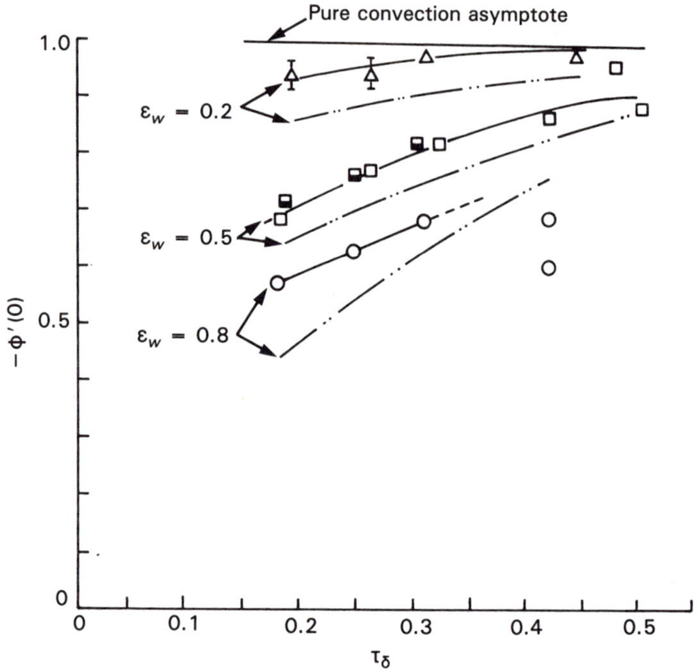

Figure 17.6.2 Variation of the temperature gradient at the wall, with the optical thickness τ_δ of the boundary layer and the surface emissivity, ϵ_w. (——) Experimental data; (– · · –) theoretical results. *(Reprinted with permission from T. Audunson and B. Gebhart,* J. Fluid Mech., *vol. 52, p. 57. Copyright © 1972, Cambridge University Press.)*

of the fluid medium and δ a characteristic boundary layer thickness. The total surface heat flux is the sum of the convective and radiative heat fluxes. The latter is reduced by absorption back at the surface. Convection is increased since this absorbed energy increases the energy conducted back again into the gas, increasing buoyancy. Therefore, depending on the fraction each mode represents in the total process, gaseous radiation may increase or decrease the total surface heat flux for a given temperature level.

Ali et al. (1984) analyzed the natural convection flow over a horizontal surface, considering the fluid as a gray, optically thick medium. Similarity was found to arise and the temperature and velocity profiles were computed. Radiation effects led to an increase in the temperature level in the flow for a heated surface facing upward.

Work has also been done on the radiation-convection interaction for nongray radiating fluids. Bratis and Novotny (1973) used limiting forms of radiation to approximate radiation band profiles in gases. Novotny et al. (1976) employed a local nonsimilarity method of analysis for nongray gases. Radiation in liquids is important in glass technology, solar ponds, and nuclear accident containment. Bankston et al. (1977) considered the total band absorption of absorbing-emitting liquids. The radiation-convection interaction in a liquid boundary layer flow adjacent to a uniform heat flux vertical surface was calculated, using local nonsimilarity methods for carbon tetrachloride. The computed effects of radiation on the surface temperature, $\phi(0, \xi)$, and on the temperature gradient at the wall, $\phi'(0, \xi)$, are shown in Fig. 17.6.3. The case of negligible thermal radiation, $\epsilon_w = 0$, is also shown for comparison. Here ξ is the local nonsimilarity variable, which varies with x, ϕ is the dimensionless local temperature, and ϕ_0 is the temperature in the absence of radiation effects. As expected, the effect of radiation increases as ϵ_w increases and also as the flow proceeds downstream.

Thermal instability. Several studies have considered this effect in horizontal and inclined fluid layers for gray fluids; see Arpaci and Gozum (1973) and Hassab and Ozisik (1979). Vertical slots have been considered by Lauriat (1980, 1982a), Balvanz and Kuehn (1980), and Hatfield and Edwards (1982). Two major effects have been predicted. Radiation delays the onset of instability in the vertical slot and also increases the heat transfer across the cavity. A differential formulation, called the modified P-1 approximation, was employed by Lauriat (1982b) for two-dimensional radiative transfer. However, it was found that this approximation underpredicts the interaction between convection and radiation. It was concluded that the approximation yields poor results for carbon dioxide and is inapplicable for water vapor.

It may be seen from the above discussion of natural convection flows with

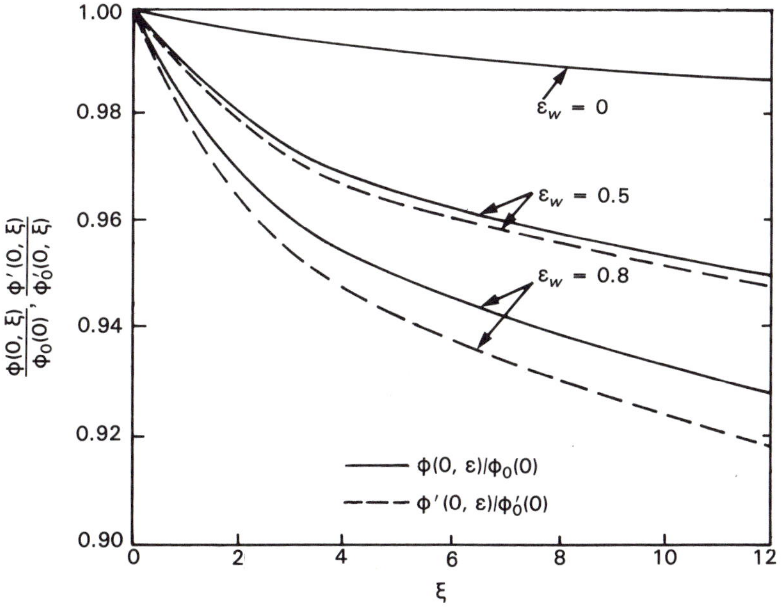

Figure 17.6.3 Effect of radiation on the wall temperature and the temperature gradient at the surface in liquid carbon tetrachloride for a uniform surface heat flux condition. *(Reprinted with permission from J. D. Bankston et al.,* J. Heat Transfer, *vol. 99, p. 125. Copyright © 1977, ASME.)*

coupled radiation effects that not much work has been done, perhaps because of the complexity arising in the formulation of radiative transport. Various approximations, considering both gray and nongray fluids, have been used to simplify the analysis. However, because of the importance of these processes in practical applications, such as furnaces, fires, and transport in natural water bodies, there is continually growing interest in natural convection with coupled radiative heat transfer.

17.7 ENTROPY GENERATION

The preceding chapters give a detailed account of the transport mechanisms, velocity and thermal fields, and resulting transport rates in a variety of buoyancy-induced flows in various configurations. In these flows the motive buoyancy force is due to the interaction of terrestrial gravity with density gradients in the fluid body. These density variations arise from gradients of temperature and concentration of chemical species. In general, for finite temperature and/or concentration differences, transport of momentum, thermal energy, and mass results. From a thermodynamic point of view, all such transport processes are irreversible.

A simple example of such an irreversible process is a small body of ice melting in warmer quiescent ambient water in an insulated container. If such a system is isolated, the phase change occurs in only one direction. Ambient thermal energy diffuses to the solid ice through the thermal boundary region in the warmer water, causing melting. After all the ice has melted, thermodynamic equilibrium is eventually established throughout the water. The system cannot, alone, return to its original state. Previous chapters indicate how to predict the rate of the ice melting and how long it will take to reach thermodynamic equilibrium.

The second law of thermodynamics quantifies the irreversibility of such transport processes. It may easily be shown that the entropy of such a system increases as melting progresses. The entropy is a maximum at equilibrium. Since the entropy of this isolated system cannot decrease, the melting process cannot reverse. The net total entropy increase is a quantitative measure of the irreversibility.

Buoyancy-induced transport always generates entropy. This arises from fluid friction and from heat and mass transfer through finite temperature and concentration differences. The same processes also generate entropy in forced convection flows. It is appropriate here to restate, in the context of entropy generation, a systematic difference between forced and buoyancy-induced convection. Consider forced convection flows with constant fluid properties. The velocity field, and hence the fluid friction, is then independent of the thermal field. However, coupling between them may arise due to the temperature- or concentration-dependent properties that influence the velocity field. Otherwise, the two entropy-generating mechanisms, fluid friction and heat transfer, are independent of each other. On the other hand, in natural convection flows, the velocity field is a consequence of temperature and/or concentration gradients. Therefore, the several mechanisms that contribute to entropy generation are directly coupled with each other.

It is often important and useful to assess heat transfer processes from a thermodynamic viewpoint. All transfer processes and devices are inherently irreversible and result in a one-way destruction of useful or available energy, sometimes called exergy. A growing awareness of energy conservation has led to interest in how much exergy is dissipated in heat transfer and how better thermodynamic efficiency may be achieved. The laws of thermodynamics are applied. The first law has established the energy equation and the second law is often not used to analyze convective transport. However, to determine the conditions under which there is a minimum loss of exergy, that is, minimum entropy generation, the second law is used. Such analysis of various thermal processes is discussed in detail by Bejan (1982).

Evaluation of the entropy production rate. The rate of local entropy generation in convective circumstances may be obtained by applying the second law to a differential control volume in fluid engaged in convective heat transfer. The second law for such an open system takes the form

$$\sum_{\text{in}} \dot{m}s - \sum_{\text{out}} \dot{m}s + \frac{q}{T} \leq \frac{\partial S}{\partial \tau} \tag{17.7.1}$$

where $\dot{m}$, s, S, q, and T are the mass flow rate, specific entropy, total entropy, heat transferred to the control volume, and absolute temperature, respectively. The left and the right sides of Eq. (17.7.1) represent the nominal entropy transfer and the actual entropy changes, respectively. The equality applies only for a completely reversible process. To determine the degree of irreversibility, $\dot{S}_{\text{gen}}$, it is convenient to rearrange Eq. (17.7.1) as

$$\dot{S}_{\text{gen}} = \frac{\partial S}{\partial \tau} - \frac{q}{T} + \sum_{\text{out}} \dot{m}s - \sum_{\text{in}} \dot{m}s \geq 0 \tag{17.7.2}$$

Entropy production in a laminar flow. The term $\dot{S}_{\text{gen}}$ represents the entropy generation rate due to irreversibility. It is zero for a reversible process. The physical effects formulated for a differential control volume are substituted into Eq. (17.7.2). The result is simplified by using the continuity and the energy equations. The following equation for the volumetric entropy generation rate $\dot{S}'''_{\text{gen}}$ results:

$$\dot{S}'''_{\text{gen}} = \frac{k}{T^2}(\nabla t)^2 + \frac{\mu}{T}\Phi \tag{17.7.3}$$

where Φ, the viscous dissipation function, is given by Eq. (2.1.4). The details of the derivation of Eq. (17.7.3), omitted here for brevity, are given in Chapter 5 of Bejan (1982). For transport described in a two-dimensional Cartesian system, Eq. (17.7.3) reduces to

$$\dot{S}'''_{\text{gen}} = \frac{k}{T^2}\left[\left(\frac{\partial t}{\partial x}\right)^2 + \left(\frac{\partial t}{\partial y}\right)^2\right] + \frac{\mu}{T}\left\{2\left[\left(\frac{\partial u}{\partial x}\right)^2 + \left(\frac{\partial v}{\partial y}\right)^2\right] + \left(\frac{\partial v}{\partial x} + \frac{\partial u}{\partial y}\right)^2\right\} \tag{17.7.4}$$

Equations (17.7.3) and (17.7.4) show that the irreversibility is due to two effects, energy conduction and viscous dissipation. If any temperature or velocity gradients are present, $\dot{S}'''_{gen}$ becomes positive and finite. It is important to consider the relative contribution of these two factors to the total entropy generation rate. In Section 2.6 it was shown that the viscous dissipation effects, when compared with conduction, may often be neglected in the energy equation. For boundary region transport, the leading terms from Φ and from the conduction term $k\nabla^2 t$ in the energy equation are $\mu(\partial u/\partial y)^2$ and $k\,\partial^2 t/\partial y^2$, respectively. Their ratio R_4 is given by Eq. (2.6.3) as

$$R_4 = \frac{\mu(\partial u/\partial y)^2}{k(\partial^2 t/\partial y^2)} = O\left[\frac{\mu u^2}{k(t_0 - t_\infty)}\right] = O\left(\Pr \frac{g\beta L}{c_p}\right) \tag{17.7.5}$$

Following Eq. (2.5.8), the term $g\beta L/c_p$ was seen to be very small for terrestrial gravity. Thus, for $\Pr \leq 1$ the ratio $R_4 << 1$. This conclusion also holds for large Prandtl number fluids since $g\beta L/c_p << 1$. Although it is common to neglect the viscous dissipation effects in the energy equation, their exclusion from the entropy generation, Eq. (17.7.4), needs more careful consideration. Comparison of the leading viscous and conduction terms of irreversibility in Eq. (17.7.4) yields

$$R_6 = \frac{(\mu/T)(\partial u/\partial y)^2}{(k/T^2)(\partial t/\partial y)^2} = O\left[\frac{\mu T u^2}{k(t_0 - t_\infty)^2}\right]$$

$$= O\left(\frac{T}{t_0 - t_\infty}\frac{\Pr g\beta L}{c_p}\right) = O\left(\frac{T}{t_0 - t_\infty} R_4\right) \tag{17.7.6}$$

where T is some characteristic absolute temperature level. From Eq. (17.7.6) it can be seen that, although R_4 may be much smaller than unity in many applications, the relative fluid friction irreversibility effect, R_6, is not necessarily negligible, compared to heat transfer irreversibilities. In many convective flow circumstances the ratio $T/(t_0 - t_\infty)$ is considerably greater than unity. Thus, in the entropy generation analysis of convection, both entropy-generating factors must be considered.

From the above discussion, two dimensionless parameters appear in the entropy generation analysis. These are the dimensionless temperature difference $(t_0 - t_\infty)/T$, related to R_1, and the ratio of entropy generation rates due to friction and heat transfer, $\dot{S}'''_{gen,f}/\dot{S}'''_{gen,h}$, related to R_4.

Other aspects. For flows in which the velocity and temperature are known at each point in the medium, the local volumetric entropy generation rate $\dot{S}'''_{gen}$ may be calculated from Eq. (17.7.3) or (17.7.4). Such a procedure is suitable when exact or finite-difference solutions of the flow equations are available, as they often are for laminar flows. On the other hand—for example, for turbulent transport—the velocity and temperature fields are frequently unknown. However, heat transfer and fluid friction data along fluid-solid interfaces are available for some such flows of practical importance. With this information, the entropy generation rate may be calculated by an integral balance in the region of transport.

Using both of these procedures, Bejan (1979) reported the entropy generation rates for several internal and external forced convection flows. As discussed earlier, in forced flows the two entropy-generating factors, fluid friction and heat transfer, are often independent of each other. Hence, an optimization procedure may be used. The surface geometry—plate length, tube diameter, and so forth—may be adjusted to yield minimum entropy generation for given values of heat transfer and flow velocities. This irreversibility minimization, in terms of constraints such as a required amount of heat transfer, heat transfer area, volume of heat exchanging device, and total pressure drop, may sometimes be applied to both internal and external forced flows.

For buoyancy-induced transport, the entropy generation rates may also be calculated by the procedures discussed above, from the general equations or from local considerations. However, then the irreversibility-generating mechanisms, heat transfer and fluid friction, directly interact. They are not independent of each other. This requires a much more complicated treatment. Nevertheless, there is sufficient information for many buoyancy-induced processes to make such determinations and optimizations therefrom.

REFERENCES

Ahmad, T., and Faeth, G. M. (1978). *J. Heat Transfer 100,* 112.

Ali, M. M., Chen, T. S., and Armaly, B. F. (1984). *AIAA J. 22,* 1797.

Arpaci, V. S. (1968). *Int. J. Heat Mass Transfer 11,* 871.

Arpaci, V. S., and Gozum, D. (1973). *Phys. Fluids 16,* 581.

Audunson, T., and Gebhart, B. (1972). *J. Fluid Mech. 52,* 57.

Balvanz, J. L., and Kuehn, T. H. (1980). In *Natural Convection in Enclosures, ASME HTD Symp.,* vol. 8, p. 55.

Bankston, J. D., Lloyd, J. R., and Novotny, J. L. (1977). *J. Heat Transfer 99,* 125.

Barcilon, V., and Pedlosky, J. (1967). *J. Fluid Mech. 29,* 673.

Bejan, A. (1979). *J. Heat Transfer 101,* 718.

Bejan, A. (1982). *Entropy Generation Through Heat and Fluid Flow,* Wiley, New York.

Bratis, J. C., and Novotny, J. L. (1973). *Prog. Astronaut. Aeronaut. 31,* 329.

Buchberg, H., Catton, I., and Edwards, D. K. (1976). *J. Heat Transfer 98,* 182.

Bühler, K., and Oertel, H. (1982). *J. Fluid Mech. 114,* 261.

Busse, F. H., and Whitehead, J. A. (1974). *J. Fluid Mech. 66,* 67.

Cess, R. D. (1964). *Adv. Heat Transfer 1,* 1.

Cess, R. D. (1966). *Int. J. Heat Mass Transfer 9,* 1269.

Chandrasekhar, S. (1961). *Hydrodynamic and Hydromagnetic Stability,* Oxford Univ. Press (Clarendon), London.

Chen, C. F., and Kirchner, R. P. (1971). *J. Fluid Mech. 48,* 365.

Chen, C. F., and Sandford, R. D. (1977). *J. Fluid Mech. 83,* 83.

Cramer, K. R. (1962). ASME Paper 62-HT-22. Also, Cramer, K. R. (1963). *J. Heat Transfer 85,* 35.

Davis, S. H. (1976). *Annu. Rev. Fluid Mech. 8,* 57.

Dorfman, L. A. (1963). *Hydrodynamic Resistance and the Heat Loss of Rotating Solids,* Oliver & Boyd, Edinburgh.

Dropkin, D., and Gelb, G. (1964). *J. Heat Transfer 86,* 203.

Dropkin, D., and Globe, S. (1959). *J. Appl. Phys. 30,* 84.

England, W. G., and Emery, A. F. (1969). *J. Heat Transfer 91,* 37.
Gebhart, B. (1963). *AIAA J. 1,* 380.
Gershuni, G. Z., and Zhukhovitskii, E. M. (1958). *Sov. Phys. JETP 34*(7), 461.
Globe, S., and Dropkin, D. (1959). *J. Heat Transfer 81,* 24.
Grodzka, P. G., and Bannister, T. C. (1972). *Science 176,* 506.
Grodzka, P. G., and Bannister, T. C. (1974). *Natural Convection in Low-g Environments,* AIAA Paper 74-156.
Hassab, M. S., and Ozisik, M. N. (1979). *Int. J. Heat Mass Transfer 22,* 1095.
Hatfield, D. W., and Edwards, D. K. (1982). *Int. J. Heat Mass Transfer 25,* 1363.
Hering, R. G., and Grosh, R. J. (1963). *J. Heat Transfer 85,* 29.
Homsy, G. M., and Hudson, J. L. (1969). *J. Fluid Mech. 35,* 33.
Homsy, G. M., and Hudson, J. L. (1971a). *J. Fluid Mech. 48,* 605.
Homsy, G. M., and Hudson, J. L. (1971b). *Int. J. Heat Mass Transfer 14,* 1149.
Hudson, J. F., Tang, D., and Abell, S. (1978). *J. Fluid Mech. 86,* 147.
Jaluria, Y. (1982). *Comput. Fluids 10,* 95.
Jaluria, Y., and Torrance, K. E. (1986). *Computational Heat Transfer,* Hemisphere, Washington, D.C.
Jhaveri, B., and Homsy, G. M. (1980). *J. Fluid Mech. 98,* 324.
Jhaveri, B., and Homsy, G. M. (1982). *J. Fluid Mech. 114,* 251.
Kelleher, M. D., and Yang, K.-T. (1967). *Appl. Sci. Res. 17,* 249.
Kreith, F. (1968). *Adv. Heat Transfer 5,* 129.
Kreith, F., and Kneisel, K. (1963). *6th Natl. Heat Transfer Conf.,* Boston, AIChE Paper 44.
Kuiken, H. K. (1970). *J. Fluid Mech. 40,* 21.
Kwon, S. S., Kuehn, T. H., and Tolpadi, A. K. (1983). ASME Paper 83-HT-100.
Lauriat, G. (1980). In *Natural Convection in Enclosures, ASME HTD Symp.,* vol. 8, p. 63.
Lauriat, G. (1982a). *J. Heat Transfer 104,* 609.
Lauriat, G. (1982b). *Proc. 7th Int. Heat Transfer Conf.,* Paper NC6.
Lehnert, B., and Little, N. C. (1957). *Tellus 9,* 97.
Liu, C. N., and Shih, T. M. (1980). *J. Heat Transfer 102,* 724.
Liu, K. V., Lloyd, J. R., and Yang, K. T. (1981). *Int. J. Heat Mass Transfer 24,* 1959.
Liu, V. K., Yang, K. T., and Lloyd, J. R. (1982). *Int. J. Heat Mass Transfer 25,* 863.
Lock, G. S. H., and Gunn, J. C. (1968). *J. Heat Transfer 90,* 63.
Lugt, H. J., and Schwiderski, E. W. (1965). *Q. Appl. Math. 23,* 130.
Lugt, H. J., and Schwiderski, E. W. (1966). *J. Atmos. Sci. 23,* 54.
Luikov, A. V. (1974). *Int. J. Heat Mass Transfer 17,* 257.
Lykoudis, P. L. (1962). *Int. J. Heat Mass Transfer 5,* 23.
Meyer, B. A., Mitchell, J. W., and El-Wakil, M. M. (1982). *J. Heat Transfer 104,* 111.
Nakagawa, Y. (1955). *Nature (London) 175,* 417.
Nakagawa, Y. (1957). *Proc. R. Soc. Ser. A 240,* 108.
Nakagawa, Y. (1960). *Rev. Mod. Phys. 32,* 916.
Nakagawa, Y., and Goroff, I. R. (1961). *Phys. Fluids 4,* 342.
Negrelli, D. E., Lloyd, J. R., and Novotny, J. L. (1977). *J. Heat Transfer 99,* 212.
Newell, A. C., Lange, C. G., and Aucoin, P. J. (1970). *J. Fluid Mech. 40,* 513.
Nielsen, C. E. (1979). Nonconvective Salt-Gradient Solar Ponds, in *Solar Energy Handbook,* Marcel Dekker, New York.
Novotny, J. L., Lloyd, J. R., and Bankston, J. D. (1976). *Thermophysics and Heat Transfer Conf.,* Boston, AIAA Paper 74-653.
Payvar, P. (1977). *Int. J. Heat Mass Transfer 20,* 431.
Poots, G. (1961). *Int. J. Heat Mass Transfer 3,* 1.
Popiel, C. O., and Bosuslawski, L. (1975). *Int. J. Heat Mass Transfer 18,* 167.
Quintiere, J. G. (1981). *Fire Safety J. 3,* 201.
Ramachandran, N., Gupta, J. P., and Jaluria, Y. (1981). *Numer. Heat Transfer 4,* 469.
Ramachandran, N., Gupta, J. P., and Jaluria, Y. (1982). *Int. J. Heat Mass Transfer 25,* 187.

Riley, N. (1964). *J. Fluid Mech. 18,* 577.
Romig, M. F. (1964). *Adv. Heat Transfer 1,* 267.
Rossby, H. T. (1969). *J. Fluid Mech. 36,* 309.
Rotem, Z., and Claassen, L. (1970). *4th Int. Heat Transfer Conf.,* Versailles, IV Paper NC4.3.
Schmidt, E., and Leidenfrost, W. (1953). *Forsch. Geb. Ingenieurwes. 19,* 65.
Senftleben, H., and Braun, W. (1936). *Z. Phys. 102,* 480.
Shah, R. K., and London, A. L. (1978). *Laminar Flow Forced Convection in Ducts,* Academic Press, New York.
Shen, S. F. (1961). *J. Aerosp. Sci. 28,* 397, 417.
Siegel, R., and Howell, J. R. (1981). *Thermal Radiation Heat Transfer,* Hemisphere, Washington, D.C.
Singh, K. R., and Cowling, T. G. (1963). *Q. J. Mech. Appl. Math. 16,* 1, 16.
Sparrow, E. M., and Cess, R. D. (1961). *Int. J. Heat Mass Transfer 3,* 267.
Sparrow, E. M., and Chyu, M. K. (1982). *J. Heat Transfer 104,* 204.
Sparrow, E. M., and Faghri, M. (1980). *J. Heat Transfer 102,* 402.
Sparrow, E. M., and Prakash, C. (1981). *Int. J. Heat Mass Transfer 24,* 895.
Spradley, L. W. (1974). *Thermoacoustic Convection of Fluids in Low Gravity,* AIAA Paper 74-76.
Stratton, J. A. (1941). *Electromagnetic Theory,* McGraw-Hill, New York.
Suwono, A. (1980). *Int. J. Heat Mass Transfer 23,* 819.
Tang, D., and Hudson, J. L. (1983). *Int. J. Heat Mass Transfer 26,* 943.
Viskanta, R. (1966). *Adv. Heat Transfer 3,* 175.
Viskanta, R. (1982). *Proc. 7th Int. Heat Transfer Conf. 1,* 103.
Zinnes, A. E. (1970). *J. Heat Transfer 92,* 528.

PROBLEMS

17.1 Consider a vertical cylindrical tank of 5 cm diameter and 1 cm height, with closed ends, rotating at N rpm about its axis of symmetry. The top and bottom surfaces are at 20 and 30°C, respectively, and the vertical cylindrical wall is made of insulating material. The tank contains water.

(a) Calculate the relevant Grashof, Rayleigh, Ekman, and Taylor numbers and estimate the ratio of the convection and rotational velocities.

(b) Determine the range of the rotation rate N for which the heat transfer is driven essentially by buoyancy effects alone.

(c) Compare the resulting Nusselt number in part (a) with that given for unstably stratified layers in Chapter 13.

(d) What rotation rate gives marginal stability?

(e) What rate of rotation results in a minimum heat transfer?

17.2 Consider the conditions given in Problem 17.1, but with the top surface heated instead.

(a) Describe the direction of the rotational effect on the flow.

(b) If the fluid is silicone of viscosity 0.65 centistokes, determine the rotation rate and the imposed temperature difference such that $A = 50$ and $\beta\, \Delta t \mathrm{Pr}\sqrt{\mathrm{Ta}} = 0.1$.

(c) Calculate the resulting Nusselt number and heat transfer rate.

(d) What would the value of Nu be in the absence of rotation?

17.3 A downward-facing cone with a vertex angle of 45° and surface temperature of 40°C is in quiescent air at 20°C. Over what range of rotational speed would buoyancy effects be appreciable?

17.4 A vertical surface 25 cm high and 25 cm wide at 30°C is in an extensive ambient of liquid mercury at 20°C. There may be a uniform horizontal magnetic induction B_0.

(a) Calculate the Nusselt number and heat transfer rate per side for $B_0 = 0$.

(b) Calculate the necessary value of the Hartmann number and the induction B_0 for a reduction of 25% in heat transfer.

17.5 For the ambient conditions t_∞ and B_0 in Problem 17.4, assume that the surface temperature varies vertically as $t_0 - t_\infty = Nx = 0.4x$, where x is in centimeters. For Hartmann number $M = 2.5\sqrt[4]{\mathrm{Gr}_L}$

(a) Determine the maximum velocity in the boundary region.

(b) Estimate the surface heat flux near the trailing edge of the surface.

(c) Calculate the relevant value of B_0.

17.6 In a materials processing experiment in an earth-orbiting space laboratory, spherical crystals of 0.1 mm diameter are growing while suspended in a large supersaturated water solution of the same material. The solidification alone would result in a temperature difference of 1°C between the water solution, at 20°C, and the growing crystals. The density ratio of the crystals and water solution is 4.2. Due to spacecraft controls and other perturbations of the experiment, random displacements of 1 mm occur. Their most probable time interval is 5 s.

(a) Calculate the Nusselt number and the heat transfer rate q between the solution and a crystal if the disturbance events are exponentially distributed.

(b) Calculate the maximum range of q over all probability distributions.

17.7 Repeat part (a) of Problem 17.6 over the crystal diameter range from 10^{-2} to 0.5 mm and explain the variations of q and Nu.

17.8 Consider the conjugate heat transfer circumstance diagrammed in Fig. 17.5.1*c*, in air, with a ceramic plate. If only the lower heater is active, at $x/L = 0.25$, estimate the maximum temperature at the surface and the fraction of input energy lost by conduction to the plate. Take $L = 1$ m and $q''_{s,2} = 400\ \mathrm{W/m^2}$, with a source height of 2 cm.

17.9 A 20-cm-high vertical surface at 40°C is in quiescent ambient gas at 20°C and 10 atm pressure. Assume Pr = 0.733 and use the properties of air.

(a) Calculate the upper limit of the absorption coefficient of the gas for negligible radiation effects.

(b) Repeat part (a) for the highest value of the coefficient to remain in the optically thin regime.

17.10 Calculate the rate of entropy production that accompanies the heat transfer processes specified in

(a) Problem 3.5
(b) Problem 4.2
(c) Problem 5.5
(d) Problem 9.6

NOMENCLATURE

a	half distance between two parallel plates, radius, species disturbance amplitude function
A	area, disturbance amplitude $(= -\int_{x_N}^{x} \alpha_i \, dG/4)$
A_c	flow cross-sectional area
$b(x)$	η/y
B, $\mathbf{B}$	scalar or vector buoyancy force
B_0	magnetic induction
$c(x)$	x-dependent function in Eq. (3.3.7)
c_p	specific heat at constant pressure
c_v	specific heat at constant volume
c'''	species rate of production
C	concentration of chemical species
C'	concentration disturbance
$\tilde{C}$	dimensionless concentration $[=(C - C_\infty)/(C_0 - C_\infty)]$
C_D	drag coefficient
d	$t_0 - t_\infty$, depth of layer, an enclosure dimension
D	species diffusion coefficient, diameter, drag
D_h	hydraulic diameter $(=4A_c/P)$
e	$C_0 - C_\infty$
E	transition parameter $[=G^*(\nu^2/gx^3)^{2/15}]$
Ek	Ekman number $(=\nu/2\Omega L^2)$
f	stream function in terms of η
F	disturbance frequency
Fo	Fourier number $(=\alpha\tau/L^2)$
Fr	Froude number
Fr_L	local Froude number, see Eq. (12.4.2)
g, $\mathbf{g}$	scalar or vector acceleration due to gravity

G	$4(\mathrm{Gr}_x/4)^{1/4}$
G_θ	$4(\mathrm{Gr}_{x,\theta}/4)^{1/4}$
G^*	$5(\mathrm{Gr}_x^*/5)^{1/5}$
G_1	$4(\mathrm{Gr}_x'/4)^{1/4}$, $G(1 + \tilde{N})^{1/4}$
Gr	Grashof number $[=(gL^3/\nu^2)(\Delta\rho/\rho) = (gL^3/\nu^2)\beta(t_0 - t_\infty)$ for Boussinesq approximation; $= g\beta(t_h - t_c)d^3/\nu^2$ in enclosures]
Gr′	Grashof number $\{=(gL^3/\nu^2)[\alpha(s, p)d^q]\}$
Gr*	flux Grashof number $(=g\beta q''L^4/k\nu^2)$
Gr_m	Grashof number, with exponential variation of $t_0 - t_\infty$
Gr_x	local Grashof number $[=(gx^3/\nu^2)(\Delta\rho/\rho) = (gx^3/\nu^2)\beta(t_0 - t_\infty)]$
$\mathrm{Gr}_{x,\theta}$	$\mathrm{Gr}_x \cos\theta$
Gr_x^*	flux local Grashof number $(=g\beta q''x^4/k\nu^2)$
$\mathrm{Gr}_{x,C}$	local species Grashof number $[=g\beta x^3(C_0 - C_\infty)/\nu^2]$
Gr_x'	Grashof number $\{=(gx^3/\nu^2)[\alpha(s, p)d^q]\}$, combined Grashof number $(=\mathrm{Gr}_x + \mathrm{Gr}_{x,C})$
Gz	Graetz number
h	heat transfer coefficient based on characteristic length L
h_x	local heat transfer coefficient
H	height, stratification length, Eq. (2.5.16)
H'	stratification length, Eq. (2.5.16)
He	Hedstrom number $(=\rho\tau_0 L^2/\mu_B^2)$
j	$t_\infty - t_r$
k	thermal conductivity
k_T	thermal diffusion ratio
K	permeability, Eq. (15.1.1)
L	size or characteristic length
Le	Lewis number $(=\mathrm{Sc}/\mathrm{Pr} \equiv \alpha/D)$
m''	mass flux rate
m'''	mass source generation rate
M	as defined in $d = Me^{mx}$, momentum flow, Hartmann number
N, n	as defined in $d = Nx^n$
$\tilde{N}$	$\mathrm{Gr}_{x,C}/\mathrm{Gr}_x$
Nu	Nusselt number based on the average heat transfer coefficient h $(=hL/k)$
Nu_x	local Nusselt number $(=h_x x/k)$
$\mathrm{Nu}_{x,C}$	local Sherwood number $(=h_{x,C}x/D)$
p	pressure
p_h	hydrostatic pressure
p_m	motion pressure
P	dimensionless pressure, perimeter
Pr	Prandtl number $(=\mu c_p/k = \nu/\alpha)$
q	exponent in density Eq. (9.1.1)
q''	heat flux
q'''	volumetric heat generation rate
$Q(x)$	local heat flow or thermal energy convection rate

Q	energy input per unit span per side in two-dimensional plume and total energy input for axisymmetric plume
Q^*	nondimensional thermal capacity parameter $[=(c''/\rho c_p x)\,\mathrm{Pr}(\mathrm{Gr}_x^*)^{1/4}]$
r	$C_\infty - C_r$, horizontal distance from the surface element to the axis of symmetry (e.g., Fig. 5.1.2)
R	gas constant, cylinder or sphere radius, also $(t_m - t_\infty)/(t_0 - t_\infty)$
R^*	$5\mathrm{Re}/(\mathrm{Gr}_x^*)^{1/4}$
Ra	Rayleigh number (=GrPr)
Ra*	flux Rayleigh number (=Gr*Pr)
Ra_x	local Rayleigh number $(=\mathrm{Gr}_x\mathrm{Pr})$
Ra_x^*	flux local Rayleigh number $(=\mathrm{Gr}_x^*\mathrm{Pr})$
Ra′	Darcy modified Rayleigh number $[=g\beta\rho_\infty K(t_0 - t_\infty)L/\mu\alpha]$
Ra_x'	local Darcy modified Rayleigh number $(=\mathrm{Ra}'x/L)$
Re	Reynolds number $(=uL/\nu)$
Re_x	local Reynolds number $(=ux/\nu)$
Ro	Rossby number $(=U_c/\Omega L)$
s	salinity, temperature disturbance amplitude function, length, specific entropy
S	dimensionless salinity, entropy, stratification parameter
Sc	Schmidt number $(=\nu/D)$
t	temperature
t_{il}	equilibrium ice-liquid temperature
t_m	medium or extremum temperature
t'	temperature disturbance
Δt	$t_0 - t_\infty$
T	absolute temperature
Ta	Taylor number $[=(\mathrm{Ek})^{-2}]$
u_m	mean velocity for internal flows
u, v, w	velocity components in x, y, z directions, respectively
$\bar{u}, \bar{v}$	average values of base flow velocity in x, y directions, respectively
u', v', w'	components of velocity disturbance in x, y, z directions, respectively
U, V, W	dimensionless velocity components in x, y, z directions, respectively
W	local buoyancy force defined in Eq. (9.3.15)
U_c	characteristic flow velocity
$\mathbf{V}$	velocity vector
x	coordinate along flow direction
X	dimensionless coordinate along flow direction
y	coordinate normal to flow direction
Y	dimensionless coordinate normal to flow direction
z	transverse coordinate

Greek symbols

α	molecular thermal diffusivity $(=k/\rho c_p)$; dimensionless wave number; entrainment function; $=\alpha(s, p)$ in saline water density relation
$\hat{\alpha}$	wave number $(=\hat{\alpha}_r + i\hat{\alpha}_i)$

β volumetric coefficient of thermal expansion
β^* species expansion coefficient
γ ratio of specific heats ($=c_p/c_v$)
δ velocity boundary layer thickness
δ_t thermal boundary layer thickness
Δ, Δ_t, Δ_C δ/L, δ_t/L, δ_C/L
ϵ perturbation parameter, emissivity
ϵ_m eddy viscosity
ϵ_t turbulent thermal diffusivity
η similarity parameter
θ angle of inclination from vertical, angle of incidence, transverse wave number
λ wavelength
μ dynamic viscosity
ν kinematic viscosity
ξ angle of inclination from horizontal, positive upward
ρ density
τ time, shear stress
τ_0 time constant for transients
$\boldsymbol{\tau}'$ stress tensor
τ_{ij} components of stress tensor
ϕ dimensionless temperature excess [$=(t - t_\infty)/(t_0 - t_\infty)$], also Eq. (7.2.40)
Φ viscous dissipation term, Eq. (2.1.3), velocity disturbance amplitude function
ψ stream function in terms of x and y
ψ' disturbance stream function in x and y
ω dimensionless disturbance frequency
$\hat{\omega}$ disturbance frequency ($=2\pi F$)
ω_1 $\omega(1 + \tilde{N})^{-3/4}$
Ω generalized frequency, angular velocity
Ω^* generalized frequency for uniform flux surface condition

Subscripts

C concentration
c critical condition
ET end of transition
f quantities evaluated at film temperature
h hydrostatic
i imaginary part
m motion, mean, ambient medium
N neutral
0 imposed surface or midplane condition
r reference condition, real part, radiation

t	thermal
TT	beginning of thermal transition
VT	beginning of velocity transition
x	at x
∞	in distant medium
δ	at edge of boundary layer

APPENDIX
A
CONVERSION FACTORS

Table A.1 Conversion from English to metric units

Quantity	To convert number of	To	Multiply by
Length	in.	cm	2.540
	ft	m	0.3048
Area	ft^2	m^2	0.0929
Volume	ft^3	m^3	0.02832
Mass	lbm	kg	0.45359
	slugs	kg	14.594
Force	lbf	newtons	4.4482
Density	lbm/ft^3	kg/m^3	16.02
Work	ft lbf	mkgf	0.1383
	hp h	mkgf	273.700
Heat	Btu	kcal	0.2520
	Btu	joules	1054.35
	Btu	ft lbf	778.26
	kW hr	Btu	3412.75
Specific heat	Btu/lbm °F	cal/g °C	1.000
	Btu/lbm °F	W s/kg °C	4186.7
Pressure	$lbf/in.^2$ (psi)	kgf/cm^2	0.070309
	psi	atm	0.068046
	psi	bars	0.068948
	psi	$dyne/cm^2$	68947.0
Surface tension	lbf/ft	N/m	14.5937

Source: From Rohsenow, W. M., and Hartnett, J. P., Eds. (1973). *Handbook of Heat Transfer,* McGraw-Hill, New York.

Table A.2 Conversion from SI to other units

Quantity	To convert number of	To	Multiply by
Pressure	N/m^2	$dyne/cm^2$	10
		bar	10^{-5}
		atm (phys)	0.9869×10^{-5}
	bar = 10^5 N/m^2	atm (phys)	0.9869
		kgf/cm^2 (atm abs)	1.0197
		mm Hg	750
Density	kg/m^3	g/cm^3	10^{-3}
Specific volume	m^3/kg	cm^3/g	10^3
Heat capacity	kJ/kg °C	kcal/kg °C	0.2388
Thermal conductivity	W/m °C	cal/cm s °C	2.388×10^{-3}
		kcal/m h °C	0.86
Viscosity	N s/m^2 (kg/m s)	g/cm s (poise)	10
		kg s/m^2	0.102
Kinematic viscosity	m^2/s	cm^2/s (stokes)	10^4
Surface tension	N/m	dyne/cm (erg/cm^2)	10^3

Source: From Rohsenow, W. M., and Hartnett, J. P., Eds. (1973). *Handbook of Heat Transfer*, McGraw-Hill, New York.

Table A.3 Conversion factors for heat flux, *q''*

To obtain ↓	Multiply number of the following units by: Btu/ft² h	W/cm²	kcal/h m²	cal/s cm²
Btu/ft² h	1	3,170.75	0.36865	13,277.26
W/cm²	3.154×10^{-4}	1	1.63×10^{-4}	4.1868
kcal/h m²	2.7126	8,600	1	2.778×10^{-5}
cal/s cm²	7.536×10^{-5}	0.2389	36,000	1

Source: From Rohsenow, W. M., and Hartnett, J. P., Eds. (1973). *Handbook of Heat Transfer*, McGraw-Hill, New York.

Table A.4 Conversion factors for heat transfer coefficient, *h*

To obtain ↓	Multiply number of the following units by: Btu/h ft² °F	W/cm² °C	cal/s cm² °C	kcal/h m² °C
Btu/h ft² °F	1	1761	7376	0.20489
W/cm² °C	5.6785×10^{-4}	1	4.186	1.163×10^{-4}
cal/s cm² °C	1.356×10^{-4}	0.2391	1	2.778×10^{-5}
kcal/h m² °C	4.8826	8600	36000	1

Source: From Rohsenow, W. M., and Hartnett, J. P., Eds. (1973). *Handbook of Heat Transfer*, McGraw-Hill, New York.

Table A.5 Conversion factors for thermal conductivity, *k*

To obtain ↓	Multiply number of the following units by				
	Btu/h ft °F	W/cm °C	cal/s cm °C	kcal/h m °C	Btu in./h ft^2 °F
Btu/h ft °F	1	57.793	241.9	0.6722	0.08333
W/cm °C	0.01730	1	4.186	0.01171	1.442×10^{-3}
cal/s cm °C	4.134×10^{-3}	0.2389	1	2.778×10^{-3}	3.445×10^{-4}
kcal/h m °C	1.488	86.01	360	1	0.1240
Btu in./h ft^2 °F	12	693.5	2903	8.064	1

Source: From Rohsenow, W. M., and Hartnett, J. P., Eds. (1973). *Handbook of Heat Transfer,* McGraw-Hill, New York.

Table A.6 Conversion factors for viscosity, μ

To obtain ↓	Multiply number of the following units by				
	lbm/ft h	lbf s/ft^2	centipoise	kgm/m h	kgf s/m^2
lbm/ft h	1	116,000	2.42	0.672	23733
lbf s/ft^2	0.00000862	1	0.00002086	0.00000579	0.2048
centipoise[a]	0.413	47,880	1	0.278	9807
kgm/m h	1.49	172,000	3.60	1	35305
kgf s/m^2	0.0000421	4.882	0.0001020	0.0000284	1

[a]100 centipoise = 1 poise = 1 g/s cm = 1 dyne s/cm^2.

Source: From Rohsenow, W. M., and Hartnett, J. P., Eds. (1973). *Handbook of Heat Transfer,* McGraw-Hill, New York.

Table A.7 Conversion factors for kinematic viscosity, ν

To obtain ↓	Multiply number of the following units by			
	ft^2/h	stokes	m^2/h	m^2/s
ft^2/h	1	3.875	10.764	38.751
stokes	0.25806	1	2.778	10^4
m^2/h	0.092903	0.3599	1	3600
m^2/s	0.00002581	10^{-4}	0.0002778	1

Source: From Rohsenow, W. M., and Hartnett, J. P., Eds. (1973). *Handbook of Heat Transfer,* McGraw-Hill, New York.

APPENDIX

B

PROPERTIES OF GASES

Table B.1 Properties of dry air at atmospheric pressure—SI units[a]

Temperature			Properties							
K	°C	°F	ρ	c_p	c_p/c_v	μ	k	Pr	h	V_s
100	−173.15	−280	3.598	1.028		6.929	9.248	0.770	98.42	198.4
110	−163.15	−262	3.256	1.022	1.4202	7.633	10.15	0.768	108.7	208.7
120	−153.15	−244	2.975	1.017	1.4166	8.319	11.05	0.766	118.8	218.4
130	−143.15	−226	2.740	1.014	1.4139	8.990	11.94	0.763	129.0	227.6
140	−133.15	−208	2.540	1.012	1.4119	9.646	12.84	0.761	139.1	236.4
150	−123.15	−190	2.367	1.010	1.4102	10.28	13.73	0.758	149.2	245.0
160	−113.15	−172	2.217	1.009	1.4089	10.91	14.61	0.754	159.4	253.2
170	−103.15	−154	2.085	1.008	1.4079	11.52	15.49	0.750	169.4	261.0
180	−93.15	−136	1.968	1.007	1.4071	12.12	16.37	0.746	179.5	268.7
190	−83.15	−118	1.863	1.007	1.4064	12.71	17.23	0.743	189.6	276.2
200	−73.15	−100	1.769	1.006	1.4057	13.28	18.09	0.739	199.7	283.4
205	−68.15	−91	1.726	1.006	1.4055	13.56	18.52	0.738	204.7	286.9
210	−63.15	−82	1.684	1.006	1.4053	13.85	18.94	0.736	209.7	290.5
215	−58.15	−73	1.646	1.006	1.4050	14.12	19.36	0.734	214.8	293.9
220	−53.15	−64	1.607	1.006	1.4048	14.40	19.78	0.732	219.8	297.4
225	−48.15	−55	1.572	1.006	1.4046	14.67	20.20	0.731	224.8	300.8
230	−43.15	−46	1.537	1.006	1.4044	14.94	20.62	0.729	229.8	304.1
235	−38.15	−37	1.505	1.006	1.4042	15.20	21.04	0.727	234.9	307.4
240	−33.15	−28	1.473	1.005	1.4040	15.47	21.45	0.725	239.9	310.6
245	−28.15	−19	1.443	1.005	1.4038	15.73	21.86	0.724	244.9	313.8

Table B.1 Properties of dry air at atmospheric pressure—SI units[a] (*Continued*)

Temperature			Properties							
K	°C	°F	ρ	c_p	c_p/c_v	μ	k	Pr	h	V_s
250	−23.15	−10	1.413	1.005	1.4036	15.99	22.27	0.722	250.0	317.1
255	−18.15	−1	1.386	1.005	1.4034	16.25	22.68	0.721	255.0	320.2
260	−13.15	8	1.359	1.005	1.4032	16.50	23.08	0.719	260.0	323.4
265	−8.15	17	1.333	1.005	1.4030	16.75	23.48	0.717	265.0	326.5
270	−3.15	26	1.308	1.006	1.4029	17.00	23.88	0.716	270.1	329.6
275	−1.85	35	1.235	1.006	1.4026	17.26	24.28	0.715	275.1	332.6
280	6.85	44	1.261	1.006	1.4024	17.50	24.67	0.713	280.1	335.6
285	11.85	53	1.240	1.006	1.4022	17.74	25.06	0.711	285.1	338.5
290	16.85	62	1.218	1.006	1.4020	17.98	25.47	0.710	290.2	341.5
295	21.85	71	1.197	1.006	1.4018	18.22	25.85	0.709	295.2	344.4
300	26.85	80	1.177	1.006	1.4017	18.46	26.24	0.708	300.2	347.3
305	31.85	89	1.158	1.006	1.4015	18.70	26.63	0.707	305.3	350.2
310	36.85	98	1.139	1.007	1.4013	18.93	27.01	0.705	310.3	353.1
315	41.85	107	1.121	1.007	1.4010	19.15	27.40	0.704	315.3	355.8
320	46.85	116	1.103	1.007	1.4008	19.39	27.78	0.703	320.4	358.7
325	51.85	125	1.086	1.008	1.4006	19.63	28.15	0.702	325.4	361.4
330	56.85	134	1.070	1.008	1.4004	19.85	28.53	0.701	330.4	364.2
335	61.85	143	1.054	1.008	1.4001	20.08	28.90	0.700	335.5	366.9
340	66.85	152	1.038	1.008	1.3999	20.30	29.28	0.699	340.5	369.6
345	71.85	161	1.023	1.009	1.3996	20.52	29.64	0.698	345.6	372.3
350	76.85	170	1.008	1.009	1.3993	20.75	30.03	0.697	350.6	375.0
355	81.85	179	0.9945	1.010	1.3990	20.97	30.39	0.696	355.7	377.6
360	86.85	188	0.9805	1.010	1.3987	21.18	30.78	0.695	360.7	380.2
365	91.85	197	0.9672	1.010	1.3984	21.38	31.14	0.694	365.8	382.8
370	96.85	206	0.9539	1.011	1.3981	21.60	31.50	0.693	370.8	385.4
375	101.85	215	0.9413	1.011	1.3978	21.81	31.86	0.692	375.9	388.0
380	106.85	224	0.9288	1.012	1.3975	22.02	32.23	0.691	380.9	390.5
385	111.85	233	0.9169	1.012	1.3971	22.24	32.59	0.690	386.0	393.0
390	116.85	242	0.9050	1.013	1.3968	22.44	32.95	0.690	391.0	395.5
395	121.85	251	0.8936	1.014	1.3964	22.65	33.31	0.689	396.1	398.0
400	126.85	260	0.8822	1.014	1.3961	22.86	33.65	0.689	401.2	400.4
410	136.85	278	0.8608	1.015	1.3953	23.27	34.35	0.688	411.3	405.3
420	146.85	296	0.8402	1.017	1.3946	23.66	35.05	0.687	421.5	410.2
430	156.85	314	0.8207	1.018	1.3938	24.06	35.75	0.686	431.7	414.9
440	166.85	332	0.8021	1.020	1.3929	24.45	36.43	0.684	441.9	419.6
450	176.85	350	0.7342	1.021	1.3920	24.85	37.10	0.684	452.1	424.2
460	186.85	368	0.7677	1.023	1.3911	25.22	37.78	0.683	462.3	428.7
470	196.85	386	0.7509	1.024	1.3901	25.58	38.46	0.682	472.5	433.2
480	206.85	404	0.7351	1.026	1.3892	25.96	39.11	0.681	482.8	437.6
490	216.85	422	0.7201	1.208	1.3881	26.32	39.76	0.680	493.0	442.0
500	226.85	440	0.7057	1.030	1.3871	26.70	40.41	0.680	503.3	446.4
510	236.85	458	0.6919	1.032	1.3861	27.06	41.06	0.680	513.6	450.6
520	246.85	476	0.6786	1.034	1.3851	27.42	41.69	0.680	524.0	454.9

Table B.1 Properties of dry air at atmospheric pressure—SI units[a] (*Continued*)

Temperature			Properties							
K	°C	°F	ρ	c_p	c_p/c_v	μ	k	Pr	h	V_s
530	256.85	494	0.6658	1.036	1.3840	27.78	42.32	0.680	534.3	459.0
540	266.85	512	0.6535	1.038	1.3829	28.14	42.94	0.680	544.7	463.2
550	276.85	530	0.6416	1.040	1.3818	28.48	43.57	0.680	555.1	467.3
560	286.85	548	0.6301	1.042	1.3806	28.83	44.20	0.680	565.5	471.3
570	296.85	566	0.6190	1.044	1.3795	29.17	44.80	0.680	575.9	475.3
580	306.85	584	0.6084	1.047	1.3783	29.52	45.41	0.680	586.4	479.2
590	316.85	602	0.5980	1.049	1.3772	29.84	46.01	0.680	596.9	483.2
600	326.85	620	0.5881	1.051	1.3760	30.17	46.61	0.680	607.4	486.9
620	346.85	656	0.5691	1.056	1.3737	30.82	47.80	0.681	628.4	494.5
640	366.85	692	0.5514	1.061	1.3714	31.47	48.96	0.682	649.6	502.1
660	386.85	728	0.5347	1.065	1.3691	32.09	50.12	0.682	670.9	509.4
680	406.85	764	0.5189	1.070	1.3668	32.71	51.25	0.683	692.2	516.7
700	426.85	800	0.5040	1.075	1.3646	33.32	52.36	0.684	713.7	523.7
720	446.85	836	0.4901	1.080	1.3623	33.92	53.45	0.685	735.2	531.0
740	466.85	872	0.4769	1.085	1.3601	34.52	54.53	0.686	756.9	537.6
760	486.85	903	0.4643	1.089	1.3580	35.11	55.62	0.687	778.6	544.6
780	506.85	944	0.4524	1.094	1.3559	35.69	56.68	0.688	800.5	551.2
800	526.85	950	0.4410	1.099	1.354	36.24	57.74	0.689	822.4	557.8
850	576.85	1,070	0.4152	1.110	1.349	37.63	60.30	0.693	877.5	574.1
900	626.85	1,160	0.3920	1.121	1.345	38.97	62.76	0.696	933.4	589.6
950	676.85	1,250	0.3714	1.132	1.340	40.26	65.20	0.699	989.7	604.9
1,000	726.85	1,340	0.3529	1.142	1.336	41.53	67.54	0.702	1,046	619.5
1,100	826.85	1,520	0.3208	1.161	1.329	43.96			1,162	648.0
1,200	926.85	1,700	0.2941	1.179	1.322	46.26			1,279	675.2
1,300	1,026.85	1,580	0.2714	1.197	1.316	48.46			1,398	701.0
1,400	1,126.85	2,060	0.2521	1.214	1.310	50.57			1,518	725.9
1,500	1,220.85	2,240	0.2353	1.231	1.304	52.61			1,640	749.4
1,600	1,326.85	2,420	0.2206	1.249	1.299	54.57			1,764	772.6
1,800	1,526.85	2,780	0.1960	1.288	1.288	58.29			2,018	815.7
2,000	1,726.85	3,140	0.1764	1.338	1.274				2,280	855.5
2,400	2,126.85	3,860	0.1467	1.574	1.238				2,853	924.4
2,800	2,526.85	4,580	0.1245	2.259	1.196				3,599	983.1

[a]Symbols and units: K, absolute temperature, degrees Kelvin; °C, temperature, degrees Celsius; °F, temperature, degrees Fahrenheit; ρ, density, kg/m^3; c_p, specific heat capacity, kJ/kg K; c_p/c_r, specific heat capacity ratio, dimensionless; μ, viscosity [for N s/m^2 (=kg/m s) multiply tabulated values by 10^{-6}]; k, thermal conductivity, MW/m K; Pr, Prandtl number, dimensionless; h, enthalpy, kJ/kg; V_s, sound velocity, m/s.

Source: From Weast, R. C., Ed. (1970). *Handbook of Tables for Applied Engineering Science*, CRC Press, Boca Raton, Fla.

Table B.2 Properties of dry air at atmospheric pressure—English units[a]

Temperature			Properties							
K	°R	°F	ρ	c_p	c_p/c_v	μ	k	Pr	h	V_s
100	180	−280	0.2247	0.2456		0.0466	0.00534	0.770	42.3	651
110	198	−262	0.2033	0.2440	1.4202	0.0513	0.00586	0.768	46.7	685
120	216	−244	0.1858	0.2430	1.4166	0.0559	0.00638	0.766	51.1	717
130	234	−226	0.1711	0.2423	1.4139	0.0604	0.00690	0.763	55.5	747
140	252	−208	0.1586	0.2418	1.4119	0.0648	0.00742	0.761	59.8	776
150	270	−190	0.1478	0.2414	1.4102	0.0691	0.00793	0.758	64.2	804
160	288	−172	0.1384	0.2411	1.4089	0.0733	0.00844	0.754	68.5	831
170	306	−154	0.1301	0.2408	1.4079	0.0774	0.00895	0.750	72.9	856
180	324	−136	0.1228	0.2406	1.4071	0.0815	0.00946	0.746	77.2	882
190	342	−118	0.1163	0.2405	1.4064	0.0854	0.00996	0.743	81.5	906
200	360	−100	0.1104	0.2404	1.4057	0.0892	0.01045	0.739	85.8	930
205	369	−91	0.1078	0.2403	1.4055	0.0911	0.01070	0.738	88.0	941
210	378	−82	0.1051	0.2403	1.4053	0.0930	0.01095	0.736	90.2	953
215	387	−73	0.1027	0.2403	1.4050	0.0949	0.01119	0.734	92.3	964
220	396	−64	0.1003	0.2402	1.4048	0.0967	0.01143	0.732	94.5	976
225	405	−55	0.0981	0.2402	1.4046	0.0986	0.01168	0.731	96.7	987
230	414	−46	0.0959	0.2402	1.4044	0.1004	0.01191	0.729	98.8	998
235	423	−37	0.0939	0.2402	1.4042	0.1022	0.01215	0.727	101.0	1,008
240	432	−28	0.0919	0.2401	1.4040	0.1039	0.01239	0.725	103.1	1,019
245	441	−19	0.0901	0.2401	1.4038	0.1057	0.01263	0.724	105.3	1,030
250	450	−10	0.0882	0.2401	1.4036	0.1074	0.01287	0.722	107.5	1,040
255.4	459.7	0	0.0865	0.2401	1.4034	0.1092	0.01310	0.721	109.6	1,051
260	468	8	0.0848	0.2401	1.4032	0.1109	0.01334	0.719	111.8	1,061
265	477	17	0.0832	0.2402	1.4030	0.1126	0.01357	0.717	114.0	1,071
270	486	26	0.0817	0.2402	1.4029	0.1143	0.01380	0.716	116.1	1,081
275	495	35	0.0802	0.2402	1.4026	0.1160	0.01403	0.715	118.3	1,091
280	504	44	0.0787	0.2402	1.4024	0.1176	0.01426	0.713	120.4	1,101
285	513	53	0.0774	0.2402	1.4022	0.1192	0.01448	0.711	122.6	1,111
290	522	62	0.0760	0.2403	1.4020	0.1208	0.01472	0.710	124.8	1,120
295	531	71	0.0747	0.2403	1.4018	0.1224	0.01494	0.709	126.9	1,130
300	540	80	0.0735	0.2404	1.4017	0.1241	0.01516	0.708	129.1	1,140
305	549	89	0.0723	0.2404	1.4015	0.1257	0.01539	0.707	131.3	1,149
310	558	98	0.0711	0.2405	1.4013	0.1272	0.01561	0.705	133.4	1,158
315	567	107	0.0700	0.2405	1.4010	0.1287	0.01583	0.704	135.6	1,167
302	576	116	0.0689	0.2406	1.4008	0.1303	0.01606	0.703	137.7	1,177
325	585	125	0.0678	0.2407	1.4006	0.1319	0.01627	0.702	139.9	1,186
330	594	134	0.0668	0.2407	1.4004	0.1334	0.01649	0.701	142.1	1,195
335	603	143	0.0658	0.2408	1.4001	0.1349	0.01670	0.700	144.2	1,204
340	612	152	0.0648	0.2409	1.3999	0.1364	0.01692	0.699	146.4	1,213
345	621	161	0.0639	0.2410	1.3996	0.1379	0.01713	0.698	148.6	1,221
350	630	170	0.0630	0.2411	1.3993	0.1394	0.01735	0.697	150.7	1,230
355	639	179	0.0621	0.2411	1.3990	0.1409	0.01758	0.696	152.9	1,239
360	648	188	0.0612	0.2412	1.3987	0.1423	0.01779	0.695	155.1	1,247
365	657	197	0.0604	0.2413	1.3984	0.1437	0.01800	0.694	157.3	1,256
370	666	206	0.0595	0.2415	1.3981	0.1452	0.01820	0.693	159.4	1,264

Table B.2 Properties of dry air at atmospheric pressure—English units[a] (*Continued*)

Temperature			Properties							
K	°R	°F	ρ	c_p	c_p/c_v	μ	k	Pr	h	V_s
375	675	215	0.0588	0.2416	1.3978	0.1465	0.01841	0.692	161.6	1,273
380	684	224	0.0580	0.2417	1.3975	0.1479	0.01862	0.691	163.8	1,281
385	693	233	0.0572	0.2418	1.3971	0.1494	0.01883	0.690	166.0	1,289
390	702	242	0.0565	0.2420	1.3968	0.1508	0.01904	0.690	168.1	1,298
395	711	251	0.0559	0.2421	1.3964	0.1522	0.01925	0.689	170.3	1,306
400	720	260	0.0551	0.2422	1.3961	0.1536	0.01945	0.689	172.5	1,314
410	738	278	0.0537	0.2425	1.3953	0.1563	0.01985	0.688	176.9	1,330
420	756	296	0.0525	0.2428	1.3946	0.1590	0.02026	0.687	181.2	1,346
430	774	314	0.0512	0.2432	1.3938	0.1617	0.02066	0.686	185.6	1,361
440	792	332	0.0501	0.2435	1.3929	0.1643	0.02106	0.684	190.0	1,377
450	810	350	0.0490	0.2439	1.3920	0.1670	0.02144	0.684	194.4	1,392
460	828	368	0.0479	0.2443	1.3911	0.1695	0.02183	0.683	198.8	1,407
470	846	386	0.0469	0.2447	1.3901	0.1719	0.02222	0.682	203.2	1,421
480	864	404	0.0459	0.2451	1.3892	0.1744	0.02260	0.681	207.6	1,436
490	882	422	0.0450	0.2456	1.3881	0.1769	0.02298	0.680	212.0	1,450
500	900	440	0.0441	0.2460	1.3871	0.1794	0.02335	0.680	216.4	1,464
510	918	458	0.0432	0.2465	1.3861	0.1818	0.02373	0.680	220.8	1,478
520	936	476	0.0424	0.2469	1.3851	0.1842	0.02409	0.680	225.3	1,492
530	954	494	0.0416	0.2474	1.3840	0.1867	0.02445	0.680	229.7	1,506
540	972	512	0.0408	0.2479	1.3829	0.1891	0.02482	0.680	234.2	1,520
550	990	530	0.0400	0.2484	1.3818	0.1914	0.02518	0.680	238.7	1,533
560	1,008	548	0.0393	0.2490	1.3806	0.1937	0.02554	0.680	243.1	1,546
570	1,026	566	0.0386	0.2495	1.3795	0.1960	0.02589	0.680	247.6	1,559
580	1,044	584	0.0380	0.2500	1.3783	0.1983	0.02624	0.680	252.1	1,572
590	1,062	602	0.0373	0.2506	1.3772	0.2005	0.02659	0.680	256.6	1,585
600	1,080	620	0.0367	0.2511	1.3760	0.2027	0.02694	0.680	261.1	1,597
620	1,116	656	0.0355	0.2522	1.3737	0.2071	0.02762	0.681	270.2	1,622
640	1,152	692	0.0344	0.2533	1.3714	0.2115	0.02829	0.682	279.3	1,647
660	1,188	728	0.0334	0.2545	1.3691	0.2156	0.02896	0.682	288.4	1,671
680	1,224	764	0.0324	0.2556	1.3668	0.2198	0.02962	0.683	297.6	1,695
700	1,260	800	0.0315	0.2568	1.3646	0.2239	0.03026	0.684	306.8	1,718
720	1,296	836	0.0306	0.2579	1.3623	0.2279	0.03089	0.685	316.1	1,742
740	1,332	872	0.0298	0.2591	1.3601	0.2320	0.03151	0.686	325.4	1,764
760	1,368	908	0.0290	0.2602	1.3580	0.2359	0.03214	0.687	334.8	1,787
780	1,404	944	0.0282	0.2613	1.3559	0.2398	0.03275	0.688	344.1	1,808
800	1,440	980	0.0275	0.2624	1.354	0.2435	0.03337	0.689	353.6	1,830
850	1,530	1,070	0.0259	0.2653	1.349	0.2529	0.03485	0.693	377.3	1,883
900	1,620	1,160	0.0245	0.2678	1.345	0.2618	0.03627	0.696	401.3	1,934
950	1,710	1,250	0.0231	0.2704	1.340	0.2705	0.03768	0.699	425.5	1,985
1,000	1,800	1,340	0.0220	0.2728	1.336	0.2790	0.03903	0.702	450.0	2,032
1,100	1,980	1,520	0.0200	0.2774	1.329	0.2954			499.5	2,126
1,200	2,160	1,700	0.0184	0.2817	1.322	0.3108			549.8	2,215
1,300	2,340	1,880	0.0169	0.2860	1.316	0.3256			600.9	2,300
1,400	2,520	2,060	0.0157	0.2900	1.310	0.3398			652.7	2,381
1,500	2,700	2,240	0.0147	0.2940	1.304	0.3535			705.3	2,459
1,600	2,880	2,420	0.0138	0.2984	1.299	0.3667			758.6	2,535
1,800	3,240	2,780	0.0122	0.3076	1.288	0.3917			867.7	2,676

Table B.2 Properties of dry air at atmospheric pressure—English units[a] (*Continued*)

Temperature			Properties							
K	°R	°F	ρ	c_p	c_p/c_v	μ	k	Pr	h	V_s
2,000	3,600	3,140	0.0110	0.3196	1.274				980.5	2,807
2,400	4,320	3,860	0.0092	0.3760	1.238				1,226.8	3,033
2,800	5,040	4,550	0.0078	0.5396	1.196				1,547.3	3,225

[a]Symbols and units: K, degrees Kelvin; °R, degrees Rankine; °F, degrees Fahrenheit; ρ, density, lb_m/ft^3; c_p, specific heat capacity, Btu/lbm R = cal/g K; c_p/c_r, specific heat capacity ratio dimensionless; μ, viscosity (for lbm/ ft multiply by 10^{-4}); k, thermal conductivity, Btu/hr ft °R; Pr, Prandtl number, dimensionless; h, enthalpy, Btu/lbm (for cal/g multiply by 0.5555); V_s, sound velocity, ft/s.

Source: From Weast, R. C., Ed. (1970). *Handbook of Tables for Applied Engineering Science*, CRC Press, Boca Raton, Fla.

Table B.3 Effect of temperature on properties of gases and vapors at atmospheric pressure[a]

Substance	Temperature °C	Temperature °F	ρ g/cm³	c_v cal/g K	k cal/s cm K	μ centipoise
Ammonia	0	32	9.56×10^{-4}	0.52	5.23×10^{-9}	9.18×10^{-3}
	20	68	8.94×10^{-4}	0.52	5.69×10^{-5}	9.82×10^{-3}
	50	122	8.11×10^{-4}	0.52	6.48×10^{-5}	1.09×10^{-2}
	100	212	7.02×10^{-4}	0.53		1.28×10^{-2}
	200	392	6.20×10^{-4}			1.64×10^{-2}
	300	572	5.12×10^{-4}			1.99×10^{-2}
Argon	−13	9	1.87×10^{-3}	0.125	3.74×10^{-3}	2.04×10^{-2}
	−3	37	1.81×10^{-3}	0.125	3.87×10^{-3}	2.11×10^{-2}
	7	45	1.74×10^{-3}	0.125	3.99×10^{-3}	2.17×10^{-2}
	27	81	1.62×10^{-3}	0.125	4.22×10^{-3}	2.30×10^{-2}
	77	171	1.39×10^{-3}	0.124	4.79×10^{-3}	2.59×10^{-2}
	227	441	9.74×10^{-4}	0.124	6.31×10^{-5}	3.37×10^{-2}
	727	1,341	4.87×10^{-4}	0.124	1.02×10^{-4}	5.42×10^{-2}
	1,227	2,241	3.25×10^{-4}	0.124	1.31×10^{-4}	7.08×10^{-2}
	1,727	3,141	2.43×10^{-4}	0.124		
Butane	0	32	2.59×10^{-3}	0.3802	3.16×10^{-5}	6.84×10^{-3}
	100	212	1.90×10^{-3}	0.4842	5.60×10^{-5}	9.26×10^{-5}
	200	392	1.50×10^{-3}	0.5865	8.70×10^{-5}	1.17×10^{-2}
	300	572	1.24×10^{-3}	0.6721	1.24×10^{-4}	1.40×10^{-2}
	400	752	1.05×10^{-3}	0.7474	1.66×10^{-4}	1.64×10^{-2}
	500	932	9.16×10^{-4}	0.8131	2.15×10^{-4}	1.87×10^{-2}
	600	1,112	8.12×10^{-4}	0.8704	2.69×10^{-4}	2.11×10^{-2}
Carbon dioxide	−13	9	2.08×10^{-3}	0.1944	3.25×10^{-3}	1.31×10^{-2}
	−3	27	2.00×10^{-3}	0.1967	3.42×10^{-3}	1.36×10^{-2}
	7	45	1.93×10^{-3}	0.1989	3.60×10^{-3}	1.40×10^{-2}
	17	63	1.86×10^{-3}	0.2012	3.78×10^{-5}	1.45×10^{-2}
	27	81	1.80×10^{-3}	0.2035	3.96×10^{-5}	1.49×10^{-2}
	77	171	1.54×10^{-3}	0.2146	4.89×10^{-5}	1.72×10^{-2}
	227	441	1.07×10^{-3}	0.2424	8.01×10^{-5}	2.32×10^{-2}

Table B.3 Effect of temperature on properties of gases and vapors at atmospheric pressure[a] (*Continued*)

Substance	Temperature °C	°F	ρ g/cm³	c_v cal/g K	k cal/s cm K	μ centipoise
	727	1,341	5.36×10^{-4}	0.2946		3.89×10^{-2}
	1,227	2,241	3.57×10^{-4}	0.3166		
Carbon monoxide	−13	9	1.31×10^{-3}	0.2489	5.30×10^{-5}	1.59×10^{-2}
	−3	27	1.27×10^{-3}	0.2489	5.49×10^{-5}	1.64×10^{-2}
	7	45	1.22×10^{-3}	0.2489	5.67×10^{-5}	1.69×10^{-2}
	17	63	1.18×10^{-3}	0.2489	5.85×10^{-5}	1.74×10^{-2}
	27	81	1.14×10^{-3}	0.2489	6.03×10^{-5}	1.79×10^{-2}
	77	171	9.75×10^{-4}	0.2493	6.89×10^{-5}	2.01×10^{-2}
	227	441	6.82×10^{-4}	0.2542	9.23×10^{-5}	2.61×10^{-2}
	727	1,341	3.41×10^{-4}			4.17×10^{-2}
	1,227	2,241	2.27×10^{-4}			5.44×10^{-2}
Ethane	0	32	1.342×10^{-3}	0.3934	4.52×10^{-5}	8.60×10^{-3}
	100	212	9.83×10^{-4}	0.4938	7.59×10^{-5}	1.14×10^{-2}
	200	392	7.76×10^{-4}	0.5947	1.13×10^{-4}	1.41×10^{-2}
	300	572	6.40×10^{-4}	0.6854	1.55×10^{-4}	1.68×10^{-2}
	400	752	5.45×10^{-4}	0.7676	2.04×10^{-4}	1.93×10^{-2}
	500	932	4.74×10^{-4}	0.8405	2.57×10^{-4}	2.20×10^{-2}
	600	1,112	4.20×10^{-4}	0.9045	3.15×10^{-4}	2.45×10^{-2}
Ethanol	100	212	1.49×10^{-3}	0.403	5.50×10^{-3}	1.08×10^{-2}
	200	392	1.18×10^{-3}	0.480	8.39×10^{-3}	1.37×10^{-2}
	300	572	9.74×10^{-4}	0.554	1.19×10^{-4}	1.67×10^{-2}
	400	752	8.28×10^{-4}	0.624	1.59×10^{-4}	1.97×10^{-2}
	500	932	7.20×10^{-4}	0.691	2.05×10^{-4}	2.26×10^{-2}
Helium	−240	−400	1.463×10^{-3}		8.43×10^{-5}	3.74×10^{-3}
	−129	−200	3.38×10^{-4}		2.22×10^{-4}	1.19×10^{-2}
	0	32	3.68×10^{-4}	1.23	3.40×10^{-4}	1.86×10^{-2}
	20	68	1.67×10^{-4}	1.24	3.55×10^{-4}	1.94×10^{-2}
	40	104	1.56×10^{-4}	1.24	3.70×10^{-4}	2.03×10^{-2}
	49	120		1.24	3.76×10^{-4}	2.06×10^{-2}
Hydrogen	−13	9	9.44×10^{-3}	3.373	3.86×10^{-4}	8.14×10^{-3}
	−3	37	9.10×10^{-5}	3.388	3.98×10^{-4}	8.35×10^{-3}
	7	45	8.77×10^{-5}	3.400	4.11×10^{-4}	8.55×10^{-3}
	27	81	8.47×10^{-5}	3.410	4.22×10^{-4}	8.76×10^{-3}
	77	171	8.19×10^{-3}	3.418	4.34×10^{-4}	8.96×10^{-3}
	227	441	7.02×10^{-3}		4.91×10^{-4}	9.94×10^{-3}
	727	1,341	4.912×10^{-3}	3.467	6.50×10^{-4}	1.26×10^{-2}
Methane	0	32	7.16×10^{-4}	0.5172	7.33×10^{-3}	1.04×10^{-2}
	100	212	5.25×10^{-4}	0.5848	1.11×10^{-4}	1.32×10^{-2}
	200	392	4.14×10^{-4}	0.6704	1.52×10^{-4}	1.59×10^{-2}
	300	572	3.42×10^{-4}	0.7584	1.96×10^{-4}	1.83×10^{-2}
	400	752	2.91×10^{-4}	0.8430	2.43×10^{-4}	2.07×10^{-2}
	500	932	2.53×10^{-4}	0.9210	2.91×10^{-4}	2.29×10^{-2}
	600	1,112	2.24×10^{-4}	0.9919	3.43×10^{-4}	2.52×10^{-2}
Nitrogen	−13	9	1.31×10^{-3}	0.2488	5.52×10^{-5}	
	−3	37	1.27×10^{-3}	0.2487	5.71×10^{-5}	
	7	45	1.22×10^{-3}	0.2487	5.89×10^{-5}	
	27	81	1.18×10^{-3}	0.2487	6.06×10^{-5}	
	77	171	1.14×10^{-3}	0.2487	6.24×10^{-5}	1.79×10^{-2}

Table B.3 Effect of temperature on properties of gases and vapors at atmospheric pressure[a] (*Continued*)

Substance	Temperature °C	Temperature °F	ρ g/cm³	c_v cal/g K	k cal/s cm K	μ centipoise
	227	441	9.75×10^{-3}	0.2490	7.11×10^{-3}	2.00×10^{-2}
	727	1,341	6.82×10^{-3}	0.2524	9.49×10^{-2}	2.57×10^{-2}
Oxygen	−13	9	1.50×10^{-3}	0.2188	5.60×10^{-3}	1.85×10^{-2}
	−3	37	1.45×10^{-3}	0.2190	5.80×10^{-3}	1.90×10^{-2}
	7	45	1.39×10^{-3}	0.2193	5.98×10^{-3}	1.96×10^{-2}
	27	81	1.35×10^{-3}	0.2195	6.22×10^{-3}	2.01×10^{-2}
	77	171	1.30×10^{-3}	0.2198	6.40×10^{-3}	2.06×10^{-2}
	227	441	1.11×10^{-3}	0.2221	7.33×10^{-3}	2.32×10^{-2}
	727	1,341	7.80×10^{-3}	0.2324	9.97×10^{-3}	2.99×10^{-2}
Propane	0	32	1.97×10^{-3}	0.3701	3.62×10^{-3}	7.50×10^{-3}
	100	212	1.44×10^{-3}	0.4817	6.26×10^{-3}	1.00×10^{-2}
	200	392	1.14×10^{-3}	0.5871	9.56×10^{-3}	1.25×10^{-2}
	300	572	9.39×10^{-4}	0.6770	1.34×10^{-4}	1.40×10^{-2}
	400	752	7.99×10^{-4}	0.7550	1.78×10^{-4}	1.72×10^{-2}
	500	932	6.94×10^{-4}	0.8237	2.28×10^{-4}	1.94×10^{-2}
	600	1,112	6.16×10^{-4}	0.8831	2.83×10^{-4}	2.18×10^{-2}

[a]Symbols and conversion factors: ρ, density in g/cm³ (for lb/ft³ multiply by 62.428, for kg/m³ multiply by 1000); c_p, specific heat, cal/g K (for Btu/lb °R multiply by 1, for J/kg K multiply by 4184.0); *k*, thermal conductivity, cal/s cm K (for W/m K multiply by 418.4, for Btu/h ft °R multiply by 241.9); μ, absolute viscosity in centipoises (for lb/ft h multiply by 2.419, for N s/m² multiply by 0.001).

Source: Reprinted with permission from R. C. Weast, Ed., *Handbook of Tables for Applied Engineering Science*. Copyright © 1970, CRC Press, Inc., Boca Raton, Fla.

APPENDIX

C

PROPERTIES OF STEAM

Table C.1 Properties of steam

Pressure, psia	Saturated vapor	32	200	400	600	800	1000	1200
		Temperature, °F						
		Viscosity, lb/h ft						
0	—	0.023	0.031	0.041	0.050	0.059	0.067	0.074
500	0.054	—	—	—	0.059	0.073	0.073	0.080
1,000	0.070	—	—	—	0.069	0.074	0.080	0.086
1,500	0.082	—	—	—	0.082	0.082	0.087	0.092
2,000	0.094	—	—	—	—	0.086	0.094	0.097
2,500	0.108	—	—	—	—	0.101	0.101	0.104
3,000	0.116	—	—	—	—	0.110	0.108	0.110
3,500	—	—	—	—	—	0.119	0.114	0.116
		Thermal conductivity, Btu/h ft °F						
0	—	0.0092	0.0133	0.0184	0.0238	0.0292	0.0347	
250	0.0211	—	—	—	0.0247	0.0206	0.0349	
500	0.0251	—	—	—	0.0260	0.0302	0.0352	
1,000	0.0316	—	—	—	0.0301	0.0314	0.0357	
1,500	0.0379	—	—	—	0.0376	0.0332	0.0364	
1,750	0.0408	—	—	—	—	0.0343	0.0368	
2,000	0.0445	—	—	—	—	0.0355	0.0372	

Source: Reprinted with permission from B. Gebhart, *Heat Transfer*. Copyright © 1971, McGraw-Hill, New York. Also from J. Keenan and F. Keyes, *Thermodynamic Properties of Steam*. Copyright © 1955, John Wiley & Sons.

APPENDIX

D

DIFFUSION OF WATER VAPOR INTO AIR

Table D.1 Values of diffusion constant and Schmidt number

Temperature, °C	Diffusion constant, D (ft^2/h)	Diffusion constant, D (cm^2/s)	$\left(\frac{\mu}{\rho D}\right)^a$
0	0.844	0.218	0.608
10	0.898	0.232	0.610
20	0.952	0.246	0.612
30	1.01	0.260	0.614
40	1.06	0.275	0.615
50	1.12	0.290	0.616
60	1.18	0.305	0.618
70	1.24	0.321	0.619
80	1.30	0.337	0.619

[a]The values of $\mu/\rho D$ were calculated using the viscosity and density of dry air. Thus the values apply only when the diffusing water vapor is very dilute.

Source: Reprinted with permission from R. C. Weast, Ed., *Handbook of Tables for Applied Engineering Science*. Copyright © 1970, CRC Press, Inc., Boca Raton, Fla.

APPENDIX

E

DIFFUSION OF GASES AND VAPORS INTO AIR

Table E.1 Values of diffusion constant and Schmidt number at 1 atm pressure

	Diffusion constant, D, ft²/h		Diffusion constant, D, cm²/s		$\left(\frac{\mu}{\rho D}\right)^a$	
Substance	0°C	25°C	0°C	25°C	0°C	25°C
H_2	2.37	2.76	0.611	0.712	0.217	0.216
NH_3	0.766	0.886	0.198	0.229	0.669	0.673
N_2	0.691		0.178		0.744	
O_2	0.689	0.80	0.178	0.206	0.744	0.748
CO_2	0.550	0.635	0.142	0.164	0.933	0.940
CS_2	0.36	0.414	0.094	0.107	1.41	1.44
Methyl alcohol	0.513	0.615	0.132	0.159	1.00	0.969
Formic acid	0.509	0.615	0.131	0.159	1.01	0.969
Acetic acid	0.411	0.515	0.106	0.133	1.25	1.16
Ethyl alcohol	0.394	0.461	0.102	0.119	1.30	1.29
Chloroform	0.352		0.091		1.46	
Diethylamine	0.342	0.406	0.0884	0.105	1.50	1.47
n-Propyl alcohol	0.329	0.387	0.085	0.100	1.56	1.54
Propionic acid	0.328	0.383	0.0846	0.099	1.57	1.56
Methyl acetate	0.325	0.387	0.0840	0.100	1.58	1.54
Butylamine	0.318	0.391	0.0821	0.101	1.61	1.53
Ethyl ether	0.304	0.360	0.0786	0.093	1.69	1.66
Benzene	0.291	0.341	0.0751	0.088	1.76	1.75
Ethyl acetate	0.277	0.330	0.0715	0.085	1.85	1.81
Toluene	0.274	0.325	0.0709	0.084	1.87	1.83

Table E.1 Values of diffusion constant and Schmidt number at 1 atm pressure (*Continued*)

	Diffusion constant, D, ft^2/h		Diffusion constant, D, cm^2/s		$\left(\frac{\mu}{\rho D}\right)^a$	
Substance	0°C	25°C	0°C	25°C	0°C	25°C
n-Butyl alcohol	0.272	0.348	0.0703	0.090	1.88	1.71
i-Butyric acid	0.263	0.313	0.0679	0.081	1.95	1.90
Chlorobenzene		0.283		0.073		2.11
Aniline	0.236	0.279	0.0610	0.072	2.17	2.14
Xylene	0.228	0.275	0.059	0.071	2.25	2.17
Amyl alcohol	0.228	0.271	0.0589	0.070	2.25	2.20
n-Octane	0.195	0.232	0.0505	0.060	2.62	2.57
Naphthalene	0.199	0.20	0.0513	0.052	2.58	2.96

[a]Based on $\mu/\rho = 0.1325$ cm^2/s for air at 0°C and 0.1541 cm^2/s for air at 25°C; applies only when the diffusing gas or vapor is very dilute.

Source: Reprinted with permission from R. C. Weast, Ed., *Handbook of Tables for Applied Engineering Science*. Copyright © 1970, CRC Press, Inc., Boca Raton, Fla.

APPENDIX

F

PROPERTIES OF LIQUIDS

Table F.1 Properties of pure water at atmospheric pressure

t, °C	ρ, kg/m^3	$\mu \times 10^3$, kg/m s	$\nu \times 10^6$, m^2/s	k, W/m K	$\beta \times 10^5$, K^{-1}	c_p, W s/kg K	Pr
0	999.84	1.7531	1.7533	0.5687	−6.8143	4,209.3	12.976
5	999.96	1.5012	1.5013	0.578	1.5985	4,201.0	10.911
10	999.70	1.2995	1.2999	0.5869	8.7902	4,194.1	9.286
15	999.10	1.1360	1.1370	0.5953	15.073	4,188.5	7.991
20	998.20	1.0017	1.0035	0.6034	20.661	4,184.1	6.946
25	997.05	0.8904	0.8930	0.6110	20.570	4,180.9	6.093
30	995.65	0.7972	0.8007	0.6182	30.314	4,178.8	5.388
35	994.03	0.7185	0.7228	0.6251	34.571	4,177.7	4.802
40	992.21	0.6517	0.6565	0.6351	38.53	4,177.6	4.309
45	990.22	0.5939	0.5997	0.6376	42.26	4,178.3	3.892
50	988.04	0.5442	0.5507	0.6432	45.78	4,179.7	3.535
60	983.19	0.4631	0.4710	0.6535	52.33	4,184.8	2.965
70	977.76	0.4004	0.4095	0.6623	58.40	4,192.0	2.534
80	971.79	0.3509	0.3611	0.6698	64.13	4,200.1	2.201
90	965.31	0.3113	0.3225	0.6759	69.62	4,210.7	1.939
100	958.35	0.2789	0.2911	0.6807	75.00	4,221.0	1.729

$\rho = (999.8396 + 18.224944t - 0.00792221t^2 - 55.44846 \times 10^{-6}t^3 + 149.7562 \times 10^{-9}t^4 - 393.2952 \times 10^{-12}t^5)/(1 + 18.159725 \times 10^{-3}t)$

$\mu = 2.414 \times 10^{-5}(10)^{247.8/(t+133.15)}$

$$k = -0.92247 + 2.8395\left(1 + \frac{t}{273.15}\right)^3 - 1.8007\left(1 + \frac{t}{273.15}\right)^2 + 0.522577\left(1 + \frac{t}{273.15}\right) - 0.07344\left(1 + \frac{t}{273.15}\right)^4$$

$\beta = \{-[18.224944 - 0.0158444t - 1.6635 \times 10^{-4}t^2 + 5.9902 \times 10^{-7}t^3 - 1.9665 \times 10^{-9}t^4 - 18.159725 \times 10^{-3}\rho(t)]\}/(1 + 18.159725 \times 10^{-3}t)\rho(t)$

$c_p = 4.1868 \times 10^3[2.13974 - 9.68137 \times 10^{-3}(t + 273.15) + 2.68536 \times 10^{-6}(t + 273.15)^2 - 2.42139 \times 10^{-8}(t + 273.15)^3]$

Source: From Kukulka, D. J. (1981). Thermodynamic and Transport Properties of Pure and Saline Water, M.S. Thesis, State University of New York at Buffalo.

Table F.2 Properties of common liquids—SI units[a]

Common name	Density, kg/m³	Specific heat, kJ/kg K	Viscosity, N s/m²	Thermal conductivity, W/m K	Freezing point, K	Latent heat of fusion, kJ/kg	Boiling point, K	Latent heat of evaporation, kJ/kg	Coefficient of cubical expansion, K^{-1}
Acetic acid	1049	2.18	0.001155	0.171	290	181	391	402	0.0011
Acetone	784.6	2.15	0.000316	0.161	179.0	98.3	329	518	0.0015
Alcohol, ethyl	785.1	2.44	0.001095	0.171	158.6	108	351.46	846	0.0011
Alcohol, methyl	786.5	2.54	0.00056	0.202	175.5	98.8	337.8	1100	0.0014
Alcohol, propyl	800.0	2.37	0.00192	0.161	146	86.5	371	779	
Ammonia (aqua)	823.5	4.38		0.353					
Benzene	873.8	1.73	0.000601	0.144	278.68	126	353.3	390	0.0013
Bromine		0.473	0.00095		245.84	66.7	331.6	193	0.0012
Carbon dissulfide	1261	0.992	0.00036	0.161	161.2	57.6	319.40	351	0.0013
Carbon tetrachloride	1584	0.866	0.00091	0.104	250.35	174	349.6	194	0.0013
Castor oil	956.1	1.97	0.650	0.180	263.2				
Chloroform	1465	1.05	0.00053	0.118	209.6	77.0	334.4	247	0.0013
Decane	726.3	2.21	0.000859	0.147	243.5	201	447.2	263	
Dodecane	754.6	2.21	0.001374	0.140	247.18	216	489.4	256	
Ether	713.5	2.21	0.000223	0.130	157	96.2	307.7	372	0.0016
Ethylene glycol	1097	2.36	0.0162	0.258	260.2	181	470	800	
Fluorine refrigerant R-11	1476	0.870[b]	0.00042	0.093[b]	162		297.0	180[c]	
Fluorine refrigerant R-12	1311	0.971[b]		0.071[b]	115	34.4	243.4	165[c]	
Fluorine refrigerant R-22	1194	1.26[b]		0.086[b]	113	183	232.4	232[c]	
Glycerine	1259	2.62	0.950	0.287	264.8	200	563.4	974	0.00054
Heptane	679.5	2.24	0.000376	0.128	182.54	140	371.5	318	
Hexane	654.8	2.26	0.000297	0.124	178.0	152	341.84	365	
Iodine		2.15			386.6	62.2	457.5	164	
Kerosene	820.1	2.09	0.00164	0.145				251	
Linseed oil	929.1	1.84	0.0331		253		560		
Mercury		0.139	0.00153		234.3	11.6	630	295	0.00018
Octane	698.6	2.15	0.00051	0.131	216.4	181	398	298	0.00072
Phenol	1072	1.43	0.0080	0.190	316.2	121	455		0.00090
Propane	493.5	2.41[b]	0.00011		85.5	79.9	231.08	428[c]	
Propylene	514.4	2.85	0.00009		87.9	71.4	225.45	342	
Propylene glycol	965.3	2.50	0.042		213		460	914	
Sea water	1025	3.76–4.10			270.6				
Toluene	862.3	1.72	0.000550	0.133	178	71.8	383.6	363	
Turpentine	868.2	1.78	0.001375	0.121	214		433	293	0.00099
Water	997.1	4.18	0.00089	0.609	273	333	373	2260	0.00020

[a]At 1.0 atm pressure (0.101325 MN/m²), 300 K, except as noted.
[b]At 297 K, liquid.
[c]At 0.101325 MN, saturation temperature.
Source: Reprinted with permission from R. C. Weast, Ed., *Handbook of Tables for Applied Engineering Science*. Copyright © 1970, CRC Press, Inc., Boca Raton, Fla.

APPENDIX

G

PROPERTIES OF LIQUID METALS

Table G.1 Properties of liquid metals at atmospheric or higher pressures

Metal (melting point, °F)	Temperature		Specific gravity (cal/g °C)	Specific heat	Thermal conductivity		Absolute viscosity	
	°F	°C			Btu/h ft °F	cal/s cm °C[a]	lbm/ft s	centipoise
Aluminum (1220)	1250	677	2.38	0.259				
	1300	704	2.37	0.259	60.2	0.249	1.88×10^{-3}	2.8
	1350	732	2.36	0.259	63.4	0.262	1.61×10^{-3}	2.4
	1400	760	2.35	0.259	64.3	0.266	1.34×10^{-3}	2.0
	1450	788	2.34	0.259	69.9	0.289	1.08×10^{-3}	1.6
Bismuth (520)	600	316	10.0	0.0345	9.5	0.039	1.09×10^{-3}	1.62
	800	427	9.87	0.0357	9.0	0.037	9.0×10^{-4}	1.34
	1000	538	9.74	0.0369	9.0	0.037	7.4×10^{-4}	1.10
	1200	649	9.61	0.0381	9.0	0.037	6.2×10^{-4}	0.923
Cesium (83)	83	28	1.84	0.060	10.6	0.044		
	150	66					3.84×10^{-4}	0.571
	250	121					2.95×10^{-4}	0.439
	350	177					2.47×10^{-4}	0.368
	400	204					2.30×10^{-4}	0.343
Lead (621)	700	371	10.5	0.038	9.3	0.038	1.61×10^{-3}	2.39
	850	454	10.4	0.037	9.0	0.037	1.38×10^{-3}	2.05
	1000	538	10.4	0.037	8.9	0.036	1.17×10^{-3}	1.74
	1150	621	10.2	0.037	8.7	0.036	1.02×10^{-3}	1.52
	1300	704	10.1		8.6	0.035	9.20×10^{-4}	1.37
Lithium (355)	400	204	0.506	1.0	24.	0.10	4.0×10^{-4}	0.595
	600	316	0.497	1.0	23	0.095	3.4×10^{-4}	0.506
	800	427	0.489	1.0	22.	0.090	3.7×10^{-4}	0.551
	1200	649	0.471				2.9×10^{-4}	0.432
	1800	942	0.442				2.8×10^{-4}	0.417

Table G.1 Properties of liquid metals at atmospheric or higher pressures (*Continued*)

Metal (melting point, °F)	Temperature °F	Temperature °C	Specific gravity (cal/g °C)	Specific heat	Thermal conductivity Btu/h ft °F	Thermal conductivity cal/s cm °C[a]	Absolute viscosity lbm/ft s	Absolute viscosity centipoise
Magnesium (1203)	1250	677	1.55	0.318				
	1301	705	1.53	0.320				
	1350	732	1.49	0.322				
Mercury (−38)	50	10	13.6	0.033	4.8	0.020	1.07×10^{-3}	1.59
	200	93	13.4	0.033	6.0	0.025	8.4×10^{-4}	1.25
	300	149	13.2	0.033	6.7	0.028	7.4×10^{-4}	1.10
	400	204	13.1	0.032	7.2	0.030	6.7×10^{-4}	0.997
	600	316	12.8	0.032	8.1	0.033	5.8×10^{-4}	0.863
Tin (449)	500	260	6.94	0.058	19	0.079	1.22×10^{-3}	1.82
	700	371	6.86	0.060	19.4	0.080	9.8×10^{-4}	1.46
	850	454	6.81	0.062	19	0.079	8.5×10^{-4}	1.26
	1000	538	6.74	0.064	19	0.079	7.6×10^{-4}	1.13
	1200	649	6.68	0.066	19	0.079	6.7×10^{-4}	0.997
Zinc (787)	600	316	6.97	0.123	35.4	0.146		
	850	454	6.90	0.119	33.7	0.139	2.10×10^{-3}	3.12
	1000	538	6.86	0.116	33.2	0.137	1.72×10^{-3}	2.56
	1200	649	6.76	0.113	32.8	0.136	1.39×10^{-3}	2.07
	1500	816	6.74	0.107	32.6	0.135	9.83×10^{-4}	1.46

[a]For W/cm °C multiply by 4.184.

Source: From R. C. Weast, Ed. *Handbook of Tables for Applied Engineering Science*, CRC Press, Boca Raton, Fla., 1970.

APPENDIX
H
DIFFUSION OF SOLUTES INTO WATER

Table H.1 Diffusion of solutes into water—dilute solutions at 20°C

Substance	Diffusion constant, D^a		Schmidt number, $(\mu/\rho D)^d$	Substance	Diffusion constant, D^a		Schmidt number, $(\mu/\rho D)^d$
	English[b]	Metric[c]			English[b]	Metric[c]	
H_2	19.8	5.13	196	H_2SO_4	6.70	1.73	581
O_2	6.97	1.80	558	NaOH	5.84	1.51	666
CO_2	6.85	1.77	568	NaCl	5.22	1.35	744
NH_3	6.81	1.76	571	Ethyl alcohol	3.87	1.00	1005
N_2	6.35	1.64	613	Acetic acid	3.41	0.88	1140
Acetylene	6.04	1.56	644	Phenol	3.25	0.84	1200
Cl_2	4.72	1.22	824	Glycerol	2.79	0.72	1400
HCl	10.2	2.64	381	Sucrose	1.74	0.45	2230
HNO_3	10.1	2.6	390				

[a]The following relationship may be used to estimate the effect of temperature on the diffusion constant: $D_1/D_2 = (T_1/T_2)(\mu_2/\mu_1)$, where T is temperature, kelvins, and μ is solution viscosity, centipoises. The diffusion constant varies with concentration because of the changes in viscosity and the degree of ideality of the solution.

[b]For English units in ft^2/h multiply by 10^{-5}.

[c]For metric units in m^2/s multiply by 10^{-9}.

[d]Based on $\mu/\rho = 0.01005\ cm^2/s$ for water at 20°C. Applies only for dilute solutions.

Source: Reprinted with permission from R. C. Weast, Ed., *Handbook of Tables for Applied Engineering Science*. Copyright © 1970, CRC Press, Inc., Boca Raton, Fla.

AUTHOR INDEX

SUBJECT INDEX